Energy Use Management

Proceedings of the International Conference

Volume II

International Conference on Energy Use Management, Tucson, Ariz, 1977.
October 24-28, 1977 — Tucson, Arizona
Marriott Hotel and Community Center

ORGANIZED BY

The Interdisciplinary Group for Ecology, Development, and Energy (EDEN), Los Angeles, California

The University of Arizona

IN COOPERATION WITH

Laboratoire D'Ecologie Générale et Appliquée, Université de Paris VII

The University of Tennessee

U.S. Department of Commerce – Energy Office

Pergamon Press Journals

American Public Power Association

Asian Productivity Organization

Technology Transfer Society

ENERGY USE MANAGEMENT

PROCEEDINGS of the INTERNATIONAL CONFERENCE

Volume II

Tucson, Arizona
October 24-28, 1977

ROCCO A. FAZZOLARE
and
CRAIG B. SMITH
Editors

PERGAMON PRESS
New York / Toronto / Oxford / Sydney / Paris / Frankfurt

Pergamon Press Offices:

U.S.A.	Pergamon Press Inc., Maxwell House, Fairview Park, Elmsford, New York 10523, U.S.A.
U.K.	Pergamon Press Ltd., Headington Hill Hall, Oxford OX3, OBW, England
CANADA	Pergamon of Canada, Ltd., 75 The East Mall, Toronto, Ontario M8Z 5WR, Canada
AUSTRALIA	Pergamon Press (Aust) Pty. Ltd., 19a Boundary Street, Rushcutters Bay, N.S.W. 2011, Australia
FRANCE	Pergamon Press SARL, 24 rue des Ecoles, 75240 Paris, Cedex 05, France
WEST GERMANY	Pergamon Press GmbH, 6242 Kronberg/Taunus, Frankfurt-am-Main, West Germany

Library of Congress Catalog Card No. 77-86455

Printed in the United States of America

PREFACE

The prospect of diminishing fossil fuel supplies and progressive degradation of the environment have focused world attention on energy needs and use. Propelled by expanding population and the quest for ever increasing physical standards of living, we find ourselves in an exponential energy growth trend. The hope has been all along that science and technology would pull us through with innovative replacements for the dwindling natural resources. By close examination of the problem, compared with oil and gas, we find that we are on the path of diminishing returns with many of the energy sources on the horizon: larger capital and energy expenditures, costly and even elusive research and development goals, and the prospect of greater environmental deterioration. It has become increasingly obvious that we must examine our energy use patterns in order to improve efficiency and curtail waste.

Thus, the idea for this conference started two years ago and was slowly nourished by a small group of individuals until more than three hundred speakers and many more interested participants made this conference a reality. It is the purpose of this conference to:

- promote efficiency in the end use of energy by providing an international forum for the exchange of information, strategies, and experiences;
- assess the potential for improvements in energy utilization by considering the state-of-the-art, technology, governmental policies, and socioeconomic impacts;
- present the conference information and findings to as wide a public as possible;
- demonstrate and disseminate information on practical solutions and approaches by providing for a commercial exhibit and a community workshop.

This is an *action-oriented* meeting conducted in a semi-workshop mode. Each session is to assess current and future directions in thinking and technology for improved energy use. The creative information generated will be prepared as special publication for wide public distribution.

This conference is dedicated to the optimistic faith that science, technology, economics and reason can be applied to integrated planning of our energy use, environment and needs, in order to provide for an equitable quality of life throughout the world.

Rocco A. Fazzolare

Craig B. Smith

ORGANIZING INSTITUTIONS AND PLANNING COMMITTEES

Conference Chairmen and Executive Advisory Board

Craig B. Smith, General Chairman
Interdisciplinary Group for Ecology, Development, and Energy (EDEN)
P.O. Box 64369, Los Angeles, California 90064

Rocco A. Fazzolare, Program Chairman
Energy Systems Engineering Program
Department of Engineering, Building 20
University of Arizona, Tucson, Arizona 85721

Chauncey Starr
President, Electric Power Research Institute
Palo Alto, California

G. Melese D'Hospital
Consultant, General Atomic Company
San Diego, California

Carlos Velez
Instituto de Investigaciones Electrica
Mexico

Pierre Zaleski
Abasssade de France

S. David Freeman
Office of the President, U.S.A.

Robert Maxwell
Chairman, Pergamon Press, Ltd.

Steering Committee

Quentin Looney, U.S. Department of Commerce
(on leave at Electric Power Research Institute)
Palo Alto, California

Richard Schoen
University of California, Los Angeles

Jorge Vieira da Silva, Université de Paris

Marc Ross, University of Michigan

William Snyder, University of Tennessee

Terry Triffet, University of Arizona

Richard L. Rudman, Electric Power Research Institute

Frederick Symonds, University of Tennessee

Robert A. Sandberg, Electric Power Research Institute

Laurence Martin, Martin & Martin,
Modesto, California

Roger Tierney
Roger Tierney Associates, Los Angeles

Jon Riffel, Southern California Gas Company

Program Committee

William Snyder, University of Tennessee

Robert Taussig, Mathematical Sciences N.W.,
Washington, D.C.

John Reagan, University of Arizona

Russell B. Spencer, Applied Nucleonics Company,
Los Angeles

G. Bruce Taylor, Applied Nucleonics Company,
Los Angeles

Ali Kettani, University of Petroleum and Minerals,
Saudi Arabia

James L. Dyer, Long Beach State University,
California

Lawrence Medlin, University of Arizona

G. James Kenagy, University of Washington

Laurence Martin, Martin & Martin,
Modesto, California

Martha Gilliland, Energy Policy Studies, Inc.
and University of Oklahoma

Industrial Advisory Board

Joseph Base, Chairman
Kaiser Aluminum and Chemical Corp.,

Paul E. Anderson, Martin Marietta Aluminum

J.R. Angell, Kaiser Aluminum and Chemical Corp.

Jack P. Caulfield, Rockwell International

Henry Hirsch, Interspace Corp.

R.H. Lindquist, Chevron Research Co.

J.L. Venturini, Bethlehem Steel Corp.

Utilities Advisory Board

Richard C. Day, Oklahoma Gas & Electric Co.

John Eilering, Commonwealth Edison

Paul Greiner, Edison Electric Institute

R.E. Lloyd, Florida Power & Light Co.

René H. Malès, Chairman
Electric Power Research Institute

Herbert Nash, Pennsylvania Power & Light Co.

International Liaison

Percy Westerlund, Sweden

J. Walderley Coelho Dias, Brazil

P. Mulás, Mexico

Olufemi Lalude, Nigeria

Ryohei Kiyose, Japan

Publications Board

Robert N. Miranda

Albert Henderson

Joanne Potenza

Pergamon Press, Inc., Elmsford, New York

Deputy Program Chairman – Student Affairs/Facilities

Barry Ganapol, University of Arizona

Deputy General Chairman Information/Registration

Kuppaswamy M.S. Iyengar
Interdisciplinary Group for Ecology, Development, and Energy (EDEN), Los Angeles, California

Conference Publicity

Marcia Untracht, Coordinator
Interdisciplinary Group for Ecology, Development, and Energy (EDEN), Los Angeles, California

James Lamb, University of Arizona

Thomas Dalby, Pergamon Press, Ltd., Oxford, England

Promotion Advisors

Robert Jackson, Department of Commerce

Tres Lee, American Public Power Association

Robert Sandberg, Electric Power Research Institute

Jerry Meyer, Pergamon Press, Australia

Jon Riffel, Public Relations Society of America

Richard Perrine
Technology Transfer Society of America

Susan Underwood
Technology Transfer Society of America

Local Registration

Charles Hausenbauer, University of Arizona

Jackie Hass, University of Arizona

Planning Staff

Katherine Little, University of Arizona

Demaris Marconi – ORTH

Local Press Relations

John Crow, University of Arizona

National and International Media

Joan Berliner
Joan Berliner Associates, Los Angeles, California

Industrial Exhibit

Roger Tierney
Roger Tierney & Associates, Los Angeles, California

Community Workshop

Nancy Smith, Coordinator
Interdisciplinary Group for Ecology, Development and Energy (EDEN), Los Angeles, California

Mary Beth Carlile, Tucson Community Representative

Roger Caldwell, University of Arizona

Guest Program and Tours

Marie-Odile Fazzolare, Coordinator, Tucson

Peggy Hetrick, Tucson

INTERNATIONAL CONFERENCE ON ENERGY USE MANAGEMENT

Highlights

Sunday, October 23

5:00 PM	Registration
6:00 PM	WELCOME COCKTAIL – Marriott Hotel

Monday, October 24

8:30 AM Opening PLENARY Session – Marriott Hotel

Panel Discussion – Efficient Energy Use – The Need and the Potential

Global Energy Trends – Chauncey Starr, Electric Power Research Institute

Energy, Economics, and Human Welfare – Sam M. Schurr, Resources for the Future

World Energy – The Resource Picture – Vincent McKelvey, Director, USGS

Energy Management in the Third World – Shigeaki Ueki, Minister of Mines and Energy, Brazil

Energy Conservation and Economic Growth – René Malès, Head, EPRI Energy Conservation Department

Daily 1:00 PM Monday through Thursday, October 24 – 27

PLENARY Sessions: World Perspectives – Marriott Hotel

Conversations with prominent leaders and personalities:

E.F. Schumacher, Intermediate Technical Development Group, London

S.D. Freeman, Energy Policy and Planning, Office of the President (Others, TBA)

Monday through Thursday

8:30 – 12:00
2:00 – 5:30 Simultaneous Technical Sessions (See attachments)

Monday, October 24

6:00 PM Mariachis and Cocktails, Marriott Hotel

Thursday, October 27

7:00 PM "A Night at Old Tucson" – A gala evening party in the desert

Friday, October 28

8:30 AM PLENARY Session – Marriott Hotel

Conference Review and Recommendations – A Panel-type Presentation of Conference Findings

2:00 PM Special Tours for Participants

PROGRAM
VOLUME I

Sessions		Session Organizers
TOPIC A – GENERAL – PLENARY SESSIONS		
A-1	Efficient Energy Use – The Need and the Potential	*Chauncey Starr, Craig B. Smith*
A-2	World Perspectives – (Conversations with Prominent Leaders and Experts)	*Rocco A. Fazzolare*
A-3	Committee on Nuclear and Alternative Energy Systems (CONAES): The Demand and Conservation Panel	*John H. Gibbons*
TOPIC B – INDUSTRIAL SECTOR		*William Snyder*
B-1	Energy Accounting and Information Systems	*Quentin Looney*
B-2	Thermal Processes	*Richard Hockett*
B-3	Fuel and Power Management	*Fred Symonds*
B-4	Process and Product Design	*Jack P. Caulfield*
B-5	Industry Viewpoints	*Allen C. Sheldon*
TOPIC C – COMMERCIAL SECTOR		*Robert Taussig*
C-1	HVAC Systems	*Kenneth Cuccinelli*
C-2	Building Envelope	*Lee Windheim*
C-3	*Building Systems	*J.M. Ayres*
C-4	New Energy Systems and Components	*Fred Dubin*
TOPIC D – RESIDENTIAL SECTOR		*John Reagan*
D-1	Building Materials and Design	*M. Simon*
D-2	Environmental Conditioning	*P.O. Fanger, James E. Woods*
D-3	Domestic Energy Management	*T.T. Woodson*
D-4	Solar Augmented Systems	*Frank Kreith, E. Kauer*
TOPIC E – TRANSPORTATION AND COMMUNICATIONS		*Russell Spencer*
E-1	Vehicle Design and Concepts	*U.L. Businaro, Nicholas DiNapoli*
E-2	Urban Transport Modes and Systems	*Margaret Fels*
E-3	Intercity Transport Operations	*Richard Strombotne*
E-4	Telecommunications	*Walter S. Baer*
TOPIC F – AGRICULTURE AND FOOD		*G.B. Taylor*
F-1	*Crop Production	*David Pimentel*
F-2	Livestock Production	*William Fairbank*
F-3	*Food Processing, Distribution, and Storage	*Dennis Heldman, R.E. Berry*

*Double Session

Sessions		Session Organizers
TOPIC G – INDUSTRIALIZATION AND DEVELOPMENT		*Ali Kettani*
G-1	Energy Supply and Needs	*Joseph Soussou*
G-2	Strategies and Approaches	*J. Vieira da Silva*
G-3	Intermediate Technology	*J.G. Krishnayya*

VOLUME II

TOPIC H – SCIENCE AND ENGINEERING		*James L. Dyer*
H-1	Thermodynamic Effectiveness	*R. Stephen Berry*
H-2	Heat and Power	*Robert Essenhigh, H. Eugene Shull*
H-3	*System Analysis, Modeling, and Simulation	*Bernard Hamel, Harry Brown*
H-4	Advanced Efficient Appliances Equipment and Devices	*Kurt Riegel*
H-5	Advanced Energy Systems	*Fritz Kalhammer, Jean Mascarello*
TOPIC I – ARCHITECTURE, LAND USE, AND URBAN PLANNING		*R. Larry Medlin, Konrad Wachsman*†
I-1	Materials and Structural Design	*Richard G. Stein*
I-2	Emerging Designs: Energy, Human Needs, and Environment	*Philip Steadman, Volker Hartkopf*
I-3	Land Use and Urban Planning	*Frithjof M. Lunde*
I-4	*Urban Systems	*Richard Schoen, Allen S. Kennedy*
TOPIC J – THE NATURAL ENVIRONMENT		*G.J. Kenagy*
J-1	Energy Flow in the Biosphere	*James Cooley*
J-2	Energy Management in Nature	*George Bartholomew*
J-3	Energy and Environment	*Marc Ross*
TOPIC K – THE ROLE OF GOVERNMENT		*Lawrence Martin*
K-1	Regulation, Incentives, and Policy	*Stacy Swor, Lawrence Martin*
TOPIC L – SOCIETAL AND ECONOMIC ASPECTS		*Martha Gilliland*
L-1	International Energy Patterns	*Richard Rudman*
L-2	Public Communication	*Jon Riffel*
L-3	Productivity and Employment	*David Huettner*
L-4a L-4b	Energy Analysis	*Malcolm Slesser, Eric Hirst*
L-5	Changing Life Styles	*Albert J. Fritsch*
TOPIC M – SUMMARY AND CONCLUSIONS		
M-1	Conference Review and Conclusions	*Topic Coordinators*
M-2	Open Forum – Conclusions	*Rocco A. Fazzolare and Craig B. Smith*

†Special Guest for Topic I.
*Double Session

PROGRAM ORGANIZATION
VOLUME I

Topic A
GENERAL PLENARY SESSIONS

Topic Coordinator: *Craig B. Smith*

Topic B
INDUSTRIAL SECTOR

Topic Coordinator: *William Snyder*

KEY: * – Double Session.
(A) – Abstract Included – See Post Conference Volumes III and IV for complete paper.
(P) – See Post Conference Volumes III and IV for complete paper.

KEY: *–Double Session.
(A)–Abstract Included–See Post Conference Volumes III and IV for complete paper.
(P)–See Post Conference Volumes III and IV for complete paper.

Topic C
COMMERCIAL SECTOR

Topic Coordinator: *Robert Taussig*

KEY: *–Double Session.
(A)–Abstract Included–See Post Conference Volumes III and IV for complete paper.
(P)–See Post Conference Volumes III and IV for complete paper.

KEY: *–Double Session.
(A)–Abstract Included–See Post Conference Volumes III and IV for complete paper.
(P)–See Post Conference Volumes III and IV for complete paper.

KEY: *–Double Session.
(A)–Abstract Included–See Post Conference Volumes III and IV for complete paper.
(P)–See Post Conference Volumes III and IV for complete paper.

KEY: *–Double Session.
(A)–Abstract Included–See Post Conference Volumes III and IV for complete paper.
(P)–See Post Conference Volumes III and IV for complete paper.

Topic F
AGRICULTURE AND FOOD

Topic Coordinator: *G.B. Taylor*

F-1 CROP PRODUCTION
Session Leader: *David Pimentel*

F-2 LIVESTOCK PRODUCTION
Session Leader: *William Fairbank*

F-3 FOOD PROCESSING, DISTRIBUTION, AND STORAGE
Session Leaders: *Dennis Heldman & R.E. Berry*

KEY: *—Double Session.
(A)—Abstract Included—See Post Conference Volumes III and IV for complete paper.
(P)—See Post Conference Volumes III and IV for complete paper.

Topic G
INDUSTRIALIZATION AND DEVELOPMENT

Topic Coordinator: *M. Ali Kettani*

G-1 ENERGY SUPPLY AND NEEDS
Session Leader: *Joseph Soussou*

G-2 STRATEGIES AND APPROACHES
Session Leader: *Jorge Vieira da Silva*

KEY: *–Double Session.
(A)–Abstract Included–See Post Conference Volumes III and IV for complete paper.
(P)–See Post Conference Volumes III and IV for complete paper.

G-3 **INTERMEDIATE TECHNOLOGY**
Session Leader: *J.G. Krishnayya*

VOLUME II

Topic H
SCIENCE AND ENGINEERING

Topic Coordinator: *James L. Dyer*

H-1 **THERMODYNAMIC EFFECTIVENESS**
Session Leader: *R. Stephen Berry*

H-2 **HEAT AND POWER**
Session Leaders: *Robert Essenhigh & H. Eugene Shull*

KEY: *–Double Session.
(A)–Abstract Included–See Post Conference Volumes III and IV for complete paper.
(P)–See Post Conference Volumes III and IV for complete paper.

KEY: * – Double Session.
(A) – Abstract Included – See Post Conference Volumes III and IV for complete paper.
(P) – See Post Conference Volumes III and IV for complete paper.

KEY: * – Double Session.
(A) – Abstract Included – See Post Conference Volumes III and IV for complete paper.
(P) – See Post Conference Volumes III and IV for complete paper.

KEY: *–Double Session.
(A)–Abstract Included–See Post Conference Volumes III and IV for complete paper.
(P)–See Post Conference Volumes III and IV for complete paper.

KEY: *–Double Session.
(A)–Abstract Included–See Post Conference Volumes III and IV for complete paper.
(P)–See Post Conference Volumes III and IV for complete paper.

Topic L
SOCIETAL AND ECONOMIC ASPECTS

Topic Coordinator: *Martha Gilliland*

KEY: * –Double Session.
(A)–Abstract Included–See Post Conference Volumes III and IV for complete paper.
(P)–See Post Conference Volumes III and IV for complete paper.

L-4b ENERGY ANALYSIS
Session Leader: *Eric Hirst*

L-5 CHANGING LIFE-STYLES
Session Leader: *Albert J. Fritsch*

KEY: *–Double Session.
(A)–Abstract Included–See Post Conference Volumes III and IV for complete paper
(P)–See Post Conference Volumes III and IV for complete paper.

OPTIMIZATION OF PROCESSES WITH FINITE-TIME THERMODYNAMICS

Bjarne Andresen*, R. Stephen Berry, and Peter Salamon

Department of Chemistry and The James Franck Institute, The University of Chicago, Chicago, Illinois 60637

ABSTRACT

Traditional thermodynamics has provided natural bounds to the use of energy, but the usefulness of these bounds has sometimes been limited because they are defined in terms of the often-unrealistic case of infinitely slow processes. Three approaches to the optimization of work-producing processes have been developed which permit the extension of thermodynamic techniques to define limits for processes operating at finite rates. The most detailed information comes from an optimal control analysis from which the most effective or efficient path can be determined. This approach can be applied to such problems as the optimal design of motion transducers and power trains and of lubricant characteristics. A second approach is based on coarse-graining or cycle-averaging, and is used to obtain optimal power outputs and optimal rates (but not the optimal path itself) for a system doing work by operating between two heat reservoirs. The third and most general approach extends the concept of the thermodynamic potential to finite-time processes. The discussion concludes with the incorporation of the thermodynamic finite-time optimization to the optimization of total economic costs, particularly through the identification of the technological change that would be most effective to enhance the economically optimal use of energy.

*Also at Chemistry Laboratory 3, University of Copenhagen, Universitetsparken 5, 2100 Copenhagen, Denmark

INTRODUCTION

The first incentive to develop a thermodynamics was to describe the transformation of heat into work in a real heat engine (steam engine) and to compare its performance with that of an idealized engine. Since this first beginning, the main emphasis of thermodynamics has shifted to other areas, in particular it has changed from the consideration of process variables to state variables. We are following the original line of thought, realizing that the dynamical quantities are of primary importance for the description of real entropy-producing machines.

The main drawback of traditional thermodynamics is that the idealized limit to which one compares engine performance parameters like efficiency, effectiveness, and power is the reversibly operating Carnot engine. All real engines must, by physical necessity, leave alone human impatience, perform in finite time periods, which introduces a whole score of unavoidable loss mechanisms (friction, temperature and pressure gradients, etc.). It is to investigate these, and eventually prescribe ways to minimize them, that we have extended the traditional thermodynamic formalism to include as many of these rate-dependent losses as possible in our idealized limit. In this way the ideal limiting process becomes a more meaningful limit to the real process, and the limits based on such a model system are more realistic bounds to the real engineering problem than are those of the reversible model.

Idealized finite-time models are not intended to include all the mechanical and thermal intricacies of the real systems. If they did, they would offer complete, deterministic mimics of the real systems, rather than limits. Rather, the limits should be defined to contain the minimum set of constraints that the process designers have to accept. The very straightforward physical interpretation of our results makes the finite-time approach to irreversible thermodynamics ideally suited for drawing general conclusions about which losses are the most important, which parameters should be changed to give the largest rewards, and how the losses vary as the rate of operation is changed. The possibility of carrying out calculations of optimal real-time processes suggests the use such analyses as criteria for allocation of research and development efforts on energy management, and on the optimization of the tradeoffs between the rate of operation and the amount of capital employed.

For optimizing the operation of any physical system, the most complete information one could have is the time sequence of operation which, given that the process operates in a fixed time, produces the maximum power, best efficiency, etc. This question can be put into mathematical terms, and usually results in extremely complicated coupled differential equations, even for very simple models. In Section 2 we outline the analytical solution we have obtained for a "step-Carnot cycle," a model for a Carnot engine with stick-slip friction. We have also computed the optimal path for the more realistic system of the expansion of an ideal gas in a cylinder with a piston, which has frictional losses and a non-vanishing heat resistance to the surrounding heat reservoirs. Such results can be applied, for example, to the determination of optimal power trains.

The less ambitious goal of determining the maximum obtainable power or efficiency without specification of the path for the process has been approached by the extension of the concept of a thermodynamic potential to processes constrained to operate at finite rates. This is carried out in the sense that differences of the finite-time thermodynamic potential set a bound to the work that can be achieved by the system. Section 3 presents algorithms for evaluating such potentials for a variety of problems. We also present a new formalism, the tricycle, by which the optimal rate for a system with two heat reservoirs and a work reservoir can be deduced from thermodynamic conservation equations under simple assumptions.

2. Optimal Path

The work produced in a general process is given by the path integral of a force $\vec{F}$ through a sequence of displacement $d\vec{s}$:

$$W = \int_{\lambda_2, t_2}^{\lambda_1, t_1} \vec{F}(\lambda,t) \cdot d\vec{s}(\lambda,t) \tag{1}$$

where the generalized force $\vec{F}$ depends on a large set of internal variables λ as well as on the time t. The total heat flow can similarly be described by

$$Q = \int_{\lambda_2, t_2}^{\lambda_1, t_1} C(\lambda,t)\, dT(\lambda,t) \tag{2}$$

Both $\vec{F}$ and the generalized heat capacity C will usually be complicated functions of λ and t. Finding the optimal path involves varying the path $\vec{s}(\lambda,t)$ until one obtains an extremum for (1),(2) or any other process variable we desire. This variation can be done, at least in principle, by the methods of optimal control theory.

In conventional thermodynamics of reversible processes, the only system parameters are masses, volumes, compositions, and heat capacities. Moving to finite rates we must also include parameters to describe relaxation rates, such as diffusion coefficients, friction constants, and relaxation times among different degrees of freedom. In order to get a manageable problem we generally assume that some of the relaxation processes are irrelevant to the process we investigate, either by being much faster (essentially following it as a slave) or much slower (undisturbed by the change) than the process. In particular, we assume here that the working fluid is always uniform, so that its internal pressure, density, and temperature are always well defined and independent of position within the fluid.

2.1 The Step-Carnot Cycle

As a first example of finding an optimal path we consider a single modification of Carnot's original reversible cycle, the step-Carnot cycle, defined this way:

a) the working fluid is an ideal gas;

b) the system operates against an external pressure $P_e(t)$ that varies discontinuously, in a manner controlled by a hypothetical machine operator, with the steps always involving an instantaneous change of P_e followed by a change in the volume V and temperature T of the working fluid, at constant P_e (but not necessarily constant internal pressure P_i);

c) the temperature, pressure and density within the working fluid are uniform before each step; for those aspects dealing explicitly with time this must also be true for every instant.

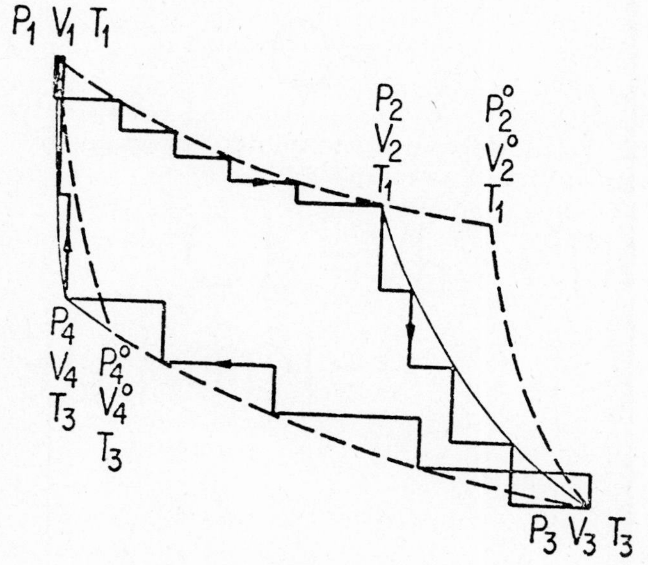

Fig. 1. The step-Carnot cycle

Figure 1 shows the ideal Carnot-cycle operating between point 1 and point 3 in dashed curves. The step-Carnot cycle connecting the same two points is indicated by the heavy stepwise line; the envelope of the adiabatic portions of the step-Carnot cycle is shown by solid curves.

The system goes through a cycle shown by the stepwise path inscribed in the reversible Carnot cycle of Figure 1. Note that although c) implies that the system follows a quasi-static path insofar as internal relaxation times are short compared with the time of any step, the path is only required to contain a finite number of points of equilibrium. This is the essential difference between the irreversible step-Carnot cycle and the reversible Carnot cycle. Along the high-temperature "isotherm," the system, in contact with a reservoir at temperature T_H, expands irreversibly, working against a constant external pressure P_e until it returns to equilibrium. Similarly, along the low-temperature "isotherm," the system, in contact with a reservoir at temperature T_L, is compressed by subjection to an instantaneous increase in P_e, until equilibrium is attained. Along the adiabatic expansion branch of the cycle, in each step the volume of the system increases until the equilibrium is reached with the external pressure. This occurs at a volume greater than that of a reversible adiabat, or at a temperature higher than that of a reversible adiabat. In other words, the stepwise adiabatic expansion has an outer envelope that is flatter than a continuous, reversible adiabat. Likewise, the stepwise adiabatic compression, branch 4, has an outer (left) envelope steeper than the curve of a reversible adiabatic compression

To complete the specification of the system, we now fix the total number of steps N. The first large stage of the problem is thus one of allocating the number of steps $N_1, \ldots N_4$ among the four branches of the cycle, so that $N_1+N_2+N_3+N_4=N$, and of determining that pressure and final volume for each step that maximizes the work done per cycle. Since the problem is no longer a continuous one, but discrete, the coupled differential equations which usually arise from an optimal control problem are replaced by a set of coupled difference equations. In Ref. 1 we solve these analytically for fixed initial and final states, V_1T_1 and V_3T_3, and find that the fractional pressure change per step, P_{n+1}/P_n remains constant along each individual branch. In addition, of course, the allocation of steps is constrained to make the branches meet at the four corners. The result, when N is large enough that the solution can be developed in powers of N^{-1}, is for the maximum effectiveness (or work per cycle)

$$\frac{W}{W_{Carnot}} = 1 + \frac{1}{2N}\frac{\sqrt{T_1}+\sqrt{T_3}}{\sqrt{T_1}-\sqrt{T_3}}\mu^2 \ln\varsigma + O(N^{-2}), \qquad (3)$$

where,

$$\gamma = C_p / C_v$$
$$\varsigma = (V_1/V_3)(T_1/T_3)^{\frac{1}{\gamma-1}} \qquad (4)$$
$$\mu = 1 - \frac{\sqrt{\gamma}}{\gamma-1}\frac{\ln(T_1/T_3)}{\ln\varsigma}$$

The maximum efficiency is achieved with a different program of steps and has the value

$$\frac{W'}{Q_1'} = \frac{T_1-T_3}{T_1} - \frac{1}{2N}\frac{T_3}{T_1}\mu^2\ln\varsigma + O(N^{-2}) \qquad (5)$$

The efficiencies and effectiveness of the two step-Carnot cycles and the reversible Carnot cycle all differ among themselves by terms of order N^{-1}.

This analysis is really intended to illustrate how an optimal finite-time process can be obtained. However we can probe further, and ask how we could use such an analysis to help optimize the design of a real reciprocating engine. First, let us suppose that the stick-slip model of the piston motion gives an adequate representation of the piston's motion, so we can argue as if there is some real counterpart to the model. Then the step-Carnot cycle optimum tells us how to allocate the steps among the four branches of the Carnot cycle, and therefore how to select the thermal properties of the lubricant, so that the system moves more easily in its high-temperature stages, in the way the optimal path requires. Second, the optimal cycle tells us how far in advance of the Carnot cycle's "Point 2" (maximum volume at the high temperature) to start the adiabatic expansion and how far beyond the Carnot "Point 4" (minimum volume at the low temperature) to carry the low-temperature isothermal compression.

2.2 Lossy Gas Expansion

A more realistic system than the step-Carnot cycle is the expansion of gas in a cylinder with a piston when

a) the working fluid is an ideal gas;

b) frictional losses between the pistons and the cylinder are

$$G = \int \alpha \frac{dV}{dt}\, dV,$$

where α is a friction constant;

c) heat flows from a reservoir at temperature T_o into the gas through a thermal conductance κ;

d) the temperature, pressure, and density within the gas are uniform at all times.

Again we wish to find that time-parametricized volume function which produces the most work on the piston shaft during the expansion from $V_1 T_1$ to V_2 in time t_o. (T_2 cannot be specified as it depends on t_o.) The optimization is carried out by setting

$$\delta(W_{rev} - G - \lambda F) = \delta\left[\int_0^{t_o}\left(P - \alpha\frac{dV}{dt}\right)dV - \lambda F\right] = 0 \qquad (6)$$

where F is the constraint imposed by energy conservation

$$C_v dT + P dV = đQ = \kappa(T - T_o)dt \qquad (7)$$

or

$$F = C_v T' + \frac{nRT}{V}V' - \kappa(T - T_o) = 0 \qquad (8)$$

The constants C_v, R, and κ are the constant-volume heat capacity, the gas constant and the heat conductance to the reservoir.

Primes denote differentiation with respect to t. The Euler-Lagrange equations (6) become

$$\frac{\kappa}{2\alpha}T' + \left(V' - \frac{\kappa}{nR}V\right)V'' = 0$$
$$\frac{1}{\gamma-1}T' + \left(V' - \frac{\kappa}{nR}V\right)\frac{T}{V} + \frac{\kappa}{nR}T_o = 0 \qquad (9)$$

Since Equations (9) are too complicated to solve analytically, we have instead solved them numerically for a wide range of loss parameters. One example of the resulting time functions are shown in Fig. 2.

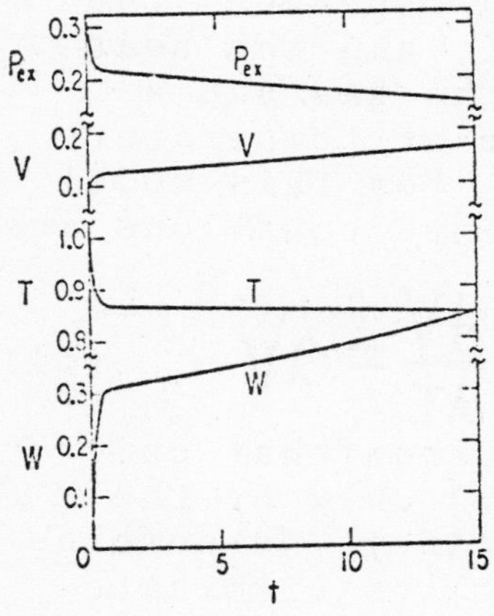

Fig. 2

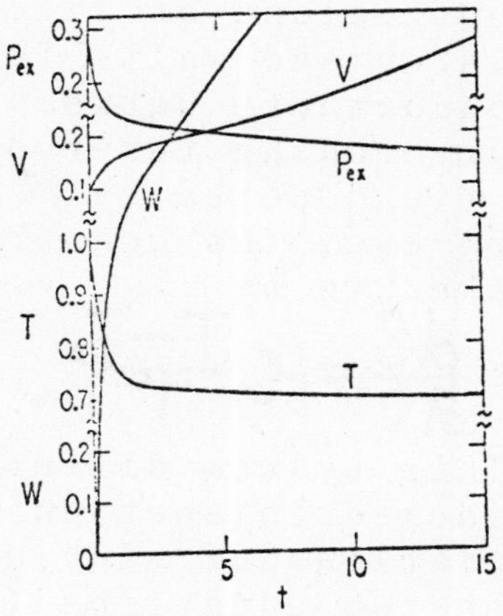

Fig. 3

Figure 2. External pressure P_{ex}, volume V, temperature T, and accumulated work W, as functions of time. Initial values and parameters: V(0)= 0.1; V'(0) = 0.1; T(0) = 1 ; W(0) = 0; T_o = the reservoir temperature = 1; γ = the heat capacity ratio = 5/3; κ/nR = 0.24; $\kappa/2\alpha$ = -0.5.

Figure 3. Optimal external pressure, volume, temperature and accumulated work; paramters as in Fig. 2, except $\kappa/2\alpha$ = -0.25.

The graph is clearly divided into an initial rapid expansion followed by a slow, almost linear volume increase. The same behavior is observed in Fig. 3 which is calculated with same initial values as Fig. 2 but twice the friction. A closer investigation of the energy flows reveals in all cases that to within about 5% the initial phase is powered solely by a decrease in internal energy (decrease in temperature), and the slow expansion by influx of heat from the reservoir. As friction is increased the latter delivers the larger fraction of the work, and at the same time the expansion becomes even more linear in time. The changes in heat flow and internal energy drop are small and largely balance each other, so that the frictional losses are taken almost exclusively from the produced work, eventually to the point where it becomes negative.

This analysis can be applied to practical problems in energy conservation, at the level of engineering design. For example, the conclusions of Section 2.2 tell us that the operation of a conventional reciprocating engine could be improved considerabley. Reciprocating engines are normally connected by connecting rods and eccentrics to driveshafts with flywheels, so that the shafts rotate at constant angular velocity. The geometry of the linkage gives the piston, and therefore the cylinder volume, a time dependence that is roughly sinusoidal. The analysis of Section 2.2 tells us that the maximum work is extracted from the piston when the volume of the cylinder follows a very non-sinusoidal path, with a rapid expansion followed by a slow expansion, and a corresponding rapid compression and a slow compression. The implication of this finding is that we could redesign the linkage between piston and driveshaft in order to make the cylinder volume vary with time in a way that mimics the ideal engine, and thereby derive more power from a given amount of fuel.

3. Maximum Work and Other Extrema of Performance

The very detailed information, the specification of optimal path, described in the previous section is unfortunately the most difficult information to obtain for finite-time processes. Thus far, it is only obtainable in mimical form for a few simple cases. However we often only want to know an upper bound to the work or efficiency as a function of the alotted time, without information about the path. If the maximum work or efficiency is adequate, then a much larger class of processes can be treated analytically. We have used two approaches, to carry out such determinations, one based on an extension of the concept of a thermodynamic potential, and one on basic conservation laws.

3.1 Thermodynamic Potentials

The change in the conventional thermodynamic potential for a process sets an upper bound to the work that can be derived from this process. We have proven[2] that a completely analogous construction can be made for any quasi-static process constrained to go to completion in a finite time. This potential is a function of state and is unique to within a constant of motion along the path between the initial and final states.

For processes which can be described by differential equations that are first order in time(this doesn't mean that the second and higher derivatives with respect to time vanish), and for which all the internal variables appearing in these equations can be derived from state functions and time, the algorithm for finding the potential $\mathcal{P}$ is as follows. Suppose for convenience that the only work is pressure-volume work, and that the path of the process is determined by the differential equation

$$đg(P,V) = g_P dP + g_V dV = 0 \tag{10}$$

This constraint can be used to transform the inexact differential forms $đW$ and $đQ$ into exact differentials and thus to give potentials for the process. We look for a function $f(P,V)$ such that $W = PdV + f\,đg$ is an exact differential, the differential of the potential that we seek. To find $f(P,V)$, we expand the expression

$$PdV + f\,đg = PdV + f(g_P dP + g_V dV) = (P + f g_V)dV + f g_P dP \tag{11}$$

Equating cross derivatives establishes the condition for an exact differential. The cross derivatives are

$$\frac{\partial}{\partial P}(P + f g_V) = 1 + \frac{\partial f}{\partial P} g_V + f\frac{\partial g_V}{\partial P} \tag{12a}$$

and

$$\frac{\partial}{\partial V}(f g_P) = \frac{\partial f}{\partial V} g_P + f\frac{\partial g_P}{\partial V} \tag{12b}$$

Setting them equal, we find a differential equation for f,

$$\left(\frac{\partial f}{\partial V}\right)_P g_P - \left(\frac{\partial f}{\partial P}\right)_V g_V + f\left(\frac{\partial g_P}{\partial V} - \frac{\partial g_V}{\partial P}\right) = 1. \tag{13}$$

When we find a solution f, we can then form the exact differential $d\mathcal{P} = PdV + f\,đg$ and obtain a potential function $\mathcal{P}$. The existence of the function f follows from a generalized version of Caratheodary's theorem[2] to which it reduces for the case $đg = đQ$ where we get $\mathcal{P} = E$ and $f = 1/T$.

One system to which we have applied this procedure is the quasi-static expansion of an ideal gas inside a cylinder equipped with a piston whose motion is exponential in time. The coefficient of friction between the piston and walls is α, and the heat conductance between the gas and a surrounding reservoir at temperature T_o is denoted by K. For this model process, the constrained equations are taken to be

$$\frac{dS}{dt} = K\frac{T - T_o}{T}, \quad \frac{dV}{dt} = aV \tag{14}$$

(This system differs from that considered in section 2.2 only by the relation defining dV/dt.) We combine the two equations (14) to eliminate time,

$$đg = (TV)\,đS - \frac{K}{a}(T - T_o)dV = 0 \tag{15}$$

From here on the procedure is exactly as outlined above, and we find

$$f = \frac{anR}{V(K - anR)} \tag{16}$$

so that the potential becomes

$$\mathcal{P} = \frac{KnRT_o}{K - anR}\ln V + \frac{aC_v nR}{K - anR} - \frac{\alpha a}{2}V^2 \tag{17}$$

This calculation is rather elementary, and by any standard is considerably easier than integrating the differential equations of the coresponding optimal control problem.

The algorithm for calculating $\mathcal{P}$ shown above can be used only when it is possible to eliminate explicit time dependence from the equations of motion; otherwise time must be included in the set of thermodynamic variables, and the potential will explicitly depend on the time interval. The general procedure has been described elsewhere.[2]

3.2 Tricycles

An alternative procedure[3] inspired by

steady-flow systems, for determining the maximum power (or average power per cycle) or efficiency of realistic systems involving two heat reservoirs and a work reservoir, or three heat reservoirs, is based on the energy conservation equation of the First Law and the entropy inequality of the Second Law. This method applies either to processes such as turbines with continuous heat flow or to the cycle averages of reciprocating processes such as Carnot-like cycles. One writes three heat-like flows, q_1, q_2, and q_3, representing the flows to the three reservoirs, each having a well-defined temperature T_i. The temperature corresponding to work reservoirs is taken to be infinity. Then one simply maximizes the expression for the efficiency, the power or any other quantity. This method has been applied successfully to the two-heat reservoir engine/heat pump with a Newtonian heat leak, quadratic (fluid-lubricated) friction (with respect to the power) and finite heat conductance to the heat reservoirs.

The natural independent variable for the system is the difference ΔT between the upper reservoir temperature and an effective temperature related to the maximum reached by the working fluid. The conditions yielding maximum power ensue from a fourth-degree equation in ΔT, which we have been able to solve analytically. A contour plot of the power function is presented in Fig. 4, and

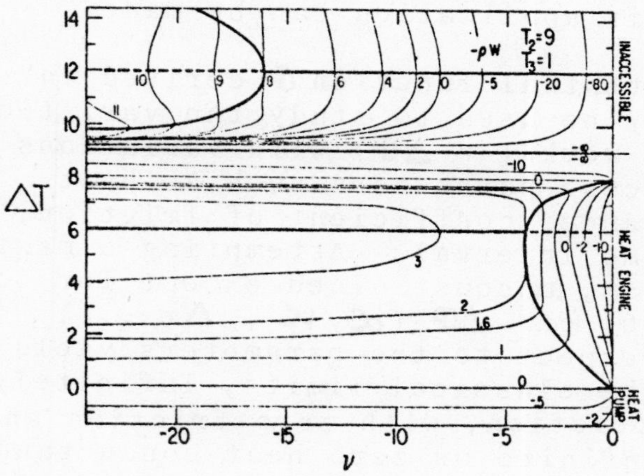

Fig. 4

Figure 4. Contour map of $-\rho W$ as a function of relative friction ($\gamma=\rho/2\alpha$) and heat exchange rate with reservoir 2 ($\Delta T=\rho q_2$) for a work-heat-heat tricycle with friction and thermal resistance. The figure is drawn for reservoir temperatures $T_2=9$ and $T_3=1$, in arbitrary units. The heavy lines indicate singularities and maxima (full line) and minimia (broken line). The region of γ near zero is the frictionally-dominated region with two real and equal maxima, and the region of larger negative γ is the resistance-dominated region with only a single maximum.

---

it seems that the solutions fall into two classes. The heat resistance is a constant $|\rho|>0$, and the friction constant is $\alpha>0$. When $|\rho|/\alpha<(\sqrt{T_{high}}-\sqrt{T_{low}})^2$, the friction dominates and the power exhibits two real and equal maxima, of course. When the inequality is reversed, the power exhibits only one maximum Analytic expressions for the power at the extrema have been derived.[3]

For the maximum efficiency, the equation is quintic, and we have not found an analytic solution to it. However, as is evident from Fig. 5,

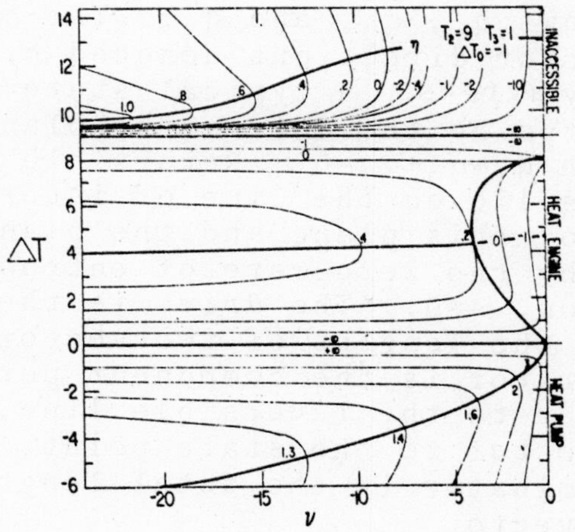

Fig. 5

Figure 5. Contour map of the efficiency as a function of relative friction ($\gamma=\rho/2\alpha$) and heat exchange rate with reservoir 2 ($\Delta T=\rho q_2$) for a work-heat-heat tricycle with friction, thermal resistance, and constant heat leak $q_0=\Delta T_0/\rho$: $\Delta T_0=-1$, which yields contours much like those of Fig. 4. However, unlike the extremal work, the extremal efficiency has a fifth root in the heat pump region of ΔT. The heavy lines indicate singularities and maxima (full line) and minimia (broken line). The extrema drawn are estimated solutions. The reservoir temperatures are the same as for Fig. 4. $T_2=9$, $T_3=1$.

the behavior of the physically meaningful roots is qualitatively similar to that for maximum power, except that the two maxima in the frictionally dominated region are unequal, and there is an additional extremum in the heat pump region.

The three quantities q_1, q_2, q_3 can be made coordinates for a three dimensional space. (See Fig. 6)

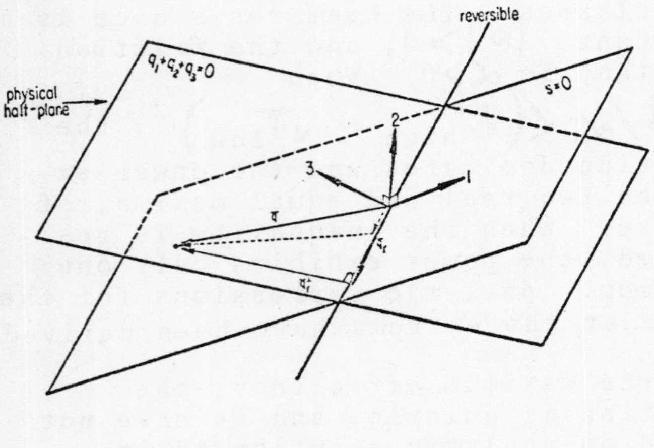

Fig. 6.

Figure 6. The space defined by the three flows q_1, q_2, and q_3. The basis vector directions denoted by the heavy arrows; a typical state is represented by the dot in the plane for which $q_1+q_2+q_3=0$. Reversible processes lie on the line of intersection of this plane and the plane defined by the zero rate of entropy production, s=0. The drive is the ratio of the lengths of two vectors: the numerator is the component perpendicular to the reversible line, of the vector to the state point, and the denominator is the total length of that vector.

Energy conservation defines one plane in this space and zero rate of entropy production by the system, continuous or cycle-averaged, defines another plane. Their intersection is a line corresponding to reversible processes. Each point in that half of the energy conservation plane lying above the zero entropy production plane defines a physical process. A quantity δ = (irreversible heat rate)/(total heat rate) which we call the drive, is a dimensionless measure of the deviation of a process point from reversibility. The value of δ depends on the choice of metric for the space, but δ is always a minimum when the efficiency η is a maximum, and $\delta \to 0$ as the efficiency approaches the efficiency of the reversible process. Thus the drive provides a quantitative measure of the irreversibility of a process.

The most immediate real system to be subjected to analysis in terms of tricycles is the Stirling engine which operates between two heat reservoirs and a work reservoir without any other exchanges with the surroundings. Otto and Rankine cycles with their continuous temperature variations can also be treated by the concept of tricycles - or conticycles in that case - after a slight generalization.

4. Applications and Prospects

By narrowing the gap between ideal reference processes and real heat processes, theory will be able to give a better understanding of the relative importance of different loss mechanisms. By using such general techniques, rather than detailed numerical simulation models, one sacrifices some accuracy in return for much simpler and more physical modeling, which emphasizes the dominant constraints and exposes the functional relationships among the characteristic parameters of the process. Using such techniques, one can predict bottlenecks in composite processes and determine the most efficient rate at which to run a thermodynamic process.

These methods can be used as powerful tools in planning the use of our energy resources. Let us explore one way this application can be made.

The potential function Φ derived in 3.1 may be used to study the variation of the work (= $\Delta\Phi$) with variations of process parameters such as heat resistance, coefficient of friction, and time interval. Attempting a maximization, unconstrained except for $\sigma_g = 0$, of $\Delta\Phi(\alpha, \kappa, \Delta t \ldots)$ with respect to the parameters yield only the classical limits, infinitely slow operation, with zero friction and with infinite or zero heat conductance. It is more interesting to optimize $\Delta\Phi(\alpha, \kappa, \Delta t, \ldots X_i \ldots)$ subject to the constraint of a fixed budget. Such a calculation reflects such extra costs as slower operation, better insulation, or better lubrication. Let $C(X_1, \ldots, X_n)$ stand for cost as a function of our process parameters $X_1, \ldots X_n$. Optimization yields

$$d(\Delta\mathcal{P}) - \lambda\, dC = 0$$

or

$$\frac{\partial \Delta\mathcal{P}}{\partial X_i} = \lambda \frac{\partial C}{\partial X_i} \tag{18}$$

for all relevant parameters X_i, where λ is a Lagrange multiplier. Note that $\partial C/\partial X_i$ is the marginal cost of the process parameter X_i and $\partial \Delta\mathcal{P}/\partial X_i$ is the marginal productivity of X_i in the generation of work, when it is generated optimally. (It will only be an approximation to suppose that the actual marginal work productivity of X_i, $\partial W/\partial X_i$, is euqal to $\partial \Delta\mathcal{P}/\partial X_i$.) Because λ is the same for all the X_i's, we can express the conditions of simultaneous thermodynamic and cost optimality for a finite-time process as

$$\frac{1}{\lambda} = \frac{\partial C/\partial X_i}{\partial \Delta\mathcal{P}/\partial X_i} = \frac{dC}{d\Delta\mathcal{P}} \tag{19}$$

In words, Eq. 19 says that for the "doubly-optimized" system, the ratio of the marginal cost of each parametric variable to the optimum marginal work productivity of that variable is a constant, the same for all the parameters, and is equal to the marginal cost of optimally-produced work.

The kind of relationship represented in (19) is of course a very general one that arises in every sort of linearized economic optimization problem[4]. The substance added in the present work is the capacity to evaluate $\Delta\mathcal{P}$ and $\partial \Delta\mathcal{P}/\partial X_i$. This capacity turns (19) from a formal truism into a useful means to test real systems for optimality. In practice, one can determine $\Delta\mathcal{P}$ and $\partial \Delta\mathcal{P}/\partial X_i$ from a knowledge of the engineeering characteristics of the system, and $\partial C/\partial X_i$ from price schedules and related data, for each X_i that one considers to be a manipulable parameter. Then, by comparing the ratios of the two derivatives, one can detect whether some parameters reflect particularly large economic efficiencies, in terms of their inefficient technological use. Obviously the identification of such parameters is a first natural step toward selecting what aspects of a process need technological improvement. Hence the application of (19) as a diagnostic tool becomes a means toward the efficient application of research and development effort.

Acknowledgement

One of us (B.A.) would like to acknowledge a travel grant from the Danish Science Foundation.

References

1. Bjarne Andresen, R. Stephen Berry, Abraham Nitzan, and Peter Salamon, "Thermodynamics in Finite Time. I. The Step-Carnot Cycle," Physical Review A. Vol.15 (in press)

2. Peter Salamon, Bjarne Andresen, and R. Stephen Berry, "Thermodynamics in Finite Time. II. Potentials For Finite-Time Processes," Physical Review A. Vol.15 (in press)

3. Bjarne Andresen, Peter Salamon, and R. Stephen Berry, "Thermodynamics in Finite Time: Extremals for Imperfect Heat Engines," Journal of Chemical Physics, Vol. 60, No. 4, 1977.

4. c.f. P.A. Samuelson, Foundations of Economic Analysis, (Harvard University Press, Cambridge, Mass., 1947), p. 60.

ENERGY CONSERVATION IN CHEMICAL REACTIONS

Some thermodynamical Aspects.

by Hon. Prof. R.A. BIDARD

C.E.N. 12, rue Partalis PARIS 8ème.

1. The energy crisis leads us to put more and more emphasis on energy conservation, and we will examine here from this general point of vue the case of chemical reactions.

It must be well understood at first that when dealing with energy, it is not sufficent to speak of heat : we must also take great care of the temperature level of that heat, as well when that heat is put into the process, as when it is an output from it. For, if the first principle of thermodynamics tells that heat and energy are equivalent, and that their sum is conservative, the second principle tells that heat cannot be totally converted in mechanical energy, but only the part $(1 - \frac{T_o}{T})$ Q of it (available energy), the rest $\frac{T_o}{T}$ Q having to go to the environment (to the heat sink), and having to be wasted.

These laws are universal, and therefore apply to chemical reactions too.

This is very well known of course, but is seems that, when chemical reactions are considered, all the implications have not been completely emphasized : fot instance we will show that the heat to be furnished theoretically to an endothermic process, even when perfect, is substantially greater than the so called "heat of reaction" of the chemical equation. You must take notice here that I have said "to be furnished to the process" and not "to be furnished to the chemical reaction" and this makes all the difference: the "process" implies not only the reaction itself, but all its environment, i.e. the thermodynamical cycke according to which that reaction is functioning and relates with the outside world: such cycle supposes occasionally the rejection of waste heat to the heat sink, even if it is as perfect as possible theoretically. And we will show that, at least when one wants to get a good energetic efficiency, such a thermodynamical cycle is absolutely necessary, as it is the only way to discard heat at the lowest possible temperature level, i.e. at the temperature T_o of th environment.

Now we must define our terms: when we speak of energetic efficiency, we speak of an efficiency quite different from the "thermal efficiency". This latter compares two quantities of heat, one beeing considered as spent, and the other beeing considered as obtained usefully; and this without any reference to the temperature levels of these heats. On the contrary, the energetic efficiency is made of the comparison of the sum of all the "available energies" that have been furnished to the process, either in the form of mechanical or electrical energy, or in the form of chemical potential energy, or finally in the form of the "available energies" of the heats furnished, compared to the sum of all the "available energies" that have been usefully extracted from the process, in any of all the forms cited above.

Now it is clear that, speaking of "energy conservation", only the "energetic efficiencies" (named sometimes "exergetic efficiencies") make any sense, because they do take care of the temperature levels of all the heats implied: As for the heat inputs it is not all the same to furnish heat at 900°C to a reaction, or to furnish it at 150°C. In the first case we could have produced electrical energy with a thermal efficiency of say 45% with that heat, and in the second case the thermal efficiency would reach only 15% may be.

And the same applies to any heat output if we intend to utilise that heat. Further more we will see that, as far as potential chemical energy is concerned, it is not correct to consider it as "heat": only the available part of the so called "heat of reaction" has to be considered (although, practically those two are often not much different).

2. This beeing settled, we must now go a little further in thermodynamics.

It is well known that chemical reactions are governed by an energy equation of the form:

$$\Delta H = \Delta G + T_o \Delta S \qquad (1)$$

This means that the energy of the reaction ΔH (the so badly called "Heat of reaction") which is to be supplied to an endothermic reaction, or to be produced by an exothermic one, has to be (for a perfect process) of two kinds, one part only $T_o\ \Delta S$ beeing heat, and all the rest ΔG beeing "available energy" of a mechanical (or electrical ... etc) kind, i.e. "noble energy" in the sense of the second law of thermodynamics. And this latter part is generally much greater than the former $T_o\ \Delta S$ one.

The best way to obtain $T_o\ \Delta S$ is to exchange heat (either positive or negative) with the environment at temperature T_o (thus this amount of heat may be minimised). This term is generally much smaller than ΔG.

To obtain ΔG things differ whether we speak of endothermic or exothermic reactions.

If we consider an endothermic reaction, we must produce and spend this amount of work: this means that, if we have only heat at our disposal, the process must imply a thermodynamical cycle able to produce and spend internally that amount of work, at the cost of an imput of external heat.

This implies too that we must the consume (Carnot second law) an amount of heat greater than ΔG, i.e. theoretically the quantity, in the best case:

$$Q = \Delta G + \int \frac{T_o}{T}\, d\, Q \qquad (2)$$

This can be much greater than ΔG if T is in low values range ! And, practically it will be still much more, as we are not able to reach the Carnot efficiency ! In any case this means that mechanical machines have to be inserted in the process.

In the exothermic case, the above means that we should produce and export (theoretically) an amount of external work almost equal to the so called "heat of reaction" (the term $T_o\ \Delta\delta$ being generally small, as already mentioned. And, theoretically too, this implies here again that mechanical machines be inserted in the process, in order to build a true thermodynamical cycle.

Not it is always possible in both cases to avoid the complication of inserting such mechanical machines, but at a certain cost.

An endothermic process may stay isobaric indeed, but we will see that, without the aid of mechanical machines, an amount of heat much greater than the theoretical one (resulting of eq (2)) will be necessary, and that either one will have to produce and export mechanical energy with the surplus, or waste it. As for the exothermic case, it is naturally always possible to produce heat instead of mechanical energy, but then this energy will be almost completely wasted too.

Finally any chemical process should imply mechanical machines to approach the maximum possible energy conservation.

3. We will now try to figure how such considerations may apply practically. We will first consider perfect processes, i.e. fully reversible ones, this permitting us to build a model, to which we could try afterwards to fit at best.

And we will consider only gaseous processes, as they are much simpler to analyse, and will not try to analyse complex processes impliing interfaces between multiple phases. It must be precised that we do not intend here to be exhaustive: the following should be considered only as a first approach to some thermodynamical aspects. So that, althought the considerations developped below apply also, in a way, and at least partially, to very complex cases, we will not consider them and will limit ourselves to thermodynamical considerations on the simpler case of gaseous components only. And, as it is always advisable to refer to a specific case, we will take as an example the combustion of H_2 and the splitting of H_2O : but any other gaseous reaction could have been chosen, and our conclusions will remain general on that respect.

4. Now we must try to figure how a perfect cycle including a chemical reaction looks like.

Actual chemical process are mostly isobaric, and that is why we will examine first the isobaric case. It is possible to represent the process in a "Temperature - Entropy" chart, providing that, for each temperature, we represent the sum of the entropies of all the flow components, be they mixed or separated.

And if we want to draw isobaric curves, we may use the convention that we represent the sum of the partial pressures of the components, even when separated. We will see that such a convention may permit useful display of the whole process.

So that we will examine first, as said, the reversible process of an isobariv, splitting of H_2O. That process must start at a temperature T_o and go back with the components O_2 and H_2 to the same temperature T_o. For the sake of simplicity we will not consider the domaine of liquid H_2O (and not more those of H_2 and O_2). These two phases situations bring some more difficulties to reach reversible cycles, but those difficulties are not tied here to any chemical situation, and are quite classical. Finally we will then suppose that we have already steam at T_o. We can draw the isobar ACD of that steam. (Figure 1).

Along A C D steam is heated to a very high temperature, where splitting begins to occur. The reaction is considered as complete in E. In G, H_2 and O_2 are supposed to be separated by a reversible process : no entropy changes occurs. (In fact, this would have to be done by reversible chemistry, or reversible cryogenics, coupled with an internal heat exchange coming back to T_G). The along G H I J K, the two gases, separated, are cooled back to T_2. Thus we come to point K. This point, as already mentioned, is not far from point A, for these two states are linked as said by the relation

$$T_o \, \Delta_A^K S = \Delta H - \Delta G$$

And it is well known that such enthalpy change ΔH and free energy change ΔG, related to room temperature and pressure, differ only by few per cents. The whole figure must then present the distorted crescentlike shape shown, owing to the fact that the internal area K A C D E I J K must be equal to ΔG, and the internal area k a A C D E I J K k equal to ΔH, (1) i.e. not much different from the former. In this figure the small term $T_o \, \Delta\delta$ is the difference of the two former, and is represented by the area aAKk.

Now we must imagine that the maximum possible of reversible internal heat exchange is realized between way up and way down.

Here we must introduce the concept of "perfect heat exchanger": such a device should exchange heat without heat losses, without temperature gaps, and without pressure losses: It is easy to see that, in such case, the two ways followed in the T.S. diagramm, by the cooled hot gas and the warmed cool gas respectively, should be "parallel" (more precisely should be superposable by a displacement parallel to O S): for, for any small exchange of heat d Q, at any temperature T, we will have the same $d\ S = \frac{dQ}{T}$ with opposite signs in both flows, and the dT will also be equal with opposite signs if no temperature gap is to build up between the two flows.

Now it is obvious that this condition cannot be fulfilled in Figure 1 owing to the shape already mentionned. The situation on this respect is represented figure 2. Points D and I are the one where the two isobars are parallel to each other. From A to D it is possible to heat the steam by heat exchange, cooling thus the products along IK. But, as the way up isobar is steeper than the way down one, there is too much heat available: it would be sufficent to cool those products along a way down line I J' K' parallel to DCA : this means that a surplus of heat H_r, measured by the area mK'J'IJKm, has to be rejected outside the process.

Further on from D to E, it is possible to recover all the heat of the products, cooling them from H to I but this amount of heat is not sufficent as, in this region, it is the way down isobar which is steeper than the way up one : one must furnish a surplus of external heat H_f, measured by the area dADEIJ'K'k'.

This area is equal (of course!) to the sum of the so called "heat of the reaction" H plus the rejected heat H_r (compare to figure 1).

$$H_f = H + H_r$$

So that the quantity of heat H_f to be furnished is, in such case, much greater than the "heat of reaction" H.

As for the quantity of heat internally exchanged, it is measured by the area k'K'IEe (if the process goes up to point E only).

If we want to maintain an overall thermal autonomy and obtain a good energical efficiency as well, we must utilise the rejected heat, and we will see that to that end, we will have to link our processes with energetical cycles, (that is to mechanical machines), as it is the only way that they may be able to reject heat only at T_o.

This may be done in different ways.

According to a first general class, we may utilise this rejected heat, still at high temperature, to produce external work, electricity for instance. But is must be understood indeed that, in the perfect isobaric case considered, there is no possible internal use of this energy inside the process. We get then no more a pure hydrogen producing process: even an endothermic chemical isobaric process would have to produce external energy as well as chemical products to remain reversible ! This is not exactly what we want: we want to spend the minimum possible amount of heat, and produce chemical goods and nothing else.

---

(1) For $Q = \int d\ Q = \int TdS$ = internal area k a A C D E I J K k = ΔH.

This is possible through a <u>second general class</u> of solutions: one may introduce a heat pump effect that may rise the temperature of that rejected heat to such a value that it could be reinjected into the process, ths replacing a great part of the former external heat input: the process will then become thermally autonomous, but will consume some work. But we will see later (see § 9) that this work may be obtained thanks to an input of heat. In any case we will need mechanical machines. A first solution of the latter case should be to remain isobaric, as supposed above, and link the process, thanks to heat exchangers (heat going either in or out) to a side thermodynamic cycle, to which all mechanical machines would belong, and in which an adequate working fluid would circulate (may be the helium of a high temperature nuclear gas reactor, as steam seems not to be an ideal fluid in such case, owing to the boiling and condensing evolutions along which much heat is to be put in or out isothermally, with no possible reversible couterpart on the chemical side).

Another solution is to utilise the gazeous products of the chemical reaction (s) themselves as working flui (s) for the thermomechanical duties as well (i.e. to flow in the mechanical machines).

Anyhow, here again, we need mechanical machines.

5. Now we may examine <u>exothermic</u> reactions: the reversible process will be the one Figure 1 but run the reverse way: we will then have an output of high temperature heat along E D and an <u>input</u> of lower temperature heat along J I. This time we could obtain the heat needed along J I as a part of the one at disposal along E D. But it would be energetically costly owing to the temperature gap between the two, and the best thing to do, here again, would be to produce energy with that heat before reinjecting it into the main cycle. And here again this could be done either in a side cycle, either in the main cycle itself, but in any case it would imply here again mechanical machines.

Finally we can conclude that any chemical process, either endothermic or exothermic, needs mechanical machines, working along a true thermodynamic cycle, to remain perfect.

6. Now it is interesting to examine in further detail what those thermodynamic cycles look like.We will limit ourselves to the case where the gaseous products of the reaction are also the working fluids in the mechanical machines in that case it is possible to draw the whole process in the same and unique T.S. diagramm.

We will describe the cycle for the endothermic reaction first, and do it in two steps: first we will suppose that we furnish only mechanical work to the process; and we will examine further the case when we have only heat at our disposal.

7. To the first step, we may have an input of work instead of heat along DE if we compress the fluid in that region, and have an output of work instead of heat along I J if we expand the fluid in said region, thus going back to the same initial total pressure. But the expansion work will be smaller than the compression one, as it takes place at lower temperature. So that we have realised two things at the same time: first a sort of very sophisticated internal heat pump, and second a zero input of high temperature heat but instead the input of an amount ΔG of work. This amount of work depends on the pressures prescribed for the products H_2 and O_2. We will suppose, for the sake of simplification that their sum is atmospheric pressure, as it is the case in Figure 1 (this could be easily modified if necessary).

We may now draw Figure 3 the whole thermodynamic cycle.

Such a cycle must have a zero internal area : no external heat is provided.

All the heating and the cooling are obtained through reversible internal heat exchange. The isobas have been shown according to Figure 1 : the way up isobars (on the left of Figure 3) have a shape analogous to A C D E G (Figure 1), and the way down isobars (on the right of Figure 3) a shape analogous to G H I J K (Figure 1).

One sees well where mechanical work is necessary: where the paths cross some isobars.

The figure has been built in such a way that internal heat exchange should be a maximum (and internal work exchange a minimum): to that end the common path follows at each point the steepest isobar of the two that cross each other at said point.

One sees that, if now no heat is to be provided at high temperature, a small amount of heat remains to be exchanged at T_o (along A B) in the form of a small compression, to take care of the term $T_o \Delta S$.

One sees also how the heat pump effect takes place: we compress, as said, at higher temperatures than we expand. But this effect is here very sophisticated as the laws Q (T) according to which heat must be extracted (by expansion) and reinjected (by compression) are very definite and not at all isothermal.

Furthermore, in that case, this heat pump effect furnishes all the high temperature heat that is necessary, thanks to its external work consumption.

8. And now the case of the exothermic reaction becomes obvious too : it is the cycle Figure 3, but run in the reverse way. Such a cycle will produce (instead of consume) the amount ΔG of work, i.e. an amount close to the "heat of reaction". In other words it is a perfect internal combustion machine that produces an amount of work almost equal to the heat content of the fuel ! This seems to violate the second principle, but it does not at all, as the second principle has nothing to do here : what we furnish is not heat, but potential chemical energy, which has an "available energy" almost equal to the so badly called "heat of the reaction".

There is no reason indeed why a perfect thermodynamical cycle would not be as perfect as a perfect fuel cell cycle: the perfect reversibility must have the same consequences for both !

On the contrary, if it was not so, we could violate the second principle in the following way : we could split H_2O chemically by thermolysis, i.e. by heat input, furnishing only the heat of the reaction, and then recombine H_2 and O_2 in a reversible fuel cell, converting thus totally the heat input in electrical energy !

Now one must not think that we could really reach a thermal efficiency of 100% (or even a little more depending on the sign the term $T_o \Delta S$): reversible combustions need extremely high temperatures, completely beyond our industrial reach !

But anyhow those considerations are of paramount importance when chemical exothermic reactions may take place at attainable tempereratures.

9. The truly autonomous perfect endothermic process.

We can easily see now how could be built a perfect endothermic process consuming only heat, with no work input nor output, and no heat outpout either (except at T_o) : the mechanical work involved in Figure 3 must be produced internally, in the cycle itself. This implies a heat input, an internal area, and a heat rejection at T_o. The solution is given Figure 4.

This figure is built supposing that all heat is available at a fixed temperature, higher than the chemical reaction temperature zone (but it could well be otherwise (see further).

The heat is provided thanks to an isothermal (i.e. reheated) expansion along F G, and the heat is rejected at T_o thanks to an isothermal (i.e. cooled) compression. These are such that they compensate exactly the difference between the mechanical work along D E and I J. The cycle is now made mechanically autonomous and receives heat only (high temperature heat in the case).

Here also it is interesting to point out that, on Figure 4, part of the mechanical machines function once more as a sort of very sophisticated heat pump, able to recycle the rejected heat of Figure 1, and reinject it into the same cycle, replacing thus a big part of the input heat represented in Figure 1. But here the heat pump functions thanks to a supplementary heat input at the maximum temperature T_1, instead of an external work input. Now, as said, it could very well be that this heat be furnished at a smaller temperature, or even with any other Q (T) law differing from the isothermal case: the shape in the T S chart would then be somewhat different, but conclusions would remain.

10. Now it must be said that these very sophisticated processes are very far from actual practice in chemistry: thus it is interesting to try to evaluate the cost we have to pay to avoid such complexity.

It must be emphasized at first that reversible chemical reactions are theoretically impossible owing to the kinetics involved.

Secondly irreversibilities are also theoretically necessary in heat exchangers (temperature gap necessary for the kinetics too). And finally mechanical machines are not perfect.

But this situation is only a good reason to try to avoid any other cause of irreversibilities.

Of course there is always, in such cases, a balance to realise between thermodynamical losses and investments in machines, heat exchangers, and so on. On that respect power plants are a good example, as they have followed secular improvements on that very question. To get a figure of merit one can compare the electrical output of power plants to the available energy input, i.e. the one of the input of heat. This "energetic efficiency" is actually better than 70%. Such a figure could be a goal for chemical processes too.

But actual practice is very far from such a figure: it is more close to 20% for a complex process than to 70%. And there are probably very good reasons to that: one of them is the price of the high temperature vessels, of the catalysators and so on : these components could not be much increased (to come closer to chemical equilibrium), the investments would be too big.

But an other one is that, generally, chemical processes do not involve internal mechanical machines and remain mostly isobaric. We have seen that in such case we should produce external energy. But generally we do not intend to do it, so that all that all that heat is wasted and has to go to the river. It is true that in many cases we try to produce steam, and to run some steam turbines with it. But even in such cases, for example in the synthesis of ammonia, these machines do not function as described above : they do not produce any external work as they should do in that case, and they only recycle internally some work, mostly for chemical purpose (high pressure in the synthesis vessel) and not for thermodynamical one.

Finally we may conclude. Actual practice is to accept very low energetical efficiencies (thermal ones are not considered here) in chemical processes. It exists means to improve substantially this situation, and we have described the general lines of some solutions. Their adoption may lead to increase the investments (specially in machinery).

But when one is faced with a sudden increase in fuel cost, one must anyhow work a new economical balance between consumptions and investments, in which consumptions have to be reduced substantially, and investments increased sufficently, so as to reach a new minimum overall cost.

R. BIDARD.

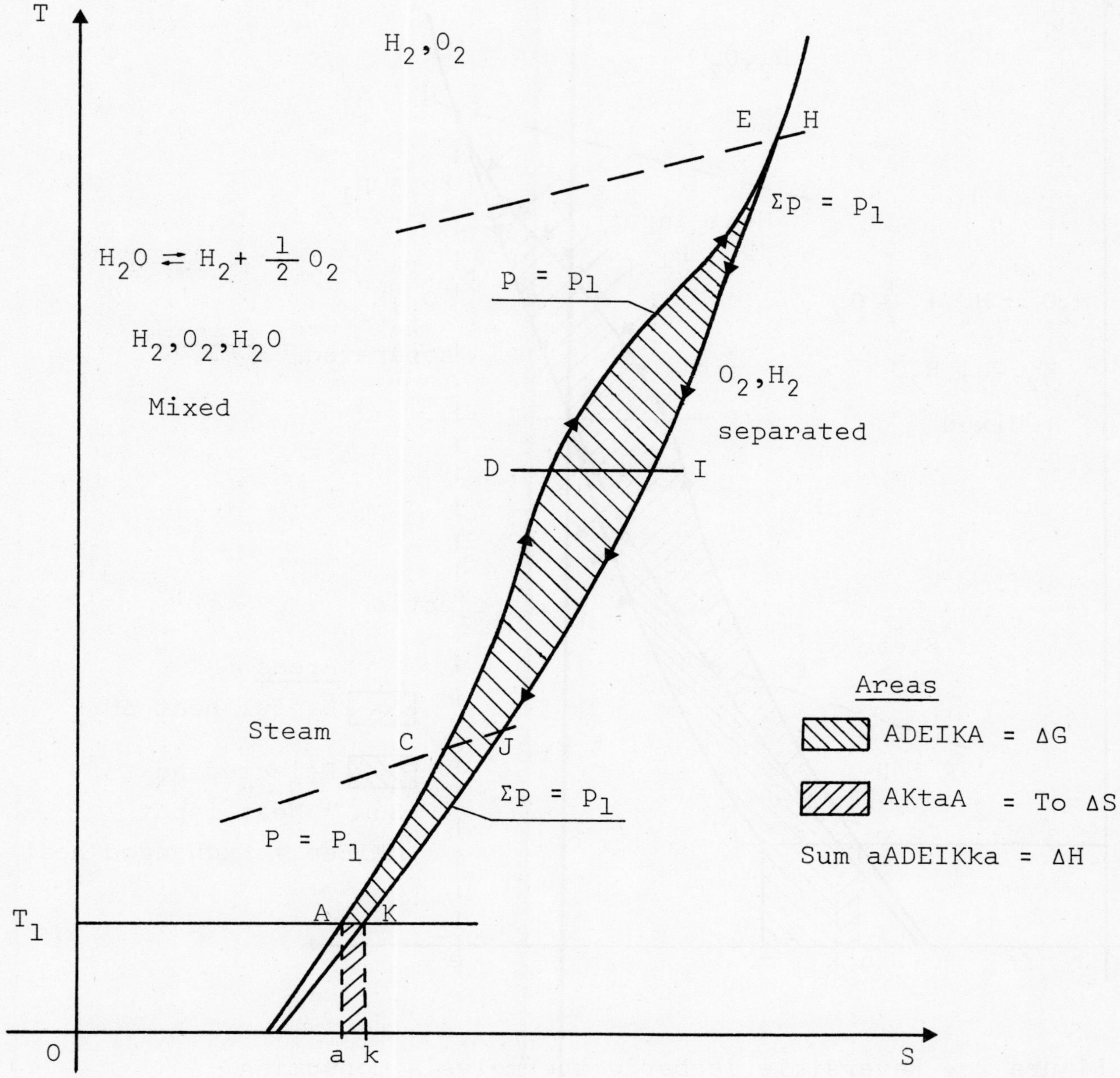

Figure 1 - Reversible isobaric thermolysis consuming and rejecting heat only.
Energies of the reaction.

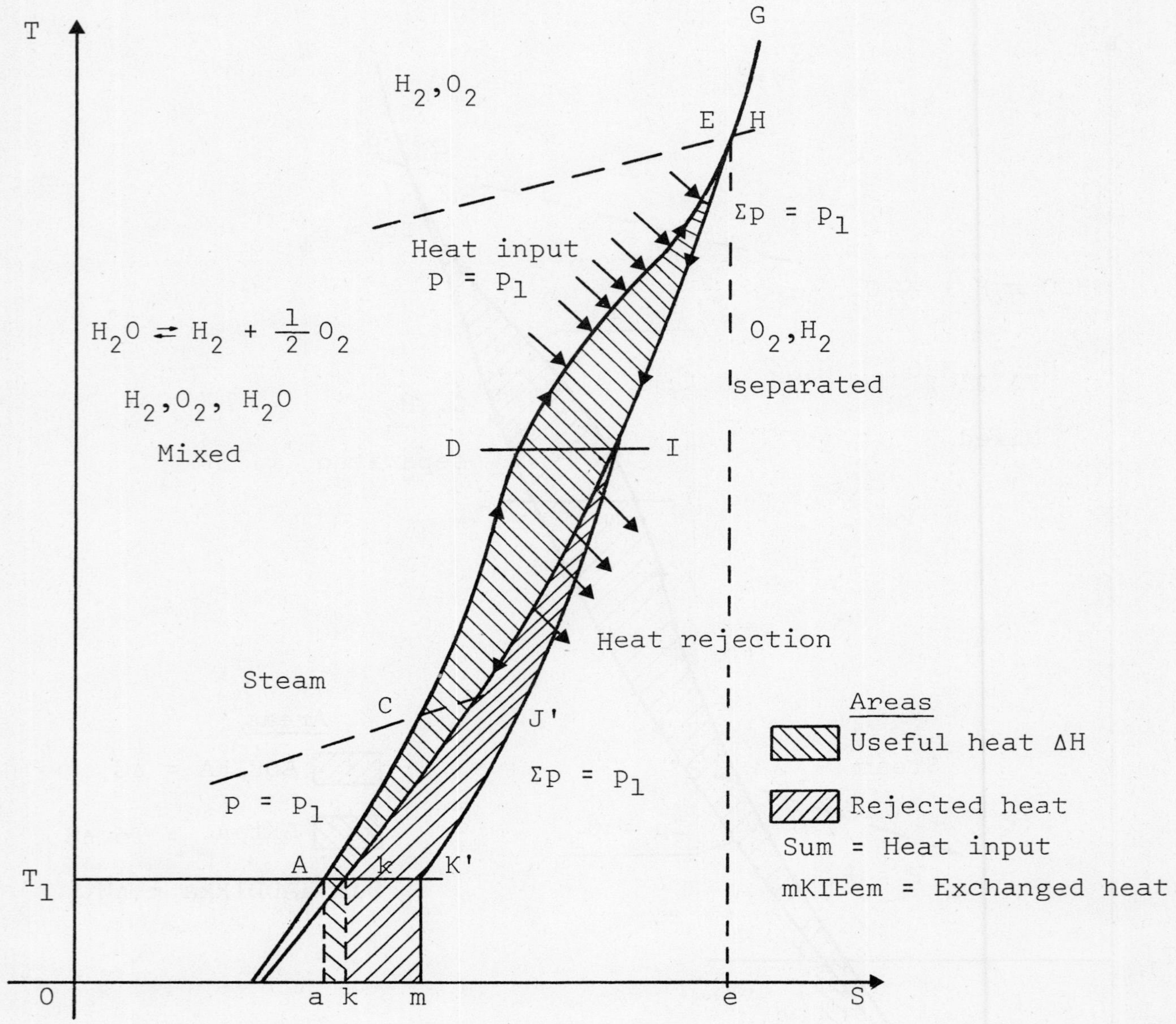

Figure 2 - Reversible isobaric thermolysis consuming and rejecting heat only.

Heat input, heat rejection, exchanged heat.

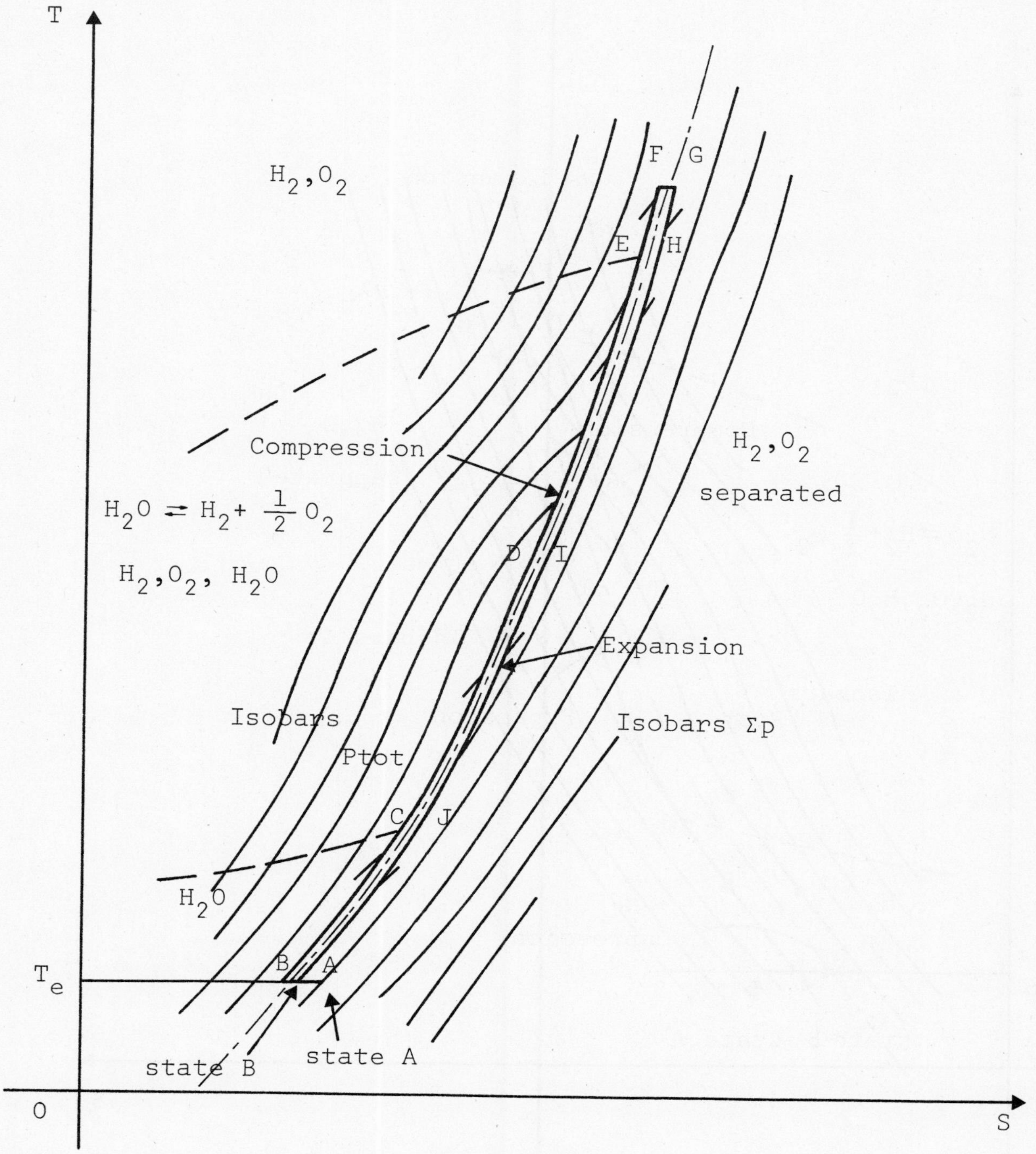

Figure 3 - Thermomechanical reversible dissociation of H_2O with work only as input and maximum internal heat exchange.

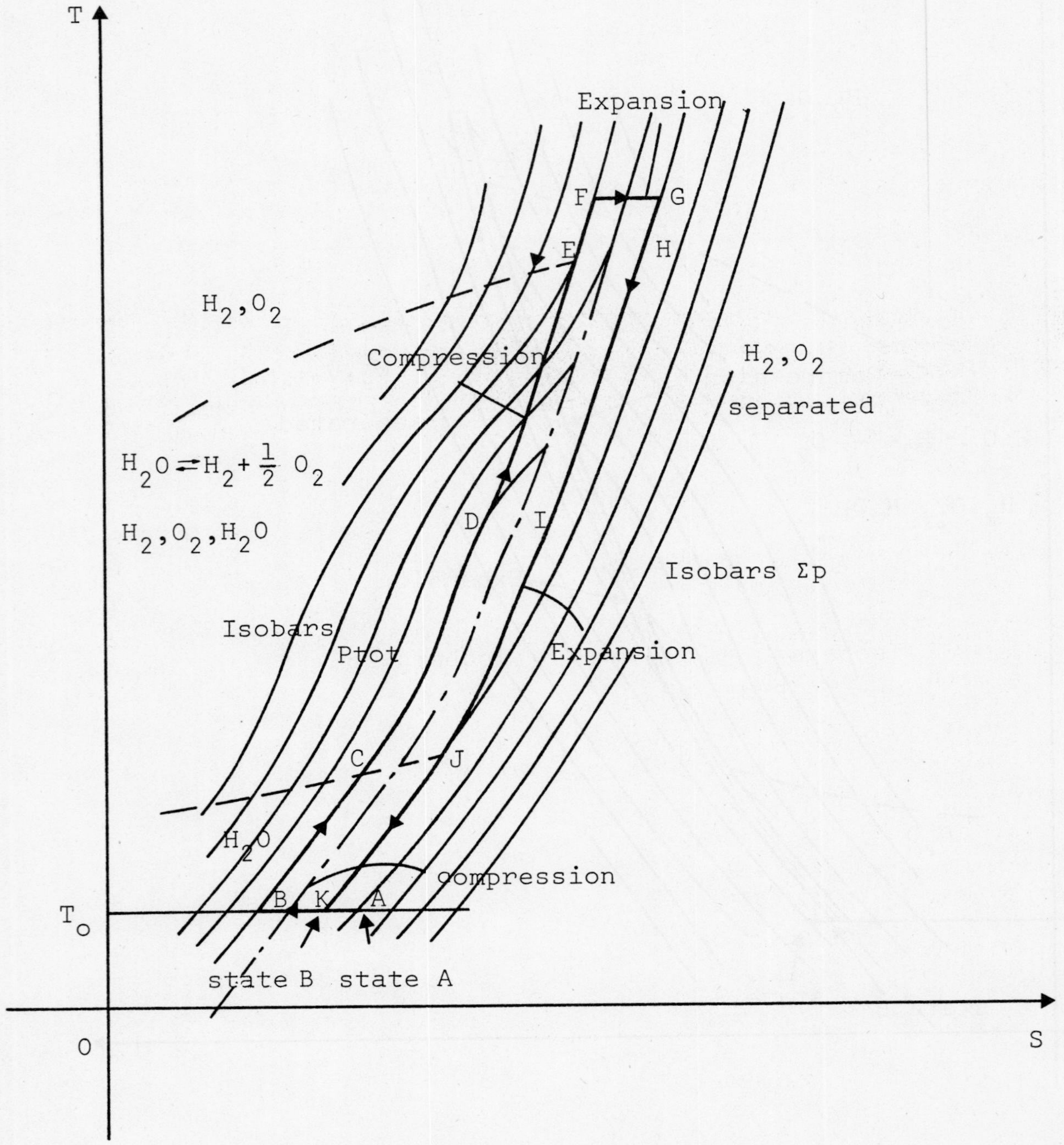

Figure 4 - Reversible thermolysis of H_2O with heat only as input and maximum internal heat exchange.

ENERGY EFFECTIVENESS MEASUREMENT FOR INTEGRATED ENERGY SYSTEMS[1]

James M. Calm, PE

Energy and Environmental Systems Division,
Argonne National Laboratory, Argonne, Illinois 60439

ABSTRACT

Performance comparison of integrated energy systems requires a measure which accounts for the diverse energy qualities of multiple inputs and outputs. A performance measure for comparison of alternative integrated systems is presented; this measure is based on consideration of dissimilar output enthalpies, varying load demands for each output, and different energy sources used as inputs. Conventional efficiency measures and the recently introduced energy effectiveness factor, Ee, are discussed. The resource utilization factor, RUF, and resource impact factor, RIF, concepts are applied to the proposed resource based coefficient of performance, RCOP.

INTRODUCTION

Concerns for depleting energy resources and growing energy demands have led to increased consideration of integrated energy systems, systems which combine energy functions to minimize resource requirements.[2] By utilizing byproduct or waste energy through cogeneration and cascading concepts, actual fuel consumption compared to independent conversion systems can be reduced. Illustrations of these systems are existing total energy systems in which heat is recovered from electrical generator engines, district heating systems utilizing low pressure steam from turbine generators, and plants which use common boilers to provide steam for heating, electrical generation, and process needs. Integrated energy systems may utilize a broad selection of conventional fuels as well as less common resources such as refuse in incineration or pyrolysis to produce usable energy for electrical generation, heating, cooling, domestic hot water heating, cooking, snow melting, or other services. Because of the integrated system sizes, they are generally better candidates for storage approaches to peak reduction and appear to offer greater potential for annual storage cycles than unintegrated systems. These factors combined with advantages of load diversity and economics of scale suggest a strong potential for energy conservation in integrated energy systems.

Performance comparison of alternative energy systems requires an evaluation function which measures the effectiveness of input resource utilization. For systems which convert energy from one form to another or transmit energy from one location to a second, a ratio of useful outputs to inputs is the generally accepted measure of efficiency, η. A simple comparison of η values, however, does not account for differences in resource and product qualities or economic values for alternative systems for which either the inputs or the outputs are not of identical form. η comparison for systems with multiple inputs or outputs is not valid for similar reasons. This paper addresses the need for a measure which affords comparison of these types of systems with primary attention to integrated energy systems, discusses several approaches, and presents an evaluation function useful in these comparisons.

SYSTEM INPUTS

In a comparison of two systems producing the same product, such as electricity, it would be insufficient to know the conversion efficiencies alone unless both used the same fuel. The same electrical generation process might utilize multiple fuels as in a steam electric plant in which some steam were derived from refuse incineration and the remainder from a fuel-fired boiler. In this case, a simple conversion efficiency might be calculated as:

$$\eta = \frac{E}{R + F} \qquad (1)$$

where: E = electrical output (expressed in thermal equivalent form),[3]

R = heating value of refuse utilized, and

F = heating value of fuel utilized.

η would not, however, be very informative as an evaluation function since it would be lowered as the portion of refuse utilized, thereby conserving fuel, was increased. This lowered η would be indistinguishable from the reduced efficiency resulting from poor combustion requiring an increase in fuel.

A simple approach to improving the η measure might be to ignore the refuse heating value in calculating the efficiency, but this leads to two objections. First, the efficiency η is a performance measure based on the first law of thermodynamics. If η were calculated as E/F, values in excess of 1.0 would be possible which would be inconsistent with the first law. This objection has been previously resolved for heat-pump and refrigeration cycles, for which the "free" ambient source energy inputs and controlled space energy removed are not counted as fuels in the performance evaluation, by use of a coefficient of performance, COP. A COP measure relates useful output to valued inputs in similar fashion to η, except that ambient heat is considered to have no value and the COP is therefore not a first-law based measure. The second objection is that the refuse should be considered as having a value. While it is generally considered a liability requiring disposal at present, as conventional fuel prices increase and refuse increases in acceptance as a fuel, refuse may be considered as a non-depleting energy source. Likewise, if the same plant as in the previous example were to use two fuels such as coal and oil, it would be inappropriate to either ignore one or treat them as equal in measuring performance. A resource weighting scheme is therefore needed.

In the development of Section 12, *Annual Fuel and Energy Resource Determination,*[4] of ASHRAE Standard 90, *Energy Conservation in New Building Design,*[5] several factors were considered for adjustment of fuel values. A Resource Utilization Factor, RUF, was defined as:

"A multiplier applied to the quantity of energy delivered to the site, which provides quantitative estimates of the resources consumed in providing that energy. This multiplier accounts for the burdens of processing, refining, transporting, converting, and delivering energy from the point of extraction to the building site."[6]

Use of the RUF adjustment would be a convenient method of considering resource consumption at the point of extraction rather than at the point of utilization.

A second factor, the Resource Impact Factor, RIF, was also considered to reflect the social value of the resources, that is to extend the measurement basis from the point of extraction to the raw resource form. The RIF was defined as:

"A multiplier applied to the fuel and energy resources required by a building project to permit evaluation of the effect on non-renewable resources resulting from the selection of fuel and energy sources giving consideration to time, location, economic, environmental, and national interest issues."[7]

Thus, when a quantity of fuel, F, is multiplied by the appropriate RUF and RIF values, the resultant FxRUFxRIF term represents the true social and resource value of the fuel required.

Conceptually, this term would be a convenient basis for evaluating effectiveness of the multiple input energy systems or for comparing alternative single input systems utilizing different fuels as follows:

$$RCOP = \frac{\text{useful outputs}}{\Sigma\ (F_j x RUF_j x RIF_j)} \qquad (2)$$

where: RCOP = the resource based coefficient of performance,

F_j = specific fuel inputs,

RUF_j = the resource utilization factor associated with F_j, and

RIF_j = the resource impact factor associated with F_j.

For a system using two fuel inputs, the RCOP would be:

$$RCOP = \frac{\text{useful outputs}}{F_1 RUF_1 RIF_1 + F_2 RUF_2 RIF_2} \qquad (3)$$

In the illustration previously used (Eq. 1) where one of the fuels is refuse, assume that the other input source is oil. The dimensionless RUF and RIF multipliers would increase the magnitude of the oil input term to reflect the resource and social values of that fuel. The refuse term, however, would be decreased to reflect its lower intrinsic value. As a result, the anomaly previously cited of an inverse relationship between the apparent efficiency and increased refuse utilization would be corrected. It is conceivable that the refuse term would be eliminated if the RIF factor for refuse were set at zero to reflect no social value or that in an extreme case the refuse term would be negative (a credit) to reflect the functional value of refuse disposal compared to energy consuming landfill disposal or disposal incineration without energy recovery. The important point for this illustration is that relative qualities or values of system inputs may be considered with this approach.

This point gains further significance for heat pumps where heat from the ambient air is utilized or for cooling systems where heat from the occupied space is rejected to the ambient. The inherent assumption is that these thermal sources have no intrinsic value and their RIF values are zero; as such the RCOP and conventional COP values are identical for simple heat pump or air conditioning systems. If, however, one wished to associate a non-zero value with ambient energy sources for environmental impact considerations or because the system under study was completely contained within a separately controlled system (such as a room air-conditioner serving an office in the core of a large warehouse with independent heating and cooling), the RIF values could be appropriately set to account for the otherwise disregarded ambient or room air energy values.

While the RUF and RIF adjustments seem valid from a theoretical viewpoint, their application requires quantitative values upon which considerable disagreement may exist. Although not identified as RUF approaches, the same concept of considering energy consumed from the point of extraction to the point of use has been incorporated in a number of recent energy cycle studies.[8,9] ASHRAE has approved a method of determing RUF values and has provided tabular values for use when locally developed values are not available.[10] Revisions to these values may be anticipated after they come into more popular usage, but the Section 12 values represent the best present data in this regard.

Accepted RIF values are not currently available, although several approaches to developing them have been considered.[7,11] While the RUF concept lends itself to empirical determination, quantification of RIF's is a more subjective venture. It seems apparent that sufficient interest in the RIF concept exists to expect a consensus formulation in the near future.

SYSTEM OUTPUTS

Having addressed the inputs of systems, attention is now focused on the outputs. It should be obvious that comparisons of systems with different outputs or different proportions of multiple outputs (different product mixes) would favor the system with the greatest output proportion of an energy form than can be efficiently produced. For example, conventional technology approaches to converting gas to heating offer higher (approximately double) efficiencies than for converting gas to electricity. For two gas systems with one producing 75% heating and 25% electricity and the second producing 50% heating and 50% electricity, the first system would almost certainly have a higher useful output to input efficiency if a unit of electricity were treated as equivalent to a unit of heating. On this basis, a comparison would be strongly biased by the produced mix.

The problem of comparing systems with different single or multiple outputs may be simply dispensed with by questioning the validity of comparing alternative systems with different outputs. For alternative system comparisons for the same application, the application defines the required outputs and the resource based coefficient of performance becomes:

$$\text{RCOP} = \frac{\Sigma\ Q_i}{\Sigma\ (F_j x\text{RUF}_j x\text{RIF}_j)} \qquad (4)$$

where: Q_i = specific outputs expressed in units consistent with F both as to dimensional units and as to period of measurement.

The validity of this approach is contingent upon acceptance of the restriction against comparing systems with non-identical outputs. This limitation may be overly restrictive in some cases as two systems designed to meet the requirements of a single application may both produce identical outputs with one or both producing an additional byproduct.[12] Simple inclusion of the additional output in the ICOP calculation would unjustifiably favor systems which produced large quantities of byproducts with high inherent conversion efficiencies, i.e. the product mix would be significantly altered thereby biasing the evaluation. Output weighting is therefore suggested.

One obvious weighting factor might be market values of the outputs. To the extent that market values are inconsistent with thermodynamic properties, which is often the case, a performance evaluation which is the subject at hand would be distorted. A weighting based on thermodynamic properties would be rather complex and would again favor those systems with large proportions of outputs with high inherent production efficiencies. A resource approach is again required.

Since the system outputs are being delivered at the point of performance measurement, the output RUF's would all be unity. This conclusion is arrived at by recalling that RUF's were introduced to adjust resource quantities measured at the point of consumption to the point of extraction. Since the point of extraction for system outputs is the system itself, a RUF adjustment is not needed for outputs.

By applying RIF adjustments to system outputs, the previous restriction against performance comparisons of systems with varying product mixes is eliminated. Equation 4 therefore becomes:

$$\text{RCOP} = \frac{\Sigma\ (Q_i x\text{RIF}_i)}{\Sigma\ (F_j x\text{RUF}_j x\text{RIF}_j)} \qquad (5)$$

where: RCOP = the resource based coefficient of performance for integrated energy

Q_i = specific outputs expressed in units consistent with F; both as to dimensional units and as to period of measurement,

RIF_i = the resource impact factor associated with Q_i,

F_j = specific fuel inputs,

RUF_j = the resource utilization factor associated with F_j, and

RIF_j = the resource impact factor associated with F_j.

As previously discussed, accepted RIF values are not presently available, but can be expected to emerge in the near future. It is probable that RIF's will initially be developed for conventional fuels only and the RCOP measure will require additional RIF determinations for both unconventional fuels and for energy forms generally recognized as in usable form rather than as raw resources. However, once consensus is achieved on the methodology of calculating RIF's, determination of the additional RIF's should be straightforward.

It should be noted that while use of the RIF adjustment affords comparison of alternative systems with nonidentical outputs, the system products should be similar if the comparison is to have validity. The approach was introduced as a method of dealing with systems producing the same basic outputs with one or more systems also producing additional byproducts not common to the others. While the RCOP measure introduces an approach to comparison of systems with totally dissimilar outputs, comparisons of such systems is still of questionable value.

Treatment of cooling as an integrated energy system output has caused some concern. While cooling may be considered an output in the product sense, it represents an energy input to the system in a thermodynamic sense. One proposal, the energy effectiveness factor Ee, has been offered which would substitute a computed "product value" for the thermal value of cooling in effectiveness measurement of integrated systems.[13] This "product value" would be calculated as the thermal value divided by a standard coefficient of performance for providing the cooling by unintegrated systems. This proposal was justified on the basis of retaining a first law consistent measure, but there are a number of problems associated with the approach.[14] First, determination of an appropriate COP to be used in the adjustment is more complicated than suggested in the Ee proposal which recommends use of the COP values set in ASHRAE Standard 90-75. These COP values are minimum rather than typical, and, in any event, are applicable only at rated conditions and not to seasonal evaluations. Second, use of the COP adjustment to "product values" would require distinction between various cooling system sizes, types, and input energy forms to select appropriate COP values. Third, heating produced by a heat-pump cycle in an integrated energy system would require a similar adjustment to be consistent. Fourth, the Ee measure would still be restricted to comparisons of systems with the same product or products even if the COP adjustment were made. Finally, the Ee proposal does not address comparisons involving differing or multiple inputs as discussed for the RCOP. The RCOP measure is not a first law efficiency, but an evaluation function relating integrated energy system outputs to resource inputs with appropriate weighting to reflect actual resource requirements and social values of the inputs. Cooling should, therefore, be treated as a system product in the same fashion as other outputs in computing the RCOP rather than treated as a thermal input.

SECOND LAW EVALUATION

A second law approach to evaluation of integrated energy systems performance would offer some interesting ideal system insights for consideration. The flows of available and unavailable energy would be evaluated with the criteria for superiority being a minimization of degradation of available energy. In this approach, the system operating with the least creation of entropy by irreversible processes would be found most attractive. The effort involved in a second law analysis[15] of a complex integrated system for even instantaneous comparisons much less for seasonal comparisons of systems would require considerably more time than the approach suggested by the RCOP. Also, a second law evaluation would provide only an ideal system bound to achievable performance and identification of system components and interfaces with the greatest potential for improvement of efficiency. It would not include fuel burdens associated with RUF's or social values of resources associated with RIF's. Given current awareness of depleting energy resources, the second law approach by itself would provide an incomplete assessment.

SUMMARY

The resource based coefficient of performance, RCOP, offers a simple evaluation measure for performance comparison of alternative integrated energy systems. Energy input quantities are adjusted by use of resource utilization factors, RUF's, and resource impact factors, RIF's, to reflect the true quantities of resources consumed as well as their social values. Energy outputs are similarly adjusted using RIF's to reflect the qualities of diverse output forms. The resultant dimensionless RCOP measure may be used for either instantaneous or seasonal comparisons of systems. This approach offers an advantage over existing performance measures in that it allows comparison of systems with diverse or multiple inputs and outputs encountered in integrated energy systems.

FOOTNOTES

[1]This paper is based on work sponsored by the Community Systems Branch, Division of Buildings and Community Systems, Office of Conservation, United States Energy Research and Development Administration.

[2]The term "integrated" also has a second meaning referring to integration of the systems to application designs to minimize energy requirements. This may be accomplished by reduction of transmission or distribution requirements thereby reducing system losses, by balancing the community or building occupancy to match system peak performance, and by community and building design to reduce or eliminate energy services needed.

[3]E, R, and F should be expressed in consistent units such as all in Btu or all in Wh; input and output rates (Btuh or W) may be utilized for an instantaneous value, but would cause η to vary with load and/or input mix.

[4]*ASHRAE Standard 90 Section 12: Annual Fuel and Energy Resource Determination*, American Society of Heating, Refrigerating, and Air Conditioning Engineers, February 1977.

[5]*ASHRAE Standard 90-75: Energy Conservation in New Building Design*. American Society of Heating, Refrigerating, and Air-Conditioning Engineers, New York, 1975. This standard includes Sections 1 to 11; based on the Standard 90 Project Committee's recommendation, the standard was published with these sections while Section 12 was processed through additional review. Future revisions of Standard 90 will include Section 12 currently published separately.

[6]Phipps, Harry H., "The RUF Concept as an Energy Measurement Tool," *ASHRAE Journal*, May 1976, pp. 28-30.

[7]Jones, Robert R., "Resource Impact Factor (RIF) Approach to Optimal Use of Energy Resources," *ASHRAE Journal*, October 1976, pp. 15-18.

[8]Baron, Seymour, "Energy Cycles: Their Cost Interrelationship for Power Generation," *Mechanical Engineering*, June 1976, pp. 22-30.

[9]Woods, J. E. and Donoso, J. E., "Energy Conversion Efficiencies for Thermal Control in Buildings," *ASHRAE Journal*, January 1977, pp. 37-41.

[10]Op. cit. *ASHRAE Standard 90 Section 12* approved by the ASHRAE Board of Directors on February 16, 1977. The RUF determination method is presented in Attachment A; regional values are tabulated in Attachment B.

[11]Weber, Stephen F., "The Effect of 'Resource Impact Factors' on Energy Conservation Standards for Buildings," National Bureau of Standards report

number NBSIR 77-1199 (R), February 1977. This study considers a variation to the RIF discussed in this paper. The variation combines the RUF and RIF into a single price multiplier (rather than quantity multipliers) determined as the ratio of the market price of resources to the social price. The social price remains a subjective estimate, although one possibility included in the study would be to set the social price at the expected substitute resource price based on depletion projections.

12
The argument against non-identical outputs meeting the same need is invalid if the outputs are measured in their final use form with secondary conversions included in the basic system analysis.

13
Coad, William J., "Energy Effectiveness Factor," *Heating/Piping/Air Conditioning*, August 1976, pp. 35-38.

14
Calm, James M., "Proposed Energy Effectiveness Factor, Ee: A Response", to be published in Heating/Piping/Air Conditioning, August, 1977.

15
For example, see: Tripp, Wilson, "Second-Law Analysis of Compression Refrigeration Systems," *ASHRAE Journal*, January 1966, pp. 49-57.

PROPER EVALUATION AND PRICING OF "ENERGY"

Richard A. Gaggioli

Mechanical Engineering Department, Marquette University, Milwaukee, WI 53233

ABSTRACT

Over the years, prominent thermodynamicists have advocated available-work analyses for properly evaluating energy-conversion processes.* Available-work is based on the Second Law, and goes back to Maxwell (1) and Gibbs (2). Unfortunately, it has not yet caught on in engineering practice, or in managerial decision-making.

It is the available-work content of a substance, not its energy content, which truly represents the potential of the substance to accomplish changes for us - to work for us. Available-work is what is called "energy" in lay terminology, and is the real commodity of value. Thus, an "energy" converter is supplied available-work in one form; the converter literally consumes (irreversibly destroys) part of it, using that part to drive the conversion processes; the remaining part is delivered in one or more converted forms. The inefficiencies, then, lie in the consumptions and in any effluent losses of available-work; only by evaluating these can proper assessments of "energy" utilization and of prospective savings be made.

Available-work is the only rational basis for evaluating (1) fuels and resources, (2) process, device and system efficiencies, (3) dissipations and their costs, (4) the value and cost of system outputs.

This paper will (i) develop the methodology of such second law analyses, which can now be made much more comprehensible as a result of recent progress in Thermodynamics; (ii) survey the results of efficiency analyses of a variety of processes, devices, systems and economic sectors; and (iii) illustrate the application of available-work methods for costing ("Thermoeconomics").

While baring many misconceptions resulting from energy analyses, the results of the efficiency analyses show great potential for alleviating "the energy problem" via conservation -- even moreso over the intermediate and long term than over the short -- and pinpoint where the opportunities are. In turn, the cost analyses show how economic-analysis decisions regarding "energy" systems can be greatly facilitated, while avoiding the misappropriations (often gross) which result from energy analyses.

The paper will be of value to engineers involved in design and in operating decisions, to managers in the private and government sectors who are involved with "energy" use and development, and to public service commissions.

INTRODUCTION

Thermodynamics -- Its Basic Implications

The basic concepts of Thermodynamics are the two commodities called Energy and, here, Available-work. The basic principles are the First Law, dealing with energy, and the Second Law, dealing with available-work.

To illustrate the basic concepts and principles, picture a conduit carrying some commodity such as electric charge, or high-pressure water, or some chemical like H_2. We could express the current by I_q (e.g., amperes = coulombs per second). Or I_v (such as gpm = gallons per minute), or I_{H_2} (such as gram-moles H_2/second). Or it could be a heat conductor carrying a thermal current I_θ. The conduit could be carrying any commodity.

---

*Different authors have presented the concept, available-work, with a variety of names (available energy, energie utilisable, exergy, essergy, potential energy, availability, ...).

Whatever the commodity might be, energy is carried concurrently with it; the rate I_E at which energy E flows is proportional to the commodity current. Thus, with charge, the rate of electric flow of energy past a cross-section of the conduit is $I_E = \phi I_q$, where ϕ is the local value of the electric potential at that cross-section.

Likewise, the hydraulic energy flow rate associated with the volumetric current I_V is $I_E = pI_V$, where p is the pressure. When a material flows and carries energy not only because of its pressure but also because of its composition, the flow of energy can be called a chemical flow: $I_E = \mu_{H_2} I_{H_2}$ where μ_{H_2} is the chemical potential.

Notice that, in each of the above examples, the proportionality factor between the commodity current and the associated energy current turns out to be the "potential" which drives the commodity current through the conduit. (Stated more precisely, the potential gradient causes the flow.)

The driving force which causes the thermal current is temperature difference; and, $I_E = TI_\Theta$. Traditionally, in science and engineering, it is I_E which has been called the rate of heat flow. It would have been better to use the word "heat" (or "heat content") for the commodity flowing with current I_Θ, but this commodity was not recognized until later, and ended up being called entropy.

Commodity Balances, and the First Law

In analysis of energy converters balances are applied for each of the relevant commodities; for examples, mass balances, energy balances, chemical compound balances, and so on. The amount of any given commodity in some container can in general be changed by either (1) transporting the commodity into or out of the container, or (2) production or consumption inside. Thus, on a rate basis {The rate of change in the amount of the commodity contained} = {The sum of all of the inlet rates} - {The sum of all of the outlet rates} + {The rate of production inside} - {The rate of consumption inside.} For steady operation the rates of influx and the productions equals the totals of the effluent rates plus consumptions.

Some commodities, like charge, cannot be produced or consumed; they are said to be conserved. The essence of the First Law is, of course, that energy is conserved; there is another aspect: the transport of any commodity has an associated energy transport.

The Potential to Cause Change for us: a Commodity

When does a commodity have the potential to cause change for us? The answer is: Whenever it is not in complete, stable equilibrium with our environment. Thus, charge has a potential to cause change when it is at $\phi \neq \phi_o$, where ϕ_o is the "ground" value; water has such potential when it is at $p \neq p_o$, where p_o is atmospheric ("ground") pressure. The useful or available power associated with the aforementioned currents are

$$P_q = \left[\phi - \phi_o\right] I_q \qquad P_V = \left[p - p_o\right] I_V$$

$$P_{H_2} = \left[\mu_{H_2} - \mu_{o,H_2}\right] I_{H_2} \qquad P_\Theta = \left[T - T_o\right] I_\Theta$$

Thus, if energy is being carried at a rate $I_E = \phi I_q$, only the portion $\left[\phi - \phi_o\right] I_q$ is available to do work, inasmuch as the current must be transmitted to ground (then carrying energy $\phi_o I_q$ to ground) in order to utilize, fully, its potential to accomplish changes for us -- its potential to work for us. (Notice that $P_\Theta = \left[1 - T_o/T\right] I_{E,\Theta}$. This is the classic result, usually derived in a complex manner from obtuse statements of the Second Law.)

I_q represents charge current, $I_E = \phi I_q$ represents energy current; $P_q = \left[\phi - \phi_o\right] I_q$ also represents a current. Namely, it is the rate of flow of the commodity called available work. (It could be symbolized by I_A; and the commodity might well simply be called work, or work content.)

Unlike energy and charge, for examples, available work is not a conserved commodity. It is the "energy" of the layman, which is the potential to cause change. In so doing, some of it is invariably consumed in every process. Consider any "energy" conversion system, such as a power plant which takes in coal and produces electricity. The rate at which available work is delivered (i.e., the electric power output) is much less than the supply rate of available work with coal (i.e., the chemical power input). Most of that difference is consumed -- used up -- driving the various processes (such as combustion, heat transfer, fluid flow) going on in the power plant; a portion is emitted in effluents (stack gases, cooling water). The losses are generally small compared to the consumptions, contrary to misconceptions which prevail as a result of first-law analyses. See Fig. 1, which will be discussed later.

driving the various processes (such as combustion, heat transfer, fluid flow) going on in the power plant; a portion is emitted in effluents (stack gases, cooling water). The losses are generally small compared to the consumptions, contrary to misconceptions which prevail as a result of first-law analyses. See Fig. 1, which will be discussed later.

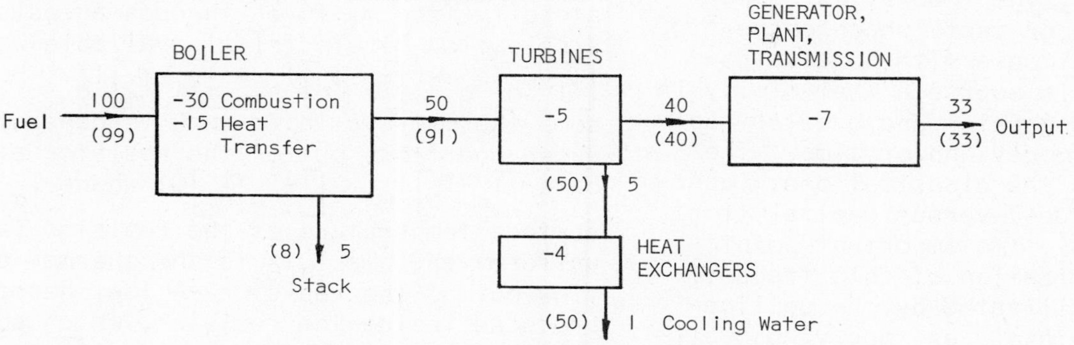

Fig. 1 Available-work (and energy) flow diagrams for a typical fossil-fired steam power plant.

The Second Law

In summary, then, it is available-work, and not energy, which represents the "capacity to cause change for us". Available-work -- which anything has when it is not in complete (stable) equilibrium with our environment and which is a commodity, which can be transferred from one thing to another -- measures the capacity to cause change for us. To accomplish changes, some is invariably used up (consumed) to make the changes occur.

The Methodology of Available-work Analyses

To improve the fuel economy of an "energy" system means to lessen the available-work consumed within the system and that lost in effluents so that, for a specified amount of product, the amount of available-work -- "fuel" -- that needs to be supplied is thereby decreased. To ascertain the consumptions and losses requires the evaluation of the rates at which available-work is transferred in and out with different commodities.

Not only is an overall analysis of an "energy" system valuable, but so is a detailed analysis which pinpoints where and to what extent the available-work consumptions occur, within the system. The procedures for making a more detailed analysis are identical to those for the overall analysis. It is simply a matter of applying balances to subsystems of the overall system. In turn, components can be broken down and analyzed further, process by process. The calculations needed to make an available-work analysis a system, device or process are the same as those for an energy analysis: Balances for the various relevant commodities -- those carrying available work -- are applied to the system (or process) of interest along with available-work balances. Also needed are the performance (input-output) characteristics of the system (or the kinetic relations for the process) and the property relationships for the various materials that may be flowing into and out of the systems (or may be involved in the process).

Concurrent thermal and chemical available-work flows. When entropy flows at a rate I_Θ, then $I_E = TI_\Theta$ and $P_\Theta = [T-T_o]I_\Theta$. When a chemical j diffuses at a rate I_j, $I_E = \mu_j I_j$ and $P_j = [\mu_j-\mu_{jo}]I_j$, where μ_{jo} is the chemical potential of j in the reference environment. When there is bulk flow of a material, carrying entropy too, $I_E = \mu_j I_j + TI_\Theta = [\mu_j+Ts_j]I_j = h_j I_j$ and $P = [\mu_j-\mu_{jo}]I_j + [T-T_o]s_j I_j = [h_j-T_o s_j-\mu_{jo}]I_j$. Reference (25) presents relationships for practical evaluation of this I_E and this P in several important special cases.

Second-law efficiency (effectiveness). In the theoretical limit, available-work supplied with any commodity can be completely transferred to any other commodity. In the case of real transformations, the degree to which this perfection is approached is measured by the so-called effectiveness (7,8,9) or second-law efficiency:

$$\varepsilon \equiv \frac{\text{available-work in product}}{\text{available-work supplied}}$$

Conventional first-law efficiencies and unit product costs defined in terms of "product" energy and "fuel" energy are generally faulty, to a degree which depends upon the kind of device or system to which they are applied. Basically, their worth is proportional to how well they approximate the effectiveness, ε.

The denominator of ε exceeds the numerator by the amount of available-work consumed plus the amount lost in effluents.

$$\varepsilon = \left[P_{out}\right] / \left[P_{out} + \dot{A}_c + \dot{A}_\ell\right]$$

For any conversion, the theoretical upper limit of ε is 100 per cent, which corresponds to the ideal case with no dissipations or losses. To approach that limit, in practice, requires greater and greater capital investment of money and/or time. The tradeoff, then, is the classical one: operating costs (for fuel) versus capital (for equipment and time). An important point here is that optimization of this tradeoff can be greatly facilitated by the application of Second Law analyses, applying available-work analyses to processes, devices and systems (7,14,16,17,19,20,21,26,27).

Costing with available-work. The cost of the product (output available work) equals the annual fuel costs, plus the annualized capital and other costs: $(\$_p/yr) = (\$_F/yr) + (\$_c/yr)$. Expressing $(\$_F/yr)$ as the product of the annual fuel consumption (A_F, Btu of available-work per year) with the unit cost of fuel (c_F,\$/Btu), $(\$_p/yr) = c_F A_F + (\$_c/yr)$. In turn, the unit cost of the product provided by the system is

$$c_p = \frac{\$_p/yr}{A_p, Btu/yr} = c_F \frac{A_F}{A_p} + \frac{(\$_c/yr)}{A_p}$$

Thus, since $\varepsilon = A_p/A_F$,

$$c_p = c_F/\varepsilon + \$_{c,annual}/A_p$$

The first term on the right reflects fuel costs; the second equals the annual capital (and other) costs per unit of system capacity. Normally, to decrease the first requires an increase in the second, and, of course, the aim is to make the investment which minimizes the total. Equations like the latter can be developed (16,19,20) to determine the unit cost of available-work at the various junctures of an energy system, including the cost of the products of a multi-purpose system. Such pricing is invaluable for the design and operation of energy systems, for technology assessment, and for decision-making. Whereas, the use of first-law analyses, and the assignment of costs to energy, leads to misconceptions, poor decisions and misappropriations. Second-law costing will be illustrated herein, under the heading THERMOECONOMICS.

TYPICAL RESULTS OF EFFICIENCY ANALYSES

Processes

Electric current. When a current flows steadily from ϕ_1 to ϕ_2 through a resistor, the dissipation of electrical available-work is $\left[\phi_1-\phi_o\right] I_q - \left[\phi_2-\phi_o\right] I_q = \left[\phi_1-\phi_2\right] I_q$. However, some thermal available-work is generally produced and flows out of the resistor at the rate $\left[T_\sigma-T_o\right] I_\Theta = \left[1-T_o/T_\sigma\right]\dot{Q}$, where T_σ is the surface temperature of the resistor (assumed uniform) and $\dot{Q} = T_\sigma I_\Theta$ is the thermal energy flow out of the resistor -- the thermal energy produced inside the resistor, which equals the dissipation of electric energy, $\left[\phi_1-\phi_2\right] I_q$.

The net consumption of available-work equals the difference between that supplied and that leaving:

$$\dot{A}_c = \left[\phi_1-\phi_o\right] I_q - \left[\phi_2-\phi_o\right] I_q - \left[T_\sigma-T_o\right] I_\Theta$$

In turn, with the relationship $\dot{Q} = T_\sigma I_\Theta = \left[\phi_1-\phi_2\right] I_q$,

$$\dot{A}_c = T_o I_\Theta = \left[T_o/T_\sigma\right]\dot{Q} = T_o/T_\sigma\left[\phi_1-\phi_2\right] I_q$$

This equation is an example of a relationship which is true in general: $\dot{A}_c = T_o\dot{S}_p$; $\dot{S}_p$ is the rate of entropy production. In the case at hand, an entropy balance on the conductor says that the entropy current out of the conductor will equal the rate of entropy production inside: $I_\Theta = \dot{S}_p$

Suppose that the resistance were an electric boiler, $T_o = 40°F$ and $T_\sigma = 212°F$. Then the electric power supply $P_q = 1200$ watts, while the thermal power input to boiling water is

$$P_\Theta = [1-T_o/T_\sigma]\dot{Q} = [1-500°R/672°R][1200 \text{ watts}]$$
$$= 308 \text{ watts} = 1048 \text{ Btu/hr}.$$

And,

$$\dot{A}_c = 1200-308 = 892 \text{ watts} \qquad \varepsilon = \frac{308}{1200} = 0.256$$

Of course, there is no energy consumed, and the first-law (energy) efficiency is 100%.

This example is typical of the second-law analysis of a process or device.

Friction. The same kind of analysis can be used to evaluate frictional processes such as flow of fluids through conduits, viscous damping, etc.

$$\dot{A}_c = T_o\dot{S}_p = \left[T_o/T\right]\dot{F} = \left[T_o/T\right] 4fL/2D\ v^2 = \left[T_o/T\right] cv^2$$

where $\dot{F}$ is the frictional rate of thermal energy production (equal to the "pump work"); for the fluid flow case, f is the "friction factor", L is the length of the conduit and D is the diameter, and v is the fluid velocity. For the viscous damping (and fluid flow) case, c is the damping coefficient (Drag force/velocity).

Relationships such as the foregoing are useful for economic analysis, such as the optimal sizing of pipe and insulation (17,28).

Heat transfer. When there is a heat flow, from high T to low, the available-work consumption is

$$\dot{A}_c = [T_1 - T_o] I_{\theta,1} - [T_2 - T_o] I_{\theta_2}$$

Since energy is conserved $T_1 I_{\theta,1} = T_2 I_{\theta,2} = \dot{Q}$. The rate of entropy production is $\dot{S}_p = I_{\theta,2} - I_{\theta,1}$. Hence, $\dot{A}_c = T_o \dot{S}_p$

$$\dot{A}_c = T_o \dot{S}_p \quad \text{and} \quad \dot{A}_c = T_o/T_2[1 - T_2/T_1]\dot{Q} = T_o/T_1[T_1/T_2 - 1]\dot{Q}$$

When there is a heat flow from a high T into the environment at T_o, the available-work loss therewith is $\dot{A}_\ell = [1 - T_o/T]\dot{Q}$.

For examples, when there is heat transfer from hot products of combustion at say T = 3540°F to water boiling at 1040°F, such as in a typical power plant boiler,

$$\dot{A}_c = \frac{500°R}{4000°R}\left[\frac{4000}{1500} - 1\right]\dot{Q} = 0.21\dot{Q}$$

$$\varepsilon = \frac{P_{\theta,2}}{P_{\theta,1}} = \frac{[1 - T_o/T_2]\dot{Q}}{[1 - T_o/T_1]\dot{Q}} = 0.76$$

For heat transfer from the same gases at 3540°F, while heating air at say 120°F, as occurs in the usual home heating furnace,

$$\dot{A}_c = 0.74\dot{Q} \qquad \varepsilon = 0.30$$

Heat transfer processes through sizeable temperature drops are very dissipative, just like electric currents with large voltage drops through resistors.

First law analyses neglect completely the available-work (fuel) consumptions due to heat transfer, and account -- erroneously at that -- only for losses. The losses from a boiler or furnace are very small compared to the consumptions inside, due to both heat transfer and combustion.

Combustion. When a fuel is burned steadily in air,

$$\dot{A}_c = P_{fuel} - P_{products} = \dot{M}_f[h_f - T_o s_f - \mu_{fo}] - \dot{M}_p[h_p - T_o s_p - \mu_{po}]$$

Evaluation of the right-hand terms shows (23, 25) that, for the conversion of chemical energy to thermal energy at atmospheric pressure,

$$\dot{A}_c \doteq 0.3P_{fuel} \quad \text{and} \quad \varepsilon \doteq 0.7$$

For combustion at higher pressures, such as in turbines and engines, the consumption is less, and the effectiveness greater; however, the gain over combustion at atmospheric pressure is virtually all lost because of additional consumptions by flow friction and losses by heat transfer.

Dissipation functions. It should be mentioned that the so-called "energy dissipation functions" or "thermal energy production" associated with fluid flows and electric currents differ from the available-work consumptions, by the factor T/T_o:

$$\dot{F} = \left[T/T_o\right]\dot{A}_c = T\dot{S}_p$$

Devices

The same methods can be applied to energy conversion devices, to evaluate $\dot{A}_c$, $\dot{A}_\ell$ and ε. Table I shows the results for some typical cases; in some instances a breakdown is given for the processes occurring in the devices. For some more or less randomly selected devices, energy losses and first-law efficiencies are also shown (in parentheses) in order to illustrate the erroneous conclusions that can be drawn therewith.

Table I gives values of device efficiencies for some typical "energy converters". The values shown are representative when the converter is operating at optimal, design-load conditions. In many instances the performance deteriorates badly at part-load; for examples, the reciprocating and gas-turbine engines, and to a lesser degree the compressors. Similarly, in most cases the performance of smaller size devices is poorer. By the way, the significant difference in efficiency between high pressure boilers and furnaces (or low-pressure boilers), even though their first-law efficiencies are practically the same, explains the emphasis on the economics of co-generation -- which has been practiced for a long time in certain industries, such as the paper industry.

TABLE 1 Available-work (and Energy) Outputs, Losses and Consumptions as a Fraction of Supply for Several Devices

Device	$\dot{A}_c/\dot{A}_s$	$\dot{A}_\ell/\dot{A}_s,(\dot{E}_\ell/\dot{E}_s)$	$\varepsilon = \dot{A}_p/\dot{A}_s,(\dot{E}_p/\dot{E}_s)$
Boiler (high-pressure steam)(23)	0.45	0.05,(0.1)	0.5,(0.9)
- Combustion	0.3		
- Heat transfer	0.15		
- Chemical effluent		0.04,(0.01)	
- Thermal effluent		0.01,(0.09)	
Furnace (comfort heating)	0.65	0.25,(0.4)	0.1,(0.6)
- Combustion	0.3		
- Heat transfer	0.35		
- Chemical effluent		0.05,(0.05)	
- Thermal effluent		0.20,(0.35)	
Engine (large, reciprocating)	0.35	0.25,(0.6)	0.4,(0.4)
- Combustion	0.20		
- Heat transfer	0.15	(0.3)	
- Exhaust		0.25,(0.3)	
Gas Turbine Engine	0.3	0.4,(0.7)	0.3,(0.3)
Steam Turbine (large)	0.1		0.9
Compressor (reciprocating)	0.3		0.7
Heat Exchangers			
- Dehumidifying and cooling coil	0.75		0.25
- Steam heating coil	0.6		0.4
Blast Furnace (24)			0.8
Open Hearth Furnace (24)			0.78
Basic Oxygen Furnace (24)			0.92

Systems

Table 2 shows the performance for some energy systems, and some whole industrial sectors, with breakdowns for devices and/or processes.

TABLE 2 Performance of Typical Systems and Sectors

System	$\dot{A}_c/\dot{A}_s$	$\dot{A}_\ell/\dot{A}_s,(\dot{E}_\ell/\dot{E}_s)$	$\varepsilon = \dot{A}_p/\dot{A}_s,(\dot{E}_p/\dot{E}_s)$
Fossil-fired Steam Power Plant (also see Fig. 1)(23)	0.55	0.06,(0.59)	0.4,(0.41)
- Boiler	0.45	0.05,(0.09)	
- Turbines	0.05		
- Condenser and heaters	0.05	0.01,(0.5)	
Co-generating Power Plant (410 kw electricity, 1130 kw steam)	0.65	0.05	0.3,(0.75)
Co-generating Power Plant (10,000 kw elec, 17,000 kw steam)	0.62	0.05	0.33,(0.75)
Equivalent, conventional	0.62	0.10	0.28
- 10,000 kw electric power	18/30	2/30	10/30
- 17,000 kw 50 psig boiler	44/70	8/70	18/70,(55/70)
All-electric Total Energy	0.65	0.06,(0.67)	0.28,(0.33)
- Power production & trans.	0.55	0.06,(0.67)	0.33
- Heat pump	0.05		

TABLE 2 (continued)	$\dot{A}_c/\dot{A}_s$	$\dot{A}_\ell/\dot{A}_s, (\dot{E}_\ell/\dot{E}_s)$	$\varepsilon = \dot{A}_p/\dot{A}_s, (\dot{E}_p/\dot{E}_s)$
Fossil-fired Total Energy	0.42	0.3, (0.4)	0.28, (0.6)
- Engine	0.37	0.3	
- Heating & cooling	0.05		
Equivalent, conventional	0.60	0.19	0.21, (0.5)
- Electricity	30/55	3/55	18/55
- Heating & cooling	30/45	11/45	3/45, (30/45)
Heating and Air-conditioning			
- Air-conditioning	0.85	0.1	0.04, (>1.0)
- Refrigerating unit (ε=0.33)	0.5	0.06	
- Compressor	0.15		
- Condenser	0.15	0.06	
- Expansion valve	0.06		
- Evaporator	0.15		
- Air-handling unit (ε=0.2)	0.3	0.04	
- Distribution (ε=0.5)	0.05		
- Heating	0.8	0.1	0.09, (0.6)
- Boiler (ε=0.35)	0.5	0.1	
- Air-handling unit (ε=0.35)	0.25		
- Distribution (ε=0.6)	0.05		
Coal Gasification (Koppers-Totzek)(25)	0.3	0.06	0.65
- Coal preparation	0.06	0.005	
- Gasifier	0.15	0.02	
- O_2 production	0.04	0.003	
- Heat recovery	0.02		
- Gas cleanup	0.035	0.03	
Economic sectors, direct fuel utilization (22,24)			
- Residential & Commercial			0.1
- Industrial			<0.15
- Iron and Steel			0.3
- Aluminum			0.12
- Transportation			0.1
- Electric Utilities			0.3

Once again, the entries in Table 2 are by-and-large for systems operating under optimal conditions - design loads, careful maintenance. For example, the effect of varying loads on the year-around performance of an air-conditioning system would reduce the overall effectiveness to 0.02 or below. For heating, the reduction, not so drastic, is to about 0.07.

Conclusions -- Specific and General

This article is not intended to conclude with proposals of specific means for improving energy systems. The aim, here is to show how to ascertain how inefficiently today's systems perform, and how to pinpoint and evaluate the true losses and consumptions. Then, needs and opportunities for improvement can be accurately assessed; the real prospective savings of "energy" that could accrue from proposed improvements -- whatever they might be -- can be rationally determined. In turn, decisions and investments (of time and/or money) can be made judiciously. Whereas, analyses made from an energy-efficiency viewpoint can be very misleading.

For example, is there hope for saving significant amounts of energy by conservation? Some have scoffed at conservation, as a significant means for relieving our energy problems -- at least over the long term. Another group has recognized the importance of conservation, but has placed most all of the emphasis on better "end-use". Thus, in the Heating and Air-conditioning sector the stress is on the prevention of losses (and gains) with better insulation, fenestration, exhaust air heat recovery, etc. Neither group is aware of how great the potential savings really are, with conservation, because they do not realize, as shown in Tables 1 and 2, how inefficient the conversion processes are. The biggest prospective savings are in the conversion processes, not in end-use. (And that is not to minimize the importance of end-use conservation.) The important point is that energy analyses recognize losses only, they do not correctly evaluate the relative importance of different losses, and they fail to recognize consumptions -- which are generally much more important than the losses! The 10 or so per cent effectiveness with which energy is utilized in this country, though improved greatly over the 1 or 2 per cent of a century ago, is very low; basically, this is encouraging inasmuch as it shows that there is great opportunity for improvement remaining. Conservation can contribute effectively to the resolution of the energy problem--even over the relatively short term, with today's technology--provided of course that capital is brought to bear.

Another important point is that the production of electricity is one of our most efficient energy conversions. The great losses commonly ascribed to the stack gases and cooling water are hardly losses at all; the actual "losses" (really consumptions) are elsewhere in the plant (primarily in the boiler), and as a fraction of input are small compared to most conversion systems. The energy in condenser cooling effluent has very little usefulness (potential to cause change). Proposals to conserve our "energy" by "making use of all that wasted energy" are misguided by the seemingly large resource in that effluent. The potential beneficial economic impact of such proposals is nil. On the other hand district steam systems, one form of co-generation, can be quite beneficial, by producing electricity with the high-temperature heat and thereby reducing it to low-temperature without the great heat transfer consumptions.

Furnaces and all-fossil total energy systems, considered to be very efficient, are very inefficient or fairly efficient, respectively.

For example, these comments have considerable negative impact on the desirability of high-Btu coal gasification and of the "Hydrogen Economy" for the purpose of distributing these synthetic fuels about, for combustion -- so highly dissipative -- in furnaces and boilers.

The results shown in Table 2 for gasification are for a state-of-the-art medium-Btu process; conversion of low- or medium-Btu to high-Btu gas is relatively more inefficient. The medium-Btu process, alone, would be useful for consumers - electrical utilities clearly are the most likely prospects - who would be willing to install their own gasification equipment locally, in order to produce a clean fuel which could be burned with minimal environmental impact. The 65% effectiveness shows the added fuel cost which must be borne: 54% greater than without the gasification. And then there is the additional capital (and other) costs. Nevertheless, this technology is certainly competitive with, if not more economic than, other methods of contending with the emissions problem.

It is evident from the foregoing information that the processes which are by far the greatest consumers of "energy" are combustion and heat transfer. Research and Development, motivated by the economic need, will develop new technology for overcoming these wastes -- directly or indirectly. Indirect improvements will come from the development of new systems combining processes -- old and new -- in novel ways. Direct improvements will be achieved with more efficient heat production, transport and transformation processes, and with more efficient chemical processes. Detailed Second Law analysis of the combustion process shows that the losses result from an imbalance between the chemical potential μ -- which is the chemical analog to the potentials p, ϕ and T -- of the reactants and of the products, just as the heat transfer losses result from the great imbalance between the temperature of combustion products and steam. Fuel cells improve the balance of the chemical potentials, and yield an electric potential energy output upon reducing μ of the reactants. But, thus far, any savings of the losses due to chemical reactions has been counteracted with other losses; and the costs of capital, and of fuel and its preparation, are unattractive. However, it is predicted here that prospective schemes for reducing the chemical potential of oxygen before it reacts with the fuel will be forthcoming, to improve performance of power stations and other combustion engines.

The intent of MHD is to capitalize on potential energy now lost in the boiler heat transfer processes. The potential energy output of the aforementioned schemes for improving combustion will capitalize not only on potential energy now lost in combustion, but also on some now lost in heat transfer, because improved combustion will lead to lower combustion temperatures and hence smaller heat transfer losses.

As mentioned earlier, energy analyses cannot locate or evaluate consumptions of "energy", because energy cannot be consumed or produced. At best, energy analyses can only determine energy losses; in so doing, however, it generally misevaluates the importance of the losses -- often, badly, as in the case of the power plant. The available-work methods for analyzing "energy" systems are the key to pinpointing the losses and consumptions, for measuring their magnitudes and resultant per cent inefficiencies, in order to determine where opportunities for improvement and conservation lie, for the purposes of decision-making for allocation of resources -- capital, R&D effort, and so on. The available-work methods, which involve exactly the same kinds of calculations as energy analyses, are the first step in evaluating the monetary costs of inefficiencies (6-9), and they also enhance the germination of prospective ideas and the quick evaluation thereof. They show that

(1) There is a need for new technology in energy conversion processes and systems.
(2) There is a great margin for improvement, which augurs well for the development of new technology.
(3) The motivation for acquiring and developing more efficient energy conversion equipment and systems will be economic -- whether socialized or free enterprise -- responding to the rising value of fuels.
(4) The key to acquiring and developing the better equipment and systems will be investment of money -- taxes or capital -- in response to the rising value of fuels.

If supply prices are held down by regulation, the investments will need to be subsidized -- tax money. (But at what subsidy level?) In turn, the required use of more efficient systems will have to be legislated. (But at what efficiency level?) Etc. The available-work costing methods to be described presently can be used to prescribe efficiencies and levels of investment, but they will yield the truly optimal results only if the real value (cost) of the energy supply is known.

THERMOECONOMICS

The use of second law methods for costing, of value in a variety of decision-making circumstances, will be illustrated in several contexts. The first context is one which is important to cost-accounting.

Cost of Co-generated Steam and Electricity

When steam is needed at moderate pressures for heating purposes less fuel resource is needed if, rather than produce the steam at the required p, it is produced at a higher p and bled down to the moderate p through a power turbine. This is so, even if the "boiler efficiency" (first-law) of a high-p boiler is no better than that of a lower-p boiler; by producing the steam at the high pressure, much of the available-work that would have been consumed by heat transfer in the lower-p boiler is saved. The amount saved is practically all delivered by the turbine shaft, say to produce electricity, inasmuch as steam turbines are so efficient. It is this saving of available-work which otherwise would have been consumed - "wasted" - which has brought co-generation so much to the fore. Figure 2 is a simplified schematic diagram of a co-generating plant.

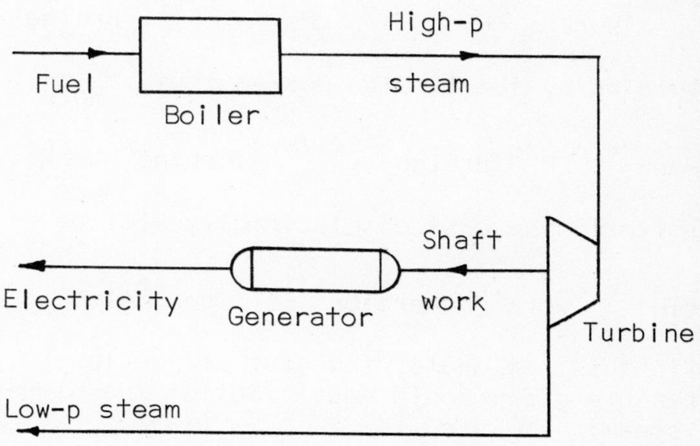

Figure 2. Simplified flow diagram of available-work, for a system co-generating steam and electricity.

When steam and electricity are co-generated, it is often critical to know how much of the costs should be attributed to each commodity. For examples, (i) if a company is co-generating, and the manufacture of one of its product is steam intensive while that of another is electricity intensive, it is certainly important to the costing of the products to know how much to charge the first one for each unit of steam and how much to charge the second for each unit of electricity. Or (ii) if a co-generating company exports excess electricity, used to service a surrounding community, it is certainly important to know how much it should charge its residential customers for electricity and how much to charge its product(s) for steam. And (iii) conversely for an electric utility which may be selling bleed steam for say comfort heating purposes in the district of a power plant.

Keenan (7) pointed out that to cost the co-generated products equitably it is necessary to recognize that their marketability and value are a consequence of their available-work not of their energy. According to the cost relationship developed earlier* the unit cost of the high-pressure steam leaving the boiler is

$$c_{HP} = c_{Fuel}/\varepsilon_{Boiler} + (\$/yr)_{Boiler}/A_{HP}$$

where A_{HP} is the annual boiler output of available-work in the steam. The turbine work output could be costed by charging it for the capital (and other) costs of the turbine and for the available-work extracted from the steam to produce the turbine work:

$$(\$/yr)_{Work} = c_{HP}[A_{HP} - A_{LP}] + (\$/yr)_{Turbine}$$

Dividing by the turbine work output, A_{Work},

$$c_{Work} = c_{HP}/\varepsilon_{Turbine} + (\$/yr)_{Turbine}/A_{Work}$$

Similarly the cost of electricity will be

$$c_{elec} = c_{Work}/\varepsilon_{generator} + (\$/yr)_{gen}/A_{elec}$$

With this rationale, the cost of the low-pressure steam would equal that of the high-p steam: $c_{LP} = c_{HP}$.

Consider the 10,000 kw electric, 17,000 kw steam power plant referred to in Table 2. A boiler cost of $10(10^6)$ dollars, amortized over 32 years at 8.5%, and accounting for income taxes (using simple straight-line depreciation), etc., leads to an annual cost of $(\$/yr)_{Boiler} = 0.8(10^6)$, while $A_{HP} = 180(10^8)$ kwh/yr assuming a load factor of ~0.7. If $c_{Fuel} = \$1.60/10^6$ Btu and $\varepsilon_{Boiler} = 0.45$ then

$$c_{HP} = \frac{1.6\$/10^6\text{Btu}}{0.45}\,\frac{3413\text{ Btu}}{\text{kwh}} + \frac{0.8(10^6)}{180(10^6)}\,\frac{\$}{\text{kwh}}$$

$$= 1.2¢/\text{kwh} + 0.45¢/\text{kwh} = 1.65¢/(\text{kwh of available-work}) = c_{LP}$$

*The unit costs of the products should be their incremental costs. Here, their average costs are used, which are the same as long as the effectiveness of the devices are relatively independent of load - which they are (23). For relationships and methods to be used when such simplifications cannot be made, see the work of Evans (e.g., **20**).

In turn,

$$c_{Work} = \frac{1.65}{0.9}\,\frac{¢}{\text{kwh}} + \frac{\$250{,}000/\text{yr}}{70(10^6)\text{kwh/yr}} = 2.2¢/\text{kwh}$$

and

$$c_{elec} = \frac{2.2¢}{0.95} + \frac{\$150{,}000/\text{yr}}{65(10^6)\text{kwh/yr}} = 2.5¢/\text{kwh}$$

For steam produced at 650 psi, and superheated to 750°F, the available-work per pound is $a = h - T_o s - \mu_o$ = 559 Btu/lbm = 0.165 kwh/lbm; for low-p steam at 50 psi, a = 338 Btu/lbm = 0.1 kwh/lbm. Thus, the costs per pound of steam are

$$c_{HP} = 1.65\frac{¢}{\text{kwh}} \cdot \frac{0.165\text{ kwh}}{\text{lbm}} = 0.27\frac{¢}{\text{lbm}}$$

$$c_{50\text{ psi}} = 1.65\frac{¢}{\text{kwh}} \cdot \frac{0.1\text{ kwh}}{\text{lbm}} = 0.17\frac{¢}{\text{lbm}}$$

Had the 50 psi steam been produced directly, in a low-pressure boiler, its unit cost would be calculated with the formula used here for c_{HP}. With the boiler's lower ε and its lower available-work output (per pound of steam produced), the unit cost turns out to be $c_{50\text{ psi}}$ = 2.2 ¢/kwh = 0.23 ¢/lbm. Comparing this with the 0.17 ¢/lbm for steam produced by co-generation explains the desirability of co-generation. (With a given fuel cost, the electricity produced by co-generation would have essentially the same cost as that produced alone, in a larger power plant. The larger plant introduces some economics "of scale", while the co-generation saves on some capital items - elimination of or reduction in size of (i) condenser, (ii) costly low-pressure turbines.)

If energy costing is used, the procedural steps would be the same as the foregoing, but with first-law efficiency rather than effectiveness (ε) and with energy (E or e) rather than available-work (A or a). The results are

$$c'_{HP} = 0.8\frac{¢}{\text{kwh energy}} = 0.27\text{ ¢/lbm}$$

$$c'_{Work} = 1.3\text{ ¢/kwh}$$

and

$$c'_{Elec} = 1.7\text{ ¢/kwh}$$

$$c'_{50\text{ psi}} = 0.8\frac{¢}{\text{kwh energy}} = 0.27\frac{¢}{\text{lbm}}$$

Comparison of the prices determined with available-work costing and with energy costing shows clearly that back-pressure steam is priced higher and electricity is priced lower with energy costing.

If energy costing is correct for 50 psi steam, then it is correct for 40 psi, for 30 psi and so on down to lower and lower pressures, and those prices would follow the upper curve in Figure 3. This would lead to the absurdity that steam at pressures like 0.5 psia would be worth about 0.2¢/lbm - almost as much as at 50 psi - while it has virtually no usefulness for heating (T=80°F, a = 0.015 kwh/lbm). There would be no buyers of 0.5 psia steam at 0.2¢/lbm - at $2 per million Btu of energy.

Available-work costing yields that the price per lbm of steam does go to zero as its usefulness goes to zero, as shown by the lower curve of Figure 3.

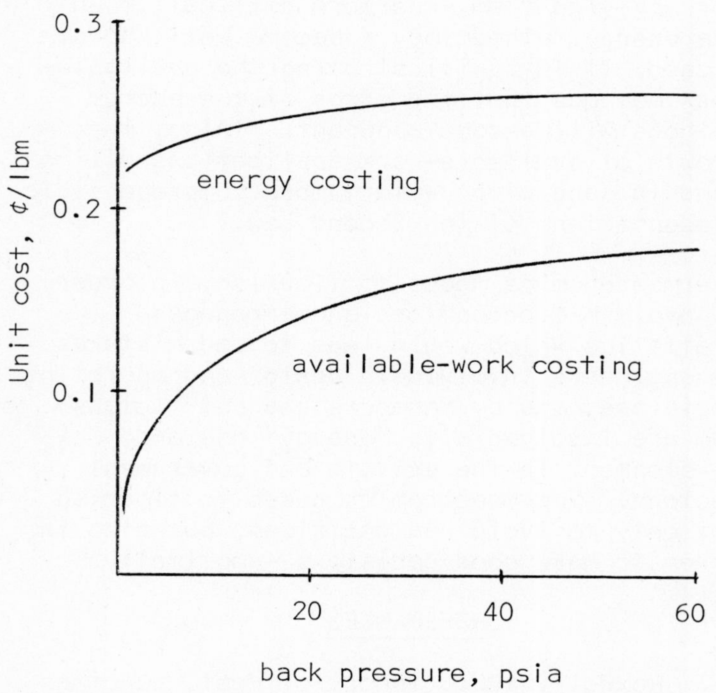

Figure 3. Cost of co-generated steam versus back pressure, with energy costing and available-work costing.

In industrial practice, energy methods are normally used, and consequently important decisions are made with misinformation from such erroneous cost accounting. In one known instance an industry has been selling excess electricity at a price significantly less than its value - at a price less than the cost to produce it - while the steam which is charged to its own products is overpriced. That is, after costing with energy methods and then marking up for a certain percent profit, the price is still lower than the cost to produce the electricity. This is true even if the electricity is considered to be a by-product.

For the case where the electricity is a by-product (and in some cases the steam may be a by-product), the costing rationale would be different from that followed above. The cost c_{LP} of available-work in back-pressure steam would be calculated as if it were produced directly in a low-pressure boiler, using the same formula as that used above for high-p steam; with the lower ε of a low-p boiler, the c_{LP} would be greater. Then

$$(\$/yr)_{Work} = c_{HP}A_{HP} - c_{LP}A_{LP} + (\$/yr)_{Turbine}$$

This yields a lower cost of electricity. The results are

$$c_{LP} = 2.2 \frac{\text{¢}}{\text{kwhr}} = 0.23 \frac{\text{¢}}{\text{lbm}}$$

$$c_{Elec} = 1.9 \frac{\text{¢}}{\text{kwhr}}$$

which are still substantially different from the numbers obtained with energy accounting.

Furthermore, there are arguments against classifying the electricity as a by-product in cases such as this - at least in the particular case being referred to. And, even if it is considered to be a by-product, should the outside consumers be the primary beneficiaries? In a free market, the question would pretty well take care of itself. In the regulated electricity market, the advocate for the company which is servicing its surrounding community might well argue that the equitable way of costing the co-generated steam and electricity is to charge them with equality, for fuel and capital devoted to producing them. Thus, since both boiler and turbine, in series, are devoted to producing shaft work and steam in parallel, the annual cost of the turbine products is

$$(\$/yr)_{products} = c_{HP}A_{HP} + (\$/yr)_{Turbine}$$

Inasmuch as it is the available-work of the products which has value, this annual cost of products equals the sum of the individual product costs, $c_{LP}A_{LP} + c_{Work}A_{Work}$. Then, setting $c_{LP} = c_{Work}$ in order to follow the advocated equality,

$$c_{LP}[A_{LP} + A_{Work}] = c_{HP}A_{HP} + (\$/yr)_{Turbine}$$

which can be solved to get the unit cost of low-pressure steam, c_{LP}. The unit cost c_{Elec} can be evaluated with the same equation used before, but with the new value of $c_{Work} = c_{LP}$. The results are

$$c_{50\ psi} = 2¢/kwh = 0.2¢/lbm; \quad c_{Elec} = 2.3¢/kwh$$

When c_{LP} is evaluated at a variety of back pressures besides 50 psi, the results look practically the same as those in Figure 3, with slightly higher steam costs with this "equality" method.

If an "equality" approach is used, but with energy taken as the commodity of value rather than available-work, the cost of steam is again too great, hardly reduced from the values shown on Figure 3. Again, with energy accounting, the price of valueless steam at very low pressures, like the aforementioned 0.5 psia, approaches 0.2¢/lbm ($2/10^6$ Btu of energy)!

Regardless which approach is appropriate — "equality", "byproduct", "extraction" (the first one used herein), or whatever — rational accounting requires that available-work not energy be taken as the commodity of value.

Before leaving the topic of co-generation, it should be mentioned that, ironically, in the last few years there has probably been more shutdown than startup of co-generating plants. This has been a consequence of (i) fuel interruptions, (ii) sky-rocketing fuel-costs, and (iii) energy accounting. If an industry must pay twice as much as a utility for fuel, even when costing electricity with energy accounting which makes it look cheaper than it really is, the apparent cost of electricity is still greater than purchased electricity. And the energy accounting makes the co-generated steam appear to be as expensive - sometimes more expensive - than direct-fired steam. So, the industry opts to shut down co-generation, to save "its" fuel, and money.

The fuel that is being diverted away from such industries, by the higher prices and by allocation, is being "reserved" for comfort heating applications - where the effectiveness of utilization is 10% compared to the 30% with co-generation!

Other Applications

References (17, 19, 26, 27, 28) among others, present additional illustrations of Thermoeconomics. References (16, 28) present elementary applications, to the optimal sizing of piping and pipe insulation, in the design of systems. Reference (19) and others by Evans et al apply Thermoeconomics to the design of desalination systems, including cases compounded with power generation. Reference (26) illustrates an application to the operation and maintenance of power plants, in particular to feedwater heaters. The methods can be applied to other components besides feedwater heaters, (e.g., Reference 27) and to design as well as operation-and-maintenance cases - to design optimally.

These applications hardly scratch the surface of the potential of available-work costing. Thermoeconomics is only in its infancy. It is first of all very important that the fallacy of energy methods become well known, in order to avoid inevitable mistakes that will otherwise arise. In turn, it is equally important that the validity of available-work methods, and the methodology for applying them - no more difficult than the energy methodology - become well known. Indeed, it is by illustrating the available-work methods that the error of the energy methods will become apparent. (Also, the growth of available-work applications will go hand in hand with better, more comprehensible presentations of the Second Law.)

Thermoeconomics needs to flourish, in order to avoid misconceptions and erroneous statistics which would lead to bad mistakes by engineers involved in design and operating decisions, and by managers and politicians who are involved with "energy" use and development in the private and government sectors. Thermoeconomics needs to flourish not only to avoid bad decisions, but also in order to make good decisions - optimally.

REFERENCES

(1) Maxwell, J.C., Theory of Heat, Longmans Green, London, 1871 (also see later editions).

(2) Gibbs, J.W., 1875; see Collected Works, vol. 1, p. 77, Yale U. Press, 1948.

(3) Gouy, M., 'Sur l'energie utilisable', Jl. Phys., 8, 504 (1889).

(4) Stodola, A., Die Dampfturbinen, Springer Verlag, Berlin, 1905 (also see later editions and the English translation, Steam Turbines).

(5) Jouguet, M., 'Le theoreme de M. Gouy', Revue Mec., 20, 213 (1907).

(6) Darrieus, G., "The Rational Definition of Steam Turbine Efficiencies", Engineering, 130, 283 (1930).

(7) Keenan, J., "A Steam Chart for Second Law Analysis", Trans. A.S.M.E., 54, 195 (1932).
(8) Keenan, J., "Thermodynamics", Wiley, 1941.
(9) Obert, E., "Thermodynamics", McGraw-Hill, 1948.
(10) Obert, E.,and Birnie, C., Losses in a 60MW Power Station, Proc. Midwest Power Conf., pp. 187 ff, 1949.
(11) Marchal, R., La Thermodynamique et le Theoreme de l'energie utilisable, Dunod, Paris, 1956.
(12) Obert, E., Concepts of Thermodynamics, McGraw-Hill, 1960.
(13) Bosnjakovic, F., Technische Thermodynamik, Steinkopf, Dresden, 1960.
(14) Gaggioli, R., "The Concept of Available Energy", Chemical Engineering Science, 16, 87-96, 1961.
(15) Gaggioli, R., "The Concepts of Thermodynamic Friction, Thermal Available Energy, Chemical Available Energy and Thermal Energy", Chemical Engineering Science, 17, 523-530, 1962.
(16) Tribus, M., and Evans, R., Thermoeconomics, UCLA Report No. 62-63, 1962.
(17) Obert, E., and Gaggioli, R., Thermodynamics, 2nd ed., McGraw-Hill, 1963.
(18) Hatsopoulos, G., and Keenan, J., Principles of General Thermodynamics, Wiley, 1965.
(19) Tribus, M., Evans, R., and Crellin, G., Thermoeconomics, Ch. 3 of Principles of Desalination, edited by K. Spiegler, Academic Press, 1966.
(20) Evans, R. and El-Sayed, Y., "Thermoeconomics and the Design of Heat Systems", A.S.M.E. Paper No. 69-Pwr-A, 1969. Also published in Trans. A.S.M.E., J. for Power.
(21) Reistad, G., Obert, E., and Gaggioli, R., "Available Energy and Economic Analyses of Total Energy Systems", Proc. Amer. Power Conf., 32, 603-611 (1970).
(22) Reistad, G., "Available Energy Conversion and Utilization in the United States", A.S.M.E. Paper No. 74/WA-PWR-1, 1974. Also published in Trans. A.S.M.E., J. of Power, 1975.
(23) Gaggioli, R., et al, Pinpointing Inefficiencies in Power Plants, Proc. Amer. Power Conf., pp. 656 ff., 1975.
(24) Gyftopoulos, E., Lazarides, L., and Widmer, T., Potential Fuel Effectiveness in Industry, Ballinger Publ. Co., 1975.
(25) Gaggioli, R. and Petit, P., "Second Law Analysis of Fuel Conversion Systems," A.C.S. Symposium Series, 21, No. 2, 56 (1976).
(26) Fehring, T. and Gaggioli, R., "Economics of Feedwater Heater Replacement", A.S.M.E. Paper No. 76/JPGC-PWR-7 (to be published in Trans. A.S.M.E., J. of Power).
(27) Gaggioli, R. and Fehring, T., "Economics of Boiler Feed Pump Drive Alternatives," accepted for presentation at the 1977 ASME-IEEE Joint Power Generation Conference, Long Beach, CA, Sept. 19-23, 1977.
(28) Gaggioli, R. and Contratto, R., "Economically Optimal Pipe and Pipe-insulation Sizing", to be submitted to A.S.M.E.

EFFECTIVE ENERGY END-USE — OPPORTUNITIES AND BARRIERS

Elias P. Gyftopoulos, Massachusetts Institute of Technology, Cambridge, Mass.
Thomas F. Widmer, Thermo Electron Corporation, Waltham, Mass., USA

ABSTRACT

The contribution of improved end-use efficiency to the resolution of the energy problem is discussed. An accelerated conservation plan is presented, and measures for overcoming barriers that inhibit the adoption of conservation are described.

INTRODUCTION

This paper discusses: 1) the significant contribution that improved end-use efficiency can make to the resolution of the energy problem, not only from the standpoint of more effective use of resources, but more importantly from that of the national economy; 2) some barriers that inhibit the adoption of conservation measures; and 3) an accelerated conservation plan.

Two well-known aspects of the energy problem are: 1) the fuels that we currently consume (especially the liquid and gaseous forms) are finite and are being exhausted at an alarming rate; and 2) the incremental capital investments for the new energy supplies are much larger than the average investments that have been made in the past to obtain comparable amounts of existing supplies.

Because of the scarcity of liquid and gaseous fuels and the high cost of new energy sources, international fuel prices have increased much faster than the prices of other commodities. Reasonable estimates indicate that the gap between relative prices of energy and other factors of production will keep widening for many decades to come. The change in relative prices raises the question: "Is it possible to reoptimize each energy-consuming task so as to achieve the same result at equal or lower overall cost, but using less energy?"

The answer to this question is an unqualified yes, based upon firm technical and economic considerations. The opportunity for reduced energy consumption per unit of product in every sector of the economy is enormous. If we do not take advantage of this opportunity, our economic well-being and security will be adversely affected.

The reoptimization of energy end-use will, of course, require long-term commitments involving significant restructuring of all sectors of society. It cannot happen automatically because of many traditional barriers and many distortions of the free market system introduced by past decisions. Though difficult, these barriers and distortions are not immovable. If boldly faced by a comprehensive energy policy, they can be largely eliminated.

TECHNICAL BASIS FOR CONSERVATION

The technical basis for energy conservation can be estimated in a most convincing fashion by means of the laws of thermodynamics. To do so, we need a yardstick that is universally applicable to all fuels and all processes, and that leads to the so-called second-law efficiency.

The necessity for such a yardstick was illustrated by several examples in a paper presented at the 1972 MIT Energy Conference (1) where the authors concluded that: "The laws of thermodynamics indicate that neither energy, nor heat, nor enthalpy nor Gibbs free energy are, in general, satisfactory yardsticks. The relevant quantity is a property called available useful work that is, in turn, uniquely related to another important property called entropy."

This yardstick has been used in several studies to evaluate the second-law efficiency of various processes (2,3). Some results are: residential and commercial space heating, 6%; residential and commercial water heating, 3%; air conditioning and refrigeration, 5%; automobile propulsion, 10%; steel production, 21%; petroleum refining, 9%; cement manufacturing, 10%; and paper production, less than 1%.

The average of the above efficiencies, weighted according to the total amount of fuel used for each purpose, is only 8.3%. Collectively, the listed end-use applications comprise almost 60% of all U.S. energy consumption. Moreover, the figure of about 8% is believed to be fairly representative of the overall energy effectiveness prevailing throughout the economy.

It is not suggested that energy efficiency will ever approach 100% for real devices or processes, even in the remote future. Nevertheless, the opportunity for savings is enormous and it is important to draw attention to the fact that no fundamental scientific barriers exist that might prevent substantial improvements in energy end-use effectiveness. Even a modest improvement from 8 to 9% efficiency, for example, represents a saving of almost 10 Quads per year (10×10^{15} Btu per year) at 1975 consumption levels, the energy equivalent of 4.6 million barrels of petroleum per day.

A convenient form of writing the second-law efficiency is given by the relation

$$\eta = C_2 E_2 / C_1 E_1$$

where E_2 is the energy required by the transformation induced by the process, C_2 is the quality of the required energy, E_1 is the total primary energy (equivalent fuel energy) consumed in the process, and C_1 is the quality of the consumed energy. Quality is an important characteristic of energy because, as is well known, an amount of energy at high temperature is more useful than an equal amount at low temperature.

Values of efficiencies can be calculated for complete processes, such as the transformation of wood into a special type of paper, or of iron ore into a special steel. The results vary widely even for slightly different types of the same product (e.g., hardened steel versus steel ingots). For this reason, efficiencies of complete processes are not a practical basis for formulating policies.

On the other hand, values of efficiencies can be determined for specific stages or functions in manufacturing, such as steam raising at specified conditions, or heat treating at a given temperature. For certain functions, the results are almost independent of the product and uniquely related to the equipment in question and its function. Thus, they can be helpful to policy makers.

Examples of functions that fall in the category just cited are: heating of stock, raising of steam, and generation of motive power or electricity. A detailed list of typical equipment and functions is given in Table 1. These classes of equipment are widely used in manufacturing and account for about 80% of all the energy consumed in industry.

TABLE 1 Examples of Process Equipment and Function

Equipment	Function
Electric Motor	Provision of motive power
Oven	Heating of stock, at T ~350°F
Heat-Treating Furnace	Heating of stock, at T ~1600°F
Dryer	Evaporation of moisture, at T ~200°F
Steam Boiler	Production of steam, at T ~350°F
Space Heater	Space heating, at T ~100°F
Water Heater	Water heating, at T ~150°F
Bottoming Engine	Generation of electricity from waste heat
Topping Engine	Generation of electricity and production of heat or steam
Slab Reheat Furnace	Heating of stock, at T ~1400°F (before forming)
Electric Furnace	Heating and melting of stock, at T ~2800°F

Each of the functions listed in Table 1 requires an amount of energy, E_2, of a certain quality, C_2. For a given level of production, the energy demand, E_2, can be readily calculated by means of standard procedures. Its quality, C_2, on the other hand, can be evaluated as follows: (i) if the function is generation of motive power or electricity, $C_2 = 1$; (ii) if the function is heating of stock at a particular temperature, T(°F), C_2 is given by the thermodynamic relation

$$C_2 = 1 - \frac{530}{T - 70} \ln \frac{T + 460}{530};$$

and (iii) if the function is steam raising or drying of stock, C_2 is given by the relation

$$C_2 = 1 - 530 \frac{\Delta s}{\Delta h}$$

where Δs and Δh are the entropy and enthalpy changes from water at atmospheric conditions to steam at the desired conditions, respectively.

The energy required by each of the functions in question is satisfied by consuming fuel, electricity, or recovered waste heat. For each fuel, the energy, E_1, is computed by using the heating value of the fuel. For electricity, the energy is the number of kWh consumed times 10,000 Btu/kWh because electricity is generated from fuels at a rate such that one unit of electricity energy requires about 3 units of fuel energy. For waste heat, the energy is found by considering changes in the energy content of the material that carries the waste heat.

Qualities, C_1, of different forms of primary energy and waste heat are as follows: for gas, distillate oil and electricity, C_1 is about unity; for residual oil and coal, C_1 is 0.98 and 0.97, respectively, because some energy is required to prepare these fuels prior to combustion; for a waste heat source at temperatures, T, C_1 is evaluated by the same formula as C_2 for a heating function.

From a number of specific examples we can ascertain that, under present practices, the efficiency ($C_2 E_2/C_1 E_1$) for many of the functions in Table 1 ranges between 0.15 and 0.25, and that, using only known technology, these efficiencies can be substantially increased.

The function-related efficiency can be evaluated either for a single piece of equipment or for several pieces collectively. It can also be applied to equipment with different types of materials (outputs) being processed and several forms of energy being supplied. The energy-quality product for each output is evaluated as discussed above and the results are additive. Moreover, the energy-quality product for each of the energies supplied is computed as above and the results are additive. The overall efficiency is the ratio of the two sums.

ECONOMIC BASIS FOR CONSERVATION

A useful perspective on the key issues of energy economics can be obtained by examining capital investment requirements for various alternative supply and conservation measures.

On the supply side, the diminishing store of readily available fossil-fuel resources has resulted in the investment of large amounts of capital per unit of energy production capacity. Apart from the Middle East where reserves are still readily accessible, most new petroleum or natural gas production facilities (e.g., U.S. outer continental shelf, North Sea, Alaska, etc.) require anywhere from \$10,000 to upwards of \$15,000 for each barrel per day of equivalent fuel energy provided. This translates into a capital demand of about \$4.5 to \$6.8 billion for every Quad per year of energy delivered. Synthetic gas and oil obtained from coal will be even more capital intensive, probably in excess of \$10 billion per annual Quad.

New coal supplies are still obtainable at a capital cost of \$1.5 billion to \$2.0 billion per Quad per year. However, the mining, processing, and combustion of coal are attended by serious environmental and safety problems that may ultimately limit the rate of coal consumption, or at least cause increases in the cost of supply. Moreover, coal does not possess the flexibility of application inherent to oil and gas. The industrial sector could undoubtedly substitute more coal for such purposes as the raising of process steam, but any major increase in our reliance upon coal will depend, for the most part, upon its greater use by electric utilities or the development of economical gasification methods.

The capital investment requirement is much higher when energy is provided in the form of electricity. For every Quad of delivered electricity per year, the capital investment in facilities for fuel supply, generation, transmission and distribution will range from $45 billion for coal-based systems to about 1-1/2 times this figure for nuclear systems. These figures are not directly comparable to the costs for coal and petroleum fuel resources, since electricity has far greater flexibility of usage than does raw fuel. Even so, the capital cost of coal-based electricity is about $15 billion per annual Quad of coal converted to electricity, or more than 8 times the capital cost of raw coal supply itself.

Despite its high capital cost, electricity occupies a unique and vital place in the spectrum of energy forms. Many tasks exist that can be performed only by energy of the highest thermodynamic grade; e.g., electricity or mechanical work. For this reason, electricity is an essential part of a balanced energy supply system. Electricity should be recognized, however, as having both special properties and high capital intensity, and therefore should not be used as a convenience fuel.

The enormous and growing capital requirements of the energy supply industry hold serious implications for the entire economy. The New York Stock Exchange (4) estimated that total capital demands of the energy supply industry would exceed $800 billion in the 1975 to 1985 period. Even with highly optimistic projections concerning economic growth and capital formation, it is unlikely that the U.S. economy will produce more than $2.7 trillion of investment capital for all purposes over that ten-year span. Assuming that the long-standing ratio between business and residential investment prevails, there will be about $1.8 trillion to be divided among all business investments for both new capacity and replacement purposes.

The prospect of allocating almost half of all business capital to energy supply investments alone is truly alarming in its implications for the economy. In the recent past, the energy industry has consumed only about one-fourth of total U.S. business capital, and even this fraction has created growing stresses in the capital markets. Unless this trend is reversed, we will soon be devoting so much of our scarce capital resources to energy production that other business needs will be denied access to sufficient investment funds.

In contrast to the picture of rising expenditures to develop diminishing fuel reserves, conservation can be put to effective use with substantially smaller commitments of capital. An indication of the capital savings potential is best seen by referring to some typical energy-conserving investment opportunities.

One of the simplest of such examples is afforded by the modular air-conditioning units sold for homes and apartments. Data published by the Committee on Energy Research of the Federal Council on Science and Technology (5) showed the following figures for energy consumption and initial cost of three commercially available room air conditioners having exactly the same cooling capacity, 5000 Btu per hour. For specific cooling capacities of 4.58, 5.8, and 8.7 Btu/watt hour, the retail prices are $120, $140, and $165, respectively. By investing only $45 additional first cost (38%), it is possible to obtain an efficiency improvement of 89%.

Since the air conditioner is likely to be used only 500 hours per year, or about 6% of the time, the energy saving bill will be 258 kWh per year. However, its usage is likely to coincide with the period of highest summer demand on the local electric utility system. Hence, the $45 increment for conservation can be viewed as a direct substitution for over 1/2 kW of utility system peak generation and distribution capacity having a value of at least $200.

Unfortunately, the benefits realized by the user do not reflect the same degree of advantage indicated by the capital cost comparison. In fact, the consumer would save only about $18 per year for 500 hours of use, yielding a gross payback of about 4 years. Even this moderately attractive return can be illusory when the ultimate consumer does not participate in the initial purchase decision (e.g., rental apartments or housing equipped by the builder rather than by the owner).

Another example is waste-heat recuperators. Recuperators can provide fuel savings of at least 25% on most high-temperature furnaces used for controlled-atmosphere metal processing. The cost of such recuperators is about $1300 for each combustion burner on a radiant tube furnace, with fuel savings amounting to about 125,000 Btu per hour per recuperator. Under normal operating schedules, this represents a capital cost investment of $1.5 billion per annual Quad of fuel saved, compared to the $6 billion per annual Quad cost of new domestic gas supply.

With the recent sharp rise in industrial gas prices, the payback period for recuperators has shortened to about 3 to 4 years, a range that is still only marginally attractive to most industrial firms whose capital budgets can barely cover essential or "mainstream" business investment needs.

Another example is the generation of electricity from waste heat. Specific engineering studies indicate that bottoming engines for generation of electricity from waste heat cost only 1/3 to 1/2 as much as equivalent new electricity supply systems.

In general, capital investments for on-site generation of electricity by various schemes are smaller than those for central power stations (Fig. 1).

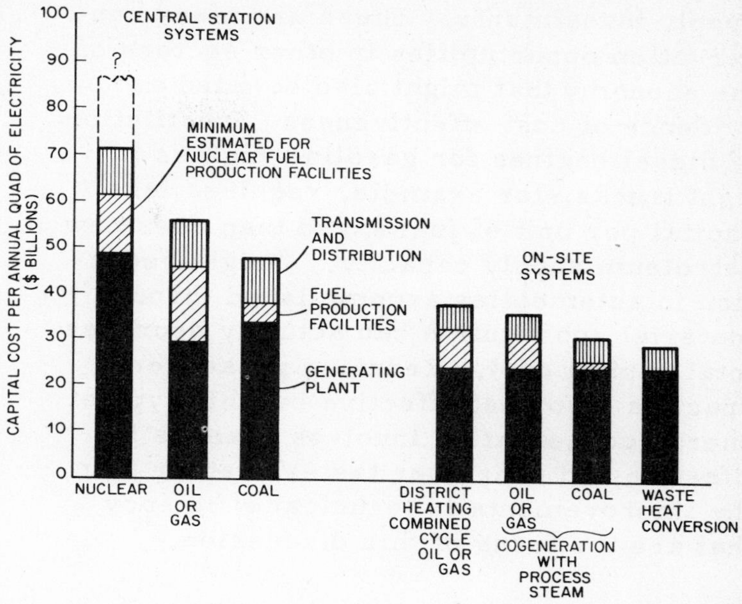

Fig. 1 Capital cost of conventional versus alternative electric generating systems, including fuel supply facility costs; load factor 0.7

Some additional examples of conservation opportunities in industry are summarized in Table 2 to illustrate typical economic tradeoffs between savings and new energy

TABLE 2 Typical examples of industrial conservation measures

Conservation Measure	Energy Saved	1985 Potential National Savings (equiv. bbl/day)	Capital Cost	
			Conservation	New Supply
Recuperators on Steel Reheat Furnaces	10^8 Btu/hr 170 tons/hr steel throughout	35,000 (petroleum or natural gas)	$480,000	10 times conservation
Cogeneration of Electricity and Process Steam in Steel Plants	40 MW electricity for 10,000 ton per day plant	14,600 (fuel for electric utilities)	$41.4 x 10^6	1.7 times conservation
Dry Quenching of Coke in Steel Industry	4 x 10^9 Btu per day of steam fuel plus 1.4 x 10^9 Btu/day coke (3000 ton/day plant)	62,000 (coal, coke and petroleum)	$9.3 x 10^6 (retrofit)	Same as conservation
			$5.3 x 10^6 (new plant)	1.7 times conservation
Reduced Current Density in Hall Process Aluminum Reduction Cells	2,160 kWh per ton of primary aluminum	70,000 (fuel for electric utilities)	$350 per annual ton of primary aluminum	1.25 times conservation
Air Preheaters on Ethylene Plant Process Heaters	4.8 x 10^{11} Btu/yr for 1400 ton per day plant	7,200 (petroleum or natural gas)	$1.14 x 10^6	2.9 times conservation
Steam Generation From Ammonia Process Heater Flue Gas	11 x 10^6 Btu/hr for 1000 ton per day ammonia plant	3,300 (petroleum or natural gas)	$130,000	6 times conservation
Power Recovery Turbines on High-Pressure Refinery Process Flow Streams	11.0 MW for 100,000 barrel per day refinery	44,000 (fuel for electric utilities)	$2.8 x 10^6	7.2 times conservation
Increased Use of Bark and Wood Wastes as Fuel in Paper Industry	10^9 Btu per day for 400 ton per day mill	97,000 (coal, gas and petroleum)	$1.7 x 10^6	1.1 times conservation

Incremental new energy supply facility costs per daily barrel equivalent:

Coal $ 4,000
Oil or Gas $15,000

Baseload Electric (1/2 coal, 1/2 nuclear): $130,000

supply investments. There are many conservation opportunities in other sectors of the economy that might also be cited as evidence of cost effectiveness. Substitution of diesel engines for gasoline engines in light trucks, for example, requires less capital per unit of fuel saved than does new petroleum supply capacity. Weight reduction in automobiles accomplished through material substitution can actually decrease total capital cost. Reducing passenger space is also cost effective but this type of energy conservation involves changes in lifestyle and consumer tastes, rather than the improvements in technical efficiency that are the focus of this discussion.

It is important to make clear the distinction between two kinds of conservation, and to dispel the popular misconception that conservation is equivalent to belt-tightening. Usually taken in response to immediate crises, belt-tightening measures tend to obscure the real and lasting benefits of conservation through improved end-use effectiveness.

Because of their economic attractiveness, capital investments in conservation might be expected to proceed at a faster rate in the industrial sector of the economy where energy users are likely to have a greater awareness of first-cost versus operating-cost tradeoffs. Recent experience does not bear out this assumption, however, and industry has not significantly outpaced other sectors in improving its energy efficiency.

BARRIERS TO CONSERVATION

Several factors contribute to the inertia of industrial energy-users against investing in energy-efficient equipment:

(1) Most energy-user companies are forced to maintain conservative debt-to-equity ratios because of uncertainty about the future availability and cost of financing. Conservation investments, therefore, do not usually command high priority in the competition for limited capital funds that must first be reserved for essential mainstream business purposes (e.g., tooling new products, expansion of capacity to meet market conditions, etc.).

(2) Criteria for investment payback are more stringent for manufacturing companies than for regulated utilities whose risks are lower. Typical values of return on investment of manufacturers, utilities, and energy suppliers are shown in Fig. 2.

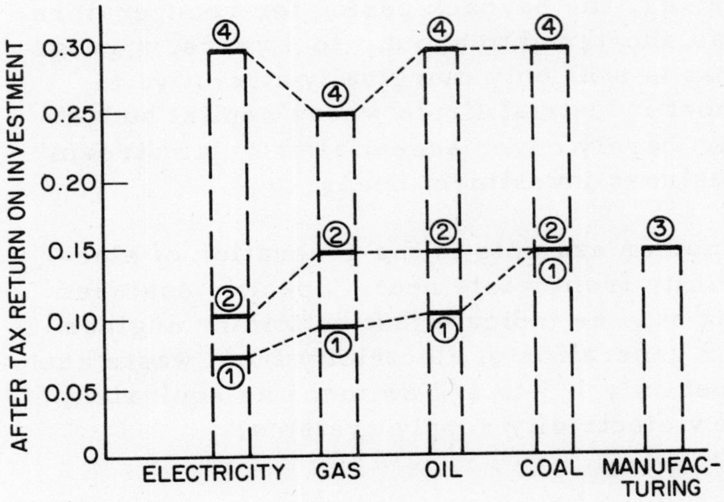

Fig. 2 Prevailing returns on investment in energy supply and manufacturing: 1 incremental investments in supply; 2 average investments in supply; 3 mainstream investments in manufacturing; and 4 energy conservation investments in manufacturing

(3) The pricing of industrial electricity and fuel is largely based on average, rather than replacement costs of supply. The difference between average and replacement costs over the past few years is shown in Fig. 3.

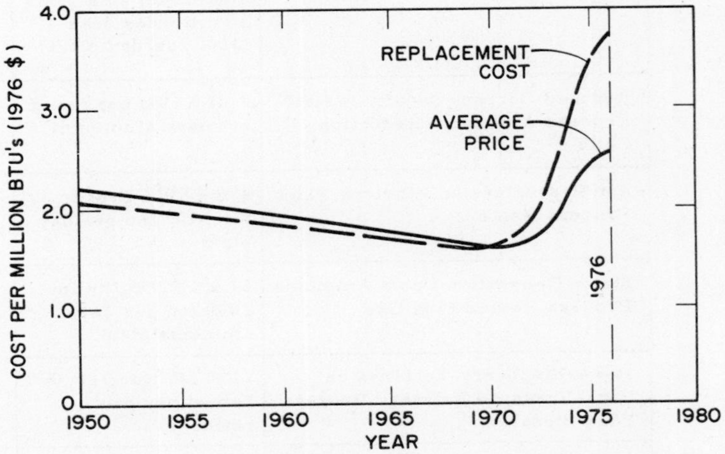

Fig. 3 Aggregate cost in constant dollars of industrial energy from 1950 to 1976, weighted according to usage of coal, oil, gas and electricity in industry

(4) Regulations regarding the generation and sale of electricity restrict the adoption of cogeneration of electricity and process steam at industrial sites.

Taken together, these factors tend to create a major distortion in the deployment of scarce resources toward the most optimum balance between new energy supply investments and energy conservation investments.

EFFICIENT GROWTH THROUGH ACCELERATED CONSERVATION

Adoption of a policy by the Government that stresses energy conservation can produce major changes in our energy usage patterns in a relatively short time without impairing economic expansion. Comparison of energy demand estimates, with and without an accelerated conservation approach, reveals the important differences that might be anticipated during the next decade.

In 1975, FEA (6) provided a ten-year forecast for U.S. energy production and usage. The effects of higher energy prices alone, according to FEA, could be expected to cut 1985 consumption from an unconstrained demand level of 123 Quads to about 107 Quads. Demand would be restricted further, to about 101 Quads, by some additional conservation measures that were not specified.

Under the FEA plan, electrification was to increase from 1.93 trillion kWh (or 25% of all energy input in 1975) to 3.35 trillion kWh (or more than 34% in 1985). Thus, with the real GNP expanding at about 3% per year (34% over the decade), total energy was to grow by 2.8% per year, and electricity by 5.5% per year. The plan was expected to produce almost no change in the distribution of energy by the end-use sectors relative to the pattern existing in 1975 (residential and commercial 37%, transportation 24%, and industry 39%).

A shift in the sources of energy was projected, with coal rising from 18% in 1975 to 22% in 1985, nuclear energy rising from 3% to almost 10% (accounting for over one-fourth of electricity generation) and oil and gas declining from 74% to 63% (the major reduction occurring in electric utility consumption of these fuels).

In order to provide a framework for evaluating these forecasts, an alternative plan has been devised that stresses conservation measures. Based upon the same growth in real GNP as assumed in the FEA plan, about 3% per year, the alternative approach postulates no substantial social changes or curtailment of living standards. This plan, referred to as an Accelrated Conservation Policy, is by no means the only plan that might be considered. However, its key elements will illustrate some of the benefits that can be realized by more effective end-use of energy.

A prime element of the alternative plan involves transfer of a major portion of the capital now marked for new energy supplies into investments aimed at energy conservation in each of the end-use sectors. A variety of specific measures is needed to accomplish this change, an important result of which will be a major reduction in the total amount of capital required for all energy investments.

The proposed plan calls for sharp curtailment in the rate of growth for electricity, the most capital intensive form of energy. It will also be necessary to implement certain specific measures aimed at improving end-use efficiencies. Some of the major actions which might be taken include:

(1) Strict enforcement of the mandatory automobile fuel economy standards already enacted by Congress.

(2) Construction of alternative electric generation capacity in lieu of 103,000 megawatts of planned central station capacity. This would be allocated as follows: cogeneration with process steam and by means of waste-heat recovery from industrial processes, 64,000 MW; generation by combined cycle district heating plants producing both electricity and heat for residential and commercial buildings, 32,000 MW; and generation from heat provided by municipal trash incinerators, 7,000 MW.

Together, these sources would contribute 24.5% of all U.S. electricity.

A variety of measures will be needed to stimulate this substantial shift away from central station utilities to the far more efficient combined cycle and waste-heat systems identified above. Possible actions include, but are not limited to, the following: (a) setting mandatory rules for purchase of surplus industrial electricity by utilities; (b) restructuring of back-up or demand charges originally designed by utilities to discourage on-site generation; (c) providing direct Government loans to industries and apartment or commercial complexes in order to finance investments in on-site generating capacity; (d) special taxes levied against industries and commercial businesses failing to take advantage of proven cogeneration opportunities; and (e) changes in Federal, state and local rules regulating utilities.

(3) Establishment of efficiency goals for all energy-intensive industrial processing equipment and systems. The efficiency discussed earlier can be used as a guide, especially for the functions listed in Table 1, which consume 80% of industrial energy.

(4) Enactment of mandatory heating, insulation, and lighting standards for all types of new residential and commercial construction.

(5) Enactment of progressively stricter efficiency standards for all major energy-consuming appliances such as water heaters, refrigerators, air conditioners, home furnaces, etc.

(6) Rapid phase-out of natural gas as a fuel for central station electricity generation or for any process steam applications in industry. This can be accomplished by either a direct ban on such use, or by charging steeply progressive taxes on gas fuel that is misused in this fashion. It is essential that sufficient gas be reserved for residential space heating and for direct-fired high-temperature industrial process heating in order to avoid excessive growth in electricity demand.

(7) Provision for direct Government loans and other economic incentives to finance the retrofitting of insulation, storm windows, improved furnaces, and other cost-effective conservation systems to existing housing. This program should be pursued on a continuing basis until every structure in the Nation has been modified to an extent commensurate with the capital cost of incremental new energy supply. It is not anticipated that this program would be completed until well beyond 1985, and less than 1 Quad was factored into projected savings for these measures over the next decade.

Collectively, these and a number of less important actions would reduce energy consumption over the next decade to 80 Quads per year, a saving of 21 Quads relative to the FEA plan. In effect, energy growth can be almost halted over the ten-year span while economic activity expands by 3% per year. Moreover, the costly electrical sector would increase to only 2.53 trillion kWh, a growth rate of 2.8% per year relative to 1975. The fraction of total energy converted to electricity, 29%, is higher than in 1975, but still well below the 34% figure projected by the FEA plan.

Distribution of energy by end-use sector differs from that of the FEA plan, with transportation accounting for only 20%, and industry rising slightly to 42%. Sources of energy would change somewhat, with nuclear fuel contributing only 7% instead of 10% of all energy. The fraction for oil and gas and hydraulic head is about the same for both plans (69%), but coal's contribution rises from 22% to 24% under the accelerated conservation alternative.

Major contributions to the 21 Quads of total energy savings are due to automobile fuel economy standards (5.6 Quads), alternative methods for electricity generation (2.9 Quads), improved efficiencies in industrial processes (3.5 Quads), and appliance efficiency standards (2.5 Quads). The remaining 6.5 Quads of savings result from improved insulation standards for new commercial and residential buildings, increased retrofitting of insulation in existing structures, small amounts of solar-assisted water and space heating, and greater efficiency in trucks due to wider use of diesel engines, improved scheduling practices, drag reductions, etc.

None of the proposed measures, except for the final stages of automobile efficiency improvement, requires the use of unproven technology. Moreover, the overall improvement represents only a modest aggregate gain in the absolute efficiency of devices and processes. In fact, the absolute efficiency of energy utilization is increased by only 2.6 percentage points to a figure of 10.9%, approximately one-third of the gain being attributable to automobile fuel-economy improvements alone.

The most striking difference between the accelerated conservation policy and the FEA plan is in the amount of capital needed to implement each of these programs. We estimate that the FEA plan would cost $648 billion, $570 billion for new energy supply and $78 billion for conservation, whereas the accelerated conservation policy would cost $318 billion, $161 billion for new energy supply and $157 billion for conservation measures.

The enormous capital savings are due in large part to sharply lower central station electricity generating capacity. This results not only from a reduction in total electrical demand, but also from the lower cost of alternative combined cycle generating equipment (e.g., cogeneration).

The conservation policy requires construction of only 45,000 MW of nuclear capacity, instead of 133,000 MW (Fig. 4). Coal consumption by electric utilities rises from about 390 million tons per year in 1975 to 460 million in 1985, rather than the 700 million tons required under the FEA plan. Over the decade, total coal consumption for all purposes expands from 660 million tons to 890 million tons, for the conservation plan, compared to 1060 million tons for the FEA plan.

Overall investment in central station electric generating capacity, excluding facilities for fuel supply, would total $391 billion under the previous FEA plan. With a policy of accelerated conservation, only $91 billion would be invested by the utilities; an additional increment of $61 billion being spend for alternative combined cycle systems in industry and in district heating and electricity projects.

The greatest dividends of an accelerated conservation policy occur in oil and gas demand. Consumption of both of these fuels declines substantially from 1975 levels, petroleum dropping from 35.9 to 32.3 Quads, and natural gas from 22.0 to 17.8 Quads. In effect, accelerated conservation measures provide the second most important energy source ten years from now (21 Quads versus 32.3 Quads for oil). By contrast, FEA projects that petroleum consumption increases from 35.9 Quads in 1975 to 42.3 Quads in 1985, while gas usage remains nearly constant (21.5 Quads in 1985).

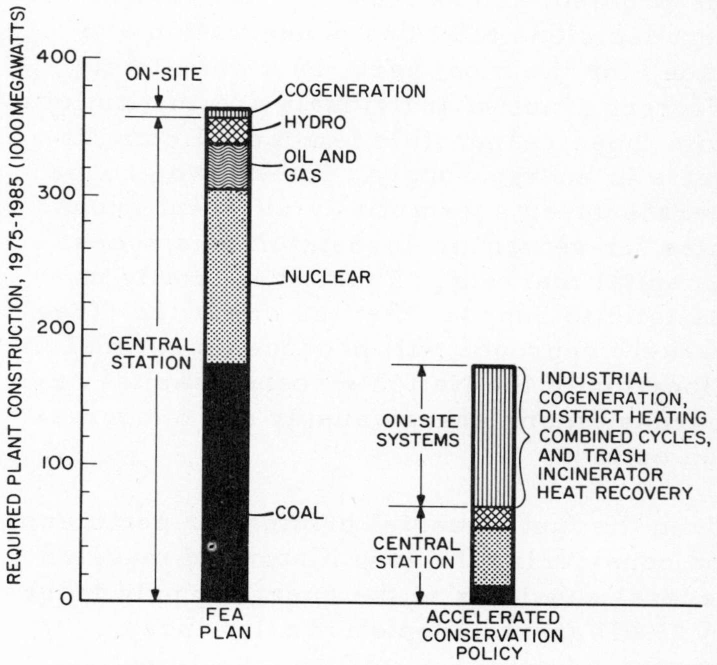

Fig. 4 Distribution of new and replacement capacity for electric generation facilities constructed over the next decade

The all-important effect on imports can be seen by comparing the above demand figures with FEA projections for maximum domestic supply under conditions of price deregulation. Accelerated conservation policies produce a net surplus or reserve of 4.5 Quads per year of natural gas by the end of the decade. Required petroleum imports are only 5.3 Quads or 2.4 million barrels per day (about one-third the present level of imports). Even if natural gas price controls were to be continued, the accelerated conservation plan would curtail usage sufficiently to almost exactly balance FEA's domestic supply forecast of 17.9 Quads for price-controlled supply. In contrast, the previous FEA plan with complete price deregulation achieves zero imports in natural gas, but still requires the import of about 6 million barrels per day of petroleum (over 13 Quads).

There are clear and compelling economic advantages to the Nation for conserving more energy. Before this can occur, however, it will be necessary to resolve the differences between two distinctly different sets of economic assessments that ultimately determine the balance between energy supply and energy conservation decisions.

The problem arises from the fact that investment decisions affecting conservation are made, for the most part, by a completely different group of individuals and institutions from those responsible for capital commitments in energy supply. These two groups operate under substantially different ground rules for return-on-investment and access to capital markets. Thus, it is totally unrealistic to suppose that the so-called "free market" approach will produce an optimal allocation of the Nation's scarce capital resources among energy supply and conservation options.

Given the fact of cartel pricing for petroleum, and considering the long history of massive Federal subsidies to the energy supply industry (R&D grants, depletion allowances, guaranteed return on utility investments, etc.), nothing resembling a free marketplace exists today. This fact has often been obscured by those who promote price deregulation as the only means for a comprehensive energy policy.

ACCELERATED CONSERVATION, THE TRANSITION AND THE LONG-TERM POTENTIAL

The proposed policy for accelerated energy conservation leans heavily upon mandatory measures to improve end-use efficiency. It is inevitable that this approach will raise arguments against tampering with the so-called "free market." Direct intervention must be considered, however, because price alone cannot provide sufficiently strong motivation for accelerated conservation.

The limitations of price as a conservation stimulus, even for the industrial sector, are illustrated by the fact that energy cost still averages well below 10% of value-added for all manufacturing. Thus, even large additional rises in fuel prices will not necessarily place overwhelming conservation pressures upon manufacturers.

The changeover of national priorities to a conservation orientation will require imaginative use of all available instruments of policy, including further mandatory measures tailored to resolve each of a variety of specific problems or barriers to effective energy usage. Overall, the goal should be to devise programs and regulations which will narrow the wide gap that now exists between criteria for investing in energy savings versus new energy supplies.

The implications of a conservation strategy for the more distant future are perhaps even more significant than for the nearer term. Present technology can clearly provide the 21-Quad incremental savings over the next 10 years. Considering that overall second-law efficiency would still be less than 11% at that point, it is almost certain that aggressive research into end-use efficiencies could advance the technology still further. A concerted effort in this area has not even begun, and the untapped potential for improvement may well exceed anything on the horizon among the various alternative energy supply options. If, for example, we were able to continue improving energy efficiencies by about one percentage point every 2-1/2 years, it would be possible to sustain an uninterrupted growth in real GNP of 3% per year for the next 3 decades, and still consume no more energy than we do today (Fig. 5). Even then, our overall end-use efficiency would be only 20%, about equal to that of the steelmaking process today.

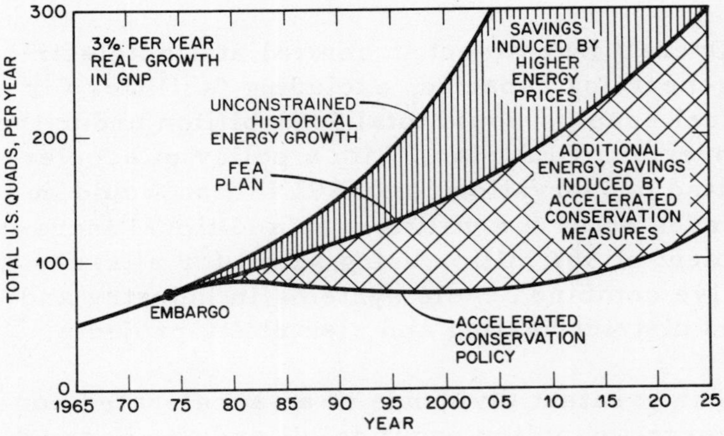

Fig. 5 Projected long-term energy consumption trends for alternative energy policies

The approximate progression of end-use efficiencies corresponding to alternative energy policies is plotted in Fig. 6. It is interesting to note that the progress that must be achieved under a long-term program of accelerated conservation is still less than that accomplished over a comparable number of decades in the improvement of electric generating plant efficiencies. The latter process, of course, has been subjected to enormous and continuing commitments of technological resources, the same prescription that is suggested here for energy end-use processes.

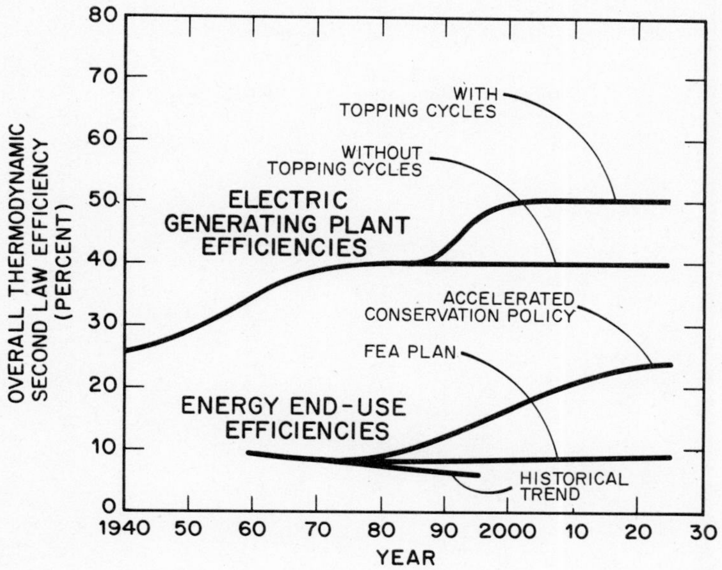

Fig. 6 Projected long-term energy end-use efficiency for alternative energy policies, and evolution of efficiency of central station plants for converting fossil fuel to electricity

Some progress has been made in overcoming the notion that the conservation of energy is synonymous with decreased economic activity. There is a growing awareness that capital investments in energy-saving devices can often yield greater dividends than comparable investments in new supply. Given appropriate stimulus, then, it is likely that the U. S. economy will make substantial progress towards more efficient end-use of energy over the next 10 years. Unfortunately, there is little appreciation of the fact that conservation can play a major role in our long-term energy future. This misconception must be changed in order that attention can be focused upon the task of developing the new conservation technology needed to insure continuing reductions in energy consumption in the period beyond 1985.

BIBLIOGRAPHY

1. J.H. Keenan, E.P. Gyftopoulos, and G.N. Hatsopoulos, The Fuel Shortage and Thermodynamics: The Entropy Crisis, MIT Energy Conference Proceedings, edited by M. Macrakis, MIT Press, 1972.

2. E. P. Gyftopoulos, L. J. Lazaridis, T. F. Widmer, "Potential Fuel Effectiveness in Industry," a Report to the Ford Foundation Energy Policy Project, Ballinger Publishing Company, 1974.

3. Technical Aspects of Efficient Energy Utilization, Summer Study, American Physical Society, 1974.

4. Toward a National Energy Policy, Capital Requirements, Mobil Oil Corporation, 1975.

5. C.A. Berg, A Technical Basis for Energy Conservation, Technology Review, February 1974.

6. National Energy Outlook, Federal Energy Administration, FEA-N-75/713, 1976.

AVAILABILITY ANALYSIS

M. V. Sussman

Tufts University, Medford, Mass., 02155

All Calories are Equal, but Some are Worth More than Others

The cost of energy depends not only on the quantity used but to a very large measure on its quality. Yet the fact that all Btu's or calories of energy do not have the same quality, is not always appreciated. The quality of energy is measurable by the amount of work that may be extracted from it. Different sources of energy contain radically different amounts of work. For example, one thousand Btu (252 Kcal) of electricity (0.29 K -Watt - hrs.) may, in theory, be converted completely into work, whereas 1000 Btu of 250 psi steam (0.87 pounds) may at best produce only 330 Btu of work, and 1000 Btu of atmospheric steam may at best surrender only 180 Btu of its energy as work.

The measure of the work available from any energy stream is given by the thermodynamic property called "*Availability*".

Availability

The availability is the same as the theoretical maximum work (i.e. *the reversible work*), that may be obtained from an energy source when the material in that source is moved by completely reversible processes into equilibrium with the atmosphere. Completely reversible processes are processes in which heat transfer occurs only when the system is at the temperature of the surroundings; mass transfer occurs only after the system has reached the concentration of the surroundings; and all mechanical work effects occur without friction and with infinitesimal force imbalance. The processes may in general be represented on Temperature - Entropy diagrams (Fig. 1) by a vertical, isentropic, path (1-2) during which the system moves reversibly into temperature equilibrium with the atmosphere, followed by a horizontal, isothermal, path (2-a) during which pressures and concentrations equilibrate reversibly with the atmosphere. An idealized set of expansion turbines for affecting these processes in a steady-flow manner is shown in Fig. 2.

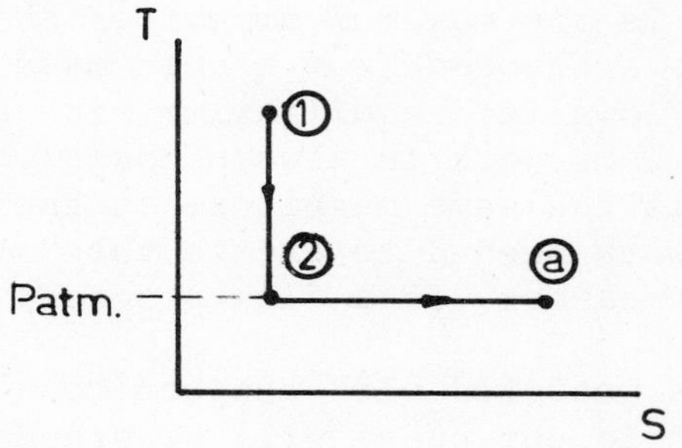

Figure 1

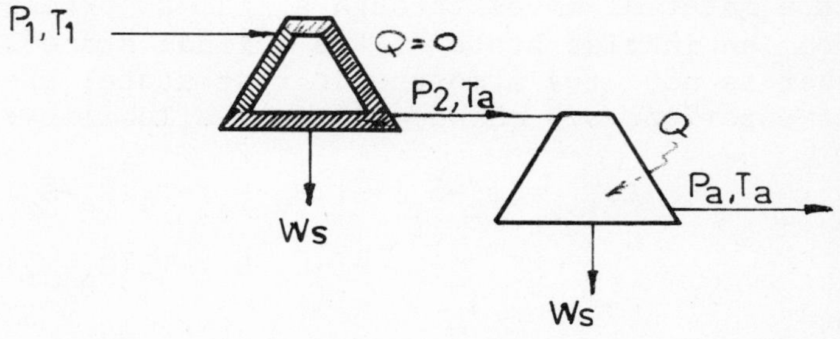

Figure 2

When we account for all the energy entering or leaving any *steady flow process*, the accounting, which is simply a statement of energy conservation, yields the simple equation:

$$\bar{Q} - \bar{W}_s = \Delta\bar{H} \qquad [1]$$

when there are no potential or kinetic energy effects. $\bar{Q}$ is the heat entering the process. $\bar{W}_s$ is the work leaving through the turbine shafts, and $\Delta\bar{H}$ is the enthalpy change of a unit mass of the material traversing the apparatus. The work output, $\bar{W}_s$, is a *maximum* when the process is *completely reversible*.

Therefore from equation [1]

$$\bar{W}_{s(reversible)} = \bar{W}_{s(maximum)}$$

$$= -(\Delta\bar{H}_{1-a} - \bar{Q}_{reversible})$$

Because all the heat effects occur reversibly at T_a we may also write:

$$\bar{W}_{s(MAXIMUM)} = -(\Delta\bar{H}_{1-a} - T_a\Delta\bar{S}_{1-a})$$

or

$$\text{AVAILABILITY} = -(\Delta\bar{H}_{1-a} - T_a\Delta\bar{S}_{1-a})$$

$$\equiv \phi \qquad [2]$$

Equation [2] is the formal definition of the thermodynamic function, AVAILABILITY, ϕ. "Availability" measures the work extractable from an energy stream, in a *steady flow process**. It is the *maximum amount of shaft work that can be extracted from a unit mass of matter as it flows into equilibrium with the atmosphere*. This work is always *positive* (or zero) and is the same magnitude as the least amount of work needed to create the matter from the atmosphere in a flow process.

ϕ is always measured with respect to a rest state, that in our usage will be standard environment at 25°C and one atmosphere. T_a is therefore 298°K, or 573°R.

If a material moves through a flow process from an initial state (1) to a final state(2), that is *not* the atmospheric rest state, then it experiences a *change,* Δ, in availability:

$$\Delta\phi_{1-2} = (\phi_2 - \phi_1) = -[(\bar{H}_a - \bar{H}_2) - T_a(\bar{S}_a - \bar{S}_2)]$$
$$-[(\bar{H}_a - \bar{H}_1) - T_a(S_a - S_1)]$$

or

$$\Delta\phi_{1-2} = (\bar{H}_2 - \bar{H}_1) - T_a(\bar{S}_2 - \bar{S}_1), \text{ or}$$

$$\Delta\phi = \Delta\bar{H} - T_a(\Delta\bar{S}) \qquad [3]$$

where the Δ signifies a *change* between an initial and any final state, not limited to the rest state of the atmosphere, as in equation [2]. However, T_a is, as before the temperature of the standard atmosphere. Notice the minus sign in equation [2] and none in equation [3].

*Another "availability" function, has been defined for *non-flow* systems:

$$\Delta B \equiv \Delta U + P_a\Delta V - T_a\Delta S + \Sigma\mu_{ia}n_i.$$

The function is also called "essergy", and "exergy". Equation [2] is more useful for engineering flow-process analysis.

Equation [2] and [3] are perfectly general and $\Delta\phi$ depends only on the initial and final states (1) and (2), and is *independent of the path* taken to get from (1) to (2). Consequently, the change in availability of a unit mass of matter flowing from state (1) to state (2) is found by evaluating $\Delta\bar{H}$ and $\Delta\bar{S}$ for that change and inserting their values in equation [3].

Availability and Gibbs Free Energy

A function closely related to, but *not* identical with the availability is the *Gibbs free energy*. A change in free energy is defined as:

$$\Delta G \equiv \Delta H - \Delta(TS)$$

Under the special circumstances of an *isothermal* process:

$$\Delta G_T = \Delta H - T\Delta S$$

If, in addition, the isotherm is at T_a:

$$\Delta G_{T_a} = \Delta H - T_a\Delta S = \Delta\phi \qquad [4]$$

or, the free energy change is equal to the availability change only for an isothermal process at T_a.

The free energy change is a measure of work extractable from *isothermal* changes of state. The availability by contrast, is completely general; it measures the work extractable from *any* change of state.

Availability Analysis of Physical Processes

a) ϕ of Saturated Steam: Example (a); Find ϕ for 250 psia saturated steam relative to liquid water at 77°F and 1 atm total pressure.

Answer: To evaluate Equation [2] we use information on the enthalpy and entropy of steam obtained from a set of "Steam Tables". (Combustion Engineering Co.)

State	(1)	(a)	
P	250	14.7	psia
$\bar{H}$	1201.1	45.04	Btu/lb
T	401	77	°F
$\bar{S}$	1.5364	0.0877	

Therefore $\phi = -[(45.04-1201.1)$
$-537(0.0877-1.5264)]$
$= +383.5$ Btu/lbm.

Notice that the enthalpy change for the steam, (the heat released by condensing the steam to liquid water at 77°F) is -1156 Btu/lb. Only 383.5 or 33% of this heat energy can, even under optimal theoretical circumstances, be converted to work.

The availability of steam is more sensitive to temperature and pressure then is the enthalpy. Table I compares the values of enthalpy, and availability for saturated steam at pressures from 1000 to 15 psia. Enthalpy drops only 5%, while availability drops 53% over this range.

TABLE 1. SATURATED STEAM

P	T	$\bar{H}$ (Ref: $H_2O(l)$ @ 77°F) Btu/lb	ϕ Btu / lb	ϕ/H	C_e $/1000 lb.
1000	544.6	1147.9	448	.38	7.17
500	467.0	1159.7	420	.35	6.72
250	401.0	1156.1	384	.32	6.14
100	327.8	1142.2	329	.28	5.26
50	281.0	1129.1	286	.24	4.58
30	250.3	1119.1	254	.21	4.06
15	213.0	1105.9	210	.18	3.36

b) ϕ of Liquid Air:

i) What is the most work that can be obtained from a kilogram of liquid air at its boiling point of -194.5°C and 1 atmosphere by allowing it to come to equilibrium with the atmosphere at 25°C?

ii) What is the least amount of work needed to liquify a kilogram of air at one atmosphere pressure starting with air at 1 atm and 25°C?

Answer: The answers to (i) and (ii) are $+\phi$ and $-\phi$ respectively, of saturated liquid ait with initial state, 1 atm and -194.5°C.

$$\phi = -[\Delta\bar{H} - T_a\Delta\bar{S}] \qquad [2]$$

To obtain this work we could use the liquid air in a special expansion engine heated by the atmosphere or we could use the liquid air as a heat sink for a Carnot engine drawing its heat from the atmosphere. The example illustrates that $\phi > 0$ even when $T_1 < T_a$.

c) $\Delta\phi$ for a Throttling Process: Compute the loss of availability resulting from the adiabatic throttling of a superheated steam, initially at 250 psia and 600°F to a pressure of 100 psia.

Answer: Throttling processes occur at *constant enthalpy*. Knowing that the pressure after throttling is 100 psia and that the enthalpy is the same as in the initial state, allows us to find the corresponding temperature* and entropy* by interpolation in a superheated steam thermodynamic property table.

STATE			
P	250	100	psia
T	600	578*	°F
$\bar{H}$	1319.0	1319.0	
$\bar{S}$	1.6502	1.7483*	

In this example we move to a final state at a higher T and P than the atmosphere. $\Delta\phi$ is given by equation [3].

$$\Delta\phi_{1\to2} = \phi_2 - \phi_1 = (\bar{H}_2 - \bar{H}_1) - T_a(\bar{S}_2 - \bar{S}_1)$$

For our example:

$$\Delta\phi = 0 - 537(1.7483 - 1.6502)$$
$$= 52.6 \text{ Btu/lb}$$

The throttling process *wastes available energy* even while *conserving enthalpy*. In this example, 12% of the available energy is lost by dropping the steam pressure.

d) $\Delta\phi$ for a Mixing Process; Ideal Solutions: The availability loss on preparing, in a flow process, *one mole of ideal solution* from *pure* ingredients at the rest state temperature is:

$$\Delta\phi_{\text{ideal solution}} = RT_a\Sigma x_i \ln x_i \qquad [5]$$

where x_i is the mole fraction of each component in the solution. Notice that equation [5], because it involves the logarithm of composition fractions (x_i), is always negative. Mixing always result in a *loss* of availability.

Equation [5] may also be used for nonideal solutions if the *activity coefficients*, γ_i, for the components are known. The equation, in its more general form is:

$$\Delta\phi_{\text{solution}} = RT_a\Sigma x_i \ln\gamma_i x_i \qquad [6]$$

where γ_i is the activity coefficient of component i, and is equal to unity in ideal solutions.

e) <u>$\Delta\phi$ for a Thermal Mixing Process with No Energy Loss</u>:

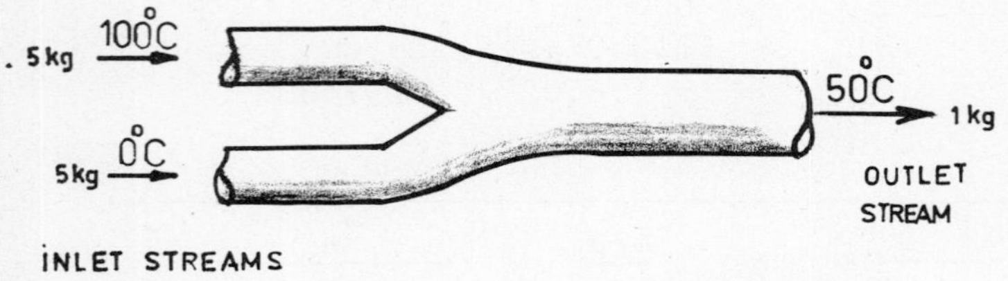

In order to produce 50°C water, a stream of water at 100°C is continuously mixed with stream of water at 0°C. What is the $\Delta\phi$ per kilogram of 50°C

$$\Delta\phi(\text{process}) = \phi(\text{outlet}) - \Sigma\, n\phi(\text{inlet streams})$$
$$= \phi_2 \quad - \Sigma\, n\phi_1$$

From Equation [2]:

$$\phi_2(\text{outlet}) = -[\Delta\bar{H}_{2-a} - T_a\Delta\bar{S}_{2-a}]$$
$$= -\ 1(298-323)$$
$$+\ 298(1)\ \ln(298/323)$$
$$= +\ 25.0 - 24.01$$
$$= 0.99 \text{ kcal./kg.}$$
$$\phi_1(\text{hot}) = -[(298-373) - 298\ \ln(298/373)]$$
$$= -[-75.0+66.90] = \frac{8.10}{\text{kcal./kg.}}$$
$$\phi_1(\text{cold}) = -[(298-273) - 298\ \ln(298/273)]$$
$$= -[25.0 - 26.11]$$
$$= 1.11 \text{ kcal/kg.}$$

Therefore:

$$\Delta\phi(\text{process}) = .99 - (.5\text{x}8.10 + .5\text{x}1.11)$$
$$= -3.62 \text{ Kcal/kg.}$$

The flow of availability in this simple mixing process is illustrated in Fig. 2.

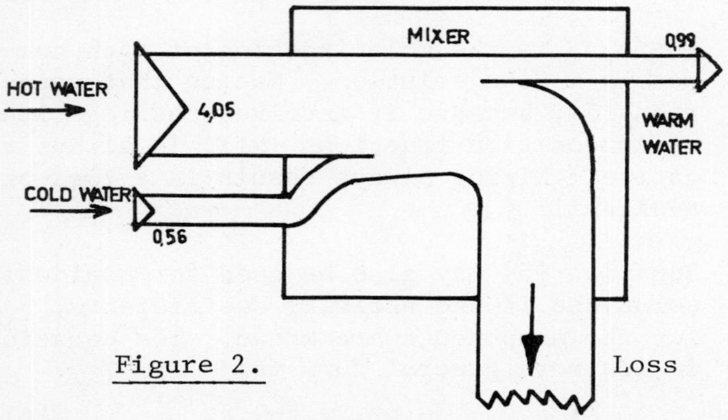

<u>Figure 2.</u>

Notice that enthalpy is conserved in this process. There are *no* heat losses. Nevertheless, 78% of the *availability is lost*, because the heat is transferred across a large temperature interval.

<u>Availability Analysis of Chemical Reaction Processes</u>

For computing ϕ of reactive chemicals, and $\Delta\phi$ of Chemical reaction processes, we define the rest state as: *the products of combustion* in the form of H_2O (liquid), CO_2 (g) SO_2 (g), etc., each *pure*, and in their normal state of aggregation at 25°C and one atmosphere. This is the same reference state used for tables of enthalpy of combustion*.

With the rest state so defined, the availability of any pure material at *25°C* and *1 atm* is the negative of its standard free energy of combustion, i.e., its standard free energy of formation** minus the standard free energies of formation of its combustion products. We will call this quantity the *STANDARD CHEMICAL AVAILABILITY* and represent it by the symbol "ϕ_0".

Table 2 lists Standard Chemical Availabilities of common compounds and elements computed as in Equation [7].

$$\phi_0(\text{of } C_xH_yO_z) = G^0 \text{ (formation)}$$
$$- xG^0_f\ (CO_2 \text{ gas})$$
$$+ \frac{y}{2}\, G^0_f(H_2O\ .\ \text{liq.}) \qquad [7]$$

If the compound or element is at a temperature and pressure other than 25°C and 1 atm, then its availability, ϕ_1, increases by the $\Delta\phi$ for moving the compound from the standard state, ($_\circ$), of 25°C and 1 atm, to the pressure and temperature of the given state, (1).

$$\phi_1 = \phi_0 + \Delta\phi_{0-1} \qquad [8]$$

<u>Example: A Combustion Reaction</u>

Methane, CH_4, is burnt with 20% excess air. The mixture of gas and air enter the burner at 25°C and 1 atm. and burns completely to form CO_2, H_2O (vapor), oxygen, and nitrogen at 1800 C. Compare the availability of the hot combustion gases with the fuel-air mixture entering the burner.

---

*Table 3-203, p. 3-156 "Chemical Engineer's Handbook". Perry, 5th Ed., McGraw-Hill (1973).

**Table 3-202, p. 3-127, Ibid.

25°C

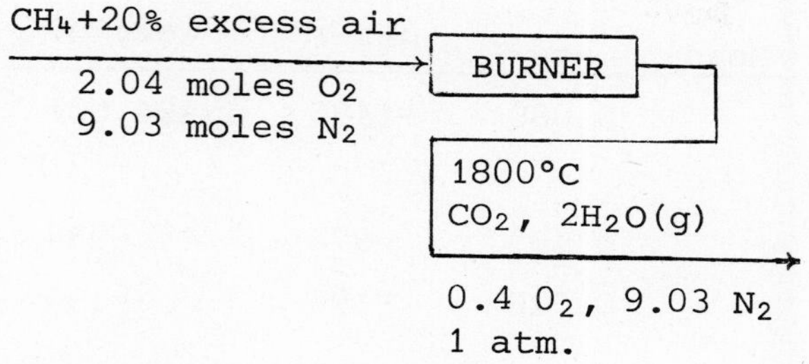

$$CH_4(g)+2O_2(g) \longrightarrow CO_2(g)+2H_2O(g)$$

C_p(mean) 25–1800°C : CO_2 = 12.95
$H_2O(g)$ = 10.25
O_2 = 8.0
N_2 = 7.9

Answer: We will do this computation in tabular forms on the basis of one gram-mole of CH_4, and assume ideal gas behavior. ($\gamma_i=1$)

Moles O_2 required= 2.0(1.20)=2.4

$$\text{Moles } N_2 \frac{\text{entering}}{\text{with air}} = 2.4\text{x} \frac{79}{21} = 9.03$$

The gases leaving the burner will contain 1 mole CO_2, 2 moles $H_2O(g)$, 0.4 moles O_2, and 9.03 moles N_2. Both the fuel and exhaust gases are mixtures. Therefore the pressures of each component is the partial pressure, p_i, in the mixture:

$$p_i = y_i P$$

Where y_i = mol fraction of component "i" in the gas mixture
P = total pressure = 1 atm.

The computations for $\Delta\phi_{0-1}$ for methane and oxygen are represented in Fig. 3, the computation paths move from the standard state, (o), to the partial pressure in the inlet streams.

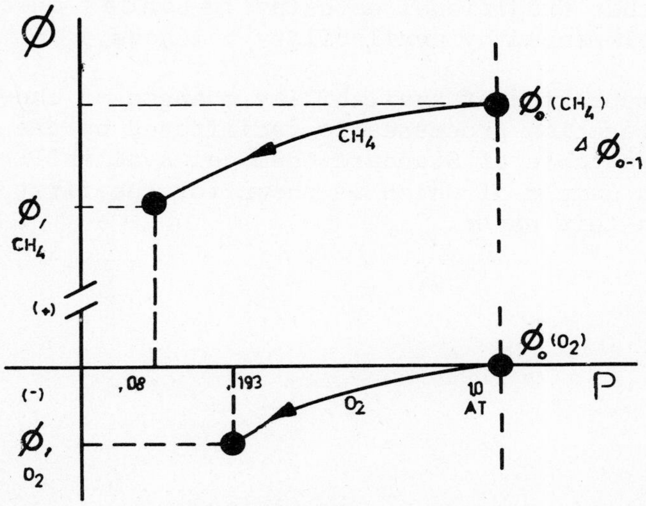

Figure 3.

$$\Delta\phi_{0-1}(CH_4) = (\bar{H}_2-\bar{H}_0)-T_a(\bar{S}_1-\bar{S}_0)$$
$$= C_p(T_1-T_0)-T_a(R\ln P_0/P_{CH_4})$$
$$= -1496 \text{ cal/gm-mol}$$

$$\Delta\phi_{0-1}(O_2) = -298(1.987)\ln 1/0.193$$
$$= -974 \text{ cal/gm-mol}$$

$$\Delta\phi_{0-1}(N_2) = -298(1.987)\ln 1/0.726$$
$$= -190 \text{ cal/gm-mol}$$

The availabilities, ϕ_1, of the components of the inlet streams are:

$$\phi_1 = \phi_0 + \Delta\phi_{0-1};$$
$$\phi_1(CH_4) = 195{,}500-1496$$
$$= 194{,}004 \text{ cal/gm-mole}$$
$$\phi_1(O_2) = 0 - 974 = -974$$
$$\phi_1(N_2) = 0 - 190 = -190$$

The total available energy in the fuel-air mixture is:

$$\Sigma\phi_1(\text{inlet}) = 194{,}004+(2.4)(-974) + (9.03)(-190)$$
$$= 189{,}951 \text{ cal/gm-mole } CH_4.$$

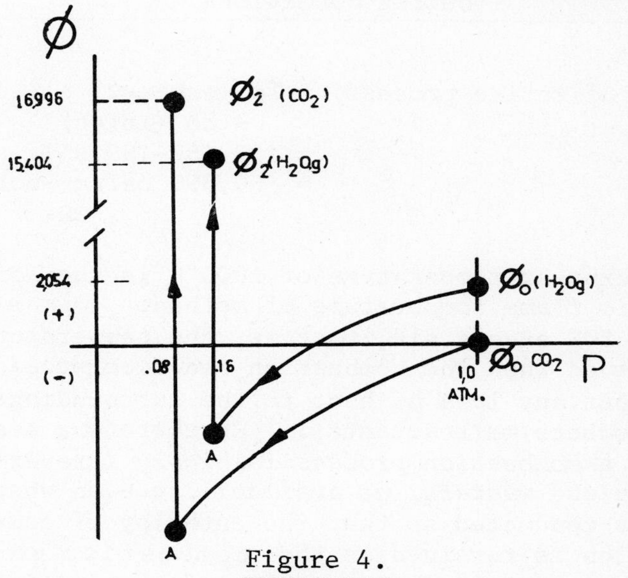

Figure 4.

Figure 4 shows convenient computation paths for evaluating $\Delta\phi_{0-2}$ of the water vapor and the CO_2 in the exhaust gas. The path move through two processes: an *isothermal expansion* from ϕ_0, at 1 atm and 298°K, to an intermediate state, at the partial pressure of the component in the exhaust stream; followed by constant pressure heating from 25°C to 1800°C. The total available energy in the hot exhaust gas stream is:

$$\Sigma\phi_2(\text{exhaust}) = 16{,}996+(2)15404 + (9.03)9{,}646 + (0.4)11{,}614$$
$$= 139{,}552 \text{ cal/gm-mol } CH_4 \text{ burned.}$$

ADIABATIC COMBUSTION of METHANE

Stream	Composition	n_i moles	T°K	ϕ_0 Table 2	C_p Mean	P_i atm	$\Delta\phi_{0-1}$	$\phi=\phi_0+\Delta\phi$	$n_i\phi$
	CH_4 (.080)	1.0	298	195,500		0.080	-1496	194,004	194,004
INLET(1)									
	O_2 (0.193)	2.4	298	0.0		0.193	-974	-974	-2338
	N_2 (0.726)	9.03	"	0.0		0.726	-190	-190	-1716
								$\Sigma n\ \phi_1$	189,951
							$\Delta\phi_{0-2}$		
EXHAUST(2)	CO_2 (.080)	1.0	2073	0.0	12.95	0.080	16,996	16,996	16,996
	$H_2O(g)$ (.161)	2.0	"	2,054	10.25	0.161	13,350	15,404	30,809
	N_2 (.726)	9.03	"	0.0	7.9	0.726	9,646	9,646	87,101
	O_2 (.032)	0.4		0.0	8.0	0.032	11,614	11,614	4,646
								$\Sigma n\ \phi_2$	139,552

$\Delta\phi_{0-1} \equiv$ Availability change as an inlet material moves from S.T.P. to the given inlet conditions.

$\Delta\phi_{0-2} \equiv$ Availability change as an outlet material moves from S.T.P. to the given outlet conditions.

$$\Delta\phi(\text{Entire process}) = \Sigma\phi\ (\text{exhaust}) - \Sigma\phi\ (\text{inlet}) = 139{,}552-189{,}951 = -50{,}399\ \text{cal/gm-mole}\ CH_4$$

The exhaust temperature of 1800°C is the *adiabatic flame temperature* of methane burned with 20% excess air, that is, the temperature achieved when the combustion reaction proceeds without any loss of heat to the surroundings. H(products)=H(reactants). We therefore see that a *combustion process is highly irreversible* and *wasteful of availability* even when it is conducted so that the enthalpy of combustion is retained in the combustion products. In this instance 27% of the initial availability is lost. Availability analysis reveals that a chemical process may be degrading energy quality while conserving quantity.

Availability Analysis of a Methane Reforming Process

An availability flow diagram for the simplified reforming process shown in Fig. 5, appears in Fig. 6. Analysis of availability flows indicates that major losses occur in the reforming furnace and in the heat exchanger. Small losses occur at every mixing junction and throttling control value. The net loss of availability is 20% with the stack gasses accounting for only 1/20 of the total loss. By contrast an enthalpy balance on the same process shows only a 9% net enthalpy loss with 1/3 of this loss in the stack gas. The availability analysis suggests that affective energy management requires redesign of the temperature ranges of the furnace and heat exchanger and tells us to ignore the stack gas losses. The enthalpy analysis erroneously directs attention to the large volume stack gas that contains a large quantity of low quality energy.

Conclusion

Proper energy management requires that we account for the quality (availability) as well as the quantity (enthalpy) of energy used. This means that traditional enthalpy balances must be supplemented by availability balances.

The computation of availability changes of chemical reaction processes is facilitated by the use of a table of Standard Chemical Availabilities, a sample of which is shown for the first time in this paper.

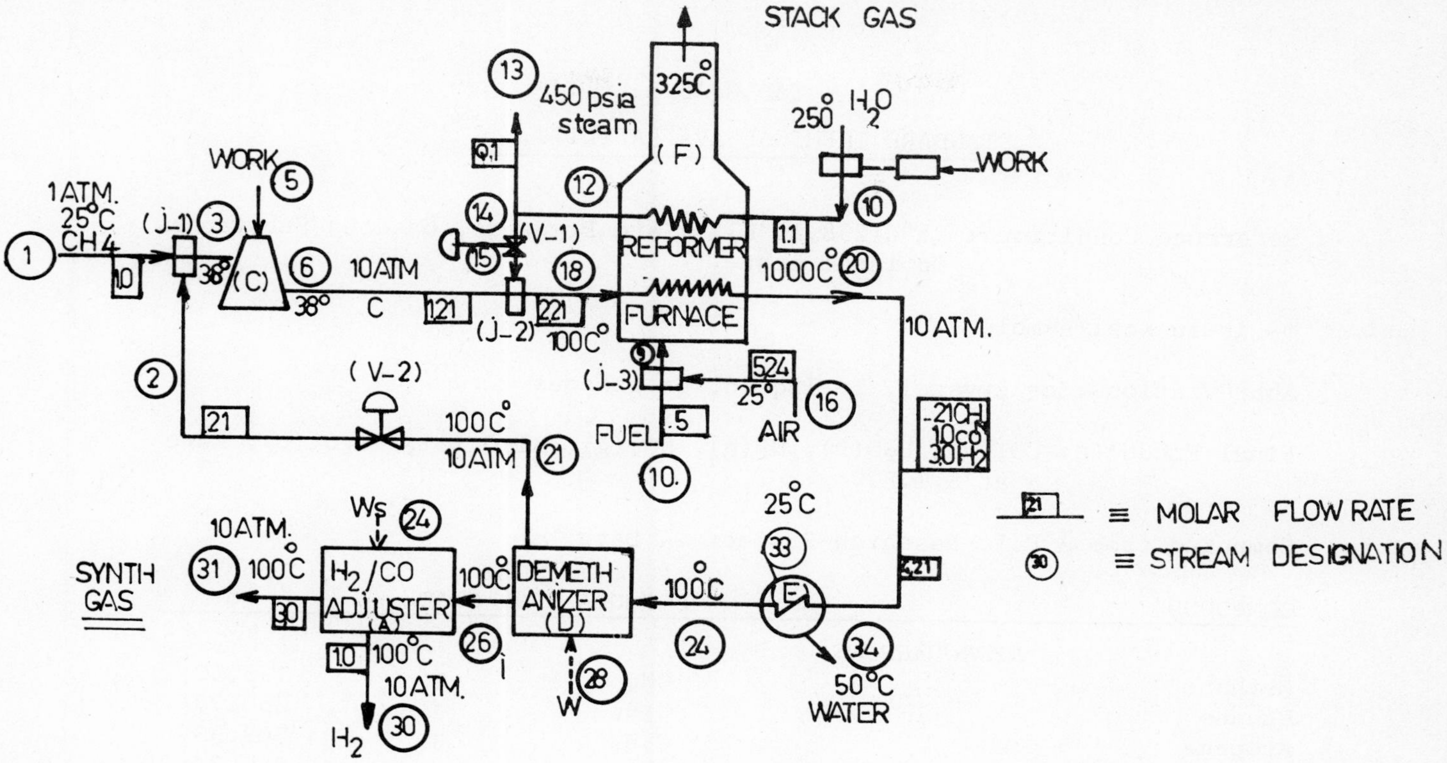

Figure 5. Methane Reforming Process: $CH_4+H_2O \longrightarrow CO+3H_2$

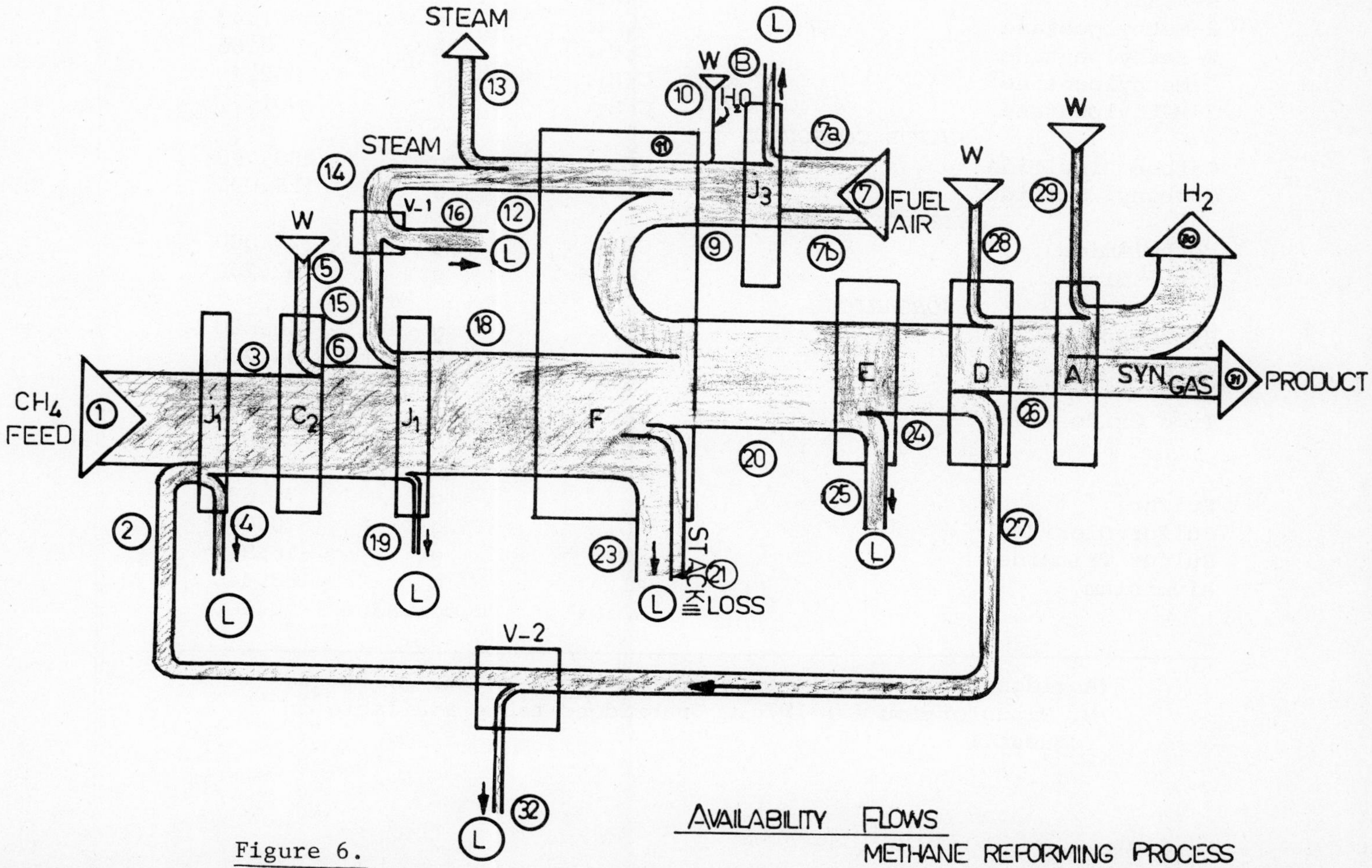

Figure 6. AVAILABILITY FLOWS METHANE REFORMING PROCESS

TABLE 2

STANDARD CHEMICAL AVAILABILITY, ϕ_0

Reference Conditions: 25°C(298.16°K), 1 atm Pressure, Gaseous Substances in ideal state.

ϕ_0 is in Kcal/g-mole.

Abbreviations: c= crystal, 1= liquid, g= gaseous

Final Products: $CO_2(g)$, $H_2O(\ell)$, $N_2(g)$, $SO_2(g)$, $Al_2O_3(c)$, $Fe_2O_3(c)$, Pure & at S.T.P.

Computed from A.P.I. Research Project 44 Data.

COMPOUND	FORMULA	STATE	ϕ_0
HYDROCARBONS			
Methane	CH_4	g	195.50
Ethane	C_2H_6	g	350.73
Propane	C_3H_8	g	503.93
N-Butane	C_4H_{10}	g	656.74
Isobutane	C_4H_{10}	g	656.19
N-Pentane	C_5H_{12}	g	809.48
N-Pentane	C_5H_{12}	ℓ	809.23
2-Methylbutane	C_5H_{12}	g	807.94
2-Methylbutane	C_5H_{12}	ℓ	807.85
2,2,-Dimethylpropane	C_5H_{12}	g	807.80
N-Hexane	C_6H_{14}	g	962.44
N-Hexane	C_6H_{14}	ℓ	961.48
2-Methylpentane	C_6H_{14}	g	961.43
2-Methylpentane	C_6H_{14}	ℓ	960.66
3-Methylpentane	C_6H_{14}	g	962.10
3-Methylpentane	C_6H_{14}	ℓ	961.27
SULFUR COMPOUNDS			
Carbon Disulfide	CS_2	g	253.750
Carbonyl Sulfide	COS	g	125.090
NITROGEN COMPOUNDS			
Methylamine	CH_5N	g	220.000
Ethylamine	C_2H_7N	g	63.200
INORGANICS			
Hydrogen	H_2	g	56.690
Carbon	C	c,graphite	94.260
Iron	Fe	c,alpha	89.550
Iron Oxide	FeO	c	30.170
	Fe_2O_3	c	0.000
	Fe_3O_4	c	26.35
Sulfur	S	c	71.680
Sulfur Dioxide	SO_2	g	0.000
Sulfur Trioxide	SO_3	g	-16.910
Aluminium	Al	c	188.440
	Al_2O_3	c,corundum	0.000

(Abridged from a table constructed by M. Viola, R. Stern, N. Papagiorgiou (c) 1976. Unabridged table available on request.)

EVALUATION OF ENERGY UTILIZATION ANALYSIS METHODS FOR INDUSTRIAL PROCESSES

Alfredo Urdaneta and Philip S. Schmidt

Department of Mechanical Engineering, University of Texas at Austin

ABSTRACT

Various methods for analyzing energy utilization in industrial processes are reviewed. These methods are categorized as non-economically- versus economically-based, and as analytical versus synthesis-oriented. It is shown that most analytical formulations based on thermodynamic availability are not well-suited as design tools; whereas, enthalpy-based design methods do not clearly delineate sources of energy dissipation. Integration of availability-based methods with economic considerations to produce a more general design approach is discussed.

INTRODUCTION

The term "energy utilization analysis" encompasses a spectrum of theoretical methods for determining the effectiveness with which energy is used in industrial processes and for designing processes to maximize this effectiveness. These methods share similar, but not necessarily identical, goals: that is, to reduce the total energy expended to perform a specified task or to minimize the total dollars expended. The extent to which the methods actually achieve these goals varies greatly. The various approaches may be defined in two broad categories. The first category focuses on analyzing the dissipation of energy within an existing process network and highlighting those points at which maximum dissipation occurs; the second category is aimed at synthesizing process networks to produce the specified product with minimum expenditure of total energy or dollars. Strong motivation exists, particularly in the energy-intensive industries, to couple the two approaches, since, as energy becomes increasingly expensive, optimum economics and optimum energy utilization more closely parallel one other.

The thermodynamic basis of the energy accounting scheme greatly influences the nature and usefulness of an energy utilization analysis. Conventionally, process design has been largely based on "First Law" energy analysis, i.e., characterization of components based only on the principle of conservation of energy. First Law analysis provides a direct economic connection, since components are sized on the basis of their energy input or dissipation, and operating costs are represented in terms of fuel or electricity used, both of which have a direct dollar equivalent. The problem with First Law analysis is that, from the standpoint of optimum energy utilization, "all BTU's are not created equal". The potential applications for a million BTU in a 2000°C flue gas stream are considerably greater than the same million BTU in a wastewater stream at 50°C. "Second Law" analysis, i.e., analysis based on the flow of thermodynamic availability through processes, does, indeed, reflect the difference in value of energy at varying temperatures, pressures, and chemical potentials. As will be pointed out later, however, Second Law analysis presents some difficult problems for use in process design and optimization. Since design decisions in the real world are economically based, it is important that energy utilization analysis methods be related to economic parameters under the control of the designer. Only in the most extremely energy-intensive processes is it likely that a design approach based exclusively on energy optimization will produce similar solutions. Thus, the relation between the two must be clarified.

The objective of this paper is to review the body of theory relating to the evaluation of energy utilization efficiency and to the use of energy analysis in synthesizing and optimizing industrial process networks. The connection between these two areas and their potential applications will be discussed, and recommendations put forth on useful directions for future work.

ANALYSIS OF ENERGY UTILIZATION EFFICIENCY AND DISSIPATION

This section will focus on methods for evaluating and interpreting energy utilization on the basis of the Second Law of thermodynamics. A simple energy balance, with its attendant efficiencies is still a useful tool for analyzing engineering systems. Such analysis, however, as pointed out above, has only limited value for evaluating the thermodynamic losses inherent in processes. Second Law analysis, on the other hand, is much more enlightening because dissipations and efficiencies measured with the help of availability concepts do provide useful insight into the mechanisms of irreversibility and the effiency with which basic energy resources are used.

Within this first category a distinction will be made between investigations that do not consider the influence of economic objectives and those that formulate the interactions between thermodynamics and economics so as to satisfy certain economic constraints.

Non-economic energy analysis methods

Some early examples of non-economic approaches will be found in the works of Birnie and Obert (1), Bruges (2), Meyer et al (3), Chambadal (4), and Gaggioli et al (5). These investigations all addressed the calculation of the thermodynamic availability of working fluids at the terminal points of individual components of power plants to obtain a detailed loss distribution and to evaluate thermodynamic effectiveness. No attempt was made to investigate improvements in power cycle performance that might be brought about by design modifications of components.

Riekert (6) proposed measuring the efficiency of energy utilization for entire chemical processes by comparing the total output of available work in all outgoing streams to the total input of availability. This analysis does not allow the elucidation of the physics of interactions between the components of the process network, an important requirement if one wishes to determine changes in component specificiations to achieve better energy utilization.

Smith (7) performed a thermodynamic analysis of a petroleum refinery unit based on an availability balance including a lost work term to account for the availability destruction associated with each piece of equipment. A major contribution of this work was the development of computer algorithms to predict the thermodynamic availability of petroleum fractions. In Smith's work, each piece of equipment is considered to be a "black box", or node, with input and output terminals. The lost work at each node is calculated by summing the availabilities of the inlet streams and subtracting the sum of the availabilities of the outlet streams. The ideal isothermal work of separation is calculated for the refinery process by considering the complete flowsheet also as a black box where the inlet is crude oil feed and the outlets are the products. Based on the same approach, the ideal actual work of production is determined from the availability of the products at the temperature and pressure at which they exit the sample flowsheet. From the total lost work for the process, the ideal isothermal work of separation, and the ideal actual work of production, one can determine the availability efficiency of separation and the standard thermodynamic efficiency.

From the point of view of engineering design, only the nodal lost work output is of value, since it allows the designer to identify the pieces of equipment and areas of the flowsheet which have the greatest losses, and hence to determine the equipment or steps in the process which could be replaced to advantage. However, this type of analysis gives no hint as to the nature of the changes that might be made.

Sweeney (8) analyzed the same refinery process by considering the energy system as a demand element in a series of interconnections and converters by which the primary energy sources supply the needs of the demand element (considered to be the crude oil undergoing the specified thermodynamic state changes). Each piece of equipment in the system is a converter in that it uses available energy to transform the thermodynamic state of the process material in some specified manner. Sweeney developed an algorithm for tracing mass, energy, and thermodynamic availability as they flow through the given process, supplied either from a utility flow, or from the process material or product flow. A computer is used to implement the algorithm and to perform a configurational analysis of the flowsheet. The program interprets the flowsheet as a graph and follows each of the intermediate products and utility flows from its entry to its exit, enumerating the alternating sequence of streams and pieces of equipment.

The thermodynamic analysis of the refinery unit is performed by integrating Smith's thermodynamic data program into the configurational analysis algorithm and developing a "nodal utilization accounting system" to categorize and quantize the use of availability at each piece of equipment in the process. Two major classifications of activity are used to quantify the availability transfer at each node: availability needed by the node to perform its specified system function, and availability supplied to the node to provide the capacity for this function. Availability interchange at each node is further subdivided by separating needed availability into three categories: discharge with no further utilization, increase due to boundary supply, and increase due to recycle supply. Supplied availability is divided into four categories: recycle to be cascaded, boundary supply to be cascaded, recycle to be discharged with no further utilization, and boundary supply to be discharged with no further utilization. System energy utilization parameters are defined based on the nodal utilization accounting system to measure the process cascade, use, and discharge of availability. Similarly to Smith, Sweeney includes the computation of the ideal isothermal work of separation and the ideal actual work of production into his efficiency analysis.

In its present form, the algorithm developed by Sweeney is computationally cumbersome and, therefore, of limited value as a design tool.

The last contribution to be discussed in this first category is the important contribution to utilization analysis of energy systems presented by Hamel and Brown (9, 10, 11). Based on thermodynamic availability concepts, Hamel and Brown define system energy utilization in terms of components categorized as discrete energy users and converters. A dimensionless energy grade function, defined as the ratio of thermodynamic availability to either internal energy or enthalpy, is used together with input-output coefficients and matrices, An energy system profile can be developed which provides an in-depth visualization of the system energy utilization. Utilization effectiveness parameters are also defined in terms of the ideal performance of the system when the transfer of energy is considered to occur reversibly, to quantify actual system utilization and the degree of departure of system performance from the ideal.

In the utilization analysis of energy systems presented by Hamel and Brown, no attempt is made to characterize individual performance parameters for discrete energy users that would permit the system designer to control the energy efficiency or energy productivity for these demand elements. Each discrete energy user is characterized in terms of system parameters by means of histogram-distributions that show the fraction of the total availability and energy that it uses, and in addition calculates average values for these distributions.

The analytical framework developed by Hamel and Brown has been applied to the energy systems of an all-electric and a gas-electric house to demonstrate the application of their approach to real systems and to gain insight into the characteristics of energy utilization in residences. Profiles are presented which show the grade level of the energy required for various functions and the mismatches between supply and demand. Improvements in system performance that might be brought about by revising internal energy exchanges or by introduction of alternative components are not discussed, and no conclusions are drawn regarding the relative merits of the different primary energy sources.

The utilization analysis formulated by Hamel and Brown is not considered applicable to some more complex industrial processes, such as refinery systems, because of the monotonic nature of the energy grade function and the difficulty of application to processes where large energy contents are inherent in the process streams. Since the components of the energy system are considered to be "black boxes" with input-output connections as the only carriers of information, little is learned concerning the character of the dissipations. The analysis method does not lead to an alternative network structure nor to specific design modifications to achieve improved energy usage.

Reistad et al (12) performed thermodynamic and economic comparisons of several total energy systems with a conventional system supplying both heating and electrical energy. An availability analysis is used to show qualitatively why total energy systems, even with relatively inefficient power plants, may be thermodynamically desirable, and to show generally where improvement efforts should be directed in the conventional systems. The economic analysis, however, is not actually coupled to availability, but rather is derived from conventional First Law considerations As in most of the cases discussed previously, the availability analysis does not include a characterization of availability dissipation as a function of relevant engineering parameters of the elements in which dissipations occur, and therefore provides only an indication of the magnitude and location of losses.

In summary, availability analysis of industrial processes, as demonstrated in the investigations described in this section, can provide a general indication of areas for potential improvement in energy utilization. These methods must be considered as "broad brush" approaches, however, and at this level are not well-suited as process design tools. To serve the needs of the designer, a more detailed characterization of availability dissipation as a function of controllable process variables, preferably economically-based, is required.

The following section describes analytical methods which incorporate some of these considerations.

Energy/economic analysis methods

Tribus and Evans (13) were the first to consider the use of available energy analysis coupled with traditional concepts of cost accounting to optimize the design of energy-using systems. In this respect, thermo-economics, as Tribus and Evans have called their methodology, forms a bridge between the methods described above, which are concerned only with the measurement and evaluation of availability dissipations and efficiency of energy utilization, and those concerned with the economic optimization of energy-using networks.

Thermo-economics uses an information-theory approach to thermodynamics to provide a single "exergy balance" equation to summarize the thermodynamics of the process. The exergy function is a general parameter which embraces most other functions that measure the work potential of energy, including the Keenan availability function, free energy, available work, and chemical potential.

Introducing exergy dissipation as a unified extensive thermodynamic measure, thermo-economics considers a complex system as made up of a number of dissipatives zones for which the relative economic value of each dissipation is estimated. A structure of internal unit costs, representing a fictitious internal economy, is introduced in order to provide for the evaluation of the overall relative economic values of the local dissipations. In each zone, exergy and material streams are identified and assigned a cost. Exergy fluxes are divided into various categories representing the several ways exergy may be transported across a boundary (e.g., thermally, mechanically, or chemically). Also, the division of exergy fluxes reflects the fact that different forms of exergy, equivalent in a thermodynamic sense, are not necessarily of equal economic value. The amortized cost of capital equipment in each zone is also determined such that an economic balance may be written in parallel with the exergy balance.

Proper substitution of exergy balance equations into the cost accounting equations leads to a system of equations that, when divided by the rate of production of the principal product, can be optimized by identifying in each zone the entropy production rate and by balancing this entropy production against capital expenditure. Thermo-economics introduces physical considerations by means of generalized resistance expressions that reflect the most important mechanism of dissipation in each zone as a simple function of the process variables.

One of the features of thermo-economics which appears as a consequence of the additive nature of entropy is the possibility of analyzing complex processes on a zone-by-zone basis.

Thermo-economics has been applied only in the evaluation and optimization of vapor-compression water desalination systems. Tribus and Evans applied thermo-economics to simple linear multizone (single product) systems under the simplified viewpoint of fixed price structure for the external economics of the plant. Subsequently, the same problem was analyzed with a variable external price structure. El-Sayed and Evans (14) extended the thermo-economic approach to stagewise as well as to interconnected multizone multiple independent variable systems. El-Sayed and Aplenc (15) applied the latter analysis to a vapor-compression desalting system, and Fruth (16) used similar concepts to perform a general economic comparison between nuclear dual-purpose plants (operating on the saturated steam cycle) and conventional dual-purpose plants.

Thermo-economic analysis produces a set of optimum dimensions or operating conditions that minimize the unit product cost. The method also provides insight into the mechanisms of dissipations occurring in real components. The input data required, however, is difficult to obtain, expecially the unit cost of exergy and material; the output is not easily interpreted in a decision making sense, and the analysis becomes extremely complicated when multiple product processes are considered.

While all of the methods described in the above sections, both economically-based and non-economic, have certain shortcomings in terms of practical applicability, it is important to emphasize again the unique contribution of Second Law analysis methods. Thermodynamic availability reflects the true intrinsic potential of different energy sources, and, unlike economics, is invariant with time for a given process. Thus, as a measure of energy utilization effectiveness, thermodynamic availability and associated measures, such as exergy, should be the fundamental basis for design of energy systems.

In the following section, some of the methods used to synthesize and economically optimize energy exchange networks will be described, and the potential for integrating Second Law analysis into these methods will be discussed.

SYNTHESIS AND ECONOMIC OPTIMIZATION OF ENERGY-USING NETWORKS

Process synthesis may be defined as the creation of a configuration of components to satisfy a given set of external constraints while minimizing some objective variable, usually total cost. Synthesis is a relatively new area of engineering research, as described by Sherwood (17) in 1963. Since, in principle, it is possible to optimize a network on the basis of any relevant variable, it is of interest in this paper to examine the approaches used in network synthesis and to discuss the potential for using these methods to optimize energy utilization efficiency. Synthesis techniques have been applied to the integration of heat exchanger networks, the design of processes requiring refrigeration, compression, expansion, and to multi-component separation by distillation. Common to all of these problems is their energy-intensiveness; hence, the synthesis of an optimum economic process network closely parallels energy optimization.

The use of synthesis-like concepts for selecting optimal heat exchanger networks was first reported by Kesler and Parker in 1969 (18).

Kesler and Parker considered the design of heat exchanger strings as a network problem made up of nodes and arcs (heat transfer devices and process streams, respectively). Their methodology allows for optimal sequencing, as well as allocation of heat duties in an arbitrary network of hot and cold streams. The problem is stated in terms of an equation that expresses the total cost of all the exchangers in the network as a function of size, inlet and outlet conditions, and cost of associated equipment and utilities to supply and remove the heat in the various streams. The resulting equation is nonconvex and strongly nonlinear, but by fractioning each stream in the network into "heat elements", the equation representing the cost of the network can be approximated with limited accuracy as a linear function of these elements. The optimal network is found by searching via alternate use of an assignment algorithm and a modified linear programming algorithm. This strategy has the objective of achieving a global economic optimum in a finite number of steps. However, the resulting network using relatively coarse heat elements specifies only approximate values of the areas for heat exchange and the costs of the equipment and utilities associated with the final network, and may not be truly optimal. Kesler and Parker's approach requires a large matrix of heat exchange possibilities, even for the coarse heat elements considered in their example, and even larger matrices would be required if smaller heat elements are used to increase the accuracy of the solution.

Based on concepts similar to Kesler and Parker's, combined with the decomposition of the network problem into two different levels of design, Kobayashi et al (19) reported a method for synthesizing optimal networks of heat exchangers by means of linear programming. The first level is the optimum design of the structure, and the second level determines the optimum total exchange duty of the network generated by the first level problem. This two level procedure is iteratively repeated until identical results are obtained on successive iterations. Kobayashi's approach produces a manually generated representation of the final network and a summary of relevant costs associated with heat exchangers and auxiliary heating and cooling equipment, and number, duty and areas of heat exchanger elements.

Unlike Kesler and Parker, Kobayashi allows each stream to be processed more than once; however, no guidance is given with respect to splitting of process streams.

Nishida et al (20) approached the synthesis problem previously attacked by Kobayashi et al by exploring analytically the necessary conditions for the optimal structure of heat exchanger systems. Considerable simplification, both in system performance description and in the objective function are needed to obtain the necessary conditions analytically.

A major contribution of Nishida was the introduction and use of a graphical representation, called the heat content diagram, to synthesize the optimal structure of the heat exchange system. On the heat content diagram, each stream is represented by a block, the area of each block corresponding to the amount of heat to be removed or added in order for the stream to reach its specified output temperature. The input and output temperatures of each stream are represented on the vertical axis of the diagram, and the capacity rate on the horizontal axis. Dividing a block horizontally corresponds to multiple heat exchange, and vertically to stream splitting.

Like Kobayashi, Nishida divided the synthesis problem into two subsystems, of which only the interior subsystem is represented by means of the heat content diagram. The assignment of heat duties within the interior network is accomplished by matching areas of the hot and cold streams to minimize total heat exchanger area. Once the synthesis of the interior subsystem is completed, synthesis of the exterior subsystem can be obtained by bringing each stream leaving the interior system to its specified temperature using auxiliary heating or cooling equipment. This does not necessarily complete the synthesis, since the value of the total interior system heat duty may not yet be optimal. At this point, a trial and error iteration with assumed values of the total heat duty of the interior subsystem may be necessary to bring the value of the total heat duty to an optimum. Only the special case where all overall heat transfer coefficients are equal and the heat capacity-mass flowrate product is the same for all the streams was treated.

Hoffman (21) in his review and evaluation of current methods of synthesizing heat exchanger networks, suggested the use of modular simulation routines to express the economic objective function for an entire process. His simulation technique involves translating the process flowsheet into an information flow diagram in which specifications concerning the flow and thermodynamic state of any stream are transferred from one module to another. All incoming information to the initial module is assumed known. This information is modified by the computations performed in the initial module to become the incoming information for the next module, and so on, in a specified calculation order. The modules are applied in a continuous sequence of design, simulation, testing, and evaluation, and the optimization process ends when all the problem constraints are satisfied.

Hoffman found that his approach, when coupled with Nishida's graphical interpretation scheme, produced near-optimal networks without excessive computer requirements. Even though the results using Nishida's method do not guarantee global optimality, the advantages stated previously make Hoffman's procedure an extremely powerful one.

A number of other investigators have contributed to the theory of optimization of heat transfer networks. While these works will not be detailed here, some worthy of mention are Rudd (22) (decomposition of complex networks into simpler subproblems), Masso and Rudd (23) (use of heuristic constraints in network structuring), and Siirola, Powers, and Rudd (24) (process-generator conceptualization).

Menzies and Johnson (25) have approached the problem of synthesizing optimal supporting equipment networks in cases where a basic processing scheme has already been formulated. The method uses a modular program system with a price-based decomposition algorithm for subprocess integration. Heuristic decision criteria are used and a Lagrangian-based decomposition technique is employed to break complete processes into sub-processes that are then optimized by branch-and-bound combinatorial optimization.

The system developed by Menzies and Johnson is limited to satisfying fixed stream temperature and pressure specifications. The system proceeds through the following synthesis stages: analysis of the given process scheme to identify a set of streams with unsatisfied temperature and pressure specifications; generation of all possible equipment networks to satisfy the stream requirements; and selection of the optimal cost network using the branch-and-bound technique. Allowed process equipment includes countercurrent heat exchangers, polytropic single stage compressors, adiabatic expanders, and adiabatic stream mixers and splitters. The program provides for multistage compression with water intercooling, and for vapor recompression condensation or reboiling by means of precoded assemblies of process equipment. An independent subprocess procedure, with some limited decision-making capability, is also included in the system to generate the equipment sequence for a conventional cascade refrigeration unit. Design considerations are, when possible, translated into programmable heuristics. The heuristics are used in setting the order and extent of unit operations and particularly in the prescreening of prospective stream matchings. Thermodynamic data is supplied to the system by means of an executive routine with a physical properties calculation package.

Thermal energy recovery is regarded as the prime consideration for estimating the basic stream prices required for process integration and decomposition. Two factors which affect the true stream value are considered: the degree of irreversibility involved in stream use, and the variation in the cost of stream usage. Both factors are accounted for by the introduction of a discount parameter that adjusts stream transfer prices.

Menzies and Johnson's methodology was tested for a high pressure ethylene separation scheme, and an optimum network, essentially identical to an existing plant, was obtained. A very important parameter associated with the problem solution was found to be the maximum entropy change to minimize exchanger irreversibilities. The efficiency of the branch-and-bound optimization procedure was found to be very dependent on the selection of the bounding problem. For real systems, the task of testing the large number of possible networks for feasibility can become computationally formidable.

The last research to be evaluated here is the work of King, Gantz, and Barnes (26) on the strategy of systematic evolutionary process synthesis. This work is very important, from the point of view of energy utilization analysis, because of the explicit use of heuristics specifically designed to reduce available energy dissipation.

The evolutionary strategy is initiated by providing a basic simple process flowsheet to accomplish the desired task. A first level of heuristics is applied to the initial network to identify a portion of the process to be modified. The portion of the process so identified is then modified through steps involving the use of second level heuristics. Two specific applications of systematic evolutionary process synthesis are presented: the synthesis of a methane liquefaction process, and the synthesis of demethanizer configurations.

The concepts and applications discussed by King et al were considered by the authors of that paper to serve as a progress report on a novel approach to the logic of process synthesis. Further search of the literature in this field has failed, however, to uncover additional research aimed at the extension of this approach to more general process design problems. Readers interested in a more extensive review of synthesis methods for chemical processes will find the works of Hendry (27) and King (28) to be helpful.

CONCLUSIONS

All of the approaches described in the above sections are, in some sense, energy utilization analysis methods. The various non-economic methods produce outputs directly measured in terms of energy or some directly associated variable. Even the methods which are primarily economic optimization techniques, are, in a very real sense, energy utilization analysis tools, since their application has been overwhelmingly in the area of energy-intensive process design.

The application of the non-economic approaches appears to be in the delineation of segments of a process requiring special attention to reduce energy inefficiencies. These methods are also useful in providing a thermodynamic basis for comparison of various components of an industrial system.

Process synthesis methods have their most obvious application in the area of detailed process design for optimum cost. It should be noted, however, that potentially these same methods could be used to optimize a system design on the basis of any process variable, including Second Law measures of energy utilization, such as availability, exergy, or energy grade function. Some of these methods do, in fact, use heuristic rules based on availability dissipation to aid in reducing the number of candidate systems for economic optimization.

It is useful to consider how the strongest points of these various approaches might be brought together to produce a process design tool based on Second Law energy utilization measures. The thermo-economic theory originally developed by Tribus and Evans appears to provide the bridge, since the thermo-economic method essentially optimizes economics of the system based on a cost per unit availability dissipation measure. Exergy dissipation is directly tied to process variables, and in the economic optimization, component design variables are taken into account. The basic problem of thermo-economics appears to lie with the application of the method to a broader range of industrial processes than the desalination applications described in the existing literature. Thermo-economics is not a simple tool. Development of the exergy cost functions in terms of relevant process variables requires a complex formulation in terms of data which is normally not readily available to the designer. Thus, it appears that this powerful tool has been under-utilized primarily because of the difficulty of translating "natural" process information into the specialized terminology which characterizes the method. How can thermo-economic analysis be made more accessible to the process engineer?

First, a valuable contribution would be the formulation of standard component "modules" to translate design and process variables into exergy (or availability) terms for a number of components common to industrial processes, such as pumps, compressors, heat exchangers, piping, and heaters. Second, capital and operating costs could be incorporated into the same modular format such that the costs per unit exergy dissipation function could be easily generated. Such a system of "building blocks" could be easily incorporated into process networks which could be manipulated using any of the various synthesis methods discussed above. It should be possible to generate a "cost flow" process diagram in which regions of high availability dissipation would stand out in dollar terms (which are much more easily understood by industrial managers and project engineers!). Both capital and operating costs could be incorporated into the cost flow. For example, the capital cost of a run of insulated piping could be amortized over the total expected mass flow through the pipe in its lifetime, so that this cost could be directly added to the costs associated with exergy dissipation in the pipe. The trade-off between investment for insulation and larger piping and operating cost of heat loss and pressure drop (as reflected in exergy dissipation) would be combined in minimizing the single cost flow measure.

The marriage of Second Law-based energy analysis methods with economic process synthesis techniques might also have another intriguing application beyond that of industrial process design. In circumstances where, for an energy-intensive process, a significant difference exists between the optimum economic process and the optimum availability dissipation process, the root cause is a distortion in the true value of energy such as that caused by unrealistic price regulation policy. Energy/economic models could conceivably be used to examine regulatory policy with a view toward eliminating price distortions. Whether government regulatory bodies would use such a rational tool if it were available is a question which the authors prefer to leave to the political scientists.

REFERENCES

(1) C. Birnie and E. F. Obert, Evaluation and Location of the Losses in a 60,000 KW Power Station. Proc. Midwest Power Conf., 11, 187-93 (1949).

(2) E. A. Bruges, Applications of Second Law Analysis, The Engineer, (Oct. 8, 1954) and (March 5, 1954).

(3) C. A. Meyer, G. J. Silvestri and J. A. Martin, Availability Balance of Steam Power Plants, J. Eng. Power, (January 1959).

(4) P. Chambadal, Availability Balance of Gas Turbines, ASME Paper 66-GT-96, (October 1965).

(5) Richard A. Gaggioli, Jae-Joon Yoon, Scott A. Patulski, Alan J. Latus, and E. F. Obert, Pinpointing the Real Inefficiencies in Power Plants and Energy Systems, Proc. American Power Conf., 37, 656 (1975).

(6) L. Riekert, The Efficiency of Energy-Utilization in Chemical Processes, Chem. Eng. Science, 29, 1613 (1974).

(7) S. V. Smith, J. C. Sweeney, H. L. Brown, B. B. Hamel, and E. D. Grossmann, A Thermodynamic Analysis of a Refinery Process, Energy Institute Report-75-3, Drexel University (June 1975).

(8) J. C. Sweeney, S. V. Smith, H. L. Brown, B. B. Hamel, and E. D. Grossmann, A Fundamental Approach to the Thermodynamic Configurational Analysis of Process Flowsheets As Applied to a Refinery Unit, Energy Institute Report 75-4, Drexel University (June 1975).

(9) B. B. Hamel and H. L. Brown, Utilization Analysis of Energy Systems, Part I: Fundamentals of Energy Utilization, Drexel University Report No. 729181, (December 1971).

(10) B. B. Hamel, H. L. Brown, R. E. Crane, and J. K. Nwude, Part II, Application of Utilization Criteria to House Energy System, Drexel University Report No. 729182, (December 1971).

(11) H. L. Brown, B. B. Hamel, B. K. Hinkle, and W. T. Schleyer, Part III, Urban Energy Systems Manufacturing Industry Data Base and Energy Maps, Drexel University Report No. 729186, (December 1971).

(12) G. M. Reistad, R. A. Gaggioli, and E. F. Obert, Available Energy and Economic Analysis of Total Energy Systems, Proc. American Power Conf., 32, 603 (1970).

(13) R. B. Evans and M. Tribus, Thermo-Economics of Saline Water Conversion, I & EC Process Design and Development, 4 (April 1965).

Also: R. B. Evans, G. L. Crellin, and M. Tribus, Thermo-Economic Considerations of Sea Water Demineralization, Chapter 2, Principles of Desalination, K. W. Spiegler, Ed., Academic Press (1966).

(14) Y. M. El-Sayed and R. B. Evans, Thermo-Economics and the Design of Heat Systems, J. Eng. Power, 27 (January 1970).

(15) Y. M. El-Sayed and A. J. Aplenc, Application of the Thermo-Economic Approach to the Analysis and Optimization of a Vapor Compression Desalting System, J. Eng. Power, 17 (January 1970).

(16) H. Fruth, Cost Determination and Comparison of Nuclear and Fossil Fueled Dual Purpose Power and Desalination Plants, First Int. Symposium on Water Desalination, Washington, D.C. (October 1965).

(17) T. K. Sherwood, A Course in Process Design, MIT Press (1963).

(18) M. G. Kesler and R. O. Parker, Optimal Networks of Heat Exchange, Chem. Eng. Progr. Symp. Ser., No. 92, 65, 111 (1969).

(19) S. Kobayashi, T. Umeda, and A. Ichikawa, Synthesis of Optimal Heat Exchange Systems -- An Approach by the Optimal Assignment Problem in Linear Programming, Chem. Engrg. Sci., 26, 136 (1971).

(20) N. Nishida, S. Kobayashi, and A. Ichikawa, Optimal Synthesis of Heat Exchange Systems -- Necessary Conditions for Minimum Heat Transfer Area and Their Application to System Synthesis, Chem. Engrg. Sci., 26 (1971).

(21) T. W. Hoffman, The Optimal Design of Heat Exchanger Networks -- A Review and Evaluation of Current Procedures, Chapter 6, Heat Exchangers: Design and Theory Sourcebook, McGraw Hill (1974).

(22) D. F. Rudd, Elementary Decomposition Theory, AIChE Journal, 14, 343 (Mar. 1968).

(23) A. H. Masso and D. F. Rudd, Heuristic Structuring, AIChE Journal, 15, 10 (January 1969).

(24) J. J. Siirola, G. J. Powers, and D. F. Rudd, Toward a Process Concept Generator, AIChE Journal, 17, 677 (May 1971).

(25) M. A. Menzies and A. I. Johnson, Synthesis of Optimal Energy Recovery Networks Using Discrete Methods, The Canadian Journal of Chem. Eng., 50, 290 (April 1972).

(26) C. J. King, D. W. Gantz, and F. J. Barnés, Systematic Evolutionary Process Synthesis, I. & E.C. Process Design and Development, 11, 271 (1972).

(27) J. E. Hendry, D. F. Rudd, and J. D. Seader, Synthesis in the Design of Chemical Processes, AIChE Journal, No. 11, 19, 1 (January 1973).

(28) J. C. King, Understanding and Conceiving Chemical Processes, AIChE Monograph Series, No. 8, 70 (1974).

THE ROLE OF THERMODYNAMIC EFFECTIVENESS IN EVALUATING COAL CONVERSION RD&D

W.C. Peters[1], T.C. Ruppel[2] and J.W. Mulvihill[3]
U.S. Energy Research and Development Administration
Pittsburgh Energy Research Center
4800 Forbes Avenue
Pittsburgh, PA 15213

ABSTRACT

There is an ever increasing worldwide interest in the conversion of lower grade fossil fuels into higher grade, more useful forms. The major effort is underway in the United States, where a number of competing processes for conversion of abundant domestic coal supplies to gaseous fuels are under active development.

The general approach to the evaluation of alternative conversion methods is a comparison of their relative merits in terms of: economics, energy efficiencies and environment. The three are inherently linked through the inefficiencies associated with a particular process which dictates to a large extent the environmental impact and costs of associated environmental control technologies. Thus, evaluation of energy efficiency takes on added importance, as environmental constraints are tightened in the future, particularly in connection with evaluating RD&D on emerging technologies. Although they are being developed and demonstrated now, they will not be extensively utilized until the 1980-1990 timeframe and beyond.

The problems with conventional (First Law) thermodynamic analysis in evaluating and developing coal conversion RD&D priorities are dicussed and compared with preliminary second law "availability" studies of gasification reactors. The gasification process is characterized with respect to the relative quantity and quality of the input/output streams. The analysis forms the basis for a discussion of the potential practical value of "effectiveness" as a parameter for optimizing coal conversion efficiency.

In addition, the "effectiveness" of coal gasification is presented in conjunction with "effectiveness" evaluations by others for present day energy conversion technologies (combustion of coal and oil to produce electricity), and end-use technologies (steel industry).

[1]Chief, Energy Conservation Branch, Environment and Energy Conservation Division

[2]Chemical Engineer, Energy Conservation Branch, Environment and Energy Conservation Division

[3]Director, Environment and Energy Conservation Division

L. BOLLE,

Chargé de Cours à l'U.C.L.

2, Place du Levant - B - 1348 - Nouvain-la-Neuve

BELGIQUE

I - INTRODUCTION

Notre civilisation industrielle a engendré un grand nombre de produits ou de procédés qui, pour être mis à notre disposition, requièrent la dépense d'importantes quantités d'énergie. L'évolution du mode de vie et des objets de consommation - cette évolution que l'on appelle souvent progrès - peut sembler s'être faite de manière continue ; c'est une simplification acceptable seulement si l'observateur qui croit découvrir cette propriété se situe loin de son domaine d'observation. En réalité, l'évolution dont je parle doit plutôt être décrite comme un ensemble de mutations, chacune d'assez petite amplitude, et dont toutes ne sont d'ailleurs pas promises à une longue durée de vie. Ces mutations se font comme par accident, sans vraie coordination et selon une séquence dans le temps qui n'est pas exempte d'arbitraire Etant donné la multiplicité de leurs causes, ces mutations présentent un caractère aléatoire certain.

Après une période d'évolution dans la facilité, apparait soudain une contrainte majeure : une source première d'évolution, malmenée, ne se prête plus docilement aux caprices de ceux qui l'épuisent. Il faudra donc économiser de l'énergie. C'est une rupture dans l'évolution connue mais non l'avènement d'un statisme oublié. Ainsi s'amorce depuis quelque temps une évolution en profondeur, saine, qui nous oblige à repenser aux sources de notre abondance, et nous motive à respecter notre patrimoine commun et premier : nos richesses naturelles.

Economiser de l'énergie, mais encore.... Nous distinguerons trois classes de moyens qui permettent d'atteindre cette fin :

-La simple récupération des énergies perdues ;
-La modification de procédés de fabrication, tendant à réduire leurs besoins ou pertes d'énergie ;
-La conception de processus globaux, intégrant plusieurs procédés classiques distincts. Ces processus intégrés se prêtent évidemment à optimisation. Ils tendent à annihiler les incohérences et gaspillages nés des composantes aléatoires de l'évolution que nous venons de décrire.

Il n'y a pas lieu d'attacher une échelle de valeur à ces différents moyens. L'un d'eux sera choisi, dans un cas déterminé, en fonction de sa faisabilité et de sa rentabilité.

La conception d'un processus global intégrera le plus souvent la production d'énergie sous les formes mécanique (ou électrique) et thermique. Selon un vocable né aux Etats-Unis, on dira qu'il s'agit alors d'un procédé à "énergie totale".

Il n'y a pas de taille caractéristique des systèmes à énergie totale. Nous en connaissons dont la puissance globale utile n'est que de quelque 200 kW ; il en est d'autres de plus de 1000 MW.

Il y a une multiplicité de moyens de réaliser ces systèmes ; la multiplicité des applications possibles - et des réalisations existantes - est tout aussi grande. Avant d'étudier pratiquement la mise en oeuvre des systèmes à énergie totale, nous rappellerons quelques points de thermodynamique qui les justifient tous.

II - CONSIDERATIONS DE THERMODYNAMIQUE

Examinons la représentation dans un diagramme T,S (température absolue - entropie) d'un cycle moteur quelconque A 1 B 2 (fig. 1).

Nous supposerons pour l'instant que toutes les transformations encourues par le fluide thermodynamique au cours de son évolution cyclique sont réversibles.

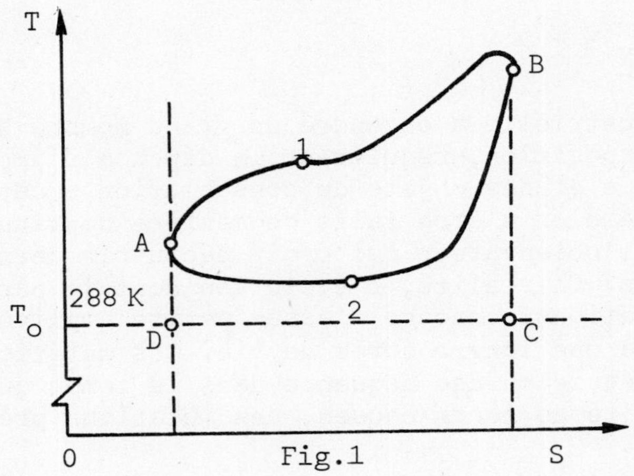

Fig.1

Pour faire suivre au fluide la transformation A 1 B (branche supérieure de la transformation cyclique), il faut lui fournir une quantité de chaleur positive Q_I que l'on peut évaluer par (1) :

$$Q_I = \oint_{A1B} T\, d\, S = \int_{S_A}^{S_B} T_I(S)\, d\, S. \qquad (1)$$

Semblablement, pour faire suivre au fluide la branche inférieure B 2 A de la transformation cyclique concidérée, il faut en extraire la quantité de chaleur positive Q_{II} donnée par :

$$Q_{II} = - \oint_{B2A} T\, d\, S = \int_{S_A}^{S_B} T_2(S)\, d\, S. \qquad (2)$$

Si, en outre, il n'y a pas de frottement dans aucun organe mécanique de transmission, la machine fonctionnant selon le cycle décrit produira un travail moteur :

$$\tau_m = Q_I - Q_{II}. \qquad (3)$$

Cette dernière expression n'est autre que l'écriture du principe de conservation de l'énergie pour la machine motrice, soumise d'une part à un influx d'énergie Q_I, et cédant d'autre part au monde extérieur les énergies τ_m et Q_{II} respectivement sous forme mécanique et thermique.

Le cycle que nous considérons est parcouru par le fluide dans le sens des aiguilles d'une montre. C'est la condition pour qu'il soit un cycle moteur. En effet, si le fluide parcourait le cycle en sens inverse, les quantités de chaleur Q_I et Q_{II} que nous avons définies seraient négatives, de même alors que τ_m : le cycle absorberait τ_m et Q_{II}, et restituerait Q_I au monde environnant. Il s'agirait d'un cycle dit récepteur de travail moteur.

Se basant sur les interprétations géométriques des intégrales définies (1) et (2), on voit que la représentation graphique de τ_m est l'aire de la fraction du plan (T,S) intérieure à la frontière A 1 B 2 .

Si le travail moteur est considéré comme seul effet utile, on définira le rendement thermique ou énergétique de la machine par :

$$\eta_t = \frac{\tau_m}{Q_I} = I - \frac{Q_{II}}{Q_I}. \qquad (4)$$

Selon ce concept, l'intensité du travail moteur produit est donc comparé à Q_I, la quantité d'énergie thermique noble fournie au fluide de cycle. On voit qu'à l'évidence, la quantité de chaleur Q_{II} doit être réduite pour accroître le rendement thermique. Elle ne peut pourtant l'être au-delà de toute limite. Dans l'exemple que nous étudions, si la branche A 1 B est fixée, la chaleur Q_{II}, dans une ambiance prise conventionnellement à 288 K (15°C), sera minimum pour le cycle limite dont la branche inférieure d'évolution est la ligne brisée BCDA. Il y correspond la valeur limite.

$$Q_{II,lim} = T_o (S_B - S_A) .$$

Les valeurs du rendement énergétique de la production de puissance mécanique sont toujours assez faibles, de l'ordre de 0,3... 0,4 en pratique. Cette constatation pessimiste devrait être corrigée, selon les cas, par les considérations suivantes :

1 . Si l'objet de l'application technique est uniquement la production de travail moteur, la comparaison de τ_m à Q_I, pour intéressante qu'elle soit, n'est pas la plus logique.

En effet, étant donné le fait que la température d'ambiance naturelle est toujours voisine de la températue conventionnelle que choisit le thermodynamicien (288K), l'entièreté de la quantité d'énergie thermique Q_I n'est pas convertible en travail moteur ; seule la partie $(Q_I - Q_{II,lim})$ l'est. Il serait donc plus logique de comparer τ_m à cette dernière quantité, définissant ainsi un nouveau rendement que nous ne nommerons que par son sigle η^* :

$$\eta^* = \frac{\tau_m}{Q_I - Q_{II,lim}} = I - \frac{Q_{II} - Q_{II,lim}}{Q_I - Q_{II,lim}}$$

Afin de pouvoir se rendre compte quantitativement des implications de cette remarque, considérons un cycle à vapeur de Rankine-Hirn, pourvu des améliorations classiques que constituent la resurchauffe de la vapeur et les soutirages en cours de détente.

Dans le condenseur, l'énergie cédée à l'ambiance vaut invariablement une fraction 0,50... 0,55 de l'énergie fournie à la vapeur dans le générateur de vapeur. Nous basant sur un exemple détaillé dans la littérature technique (2), nous relevons, par kg de vapeur passant au condenseur :

$$Q_I = 4\ 200 \text{ kj/kg},$$

$$Q_{II} = 302\ (S_B - S_A) = 2238 \text{ kj/kg}.$$

Q_{II} est calculé en supposant une condensation isobare à 29°C (0,04 bar) d'une vapeur possédant un titre initial de 0,92 ; la chaleur de condensation à 29°C est 2433 kj/kg. Le rapport des chaleurs fournies et rejetées vaut ici :

$$\frac{Q_{II}}{Q_I} = 0,533$$

et la quantité de chaleur $Q_{II,lim}$ est :

$$Q_{II,lim} = 288\ (S_B - S_A) = 2134 \text{ kj/kg}.$$

D'où l'on déduit les valeurs :

$$\eta_t = 0,467,$$

$$\eta^* = 1 - \frac{104}{2066} = 0,950 .$$

Ces résultats montrent que si la perte d'énergie non récupérable est très importante au condenseur, au contraire, la perte d'énergie convertible en travail moteur y est très réduite.

Je n'aurais garde de laisser le lecteur sur cette dernière impression. Ce serait, cette fois , pêcher par excès d'optimisme. En effet, le rendement η^* et nos dernières considérations sont destinées à mettre en évidence les perte à la source froide seulement, dues à l'irréversibilité du transfert de chaleur et à l'anisothermie de cette source, dans la pratique industrielle. Ces mêmes raisonnements faits à propos de la source froide doivent l'être aussi à propos de la source chaude.

Continuons donc notre analyse sur base du même exemple. Nous supposerons que le générateur de vapeur auquel nous avons fait allusion ci-dessus est classique, à combustible fossile. La source d'énergie est un système (dont les caractéristiques seront affectées de l'indice I) composé de combustible et d'air comburant, supposés initialement à la pression de 1 bar et à la température T. La source chaude proprement dite est les gaz comburés. Ces derniers, ramenés à la températue initiale T et à la pression de 1 bar, constituent un nouveau système (indice II). L'on sait que la chaleur de réaction ou d'oxydation du combustible est égale à la différence des enthalpies des systèmes I et II : on l'appelle encore pouvoir calorifique supérieur ($C_{p,s}$). Nous pourront écrire :

$$C_{p,s}\ (T) = H_I\ (T) - H_{II}\ (T).$$

Pour apprécier la qualité des échanges thermiques entre la vapeur et la source chaude, et les qualités intrinsèques de cette source, il nous faut maintenant évaluer la fraction de la chaleur de réaction dégagée par l'oxydation du combustible qui soit convertible en travail moteur. Nous nous placerons ici encore dans des circonstances idéales, en faisant l'hypothèse que la combustion se fait sans pertes thermiques vers l'ambiance et sans travaux de frottement visqueux décelables au sein du mélange polyphasique. Considérons l'équation énergétique du travail moteur, relative à une transformation élémentaire quelconque d'un système produisant du travail moteur :

$$d\,\mathcal{T}_m = dQ - d\ (K + gZ) - dH,$$

où d Q est la chaleur reçue par le système, d (K + gZ) l'accroissement des énergies cinétiques et potentielle du système après transformation et où d H est l'accroissement d'enthalpie. Cette relation peut encore s'écrire :

$$d H = - d \tau_m - d(K + g Z) + d Q,$$

et nous l'interpréterons de la manière suivante (3) : l'accroissement (positif ou négatif) d H sera convertible en travail moteur pour la partie ($- d \tau_m - d (K + gZ)$) évidemment, puisque ce sont les termes d'énergie mécanique,et pour une fraction de d Q. Pour apprécier quelle est cette fraction, il faut tenir compte de la températue T du fluide au moment où il subit l'action calorifique d Q ; le travail maximum réalisable au moyen de cette action calorifique sera obtenu par un cycle de Carnot, opérant entre les températues T et T_o, et vaudra :

$$\frac{T - T_o}{T} \, d Q.$$

Il en résulte que la récupération maximale de d H sous forme mécanique sera :

$$d H_r = - d \tau_m - d (K + gZ) + \frac{T - T_o}{T} \, d Q$$

$$d H - \frac{T_o}{T} \, d Q$$

$$= d H - T_o \, d S,$$

la dernière égalité n'étant valable que pour une transformation réversible. Si l'on introduit la fonction E, l'exergie du fluide, définie par :

$$E = H - T_o S,$$

on voit que l'on a :

$$d H_r = d E,$$

ce qui élucide le sens physique de l'exergie.

Si l'on applique cette notion au cas de la combustion, on notera que l'exergie du combustible, ou mieux, l'exergie libérée par la combustion, vaut :

$$E_I - E_{II} = H_I - H_{II} - T_o \, (S_I - S_{II}).$$

L'énergie libérée par la combustion est toujours inférieure à $C_{p,s}$ (p_o, T_o). Pourtant, pour les combustibles solides et liquides, le terme T_o $(S_I - S_{II})$ est très petit, et l'on a :

$$E_I - E_{II} \simeq C_{p,s}.$$

Pour le gaz naturel, on a approximativement :

$$E_I - E_{II} \simeq 0{,}90 C_{p,s}.$$

Malheureusement, pour une combustion isobare, la libre expansion des gaz chauds leur fait encourir une diminution d'exergie valant environ 30% de l'exergie du combustible. D'autre part, l'irréversibilité du transfert de chaleur entre les gaz chauds et la vapeur suscite, ainsi que le montre un calcul détaillé, une nouvelle perte d'environ 25% de l'exergie libérée par la combustion.

Le rendement exergétique, que l'on peut définir comme le rapport de τ_m à la fraction maximale convertible en travail moteur de l'énergie contenue dans le combustible,

$$\eta_{ex} = \frac{\mathcal{T}_m}{E_I - E_{II}} ,$$

sera finalement du même ordre de grandeur que le rendement thermique. Cette coincidence est pur fruit du hasard. La considération simultanée de ces deux rendements, pour un cycle à vapeur, met en lumière deux conclusions essentielles : les pertes aux sources froides et chaudes, quoique de nature différente, sont toutes deux très intenses. Mais la perte à la source froide est principiellement inévitable, tandis que les transferts à la source chaude sont thermodynamiquement fort imparfaits et théoriquement très perfectibles. Les procédés de resurchauffage et de soutirages améliorent le cycle selon deux points de vue que nous avons distingués.

2 . Si l'on attribue une valeur intrinsèque à la production de calories à relativement basse température, on parvient à accroître de manière déterminante le rendement thermique. Si l'on ne prend en considération que des transformations théoriques (réversibles), ce qui fut notre propos jusqu'à présent, on notera que ce rendement vaut d'ailleurs toujours l'unité, puisque la chaleur Q_{II} est ici considérée comme utile :

$$\eta_{e,t} = \frac{\mathcal{T}_m + Q_{II}}{Q_I} = 1$$

La nécessité de rendre valorisables les calories Q_{II} impose souvent de modifier, en le relevant, le niveau de température de la source froide du cycle moteur. Les pertes d'énergie et d'exergie, comptabilisées au point de vue du seul cycle moteur, deviennent plus importantes. Le rendement thermique η_t du cycle de production d'énergie mécanique décroit. C'est par ailleurs ce rendement qui dicte la répartition de l'énergie Q_I entre celles $\mathcal{T}_m$ et Q_{II}. Le facteur de répartition chaleur/travail a pour valeur, dans un système idéal :

$$\frac{Q_{II}}{m} = \frac{1}{\eta_t} - 1, \tag{5}$$

puisque :

$$Q_{II} = Q_I - \mathcal{T}_m = \frac{\mathcal{T}_m}{\eta_t} - \mathcal{T}_m .$$

Le caractère irréversible de certaines transformations du fluide parcourant le cycle moteur (hormis les échanges calorifiques incontrôlés avec l'ambiance) ne modifie pas l'expression formelle du facteur de répartition et n'a pas d'influence sur la valeur unitaire du rendement global d'un système à énergie totale (4). Ces irréversibilités, diminuant toutefois la valeur numérique de η_t, modifient la valeur du coefficient de répartition.

D'autre part, dans le générateur de vapeur, l'énergie transmise au fluide de cycle n'est qu'une fraction η_G de la chaleur disponible dans le combustible, et une seule fraction η_c de Q_{II} est transmise au demandeur de calories. En outre, seule la fraction η_m de la puissance transmise par le fluide de cycle à l'organe moteur de la machine motrice est transmise utilement (η_m est le produit des rendements mécanique du turbogroupe, de la conversion électrique et de transport de l'énergie thermique avant sa conversion en énergie mécanique). En pratique, l'expression du rendement énergétique d'un système à énergie totale s'écrit finalement :

$$\eta_{e,t} = \frac{\eta_m \mathcal{T}_m + \eta_c Q_{II}}{Q_I/\eta_G} = \eta_G \left(\eta_m \eta_t + \eta_c \frac{Q_{II}}{Q_I}\right). \tag{6}$$

Le facteur de répartition chaleur/travail (φ) devient :

$$\varphi = \frac{\eta_c Q_{II}}{\eta_m \mathcal{T}_m} = \frac{\eta_c}{\eta_m \eta_t} \frac{Q_{II}}{Q_I} \tag{7}$$

Les valeurs du rendement énergétique $\eta_{e,t}$ se trouvent être comprises, en pratique, dans l'intervalle 0,75... 0,90.

III - APPLICATIONS DU PRINCIPE DE L'ENERGIE TOTALE.

Nous classerons ces applications en fonction du type de machine produisant du travail moteur, et terminerons par quelques considérations relatives au chauffage urbain.

A . Les turbines à vapeur.

Ces machines sont certes les plus répandues pour la production d'énergie mécanique, en général convertie immédiatement sous forme électrique par turboalternateur appartenant à la même ligne d'arbre. Parfois, une turbine à vapeur entraine directement un turbo-compresseur ou une turbo-pompe, mais cette éventualité pourra tout aussi bien être couverte par les propos qui vont suivre.

En cas de production exclusive de travail moteur, la turbine est toujours à condensation. En cas de production simultanée de travail moteur et de chaleur, deux variantes fondamentales sont possibles, qui sont la turbine à contrepression et la turbine à condensation, avec un ou plusieurs prélèvements de vapeur. Des prélèvements de vapeur peuvent aussi être opérés sur des turbines à contrepression, si des actions calorifiques à deux températures différentes (et en proportions convenables) sont indispensables. Les systèmes pratiques comportent en outre de nombreuses variations. Les principales sont : possibilité de détente directe et désurchauffe de la vapeur vive en cas de demande de chaleur accrue ; turbine à contrepression avec soutirage de vapeur réglable de manière à assurer une température de l'eau d'alimentation du générateur de vapeur qui soit constante en dépit de besoins calorifiques variables (ce procédé permet l'accroissement du facteur de répartition chaleur/travail) ; turbine à un corps haute pression (HP) et deux corps basse pression (BP) en parallèle, l'un à condensation, l'autre à contrepression ; turbine à admission intermédiaire de vapeur moyenne pression générée par récupération de calories perdues par ailleurs.

Dans les cas que nous venons d'énumérer, il y a entorse au principe de l'énergie totale dès que la turbine fonctionne pour partie selon un cycle à condensation (sans récupération de la chaleur cédée). Des solutions hybrides sont souvent rendues nécessaires par le défaut d'adéquation entre les besoins en énergie électrique et en énergie thermique.

La vapeur d'échappement d'une turbine est un excellent fluide caloporteur à courte distance. Porteuse de sa chaleur de condensation, elle peut fournir une action calorifique importante à température constante, avec un excellent coefficient de transmission calorifique. Dans des circonstances normales, on prélève à la turbine de la vapeur possédant une surchauffe de 10... 12°C par rapport à la température (de saturation) à laquelle sera produit l'effet calorifique.

Le transport de vapeur à grande distance n'est pas à promouvoir : en effet, il faut assurer alors une plus grande surchauffe à la vapeur prélevée ; les pertes de charge du circuit seront importantes, du fait du débit volumique élevé de la vapeur ; il faut prévoir un système de purge et éventuellement de collecte des condensats "vagabonds" ; enfin, l'investissement du circuit de distribution de la vapeur est élevé, étant donné le gros diamètre des conduites.

Pour le transport de calories à grande distance, on adoptera plutôt l'eau comme fluide caloporteur, cette eau étant réchauffée au voisinage immédiat de la turbine dans un condenseur de vapeur d'échappement ou de prélèvement à 1 ou quelques bars.

Les installations industrielles de production simultanée d'énergie mécanique (ou électrique) et thermique ont le plus souvent une puissance inférieure à 100 MWth. Le choix optimum de l'état initial de la vapeur dépend du facteur de répartition et de l'option prise pour le dimensionnement de l'installation. En tout état de cause il existe (5) une limite inférieure à la surchauffe de la vapeur haute pression, se situant vers :

160°C	à	25 bars,
180°C	à	50 bars,
200°C	à	100 bars,

Même si les conditions à l'échappement de la turbine étaient favorables, il n'est pas souhaitable de produire au générateur de la vapeur à trop faible surchauffe ; cette vapeur aurait un volume massique trop faible, ce qui entraînerait, surtout pour des groupes de faible puissance, la nécessité de prévoir des ailettes de détente HP très courtes. Le rendement de la détente en serait fortement détérioré.

Le dimensionnement et le choix du type d'installation à énergie totale dépendra essentiellement :

- De la valeur moyenne et des fluctuations du facteur de répartition chaleur/travail (φ) ;
- De l'option prise au départ : soit un fonctionnement en marche autonome pour la production d'énergie électrique, soit un fonctionnement en parallèle sur le réseau (6).

Dans le premier cas, la demande de puissance électrique dicte le réglage du groupe. Nous avons énuméré ci-dessus plusieurs moyens de rendre compatibles les puissances calorifiques et électrique produites instantanément. Toutefois, l'installation se complique fort, les frais de premier investissement et les frais de fonctionnement s'accroissent sensiblement si la fraction de la vapeur parcourant un cycle à condensation (basse températue) est élevée, ce qui est le cas lorsque les besoins de chaleur sont déficitaire par rapport aux possibilités du groupe. Rappelons encore qu'une turbine à contrepression constitue un investissement beaucoup moins coûteux qu'une turbine à condensation.

La seconde option possible est plus efficace et plus simple à mettre en oeuvre. Dans ce cas, la puissance électrique générée par le groupe suit les fluctuations de la demande de chaleur, le débit de vapeur étant adapté en conséquence. Le réseau extérieur de distribution d'électricité intervient soit pour fournir le complément de puissance électrique indispensable, soit pour absorber l'exédent produit. Cette dernière solution sera en général la plus rentable, tant du point de vue économique qu'énergétique, à condition que le propriétaire du groupe à énergie totale puisse négocier un contrat d'achat et de vente de puissance électrique acceptable avec les sociétés de production ou de distribution.

A titre d'exemple et pour présenter certains ordres de grandeur, considérons une application pure du principe de la contrepression, sans pertes calorifiques à l'échangeur principal de chaleur ($\eta_c = 1$). La puissance thermique nominale exigée est 15MWth. Cette action calorifique est à produire à 110°C. On suppose un écart de 10°C à l'échangeur. La température de condensation de la vapeur doit donc être 120°C, ce qui correspond à une pression de 2 bars. Envisageons deux qualités de vapeur HP.

	1er cas	2ème cas
Pression de la vapeur vive :	25 bars	75 bars
Température de surchauffe :	380°C	480°C
Rendement isentropique de la détente :	0,84	0,84
τ_m de la détente :	460kJ/kg	660kJ/kg
Surchauffe de la vapeur à l'échappement de la turbine :	15°C	9°C
Chaleur de condensation à 2 bars :	2198kJ/kg	2198kJ/kg
Facteur de répartition chaleur/ travail (φ)	4,77	3,33
Débit nominal de vapeur :	24,6 T/h	24,6 T/h
Puissance nominale à l'accouplement de la turbine :	3,1 MW	4,5 MW

Ces valeurs supposent $\eta_m = \eta_c = 1$. Si l'on choisit les valeurs plus réalistes $\eta_m = 0,94$, $\eta_c = 0,98$ et si l'on adopte pour le générateur de vapeur $\eta_G = 0,91$, les facteurs de répartition se modifient quelque peu :

$\varphi_1 = 4,92$ $\qquad\qquad$ $\varphi_2 = 3,44.$

Les rendements globaux des deux systèmes envisagés sont très voisins ; ils valent respectivement :

$\eta_{e,t,1} = 0,880$ $\qquad\qquad$ $\eta_{e,t,2} = 0,882.$

B - Les turbines à gaz.

Les rejets thermiques des turbines à gaz sont de nature très différente de ceux des turbines à vapeur. Ils sont à haute température et sont constitués par un mélange gazeux à faible capacité calorifique dont la fraction condensable est très minime.

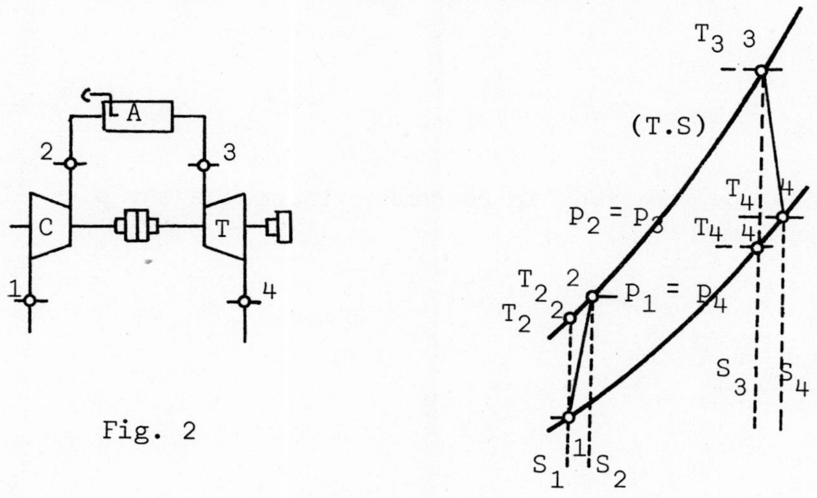

Fig. 2

On sait que le rapport de compression d'une turbine à gaz fonctionnant selon un cycle simple (fig. 2) peut être choisi de manière à rendre maximum soit le travail moteur produit, soit le rendement thermique de la machine.

En supposant que les gaz comburés sont assimilables à de l'air, que les phases de compression et de détente sont des évolutions polytropiques de gaz parfaits à chaleur spécifique constante, et qu'il n'y a pas de pertes de charge dans la chambre de combustion, on trouve aisément les expressions du travail moteur et du rendement :

$$\tau_m = c_p\, T_1 \left[y\,(1 - x^{-\eta_T}) - (x^{1/\eta_c} - 1) \right] \qquad (8)$$

$$\eta_t = \frac{y\,(1 - x^{-\eta_T}) - (x^{1/\eta_c} - 1)}{y - 1 - (x^{1/\eta_c} - 1)} \qquad (9)$$

où η_T et η_c sont les rendements polytropiques internes de la turbine et du compresseur, c_p est la chaleur spécifique de l'air. En outre, on a posé :

$y = T_3/T_1$

et

$$x = \left(\frac{P_2}{P_1}\right)^{\frac{k-1}{k}}$$

k est le rapport des chaleurs spécifiques à pression constante et à volume constant (k = 1,4 pour l'air).

On peut vérifier que le rendement η_t est continûment croissant avec y, donc avec la température d'admission des gaz comburés dans la turbine. L'optimisation de la turbine à gaz fixera le paramètre x, l'image du rapport de compression.

Pour des machines récentes, où la température T_3 peut être voisine de 900°C si les gaz comburés sont propres, et où les rendements polytropiques internes η_T et η_c sont voisins de 0,90, on obtient un rendement thermique de l'ordre de... 0,30 et un rapport de compression de...9... pour l'optimisation du travail moteur. L'expression théorique donnant ce rapport optimal est :

$$\frac{1}{x^{\eta_c}} + \eta_T = y\, \eta_c\, \eta_T \qquad (10)$$

Le rapport de compression rendant maximum le rendement thermique est plus élevé, ainsi que le montre son expression théorique :

$$\frac{1}{x^{\eta_c}} + \eta_T = \frac{y\, \eta_c\, \eta_T}{1 - \eta_t} \qquad (11)$$

Dans le cas évoqué ci-dessus, il serait de l'ordre de 20. Il est à noter toutefois qu'au voisinage de l'optimum, l'accroissement de η_t est très faible ; il y aura donc toujours un avantage économique à adopter, même si l'optimum de rendement est le but visé, un rapport de compression un peu inférieur à la valeur tirée de (11).

La médiocrité du rendement thermique d'une turbine à gaz fait en sorte qu'il est rarement économique de l'utiliser dans des installations statiques selon le cycle simple représenté à la fig. 2. On peut améliorer ce cycle en y greffant un récupérateur de chaleur transférant vers l'air à la sortie du compresseur une partie des calories perdues à l'échappement. Malheureusement, l'échangeur-récupérateur sera toujours volumineux et coûteux, et source de pertes de charge non négligeables, ce qui diminuera la puissance disponible. Le rendement sera ici de l'ordre de 0,35.

On peut diminuer dans une proportion considérable (30...40%) l'extension de la surface d'échange en transférant les calories perdues à de la vapeur qui pourra produire elle aussi du travail moteur, ou éventuellement sera considérée comme vapeur de procédé. Voici donc la turbine à gaz insérée dans un système à énergie totale. Les cycles moteurs réalisables à partir de vapeur générée dans une chaudière de récupération n'auront que de modestes performances. Aussi limite-t-on souvent la simple récupération aux cas de production de vapeur de procédé. Par contre, si l'on fournit un appoint de combustible, le mélange comburant étant les fumées d'échappement de la turbine à gaz, on n'est plus limité d'aucune manière. Ce dernier cycle combiné est la conjonction d'un cycle à vapeur classique (éventuellement à contrepression ou à prélèvement) et d'un cycle de turbines à gaz sans aucune perte à l'échappement. On en voit l'avantage évident. On comprendra aussi que la turbine à gaz soit toujours conçue, dans ce cas, pour produire le maximum de travail moteur.

Une combustion d'appoint dans la chaudière de récupération est toujours avantageuse, et cela à un double titre : la combustion intensifie fortement la fraction de la puissance échangée par rayonnement plutôt que par convection, ce qui permet de réduire la surface d'échange par unité de puissance transmise. Qu'il y ait combustion d'appoint ou non, pourvu que l'on ne doive pas apporter d'appoint d'air frais, les pertes à l'échappement de la chaudière de récupération sont fixes ; ces pertes, rapportées à l'unité de puissance transmise, seront donc d'autant plus faibles que la puissance injectée par le combustible d'appoint sera grande.

Au passif du procédé, il faut noter que la turbine à gaz ne s'accomode que de combustibles de très bonne qualité, donc chers, ce qui annihile parfois l'avantage économique du système. D'autre part, le procédé ne supprime pas le problème de la récupération des calories à basse température : il est malaisé d'obtenir des températures de sortie des gaz d'une chaudière de récupération avec post-combustion qui soient aussi basses que dans les générateurs de vapeur classiques équipés d'un échangeur réchauffeur d'air (8).

Un exemple de cycle combiné où la puissance produite par la vapeur prédomine est la centrale de Vitry, qui comporte deux tranches de 320MW dont la répartition de puissance est pour chacune :

$$TG = 70\ MW \quad ; \quad TV = 250\ MW$$

Les turbogroupes à vapeur peuvent fonctionner en marche autonome ou combinée ; dans le premier cas, le réchauffage de l'air se fait par soutirage de vapeur (cycle CESAS). En marche combinée, le rendement global de la centrale de Vitry dépasse 42% (7).

L'autoproduction industrielle d'électricité par turbine à gaz est une solution éprouvée. La puissance des groupes est dans la gamme 10...20 MW. La vapeur produite est utilisée le plus souvent comme caloporteur. Lorsque la vapeur n'est pas destinée à produire du travail moteur, et lorsqu'il n'y a pas de combustion d'appoint, le facteur de répartition chaleur/travail dépendra de la qualité de la récupération. Sa limite supérieure est donnée par :

$$\varphi_{max} = \frac{T_4 - T_1}{T_3 - T_4} = \frac{(y - 1) - y\,(1 - x^{-\eta_T})}{y\,(1 - x^{-\eta_T}) - (x^{1/\eta_c} - 1)}$$

Avec comme plus haut, $\eta_T = \eta_c = 0{,}90$, $y \cong 4$, on trouve :

$$\varphi_{max} \cong 1{,}77.$$

Un écart de température de 120°C entre l'ambiance et la sortie de la chaudière de récupération réduit d'un tiers environ le facteur de répartition, qui devient :

$$\varphi \cong 1{,}15.$$

On envisage actuellement une récupération très complète de l'enthalpie des fumées à l'échappement, lesquelles, grâce à un système nouveau actuellement à l'essai à Gaz de France, pourraient être refroidies jusqu'à environ 45°C, si le combustible est du gaz naturel. On récupèrerait ainsi une grande partie de la chaleur de condensation de la vapeur d'eau générée par la combustion de l'hydrogène. Les gaz sont refroidis par aspersion d'eau pulvérisée par des gicleurs. L'eau chaude est récoltée vers 50°C et peut trouver des utilisations telles que chauffage de bureaux, préchauffage de bains, eau chaude sanitaire (9).

C . Les moteurs à combustion interne.

Les moteurs à allumage spontané fonctionnent au fuel ou au gaz naturel. Les fuels lourds ne peuvent être utilisés que pour de gros moteurs lents, car ils nécessitent une grande chambre de combustion.

Le rendement thermique de la production d'énergie mécanique par moteurs Diesel et Diesel-gaz est d'environ 38%, ce qui est notablement plus élevé que pour les turbines à gaz. Ces moteurs conviennent bien à la constitution de groupes à énergie totale. La conception des chaudières de récupération pour moteurs Diesel ne sera pas très différente de celles pour turbines à gaz. Une combustion d'appoint pourra être pratiquée.

La récupération des calories perdues lors d'un cycle Diesel est toutefois plus partielle et moins aisée que pour une turbine à gaz. En effet, 40% environ de la chaleur résiduelle est évacuée soit au réfrigérant d'huile, soit par l'eau de refroidissement, soit par pertes pariétales (convection et rayonnement tnermique). L'eau de refroidissement et l'huile ne peuvent utilement rétrocéder leurs calories qu'à fort basse température. Enfin les gaz d'échappement sont à une température quelque peu plus basse que ceux de la turbine à gaz. Par comparaison, les gaz d'échappement d'une turbine à gaz portent 98 à 99% de la chaleur résiduelle du cycle (5), p. 179.

Le prix à l'investissement d'un moteur Diesel, par unité de puissance, varie très peu avec la dimension du moteur. Cet avantage permet de modulariser les installations à énergie totale conçues selon ce principe. Pour la turbine à gaz, au contraire, le prix d'investissement par unité de puissance varie du simple au double si l'on compare des machines de 10 et 3 MW, respectivement. Dès lors, le seuil de rentabilité pour l'installation d'un système à énergie totale se situe à 10... 15 MWe pour une turbine à gaz, et environ 1 MW seulement pour les groupes Diesel.

Selon le degré de récupération de la chaleur, le facteur de répartition sera plus ou moins élevé, mais demeurera toujours inférieur à ce qu'il peut être dans des systèmes à turbine à vapeur ou à gaz. Les rendements globaux que l'on peut escompter pour les groupes Diesel sont normalement compris dans l'intervalle 0,65... 0,80.

D . Le chauffage urbain (extrait de (10)).

Dans nos pays occidentaux, les plus gros consommateurs d'énergie primaire sont les sociétés productrices d'électricité (secteur très concentré) et la multitude des consommateurs devant assurer le chauffage de locaux (secteur parfaitement essaimé). La nécessité d'une reconversion des mentalités est reconnue ; mais le premier bon pas reste toujours à faire en Belgique. La chaleur doit être distribuée, en zone d'habitation dense ; encore faut-il que la source de chaleur soit judicieusement choisie. Sur ce point, il n'y a pas de controverse possible : les productions d'électricité et de chaleur par groupes à vapeur doivent être combinées.

En examinant certaines statistiques, on peut croire qu'une habitation moyenne (maison, villa, appartement, studio,...), bien isolée et chauffée raisonnablement ne devrait pas consommer pour son chauffage plus de $C = 125 \cdot 10^6$ kJ par an (ce qui correspond approximativement à 30 Gcal/a et à une consommation de fuel de l'ordre de 5 000 kg/a). Cette consommation se répartit sur plusieurs mois, suivant un diagramme qui dépend de l'évolution de la température atmosphérique et de l'occupation du logement. Chaque installation doit avoir une puissance déterminée (puissance installée) qui s'identifie en pratique avec la puissance de pointe. Le temps d'utilisation est la durée en heures pendant laquelle l'installation pourrait fournir l'énergie C si elle fonctionnait constamment à la puissance de pointe. Admettons que ce temps soit U = 2000 h. La puissance de pointe de toutes les installations supposées les mêmes doit donc être :

$$p = \frac{C}{3600\ U} = \frac{125 \cdot 10^6}{3600 \cdot 2000} = 17{,}4 \text{kJ/s ou kW},$$

ce qui correspond à environ 15 000 kcal/h.

On pourra trouver aux figures 3 à 6 des représentations schématiques d'installations motrices à vapeur, classiques et nucléaires, ainsi que les évolutions correspondantes de la vapeur représentées dans un diagramme (h,s).

Nous allons montrer quelles modifications apporter à ces cycles pour combiner production de travail moteur et de chaleur.

a) Cycles utilisant un combustible fossile.

Considérons une installation alimentée par un combustible fossile (charbon, pétrole,...) comportant une turbine ayant à l'entrée les caractéristiques de vapeur : 140 bars, 540/540°C (fig. 4). Concevons cette machine pour une pression de sortie de 2,75 bars (pression du soutirage IV de la fig. 4). Nous utiliserons la vapeur d'échappement, encore surchauffée (238°C) pour réchauffer l'eau d'une distribution de chaleur jusqu'à une température qui peut au moins atteindre 131°C. Il n'y a plus d'intérêt de conserver les soutirages dont le but était de diminuer la chaleur perdue à la source froide ; l'eau alimentaire sera cependant à 131°C (état liquide saturé sous la pression 2,75 bars).

Proposons nous de faire passer par la turbine un débit de vapeur tel qu'au moyen de la vapeur d'échappement on puisse alimenter une distribution de chaleur de 1000 $.10^6$ kJ/h. Nous admettrons que les pertes du réseau d'eau chaude soient de 10%. Il faudra dès lors prévoir au départ une puissance de chauffage de 1100 . 10^6 kJ/kg.

Chaque kg de vapeur d'échappement en passant à l'état liquide saturé sous la pression de 2,75 bars (dont l'enthalpie est h'_{IV} = 548,8 kJ/kg) pourra donc céder une chaleur :

$$Q_2 = h_{6,IV} = 2940 - 548,8 = 2391 \text{ kJ/kg.}$$

Cependant à cause des pertes de la distribution, la partie utile de cette chaleur ne sera que :

$$Q'_2 = Q_2 \frac{1000}{1100} = 2173,7 \text{ kJ/kg.}$$

Par ailleurs le travail τ_m effectué par chaque kg de vapeur dans la turbine ressortit à :

$$\tau_m = h_3 - h_4 + h_5 - h_6 = 988,3 \text{ kJ/kg,}$$

Tandis que l'action calorifique Q_1 à fournir par kg d'eau dans le générateur s'élève à :

$$Q_1 = h_3 - h' + h_5 - h_4 = 3379,5 \text{ kJ/kg.}$$

Si η = 0,991 est le rendement mécanique de la turbine, et si les auxiliaires (y compris les pompes de la distribution) demandent 5% du travail de la machine, on aura :

$$\tau - \tau' = \eta_a \tau = \eta_a \eta \tau_m = 930,4 \text{kJ/kg.}$$

Si le rendement du générateur de vapeur est η_G = 0,940 on trouve pour le rendement total de l'installation (pertes de la distribution comprises) :

$$\eta_{e,t} = \frac{\tau - \tau' - Q'2}{Q_I} \eta_G = \frac{930,4 + 2173,7}{3379,5} 0,940 = 0,863.$$

Le débit de vapeur requis pour la distribution étant :

$$G_v = \frac{1100 .10^6}{2391,1} = 460 .10^3 \text{ kg/h ou } 127,8\text{kg/s,}$$

La puissance nette de la machine est :

$$N - N' = G_v (\tau - \tau') = 127,8 . 930,4 = 118\ 905 \text{ kW.}$$

En dehors du réseau de distribution de chaleur, cette application demande un générateur de vapeur, une turbine et des échangeurs de chaleur en remplacement du condenseur de la turbine. Celle ci ne possède plus de corps BP, ni de poste d'eau (sauf une bâche alimentaire, qu'il est pratique de conserver dans le circuit). Toute la détente se produit dans le domaine de la surchauffe, ce qui élimine les difficultés provenant de la présence d'une phase liquide dans la vapeur, au moment même où le volume massique du fluide devient considérable. Enfin, comme il n'y a plus de source froide à la température de l'atmosphère, on libère l'installation des sujétions qui entraine une prise d'eau à la rivière ou le recours à un réfrigérant atmosphérique.

Il est certain que l'utilisation d'une centrale ainsi conçue est mauvaise puisqu'on l'arrête pendant 3 ou 4 mois par an. Compte tenu du rendement obtenu, d'une certaine simplification de l'installation, de l'accroissement du coût de l'énergie primaire, d'une certaine diminution de la pollution,... Il est difficile de croire, même dans le cadre d'une tarification avantageuse pour les abonnés et d'un amortissement raisonnable, que la méthode de chauffage

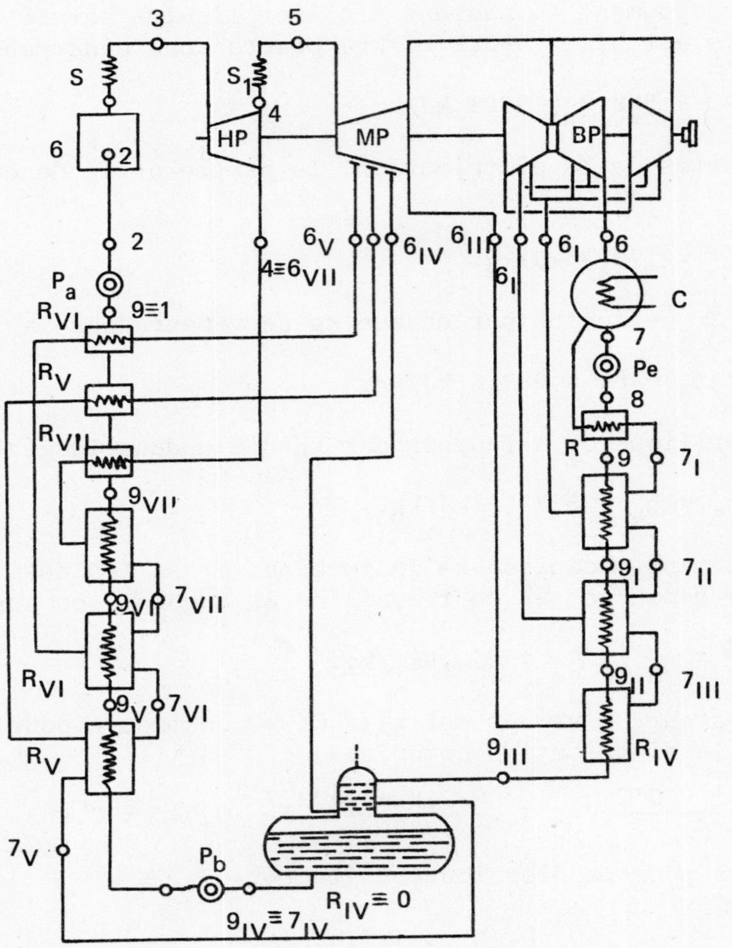

Fig. 3 - Schéma d'un cycle classique à vapeur.

La pompe P introduit l'eau dans le générateur de vapeur G (état 1) d'où elle sort (en 3) à l'état de vapeur surchauffée. Elle se détend ensuite dans une turbine réalisée en trois corps celui haute pression HP, celui moyenne pression MP et celui basse pression BP supposé subdivisé en trois flux. Entre le corps HP et MP a lieu une resurchauffe de la vapeur dans l'appareil S_1 (états 4 et 5). Le condenseur C maintient à la sortie de la turbine (en 6) un vide de l'ordre de 0,04-0,05 bar. L'eau condensée revient au générateur par l'intermédiaire d'une série de réchauffeurs R, recevant d'autre part de la vapeur soutirée à la turbine aux points 6_I à 6_{VII}. Les trois premiers réchauffeurs sont dits à basse pression ; celui R_{IV} est à mélange et est installé au-dessus de la bâche d'eau ; les derniers réchauffeurs sont à haute pression (à cause de la pompe de bâche P_b). L'ensemble des réchauffeurs constitue le poste d'eau.

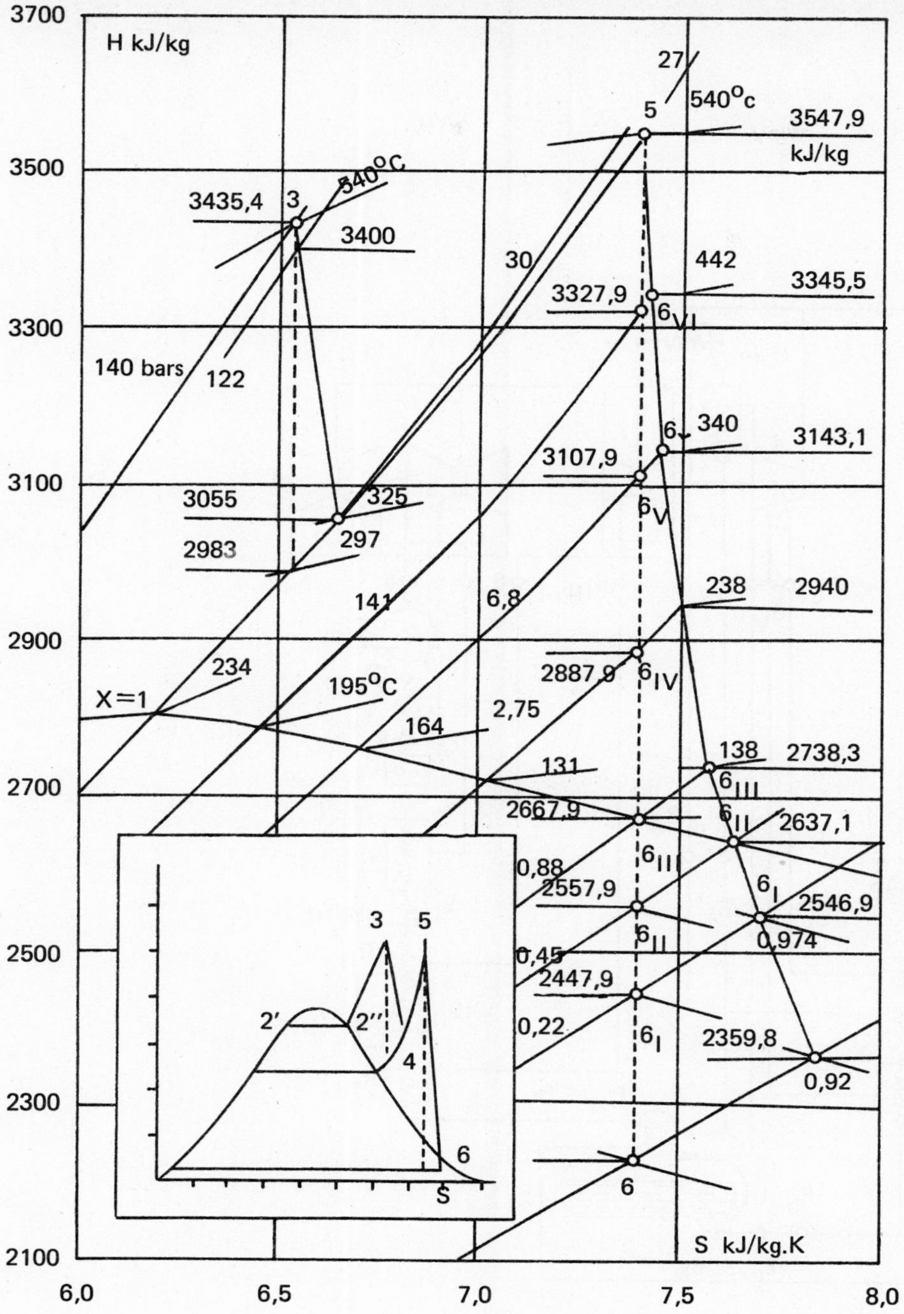

Fig. 4 - Diagramme de Mollier correspondant à l'installation de la Fig. 3.

La pression initiale de la vapeur à l'état 3 est 140 bars et la température de surchauffe est 540°C . Après la détente (3,4) a lieu la resurchauffe à la température 540°C (transformation 4,5). A partir de l'état 5 la vapeur est détendue jusqu'à la pression du condenseur. Les points où sont effectués les 7 soutirages sont répartis le long de la ligne de détente. On dit que les caractéristiques de ce cycle sont :

140 bars/540°C/540°C/VII

Les valeurs indiquées étant respectivement la pression à l'entrée de la turbine, la température de surchauffe, celle de resurchauffe et enfin (en romain) le nombre de soutirages.

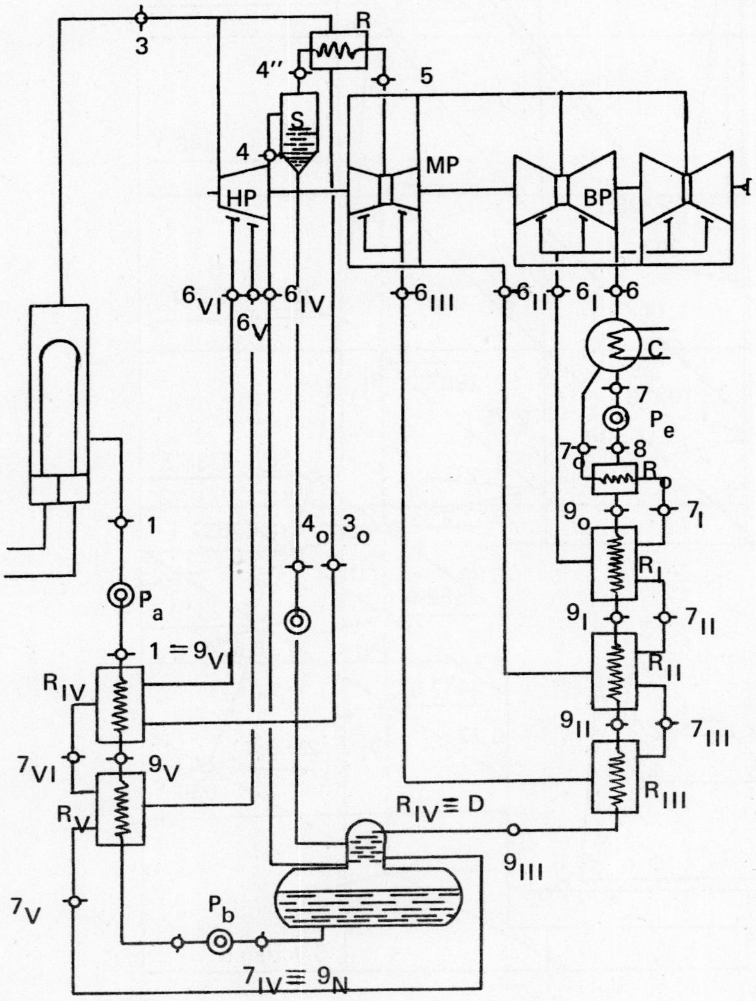

Fig. 5 - Schéma d'un cycle à vapeur alimenté par un réacteur PWR (réacteur à eau surpressée)

La vapeur est produite dans le générateur G parcouru d'autre part par l'eau surpressée d'un circuit primaire, au moyen duquel la chaleur dégagée au réacteur est tranférée au fluide parcourant le cycle moteur. Les températures d'entrée et de sortie du fluide primaire dans G étant respectivement de l'ordre de 320... 325°C et 280 ...285°C, la pression de la vapeur produite peut atteindre ... 60 ... bars. La turbine comprend un étage HP à simple flux un corps MP à double flux et deux corps BP chacun à double flux. Entre les corps HP et MP la vapeur est séchée dans le séparateur S, puis surchauffée au moyen de la vapeur vive, prélevée au générateur de vapeur. Six soutirages sont prévus ; celui R_{IV} fonctionne par mélange et est monté comme dans l'exemple de la fig. 3 au dessus de la bâche alimentaire ; c'est dans ce réchauffeur que l'eau déposée dans le séparateur S rejoint le cycle. Enfin la vapeur condensée dans le surchauffeur R est réintroduite dans le cycle au réchauffeur R_{VI}.

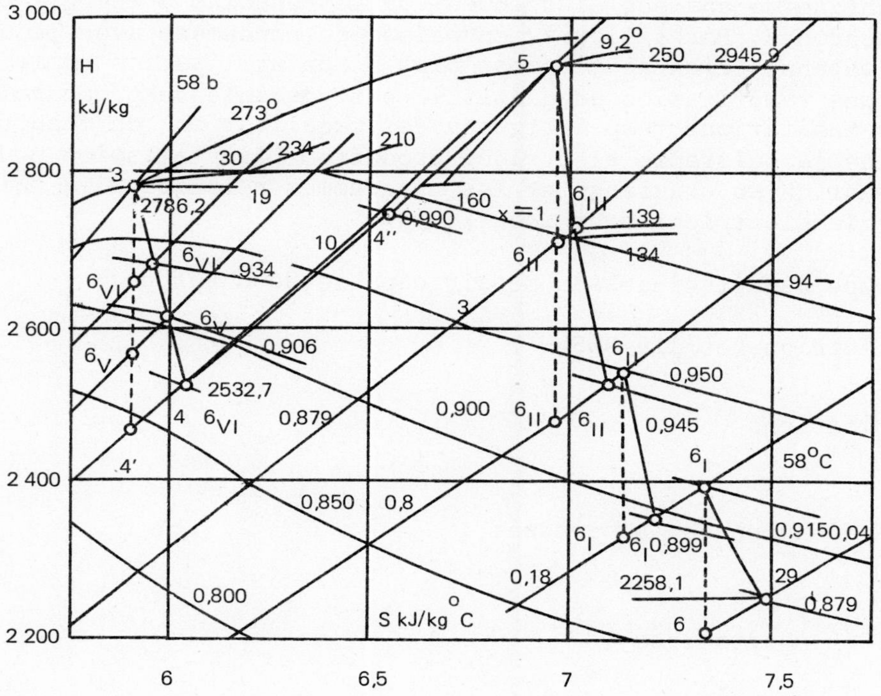

Fig. 6 - Diagramme de Mollier correspondant à l'installation de la fig. 5.

Comme on le voit la plus grande partie de la détente de la vapeur se fait dans le domaine des vapeurs saturées. En appliquant ce que nous avons expliqué dans la légende de la fig.4, on peut conclure que les caractéristiques du cycle sont :

58 bars/273°C/250°C/VI.

Ce qui donne :

$$\tau_m = 1261,0 \text{ kJ/kg.}$$

Compte tenu des pertes mécaniques (estimées à 0,9%), le travail utile à l'accouplement ressortit à :

$$\tau = (1 - 0,009)\ \tau_m = 1249,7 \text{ kJ/kg}$$

Dans une installation nucléaire, les auxiliaires, qui comprennent en plus les pompes de circulation du circuit d'eau primaire, ont une consommation que nous estimerons à 7%. Il vient donc :

$$\tau - \tau' = (1 - 0,07) = 1162,2 \text{ kJ/kg ;}$$

le débit de vapeur arrivant au condenseur s'élève dès lors à :

$$G_v = \frac{1000 \ . \ 10^3}{1162,2} = 860,44 \text{ kg/s ou } 3097,5 \text{ t/h.}$$

et celui à l'entrée de la machine :

$$G'_v = G_v\ (1 + \sum X_i + Y + Z) = 1657,5 \text{ kg/s ou } 5\ 967 \text{ t/h.}$$

ainsi conçue ne puisse laisser une certaine rentabilité.

Dans le procédé que nous venons d'étudier, l'énergie électrique produite est liée d'une manière très étroite à la quantité de chaleur distribuée. Or les besoins d'énergie électrique et de chaleur ne sont pas synchrones. Par ailleurs, en raison du rendement avec lequel l'énergie électrique produite est obtenue, rien ne s'oppose plus à son utilisation comme moyen de chauffage. Or, à la faveur d'une tarification adéquate, il est possible de promouvoir le chauffage electrique pour les habitations trop éloignées des canalisations du chauffage urbain, jusqu'à utilisation complète de la puissance électrique produite. L'installation peut alors être considérée comme ne servant qu'au chauffage et les diagrammes relatifs à la puissance appelée en chaleur ou en énergie électrique sont identiques.

Dans cette optique, le groupe décrit ci-dessus serait capable de chauffer :

- Au moyen de sa puissance électrique de 118 905kW :

$$\frac{118\ 905}{17{,}4} = 6834 \text{ habitations}$$

- Au moyen de la distribution d'eau chaude surpressée :

$$\frac{1000 \;.\; 10^6}{3600 \;.\; 17{,}4} = 15\ 964 \text{ habitations ,}$$

Soit au total 22 798 logements.

b) Cycle alimenté par un réacteur PWR

Supposons que le cycle représenté par le diagramme (h,s) de la fig. 6 soit utilisé dans une installation d'une puissance de 1000 MW. Pour ce cycle, X_i (i = I, ... VI) sont les débits prélevés à la turbine aux points de soutirage, Y est le débit d'eau en provenance du séparateur S et Z est le débit de vapeur vive passant au surchauffeur. Dans l'exemple de la fig. 6 ces débits, rapportés au débit entrant en 6 au condenseur, valent :

$X_I = 0{,}0803$ $\qquad$ $X_V = 0{,}1206$

$X_{II} = 0{,}0810$ $\qquad$ $X_{VI} = 0{,}1037$

$X_{III} = 0{,}0905$ $\qquad$ $Y = 0{,}1581$

$X_{IV} = 0{,}1421$ $\qquad$ $Z = 0{,}1500.$

Il en résulte que par kg de vapeur à la sortie des corps BP, le débit à l'entrée du corps HP et celui au générateur de vapeur G sont respectivement :

$$1 + \sum X_i + Y = 1{,}776 \text{ kg} \qquad \text{et} \qquad 1 + \sum X_i + Y + Z = 1{,}926 \text{ kg.}$$

Par kg de vapeur arrivant au condenseur, le travail effectué sur le rotor de la turbine est obtenu par la formule :

$$\tau_m = (1 + Y)(h_3 - h_4) + (h_5 - h_6) + \sum_{IV}^{VI} X_K (h_3 - h_{6,K})$$
$$+ \sum_{I}^{III} X_K (h_3 - h_4 + h_5 - h_{6,K}) ,$$

Proposons nous de prélever au soutirage III, à la pression de 3 bars une quantité de vapeur suffisante pour alimenter un chauffage urbain d'une puissance nette de 1000.10^6 kJ/h. A cause des pertes calorifiques des canalisations il faut prévoir au départ de la centrale une action calorifique de $1100 \, . \, 10^6$ kJ/h.

A l'état 6_{III} la vapeur possède une enthalpie $h_{6,III} = 2736{,}6$ kJ/kg, alors que pour le liquide saturé sous la pression de 3 bars on a : $h'_{III} = 561{,}4$ kJ/kg. Le prélèvement proposé réduit donc le débit de vapeur dans la partie de la turbine en aval du point 6_{III}, y compris celui du soutirage III (qui ne doit réchauffer que l'eau revenant du condenseur et celles des réchauffeurs I et II) de :

$$\frac{1100 \, . \, 10^6}{3600\,(h_{6,III} - h'_{III})} = \frac{1100 \, . \, 10^6}{3600\,(2736{,}6 - 561{,}4)} = 140{,}47 \text{ kg/s}$$

et diminue la masse de vapeur arrivant par seconde au condenseur de :

$$G_v = \frac{140{,}47}{1 + X_I + X_{II} + X_{III}} = \frac{140{,}47}{1{,}2518} = 112{,}2 \text{ kg/s}$$

soit 13% de G_v. Ce prélèvement ne perturbe guère la veine fluide dans la partie BP de la machine, où les débits volumiques sont considérables et même quelque peu excessifs, au point qu'à la sortie de la turbine la vitesse axiale de la vapeur à pleine charge peut atteindre 280...300 m/s.

Calculons par ailleurs la perte de puissance subie par la turbine à cause de ce prélèvement. Pour les états intervenant dans ce calcul les enthalpies sont :

$h_{6,III} = 2736{,}6$ kJ/kg, $h''_{6,II} = 2540{,}6$ kJ/kg, $h_{6,II} = 2552$ kJ/kg,

$h''_{6,I} = 2358$ kJ/kg, $h_{6,I} = 2398$ kJ/kg, $h_6 = 2260$ kJ/kg.

Par kg de vapeur arrivant au condenseur, la vapeur qui se détent au-delà du point 6_{III} effectue un travail :

$$\mathcal{T}_m = (1 + X_I + X_{II})\,(h_{6,III} - h''_{6,III}) + (1 + X_I)\,(h_{6,II} - h''_{6,1}) + h_{6,1} - h_6$$

$$= 1{,}1613 \, . \, 196 + 1{,}0803 \, . \, 194 + 138 = 575{,}2 \text{ kj/kg,}$$

ce qui correspond à un travail net :

$$\mathcal{T} - \mathcal{T}' = \eta\eta_a \, \mathcal{T}_m = 0{,}991 \, . \, 0{,}93 \, . \, 575{,}2 = 530{,}1 \text{ kJ/kg.}$$

Il en résulte que la perte de puissance inhérente à la diminution du débit ΔG_v est de :

$$\Delta G_v\,(\mathcal{T} - \mathcal{T}') = 112{,}214 \, . \, 530{,}1 = 59\,484 \text{ kW ou } 59{,}5 \text{ MW,}$$

soit 5,95% de la puissance de la machine sans prélèvement.

L'eau alimentaire sortant du réchauffage 6_{VI} est à la température 234°C et possède une enthalpie $h_1 = 1008{,}4$ kJ/kg ; après la pompe alimentaire son enthalpie est de l'ordre de 1018 kJ/kg. La chaleur communiquée à la vapeur dans le générateur par seconde vaut dès lors :

$$G'_v\,(h_3 - h_2) = 1657{,}47\,(2786{,}2 - 1018) = 2930{,}75 \, . \, 10^3 \text{ kJ/s ou kW.}$$

Comme la machine a une puissance électrique de (1000 - 59,5) 10^3 kW et une puissance nette de chauffage de 1000 . 10^6/3600 = 277,78 .10^3 kW, le rendement total de l'installation devient :

$$\eta_{e,t} = \frac{(1000 - 59,5)\ 10^3 + 277,78\ .\ 10^3}{2930,75\ .10^3} = 0,416,$$

alors que sans prélèvement il était 0,341.

L'augmentation du rendement énergétique est considérable pour une faible baisse de production de puissance électrique.

Le rendement énergétique ainsi obtenu pour une centrale nucléaire est comparable à celui que nous avons cité pour une centrale électrique utilisant un cycle combiné avec turbines à gaz et à vapeur.

IV - CONCLUSIONS

Nous avons examiné les contributions que peut apporter la thermodynamique au problème de l'économie d'énergie, en ce qui concerne les systèmes intégrés de production de chaleur et de travail moteur. Les solutions qui ont été discutées ne demandent la mise en oeuvre d'aucune technologie inconnue et, si elles ne le sont déjà, pourraient être appliquées à relativement berf délai. Etant donné la grande densité de population et d'activités industrielles de notre pays, le problème du transport de calories ne devrait pas être trop aigu en Belgique.

Les principes sont clairs et admis ; de nombreuses possibilités d'application sont reconnues ; la technologie indispensable est acquise. Il manque peut-être certaines études d'optimalisation de systèmes à énergie totale, pour parfaire encore les preuves de leur rentabilité. Mais pour voir se multiplier les systèmes que nous défendons, il ne manque qu'un élément, fondamental toutefois : la volonté commune des "décideurs" d'entreprendre une politique concertée d'investissements qui soit à la mesure de déclarations d'intentions souvent faites : préserver l'équilibre de la balance des paiements du pays, économiser les ressources naturelles de la planète.

BIBLIOGRAPHIE

(1) A. HOUBERECHTS : La thermodynamique technique, tome I, Vander,Louvain (1975), p.64.

(2) A. HOUBERECHTS : La thermodynamique technique, tome II, en préparation.

(3) A. HOUBERECHTS : L'exergie, EPE, 7 ,(1971), 37-67.

(4) J. MARTIN : La production combinée d'électricité et de chaleur, notions de thermodynamique : l'énergie totale, énergie n°207 (1974), 79-92.

(5) R. M. E. DIAMANT : Total Energy. Pergamon Press, Oxford (1970).

(6) F. MEUNIER : La production combinée d'électricité et de chaleur. Point de vue d'un bureau d'études. Energie, n°207 (1974), 95-125.

(7) H. PFENNINGER : Centrale électrique combinée à turbines à vapeur et à gaz. Revue Brown Boveri, 60 , (1973), 389-397.

(8) E. AGUET : Evolution actuelle des cycles combinés dans le contexte de l'environnement et de la pénurie de combustibles. Communication n° 46 aux journées d'étude des "Centrales Electriques Modernes" (oct. 1974), 12p.

(9) ANON : Récupérateur à condensation derrière chaudière à gaz. Revue générale de thermique, n° 133 (1973), 81.

(10) A. HOUBERECHTS : Cycles thermiques moteurs, problèmes de chauffage, énergie totale et thermopompage. Conférence prononcée à la tribune de l'U. I. Lv. A paraître.

UTILISATION D'UN CYCLE BINAIRE EAU-AMMONIAC DANS LES CENTRALES ELECTRIQUES (POWER PLANT INCLUDING AN AMMONIA BOTTOMING CYCLE)

Jean Fleury

Electricité de France - Direction des Etudes et Recherches - 6, quai Watier - 78400 Chatou

ABSTRACT

Settlement of power plants has become to day a difficult problem because it is bound by economical, ecological and sociological coercions. Every freed parameter is there fore particularly interesting. So, air cooling increases considerably the number of possible places to install power plants and permit to bring them more easily nearer to the consumption centers. This results in a reduction of the high voltage grid and in the suppression of heating of the water in rivers or coast zones, things which are welcome through the population.

Bottoming cycles are well adapted to air cooling systems. In addition they bring to them further advantages which promote their development.

The object of this paper is to present the ammonia bottoming cycle foreseen at "Electricité de France" and demonstrate the further advantages it brings to the air cooling systems associated with classical steam cycle power plants.

CONSIDERATIONS AYANT CONDUIT A ENVISAGER LES CYCLES BINAIRES

Parmi toutes les conditions à remplir par un site destiné à recevoir une centrale nucléaire, celles concernant la réfrigération sont de loin les plus contraignantes. Elles nécessitent de disposer de ressources locales en eau minimales, ce minimum étant fonction de la puissance qu'on désire installer et du type de réfrigération choisi :

Réfrigération par circuit ouvert (sur mer ou sur rivière)

Dans ce mode de réfrigération toute la charge thermique est évacuée dans un fort débit d'eau : 40 m^3/s pour un PWR 900 MW. Ce débit élevé permet toujours une dilution correcte des effluents.

Avantages : coût acceptable - simplicité du matériel.

Inconvénients :
- limitation des sites possibles,
- éloignement des centres de consommation,
- échauffement des eaux.

Réfrigération humide

Dans ce procédé, l'évaporation d'une partie du débit d'eau de circulation absorbe environ 80 % de la charge thermique. Les besoins minimums en eau sont de 1,6 m^3/s pour un PWR 900 MW, dont 0,6 m^3/s sont évaporés.

Avantages : plus grande liberté du choix des sites.

Inconvénients :
- nuisance esthétique des tours,
- bruit dans le cas du tirage induit,
- panaches de vapeur d'eau.

Réfrigération sèche

Ici, toute la charge thermique est transmise à l'atmosphère par un échangeur à surface. Il n'y a pas de consommation d'eau pour la réfrigération mis il subsiste néanmoins un besoin d'au moins 1,5 m^3/s pour la dilution des effluents (PWR 900 MW). Par contre dans le cas des surrégénérateurs le débit d'eau de dilution n'est plus nécessaire mais il subsiste la nécessité d'un débit d'eau d'appoint de l'ordre de 1 m^3/s notamment pour les pertes d'eau au niveau de la turbine.

Avantages : la réfrigération sèche permet une grande liberté dans le choix du site excepté dans une région désertique.

Inconvénients :

- coût très élevé ; le coût du kW installé s'en trouve augmenté d'environ 12 % par rapport au réfrigérant humide ;
- on dispose de peu d'expériences : les techniques ont été testées sur des unités ne dépassant pas 200 MW ;
- les effets sur l'atmosphère sont mal connus. On craint notamment la formation de panaches de grandes dimensions à une altitude relativement élevée;
- ils posent de graves problèmes vis-à-vis de la turbine à vapeur entraînant des indisponibilités en saison chaude pouvant atteindre plus de 200 h par an. En effet, les variations de la température sèche de l'air sont beaucoup plus importantes que celles de la température humide et cela entraîne des variations de la pression d'échappement qui dépassent les limites autorisées par les constructeurs. (Voir annexe 1).

Cycles binaires

Il semble bien qu'à terme les problèmes d'environnement conduisent à avoir recours à la réfrigération sèche. Aussi a-t-on recherché une solution de réfrigération sèche qui ne présente pas les inconvénients énumérés ci-dessus.

Une solution qui paraît être intéressante est celle du cycle binaire dont un exemple est schématisé ci-dessous :

L'utilisation d'un cycle binaire dans lequel le fluide secondaire a une densité de vapeur très importante permet de résoudre de façon satisfaisante les problèmes énoncés ci-dessus. Le cycle binaire est schématisé sur la Fig. 1

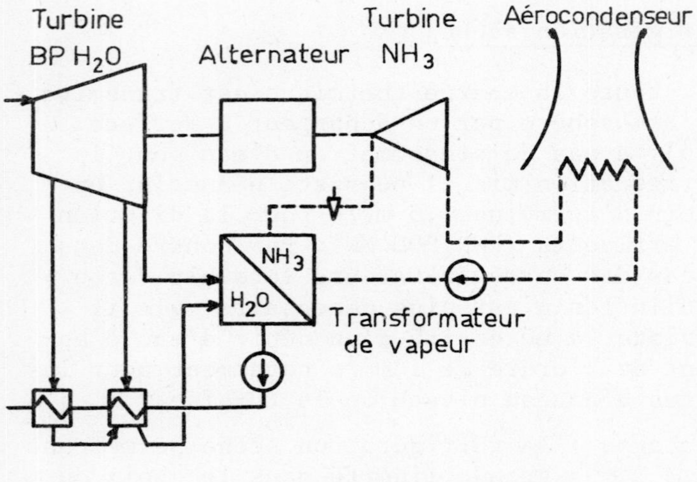

Fig. n° 1 - Cycle binaire H_2O/NH^3

Le principe du cycle binaire consiste à arrêter la détente de la vapeur d'eau à un niveau de pression supérieur à la pression de condensation habituelle et de la condenser alors dans un "transformateur de vapeur" qui en utilise la chaleur latente de condensation pour évaporer un fluide secondaire très dense. Le fluide secondaire est ensuite envoyé dans une turbine où il se détend en produisant du travail puis il est recondensé dans un aérocondenseur.

Les avantages que l'on espère obtenir avec un tel cycle sont les suivants :

- suppression du risque de gel dans l'aéro-réfrigérant ;
- suppression des indisponibilités d'été (voir annexe 1) ;
- réduction des surfaces d'échappement basse pression de la turbine à vapeur d'eau. Ceci permet de reculer la limite maximale en puissance unitaire des groupes et du même coup de revenir à la vitesse de 3000 t/mn au lieu de 1500 t/mn :
- légère augmentation de rendement du cycle car l'utilisation d'un changement de phase pour l'évacuation de la chaleur vers l'atmosphère permet une réduction des irréversibilités thermiques ;
- variabilité de la puissance nette de la centrale en fonction de la température de l'air d'où une meilleure utilisation de la source froide (voir annexe 2) ;
- enfin, possibilités accrues pour le transport et l'utilisation à distance de la chaleur rejetée par la centrale.

CHOIX DU FLUIDE SECONDAIRE

Sélection des fluides possibles

Le fluide secondaire, aux températures de fonctionnement envisagées, doit se trouver dans des états physiques convenables :

- dans les conditions de fonctionnement il doit toujours garder une pression de condensation supérieure à la pression atmosphérique ce qui, si l'on admet une température d'air minimale de -10°C et une approche à l'aérocondensateur de 25°C peut se traduire par la condition suivante :

  Température de condensation T_K à 1 bar $\leqslant$ 15°C

- pour limiter les irréversibilités thermodynamiques au bouilleur, il faut rester nettement sous-critique. En admettant a priori que la pression de coupure sur la vapeur d'eau se situe autour de 300 m bar, soit 70°C, et si on prend un écart de température au bouilleur de 10°C, la température du fluide secondaire ne dépasse jamais 60°C. Il paraît donc raisonnable d'exiger une

température critique du fluide secondaire supérieure à 90°C ;
- pour éviter tout risque de gel de l'aéro-condenseur on impose que la température de congélation à la pression atmosphérique soit inférieure à -20°C ;
- il faut par ailleurs pouvoir disposer de caractéristiques thermodynamiques assez précises du fluide envisagé.

Ces diverses conditions ont permis de dresser sur le tableau 1 une liste de fluides envisageables

On suppose que le cycle a les caractéristiques suivantes :

Pression de coupure H_2O	300 m bar
Température de coupure H_2O	69°C
Pincement au bouilleur	10°
Température ébullition secondaire	59°C
Puissance thermique du bouilleur	2700 MW

La turbine secondaire, à 50 % de degré de réaction, est supposée tourner à 3000 t/mn et constituée en un seul coprs du type diabolo à deux échappements.

Tableau 1 : fluides secondaires envisageables pour un cycle binaire

Fluide	Formule	Temps de condensation à 1 atm °C	Temps critique °C	Temps de congélation à 1 atm °C
Ammoniac	NH_3	- 33,6	132	- 77,9
Fréon 12	$C\,cl_2\,F_2$	- 19,8	111,5	-155
Fréon 21	$CH\,cl_2\,F$	8,9	178,5	-135
Fréon 22	$CH\,cl\,F_2$	- 40,8	96	-160
Fréon 114	$C_2\,cl_2\,F\,4$	3,5	145,7	- 94
Chlorure de méthyle	$CH_3\,cl$	- 24	143,8	- 97,6
Propane	$C_3\,H_6$	- 42,1	96,8	-187,7
Isobutane	$C_4\,H_{10}$	11,7	134	-159,4
n Butane	$C_4\,H_{10}$	- 0,5	152,8	-138,3

Etude comparative des divers fluides retenus

Hypothèse de travail :

Pour faciliter la comparaison, on a admis a priori que le cycle secondaire serait un cycle de Rankine sans surchauffe :

On impose que la vitesse périphérique de la turbine au rayon moyen ne dépasse pas 0,8 C_s (C_s = vitesse du son dans le fluide considéré):

On admet un angle des distributeurs de 26°. On peut alors dresser le tableau II qui montre à l'évidence que l'ammoniac, sur le

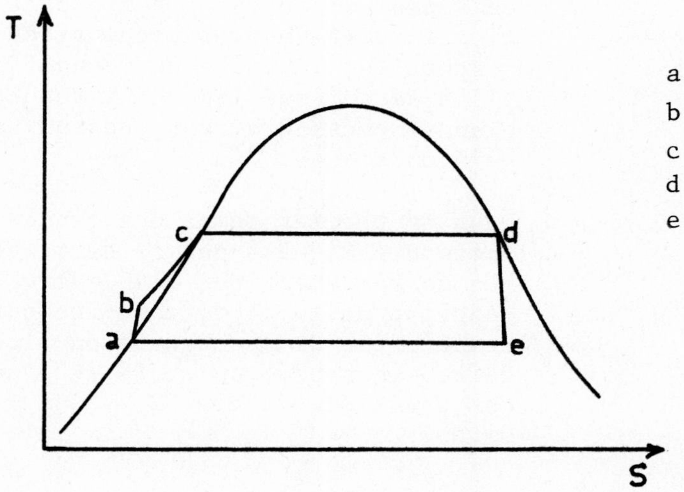

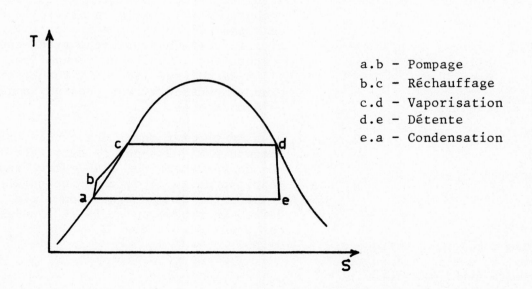

a.b - Pompage
b.c - Réchauffage
c.d - Vaporisation
d.e - Détente
e.a - Condensation

plan du dimensionnement du tracé de la veine de turbine, se place en tête avant le propane et le fréon 22.

De plus il a un prix modéré en comparaison du fréon 22. Enfin c'est un produit largement industrialisé et les technologies nécessaires pour son utilisation sont maintenant bien connues.

Tableau 2

		R 12 C Cl_2 F_2	R 22 CH Cl F_2	R 114 C_2 Cl_2F_4	C_3 H_8	ISO C_4 H_{10}	N C_4 H_{10}	NH_3
Δh au bouilleur	kJ/kg	114	144	114	263	278	322	995
Débit massique de fluide secondaire q_m	kg/s	23700	18750	23700	10260	9720	8380	2710
Vitesse du son C_s	m/s	144	178	120	252	234	212	430
Δh turbine	kJ/kg	12,6	10,5	8,1	27,8	20,9	23,3	75
volume spécifique v	m^3/kg	0,0193	0,0159	0,0457	0,0350	0,0805	0,1185	0,0862
surface nécessaire S	m^2	9,05	4,77	25,7	4,07	9,53	13,3	1,55
débit volumique q_v	m^3/s	457,4	298	1083	359	780	993	233
vitesse périphérique moyenne U	m/s	115,2	142,4	96	202	187	170	344
diamètre moyen D_m	m	0,73	0,91	0,61	1,29	1,19	1,08	2,19
hauteur d'ailetage pour une turbine diabolo h	m	1,97	0,83	6,7	0,5	1,27	1,96	0,112

Enfin la considération des propriétés de toxicité et d'inflammabilité a joué un rôle déterminant dans le choix définitif du fluide secondaire.

La classification des Underwriter's Laboratoires attribue des indices pour la toxicité allant en décroissant pour une toxicité croissante :

	Toxicité	Limites d'inflammabilité % en volume dans l'air
NH_3	2	16 à 25
Fréon 22	5 a	non
Propane	5 b	2,3 à 7,3
Isobutane	5 b	1,8 à 7,4
N. Butane	5 a	1,6 à 6,5

Finalement, bien que modérément toxique et inflammable, l'ammoniac a été retenu car c'est lui qui donne la turbine secondaire la plus compacte et la mieux adaptée aux conditions de fonctionnement envisagées.

CHOIX DU CYCLE SECONDAIRE

Une seconde étape a été le choix des caractéristiques du fluide secondaire et en particulier la détermination de la pression de coupure P_c à laquelle on arrête la détente en vapeur d'eau, et d'autre part la définition de certaines particularités comme la surchauffe, le poste de réchauffage, etc... Il va de soi que les choix résultent aussi bien de considérations thermodynamiques qu'économiques.

Dans un premier temps une étude purement thermodynamique a permis de tracer les courbes de rendement net d'un cycle binaire en fonction de la pression de coupure et de la température de surchauffe du fluide secondaire. Le schéma du cycle le plus général est donné sur la Fig. 2

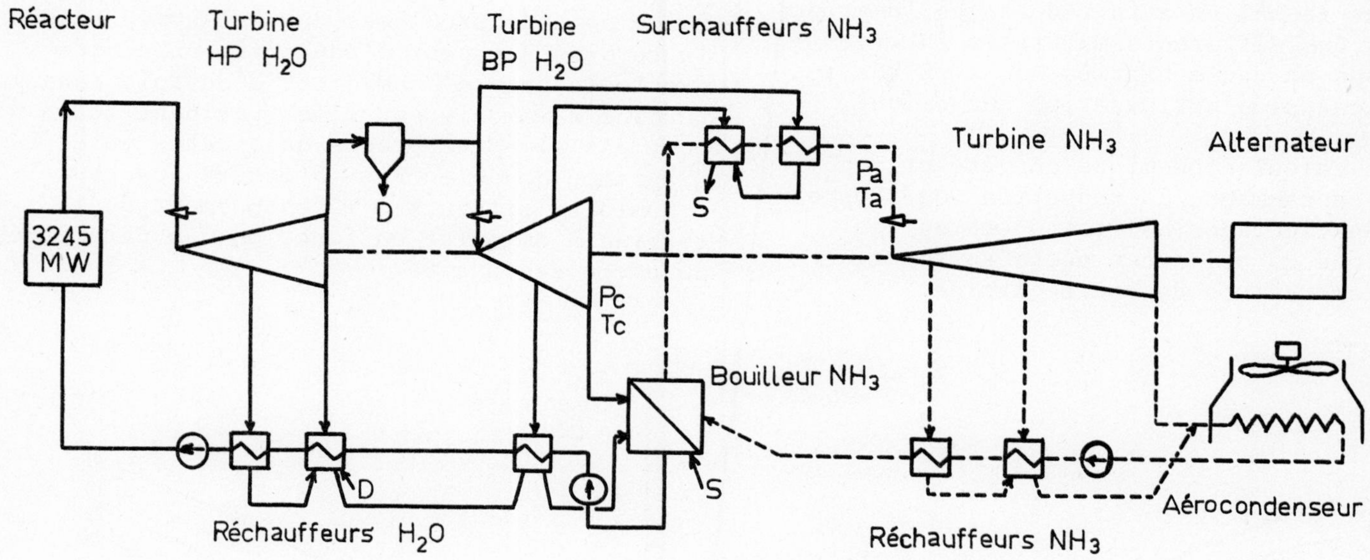

Fig.n° 2 - Cycle binaire sur PWR 1300 MW

Le diagramme de la Fig. 3 donne la variation du rendement net du cycle binaire en fonction de la température à l'admission de la turbine à ammoniac. La courbe en trait plein concerne le cas où le flux secondaire n'est pas surchauffé. Les courbes A, B et C

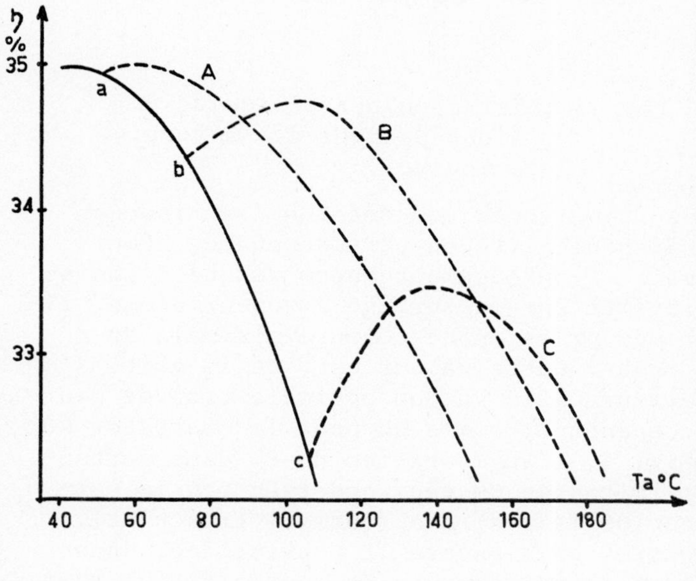

Fig. n° 3 - Rendement net d'un cycle binaire à NH_3 en fonction de la température d'admission à la turbine secondaire (cas d'un PWR 1300 MW).

concernent des cycles secondaires avec surchauffe avec une pression à l'admission de la turbine secondaire correspondant à la pression de saturation définie aux points a, b et c. Ce diagramme montre à l'évidence que sur le plan purement thermodynamique on a intérêt d'une part à avoir une pression de coupure faible et d'autre part à ne pas faire de surchauffe.

Par ailleurs, le choix d'une pression de coupure faible conduit à ne supprimer qu'un seul réchauffeur sur la partie vapeur d'eau lequel devrait en toute rigueur être remplacé par un réchauffeur sur le cycle secondaire. Cependant, dans un premier temps, et dans un souci de simplification, on ne prévoira pas de réchauffeur sur l'ammoniac, ce qui aura, bien sûr pour conséquence une légère perte de rendement.

Le fait de travailler sans surchauffe et sans réchauffeur étant acquis, l'étude du rendement net en fonction de la pression de coupure P_c a été reprise, ceci en tenant compte d'hypothèses précises pour les écarts de température dans les échangeurs, les rendements, les pertes de charge, etc...

Un programme de calcul thermodynamique a été mis au point.

Les résultats sont donnés sur le diagramme de Fig.4 . On voit que pour des valeurs raisonnables des paramètres les plus importants:

approche à l'aérocondenseur = 28°
ΔT au bouilleur = 8°

l'optimum de la pression de coupure se situe

entre 200 et 300 m bar.

Enfin, à partir du programme thermodynamique dans lequel on a introduit les fonctions de prix des différents matériels intervenant dans un cycle binaire, on a pu réaliser un calcul d'optimisation économique.

Dans ce calcul économique ont été pris en compte, notamment, la réduction (ou suppression) des indisponibilités d'été et les variations de puissance nette en fonction de la température de source froide.

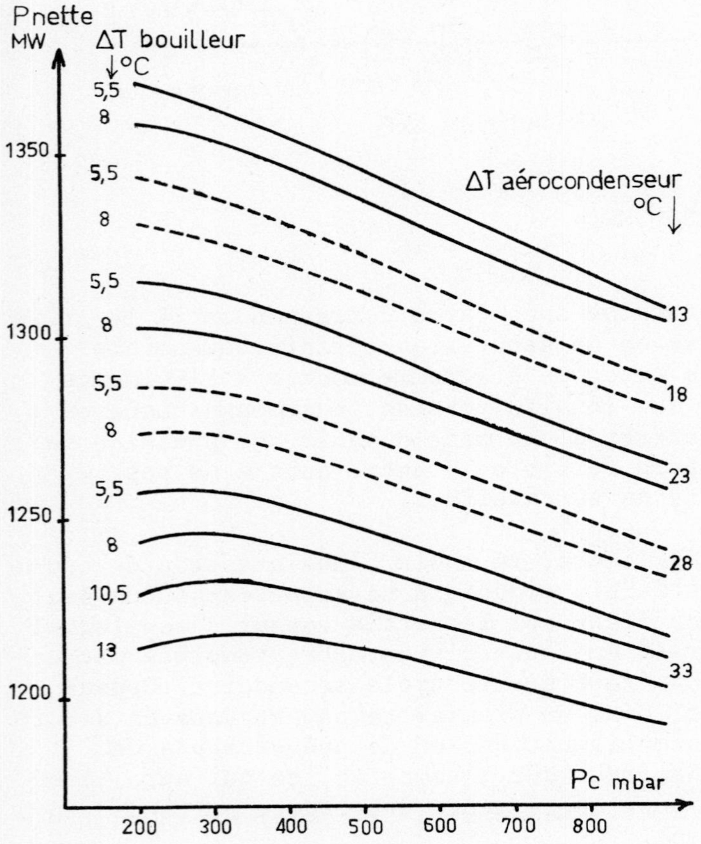

Fig.4 : Puissance nette d'un PWR 1300 MW à cycle binaire à NH_3

Pour le cas étudié d'un cycle binaire à ammoniac associé à une centrale PWR 1300 MW, l'optimisation a conduit aux valeurs suivantes pour les principaux paramètres du cycle:

Pression de coupure	: P_c = 200 m bar
Pincement au bouilleur	: ΔT = 8°C
Approche à l'aérocondenseur	: A = 28°C

<u>Remarques</u> :

Deux considérations viennent en réalité limiter les possibilités de choix de la pression de coupure P_c :

La première est que plus on choisit P_c élevée moins on a de problèmes dus à l'humidité dans la turbine à vapeur d'eau. Mais en contrepartie le taux d'"humidité" s'accroît dans la turbine à ammoniac et il est évident qu'on ne peut pas aller trop loin dans cette voie. Le diagramme donné en Fig. 5 montre les taux d'humidité atteints à l'échappement de la turbine à ammoniac en fonction des paramètres du cycle secondaire.

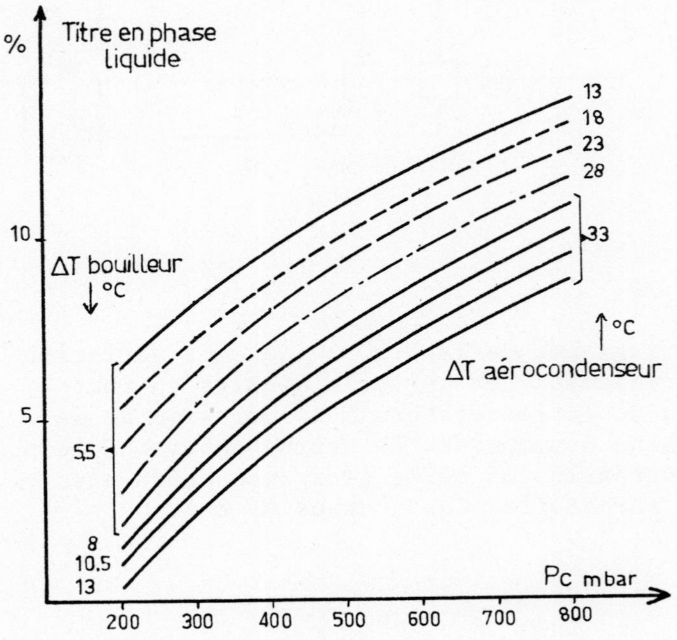

Fig. 5 : Titre en phase liquide à l'échappement de la turbine à ammoniac.

La deuxième remarque concerne les risques d'indisponibilité en période chaude. Pour n'avoir rigoureusement aucun risque d'indisponibilité sur la turbine à vapeur d'eau, il faut une pression de coupure minimale de 340 m bar. Cette valeur est, on le voit, supérieure à la valeur optimale trouvée pour P_c. Cependant, comme on peut le constater sur la Fig. 4, l'augmentation de P_c dans certaines limites ne dégrade que très peu le rendement de la centrale (optimum très plat). En outre, on a encore la possibilité, dans les cas de température de source froide très élevée, de by-passer une partie du débit de la turbine à ammoniac ce qui permet d'en réduire la température d'admission et par conséquent d'abaisser la pression de coupure P_c.

CONCLUSION

L'utilisation d'un cycle binaire à ammoniac permet d'utiliser le mode de réfrigération atmosphérique sec. Il en résulte des avantages considérables vis-à-vis des problèmes d'environnements qui sont ceux de la réfrigération atmosphérique sèche en général, à savoir :

- l'absence de besoins en eau (hormis pour l'appoint et la dilution des effluents) permet d'augmenter considérablement le nombre de sites possibles d'implantation des centrales. Ainsi on pourra plus aisément choisir des sites discrets et éloignés de zones touristiques ou protégées ;

- les centrales électriques pourront plus facilement être rapprochées des centres de consommation ce qui permettra de réduire la longueur des lignes de transport.

De plus les centrales fonctionnant avec un cycle binaire présentent des avantages considérables par rapport aux centrales conventionnelles à aéroréfrigérant sec :

- elles ne présentent aucun risque de gel de l'aéroréfrigérant ;
- elles permettent de réduire considérablement les dimensions des corps basse pression des turbines à vapeur d'eau. Cette particularité autorise de revenir à la vitesse de rotation de 3000 tours/mn au lieu de 1500 tours/mn et permet de reculer notablement la limite de puissance unitaire des groupes ;
- elles permettent d'utiliser au mieux les basses températures d'air dont on dispose en hiver, en procurant un gain de puissance important : 0,5 % par degré centigrade ;
- enfin elles ne sont pas sujettes aux indisponibilités en été.

Tous ces avantages, et notamment les deux derniers, se traduisent financièrement par des gains importants qui rendent le bilan économique favorable au cycle binaire, comparé à celui de la réfrigération sèche conventionnelle.

Par ailleurs, à l'époque actuelle où l'on se préoccupe de plus en plus d'utiliser la chaleur rejetée par les centrales électriques, l'ammoniac qui assure le transport de la chaleur rejetée par la centrale vers l'aérocondensateur, est un excellent support pour transporter cette chaleur vers des utilisateurs plus éloignés.

ANNEXE 1

FONCTIONNEMENT D'UNE TURBINE A VAPEUR D'EAU A PRESSION D'ECHAPPEMENT VARIABLE

Pressions d'échappement inférieures à la pression nominale

La chute enthalpique dans la turbine croît lorsque la pression d'échappement P_K diminue. Pour une certaine valeur de P_K il se produit un blocage sonique dans l'ensemble constitué par la dernière roue, le diffuseur et le fond d'échappement, par suite de l'augmentation du débit volumique de la vapeur. A partir de ce moment, si l'on continue à baisser la pression P_K, la pression à l'aval du dernier étage ne varie plus et la chute enthalpique n'augmente plus.

Cependant, le fonctionnement à des valeurs de P_K faibles ne présente pas de risques pour la machine.

Pressions d'échappement supérieures à la pression nominale

Lorsque, à débit massique constant, la pression de condensation P_K augmente, le débit volumique à l'échappement de la dernière roue diminue et les triangles de vitesses se trouvent désadaptés. Il en résulte ce qui suit :

- réduction de la chute enthalpique dans la machine,
- diminution du rendement dans les derniers étages,
- déviation de l'écoulement de vapeur vers la périphérie avec apparition d'une recirculation en pied d'aubage ce qui produit un échauffement dans cette zone du fait des "pertes par ventilation" (voir Fig. A.1.1),
- accroissement de l'érosion par les gouttes aux extrémités d'ailettes dû à l'augmentation de granulométrie et à la désadaptation des triangles de vitesses.

Tous ces effets expliquent que la courbe de puissance nette en fonction de la température de condensation T_K présente l'allure indiquée sur la Fig. A.1.2.

Pratiquement les turbines actuelles peuvent fonctionner avec une pression de condensation variant au maximum dans un rapport 1 à 2.

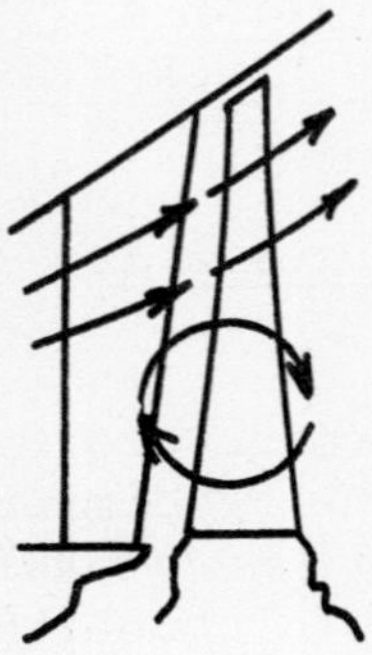

Fig. A.1.1

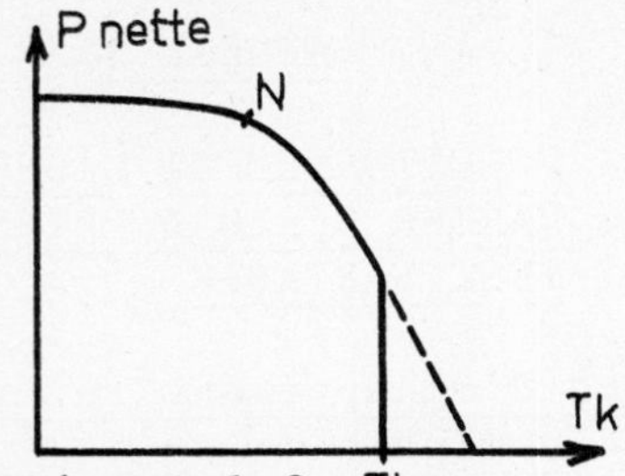

Fig. A.1.2

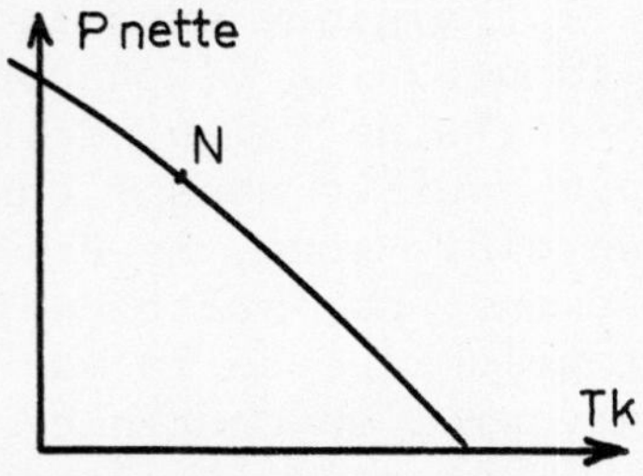

Fig. A.2.1

CAS DU CYCLE BINAIRE

Dans ce cas, la présence de la turbine secondaire amortit au niveau de l'échappement de la turbine à vapeur d'eau les effets des variations de la température de l'air. On peut illustrer ceci par un exemple :

Pour une température d'air ambiant variant de −5°C à 35°C, on observe, dans le cas d'une centrale conventionnelle sur aéroréfrigérant sec, une pression de condensation variant de 28 m bar à 230 mbar alors que en cycle binaire la pression de coupure P_c varie de 160 à 325 m bar (pression de coupure nominale : 200 m bar).

On a donc pu ainsi limiter les variations relatives de pression à l'échappement de la turbine à vapeur d'eau. Quant à la turbine à ammoniac, compte tenu des ses dimensions réduites et des faibles vitesses, elle supporte bien les variations de volume massique de la vapeur d'ammoniac.

ANNEXE 2

VARIABILITE DE LA PUISSANCE NETTE D'UNE CENTRALE A CYCLE BINAIRE EN FONCTION DE LA TEMPERATURE AMBIANTE

On a vu en annexe 1 que la courbe de puissance nette d'un groupe à vapeur classique en fonction de la température de source froide avait l'allure indiquée sur la Fig. A.1.2. Il est donc impossible de bénéficier pleinement des basses températures d'hiver. De plus, pour $T_K > T_i$ la puissance devient nulle (indisponibilité).

Avec les cycles binaires, par contre, les indisponibilités sont supprimées et la courbe de puissance nette ne présente pas le maximum du cas précédent, ou plutôt, ce maximum est reporté vers des températures très basses (Fig. A.2.1)

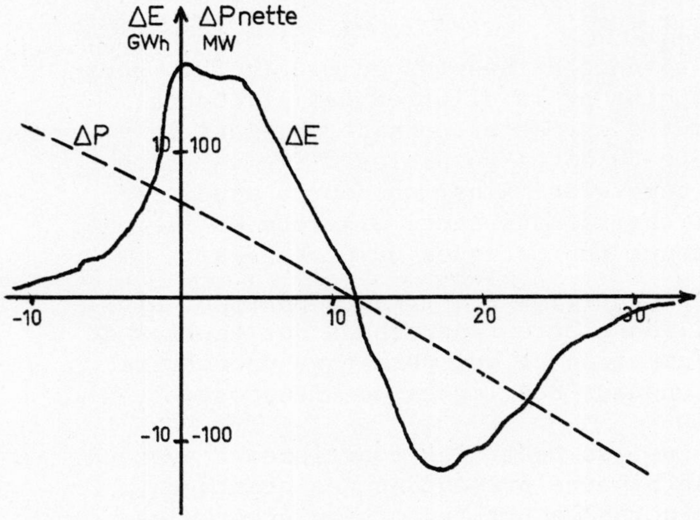

Fig. A.2.2 - Ecarts de puissance et d'énergie produite : cas du cycle binaire

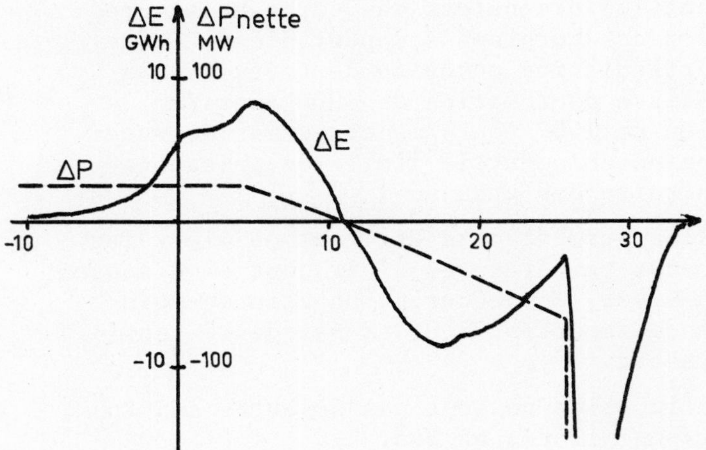

Fig. A.2.3 - Ecarts de puissance et d'énergie produite : cas de l'aéroréfrigérant sec.

Les Fig. A.2.2 et A.2.3 indiquent les variations de puissance et d'énergie produites dans l'année par une tranche PWR 1300 en fonction de la température ambiante. Dans le cas du cycle conventionnel on constate une perte de 12 GWh par rapport à un fonctionnement à puissance nominale continue, ce qui, actualisé sur 20 ans représente 28 MF. Dans le cas du cycle binaire on trouve un gain de 40 GWh, ce qui, actualisé sur 20 ans représente 72 MF.

TOTAL ENERGY SYSTEMS

L. Bolle
Charge de Cours
Universite Catholique de Louvain

ABSTRACT

There are three ways to promote energy conservation:
- the recovering of waste heat;
- the reduction of the consumption or of the losses of energy by modification of the concerned industrial processes;
- a better integration of unit operations, processes and power plants.

Total energy systems belong to this third class. Optimization of deeply integrated installations can insure the most attractive amount of energy savings.

As far as the production of mechanical work only is concerned, the energetic efficiency of plants cannot be higher than ...0.60... for thermodynamical reasons, the actual value being approximately 0.30...0.40. The main energy loss is at the cold source of the thermal cycles, although the plants are nearly perfect at this side (irreversibilities reduced to a few °C). The strongest imperfections are located at the hot source, where their effect may be characterized by the so-called "exergetic efficiency". A detailed analysis of fossil fired steam power plants show that the potential work which is involved in the primary energy consumed and which would be generated if all thermodynamic transformations would be reversible, is reduced by an amount of 30% due to the free expansion and the dilution of the smokes during combustion and of 25% due to heat transfer irreversibilities at the boiler.

The more drastic way to reduce energy losses at the cold source is not to improve the exergetic efficiency of the hot source, but to use the waste heat at the cold source of power plants for several purposes: it is the aim of the total energy systems.

In the text of one conference, we then emphasize the following systems:
- steam cycles: back pressure turbines and throwing-out of vapor.
- gas turbines: waste heat recuperators; combined gas and vapor cycles, with either low or high power ratio between them.
- diesel and diesel-gas engines: waste heat recuperator.

The essential properties and advantages, the thermodynamic efficiency and heat/work ratio are given for each system. Some numerical examples are presented. More attention is paid to district heating systems in connection with large steam power plants.

THE POTENTIAL OF RESIDENTIAL TOTAL ENERGY SYSTEMS

Eugene E. Drucker
Syracuse University

ABSTRACT

The feasibility of small, packaged total energy systems will be examined, for application to residential dwellings. It is believed that systems of the order of 5 to 10 KW size, utilizing mass produced gas engines of the simple reciprocating variety can be produced for sums in the order of 100-200 $/KW. Central station electrical capacity costs at least 5 times that amount. Waste heat from the engine can be used for space and hot water heating. Both the heating and the electrical loads of a typical residence can be satisfied with the same basic gas consumption that homes now require for heating alone.

The necessary load balancing and thermal storage to make the system work will be examined, as well as the economic feasibility of the scheme.

A diesel engine can replace the gas engine in areas which now use oil predominately, with the same order of energy savings.

GRID CONNECTED COGENERATION ALTERNATIVES FOR CLARK UNIVERSITY*

Robert L. Goble
Program on Science, Technology, and Society
Clark University, Worcester, MA 01610

Sander E. Nydick
Thermo Electron Corporation
Waltham, MA 02154

ABSTRACT

Cogeneration offers great promise both as a means of energy conservation and as a source of new electrical generating capacity. Its promise can be fully realized, however, only if cogeneration installations interchange electricity with the utility grid. In this paper we compare cogeneration alternatives for a test community, Clark University, and show that the choice of system size and type is sensitive to the terms of interchange and the price schedule.

The Clark campus is located in Worcester, Massachusetts. It is occupied by about 2500 people during the day and 1200 at night. In size and energy use it is typical of many New England facilities, commercial and residential, as well as colleges and universities. The generators compared, steam turbines, gas turbines, and diesels, are all commercially available. These systems have different operating characteristics, including a large difference in the amount of electricity produced per unit heat output. We considered a wide range of heat output capacities and so included systems which export large amounts of electricity and systems designed to be totally independent of the utility. If the electric utility will purchase electricity for a reasonably high price, one still less than its marginal costs, we find that Clark will select a system which will export a substantial amount of electricity. Lower prices, predictably, destroy the incentive to choose a system sized to export very much electricity. This is particularly true for a community such as Clark, in which the cogeneration system is to be retrofitted to an existing boiler facility which will provide as much heat as necessary. The sensitivity of the choice of system electrical capacity to interchange prices implies that any regulation which encourages or discourages interchange will affect strongly the future development of cogeneration.

---

* Research supported in part by ERDA contract #.

ADVANCED AND DECENTRALIZED INSTALLATIONS FOR COMBINED PRODUCTION OF HEAT AND POWER IN DENSELY POPULATED AREAS

P.H.H. Leijendeckers
Van Heugten, The Netherlands

ABSTRACT

The rapid growth in the consumption of energy in the last few decades has considerably reduced the length of time before our supplies of fossil fuels run out.

The exhaustion of certain fuel deposits is already in sight for the present world population. This confronts us with the necessity of developing new sources to replace them; or of giving the fuel deposits we have at the moment a longer lease of life by using them more efficiently. The application of the last strategy will mean that not only will more time be gained for the development of new sources, but at the same time the way will be prepared for a future society in which definite limits imposed on the consumption of energy will be recognized.

It is of course highly probable that in any future alternative systems for environmental, technical and economical reasons, also energy can not be made available in unlimited quantities.

These considerations make an efficient use of energy, therefore, of great fundamental importance: and both in the short and medium term.

For the realization of a more efficient use of energy two important possibilities can be recognized, i.e.:
- a reduction in the end use of energy, as regards heating, lighting, transport etc.;
- an improvement in the efficiency of the processes which convert fuels into the end products e.g. heat, electricity and their transportation.

The first possibility can be implemented mainly by the consumer himself taking measures: such as improved forms of insulation in the house and better appliances (refrigerators etc.). The second possibility will demand above all measures in the field of technology, town and country planning as well as regulations on the part of local authorities and the Government, in behalf of supplying energy to larger conglomerates of consumers (industries, urban residential areas etc.). These in turn will effect the infrastructural system within society and the choice of the systems to be used. For these reasons this second category forms the subject of this paper.

A closer look at the possibilities for improving the efficiency of fuel conversion into useful end products makes it obvious that the methods that have been used for years offer many technical solutions which it is true up until now have not or hardly ever been used.

The problems, therefore, are not to be found primarily in the field of technical research or in industrial product development, but rather in those technical governmental changes introduced into the energy supply system that make the application of more efficient systems possible.

A primary point is that the end products will, from a production point of view, have to be grouped far more closely together. Futhermore the production processes will have to take place closer to the consumers to reduce transport losses. In concrete terms this means the establishing of small scale supply units built in direct proximity to the consumers to provide them with combined electricity and heat (total energy system). The next important point is that the production of the end products must be more closely co-ordinated. One way of doing this is the creation of storage facilities to deal with temporary surpluses caused by fluctuations in demand. Another is the application of conversion systems which, by maintaining a high conversion rate, permit a production interchange of electricity and heat to a large extent. Additional possibilities are heat pumps and bottoming units for the production of electricity.

If the conversion centers are sited close to the areas of consumption, it seems necessary that only a very pure fuel can be considered for use if the environment quality of the area is to be maintained. This means that as far as the supply to urban residential areas of basic fuels is concerned gas (natural gas, synthetic gas or hydrogen) must be regarded as a viable option.

The result of all these considerations is an energy supply system characterized by the following structure:

1. For the supply of basic fuel a coarse-meshed nationwide gas distribution network with supply and conversion stations near to the centers of consumption, the gas coming from huge national production centers.

2. Close to the centers of consumption fuel conversion stations for the combined production of the end products heat and electricity.

3. Within the centers of consumption a distribution system for heat and electricity. Heating and electrical connections direct to individual homes. Heat pumps and energy accumulators must be installed to maintain a national balance between the supply and demand for the end products.

Such a concept differs greatly from what is the norm at the moment: a huge nationwide centralized conversion of fuel into electricity which then has to be distributed over large distances, followed by a separate and completely decentralized conversion of fuel into heat by the individual consumer. Sweeping changes are, therefore, necessary if this new concept is to be introduced. The mere fact, however, that in the new process the amount of fuel consumed in the conversion into the end product can be halved if the end consumption remains constant, makes a close study of the necessary changes desirable.

It is opportune at this point that we should bear in mind that if no limits are imposed on the present rate of energy consumption, this to say the least will result in large-scale fundamental changes within society. We have only to think of the consequences which could arise from the planning nationwide of an increasing number of large electric (fossile or nuclear) power stations and high-voltage distribution systems, not to mention other consequences. The question facing us is why up to this point so little attention has been devoted to these other possibilities, while at the same time the changes being introduced as a result of increasingly larger-scale central generating systems and distribution networks with their detrimental effect on the countryside and the environment generally meet with little or no opposition. One of the explanations for this lack of interest can certainly be attributed to the inbuilt inertia within our society: the flywheel effect prevalent in the present system in the form of administrative procedures and erroneous criteria used for judging the economy: above all the short term vision of economic returns. But certainly also the lack of conviction that fundamental changes are necessary.

All this leads to great resistance towards accepting new possibilities and other solutions and strengthens the pre-occupation with the supposed drawbacks of the solutions outlined earlier. People should, however, realize that the whole course of human civilization has been characterized by changes and developments. Changes, therefore, constitute no harm in themselves. Only by consciously choosing and in advance influencing certain adaptations can we bring order and direction in the process of development.

fig. 0

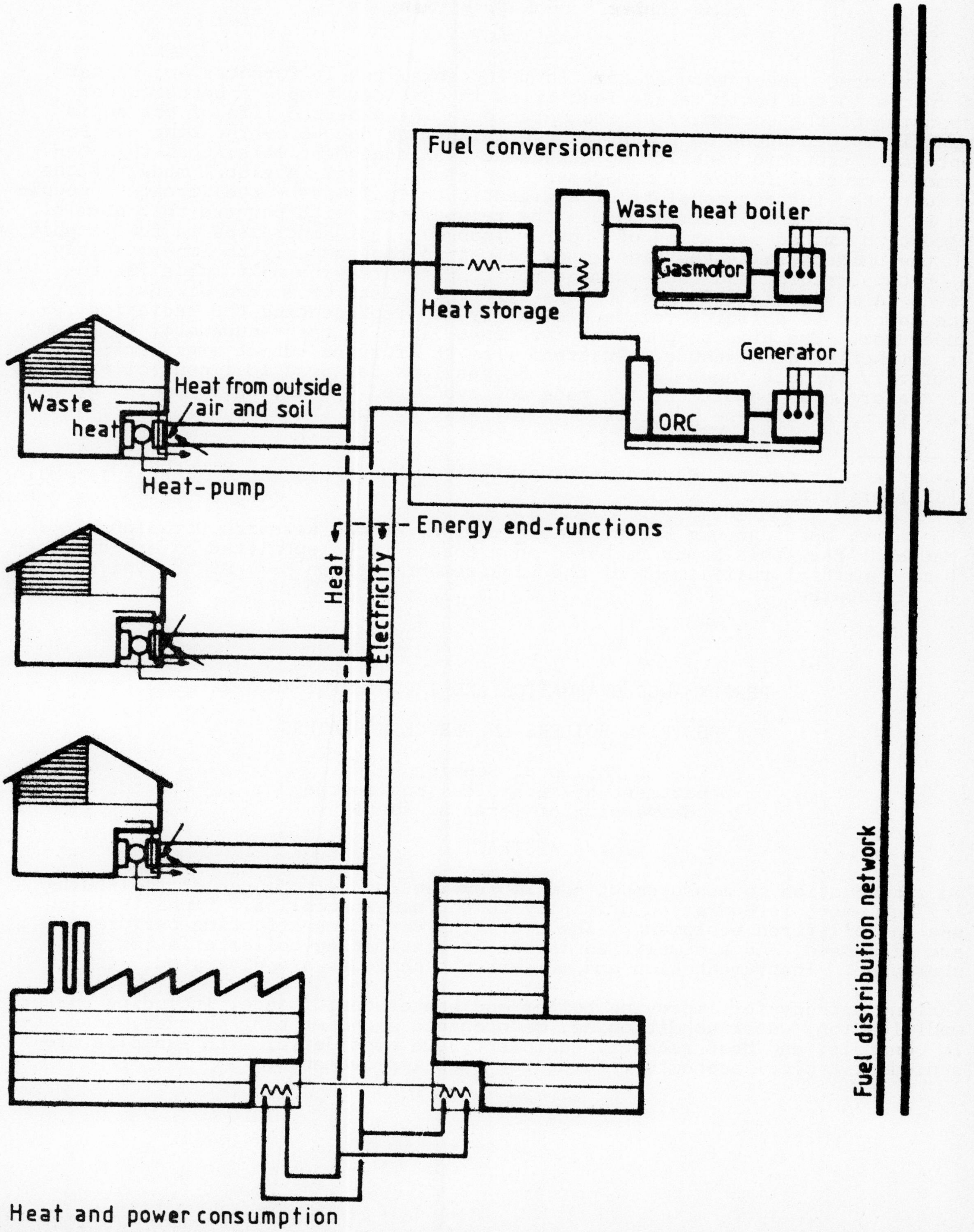
Fuel conversioncentre
Waste heat boiler
Gasmotor
Heat storage
Generator
ORC
Heat from outside air and soil
Waste heat
Heat-pump
Energy end-functions
Heat
Electricity
Fuel distribution network
Heat and power consumption

A DYNAMIC MODEL OF A RECUPERATED INDUSTRIAL FURNACE

W.M. Rohrer[1] and D.P. Birmingham[2]

ABSTRACT

The expanded use of recuperators in melt tanks, remelt furnaces and reheat furnaces in the basic metals industries in contingent upon a better understanding of the economics of recuperation. The expected life of the equipment is so dependent on potential thermal damage due to overheating and the potential economic benefits so dependent upon equipment life, that this becomes a crucial factor in management decision making. A global model of the recuperated furnace consists of a gigantic thermal mass - the furnace - coupled to a negligible thermal mass - the recuperator, with burners that permit almost unlimited variation of input. Thus can rapid increases in fuel supply to the burners raise the flue gases almost instantaneously to impermissibly high temperatures and cause damage to recuperators with no furnace interaction. A more detailed model of the system consists of a zero dimension furnace coupled to a finite volume element system representing the radiation recuperator. The time response of the inner liner of the recuperator is shown as a function of distance downstream from the furnace output with various arbitrary furnace input functions. Suggestions for practical control of recuperator input conditions are made and evaluated. Measured responses of test furnaces will be presented if available at the time.

[1]Associate Professor, Mechanical Engineering, University of Pittsburgh, Pittsburgh, PA

[2]Engineer, Westinghouse Electric Corporation, Atomic Research Division, Madison, PA. This paper is based on a thesis to be submitted by Mr. Birmingham in partial fulfillment of the requirements for an M.S. at the University of Pittsburgh.

PERFORMANCE EVALUATION AND IMPROVEMENT IN INDUSTRIAL BOILERS AND PROCESS HEATERS

Philip S. Schmidt
Department of Mechanical Engineering
University of Texas at Austin

ABSTRACT

An introduction to measurement and improvement of performance of fuel-burning equipment is addressed primarily to the non-specialist. Emphasis is on gas and oil-fired equipment. The operating variables affecting performance are discussed, and a simplified method for estimating boiler efficiency is presented. Instrumentation and measuring techniques are discussed.

Various methods for improving boiler and heater performance, including burner modification, water conditioning, maintenance and operating changes, automatic controls, and heat recovery equipment, are considered, with examples presented of typical economic returns for such improvements.

STATE ENERGY FLOW PATTERNS

R. B. Kidman and R. J. Barrett

Los Alamos Scientific Laboratory, University of California
Theoretical Division, Los Alamos, New Mexico 87545

ABSTRACT

Highly visual and self-explanatory 1975 energy flow diagrams are presented for each of the 50 states and for the entire United States. Each diagram illustrates the energy produced and how it is consumed or lost. The diagrams are meant to serve as a convenient and useful reference (or starting point) for consideration of energy-related problems.

INTRODUCTION

A convenient and simple display of the nation's energy system could prove invaluable to individuals who must have the proper perspective to understand our energy-related problems and to evaluate the proposed solutions. With this in mind, we have diagrammed 1975 energy flow patterns for each of the 50 states and for the entire United States. The diagrams illustrate the energy produced and how it is stored, imported/exported, consumed, or lost. With only a minimum of time invested in reviewing these diagrams, policy makers and private citizens can obtain an understanding of the current energy situation that, in turn, should help them to eveluate energy-related questions or, at least, to identify further information required for informed discussions.

In the following sections we briefly explain the diagram structure, methodology, and data. If these sections fail to resolve any difficulty, please call the authors at 505-667-7671 (commercial) or 8437671 (FTS).

DESCRIPTION OF DIAGRAM STRUCTURE

Each diagram illustrates the production of various kinds of energy and the subsequent movement, storage, import/export, and consumption of this energy within the various sectors of society. We made the widths of the flow lines proportional to the amounts of energy that flow through each path, so that one can get an immediate intuitive feeling for the relative magnitude of the various energy uses.

The widest parts of the energy flow lines, near the left-hand side of each diagram, represent the total source of each kind of energy available for export, consumption, and storage. The total source comes from production, imports, and storage. Any Flows beginning or ending with an arrow are net flows. The net flows to storage represent the net change in storage that occurs at the production, transportation, and consumption facilities. From this point, energy is shown as flowing to the various consuming sectors. The energy proceeding to "electricity generation" is converted to electric energy, which is then exported or distributed among the residential, commercial, and industrial sectors. The energy lost in converting to and transmitting electric energy proceeds to lost energy. Each diagram is completed by showing the amount of energy that is actually used and lost by each consuming sector.

Each diagram also contains a summary legend that gives the following information: energy units used [all energy types were converted to trillions of British thermal units (tBtu)], the state's population (1), the net energy exported or imported, and the total energy produced and consumed. Note that energy that is stored is considered not consumed, and energy from storage is not part of production.

The procedures, conventions, assumptions, and data sources used to compute the energy flow patterns are described below.

Electricity Generation

We obtained the nonnuclear electricity production for each state from Federal Power Commission (FPC) data (2), and electricity production from nuclear plants mostly from Nuclear Regulatory Commission (NRC) data (3) (see Sec. C below). Electricity consumed by the various sectors was also obtained from FPC data (4). Electricity consumed in the FPC's "other" sector was added to the commercial sector.

The amount of electricity imported or exported was computed as follows: If electricity production minus 9.55% for transmission loss is greater than the state's total electricity comsumption, the difference is shown as an export of electric energy. If production minus transmission loss is less than the state's total consumption, the difference is shown as an import of electric energy. It was assumed that Alaska and Hawaii did not export or import any electricity.

Electricity consumed or exported is subtracted from the energies input to electricty generation to obtain the amounts of energy lost in generating and transmitting electricity.

Electric energy units, kilowatt-hours (kWh), were converted to British thermal units (Btu) at the rate of 1 kWh = 3413 Btu.

Hydroelectric

The amount of electrical energy generated from hydropower was taken from FPC data (2). For some states, a small amount of electricity production from other sources (geothermal, wood, and waste) is included in this figure. When this occurs, the label is changed to "Hydro + Other." Because no attempt was made to associaate any generation losses with this energy, the only losses this energy suffers are transmission losses and inefficiencies during consumption.

Nuclear

The amount of nuclear energy imported for electricity generation was taken to be the gross thermal energy released in reactors. Except for Washington, Oregon, and California, these data were obtained from the NRC (3). For the three exceptions, the larger FPC nuclear electricity generation numbers (2) were converted back to gross thermal energies. A Bureau of Mines (BOM) national average conversion factor of 10 660 Btu/kWh (Ref. 5, December 1975) was used for Washington and Oregon. The NRC (3) California numbers were used to compute a conversion factor for the FPC California number.

Table I shows the amount of uranium produced in the US (6) and some estimates of its energy content. The "High tBtu" column refers to the energy content if all the U-235 and U-238 were fissioned, that is, if breeder reactors were operating. We obtained this estimate by multiplying the number of U-235 and U-238 atoms by 215 MeV (the energy released in a fission) and converting the result to Btu's. The "Low tBtu" column refers to the energy that thermal reactors can extract. We obtained this estimate by assuming that natural uranium is 0.7% U-235, that 0.2% of the U-235 is lost during fluorination, that 32% of the U-235 is removed as tailings during enrichment (in going from 0.7 to 3.2% enrichment), that 1% of the U-235 is lost during fuel rod fabrication, that 72% of the U-235 loaded into reactors is burned, and that one-third of the fission energy released comes from U-238. These assumptions, which typify the present day thermal reactor fuel cycle, indicate that only 0.508% of the uranium energy present in the yellowcake is actually extracted by thermal reactors. Because the heat value of the uranium exported is conditional and is a large amount, it was not included in the energy flow diagrams. The nuclear energy represented by the higher export numbers is not lost but is being accumulated for recycle or for use in breeder reactors.

Natural Gas

The production of dry natural gas was computed from BOM data (7) by subtracting from gross withdrawals the volume of gas returned to the earth for "repressuring" and the volume shrinkage caused by the extraction of natural gas liquids. Natural gas liquids (NGL) are included in "Oil and NGL" production (see Sec. F below). The Illinois extraction shrinkage is larger than its gross withdrawals because much of the imported gas is processed for NGL in Illinois. Consequently, the gas production figure for Illinois is negative.

Natural gas imports/exports and storage changes were obtained directly from BOM data (7). Because only one BOM figure is given for the combined imports of Maine, New Hampshire, and Vermont, separate figures for these states were deduced from their storage and consumption figures. Residential and electric-utility consumption figures were also taken directly from the BOM (7). Commercial consumption includes the BOM's "commercial" category plus "other consumers." Natural gas consumed in transportation is simply "pipeline fuel." Industrial consumption includes the BOM "industrial" category plus "plant and lease fuel," gas "vented and flared" and "unaccounted for." Consumption figures for New Hampshire, Vermont, and Maine were taken from American Gas Association (AGA) data (8) and renormalized to equal the BOM total for those three states.

All gas volumes were converted to Btu's according to the AGA (8) conversion factors shown in Table II.

Coal

When converting coal tonnage to Btu's, one should be aware of the variations in the heat content of coal from different mines. The FPC has compiled a comprehensive data base (9) covering the delivery of fuel to electric utilities. The coal deliveries to the utilities in a state can be appropriately weighted with their heat values to obtain a single conversion factor for the coal delivered to the electric utilities in the state. Also, coal deliveries from the mines in a state to electric utilities can be appropriately weighted with their heat values to obtain a single state conversion factor for the coal production used at electric utilities. Because electric utilities use most of the coal, the above conversion factors should adequately describe all coal consumption and production, respectively, in the state. These state coal conversion factors are shown in Table II. (If one would choose to follow the distribucion of coke, which is not done here, the above conversion factors may not be appropriate.)

Coal production figures were obtained from the BOM (10). Coal storage at the mines is tabulated by the BOM (11) according to their producing districts. If a producing district coincides with a single state, there is no problem in obtaining the state's coal mine storage. If a state is only part of a district, or if a district is only part of a state, the storage values of the affected districts are added and then reallocated among the constituent states according to their production figures. We converted the above tonnages to Btu's using the production conversion factors in Table II (12-15).

The amount of coal consumed for electricity generation was taken from FPC data (2). Data for coal consumed by commercial and industrial sectors were provided by the BOM (11). The "coke gas plants" and "all others" categories of that reference were assigned to the industrial sector, whereas the "retail dealers" category was assumed to be the commercial sector. Several small consumption categories, which were tabulated by the BOM according to their producing districts, were broken down by state in the same manner as was done above for coal storage. The following small categories were added to the industrial sector: "railroad fuel," "coal used at mines and sales to employees," 61.6% of "destinations not revealable," and "vessel fuel," which was distributed to the states according to their individual Great Lakes coal movements compared to the total Great Lakes coal movements. The commercial sector received 5.2% of "destinations not revealable" with the rest assumed to be already accounted for by the FPC (2) electricity generation consumption. The residential and transportation sectors were assumed to use negligible amounts of coal. The differences between coal delivered to (9) and coal consumed (2) by the electric utilities was assumed to go into storage. We converted all of the above tonnages to Btu's using the consumption conversion factors in Table II. The sum of the mine and utility storage numbers gives the total coal storage.

Anthracite production (16) was added to Pennsylvania's coal production figure. Of the anthracite distributed to the various states, 41.7% was added to the commercial sector, 29.3% was added to the industrial sector, and the rest was assumed to be already accounted for by the FPC (2) electricity generation consumption. The anthracite conversion factor was assumed to be 12 700 Btu/lb.

The coal energy needed to balance production, storage, and consumption was assumed to be exported (or imported).

Oil and NGL

This category encompasses various liquid fuels, each having a different heat content (Table III). Care was taken to ensure that an overall balance of production, imports/exports, storage changes, and consumption was obtained for each fuel on a national level. Because consumption figures are the least accurate, they were renormalized to yield the correct national balance. The renormalization procedure is similar to that used by the BOM in arriving at their consumption figures (17-19).

Production figures represent the sum of crude petroleum produced at wells (Ref. 5, March 1976) and NGL produced at gas processing plants (Ref. 5, January 1976). However, for the latter, several states had to be disaggregated in proportion to the natural gas extraction losses given in the BOM natural gas annual (7).

Storage figures include changes in stocks of crude petroleum, unfinished oils, refined products, liquefied petroleum gases (LPG), other NGL, and fuel oils stored at electric generating plants. The amount of fuel oil stored at electric generating plants was taken from Table 13a of Ref. 18. All other storage data were obtained from the FEA (20). The FEA data for refined products did not include all of the storage quoted by the BOM in Ref. 5, December 1975. Consequently, we adjusted the changes in storage of refined products (including fuel oil at electric plants) to be consistent with the BOM figures. Import/Export figures for individual states and regions were deduced from production, storage, and consumption figures. Imports and exports for the US as a whole were obtained from the BOM (21).

Oil and NGL are consumed in the form of nearly 20 separate fuels (and nonfuels). To convey the relative importance of each fuel, we have shown the percentage used in each category in relation to the total national consumption of oil and NGL. To as great an extent as possible, the percentages shown are consistent with the FEA's FEDS data base (22).

Motor Gasoline 39.3%: These data are available from the Federal Highway Administration. Table MF-21 of Ref. 23 contains a breakdown of highway vs non highway consumption of motor fuels, which includes motor gasoline and special fuels. Special fuels consist mainly of diesel fuel and some LPG, both of which are accounted for under other consumption categories. The "highway" figures (reduced by the "special fuel" consumption from Table MF-25 of Ref. 23) are all attributed to the transportation sector, along with the "marine" usage from Table MF-24 (23). Commercial consumption consists of "public nonhighway," "agriculture," "construction," and "miscellaneous."

Distillate Fuels (including diesel) 18.5%: These data were derived primarily from a BOM publication (18). Consumption in the transportation sector includes distillate fuel oil for "railroads," "bunkering," and "military" in addition to the "on-highway" use of diesel. Residential heating figures were taken from the FEDS data base (22). Commercial includes the rest of the "heating" category plus the "off-highway" uses of diesel. Industrial consumption encompasses "industrial," "oil company," and "other." Electric utility consumption, quoted directly from Ref. 18, includes the consumption of some jet fuel. The electric consumption figures for some states are reduced to reflect changes in storage.

Residual Fuel Oil 17.1%: These data were taken from Ref. 18. Industrial consumption includes "industrial," "oil company," and "miscellaneous." Transportation includes "bunkering," "railroad," and "military." All of the "heating" is allocated to the commercial sector. Electric utility consumption is quoted directly. The electric consumption figures are reduced to reflect changes in storage.

Aviation Gasoline and Jet Fuel 6.5%: The figures for aviation gasoline consumption and civilian consumption of jet fuel were obtained from a proprietary source. The data for military consumption of jet fuel were obtained from the FEDS data base (22). Consumption of jet fuel for electrical generation was accounted for under distillate fuels.

LPG, Ethane, and Other NGL 6.2%: These figures were obtained from a BOM publication (17). Commercial consumption includes 10% of "household/commercial" and "miscellaneous" and 25% of "internal combustion." Residential consumption consists of 90% of "household/commercial," and transportation consumption includes 25% of "internal combustion." Industrial consumption includes "industrial", "utility gas," "chemical and rubber," and 50% of "internal combustion." Figures for "chemical and rubber" consumption are given by Petroleum Allocation for Defense (PAD) district only. A state-by-state breakdown for PAD-III was obtained from BOM (24). Breakdowns for other districts were calculated on the basis of the state's share of the chemical and synthetic rubber industry, as determined from the 1972 census of manufacturers (25).

Asphalt and Road Oil 3.1%: The principal source of data is the BOM (19). Transportation consumption includes all road oil and "asphalt for paving." Commercial consumption includes "asphalt for roofing" and "all other."

Petrochemical Feedstocks 2.1%: These were all allocated to the industrial sector. State-by-state breakdowns were done in proportion to the use of LPG and ethane for "chemical and rubber."

Petroleum Coke 1.7%: According to Ref. 5, May 1976, over 70% of all petroleum coke is consumed at refineries. We have allocated all petroleum coke consumption to the industrial sector. State-by-state breakdowns were done in proportion to the refinery usage figures from Ref. 5, May 1976. When necessary, state- by-state breakdowns were done on the basis of refining capacity (26).

Kerosene 1.0%: Data are from Ref. 18. Kerosene for "heating" was allocated to the residential sector, and "other" consumption was allocated to the industrial sector.

Special Naphthas and Miscellaneous 1.0%: This category includes paint thinners, cleaners, solvents, absorption oils, medicinals, and insecticides. They were allocated to the residential sector according to state population.

Lubricants and Wax 1.0%: The Census Bureau publishes a semiannual survey of lubricant sales (27). Our national total for lubricants was allocated to the industrial and transportation sectors in proportion to the Census Bureau's figures on consumption of oils and greases. State-by-state breakdowns were made in proportion to the Census Bureau's figures for total sales of lubricating oils. Wax consumption, which is a small portion of this category, falls mostly in the paper and candle industries. State-by-state breakdowns for wax consumption were done in the same manner as for lubricants.

Crude Petroleum 0.1%: Crude petroleum is included in the consumption of residual and distillate fuels.

Still Gas 2.4%: This category represents the net consumption of still gas in the refining cycle. This was computed as the difference between the total heat content of all refinery feedstocks (plus the still gas used as fuel) and the total heat content of all refined products produced. This energy is passed through the industrial sector to lost energy. Refinery fuels other than still gas are accounted for separately.

Used vs Lost Energy

The "softest" numbers in this report are the fractions of consumed energy that are assumed to do useful work for the various sectors. Even if current numbers, carefully collected, were available there is still the question of whose definition of efficiency should be used. For example, one person may figure the useful energy of a toaster to be the energy absorbed by the bread; another may figure it to be the total heat produced since that which is not absorbed by the bread is used to heat the home; and a third may partially agree and explain that the usefulness of the home heating portion depends on the seasons and the house and will require a comprehensive data collection effort.

Thus, to convey some idea of the energy that performs useful work, the following arbitrary efficiencies have been adopted for the various sectors: transportation, 25%; residential, commercial, and industrial, 75%. However, certain uses of energy are treated uniquely. Energy used to produce, transport, or process energy is passed through the appropriate sector to lost energy. This includes natural gas used as pipeline fuel; natural gas that is vented, flared, or used in the field and processing plants; unaccounted for gas; and all energy lost in the petroleum refinery cycle. Nonenergy uses of petroleum products are passed through the appropriate sectors to used energy. These include lubricants, wax, asphalt, road oil, special naphthas, miscellaneous products, petrochemical feed-stocks, and LPG used in the chemical and synthetic rubber industries. The efficiency factors are then applied to what remains in each sector.

THE DIAGRAMS

We show a different energy scale for each of the energy flow diagrams so that we can accurately convey the relative magnitude of energy flows within the state. Otherwise, for example, if South Dakota and California had the same energy scale, all flows in South Dakota would essentially be thin lines.

Flow lines <0.05 tBtu are not shown. The fact that all numbers shown have one place to the right of the decimal point is not to be construed as a statement of accuracy -- it is simply a choice for uniformity of presentation.

The US diagram was produced by adding the constituent state contributions for each component of the diagram.

ACKNOWLEDGMENTS

We gratefully acknowledge Dolly M. McClellan for her assistance in reducing and manipulating the data.

REFERENCES

1. "Population Estimates and Projections," Series P-25, No. 615, U. S. Department of Commerce, Bureau of Census (November 1975).

2. Federal Power Commission News Release No. 22686 (20 October 1976).

3. "Operating Units Status Report," Nuclear Regulatory Commission report NUREG-0020-1 (January 1976).

4. Federal Power Commission News Releases 21499 (20 June 1975), 21578 (23 July 1975), 21619 (6 August 1975), 21710 (8 September 1975), 21771 (2 October 1975), 21844 (31 October 1975), 21978 (9 December 1975) 22052 (12 January 1976), 22155 (18 February 1976), 22217 (17 March 1976), 22315 (23 April 1976).

5. U. S. Department of the Interior, Bureau of Mines, Mineral Industry Surveys: Crude Petroleum, Petroleum Products and Natural Gas Liquids, Monthly, Washington, D. C.

6. "Statistical Data of the Uranium Industry," United States Energy Research and Development Administration, Grand Junction Office report GJO-100(76) (1 Jan. 1976).

7. U. S. Department of the Interior, Bureau of Mines, Mineral Industry Surveys: Natural Gas Production and Consumption: 1975, Washington, D. C. (October 4, 1975).

8. "Gas Facts: 1975," American Gas Association, Arlington, Va. (October 1976).

9. "Monthly Reports of Cost and Quality of Fuels for Steam-Electric Plants (Form 423) of U. S. Electric Utilities," Federal Power Commission, Washington, D. C. (1976).

10. Weekly Coal Report No. 3090, U. S. Department of the Interior, Bureau of Mines, Washington, D. C. (December 1976).

11. "Mineral Industry Surveys: Bituminous Coal and Lignite Distribution Calendar Year 1975," U. S. Department of the Interior, Bureau of Mines, Washington, D. C. (3 Dec. 1976).

12. "Historical Fuels and Energy Consumption Data, 1960-72, United States by States and Census Districts West of the Mississippi," U. S. Department of the Interior report IC 8705 (1976).

13. "Historical Fuels and Energy Consumption Data, 1960-72, United States by States and Census Districts East of the Mississippi," U. S. Department of the Interior report IC 8704 (1976).

14. "Steam-Electric Plant Factors," 1975 edition, National Coal Association (January 1976).

15. 1976 Keystone Coal Industry Manual, (McGraw-Hill, New York, 1976).

16. "Mineral Industry Surveys: Coal - Pennsylvania Anthracite in 1975," U. S. Department of the Interior, Bureau of Mines, Washington, D. C. (10 November 1976).

17. U. S. Department of the Interior, Bureau of Mines, Mineral Industry Surveys: Sales of Liquefied Petroleum Gases and Ethane in 1975 (October 2, 1976).

18. U. S. Department of the Interior, Bureau of Mines, Mineral Industry Surveys: Sales of Fuel Oils and Kerosine in 1975 (September 17, 1976).

19. U. S. Department of the Interior, Bureau of Mines, Mineral Industry Surveys: Sales of Asphalt in 1975 (July 19, 1976).

20. Chan Stalvey, Federal Energy Administration, Washington, D. C., personal communication, Februrary 3, 1977.

21. Robert J. Schmer, Bureau of Mines, Washington, D. C., personal communication, February 3, 1977.

22. J. P. Galliker and R. Fuller, "Documentation of the Federal Energy Data System," Federal Energy Administration working paper 77-WPIA-03 (1977).

23. U. S. Department of Transportation, Federal Highway Administration, Highway Statistics (1975 data not yet available in published form).

24. Leonard Fanelli, Bureau of Mines, Washington, D. C., personal communication, Janaury 1977.

25. United States Department of Commerce, "1972 Census of Manufacturers," Bureau of the Census report MC72(2)-28 (September 1974).

26. U. S. Department of the Interior, Bureau of Mines, Mineral Industry Surveys: Petroleum Refineries in the United States and Puerto Rico January 1, 1976 (July 14, 1976).

27. U. S. Department of Commerce, Bureau of Census, Current Industrial Reports: Sales of Lubricating and Industrial Oils and Greases: 1975 (September 1976).

TABLE I

HEAT CONTENT OF URANIUM PRODUCTION

State	Tons of U_3O_8	High tBtu Breeder Reactors)	Low tBtu (Thermal Reactors)
New Mexico	5500	333300	1693
Wyoming	3700	224220	1139
Colorado, Texas, Utah, Washington	3100	187860	954
	12300	745380	3786

TABLE II

CONVERSION FACTORS[a]

	Coal (BTU/lb)		
	Consumption	Production	Gas (BTU/Cu ft)
Alabama	11581.7	11807.5	1029
Alaska	11315.4[12]	11315.4[12]	1005
Arizona	10544.8	10824.7	1052
Arkansas	14380.0[13]	11030.4	997
California	12724.7[12]		1057
Colorado	9903.9	10516.8	913
Connecticut	11933.0[14]		1005
Delaware	12270.8		1020
Florida	11546.3		1043
Georgia	11880.9	13500.0[15]	1027
Hawaii			947
Idaho	10790.0[12]		1055
Illinois	10134.7	10710.3	1026
Indiana	10614.6	10811.7	990
Iowa	10193.5	9657.0	1008
Kansas	9978.4	10157.7	984
Kentucky	10740.5	11326.5	1008
Louisiana			1037
Maine	13400.0[14]		1024
Maryland	12161.4	11344.5	1013
Massachusetts	12173.6		1004
Michigan	11833.2		1012
Minnesota	9027.0		1001
Mississippi	11627.7		1023
Missouri	10753.4	9514.8	1006
Montana	7979.5	8987.2	1021
Nebraksa	10476.8		994
Nevada	11193.9		1067
New Hampshire	13350.5		1010
New Jersey	12700.6		1031
New Mexico	8924.4	8966.5	1064
New York	12024.5		1015
North Carolina	11894.2		1018
North Dakota	6672.2	6515.2	1001
Ohio	10959.8	10910.5	1023
Oklahoma	11330.0[14]	11952.2	1015
Oregon	10790.0[12]		1039
Pennsylvania	11776.0	11727.0	1025
Rhode Island			1014
South Carolina	12080.2		1024

TABLE II (cont)

South Dakota	6308.1		1000
Tennessee	10991.4	11563.0	1031
Texas	6551.5	6551.5	1026
Utah	11824.9	11910.7	950
Vermont	12872.0		1008
Virginia	11965.1	11803.1	1019
Washington	8100.0	8100.0	1042
West Virgina	11610.4	11975.1	1037
Wisconsin	10635.4		1020
Wyoming	8312.9	9166.5	934

[a]Superscript numbers refer to references cited at end of text.

TABLE III

HEAT CONTENT VALUES FOR LIQUID FUELS[a]

Fuel Type	Heat Content (10^6 BTU/42 gal barrel)
Crude oil	5.800
Motor gasoline	5.248
Aviation gasoline	5.248
Jet fuel (naphtha)	5.355
Jet fuel (kerosene)	5.670
LPG and ethane	4.011
Other NGL	4.620
Distillate fuel oil	5.825
Residual fuel oil	6.287
Petrochemical feedstock[b]	5.825
Asphalt	6.636
Road Oil	6.636
Petroleum coke	6.024
Kerosene	5.670
Lubricants	6.065
Wax	5.537
Special naphthas	5.248
Miscellaneous[b]	5.825
Still gas	6.000

[a]Reference 5, December 1975.

[b]Assumed equal to distillate fuel oil.

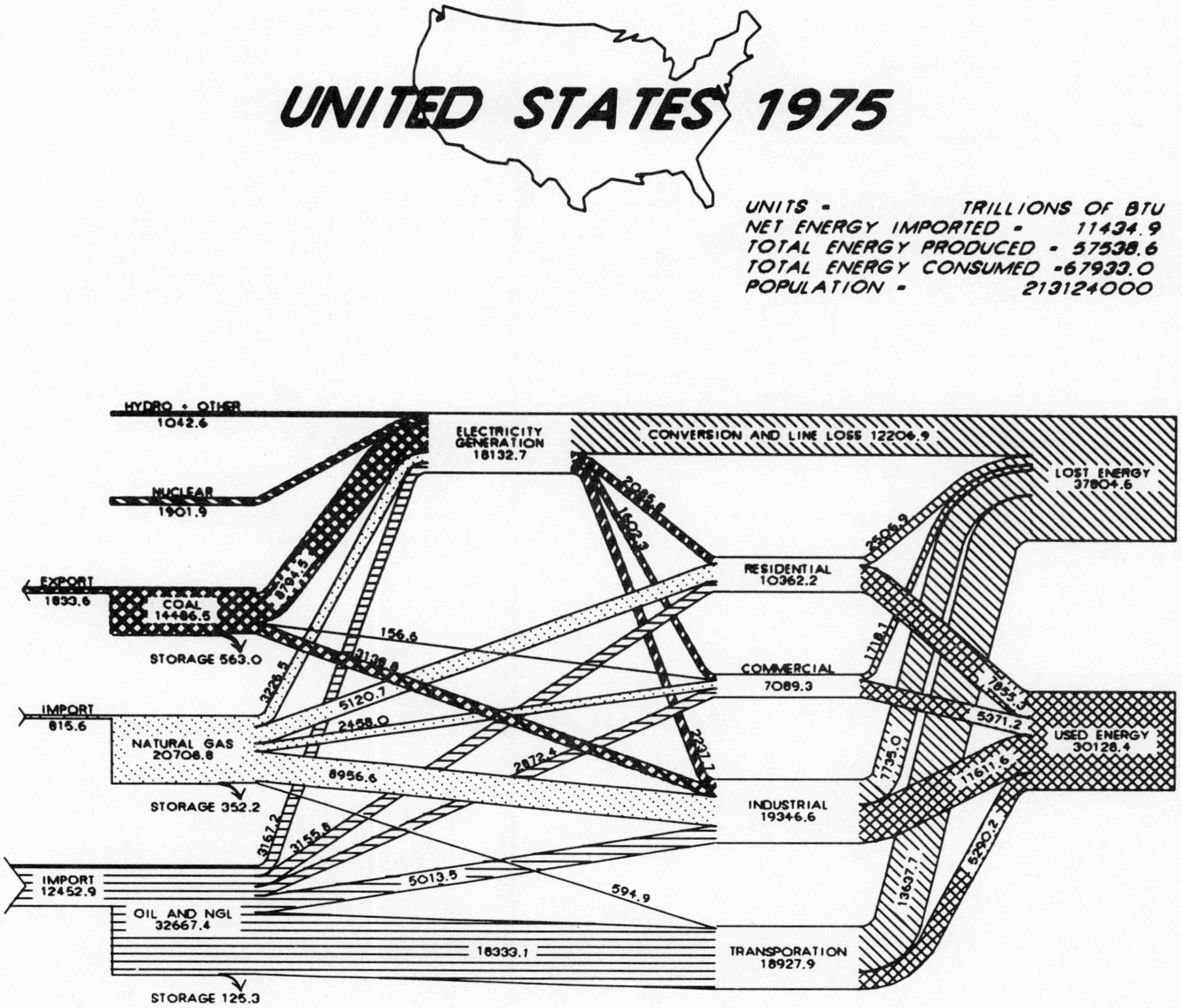

LOS ALAMOS SCIENTIFIC LABORATORY

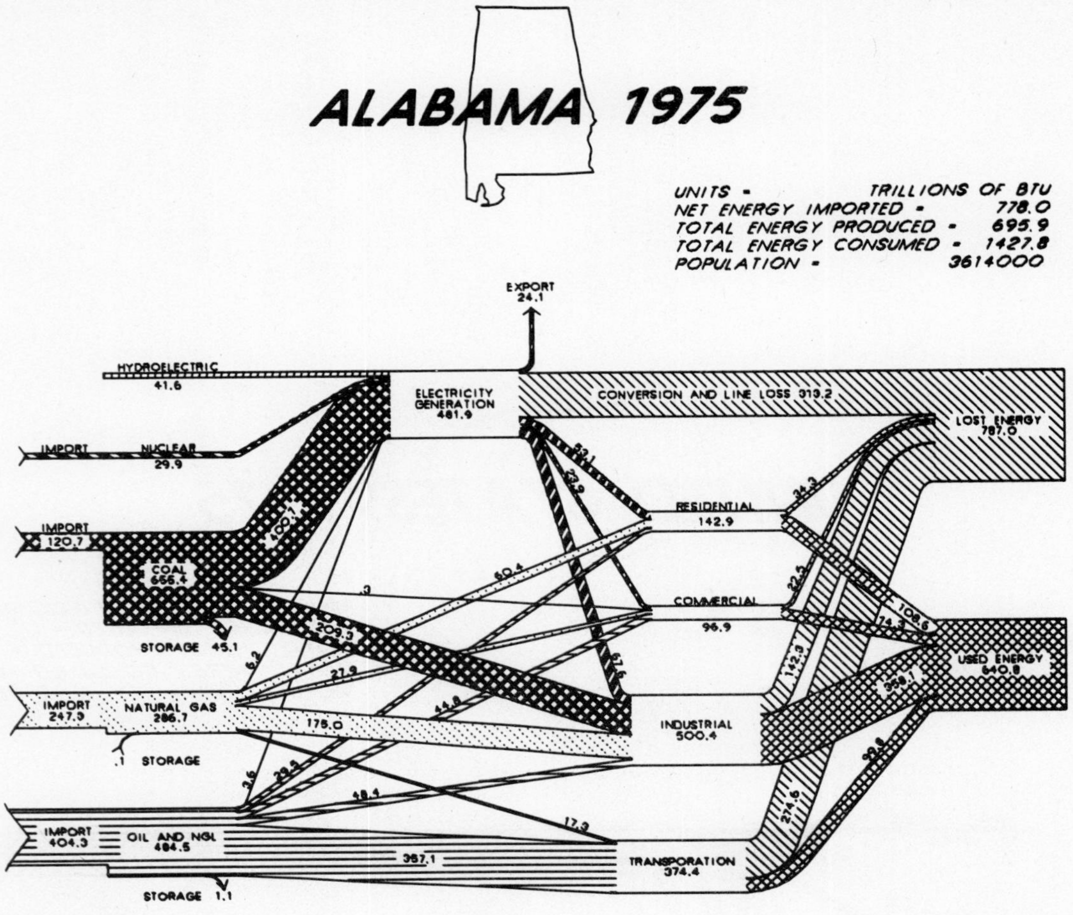

LOS ALAMOS SCIENTIFIC LABORATORY

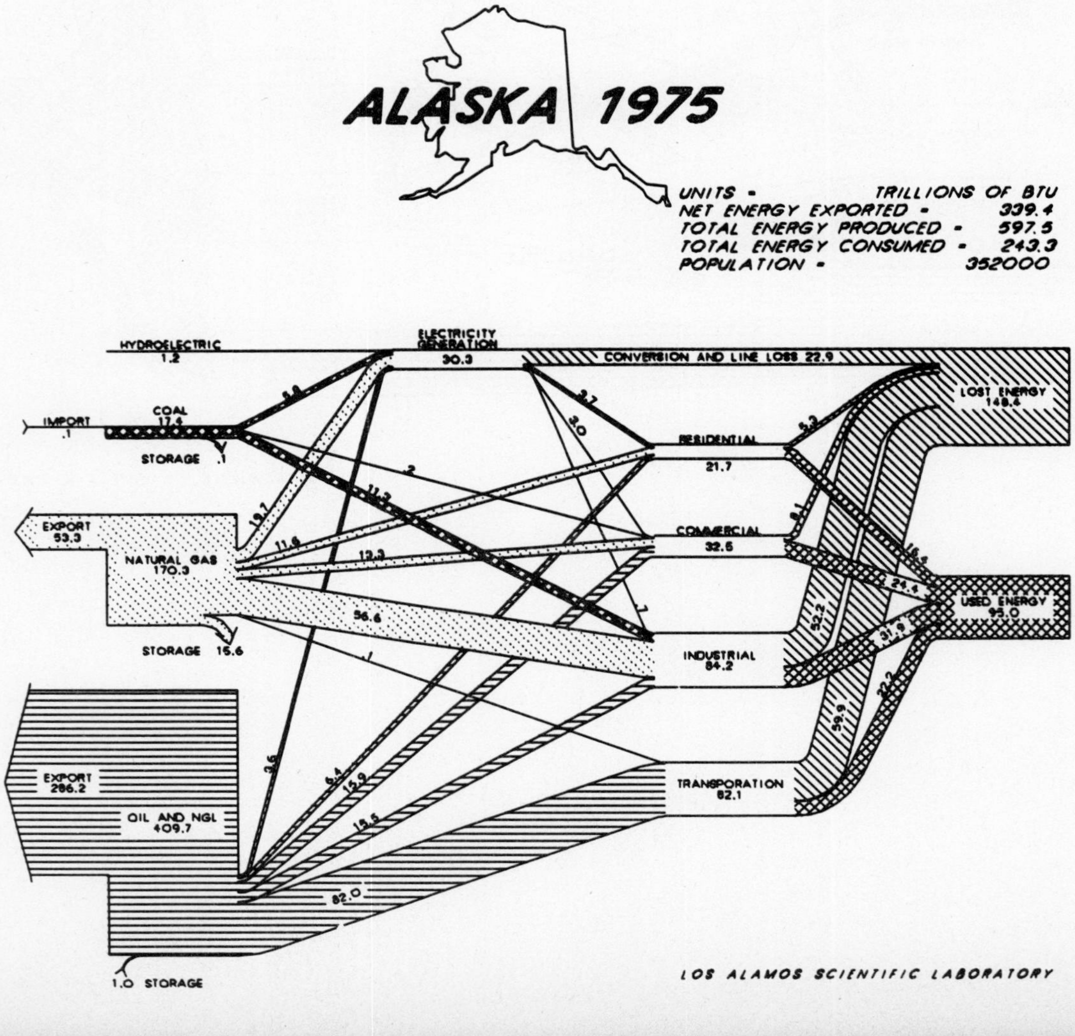

LOS ALAMOS SCIENTIFIC LABORATORY

ARIZONA 1975

UNITS - TRILLIONS OF BTU
NET ENERGY IMPORTED - 415.4
TOTAL ENERGY PRODUCED - 179.9
TOTAL ENERGY CONSUMED - 597.9
POPULATION - 2224000

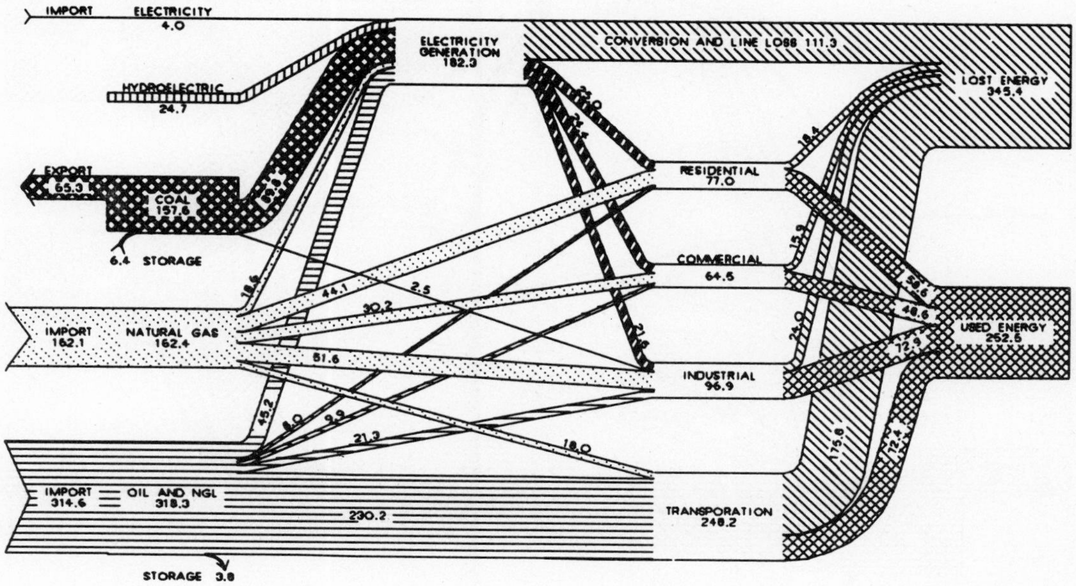

LOS ALAMOS SCIENTIFIC LABORATORY

ARKANSAS 1975

UNITS - TRILLIONS OF BTU
NET ENERGY IMPORTED - 447.5
TOTAL ENERGY PRODUCED - 234.9
TOTAL ENERGY CONSUMED - 678.0
POPULATION - 2116000

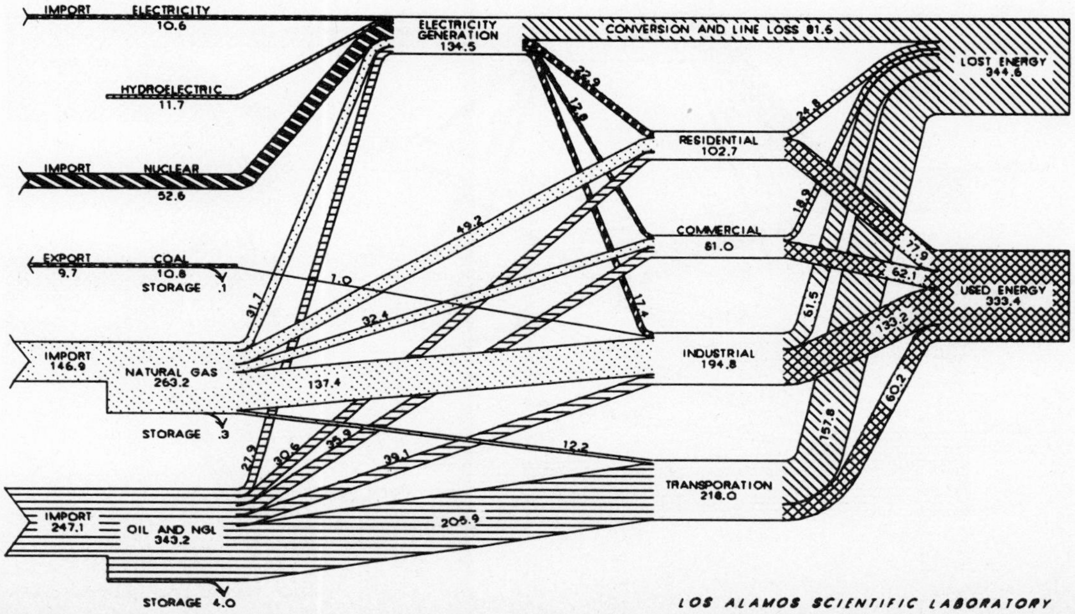

LOS ALAMOS SCIENTIFIC LABORATORY

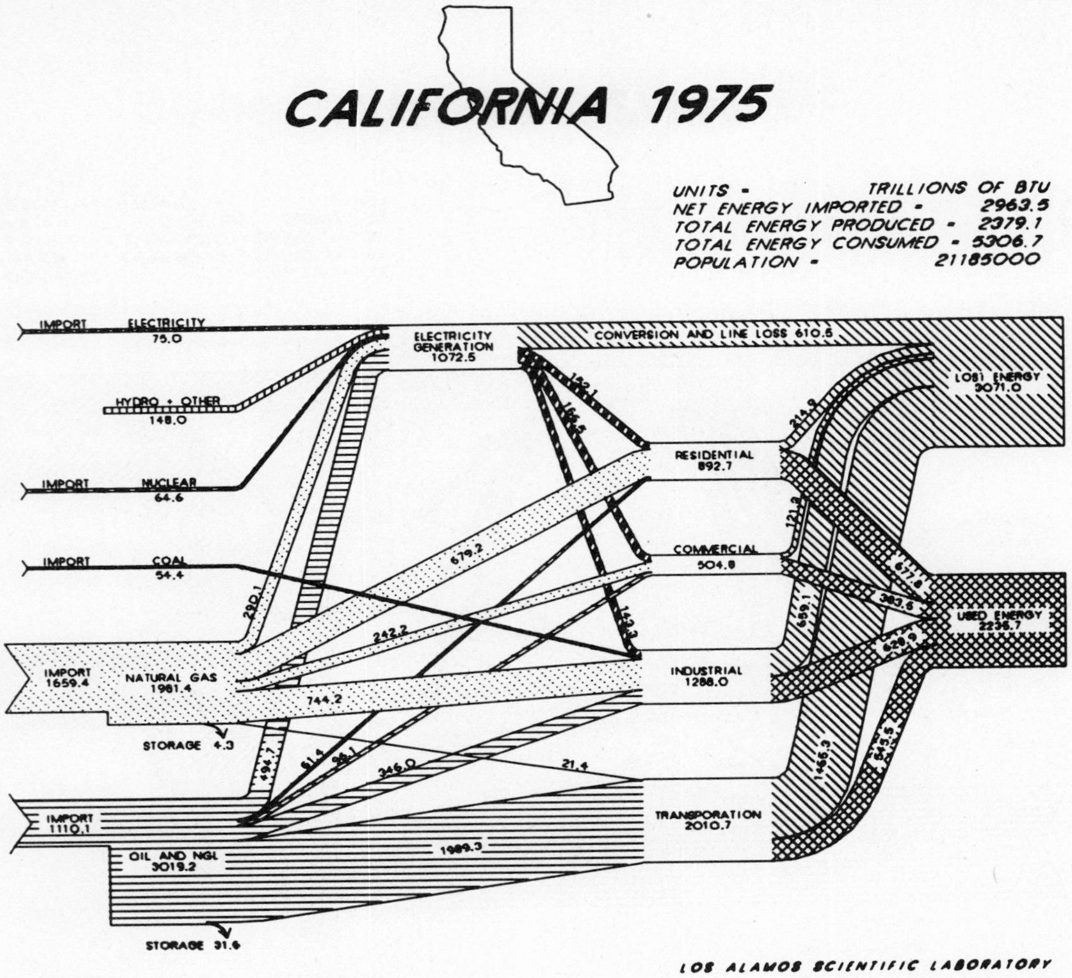
CALIFORNIA 1975
UNITS - TRILLIONS OF BTU
NET ENERGY IMPORTED - 2963.5
TOTAL ENERGY PRODUCED - 2379.1
TOTAL ENERGY CONSUMED - 5306.7
POPULATION - 21185000
IMPORT ELECTRICITY 75.0
HYDRO + OTHER 148.0
IMPORT NUCLEAR 64.6
IMPORT COAL 54.4
IMPORT 1659.4
NATURAL GAS 1961.4
STORAGE 4.3
IMPORT 1110.1
OIL AND NGL 3019.2
STORAGE 31.6
ELECTRICITY GENERATION 1072.5
CONVERSION AND LINE LOSS 610.5
LOST ENERGY 3071.0
RESIDENTIAL 892.7
COMMERCIAL 504.8
INDUSTRIAL 1288.0
TRANSPORATION 2010.7
USED ENERGY 2235.7
679.2
242.2
744.2
290.1
494.7
346.0
21.4
1989.3
1466.3
LOS ALAMOS SCIENTIFIC LABORATORY

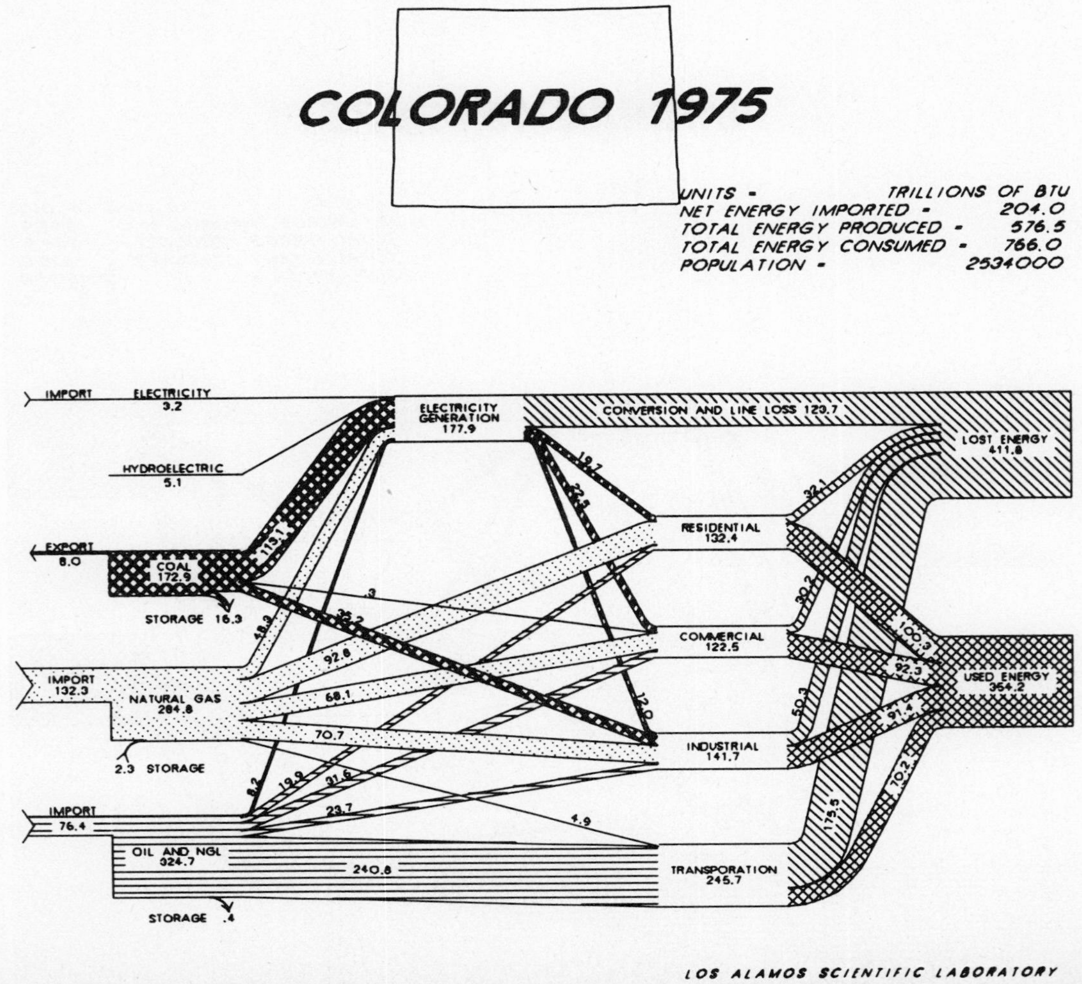
COLORADO 1975
UNITS - TRILLIONS OF BTU
NET ENERGY IMPORTED - 204.0
TOTAL ENERGY PRODUCED - 576.5
TOTAL ENERGY CONSUMED - 766.0
POPULATION - 2534000
IMPORT ELECTRICITY 3.2
HYDROELECTRIC 5.1
EXPORT 8.0
COAL 172.9
STORAGE 16.3
IMPORT 132.3
NATURAL GAS 284.8
2.3 STORAGE
IMPORT 76.4
OIL AND NGL 324.7
STORAGE .4
ELECTRICITY GENERATION 177.9
CONVERSION AND LINE LOSS 129.7
LOST ENERGY 411.8
RESIDENTIAL 132.4
COMMERCIAL 122.5
INDUSTRIAL 141.7
TRANSPORATION 245.7
USED ENERGY 354.2
115.1
92.8
68.1
70.7
23.7
4.9
240.8
175.5
LOS ALAMOS SCIENTIFIC LABORATORY

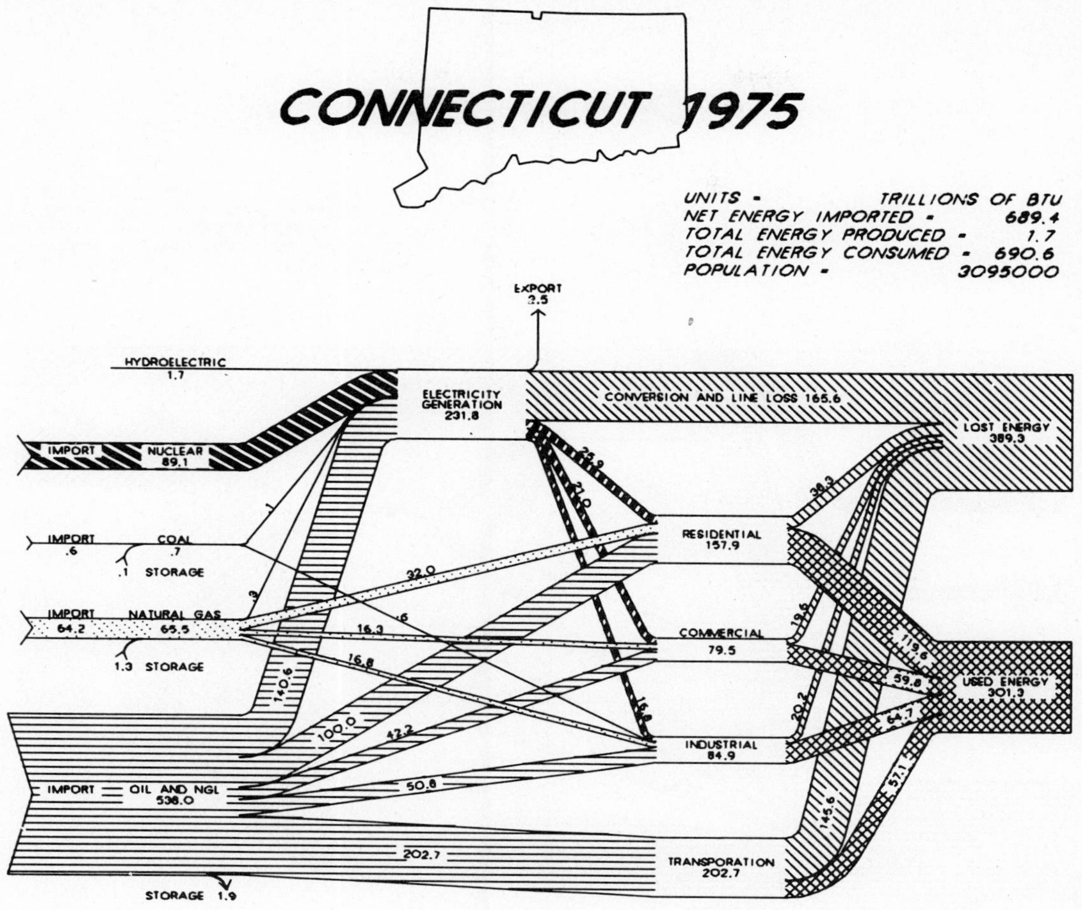

LOS ALAMOS SCIENTIFIC LABORATORY

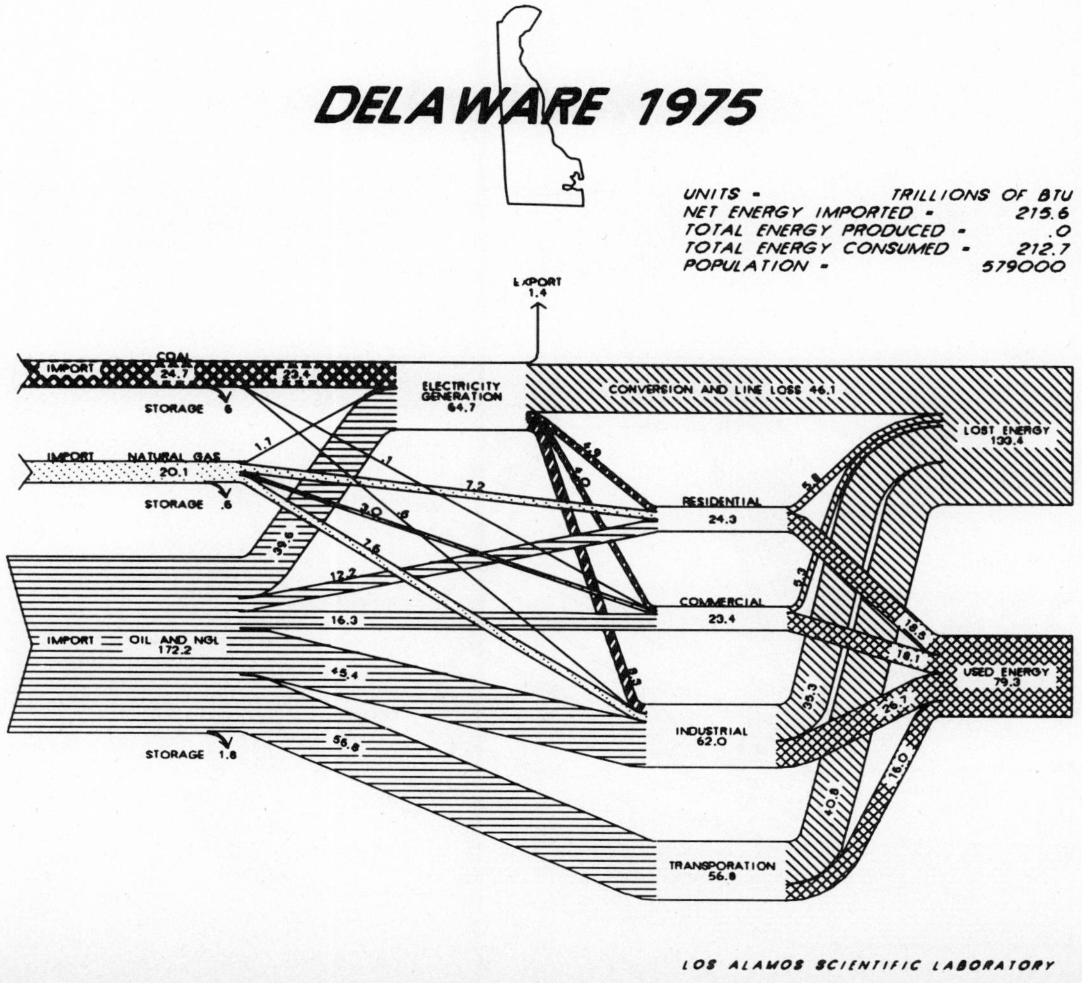

LOS ALAMOS SCIENTIFIC LABORATORY

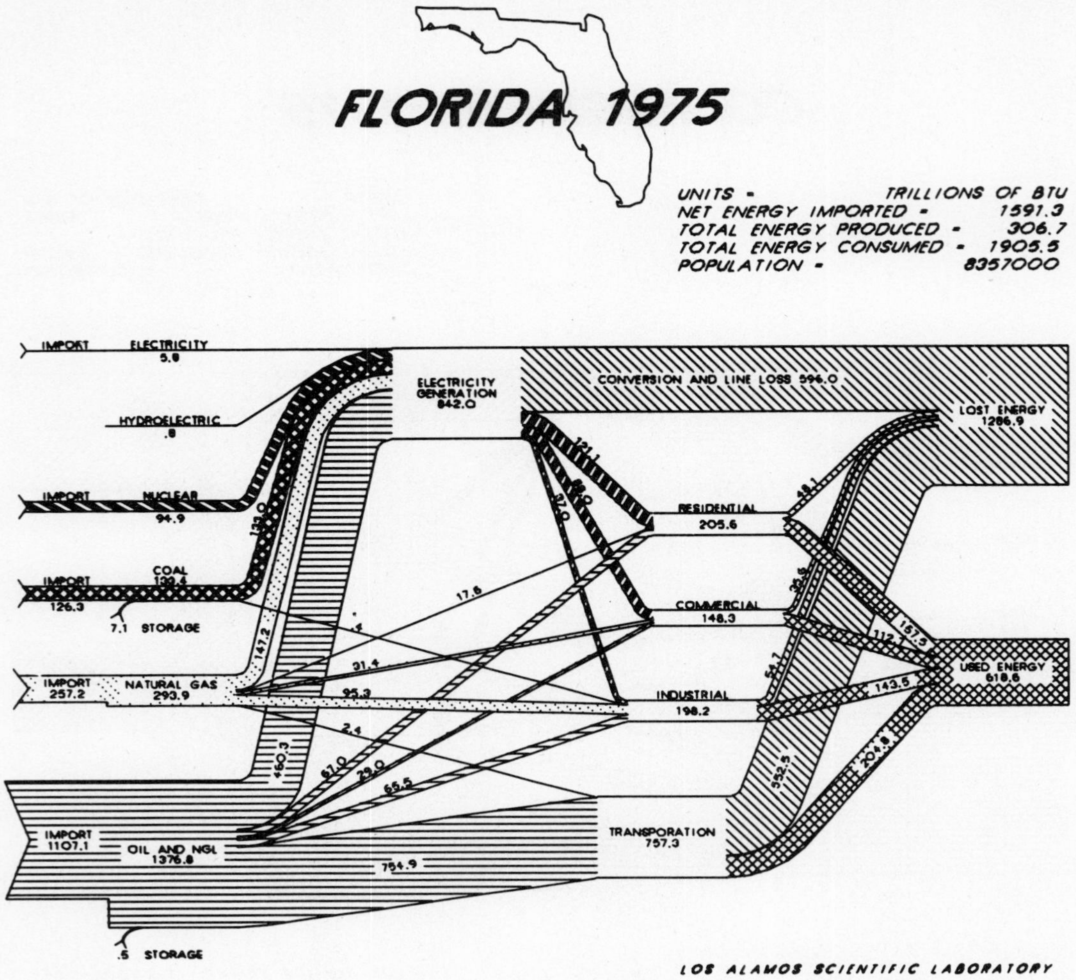
FLORIDA 1975
UNITS - TRILLIONS OF BTU
NET ENERGY IMPORTED - 1591.3
TOTAL ENERGY PRODUCED - 306.7
TOTAL ENERGY CONSUMED - 1905.5
POPULATION - 8357000
IMPORT ELECTRICITY 5.8
HYDROELECTRIC .8
IMPORT NUCLEAR 94.9
IMPORT 126.3 COAL 130.4
7.1 STORAGE
IMPORT 257.2 NATURAL GAS 293.9
IMPORT 1107.1 OIL AND NGL 1376.8
.5 STORAGE
ELECTRICITY GENERATION 842.0
CONVERSION AND LINE LOSS 594.0
LOST ENERGY 1286.9
RESIDENTIAL 205.6
COMMERCIAL 148.3
INDUSTRIAL 198.2
TRANSPORATION 757.3
USED ENERGY 618.6
754.9
LOS ALAMOS SCIENTIFIC LABORATORY

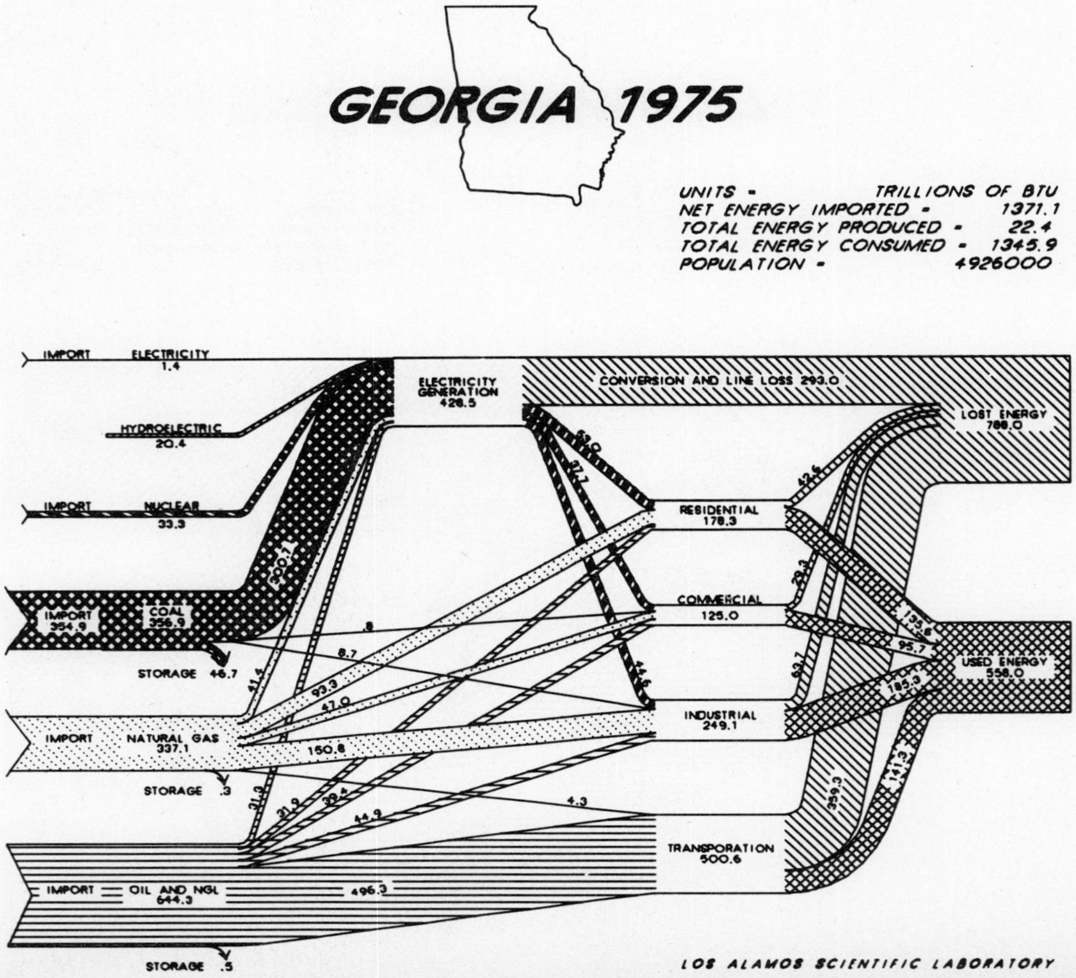
GEORGIA 1975
UNITS - TRILLIONS OF BTU
NET ENERGY IMPORTED - 1371.1
TOTAL ENERGY PRODUCED - 22.4
TOTAL ENERGY CONSUMED - 1345.9
POPULATION - 4926000
IMPORT ELECTRICITY 1.4
HYDROELECTRIC 20.4
IMPORT NUCLEAR 33.3
IMPORT 354.9 COAL 356.9
STORAGE 46.7
IMPORT NATURAL GAS 337.1
STORAGE .3
IMPORT OIL AND NGL 644.3
STORAGE .5
ELECTRICITY GENERATION 426.5
CONVERSION AND LINE LOSS 293.0
LOST ENERGY 788.0
RESIDENTIAL 178.3
COMMERCIAL 125.0
INDUSTRIAL 249.1
TRANSPORATION 500.6
USED ENERGY 558.0
150.8
496.3
LOS ALAMOS SCIENTIFIC LABORATORY

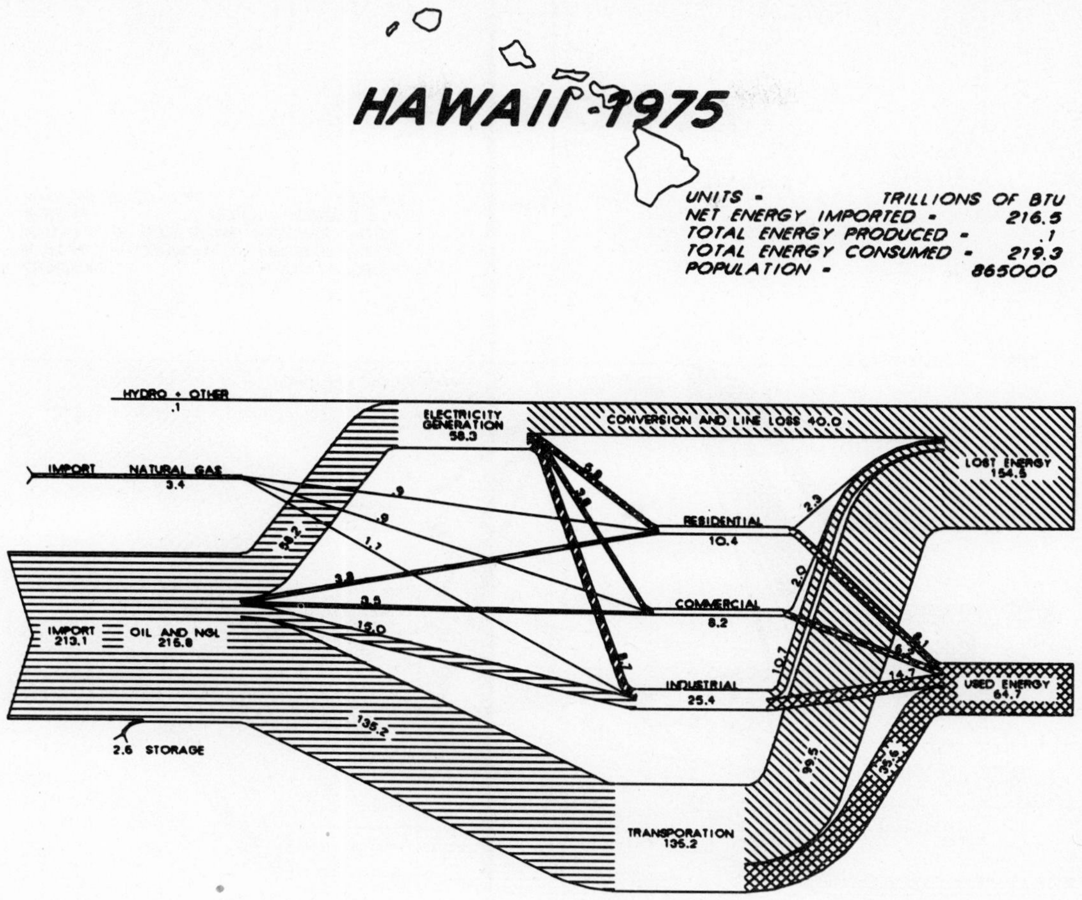

LOS ALAMOS SCIENTIFIC LABORATORY

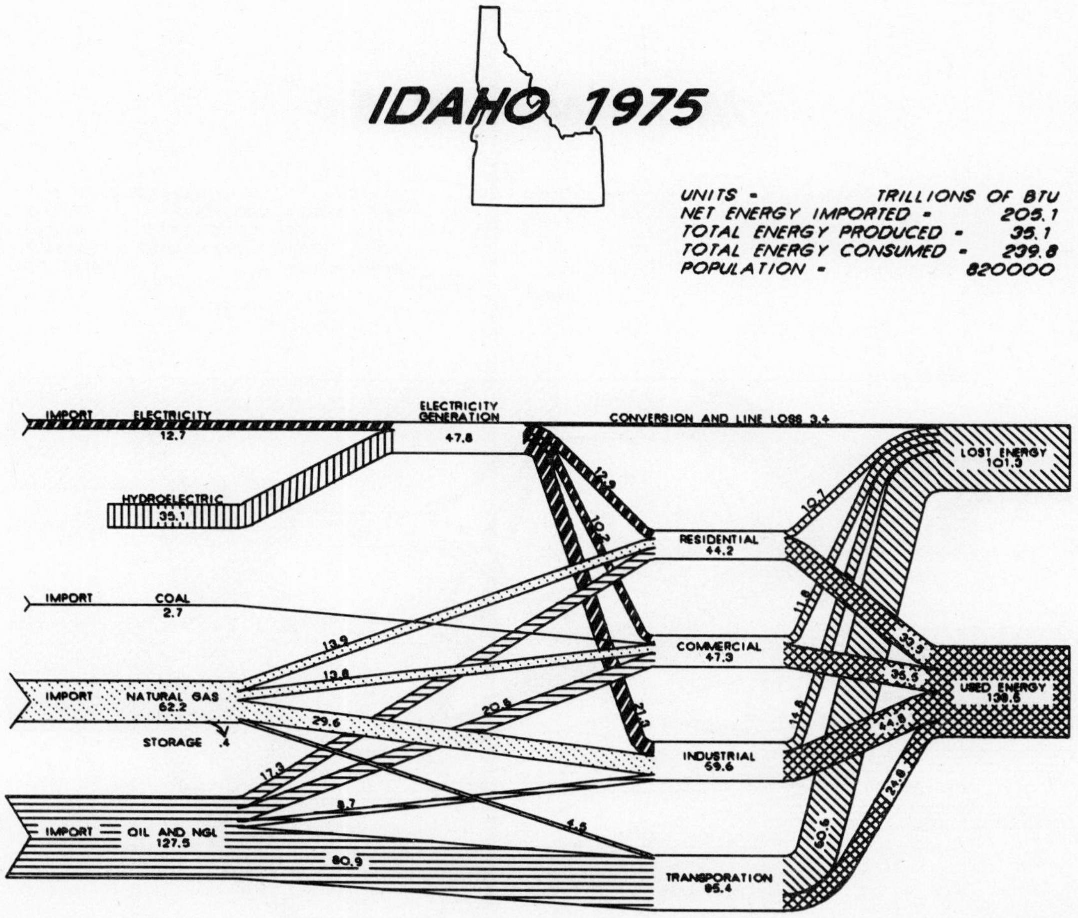

LOS ALAMOS SCIENTIFIC LABORATORY

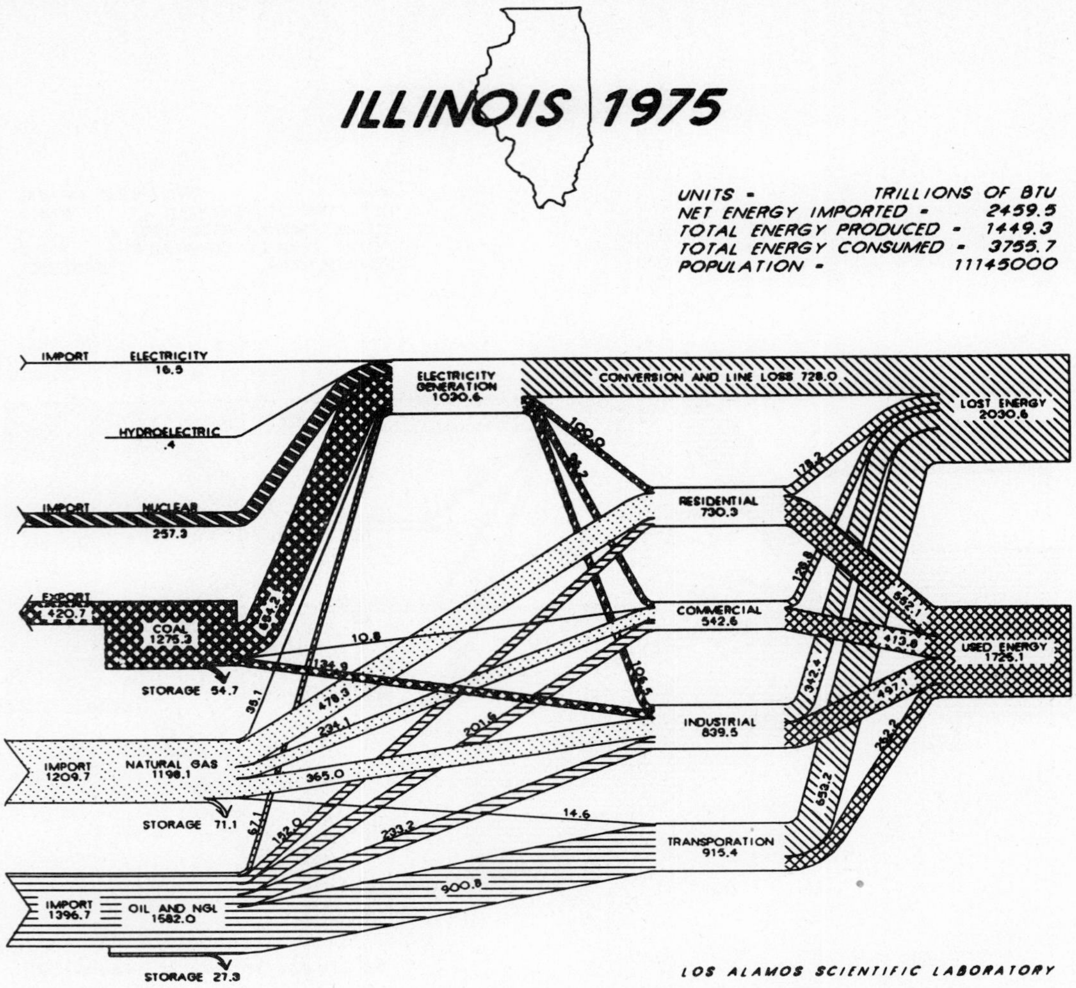
ILLINOIS 1975
UNITS - TRILLIONS OF BTU
NET ENERGY IMPORTED - 2459.5
TOTAL ENERGY PRODUCED - 1449.3
TOTAL ENERGY CONSUMED - 3755.7
POPULATION - 11145000
IMPORT ELECTRICITY 16.5
HYDROELECTRIC .4
IMPORT NUCLEAR 257.3
EXPORT 420.7
COAL 1275.3
STORAGE 54.7
IMPORT 1209.7
NATURAL GAS 1196.1
STORAGE 71.1
IMPORT 1396.7
OIL AND NGL 1582.0
STORAGE 27.3
ELECTRICITY GENERATION 1030.6
CONVERSION AND LINE LOSS 728.0
LOST ENERGY 2030.6
RESIDENTIAL 730.3
COMMERCIAL 542.6
INDUSTRIAL 839.5
TRANSPORATION 915.4
USED ENERGY 1725.1
LOS ALAMOS SCIENTIFIC LABORATORY

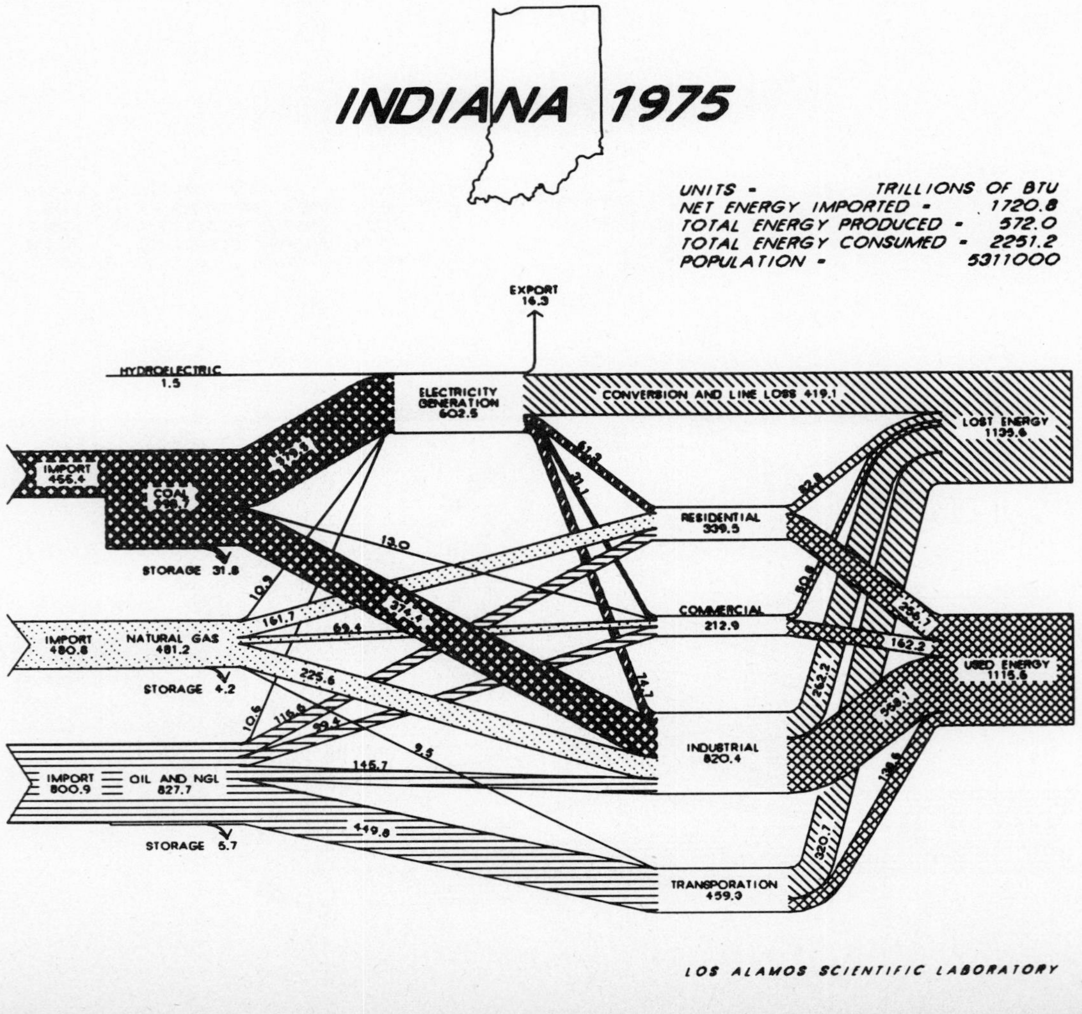
INDIANA 1975
UNITS - TRILLIONS OF BTU
NET ENERGY IMPORTED - 1720.8
TOTAL ENERGY PRODUCED - 572.0
TOTAL ENERGY CONSUMED - 2251.2
POPULATION - 5311000
EXPORT 16.3
HYDROELECTRIC 1.5
IMPORT 456.4
STORAGE 31.8
IMPORT 480.8
NATURAL GAS 481.2
STORAGE 4.2
IMPORT 800.9
OIL AND NGL 827.7
STORAGE 5.7
ELECTRICITY GENERATION 602.5
CONVERSION AND LINE LOSS 419.1
LOST ENERGY 1135.6
RESIDENTIAL 339.5
COMMERCIAL 212.9
INDUSTRIAL 820.4
TRANSPORATION 459.3
USED ENERGY 1115.6
LOS ALAMOS SCIENTIFIC LABORATORY

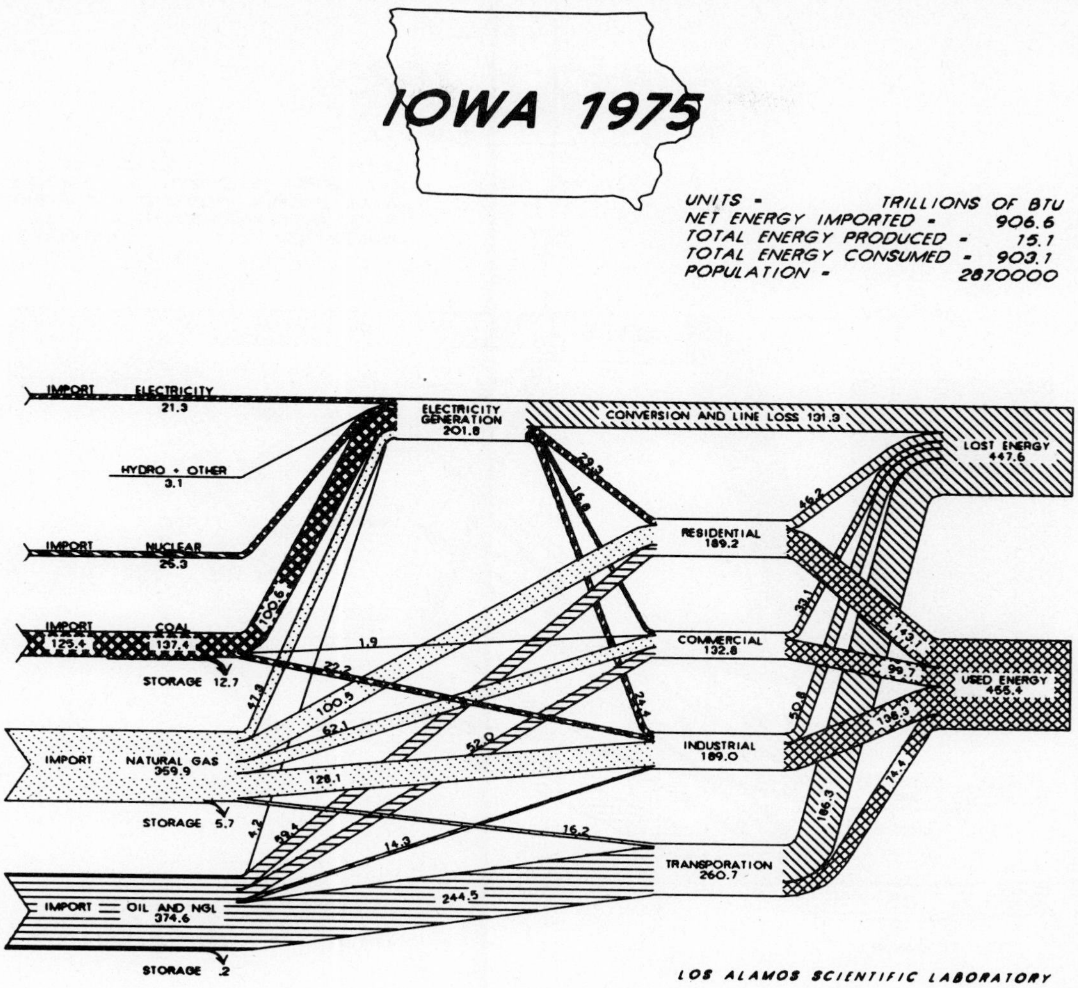
IOWA 1975
UNITS - TRILLIONS OF BTU
NET ENERGY IMPORTED - 906.6
TOTAL ENERGY PRODUCED - 15.1
TOTAL ENERGY CONSUMED - 903.1
POPULATION - 2870000
IMPORT ELECTRICITY 21.3
HYDRO + OTHER 3.1
IMPORT NUCLEAR 26.3
IMPORT COAL 125.4 137.4
STORAGE 12.7
IMPORT NATURAL GAS 359.9
STORAGE 5.7
IMPORT OIL AND NGL 374.6
STORAGE .2
ELECTRICITY GENERATION 201.8
CONVERSION AND LINE LOSS 131.3
LOST ENERGY 447.6
RESIDENTIAL 189.2
COMMERCIAL 132.8
INDUSTRIAL 189.0
TRANSPORATION 260.7
USED ENERGY 455.4
100.5
1.9
22.2
47.8
100.5
67.1
128.1
52.0
4.2
59.4
14.3
16.2
244.5
29.3
16.8
24.4
46.2
33.1
143.1
99.7
50.8
138.3
186.3
74.4
LOS ALAMOS SCIENTIFIC LABORATORY

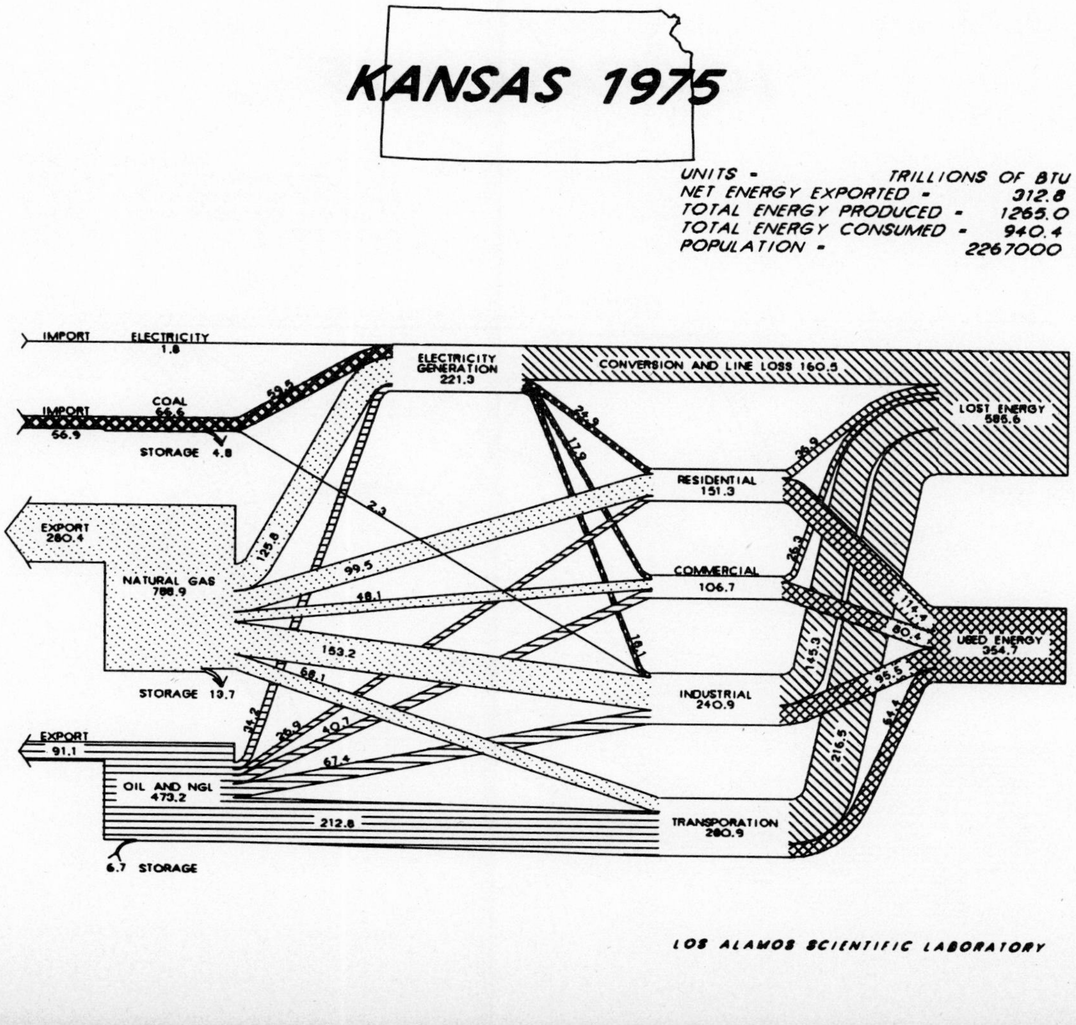
KANSAS 1975
UNITS - TRILLIONS OF BTU
NET ENERGY EXPORTED - 312.8
TOTAL ENERGY PRODUCED - 1265.0
TOTAL ENERGY CONSUMED - 940.4
POPULATION - 2267000
IMPORT ELECTRICITY 1.8
IMPORT COAL 66.6
56.9
STORAGE 4.8
EXPORT 280.4
NATURAL GAS 788.9
STORAGE 19.7
EXPORT 91.1
OIL AND NGL 473.2
6.7 STORAGE
ELECTRICITY GENERATION 221.3
CONVERSION AND LINE LOSS 160.5
LOST ENERGY 585.6
RESIDENTIAL 151.3
COMMERCIAL 106.7
INDUSTRIAL 240.9
TRANSPORATION 280.9
USED ENERGY 354.7
59.5
125.8
2.3
99.5
48.1
153.2
68.1
34.2
26.9
40.7
67.4
212.8
24.9
17.9
18.1
36.9
35.3
116.4
60.4
145.3
95.5
215.5
64.4
LOS ALAMOS SCIENTIFIC LABORATORY

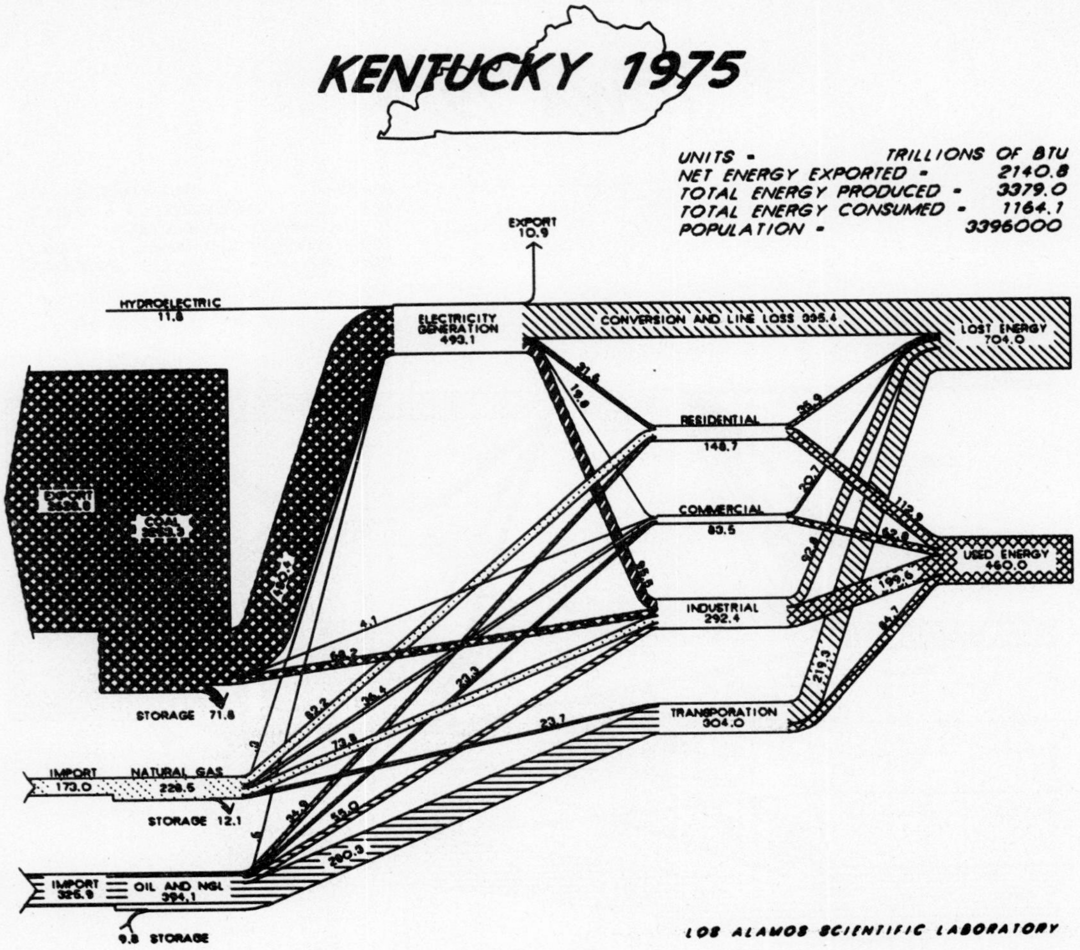
KENTUCKY 1975
UNITS - TRILLIONS OF BTU
NET ENERGY EXPORTED - 2140.8
TOTAL ENERGY PRODUCED - 3379.0
TOTAL ENERGY CONSUMED - 1164.1
POPULATION - 3396000
EXPORT 10.9
HYDROELECTRIC 11.8
ELECTRICITY GENERATION 493.1
CONVERSION AND LINE LOSS 335.4
LOST ENERGY 704.0
RESIDENTIAL 148.7
COMMERCIAL 83.5
INDUSTRIAL 292.4
TRANSPORATION 304.0
USED ENERGY 460.0
STORAGE 71.8
IMPORT 173.0
NATURAL GAS 228.5
STORAGE 12.1
IMPORT 326.9
OIL AND NGL 394.1
9.8 STORAGE
LOS ALAMOS SCIENTIFIC LABORATORY

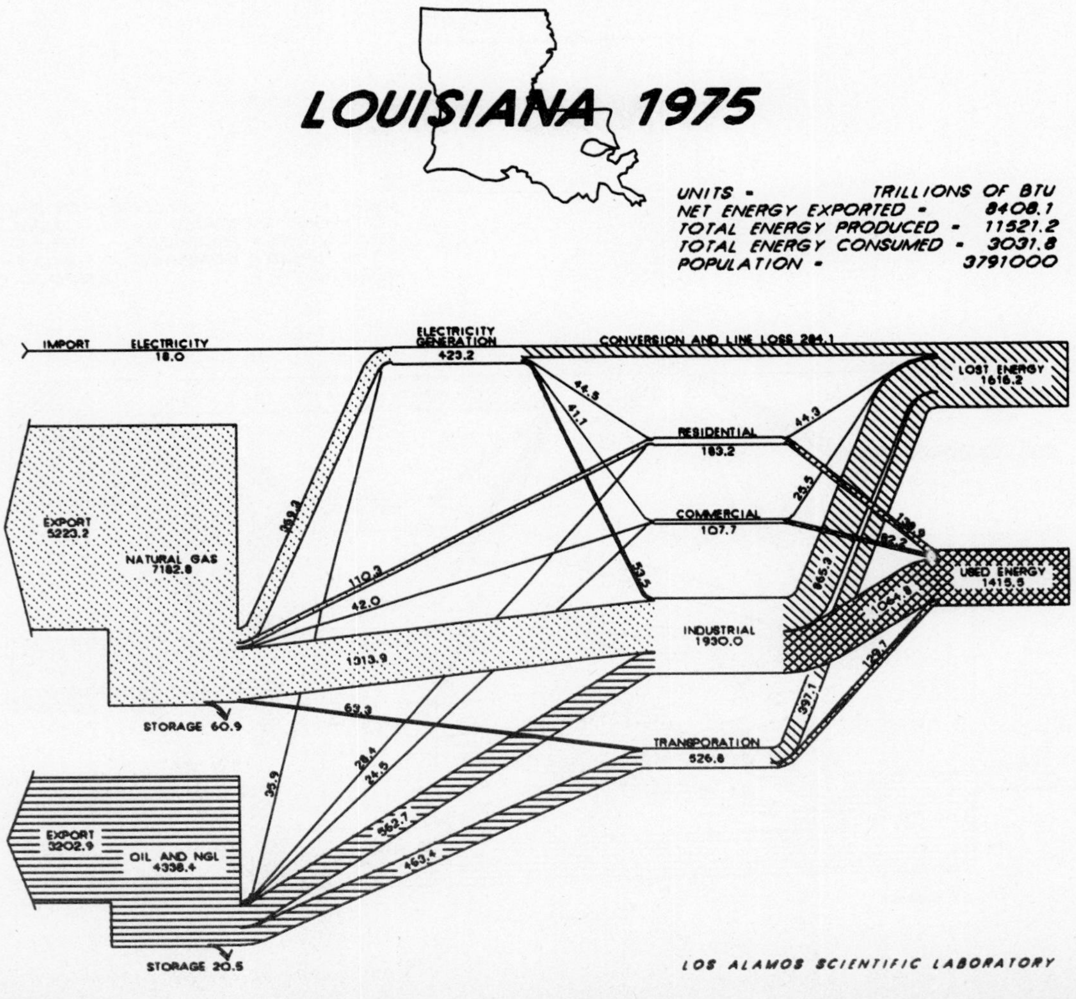
LOUISIANA 1975
UNITS - TRILLIONS OF BTU
NET ENERGY EXPORTED - 8408.1
TOTAL ENERGY PRODUCED - 11521.2
TOTAL ENERGY CONSUMED - 3031.8
POPULATION - 3791000
IMPORT ELECTRICITY 18.0
ELECTRICITY GENERATION 423.2
CONVERSION AND LINE LOSS 284.1
LOST ENERGY 1616.2
RESIDENTIAL 183.2
COMMERCIAL 107.7
INDUSTRIAL 1930.0
TRANSPORATION 526.8
USED ENERGY 1415.5
EXPORT 5223.2
NATURAL GAS 7182.8
1313.9
STORAGE 60.9
EXPORT 3202.9
OIL AND NGL 4336.4
STORAGE 20.5
LOS ALAMOS SCIENTIFIC LABORATORY

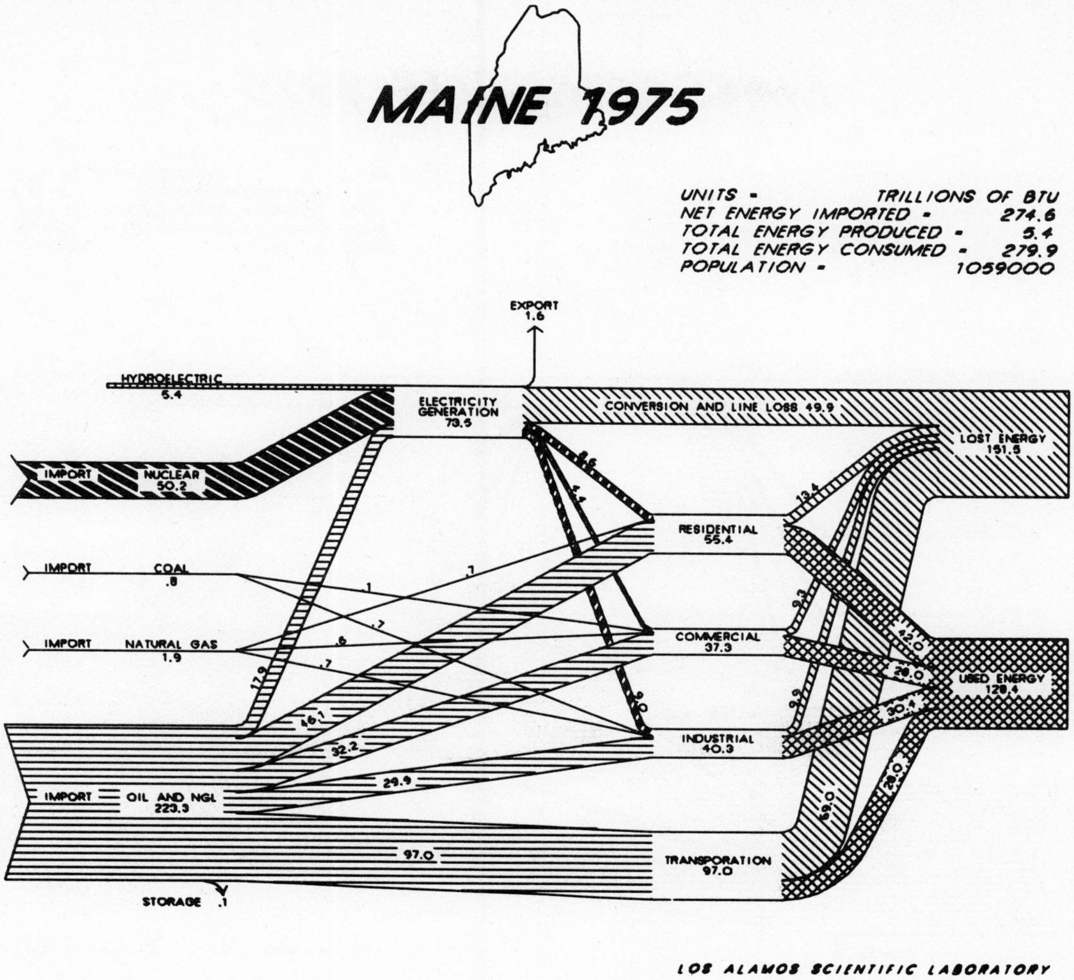
MAINE 1975
UNITS - TRILLIONS OF BTU
NET ENERGY IMPORTED - 274.6
TOTAL ENERGY PRODUCED - 5.4
TOTAL ENERGY CONSUMED - 279.9
POPULATION - 1059000
EXPORT 1.6
HYDROELECTRIC 5.4
ELECTRICITY GENERATION 73.5
CONVERSION AND LINE LOSS 49.9
LOST ENERGY 151.5
IMPORT NUCLEAR 50.2
IMPORT COAL .8
IMPORT NATURAL GAS 1.9
IMPORT OIL AND NGL 223.3
RESIDENTIAL 55.4
COMMERCIAL 37.3
INDUSTRIAL 40.3
TRANSPORATION 97.0
USED ENERGY 128.4
STORAGE .1
97.0
29.9
32.2
46.1
17.9
LOS ALAMOS SCIENTIFIC LABORATORY

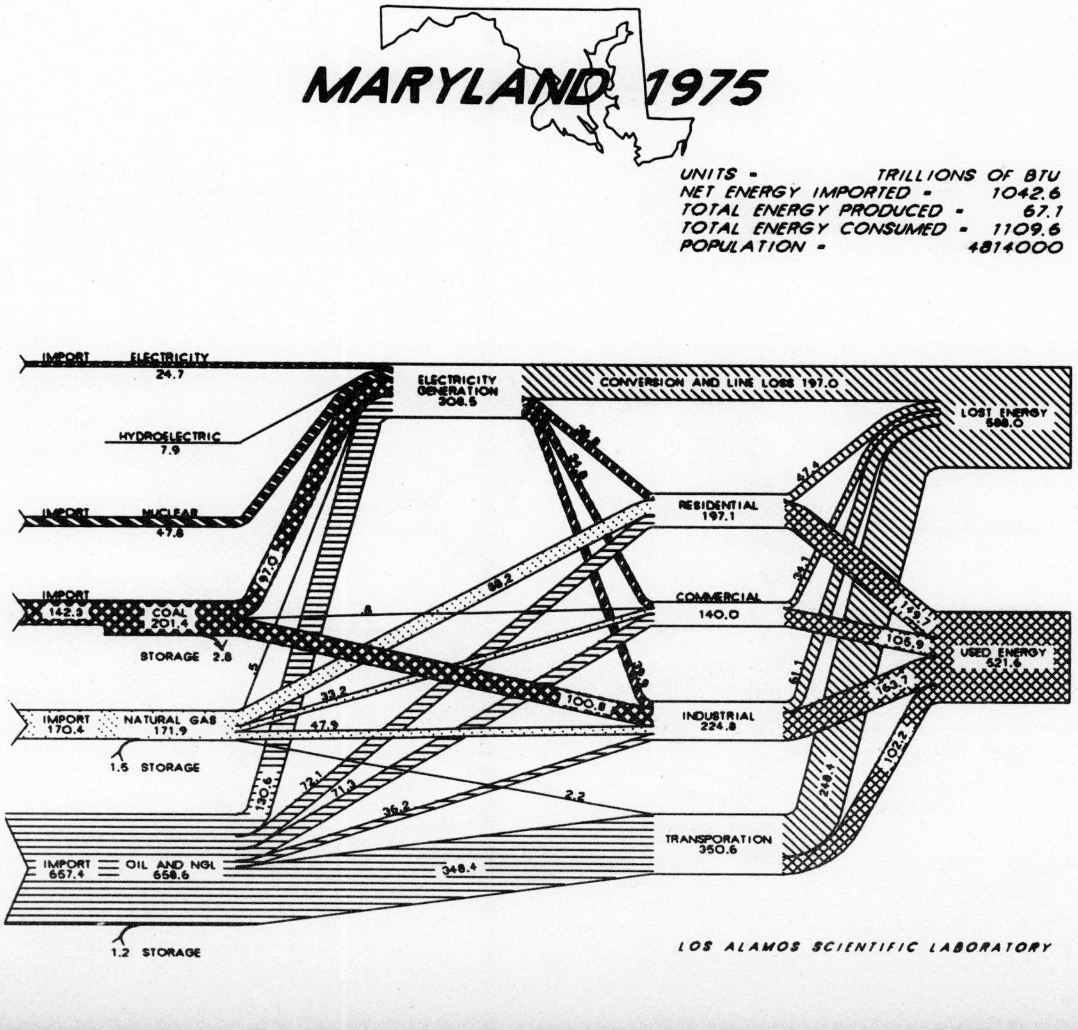
MARYLAND 1975
UNITS - TRILLIONS OF BTU
NET ENERGY IMPORTED - 1042.6
TOTAL ENERGY PRODUCED - 67.1
TOTAL ENERGY CONSUMED - 1109.6
POPULATION - 4814000
IMPORT ELECTRICITY 24.7
HYDROELECTRIC 7.9
IMPORT NUCLEAR 47.8
IMPORT 142.9 COAL 201.4
STORAGE 2.8
IMPORT 170.4 NATURAL GAS 171.9
1.5 STORAGE
IMPORT 657.4 OIL AND NGL 658.6
1.2 STORAGE
ELECTRICITY GENERATION 306.5
CONVERSION AND LINE LOSS 197.0
LOST ENERGY 588.0
RESIDENTIAL 197.1
COMMERCIAL 140.0
INDUSTRIAL 224.8
TRANSPORATION 350.6
USED ENERGY 521.6
100.8
47.9
33.2
36.2
2.2
348.4
130.6
LOS ALAMOS SCIENTIFIC LABORATORY

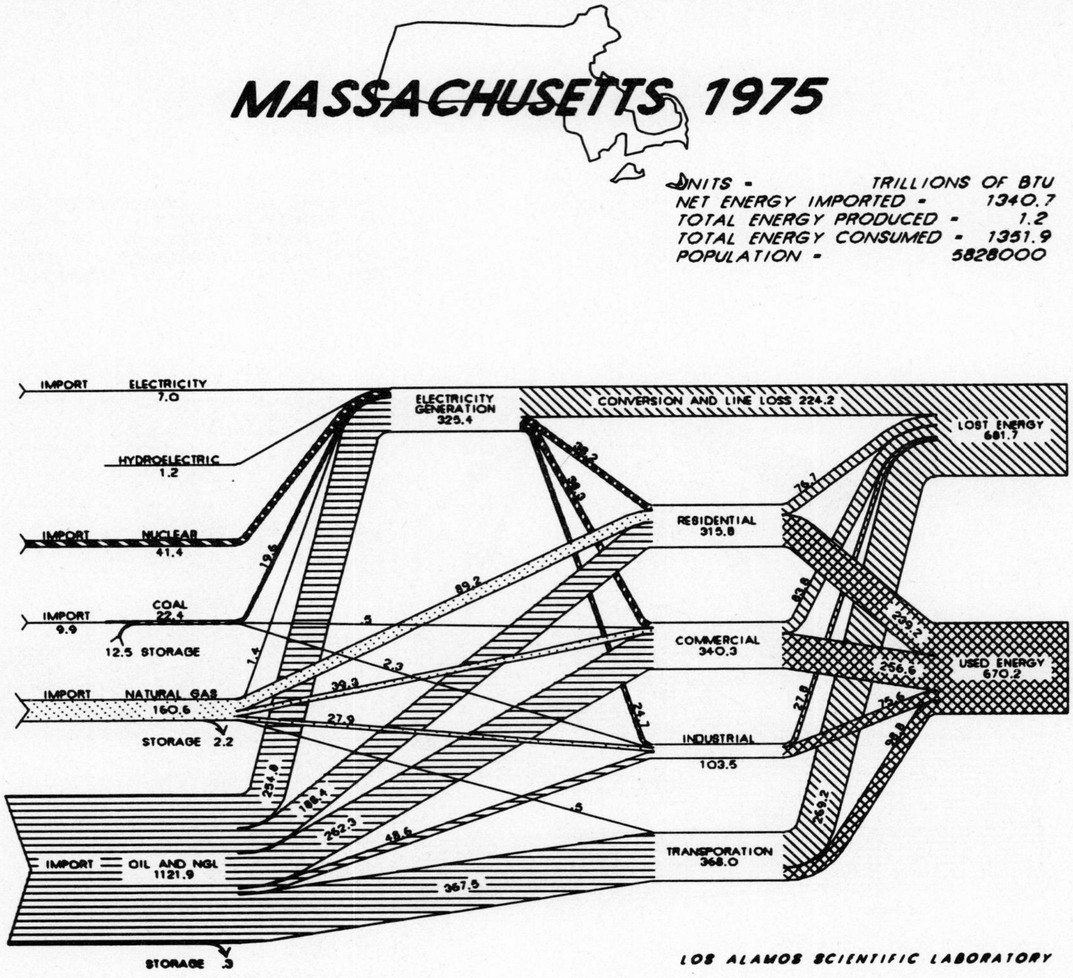
MASSACHUSETTS 1975
UNITS - TRILLIONS OF BTU
NET ENERGY IMPORTED - 1340.7
TOTAL ENERGY PRODUCED - 1.2
TOTAL ENERGY CONSUMED - 1351.9
POPULATION - 5828000
IMPORT ELECTRICITY 7.0
HYDROELECTRIC 1.2
IMPORT NUCLEAR 41.4
IMPORT 9.9 COAL 22.4
12.5 STORAGE
IMPORT NATURAL GAS 160.6
STORAGE 2.2
IMPORT OIL AND NGL 1121.9
STORAGE .3
ELECTRICITY GENERATION 325.4
CONVERSION AND LINE LOSS 224.2
LOST ENERGY 681.7
USED ENERGY 670.2
RESIDENTIAL 315.8
COMMERCIAL 340.3
INDUSTRIAL 103.5
TRANSPORATION 368.0
LOS ALAMOS SCIENTIFIC LABORATORY

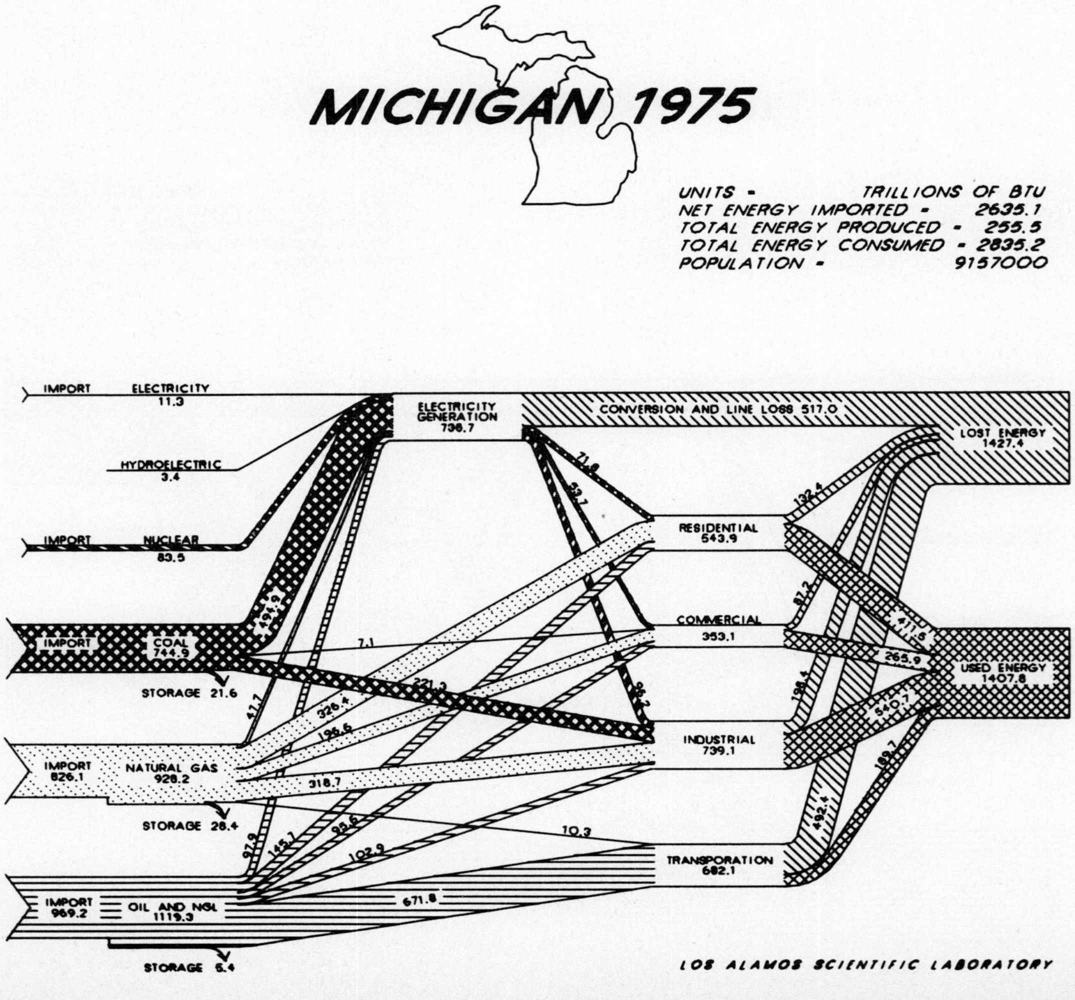
MICHIGAN 1975
UNITS - TRILLIONS OF BTU
NET ENERGY IMPORTED - 2635.1
TOTAL ENERGY PRODUCED - 255.5
TOTAL ENERGY CONSUMED - 2835.2
POPULATION - 9157000
IMPORT ELECTRICITY 11.3
HYDROELECTRIC 3.4
IMPORT NUCLEAR 83.5
IMPORT COAL 744.9
STORAGE 21.6
IMPORT 826.1 NATURAL GAS 928.2
STORAGE 28.4
IMPORT 969.2 OIL AND NGL 1119.3
STORAGE 6.4
ELECTRICITY GENERATION 736.7
CONVERSION AND LINE LOSS 517.0
LOST ENERGY 1427.4
USED ENERGY 1407.8
RESIDENTIAL 543.9
COMMERCIAL 353.1
INDUSTRIAL 739.1
TRANSPORATION 682.1
LOS ALAMOS SCIENTIFIC LABORATORY

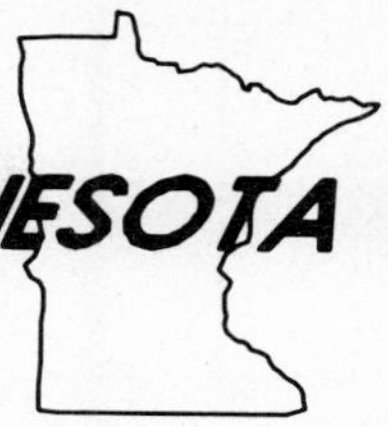

MINNESOTA 1975

UNITS - TRILLIONS OF BTU
NET ENERGY IMPORTED - 1213.4
TOTAL ENERGY PRODUCED - 2.5
TOTAL ENERGY CONSUMED - 1193.7
POPULATION - 3926000

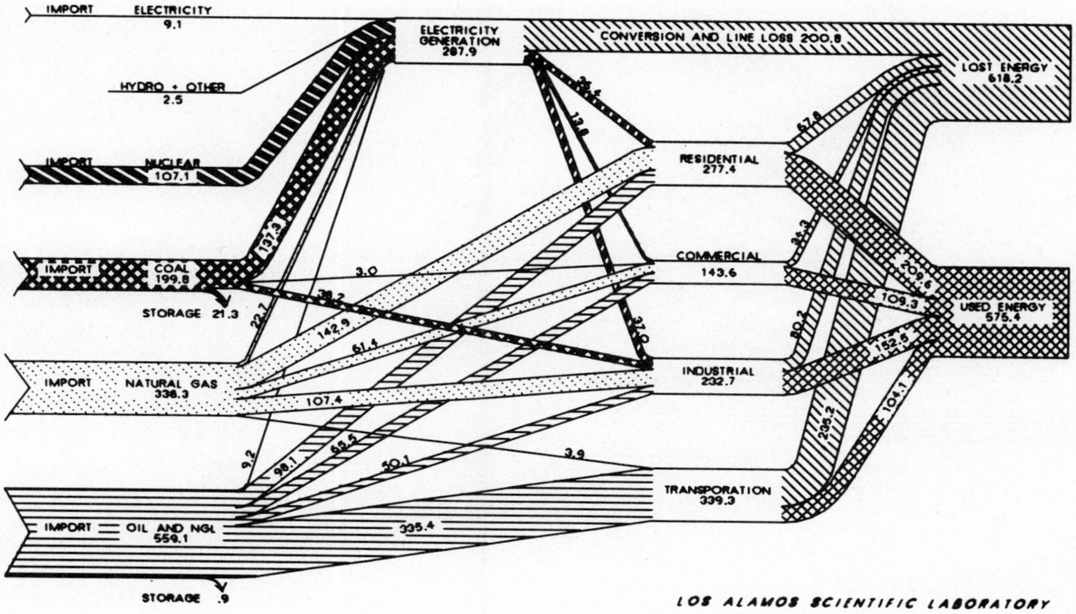

LOS ALAMOS SCIENTIFIC LABORATORY

MISSISSIPPI 1975

UNITS - TRILLIONS OF BTU
NET ENERGY IMPORTED - 352.0
TOTAL ENERGY PRODUCED - 359.6
TOTAL ENERGY CONSUMED - 696.1
POPULATION - 2346000

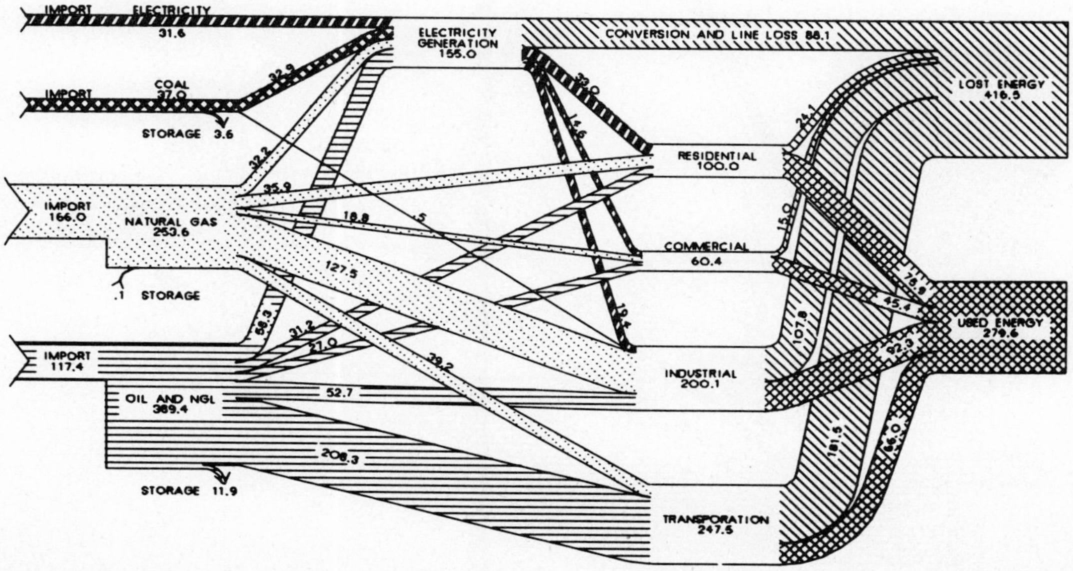

LOS ALAMOS SCIENTIFIC LABORATORY

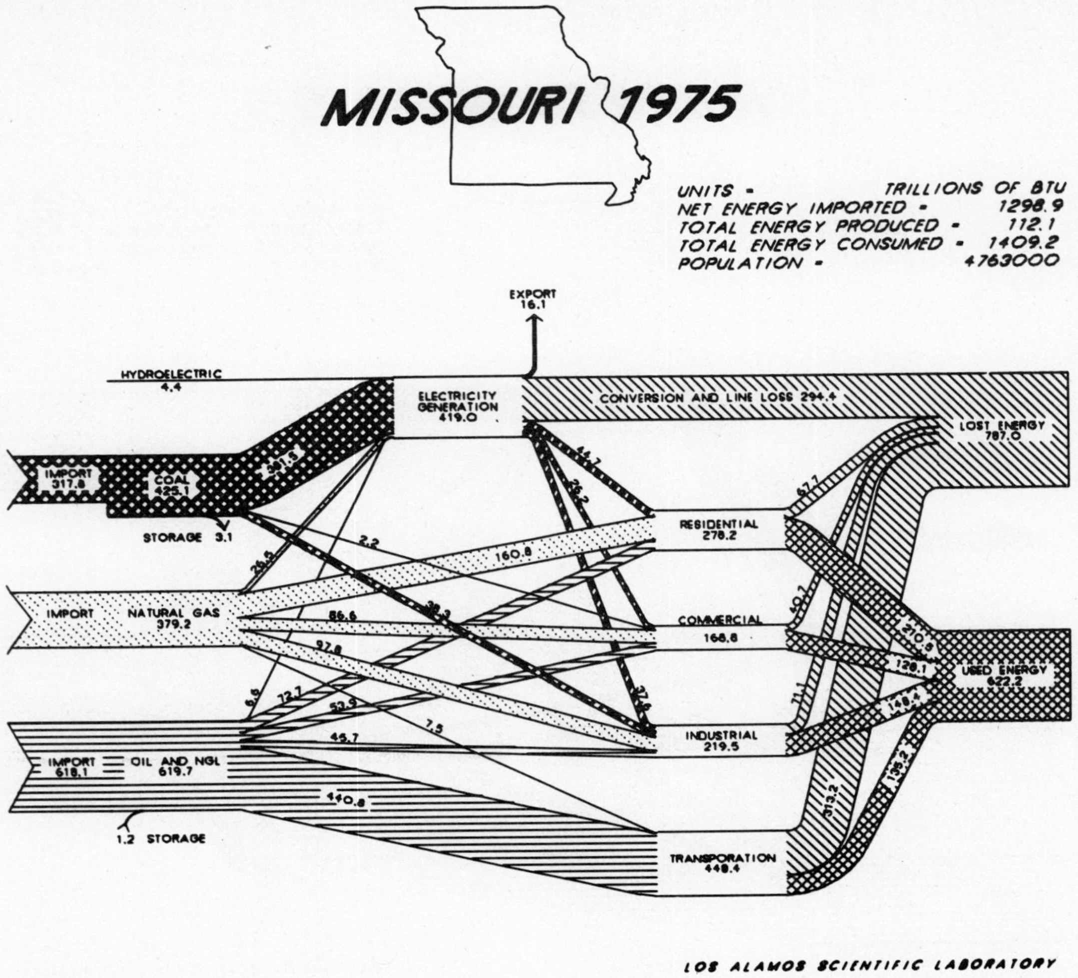
MISSOURI 1975
UNITS - TRILLIONS OF BTU
NET ENERGY IMPORTED - 1298.9
TOTAL ENERGY PRODUCED - 112.1
TOTAL ENERGY CONSUMED - 1409.2
POPULATION - 4763000
EXPORT 16.1
HYDROELECTRIC 4.4
ELECTRICITY GENERATION 419.0
CONVERSION AND LINE LOSS 294.4
LOST ENERGY 787.0
IMPORT 317.8
COAL 425.1
STORAGE 3.1
RESIDENTIAL 278.2
IMPORT NATURAL GAS 379.2
COMMERCIAL 168.8
USED ENERGY 622.2
INDUSTRIAL 219.5
IMPORT 618.1
OIL AND NGL 619.7
1.2 STORAGE
TRANSPORATION 448.4
LOS ALAMOS SCIENTIFIC LABORATORY

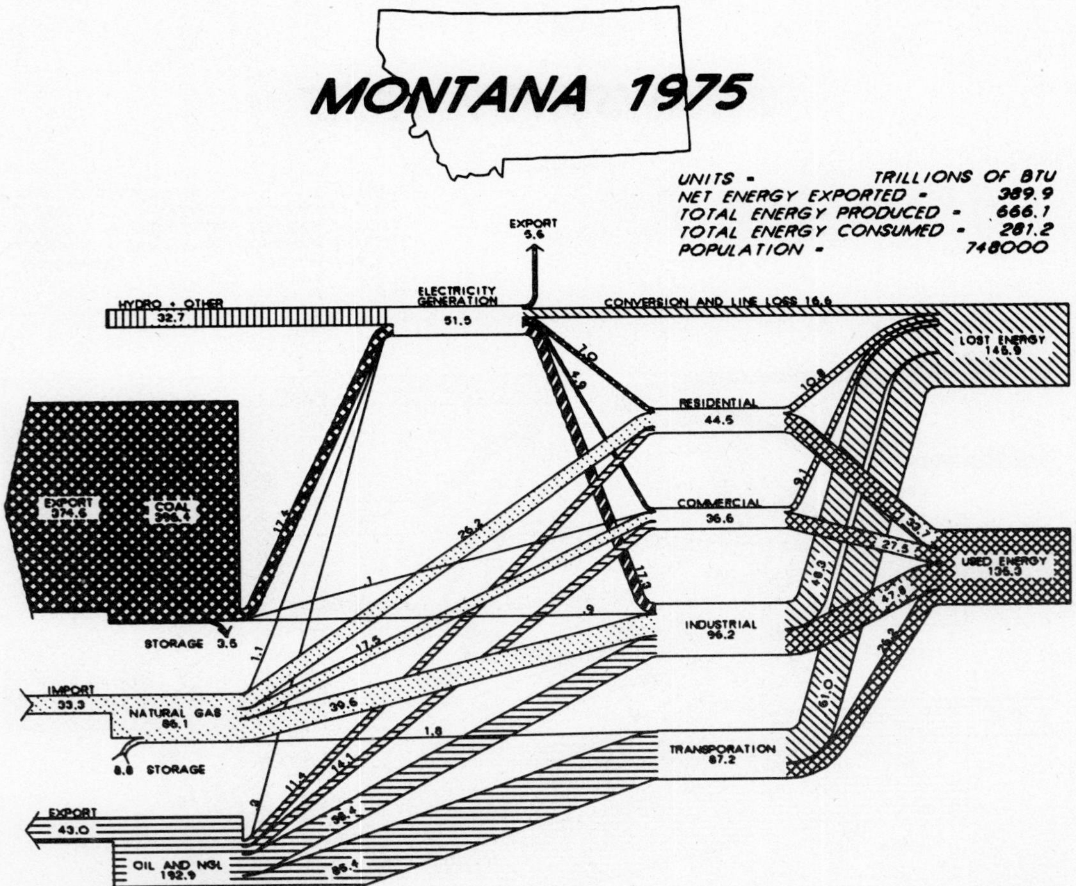
MONTANA 1975
UNITS - TRILLIONS OF BTU
NET ENERGY EXPORTED - 389.9
TOTAL ENERGY PRODUCED - 666.1
TOTAL ENERGY CONSUMED - 281.2
POPULATION - 748000
EXPORT 5.6
HYDRO + OTHER 32.7
ELECTRICITY GENERATION 51.5
CONVERSION AND LINE LOSS 16.6
RESIDENTIAL 44.5
COMMERCIAL 36.6
INDUSTRIAL 96.2
TRANSPORATION 87.2
STORAGE 3.5
IMPORT 33.3
NATURAL GAS 85.1
8.8 STORAGE
EXPORT 43.0
OIL AND NGL 192.9
STORAGE .3
LOS ALAMOS SCIENTIFIC LABORATORY

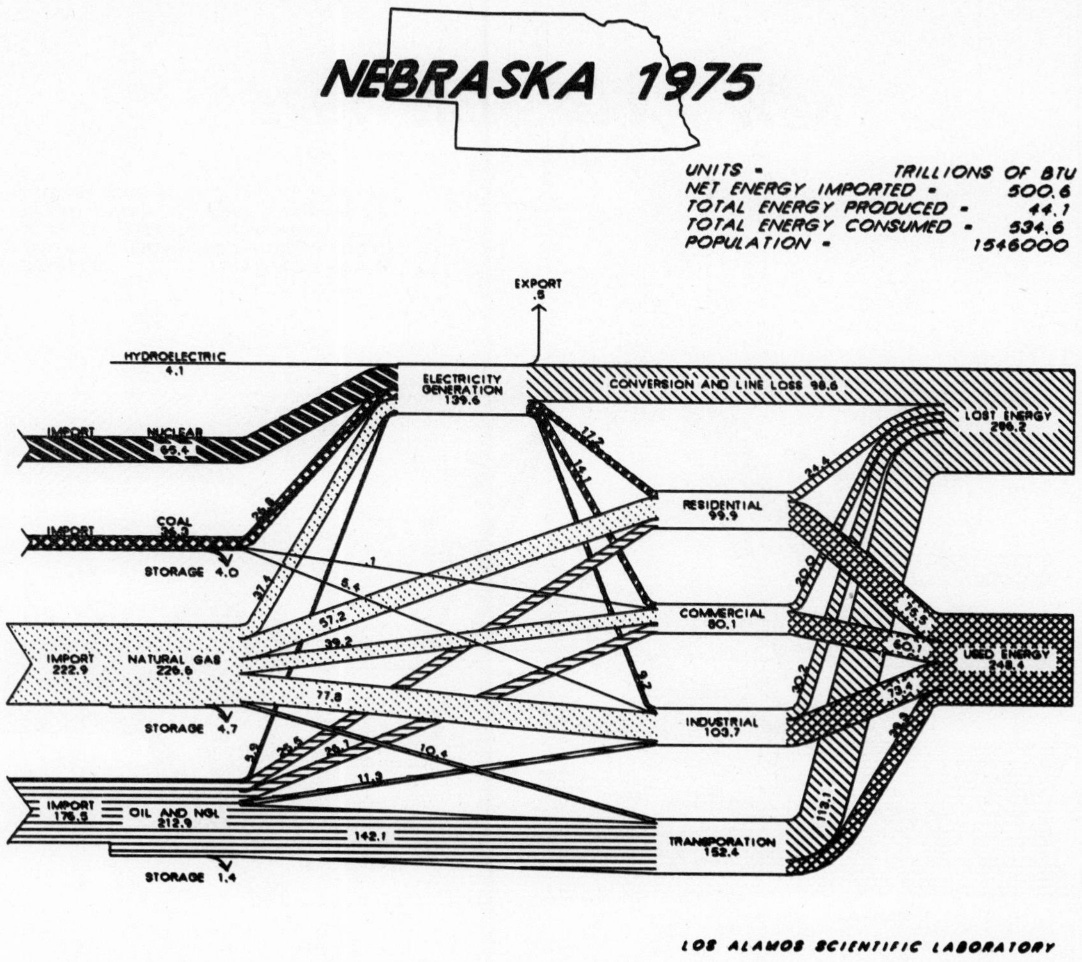
NEBRASKA 1975
UNITS - TRILLIONS OF BTU
NET ENERGY IMPORTED - 500.6
TOTAL ENERGY PRODUCED - 44.1
TOTAL ENERGY CONSUMED - 534.6
POPULATION - 1546000
EXPORT .5
HYDROELECTRIC 4.1
ELECTRICITY GENERATION 139.6
CONVERSION AND LINE LOSS 98.6
LOST ENERGY
IMPORT NUCLEAR
IMPORT COAL
STORAGE 4.0
RESIDENTIAL 99.9
COMMERCIAL 80.1
IMPORT 222.9 NATURAL GAS 226.6
USED ENERGY
STORAGE 4.7
INDUSTRIAL 103.7
IMPORT 176.5 OIL AND NGL 212.9
142.1
TRANSPORATION 152.4
STORAGE 1.4
LOS ALAMOS SCIENTIFIC LABORATORY

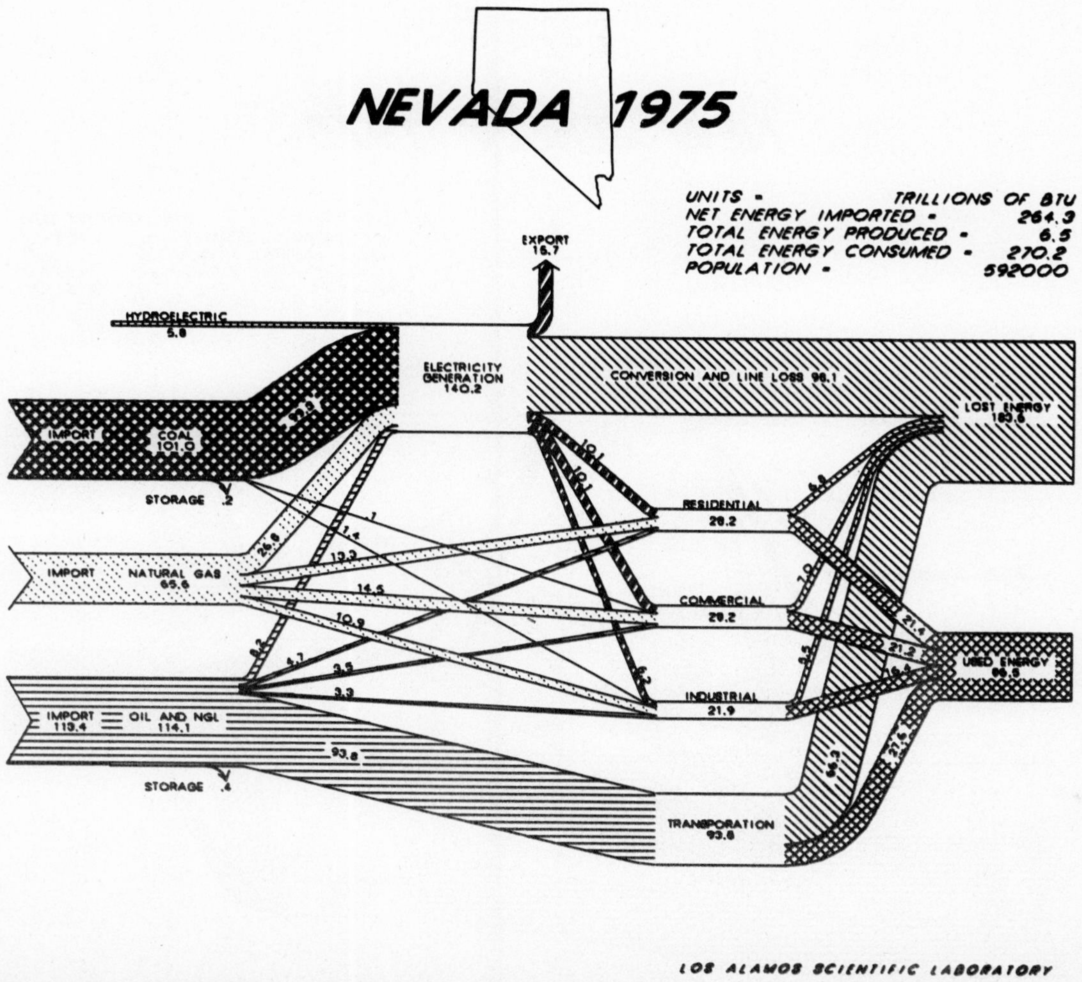
NEVADA 1975
UNITS - TRILLIONS OF BTU
NET ENERGY IMPORTED - 264.3
TOTAL ENERGY PRODUCED - 6.5
TOTAL ENERGY CONSUMED - 270.2
POPULATION - 592000
EXPORT 15.7
HYDROELECTRIC 5.8
ELECTRICITY GENERATION 140.2
CONVERSION AND LINE LOSS 98.1
LOST ENERGY 183.5
IMPORT COAL 101.0
STORAGE .2
RESIDENTIAL 26.2
IMPORT NATURAL GAS 65.6
COMMERCIAL 29.2
USED ENERGY
INDUSTRIAL 21.9
IMPORT 113.4 OIL AND NGL 114.1
93.8
STORAGE .4
TRANSPORATION 93.8
LOS ALAMOS SCIENTIFIC LABORATORY

NEW HAMPSHIRE 1975

UNITS - TRILLIONS OF BTU
NET ENERGY IMPORTED - 181.0
TOTAL ENERGY PRODUCED - 3.7
TOTAL ENERGY CONSUMED - 179.3
POPULATION - 818000

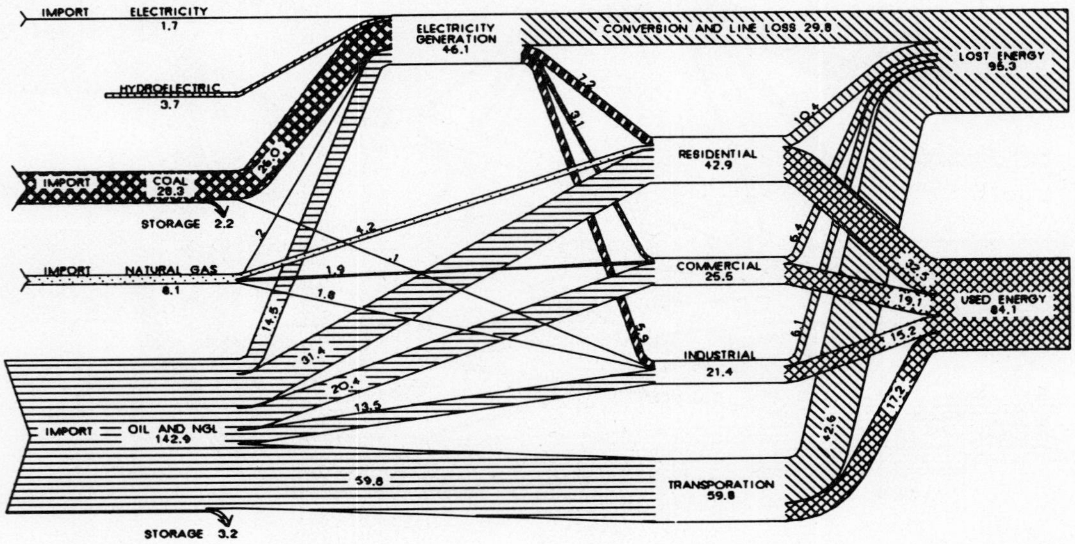

LOS ALAMOS SCIENTIFIC LABORATORY

NEW JERSEY 1975

UNITS - TRILLIONS OF BTU
NET ENERGY IMPORTED - 1690.2
TOTAL ENERGY PRODUCED - .9
TOTAL ENERGY CONSUMED - 1685.8
POPULATION - 7316000

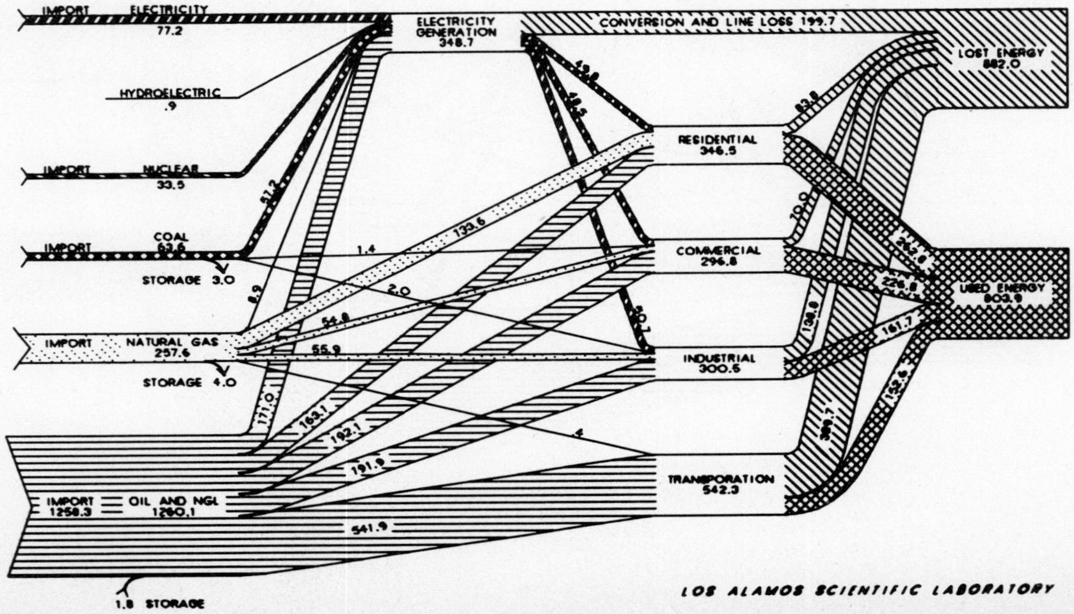

LOS ALAMOS SCIENTIFIC LABORATORY

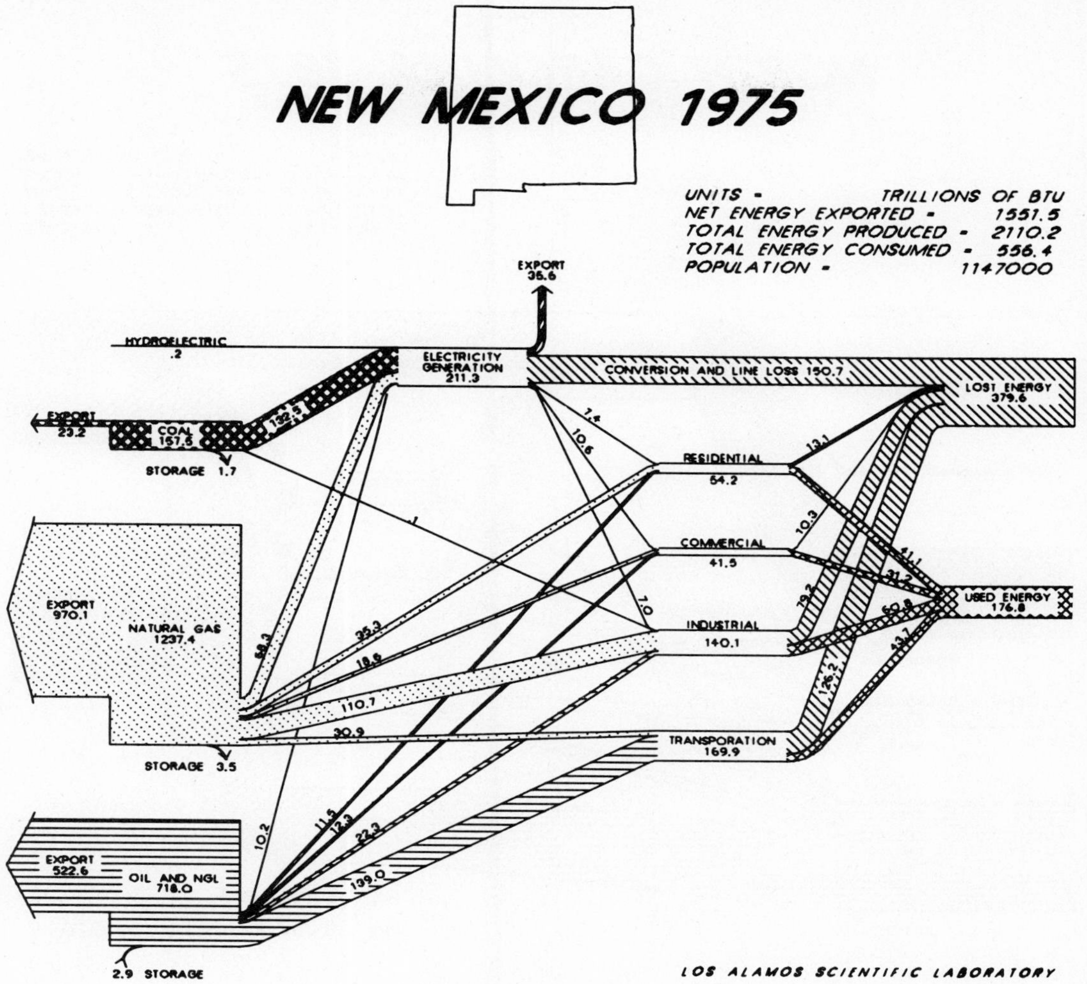
NEW MEXICO 1975
UNITS - TRILLIONS OF BTU
NET ENERGY EXPORTED - 1551.5
TOTAL ENERGY PRODUCED - 2110.2
TOTAL ENERGY CONSUMED - 556.4
POPULATION - 1147000
EXPORT 36.6
HYDROELECTRIC .2
ELECTRICITY GENERATION 211.3
CONVERSION AND LINE LOSS 150.7
LOST ENERGY 379.6
EXPORT 23.2
COAL 157.6
132.5
STORAGE 1.7
RESIDENTIAL 54.2
COMMERCIAL 41.5
INDUSTRIAL 140.1
TRANSPORATION 169.9
USED ENERGY 176.8
EXPORT 970.1
NATURAL GAS 1237.4
STORAGE 3.5
EXPORT 522.6
OIL AND NGL 718.0
2.9 STORAGE
LOS ALAMOS SCIENTIFIC LABORATORY

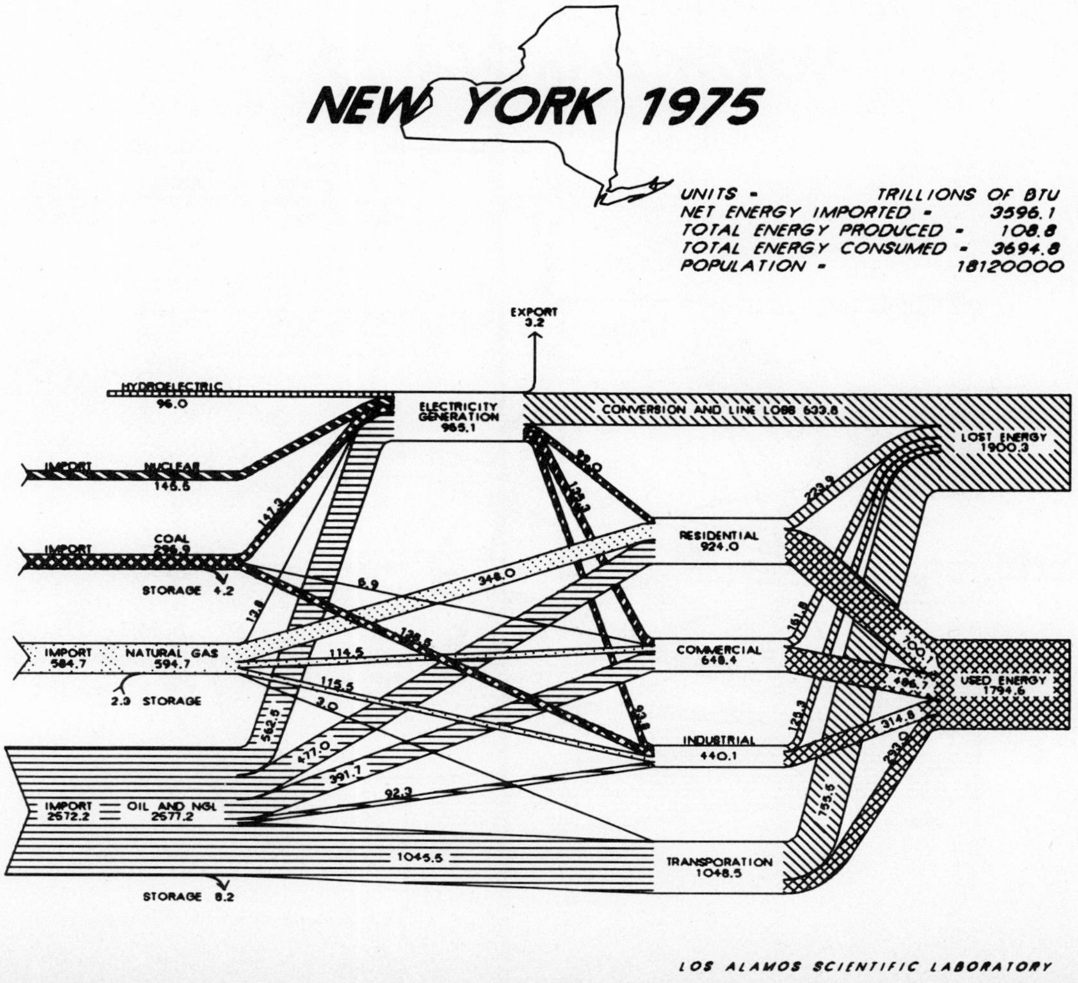
NEW YORK 1975
UNITS - TRILLIONS OF BTU
NET ENERGY IMPORTED - 3596.1
TOTAL ENERGY PRODUCED - 108.8
TOTAL ENERGY CONSUMED - 3694.8
POPULATION - 18120000
EXPORT 3.2
HYDROELECTRIC 96.0
ELECTRICITY GENERATION 985.1
CONVERSION AND LINE LOSS 633.8
LOST ENERGY 1900.3
IMPORT NUCLEAR 146.5
IMPORT COAL 296.9
STORAGE 4.2
RESIDENTIAL 924.0
COMMERCIAL 648.4
INDUSTRIAL 440.1
TRANSPORATION 1048.5
USED ENERGY 1794.6
IMPORT 584.7 NATURAL GAS 594.7
2.3 STORAGE
IMPORT 2672.2 OIL AND NGL 2577.2
1045.5
STORAGE 8.2
LOS ALAMOS SCIENTIFIC LABORATORY

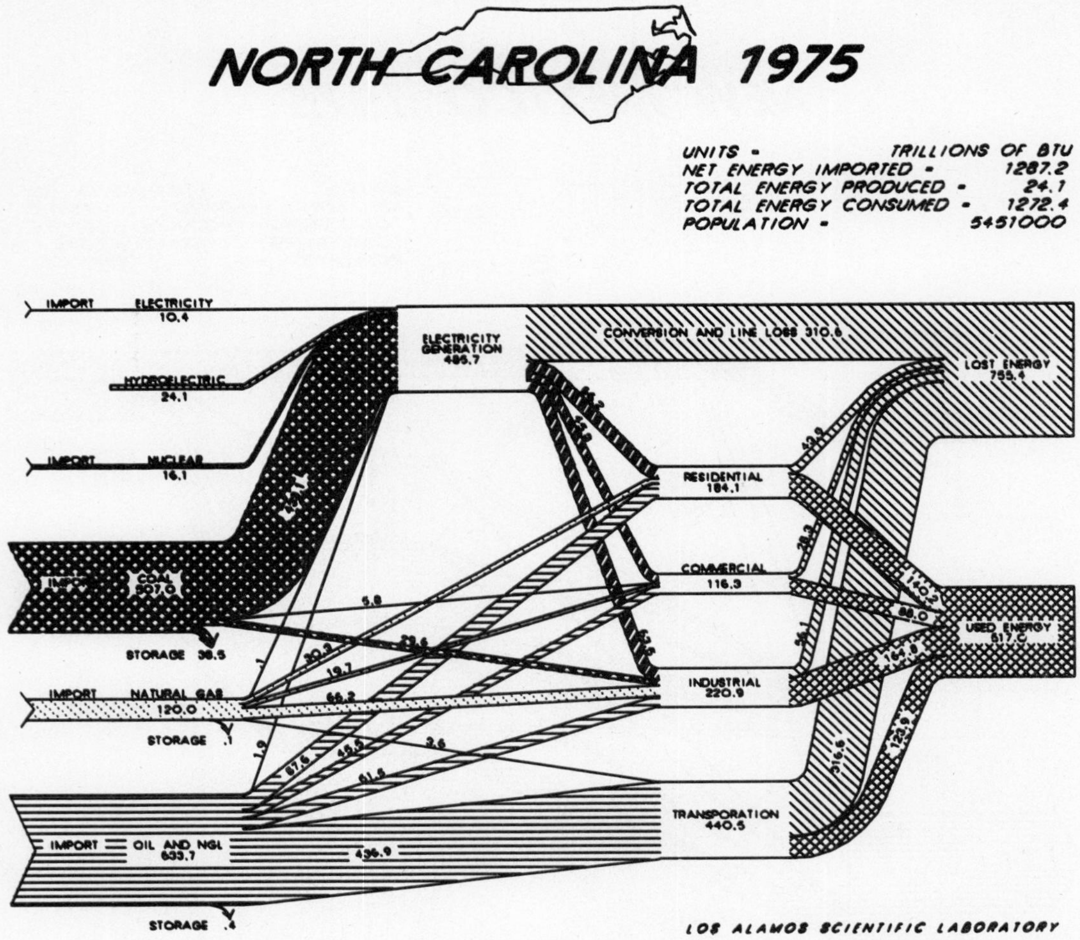
NORTH CAROLINA 1975
UNITS - TRILLIONS OF BTU
NET ENERGY IMPORTED - 1287.2
TOTAL ENERGY PRODUCED - 24.1
TOTAL ENERGY CONSUMED - 1272.4
POPULATION - 5451000
IMPORT ELECTRICITY 10.4
HYDROELECTRIC 24.1
IMPORT NUCLEAR 16.1
IMPORT COAL 507.0
STORAGE 36.5
IMPORT NATURAL GAS 120.0
STORAGE .1
IMPORT OIL AND NGL 633.7
STORAGE .4
ELECTRICITY GENERATION 485.7
CONVERSION AND LINE LOSS 310.6
LOST ENERGY 755.4
RESIDENTIAL 184.1
COMMERCIAL 116.3
INDUSTRIAL 220.9
TRANSPORATION 440.5
USED ENERGY 517.0
436.9
LOS ALAMOS SCIENTIFIC LABORATORY

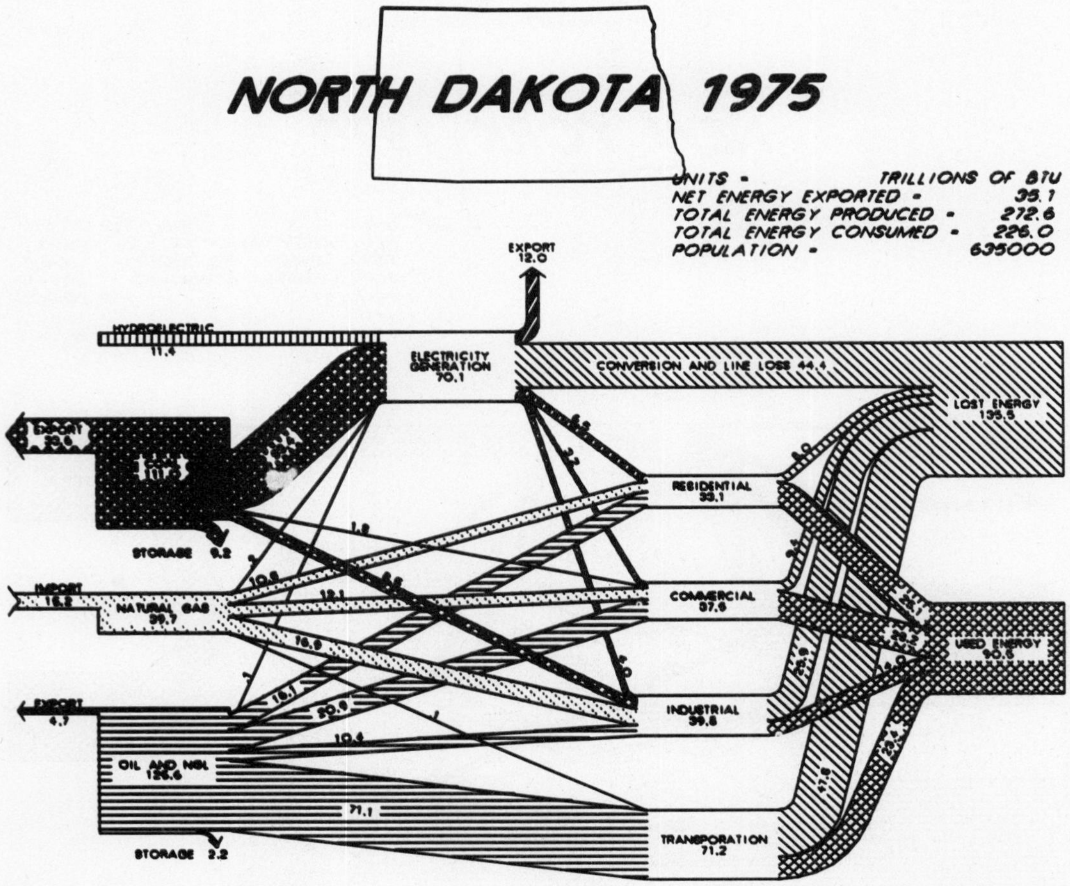
NORTH DAKOTA 1975
UNITS - TRILLIONS OF BTU
NET ENERGY EXPORTED - 35.1
TOTAL ENERGY PRODUCED - 272.6
TOTAL ENERGY CONSUMED - 226.0
POPULATION - 635000
EXPORT 12.0
HYDROELECTRIC 11.4
STORAGE 9.2
IMPORT
NATURAL GAS
EXPORT 4.7
OIL AND NGL 126.6
STORAGE 2.2
ELECTRICITY GENERATION 70.1
CONVERSION AND LINE LOSS 44.4
LOST ENERGY 135.5
RESIDENTIAL 33.1
COMMERCIAL 37.6
INDUSTRIAL 39.8
TRANSPORATION 71.2
USED ENERGY 90.5
71.1
LOS ALAMOS SCIENTIFIC LABORATORY

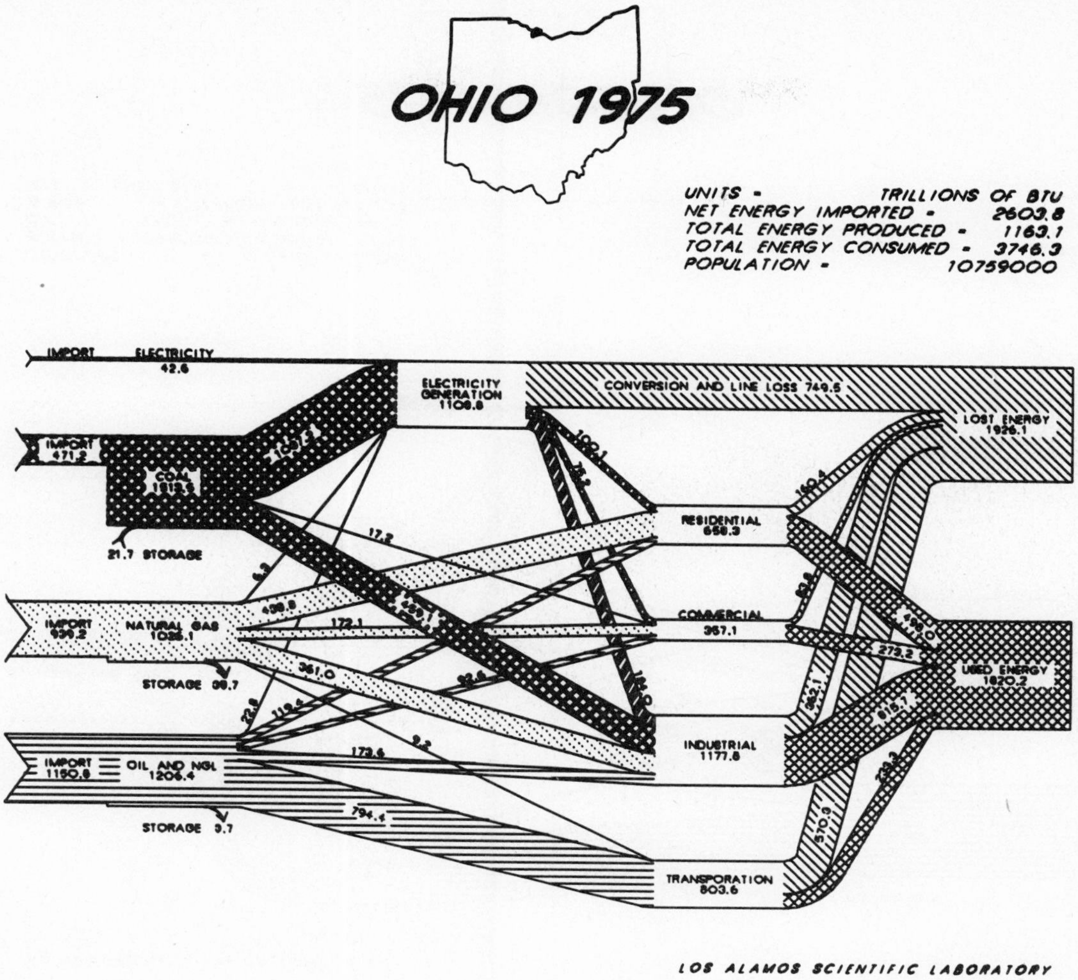
OHIO 1975
UNITS - TRILLIONS OF BTU
NET ENERGY IMPORTED - 2603.8
TOTAL ENERGY PRODUCED - 1163.1
TOTAL ENERGY CONSUMED - 3746.3
POPULATION - 10759000
IMPORT ELECTRICITY 42.6
ELECTRICITY GENERATION 1106.8
CONVERSION AND LINE LOSS 749.5
LOST ENERGY 1926.1
21.7 STORAGE
STORAGE 3.7
RESIDENTIAL 658.3
COMMERCIAL 357.1
INDUSTRIAL 1177.8
TRANSPORATION 803.6
USED ENERGY 1820.2
173.4
794.4
LOS ALAMOS SCIENTIFIC LABORATORY

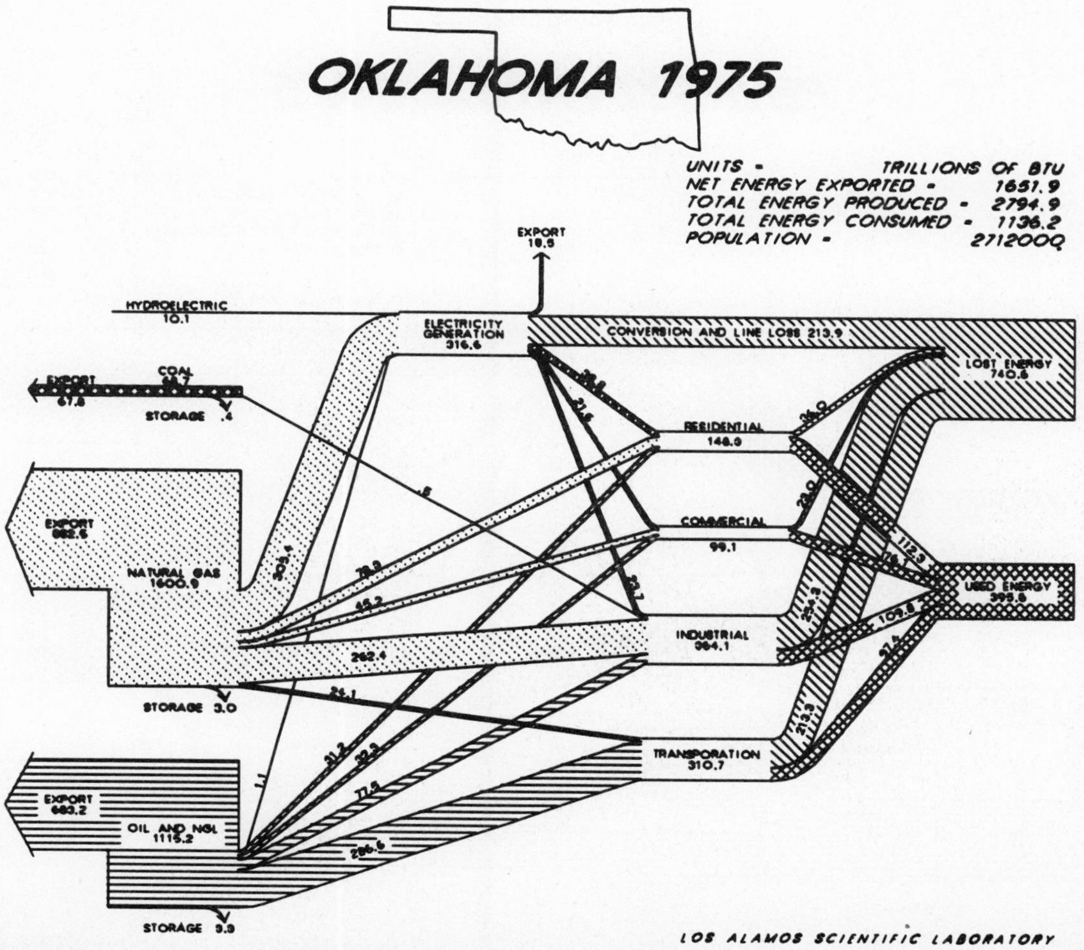
OKLAHOMA 1975
UNITS - TRILLIONS OF BTU
NET ENERGY EXPORTED - 1651.9
TOTAL ENERGY PRODUCED - 2794.9
TOTAL ENERGY CONSUMED - 1136.2
POPULATION - 2712000
EXPORT
HYDROELECTRIC 10.1
ELECTRICITY GENERATION 316.6
CONVERSION AND LINE LOSS 213.9
LOST ENERGY 740.6
COAL
STORAGE .4
RESIDENTIAL 146.9
COMMERCIAL 99.1
INDUSTRIAL 364.1
TRANSPORATION 310.7
USED ENERGY 395.6
NATURAL GAS 1600.9
STORAGE 3.0
EXPORT 683.2
OIL AND NGL 1115.2
STORAGE 3.9
262.4
LOS ALAMOS SCIENTIFIC LABORATORY

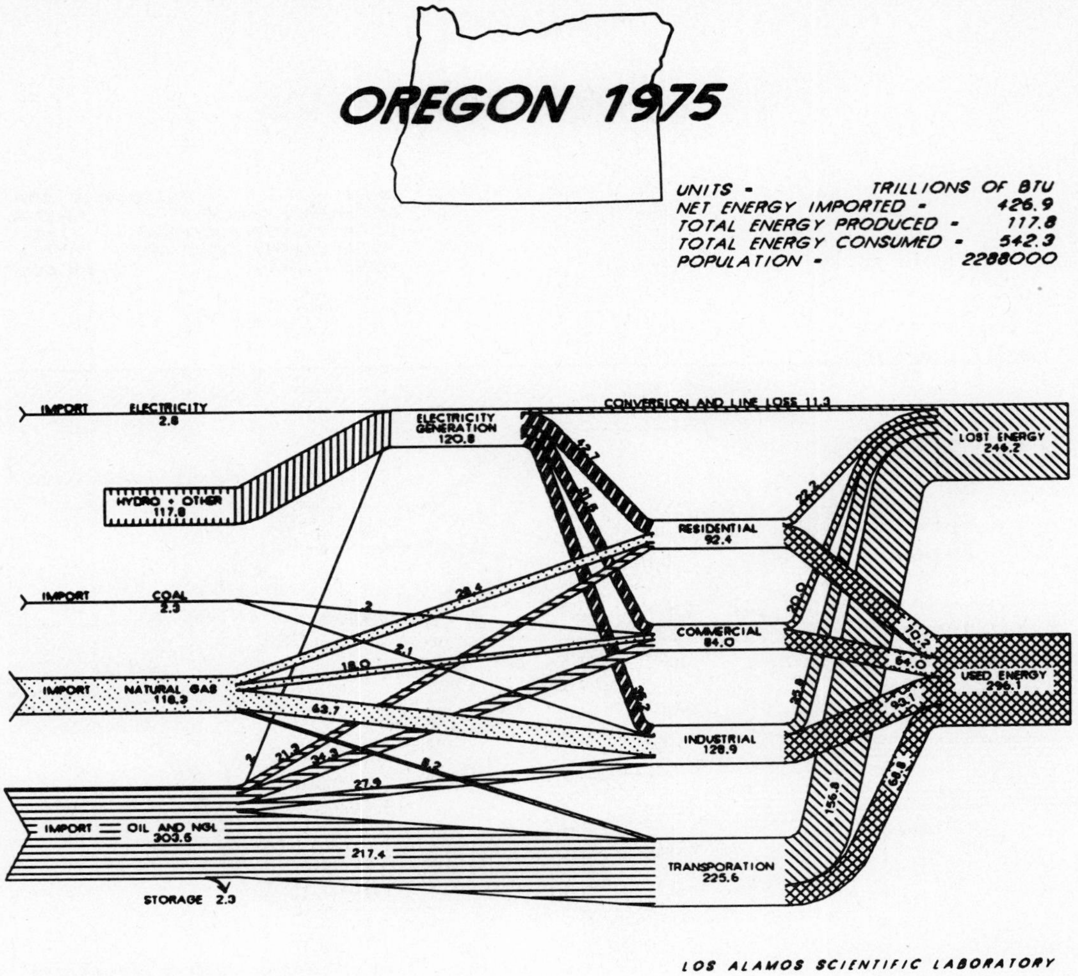
OREGON 1975
UNITS - TRILLIONS OF BTU
NET ENERGY IMPORTED - 426.9
TOTAL ENERGY PRODUCED - 117.8
TOTAL ENERGY CONSUMED - 542.3
POPULATION - 2288000
IMPORT ELECTRICITY 2.8
HYDRO + OTHER 117.8
IMPORT COAL 2.3
IMPORT NATURAL GAS 118.9
IMPORT OIL AND NGL 203.6
STORAGE 2.3
ELECTRICITY GENERATION 120.8
CONVERSION AND LINE LOSS 11.3
RESIDENTIAL 92.4
COMMERCIAL 84.0
INDUSTRIAL 128.9
TRANSPORATION 225.6
LOST ENERGY 246.2
USED ENERGY 296.1
217.4
27.9
63.7
18.0
2.1
29.4
70.2
64.0
LOS ALAMOS SCIENTIFIC LABORATORY

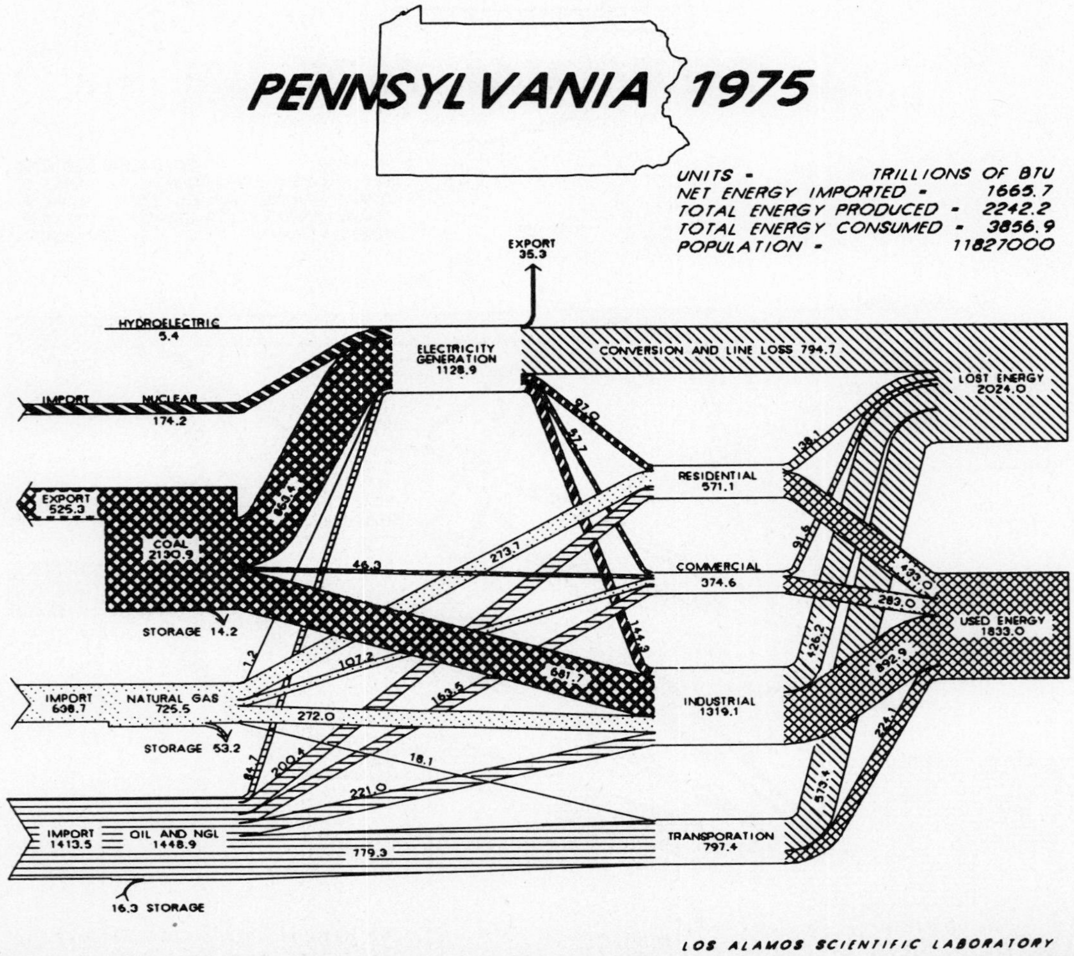
PENNSYLVANIA 1975
UNITS - TRILLIONS OF BTU
NET ENERGY IMPORTED - 1665.7
TOTAL ENERGY PRODUCED - 2242.2
TOTAL ENERGY CONSUMED - 3856.9
POPULATION - 11827000
EXPORT 35.3
HYDROELECTRIC 5.4
IMPORT NUCLEAR 174.2
EXPORT 525.3
COAL 2130.9
STORAGE 14.2
IMPORT 638.7 NATURAL GAS 725.5
STORAGE 53.2
IMPORT 1413.5 OIL AND NGL 1448.9
16.3 STORAGE
ELECTRICITY GENERATION 1128.9
CONVERSION AND LINE LOSS 794.7
RESIDENTIAL 571.1
COMMERCIAL 374.6
INDUSTRIAL 1319.1
TRANSPORATION 797.4
LOST ENERGY 2024.0
USED ENERGY 1833.0
97.0
46.3
273.7
107.2
163.5
681.7
272.0
18.1
221.0
200.4
779.3
138.1
91.6
433.0
283.0
426.2
892.9
573.4
224.1
LOS ALAMOS SCIENTIFIC LABORATORY

RHODE ISLAND 1975

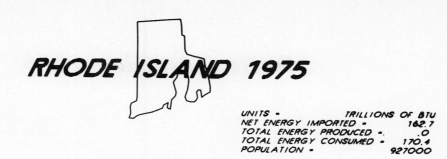

UNITS - TRILLIONS OF BTU
NET ENERGY IMPORTED - 162.7
TOTAL ENERGY PRODUCED - .0
TOTAL ENERGY CONSUMED - 170.4
POPULATION - 927000

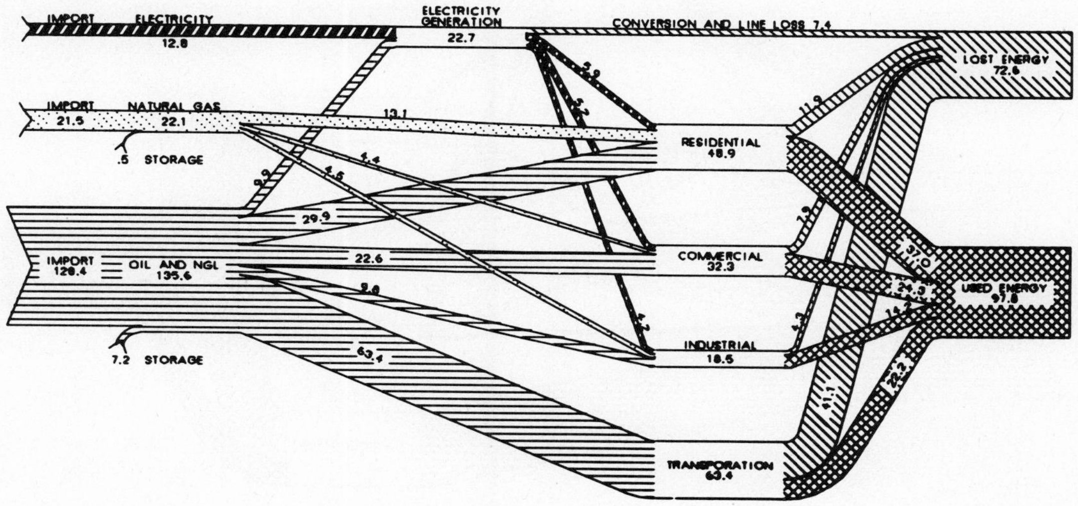

LOS ALAMOS SCIENTIFIC LABORATORY

SOUTH CAROLINA 1975

UNITS - TRILLIONS OF BTU
NET ENERGY IMPORTED - 798.9
TOTAL ENERGY PRODUCED - 14.9
TOTAL ENERGY CONSUMED - 812.9
POPULATION - 2818000

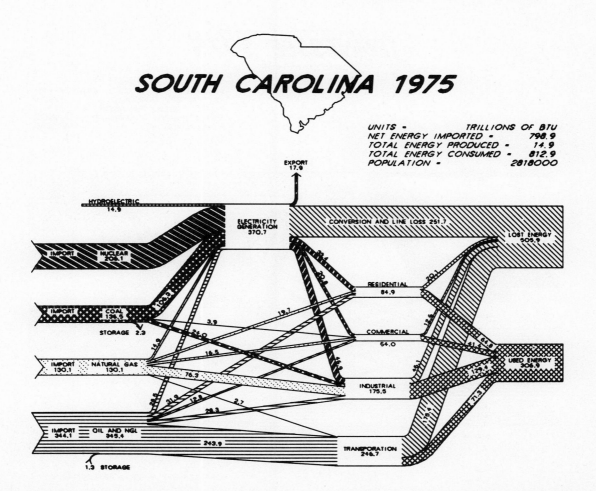

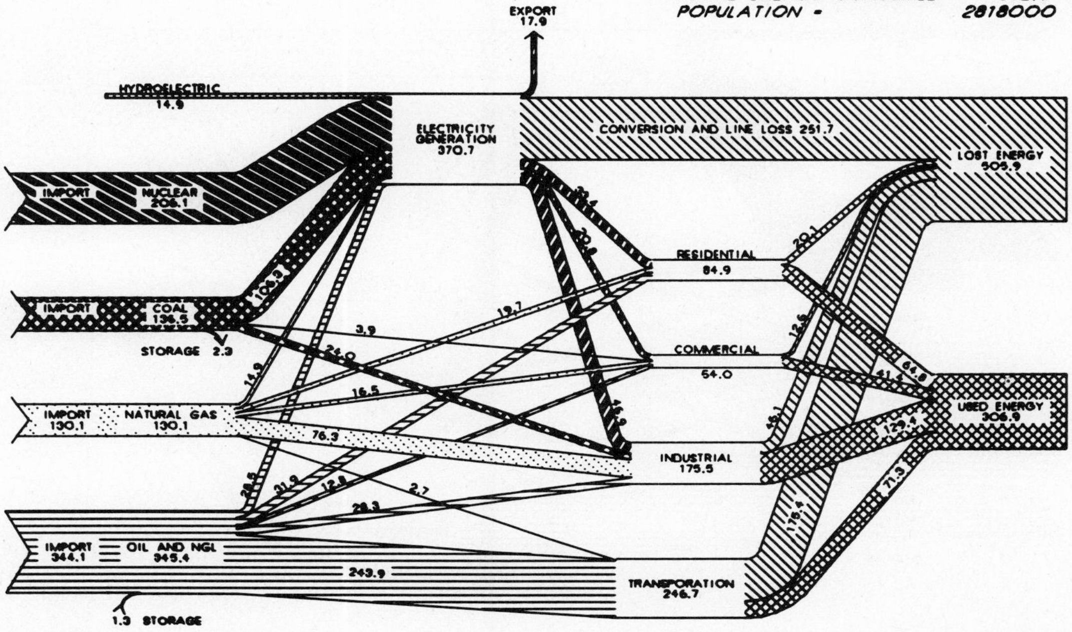

LOS ALAMOS SCIENTIFIC LABORATORY

SOUTH DAKOTA 1975

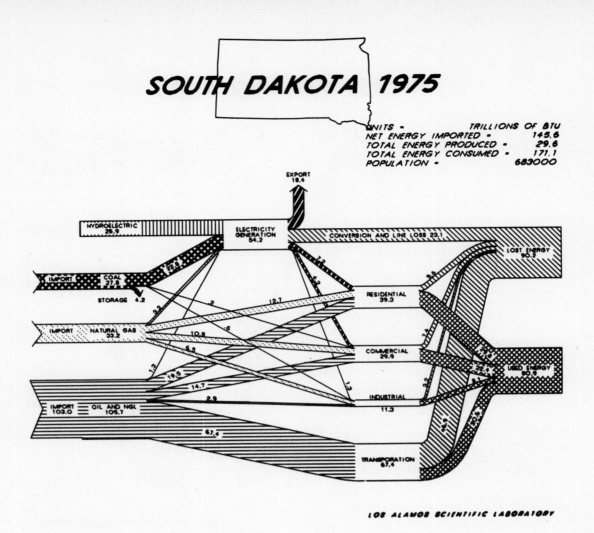

UNITS - TRILLIONS OF BTU
NET ENERGY IMPORTED - 145.6
TOTAL ENERGY PRODUCED - 29.6
TOTAL ENERGY CONSUMED - 171.1
POPULATION - 683000

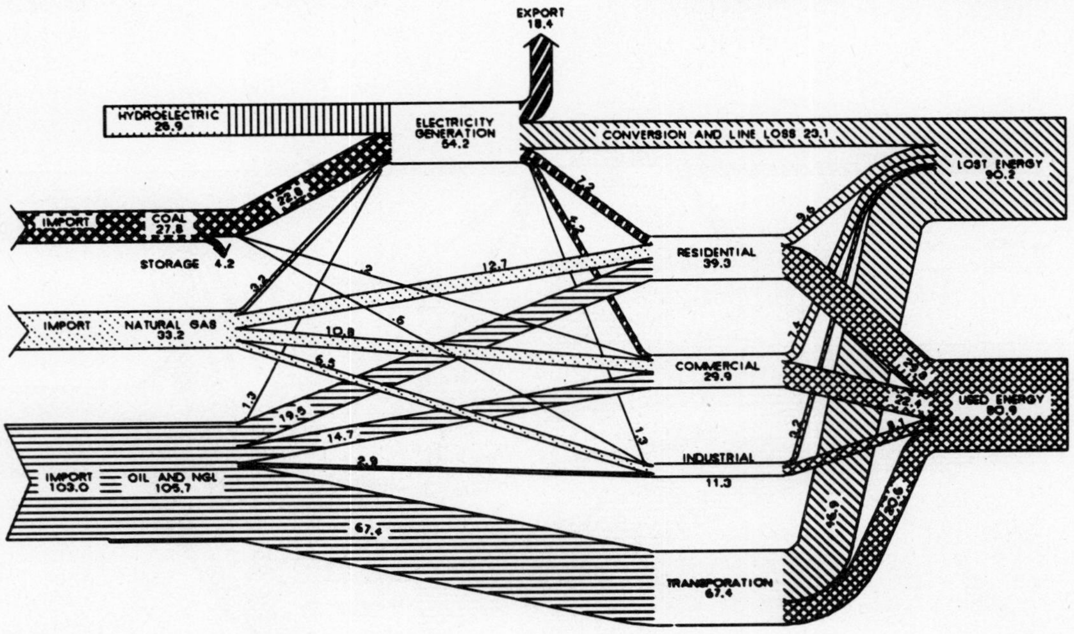

LOS ALAMOS SCIENTIFIC LABORATORY

TENNESSEE 1975

UNITS - TRILLIONS OF BTU
NET ENERGY IMPORTED - 1192.4
TOTAL ENERGY PRODUCED - 234.7
TOTAL ENERGY CONSUMED - 1292.0
POPULATION - 4188000

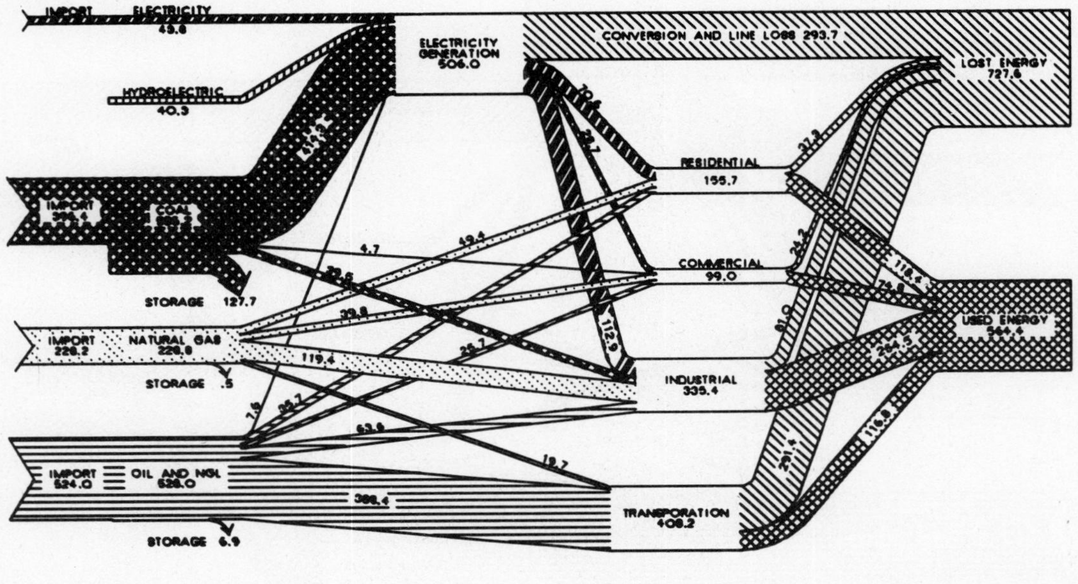

LOS ALAMOS SCIENTIFIC LABORATORY

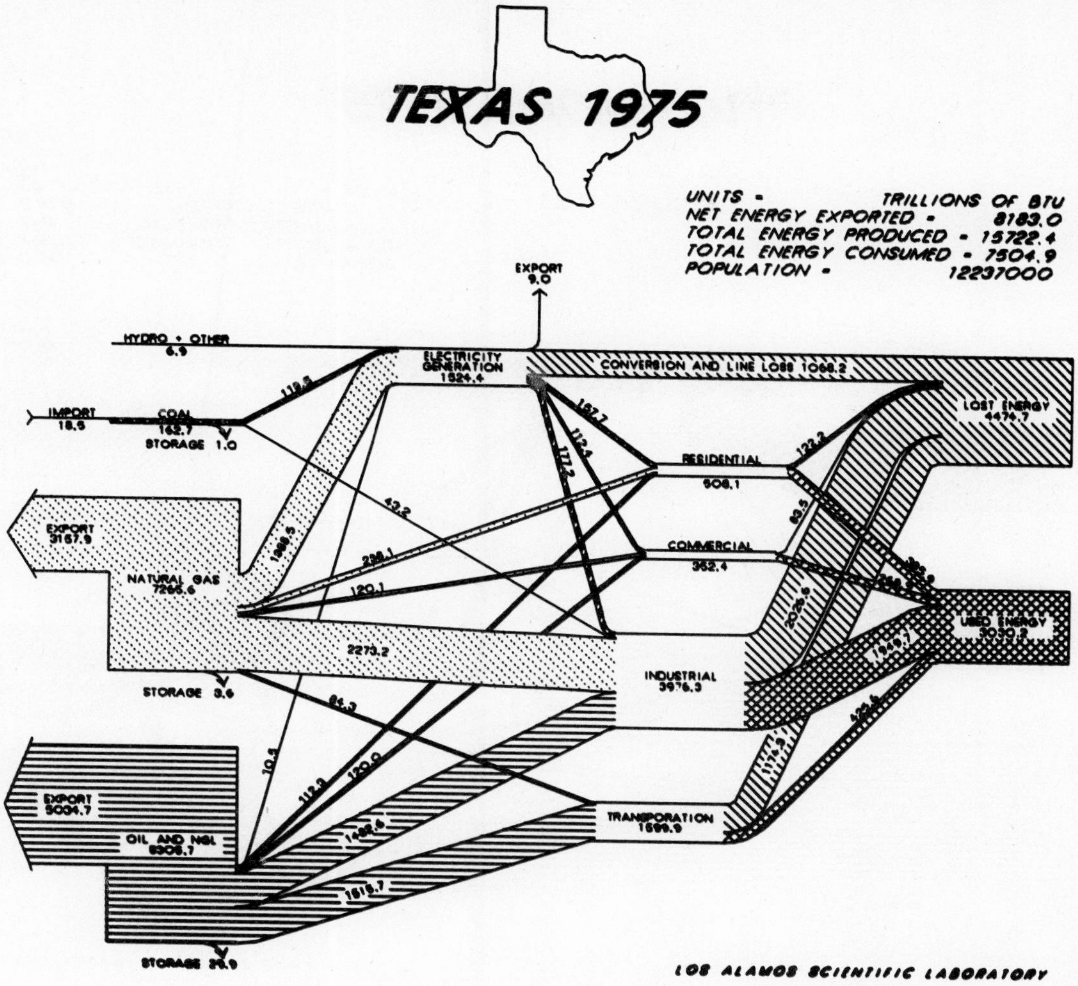
TEXAS 1975
UNITS - TRILLIONS OF BTU
NET ENERGY EXPORTED - 8183.0
TOTAL ENERGY PRODUCED - 15722.4
TOTAL ENERGY CONSUMED - 7504.9
POPULATION - 12237000
EXPORT 9.0
HYDRO + OTHER 6.9
ELECTRICITY GENERATION 1524.4
CONVERSION AND LINE LOSS 1066.2
IMPORT 18.5
COAL 162.7
STORAGE 1.0
LOST ENERGY 4474.7
RESIDENTIAL 908.1
EXPORT 3167.9
NATURAL GAS 7295.6
COMMERCIAL 352.4
USED ENERGY 3030.2
2273.2
STORAGE 3.6
INDUSTRIAL 3976.3
EXPORT 5004.7
OIL AND NGL 8905.7
TRANSPORATION 1599.9
STORAGE 25.9
LOS ALAMOS SCIENTIFIC LABORATORY

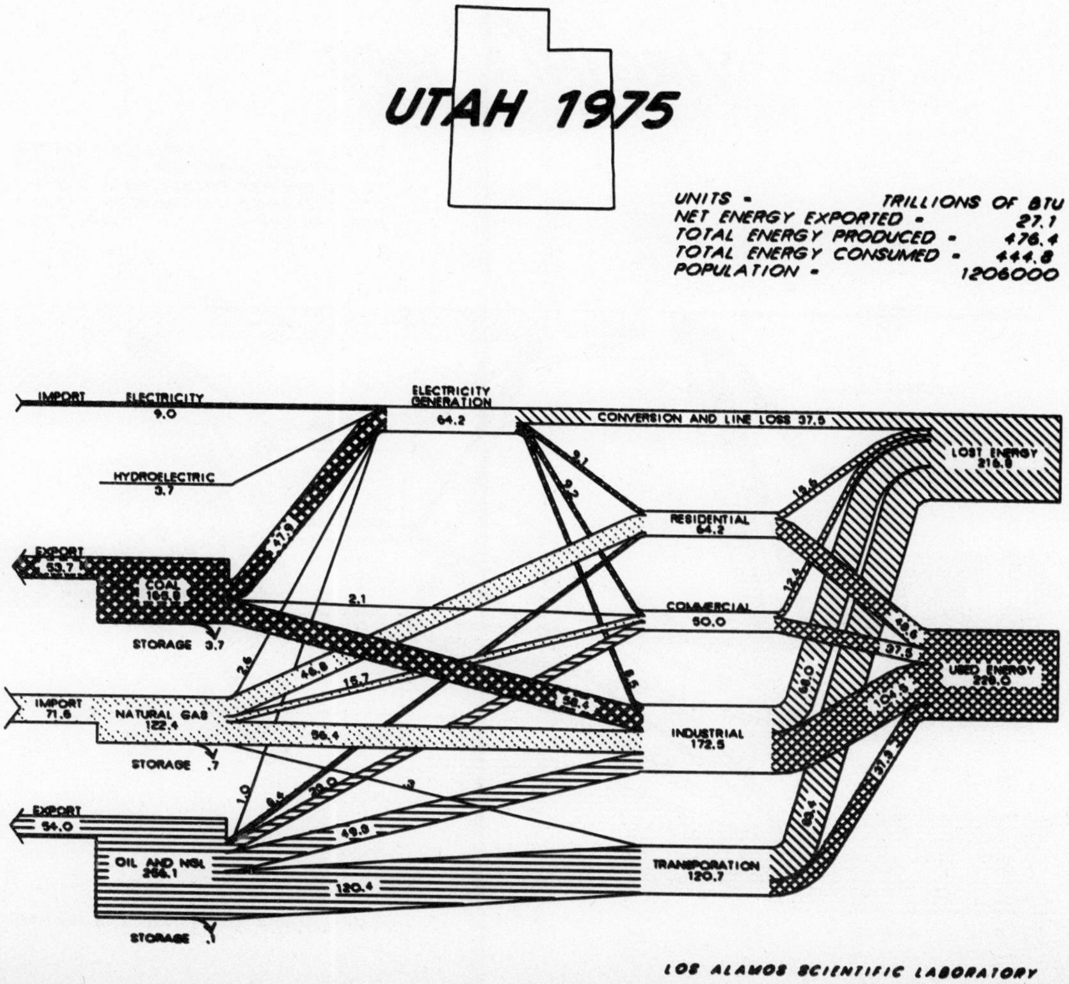
UTAH 1975
UNITS - TRILLIONS OF BTU
NET ENERGY EXPORTED - 27.1
TOTAL ENERGY PRODUCED - 476.4
TOTAL ENERGY CONSUMED - 444.8
POPULATION - 1206000
IMPORT ELECTRICITY 9.0
ELECTRICITY GENERATION 64.2
CONVERSION AND LINE LOSS 37.5
HYDROELECTRIC 3.7
LOST ENERGY 215.8
RESIDENTIAL 64.2
EXPORT 53.7
COAL 169.9
COMMERCIAL 50.0
STORAGE 3.7
USED ENERGY 229.0
IMPORT 71.6
NATURAL GAS 122.4
INDUSTRIAL 172.5
STORAGE .7
EXPORT 54.0
OIL AND NGL 266.1
TRANSPORATION 120.7
120.4
STORAGE .1
LOS ALAMOS SCIENTIFIC LABORATORY

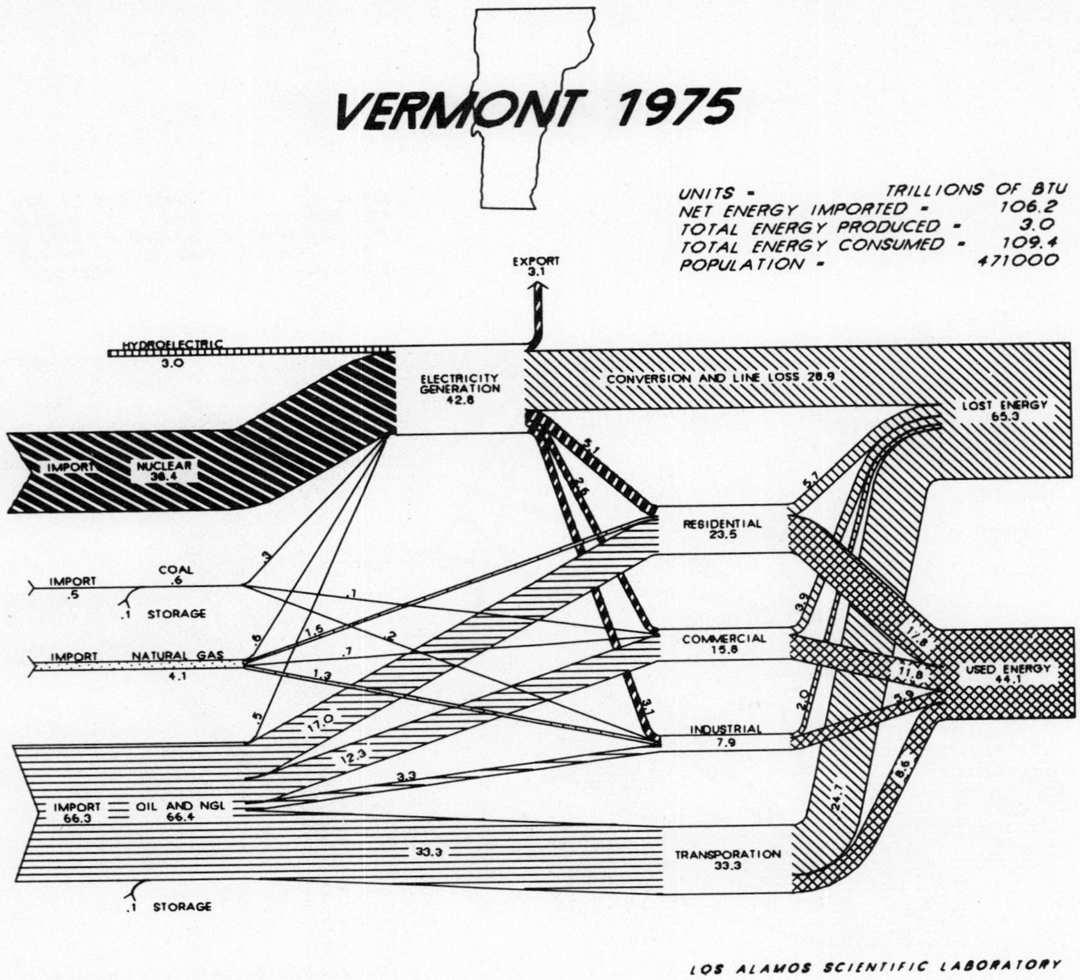

VERMONT 1975
UNITS - TRILLIONS OF BTU
NET ENERGY IMPORTED - 106.2
TOTAL ENERGY PRODUCED - 3.0
TOTAL ENERGY CONSUMED - 109.4
POPULATION - 471000
EXPORT 3.1
HYDROELECTRIC 3.0
ELECTRICITY GENERATION 42.8
CONVERSION AND LINE LOSS 28.9
LOST ENERGY 65.3
IMPORT NUCLEAR 38.4
IMPORT .5 COAL .6
.1 STORAGE
IMPORT NATURAL GAS 4.1
RESIDENTIAL 23.5
COMMERCIAL 15.8
INDUSTRIAL 7.9
USED ENERGY 44.1
IMPORT 66.3 OIL AND NGL 66.4
TRANSPORATION 33.3
33.3
.1 STORAGE
LOS ALAMOS SCIENTIFIC LABORATORY

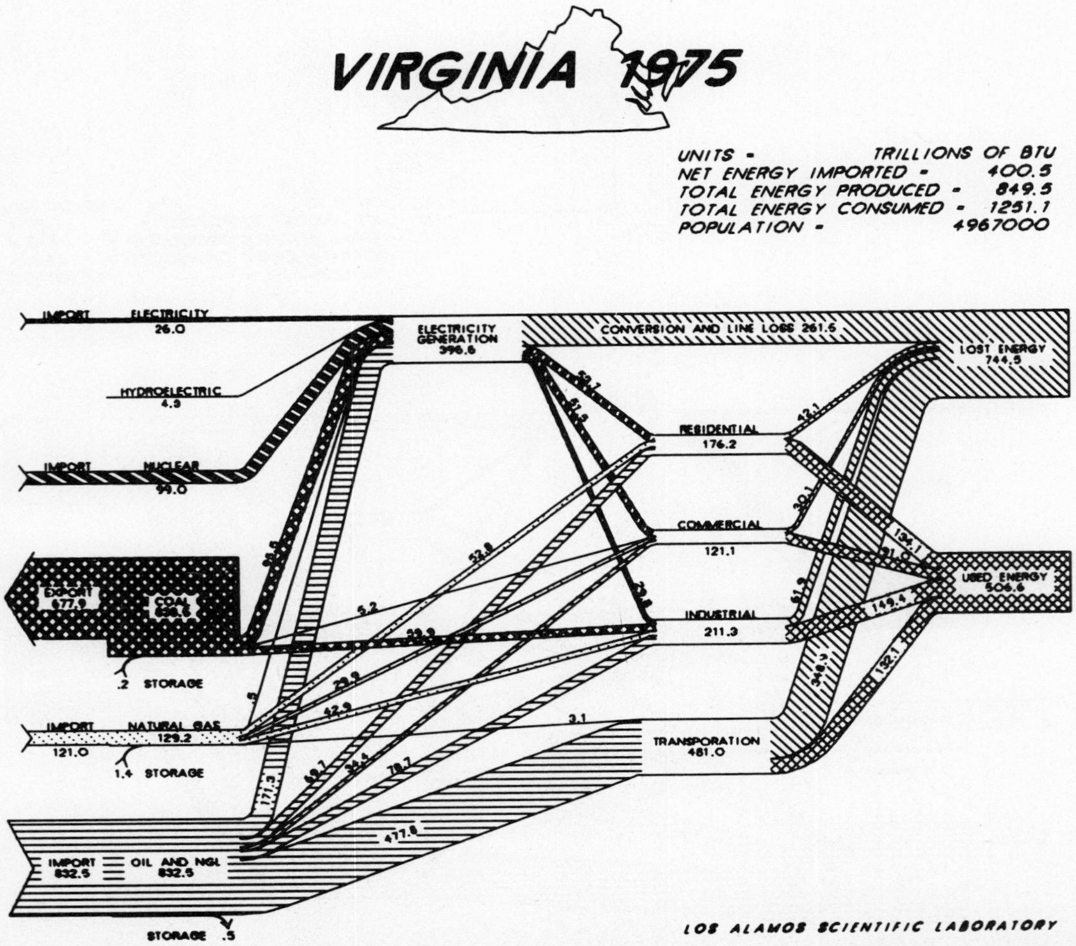

VIRGINIA 1975
UNITS - TRILLIONS OF BTU
NET ENERGY IMPORTED - 400.5
TOTAL ENERGY PRODUCED - 849.5
TOTAL ENERGY CONSUMED - 1251.1
POPULATION - 4967000
IMPORT ELECTRICITY 26.0
ELECTRICITY GENERATION 396.6
CONVERSION AND LINE LOSS 261.6
LOST ENERGY 744.5
HYDROELECTRIC 4.9
IMPORT NUCLEAR 99.0
RESIDENTIAL 176.2
COMMERCIAL 121.1
INDUSTRIAL 211.3
USED ENERGY 506.6
.2 STORAGE
IMPORT NATURAL GAS 129.2
121.0
1.4 STORAGE
TRANSPORATION 481.0
477.8
IMPORT 832.5 OIL AND NGL 832.5
STORAGE .5
LOS ALAMOS SCIENTIFIC LABORATORY

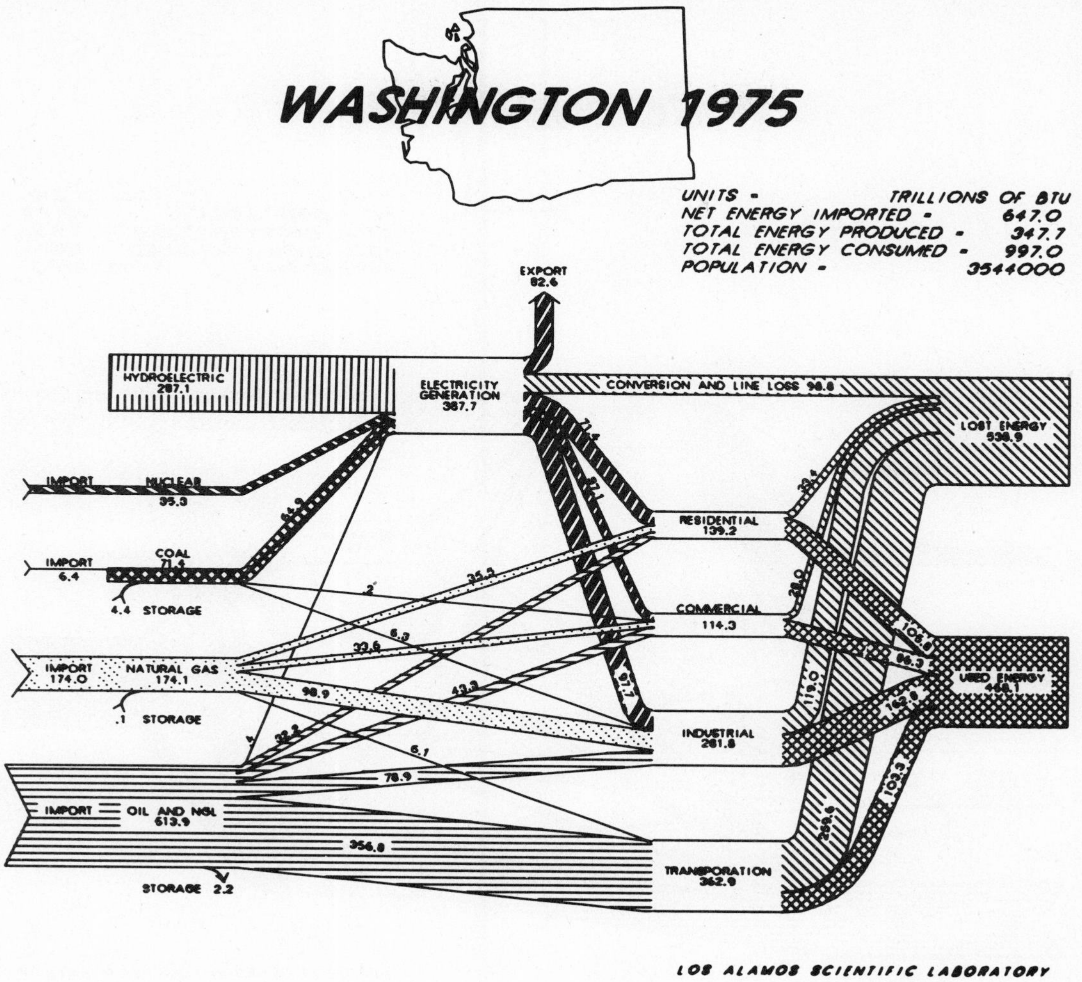
WASHINGTON 1975
UNITS - TRILLIONS OF BTU
NET ENERGY IMPORTED - 647.0
TOTAL ENERGY PRODUCED - 347.7
TOTAL ENERGY CONSUMED - 997.0
POPULATION - 3544000
EXPORT 82.6
HYDROELECTRIC 287.1
ELECTRICITY GENERATION 387.7
CONVERSION AND LINE LOSS
LOST ENERGY
IMPORT
NUCLEAR 35.3
IMPORT 6.4
COAL 71.4
4.4 STORAGE
IMPORT 174.0
NATURAL GAS 174.1
.1 STORAGE
IMPORT
OIL AND NGL 613.9
STORAGE 2.2
RESIDENTIAL 139.2
COMMERCIAL 114.3
INDUSTRIAL 261.8
TRANSPORATION 362.9
USED ENERGY
LOS ALAMOS SCIENTIFIC LABORATORY

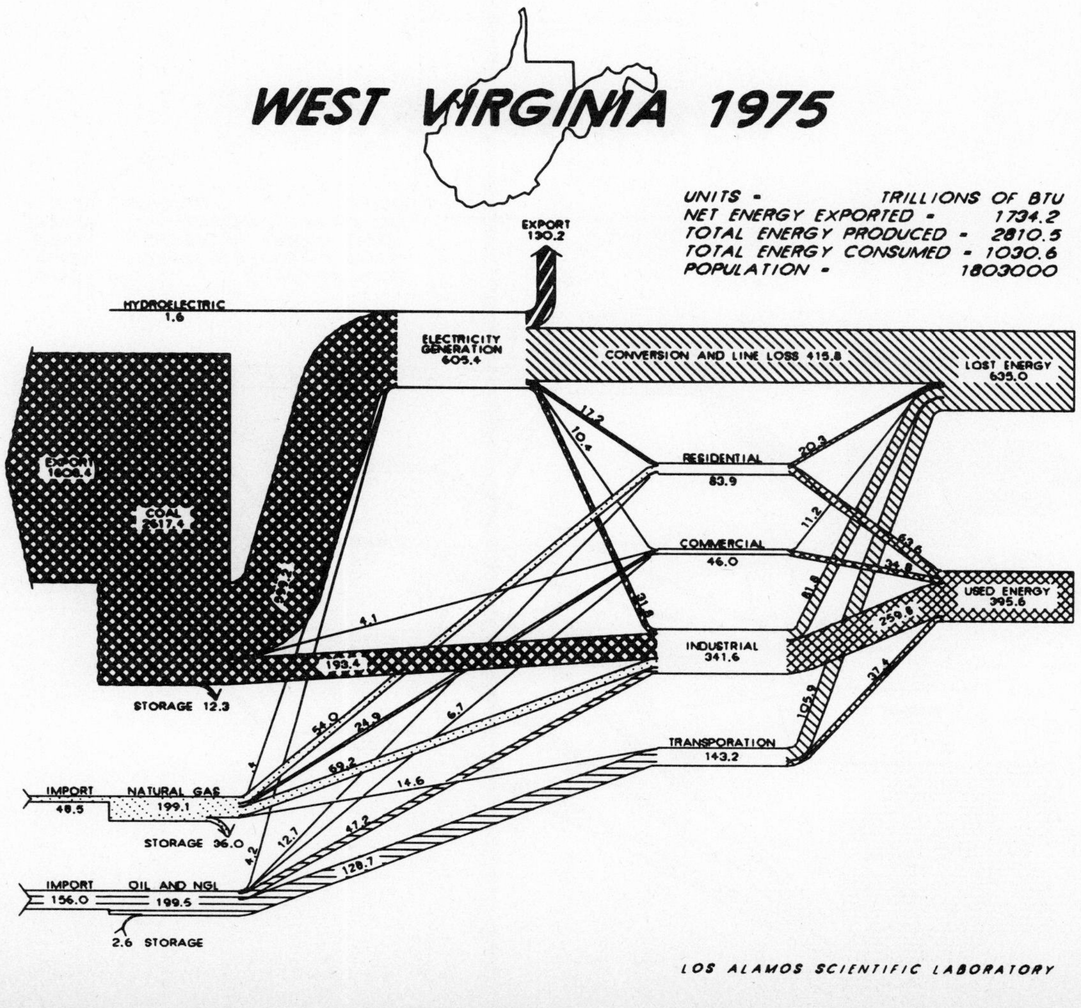
WEST VIRGINIA 1975
UNITS - TRILLIONS OF BTU
NET ENERGY EXPORTED - 1734.2
TOTAL ENERGY PRODUCED - 2810.5
TOTAL ENERGY CONSUMED - 1030.6
POPULATION - 1803000
EXPORT 130.2
HYDROELECTRIC 1.6
ELECTRICITY GENERATION 605.4
CONVERSION AND LINE LOSS 415.8
LOST ENERGY 635.0
EXPORT
COAL
STORAGE 12.3
RESIDENTIAL 83.9
COMMERCIAL 46.0
INDUSTRIAL 341.6
TRANSPORATION 143.2
USED ENERGY 395.6
IMPORT 48.5
NATURAL GAS 199.1
STORAGE 36.0
IMPORT 156.0
OIL AND NGL 199.5
2.6 STORAGE
LOS ALAMOS SCIENTIFIC LABORATORY

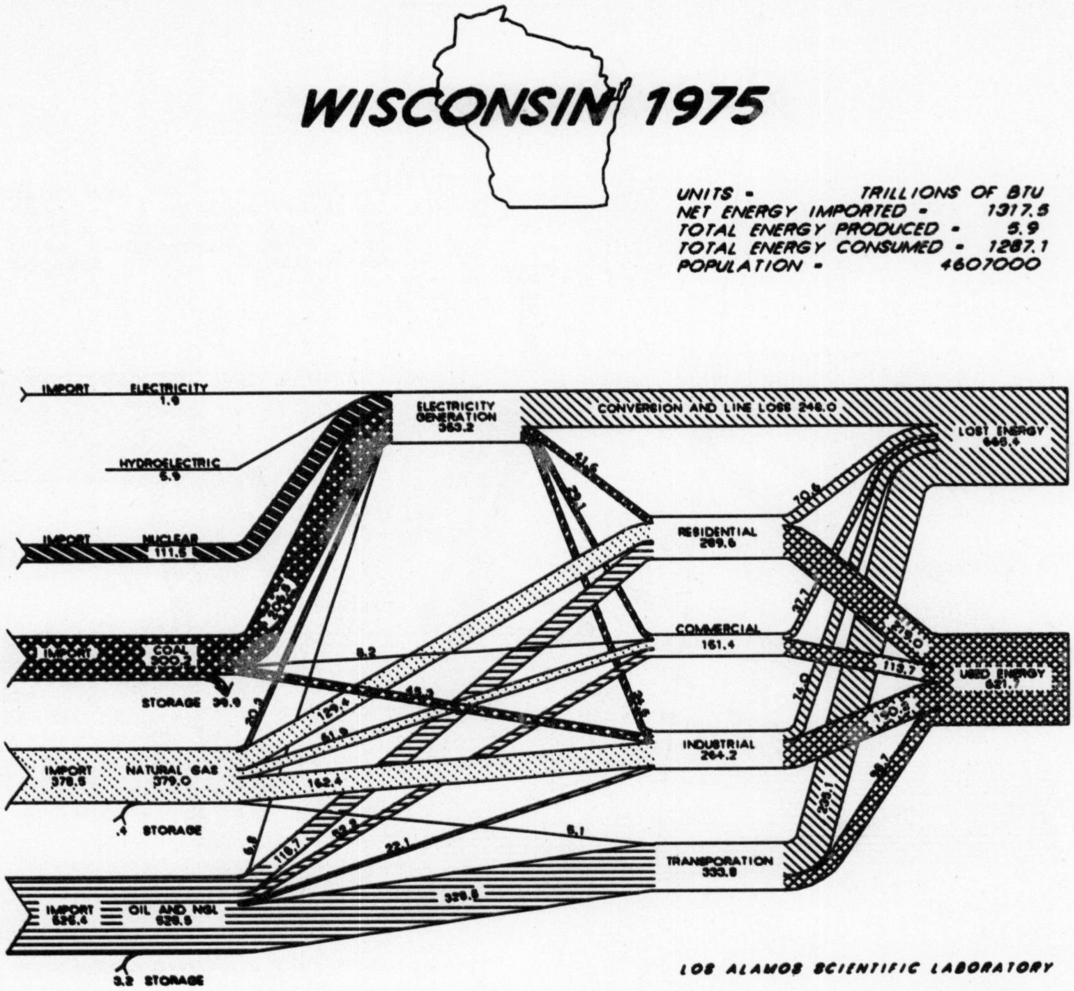
WISCONSIN 1975
UNITS - TRILLIONS OF BTU
NET ENERGY IMPORTED - 1317.5
TOTAL ENERGY PRODUCED - 5.9
TOTAL ENERGY CONSUMED - 1287.1
POPULATION - 4607000
IMPORT ELECTRICITY 1.9
HYDROELECTRIC 5.9
IMPORT NUCLEAR 111.5
IMPORT COAL
STORAGE
IMPORT 378.5 NATURAL GAS 379.0
.4 STORAGE
IMPORT OIL AND NGL
3.2 STORAGE
ELECTRICITY GENERATION 363.2
CONVERSION AND LINE LOSS 246.0
LOST ENERGY
RESIDENTIAL
COMMERCIAL 161.4
INDUSTRIAL 264.2
TRANSPORATION 333.8
USED ENERGY
LOS ALAMOS SCIENTIFIC LABORATORY

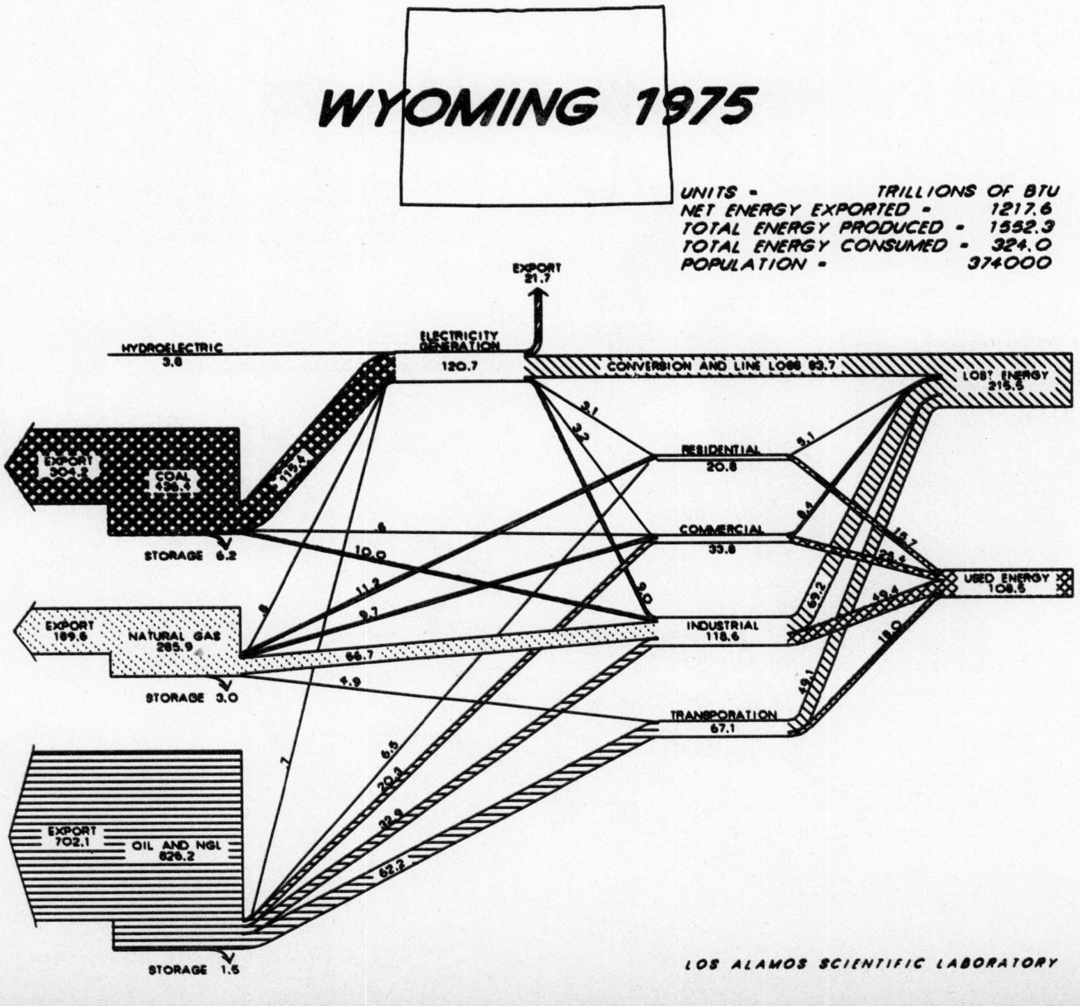
WYOMING 1975
UNITS - TRILLIONS OF BTU
NET ENERGY EXPORTED - 1217.6
TOTAL ENERGY PRODUCED - 1552.3
TOTAL ENERGY CONSUMED - 324.0
POPULATION - 374000
EXPORT 21.7
HYDROELECTRIC 3.8
ELECTRICITY GENERATION 120.7
CONVERSION AND LINE LOSS 83.7
LOST ENERGY 215.5
EXPORT 504.2
COAL
STORAGE 6.2
EXPORT 189.6
NATURAL GAS 265.9
STORAGE 3.0
EXPORT 702.1
OIL AND NGL 826.2
STORAGE 1.5
RESIDENTIAL 20.8
COMMERCIAL 33.8
INDUSTRIAL 118.6
TRANSPORATION 67.1
USED ENERGY 108.5
LOS ALAMOS SCIENTIFIC LABORATORY

O A S I S

A COMPUTER PROGRAM FOR SIMULATION AND OPTIMIZATION OF CENTRAL PLANT PERFORMANCE*

Veronika A. Rabl

Energy and Environmental Systems Division
Argonne National Laboratory, Argonne, Illinois 60439

and

Zulfikar Cumali, Paul K. Davis, and Aydin Ulkucu

Consultants Computation Bureau
594 Howard Street, San Francisco, CA 94105

ABSTRACT

This paper describes a computer program which was developed for the purpose of simulation and optimization of central plants. Since the program was designed to be applicable to a diverse set of system configurations and a large number of runs was anticipated, the primary emphasis was placed on flexibility, modularity, short run times, and ease-of-use. The program simulates the quasi-steady state operation of a user-defined component configuration for a chosen period of time, generally in hourly increments. Plant operation is simulated in accordance with either:

a) user-defined operating rules,

b) weighted resource energy minimization, or

c) operating and maintenance cost minimization.

A library of component subroutines is provided within the program and contains performance and economic data for several major components, e.g., prime movers, generators, boilers, chillers, etc. This data can be readily modified by the user. The library is being continuously updated and expanded as new data and systems emerge.

INTRODUCTION

Energy shortages experienced in recent years led to a recognition that energy conservation is one of the most important means of extending the availability of scarce natural resources. Within the search for new energy conserving technologies certain old concepts were revitalized and expanded. The concept of integrated energy systems belongs to this category. Total energy systems that utilize waste heat from prime movers were built and operated at varying levels of success in the past. In those systems, the extent of integration was generally limited to generation of electricity with simultaneous utilization of waste heat or extracted steam for heating. Most applications fell within the range of 100 kW_e to a few MW_e with customers ranging from housing complexes through large hospitals and commercial establishments.

The concept of integrated energy systems represents a significant evolution both in size and in scope of this approach. The on-site systems, as viewed at present, are also considered for applications where all or part of energy related services are supplied to communities as large as several hundred MW_e of equivalent electric demand. The target communities thus may be as diverse as a municipal or suburban business

*Work was supported by the Community Systems Branch, Division of Buildings and Community Systems, Office of the Assistant Administrator for Conservation, Energy Research and Development Administration.

district, a farm community, or a multi-zoned Planned Unit Development, to name only a few. The services considered reflect the diversity of the customers; they may range from traditional electricity supply and environmental conditioning through transportation, communications, snow removal, trash collection, etc. In general, not all services have to be supplied by the on-site integrated energy system; for example, in some cases it may be neither energy conserving nor economical to generate electricity. The system components may also be very different from those utilized in the traditional total energy systems; a number of new or improved types of equipment are being developed for the near term and future markets, such as fuel cells, improved turbines, new cycles, solar, and storage technologies. A number of technologies, existing, emerging, and advanced, will be subjected to an evaluation of their performance within an integrated energy system.

The task of selecting a system that provides the best match, both in energy and economic sense, to a community requires tools capable of evaluating the interaction with the community demand as well as that among the closely coupled components. Due to the complexity of the problem, the analysis would largely rely on a suitable computer program. Such a program would have to be highly modularized to afford maximum flexibility in configuring the system. It would also have to allow the user to readily modify the library component routines (to account for variations among specific components) as well as an adequate capability for describing the operating strategies and controls applicable to the plant. Determination of an optimal operating strategy would allow the user to evaluate the full potential of the selected system. A computer program that attempts to satisfy these requirements has been developed and is described below.

COMPUTER PROGRAM DESCRIPTION

The computer program is designed to simulate and optimize the operation of a user-configured and sized plant. The simulation is performed on a quasi-steady basis, i.e., steady state operation is assumed for each simulation time increment, which is generally on the order of one hour. Plant operation is simulated according to either:

a. user defined operating rules,

b. weighted resource energy minimization, or

c. operating and maintenance cost minimization.

The program offers high flexibility in configuring the plant. A library of generic component subroutines is provided with the program; the performance curves are based on data averaged over several manufacturers and a range of sizes. The user specifies the size of each piece of equipment installed in the plant and connects equipment inputs and outputs to each other as appropriate for the plant to be simulated. To the extent possible, an attempt is made in the program to prevent the user from making inappropriate connections, such as routing the steam from the boiler into the diesel engine. In some cases, the component operation may depend, in addition to the load imposed on that component, on certain parameters, such as ambient air temperature. The source of such information is also specified by the user - either explicitly or by default. Based on the plant structure and required output information, the user, while configuring the system, selects the types and grades of energy flowing into, within, and out of the plant.

After the plant has been configured, its performance and economics can be analyzed. This analysis requires specification of loads on the plant, instructions for operating the plant, and pertinent economic data. The loads are given at hourly or other user selected intervals. Instructions for plant operation presently consist of an allocation of load among components installed for the purpose of satisfying that particular load, e.g., a bank of electric generators to satisfy the electric load or several chillers to satisfy the cooling load. Different load splits can be specified for different simulation time periods. Operating strategies do not have to be provided if an optimization run is selected. In such cases, the program determines the optimal operating procedures and supplies the rules (for each hour) to the simulation program.

The level of detail of output information from the program can be selected by the user. Available reports include a description of plant configuration, parameters of equipment input/output curves, and simulation results. The reports summarize, at user specified intervals, the amounts of energy supplied to the plant, transferred within the plant, and output by the plant. Component operation is described in terms of total operating time and corresponding part loads. Maxima and minima of quantities are indicated. Economic reports, if requested, provide information on costs of equipment, plant operation and maintenance, and life cycle costs.

The structure of the computer program is shown in Fig. 1. It can be seen that the code, in fact, consists of several subprograms:

a. Input Processor - to process all input data and commands given in the System Definition and Control Language;

b. Plant Simulation - to process the configuration, determine the order of simulation, identify information loops, and execute the simulation;

c. Plant Optimization - to determine optimal operating strategies and storage policy;

d. Economic Analysis - to estimate the costs of operation and construction of the plant; and

e. Reports Program - to print the appropriate output information.

The input into the program consists of two parts:

a. The user provides a data file containing the values of required

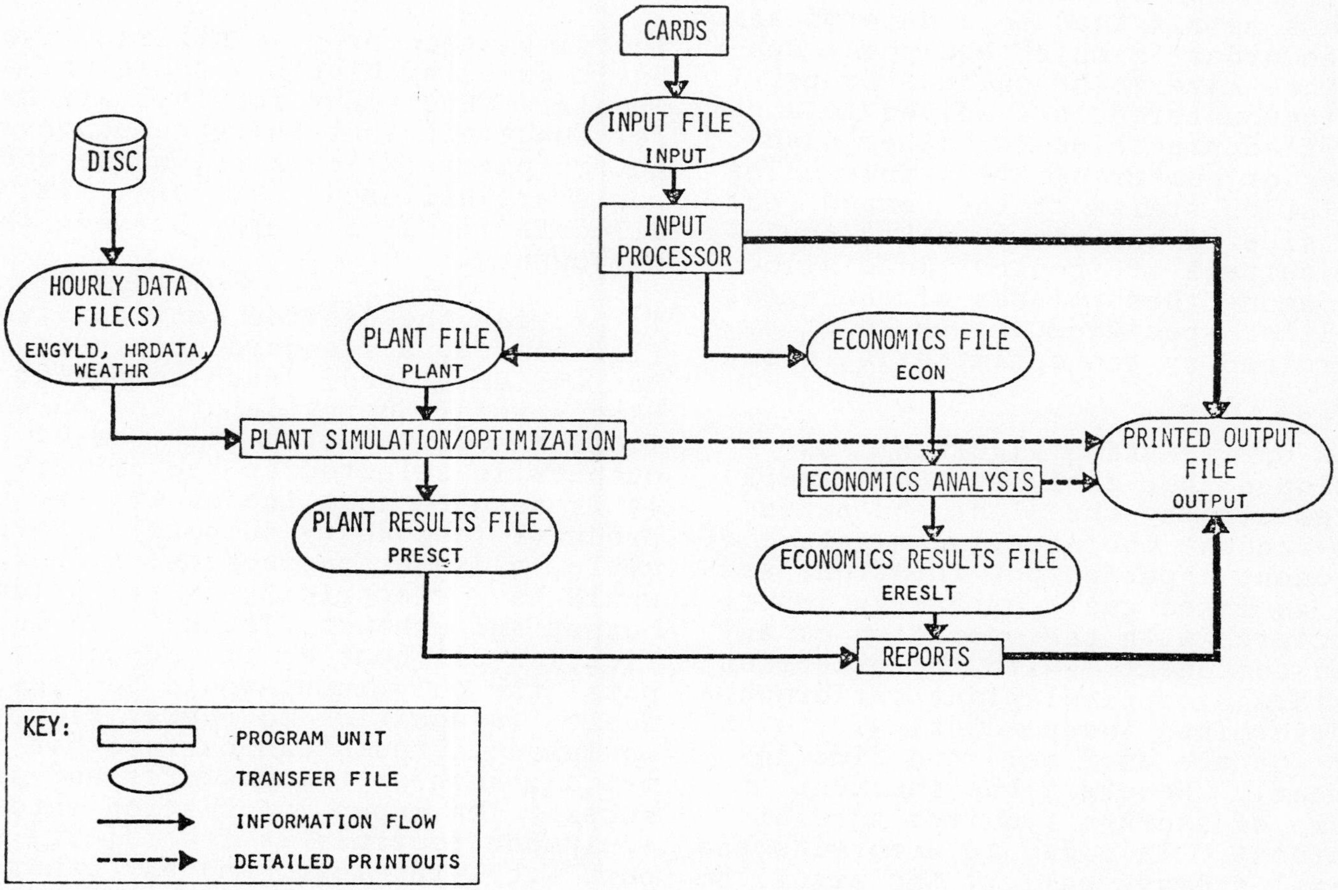

Fig. 1. Computer program structure

plant outputs in hourly or other specified time increments. There is no restriction on energy forms - they may include electricity, various grades of heat, chilled water, fuels, etc. The simulation can utilize all data on the file or any specified subset thereof.

b. The second part of the user input utilizes the input language (System Definition and Control Language) to describe the data file, plant configuration, operating strategies, and economic data. The operating rules can be omitted if the optimization feature of the program is to be utilized. The input language also allows the user to specify what portions of the data file are to be used for the particular run, to modify default parameters, and to select output reports.

The simulation program sets up a relation matrix, based on the user-specified configuration. The entries in this matrix lead to a determination of the order in which equipment subroutines have to be called. Loops, when encountered, are solved to a user specified precision in either quantities or number of iterations. The simulation starts at the demand (output) side of the plant. When equipment outputs are pooled, load allocation among these pieces of equipment is either specified in the input or determined by the optimization program.

A new optimization algorithm was developed that analyzes the network represented by the plant and takes into account the non-linear nature of component input/output relations as well as fixed costs or energy inputs associated with the operation of any given component. With the exception of storage, optimal plant performance is determined independently for each hour (or the user selected time increment). However, the inherent nature of storage requires special treatment. In order to determine the optimal storage policy, the algorithm "looks-ahead" for a certain time period and determines the policy for each hour based on the anticipated profile of demands during the look-ahead period. The "forecast" is repeated at intervals smaller than the forecasting period and the policy is adjusted accordingly.

The economic analysis does not have to be performed in conjunction with the simulation or energy optimization of the plant. Economic reports appears only if economic analysis is called for in the input and parametric runs may be requested.

The program is operational on a CDC 7600 computer; its conversion to an IBM compatible version is anticipated. The full year run (8760 hours) for a sample plant, consisting of diesel engine/generators with heat recovery, boilers, and chillers, requires less than 20 cpu seconds and costs less than $4.-. More complex systems with a large number of components and with storage optimization may cost as much as $20.-. The indicated figures apply to government users; commercial users are likely to experience higher costs.

SIMULATION PROCEDURE

The simulation program utilizes two basic building blocks - equipment and "pools." Equipment routines are transfer functions containing component input/output relationships, while the pools are utilized to record, mix, and allocate flows of energy between the components.

To provide the desired configurational flexibility, a standard component format was developed. Each component can have several energy input and output flows associated with it. One of the outputs is selected as the primary one; it is assumed that the others are byproducts (secondary outputs). For example, a diesel engine/generator set would have electricity as its primary output and exhaust, jacket, and lubrication oil heat as its secondary outputs; the only input would be diesel fuel. In addition to energy flow, equipment outputs can carry other information, such as temperatures and pressures. The input information into the equipment routines is its primary output, i.e., the load, and any other parameters the particular routine requires to calculate the amount and quality of input energy and byproducts.

The user can define a number of pools for the plant reflecting the types and qualities of energy flowing through the system. Several types of pools can be distinguished:

a. load or plant output - these receive information from the data file;

b. external resource - these calculate the amount of fuel or other resource inputs into the plant, based on the simulation;

c. internal pools - these record and allocate energy flows among equipment in the plant.

Each input and output of an equipment is connected to a pool. Several components can have their input and/or output associated with the same pool. After a component has been simulated, the output information, i.e., amounts and qualities of required inputs and produced byproducts, is routed to the appropriate pools. The pools thus carry the information on the load imposed on the components upstream. The byproducts are assigned to satisfy the requirements to the extent possible, and the remaining load, if any, is allocated to the equipment whose principal outputs are produced into the pool under consideration. This allocation is either specified by the user or determined by the optimization program.

An extensive library of component subroutines will be built for the program. It will contain equipment performance and economic data to be used as default values for the simulation. Essentially all parameters are accessible through the input language and are readily modifiable by the user.

SYSTEM DEFINITION AND CONTROL LANGUAGE

The entire plant structure and all data used for the simulation and optimization is fully controlled by the user through the input language. The language is conversational and allows for a significant amount of freedom in formatting of the input by being largely free of fixed fields. The user can caption the analysis, and may also provide an identifying name to any group of identical equipment, as well as to energy types, units-of-measure, etc.

The entries in the language are grouped in commands identified by distinct keywords. The command begins with either the command keyword or a user-defined name followed by this keyword. If data entries are required, the command continues with a grouped series of data keywords followed by the data items. Some of the features of the language are illustrated below.

Each piece of equipment (or group of identical pieces) is described to the program, related to a component subroutine, and its inputs and outputs are associated with pools. The command keyword EQUIPMENT is used to introduce this type of data. For a diesel engine/generator set this command might be provided in the following form:

```
DSLGEN75  EQUIPMENT  TYPE = DIESL
                     SIZE = 75 KW
                     NUMBER = 2
                     AVAILABLE = 2
          INPOOLS =  OIL
          OUTPOOLS = ELECT, EXHAUST,
                     JACKET, LUBE .
```

Here, for clarity, the command keyword is underlined with a solid line and data keywords with dashed lines. DSLGEN75 is a user-given name, all other entries, other than delimiters, are data items. The data item DIESL refers to a library subroutine that simulates diesel engine/generator sets with heat recovery, NUMBER is a data keyword for the number of installed pieces of this capacity, and AVAILABLE specifies the number available for operation at any given time. The data entries for INPOOLS and OUTPOOLS are ordered sets of pool names assigned to the inputs and outputs of this component. The pool names are recognized by the program if they belong to the default set of pools, or had been specified previously in the POOL command, e.g.,

```
ELECT POOL
```

The unit KW or KWATT is recognized by the program. A unit of measure definition command is available to define new units, e.g.,

UNITS TON = 12000 BTUH .

Here the new unit, TON, had been defined in terms of a recognized unit, BTUH, and became available for use in the appropriate data items.

The data file for the plant contains the values of the required plant outputs and environmental data needed by the simulated components. This file can be described to the program in the following manner:

HOURLY DATA FILE = ENGYLD

ITEMS = HEATLOAD, COOLLOAD, ELECTRIC, AIRTEMP, HUMIDITY, HOTWATER .

The file name selected is ENGYLD; the data items for each time increment are described by user defined names in the order they appear on the file and represent the heating load, cooling load, electric load, dry bulb temperature, humidity ratio, and domestic hot water load, respectively. If the values are not given in the default units, a conversion factor can be specified in the corresponding POOL command, e.g., if the electric load is given in KW, the appropriate command would be:

ELECTRIC POOL, CONVERSION = 1 KW .

If the cooling load can be satisfied by compression and/or absorption chillers the user can specify how this load is to be split. This is accomplished by a successive series of commands. The first command points to a pool when the load is to be allocated.

OPERATE POOL = COOLLOAD .

The set of commands that follow specify the load range and allocation within that range, e.g.,

RANGE CUTOFF = 60 TON
EQUIP1 = COMPCHLR 1,1.,0.

RANGE CUTOFF = 90 TON
EQUIP1 = COMPCHLR 1,.6,0.
EQUIP2 = ABS-CHLR 1,.4,0.

RANGE CUTOFF = 130 TON
EQUIP1 = COMPCHLR 1,.5,0.
EQUIP2 = ABS-CHLR 1,.5,0.

Three load ranges have been selected in the preceding example, 0 to 60 ton, 60 to 90 ton, and 90 to 130 ton. The equipment names, COMPCHLR and ABS-CHLR, which will have been defined in EQUIPMENT commands, are followed by three numbers - the first is the number of pieces assigned, the second is the fraction of the load to be carried by that component after the minimum output, indicated by the third number, had been applied towards the load. The OPERATE command can be omitted if the optimization program is called for, e.g.,

OPTIMIZE BASIS = ENERGY .

The program would then run the plant so as to minimize at each hour the fuel energy input into the system.

RUN commands are provided for program selection and specify the desired analysis, e.g.,

RUN PLANT , and

RUN ECONOMICS .

If no RUN command is encountered the program merely scans the syntax of the input.

A series of economic commands is available to specify the cost structure for the plant itself and for the resources utilized.

The examples presented above should be treated as such; their purpose is to illustrate rather than fully describe the language features. In fact, some of the indicated entries may not even appear in the input - default specifications would be substituted in those cases.

CONCLUSIONS

It is believed that the computer program described above will provide a qualified user with a capability for analysis of a large number of different central plants. It combines a great amount of flexibility in plant configuration with low running costs. At the time this paper is being written, several features are being added to the program, while some of the others are being modified. The library of component subroutines is being expanded as new data becomes available. In the future, the program will be supplemented by a simulation capability for distribution systems and by a program to calculate service and energy demands of the community served by the plant. An extensive testing and validation phase is in preparation.

ACKNOWLEDGMENTS

The authors would like to thank James M. Calm for helpful comments and reviews of the program and its documentation.

EFFICIENT USE OF ENERGY IN A SOLAR BUILDING

M. Ucar, Research Associate; E. Drucker, Professor; J. LaGraff Associate Professor; College of Engineering, Syracuse University, Syracuse, New York

ABSTRACT

A generalized dynamic computer program (SYRSOL) has been developed at Syracuse University under an ERDA grant, for the mathematical simulation of the thermal behavior of multi-zone solar heated buildings. SYRSOL stands for Syracuse University Solar Building Energy Analysis Program. The system modeled employs a series of water-to-air heat pumps connected in a closed loop flat plate water cooled solar collector, a water storage tank, and a coolant tower. Weather data are represented by sinusoids, which save programming and computing time.

Many simulation runs were made with SYRSOL on a school and an office building in several cities. The following are some of the conclusions that can be drawn from the results of the simulation runs. The use of sinusoidal temperature functions and monthly average cloud cover modifier constants is realistic and accurate. The decentralized water-to-air system is indeed a very good choice for commercial buildings. The addition of small collector areas (about 10% of the floor area) produce a system that is even more attractive and highly energy conserving. Single glazing black coated flat plate collectors are very well suited to the system described and provide seasonal utilization efficiencies in excess of 50%. Utilization efficiency drops with increasing collector area. It is feasible to use the solar collectors for desiccant dehumidification of ventilation air in the summer.

Temperature functions and cloud cover modifiers are presented in the form that can readily be used by researchers and designers.

INTRODUCTION

Closed liquid loop decentralized heat pump systems offer many practical advantages in HVAC applications, especially in commercial buildings with a large number of individual spaces with diverse loads. These systems utilize unitary water-to-air heat pumps, typically from 2 1/2 to 5 tons capacity, and are also known as electro-hydronic systems. Operating about a single water loop with temperatures near building ambient, they offer the flexibility of providing heating or cooling in a space at any time, and therefore simultaneous heating and cooling in the building. Since in the heating mode, the heat pumps remove heat from the loop while in the cooling mode they deliver heat to the loop, the system is inherently and strongly energy conserving. They can be simply turned off when not needed, without causing any unbalanced problems. Interior spaces of large buildings often require cooling year-around due to internal heat sources and the lack of exposed cold wall surfaces. Also, at some times of the year, opposite exposured of the perimeter zones may have different heating or cooling requirements due to solar loads or winds. The ability to transfer the excess heat from one zone to another via the liquid loop offers significant energy savings. In addition, the excess heat from the cooled zones can be used to supply domestic hot water, heat swimming pools, etc.

When there is a net heating load in the building auxiliary energy must be added to the loop which is typically kept in a temperature range of about 10°C (50°F) to 40°C (104°F). In this temperature range flat plate solar collectors provide an attractive auxiliary energy source. In fact, with these low temperature requirements, it is possible to obtain seasonal collection efficiencies in excess of 50% with single glazed collectors in northern climates. There are, of course, also capital cost and efficiency advantages for the storage part of the solar system.

An algorithmic dynamic computer program has been written for determining the thermal performance of two typical commercial building types with heat pumps and solar auxiliary heat (Refs. 1,2,3,4). Weather functions have been generated for thirteen U.S. cities based on long term weather data to simplify computational procedures. An additional operating mode has also been added. This is solar dehumidification in the summer to lower the energy required to satisfy the cooling load. The excess solar energy in the summer months cannot help the heat pump cooling operation directly. However, if this excess energy is used to regenerate a desiccant dehumidification system, potential energy savings are apparent.

This paper is essentially an overview of our observations and conclusions on the solar-assisted unitary (i.e., decentralized) water-to-air heat pump system. After four years of in-depth study of this particular system, as it applies to commercial buildings, the authors are convinced that it is a very effective, energy-conserving, versatile, demand-responsive and economical system. Therefore, the authors believe that at this point the results should be desseminated widely to researchers and designers. Technical details and analyses are kept to a minimum to emphasize the fundamental aspects of the conclusions. The reader may refer to earlier publications for greater detail.

ANALYSIS

The methodology of the research is essentially a parametric study of a basic solar supplemented space conditioning system, using mathematical modeling and a high speed computer. Two representative commercial building types have been modeled, a two-story school and a five-story office building. Details of the models appear in previous reports and papers (Refs. 1-7). Load calculations are performed on an hourly basis, for from 4 to 11 separate zones in the buildings. State-of-the-art flat plate collectors and water-to-air heat pumps were simulated. Realistic schedules for occupancy, ventlation, lighting, hot water consumption and night set back of temperature were incorporated into the computer program. Average weather data for each of 13 geographic areas was prepared using a set of generalized weather functions (Ref. 2) developed for the program. These functions give dry bulb temperature, dew point temperature and cloud cover. From the cloud cover function (expressed as a cloud-cover modifier) and the theoretical insolation, the actual insolation can be computed for any time, location and collector orientation. Heat and moisture balances are taken in the zones of the building and heat pump operation simulated for heating or cooling, including sensible and latent components. Auxiliary electric heat, solar heat, and cooling tower heat rejection are utilized when necessary to balance the total energy and moisture flows in the building. Domestic hot water production is simulated, using a water-to-water heat pump and a topping electric heater.

The total amount of energy collected and used in the various categories is summed and appears as output information. The two primary outputs are the yearly total consumption of electricity, and the yearly auxiliary electric heat used. The main parameters are the collector area, water storage volume and climate. Other parameters include collector inclination, collector orientation, heat pump coefficient of performance, lighting levels and ventilation heat recovery. A number of other parameter variations have been tested that concern cloud cover, dry bulb temperature and the number of days per month which are actually simulated.

DOMESTIC HOT WATER

The existence of a larger water storage tank in the building HVAC system described in this paper made the possibility of its use as a source of energy for domestic hot water (DHW) of interest. Heating water from this source which draws upon both solar inputs and heat rejected from some building zones has obvious energy conserving advantages.

Two approaches to the production of DHW were evaluated in the simulation program and compared to the energy of a direct resistance heating mode. The first scheme involved the use of a water-to-water heat pump operating between the storage tank source and a smaller DHW tank. The performance of this heat pump was based on actual data for a commercial five-ton water chiller given as a function of evaporator entering water temperature from the storage tank. DHW temperature circulating through the condenser was limited to 120°F (48.9°C) for the heat pump chosen. Temperature topping to 140°F (60°C) was accomplished by resistance heating. A characteristic water schedule was assumed for each building (1600 gallons/day for the school; 735 gallons/day for the office building). More descriptive details of this system can be found in Reference 6.

The second approach to DHW heating involved direct solar heating of the water (through a heat exchanger) in the smaller tank during that portion of the year when solar collection capacity exceeds the auxiliary energy needs of the building. The example used for a representative comparison was for the Kay Office Building in New York City using 8000 ft^2 (743 m^2) of collector area (6 months of excess collection capacity). Table 1 summarizes the electrical energy needs of each system. The resistance heating energy use and the combined heat pump/resistance energy are essentially independent of the geographical location and solar system size. The system using some direct solar heating, however, does depend on the location and the collector size in that this determines the portion of the year when excess collection capacity exists.

The energy reduction of each system is significant. The cost effectiveness of an investment in a water-to-water heat pump would be very attractive compared to resistance heating (2-3 year pay back, see Ref. 6). The savings of direct solar heating during the summer months of excess collection capacity would also certainly improve the cost effectiveness of the solar installation.

SOLAR COLLECTORS

Three representative solar collectors were selected for evaluation in the system model discussed in this paper. One was a moderate performance single glazed, non-selective flat plate collector (Revere). Another was a higher performance flat plate with a selective surface and single glazed (Lennox/Honeywell). The third was a moderately focusing (10X) tracking concentrating collector (Northrup). In the temperature range of operation appropriate to water source heat pump systems (generally less than a ΔT of 50°F [10°C]) the difference in efficiencies are not too great and economics would be the dominant parameter of design choice. In this low temperature range it is worth noting that single glazed collectors usually have a higher efficiency than double glazed collectors. It is also of interest to note the poorer performance of the concentrating collector at low fluid temperatures. The schematic of the basic system modeled is shown in Fig. 1.

WEATHER DATA

Weather data are one of the most important elements of building energy use simulation whether the building has solar collectors or not. The measured weather data provided by the U.S. Wea-

TABLE 1

Electrical Energy (KWH) Requirements
Domestic Hot Water

	Resistance Only	Heat Pump & Resistance (% Reduction)	Heat Pump-Resistance & Solar Direct (% Reduction)
Office	65,000	23,000 (65%)	11,700 (82%)
School	129,000	55,000 (57%)	

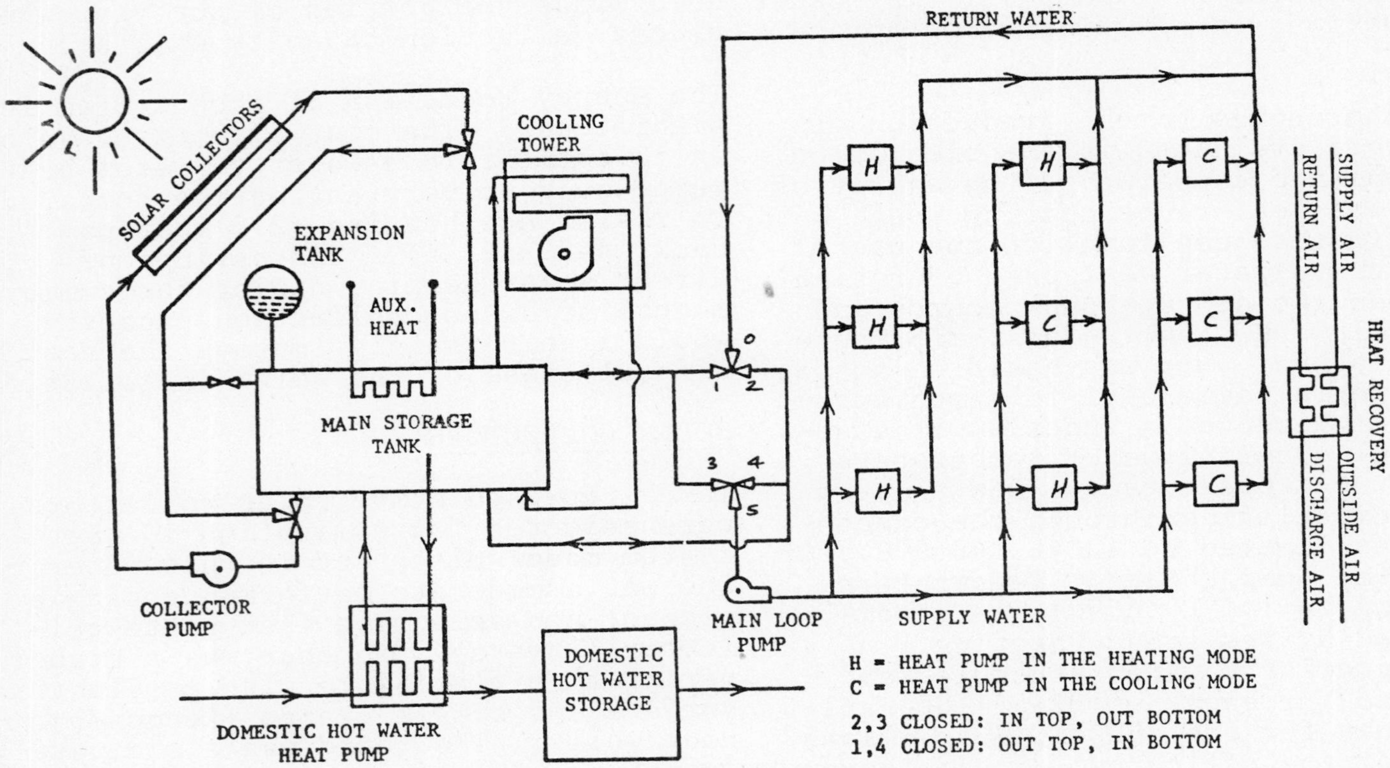

Fig. 1. Schematic of Typical Solar Assisted Closed Loop Heat Pump System Model

ther Bureau is not in a very convenient form for use in building simulation. Therefore, it is necessary to process this data in some way to formulate a weather model that represents typical weather for a given location on a continuous or hour-by-hour basis. The conventional method of representing weather with winter and summer extreme temperatures is no longer valid because it is energy-wasteful and also it gives no idea of the average hourly conditions.

The approach taken by the present study to the representation of weather data is fundamentally different from others, and also very powerful Sinusoidal functions are used for dry bulb and dew point temperatures, and monthly average constants for cloud cover and a constant wind speed. The generalized weather functions, as this approach has been called, provide an alternative to magnetic weather tapes.

GENERALIZED TEMPERATURE FUNCTIONS

The derivation of these functions have been discussed in detail in an earlier ASME paper (Ref. 2). Here only their advantages will be discussed and the most up-to-date version be presented.

The generalized temperature functions have the following advantages over the weather tapes:

1) They represent many years' average experience of weather data.

2) They can be derived once and for all.

3) They make programming easier more effective as well as faster.

4) They eliminate the need for computer tape drives, thus enabling the user to run his/her program on very small computer installations (such as mini-computers) in the absence of a powerful computer system.

5) They reduce the residence time of the program in the computer core, thus providing much shorter turn-around time.

6) They generally reduce the cost of computing.

7) Weather functions, once derived, cost little to use. These functions are very concise and easily communicated equivalents of costly magnetic tapes. Weather tapes would have cost over one thousand dollars for the 13 cities for which we have generated the functions.

8) Weather functions are simply more convenient.

9) They are very versatile. One can simulate only several days a month

without any change in the program or the weather routine.

10) They are very powerful. One can easily perform many kinds of studies of building performance and energy use.

CLOUD COVER DATA

Cloud cover data are naturally very important in the simulation of solar buildings, and are presented by "Cloud Cover Modifiers". For a given city 12 cloud cover modifiers are computed by dividing the measured monthly total insolation by the ideal clear sky total. The clear sky insolation intensities are predicted by the computer program SYRSOL within less than 1% inaccuracy.

WIND SPEED

The wind speed affects energy calculations through the outside wall surface heat transfer coefficient and infiltration. In the recent past commercial buildings have been considered sealed structures with very large forced ventilation rates. Therefore, infiltration was considered wind independent in our own analysis. The wind speed was treated as a constant.

However, with today's necessity and tomorrow's requirements of much lower ventilation rates, infiltration may gain more significance in the energy consumption of a building. These two points will be studied in more detail in the future.

MAJOR CONCLUSIONS OF THE STUDY

1) For commercial buildings in the 100,000 ft^2 area class, collector areas equal to 5 to 15% of the building area only are feasible. In the case of very large hot water demand, larger collectors may be justified.

2) Very large water storage volumes are cost effective. Ratios as high as 40 lb. of water per square foot of collector are economic in moderate climates and 60 in areas with colder climates.

3) Since collector cost increases almost linearly with size, but the rate of increase of useful energy collected decreases sharply with size, the cost/benefit ratio increases with collector size. There is no optimum collector size from an economic point of view, except perhaps for trivially small collectors.

4) The attractiveness of solar assisted heat pump systems for heating is greatest in the colder climates, where despite lower collector efficiencies, the total amount of solar energy actually is maximized.

5) The solar assisted heat pump system virtually removes the necessity for auxiliary heat at very low collector area to floor area ratios.

6) The system, with large storage volume, is capable of maintaining space conditions over periods of several cloudy days. The yearly performance of the system does not change appreciably when random cloud cover sequencies are substituted for a uniform cloud cover equal to the average of the random sequences.

7) The system performance is not highly dependent upon collector orientation or angle of inclination.

8) For a system with collector sized for commercially available heat pumps, an improvement in the COP of the heat pumps does not reduce significantly the total electrical energy consumption. To achieve an overall reduction, the collector area must increase.

9) For the purpose of generating domestic hot water, a water-to-water heat pump with storage water as the source, is highly cost effective relative to electric resistance heating.

10) It is essential to do multi-zone analysis of a large building since there may be appreciable amounts of heating and cooling done simultaneously even though the net thermal load on the building may be small or zero.

11) Dry bulb and dew point temperature data for any particular city may be accurately represented by sinusoidal temperature functions (with harmonics). These functions predict average monthly temperatures with an average error of less than 0.5 F.

12) Using empirical single glazed collector efficiency data, simulation runs for a whole system produce seasonal efficiencies for heat collection of the order of 60%. That this could be achieved in a Northern climate illustrates the effectiveness of the low temperature water loop system.

13) Ventilation heat recovery systems can save a considerable amount of energy only if proper control logic is used.

14) Desiccant dehumidification systems, with solar heat regeneration, reduce the required electrical energy for heat pump cooling. Secondarily, dehumidification permits a higher interior dry bulb temperature for a given comfort level.

ACKNOWLEDGEMENTS

The authors gratefully acknowledge the financial support of the Solar Heating and Cooling Branch of Energy Research and Development Administration (ERDA) in the conduct of the research described in this paper.

REFERENCES

(1) Ucar, M., et al: "Thermal Simulation of a Building with Solar Assisted Closed Liquid Loop Unitary Heat Pumps", ASME Publication 76-WA/Sol-23, 1976.

(2) Card, W. H., et al: "Generalized Weather Functions for Computer Analysis of Solar Assisted HVAC Systems", ASME Publication 76-WA/Sol-20, 1976.

(3) Ucar, M., et al: "Computer Model for a Solar Assisted, Closed Liquid Loop Heat Pump System", ASHRAE Transactions, Vol. 82, Part I, 1976.

(4) Drucker, E. E., "Commercial Building Solar Heat Pump Systems", ERDA Workshop on "Solar Energy Heat Pump Systems for the Heating and Cooling of Buildings", The Pennsylvania State University, University Pk., Pa., 12-14 June, 1975.

(5) Ucar, M., "Solar Building Energy Use Analysis", Dissertation, Syracuse University, Syracuse, N.Y., 1976.

(6) Drucker, E. E., et al: "Commercial Building Unitary Heat Pump System with Solar Heating", Final Report to ERDA, Syracuse University, Syracuse, N.Y., 31 October, 1975.

(7) Drucker, E. E., et al: "Commercial Building Unitary Heat Pump System with Solar Heating", Semi-annular Report to NSF, Syracuse University, Syracuse, N.Y., 31 December, 1974.

MODELS AND METHODS FOR THE ANALYSIS AND OPTIMIZATION OF SOLAR ENERGY SYSTEMS

R. Bruno and R. Kersten
Philips, Aachen, Germany

ABSTRACT

Several methods of calculating the dynamic behaviour of solar energy systems have been investigated. These methods can be classified into three categories; the finite element approach, the component-'black box' approach, and simplified methods. In each case special attention has been concentrated on the reduction of computer time and maintenance of calculation accuracy. In the finite element approach it is shown that, in general, only a one dimensional treatment of both acyclic and recyclic circuits is necessary. It is shown that the component-'black box' approach can be solved by either an exact integration over an elementary time period or by cyclic convergence (modular method). From the exact solution it is found that the number of equations to be solved may be reduced by using a lumped circuit separation model. Both of these approaches require hour by hour weather and load data for calculations. Two types of simplified methods have been considered, the first is based on the lumped circuit separation equations and uses an analytic form for the weather and load data while the second is an empirical method. It is shown that in going from the exact finite approach to the former simplified method computation time can be reduced by over three orders of magnitude and accuracy is reduced by about 5/1000 for all energies calculated. It is indicated that the empirical simplified method provides the basis for a system's optimization procedure.

POWER AND ENERGY MANAGEMENT SIMULATION (PEMS) COMPUTER PROGRAM

David W. Galehouse, P.E.
Vice President-Engineering
Automated Construction Technology, Inc.
Dayton, Ohio

ABSTRACT

A computer program developed to simulate the impact of commercially available power and energy management systems on a facility's electric profile and electric utility cost is described. Both hardware- and software-based systems can be simulated including instantaneous, ideal rate and predictive electrical demand limiting algorithms. The program uses existing recording demand meter data, a description of the power and energy management system to be simulated, and programmed, duty-cycled and sheddable load tables as input data. Output includes overall results such as reduction in kilowatt-hours, kilowatts of demand and load factor and statistical data for each controlled load such as number of times shed and total off-time. Several runs are included to illustrate the use of the program in cost/benefit analyses on representative recent projects.

SOME PROBLEMS IN ACHIEVING EFFECTIVE MACRO-ENERGY MANAGEMENT

W. Murgatroyd
Imperial College, London

ABSTRACT

A short review is given of predicted energy supplies and demands showing the need for and possible impact of better energy management.

A typical energy system is described, with special reference to electrical energy, as well as the two major inputs - fuel and capital - which form the basis of energy management optimization. The system components from the generator to the final appliance are dicussed in terms of technical efficiency and cost, and the importance is stressed of achieving an overall optimization of resources across the system. The difficulties of achieving this are outlined, and the roles which can be played by legislation, tariffs, fiscal measures etc. are analyzed. The influence of changes in the capital/fuel input balance is stressed, together with the roles of energy storage in responding to these changes.

Examples are given of non-optimum use of resources in the industrial sector.

PIES: THE NATIONAL ENERGY MODEL OF FEA

Paul H. Randolph
Federal Energy Administration
Washington, D.C.

ABSTRACT

The purpose of the Project Independence Evaluation System (PIES) is to forecast the state of the energy supply and demand of the nation. PIES makes this forecast by taking a snapshot of the nation's energy economy on an average day at the beginning of 1980, 1985 and 1990. The model predicts energy consumption in each of ten regions for eight fuel products. In addition, the model determines the sources of these fuels, methods of converting these fuels into energy forms suitable for final consumption, and the transportation of fuel or energy from one part of the country to another.

The model assumes a competitive economic structure with upward sloping supply curves and downward sloping demand curves. Within this framework the model describes a static market equilibrium of the energy system. No attempt is made to describe the trajectories by which this equilibrium is achieved.

This paper will examine the mathematical structure of the PIES model, along with some of its energy projections and their implications.

A MORPHOLOGICAL MODELLING ON LONG-TERM ENERGY SYSTEM

Tadashi AOYAGI, Atsuyuki SUZUKI and Ryohei KIYOSE

Department of Nuclear Engineering, Faculty of Engineering
University of Tokyo, Bunkyo, Tokyo 113, Japan

INTRODUCTION

One of the most important issues of long-term energy problem is on R & D planning of advanced energy technologies which supply virtually unlimited energy but require a great deal of capital investment. The need of advanced energy technology R & D rests primarily upon energy consumption level in the future. As well known, energy consumption level is heavily related to economic productivity, gross national product. A part of gross national product goes to the expenditure on advanced energy technology R & D. A question is how much, out of gross national product, should be invested onto advanced energy technology R & D. This study is concerned with such an energy R & D planning problem.

To draw a general conclusion from generic statement and quantification of the problem, one must take into account a lot of factors relevant to the problem and specify interrelationships between the factors. In actual, however, it is of great difficulty since the spectrum of those factors is widely broadening over technological, economic and societal aspects and the quantitative specification of interrelationships between the factors is associated with much ambiguity.

From this consideration, this study is not aimed at providing comprehensive interpretation on long-term energy R & D problem but rather at understanding morphological relations between some key factors among all others; the key factors treated in this work are energy consumption level, capital investments of energy and non-energy sectors, and gross national product.

BASIC EQUATIONS

For simplicity, suppose the gross national product is distributed into only aggregated totals of consumption goods and capital goods, and consider the following most simplified macro-economic model.

$$G = C + K + I + U, \tag{1}$$

$$G = a(F + E), \tag{2}$$

$$\dot{K} = \alpha G - \beta K; \quad 0 < \alpha < 1, \ \beta > 0, \tag{3}$$

$$F = \gamma I, \tag{4}$$

$$E = \varepsilon \int_0^t U \, dt', \tag{5}$$

where, G = gross national product,
C = consumption goods,
K = productive capital of non-energy sector,
I = annual investment on conventional energy technology (CET),
U = annual investment on advanced energy technology (AET),
F = annual energy consumption, the supply source of which is CET,
E = annual energy consumption, the supply source of which is AET, and
$\alpha, \beta, \gamma, \varepsilon$ and a = exogenous parameters.

The meaning of Eq. (1) is clear because of the assumption that gross national product, gross national income and gross national expenditure are all identical. The coefficient a in Eq. (2) is constant if industrial structure and style of life do not change significantly. If industrial structure shifts to use energy more efficiently, it increases the value of a. Equation (3) is one of the expressions of capital investment function based on macro-economic theory. The assumption that the energy F from the CET can be simply purchased by paying I is made to set out Eq. (4).

Equation (5) is valid if the AET is defined as the technology which is extremely capital-intensive and generates unlimited energy with infinite service life.

Now, it is assumed for simplification that the energy consumption F is kept constant at the present level F_0 because of the limitedness of conventional energy resource, and that the amount of consumption goods at any point of time must exceed the initial level C_0 because the society prefers to behave so that individual income may become higher. Under these assumptions, Eqs. (1) to (5) are reformed into

$$0 \leq U \leq aE - C_0 - \dot{P} - I_0' ; \quad I_0' = (1/\gamma - a)F_0 , \tag{1'}$$

$$\dot{P} = \alpha a E - \beta P \; ; \quad P = K - \alpha a F_0/\beta , \tag{3'}$$

$$\dot{E} = \varepsilon U . \tag{5'}$$

Taking P and E as state variables and U as a control variable, our problem is reduced to a linear control problem with the admissible control region depending on the state variables.

EXTREME CASES

There are two extreme cases on the control variable; one is the case that $U = 0$ and the other, $U = aE - C_0 - \dot{P} - I_0'$.

(1) $U = 0$.

In this case, nothing is invested onto the AET and the energy consumption level is not expected to increase, being stabilized at the initial level $E_0 + F_0$. The solution of Eqs. (3') and (5') is easily obtained and illustrated in Fig. 1 where the initial condition is treated parametricly.

(2) $U = aE - C_0 - \dot{P} - I_0'$.

This is the case where the annual investment on the AET is taken up to the utmost being subject to the constraints on consumption goods, $C \geq C_0$. By taking the linear transformation:

$$P' = P - \alpha(C_0 + I_0')/\beta ,$$

$$E' = E - (C_0 + I_0')/a ,$$

Eqs. (3') and (5') are reduced to

$$\begin{pmatrix} \dot{P}' \\ \dot{E}' \end{pmatrix} = \tilde{T}\begin{pmatrix} P' \\ E' \end{pmatrix} : \quad \tilde{T} = \begin{pmatrix} -\beta , & a\alpha \\ \varepsilon\beta , & a\varepsilon(1-\alpha) \end{pmatrix} . \tag{6}$$

The general expression of the solution of Eq. (6) is as follows.

$$P' = A_1 e^{\lambda_1 t} + A_2 e^{\lambda_2 t} , \tag{7}$$

$$E' = B_1 e^{\lambda_1 t} + B_2 e^{\lambda_2 t} , \tag{8}$$

where, A_1, A_2, B_1 and B_2 = constants,

λ_1 and λ_2 = eigenvalues of the matrix $\tilde{T}$, given by

$$\lambda_i^2 - (a\varepsilon(1-\alpha) - \beta)\lambda_i - a\varepsilon\beta = 0; \quad i = 1, 2. \tag{9}$$

It is worth noticing that Eq. (6) possesses the following topological properties.

(i) The necessary conditions for stabilization, $\dot{P}' = 0$ and $\dot{E}' = 0$, are

$$E'/P' = \beta/a\alpha > 0 , \text{ and} \tag{10}$$

$$E'/P' = -\beta/a(1-\alpha) < 0 , \tag{11}$$

respectively.

(ii) The point $(P', E') = (0, 0)$ which satisfies both $\dot{P}' = 0$ and $\dot{E}' = 0$ is a sort of singular point, and since $\lambda_1\lambda_2 < 0$ (suppose here $\lambda_1 > 0$ and $\lambda_2 < 0$, for convenience), that singular point is a saddle point[1].

(iii) At the asymptotic state $t \to +\infty$,

$$E'/P' = B_1/A_1 = (\lambda_1 + \beta)/a\alpha > 0 , \tag{12}$$

and at $t \to -\infty$,

$$E'/P' = B_2/A_2 = (\lambda_2 + \beta)/a\alpha < 0 . \tag{13}$$

(iv) The topological relationship between the four straight lines, given by Eqs. (10) to (13) is as,

$$(\lambda_1 + \beta)/a\alpha > \beta/a\alpha > 0$$

$$> (\lambda_2 + \beta)/a\alpha > -\beta/a(1-\alpha) . \tag{14}$$

By making use of these properties, Fig. 2 represents the trajectories in the $P'-E'$ and $K-E$ spaces on Eqs. (7) and (8) for various initial conditions. It is observed from Fig. 2 that the two asymptotes separate the space into four regions, in each of which the trajectories behave in different ways, and that, in the long distance of future, both the productive capital and the energy consumption level are either increasing or decreasing, depending on the initial condition but the ratio of the changing rates is rather fixed by Eq. (12). These observations throw some light upon energy policies. For instance: The right-hand-side of Eq. (12) can be one of the indices in making the policies, or the difference between the right-hand-sides of Eqs. (10) and (11) can be another index to represent the expectation of economic growth.

OPTIMUM CASE

Despite the fact that a variety of optimization criteria can be considered, the following optimization problem is now posed as an illustrated example;

$$maximize,\ J=\int_0^T (aE - \dot{P} - U)\cdot Edt, \tag{15}$$

subject to Eqs. (1'), (3') and (5').

The objective function is an expression of the case where the society prefers to increase both the consumption goods and the energy supplied by the AET.

Owing to the maximum principle formulation, the Hamiltonian function, minimized by U, must be constant at any point of time.

$$H = \min_U [X + \Gamma U] = \text{constant}, \tag{16}$$

$$X = (a(1-\alpha)E + \beta P)(v - E) + (a\alpha E - \beta P)\psi_1 - (C_0 + I_0')v, \tag{17}$$

$$\Gamma = (E + \varepsilon\Psi_2 - v), \tag{18}$$

$$\text{where, } \dot{\Psi}_1 = \beta(E + \Psi_1 - v), \tag{19}$$

$$\dot{\Psi}_2 = \cdot\beta P + 2a(1-\alpha)E - a\alpha\Psi_1 - a(1-\alpha)v - U, \tag{20}$$

$$v = \begin{cases} E+\varepsilon\Psi_2; & U=a(1-\alpha)E-\beta P-C_0-I_0', \\ 0 & ; \text{ otherwise.} \end{cases} \quad (2)$$

It follows from Eq. (16) that the optimal control is possibly composed of

$U = 0$; for $\Gamma \geq 0$,

$U = a(1-\alpha)E - \beta P - C_0 - I_0'$; for $\Gamma \leq 0$,

and/or

$U = U_s$; for $\Gamma = 0$.

The control U_s, satisfying $\dot{\Gamma} = \dot{X} = 0$, is called a singular control and it exists under the condition:

$$E/P = \theta; \quad \theta = \beta(1 \pm 1/\sqrt{1-\alpha})/a\alpha, \tag{22}$$

in the form of,

$$U_s = \theta(a\alpha E - \beta P)/\varepsilon. \tag{23}$$

The synthesis of an optimal control over the interval $[0, T]$ is done for a given initial condition: (P^0, E^0) and for a given end-point condition: $f(P^T, E^T; T) = 0$. An example is illustrated in Fig. 3, where the initial condition is roughly the present state of Japan and yet the end-point condition is fixed just for illustrating purpose.

CONCLUDING REMARKS

A model on long-term energy system was discussed with emphasis upon not quantitative but qualitative understanding on the morphological structure of the energy system: in particular, the relationships between the productive capital of non-energy sector, P, the capital investment of energy sector, U and the energy availability based on the AET, E. The salient feature of the model is that, without numerical specification on the exogenous parameters included in the mathematical formulation, it can look into the question how each of the factors behaves in the future.

There, however, are lots of simplifications assumed in the model. Even for the relationships between the above factors, the mathematical expression is not unique but rather many other myths are possible. The comparison to another formulation is required to observe more practical informations. This study, therefore, is to be regarded as an attempt of morphological modelling on long-term energy system.

ACKNOWLEDGMENTS

The authors would like to express their sincere thanks to Professors Y. Yamamoto and T. Kihara of the University of Tokyo for their continual guidance and encouragements for many years. Thanks are also due to Miss K. Shimada for her patience in typing this manuscript.

REFERENCES

1. Pontryagin, L. S., "Ordinary Differential Equations", translated in Japanese by T. Kimura and K. Chiba, pp. 105-110, Kyoritsu Shuppan, Tokyo. 1963.

2. Leitmann, G., "An Introduction to Optimal Control", pp. 88-92, McGraw-Hill, New York, 1966.

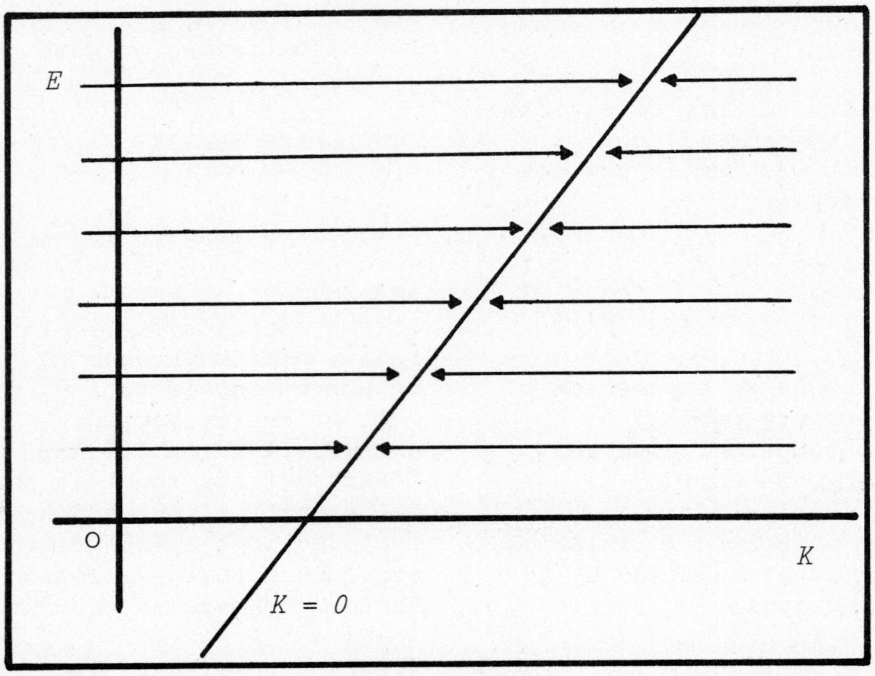

Figure 1: Phase Trajectories for $U=0$.

Notes : (1) While the trajectories depend on the initial condition, each of them, being parallel to the axis K, goes to the state: $\dot{K} = 0$ at the limit $t \to \infty$.

(2) The intersection between the asymptotic state $\dot{K} = 0$ and the axis K satisfies $P = 0$ $(K = a\alpha F_0/\beta)$.

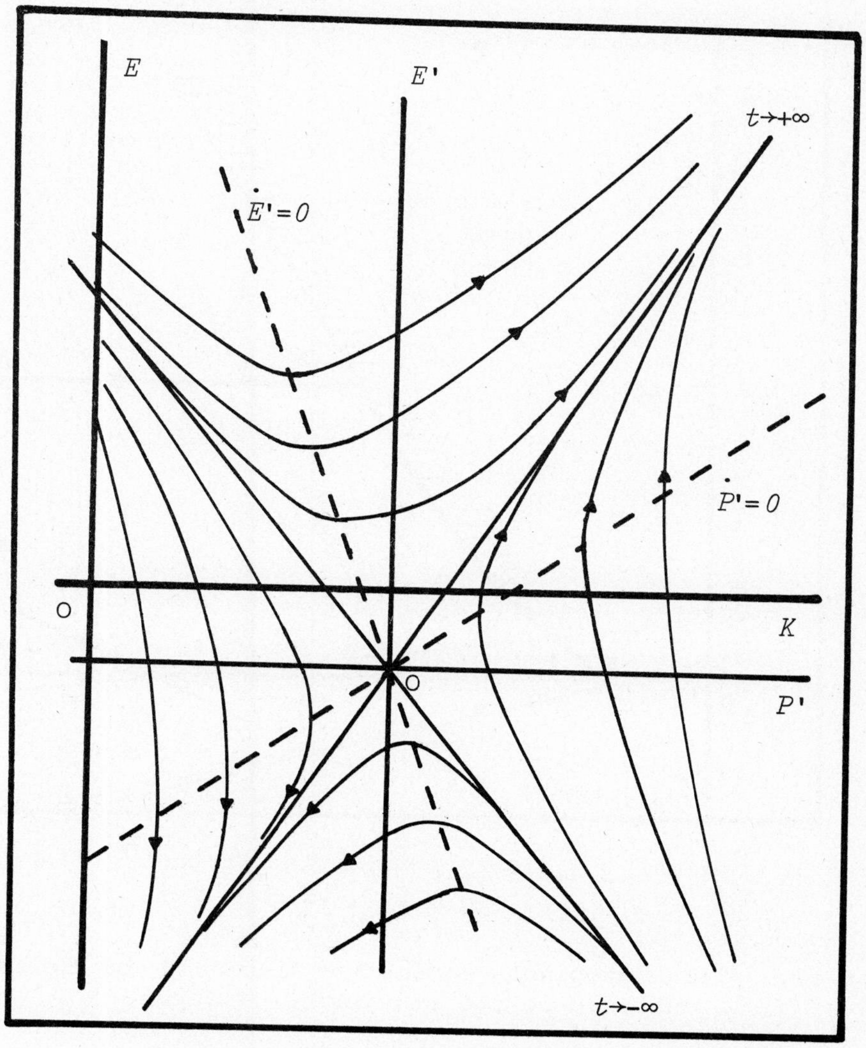

Figure 2: Phase Trajectories for $U = \alpha E - C_0 - \dot{P} - I_0'$.

Notes : (1) The differential properties of the trajectories are based on the topological relationships to the asymptotic states $t \to +\infty$ and $t \to -\infty$, and to the stabilized states $\dot{P}'= 0$ and $\dot{E}'= 0$. If and only if a state is located at the subregion of $\dot{P}' \geq 0$, surrounded by the straight lines: $t \to +\infty$ and $\dot{E}'= 0$, it satisfies $\dot{P}'> 0$ and $\dot{E}'> 0$.

(2) As clear from the linear transformation from $(\dot{P}, \dot{E})$ into (P', E'), if $\dot{P}'= \dot{E}'= 0$ then $\dot{P} = \dot{E} = 0$ (also $\dot{K} = 0$).

(3) The point $(P', E') = (0, 0)$ is represented by $(\alpha(C_0+I_0)/\beta, (C_0 + I_0)/\alpha - F_0)$ in the $K - E$ space.

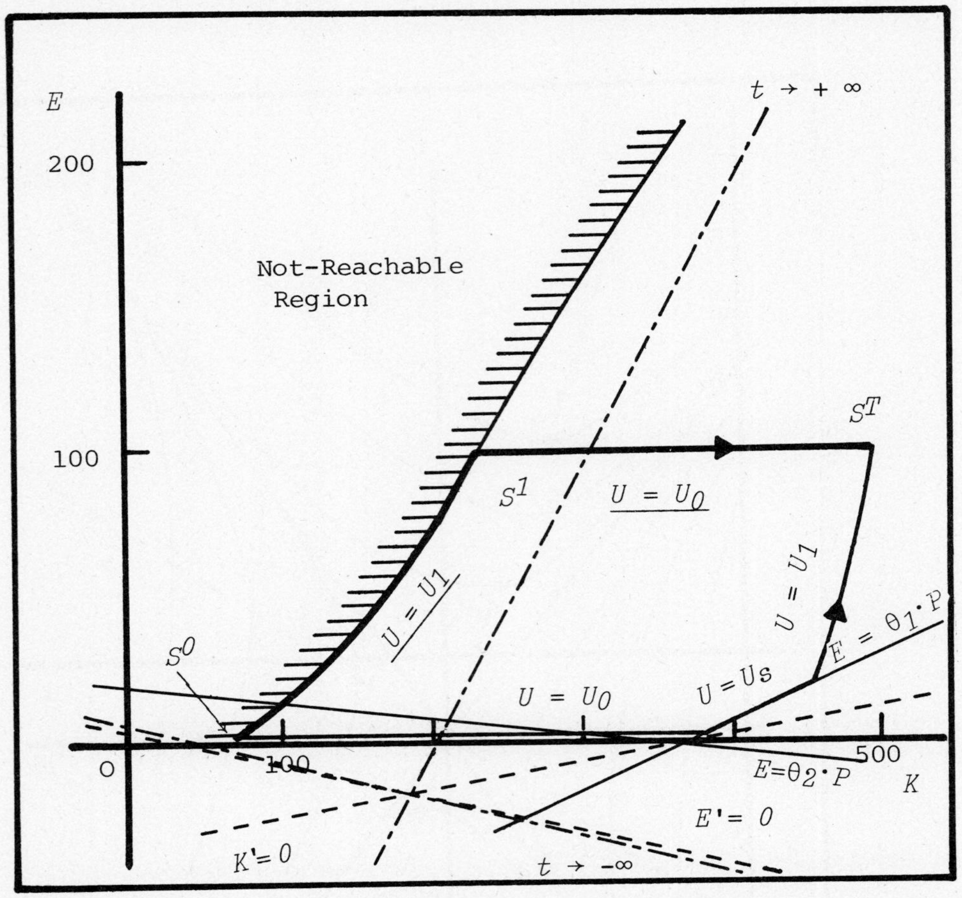

Figure 3: Optimal Trajectory for Initial State S^0 (70.1, 0.5) and Final State S^T (500, 100); $S^0 \to S^1 \to S^T$.

Notes : (1) The exogenous parameters included in the model are fixed from the analysis on the statistics of Japan.

$a = 0.454$ (¥ $10^{13}/10^{14}$ kcal)	$C_0 = 6.986$ (¥ 10^{13}/yr)
$\alpha = 0.604$ (—)	$I_0 = 0.635$ (¥ 10^{13}/yr)
$\beta = 0.025$ (1/yr)	$K^0 = 70.1$ (¥ 10^{13})
$\gamma = 52.6$ (10^{14} kcal/¥ 10^{13})	$F_0 = 33.4$ (10^{14} kcal/yr)
$\varepsilon = 1.297$ (10^{14} kcal/yr/¥ 10^{13})	$E^0 = 0.5$ (10^{14} kcal/yr)

The monetary terms are in a unit of 1975 Yen.

(2) The initial condition S^0 (K^0, E^0) given above corresponds roughly to the 1975 level. The final state S^T K^T, E^T) is fixed in such a way that $K^T \simeq 7 \times K^0$, and $E^T + F_0 \simeq 4 \times (E^0 + F_0)$.

(3) The optimal trajectory is determined as $S^0 \to S^1 \to S^T$, where, for $S^0 \to S^1$, $U = U_1 = a(1-\alpha)E - \beta P - C_0 - I_0'$, and for $S^1 \to S^T$, $U = U_0 = 0$.

(4) The sequence: $U = U_0$, $U = U_s$ and then $U = U_1$ is also able to reach S^T but not optimal. The singular control surface is given by,

$E = \theta_1 \cdot P$ or $E = \theta_2 \cdot P$,

where, $\theta_1 = \beta(1 + 1/\sqrt{1-\alpha})/a\alpha = 0.236$ (10^{14}kcal/yr/¥ 10^{13})

$\theta_2 = \beta(1 - 1/\sqrt{1-\alpha})/a\alpha = -0.054$ (10^{14}kcal/yr/¥ 10^{13}).

THE EFFECT OF ENERGY CONSERVATION ON ECONOMIC GROWTH

Timothy Jordanides

Department of Electrical Engineering
California State University,
Long Beach, California 90840

ABSTRACT

A model of petroleum utilization in the United States is formulated and used to investigate the effects of energy conservation as well as different tax incentive and R & D funding policies on economic growth.

INTRODUCTION

The shortage of energy that first occurred in late 1973 and early 1974 and subsequent events made apparent to the public the dependence of our modern technological society on that commodity. A recently announced national energy policy has placed substantial emphasis on the conservation of energy. What is perhaps not so generally well emphasized is the importance of conservation of petroleum in the light of rising imports and dwindling domestic supplies. Petroleum and natural gas constituted the source for 77 percent of all energy consumed in the United Stated in 1973 [1].

There are, of course, certain characteristics of petroleum that make it quite different from other sources of energy. Among these, it is the most portable such source. Indeed, a substantial amount of ground transportation and all of air transportation is possible today only if oil is available. However, a consequence of this is that the economic principle of substitution via the price mechanism simply will not suffice for regulating the consumption of petroleum, at least not as clearly as it would for other commodities.

Another important aspect of petroleum that deserves attention is its finite nature. Several researchers [1]-[3] have pointed with convincing arguments to the finiteness of most energy forms, having particularly singled out petroleum. And there is no doubt that conservation of petroleum will receive more careful attention in the next 10 years compared to the last decade. In this research, we use a model for petroleum utilization developed previously [4], to show the importance of conservation as one of the components of the policy (control) variables. Conservation is usually loosely defined as turning off unnecessary lights, car pooling, etc. Of course the above are very cost effective, have the least adverse impact on economic growth and have the immediate effect of saving money to the individual involved, which means that they can be successfully moved into the public sector. But the complexities and dimensions of conservation are far greater than plain curtailment.

An attempt is made here to analyze conservation of energy from the perspective of the energy consumption/GNP ratio. The feeling is here that the significance of the energy consumption vs. the total output per capita in constant dollars is far greater than what it has been accorded in the literature. One component of this statistical ratio signifies efficiency of energy use. Other components of the ratio deal with nonenergy uses of petroleum. The impact of increased efficiency on ecomonic growth is clearly positive. The technologies involved in getting more benefits out of a unit of energy will aid economic growth in creating new jobs and markets.

SYSTEM ANALYSIS

The mechanics of the formulation of our model are based on Forrester's system dynamics approach [5]. The rationale for the selection of variables, coefficients, and term-by-term construction of the equations is based on economic and geologic literature, oil company reports and policies, and the examination and scrutiny of the events during 1973-1975, concerning energy in general and oil in particular. Our model is dynamic; it does not merely yield "snapshots" of supply and demand. Further, we include the finite nature of petroleum pools [2], [13] and allow nonlinear relationships among the variables as warranted.

There is considerable documentation in the literature [1], [2] that the average cost of petroleum increases exponentially as the depth of drilling increases both on shore and off shore. Consequently, the task of the analyst who attempts to model the behavior of the cost of domestic petroleum in the years to come is quite formidable. In this study, in attempting to quantify the dynamics of cost, the following relationship was established between fraction of deposit remaining in the ground $FR(k)$ versus actual cost $AC(k)$,

$$AC(k) = \frac{AC(1)}{FR^2(k)} .$$

The above relationship, as reflected in (13) below, is an estimate between cost and depletion factors. As major discoveries occur, the cost curve must be adjusted downward.

Our model in block diagram form is shown in Fig. 1. There are five level variables: domestic supply; inventory; capital investment for the production, exploration, and refining of petroleum; capital investment for research and development of new sources of energy; and substitution fraction. Each level variable is increased or decreased by its associated rates of flow. For example, the sum of SIR (supply import rate) and DSR (domestic supply rate) increases inventory, while DE (demand) decreases it. The investment rate for the production, exploration and refining of petroleum (PERIR) increases the level of investment CIPER, while PERDR decreases it, representing an obsolescence factor (equipment which is removed from active use each year). The circles in the diagram are the auxiliary variables of the system. The description and classification for all variables in the model are given in Table II. The variable population (P) is regarded as an exogenous input. As shown in Fig. 1, DE decreases DSFR which increases COST, which in turn impacts on PERIR and SF.

TABLE II
MODEL VARIABLES

Variable	Classification*	Description
DS	$x_1(A)$	Domestic supply
IA	x_2	Inventory
CIPER	$x_3(B)$	Capital investment for the production, exploration, and refining of petroleum
CIRD	$x_4(C)$	Capital investment for research and development of new sources of energy
SF	$x_5(D)$	Substitution fraction
DSR	$x_6(E)$	Domestic supply rate
TXI	u_1	Tax incentives
RDIR	u_2	Investment rate for research and development for new sources of energy
PERIR	z_1	Investment rate for production, exploration, and refining of petroleum
PERDR	z_2	Depreciation rate for production, exploration and refining of petroleum
DE	$z_3(F)$	Demand
RDDR	z_4	Depreciation rate for research and development
DSFR	z_5	Domestic supply fraction remaining
AC	z_6	Cost of petroleum
SIR	w_1	Supply import rate
P	w_2	Population

* x = state variable, u = control variable, z = intermediate variable, w = exogenous variable.

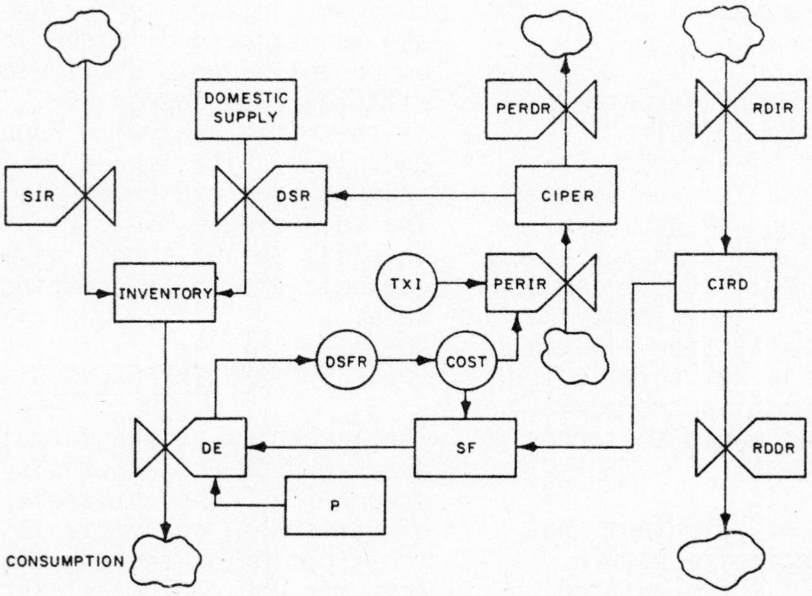

Fig. 1. Petroleum utilization in United States.

The equations for the model are

$$DS.K = DS.J + DT(-DSR.JK) \quad (2)$$

$$IA.K = IA.J + DT \cdot (DSR.JK + SIR.JK - DE.JK) \quad (3)$$

$$CIPER.K = CIPER.J + DT \cdot (PERIR.JK - PERDR.JK) \quad (4)$$

$$CIRD.K = CIRD.J + DT \cdot (RDIR.JK - RDDR.JK) \quad (5)$$

$$SF.K = SF.J + C_1[CIRD.J - 1] + C_2[C_3AC.J - SF.J] \quad (6)$$

$$DSR.JK = DSRI(e^{-DT/20}) \cdot (CIPER.J/CIPERI) \quad (7)$$

$$PERIR.JK = \alpha x AC.J \times TXI.J \quad (8)$$

$$PERDR.JK = \theta x CIPR.J \quad (9)$$

$$RDDR.JK = \delta x CIRD.J \quad (10)$$

$$DE.K = [1 - SF.K] \times DEI \times P.K/PI \quad (11)$$

$$DSFR.K = DS.K/DSI \quad (12)$$

$$AC.K = ACI(DSFR.K)^2 \quad (13)$$

The letters, *J*, *K*, and *L* following a decimal point are time step indicators; they refer to past, present, and future points in time, respectively. Also, *JK* and *KL* refer to rate variables during intervals from *J* to *K* and from *K* to *L*, respectively. Each time interval is taken to be one year. An *I* following a variable indicates the value of the variable in 1973 (all initial data refer to 1973 as these are the most recent data available). The description for the variables and their notation is given in Table II.

Equations (2)-(6) describe the five level variables in the model, (7)-(10) specify the rate variables, and the rest of the equations define the auxiliary variables. In (6), the two factors involved in the substitution are considered to be the cost of oil and the fruition of the development of alternate sources of energy. The term $C_2[C_3AC.J - SF.J]$ also contains the conservation factor in an aggregate fashion; as the price goes up people will tend to use less, some of the difference going to alternate forms of energy and some credited to conservation. The term $C_1[CIRD.J - 1]$ was constructed to reflect the time lag involved in the development of other sources of energy. With C_1 taken as 0.01, the effect of CIRD on SF will be minimal for the first five years (about 0.06 for t = 5). On the other hand, as CIRD accumulates, its effect becomes stronger. This is somewhat simpler in modeling time lags than increasing the order of the system by introducing another state variable.

Equation (7) expresses the behavior of domestic supply rate (DSR) based on both economic and geological considerations. It is well documented in the literature [13] that the production of oil in the United States has just about reached its peak and further improvements or even maintaining the present rate will require progressively higher rates of investment.

Equation (8) describes the behavior of the investment rate for the production, exploration, and refining of petroleum (PERIR). The decision to aggregate the outlays for production, exploration, and refining was made because it was realized that the actual (retail) cost of oil is the dominant factor for all three. As depicted in Fig. 1, we assumed for our model that price and cost are the same,[1] AC (actual cost) specified in (13). Usually, PERIR, private money (oil company) in its entirety, is based on total revenue. The ratio (PERIR)/(TR) is surprisingly close to 0.10 in many oil company annual reports [20], [21]. Now if we notice that

$$\Delta TR = DE \times \Delta AC$$

where TR is the total revenue for a ΔAC of \$0.001 per gallon, a ΔTR of $\$2.345 \times 10^8$ per year is generated based on total 1973 consumption of 5.584×10^9 barrels.[2] If the above ΔTR of $\$2.345 \times 10^8$ is considered extra profitability, the importance of the AC term is apparently demonstrated.

Finally, the oil companies aggregate the production and refinery expenses as capital expenses, which makes our aggregation quite realistic. The investment rate for research and development for new sources of energy admittedly has a rather weak input from the oil companies. The decision as to how much to spend each year is made by the government. Consequently, RDIR is treated as a control variable in the model.

---

[1] Average cost and price are not, of course, in general identical. However, average cost is indicative of price, and since the model here is highly aggregated, it is reasonable not to distinguish between the two. One must, however, be aware that the two are the same only under conditions of perfect competition and short-term equilibrium.

[2] 1 barrel contains 42 gallons.

Equations (9) and (10) designate the depreciation and obsolescence of capital equipment in the respective level variables.

The depreciation rate for research and development for new sources of energy is considered higher than equipment depreciation and obsolescence. Consequently, the values for δ and θ are 0.05 and 0.04, respectively. These are equivalent to discard rates of 5 percent and 4 percent per year, which are equivalent to an average life of 20 and 25 years, respectively.

In (8) the parameter α describes the propensity of the oil companies to spend a certain percentage of their revenues for the production, exploration, and refining of petroleum. In the past that percentage has been about 0.10 [20], [21]. However, the indication is [22] that this percentage will increase in the future. For example, the Alaska pipeline, already under construction, will require far greater amounts of capital than previously spent on similar projects. With PERIR and AC normalized to 1, the value of α is taken as 0.18.

The above system of equations, which describes the utilization of petroleum in the U.S., will now be rearranged in state-variable form. The state vector $\overline{X}(k)$ is defined as

$$\overline{X}(k) = \begin{bmatrix} DS(k) \\ IA(k) \\ CIPER(k) \\ CIRD(k) \\ SF(k) \\ DSR(k) \end{bmatrix} \triangleq \begin{bmatrix} x_1(k) \\ \cdot \\ \cdot \\ \cdot \\ x_6(k) \end{bmatrix} = x(k)$$

where the change to lower case letters is a matter of convention in state-variable representations, and the control vector is specified by

$$\overline{U}(k) = \begin{bmatrix} TXI(k) \\ RDIR(k) \end{bmatrix} \triangleq \begin{bmatrix} u_1(k) \\ u_2(k) \end{bmatrix} = u(k)$$

The remaining variables are the auxiliary variables $\overline{Z}(k)$ and the exogenous parameters $\overline{W}(k)$ defined by

$$\overline{Z}(k) = \begin{bmatrix} PERIR(k) \\ PERDR(k) \\ DE(k) \\ RDDR(k) \\ DSFR(k) \\ AC(k) \end{bmatrix} \triangleq \begin{bmatrix} z_1(k) \\ \cdot \\ \cdot \\ \cdot \\ z_6(k) \end{bmatrix} = z(k)$$

and

$$\overline{W}(k) = \begin{bmatrix} SIR(k) \\ P(k) \end{bmatrix} = \begin{bmatrix} w_1(k) \\ w_2(k) \end{bmatrix} = w(k)$$

respectively. Equations (2)-(13) now assume the form

$$x_1(k+1) = x_1(k) - x_6(k) \tag{14}$$

$$x_2(k+1) = x_2(k) + x_6(k) + w_1(k) - z_3(k) \tag{15}$$

$$x_3(k+1) = x_3(k) + z_1(k) - z_2(k) \tag{16}$$

$$x_4(k+1) = x_4(k) + u_2(k) - z_4(k) \tag{17}$$

$$x_5(k+1) = x_5(k) + c_1[x_4(k) - 1] + c_2[c_3 z_6(k) - x_5(k)] \tag{18}$$

$$x_6(k+1) = x_6(1)e^{-k/20}[x_3(k)/x_3(1)] \tag{19}$$

$$z_1(k) = \alpha \cdot z_6(k) u_1(k) \tag{21}$$

$$z_2(k) = \theta \cdot x_3(k) \tag{22}$$

$$u_2(k) = \gamma(k) x_4(k) \tag{23}$$

$$z_3(k) = [1 - x_5(k)] z_3(1) \frac{w_2(k)}{w_2(1)} \tag{24}$$

$$z_4(k) = \delta x_4(k) \tag{25}$$

$$z_5(k) = x_1(k)/x_1(1) \tag{26}$$

$$z_6(k) = z_6(k)/z_5{}^2(k) \tag{27}$$

$$w_2(k+1) = r w_2(k) \tag{28}$$

where use has been made of the fact that DT = 1.

Substitution of the z_i from (21), (22), (24)-(26), and (27) into (14)-(19) permits the system to be written as

$$x_1(k+1) = x_1(k) - x_6(k) \tag{29}$$

$$x_2(k+1) = x_2(k) + x_6(k) + w_1(k) + z_3(1) \frac{w_2(k)}{w_2(1)} x_5(k) - z_3(1) \frac{w_2(k)}{w_2(1)} \tag{30}$$

$$x_3(k+1) = (1 - \theta) x_3(k) + \alpha u_1(k) \frac{x_1(1)^2}{x_1(k)^2} \tag{31}$$

$$x_4(k+1) = (1 - \delta) x_4(k) + \gamma(k) x_4(k) \tag{32}$$

$$x_5(k+1) = (1 - c_2)x_5(k) + c_1x_4(k) + c_2c_3\frac{x_1(1)^2}{x_1(k)^2} - c_1 \quad (33)$$

$$x_6(k+1) = x_6(1)e^{-k/20}x_3(k). \quad (34)$$

This is in the standard state-variable form

$$x(k+1) = f(x(k), u(k), w(k)).$$

ENERGY CONSERVATION POLICY

Possible means to encourage conservation have been discussed in Congress [24], but the actual formulation of energy conservation policy has been very slow. One reason is perhaps that policy formulation, like the formulation of the objective function in an optimization problem involves the examination of several goals, often conflicting with each other, and finally arriving at the "best" compromise, in some sense. The optimal solution or policy need not be unique, as often is the case in an engineering system; rather, a set of alternatives ought to be clearly identified.

Two classes of control or policy variables have been identified in this study, dealing with the two components of energy conservation, as outlined above, curtailment and increased efficiency. Various measures such as price increases, tax increases, etc., tend to discourage consumption of energy, while measures such as tax subsidies, low interest loans, etc., tend to encourage better efficiencies and will have a more lasting effect on future growth. Indeed, the issues involved in energy conservation are social, psychological, institutional and physical consequently the task of formulating policy becomes formidable and highly complex.

One area which we have just begun to examine at the Energy Systems Laboratory at California State University, Long Beach is the application of input-output techniques to the analysis of energy conservation. These techniques have proven successful in other areas of economic analysis, and the feeling here is that the cause-effect perspective of input-output techniques provides a detailed framework for quantizing the effect of energy flow on the output of each sector of the economy.

CONCLUSIONS

In the model presented above, we have brought together certain geologic and economic considerations which are unique to the case of petroleum utilization vis-a-vis other energy forms. The unique features of our model which distinguish it from its predecessors, and which we judge are significant steps toward more complete and realistic modeling of petroleum utilization, are 1) its dynamic nature, 2) the state variable format which makes clear the distinction among state, control, and auxiliary variables, and 3) the introduction of nonlinear relationships among the variables as especially warranted by our treatment of the geologic and economic factors noted above.

As with any model of the Forrester type, there is uncertainty regarding the numerical values assigned to the coefficients and exponents in the equations. While we are readily able to parameterize our simulations, neither we nor anyone else can be certain of what population growth or the cost of synthetic oil will do over the next ten or even five years. Such models thus are useful in indicating gross behavior and trends that can result as a consequence of existing (current) conditions. The parameterization provides the "envelope" for this behavior.

Our results do not (nor are they intended to) provide a "master policy" for reaching certain national objectives or goals by a fixed deadline. Rather, in classifying the policy tools available to the policy maker into two categories, based on their effect on increasing the domestic supply or the development of alternative energy sources, we provide a basis for weighing certain choices and trade-offs in considering a variety of energy policies regarding petroleum.

The thrust of our work here has been to present and explore the consequences of some aspects of the petroleum supply-demand-utilization picture which have not been given adequate consideration previously, and to do so in the framework of nonlinear, dynamic systems with control inputs. We judge our effort to be one more step in a series of steps that have already been taken and that yet remain to be taken in understanding and ultimately being able to utilize petroleum and other sources of energy prudently.

REFERENCES

[1] E. Cook, "The economy of energy and materials," manuscript prepared for the "Earth 2020" Faculty Institute sponsored by NASA, San Diego, CA, Summer 1974.

[2] M. K. Hubbert, "Energy Resources: A report to the Committee on Natural Resources," National Academy of

Sciences--National Research Council, Publ. 1000-D, Washington, D.C., 1962.

[3] A. D. Zapp, "Future Petroleum Producing Capacity of the United States," U.S. Geological Survey Bulletin 1142-H, 1962.

[4] T. Jordanides and J. Meditch, "Some Aspects of Modeling Petroleum Utilization in the United States," IEEE Transactions on Systems, Man and Cybernetics, vol. SMC-7, No. 4, April 1977.

[5] J. Forrester, World Dynamics, Cambridge, MA: Wright-Allen Press, 1971.

[6] K. C. Hoffman, "The United States energy system--a unified planning framework," Ph.D. dissertation, Systems Engineering Dept., Polytechnic Institute of Brooklyn, June 1972.

[7] Policy Study Group of the MIT Energy Laboratory, "Energy Self-Sufficiency: An Economic Evaluation," Technology Review, vol. 76, no. 6, pp. 28-58, May 1974.

[8] W. E. Mooz and C. C. Mow, California's Electricity Quandry: Estimated Future Demand, Rand Corporation, Report no. R-1084-NSF/CSRA, Santa Monica, CA, Sept. 1972.

[9] W. Leontief, The Structure of the American Economy, 1919-1939, Oxford U. Press, New York, 2nd Edition, 1951.

[10] A. P. Carter, "Applications of input-output analysis to energy problems,' Science, vol. 184, no. 4134, April 1974.

[11] A. Avamides and J. Cross, "NPC Analysis of oil and gas supply," in Energy Modeling, Milton Searl, Editor, Resources for the Future, Inc., Washington, D.C., pp. 304-315, 1973.

[12] Federal Energy Administration, "Project Independence report," Washington, D.C., November 1974.

[13] M. K. Hubbert, "Energy Resources," in Resources and Man, P. Cloud, Ed., W. H. Freeman and Co., San Francisco, 1969.

[14] E. J. List, "Energy use in California: Implications for the environment," California Institute of Technology, Environmental Quality Laboratory report no. 3, December 1971.

[15] D. R. Limaye, "TERA--the total energy resource analysis model," Gas Magazine, November 1972.

[16] D. Meadows, et al., Limits to Growth, New York: Universe Books, 1972.

[17] H. Landsberg, Resources in America's Future, Resources for the Future, Inc., Washington, D.C., 1963.

[18] ________, "Learning from the past: RFF's 1960-1970 energy projections," in Energy Modeling, Milton Searl, Ed., Resources for the Future, Inc., Washington, D.C., pp. 416-436, 1973.

[19] National Economic Research Associates, Inc., Energy Consumption and Gross National Product in the United States: An Examination of a Recent Change in the Relationship, New York: March, 1971.

[20] Exxon Corporation 1973 Annual Report, Exxon Corporation, 1251 Avenue of the Americas, New York, NY.

[21] Standard Oil Company of California 1973 Annual Report, Standard Oil Company of California, 225 Bush Street, San Francisco, CA.

[22] T. Jordanides, "Optimization of U.S. energy policy: Objectives and alternatives," Ph.D. dissertation, SEOR Rept. no. 75-4, School of Engineering, University of California, Irvine, June 1975.

[23] T. Jordanides and J. S. Meditch, "On the management of U.S. energy resources," Prod. of Int. Conf. on Cybernetics and Society, San Francisco, CA, pp. 269-271, Sept. 1975.

[24] "Conservation and Efficient Use of Energy," PTS 1-4, Hearings before House Subcommittee on Science and Astronautics, May 1973.

DYNAMIC NET ENERGY SIMULATION MODEL OF THE EEC ENERGY TRANSFORMATION SYSTEM: PHYSICAL VIABILITY OF ECONOMIC CHOICES

Malcolm Slesser

Head, Systems Analysis Division, EURATOM, Ispra, Italy

ABSTRACT

Evaluation of net energy is sensitive to the system boundary chosen. However by simulating an entire national or international energy transformation system in a dynamic model, the system boundary problem is resolved. The model built to analyse the net effect of the EEC's energy transition programme is briefly described and some of the results presented.

INTRODUCTION

Net energy was the topic of a meeting convened by Stanford University, under NSF auspices, in August 1975 (1). Almost every shade of opinion was represented, and perhaps for this reason, one outcome of the meeting was a reluctance to formulate any algorithm for net energy. In many ways this was a sad outcome, yet the reasons were not far to seek, and were principally three in number: there was no agreement as to the virtue of net energy as a measure, whether of energetic or economic efficiency; there was no agreement on what constituted the correct system boundary; there was no agreement on how to handle the matter of energies of different quality, such as coal and electricity.

A few months later the Energy Studies Unit of the University of Strathclyde was commissioned by the EEC to make a dynamic net energy model of the EEC energy transformation system. The objectives were quite straight forward: to assess, as a function of time the net amount of fuel entering the demand sector (that is to say the non-energy sector) of the EEC economy, the total amount of energy resource needed to furnish this fuel, and the energy content of the feedback from the demand sector to the energy sector in order to sustain the growth and change in the energy sector. This paper explains the methodology used, and how, to a great extent, we were able to solve the system boundary problems raised by the Stanford meeting. The paper concludes with a summary of the results obtained and their significance.

System Boundary

Though our aim was to produce a dynamic model, it was not intended that it should be entirely endogenous, such as the Club of Rome sponsored World 3 model of Meadows and associates (2). Rather it was to be dependant on an endogenous economic scenario. The assumptions taken were the EEC's stated objectives for energy supply and mix in the year 1985 (3), starting from a known energy mix in the year 1973. The point of origin of our work was to disentangle the numbers given, for they were reported at different system boundaries, notwithstanding the fact that they are officially added as if they were at the same system boundary. Table 1 lists the stated objectives, the relevant system boundaries as indicated in Fig. 1, and the actual energy totals using the outer system boundary (energy resource in the ground) and the inner system boundary (fuel flowing into the demand sector). It is part of our contention that energy figures can only be summed when they are measured at the same system boundary.

The Model

The model used was a 'top-down' model, that is to say each energy sector was independantly modelled. Thus, if we treat each sector for the moment as a black box, each was furnished with three inputs: the energy resource in question, thermal inputs from other sectors or from the demand sector of the economy (all summed as heat) and electricity from other sectors or from the demand sector (all summed

TABLE 1 EEC Energy Use in 1973 According to System Boundary: MTCE (thermal)

Energy source	1973 official figure	System boundary measured (Fig. 1)	At system boundary 1	At system boundary 6
Solid fuel	315	3 and 5	315	178
Oil	886	3	874	650
Natural gas	167	3	165	128
Hydro- and geothermal	35	6	renewable	as electricity
Nuclear	20	6	145	as electricity
Electricity	nil		nil	130
Total	1,423		1,499	1,086

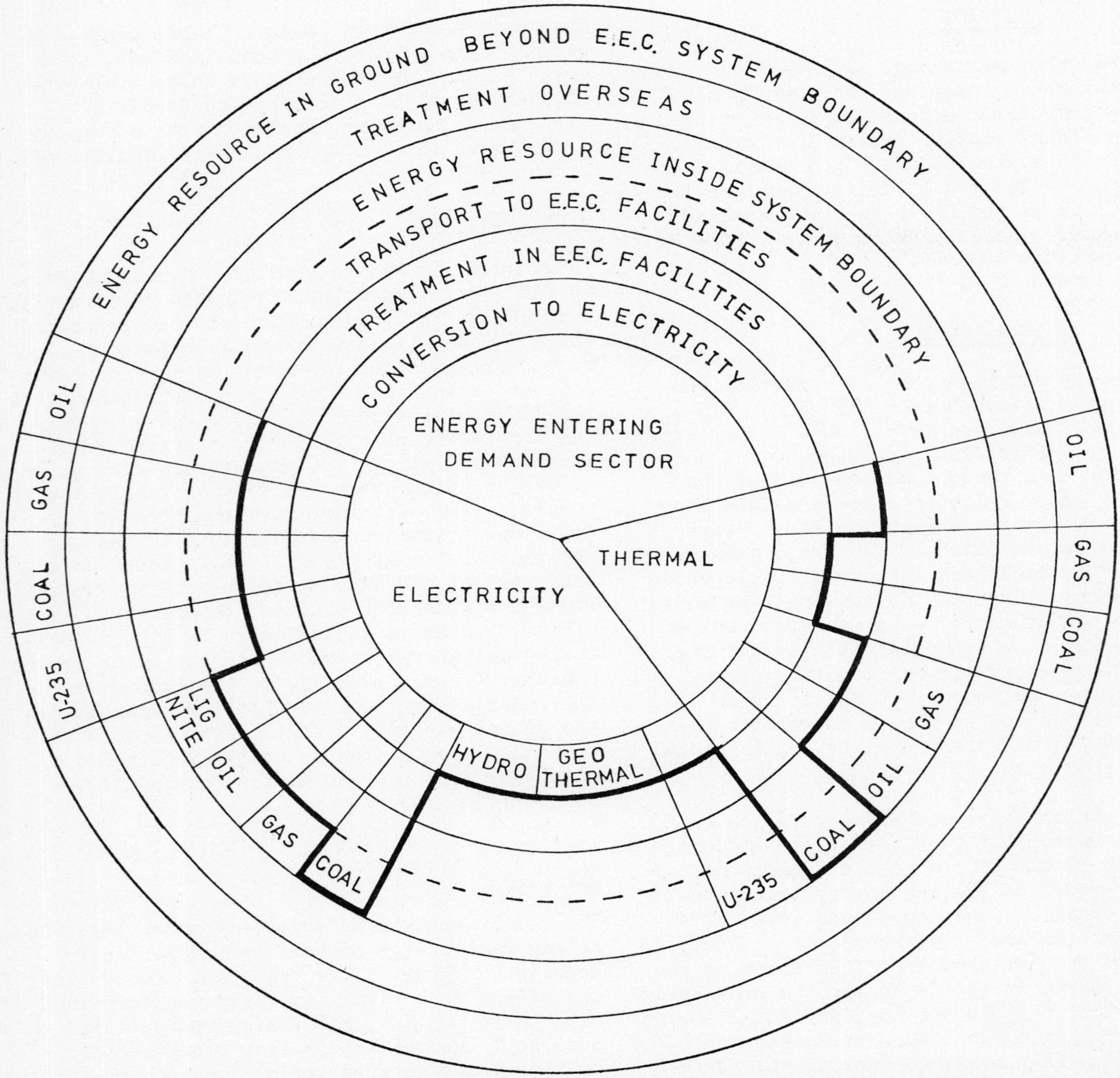

Fig. 1. System boundary diagram EEC energy transformation system

as mwh electric). Each black box had two outputs: thermal energy or electrical energy, the former summed as terajoules, the latter as mwh electric. The net amount of these two, that is after all demands by the black boxes have been met, passes to the demand sector, and constitutes the fuel supply, in two types, heat and electricity.

Within the black boxes each energy sector was modelled using the system simulation language 'Dynamo'. Dynamo uses 'levels' as a conservation device, and expresses change in terms of rates. The simple algorithm is:

$$\text{Level at next time interval} = \text{Level now} + (dt * \text{rate of change})$$

Energy was thus treated as a level, and power as a rate. Of course other essentials like the total number of power stations or refinery capacity were also treated as levels or auxiliaries. Rates of construction were computed to match the EEC objectives, duly modified to the appropriate system boundary. Figure 2 exhibits the hard coal sector, and one can see many feedback loops controlling the net output of the coal sector or black box. This sector is in fact one of the simplest, the nuclear sector being the most complex.

It was found that the simulation was satisfactory at a time interval, dt, of one month. Figure 3 indicates how the sectors interlink.

Electricity

A crucial computation in the model was the gross energy requirement (GER) for electricity, using GER in the sense defined in the IFIAS report on energy analysis methodology (4). Where electricity is internally generated and used it may have one GER value, but where it enters the international grid, one cannot identify electricity from Italy or Scotland, or whether from nuclear sources or lignite. It was essential therefore to have an instantaneous value of the GER for electricity in the whole of the EEC. The value of a dynamic simulation model is that this is readily generated, and the only inaccuracy is in assuming that the GER for electricity used in this month's operations is based upon last month's primary fuel requirements. In sensitivity testing it was found that whatever the initial value chosen for the GER of electricity, the model settled down to its own inherently consistent value within six months of initiation. Of course, where electrical and thermal inputs could be kept completely separate, that was done.

Data

No algorithm for net energy was used. Rather energy transformation systems were analysed for the energy flows within each system. Data sources were many and varied, and all have been published. Of particular value was the energy analysis by the Colorado State Energy Institute (5) at Boulder and the energy analysis of pressurised water reactors made by the Institute for Energy Analysis (6) at Oak Ridge. In addition it was necessary to compare these figures with those produced from known European sources. Only one internal energy analysis was carried out, and that was for North Sea oil production, where the main factor is the energy required for investment in the system (7).

Nuclear Energy

In the literature there are three current conventions for dealing with the resource equivalent of nuclear energy:

1. Ascribing to nuclear electricity an energy equivalent equal to the current average fossil energy requirement for electricity. This method, used in all official energy statistics except UN, does not reflect the resource element of nuclear fuel. (UN statistics treat all electricity in terms of its thermal equivalent.) This method cannot be used to establish net energy.

2. Treat uranium as a non-fuel.

3. Ascribe to nuclear electricity the amount of uranium 235 consumed in its production and convert that into its equivalent potential fissile heat.

It is a convention in energy analysis that energy containing products of a system should be credited to the energy resource consumed by the system. Since the energy containing products can only yield energy under certain conditions of investment and technology, no credits are actually introduced into the model. Instead the cumulative production of fissile plutonium and uranium plus unconsumed uranium 238 are summed, and become an output of the model. This is an important advantage of a dynamic model. If the proposed European nuclear programme goes forward, the amount of by-product uranium 238 in storage by 1985 is enough, given breeder reactors, to satisfy the entire European energy needs for the next century. This, however, is only true in

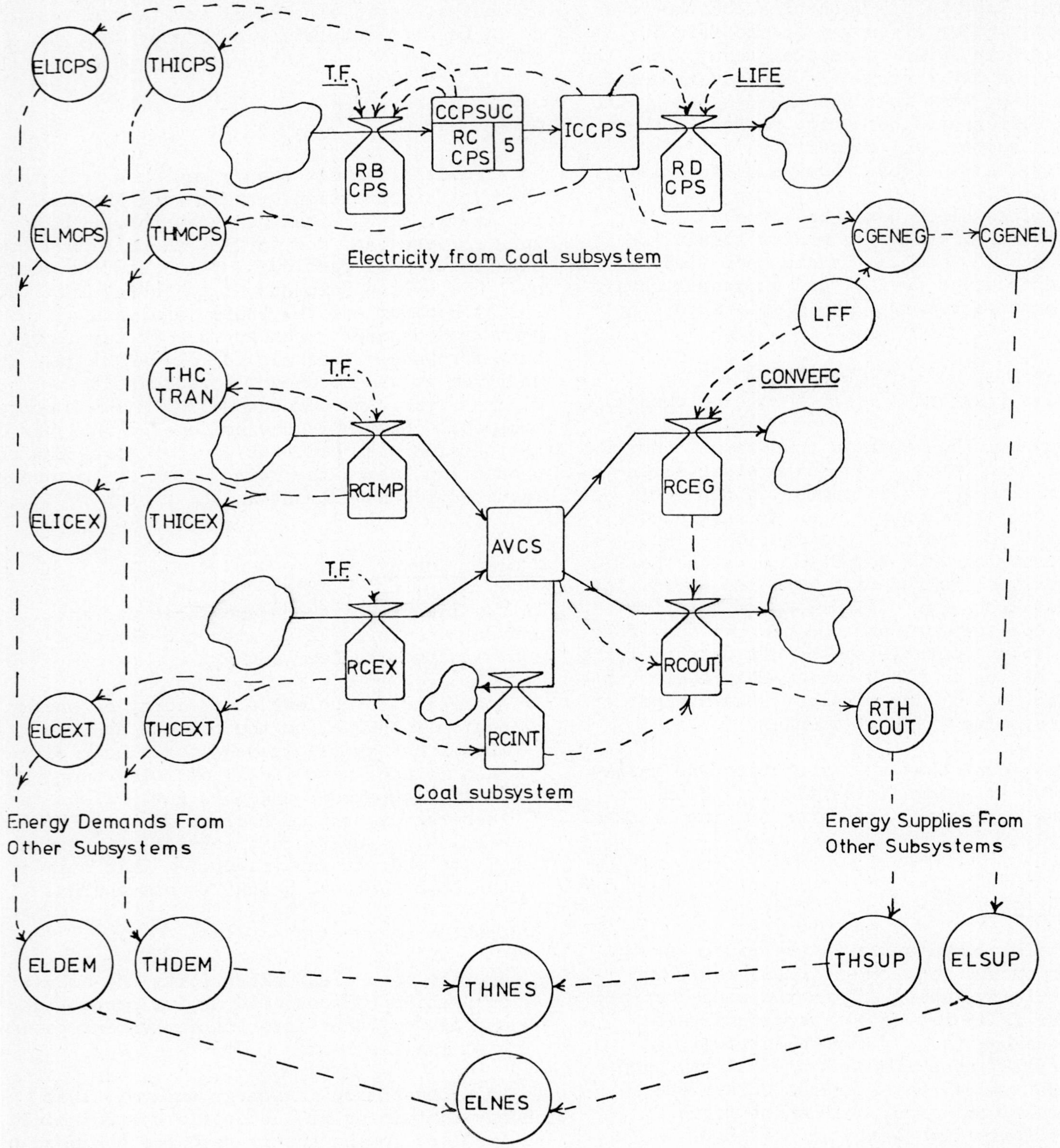

Fig. 2. Coal and electricity from coal subsystems

principle. Until the breeder system is introduced into the model, and the entire fuel and investment cycle incorporated, no-one can say how long it will be before the breeder system is a net energy producer.

Output

In a paper of this brevity it is impossible to list the range of outputs yielded by a dynamic model. The net energy picture is summarised in Fig. 3 from which it may be seen that if uranium is treated as a fuel (convention 3) the system GER rises from 1.38 in 1973 to 1.52

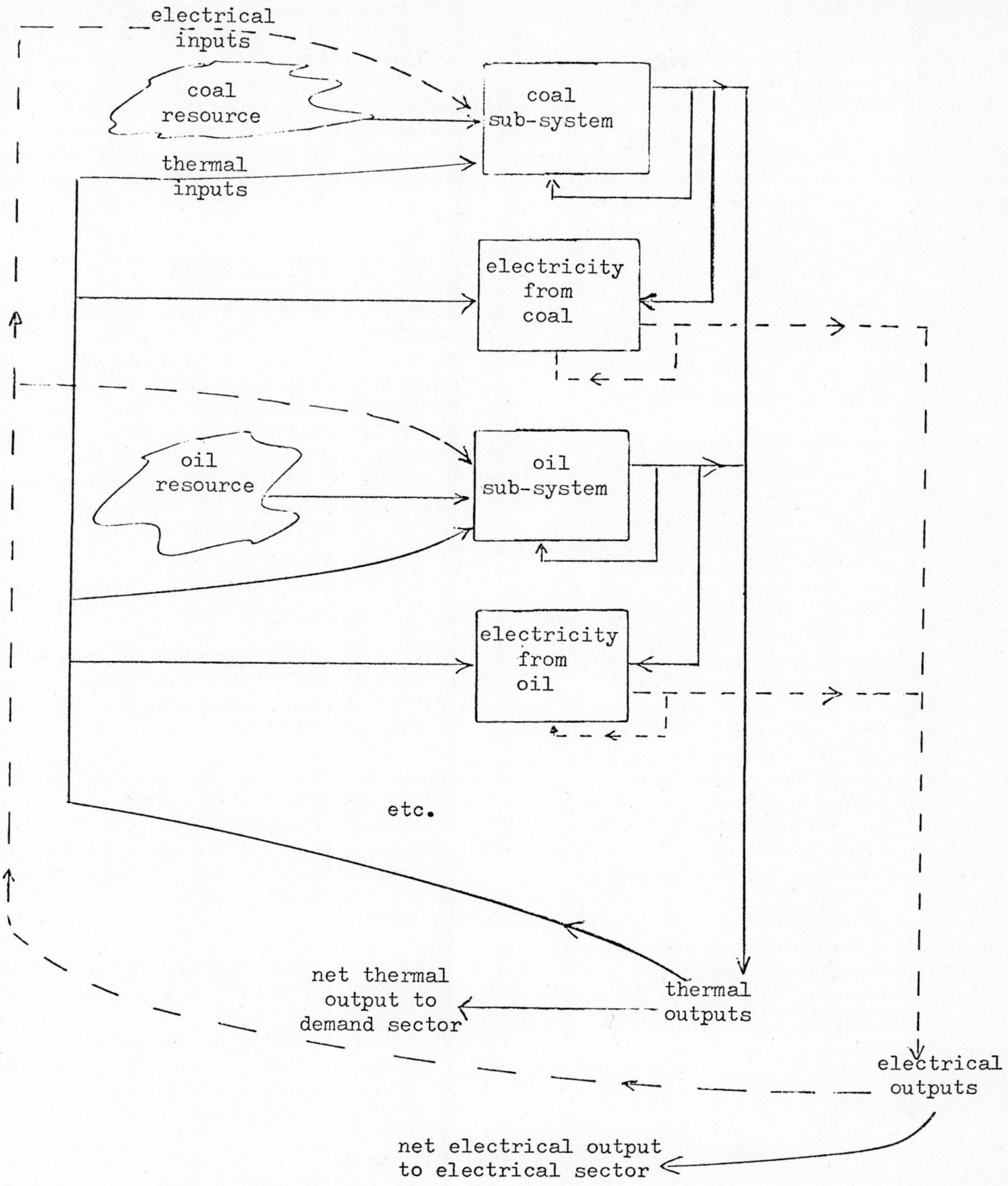

Fig. 3. Two elements of a national energy transformation system

in 1985, while the average electrical GER rises from 3.0 in 1973 to 3.6 in 1985, notwithstanding improvements in technology or load factor. On the other hand if the policy is to consider uranium as a non-fuel, and hence essentially a fossil fuel extender, the system GER starts at 1.25 and after a short rise due to the high rate of nuclear reactor construction, falls away to 1.17 in 1985. While on the face of it this suggests a great deal of support for the virtues of nuclear power, one must remember that if eventually nuclear power became the dominant energy source, then this fiction would simply lead to false conclusions, such as we have seen in economics where air and water were for long treated as free common goods.

Nuclear energy technology, however, creates its own powerful message. By 1985 the accumulated energy content of plutonium, U235 in the tailings and U235 in spent fuel will amount to 43,000 TJ or roughly 1.2 x 10^9 tons of coal equivalent, and very many times that if stock-piled U238 is included.

A dynamic model can very conveniently accumulate total radioactive wastes, taking into account decay. Strontium 90 in storage would amount to some 700 M curies by 1985 while Caesium 137 would amount to almost a 1000 Mcuries. Models of this type can conveniently handle many variants on a theme. It could, for example, separately accumulate long and short life actinides, or deal with the level of wastes in a breeder programme aimed at, inter alia, burning up wastes.

An exceptionally interesting output is the amount of energy recycled from the demand sector to the energy transformation sector. Largely this is an investment and maintenance supply. Where, as in the EEC situation there is a move to new and expensive technologies, this investment rises from about 2.8% of the fuel entering the demand sector in 1973 to a peak of 3.1% in 1983. Under an earlier EEC energy scenario now abandonned, this investment would have amounted to 4.4%. This figure may seem small, and some analysts have regarded it as trivial. However, if we apply to such figures current energy intensity data (energy per unit money cost) such as generated by Chapman (8) and Herendeen (9), then we find that even the lower figure amounts to an investment rate in the energy transformation sector of the EEC of 31 x 10^9 US dollars (1973) a year by 1983. Such figures are very close to estimates made by more traditional econometric methods, but are in this instance arrived at easily as a by-product of the model.

CONCLUSION

By creating a dynamic simulation model of a complete energy transformation system, as opposed to a static analysis of one element of such a system, the system boundary problems inherent in the static model are removed, and no assumptions need be involved. Such a model can generate a wealth of time-variant by-product information. It is still too early to state whether the net energy flow (fuel) entering the demand sector is a more useful number than the primary energy consumption required to meet that flow, but at this stage this much may be said. Firstly, there are quite reasonable theoretical grounds for thinking net energy is the more important of the two numbers so far as economic computations are concerned. For example, the net energy per unit of GNP in the EEC turns out to be a more consistent figure than when reckoned in gross energy terms. Secondly, if energy resources are important to our reckoning, then it is important to consider the energy resource consequences of any given economic scenario.

ACKNOWLEDGEMENTS

This paper is presented under the auspices of the Euratom Joint Research Centre of the EEC, at Ispra, Italy, and is based on work done in collaboration with R. Veitch, A.R. Gloyne and R. Peckham at the Energy Studies Unit, Strathclyde University, in 1976. The work was carried out under contract 204-75-10-EC1-UK on behalf of the European Commission. The results and conclusion do not necessarily reflect those of the Commission.

REFERENCES

(1) Report of NSF-Stanford Workshop on net energy analysis: T.J. Connolly and J.R. Spraul, eds., Stanford University, California, 1975.

(2) Meadows, D.L., et al, The Limits to Growth Universe Books, NY, 1972.

(3) Commission of the European Communities, Prospects of Primary Energy Demand on the Community (1975-1980-1985), Brussels 1976.

(4) Energy Analysis, Report No. 6, IFIAS, Ulriksdals Slott, S-17171 Solna, Sweden.

(5) Net Energy Analysis: an energy balance study of fossil fuel resources, Energy Research Institute of Colorado, Golden, Colorado, USA, April 1976.

(6) Rotty, R.M., Perry, A.M. and Reisler, D.B., 'Net energy from nuclear power', Institute for Energy Analysis, Oak Ridge Associated Universities, IEA report 1975.

(7) Energy requirement of the Forties Field, North Sea, E. McLeod, B.Sc. thesis, Energy Studies Unit, Strathclyde University, Glasgow, 1976.

(8) Chapman, P., Energy analysis of the census of production, Open University Energy Research Group report ERG006, 1975, Open University, England.

(9) Herendeen, R.A., An energy input-output matrix for the USA, 1963; User's Guide, 1973, Centre for Advanced Computation, University of Illinois at Urbana-Champaign, Urbana, Illinois, USA.

MODELING RESIDENTIAL HEAT LOAD FROM EXPERIMENTAL DATA: THE EQUIVALENT THERMAL PARAMETERS OF A HOUSE

Robert C. Sonderegger
Center for Environmental Studies, Princeton University
Princeton, New Jersey, 08540

ABSTRACT

Two different approaches to the determination of heating and cooling loads in buildings are compared: deterministic models and methods based on equivalent thermal parameters (ETP's) of a building. While the former are algorithms for the determination of all heat transfer through the shell of any building, the latter is a data oriented approach that infers the ETP's of a particular building by multiple correlation of load data and weather. The ETP method is convenient to provide a rank ordering of different houses by their energy performance and to assess the overall effects of retrofits on a house. Like deterministic methods, it can also predict accurate heating loads as a function of weather.

A convenient set of ETP's is established for a gas heated residential townhouse. Multiple step regressions with weather, of gas consumption data with and without auxiliary electrical heat yield an estimate of the first ETP, the furnace field efficiency, of 76.5%. Except for the "equivalent solar window area", the other ETP's obtained from data regressions reproduce satisfactorily the results of detailed calculations. The overall effect of retrofits is tested by comparison with the ETP's obtained from a pilot study with no attic insulation, performed before caulking of the house shell.

INTRODUCTION

A large number of methods have been developed to predict the heating/cooling load or the indoor temperature of a building in response to weather and usage profiles (Ref. 1-2). Most of these methods involve computer oriented algorithms that balance all heating and cooling terms caused by equipment, appliances, outdoor weather and other factors. The time delays caused by heat conduction through walls greatly add to the complexity of these algorithms. While many of the resulting large computer packages have been increasingly successful in predicting energy requirements for buildings of widely different construction and location, they also are inherently deterministic: once all input data concerning building construction, weather and operating schedules are fed into the computer, the predicted temperatures and/or heating or cooling loads are fixed. If this output does not agree with measured data to the user's satisfaction, there is no obvious way to correct the prediction.

A different approach, using equivalent thermal parameters (ETP's) of a building, has been proposed (3-4): Instead of telling the computer how the building is built and asking it for the heating load, one tells the computer the measured heating load and asks it for the building parameters -- parameters (ETP's) describing what the building is like. The same can be done using cooling load and/or indoor temperature as input instead of heating load. A convenient set of ETP's is 1) total heat loss rate per unit temperature difference, 2) equivalent solar window area, 3) constant heat loss rate to the ground, 4) equivalent thermal mass and 5) furnace field efficiency. Why the furnace efficiency is included as a building feature (rather than as an equipment standard) and the definition of the other ETP's is the main subject of this paper.

Such a set of ETP's for a particular building can thereafter be used to estimate the same building's thermal performance in other weather situations; or, it can quantify the building's performance before and after retrofits. In many cases, the ETP's are best thought of in analogy to measured "miles per gallon" for a car. It is important to recognize the distinctions in the philosophy and the applicability between the deterministic and the ETP methods. Once the actual data on heating/cooling load and indoor temperature of a building exist, it is comparatively easy to obtain the building's ETP's. This can be done without knowing anything else about the building. Specifically, no floor plans or other construction data are needed. In fact, to a certain extent, we get them as output, in the form of ETP's. To be able to catalogue the energy requirements and the thermal response of any house by a quite general set of three or four numbers should look attractive to policy makers, real estate agents and homeowners. The same set of numbers also facilitates a meaningful assessment of the overall effects of retrofits on a house. It will be shown that one can obtain the ETP's of a house cheaply enough to make them useful.

Deterministic methods, in turn, can tell you what (almost) any building does anywhere, not only the one on which you have load data; they are especially useful when the building exists only on the drawing board. The price for this generality is a large and complex program and considerable paper work on input sheets, listing everything from the composition of each wall layer to the efficiency of the heating system.

There are also hybrid approaches like estimating a set of ETP's from load data predicted by a deterministic computer package (5). In this paper, in turn, we will compare the ETP's obtained from measured load data with what is calculated by a very detailed accounting of all heat transfer mechanisms between the house and the environment.

The ETP method has been applied to experimental buildings (3). This paper describes a similar approach, including the contribution by the sun and internal heat sources, applied to an actual occupied townhouse. The fuel burned by the thermostatically regulated, central gas furnace in this townhouse, the outside temperature, the solar flux and the electric consumption are continuously monitored. The house's ETP's obtained from correlation of this data are discussed in view of their physical interpretation and compared to what is calculated using detailed floor plans and construction data. The amount of furnace gas "saved" when auxiliary electric heating is provided, gives a direct estimate of the furnace field efficiency, independent of design calculations. A similar experiment, with a free-floating inside temperature (as opposed to a thermostatically controlled indoor temperature) was also performed and is described elsewhere (6).

The experiment described in this paper is part of the ongoing project on energy conservation carried out by the Center for Environmental Studies of Princeton University at Twin Rivers, New Jersey, a 3000 dwelling, Planned Unit Development (7). The experiment, outlined in the following section, was carried out in a three-bedroom, two-story wood-frame townhouse, occupied by the author and contiguous to other identical units within the same block.

A SIMPLE HEAT LOAD MODEL

A general model relating the change in time of the inside temperature to the inside-outside temperature difference and the heat added by furnace, sun and appliances is described in (6). When the inside temperature is held constant by a thermostat, the resulting model equation is considerably simplified to a simple energy balance of the heat gains provided by the furnace, the sum of all interior heat sources, the heat conduction through all house boundaries, the air infiltration and the sun.

$$\varepsilon G = -(E+P+B+L) + H(T-TO) - AS \quad (1)$$

- G is the rate of energy released by the furnace gas combustion [Watt];
- ε is the furnace field efficiency;
- E,P are the rates of energy released by all electrical appliances, including special resistance heaters introduced for experimental purposes and the people in the house [Watt];
- B is the rate of "constant" basement and neighbor heat gain [Watt];
- L is the latent heat gain of the

humidifier (usually negative) [W];
H is the equivalent heat transfer per unit inside-outside temperature difference [Watt/°C];
T is the average indoor temperature;
TO is the outdoor temperature [°C];
A is the equivalent solar window area [m^2];
S is the solar flux impinging on the south wall of the house [Watt/m^2].

The relatively light construction of the outside facing walls and the constancy of the indoor temperature make the inclusion of a heat storage term unnecessary. The same is not true when we let the indoor temperature float (disabling the thermostat) (6). Then, the considerable thermal masses of interior walls and floors and of the cinder block firewalls enter into action. To be sure, these masses always store heat, but little of it is released and little additional heat is absorbed as long as we keep the indoor temperature constant.

We will consider the simple model we outlined only for situations where the right hand side of eq. 1 remains positive. Warm outdoor temperatures or high solar or appliance heat gains that can cause the furnace to shut off for a long period of time have the effect of a transition to the free-floating indoor temperature case.

Equation 1 expresses the fundamental energy balance in a house at the most elementary level. Basically, it is an instantaneous degree-day model with a reference temperature that varies with the sun and the internal loads. The parameters ε, B, H and A constitute a set of equivalent thermal parameters for this house. The field furnace efficiency, ε, is defined as the ratio of the portion of combustion energy useful to heating the house and the total energy released by combustion. The "constant" basement and neighbor heat gain, B, includes all those heat transfers that vary very slowly when compared to the variations in outside temperature and solar flux, e.g. heat gain from the ground (usually negative) and from the neighboring houses. The non-negligible dependence of B on the neighbor temperature is a peculiarity unique to rowhouse construction. The equivalent heat transfer constant, H, is the sum, per unit inside-outside temperature difference, of the rates of heat conduction through the walls, the windows and the attic, and the rate of heat convection caused by air infiltration.

The exposed walls of this townhouse face approximately south and north. For most of the winter season, no direct sunlight reaches the northern side. Therefore, for the particular case of our south-facing townhouse (and possibly for a detached house with no windows on either east or west wall), we can define the equivalent solar window area, A, as the area of the 100% transparent, perfectly insulated opening in the south wall that would allow for the same degree of indoor solar heating as what is actually attained. It gives the net effect of the sun shining through the windows, as well as heating the opaque outside walls and roof (7).

Both the equivalent solar window area, A, and the constant heat gain, B, are seasonally variable: they do not quite fit the description of "ideal" equivalent parameters unique to the home. However, it is shown in the appendix that the seasonal dependence is relatively easy to estimate, without the need for separate data for every month of the year.

The model described by eq. 1 can now be applied to experimental data. We measure (in our case, every 20 minutes) the gas consumption, the energy released by electric appliances and people, the indoors and outdoors temperatures and the solar flux impinging on the south wall. From this data we will infer, by multiple step regression analysis, the equivalent thermal house parameters ε, H, A and B.

REGRESSIONS AND FURNACE EFFICIENCY

The experiment presented in this paper was carried out during 7 days of last winter's natural gas crisis (January 1977). Figure 1 shows the most relevant of the collected data. The indoor temperature is obtained as an average of 12 thermistor probes distributed over the two living space floors. The abnormally high outdoor temperature excursion on January 28 indicates a heavy winter storm on that day. During the first 47 hours of the experiment, two portable electrical heaters equipped with fans placed upstairs and downstairs provided a constant 2.92 kW heat source assisting the furnace in heating the house.

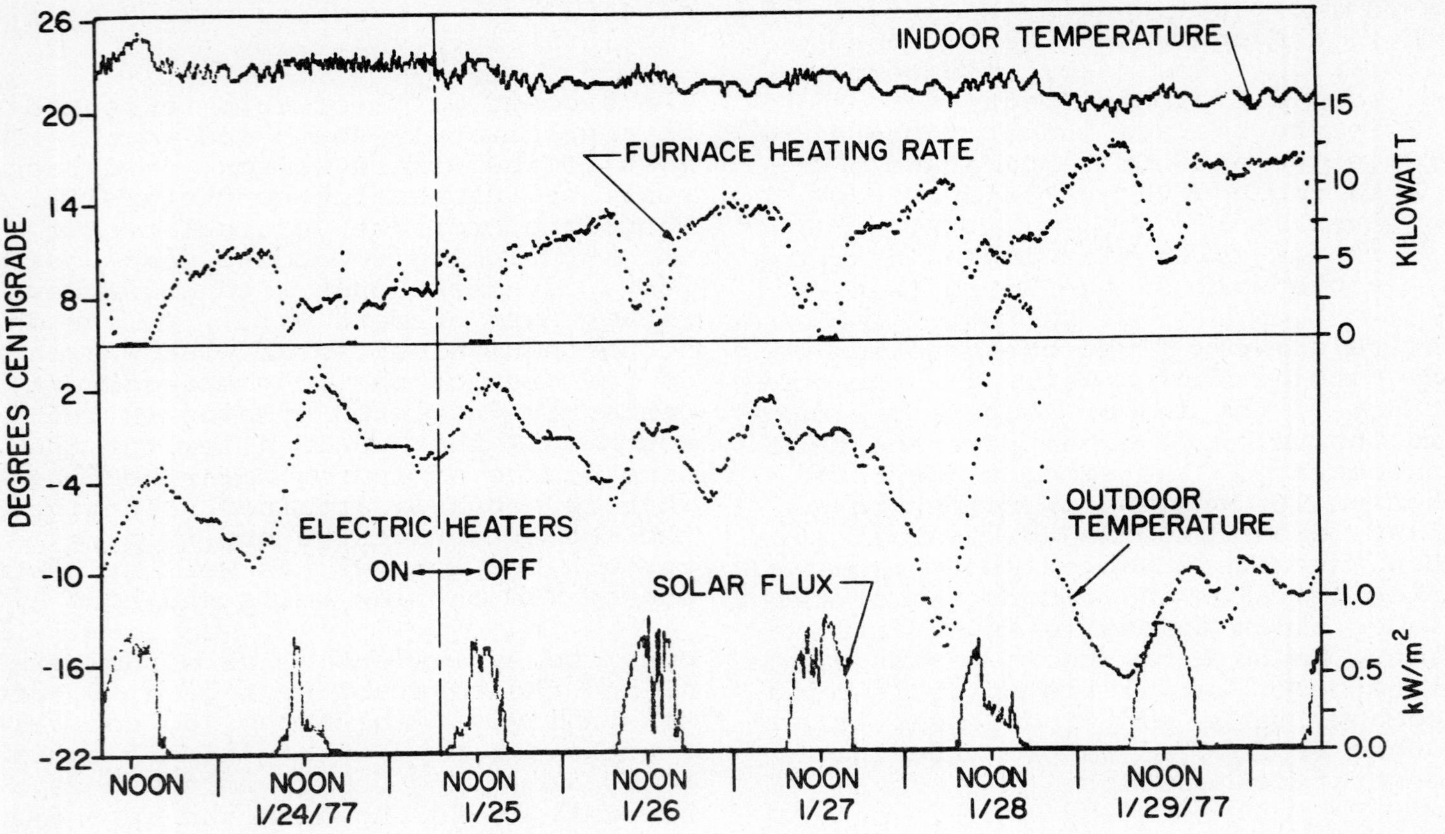

INDOOR AND OUTDOOR TEMPERATURE, FURNACE HEATING RATE AND SOLAR FLUX ON SOUTH FACING WALL

FIG. 1

The equation used for the regression of these data is of a form consistent with the original model in eq. 1:

$$G = -\underline{c} + \underline{b}(T-TO) - \underline{a}S \qquad (2)$$

$\underline{c}$, $\underline{b}$, $\underline{a}$ are the regression coefficients;

G, T, TO and S conform to the nomenclature below eq. 1.

Two separate regressions were carried out: one for the data recorded while the electric heaters were operating, and one after they were shut down. The resulting coefficients and the R-squared's are shown in Table 1. The numbers in parentheses under the coefficients indicate the t-statistics (11) of the estimates.

As one would expect, the coefficient in inside-outside temperature difference, $\underline{b}$, changes little after heater shutdown. The change in the solar coefficient, $\underline{a}$, is a statistical artifact discussed in a later section.

The dramatic change in the intercept, $\underline{c}$, between the two experiment periods reflects the drop in electrical heat gain after heater shutdown and gives us a measure for the furnace field efficiency, ε: The relation between the drop in the intercept in eq. 2, $\Delta\underline{c}$, and the concurrent drop in the constant "energy" term in eq. 1, $\Delta(E+P+B+L)$, is

$$\Delta\underline{c} = \frac{\Delta(E+P+B+L)}{\varepsilon} \qquad (3)$$

TABLE 1 Regression Coefficients During and After Electric Heating

		$\underline{c}$	$\underline{b}$	$\underline{a}$	R^2
Electric Heaters	on	3.399 kW	279 W/°C (12.6)	4.35 m^2 (9.7)	0.67
	off	-0.519 kW	295 W/°C (28.4)	6.35 m^2 (24.4)	0.80

From Table 1 we know $\Delta\underline{c}$ to be 3.918 kW. The change in (P+B+L) before and after heater shutdown is extremely small (a mere 81 Watts), while the much larger drop in electrical heat, ΔE, is very well determined by direct measurements. Thus we obtain a reliable value of 2.996 kW for Δ(E+P+B+L). Solving eq. 3 for ε, we obtain an estimate of the field furnace efficiency of

$$\varepsilon = \frac{2.996 \text{ kW}}{3.918 \text{ kW}} = 0.765 \pm 0.030 \qquad (4)$$

The indicated error margin reflects the uncertainty in the computation of Δ(E+P+B+L) and $\Delta\underline{c}$. Again we should point out that this estimate of ε results simply from looking at how much gas heat we "saved" by turning on a known amount of electric heat. In a more traditional approach one would calculate how much heat is needed to keep the house at the recorded indoor temperature, given the outdoor weather, and divide the heat load thus obtained by the recorded heat rate supplied by the furnace. However, such heat load calculations are theoretical and based on many sometimes unwarranted assumptions, like the rate of air infiltration. The lack of need for a heat load calculation in the method presented in this paper eliminates an important source of inaccuracy.

REMARKS ABOUT FURNACE EFFICIENCY

The furnace field efficiency obtained in the previous section appears somewhat high compared to what is usually assumed and measured. Earlier measurements of the same type of gas furnace yielded a laboratory plenum efficiency of 75%-80% (Ref. 8). But in a real life setting, under partial load, one would expect this number to drop significantly.

A major cause for our estimate to be a high 76.5% is attributed to our including the basement, where the furnace is located, as part of the living space. This is perfectly legitimate for those houses with finished basements (a sizeable and growing number in Twin Rivers); it is less legitimate when the basement is used mainly as a storage area. To unravel this ambiguity we must go back to a proper definition of furnace field efficiency. To say that the field efficiency of the furnace is 77% is equivalent to saying that 23% of the chemical energy released by natural gas combustion are lost, either by direct exhaustion to the outside (e.g. through the stack) or by combustion inefficiencies. The remaining 77% are "useful" heat, keeping the living space of the house warm, basement included. The definition of efficiency becomes fuzzy as soon as we decide to exclude a room or the basement from the living space. Then the portion of furnace heat reaching that room through the heating system is effectively lost, decreasing the actual furnace efficiency. For instance, excluding the basement would drop the furnace efficiency in our townhouse to about 66%. A mere decision by the occupants, without any physical change, can decrease the effective furnace efficiency! Clearly the furnace efficiency alone is a poor measure of efficient energy use and must be defined more strictly, since a small family can effectively reduce it, simply by not using some space of their home.

For the purposes of this paper we will stick to the "full use" definition of furnace field efficiency that draws the house boundaries around all rooms <u>and</u> the basement. When striving for efficient energy use, though, we must not only ask for a high furnace field efficiency, but also for high equivalent thermal parameters A and B, a low <u>H and</u> a high occupancy of the living space. Retrofitting existing houses and their heating systems will save energy; inducing families to buy smaller homes would save even more.

OTHER ETP'S OBTAINED FROM REGRESSION

The regression coefficients $\underline{c}$, $\underline{b}$ and $\underline{a}$ listed in Table 1 can be used in two ways: to predict the furnace heating rate in a given weather situation and to evaluate the equivalent thermal parameters A, B and H. Figure 2 gives an idea of the predictive value of the regression coefficients. While discrepancies between predicted and measured furnace heating rates are clearly present, the general trend is well reproduced. Some of these discrepancies can be related to sudden stirrings of the air around the thermostat or to temporary surges in interior heat gains like the double spike on the evening of January 24, caused by the cooking of a larger dinner and the presence of a guest. The correlation of the discrepancies with wind speed and/or wind direction is inconclusive. The wide range of the weather variables and of the internal heat gains during the ex-

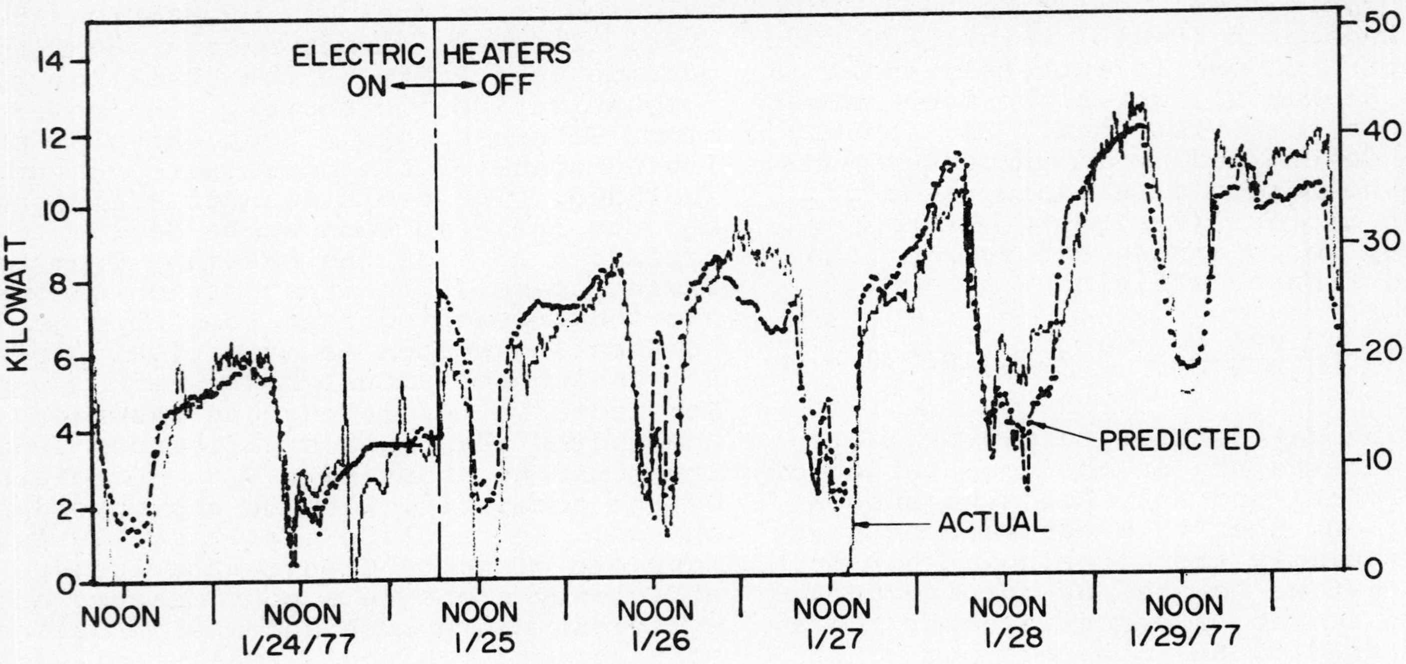

ACTUAL AND PREDICTED FURNACE HEATING RATE
FIG. 2

periment makes one confident that the set of regression coefficients c, b and a can be used as reliable predictors for the furnace gas requirements over most of the heating season, provided the proper adjustments are made for changes in internal appliance loads and solar height, affecting c and a, respectively. How this is done will become clear from what follows.

A consistency check of our simple model results from comparing the ETP's B, H and A obtained from the regression coefficients c, b and a to what we would expect these parameters to be from theoretical calculations. The relation between regression coefficients and ETP's is obtained through comparison of eqs. 1 and 2:

$$B = \varepsilon \underline{c} - (E+P+L) \quad (6a)$$

$$H = \varepsilon \underline{b} \quad (6b)$$

$$A = \varepsilon \underline{a} \quad (6c)$$

Table 2 displays the three parameters B, H and A obtained from the regression coefficients, along with what is calculated by design load methods in the appendix. The most striking feature is the excellent agreement between the estimated and calculated values of H. The calculation of H is part of what is commonly referred to as a static heat load calculation. It entails the separate computation of the heat transfers per unit temperature difference for each wall, each window, the attic, the front door, etc. Also included is the contribution of air infiltration, measured during the experiment to be an average of 0.5 exchanges per hour. It is this measurement that eliminates the largest source of error in a standard heat load calculation and makes the good agreement of estimated and calculated H plausible. The slight increase in the calculated values of H after heater shutdown is due to a higher average wind speed of 4.8 m/s (10.7 mph), up from 1.8 m/s (4.1 mph). The discrepancy

TABLE 2 ETP's Obtained From Regressions and From Design Load Calculations

ε = 76.5%	Heaters on		Heaters off	
	Regressed	Calculated	Regressed	Calculated
B [kW]	-1.14	-1.03	-1.14	-1.02
H [W/°C]	213	213	226	220
A [m^2]	3.33	5.98	4.86	5.98

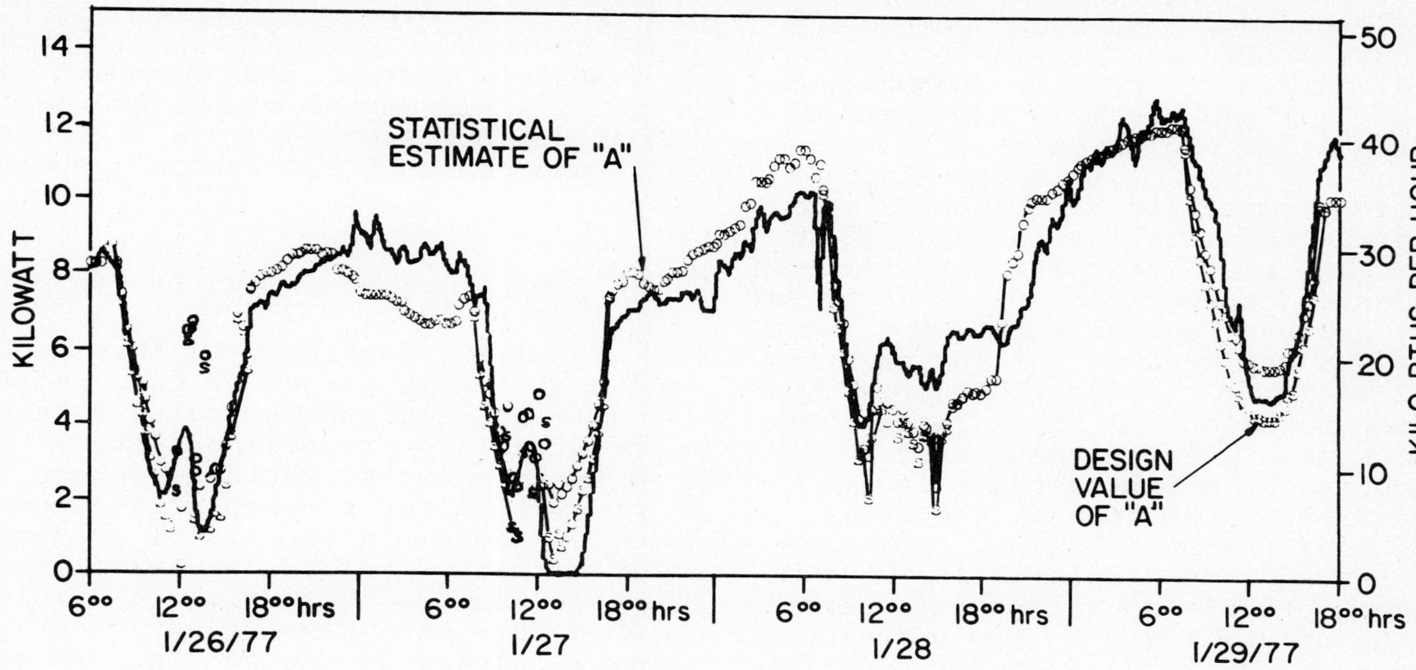

ACTUAL FURNACE HEATING RATE AND ITS PREDICTION USING DIFFERENT EQUIVALENT SOLAR WINDOW AREAS "A"

FIG. 3

between regressed and calculated values of the solar window area A is a statistical artifact. As can be seen in Fig. 1, the solar flux is rising and falling most of the time, while rarely on a steady peak level. Consequently the regression algorithm tends to "match the slopes" of the rising and falling solar flux to the decreasing and increasing furnace operation, at the expense of the few poorly matched peak points. The result is a chronic underprediction of the equivalent solar window efficiency, reflected in the underprediction of the drop in furnace operation during daytime, visible in Fig. 2. Setting A to the design value of 5.98 m^2 computed for this season results in a visibly better matching between predicted and actual data, as shown in Fig. 3. The estimate of A is even further off target in the first experiment period because the combined effect of the auxiliary heaters and the sun made the furnace shut off completely for almost six and a half hours, on January 23. The necessary exclusion of such points from the regression analysis takes away even more "solar peak points", further underestimating A. Since we cannot make the sun shine as a square wave, this kind of experiment is not a good estimator of the equivalent solar window area A. A better estimate for A is provided by experiments based on free-floating temperature, where thermal storage plays a major role, literally "saving" the transient solar energy for experimental detection and correlation (6).

EVALUATING THE EFFECTS OF RETROFITS

A five day long pilot study similar to

TABLE 3 Regression of Pilot Study and Regressed and Calculated ETP's

Regression Coefficients		ETP's:	Regressed	Calculated
$\varepsilon = 0.765$	c = 1.273	B [kW]	0.38	0.04
$R^2 = 0.85$	b = 465 W/°C (29.5)	H [W/°C]	356	327
	a = 3.45 m^2 (10.8)	A [m^2]	2.64	6.32

this experiment had been conducted 9 months earlier in April 1976 in the same house. In the time span between that study and the main experiment a series of retrofits were performed on the house. Wall joints and window frames were sealed and caulked, reducing air infiltration rates by about 30%. The most important difference in the house between that pilot study and now, however, was an uninsulated attic. This was achieved by rolling back the 0.52 W/(°Cm^2) (R-11) fiberglass insulation batts on the attic floor. The anticipated effect, calculated in the appendix, is a 54% increase in the H-parameter of the house. The regression coefficients obtained from this pilot study are displayed in Table 3. Since there was no intermittent operation of auxiliary electrical heaters, no direct estimate of the furnace field efficiency is available from that pilot study. Using ε=76.5% obtained from the main experiment, the equivalent thermal parameters B, H and A are evaluated and also listed in Table 3, along with what is expected from calculation.

Just as in the main experiment, there is fairly good agreement between calculated and regressed H, the equivalent rate of heat transfer per unit temperature difference. Comparing the 356 Watt/°C obtained without attic insulation to the 226 Watt/°C obtained with insulation provides direct evidence supporting the benefits of this easy retrofit. Caulking and sealing may also have helped, but it is difficult to detect it this way because the higher "leakiness" of the house in that pilot study is offset by the much colder weather in the main experiment (-5 ± 6 °C vs. $+7 \pm 4$ °C). That the equivalent solar window area is again underestimated comes as no surprise for the reasons stated earlier. Compared to the main experiment, the discrepancy in the constant heat gain parameter is larger, but so is the uncertainty in the determination of the electrical heat gain E, the latent load L and the neighbor temperature. Actually, the discrepancy is only slightly larger than the uncertainty.

It is through simple experiments of the type described in this paper that the equivalent thermal parameters of a house can be determined before and after different kinds of retrofits. Though the procedures we outlined were applied to a rowhouse, they are just as valid and easier to perform for detached houses, where B, the constant heat transfer, would depend on the average indoor-outdoor temperature difference only, and not on the neighbor temperature.

A SIMPLE KIT TO DETERMINE ETP'S

The experiment described in this paper made use of a large number of sensors and a relatively sophisticated data acquisition and processing system. Most of the results could have been obtained with a far simpler arrangement, described here: a thermistor or similar temperature recording device located in a shady spot outside, a simple relay on the gas valve or any other similar mechanism on the heating and cooling plant, a solar flux meter (even a sophisticated temperature-compensated unit can cost as little as $150) and a few precise mercury thermometers. Using an appropriate power supply unit, the thermistor, the solar flux meter and the furnace relay can be monitored by a multi-channel chart recorder. A constant setting of the thermostat and periodic readings of the indoor temperature in strategic spots and of the electric utility meter should provide all remaining necessary information. In a rowhouse setting, additional visits to the neighbors may be necessary. Data reduction could be carried out manually with relatively little effort, if hourly averages of the readings are taken. Ordinary least square regressions in three variables necessary for the data analysis can be carried out on any desk (or some pocket) calculator. If ETP's were to be determined on a commercial basis, a $20 fee per visited house should cover the capital cost (12). An additional $80 should cover the 3-4 hours of work for a trained technician to install and to remove the kit and to reduce the collected data.

CONCLUSION

The concept of equivalent thermal parameters (ETP's) for a house has been presented and applied to actual data gathered from an occupied townhouse. A convenient set of ETP's is found to be the equivalent rate of heat transfer per unit inside-outside temperature difference, H, the equivalent solar window area, A, the rate of constant heat transfer, B, and the furnace field efficiency, ε. All ETP's are experimentally determined through regression analysis of the data. The efficiency, ε, is obtained by comparison of two separate runs, one with straight gas furnace heat and one with auxiliary electrical heat. The difference in the regression intercepts represents the gas heat "saved"; dividing the known auxiliary electric heat by the saved gas yields our estimate of the 76.5% field efficiency of this furnace, independent of design heat load calculations. With this number, all other ETP's can be derived from the regression coefficients and compared with standard engineering calculations applied to that house. The correspondence is better than some uncertainties in the calculation, with the exception of the equivalent solar window area A, chronically underestimated by this type of experiment. Only a sequence of uniformly sunny days, cold enough never to turn the furnace off completely, or a free-floating indoor temperature experiment can give a more reliable estimate of A.

The effect of attic insulation on the H-parameter of this house is also shown. It confirms the about 35% reduction in H anticipated from desing calculations. However, other investigations at our lab indicate that the situation may be more complex than what this result implies.

The concept of ETP's lends itself to a variety of purposes; for instance, establishing a rank-ordering of different houses according to their energy efficiency, assessing the effect of retrofits and predicting the energy use of a house in a given weather situation. Although in this paper the concept has been applied to the heating season only, it is clearly generalizable to the cooling and intermediate seasons. A prototype of a low cost experiment kit aimed at determining the ETP's of any house is also suggested.

APPENDIX

Synchronization of Furnace Operation

Controlled by the thermostat located in the downstairs living room, the gas furnace operates in an on-off mode, asynchronous to all other data sampling. Figure 4 shows the qualitative furnace operation covering three separate firings. The furnace fires more often and for longer periods if the outdoor temperature drops or if the sun is down. Two furnace variables are involved: the on-time, g, (the time the gas valve is open) and the cycle time, c, between two successive firings. A meaningful smoothing into one furnace variable is provided by the fractional on-time, f, defined as the share of on-time in the cycle time between consecutive firings:

$$f_i = g_i / c_i \tag{7}$$

The index i stands for the i-th firing. An f-value of 0 means that the furnace is totally shut off, while an f-value of 1 indicates full power operation. Figure 5 shows on-times, cycle times and fractional on-times over the whole ex-

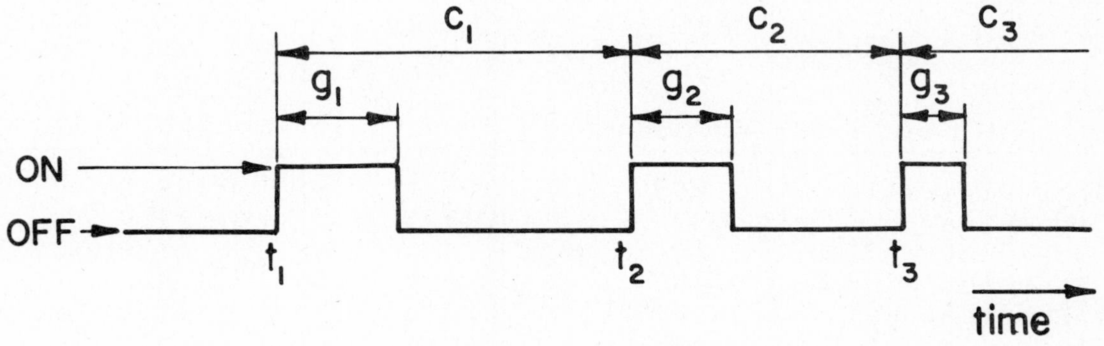

INTERMITTENT FURNACE OPERATION: FIRING TIMES (t_i), ON TIMES (g_i), CYCLE TIMES (c_i)

FIG. 4

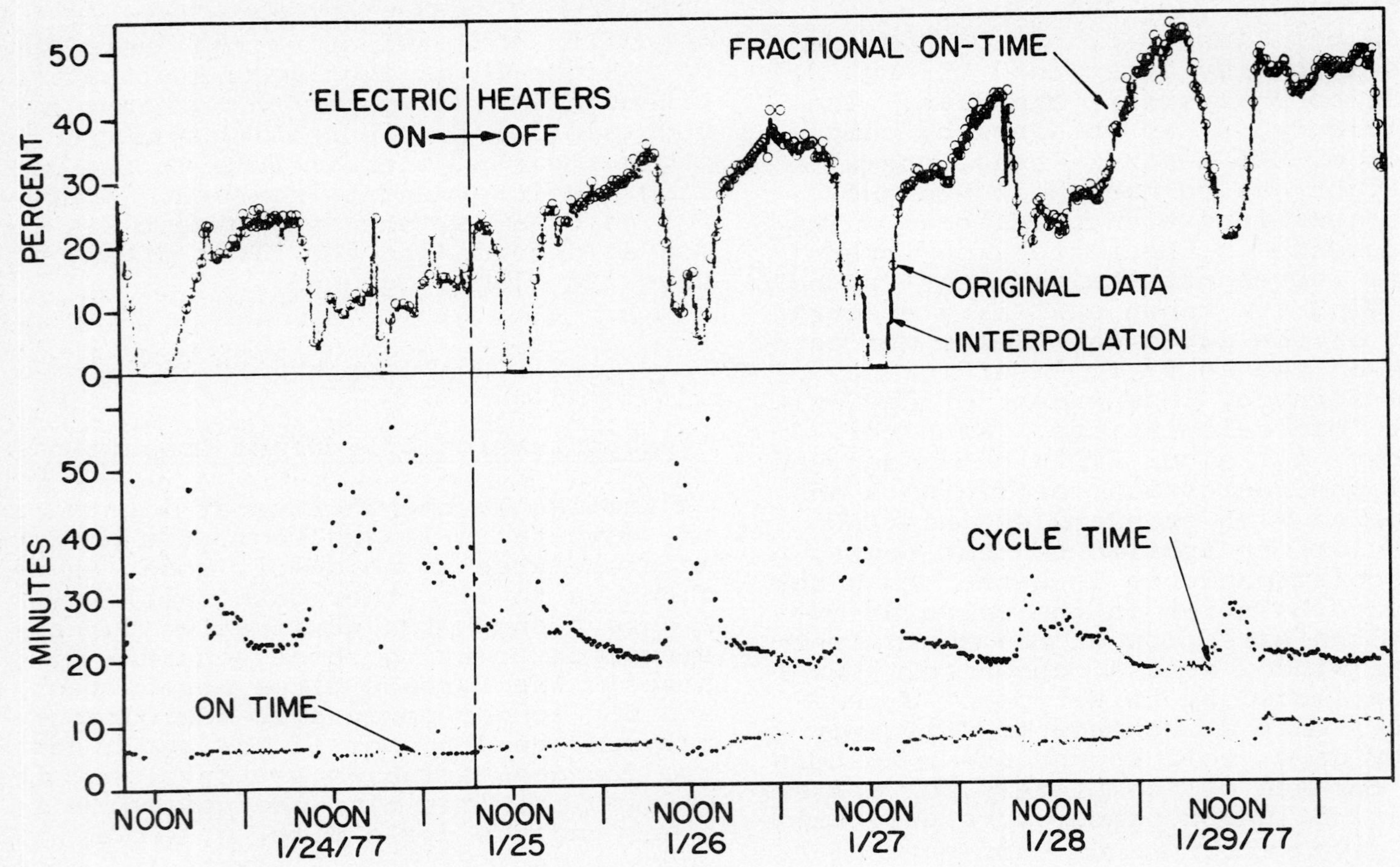

FURNACE OPERATION: ON TIME, CYCLE TIME AND FRACTIONAL ON TIME (ORIGINAL DATA AND INTERPOLATION)

FIG. 5

TABLE 4 Heat Load Calculation for 3-Bedroom Wood-Frame Townhouse

Contribution from	U [W/(°Cm^2)]	A [m^2]	UA [W/°C]
Outside Walls	0.556	63.3	35.2
Front Door	3.07	1.9	5.7
Double Pane Windows[1]	4.26 (3.98)[2]	11.67	49.7 (46.4)[2]
Single Pane Patio Door[1]	6.07 (5.45)[2]	5.58	33.9 (30.4)[2]
Ceiling: Insulated and Bare	0.488 (3.50)[3]	70.9 (67.4)[3]	34.6 (236.2)[3]
Roof[4]	3.95	84.2	332.8
Attic: Ceil. & Roof in Series	—	—	31.3 (138.2)[3]
Air Infiltration	0.5 ex/hr	V=382m^3	64.3
Total Equivalent Heat Transfer Constant			H = 220.1 (213.3)[2] (327.0)[3]
1. & 2. Floor Firewalls	0.965	104.2	HN=100.6
Basement Firewalls	1.87	49.1	HB= 91.9
Ab. Grade Basement Walls	2.17 (0.553)[5]	6.0	HW= 13.0 (3.3)[5]
Ab. Grade Basement Windows	5.45	0.77	4.2
Bel. Grade Basement Walls	0.566 (0.286)[5]	22.9	HG= 13.0 (6.6)[6]
Basement Floor	0.095	67.4	6.4

(1) 80% glass area, metal sash.
(2) 2.2 m/s (5 mph) wind, instead of 4.5 m/s (10 mph).
(3) Attic insulation rolled back.
(4) Includes 3 exchanges per hour attic ventilation (measured).
(5) Insulated with 0.52 W/(°Cm^2) (R-11) insulation.
(6) Insulated with 0.63 W/(°Cm^2) (R-9) insulation.

perimental period. Notice how the cycle times become much larger and the on-times somewhat shorter during daytime, both concurring in reducing the corresponding fractional on-times. Even in very cold weather the furnace rarely works harder than 50% of its full power.

A four-point Lagrange Interpolation was used to synchronize the irregularly spaced f-points (circles in Fig. 5) with the 20-minute sampling intervals of all other data. The result is the continuous curve in Fig. 5. When the furnace is off for a long time, the interpolation may yield negative values. All such instances are excluded from the analyses in this paper and the corresponding f-values set to zero in all figures. When we multiply the fractional on-times f by the heat content of natural gas and by the measured gas flow when the furnace is on, we obtain what we called G:

$$G = f \cdot 24.03\,\text{kW} = f \cdot 82{,}000\,\text{Btu/hr} \qquad (8)$$

G was defined in the text as the heat released by furnace gas combustion.

Heat Load Calculations

The floor plans of the townhouse and the detailed breakdown of all contributions in the steady-state heat load calculation are given in (7), using heat conductances from (10). A summary is given in Table 4.

The heat transfer rate B is calculated as the sum of the heat transfers to the ground and to the neighbors. The conduction to the ground per average basement-outdoor temperature difference was calculated following a method developed by Frank Sinden, of our lab (9), based on an exact steady-state solution of the three-dimensional Fourier Equation. The thermal mass of the earth smoothes the fluctuations of outdoor temperature. Thus the time average, over the whole experiment, of the outdoor temperature was taken to estimate the heat transfer to the ground.

The heat flux by conduction and air infiltration through the above grade basement walls fluctuates with the daily outdoor temperature. However, it can be shown that the amplitude of the fluctuating portion of this heat flux is reduced by a factor of about three by the considerable thermal mass of all basement walls and the floor. Thus only about one third of the fluctuating heat flux reaches the first floor, directly influencing the thermostat: only that third is "attributed" to H by the regression, while the other two thirds are attributed to the "constant" heat transfer rate B. The exact proportions can not be well established in this experiments. As a compromise we include all conductive basement heat transfer into the B parameter, while lumping all air infiltration (including the basement contribution) into the H parameter. Thus the heat gain B is calculated as

$$B = (HG+HW)(\overline{TO}-\overline{TB}) + HB(\overline{TN}-\overline{TB}) + HN(\overline{TN}-\overline{T}) \qquad (9)$$

HG, HW, HB, HN are defined in Table 4; $\overline{TO}$, $\overline{TB}$, $\overline{TN}$, $\overline{T}$ are the time averages, over the full experimental periods, of the outdoor, basement, neighbor and indoor temperatures.

In the pilot study, all basement walls were bare. In the first period of the main experiment, 43% of the above grade wall and half of the below grade walls were insulated. The remaining portions of both were insulated after 2 of the 5 days of the second period. Using the temperature values listed in Table 5, we compute B to be

$$B = -1.030\ \text{kW} \quad \text{(heaters on)} \qquad (10a)$$
$$B = -1.024\ \text{kW} \quad \text{(heaters off)} \qquad (10b)$$
$$B = +0.043\ \text{kW} \quad \text{(pilot study)} \qquad (10c)$$

TABLE 5 Temperature Time Averages

All °C	$\overline{TO}$	$\overline{TN}$	$\overline{TB}$	$\overline{T}$
H'ters on	-3.9±3.6	19.7	19.6	23.0
H'ters off	-5.6±6.3	19.3	21.3	21.3
Pilot St.	+6.9±3.6	22.8	19.9	20.3

TABLE 6 Time-Averaged Heat Gains

All kW	E[1]	P[2]	L[3]
Heaters on	4.095	0.155	-0.506
Heaters off	1.180	0.125	-0.563
Pilot Study	0.568	0.142	-0.118

(1) Includes all electric consumption, except what makes up for the hot water that "goes down the drain".
(2) Obtained from people presence times, using 117 Watts (400 Btu/hr) per person.
(3) Includes latent heat of water to humidifier and to plants, at .687 kWhr/kg.

Equivalent Solar Window Area A

This parameter is the result of a contribution from the transparent windows (about 80% of the total A, in our townhouse with south facing double pane windows making up 17% of the total south wall surface) and a contribution from the opaque walls. Using the sol-air temperature concept, the equivalent solar window area A is defined as

$$A = \varepsilon_s W + H'(\alpha/h) \qquad (11)$$

ε_s is the average net transmissivity of the window glass to solar radiation (0.73 measured in January, 0.58 in April)*;

W is the net transparent glass area of the south-facing windows ($6.39\ m^2$ & $5.98\ m^2$);

α is the absorptivity of the outside opaque walls to solar radiation;

h is the outside film coefficient ($\alpha/h = 0.035\ (°Cm^2)/W$ from (10));

H' is the heat conduction per unit temperature difference through the opaque outside walls and the attic ($37.2\ W/(°Cm^2)$ & $82.1\ W/(°Cm^2)$).

While the net glass area W can be determined quite easily, the net transmissivity ε_s of glass to solar radiation is a complicated function of solar altitude, time of day, glass properties, number of window panes and more. For this experiment ε_s was determined experimentally with the use of two solar flux meters placed inside and outside the window.

ACKNOWLEDGEMENT

The author wishes to thank ERDA and NSF-RANN for their research support.

*All value pairs in this nomenclature refer to the main experiment in the first number and to the pilot study in the second.

REFERENCES AND FOOTNOTES

(1) D.G. Stephenson and G.P. Mitalas: Cooling Load Calculations by Thermal Response Factor Method, ASHRAE Trans. 73(1), III.1.1 (1967).

(2) T. Kusuda: NBSLD, The Computer Program for Heating and Cooling Loads in Buildings, NBS Bui. Sci. Ser. 69 (1976).

(3) T. Kusuda, T. Tsuchiya and F.J. Powell: Prediction of Indoor Temperature by Using Equivalent Thermal Mass Response Factors, Proc. 5. Symp. on Temperature, NBS (1971).

(4) K. Kimura and H. Ishino: Air Conditioning Load Calculations by the Equivalent Mass Weighting Factor Method for the Computerized Control, Proc. of the Japan. Arch. Soc., Kyushu Meeting (Oct. 1972).

(5) R. Muncey, J. Spencer and C. Gupta: Method for Thermal Calculations Using Total Building Response Factors, NBS Bui. Sci. Ser. 39 (1971).

(6) R. Sonderegger: Diagnostic Tests Determining the Thermal Response of a House, ASHRAE Meeting, Atlanta (1978).

(7) R. Socolow and R. Sonderegger: The Twin Rivers Program on Energy Conservation in Housing: Four-Year Summary Report, Center for Envir. Studies Rep. #32, Princeton U. (1976). Excerpts in ASHRAE Trans. 83(1) (1977).

(8) M. Nowotarski and D.T. Harrje: Warm Air Furnace Non Steady Effects, Center for Envir. Studies Report (Number to be det.), Princeton U. (June 1977).

(9) F.W. Sinden: Conductive Losses From Basements, Center for Envir. Studies Note #4, Princeton U. (1976).

(10) ASHRAE Handbook of Fundamentals (1972).

(11) The t-statistic is defined, here, as the ratio of an estimated regression coefficient and the standard error of the estimate.

(12) Based on $1,000 capital cost (incl. calculator), 5 houses per heating season, 10 years projected kit life.

POSSIBLE DEVELOPMENTS OF THE ENERGY SUPPLY SYSTEM IN THE FEDERAL REPUBLIC OF GERMANY

T. Bohn
University of Essen
Germany

ABSTRACT

An analysis of the future energy demand has been performed based on the foreseeable economic development of individual consumer sectors. Based on the specific situation of Germany regarding resources and constraints of the economic and the environment, options are given for an optimised future energy supply system. Possibilities of energy savings, rational energy utilisation and new energy technologies are taken into account.

A METHODOLOGY FOR EVALUATION AND USE OF INTERMEDIATE TERM METEOROLOGIC FORECASTS IN ADJUSTING TO ENERGY CRUNCHES CAUSED BY EXTREME WEATHER CONDITIONS

Donald R. Davis
Assistant Professor, Departments of Systems and Industrial Engineering and of Hydrology and Water Resources
University of Arizona, Tucson Arizona 85720

Roman Krzysztofowicz
Research Associate, Department of Hydrology and Water Resources
University of Arizona, Tucson Arizona 85720

ABSTRACT

The newspapers quoted the utility spokesman as saying that they thought the cold weather was coming and if they had made adjustments and curtailments a week in advance the shortage would not have been so severe, but since the forecast might not have been right they took no action.

The effectiveness of using intermediate term meterologic forecasts to make advance adjustments in an area's mix of energy sources so as to mitigate the effects of sudden heavy demand for energy supplies due to unseasonably severe weather is made difficult by the uncertainty inherent in the weather forecast. A methodology is developed for comparing the effectiveness of the plans and stand by procedures, such as curtailments, which are designed to be activated during weather severities, such as cold snaps.

The methodology is based on a systems model, which is termed the weather severity forecasting-response model, and consists of a meterologic network, a forecast model, a decision model and the action response. These are mathematically modeled in a manner which allows evaluation a weather severity forecasting-response system. Such evaluation depends on the accuracy of the forecasts and their timelessness as well as on the costs and losses involved in taking or not taking action.

The actual value of a particular weather severity mitigation procedure is compared with the optimal value that could be obtained if the uncertainty in the forecast subsystem were handled in an optimal manner. The optimal value is in turn compared with the potential value which is based on perfect forecasts and response. Efficiencies of the overall weather severity forecast-response system can then be calculated as well as the efficiency of the forecasting subsystem and the response subsystem.

The costs involved in activating some weather severity procedures, such as curtailments to industry, may be measured in economic terms, but the costs in other actions such as school closings and curtailments to residential areas involve social, psychological and economic costs. Multiobjective techniques are introduced to handle these noncommensurate costs.

An example of the use of this methodology, for analyzing the trade off between forecast accuracy and lead time is presented showing that for some situations, a less accurate forecast subsystem with a long lead time may be preferable to a more accurate forecast with a shorter lead time, providing the uncertainties in the forecast are considered in determining the sequence of actions to be taken in response to a forecast of extreme weather conditions.

EFFECTS OF ALTERNATIVE TECHNOLOGIES ON THE FINAL ENERGY DEMAND SECTOR (A SIMULATION MODEL APPROACH)

R. Heckler, R. Patzak, H. Reents, K. Schmitz
KFA, Jülich, Germany

ABSTRACT

Almost two thirds of the primary energy consumption of the FRG are being dumped unused as heat into the environment. The majority of 58% of these losses occur in the demand sector. The remaining losses in the amount of 42% arise from the conversion and distribution sector.

In order to achieve a uniform evaluation of alternative measures for the improvement of this situation as well as the demonstration of its effects on the synoptic system man-energy-environment, an energy model has been built in the research group of Systems Research and Technological Development (STE) of the Nuclear Research Center in Jülich. This model is suitable for the simulation of technological, structural as well as measures of price policy in the sectors of energy production and use. By taking into account the interdependencies between energy economy and the other branches of economy on the one hand and the environment on the other hand, it demonstrates the effects of alternative measures with regard to the development of

- energy demand with special regard to the final energy demand
- emission of pollutants
- demand for capital

Starting from the presentation of the energy demand structures in the basic (useful) energy, as well as the final energy sectors and from the characterization of the decisive factors of the energy demand, the lecture will explain the simulation model for the Federal Republic of Germany. Herein the conventional method of estimating the future energy demand by means of the final energy demand, has been abandoned as the basic energy demand has been made the basis of the model calculations. Thereafter, the effects on the energy economy, the remaining sectors of the macro economy, as well as the environment are being depicted by means of selected technological and structural measures in the demand sectors: industry, residential, commercial consumption and transport.

For the presentation of different scenarios, different strategy calculations are planned for the sectors of

- industry
    - enforced use of "direct reduction - electric furnace" for steel production instead of the conventional methods of steel production "blast-furnace-oxygen blast converter"
    - stagnation of growth in selected sectors of basic industry
- residential and commercial consumption
    - introduction of alternative heat pump systems
    - warm water heating by means of solar energy
    - improved insulation of buildings
- transport
    - substitution of carburetor fuels by methanol and diesel fuels
    - transfer from individual traffic to public transportation
- environmental protection
    - desulfurization of residential heating
    - effects of introduction of district heating

ENERGY DEVELOPMENT DECISION-MAKING: MINIMAX VERSUS EXPECTED VALUE CRITERION

Jean E. Weber
Department of Management

Lucien Duckstein
Department of Systems & Industrial Engineering

Donald R. Davis
Department of Hydrology & Water Resources

The University of Arizona
Tucson, Arizona 85721

ABSTRACT

Energy decisions, which are essentially made under various uncertainties on the state of nature, can be approached from either an expected value or a minimax viewpoint. Several examples are used throughout the study to illustrate this dichotomy. Formally, a decision problem includes a decision or action space (alternate sources of energy), a reward or loss space (monetary and social losses due to wrong decision), and probability distributions mapped on the reward space (probability of shortage). The decision-maker may choose such a probability distribution according to a Bayes or expected value viewpoint or a minimax viewpoint, which may itself be expected value or absolute. The Bayes viewpoint is appropriate when (a) the decision-maker can specify a prior distribution which is an accurate representation of the states of nature, and (b) a repetitive situation exists in which many large and small individual losses average out to the expected value without unacceptable consequences (many minor incidents). An expected minimax, which minimizes maximum expected loss, requires a repetitive situation based on the least favorable distribution of the state of nature; an absolute minimax, on the other hand, minimizes maximum possible loss. Minimax criteria are appropriate when a very serious loss, such as a major energy shortage, cannot be tolerated.

Actual decisions appear to be a mix of the two viewpoints, which require different types of data. For this reason and for a better understanding of decision-making in energy management, it is important to identify which approach is to be used in a given problem.

HEAT RECOVERY FROM WASTE WATER IN DOMESTIC BUILDINGS

Karl Friedrich Ebersbach

Forschungsstelle für Energiewirtschaft München

INTRODUCTION

Heat recovery is one of the important ways leading to a rational use of energy. In many industries techniques of heat recovery have been applied to a certain extent for a couple of years. For private households, however, such techniques do not yet exist. In the last few years the increasing necessity of saving energy induced activities in dealing with problems and possibilities of recovering waste heat from households. Already special techniques for this kind of heat recovery are being developed.
In the Federal Republic of Germany this topic was dealt with especially in the framework oftwo R + D - projects sponsored by the Bundesministerium für Forschung und Technologie. The task of the first one is to get detailed information of the structure of energy demand and energy consumption of private households, to work out the technical requirements on heat recovery plants and to develop basic designs of heat recovery plants.
Within the second one, jointly accomplished by Rheinisch Westfälisches Elektrizitätswerk AG,AEG-Telefunken and Forschungsstelle für Energiewirtschaft, a pilot plant was constructed and has been running in a six-family-house since December 1976.
This paper will give an account of the problems, the results and the operating experiences from the two projects.

HOT WATER CONSUMPTION OF PRIVATE HOUSEHOLDS

In 1975 private households in the FRG had a final energy consumption of 520 TWh or $64 \cdot 10^6$ tce being 27,4 % of the final energy consumption of the FRG in total. About 80 % are used for space heating, 10 % for hot water generation and another 10 % for the remaining energy applications. Electrical energy for hot water generation is used in 46 % of the households. For covering the hot water demand of washing machines and dishwashers these machines are gereally equipped with a particular heating system. A full-electric household of four persons needs for

Hot water (bathroom and kitchen)	2000 kWh/a
Washing machine	500 kWh/a
Dishwasher	600 kWh/a
Total	3100 kWh/a.

About 2300 kWh/a are dissipated in waste water and can be used for heat recovery.

Very important for developing and designing of waste heat recovery systems is the basic knowledge of the daily curves of hot water demand and waste heat output. Therefore, these quantities have been measured in a lot of houses over a period of about two years. In addition, the consumption of cold water was measured too.

The evaluation of these

measurements showed a significant dependence of the daily consumption and its behaviour on the day of the week. This is to be seen in Fig. 1, showing the hot water consumption of a 45-family-house accumulated over 24 hours of the day. While the curve characteristics of the days from Monday to Thursday are nearly the same, the Friday consumption is significantly higher, but is exceeded clearly by the Saturday and Sunday values.
The utilization of the washing machines is shown to be also dependent on the week-day. Nearly 40 % of the washing-processes take place on Monday; with the exception of Sunday, the remaining days show nearly the same washing frequency (see Fig. 2).

The cold water consumption is essentially influenced by toilet flushing the daily curves of which can be seen in Fig. 3 for the different days of the week.
The water consumption in total (hot and cold) is shown in Fig. 4 as an average balance of daily water consumption of a 4-persons-family, equipped with washing-machine and dishwasher.

BASIC DRAFT ON HEAT RECOVERY

If there is heat recovery from waste water you have to take into account that cold and warm waste water are drained off together. A separation is technically hardly to achieve and too expensive. Waste water from water-closets has to be drained off with a separate sewer. The average temperature of the resulting waste water is about 20 to 25 °C. To use heat at this temperature level calls for a heat-pump. Further more, there has to be a waste water storage and a fresh water storage to compensate the peaks and to bridge over the temporal displacement of waste water drained off and heat demand.

Contamination and chemical aggressivity require special materials and special constructed heat exchangers. There are experiences concerning these problems in industry. But these experiences can be applied only to a certain part. First of all heat recovery plants in domestic buildings have to run reliable many years without enduring supervision and maintenance.

First experiences about these problems are just gathered with the before mentioned pilot-plant in Essen. The heat pump that takes heat from the waste water runs as long as the warm water storage is full charged, or the waste water storage completely discharged. In that case, the waste water may become undercooled, that means the waste water temperature falls below the temperature of the fresh-water feed. The result is a declining coefficient of performance.
Because of the technical qualities of modern heat-pumps and refrigerants utilized the temperature of the warm water is limited to ~55 °C.
Fig. 5 shows the annual balance of a heat recovery plant, which has been designed for a four-person household Without heat recovery there would be a demand of electrical energy of 3700 kWh/a (3100 kwH/a for washing machine, dishwasher and demand for warm water with 600 kWh/a dissipation of the warm water storage) with heat recovery the demand is reduced to 1950 kWh/a.
From that amount the heat pump requires 1450 kWh/a. The remaining part has to cover the demand of mechanical energy (to run the washing machine and the dishwasher) and to cover the demand of this machines to generate higher water temperatures. From the waste water 2050 kWh/a can be recovered.

DESCRIPTION OF THE PILOT-PLANT IN ESSEN

The result of investigations of demand and consumption was the fact that with increasing number of provided households the maximum of the warm water demand per person declined strongly. The values for the hourly maximums are:

3	person household	21 l/pers.h
10	person household	12 "
15	person household	8,5 "
45	person household	4 "

and for a big plant with 324 households only 3 l/pers.h.

The result of this declining is, that with increasing number of supplied households, the heat recovery plants can be constructed at lower specific costs.

This plant has been designed for supplying a 6 family-house with the following consumption figures:

total water consumption	150 l/pers.
warm water consumption	40 l/pers.
consumption for washing machine and dishwasher	20 l/pers.

The schematical construction of the plant is shown in Fig. 6. First of all the waste water is drained into a cleaning tank to keep back the rough particles and then drained into the storage-tank. If there is a demand it will be conducted into the vaporizer of the heat-pump and gives off its heat. The condenser of the heat pump transfers heat to a intermediate-water-storage which is followed by an electrical heating unit for complementation and reserve.

The results which have been achieved up to now show that the heat pump works with a coefficient of performance of $\varepsilon = 2$.

From the inlet temperature in the storage-tank (22 - 30 °C) the waste water gets cooled to $\approx$ 10 °C. The temperature of the process water amounts to 45...50 °C.

The target of the study is beside the measuring of energy flows and coefficients of performance to investigate following questions.

1. Investigation of the contamination of various heat exchangers

2. Integration of the cleaning-storage and the storage-tank, if possible with waste water heat exchanger

3. Integration of the fresh water heat exchanger and fresh water storage.

Point 2 and 3 cause a considerable simplification of the technology of these plants.
Actually with this plant only warm waste water is used and thus warm water produced. In future, a intensive investigation will be necessary how to use other waste heat resources (e.g. refrigerator, freezer) and their utilisation for other purposes (e.g. space heating).

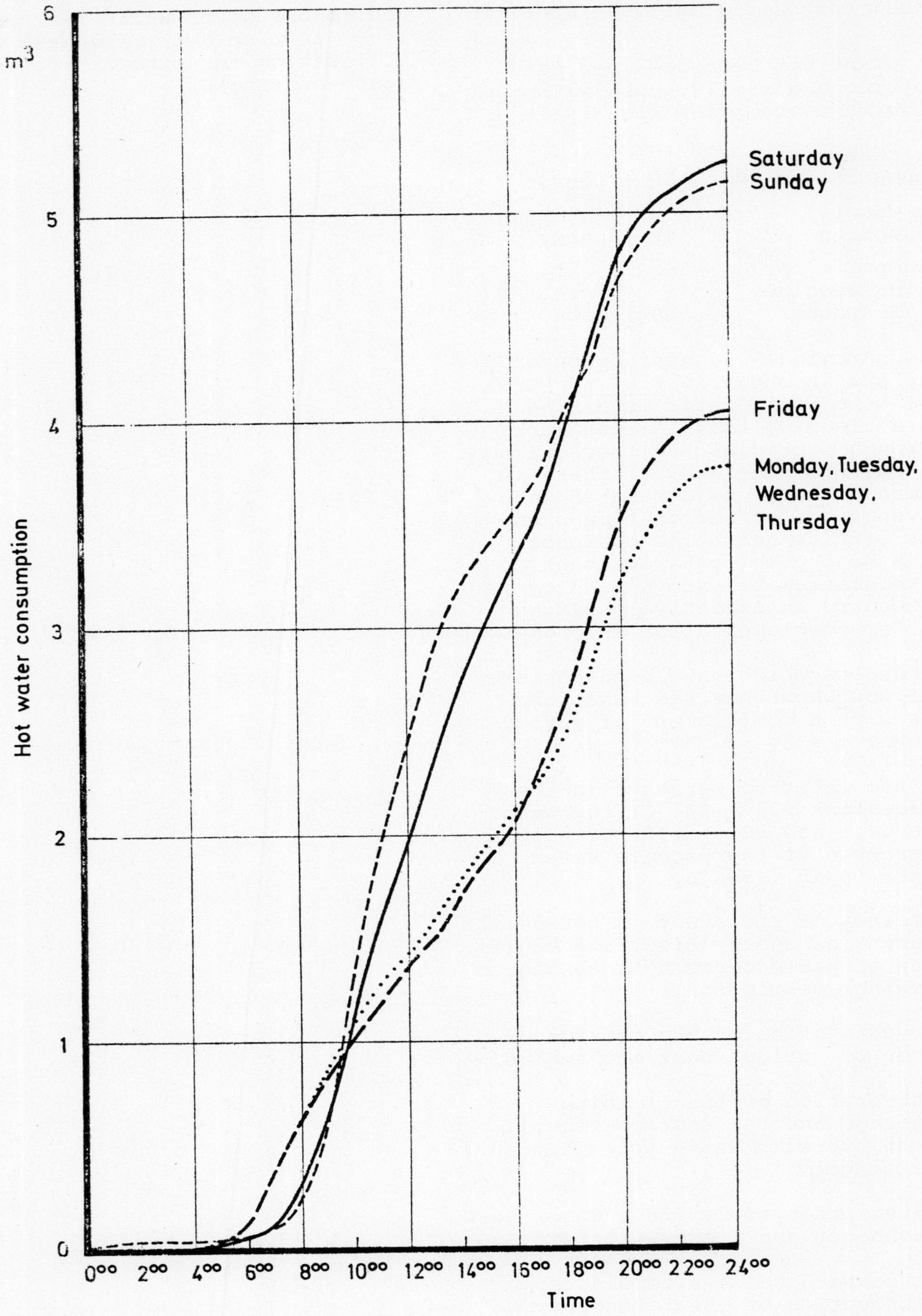

FfE	Daily Cumulative Consumption of Hot Water (45 - Family - House, 106 Persons)	Fig. 1

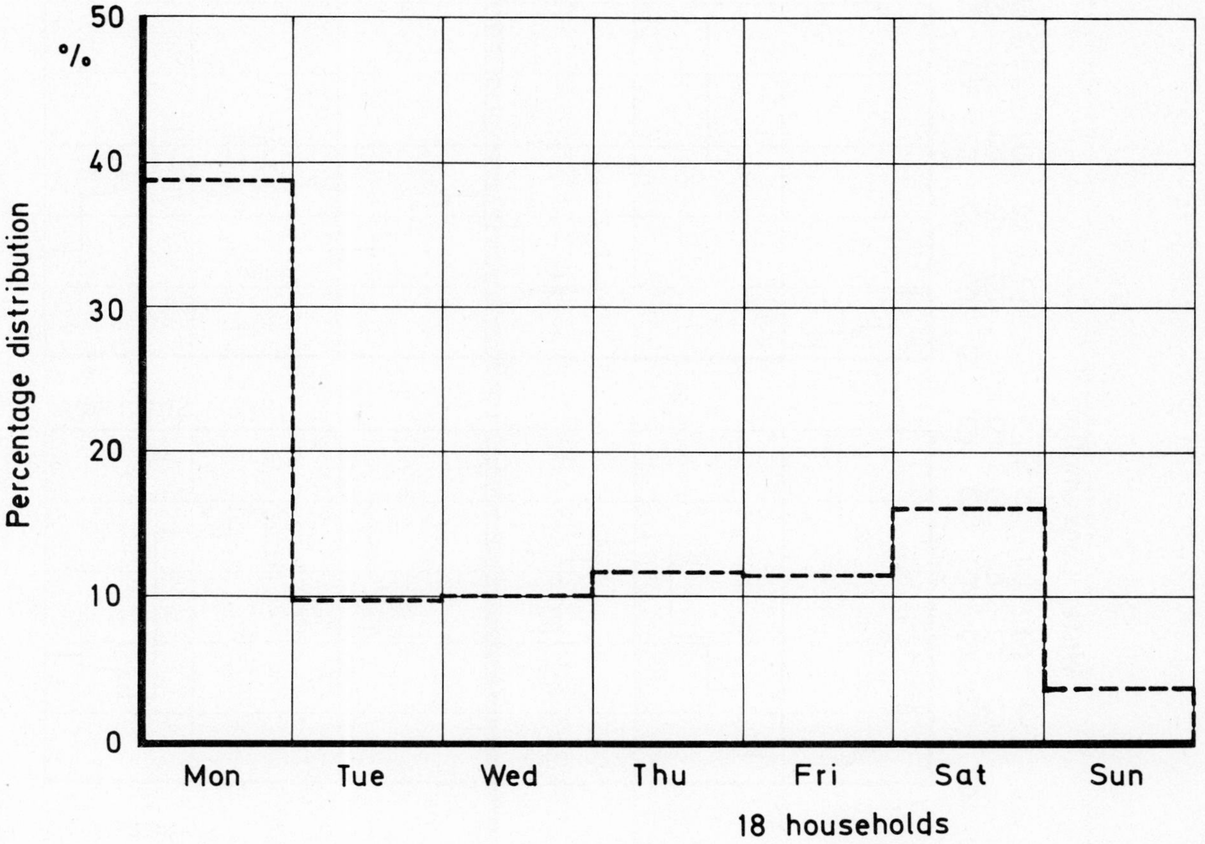

100 % ≙ 496 washing - processes

number of washing - processes per week = 100 %

FfE	Percentage Distribution of Washing - Processes on the Different Week-Days	Fig. 2

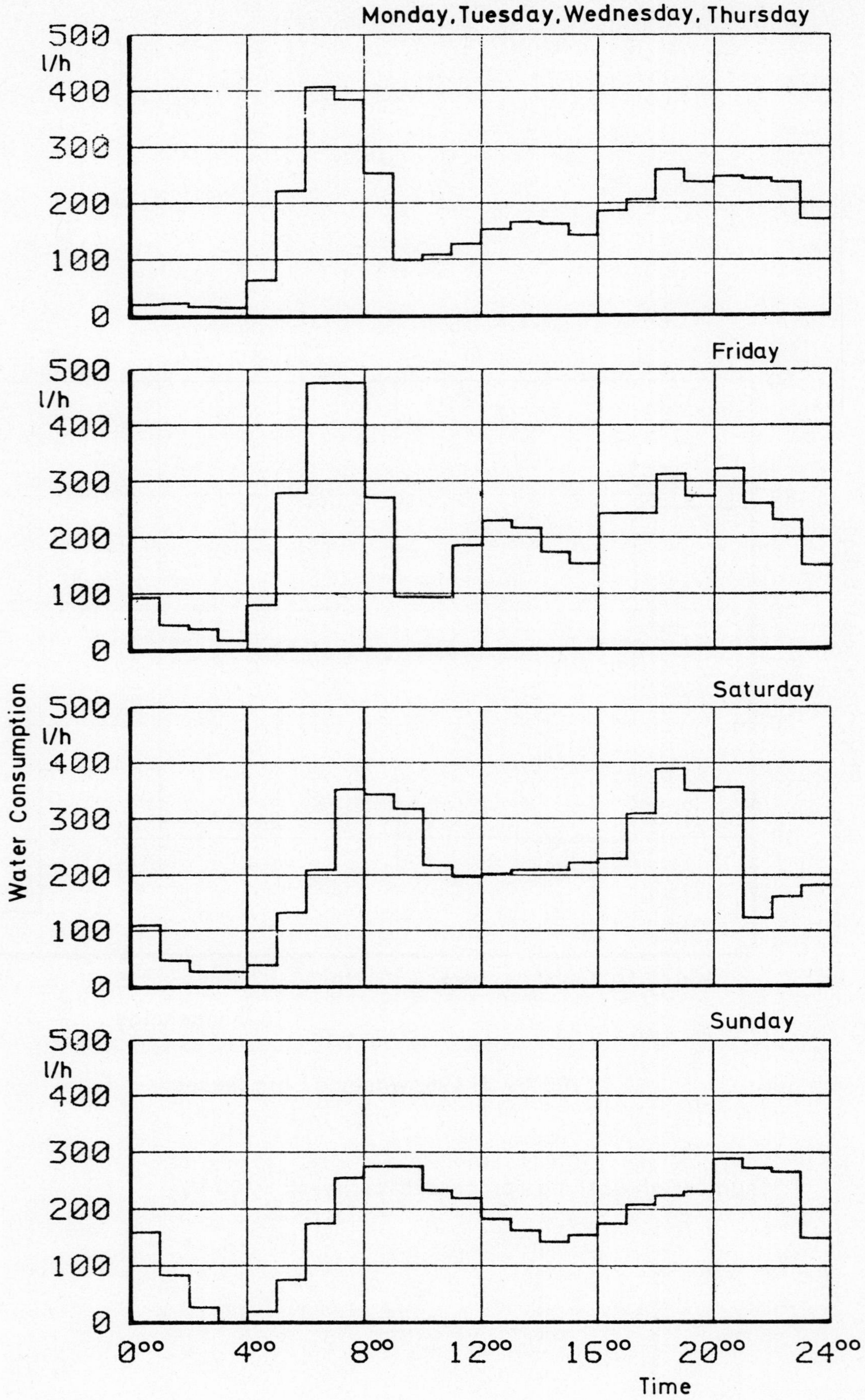

FfE	Consumption for Water - Closets (45 Family - House, 106 Persons)	Fig. 3

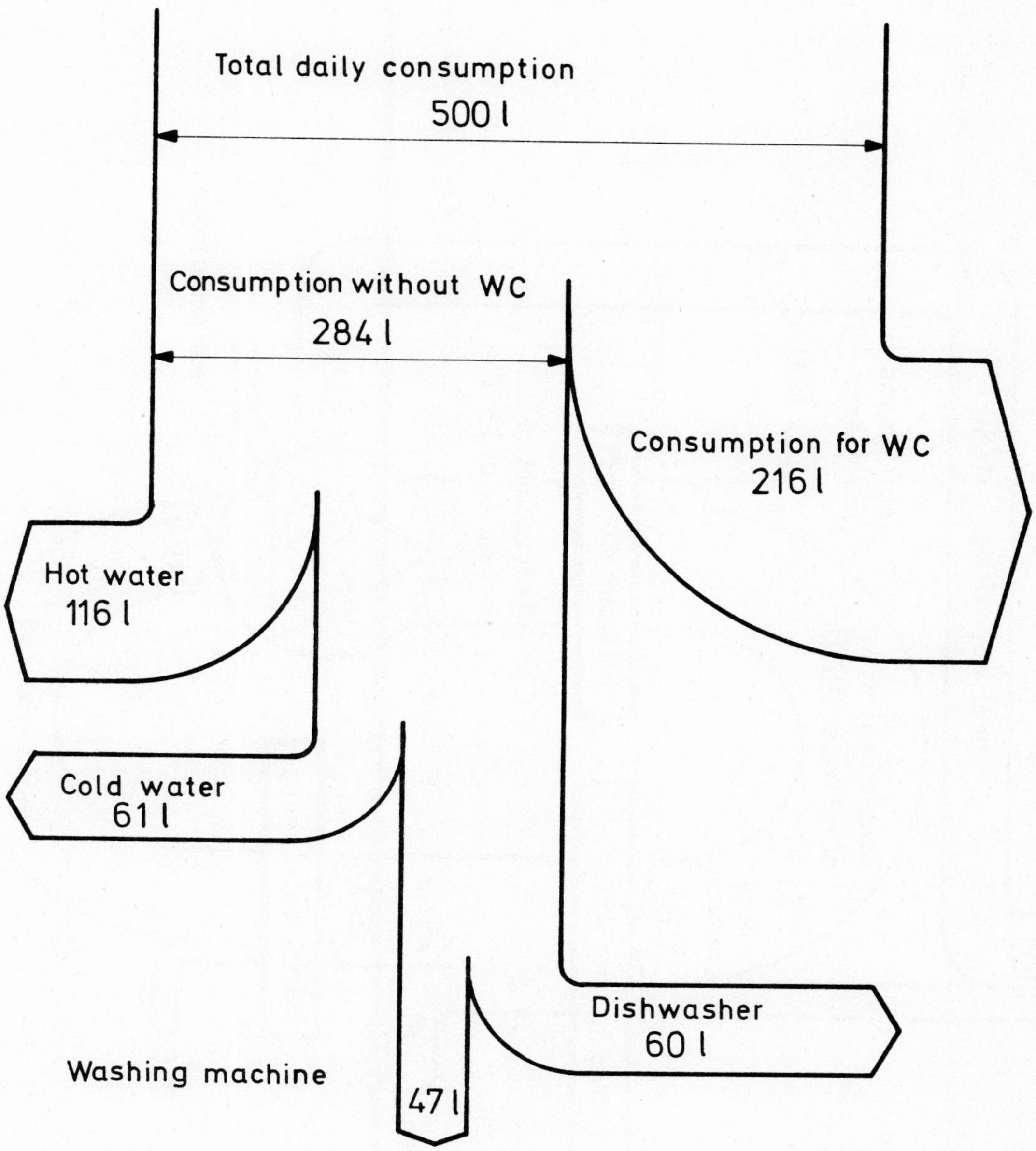

(Figures represent average daily consumption)

FfE	Flowdiagram of Water Consumption in Households	Fig. 4

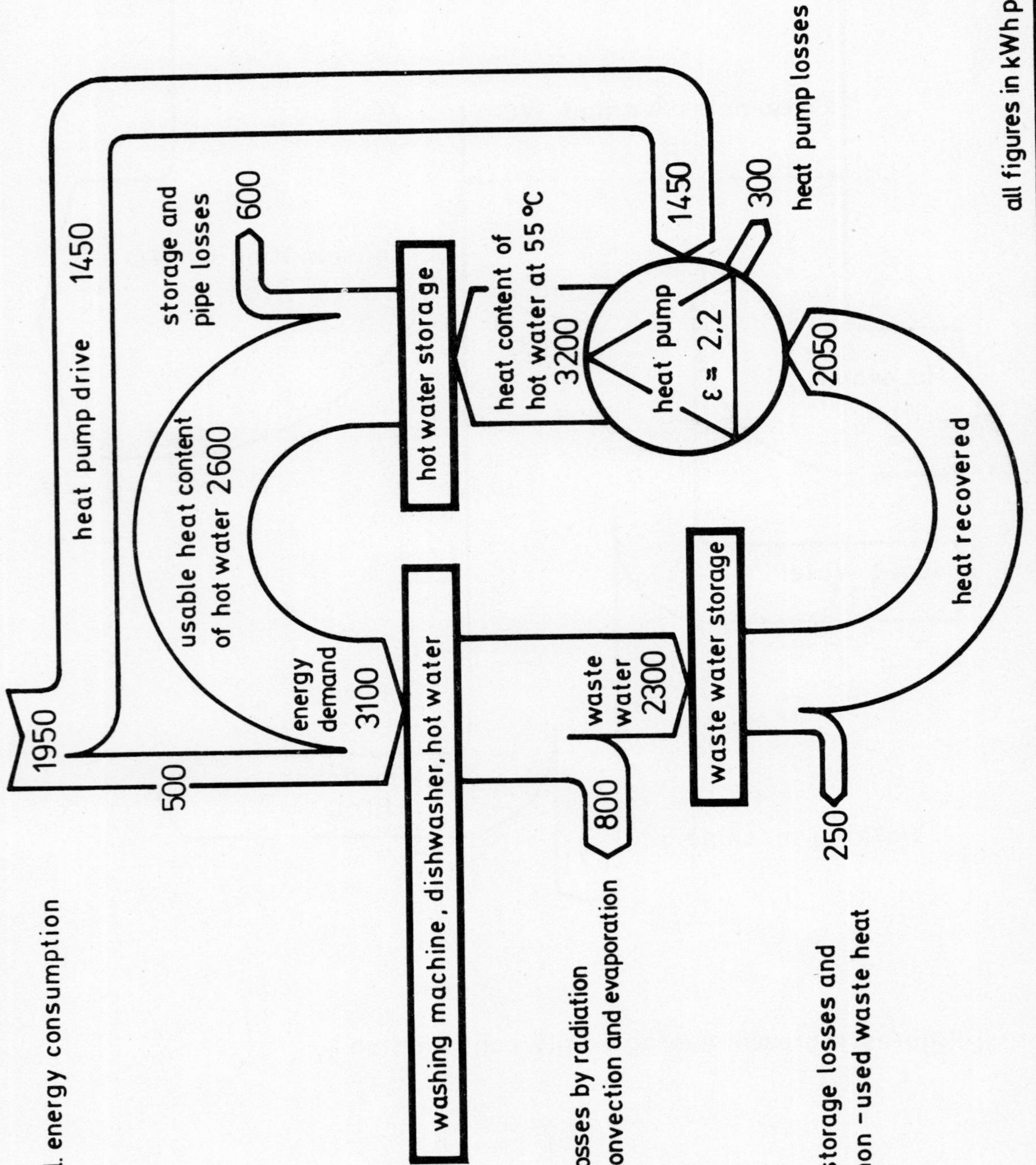

Fig. 5 Energy Balance of Heat Recovery from Waste Water in Households

Ff E

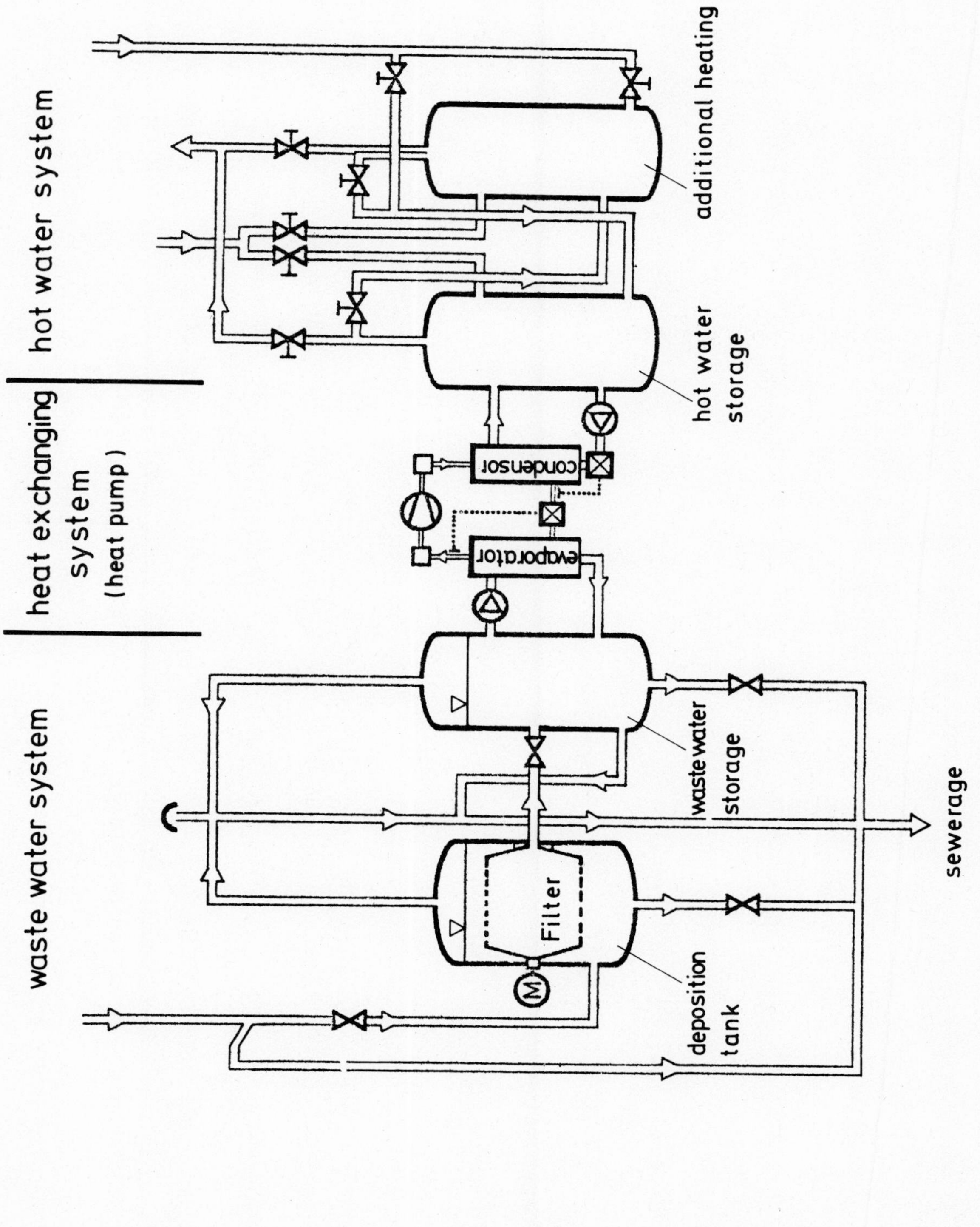

Fig. 6 Waste Heat Recovery Plant for a Six - Family - House FfE

DEVELOPMENT OF AN ENERGY CONSERVING ELECTRODELESS FLUORESCENT LAMP (LITEK)

Donald D. Hollister

Lighting Technology Corporation, Fullerton, Calif.

ABSTRACT

The status of the LITEK lamp development program is reviewed. Technical operation of the device is outlined. The results of life cycle cost effectiveness analyses are presented and the energy conservation potential of the LITEK lamp is estimated.

INTRODUCTION

Edison's original electric lamp patent issued January 27, 1880, and this lamp is considered by many to be the prominent invention of its century, for eventually, at the flip of a switch, it brought the gaslight era to an end and extended everyone's useful day to a full twenty-four hours. The spectacular convenience of the screw-in Edison lamp remains the signal feature upon which the incandescent lamp's strong continued popularity is based.

The fluorescent lamp initially was marketed in the 1940s. During the following twenty years phosphor color rendering characteristics and process technology improvements combined with demonstrated operating cost effectiveness to enable the widespread propagation of the fluorescent lamp in the commercial and industrial sectors. Significant residential applications,however, failed to materialize. Both the fixture, which is necessary for fluorescent lamp operation, and its installation are relatively costly; and, until only recently, the color rendering capabilities of commercial fluorescent lamps did not yield results amenable to interior decoration.

Lighting accounts for about five percent of the United States' total energy consumption, equivalent to about two million barrels of oil per day. Incandescent lamps, which typically have luminous efficacies of about 15 lumens per watt, use about 55 percent of the nation's total lighting energy consumption, roughly the equivalent of 1 million barrels of oil per day. Fluorescent and HID lamps, on the other hand, are two to five times more efficient than incandescent lamps, the fluorescents having overall luminous efficacies (including electrode and ballast losses) up to about sixty lumens per watt.

Thus, considerations of overall usage and relative lamp efficiency strongly indicate that energy conservation efforts in lighting should initially be directed toward the incandescent lamp.

TECHNICAL BACKGROUND

The LITEK lamp,shown in Fig.1, is intended to combine the convenience of the incandescent lamp with the efficiency of a fluorescent lamp to achieve an overall three-to-four fold decrease in lighting energy usage. The LITEK lamp is similar in size and shape to a conventional incandescent lamp, it has no hot filament. Instead, it has a relatively cool magnetic coil that is energized by an electronics package contained in the lower portion of the lamp.The coil produces a magnetic field that excites mercury vapor contained within the lamp resulting in emission of ultraviolet light from an electrodeless arc discharge. Ultraviolet light causes the phosphor layer to fluoresce and emit visible light. The overall luminous efficacy of the LITEK lamp, when treated as a complete system, presently is about fifty lumens per watt of input power. Identified sources of inefficiency in the present lamp electronics design suggest that the market ready device will perform at the overall rate of sixty lumens per watt.

Electrodeless discharges in gases are not new; in fact the first public dis-

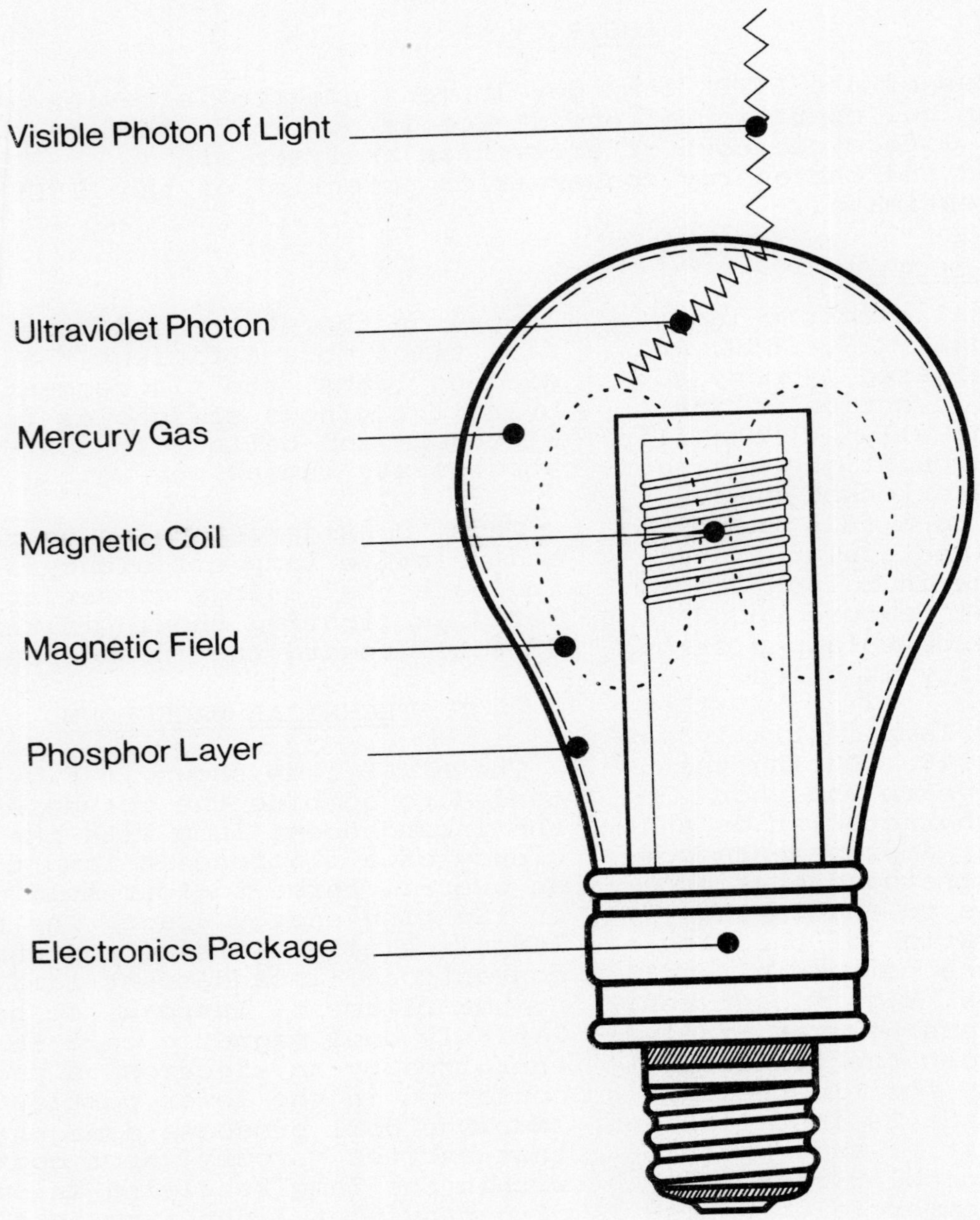

Fig. 1. LITEK Electrodeless fluorescent lamp

closure (1) of the observation of this phenomenon dates back nearly a century to the year 1884, while the initial attempt to provide a physical (i.e., mathematical) description of the phenomenon was made seven years later (2,3) in 1891.

The existence of more than one kind of electrodeless discharge was recognized nearly fifty years ago. This is evidenced by the famous Thomson-Townsend dispute (4,5) which finally was resolved (6) in 1929 by McKinnon with the disclosure that under certain conditions the "electrodeless" discharge is electric in origin, and under other conditions it is magnetic (i.e., induced) in origin. The status of early knowledge was best summarized by Babat (7), who appropriately defined the electric electrodeless discharge as an "E-Discharge", and the magnetically induced electrodeless discharge as an "H-Discharge". It is necessary that the differences between the "E-Discharge" and the "H-Discharge" be recognized as fundamental because the applications to which an electrodeless discharge can be put successfully depend entirely upon these differences and how such differences are exploited.

In the "E-Discharge" an electric field is impressed on a gas by means of external electrodes as in Fig. 2. An electric discharge is maintained in this configuration by continuing the discharge conductive current by means of displacement currents through the dielectric discharge chamber walls to the external electrodes. An equivalent circuit for the "E-Discharge" is presented in Fig.3, where the effects of dielectric currents are denoted by capacitors in series with the discharge.

In the "H-Discharge" the electric field is induced in the discharge gas by the transformer action of a high frequency magnetic field, as shown in Fig. 4. The discharge is maintained in this configuration entirely by induction, and the discharge current flows in closed loops about the central axis of this system. The equivalent circuit of the "H-Discharge" is shown in Fig. 5, where the discharge is presented as a single-turn secondary winding of the induction coil, which has been shorted through its own resistence.

It is of considerable importance to note that the "E-Discharge" can and does occur in a rarefied gas located within the volume of a solenoid carrying high frequency current when the axial electric field of the solenoid (roughly Ldi/dt per unit length) has a value which exceeds that of the induced electrical field in the rarefied gas, as was emphatically demonstrated (5) by Townsend and Donaldson. Or, the two discharge types can coexist (8) simultaneously, in the same apparatus. Thus, electrodeless discharges can be formed electrostatically (i.e., capacitively), or by a time-varying magnetic field (i.e., magnetically), or they can result from a combination in nearly any proportion of these two mechanisms (9).

The merits and demerits of the electrodeless discharge as a means for transferring large amounts of power to high-pressure gas were treated in detail by Hollister (10), who showed that in high-power applications, the "E-Discharge" is considerably inferior to the "H-Discharge". In this study it was determined that scaling laws exist for the "H-Discharge" and that under conditions of thermal equilibrium, the "H-Discharge" is indeed an arc type of electrical discharge. This is in contrast with the low-pressure "ring" discharge of Thomson (2,3,4) and with Townsend's(5) capacitive, low pressure, glow discharge. It had earlier been reported (11) that scaling laws exist for the "electrodeless arc", as the thermal "H-Discharge" is now called.

Subsequent theoretical studies (12) demonstrated that the existence of the electrodeless arc discharge is <u>independent</u> of the outer boundary condition (i.e.,the discharge vessel wall), hence that it is <u>not</u> necessary to "stabilize" (13) the discharge by convective mass throughflow. Thus, it was only after the year 1969 that a non-flow stabilized (i.e.,sealed-off) high pressure electrodeless arc discharge was possible; and it was only after the year 1970 that feasibility was established for the application of this physical knowledge toward the development of an electrodeless arc high intensity light source. The first public disclosure (14) of these principles in the high intensity lighting application was subsequently made in February, 1971, with full technical disclosure following in 1973 (15) and 1975 (16).

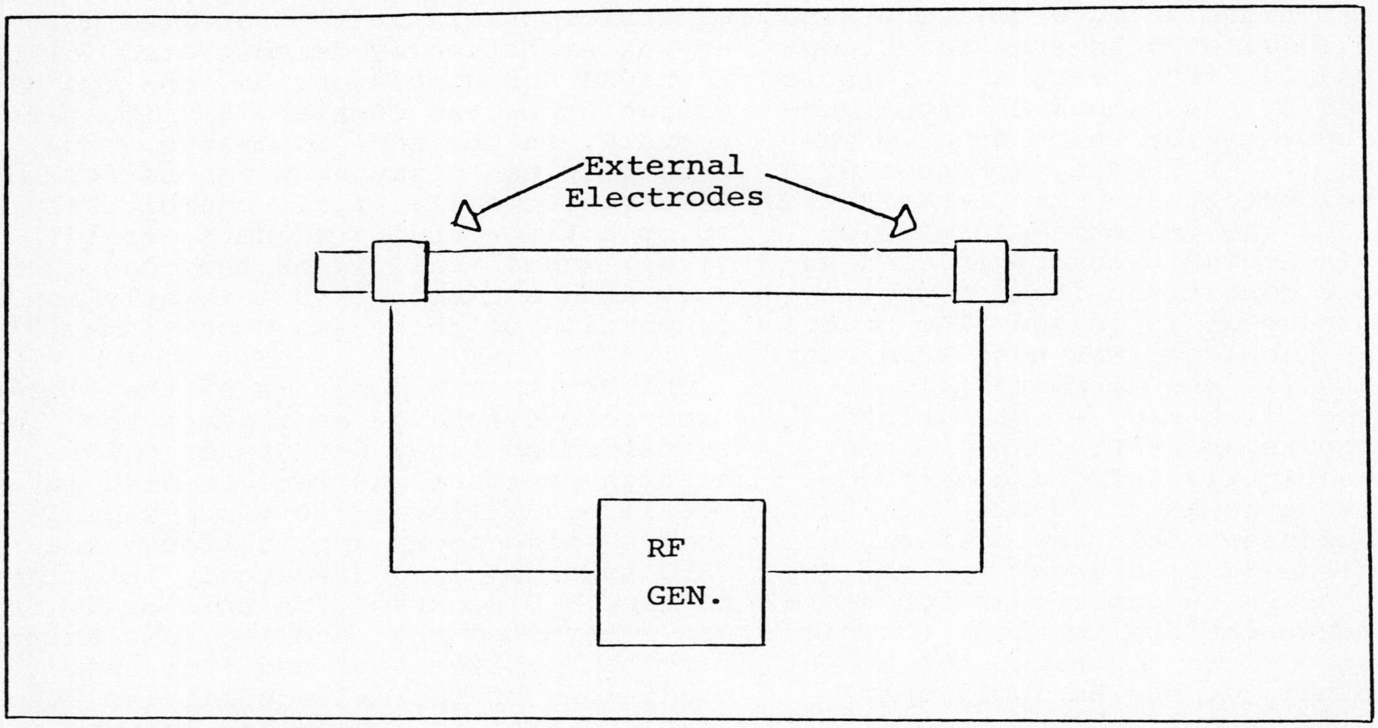

FIGURE 2
E-DISCHARGE CONFIGURATION

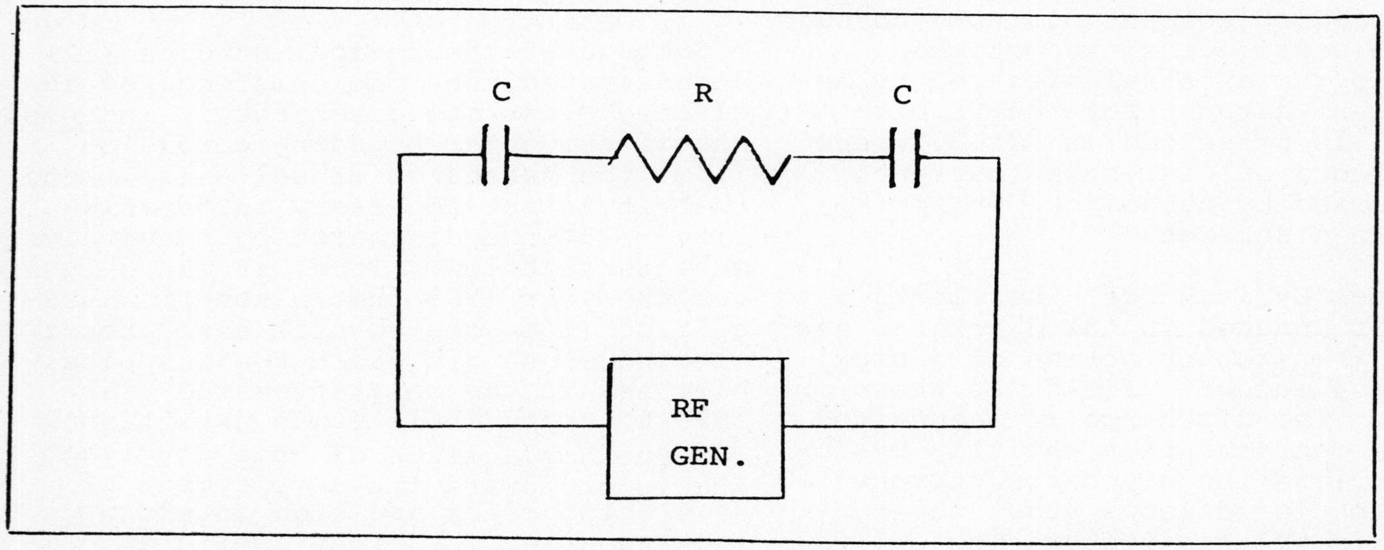

FIGURE 3
EQUIVALENT CIRCUIT OF E-DISCHARGE

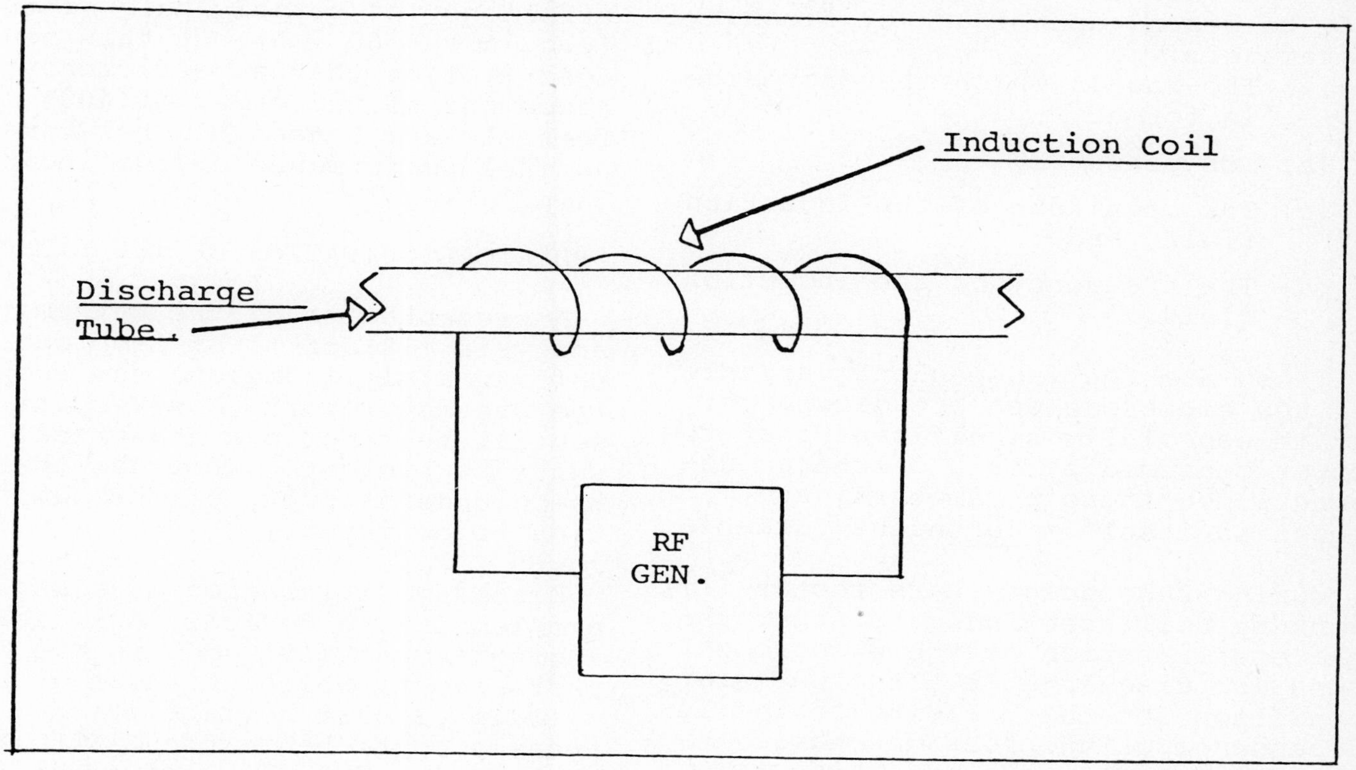

FIGURE 4

H-DISCHARGE CONFIGURATION

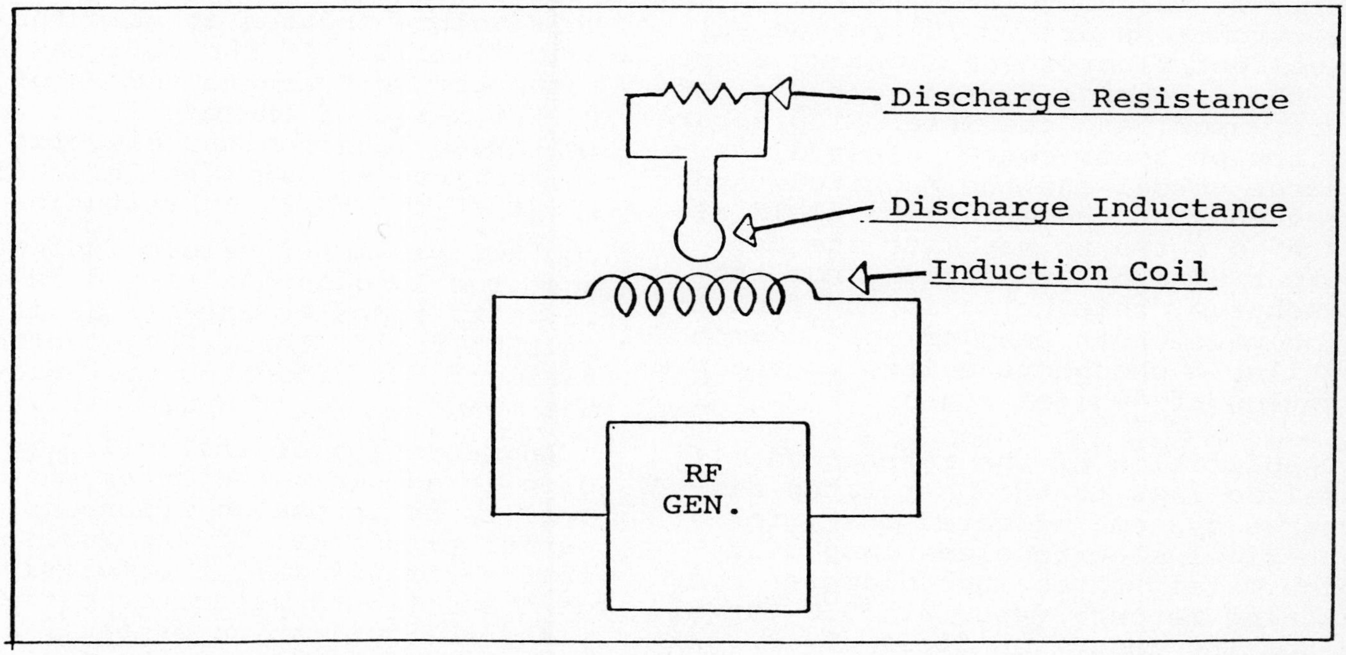

FIGURE 5

EQUIVALENT CIRCUIT OF H-DISCHARGE

The operating state of the thermally equilibrated, high pressure electrodeless arc has been found (11,12,14,15, 16) to depend upon four key discharge parameters.

(a) The gas in which the discharge is formed,

(b) Its pressure,

(c) The magnitude of the induction field, and

(d) The frequency of the induction field.

These are the independent variables of the electrodeless arc discharge. It is especially significant that the power consumed by this discharge depends upon these parameters, and hence is itself a dependent variable.

During subsequent years,recent (17) work by Hollister indicated that thermal equilibration of the electrodeless arc discharge is not a necessary condition for the validity of the independence of the four discharge parameters. Thus, it was predicted that a low pressure, non-equilibrated electrodeless arc would follow the same physical rules as does the high pressure, equilibrated electrodeless arc discharge, and that efficiency in power transfer to this discharge would depend on these parameters as well. It was postulated that such an electrodeless arc would provide an excellent fluorescent lighting source if the discharge forming gas was mercury vapor, and the internal pressure of the phosphor coated glass discharge vessel and the magnitude and frequency of the induction field all were adjusted to maximize the net output of $\lambda 2537$ radiation from the discharge, this being in the excitation wavelength band of most common available phosphors suited to the production of "white" light.

Application of the referenced (11) scaling laws to the postulated discharge system indicated that, for a spherical electrodeless lamp of approximately three inch diameter containing mercury vapor at 50°C, an alternating induction field of approximately one gauss strength at several tens of megacycles would dissipate approximately twenty watts of rf power. Assuming a net 50% phosphor efficiency, such a discharge could yield a light output up to 115 lumens per watt of dissipated power. Real world effects, mainly in terms of the obtainable electronic efficiency of rf power supply portend an obtainable net efficiency of roughly half of this (e.g. about 60 l/W) but this nevertheless implies that an efficiency enhancement of the order of 400% is both feasible and obtainable relative to the nominal performance of the incandescent lamp.

PHYSICAL OPERATION OF THE LITEK LAMP

A presentation of the mathematical analysis (18) of LITEK lamp operation unfortunately is beyond the scope of the present report. However, it is essential that certain analytical results be presented in order that a basic understanding of the lamp's operation be conveyed.

Discharge initiation is always a problem in arc devices. Auxiliary starting equipment (19) adds to the total parts count, which necessarily reduces overall circuit reliability and increases cost; hence the LITEK lamp has been designed to be self-starting.

This has been accomplished by noting that the voltage drop across the induction coil $L(dI/dt)$ effectively produces a moderately strong longitudinal electric field E_z within the discharge volume of the lamp. Likewise, the time rate of change of magnetic flux through this volume induces an azimuthal electric field E_ϕ. If the coil has N/ℓ turns per unit length and is of radius R, the ratio of the magnitudes of the impressed longitudinal electric field to the induced azimuthal electric field is $|E_z/E_\phi| \simeq 2\pi RN/\ell$. Substitution of nominal parameter values indicates that the longitudinal field is twenty to fifty times as intense as is the azimuthal field. Thus, the electrostatic field of the induction coil provides the means for discharge initiation.

Augmentation of the coil's voltage drop is accomplished by arranging the coil to be in pseudo resonance with a series capacitor. If the quality factor of the circuit is Q,the relative phases of the coil and capacitor currents yield multiplication of their respective voltage drops by the factor Q. The augmented electrostatic field thus produced has been found to be a completely reliable means for the production of initial ionization in the LITEK lamp via a capacitive E-Discharge.

As stated earlier, the electrostatic mechanism generally does not enable appreciable power transfer to an electrodeless discharge. Principally, this is because the capacitance between the induction coil and the discharge volume usually is quite small and the resulting capacitive reactance is sufficiently large to limit severely any plasma current, hence to prevent significant power transfer to the discharge.

However, no such constraint exists for current responding to the induced azimuthal electric field, and, provided that sufficient pre-ionization is produced capacitively, azimuthal currents rapidly form closed loops about the induction coil axis in a smooth transition from the inefficient E-discharge to a highly efficient H-discharge, enabling the coupling of power into the azimuthal discharge. This immediately reduces circuit Q to a relatively small value, where the efficiency of power transfer is proportional to ΔQ.

The actual resistance of the induced plasma torus thus formed is quite small, but effectively this resistance forms the secondary winding of a transformer and therefore reflects reciprocally into the circuit's primary winding. Thus, a small coupled resistance reflects as a large resistance in series with the induction coil, and limits the primary current to a value which can be predetermined. This self-ballasting effect enables the LITEK lamp to be run directly off the line as a self-contained unit.

The Langevin equation appropriate to the LITEK lamp's plasma was expanded to derive a discharge resistivity matrix. This was inverted to yield an expression for the appropriate conductivity tensor, which, in turn, was examined term by term. For the conditions typical of LITEK lamp operation the conductivity elements parallel and perpendicular to the incident magnetic field were found to be equal, and the Hall term was small, as were dielectric effects at the frequency of operation. Thus, dielectric, cross field, and Hall effects are negligible, and a nominal value for the real, scalar electric conductivity of the discharge is of the order of 10-20 mhos per meter.

Existence of a well-behaved conductivity being assured, the field equations were manipulated to yield a diffusion equation for the magnetic induction, the solution of which was in terms of modified Bessel functions. Expressions for the induced current density and electric field were obtained. Application of Poynting's theorem then produced closed form expressions for total discharge dissipation and plasma-augmented coil inductance which exhibited reasonable agreement with measured values and were self-consistent within about 18%. Finally, the coupling coefficient between the induction coil and its coaxial plasma was determined, significantly, to be independent of the number of turns of the induction coil.

In actuality, the LITEK lamp is a system composed of an electronic radio-frequency power supply and its load, which is an induced electrodeless arc discharge. Thus, a "bad" lamp will perform poorly even with a "good" power supply, and conversely.

It was found that the most effective circuit configuration is a simple Clapp oscillator (20). Lamp configuration was determined not to be critical with respect to external dimensions, and for a combination of convenience and cosmetic reasons, the present LITEK lamp is similar to the standard A-line incandescent shape. Spherical and "exotic" lamp shapes have been developed which exhibit performance levels essentially identical to the standard.

Performance of the system currently is at a level between 45 and 50 lumens per watt of input AC power with an efficiency spread of roughly ±5% from unit to unit, as shown in Fig. 6.

Present development efforts, consisting mainly of optimizing the mercury-indium amalgam employed in the bulb and reducing the present discrete component circuit to a more efficient (and less costly) hybrid configuration, are expected to yield an overall performance improvement of about 25%. The following nominal specifications are expected to apply to the market-ready LITEK lamp:

a. Input voltage: 120 volts @ 60 Hz, ±10 volts.

b. Total input power level: 25 watts approximately.

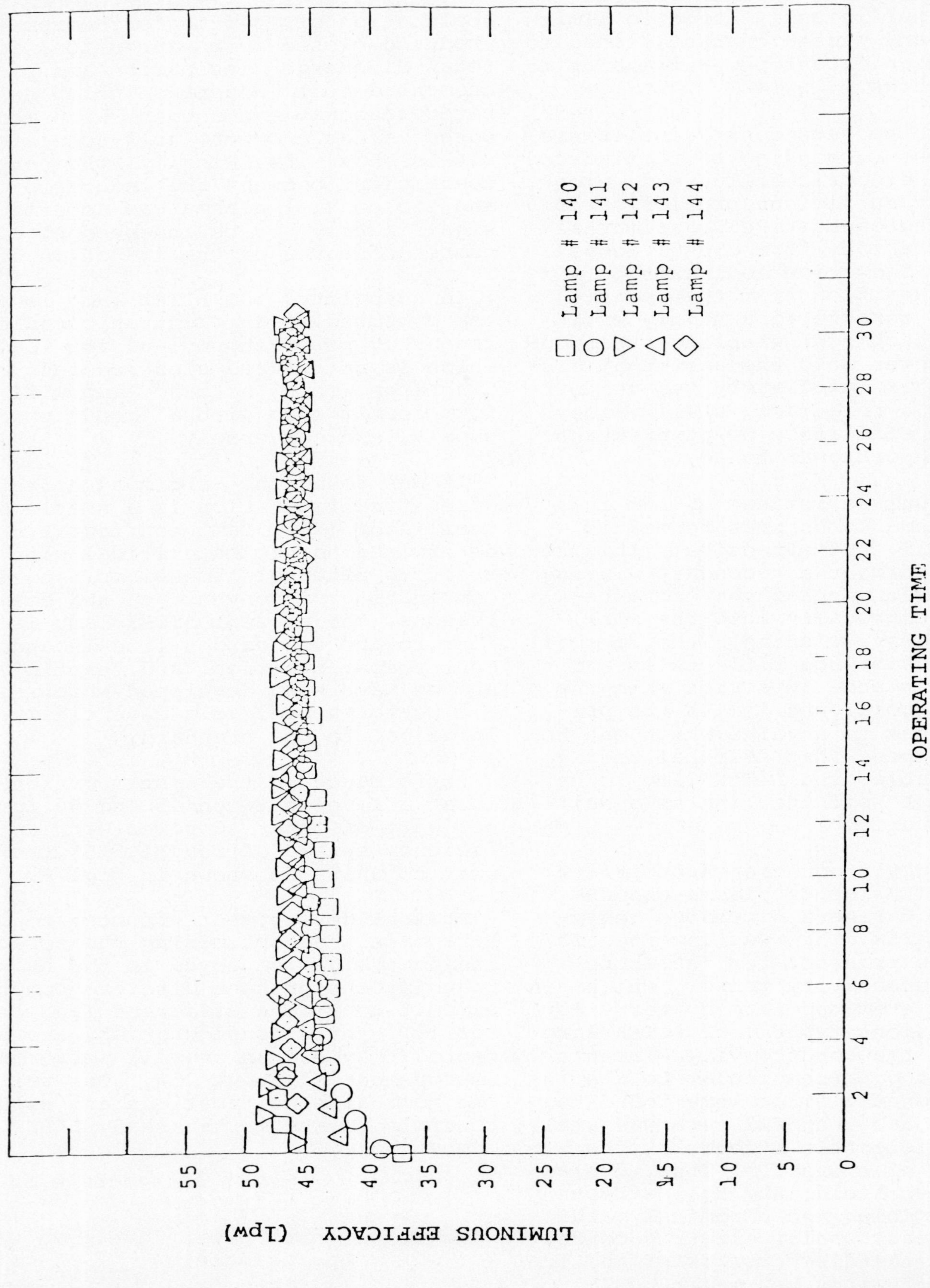

FIGURE 6 LITEK LAMP STABILIZATION

c. Operating frequency: 13.56 MHz, ±6.78 KHz for the complete system.

d. Overall circuit efficiency: 70% minimum, 60 lumens per AC watt.

e. Inband radiation: Less than 25 microvolts/meter at 1000 feet (with plasma load).

f. Spurious radiation: Less than one microvolt/meter at 1000 feet (with plasma load).

g. Lamp base: Use standard Edison medium skirted base plus a small additional section for heat sinking.

h. Ambient air operating temperature: +40° to +100°F.

END USE ENERGY MANAGEMENT: COST EFFECTIVENESS

The LITEK electrodeless fluorescent lamp offers four major advantages:

1. Efficiency-The LITEK lamp uses approximately 25% as much energy as a standard incandescent lamp for an equivalent level of illumination.
2. Long Life-The LITEK lamp employs no electrodes or filaments which are the normal failure mechanisms in existing incandescent or fluorescent lamps.
3. Edison-Socket Compatibility-The design of the LITEK lamp permits direct screw-in replacement of standard incandescent lamps without wiring modifications or ancillary starters, ballasts or power supplies.
4. Life-Cycle Cost Effectiveness-The LITEK lamp can repay its initial cost in only a few months of normal use through reduced expense of energy usage.

According to Nuckolls (21) many factors other than initial cost interrelate to determine the environmental (and economic) impact of any lighting system. Full environmental cost should include an evaluation of light output and consumption in terms of effectiveness, energy use, purchase cost (including investment interest, taxes, and insurance), installation labor and materials, and the system's impact on other aspects of the environment. Comparisons of alternate illumination methods should be used to determine the optimal system.

In the incandescent lamp replacement market many of the above factors do not directly apply. Fixtures already exist and the environmental impact of lighting is essentially predetermined. To the user, the basic benefit of LITEK lamp replacement of incandescent lamps must be the reduced cost of lighting which results from the substantially greater energy efficiency of the LITEK lamp. On the national level, however, the apparent benefits are reversed and a substantial nationwide energy saving would follow truly widespread LITEK lamp usage, with the result that additional user dollars may be diverted from business overhead into increased profits or as an offset to other energy-related cost increases.

The most rapid utilization of the LITEK lamp is most likely to take place within the commercial sector. This is because the non-industrial businessman purchases electric energy at the commercial rate, which in general is a higher rate than that paid by either the residential or the industrial user. While the actual per kilowatt-hour cost usually varies in a somewhat complicated inverse relationship to the net energy usage, the typical commercial rate will average up to 50% higher than the typical residential rate within a given community. It is axiomatic that those communities which have ample energy resources pay substantially less for energy than those communities which are energy-short. Likewise, electricity costs more in areas where the peak demand is high than it does in areas with a relatively constant demand. As a result, locations such as New York City are characterized by extremely high energy costs, while cities such as San Francisco or Seattle, which heretofore had considerable hydroelectric energy resources available, are accustomed to paying 50% to 60% less.

However, a certain relativity exists. Energy rates are rising everywhere, and today's New York rate is merely the harbinger of the local rate in a few years. In this presentation, therefore, the New York rate has been used as a portent of things to come.

The ultimate market for the LITEK lamp is equal to the number of sockets into which the lamp can be placed. In the United States this number has been estimated by Stanford Research Institute to be 2.75 billion, 500 million of which are in the non-residential sector. Segmentation of about 94% of the total market is according to the following breakdown of principal socket locations:

Major Markets		Millions of Sockets
Residential		2,250
Industrial		80
Street, highway lighting		4
Commercial indoor:		
Office buildings	35	
Retail establishments	40	
Schools	40	
Health facilities	15	
Other	60	
Subtotal		190
Commercial outdoor		35
Other miscellaneous		29

The full energy conservation potential of the LITEK lamp will only be realized if its propagation into many of the above identified sockets is assured. However, an initial assessment of the marketability of the LITEK lamp must be based on a firm foundation. A reliable measure of the LITEK lamp's marketability must first be quantified to assess the true commercial merit of the LITEK lamp. Such a measure is provided by the life cycle cost effectiveness of the LITEK lamp in its projected market ready form. To this effect, the life cycle cost effectiveness of the LITEK lamp has been evaluated in comparison with an equivalent light output incandescent lamp in an equivalent installation.

The study does not provide annualized results because the year is considered to be too arbitrary a base period from the user's point of view. Rather, an amortization period T exists for every installation after which the LITEK lamp has paid for itself through reduced energy costs. Employing this base period, one needs only to determine the daily, weekly, monthly and annual usage in the particular installation under consideration to provide an appropriate comparative basis in time.

Incandescent lamp performance and cost data were obtained from recent manufacturers' sales and pricing publications. LITEK lamp performance was based upon the projected luminous efficacy of the market ready device of 60 lumens per watt of total input power. The general amortization method is described in Fig. 7. For brevity, only those data related to 40 and 100 watt comparison lamps, where a total energy rate of \$0.10 per KWH exists, are presented in Tables I and II. The effects of maintenance (i.e., replacement) do not appear in the 100W commercial sector example because the LITEK lamp amortizes before the comparison 2500 hour lifetime incandescent lamp burns out. In these examples the "worst case" of LITEK lamp initial cost of \$10.00 was selected to maintain a degree of realistic conservatism in interpretation of the analysis.

The amortization period T was determined comparatively on the basis of equality of total costs for use of a single LITEK lamp and n incandescent lamps under the constraint that $T=(n-1)L+t$, where $(n-1)$ is the integral number of rated incandescent lamp lifetimes L in T, and t is a residual use period. The analysis was performed numerically. In cases where the crossover point occurred at a required relamping time, the analysis included the cost of the final lamp, and the amortization period was recomputed on the basis of inclusion of that cost. Because of the normally small cost of incandescent lamps, this inclusion does not appreciably affect the computation of life cycle costing for most cases treated.

However, when maintenance costs are included in the relamping scenario, as in certain burdened commercial installations, the effective relamping expense is appreciable and the resultant amortization period is relatively short.

In the past, the high cost of incandescent lamp maintenance (i.e., lamp replacement) led to the introduction of several types of long life incandescent lamp. Additionally, current OSHA rulings (mandating, for example, a two-man crew to operate a ladder during the replacement of a ceiling mounted lamp) are responsible for the increased proliferation of long life incandescent lamps in the commercial marketplace. However, the long life incandescent lamp is notoriously inefficient, and in

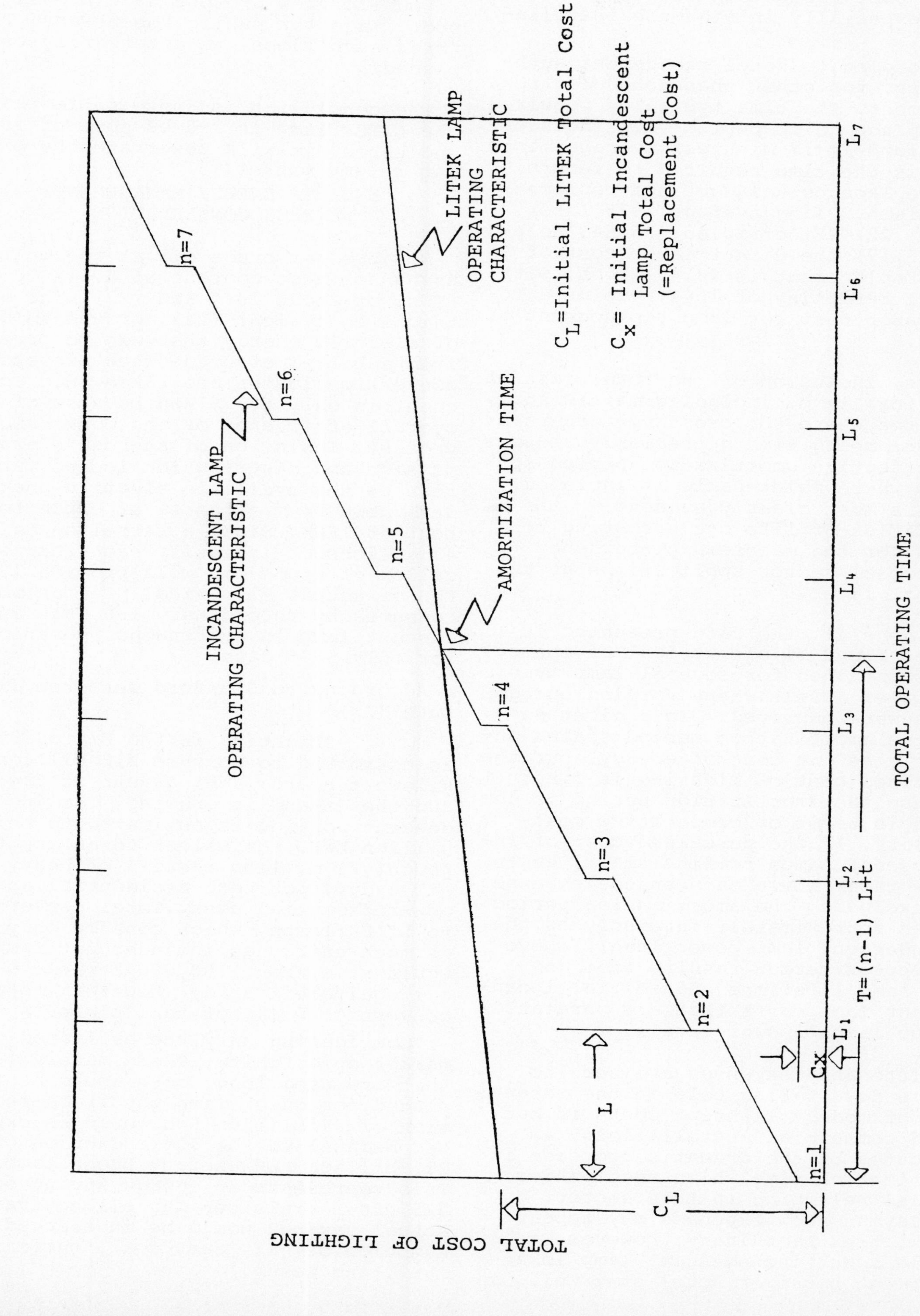

FIGURE 7 AMORTIZATION METHOD

some areas the rising cost of energy has overtaken maintenance as the principal cost element in lighting economics, expecially in high-use installations.

According to hotel management authorities, for other than scheduled replacement, the time typically required to replace an in-public-view incandescent lamp is 17 minutes. Averaged into this is the time required to secure and replace both ladder and replacement lamp. Using average 1975 labor rates (22) (Electrician: \$6.84, Helper: \$5.17) the direct labor cost of lamp replacement is \$3.40 which, with normal burdening of 100%, yields a total labor cost for lamp replacement of \$6.80.

Thus, inclusion of the total cost of the final lamp, including maintenance, in cases where the crossover occurs at the relamping time appreciably reduces the effective amortization period in cases where maintenance is included. This is most clearly evident in the tabulation of life cycle costing results for the maintenance-burdened commercial sector applications at the 40 watt level.

In general, the data presented in Tables I and II are typical of the results obtained for several lamp types and power levels,where varying energy rates were employed. In a given sector the amortization period scales inversely as the cost of energy, but the amortized cost of lighting is fixed because the amortization period is defined in terms of comparative cost equality. If the purchased cost of incandescent lamps remains small, as in those cases where maintenance expenses are excluded, the amortization period varies approximately inversely as the incandescent lamp power level, where the approximation results from unequal lamp lifetimes and initial incandescent lamp costs that are unrelated to the lamps' power levels.

Maintenance expenses are seen to play a devastating role in the retention of modest lighting costs in burdened commercial installations, as indicated by the dramatic decrease in amortization time which accompanies initial relamping in this sector. To the layman such expenses may appear somewhat extraordinary; however, a single spent incandescent lamp in an apartment house or hotel stairwell can lead to personal injury litigation and damage awards of astronomical proportions. Moreover, burned-out light bulbs in public view present an extremely poor image for public businesses. Therefore, such lamps are necessarily replaced.

The conclusion is inescapable that the LITEK lamp is indeed cost effective and hence, it will penetrate the existing relamp market.

END USE ENERGY MANAGEMENT: ENERGY CONSERVATION

A barrel of crude oil at the wellhead has a net heat content of 5.6×10^6 BTUs, or about 1640 KWH (23). The electrical equivalent (24), or the amount of electric energy that can be produced from a barrel of crude when process and conversion losses are taken into account is only 601.5 KWH because of the overall efficiency of the conversion process. Inclusion of aggregate transmission and distribution losses further reduces the available electric energy derivable from a barrel of crude to about 580 KWH. Since a barrel contains 42 gallons it is easily seen that one gallon of crude oil will provide 13.81 KWH of actual electrical use, an amount consumed by an ordinary 100 watt incandescent lamp in continuous operation for only 5.75 days.

According to Stanford Research Institute (25):

"Lighting in the United States is estimated to consume 418 billion kilowatt-hours (KWH) a year at present, and the demand is growing. The incandescent portion is estimated to be 178 billion KWH, equivalent to an input of roughly 1 million barrels of heavy fuel oil a day, for both residential and non-residential uses. Total conversion to LITEK lamps, which consume only 30% as much energy as incandescent lamps, would save about 700,000 barrels of oil equivalent a day; however, total conversion will not take place.

"Considering only the estimated total market opportunity, where mainly lamps that are used long hours would be replaced for quick lamp payoff, a reduction of 87 billion KWH could be expected because of the 70% reduction of the 125 billion KWH used by those lamps. This represents an energy saving of 425,000 barrels per day oil equivalent. Actual savings would be in reduced consumption of oil, gas, coal, nuclear fuel, and water."

TABLE I Cost Effectiveness Analysis
Comparison: 100 Watt Incandescent

	Residential	(No Maintenance)	(Including Maintenance)
Amortization Time (hours)	1247.06	1238.55	1238.55
Total Cost ($)	13.47	13.06	19.86
Number Incandescent Lamps (n)	2.00	1.00	1.00
Total Incandescent Lamp Cost ($)	1.00	0.67	7.47
Total Incandescent Energy Cost ($)	12.47	12.39	12.39
LITEK Energy Cost ($)	3.47	3.06	3.06
LITEK Energy Cost Savings ($)	9.00	9.33	9.33
KWH Savings	90.00	93.30	93.30
BBL Oil Savings	.15	0.16	0.16

TABLE II Cost Effectiveness Analysis
Comparison: 40 Watt Incandescent

	Residential	(No Maintenance)	(Including Maintenance)
Amortization Time (hours)	2744.62	2629.99	574.37
Total Cost ($)	12.06	11.82	17.20
Number Incandescent Lamps (n)	2.00	2.00	2.00
Total Incandescent Lamp Cost ($)	1.08	1.30	14.90
Total Incandescent Energy Cost ($)	10.98	10.52	2.30
LITEK Energy Cost ($)	2.06	1.82	0.40
LITEK Energy Cost Savings ($)	8.92	8.70	1.90
KWH Savings	89.20	87.00	19.00
BBL Oil Savings	0.15	0.15	0.03

The energy saving potential of the LITEK lamp in the commercial sector is easy to estimate. In this sector 375 million incandescent sockets have been identified as readily accessible, where the average socket load is 75 watts. The individual socket energy savings potential is presented in Table III as a function of daily usage, where the results of use of a 75 watt incandescent lamp are compared with those of a single LITEK lamp on an annual basis.

National energy conservation potential is presented in Table IV as a function of percent penetration into the identified commercial sector. For the sake of consistency, lamp usage is equivalent to 8.58 hours per day listed in the previous tabulation.

The obvious conclusion that, because of the relatively high energy cost within the commercial sector, and due to continuing OSHA and union induced maintenance expense increases, nearly 100% penetration of the commercial marketplace is not unlikely, emphasizes that the above energy saving potential is significant, especially when due account is taken of the air conditioning energy use synergism(26).

The present domestic residential market opportunity consists of some 500 million units out of a current total of 2.25 billion. On the average, the "typical" domestic lamp is of the 75 watt size and burns about three hours per day. On this basis the total annual energy saving potential of the LITEK lamp within the identified residential market segment is 31 billion KWH, roughly the equivalent of 145,000 barrels of oil per day.

However, the residential sockets which initially are targeted by the LITEK lamp are not "typical" within the above definition because the amortization period for such lamps may not be sufficiently short to inspire significant purchase of the LITEK lamp.

The LITEK lamp will first replace high-wattage incandescent lamps in high-use applications in regions where energy rates are high. Later,as economies of scale reduce the lamp's price, and energy rates continue to rise, significant penetration into the so-called "typical" applications would be expected.

Consistent with the results of the life cycle cost effectiveness study,the residential market is not expected to be penetrated to a significant extent until the year 1980. The number of residential sockets projected into 1980 is 2.44 X 10^9, an increase of 190 million above the 1976 value.The median daily usage can be expected to remain at about 3 hours, thus roughly half of these sockets will operate for more than three hours, and of these, one fourth are estimated to operate for six hours or more yielding a total of 305 million high-use residential sockets to form the basis of the accessible residential market opportunity in 1980.

Assuming that the lamps replaced by LITEK are of the 100 watt size, the annual usage of 66.8 billion KWH, would be reduced by 50.1 billion KWH to 16.8 billion KWH, a net savings of the energy equivalent of 237,000 barrels of oil per day. The remaining 237 million accessible sockets would be expected to be "typical", hence their replacement would save 14.6 billion KWH per year, or about 69,000 barrels of oil per day.

The energy savings potential of the LITEK lamp in the identified accessible market sectors is summarized as follows. The estimate by SRI of an energy savings potential of the equivalent of 425,000 barrels of crude per day was based on national averages of recent energy use. The present basic analysis, which projected the market into 1980, combines direct commercial applications with "typical" residential applications to yield a prediction of an energy savings potential of the equivalent of 457,000 barrels of crude per day, in close agreement with the projection of SRI. Closer examination of the marketplace, based on the results of the life cycle cost effectiveness study presented herein, indicated that high-use, high wattage residential socket installations will be relamped by the LITEK lamp much more rapidly than will be typical residential sockets, with a corresponding increase in energy savings potential as the immediate result. Similarly, the analysis indicated that far fewer of the typical sockets will be relamped than the number predicted by SRI. Moreover, the air conditioning load that has been identified as a concomitant of lighting was not included in SRI's estimate of total lighting energy consumed, hence the reduction

TABLE III ; 75 Watt Incandescent/LITEK Conservation Comparison

Daily Usage (Hrs/Day)	75 W Incan. KwH/Year	BBL Oil Per Lamp Per Year	Equivalent LITEK BBLS Use	Net LITEK Savings BBL/Year
2	55	0.098	0.025	0.074
4	110	0.188	0.047	0.141
6	164	0.285	0.071	0.214
8	219	0.375	0.094	0.281
8.58	235	0.405	0.101	0.304
10	274	0.473	0.118	0.355
12	329	0.570	0.143	0.428
14	383	0.660	0.165	0.495
16	438	0.758	0.190	0.569
18	493	0.848	0.212	0.636
20	548	0.945	0.236	0.709
22	602	1.035	0.259	0.776
24	657	1.133	0.283	0.850

TABLE IV: Energy Conservation Potential in Commercial Sector

Penetration (Percent)	Total Sales (Millions) Of Units	Oil Energy Equivalent Savings		
		Direct BBL/Year	BBL/Day	Incl. A/C BBL/Day
1.0	3.75	1.14×10^{6}	3.12×10^{3}	4.68×10^{3}
2.5	9.38	2.84×10^{6}	7.79×10^{3}	1.17×10^{4}
5.0	18.75	5.69×10^{6}	1.56×10^{4}	2.34×10^{4}
7.5	28.13	8.55×10^{6}	2.34×10^{4}	3.51×10^{4}
10.0	37.50	1.14×10^{7}	3.12×10^{4}	4.68×10^{4}
50.0	187.50	5.69×10^{7}	1.56×10^{5}	2.34×10^{5}
100.0	375.00	1.14×10^{8}	3.12×10^{5}	4.68×10^{5}

in this energy requirement which accompanies LITEK lamp usage was neglected in the SRI estimate.

Inclusion of these corrections suggests that the energy saving potential of the LITEK lamp could be as much as the energy equivalent of 774,000 barrels of crude per day. The various analytical results are presented in Table V.

ACHIEVEMENT OF THE CONSERVATION POTENTIAL OF THE LITEK LAMP

Stanford Research Institute has estimated that of the 2-3/4 billion live incandescent lamp sockets in the United States 875,000,000 high use sockets should be immediately accessable to the LITEK lamp at estimated selling prices in the range of $7.50 to $10.00. This, SRI estimates, would result in a savings of our nation's energy resources equivalent to 425,000 barrels of oil per day. More recent studies have indicated that lower prices and a deeper penetration of this available market can be projected with reasonable confidence because of advances in technology and higher electric power costs. These studies indicate that the potential energy savings could run as high as 774,000 barrels of oil per day. Averaging the high and the low estimated energy conservation potential figures yields a nominal oil equivalent energy savings potential equivalent to 600,000 barrels of oil per day.

The total Alaskan oil reserves (both onshore and offshore) have been found to amount to about 10 billion barrels of crude. The Alaskan pipeline can carry up to 1.2 million barrels of crude per day. At that rate all identified Alaskan reserves will be totally depleted in 22.83 years.

If it is assumed that strong conservation measures will become matters of National priority, it is not unlikely that the depletion rate of the Alaskan reserves will be reduced by mandate. The relative importance of the LITEK lamp will then increase.

For example, if an attempt is made to conserve the presently identified Alaskan oil reserves (i.e., to stretch them out) for about forty years, the Alaskan pipeline will inject some 600,000 barrels of oil per day into the energy economy. Properly implemented, it has been shown that the LITEK lamp's propagation could yield the conservation of the energy equivalent of about 600,000 barrels of oil per day. Thus, the two projects could be considered to be roughly equivalent in terms of their respective impact on energy reserves. Also because of the life cycle cost effectiveness of the LITEK lamp, it will save money for its users. This can be an effective hedge to the inflationary economic forces created by rising energy costs.

The major U.S. lamp producers have a combined production capability of approximately 5,000,000 lamps per day. If these producers can be encouraged to redirect a mere 5% of their production resources to the manufacture and sale of the LITEK lamp, satisfaction of the SRI identified market need could occur by 1985 or sooner. This will result in daily energy savings of more than 600,000 barrels of oil per day.

It is recognized that the maximum energy conservation potential of the LITEK lamp cannot be achieved unless the production capabilities, financial resources, and experience which are already abundantly available in the U.S. and abroad can be effectively marshalled to the task of producing and distributing this new light source.For this reason, it is proposed to broadly license this technology on reasonable terms to any company or organization which can quickly and effectively address this international need.

In the interim, every effort will be made to accelerate the propagation of the LITEK lamp. Each year that can be shaved from the total time required for its widespread implementation could mean the conservation of the energy equivalent of almost a quarter billion barrels of crude within the United States alone. On the international scale this could almost double.

TABLE V

ENERGY SAVINGS POTENTIAL OF THE LITEK LAMP

In Accessible Market Sectors

(In units of Barrels of oil energy equivalent per day)

A. STANFORD RESEARCH INSTITUTE:		425,000
B. LITEK, Basic Analysis		
Commercial	312,000	
"Typical" Residential	145,000	
Total		457,000
C. LITEK, Results of Life Cycle Cost Effectiveness Study		
Commercial, direct	312,000	
Commercial, A/C synergism	156,000	
High use residential	237,000	
"Typical" residential	69,000	
Total		774,000

REFERENCES

1. Hittorf, W., Wied. Ann 21, 137 (1884)
2. Thomson, J. J., Phil Mag. S. 5, 32, 321 (1891).
3. Thomson, J. J., Phil Mag. S. 5, 32, 445 (1891).
4. Thomson, J. J., Phil Mag. S. 7, 4, 1128 (1927).
5. Townsend, J. S. & Donaldson, R. H., Phil Mag. S. 7, 5, 178 (1928).
6. MacKinnon, K. A., Phil Mag. S. 7, 8, 605 (1929).
7. Babat, G. I., J. Inst. Elect. Engr. (London),94, 27 (1947).
8. Tykocinski-Tykociner, J., Phil Mag., S7, 13, 953 (1932).
9. Hollister, D. D., IEEE Transactions, AP-13, 134 (1965).
10. Hollister, D. D., An Investigation of the High Pressure Electrodeless Arc in Air, Technical Report, AFFDL-TR-68-160, Wright Patterson AFB, Ohio 45433, February, (1969).
11. Hollister, D. D., Physics Letters, 27A, 672 (1968).
12. Hollister, D. D., The Thermal Electrodeless Arc in Air, presented at the 1st Symposium on the High Pressure Arc, 22nd Gaseous Electronics Conference, Gatlinburg, Tennessee (1969).
13. Reed, T. B., J. Appl. Phys. 32, 821 (1960).
14. Hollister, D. D., A Xenon Lamp with Two Less Electrodes, Electro-Optical Systems Design Mag., February (1971) pg 26.
15. Hollister, D. D., U.S. Patent No. 3,763,392, Oct. 2, 1973.
16. Hollister, D. D., U.S. Patent No. 3,860,854, Jan. 14, 1975.
17. Hollister, D. D., U.S. Patent No. 4,010,400, March 1, 1977.
18. Hollister, D. D., (to be published).
19. Anderson, J. M., U.S. Patent No. 3,521,120, July 21, 1970.
20. Clapp, J. K., Proc. IRE,36,356 (1948);See also Gouriet, G. G., Wireless Engineer, April,(1950) pp 105-112.
21. Nuckolls, James L., Interior Lighting for Environmental Designers, Wiley-Interscience, New York (1976).
22. World Almanac and Book of Facts, Newspaper Enterprise Association, New New York, (1977), pg 140.
23. House Subcomittee on Energy, reported in World Almanac and Book of Facts, Newspaper Enterprise Assoc., New York, (1977), pg 136.
24. Thirring, Hans, Energy for Man, Harper and Row, New York, (1976) pg 25.
25. Business Development Strategy and Market for a New General Purpose Lamp, Stanford Research Institute, December (1976).
26. Lighting and Thermal Operations, Guidelines, FEA Office of Conservation and Environment, Wash. D.C., pg 1.

THE REFRIGERATOR AND NEW TECHNOLOGIES

W. David Lee

Arthur D. Little, Inc., Cambridge, Massachusetts

ABSTRACT

Technological developments could reduce the energy consumption of household refrigerators by 60% or more. The over 80 million refrigerators now in operation account for approximately 1.4×10^{15} Btu of primary energy consumption per year. Energy savings equal to approximately 1% of the entire nation's energy consumption could be achieved through improved refrigerator design.

Key areas identified for technological developments are: improved insulation systems, including the door closure area; improved compressor pumps; and advanced control systems. Approaches for reducing energy consumption through design changes are discussed in this paper which is based on work performed by Arthur D. Little, Inc. under contract to the Federal Energy Administration.*

INTRODUCTION

Consumption of electricity by household refrigerators is a significant portion of the nationwide energy consumption. Of the estimated 80 million refrigerators presently in use in the United States, each consumes an average of 4.5 kwh of electric power per day. In total this amounts to energy consumption of 1.4×10^{15} Btu per year at the power plant.

The energy consumption of the refrigerator can be partitioned into three areas:

Refrigeration load (heat from walls, food, door openings): 65%

Heaters (antisweat and defrost): 20%

Auxiliaries (fans and timers): 15%

Examination of the energy partitioning within a refrigerator can point to the areas in which technological advances could significantly reduce energy consumption.

HEAT LOAD

Heat enters the refrigerated volume through the walls and through the door openings. Typically, a heat load of 300 Btu per hour is gained through the walls, while on average another 100 to 200 Btu per hour is admitted through door openings and food admission.

Increased insulation thickness will reduce the thermal load, but will also reduce the available refrigerated volume. The majority of domestic refrigerators sold in the United States have reached the limit of acceptable outer dimensions, and an additional insulation thickness will necessarily decrease the storage volume. The effect of adding inches of urethane foam to a simple cube geometry is shown in Fig. 1. A point of diminishing returns is reached at which added insulation reduces the available storage volume at a faster rate than it reduces the energy consumption. The optimum insulation thickness for any given model will be a function of the refrigerator geometry and the refrigeration unit efficiency.

*Federal Energy Administration Contract No. CO-04-50228-00, "Energy-Saving Options for Refrigerators and Water Heaters; Volume 1, Refrigerators"

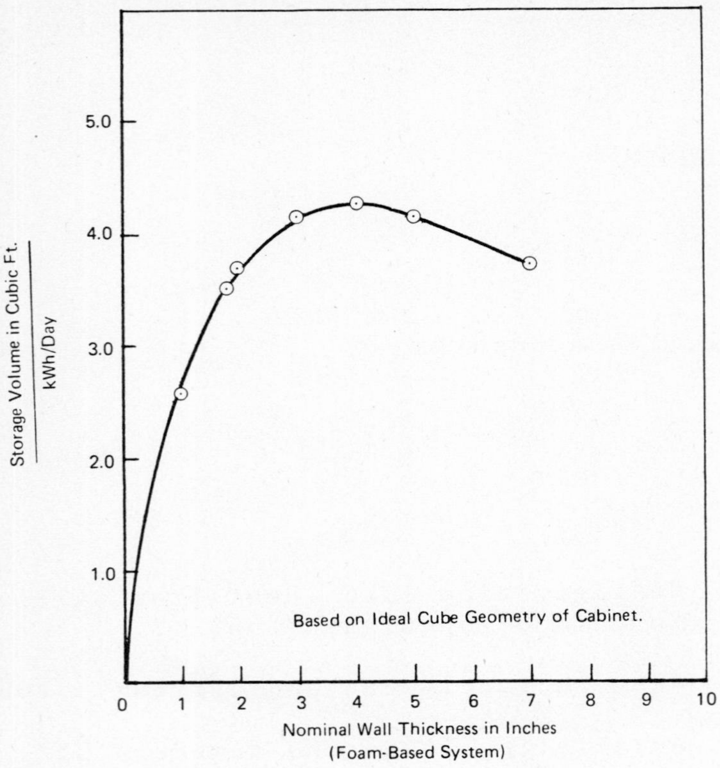

Fig. 1. Energy factor $\frac{\text{storage volume}}{\text{kwh/day}}$ vs. nominal insulation system thickness

Energy savings can be achieved with better quality insulating materials. Table 1 shows the effective conductivities of three insulation materials. The conductivity of the gas contained in the insulation (also shown) constitutes a major portion of the conductivity of the insulation; in order to achieve conductivities lower than that of polyurethane foam (which is used in most new refrigerators), a gas with a conductivity lower than freon must be used or the system must be evacuated.

TABLE 1 Effective Thermal Conductivity of Insulation

Insulation	Thermal Conductivity $\frac{\text{Btu-inch}}{\text{hour-ft}^2}$	Gas Thermal Conductivity
Fiberglass	.24	.18 Air
Polyurethane Foam	.13	.07 Freon
Powder	.01	.001 Evacuated

Evacuating the insulation system will reduce the gas conductivity, but precautions must be taken to also eliminate or minimize the radiation component of heat trasnfer between the inner and outer walls of the refrigerator. The evacuated powder insulation system nearly eliminates gas conduction, using opaque powder to minimize the radiation component.

In current refrigerator models, heat leaks still occur at the door closure area where the inner and outer walls of the refrigerator must join. Figure 2 shows the composite conductivity of the insulating material (the lower limit of the cross-hatched band) together with the effect of the door closure area (within the band). As shown, the door closure area raises the effective conductivity of a evacuated powder insulation to nearly that of the polyurethane foam insulating system for thick-walled systems. A further problem with the evacuated powder insulation system is that special stainless steel or other vacuum-tight material will be required. If evacuated powder insulation systems can be developed that are cost competitive with thick-walled polyurethane systems, energy savings in excess of 40% may be realized from this change alone.

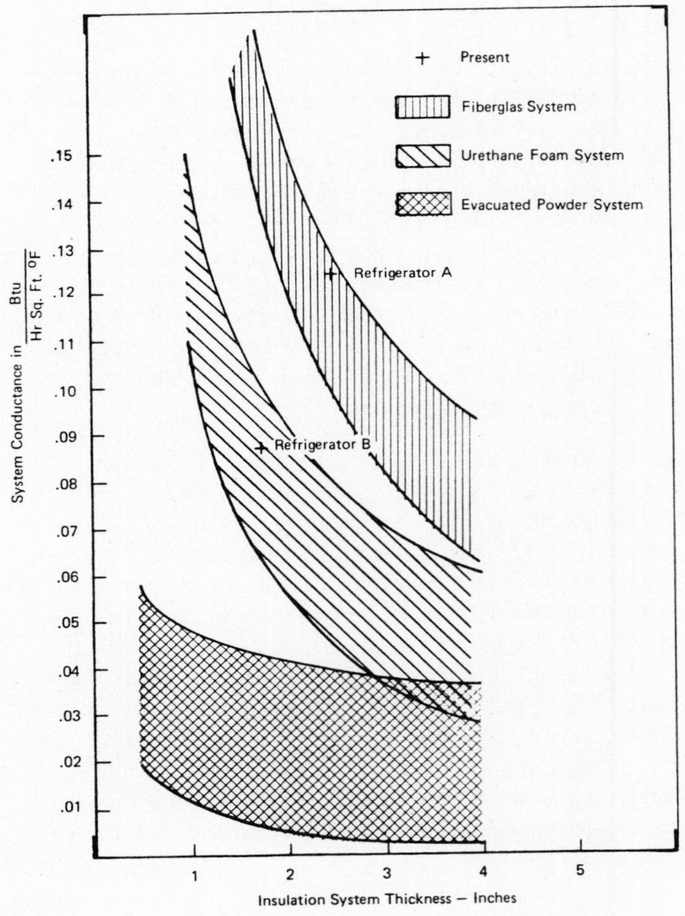

Fig. 2. Refrigerator insulation systems

For either the urethane foam or the powder insulation system, improvements in the door closure area will be required. These may include double-seal gaskets.

REFRIGERATION UNIT

Figure 3 shows a schematic of a typical refrigeration unit for domestic refrigerators. The compressor provides high-pressure gas to the condenser, which cools the gas (freon) to a liquid by either free convection or using a fan; at this point, a capillary tube expands the high-pressure liquid to low-pressure gas and liquid mixture, which reduces the gas temperature to approximately -20° F. This cold freon enters the evaporator coil, which cools the air circulated by a fan within the refrigerator. Key areas for energy savings are the compressor, heat exchangers, and the fans and their motors.

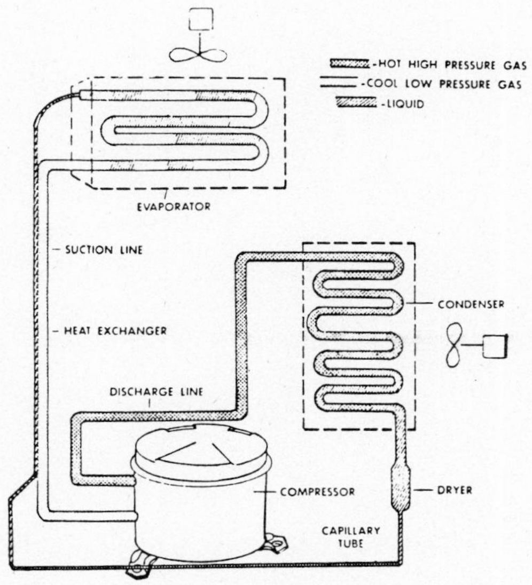

Fig. 3. Schematic of a typical refrigeration unit

Table 2 summarizes the overall compressor efficiency, separating the pump from the motor efficiency. The efficiencies of presently available pumps are relatively low because of the difficulty of making high-efficiency small compressors, which have large rubbing surfaces relative to the amount of work they perform on pumping refrigeration fluid. As the units become larger, the mechanical efficiencies generally increase. Nevertheless, some energy savings can be achieved, and we anticipate that future pump efficiencies can reach about 70% for reciprocating and rotary compressors.

TABLE 2 Compressor Efficiencies

Type	Motor Efficiency %	Pump Efficiency Present %	Pump Efficiency Future %
Rotary	80	47	70
Reciprocating	80	52	69

Heat exchangers typically operate with a 25 - 30° F temperature difference between the air and the freon. Increasing the heat exchanger size will reduce the temperature difference and increase the overall system efficiency. Increased surface area and improved finning may yield a 5% energy savings.

Fans used in today's domestic refrigerators, which account for approximately 16% of the refrigerator energy consumption, can be improved by the use of better blade design and more efficient fan motors. We estimate that an additional 8% energy savings can be achieved by improved fan and motor assemblies.

CONTROLS

In addition to the improvements in the refrigeration unit, improved unit controls can offer substantial energy savings in the refrigerator. In a typical automatic defrost refrigeration, the airflow from the cooling coil is split between the freezer and the fresh-food compartment as shown in Fig. 4. The refrigeration unit is actuated by a thermostatically-controlled switch located in the fresh-food compartment. The actual distribution of airflow between the two compartments is controlled by a manually set damper. The freezer is without active thermostat control, depending for thermal control on the thermostat in the fresh-food compartment to call for cooling. Some manufacturers offer an automatic damper in the fresh-food compartment and relocate the thermostat switch in the freezer compartment. This provides active control in both compartments.

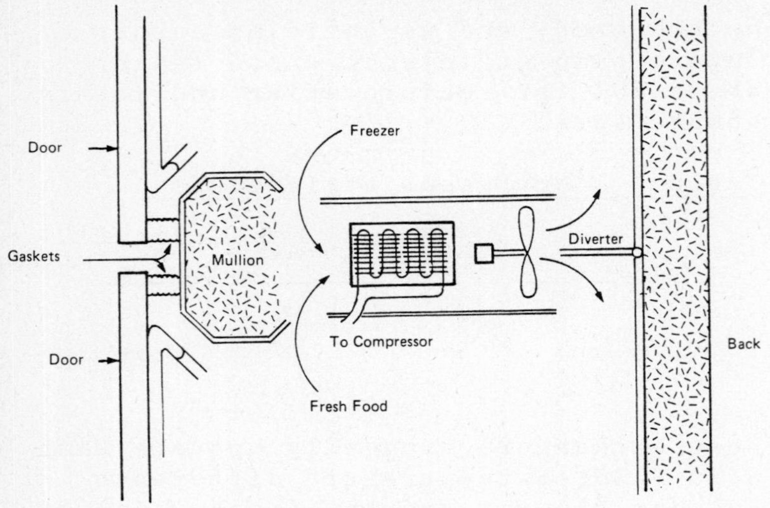

Fig. 4. Evaporator Design

Future controls may provide the separate compartment control, as well as reduce the energy consumption of the refrigerator by adjusting the refrigeration unit operation to meet optimally the refrigeration requirements. Variable speed compressors and techniques designed to adjust the refrigeration separately to the 0° freezer and the 35° fresh-food compartment can be envisioned in the future. It is estimated that operation of the refrigeration unit independently for the two compartments and optimization of the refrigeration unit for each compartment spearately could save 15% of the energy consumption.

Improved technology in insulation systems combined with improved refrigeration units, and advanced controls may someday result in an automatic defrosting refrigerator with energy consumption only 40% that of present day refrigerators.

THE ANNUAL CYCLE ENERGY SYSTEM CONCEPT AND APPLICATION*

John C. Moyers and Eugene C. Hise

Oak Ridge National Laboratory, Oak Ridge, Tenn. 37830

ABSTRACT

The Annual Cycle Energy System (ACES), under development at ERDA's Oak Ridge National Laboratory, promises to provide space heating, air conditioning and water heating at a significantly lower expenditure of energy than conventional space conditioning and water heating systems. The ACES embodies heat pumping, thermal storage and, where climate dictates, solar assistance. The concept is described, along with variations in design that permit flexibility to maximize energy conservation or to provide load management capabilities. Installations that exist or are under construction are described and variations that are incorporated to meet specific objectives are discussed.

INTRODUCTION

The residential sector of the U.S. economy is an important one from the standpoint of energy conservation, accounting for about one-fifth of total energy use and about one-third of total electricity sales. Within the home, three of the largest energy consuming functions account for approximately three-quarters of the total energy consumption.

The Annual Cycle Energy System (ACES), under development at ERDA's Oak Ridge National Laboratory since late 1974, provides these three services at an expenditure of energy that is significantly lower than conventional systems. In addition, the ACES presents the electric utility with a much flatter wintertime load profile than the air-to-air heat pump and permits managed operation to reduce the utility's peak air conditioning load.

CONCEPT

The ACES is an integrated system designed to provide efficiently the three largest energy requirements of residences and buildings — space heating, space cooling, and water heating — through the use of a heat pump, thermal storage, and auxiliary sources and sinks.

The heat pump is the conventional vapor compression cycle and uses commercially available compressors, heat exchangers and other refrigeration components.

The thermal storage is provided for both the high and low temperature sides of the heat pump. The high temperature storage consists of at least a domestic hot water storage tank and can, if advantageous, also store hot water for space heating. The low temperature storage is a

*Research sponsored by Energy Research and Development Administration and Department of Housing and Urban Development, under Union Carbide Corporation's Contract with the Energy Research and Development Administration.

fixed volume of water and serves as the heat source for the heat pump and also serves as the cold source for air conditioning. During the heating season, heat is pumped from the low temperature storage to supply space heating and domestic water heating. The bulk of this heat is taken from the heat of fusion of water so that over the heating season the water is nearly all turned to ice. This ice is stored and used for air conditioning in the summer. Thus the compressor operates to supply space and water heating but does not operate to supply air conditioning. To state the advantage of the thermal storage in simplistic terms: the system always uses both ends of the heat pump. Whenever the compressor operates, the heat is supplied to space and water heating and the cooling effect is stored in the form of ice to be used for space cooling. This is as opposed to the conventional heat pump which always discards either the heating or cooling effect.

If the annual amount of heating required were precisely balanced by the annual amount of cooling required the system could make ice all winter, melt it all summer, and come out even every fall. There is a broad band of climate across the central USA that comes remarkably close to this balance, but of course the vagaries of climates, buildings, and occupancy preclude a continuing balance and auxiliary heat sources and sinks are required.

In a northern climate where the heating requirement clearly exceeds the cooling, a heat source is required. The favored source is a simple solar collector — simple because it needs to collect heat at a temperature only high enough to melt ice. In many locations it need not be glazed or insulated. This system would be managed over the year in such a way as to collect solar energy for part of the heating season and then create enough ice to supply the air conditioning.

Conversely, in a southern climate where the cooling load clearly exceeds the heating load a heat sink is required. The favored heat sinks are the conventional fan-cooled outdoor condenser or the simple solar collector mentioned above. This system would be managed over the year so that all space heating and water heating is pumped from the low temperature storage and the resulting ice stored for air conditioning. When that ice is exhausted, the compressor is permitted to run at night, when the heat rejection is most efficient, to store cooling for the following day.

What are the advantages of adding the complexities of thermal storage and auxiliary heat sources and sinks to a heat pump? There is first the obvious advantage that both the heating and cooling effect of the heat pump are used simultaneously which can result in almost free air conditioning. Secondly, the domestic hot water is heated with about 40% of the energy required by a conventional electric heater. Less obvious advantages are that, because of the mode of operation, the heating coefficient of performance (COP) is higher than that of a current air-to-air heat pump, the capacity is constant, and the system is eminently suited to load management.

VARIATIONS ON THE THEME

The three previously described components, heat pump, thermal storage, and auxiliary heat sources and sinks, can be assembled in many ways to fit a wide variety of climates, buildings, occupancies, and requirements. In the center is the building and climate that results in a close balance between the annual total heating and cooling requirements such that all of the heating can be supplied by making ice and all of that stored ice used to supply cooling. This configuration has the maximum thermal storage, the minimum auxiliary heat sources and sinks, and a maximum annual COP.

Departing from center in the direction of a higher heating-to-cooling ratio (moving north), the amount of ice that could be formed in the heating season exceeds the amount required for cooling and so the thermal storage is sized to supply the cooling and the auxiliary heat source is increased to melt the unneeded ice.

Departing from center in the direction of a higher cooling-to-heating ratio (moving south), the thermal storage is sized to supply the heating requirements and to store all the ice made, and an auxiliary heat sink is supplied to provide heat rejection for the additional space cooling.

Yet another variation responds not to climate but to the electric utility need for load management. This configuration has both high and low temperature thermal storage and both an auxiliary heat source and an auxiliary heat sink. During the heating season all possible solar energy is collected to keep the low temperature storage melted and the compressor is operated off-peak to supply immediate heating and to store heat for the on-peak hours. During the cooling season, the compressor is operated off-peak to store cooling for the on-peak hours. The heat that is useful for domestic hot water is stored and the excess rejected to the auxiliary heat sink.

Examples of each of the configurations are in existence, under construction, or firmly planned.

Existing and Planned Installations

Four ACES installations exist, are under construction, or are planned at the present time. These include three single-family residences, two of which are designed for full annual cycle operation, and a health care facility of much larger size.

The Knoxville house. This installation (Ref. 1) is located at the Tennessee Energy Conservation in Housing site on the University of Tennessee Institute of Agriculture campus near the Knoxville municipal airport. Two other single-family residences, a solar house and a control house, are adjacent to the ACES house. The solar house incorporates both active and passive solar heating and the control house can be heated by an air-to-air heat pump or by electric resistance heaters. The floor plan and insulation level for the three houses are identical, allowing direct comparison of energy consumption by the various space conditioning and water heating systems.

An artist's rendition of the installation is shown in Fig. 1. The 70.8 m^3 (2500 ft^3) insulated ice bin, sized for full annual cycle operation, contains an array of ice-forming coils having 33 cm (13 in.) horizontal and vertical spacing. An anti-freeze methanol water solution is circulated through the coils to transport heat from the bin to the heat pump evaporator during space and domestic water heating operation or directly from the heating/cooling fan coil during air conditioning operation. The heat pump mechanical package contains the refrigerant compressor, a desuperheater and a condenser for domestic water heating, a space heating condenser, an evaporator, and pumps and valves to provide the desired routing of domestic water and anti-freeze solution.

A radiant/convector coil is located on the south-facing vertical wall between roof sections to serve as a heat source or sink for balancing the system against year-to-year variations in climate. In the case of a severe winter, when more ice is being formed than the bin will hold or will be needed for the next summer's air conditioning, the radiant/convector coil is used to collect solar energy and to convectively absorb heat from the outside air during warm hours in late winter. Because the collected heat is usable at relatively low temperature for melting ice, no insulation or glazing is used in the collector coil. The coil itself is constructed of 12.7 mm (1/2 in.) outside diameter aluminum tubing with co-extruded diametrically opposed fins, such that the tube and fin present a 7.6 cm (3 in.) wide collection surface. The tubing is mounted to present a venetian blind effect, with each collection surface inclined at 60° from the horizontal, allowing free air circulation around the tubing.

In the case of an unusually hot and long summer, the stored ice will be melted before the end of the air conditioning season. The system accommodates to this situation through night operation of the heat pump to freeze ice for the next day's cooling, disposing of the heat through the radiant/convector coil to the night air and sky. This effectively shifts the major electrical consumption for

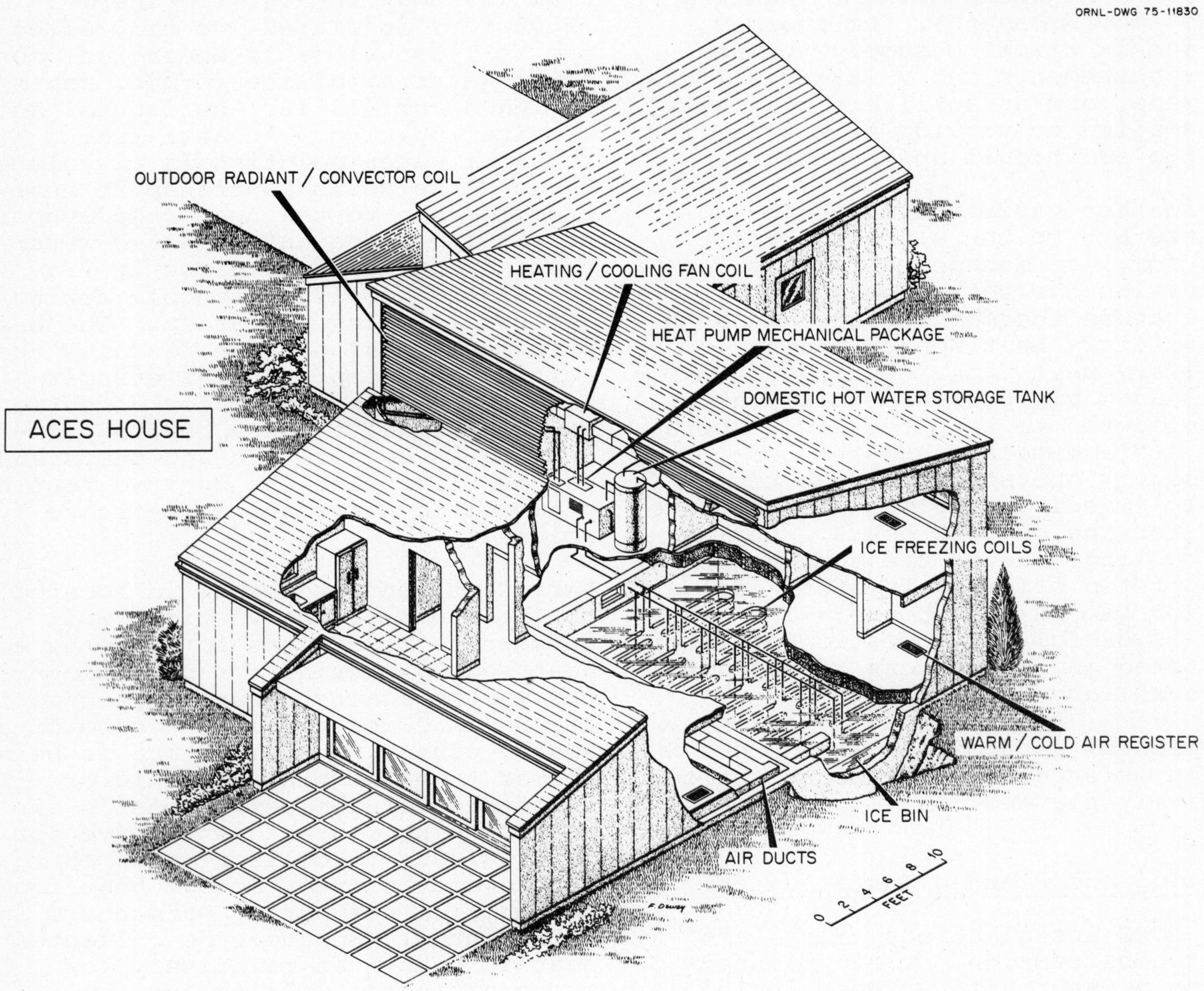

Fig. 1. The Knoxville demonstration house.

air conditioning from the utility's daytime peak-load hours to the night low-load hours.

The ACES house and the control house are unoccupied and identical interior temperature are maintained. A loads simulation package is installed in each to impose heat input and hot water consumption representing appliances and occupancy by a family of four.

Data on weather, system operation, and electrical consumption are automatically recorded hourly on cassette tapes for subsequent analysis.

The Richmond house. This installation, also sized for full annual cycle operation, is located in the Bretton Woods Subdivision in Richmond, Virginia. The project, accomplished primarily with private funds, was designed and installed by a consortium consisting of a builder, a heating, ventilating and air conditioning contractor, a consulting engineer, and equipment and materials manufacturers and suppliers.

The house, a two-story colonial style structure having 167 m^2 (1800 ft^2) of finished floor area plus a basement, is shown in Fig. 2. A free standing 70.8 m^3 (2500 ft^3) ice storage bin, of wood frame and plywood construction with vinyl lining, is located in the basement. The ice-forming coils are fabricated from 12.7 mm (1/2 in.) outside diameter

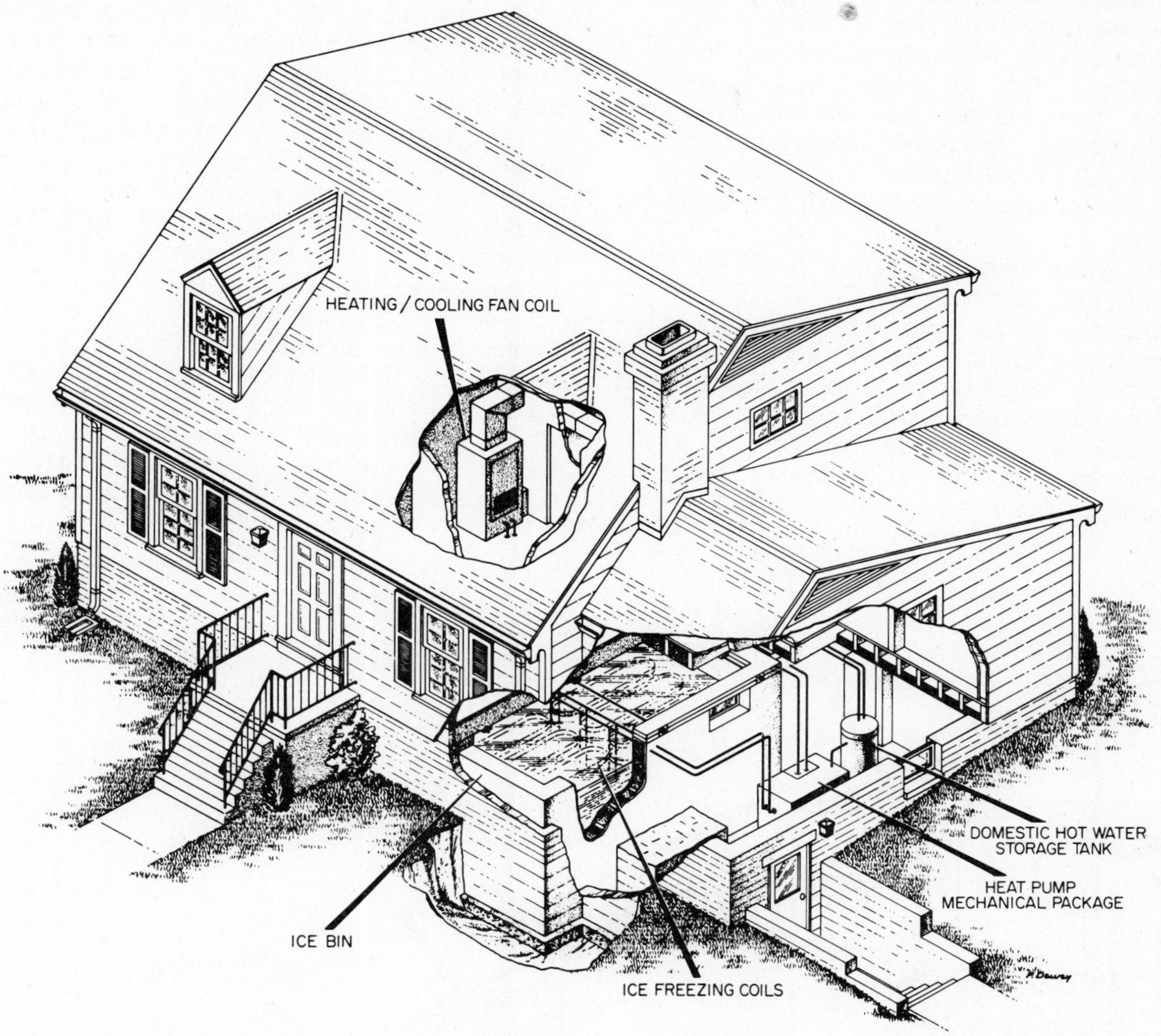

Fig. 2. The Richmond demonstration house.

copper tubing spaced on 35.6 cm (14 in.) centers.

Anti-freeze solution circulates through the ice-forming coils to transport heat from the bin to the heat pump evaporator for space heating or domestic water heating or from the space cooling coil in the air ducting to the bin for air conditioning.

Three options are provided for condensing the refrigerant. In the space heating mode, a separate coil in the air ducting serves as the condenser. A water-cooled condenser is provided for domestic water heating. An outdoor fan-blown air coil is installed to serve as the condenser for nighttime heat rejection if all of the ice has melted and cooling is required. No provision for wintertime ice melting is made; in the unlikely event that the ice bin capacity is reached before the end of the heating season, electric resistance heaters are provided in the air ducting for finishing the heating season.

The outdoor air coil was selected for this installation because performance calculations did not show the need for solar supplementation except on rare occasions and because the architecture of the house did not lend itself to the use of a solar panel.

<u>The Philadelphia house</u>. This installation, a joint venture of Philadelphia

Electric Company and ERDA, differs from the two preceding residential installations in that it is designed with load management as a major objective and will not operate on an annual cycle. It is the first installation to use an ice-maker heat pump instead of in-tank ice-forming coils (Ref. 2). Figure 3 shows a typical installation of such a system.

The house is a story-and-a-half dwelling having 148 m^2 (1600 ft^2) of living space. The indoor unit of the ice-maker heat pump, located in the utility room, contains the compressor, the air heating and cooling coils, the domestic water heating coil, and the air blower. The ice-maker heat pump outdoor unit sits over the buried ice storage bin and contains the ice-freezing plates and a fan-blown air coil. The 14.2 m^3 (500 ft^3) storage bin, of insulated cast concrete, is sized to provide source heat to the heat pump for approximately ten days of water and space heating. An uninsulated, unglazed, solar panel made of extruded polypropylene is located on the roof of the attached garage and will provide sufficient low temperature solar energy to prevent freeze-up of the bin.

Summer air conditioning will be provided by nighttime compressor operation, using the outdoor air coil as the condenser, to maintain ice in the bin. Ice water from the bin will be circulated through the cooling coil

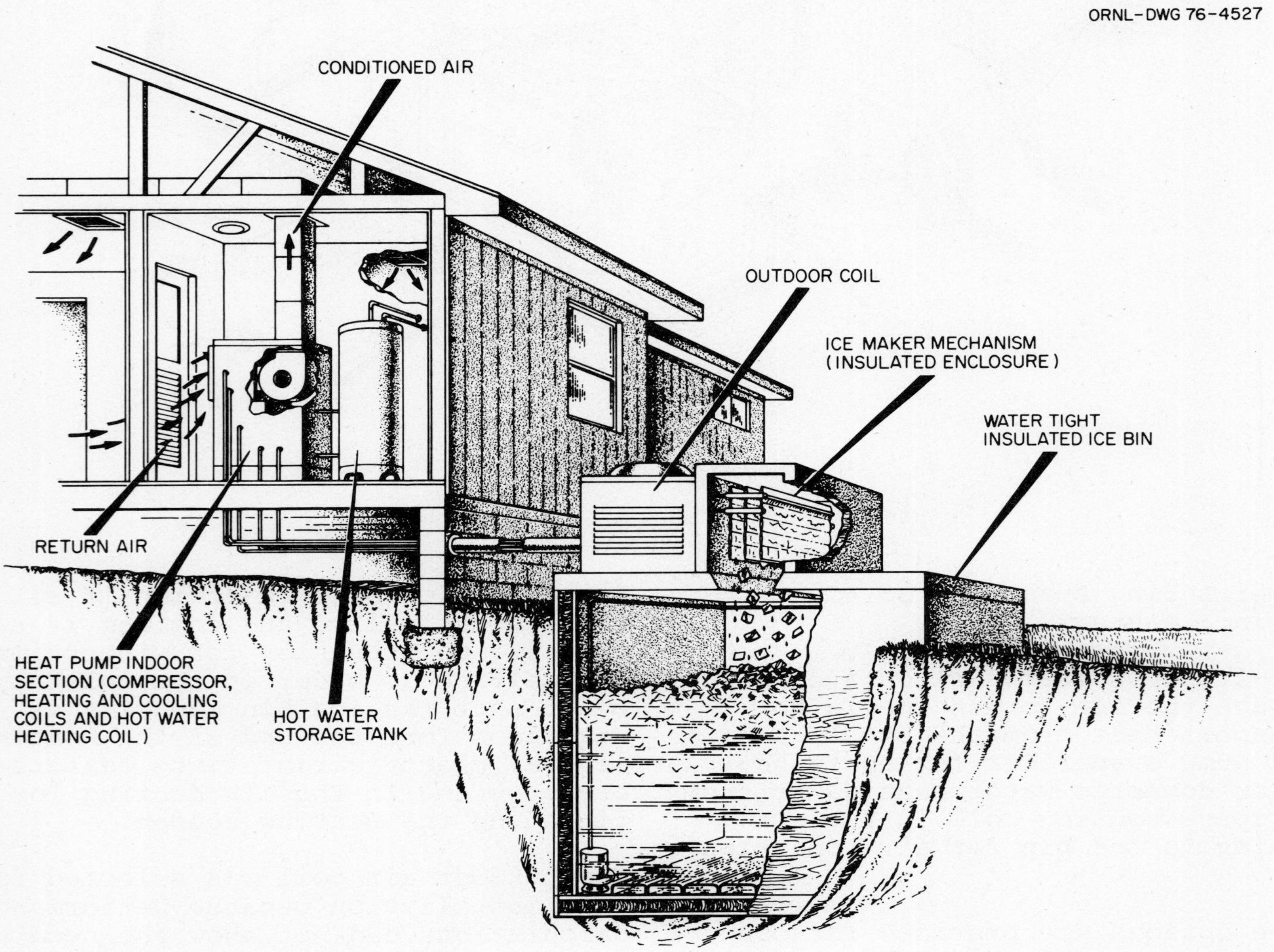

Fig. 3. Typical ice-maker heat pump installation.

in the air ducting. In this way, most of the air conditioning electrical consumption occurs during the utility's off-peak hours.

The Wilmington nursing home. This installation, now under construction by the Veterans Administration, is a 60-bed nursing home located in Wilmington, Delaware (Ref. 3 and 4). It constitutes the first large installation of an ACES and is intended both to conserve energy and to provide load management by using power when the utility's load is low. Construction was started in July 1976 and will be finished in early 1978.

At the time the decision was made to use an ACES to provide space heating and cooling for the facility, the design of the nursing home was nearly completed and included a two-pipe system feeding individual room fan coil units. A conventional steam-to-hot-water heating and mechanical air conditioning central plant was to be used. The only change of consequence to the nursing home design in adapting it to the ACES was an increase by one model size of the fan coils. This change was required because the ACES provides 40.6°C (105°F) water for heating instead of the originally planned 82.2°C (180°F) water.

The ACES will be contained in a 12.2 m by 15.2 m (40 ft by 50 ft) structure, termed the "Energy Bank" by the Veterans Administration, adjacent to the nursing home. The 623 m^3 (22,000 ft^3) concrete storage bin forms the basement of the structure and contains the ice-forming coils. The mechanical equipment, installed above the bin, consists of the compressors, a condenser, a brine cooler, a water chiller, and a fan-blown outdoor air coil that can serve as an evaporator or a condenser. The roof of the ice bank structure is the solar collector/radiator, formed from black anodized aluminum extrusions having integral axial fins and tubing.

This system is ingenuously arranged to permit several modes of heating or cooling operation. These modes, selected by a mini-computer to provide the most economical operation at given load and weather conditions, include:

1. Space heating by air source heat pumping during mild weather.
2. Space heating by bin source heat pumping, with or without solar energy collection.
3. Space cooling by melting ice formed during heating season.
4. Space cooling by heat pump operation with heat rejection via the outdoor air coil and radiator.
5. Space cooling by night heat pump operation and day ice melting.

The storage bin is not of adequate capacity for full annual-cycle operation. The stored ice is sufficient to supply the cooling needs into mid-July. Thus, more than two months of cooling will be supplied at a COP of 26. The remaining cooling will be accomplished through night heat pumping to form ice and day melting of ice as required, thereby shifting the heavy power requirement to off-peak hours.

PERFORMANCE OF THE DEMONSTRATION HOUSE

The Knoxville demonstration house was complete and formally dedicated July 30, 1976. The data acquisition system became operational in December 1976. Modification, maintenance, and the usual run of new system failures have prevented the continuous mechanical and the data acquisition system operation necessary for a full season demonstration. However, the systems have operated for several relatively short periods, some of which were during our coldest winter on record. Operation in the space heating, water heating and cooling modes has provided data for heat balances and component evaluation.

The ACES is intended to provide three services to a building, namely space heating, space cooling, and domestic water heating. These three functions are so integrated both in equipment and in time that we normally analyze system performance on the basis of the annual COP for the combined functions. Since we are not accustomed to thinking about conventional systems on an aggregate annual basis it becomes necessary, for comparison, to either estimate the annual aggregate COP for the conventional system or to break the ACES up into individual function COP's.

Table 1 compares a conventional system of resistance heat, air conditioner, and resistance water heater; a system with heat pump and resistance water heater; and ACES on both the function COP and the annual system COP. The annual energy consumption for water heating, space heating and space cooling are those calculated for the Knoxville demonstration house. The function COP's shown for ACES are the actual performance values measured when the system was operating to serve that function only. The COP's for the heat pump and air conditioner are the calculated seasonal COP's for typical equipment in this climate. The annual system COP's are calculated from the annual requirements and the function COP's.

CONCLUSIONS

The ACES system, through a combination of heat pumping, thermal storage, and solar assistance, provides an integrated approach to meeting space heating, water heating and air conditioning requirements with significantly reduced energy consumption. The concept offers flexibility in design and operation to meet alternative objectives of energy conservation and management of loads imposed on the utility.

The four demonstration projects should provide proof of the system's capabilities and serve to indicate the latitude that is available to the designer.

REFERENCES

1. Hise, E. C. et. al. (October 1976) Design Report for Annual Cycle Energy System Demonstration House, ORNL/CON-1, Oak Ridge National Laboratory.
2. Fischer, H. C. (June 1976) The Development of the Ice-Maker Heat Pump, Paper prepared for presentation at the Annual Meeting of ASHRAE, Seattle, Washington.
3. Heat pump's ice storage cuts heating/cooling costs, J. Architectural Record 160, 133-136 (1976).
4. R. A. Biehl and R. G. Werden, The energy bank, J. Heating/Piping/Air Conditioning 49, 53-60 (1977).

TABLE 1 Relative System Performance

Type of System	Function COP			Annual System COP
	Water Heating	Space Heating	Space Cooling	
Resistance heat, A/C, Elec. water heater	1	1	2	1.2
Heat pump, elec. water heater	1	2	2	1.5
ACES	2.9	2.6	12	3.6
Annual energy delivery to function	17.1 GJ (16.2 x 10^6 Btu)	18.2 GJ (17.2 x 10^6 Btu)	15.9 GJ (15.1 x 10^6 Btu	

APPLICATION OF HIGH SEASONAL PERFORMANCE FACTOR TECHNOLOGY TO A GAS HEAT PUMP

PAUL F. SWENSON AND PAUL E. CHAMBERLIN, P.E.

CONSOLIDATED NATURAL GAS SERVICE COMPANY, INC.
PITTSBURGH, PENNSYLVANIA

ABSTRACT

A joint ERDA*/Consolidated Natural Gas Company project concerning the design and development of a Rankine/Rankine thermally-activated heat pump is reviewed. The design has focused on achieving a high seasonal performance factor, especially for small commercial and industrial HVAC** applications. Equipment design point has been secondary. The paper reviews some of the problems addressed when seasonal performance is a primary design requirement. The approach to these problems and the methods employed in forecasting energy consumption and economic performance are discussed. The results obtained to date are reviewed.

1. INTRODUCTION

1.1 Background

ERDA and Consolidated Gas have been jointly pursuing the development of a Rankine/Rankine thermally-activated heat pump. The design has focused on achieving a high seasonal performance factor, especially for small commercial and industrial HVAC applications in the northern half of the United States. Equipment design-point operation has been secondary. This system has been designated the High Seasonal Performance Factor Gas Heat Pump, abreviated henceforward as the HSPF Gas Heat Pump.

A Breadboard Demonstrator unit is now under test. It represents a small commercial roof-top heating/cooling unit. Nameplate heating and cooling capacities are 10-1/2 and 7-1/2 tons, respectively. The experimental portion of the project is being managed and executed by Mechanical Technology, Incorporated, of Latham, New York.

1.2 Objectives

The first objective of this paper is to provide an outline of the steps being taken to achieve a high level of seasonal performance. Next to performance testing of the demonstrator, the most important step being taken within the program is to model, in detail, as-installed seasonal energy consumption and economic performance of the HSPF system. The second objective of this paper, then, is to review both the methodologies and the results obtained from the modeling effort.

Accurate translation of the technical performance of the gas heat pump to a measurement of it's economic attractiveness to own and operate required the examination of the energy requirements of complete buildings. The satisfaction of this requirement is also reviewed in this paper.

1.3 Definitions

The following definitions are useful in considering the performance of

---

*Energy Research and Development Administration, USA
**Heating/Ventilating/Air Conditioning

space-heating and space-cooling equipment, and are used herein.

Coefficient of Performance (COP):

$$\text{COP} = \frac{\text{Heat or coolth delivery rate to the conditioned space.}}{\text{Energy consumption rate at the site.}}$$

COP, by accepted practice, is applied to equipment that heats and cools or cools only. It is an instantaneous value, and it must be referenced to the conditions under which the equipment is operating in order to be adequately defined.

The most common practice is to omit the energy consumption due to heat or coolth delivery means in the COP calculation. Thus, in the case of a warm air furnace, the energy consumption of the conditioned air blower is usually not included. To include it would require the mixing of electric and fossil fuel energy units in the denominator. Equipment manufacturers frequently omit additional auxilary energy requirements, such as pumps and condenser fans, in publishing equipment specifications. Heat pump manufacturers do not show the impact of using supplemental resistance element heaters as is necessitated by very cold weather, and they do not employ a consistent means of accounting for the energy consumed in defrosting the "outdoor" heat exchanger in air-to-air type heat pumps.

Seasonal Performance Factor:

$$\text{SPF} = \frac{\text{Heat or coolth delivered to the conditioned space over a season.}}{\text{Energy consumed at the site over a heating } \underline{\text{or}} \text{ cooling season.}}$$

SPF is a function of extensive quantities, and therefore is not dependent on a referenced set of operating conditions. However, for heat pumps especially, SPF must be referenced to a geographic weather area. A "standard" year or a specific year must be identified, in order to provide a complete definition.

SPF, of all the performance-related parameters, is the most effective "handle" for the net energy efficiency of heating (and cooling) equipment. As is the case for COP however, the value of an SPF rating is dependent upon whether or not all of the energy consumption modes are incorporated in the definition.

The following definitions of electric and gas-energized heat pumps will be necessary to an appreciation of the subsequent sections.

Heat Pump

Any heat pump can be defined as a device which transports thermal energy from a low temperature medium (the heat source) to a high temperature medium (the heat sink). For heat pumps employed in space conditioning applications, the heat source is usually below room temperature, and the heat sink is above room termperature. The source most commonly employed is outdoor air, and the sink will usually be a warm-air distribution system.

The thermal energy delivered at the output end, which is the hot end, of a heat pump is indistinguishable from heat provided at that temperature by other means, such as direct combustion of fuels or electric resistance heaters. The advantage of using a heat pump is that it usually takes significantly less energy to pump the heat than it would take to provide the heat by direct conversion.

The energy advantage of a heat pump is greatest when the heat is pumped over a small temperature difference. In this sense, it is just like a water pump. More water can be pumped with a given amount of energy when the head rise is small. In terms of the 2nd law, the greater the temperature ratio between the heat source and sink, the greater the entropy change per unit of heat pumped, and hence the more work needed to move the heat.

Gas (Heat-Actuated) Heat Pump

Gas heat pumps differ from electrically actuated heat pumps in two ways:

1. The energy form used to pump the heat.

2. Utilization of the waste heat associated with the heat pump actuator.

The electrically-actuated heat pump has an advantage with respect to the first point. Electricity is a more convenient form of energy to use for heat pump actuation, and in practice this means a simpler and less expensive actuator. The gas heat pump wins an important advantage on the second point. The actuator or prime mover's waste heat (so called) is available at the site, and in a properly conceived system is largely available as a supplement to the pumped heat. The largest portion of waste heat associated with the electric heat pump is rejected at the central generating station. It has usually not been convenient or practical to distribute this heat for useful application away from the station site.

Where the source-to-sink temperature ratio is great, the large amount of work needed to pump a unit of heat will produce a correspondingly large amount of waste heat per unit of heat pumped. So for relatively large source-to-sink temperature ratios, gas heat pumps may do better because most of the waste heat is utilized. This leads in turn, to consideration of a gas heat pump for space heating in the northern part of the country, rather than in the South. Circumstances may favor the gas heat pump for industrial and commercial space heating because these applications do not usually need any space heat at all during mild weather when the source (outdoor air) to sink (indoor air) temperature ratio is small.

2. MATCHING EQUIPMENT DESIGN WITH SPACE CONDITIONING REQUIREMENTS

2.1 Seasonal Demand

Figure 1 shows the annual distribution of heating load occurrence and outdoor ambient temperature for Pittsburgh, Pennsylvania. Heating load occurrence is seen to peak at -1°C (30°F) outdoor ambient temperature. Somewhat over two-thirds of the annual heating load occurrence falls between -8°C and +7°C (15°F and 45°F).

Likewise, the majority of the space cooling load for residential and small commercial buildings falls between +22°C and +30°C (72°F and 86°F).

For both cases, the heat-pump equipment design-point falls outside (or at the edge) of the range of maximum load. The ARI* designated capacity and performance test-points for heat pumps fall at 22°C outdoor ambient temperature (45°F) for heating service, and at +35°C (95°F) for cooling service. Performance of a heat pump system at the ARI design-points may not represent an accurate indicator of performance over a complete heating or cooling season.

2.2 Designing for High Seasonal Performance Factor

Achievement of a high Seasonal Performance Factor requires a high Coefficient of Performance coinciding with those operating conditions corresponding to the occurrence of the majority of the seasonal (heating or cooling) load. As explained above, these conditions do not bracket the designated performance test-points.

The HSPF Gas Heat Pump is an air-to-air heat pump. This means that the heat source and heat sink are the outdoor air and indoor conditioned air, respectively, during heating service. The roles of indoor and outdoor air are interchanged for cooling service. The temperature difference over which the heat pump operates is largely a function of outdoor temperature. In turn, the heat pump temperature difference is an important determinant of heat pump operating parameters. The key components of the heat pump, in turn, are designed to perform best at selected values of operating parameters.

The foregoing indicates the nature of the conflict which must be resolved in selecting design operating parameters. Published performance data for space

*American Refrigeration Institute

conditioning equipment is invariably at the designated test-points. Current practice in the equipment industry is to base advertised performance on the test-points as well. And yet, as explained above, optimal seasonal performance, as measured by SPF, will most likely require a design based on different values of operating parameters.

For the HSPF Gas Heat Pump a large difference exists between SPF-optimum and test-point optimum operating parameters. This difference is reflected in the design requirements for each component, as well as for the system. The demonstrator unit now under test for ERDA and Consolidated was designated for optimal seasonal performance. Test-point performance was only considered in terms of achieving name-plate capacities for heating and cooling, and with reference to design limits for component stresses, temperatures, etc.

2.3 Design Features of the HSPF Gas Heat Pump

Figures 2 and 3 are schematic diagrams of the HSPF Gas Heat Pump for heating service and cooling service, respectively.1/ The configuration shown was selected for both the Breadboard Demonstrator of the present program, and also for the initial product concept. The initial product concept is a unitary HVAC package for small commercial/industrial application. The schematic diagrams utilize the following key for depicting the fluid stages.

Table 1.

Key To Fluid States/Figures 2 and 3

———	Liquid refrigerant or water.
— -- —	Liquid/vapor mixture.
------	High pressure vapor or gas.
— — —	Low pressure vapor or gas.

The reversible refrigeration loop is similar to that employed in electric heat pumps. The expansion valve (XV) and reversing valves (4-way) are shown. The steam cycle is far simpler than that employed in central electric generating facilities. Steam is delivered to the turbine at 21 Kg/Cm^{2}Abs (300 psia) and 590°C (1100°F). A single reheat and regeneration is employed. The operation of the system will employ several important features which contribute to performance on a seasonal basis as follows:

1. Phased heating of the conditioned air: First, the pumped heat is transferred to the return conditioned air, and secondly the steam turbine's exhaust is used to heat the air. This significantly reduces the head pressure requirement of the heat pump compressor. Second-law considerations show that the subsequent reduction in work per unit heat being pumped is a favorable trade-off for the resulting elevation in the turbine's back-pressure.

2. Direct heating of conditioned air with steam, by-passing the turbine at very low outdoor temperatures: The by-passed steam heats the conditioned air, again after it is heated by pumped heat. By this method, useful heat pumping can coexist with heating by means of by-passed steam. This technique, coupled with that of item 1. (above) will lead to useful heat pumping down to about -15°C with single-stage centrifugal compression. Note that heat pumping alone is optimized for an outdoor temperature of 0°C (32°F).

3. Fully-modulated steam cycle operation: The full utilization of this technique can only be accomplished when both the gas heat pump compressor and the air-moving fans are speed-coupled to the prime mover. An advantage inherent in this approach is that at low-capacity operation, air-moving power requirements do not destroy the basic energy-saving advantages of modulation. The only control inputs are to the burner and to the turbine throttle. Modulation is

an important contribution to high COP operation at less-than-full load operation.

4. Phased cooling of system condensers (cooling operation): This is the second law analog of item 1., except as employed during cooling operation. During cooling, heat being rejected to the surroundings from the heat pump (air conditioning) circuit is given first access to cooling air.

Figure 4 illustrates how pumped heat, steam turbine exhaust and direct steam injection contribute to off-setting the heating load as a function of outdoor ambient temperature. Pumped heat is the most efficient heat form from the viewpoint of energy consumption and economics as well. Pumped heat is available in greatest amount where the incidence of annual heating load is greatest.

2.4 Estimated System Performance

Figures 5, 6, 7 and 8 are maps which show COP as a function of both outdoor ambient temperature and load factor. The data shown on these maps corresponds to the 2-dimensional COP tables employed to define equipment performance in the analysis.

Figure 5: Heating Service, HSPF Gas Heat Pump.

Figure 6: Cooling Service, HSPF Gas Heat Pump.

Figure 7: Heating Service, Electric Heat Pump.

Figure 8: Cooling Service, Electric Heat Pump.

The COP values in the maps have been corrected for all auxiliary power consumption, defrost cycles, the effects of operating at partial load, and the use of supplemental heat when needed. Note that load is referenced to nameplate rating for cooling capacity, per industry practice.

Maximum capacity at a given outdoor temperature is shown on the maps as a dashed line.

It is apparent that the electric heat pump COP is best when it operates at design capacity; efficiency falls off rapidly as demand varies from that capacity. For both the HSPF Gas Heat Pump and the electric heat pump for heating service, operating below design capacity introduces losses which in turn reduce COP. Operating above capacity requires less-efficient supplemental heat, which also reduces COP. The HSPF unit is not nearly as sensitive, however, due to deliberate component design decisions. For cooling service, the effect of designing for less-than-full-load becomes especially evident for the HSPF unit. Although it would appear from these graphs that the electric heat pump COP values are higher than those for the gas heat pump, these charts are based only on on-site energy consumption. When the total energy consumed by the electric heat pump at both the site and the power-generating station is considered -- thus equating total quantities of energy consumed -- the gas heat pump will be shown to consume less total energy in terms of energy resources.

3. FORECASTING SEASONAL PERFORMANCE

3.1 Introduction

The implementation of new energy-conserving equipment naturally involves study and interpretation, of the complete ownership economics of the new device before and during market introduction. Energy conservation alone will not sell a new product. The economic incentive must be visible within the context of the intended application.

The initial phase of the Operating Cost and Investment Analysis Study2/ examined one size of equipment -- 7-1/2 tons nameplate cooling capacity -- as installed in each of four (4) types of buildings in each of ten (10) cities. The analysis compared cost-of-ownerships between the HSPF Gas Heat Pump and two (2) competing types of commercial HVAC systems in terms of return-on-investment using a discounted cash flow method and two types of pay-back calculations.

The competing types of HVAC systems are:

1. Combination gas-fueled heating/electric-compression cooling, year-round air-conditioning (YAC) unit.

2. Electric heat pump (EHP).

Both types are single-package, roof-mount designs. The gas heat pump, as previously mentioned, is also to be a single-package, roof-mount design.

The amounts of energy involved to maintain comfort conditions in a structure must be accounted for in two ways:

1. The first is concerned with the amount of energy that must be supplied by a heating or cooling system for the purpose of offsetting heat losses or gains between the structure and its surrounding environment. This is termed building energy requirements.

2. The second is related to the amount of energy that must be supplied to the heating or cooling system in order to match the loads impressed by the structure, that is heat losses or gains.

To measure the latter energy accurately, the structure cannot be viewed as playing a passive role supplying a uniform stimulus to which the different heating and cooling systems respond. The efficiencies of the heating and cooling systems responses to the energy requirements of the structure are principal variables to be measured. Their accuracy is initially dependent upon the structure energy requirement data. As will be explained, the methodology used reflects this philosphy.

3.2 Study Approach

The study examines seasonal energy consumption, operating costs and projected owning costs of the Demonstrator HSPF Gas Heat Pump, now under development, with comparative equipment expected to be concurrently marketed in the future. A principal input is the COP data shown in Figure 5 through 8 (above). It was considered essential to determine whether the gas heat pump operating costs are sufficiently lower than those of comparative equipment to attract buyers to a higher initial cost unit on the basis of ownership economics alone. Estimated representative installed costs of the equipment considered to be competitive and comparable to the gas heat pump were obtained from a Pittsburgh-area HVAC contractor. The initial and installed cost of the HSPF Gas Heat Pump are, of course, unknown. Therefore, premium or higher initial costs of the gas heat pump were calculated in a range of from plus 10% to plus 50% (in 10% increments) relative to the combination heat/cool unit. The initial cost premium of the gas heat pump, relative to the electric heat pump was calculated at only one point, 30%. The difference in the economic analysis treatment was due to the study objective being primarily to examine the comparison of the gas heat pump versus the combination heat/cool unit, with a secondary objective being the comparison with the elctric heat pump. Resource energy requirements, as such, are not evaluated in the study.

3.3 Methodology

The method of the study is based on computer simulations of the buildings and the equipment. Also, a computerized utility rate computation program and an economic analysis program were used. These programs are:

1. AGA* E CUBE Energy Requirements Program, Version 2.0.

2. AGA E CUBE Equipment Selection and Energy Consumption Program

3. Meriwether** Monthly Utility Cost Program

4. CNG*** General Investment Analysis Program

---

*American Gas Association

**Ross F. Meriwether and Associates

***Consolidated Natural Gas Company

A brief description of each program will be given later.

3.4 Study Description

Four representative building types were chosen. These are a rest home, an office building, a retail store and a light manufacturing building. ASHRAE3/ methods were applied to calculate the loads. In order to hold the building cooling requirement to a fixed equipment capacity (7-1/2 tons) in each of the ten (10) location cities, the radiant solar loads and the opaque wall and roof heat transfer factors were varied. Building areas and interior loads were maintained constant. This was done to enable direct comparison of buildings in different climatic zones furnished with identical space conditioning equipment as well as comparison of various types of equipment in a given building. The heating loads, reflecting the design temperature conditions of each city were allowed to vary. The construction configurations of the simulated buildings incorporated advanced U-factors which are expected to be in common use by 1984 which was the starting date for the fifteen year study period.

As mentioned, a prime study objective is the estimation of energy savings and economic incentive of the HSPF Gas Heat Pump relative to competitive equipment expected to be concurrently marketed. Therefore, the present efficiencies (coefficients of performance) of the existing commercial equipment marketed are adjusted to reflect future improvements expected.

3.5 Weather Data

The hourly weather data used in the computer program, which originated from the National Climatic Center*, were for typical years chosen as having degree days close to those long-term averages on the records of the gas utilities operating in each of the ten cities. The identical weather data is used by the AGA E CUBE program nationwide.

3.6 Equipment Size

The 7-1/2 ton cooling capacity was used since this is expected to be one of the optimum sizes for initial market penetration of the HSPF Gas Heat Pump.

3.7 Program Descriptions

3.7.1 Energy Requirement Program

This program uses design point values for nine components of thermal load (heating and cooling points) and the base component electric and process loads and distributes them over each hour of the year in accordance with the dry-bulb and dew-point variations, solar and cloud-cover variations, and building use and operation schedules, for the various types of operational days.

3.7.2 Equipment Program

This program determines the actual energy consumed by the various pieces of equipment used to meet the hourly energy requirements as developed by the previous program. The equipment performance characteristics are input in the form of two-dimensional coefficient of Performance (COP) tables. The COP maps, previously described in Section 2.4 (Figures 5, 6, 7 and 8), are graphic representations of the tables.

3.7.3 Monthly Utility Cost Program

The Meriwether Monthly Utility Cost (MUC) Program uses all cost factors of actual rates, converted to dollars, including gas and fuel adjustments, taxes and surcharges. All rate characteristics such as minimum bill, energy consumption, and electric demand steps are accounted for. Except for values unitized by square foot of floor area, all output values are in dollars per month and annual costs.

3.7.4 General Investment Analysis Program

The information, pertinent to this study, which the economic analysis program creates, includes a Detailed Investment Profile and Book Expense

*National Oceanic and Atmospheric Administration, Environmental Data Service, National Climatic Center, Ashville, North Carolina.

Profile for each of up to four (4) alternatives.

For each year of the analysis, the profiles consist of the Investment Factors and Book Expense Factors, respectively, and the Discounted Cash Flow (Savings) between the designated Base Investment (lowest initial cost alternative) and the other alternative investments.

The program was enhanced to printout the Discounted Cash Flow Pay-back Period (Years) and the Simple Pay-back Period (Years). The facility to inflate selected operating costs is also provided and utilized.

3.8 Relation of Technical Performance and Economic Performance

As previously mentioned, translation of the technical performance of the gas heat pump into a measurement of its economic attractiveness requires the examination of the total building utility requirements. This is due to the nature of gas and electric utility rate structures. Not only must actual rates be used -- with all adjustments, taxes and surcharges -- but since the building base load and equipment loads cannot be easily separated (and are usually handled as a single utility account) the whole building energy consumption is measured and then the appropriate rates are applied. The economic measurements then, are the total monthly and annual charges. The differential costs of buildings equipped with comparable, competitive equipment furnish the economic effectiveness comparisons needed.

3.9 Key Assumptions and Considerations

Before explaining the study results, review of the Key Assumptions and Considerations may be helpful. These are as follows:

1. The data presented are confined to equipment selected to satisfy a building with a 7-1/2 ton cooling capacity.

2. Building model load calculations followed accepted ASHRAE methods.

3. Constant inside temperature and advanced design standards with respect to insulation and double glazing were considered.

4. Building internal loads were modeled by 24-hour profiles for each day-type.

5. Comparative performance criteria corresponding to advanced gas direct-fired heating and advanced electric- compression cooling equipment were used.

6. For both cooling and heating, equipment performance was governed by two-dimensional COP tables. (The variables are outside air temperature and relative load factor.)

7. The COP tables (where applicable) corresponding to all equipment options accounted for degradation of efficiency due to defrost cycle, cycling, supplemental heating, power to drive auxiliaries, and compressor case losses.

8. Initial equipment and annual utility costs were escalated to the beginning of the fifteen-year study period, January 1, 1984, and an assumed national inflation factor was applied to utility costs during the study period. The basis for the rates used are commercial gas and electric utility tariffs in effect in 1976 including purchased gas adjustments, fuel cost adjustments, taxes and surcharges, adjusted by excalation factors to assumed 1984 levels.

9. Estimated installed costs for competitive equipment were obtained from an HVAC contractor in the Pittsburgh area. These costs were adjusted to 1984 by application of a uniform inflation factor.

10. Variations for gas heat pump first cost premium (relative to an electric cooling with gas heating year-round air-conditioning unit) in the range of 10% - 50% were applied and the investment criteria calculated accordingly.

3.10 Standards for Comparison

The Standards for Comparisons were:

1. Seasonal Performance Factors -- The measurement of each equipment type's efficiency for heating and cooling at the point of energy delivery. Refer to Section 1.3 for a complete definition.

2. Utility Costs -- The measurement of the total annual costs for all of the gas and electricity used for each building type, unitized by square foot or floor area -- not the heating and cooling equipment only.

3. Investment criteria - The economic equipment comparisons made used three investment criteria:

   A) Discounted cash flow rate of return, per cent.

   B) Discounted cash flow payback, years.

   C) Simple payback, years.

4. RESULTS

4.1 Economic Comparisons - Gas Heat Pump vs Combination Heat/Cool Unit

Only the extreme points -- 10% Premium and 50% Premium - of the range of Initial Cost assumptions are shown on the Figure 9 chart. The economic measurement is discounted cash flow rate-of-return per cent.

In regard to economic comparisons, Los Angeles is the exception with respect to the other nine (9) cities. The reasons for the relatively low economic advantage shown for the gas heat pump are because Los Angeles evidenced the best SPF for the combination heat/cool unit on cooling service of all the ten (10) cities studied. And, although the Los Angeles combination heat/cool unit on heating service had a relatively low SPF, the effect was minimized due to a relatively low gas rate. Also, the intergrated cooling and heating loads are among the lowest of the ten (10) cities.

As mentioned previously, economic performance was studied with respect to two additional investment criteria. These next two charts (Figures 10 and 11) show the comparative economics for The Discounted Cash Flow Pay-Back Period (years) and The Simple Pay-Back Period (years).

4.2 Conclusions

In conclusion, the key observations of the study were:

1. The heating seasonal performance or efficiency of the gas heat pump approached twice that of advanced direct-fired gas heating equipment.

2. The cooling seasonal performance or efficiency of the gas heat pump is more than twice that of present-day direct-fired absorption-type cooling units.

3. Annual operating or utility costs for the gas heat pump were shown to be significantly lower than the combination heat/cool units in all study cities.

4. Annual operating or utility costs for the gas heat pump were shown to be less than those of the electric heat pump in most of the study cities.

5. The effect of electric heating rates on the competitive position of the gas and electric heat pumps is substantial. (The initial cost premium of the gas heat pump relative to the electric heat pump was assumed to be 30% for these calculations.)

6. On economic criteria, the projected advanced design electric heat pump appears to be a better investment than a combination heat/cool unit in seven of the ten cities studied. This condition is introduced by the effect of special electric rates in the other

three cities rather than due to efficiency or initial cost differentials.

7. A direct economic comparison of the gas heat pump with the projected advanced design electric heat pump indicates that the gas heat pump has a competitive advantage in seven of the ten cities studied.

8. The less-than-full load performance for all equipment options was a more important determinant of SPF then designated test-point performance.

9. SPF values for the HSPF Gas Heat Pump were uniformly higher than test-point COP values for all installation cases included in the study.

10. The attractive economic performance of the HSPF Gas Heat Pump is principally a result of optimizing for seasonal efficiency (SPF) rather than test-point COP.

REFERENCE LIST

1. Swenson, P. F. and Rose, R. K. Development of The High Seasonal Performance Factor Gas Heat Pump. Submitted for publication with the 12th IECEC Session on Energy Conversion, August 30, 1977.

2. Consolidated Natural Gas Service Co., Inc., Marketing Department. Operating Cost & Investment Analysis Study for the Prototype ERDA/MTI Gas Heat Pump. Report to ERDA and Consolidated Natural Gas Service Co. under ERDA Contract #EY-76-C-02-2883. Pittsburgh, January, 1977.

3. American Society of Heating, Regrigerating and Air-conditioning Engineers, Inc., 345 East 47th Street, New York, New York 10017, 1972 Handbook of Fundamentals.

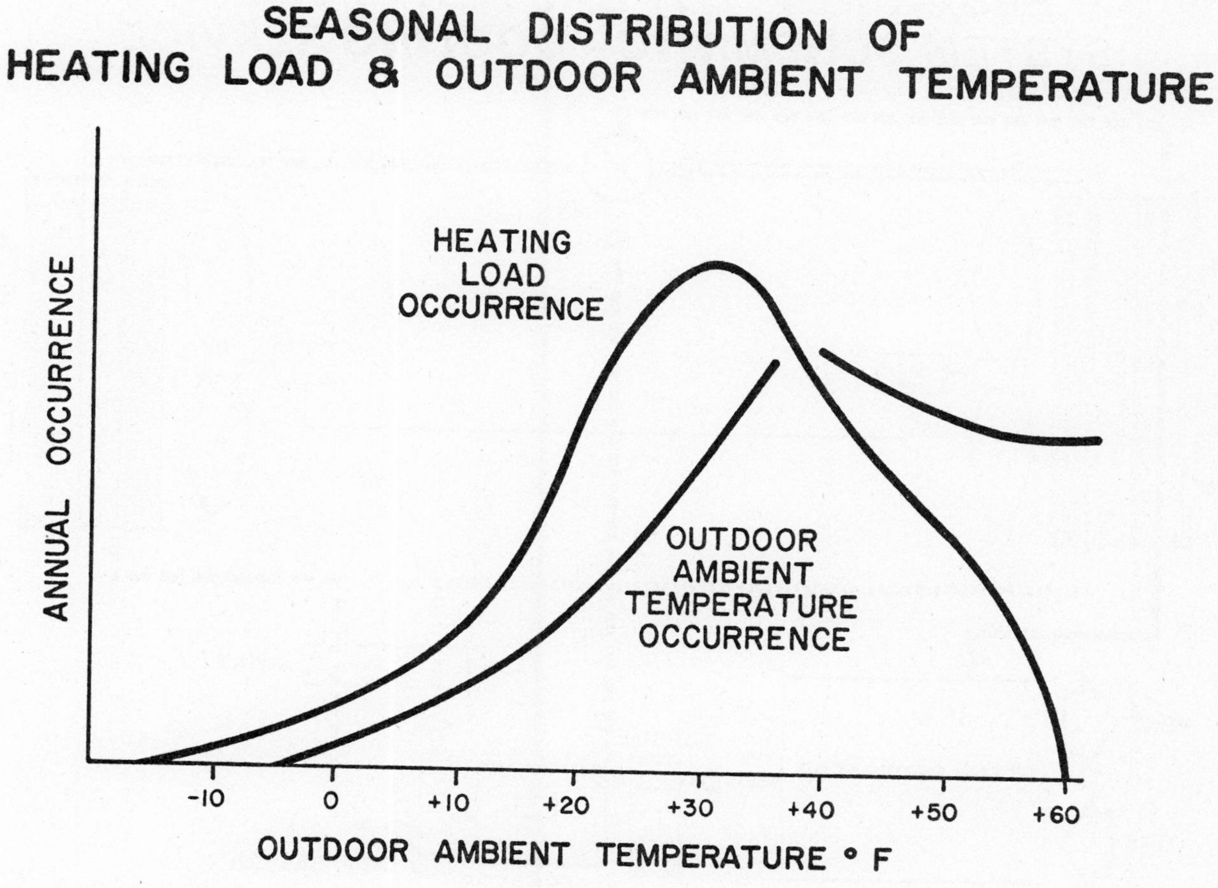
SEASONAL DISTRIBUTION OF
HEATING LOAD & OUTDOOR AMBIENT TEMPERATURE
HEATING
LOAD
OCCURRENCE
OUTDOOR
AMBIENT
TEMPERATURE
OCCURRENCE
ANNUAL OCCURRENCE
-10
0
+10
+20
+30
+40
+50
+60
OUTDOOR AMBIENT TEMPERATURE ° F

Fig. 1

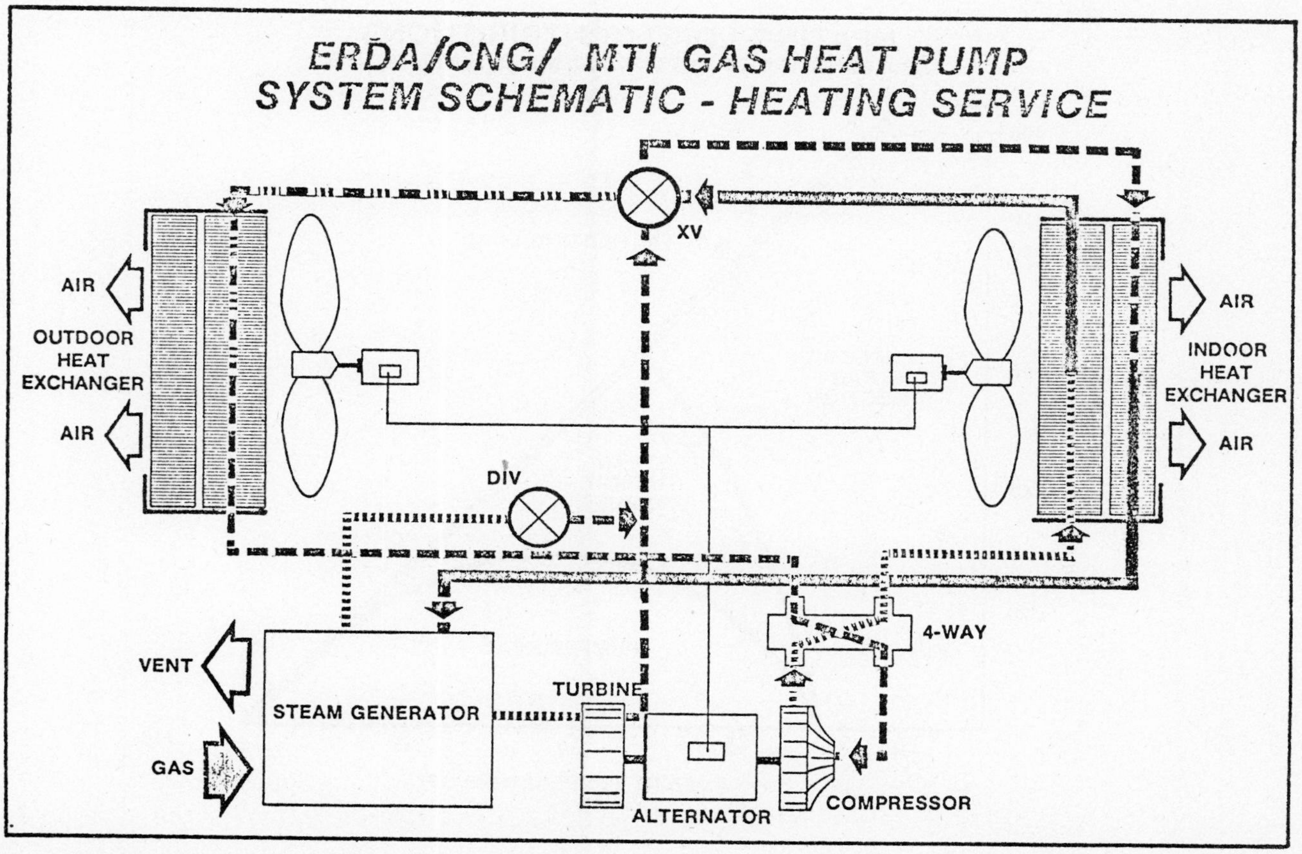
ERDA/CNG/ MTI GAS HEAT PUMP
SYSTEM SCHEMATIC - HEATING SERVICE
XV
AIR
OUTDOOR
HEAT
EXCHANGER
AIR
AIR
INDOOR
HEAT
EXCHANGER
AIR
DIV
4-WAY
VENT
STEAM GENERATOR
GAS
TURBINE
ALTERNATOR
COMPRESSOR

Fig. 2

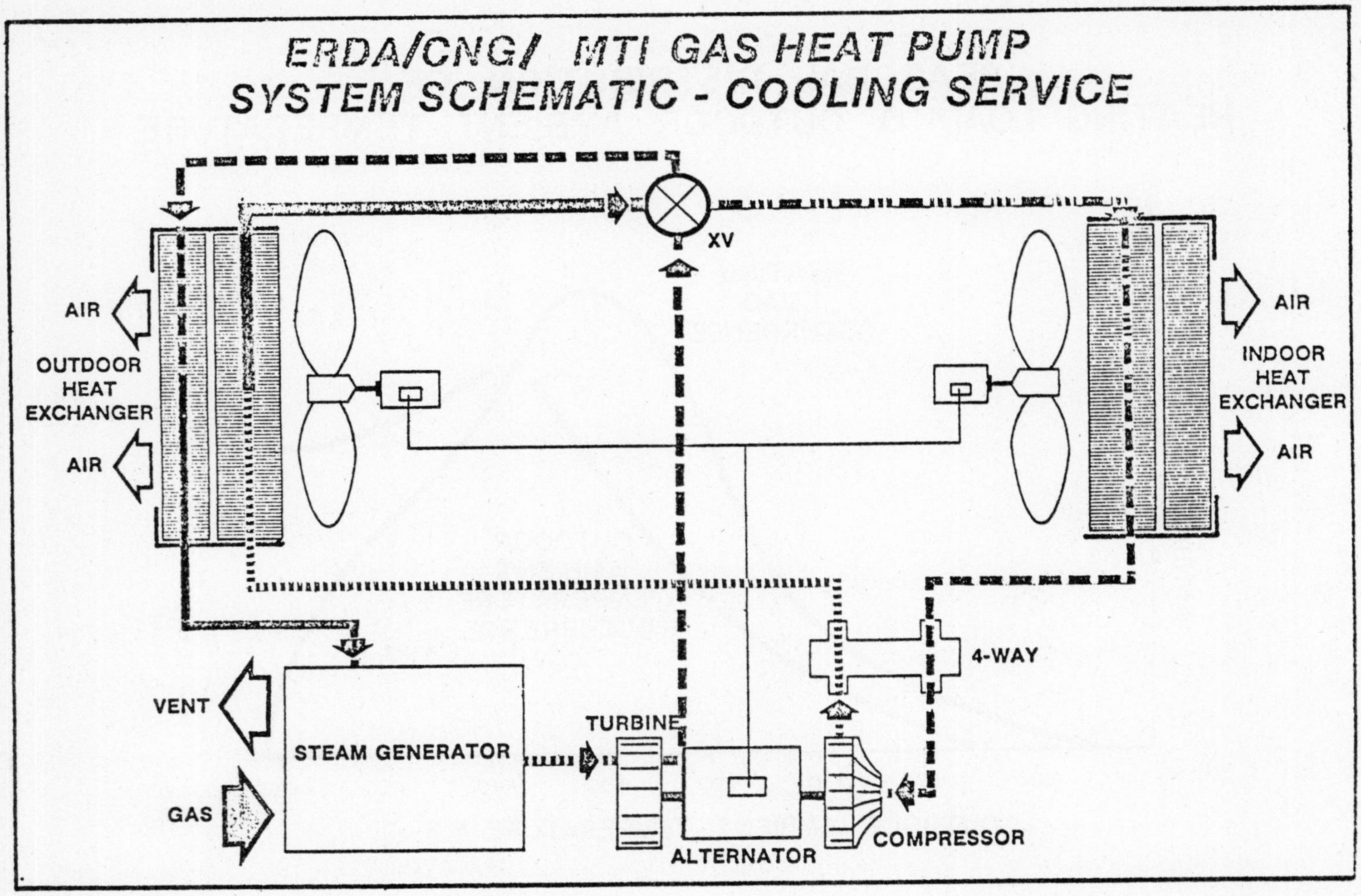
ERDA/CNG/ MTI GAS HEAT PUMP
SYSTEM SCHEMATIC - COOLING SERVICE
XV
AIR
OUTDOOR
HEAT
EXCHANGER
AIR
AIR
INDOOR
HEAT
EXCHANGER
AIR
4-WAY
VENT
STEAM GENERATOR
GAS
TURBINE
ALTERNATOR
COMPRESSOR

Fig. 3

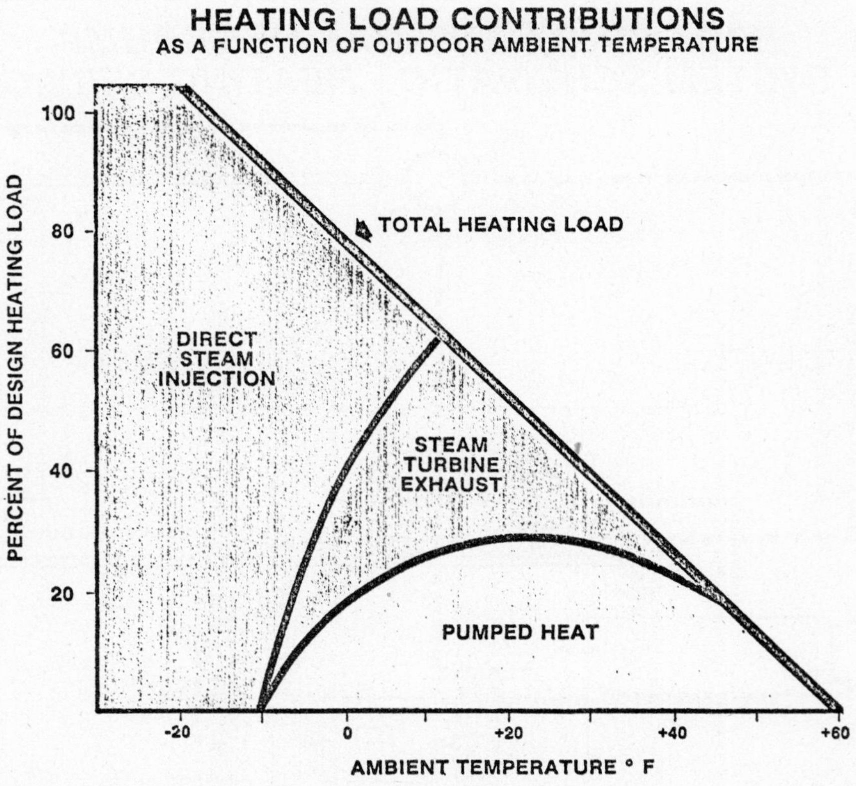
HEATING LOAD CONTRIBUTIONS
AS A FUNCTION OF OUTDOOR AMBIENT TEMPERATURE
PERCENT OF DESIGN HEATING LOAD
100
80
60
40
20
TOTAL HEATING LOAD
DIRECT
STEAM
INJECTION
STEAM
TURBINE
EXHAUST
PUMPED HEAT
-20
0
+20
+40
+60
AMBIENT TEMPERATURE ° F

Fig. 4

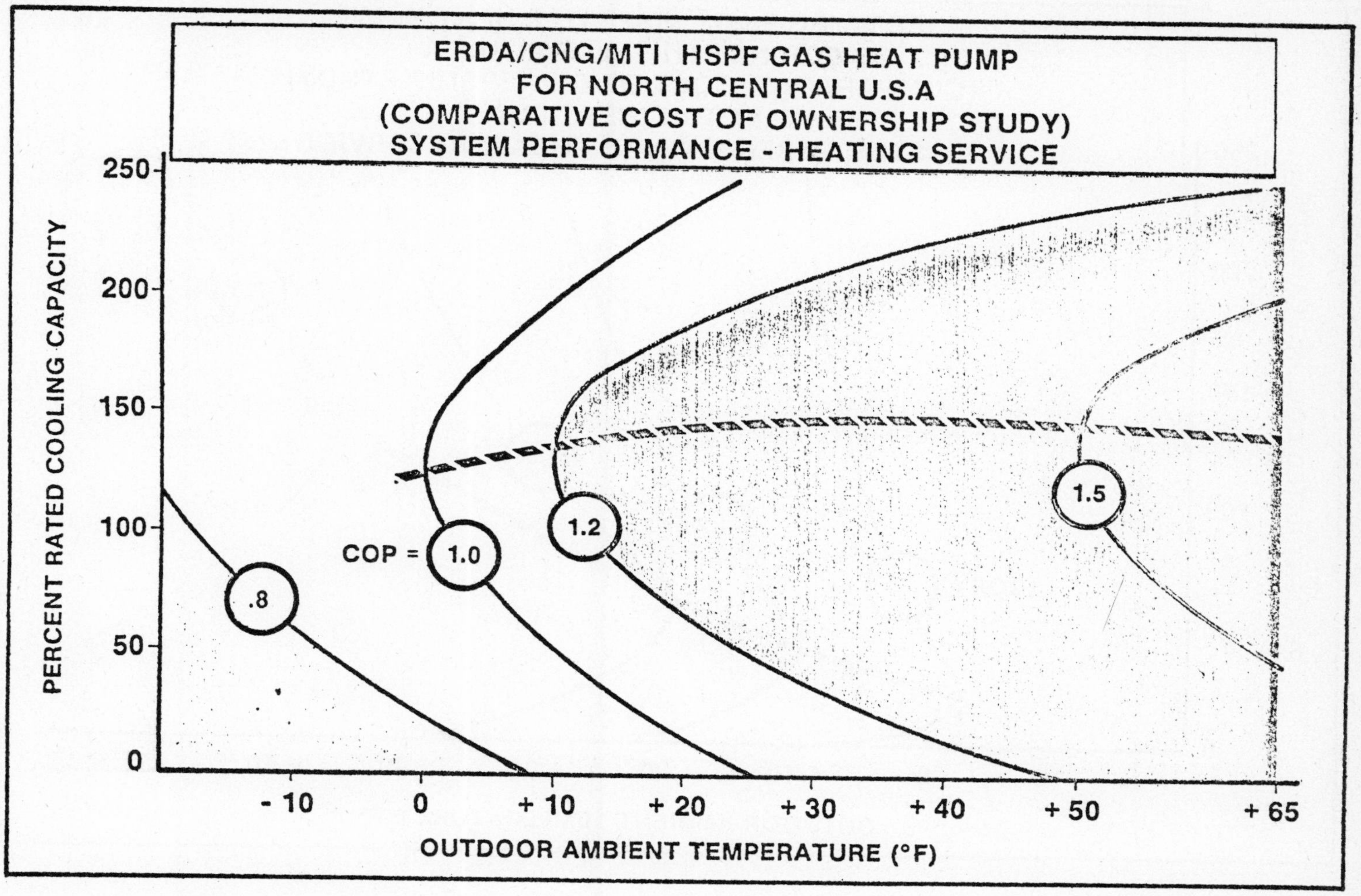
ERDA/CNG/MTI HSPF GAS HEAT PUMP
FOR NORTH CENTRAL U.S.A
(COMPARATIVE COST OF OWNERSHIP STUDY)
SYSTEM PERFORMANCE - HEATING SERVICE
PERCENT RATED COOLING CAPACITY
250
200
150
100
50
0
COP = 1.0
.8
1.2
1.5
- 10
0
+ 10
+ 20
+ 30
+ 40
+ 50
+ 65
OUTDOOR AMBIENT TEMPERATURE (°F)

Fig. 5

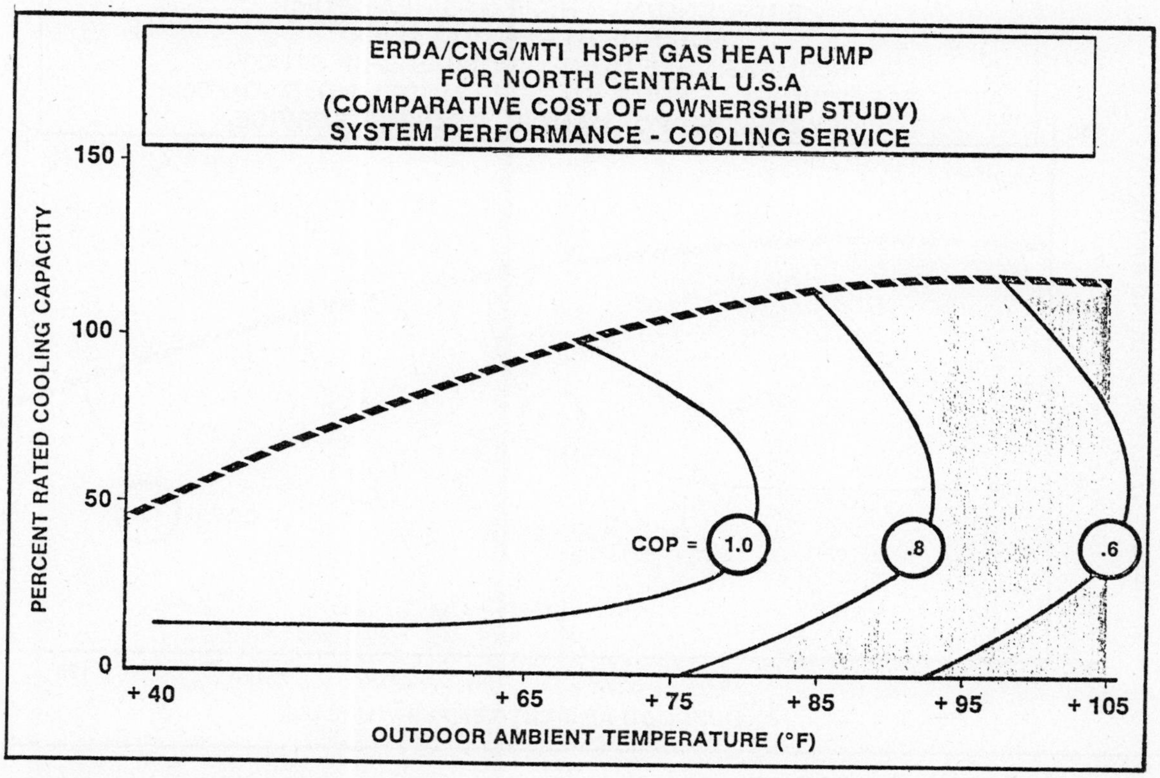
ERDA/CNG/MTI HSPF GAS HEAT PUMP
FOR NORTH CENTRAL U.S.A
(COMPARATIVE COST OF OWNERSHIP STUDY)
SYSTEM PERFORMANCE - COOLING SERVICE
PERCENT RATED COOLING CAPACITY
150
100
50
0
COP = 1.0
.8
.6
+ 40
+ 65
+ 75
+ 85
+ 95
+ 105
OUTDOOR AMBIENT TEMPERATURE (°F)

Fig. 6

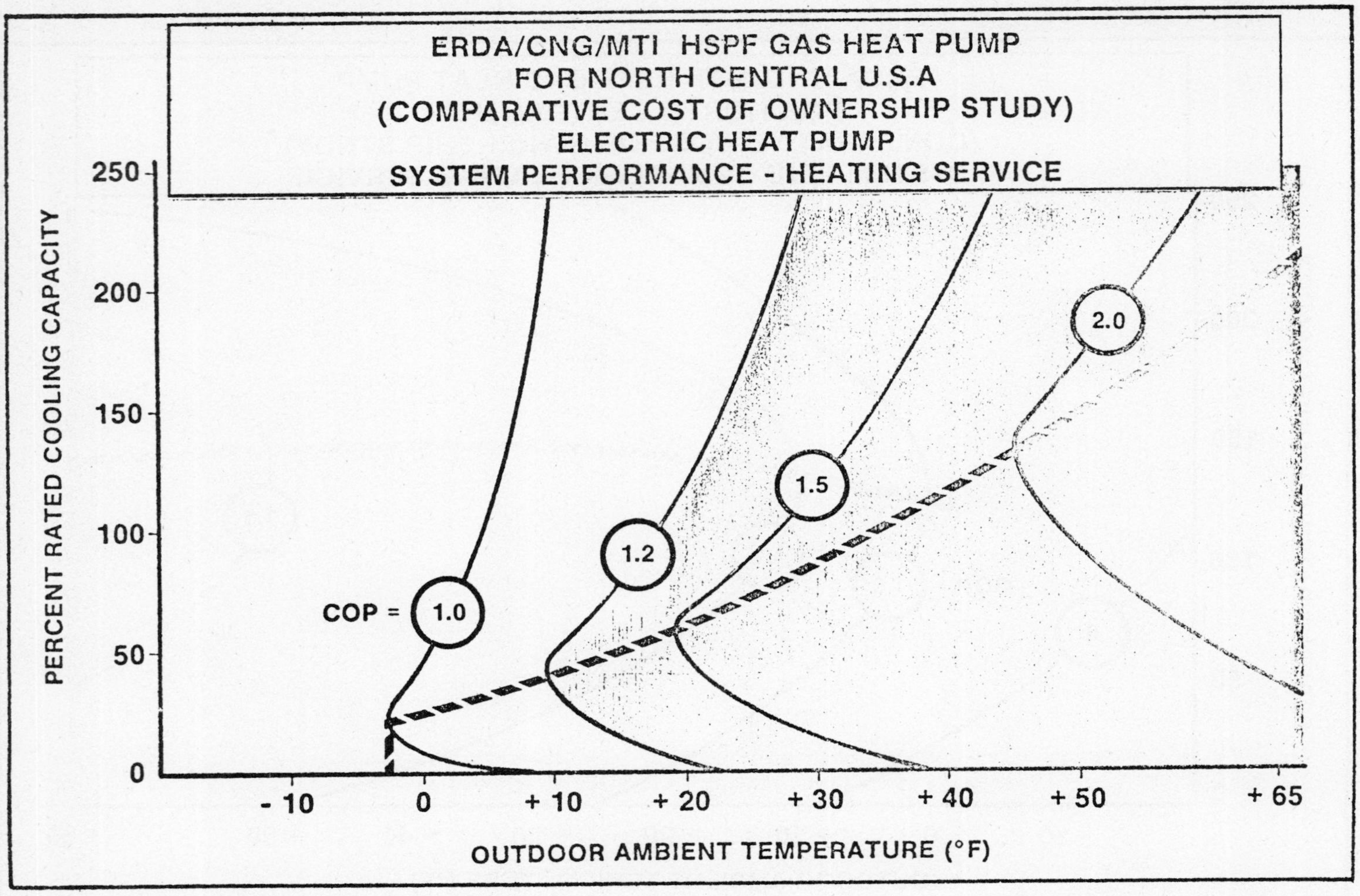
ERDA/CNG/MTI HSPF GAS HEAT PUMP
FOR NORTH CENTRAL U.S.A
(COMPARATIVE COST OF OWNERSHIP STUDY)
ELECTRIC HEAT PUMP
SYSTEM PERFORMANCE - HEATING SERVICE
PERCENT RATED COOLING CAPACITY
250
200
150
100
50
0
COP =
1.0
1.2
1.5
2.0
- 10
0
+ 10
+ 20
+ 30
+ 40
+ 50
+ 65
OUTDOOR AMBIENT TEMPERATURE (°F)

Fig. 7

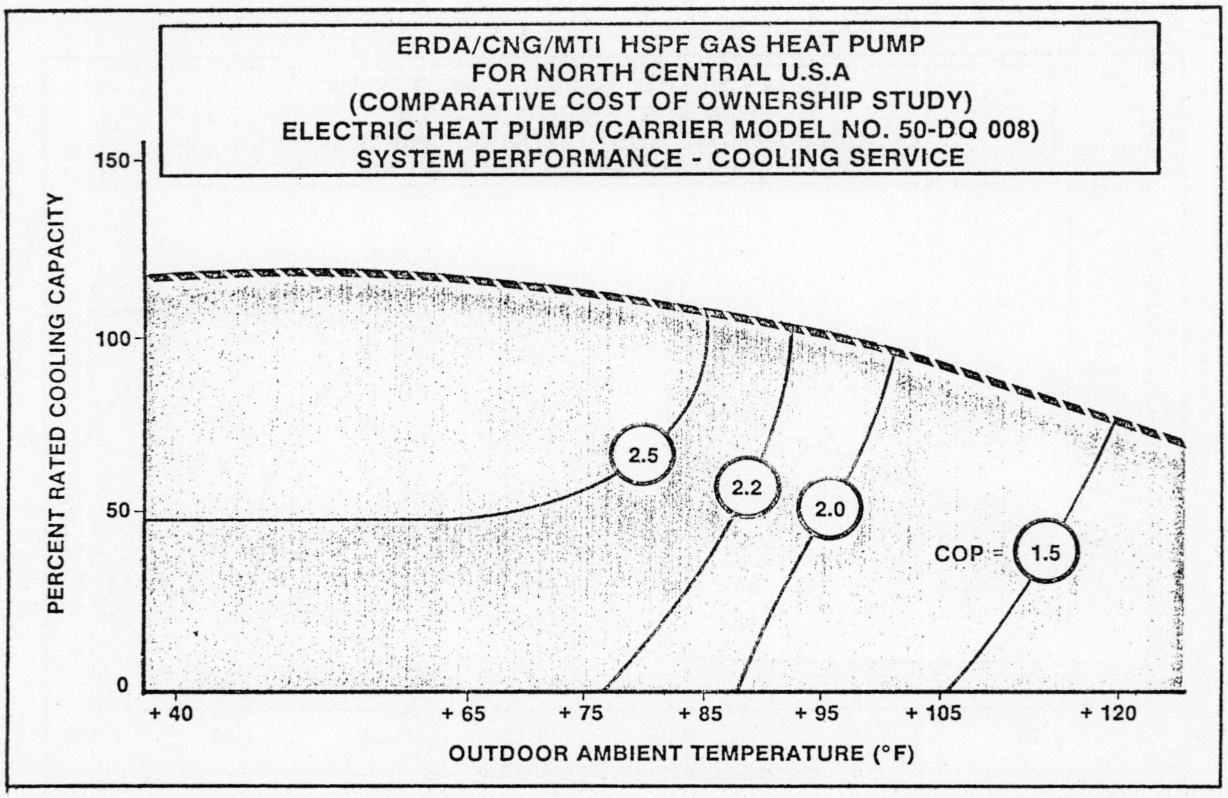
ERDA/CNG/MTI HSPF GAS HEAT PUMP
FOR NORTH CENTRAL U.S.A
(COMPARATIVE COST OF OWNERSHIP STUDY)
ELECTRIC HEAT PUMP (CARRIER MODEL NO. 50-DQ 008)
SYSTEM PERFORMANCE - COOLING SERVICE
PERCENT RATED COOLING CAPACITY
150
100
50
0
2.5
2.2
2.0
COP =
1.5
+ 40
+ 65
+ 75
+ 85
+ 95
+ 105
+ 120
OUTDOOR AMBIENT TEMPERATURE (°F)

Fig. 8

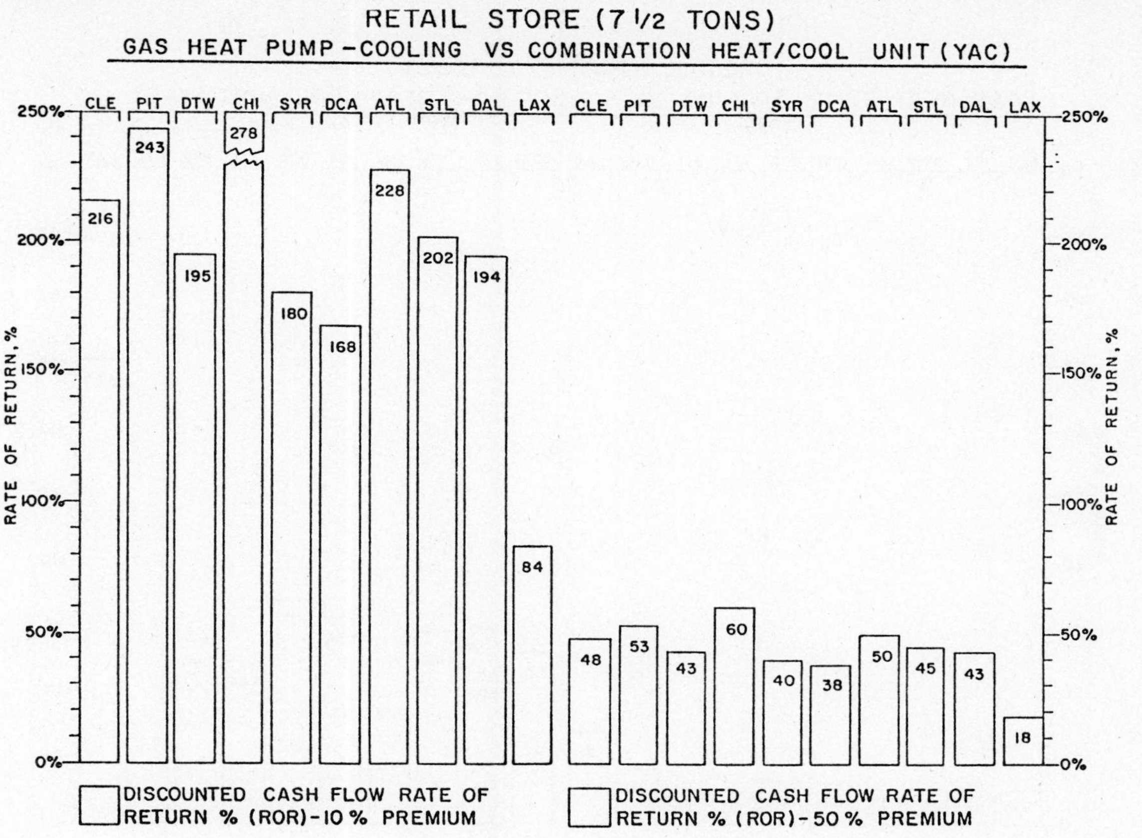
RETAIL STORE (7 1/2 TONS)
GAS HEAT PUMP - COOLING VS COMBINATION HEAT/COOL UNIT (YAC)
CLE PIT DTW CHI SYR DCA ATL STL DAL LAX CLE PIT DTW CHI SYR DCA ATL STL DAL LAX
RATE OF RETURN, %
250%
200%
150%
100%
50%
0%
216 243 195 278 180 168 228 202 194 84
48 53 43 60 40 38 50 45 43 18
DISCOUNTED CASH FLOW RATE OF RETURN % (ROR)-10 % PREMIUM
DISCOUNTED CASH FLOW RATE OF RETURN % (ROR)-50 % PREMIUM

Fig. 9

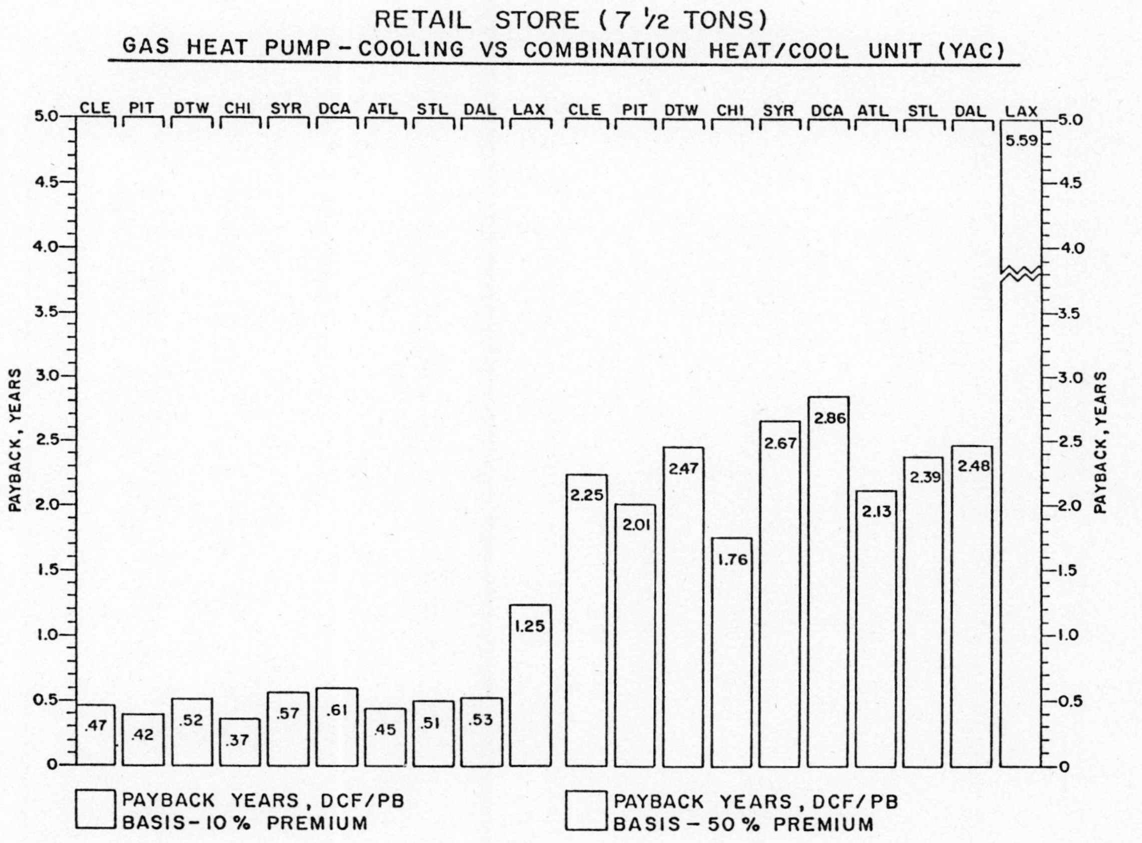
RETAIL STORE (7 1/2 TONS)
GAS HEAT PUMP - COOLING VS COMBINATION HEAT/COOL UNIT (YAC)
CLE PIT DTW CHI SYR DCA ATL STL DAL LAX CLE PIT DTW CHI SYR DCA ATL STL DAL LAX
PAYBACK, YEARS
5.0
4.5
4.0
3.5
3.0
2.5
2.0
1.5
1.0
0.5
0
.47 .42 .52 .37 .57 .61 .45 .51 .53 1.25
2.25 2.01 2.47 1.76 2.67 2.86 2.13 2.39 2.48 5.59
PAYBACK YEARS, DCF/PB BASIS-10 % PREMIUM
PAYBACK YEARS, DCF/PB BASIS - 50 % PREMIUM

Fig. 10

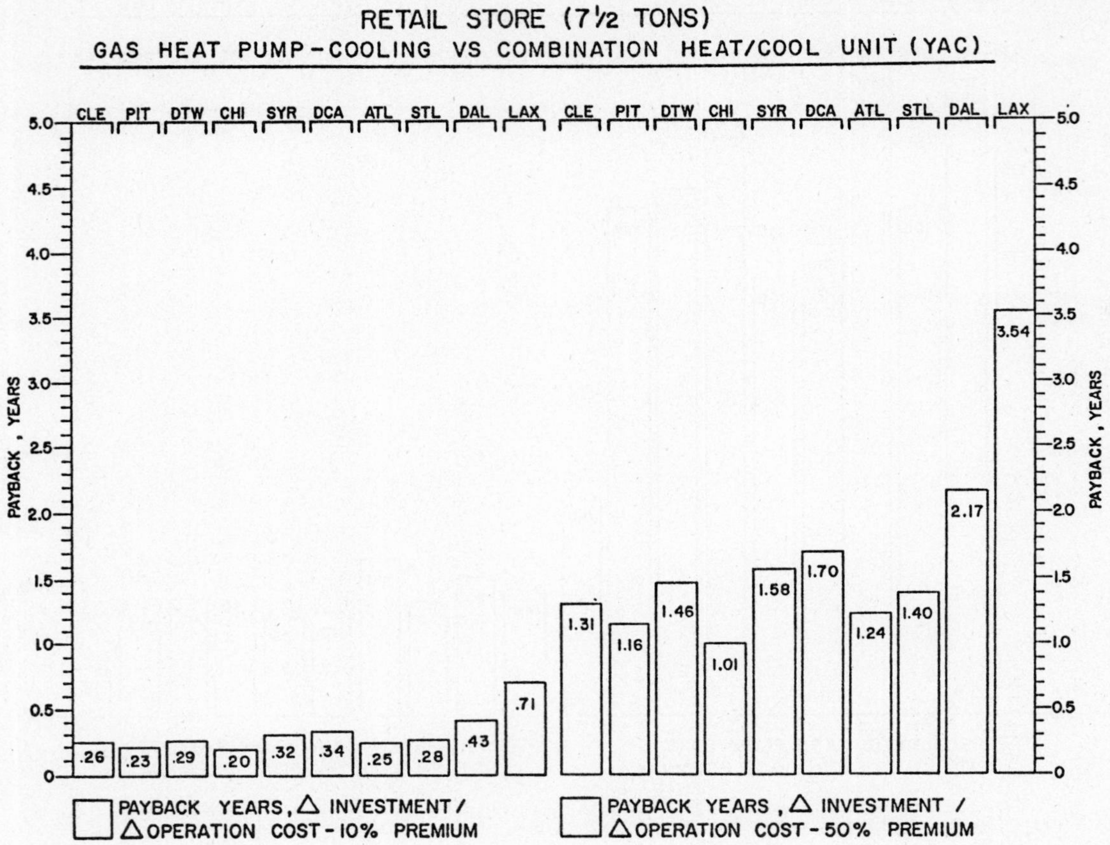
RETAIL STORE (7½ TONS)
GAS HEAT PUMP-COOLING VS COMBINATION HEAT/COOL UNIT (YAC)
CLE PIT DTW CHI SYR DCA ATL STL DAL LAX CLE PIT DTW CHI SYR DCA ATL STL DAL LAX
PAYBACK, YEARS
.26 .23 .29 .20 .32 .34 .25 .28 .43 .71
1.31 1.16 1.46 1.01 1.58 1.70 1.24 1.40 2.17 3.54
PAYBACK YEARS, △ INVESTMENT / △ OPERATION COST-10% PREMIUM
PAYBACK YEARS, △ INVESTMENT / △ OPERATION COST-50% PREMIUM

Fig. 11

ASSESSMENT OF ELECTRIC AND ELECTRIC-ASSISTED TECHNOLOGIES IN RESIDENTIAL HEATING AND COOLING APPLICATIONS

J.G. Asbury, R.F. Giese, R.O. Mueller, and S.H. Nelson

Energy and Environmental Systems Division
Argonne National Laboratory, Argonne, Illinois U.S.A.

ABSTRACT

Total costs of supplying residential heating and cooling services have been estimated for a number of electric and electric-assisted technologies. The analysis entailed the hourly simulation of device-specific loads over an annual cycle and the use of the Argonne electric utility cost allocation model. Results are presented for two representative utility service areas.

In the service area supplied by a winter peaking utility, the lowest cost space heating technologies were found to be storage-augmented resistance systems and heat pump systems augmented either by storage or by an oil furnace. Heat pumps were the most economical heating systems in the service area supplied by the summer peaking utility. Storage air conditioning was cost-effective in the summer peaking service area, storage hot water heaters in both service areas. The most cost-competitive solar energy application is domestic hot water heating.

INTRODUCTION

A number of electric and electric-assisted technologies are available for residential heating and cooling applications. These technologies include: direct resistance water and space heating, electric storage heating, electric heat pumps, storage heat pumps, electric-assisted solar heating, and bivalent heating systems. In this paper, we present the results of a recent study at Argonne National Laboratory to evaluate the relative economics of each of these technologies (1).

STUDY METHOD

The method used to evaluate the individual technologies was to calculate utility and customer capital and operating costs for each technology for several utility service areas. The two service areas for which detailed results are presented here were selected to illustrate the important factors affecting the overall cost of each of the technologies.

In each service area, the individual heating/cooling systems were sized to meet the requirements of a standard 1500 ft^2, well insulated, detached single family dwelling unit. The Argonne cost-of-service model, SIMSTOR, was used to calculate the utility costs of meeting the system loads. The model uses hourly utility load and tri-hourly weather data and heating/cooling system performance characteristics to generate load profiles over a full annual (8760 hour) cycle. It then calculates the utility capital and fuel costs to meet device-specific incremental loads. SIMSTOR incorporates a load dispatch model and observes operating constraints such as scheduled outages and the cycle time of each generating unit. It calculates transmission and distribution costs as well as generating costs. Because SIMSTOR uses an equilibrium method to solve for optimum plant capacity and mix, the estimated utility costs represent long-run marginal costs to meet the individual heating/cooling system loads.

In order to value units of capital consistently on both sides of the electric meter, one set of system cost comparisons was made with heating/cooling system capital costs calculated on the basis of the utility capital recovery rate. This accounting procedure is conceptually equivalent to assuming utility ownership of the heating/cooling system. Another set of comparisons were made under an assumed customer cost of money equal to present mortgage rates less an effective income tax credit.

Levelized annual fuel costs were calculated under an assume 0.0% real rate of escalation. Because initial-year fuel is valued at full marginal cost, this is not expected to understate fuel costs over the lifetime of the heating/cooling system.

UTILITY SERVICE AREAS

Data describing the service areas and two utility systems are presented in Tables 1 and 2. The plant-type percentages in Table 2 refer to new plant (being added at the margin) in each service area. Service Area A is supplied by a winter-peaking utility; Service Area B by a summer-peaking utility. The hourly load data, representing the reference loads to which the device-specific loads were added, were obtained directly from the two utility companies. Tri-hourly temperature and hourly insulation for weather stations located in each service area were obtained from the National Climatic Center, U.S. Department of Commerce.

TECHNOLOGIES

Supply costs for a number of electric and electric-assisted technologies were calculated.

Direct Resistance Heating

Resistance space heating and water heating applications were evaluated. For water heating, a standard 52 gallon tank was assumed. This tank, equipped with 4.5 kW heaters, has a recovery rate of 18 gal/hr (100°F temperature rise) and is considered adequate for a family of three or four. Only central furnace space heating systems were evaluated. The systems were sized on the basis of the design-day loads given in Table 1. Capacity and consumption values were increased by 10% to account for duct losses.

Air Conditioning

Air conditioning system requirements were estimated from a sizing criterion incorporating design-day dry-bulb and wet-bulb temperatures. The design criterion in Btu/hr was:

$$13{,}000 + 450\ (T_D - 73.5) + 585\ (T_w - 63.7).$$

Heat Pumps

The heat pumps were sized to the summer design-day air conditioning load. This yielded a 2.5 ton capacity in each service area. The resulting balance point usually occurred at an outdoor temperature of about 23°F.

Electric Storage Systems

Storage-augmented versions of heating, cooling, and hot water systems were examined (2). Storage capacities corresponding to the design-day loads were calculated for storage durations (discharge periods) of eight hours for resistance space heating and sixteen hours (120 gal.) for hot water heating. An earlier study indicated that these discharge periods yield the largest net benefits relative to conventional direct resistance systems.

The storage air conditioning system is a central ice-making system similar to ones developed by A.O. Smith Company and by Carrier Corporation. The eight-hour discharge period corresponds to a tank capacity of 240 gallons.

The storage requirement for the heat pump was set equal to the electric resistance energy requirement of the heat pump system on the winter design-day. Because the heat pump system operates in a nearly fully resistance mode on the design-day, the storage requirement is approximately equal to that for a central furnace system.

Bivalent Systems

An alternative to the use of storage in space heating application is the incorporation of a gas- or oil-fired backup unit in the central furnace system (3). The backup unit, ideally under the direct control of the utility, is switched on during the peak load hours, thereby substituting for utility peak capacity and fuel requirements.

When used in conjunction with the electric heat pump systems, the auxiliary fuel unit was operated in one of two modes. Under the first mode, the auxiliary unit was operated so as to simulate and exactly substitute for the heat pump's electric resistance backup. Under the second mode of operation, the auxiliary unit was switched on during those periods when normally either resistance backup and/or utility peakers would have been used.

Solar Systems

The most cost-effective residential applications of solar energy are water and space heating. However, as described in an earlier paper, active solar energy and conventional electric utility systems are a poor technological match (4). Forcing the interface of these two technologies results in impossibly low solar collector breakeven costs, rendering the question of the appropriate solar collector size a moot point. Somewhat arbitrarily, therefore, we sized the solar systems to provide about one-half (55%) the annual heating load in both hot water and space heating applications. Parametric analyses of collector versus storage capacity tradeoffs then indicated that optimal storage capacity is approximately equal to the average daily output of the collector system. This relationship held for hot water, solar/heat pump, and solar/resistance heating applications. The parametric studies were performed using the TRANSYS 8.1 computer code developed by Beckman and Duffy.

Solar/electric resistance heating simulations were run in two modes. In the first mode, the backup energy supply is simulated as an auxiliary electric furnace which cycles on as required to augment the solar input to the air flow stream. This is the approach currently adopted in conventional solar system designs. Auxiliary loads are subsequently averaged over hourly time intervals. To take advantage of the lower costs of off-peak electricity, the second mode involves the simulation of a storage resistance heater for supplying auxiliary energy requirements. Under this control strategy, perfect information is assumed for the following day's insolation and building load. Conceptually, this mode is approximately equivalent to placing the resistance heaters in a solar storage water tank that can maintain perfect temperature stratification.

In the solar/heat pump simulations, the solar energy is first input to a storage reservoir on the cold side of the heat pump. The solar energy is then removed from the reservoir by the heat pump and delivered to the building load. The second phase heating is provided by the heat pump drawing from ambient air, and the final phase by direct electric resistance heating.

A summary of the costs and capacities of each of the above technologies is presented in Table 4.

STUDY FINDINGS

Tables 5 and 6 summarize the supply costs of the services provided by each of the heating and cooling technologies.

Annual utility capital costs were calculated using the capital recovery rates shown in Table 4. These rates correspond to current utility accounting practices and incorporate a large (≃ 6%) inflation component in the cost of both bond and equity money. For consistency, fuel costs, which were assumed to have a 0% real rate of escalation, were inflated at the same 6% rate implicit in the capital recovery factor and were discounted by the same (11%) discount rate. The resulting annual fuel levelization factor was equal to 1.77. The annualized customer capital costs presented in Tables 5 and 6 were calculated according to the two accounting schemes described above. The first employs utility capital recovery factors; the second corresponds to a recovery rate equal to an effective mortgage rate of 7% (9% less a 2% income tax credit) plus straight-line depreciation.

The utility costs shown in Tables 5 and 6 represent the per-customer cost of supplying incremental heating or cooling loads. For each space heating technology, utility costs are based upon the addition of 1000 customers; air conditioning, 2000 customers; hot water heating, 10,000 customers. Each of these load increments corresponds to approximately 10 MW peak demand. Supply costs for each technology were calculated independently.

The utility costs, expressed on a cents per kilowatt hour basis, indicate the savings available to storage and bivalent systems through the transferring of on-peak demand to off-peak periods. In addition to power savings, storage systems allow substitution of low cost base load energy for relatively high cost intermediate and peaker energy thus resulting in a minimum total utility cost within a technology category. Heat pumps and solar systems on the other hand have a utility cost advantage over resistance systems due to their ability to decrease electric utility energy requirements. In both cases the reduced utility costs are coupled with increased customer capital costs.

Deferral of auxiliary energy requirements to off-peak periods in Service Area A provides a substantial economic edge for the solar storage space heating system over the solar resistance system and the solar assisted heat pump. Figure 1 shows the cost advantage in Service Area A of the solar storage resistance system over the solar resistance system for a range of collector areas. However, neither solar system is cost competitive relative to storage resistance heating. In the service area supplied by the summer peaking utility, all of the solar space heating technologies provide for substantial utility fuel saving through the displacement of peak and intermediate fuels.

Solar storage water heating is the most competitive of all the solar technologies evaluated; total costs equal storage resistance water heating total costs for collector costs of approximately $\$40/m^2$ in Service Area A and $\$70/m^2$ in Service Area B.

For Service Area A, supplied by the winter peaking utility, the lowest cost space heating systems were found to be storage-augmented resistance systems and heat pump systems augmented by storage or by an oil furnace, with the ripple controlled bivalent heat pump holding a 15-45% advantage in total cost. This cost advantage results from the relatively small customer capital cost penalty ($200) for substituting an oil furnace for a resistance furnace as a backup to the heat pump, together with the large utility cost savings ($1200) of shifting heat pump demand off-peak.

For Service Area B, supplied by the summer peaking utility, conventional heat pumps were the most economical. In this case the entire heating season is off-peak, greatly reducing the benefits of storage or bivalent systems. The storage and bivalent heat pump systems suffer a 10% cost penalty relative to conventional heat pumps and are more complicated technologies.

Storage air-conditioning is very cost effective in the summer peaking service area while conventional air conditioners hold a cost advantage in the winter peaking service area. This is due to the large coincident demand component (90%) and low load factor (10%) of the conventional air conditioners.

Storage hot water heaters which have high coincident on-peak demand in both service areas are also cost-effective.

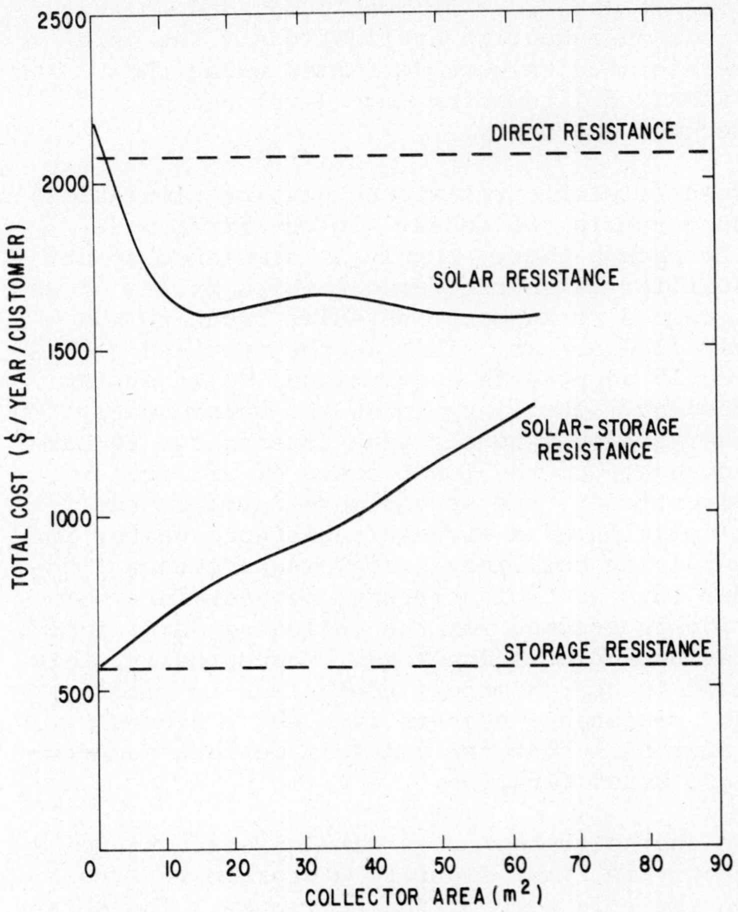

Fig. 1. Total supply costs of solar space heating systems as a function of collector area, utility Service Area A.

REFERENCES

1. Work supported by the U.S. Energy Research and Development Administration, Division of Energy Storage Systems. The results presented here are more fully documented in "Assessment of Energy Storage Technologies and Systems, Phase II," ANL/ES-64, Argonne National Laboratory (forthcoming).

2. J.G. Asbury, R.F. Giese, R.O. Mueller, and S.H. Nelson, "Commercial Feasibility of Thermal Storage in Buildings for Utility Load Leveling," Proceedings of the American Power Conference, April 18-20, 1977, Chicago, Illinois.

3. See, for example: "FMP1 Fuelmaster, Heat Pump System for Use with Gas, Oil, and Electric Furnaces," Lennox, Engineering Data, Heat Pumps, Matched Remote Systems, (November 1974).

4. J.G. Asbury and R.O. Mueller, "Solar Energy and Electric Utilities: Should They Be Interfaced," Science, 195, 445-450 (February 4, 1977).

Table 1. Utility Service Area Characteristics

Service Area	Location	Ratio Winter to Summer Peak	Climate: Annual Heating (degree days)[a]		Annual Cooling (degree days)[a]		Annual Insolation[b] (kWh/m^2 - yr)		Design Day Temperatures[c]	
			Average	1975	Average	1975	Average	1975	Heating	Cooling
A	Northeast	1.24	5621	5307	661	797	1390	1210	6	88
B	Mid-Atlantic	.62	5010	4768	940	1156	1510	1420	13	92

[a]Based on 65°F reference temperature.

[b]Horizontal surface.

[c]ASHRAE 99% heating design temperature and 2½% cooling design temperature.

Table 2. Utility System Characteristics

Service Area	Generating Plant Type (%): Base		Intermediate		Peak	Reserve Margin	T&D Loss Factor (%)[a]		
	Nuclear	Coal	Coal	Combined Cycle	Gas Turbine		Transmission	Primary Distribution	Secondary Distribution
A	100	0	0	100	100	22	5.2	2.8	3.5
B	75	25	75	25	100	18	4.5	4.0	4.5

[a]Average line losses over year include a constant 0.5% core loss and a variable I^2R loss.

Table 3. Utility Plant Characteristics and Costs

Plant Type	System Costs			Accounting Factors		Unit Operating Characteristics				
	Capital ($/kW)	Fuel ($/MWh)	Operating ($/MWh)	C.C.I.F.[a]	Capital Recovery Factor (%)	Scheduled Outage (wks/yr)	Forced Outage Rate (%)	Min. Cycle Time (hrs)	Cycling Costs ($/cycle/MW)	Min. Operate Load (%)
Base										
Nuclear	450	5	1	1.44	17	9	11	12	72	40
Coal	350	9.5	1	1.37	17	5	8	12	126	40
Intermediate										
Coal	300	12	2	1.25	17	5	8	6	35	30
Combined Cycle	250	17	3	1.12	17	4	4	6	70	30
Peak										
Gas Turbine	150	31	5	1.05	20	4	9	1	20	30
Transmission	150			1.00	17					
Distribution										
Primary	100			1.00	17					
Secondary	100			1.00	17					

[a]Construction Compound Interest Factor.

Table 4. Technology Capacities and Costs for Service Areas A and B

System	Cost Algorithm ($)	Capacity: Power A	Capacity: Power B	Capacity: Energy Storage A	Capacity: Energy Storage B	Cost ($) A	Cost ($) B
Space Heating							
Resistance							
Direct	800 + 20/kW	11.6 kW	11.0 kW	-	-	1035	1020
Storage	1020[a] + 20/kW + 8/kWh	15.5 kW	13.1 kW	124 kWh	64 kWh	2325	1795
Bivalent	2260[a] + 20/kW	11.6 kW	11.0 kW	-	-	2490	2480
Heat Pump							
Conventional	1150[b,c] + 20/kW[c]	11.6 kW	11.0 kW	-	-	1380	1370
Storage	1370[a,b] + 20/kW + 8/kWh	13.5 kW	6.5 kW	108 kWh	52 kWh	2505	1915
Bivalent 1	1590[b,d]	-	-	-	-	1590	1590
Bivalent 2	1730[a,b,d]	-	-	-	-	1730	1730
Solar							
Resistance	2000[e] + 150/m^2 + 1/gal.	31 m^2	28 m^2	600 gal.	550 gal.	7250	6750
Storage Resistance	2000[e] + 150/m^2 + 1/gal.	31 m^2	28 m^2	1000 gal.	1000 gal.	7650	7200
Heat Pump	2350[b,e] + 150/m^2 + 1/gal.	25 m^2	25 m^2	500 gal.	500 gal.	6600	6600
Air Conditioning							
Conventional	200 + 330/ton	2.5 ton	2.5 ton	-	-	1025	1025
Storage	740[a] + 330/ton + 16/kWh	2.9 ton	2.9 ton	25 kWh	25 kWh	2095	2095
Water Heating							
Resistance							
Direct	3.15/gal.	4.5 kW	4.5 kW	52 gal.	52 gal.	165	165
Storage	3.15/gal + 140	4.5 kW	4.5 kW	120 gal.	120 gal.	520	520
Solar							
Resistance	660[f] + 150/m^2 + 2/gal.	5.7 m^2	5.7 m^2	112 gal.	112 gal.	1740	1740
Storage Resistance	860[f] + 150/m^2 + 2/gal.	5.7 m^2	5.7 m^2	112 gal	112 gal.	1940	1940

[a]Includes $140 for ripple control unit.

[b]Includes $350 for incremental cost of heat pump over air-conditioner cost.

[c]Includes cost of resistance backup furnace.

[d]Includes $1240 for cost of oil furnace, oil tank and chimney.

[e]Includes $1000 for pipes and controls plus $1000 for electric furnace backup system.

[f]Includes $500 for pipes and controls.

Table 5. Cost of Supply, Service Area A

System	System Characteristics	Utility Costs ($/yr/customer)					Utility Cost (¢/kWh)	Customer Cost ($/yr/customer)			Total Cost ($/yr/customer)	
		Capital		Variable				Capital				
		Gen	T&D	Fuel	Other	Total		Util. Rate	Mort. Rate	Fuel	Util. Rate	Mort. Rate
Space Heating												
Resistance												
Direct	Central Electric Furnace	868	817	224	35	1940	9.0	240	140	-	2180	2080
Storage	8 hour Central Storage	214	0	145	5	365	1.7	535	320	-	900	685
Bivalent	Ripple Controlled Oil Furnace Backup	396	0	160	23	580	2.9	550	340	45	1175	965
Heat Pump												
Conventional	2.5 ton (SPF = 2.06)	613	741	104	11	1470	14.1	325	200	-	1795	1670
Storage	8 hour Resistance Storage	236	0	69	12	307	2.9	575	350	-	880	655
Bivalent 1	Oil Furnace Backup	271	164	73	18	527	6.4	370	240	65	960	830
Bivalent 2	Ripple Controlled Oil Furnace	157	0	61	12	230	3.0	400	260	80	710	570
Solar												
Resistance	$31m^2$ (55% solar)	353	532	27	15	927	9.6	1230	730	-	2160	1660
Storage Resistance	$31m^2$ (55% solar)	158	0	13	-16	155	1.6	1300	770	-	1455	925
Heat Pump	$25m^2$ (SPF ≃ 2.3)	284	450	54	28	816	9.0	1120	660	-	1940	1480
Air Conditioning												
Conventional	2.5 ton Heat Pump	37	0	46	11	93	4.3	255	175	-	350	270
Storage	8 hour Ice Storage	1	0	28	11	35	1.6	475	310	-	510	345
Water Heating												
Resistance												
Direct	52 gallon tank	156	72	75	20	324	5.5	40	30	-	365	355
Storage	120 gallon tank (16 hour)	24	0	53	3	80	1.4	125	80	-	205	160
Solar												
Resistance	$5.7m^2$ (55% solar)	84	41	33	11	169	6.5	300	175	-	470	345
Storage Resistance	$5.7m^2$ (55% solar)	28	0	13	0	41	1.6	325	190	-	365	230

Table 6. Cost of Supply, Service Area B

System	System Characteristics	Utility Costs ($/yr/customer) Capital Gen	Utility Costs Capital T&D	Utility Costs Variable Fuel	Utility Costs Variable Other	Utility Costs Total	Utility Cost (¢/kWh)	Customer Cost ($/yr/customer) Capital Util. Rate	Customer Cost Capital Mort. Rate	Customer Cost Fuel	Total Cost ($/yr/customer) Util. Rate	Total Cost Mort. Rate
Space Heating												
Resistance												
Direct	Central Electric Furnace	-61	0	672	197	807	4.2	235	140	-	1040	945
Storage	8 hour Central Storage	22	0	299	50	370	1.9	415	245	-	785	615
Bivalent	Ripple Controlled Oil Furnace Backup	123	0	244	84	451	2.9	545	340	20	1015	810
Heat Pump												
Conventional	2.5 ton (SFP = 2.02)	-125	0	271	116	260	2.7	325	200	-	585	460
Storage	8 hour Resistance Storage	-123	0	236	110	222	2.3	440	270	-	660	490
Bivalent 1	Oil Furnace Backup	-121	0	204	104	187	2.5	370	240	60	615	485
Bivalent 2	Ripple Controlled Oil Furnace	43	0	83	37	164	2.3	400	260	70	635	495
Solar												
Resistance	$28m^2$ (55% solar)	24	0	114	43	181	2.0	1150	680	-	1330	860
Storage Resistance	$28m^2$ (55% solar)	-19	0	121	33	135	1.5	1220	720	-	1360	860
Heat Pump	$25m^2$ (SPF ≃ 2.4)	-96	0	204	91	199	2.4	1120	660	-	1320	860
Air Conditioning												
Conventional	2.5 ton Heat Pump	179	266	195	42	681	23.3	255	175	-	935	855
Storage	8 hour Ice Storage	16	0	35	10	61	2.1	475	310	-	535	370
Water Heating												
Resistance												
Direct	52 gallon tank	82	49	139	35	305	5.2	40	30	-	345	335
Storage	120 gallon tank (16 hour storage)	3	0	76	12	91	1.5	125	80	-	215	170
Solar												
Resistance	$5.7m^2$ (60% solar)	8	0	71	20	99	4.7	300	175	-	400	275
Storage Resistance	$5.7m^2$ (60% solar)	2	0	28	5	35	1.6	325	190	-	360	225

ENERGY STORAGE ON ELECTRIC UTILITY SYSTEMS

James R. Birk and Joseph W. Pepper

Electric Power Research Institute, Palo Alto, California

ELECTRICITY DEMAND IN THE YEAR 2000

The U.S. energy supply/demand scenario presently adopted for EPRI's research and development planning process projects a base electricity demand of 7500 billion kWh by the year 2000 (1). The analysis is based on a projection of the historical relationships among energy consumption, employment, and GNP. The presumption is made that continued economic growth is necessary if the nation's societal goals are to be achieved and that the historic tie of energy use to employment will continue.

Central to this projection of energy requirements for the year 2000 are two assumptions with far-reaching implications for energy consumption in the residential, commercial, and industrial end-use sectors. The 7500 billion kWh energy consumption level assumes that the nation will have reasonable success with conservation efforts and achieve at least a 17% reduction from electricity requirements projected on the basis of normal productivity growth to the year 2000. The assumption that a 17% savings in electricity use is possible is based on a quantitative survey of conservation opportunities in industry, commerce and residences (2). It is a conservative (naturally!) estimate that categorizes the short-term (0-2 years), near-term (2-10 years), and mid-term (10-25 years) energy conservation potential of new or improved practices, devices, and systems ranging from operational or housekeeping changes to major investment in process or equipment changes. In estimating the national potential the relative importance of each of the three major end-use sectors has been recognized. Many of these end-use conservation practices, devices, and systems are being discussed in other sessions this week.

The second major assumption implicit in the projection of energy use in the year 2000 is that the U.S. will enjoy a greatly reduced dependence on oil and natural gas for the generation of electricity. EPRI's planning scenario projects that the electrical energy from oil/gas will decrease from 600 billion kWh in 1985 to 400 billion kWh in 2000. If, however, oil/gas usage had grown at the same rate (5.6%) as the overall usage of electricity, then the 2000 oil/gas requirement would have been 1360 billion kWh. Appropriate substitutes for oil and gas must provide 960 billion kWh in 2000. As in the case of the 400 billion kWh of oil/gas energy, it is also reasonable to assume that the 960 billion kWh of oil/gas substitute will be used to meet the peaking and intermediate cycling needs of the electric utility industry. The base load requirements will be met by coal- and nuclear-derived energy.

Estimates show that solar thermal, hydro, synthetic fuel from coal, and coal (intermediate duty plants) could supply about 600 billion kWh of the displaced oil/gas derived energy (3). This leaves some 360 billion kWh of off-peak energy that must be supplied to substitute for oil/gas through energy storage and/or load management.

AVAILABILITY OF OFF-PEAK ENERGY

The Public Service Electric and Gas (PSE&G) of New Jersey Energy Storage Assessment Study indicates that the practical upper unit of on-peak energy that could be supplied from available off-peak energy is 5% and 3% for weekly and daily cycle operation, respectively (4). This analysis accounts for probable device inefficiences, off-peak energy not available on a consistent basis, and probable percentage of generation capacity being used as base load. Using 5% as the practical upper limit, 375 billion kWh (5% of 7500 billion kWh) of on-peak energy could be supplied from available off-peak energy. About 55 billion kWh are likely to be available from pumped hydro systems installed before 1985. Thus, 320 billion kWh is the maximum practical upper limit of on-peak energy that could be provided by new methods of energy storage and/or load management. The primary sources of this off-peak energy are coal, hydro, geothermal, and nuclear. Some nuclear power plants will almost constantly be serving as primary baseload. Coal-derived energy will thus be called on to fulfill a large portion of baseload demands, as well as a portion of the intermediate cycling needs.

Thus we have seen that the on-peak energy demands met from energy storage and/or by load management as a result of the availability of off-peak energy (375 billion kWh) is sufficient (within the accuracy of this analysis) to displace the oil/gas derived electrical energy (360 billion kWh) for which there is no other logical alternative energy source. Our conclusion is that energy storage and load management must penetrate to their maximum potential if oil/gas use by the electric utility industry is to be significantly reduced.

ENERGY STORAGE OR LOAD MANAGEMENT

To this point energy storage and load management have been grouped together in terms of their implications for provision of on-peak energy. Both concepts involve energy management. However, from the utilities' point of view objectives of these concepts are different in that energy storage (on the utility side of the meter) is supply management, while load management (including customer energy storage) is demand management. Load management has as the primary objective decreasing peak loads while energy storage has as the primary objective using baseload energy to supply peak demand. In this context the end-use technologies that you are hearing about this week (including customer energy storage) can be classed as load (demand) management.

The analysis of load management and energy storage involves complex tradeoffs that are not fully understood. Instead of carrying out a detailed analysis we have instead made several reasonable assumptions which will have to be refined in any future work.

Let us assume that energy storage will provide roughly 60% of the on-peak energy being shifted from off peak and that load management will shift the remaining 40% away from the peak. Further, the PSE&G Energy Storage Assessment showed that of the total off-peak energy, 55% is available on a daily basis and 45% is available on the weekends. Using typical duty cycles from the PSE&G report allows an estimate of capacity requirements to be made from the energy being shifted for the daily and weekly energy storage devices.

Sample Calculation: (1) on-peak energy demands - 375 billion kWh, (2) energy demands met by energy storage (60% of #1): 185 billion kWh, (3) energy demands met by weekly energy storage (45% of #2): 85 billion kWh, (4) capacity requirement to meet energy demands of #3: 85 billion kWh/3000 hrs/year = 28 GW.

Table 1 shows the resulting energy storage load management mix in the year 2000.

We will be hearing in other sessions this week about the technological options for load management. Let us turn briefly to a review of the possibilities for utility energy storage and an estimate of the likely market for these technologies.

TABLE 1 Assumed Energy Storage/Load Management Mix in 2000

Equipment	Generation (Billion kWh)	Capacity (GW)
Daily Energy Storage	100	100
Load Management	135	--
Pumped Hydro(1)	55	19
Weekly Energy Storage(2)	85	28
Total	375	147

(1) installed before 1985
(2) installed after 1985 (including conventional pumped hydro)

ENERGY STORAGE TECHNOLOGIES

Hydroelectric pumped storage is the only method presently in use for storing energy (other than primary fuels) on electric power systems. Pumped hydro installations are by now well established as operationally and economically desirable components of a number of electric power systems around the world. In the U.S., the installed pumped hydro generating capacity at present is nearly 9 GW* with an additional 5 GW under construction (5). However, because of geologic/geographic and environmental constraints for siting of reservoirs, ultimate pumped hydro capacity is unlikely to exceed 33 GW in the U.S. This must be compared with an estimated year 2000 potential for utility energy storage of 147 GW (see Table 1). This potential is responsible for the increasing interest of U.S. electric utilities in new ways of storing energy on their power systems.

Many advanced concepts for storage of energy on the large scale desired by electric utilities have been proposed in recent years, and several concepts are now under active study and development. Their technical and economic prospects have been reviewed repeatedly (6-10), most recently in the PSE&G study referenced previously. The general conclusion from these analyses is that a rather limited number of the proposed concepts have probable or potential technical and economic feasibility. The most important characteristics of these technologies and their applications are summarized in Table 2 (adapted from Reference 11) at the end of this section.

* 1 GW = 1 x 10^6 kW

Underground Pumped Hydro (UPH). This method of energy storage has prospective operating and economic characteristics that are similar to those of conventional pumped hydro. However, UPH should face fewer siting constraints because only one surface reservoir is required, and its size can be much smaller due to the much larger difference in the elevation of upper and lower reservoirs. Thus, UPH must be considered a prime contender for future use by electric utilities. Excavation techniques and machinery components adaptable for UPH are essentially at hand, and several feasibility studies and conceptual designs have been carried out. At least one U.S. utility has tentative plans for a commercial project.

The major barriers to initiation of UPH projects appear to be remaining uncertainties regarding underground reservoir cost and stability with pressure cycling, optimum system configurations as a function of geological factors, and availability of advanced pump/turbine equipment for high pressure heads. The U.S. Energy Research and Development Administration (ERDA) and the Electric Power Research Institute (EPRI) are in the process of initiating a preliminary engineering design study that should result in a reduction of these uncertainties and the development of a general analysis and design methodology for UPH.

Compressed Air Storage (CAS). These storage systems also can be considered near-term technology. In CAS, the compressor of a combustion turbine can be uncoupled from the turbine itself. During utility system off-peak periods, the compressor is driven by the generator (which now acts as an electric motor) and charges a cavity with compressed air. The stored air is subsequently discharged during peak periods and mixed with fuel to power the turbine which now drives the generator and supplies power to the electric grid.

Potential advantages of CAS over UPH include a wider choice of reservoir types (solution-mined salt caverns, caverns excavated in hard rock, and naturally occurring aquifers). Thus, CAS may have yet broader applicability. Also, economically viable sizes of CAS installations are likely to be smaller; this will reduce construction time and financial risk. Another possible advantage is that the compressor-turbine-generator combination could be used to generate power (although at a greatly reduced rate) even after exhaustion of the stored air. On the other hand, dependence of CAS on oil or gas may become a serious restriction for its future use by utilities.*

Machinery components adaptable to CAS systems are commercially available, and existing techniques for construction of hard rock and salt caverns should be applicable. Several feasibility studies have been carried out in the U.S. and other countries, and conceptual plant designs exist. A 290 MW, two-hour discharge CAS system using solution-mined salt caverns is under construction in Huntorf, Germany and scheduled to begin operation on the power system of the Nordwestdeutsche Karftwerke later in 1977. In the U.S., the major barriers to utility commitments appear to be inadequate knowledge regarding behavior of caverns under pressure and temperature cycling, lack of information on optimum system configurations and operation (especially for aquifer air storage), and cost uncertainties. As in the case of UPH, ERDA and EPRI are planning preliminary design studies that should reduce these barriers by making analysis and design methodologies available to interested electric utilities.

<u>Sensible Heat Thermal Energy Storage (SHTES)</u>. Systems using the sensible heat capacity of working fluids (such as water/steam mixtures or oil) to store energy are technically feasible today. SHTES systems would be physically integrated with thermal power plants and thus face few, if any, siting constraints. Also, retrofit of existing power plants with thermal energy storage seems technically feasible -- particularly for systems using turbine feedwater storage. Below-ground storage in steel-lined caverns may be attractive but definitive engineering and cost analyses are still lacking. A recently proposed hot oil storage concept is attracting some interest because of the potentially attractive economics (13).

Because the current information base is inadequate for a confident assessment of sensible-heat thermal energy storage systems for electric utilities, ERDA and EPRI will be supporting engineering design studies and cost analyses, with the objective to identify the most promising systems. These analyses also will address the system reliability and availability questions arising from integration of thermal energy storage systems with the steam supply system of base load power plants.

<u>Batteries</u>. In secondary or "storage" batteries, the conversion from electrical to chemical energy (charging) and the reverse process (discharging) is performed by way of electrochemical reactions. By virtue of the electric form of input and output energy, compactness,and the modular characteristics common to electrochemical devices, batteries seem to offer the only economic method of storing energy in dispersed locations while retaining the possibility of reconverting the stored energy to electricity with high efficiency. A number of different electrochemical systems have been developed, or offer prospects for development, into practical storage batteries for utility applications. The characteristics of candidate battery systems

* "Adiabatic CAS systems as described by Glendenning (12) would store the heat of air compression and require little or no fuel on discharge. However, this interesting concept is only in a very early state of development.

have been reviewed extensively in recent years (14-16); we will discuss only those considered the most promising at this time.

Sodium-sulfur batteries were first investigated by researchers at the Ford Motor Company. The key material in this battery is β-alumina, a solid ceramic electrolyte that becomes sufficiently conducting at temperatures above $200^{o}C$. At the $250^{o}C$-$330^{o}C$ operating temperature, the reactants and most of the possible discharge products of the sodium-sulfur battery are liquid. Since liquids do not have a structural "memory", there is no possibility for development of the cumulative, deleterious changes that tend to limit the life of more conventional batteries which use solid electrodes. As a result of worldwide progress in sodium-sulfur battery development over the last several years, there seem to be no fundamental processes that would prevent attainment of the goal of at least 2000-3000 cycles for utility applications.

Major efforts to develop lithium-sulfur batteries for utility energy storage have been underway in the U.S. for a number of years at the Argonne National Laboratory (ANL) and at Atomics International. This battery type uses fused salt eutectics as electrolyte and operates in the range of $400^{o}C$-$450^{o}C$. Considerable progress has been made through development of solid iron sulfide positive electrodes and lithium alloy negative electrodes. However, the basic capability for truly long cycle life and low cost remains to be established for this battery type. The progress of the ANL program was reviewed in several earlier papers (17,18); clearly, this program represents an extensive and systematic effort to develop commercially viable batteries.

The zinc-chlorine battery is attracting considerable interest as a candidate for utility energy storage, in good part because of the development of a promising approach to chlorine storage by Energy Development Associates (EDA) in the U.S. In the EDA approach, the chlorine liberated during charging of the battery is stored in the form of ice-like chlorine hydrate, which represents a fairly compact, non-corrosive and safe form of chlorine storage. Technical progress on all aspects of the zinc-chlorine battery has been quite encouraging, and the technical feasibility of the EDA concept is presently being tested on the level of laboratory prototype batteries of up to 20 kWh capacity. Under a major contract with EPRI, EDA has begun to scale up the zinc-chlorine battery; their present goal is development and fabrication of a 1 MW/5 MWh battery by 1979. If this phase is sufficiently successful, the next step could be demonstration of a full-size (for example, 20 MW/100 MWh) battery plant on an electric utility system before 1985.

Battery installations of this size will represent a very large extrapolation from the current state of technology for the advanced batteries discussed above. To minimize the associated technical, financial and safety risks, the strategy if EPRI and ERDA is that candidate battery systems will be characterized and evaluated thoroughly on the level of modules with storage capacities of approximately 5-10 MWh. A battery energy storage test facility is presently being designed for this purpose. This "BEST" Facility, described in a paper by Casazza et al. (19), should be in operation by 1979, in time to accept the first advanced battery system to reach this stage of development.

Redox batteries -- in which the positive and/or negative active materials are dissolved in the electrolyte -- have been proposed for large-scale energy storage. The potential advantage of this approach (compared with more conventional battery designs) is that external reactant storage in tanks tends to result in relatively low capital costs for the storage-related part of capital costs. This characteristic might qualify redox batteries for accumulating and storing energy over longer periods -- for example, weekends -- than can be handled economically by conventional batteries. Because the development of redox batteries is at an early stage, research and development over three to five years are likely to be required before the true potential of this new battery type for large-scale energy storage can be assessed.

Hydrogen Energy Storage Systems. Next to batteries, hydrogen energy storage represents the best-known example of a secondary or cyclic chemical energy storage system. Several approaches have been proposed and explored for each of the required subsystems -- hydrogen generation, storage, and reconversion -- which can be combined in various ways into overall energy conversion and storage systems. Numerous surveys, and a series of papers (20,21) attest to the widespread interest in hydrogen, including its use as an energy storage medium.

Hydrogen generation by water electrolysis is an established process but currently available technology is handicapped by modest efficiency and high capital costs. Research and development currently underway in several countries should result in improved efficiency (up to 90%-95%) and reduced capital cost of electrolyzers. The development of a thermochemical process for water splitting could eliminate need for intermediate power generation and use of electrolyzers for hydrogen generation. However, recent analyses (22) indicate that even the best thermochemical processes proposed so far are unlikely to result in less expensive hydrogen than electrolysis.

Hydrogen storage, the second major subsystem of hydrogen energy storage systems, can take several different forms. Storage of compressed hydrogen is technically feasible now, but the economics are not fully established because of uncertainties regarding the lowest-cost containment methods and materials. Storing hydrogen as a cryogenic liquid or chemically bound in metal hydrides is technically feasible and will result in more compact storage systems. However, cryogenic storage of hydrogen carries a significant efficiency penalty that appears unacceptable for large-scale energy storage on utility systems. The outlook is better for metal hydride storage, but development efforts over a three- to five-year period are still required to establish the probable technical and economic characteristics of this method for hydrogen storage.

Reconversion of hydrogen to electric energy can be done in fuel cells or in combustion-based devices (gas-fired boilers or gas turbines). The fuel cell approach offers potential or high efficiency, with 60% as a realistic target for pure hydrogen fuel. Although a commercial technology is not now available, much of the fuel cell technology currently being developed in the U.S. for electric power generation will be applicable to fuel cell systems operating on pure hydrogen. Probable technical and economic characteristics, and the first generation of a commercial fuel cell technology, are expected to be established within the next three to four years.

Complete hydrogen energy storage systems consist of various combinations of hydrogen generation, storage, and reconversion. A small experimental electrolyzer/hydride storage/fuel cell system has been operated in the U.S. by PSE&G; a much larger prototype test facility is being designed at Brookhaven National Laboratory. Generally, the technical feasibility of this and similar systems is likely to become established in the coming few years. A more difficult and as yet unresolved question is whether any of the conceivable hydrogen storage systems configurations and applications can become economically competitive with further development.

Closed Loop Chemical Systems. Other recently proposed concepts for chemical conversion and storage of energy are based on closed-loop chemical reaction systems. Such systems would be thermally coupled to high temperature heat sources to achieve an energy-absorbing chemical change. The absorbed energy, now in chemical form, would be storable and transportable, possibly over significant distances. At the point of consumption, the reaction would be allowed to proceed in the reverse direction, with evolution of heat at a somewhat lower temperature. To be suitable, the forward and reverse reactions must be readily reversible and must occur at useful temperatures. One of the most promising concepts appears to be the ADAM-EVA system ($CH_4 + H_2O \leftrightharpoons CO + 3H_2$) that is currently under development in Germany for efficient and economic transport of nuclear heat to industrial and residential users.

Superconducting Magnetic Energy Storage (SMES). Storage of energy as permanent currents flowing through superconducting coils of large electromagnets is fundamentally attractive in that this appears to be the only practical method of storing electricity per se. Although a substantial amount of conceptual design and small-scale experimental work on SMES has been carried out (in the U.S. at the Los Alamos Scientific Laboratory and at the University of Wisconsin), the concept is still in an early stage of development.

The critical issues are probable capital cost of SMES and the integration of the very large installations (>10,000 MWh) needed to make the economics attractive. Even if these issues can be resolved favorably through continued study and experimentation, SMES will be faced with the barriers represented by the substantial financial risks and long time periods that are typical for large underground construction projects.

Flywheel Energy Storage

Several flywheel energy storage systems, storing electrical energy as kinetic energy of a rotating mass, have been proposed for either electric utility or transportation applications. The state-of-the-art in flywheel energy storage systems is

Table 2

Projected Characteristics and Status of Some Energy Storage Systems[a]

Type	Round Trip Efficiency (%)	Capital Costs[b] C_p ($/kW)	C_s ($/kW-hr)	Energy Density ($kW\text{-}hr/ft^3$)	Development Stage	Potential Application
Mechanical						
Pumped Hydro	67-75	150-200	2[c]-15	0.04[d]	Existing application; engineering studies for underground	Central energy storage for peak shaving and load leveling
Compressed air-gas turbine system	65-75[e]	100-200	4-20	0.1-0.5	First commerical demonstration 1977	Central energy storage for peak shaving and load leveling; reserve generating capacity
Flywheels	70-85	70-100	100-300	0.5-2	Initial development	Distributed energy storage; power factor correction; emergency generating capacity
Thermal						
Steam (pressure vessel)	70-80	150-250	30-70	up to 1	Historical installations; engineering studies of modern systems	Central energy storage, integrated with baseload steam generation
Hot Oil	65-80	150-250	10-20			
Batteries						
Lead-Acid	60-75	60-100	70-80[f]	1-2	State-of-the-art	Distributed energy storage for daily peak shaving;
Advanced aqueous	60-75	60-100	30-40[f]	1-3	Small prototypes	
High-temperature	70-80	60-100	30-50[f]	2-5	Laboratory cells	
Redox	60-70	100-200	15-35[f]	0.5-2	Conceptual and laboratory studies	
Chemical						
Hydrogen (electrolysis plus fuel cell)	35-55	400-600	5-30	N.A.[g]	Advanced development of subsystems	Central energy storage with distributed generation; combined gas/electric energy systems
Reaction systems (closed loop) CH_4+H_2O $CO+3H_2$	?	?	?	N.A.	Conceptual studies and initial development	Conversion, storage, and transport of nuclear and solar energy
Electromagnetic						
Superconducting magnets	80-90	50-60	30-200	0.5-1	Concept; key components under development	Central energy storage and system stabilization (large-scale only)

[a]Adapted from (9)

[b]Total storage system capital cost is given by $C_t = C_p + t_{max} \times C_s$ where t_{max} is the maximum period for which the storage system can be discharged at its rated power.

[c]Assuming one existing reservoir (lake).

[d]Assuming 3000-ft head.

[e]Efficiency with respect to recovery of stored energy.

[f]Including cost of installation (estimated at 10-20/kW-hr).

[g]Not applicable.

best summarized by stating that low energy density systems of a few kilowatt-hours capacity have been constructed and are being applied with great promise for success in transportation applications. Advanced, high energy density systems have been proposed and certain critical components such as the wheel itself have been built and tested on a laboratory scale. To date, a detailed system design study in which the wheel is integrated into a total, reliable system is yet to be carried out.

ECONOMICS OF ENERGY STORAGE

Electric utilities will use energy storage only if it is competitive with conventional ways of generating peaking or cycling power. Competitiveness implies that energy storage systems must meet utility standards for operating life, reliability, maintainability and safety. It also implies that the total cost of electricity generated by an electric power system using energy storage must be less than the electricity supplied by the same power system if it used only conventional generating equipment.

The energy storage methods described in the preceding sections appear to have potential for meeting the technical and operating criteria for utility equipment, but their economic competitiveness is less certain. A first step in the economic assessment of utility energy storage is to compare candidate technologies in terms of their prospective capital costs.

A breakeven cost analysis, based on the estimates of power-related costs in Table 2 gives the following targets for the energy-related costs of the various energy storage technologies. (For flywheels and batteries a \$50/kW cost credit was assumed to allow for the potential of credits for deferral of T&D expenditures, reduce spinning reserves, improve load following capability or to increase system reliability.) A comparison of these targets with the projected energy-related costs in Table 2 gives an indication of likely application of these technologies.

UPH appears to be a good prospect for weekly, daily and peaking applications, although the relative benefits in the peaking model (as evidenced by the difference between the target and projected kWh costs) are small compared to the 20-year life batteries. CAS has potential in all three modes, but is not the clear economic choice that UPH is. SHTES has potential for application in the weekly and daily modes, but not for peaking applications.

The advanced batteries (20-year life) are clear economic choices for the peaking application, with some potential for the daily and weekly modes. The 10-year life batteries (lead acid) will have application only if capital cost credits significantly greater than \$50/kW can be justified.

TABLE 3 Range of Target, Energy-Related, Capital Costs (\$/kWh)

	Peaking (2-hr discharge)	Daily Cycle (5-hr discharge)	Weekly Cycle (15-hr discharge)
UPH	<20	30-40	<50
CAS	<20	5-25	15-20
SHTES	<5	15-35	35-45
Battery (10-yr life)	15-35	15-25	<15
Battery (20-yr life)	70-90	<65	<55
Hydrogen	-	-	15-30
SMES	45-50	45-50	<50
Flywheel	75-90	60-65	<55

Hydrogen storage begins to show potential only for 15-hour (or longer) discharge times. Flywheels would appear to have application only for the peaking mode, and there they must compete with the advanced batteries. No clear picture emerges for SMES, but this reflects only the extremely uncertain nature of SMES cost projections at this stage of development.

POTENTIAL MARKET FOR ENERGY STORAGE

If 185 billion kWh in 2000 comes from new energy storage systems installed after 1985 and a 50/50 mix between peaking (1000 hours/year, 5 hours/day) and intermediate (3000 hours/year, 10 hours/day) duty will be required of energy storage, then an installed capacity of some 100 GW peaking duty energy storage and 28 GW of intermediate duty energy storage can be supported by the year 2000 (Table 1).

If we assume batteries would supply the peaking demands, about 500 x 10^6 kWh of battery energy storage capacity would have to be installed by the year 2000. This means about 33 x 10^6 kWh of additional battery capacity per year between 1985 and 2000. This is about the same size as the existing SLI (Starting, Lighting and Ignition) Battery market. At a price of $30/kWh, the resulting market is about one billion per year.

If we assume that the intermediate demands will be met by a combination of pumped hydro (conventional or underground), compressed air storage, and sensible heat thermal energy storage, most of which are site-related technologies, are there geologic constraints precluding 28 GW by the year 2000? EEI data for conventional pumped hydro capacity in operation, under construction, or planned for development, predicts about 19 GW by 1985. FPC identifies another 14 GW of conventional pumped hydro capacity as having the potential for development by 1993 (we have assumed this 14 GW would be developed by 2000).

Estimates of suitable geology combined with major load centers show roughly 30% of U.S. utilities (on a peak MW basis) as having access to all of the geologies required for the advanced intermediate duty technologies. 50-60% of U.S. utilities would appear to have access to at least one of the required geologies. This is many times the capacity estimated previously so that these technologies are not site-limited. The major limiting factor for the introduction of these technologies will be construction time. The intermediate duty advanced storage technologies are therefore probably competing for a market of some 14 GW of capacity (which excluded new conventional pumped hydro) to be added between 1985 and 2000 or about 1.5 GW/year from 1990 to 2000. At an average size of 300 to 700 MW this will require two to fiue plants to be constructed/year. At an average cost (10 hours of storage) of $400/kW this represents a potential market of some $600 million/year.

SUMMARY

Based on the EPRI R&D planning scenario projection of electricity use in the year 2000, we have postulated a possible mix of energy storage and load management (end use technologies and practices) that is consistent with the postulated oil/gas usage, alternative sources of intermediate/cycling energy, and availability of off-peak energy. While this assumed mix is only one of several plausible scenarios, we feel that any reasonable mix would call for load management, peaking storage (daily), and intermediate storage (weekly). Such a mix is required to satisfy demand while taking maximum advantage of off-peak energy.

We reviewed briefly the prospects for the major energy storage technologies from both a technical and an economic point of view and formulated an estimate of the market potential for the advanced energy storage technologies.

The net result of deploying energy storage and load mangement to supply about 375 billion kWh in the year 2000 will be a direct reduction in oil/gas usage. The electric utility's reliance on petroleum will be decreased by roughly 1-3/4 million barrels per day.

REFERENCES

1. Starr, C., "Electricity Needs to the Year 2000," EPRI, Presented to the Subcommittee on Energy Research, Development and Demonstration; House Committee on Science and Technology, Washington, DC, February 26, 1976.

2. "Efficient Electricity Use," Craig Smith, editor, Pergammon Press, Inc., New York, April 1976.

3. Birk, J. R. and Pepper, J. W., "Reducing Oil Requirements in the Electric Utility Industry: The Need for Energy Storage," Presented at the Fall Electrochemical Society Meeting, Atlanta, GA, October 1977.

4. Schneider, T. R. et al., "An Assessment of Energy Storage Systems Suitable for Use by Electric Utilities," Final Report, Volume II, EPRI RP225, Prepared by Public Service Electric and Gas Company, Newark, NJ, July 1976.

5. Loane, E., "An Assessment of Energy Storage Systems Suitable for Use by Electric Utilities," Final Report, Volume III, EPRI RP225, prepared by Public Service Electric and Gas Company, Newark, NJ, July 1976.

6. Rosengarten, W. E., A. J Kelleher, O. D. Gildersleeve, "Wanted: Load-Leveling Storage Batteries," Presented at Fall Meeting of the Electrochemical Society, Miami Beach, FL, October 1972.

7. Lewis, P. A. and J. Zemkoski, "Prospects for Applying Electrochemical Energy Storage in Future Electric Power Systems," Presented at IEEE International Convention, New York, March 1973.

8. Fernandes, R. A., O. D. Gildersleeve and T. R. Schneider, "Assessment of Advanced Concepts in Energy Storage and Their Application on Electric Utility Systems," Trans. 9th World Energy Conference, Detroit, MI, 1974.

9. Kalhammer, F. R., "Energy Storage: Applications, Benefits and Candidate Technologies," Proc. Symposium on Energy Storage, The Electrochemical Society, Princeton, NJ, 1976.

10. Gardner, G. C et al., "Storing Electrical Energy on a Large Scale," CEGB Research, May 1975, p. 12.

11. Kalhammer, F., Birk, J., and Pepper, J., "Storage of Energy on Electric Power Systems," paper No. 5B.27, WELC, Moscow, June 1976.

12. Glendenning, I., "Storing the Energy of Compressed Air," CEGB Research, May 1975, p/ 21.

13. Nicholson, E. W., and R. P. Cahn, "Storage in Oil of Off-Peak Energy from Large Power Stations," Proceedings 11th Intersoc. Energy Conversion Conference, Stateline, NV, August 1976.

14. Douglas, D. L., "Batteries for Energy Storage," Presented at Sympos. Energy Storage, Div. Fuel Chem., American Chemical Society Meeting, Atlantic City, NJ, September 1974.

15. Binder, H. and G. Sandstede, "Forschungsrichtungen bei Batterien und Brennstoffzellen im Hinblick auf die zukünftige Energieversorgung," Chem-Ing-Tech. 47; 51 (1975).

16. Yao, N. P. and J. R. Birk, "Battery Storage for Utility Load Leveling and Electric Vehicles: A Review of Advanced Batteries," Proc. 10th Intersoc. Energy Conversion Eng. Conference, University of Delaware, August 1975.

17. Walsh, W. J., "Electrical Power Generation Using Li-S Batteries," Paper No. 5B.15, WELC, Moscow, June 1977.

18. Shimotake, H., and P. Nelson, "Development of Li-Al/FeS_x Cells for Load-Levelling Application," Paper No. 5B.19, WELC, Moscow, June 1977.

19. Casazza, J. and S. Mallard, "National Facility for Testing Utilities Energy Storage Systems - The Battery Energy Storage Test (BEST) Facility," Paper No. 5B.20., WELC, Moscow, June 1977.

20. Mas, L. et al., "New Perspectives for Electrical Energy Storage by Water Electrolysis," Paper No. 5B.29, WELC, Moscow, June 1977.

21. Bockris, J. O'M., "Electrochemical Aspects of Hydrogen Economy," Paper No. 5B.31, WELC, Moscow, June 1977. Funk, J., "A Technoeconomic Analysis of Large Scale Thermochemical Production of Hydrogen," Final Report, EPRI Project RP467, Prepared by University of Kentucky, Lexington, Kentucky, and Combustion Engineering - Lummus, Bloomfield, NJ, December 1976.

22. Funk, J., "A Technoeconomic Analysis of Large-Scale Thermochemical production of Hydrogen," Final Report, EPRI Project RP467, Prepared by University of Kentucky, Lexington, Kentucky, and Combustion Engineering-Lummus, Bloomfield, NJ, December 1976.

PLANNING FOR THE FUTURE ROLES OF HIGH TEMPERATURE NUCLEAR ENERGY

P. Fortescue and R. N. Quade

General Atomic Company, San Diego, California

"Shape your heart to front the hour,
But dream not that the hour will last."

Alfred Tennyson

INTRODUCTION

Thus far, the main thrust of nuclear energy development has been directed almost exclusively towards its use for large-scale electrical power production, and its worth has been judged principally by economic competitiveness with equivalent fossil-fueled plants, on the basis of current or near future fuel cost estimates. While a healthy diversity of views on detailed means has always existed, the relative simplicity of the ends sought at least rendered the formulation of appropriate developmental programs a relatively straightforward task.

Now, however, stimulated by an unexpectedly early manifestation of the consequences of dwindling energy resources, and by increasingly strident concern with the total environmental impact involved in any technological progress, the whole role of nuclear energy and the criteria proper to its evaluation are in the process of worldwide reexamination. The problem is compounded by the fact that short-term goals can be in conflict with long-term goals, if the former are not selected in advance for compatibility with the latter.

The last consideration has always ranked high among the motives for advocacy of high temperature gas-cooled reactors (HTGR), for their full reward is an essentially long-term one, being greatly enhanced by the development of uses offering a special premium on high temperature, and by an eventual fuel scarcity, which places proper emphasis on its conservation.

This paper elaborates on this thesis, with particular reference to a progression of goals, applying modifications of the HTGR to a succession of tasks both of value in their time and essential to realization of the ultimate ends envisaged.

Since the diversity of the considerations is so wide, including not only a number of different but related tasks but also the means for sustaining the nuclear fuel supply, it is first necessary to outline the overall energy strategy implied in the particular selection of these tasks.

We turn first to the most basic consideration, which is the assurance of a sufficient fissile material supply to sustain the benefit of nuclear power over the long time period involved.

Discounting the conversion of fertile-to-fissile material, as yet only theoretically possible by hybrid fusion plant or accelerator beams, it is already generally agreed that some assistance from fast breeder reactors (FBR) is essential to secure the industry durability necessary to justify the massive monetary outlays involved.

However, the breeding capabilities of the FBR are presently far better established than its performance as a practical and reliable source of energy for universal application. Furthermore, assurance will require far more than the initial seeming success of a single demonstration plant, for a lengthy period of practical operation is essential to establish the all-important questions of reliability and maintenance. This consideration generally requires planning to be initiated far earlier than might be otherwise suggested by the anticipated timing of needs.

The breeder contribution considered in this paper, therefore, is confined to its use as a supplier of fissile makeup for the advanced converter-type reactors actually used for the designated practical tasks. More specifically, by breeding U-233 from thorium-blanketed FBRs, rather than Pu from uranium, we can utilize the very high conversion ratio attainable by use of the U-233-Th cycle in the associated duty reactors. The correspondingly reduced fissile feed requirements thereby greatly increase the number of such reactors which can be serviced by a single breeder, which, in turn, is an essential condition to any scheme relying on the thermal reactor for the bulk of ultimate practical applications.

Clearly, we also require a sufficient fast breeder conversion ratio to provide a worthwhile fissile production. For this reason, the very high conversion ratio characteristic of the gas-cooled breeder becomes of particular attraction. Since this subject has already been covered quantitatively (Ref. 1), it will suffice here to indicate only that we believe that the nuclear fuel requirements for the applications to be discussed could be indefinitely sustained by the fissile output of fast breeders of far less total power output.

Satisfied on the adequacy of sufficient nuclear fuel resources, either by the foregoing remarks or by belief in the likelihood of substantial further ore resource discovery, we pass now to the principal subject in hand, which is the most effective uses of nuclear energy and the course of development most likely to realize this, having regard both to ultimate ends and to the meeting of earlier targets.

EFFECTIVE USES AND DEVELOPMENT OF NUCLEAR ENERGY

We are here more particularly concerned with the uses and development of the HTGR-type of reactor (in either prismatic or pebble-bed form), not because it is considered a universal best prospect, but because some of the applications to be considered do not merely benefit from this high temperature but, more importantly, demand it. Also, (although this attribute is not peculiar to the HTGR) a high conversion ratio capability or, alternately, a low total fuel ore supply need is essential to assurance of long extended benefit.

Electricity Production

Following the general principle of step-by-step introduction of targets which are both realistically useful for their time, yet in furtherance of later goals, the first effective usage for the HTGR clearly lies in the field of electric power production, using traditional steam plant equipment.

Since required steam temperatures are already limited by such plants, the HTGR temperature capabilities thus are not fully exploited, but nevertheless confer the advantage over competitive LWR reactors by allowing retention of steam cycles and machinery already highly developed for fossil fuel.

The feature of a much lower total ore requirement is also gained, together with an altogether new step towards inherent safety added by the use of integral concrete vessel containment.

The next step foreseen in the pure electric generation field, presently only in the study area, is the harnessing of the HTGR to the gas turbine, thereby not only extracting benefit from the high temperatures of which this reactor core is already capable, but also eliminating the input heat exchanger problem, hitherto the prime impediment to the closed-cycle gas turbine.

Initially, two avenues of benefit are under pursuit here. Firstly, there is the prospect of a pure gas turbine system offering plant simplicity, the chance of capital cost reduction, and the freedom from the necessity of water cooling conferred by a high reject heat temperature. Alternately, by addition of an independent waste heat power generation system to this same plant, the opportunity is afforded for substantial efficiency gain, over 50% being in prospect. It is worth noting here that the achievement of this kind of efficiency by upward extension of temperature limits (as envisaged by pursuit of ceramic topping turbines or even MHD devices) would, by contrast, involve far more exacting development than its achievement here by attention to the bottom end of the cycle.

Nonelectric Applications

At this point, it is necessary to consider the energy situation more generally, for the long-term energy need also includes a large fraction for mobile application, not yet foreseeably met electrically, and also uses such as space heating, for which high-grade energy is thermodynamically wasteful.

Though it may be argued that the time available before our transport stops for want of fuel is far off, there is nevertheless even now an urgent need to plan for this situation, for proper preparation and effective demonstration of the measures required are far more time consuming than the public appreciates. Furthermore, time seemingly so easily bought now by running on the fat of of our remaining fossil reserves will be dearly paid for by subsequent generations deprived of the vast asset of a natural chemical repository earlier squandered as mere fuel.

Posterity could justifiably regard such action as the burning of the world's furniture to keep a heedless generation warm.

Some, recognizing transmission advantages of chemical energy and the benefits of a freer choice in the energy fraction used for electric production, would even predict that eventually all nuclear power will be conveniently stored in chemical form. Be this as it may, it is clearly important even now to include in our plans the consideration that at least a major part of the near future uses of nuclear energy will be directed to the synthesis of nonpolluting chemical fuels.

In this presently less well-defined area, which we may collectively term process heat applications, it is particularly important to adopt a step-by-step approach, beginning with targets well within grasp. An obvious first step in such a progression is provision of nuclear steam supply sources for existing industrial processes, which presently use fossil fuel for this purpose.

The high-temperature capabilities of the HTGR, while not fully exploited, nevertheless extend the scope of the first stage to include the possibility of a nuclear steam supply package as an alternative to an otherwise mandatory coal-fired boiler replacement, which faces many electric generation plants presently operating on oil or gas. Extended uses for steam, such as facilitating oil well extraction, and new ones that might be involved in the recovery of shale oils are also clearly within this scope.

In contemplating the next crucial step, that of combining a nuclear heat source with a chemical process specifically directed towards realization of ultimate objectives, the following points are among the dominant issues.

Considering that the freeing of hydrogen, ultimately from water, is likely to constitute the key process in any form of fuel synthesis, a first task is sought which, while still centered around the production of free hydrogen, presents this in a form both simpler of realization and more assured of immediate use.

The simplest step could be to make hydrogen from methane, utilizing current processes but substituting a nuclear heat source for the furnaces now fired with natural gas. This substitution would allow a 40% reduction in natural gas usage while producing the same amount of hydrogen for ammonia and other chemical products.

The gasification of coal by hydrogeneration provides a logical extension to this work, representing a task not only presently desirable in its own right, but also affording an immediate local use for the hydrogen produced.

The particular process envisaged here is based on gas production by hydrogeneration to methane with free hydrogen, derived from a reforming reaction of part of the product methane with steam. This latter reaction, being endothermic, is the part involving nuclear heat addition.

The point favoring the success of first steps is that the hydrogen here is essentially derived indirectly from water by way of a catalyzed steam-methane reforming reaction, requiring less and lower-grade energy and a simpler technology than that required for direct water splitting. Oxidation of part of the carbon supply, in essence, here aids by providing part of the energy for freeing the hydrogen.

It indeed seems likely that so long as carbonaceous sources hold out, they will continue to provide a fundamental easement of the basic task of producing free hydrogen. The concept of combining coal and nuclear power appears extremely promising as a key ingredient in our energy future. The practical application of this concept is described in the section, "NUCLEAR-CHEMICAL ENERGY CENTER," in this paper.

The next step, direct water splitting, is therefore to be regarded as one aimed at preparation for a future when carbonaceous sources begin to dry up, or more exactly, when their full value for other purposes becomes apparent in their cost. In this field, two avenues are apparent: firstly, simple electrolysis and, secondly, a cascade of thermochemical cycles aimed at reducing the very high temperature level of the necessary input heat that would be required to dissociate water directly.

Since electrolysis is already available as a means of hydrogen production, the whole purpose of seeking other means is primarily concerned with the efficiency, and more particularly with improvement of the basic losses in the heat engine, necessary for intermediate electric power production, and in the electrolytic cell. The performance of this route, nevertheless, serves as a useful yardstick by which to judge alternatives, but it is worth noting that this measure may well change with future development.

An illustration of the orderly progression of the HTGR development is shown in Fig. 1.

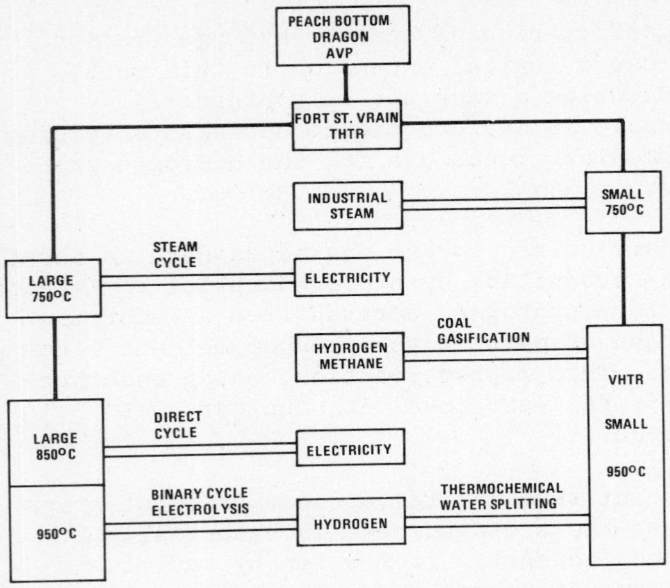

Fig. 1. Progression of HTGR development

The technical base is established by the first generation helium-cooled reactors in the U.S., Great Britain, and Germany, with outputs ranging from 50 to 100 MW(t). The second generation of Fort St. Vrain (FSV) in the U.S. and Thorium High Temperature Reactor (THTR) in Germany at 840 and 750 MW(t), respectively, are in the startup phase (FSV) or in construction (THTR). At this point, the development splits into two paths, characterized by the end product (electricity vs. process heat) and plant size [3000 MW(t) vs. 850 MW(t)]. On the electrical side, the development proceeds from the steam cycle to the closed-loop gas turbine cycle, with a progressively increasing core outlet temperature. On the process heat side, the process changes from steam producer to syngas or high Btu gas, with coal or heavy oil as a feedstock, to thermochemical water-splitting, using no carbon. Here, also, the temperature is increasing. Hydrogen production from water can be achieved by either route.

Finally, there are certain accessory systems on which research will be greatly stimulated by any progression to an economy based on wide use of synthetic fuel. Prominent among these are the fuel cell and the thermally driven heat pump. Interest in the former as a means for local electric energy generation (in principle escaping the efficiency limitations of the heat engine and also offering simplicity) will obviously be immensely stimulated by any future schemes involving chemical power transmission.

The importance of the thermal heat pump in this scenario lies in the fact that in combining a heat engine with a heat pump all the inevitable large losses in the former become usefully used in the supply of the desired low temperature heat. With locally available chemical fuel, the heat pump is also at once freed from the handicap of its present need for an eightfold more expensive electric energy source, a relaxation essential to its wide usage.

NUCLEAR-CHEMICAL ENERGY CENTER

To illustrate how the HTGR can effectively participate in the nuclear energy field for the intermediate time frame (1990 and onward), an example of a nuclear-chemical energy center utilizing coal as a feedstock has been developed. A schematic of the basic process, coal solution hydrogasification, is shown in Fig. 2 and described more fully in Ref. 2. Experimental work on the hydrogasification step is currently being conducted under a program being conducted by General Atomic and Stone and Webster Engineering Company. The process uses coal as the raw material, puts it into solution by adding hydrogen under conditions of moderate temperature and pressure, and hydrogasifies the coal solution, by adding additional hydrogen at higher temperature and pressure. Sulfur and nitrogen are removed as part of the hydrogen addition processes. Ash is taken out in the liquefaction step before pipelining to the site. A portion of the methane produced is used in conjunction with steam as the feedstock for a steam methane reformer, which is heated by hot helium from the reactor. Resulting hydrogen-rich gas is then further purified in conventional hydrogen purification equipment.

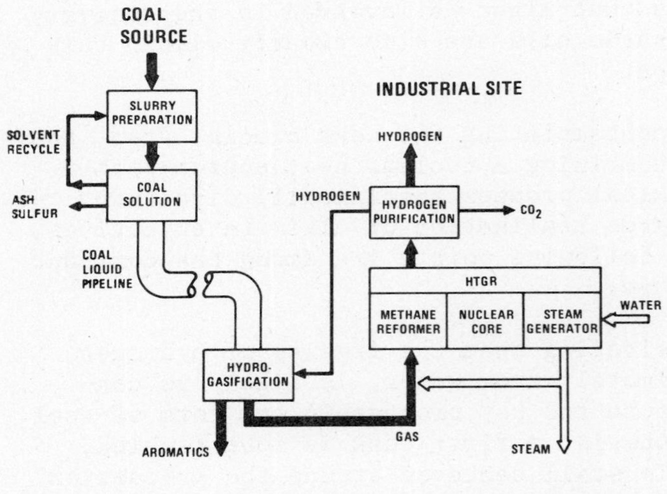

Fig. 2. Coal solution hydrogasification process

Figure 3 shows how the nuclear-chemical energy center can be integrated to serve the needs of various users. The basic HTGR process plant receives coal liquids, uranium in the form of fuel elements, and water at the plant site. The process plant output products shown in the figure are then sent to the users. Aromatics produced, primarily benzene, would

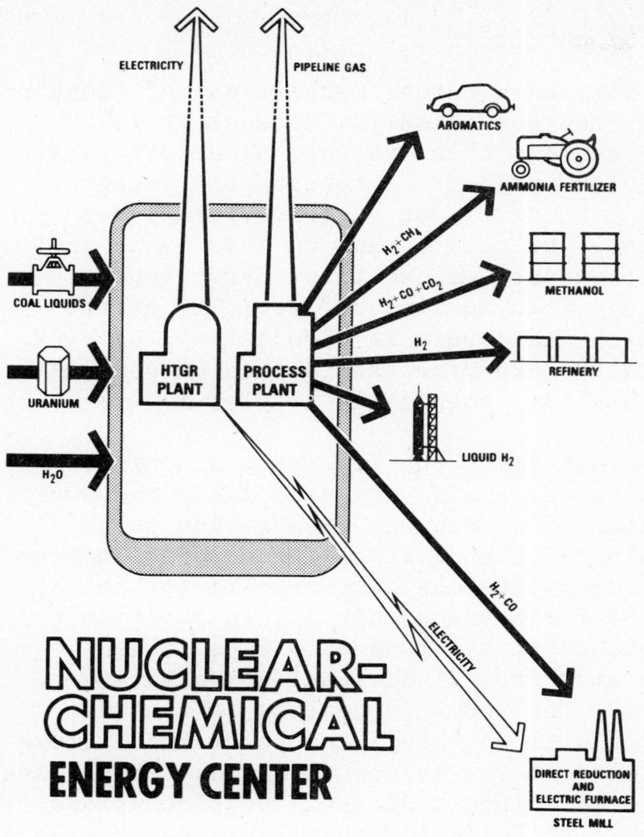

Fig. 3. Nuclear-chemical energy center

be shipped off-site via pipeline or tank car. Liquid hydrogen could also be shipped via special truck or tank car. The direct reduction plant could include an electric furnace for the downstream production of steel. A portion of the HTGR electric output then would be directed toward that end. Each of the hydrocarbon users requires a particular gas composition peculiar to his work. The conditioning equipment for the gas could be located on the energy center plant site and the special gas shipped via individual pipelines to each user. It is also possible to produce a basic gas at the HTGR plant and pipe it to the customer's site, where special conditioning equipment would convert it to the desired product.

Reasonable plant sizes, commensurate with today's technology, are shown for these users. Table 1 gives a breakdown of customer facilities, the approximate output quantity, and the quantities of the gas sent to them. These values are just matched by a 3000 MW(t) HTGR. If the nearby industrial demand only is considered, i.e., no pipeline gas or co-generated grid electricity, the HTGR size is reduced to approximately 1800 MW(t). At this point, an HTGR the size of Fort St. Vrain or the German THTR [850 MW(t)] should be given serious consideration, with a twin plant installation. This arrangement would give a very high degree of availability. A representative heat balance on the nuclear-chemical energy center shows that the nuclear heat represents 35% of the heat input, with the remainder being supplied from the liquid coal. The approximate overall thermal efficiency of the complex, including electric power production, is 66%. This value is obtained by dividing the heating value (HHV) in the products by the heat content of the coal liquid plus the nuclear heat.

TABLE 1 Plant Mass Balance

	Energy Center Product*						
Final Product	Elec. MW	CH_4	H_2	CO	CO_2	C_6H_6 (Bbl/day)	Customer Plant Capacity
		Million SCF/Day					
Oil refinery	46		72				160,000 bbl/day
Ammonia	49	15	71	7	13		1,500 tons/day
Aromatics						13,000	13,000 bbl/day
Methanol	30		103	10	28		1,500 tons/day
Liquid H_2	11		11				30 tons/day
Reduced pellets		5	103	10			1,170,000 tons/yr
Finished steel	110	5	103	10			1,000,000 tons/yr
Pipeline gas		166	20				186 MMSCFD
Grid electricity	264						264 MW
Total	510	191	483	37	41		

*Energy Center Input:
Liquid Coal - 4.27 x 10^6 tons/yr or 67,000 bbl/day
Uranium - 91.3 tons U_3O_8/yr (recycle)
H_2O - 2.15 x 10^6 gal/day (H_2 production only)

The raw materials sent to the plant, uranium fuel and water, are relatively insensitive to the plant location. The nuclear fuel has high energy density, and the water is usually locally available or easily transportable. The same cannot be said of coal in solid form. A process using the coal in liquid form, which can be readily and cheaply transported via pipeline from a coal liquefaction plant located near the coal source, has definite advantages. For this process, raw materials would not be a dominant factor in selection of plant location. Some of the output products, electricity, pipeline gas, and the liquid, can be considered relatively insensitive to plant location, but the reducing gas cannot. High volume per unit heating value of this gas makes transportation over long distances impractical.

The petrochemical, refining, and fertilizer industries are located primarily in the Gulf Coast area of the U.S. Undoubtedly the early abundance of low-cost oil and gas in this area was a major factor in locating there. Rising natural gas and oil prices, impending scarcity, and concentration of the petrochemical industry makes this portion of the country a reasonable location choice for at least the early nuclear process-heat plants.

HTGR Heat Source

The nuclear heat source that can be used with the center is shown in Fig. 4. The use of an isolation helium loop between the core and the process offers considerable flexibility for future application. The design and supporting development of a single nuclear heat source can be undertaken and matched with a variety of either chemical or direct heat processes as the industrial market demands and as process technology is developed.

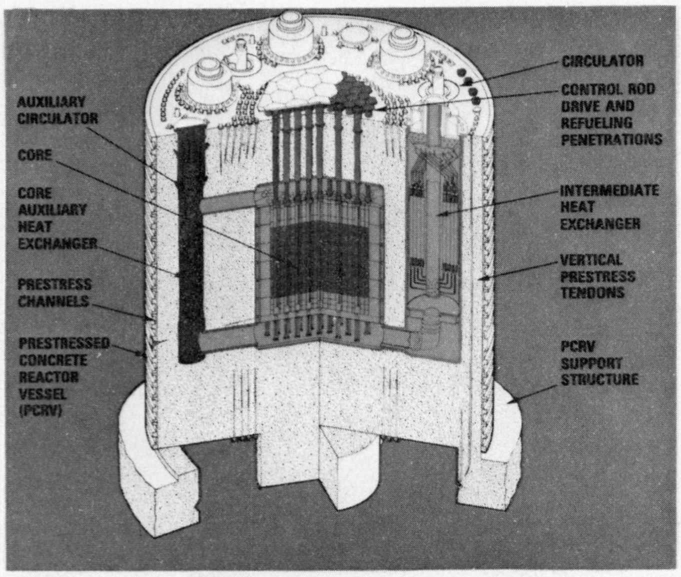

Fig. 4. HTGR process heat source

General Conclusions

Thus far, an eventual massive use of fission energy has been assumed, because it is believed that this, as yet, alone offers a realistic prospect of meeting the energy needs even of a most conservation-minded future. Our real option on this basis thus is not whether we use this energy source, but only when we will use it. The extent to which its advent is planned may, however, crucially determine the balance of penalty and benefit. Nothing is completely free.

In the meantime, the important practical question is, how is the very large endeavor necessary for adequate preparation to be sustained? The desirability of planned progression is obvious, but support for this will be forthcoming only if, in addition to acceptance of the long-term ends, these steps also offer individual commensurate reward. The former condition is the most difficult to satisfy, because of the extreme diversity of the issues involved, which range from highly important, real considerations to the pure fallacies inevitably associated with the necessary widespreading of debate.

The real risk in fact may lie rather in the inactions presently countenanced, if this serves but to delay inevitable choices until they are later forced too fast upon us by an urgency which can only grow.

REFERENCES

1. P. Fortescue, Assurance of a durable nuclear industry, Nuclear Engineering International, October (1976).

2. R. N. Quade, Hydrogen production from coal using a nuclear heat source, 1st World Hydrogen Energy Conference Proceedings, Miami Beach, Florida (March 1 - 3, 1976); [General Atomic Report No. GA-A13769, (February 26, 1976)].

FUEL CELL SYSTEMS FOR DISPERSED GENERATION OF ELECTRIC POWER

E. A. Gillis
Electric Power Research Institute, Palo Alto, California

L. J. Rogers
U.S. Energy Research and Development Administration, Washington, D.C.

ABSTRACT

Fuel cell power plants are an emerging option for electric utility power generation that show promise to favorably impact the nation's energy supply. The fuel cells' key features are high efficiency independent of size, constant generation efficiency over a wide load range, and environmental acceptability. These features permit fuel cells to be installed close to the load (which increases probability of waste heat utilization), to be installed in incremental modules to match load growth, and to be an effective tool for load management. This paper describes a program underway to evaluate a 4.8 MW fuel cell module in an electric utility system, and the potential role for this new generation option in the utility industry.

INTRODUCTION

Electric utilities in the United States are in the dilemma of meeting the need for increased generating capacity within the constraints imposed by the need for energy conservation, costs saving, and environmental protection. The problem is compounded further by the normal variations in the daily and seasonal load profile which require that a spinning reserve be maintained for immediate response to an increase in power demand. The need for a spinning reserve forces the machine to be operated at some level of inefficiency which is easily translatable into a quantified energy loss. The fuel cell generator, however, performs at maximum efficiency over an operating range of 20% to 80% of rated capacity. Application of the fuel cell for intermediate and peaking loads, and for spinning reserve, would result in substantial reductions in energy consumption. The fuel cell generator is also an attractive option for increasing capacity at existing urban sites since the level of pollutants emitted is negligible and no impact on the environment can be expected.

The first commercial hardware planned, the FCG-1 Generator, nominally rated at 27 MW, consists of six modules, each nominally rated at 4.5 MWac. The goals for the FCG-1 power plant can be summarized as follows:

Commercial Introduction: 1980
Fuel: Naphtha, Clean Coal Fuels
Installed Costs: $250/kW (1975 $)
Heat Rate, Full Load/(Part Load): 9300/(9000) Btu/kWh
Life: 20 years (stack refurbished at 40,000 hours)

BACKGROUND

The development of fuel cell generators for electric and gas utility applications was initiated in the early 1960's. First efforts involved design of a family of on-site power plants capable of generating electricity efficiently, cleanly and economically from natural gas delivered by pipe to the point of electrical demand. This, termed the TARGET program, resulted in field installation and test of 65 experimental power plants which were operated for a total of over 200,000 hours in this country, Canada and Japan. The results of this demonstration encouraged studies of the feasibility of fuel cells of higher ratings as dispersed generators in electric utility systems. A comprehensive study of the engineering and economic viability of electric utility fuel cell generators was conducted in 1972 by United Technologies Corporation and a group of utilities under the auspices of the Edison Electric Institute. The results of this study included a preliminary specification for a 27 MW fuel cell generator, and an estimate of the cost and schedule requirements to proceed with the development of this generator for initial delivery to the utilities before 1980. A program jointly sponsored by United Technologies Corporation and nine electric utilities was initiated in 1973. The desired output of the program, demonstration of a 1 MW pilot plant in an electric utility, has been achieved.

It was realized, as the pilot plant program progressed, that the 1 MW test would not provide a basis for the utilities to confirm the provisional orders which supported the effort. It was concluded that the support required to gain more experience with large fuel cell generators in the utility environment, prior to production and deployment, was beyond the normal capability of the private sector. Simply, the investment required was too high and the payback period was too long. It became evident that government and utility industry support were essential if the conservation and economic benefits of the fuel cell were to be gained. This resulted in initiation of the 4.8 MW Fuel Cell System Demonstrator Program to design, fabricate and test one module of the FCG-1 power plant.

Demonstrator Program

On June 30, 1976, a contract was executed between the U.S. Energy Research and Development Administration (ERDA), the Electric Power Research Institute (EPRI), and United Technologies Corporation (UTC) for design and fabrication of the 4.8 MW (4.5 MWac net) Fuel Cell System Demonstrator. The cost of the two-year program, $42 million dollars, was to be shared by the participants as follows: ERDA - $25 million; EPRI - $5 million; UTC - $12 million. The effort was structured to result in the delivery of the demonstrator hardware during July 1978. Concurrent with the fabrication program, a Request for Proposal was issued in February 1977 to secure a utility site for implementation of the power plant in a normal operational mode, i.e., to share in the dispatch mix of a generating facility. The host utility has been selected and site preparation has been initiated. The costs of implementation will be shared by ERDA, EPRI and the host utility.

FCG-1 CHARACTERISTICS

The FCG-1 is designed for dispersed power plant applications, that is, small remote dispatched generators dispersed throughout the utility system in close proximity to the load. The concept of dispersed generation offers certain benefits, such as:

- Reduced transmission and distribution energy losses and deferred capacity expansion
- Reduced construction lead time through use of factory-assembled power plant modules
- Improved system reliability due to the inherent redundancy of many small, dispersed power plants.

The fuel cell, however, has certain characteristics that make it better suited to the dispersed generator application than diesel or turbine engine driven generator alternatives, with additional benefits to the utilities:

- Environmental acceptability - low pollutant and noise emission levels increase the number of sites available for dispersed generators.
- Efficiency - high efficiency in small power plant sizes permits incremental additions to meet load growth without sacrificing efficiency.
- Flat Heat Rate - higher efficiency at part load provides efficient spinning reserve capacity and economic dispatch of other power plants.

- Waste Heat Availability - energy is available as steam and hot water in quantities approximately equal to electrical output.

The performance that is expected of the 4.8 MW FCG-1 Module Demonstrator will be described below in terms of these characteristics.

Emission	Pounds per Million Btu Heat Output
NO_x	2×10^{-2}
SO_2	3×10^{-5}
Particulates	3×10^{-6}
Smoke	None
Noise:	Maximum 55 db(A) 100 feet from power plant perimeter

The emission level of pollutants shown above are at least an order of magnitude lower than conventional generators fired from natural gas. Within the power plant the naphtha fuel is first desulfurized (to protect the reformer catalyst), reacted with steam in reforming and shift conversion processes to form a gaseous mixture of hydrogen, carbon oxides, methane and water. Most of the hydrogen is electrochemically oxidized in the fuel cell; the hydrogen depleted mixture exiting the fuel cell is utilized in a burner to provide reformer process heat. The low adiabatic flame temperature of this hydrogen-depleted gas is instrumental in minimizing NO_x emissions.

The primary noise sources are the various ancillary components, such as pumps, fans, compressors, etc., needed for processing and controlling reactant streams. For lower noise emission levels, additional attenuation material would be required for the fuel cell and the output magnetics.

Efficiency: The heat rate curve for the 4.8 MW FCG-1 module is shown in Fig. 1. The sudden change in heat rate at approximately one-third load results from a turbocharging system. This pressurizes the power plant to approximately three atmospheres to reduce the size and cost of plumbing and heat exchanger components. The air required by the fuel cell and the reformer burner is compressed by utilizing the energy available in the hot reformer flue gas. As the power plant output is reduced, the flue gas alone (which is proportional to output) will not supply sufficient energy to drive the turbine below about 70% power, so additional fuel must be provided. At low loads, a separate low flow turbocompressor is used to maintain low fuel consumption for spinning reserve. Transition between low and high flow turbocompressors (and minimum to rated output electrical power) takes less than fifteen seconds. In installations of more than one module, the modules can be unbalanced by the dispatcher to yield a flat heat rate for the installation. This compares to 50% rated power heat rates of 10,000-12,000 Btu/kWh for diesel and 15,000 Btu/kWh for combustion turbine generators.

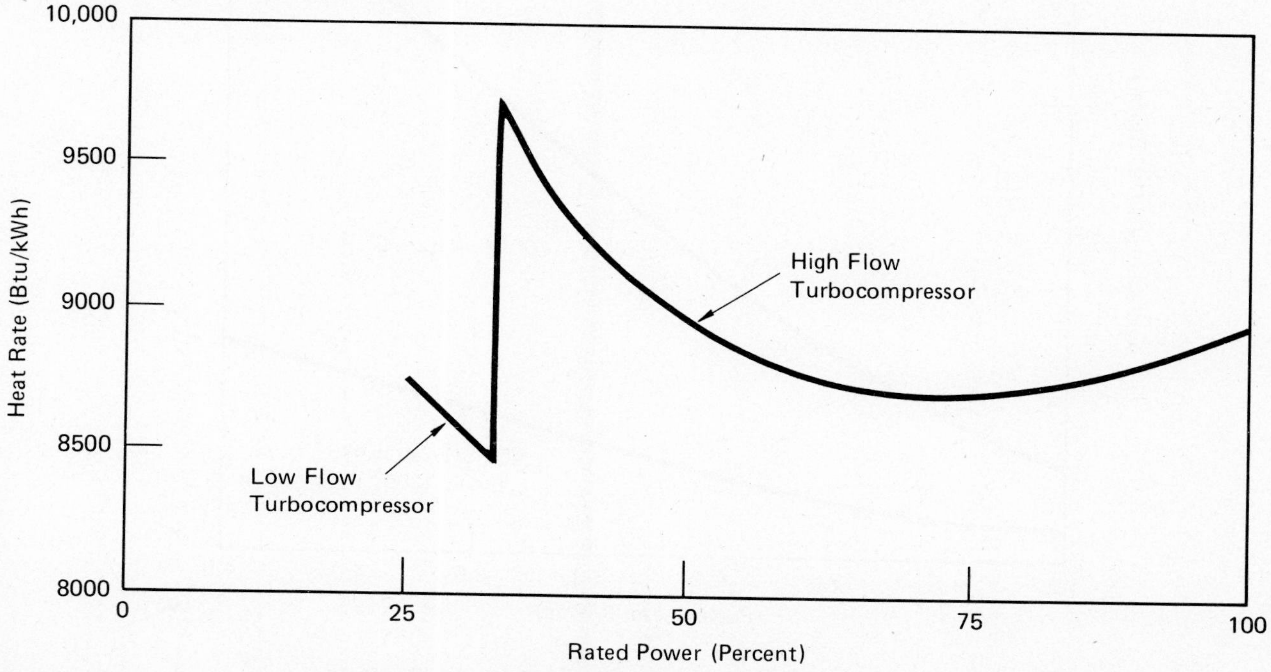

Fig. 1. FCG-1 Heat Rate

The fuel cell and fuel processing sections of the power plant are thermally coupled to maximize efficiency. The heat generated by electrochemical losses in the fuel cell raises steam which, in turn, is used in the fuel processing section. Steam in excess of that required by the fuel processing reactions is available for other uses as described in the next paragraph.

Waste Heat Recovery: There are two primary sources of waste heat from the power plant: First, the excess stack coolant (steam) that is not required by the fuel processor. Second, the latent heat of the steam that is the byproduct of the electrochemical reaction. In the 4.8 MW module, the stack coolant is available as 60 psig (approximately 300°F) saturated steam. The byproduct water is condensed within the fuel cell system to provide the makeup water for the fuel processing section. The condenser cooling water is available at 160°F. Figure 2 shows the total power plant waste heat as a function of power plant output. The heat availability is shown for a 20°F inlet to exit coolant temperature differential (i.e., 140°F coolant inlet). Slightly more heat is available at lower condenser return temperatures.

Two points concerning waste heat recovery from the 4.8 MW FCG-1 module should be made: First, the heat exchanger equipment required to collect the waste heat is an integral part of the power plant so its performance is not changed nor capital cost increased to implement waste heat utilization. If the waste heat is not utilized, it will be vented to the atmosphere via dry cooling towers. Second, the total efficiency of the power plant is 73%; 37% efficient in electrical output plus 36% in thermal energy output.

Load Management: The fuel cell power plant is an effective tool for load management of the utilities' cyclic intermediate and peaking generation capacity. This is due to the fuel cell's flat heat rate and fast response capability. The fuel cell can be dispatched to follow the utility system load fluctuations, with other intermediate or peaking capacity dispatched at their optimum operating points. The same characteristics permit the fuel cell to provide spinning reserve capacity efficiently. The value to the utility of the FCG-1 load following and spinning reserve capability has been estimated under EPRI Research Project 729-1, "Economic Assessment

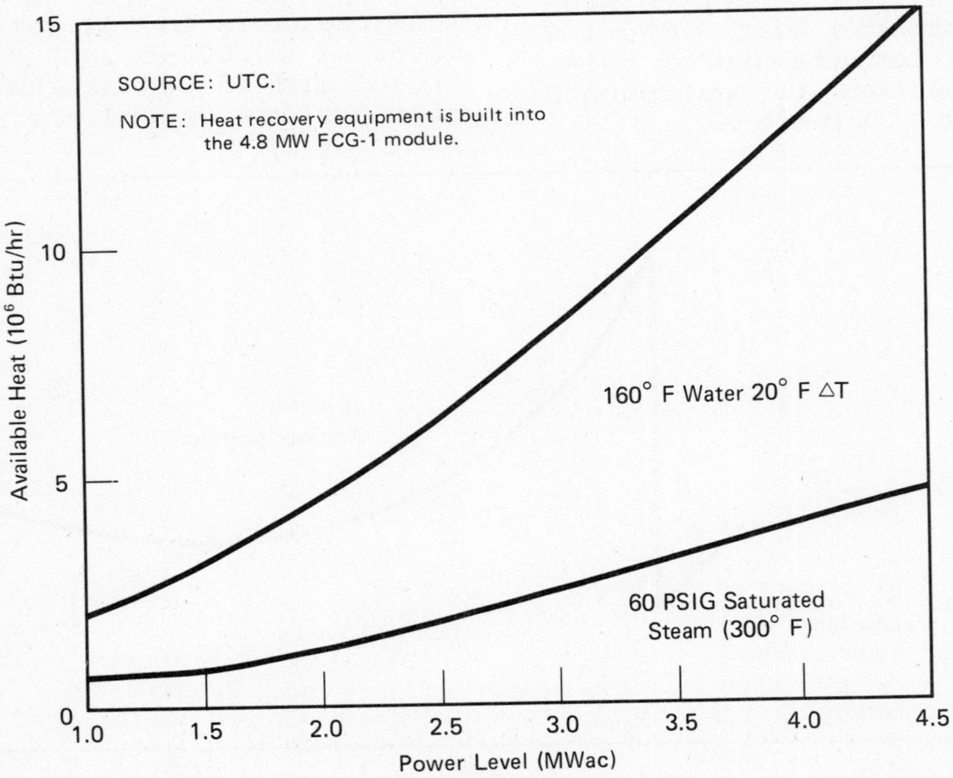

Fig. 2. FCG-1 Waste Heat Availability

of the Utilization of Fuel Cells in Electric Utility Systems." This assessment, conducted by Public Service Electric and Gas Company, Newark, New Jersey, estimated the value as a credit of at least $20/kW for the FCG-1.

Benefits

The economic benefits of dispersed fuel cell generators are being assessed for several electric utility system scenarios. EPRI project RP729-1 assessed the role of dispersed fuel cells in a typical large investor-owned utility. The role of fuel cells in small municipal and rural utilities is being assessed under RP919 by Burns and McDonnell, Inc., Kansas City, Missouri. The methodology used in both assessments is similar: Expansion plans are prepared for each utility for the years 1980-2000. The optimum mix (and location) of generating capacity is then determined by an iterative process for two cases; installation of conventional generators only, or installation of the optimum mix of fuel cells and conventional generators. The costs of generating and distributing power are calculated for each expansion plan based on minimum cost dispatch of installed capacity over daily and weekly load profiles. The optimum mix is, of course, defined by the point of minimum total costs. The various credits and penalties for dispersed fuel cells can then be assessed by comparing the capital and operating costs over the period for the expansion with and without fuel cells. The results of RP729-1 are shown in Table 1.

TABLE 1. Credits for Dispersed FCG-1

Description	Credit
Production cost including load following versus oil fired combustion turbine or intermediate capacity	$20-80/kW
Reduced reserve requirement	$16-42/kW
Reduced T&D	$29-66/kW

(Source: RP729-1 prepared by Public Service Electric & Gas, NJ)

The production cost credit represents the spinning reserve and load following credits described above under load management. Credits for reduced reserve reflect the improved system reliability due to many small, redundant generators installed on the system. Transmission and distribution credits represent capital cost savings by deferring T&D expansion and the savings due to reduced energy losses that accrue due to siting the power plant near the load.

It must be noted that the credits for reduced reserve and T&D were not included in the optimum mix calculations as these credits are very dependent on the characteristics of a particular utility;i.e., generation mix existing in 1980, geographic location, environmental constraints, land use constraints, etc. For these reasons, a specific utility may not derive all the benefits represented by the total of the three different credits.

Waste heat utilization is expected to provide even larger credits for dispersed fuel cells. Fuel cells operating in a cogeneration mode are being assessed by ERDA for onsite power in residential, commercial an industrial applications. EPRI will assess the role of Dual Energy Utilization Systems (DEUS) in electric utilities. By virtue of its efficiency, emissions, and modularity characteristics, the fuel cell is a most attractive candidate for DEUS and cogeneration implementation.

PROGNOSIS

The fuel cell power plant is the only viable near-term option that satisfies the stringent criteria for efficient dispersed power plants for the electric utilities. It is a more efficient conversion device since it is not carnot-limited; its waste heat can be utilized for cogeneration of process steam and district heating and it can be utilized as a load management tool. It has siting flexibility since it is non-polluting and, therefore, is not restricted by environmental concerns. It is capital conserving because of the short lead time between identification of need for increased capacity and installation and because of the reduction in transmisson and distribution costs. The technology is about to be demonstrated and the only deterrent to early commercialization and extensive penetration appears to be the availability of liquid fuels for power generation and the economics associated with initiating production of a new power plant option. Rational analysis leads one to the conclusion that the near-term future for fuel cell power plants is excellent.

HYDROGEN FUEL - A TECHNIQUE FOR ENERGY USE

Derek P. Gregory

Institute of Gas Technology
3424 South State Street
Chicago, IL 60616 USA

ABSTRACT

"Hydrogen energy" is a well recognized concept in which a raw energy source, such as coal, nuclear or solar, is used to split water, and the resulting hydrogen is used as the means of transmitting, storing, and utilizing energy. The cleanliness of hydrogen combustion and the economics of hydrogen delivery have been the primary advantages claimed. "Hydrogen energy" is not in use today primarily because of economic problems rather than technical shortcomings. Hydrogen is more expensive to produce from fossil fuels than are the synthetic hydrocarbons, and is more expensive to handle in aircraft and vehicle refuelling systems than conventional hydrocarbons. Its potential, however, as a residential fuel in novel appliances including catalytic combustors, fuel cells, and heat pump systems, is noteworthy because in these applications, together with in transportation applications, its higher end-use efficiency can make it compare well with electricity. The paper concludes that the time for "hydrogen energy" use has not yet come, but that it is an option that must be kept open for the future, especially when compared to other ways of utilizing the non-fossil energy sources.

INTRODUCTION

"Hydrogen energy" or the "hydrogen economy" is a concept that has become fairly well recognized in the last few years and usually means an energy-delivery system in which hydrogen is used to move energy from a nuclear or solar source to a variety of users, mainly those who are interested in using hydrogen as a fuel. Much has been written, many studies completed, and some experimental work commenced toward achieving the overall goal of implementing a significantly large use of hydrogen as a so-called "energy **vector.**" (1,2)

The economic, and sometimes the technical, justification for the use of hydrogen energy is often somewhat obscure for two principal reasons. First, the economics or efficiencies of some of the individual steps in the "hydrogen economy" - even the basic hydrogen-production step - are often disappointing and apparently unattractive; and second, there is always a tendency to compare hydrogen-energy economics with those of <u>today's</u> fossil fuels - an alternative that certainly will not be open to us 20 years from now when fossil fuel prices and availability will be radically different.

It is important, therefore, to put hydrogen energy into proper perspective, not only to compare it with alternatives but also to learn to consider the complete, overall energy system from raw energy source to useful energy output. This turns out to be a formidable task because there are so many options, optimizations, and compromises.

ADVANTAGES OF HYDROGEN ENERGY

There are three main driving forces toward considering hydrogen as an alternative fuel: The effectively infinite materials resource base (water); The cleanliness with which it can be burned (nitrogen oxides are the only possible contaminant); The superior economics of transmission, distribution, and storage when compared to electricity (this assumes that much of the present natural gas system can be used for hydrogen). In this paper, we are primarily concerned with the third of these advantages.

The long-distance transmission of hydrogen can, in principle, be carried out in pipelines similar to today's natural gas system. Economic studies (3) have shown that the cost of moving energy in such a system is of the order of 3.5 to 5.5 cents/10^6 Btu/100 miles compared to 100-600 cents/10^6 Btu/100 miles for overhead electricity transmission. The overall local distribution costs of natural gas are considerably lower than for electricity, so by analogy it has been inferred (1) that hydrogen distribution would be correspondingly lower (Table 1).

If these assumptions are correct, then, the primary justification for the use of hydrogen will be dependent upon its end use characteristics and on the economics of its production.

To date, most hydrogen energy research has concentrated on its production, because the known economics suggest that it will be considerably more expensive, on an energy basis, than both electricity and fossil fuels. Production costs for hydrogen from many different sources were estimated by Pangborn, et al.(4) in 1976, and are summarized in Table 2. Prospects for the reduction of both electrolytic and thermochemical hydrogen production costs are good, based upon anticipated developments in technology. The costs quoted in Table 2 for thermochemical hydrogen must be considered speculative, since no process plant has been designed yet.

END USE TECHNOLOGY - HIGHLIGHTS

Before we consider the long-range future of hydrogen, it is worth noting that even today hydrogen is one of the largest volume chemical commodities in use. The three largest consumers of hydrogen are ammonia synthesis, oil refining, and methanol synthesis, in that order. Together they comprise more than 95% of the current 2200 billion SCF annual hydrogen demand in the United States. The source of most of this hydrogen is the steam reforming of natural gas or, in the oil-refinery case, a by-product of catalytic reforming. If a non-petroleum alternative source of hydrogen could be found today, considerable savings of natural gas and imported oil would be possible.

TABLE 1 Relative Prices of Delivered Energy (Source: 1970 Data From the FPC and A.G.A.)

	Electricity	Natural Gas	Electrolytic Hydrogen
	$/million Btu		
Transmission	0.61	0.20	0.52*
Distribution	1.61	0.27	0.34

* Assuming pipeline hydrogen, at $3.00/million Btu used for compressor fuel in optimized pipelines compared to natural gas at $0.25/million Btu.

TABLE 2 Relative Production Price of Hydrogen from Various Sources

Process	Feedstock Type	Feedstock Price	Hydrogen Price $/$10^6$ Btu
Steam Reforming	Natural Gas	$0.6-0.65/$10^6$ Btu	1.4-1.5
Partial Oxidation	Residual Oil	$9/barrel	3.0-3.25
Electrolysis	Electricity	10-15 mill/kWhr	6.5-8.0
Coal Gasification	Coal	30¢/10^6 Btu	2.17
Thermochemical	Nuclear or Solar Heat	?	4.50 approx.

Hydrogen is also used as a direct reductant for iron and steel and as a feedstock for conversion of coal and shale to more useful gaseous and liquid fuels.

Hydrogen has been considered as a possible replacement for natural gas, either as an additive to existing gas systems or as a complete replacement. It is believed (5) that up to 10-12% by volume (only about 3-4% by energy) could be added to natural gas before problems would arise at the consumers' burners. This, however, represents a huge energy "market" of almost one quad (10^{15} Btu) per year. If complete replacement of natural gas was contemplated, the energy market for hydrogen is about 20 quads, but major conversions to every residential and industrial gas burner would have to be carried out.

Hydrogen has also been considered extensively as a transportation fuel. (6) In the land-vehicle application, the almost non-polluting aspects of hydrogen act as the incentive, while economics, on board storage, and fuel distribution are the problems. In aircraft use, the light weight of hydrogen acts as the incentive, while fuel distribution and economics become the problems.

Since the whole subject of hydrogen end use is too broad to cover in detail in this paper, the techno-economic aspects of five specific applications will be addressed. These are:

- Chemical Feedstocks
- Residential Fuel
- Fuel for Fuel Cells and Heat Pumps
- Automobile Fuel
- Aircraft Fuel

CHEMICAL FEEDSTOCK FOR "SURPLUS" ENERGY

A recent study (7) carried out by IGT for the Electric Power Research Institute surveyed this hydrogen market and addressed the possibility of supplying it with hydrogen from off-peak nuclear or coal-based electric power. It was somewhat surprising to find that, today, there really is not a sizable supply of low-cost off-peak electricity; most of the peaking capacity is supplied by low-capital-cost, high-fuel-cost plants. However, this will change in the future as nuclear plants take on a larger share of generating capacity; an economically attractive use for off-peak power could stimulate a different generating mix than is in place today. One interesting existing potential supply of low-cost interruptible power is the spinning reserve that every power pool must keep operating, though not generating, to meet emergencies within seconds of a major plant failure. It is difficult to estimate the cost at which this reserve of power could be made available on an interruptible basis, but it is thought to be on the order of 5 mill/kWhr. Again, the absolute quantity is limited to 3% to 5% of the on-line generating capacity.

Another potential source of inexpensive off-peak electricity is hydropower systems that are not endowed, geographically, with large water-storage capability. Such systems are built with generators sized to meet maximum demands, but in periods of lesser demand water must be allowed to flow through spillways, representing a potentially low-cost source of power if seasonal use could be found for it.

If off-peak power is available for a relatively short period each day (say 8 hours), or for a short fraction of the year on a seasonal basis, the cost of hydrogen produced by electrolysis becomes quite high because of the need to pay for an electrolyzer that is standing idle for long periods of time. Additionally, intermittent production coupled with constant use requires the incorporation of hydrogen storage - with resultant penalties in cost and efficiency. Figure 1 shows the manufacturing cost of electrolytic hydrogen versus plant factor at 5 and 15-mill power for a presently available Lurgi electrolyzer. Note that hydrogen from full-time operation at 15-mill power (i.e., base-load operation) is cheaper than that from 5-mill (off-peak) power at 30% (7 hours/day) operation even without storage costs, which are also heavily dependent on duty cycle.

Since firm data are not obtainable for the cost and availability of off-peak power, the data in Fig. 2 have been prepared parametrically to show the cost of electrolytic hydrogen as a function of power cost for three different plant factors. The electrolyzer in this case is an advanced-technology unit predicted by General Electric to be available by 1985. For comparison purposes the figure also shows the feedstock prices that would produce hydrogen by conventional means at comparable prices.

As an example, a 50% plant factor at 5 mill/kWhr power gives a hydrogen price of $2.00/1000 SCF ($6.15/million Btu). Hydrogen can be produced at this price from natural gas at $3.80/million Btu; from naphtha at 42¢/gal; from heavy oil at $20/bbl; or from coal at $60/ton, in conventional, large-capacity plants.

There is considerable incentive for the hydrogen-using industry to maintain a close watch on the relative rates of movement of electric-power costs (especially on an off-peak or interruptible basis), on conventional feedstock costs and availability, and on advances in electrolyzer design that will tend to reduce capital costs and increase production efficiencies.

HYDROGEN AS A RESIDENTIAL FUEL

With estimates now being made for hydrogen-production costs in the $4 to $8/million Btu range, there seems little incentive today to consider replacement of natural gas with hydrogen for residential use. A simple calculation (1) of the estimated delivered cost of hydrogen, based upon typical natural gas transmission and distribution costs and escalated to account for the different physical properties of hydrogen, yields delivered-hydrogen costs in the $4.85 to $8.85/million Btu range. Typical residential sales prices for 1976 for natural gas were in the $1.00 to $3.70/million Btu range, with an average of $1.80/million Btu.

When comparisons are made with electricity as the alternative future residential-energy supply, a somewhat different picture emerges because the 1976 typical average delivered-electricity cost to residential users was 3.6¢/kWhr or $10.56/million Btu and rising rapidly. This, however, is an unfair comparison because the end-use efficiency of hydrogen is assumed to be considerably less than that of electricity. Thus, before we can reasonably begin to compare hydrogen with other energy carriers, and in particular with electricity, we must go through a detailed total-system analysis taking into account the cost and efficiency of each conversion, transmission, and utilization step. When we do this, the efficiency of the end-using appliance becomes significantly important to the calculation.

It is in this respect that the prospects for catalytic combustion of hydrogen in pilotless, unvented appliances become significant. Research efforts have been reported (8) in which such devices are being developed. Because of the peculiar combustion characteristics of hydrogen, it is possible to

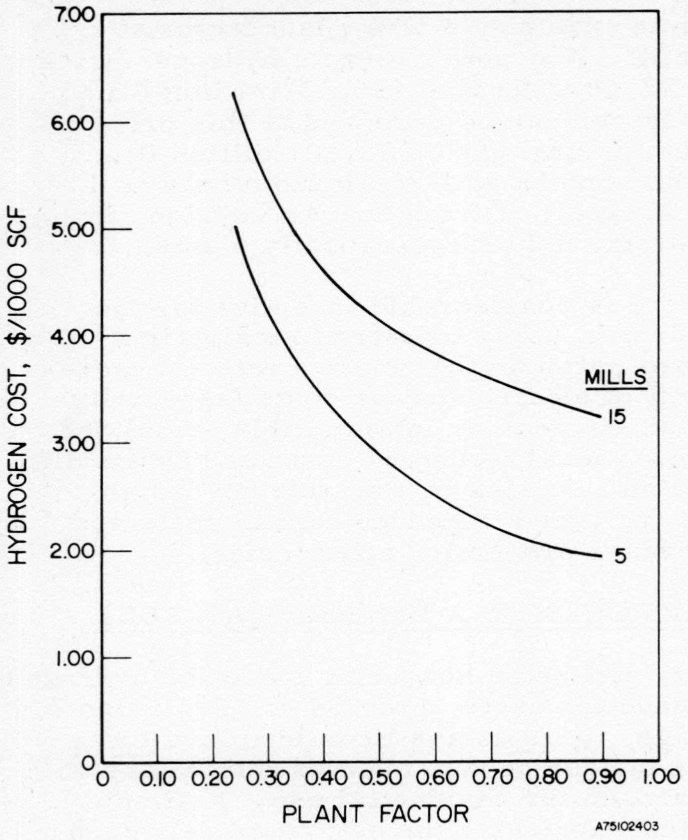

Fig. 1. Effect of plant factor on hydrogen-production costs for Lurgi electrolyzer system

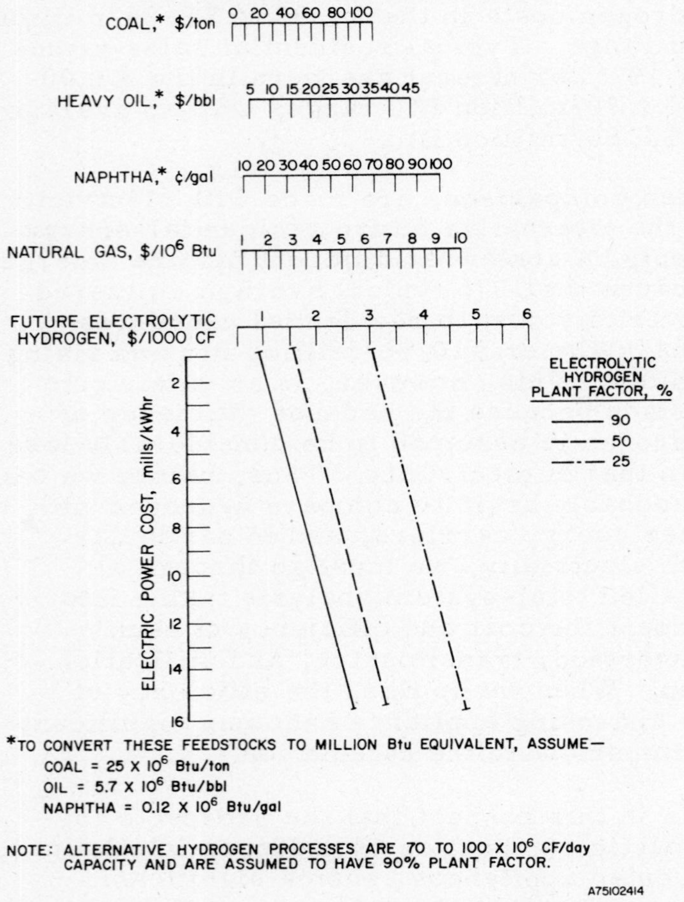

Fig. 2. Cost of electrolytic hydrogen, using "1985 technology" predicted by General Electric Co., with comparisons to conventional hydrogen-production costs

sustain smooth and controlled catalytic combustion of a hydrogen-air mixture at low temperatures - considerably below those at which a conventional flame will initiate. Because of this low combustion temperature (typically 200^0 to 800^0 F) no nitrogen oxides are formed, and no other exhaust product other than water is produced. Thus, provided humidity levels are maintained, a space or water heater can be operated without a flue (and without a pilot). In this way 100% of the high heating value of hydrogen is discharged into the house, compared with a 50% to 70% efficiency for a flued natural gas heater. Moreover, the energy losses due to free convection of room air through the conventional draft diverter, even when the furnace is not operating, are eliminated, and the initial construction cost of a chimney is not needed.

The efficiency merits of these unvented appliances have been discussed by Sharer and Pangborn, (8) who reported the results of bench-scale catalytic water heaters and space heaters. They obtained 75% efficiency in the water heater, pointing out that all of the heat released in the space heater could be delivered to the room air. However, since it might reasonably be argued that the high heating value of hydrogen can never be put to use in space heating, they used an efficiency figure of 84% corresponding to the ratio of the low to the high heating value of hydrogen, implying that nowhere in the indoor environment is water actually condensed. They then showed a single comparison of the overall efficiency of nuclear-heat-to-useful-energy in a number of systems shown in Table 3.

TABLE 3 Efficiencies for Various Energy Systems

System	Nuclear Heat to Fuel	Transmission Storage, Distribution	End Use	System Efficiency
		%		
H_2, today (electrolysis)	29 X 77*	95	65	14
Electricity, today	29	95	95	26
H_2, future (electrolysis)	45 X 95*	95	70(flame) 84(catalytic)	30 36
Electricity, future	45	90†	95	38
H_2, future (thermo-chemical)	55	95	70(flame) 84(catalytic)	37 44

* With current technology, proved nuclear-reactor/steam-turbine systems can generate electricity at 29% efficiency; electrolysis is 77% efficient. Futuristic values of 45% electrical generation and 95% electrolytic efficiency are expected.

† Due to the remote locations of nuclear power plants, line losses will be greater, causing a decrease in efficiency.

THE FUEL CELL AND THE HEAT PUMP

Two new end-use appliances are emerging from the development stage to commercial application that will have a considerable effect on overall system efficiency. The electric heat pump is already available, while engine-driven heat-pump technology is still in development. (9) Fuel cells are likely to become available commercially for onsite use within the next ten years. (10) The interaction of these devices with hydrogen energy will be interesting.

Fuel cells are presently under intensive research and development by the gas and electric utilities, by the United Technologies Corporation, and by the U. S. Energy Research and Development Administration. Two major fuel-cell types of interest are those using acid and molten-salt electrolytes. Each type requires pretreatment of conventional fuels to produce a hydrogen-rich gaseous feed to the electrochemical part of the system; the primary differences in pretreatment are due to each cell's ability to tolerate CO and CO_2. According to the Power Systems Division (PSD) of United Technologies Corp., (11) fuel-cell systems will exhibit overall efficiencies of about 40% in sizes of 40 kW and above and will have a capital cost in the $250 to $350/kW range. If pure hydrogen is available as a feed, the fuel pretreatment part of the system could be eliminated with resultant savings both in efficiency and capital cost. Fickett (12) estimates a cost savings of 25% and an increase in fuel-cell system efficiency to 49% by taking this step, pointing out that, with pure hydrogen, alkaline electrolyte systems having a capital cost between $100 and $150/kW can also be considered.

Whether or not hydrogen is considered as a fuel-cell feed, the inherent capability of an onsite fuel cell to supply heat in addition to electricity makes a significant impact on its overall efficiency. PSD (11) shows, as an example, that about 42% of the sensible "waste" heat from a 40% efficient fuel cell can reasonably be expected to be utilized, raising the overall system efficiency from 40% to 60% (Fig. 3).

The heat pump integrated with the fuel cell can have an even more dramatic effect on overall efficiency. The efficiency, or coefficient of performance (COP), of a heat pump depends markedly on the conditions under which it is operated, so it is impossible to generalize. PSD shows a sample system (Fig. 4) in which a fuel-cell/heat-pump combination has an apparent overall efficiency (or system COP) of 107% and a fuel input of 2.5 kWhr, producing a useful electrical-plus-thermal output of 2.69 kWhr.

If such a system were to be fed with hydrogen rather than a conventional fuel, a fuel-cell electrical efficiency of 50% would be achieved. Keeping the same assumptions used in the earlier example (a 75% waste-heat recovery efficiency, a 9% coincidence factor in waste-heat demand and supply, a heat-pump COP of 2.5, and a 4.0 to 1 ratio of thermal to electrical requirements), the data shown in Fig. 5 are calculated. This results in an apparent overall system efficiency (or COP) of 122%, significantly greater than the 107% for the hydrocarbon-fed example.

Further optimization would clearly have to be carried out if a fuel-cell/heat-pump combination is to be integrated with catalytic ventless appliances to meet some of the thermal requirements. In the case of hydrocarbon fuels some of the pretreated fuel, a hydrogen-rich gas suitable for feeding to the fuel cell, would be led off to feed a series of catalytic ventless appliances, while the remainder would feed the fuel-cell/heat-pump system. It has already been shown that catalytic water-heater and space-heater burners can operate satisfactorily on reformed natural gas (78.8% H_2, 20.2% CO_2, 0.33% CO) with minimal and acceptable emissions of CO and NO_x.

If a pure hydrogen-fuel feed is available, some of this would be supplied to catalytic burners and some to the fuel-cell/heat-pump system. In both the hydrogen and the hydrocarbon feed cases, the optimum split between sizing the fuel-cell/heat-pump system and the catalytic appliances is very complex and depends on many factors, including the type and location of the dwelling and the number and type of the gas appliances desired. To the author's knowledge, no treatment of such an optimization has yet been carried out.

HYDROGEN AS AN AUTOMOBILE FUEL

The conversion of conventional gasoline engines to operate on hydrogen has been successfully carried out by a number of researchers, and appears to be only a minor problem in the overall concept. Today there are several demonstration vehicles operating on hydrogen, using either cryogenic or metal-hydride storage. Pangborn et al. (13) surveyed the use of hydrogen as a vehicle fuel in comparison

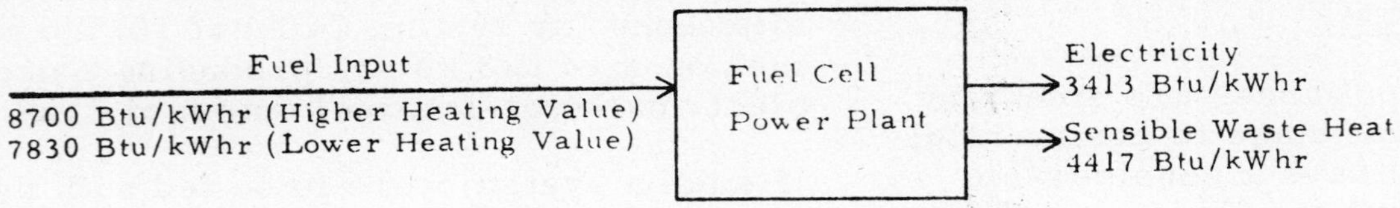

Fig. 3. Waste heat availability from a fuel cell operating at part-load (Source: United Technologies Corp.)

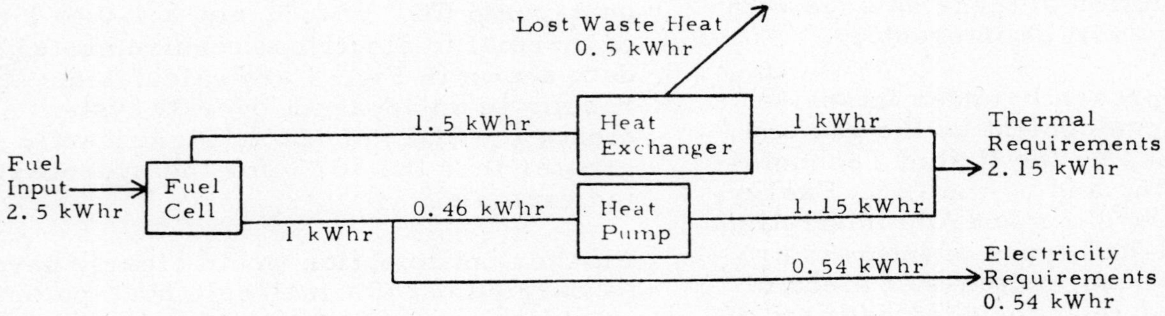

Fig. 4. Annual basis energy balance for a fuel-cell/heat-pump system, based on 1 kWhr of electricity production, hydrocarbon fuel input (Source: United Technologies Corp.)

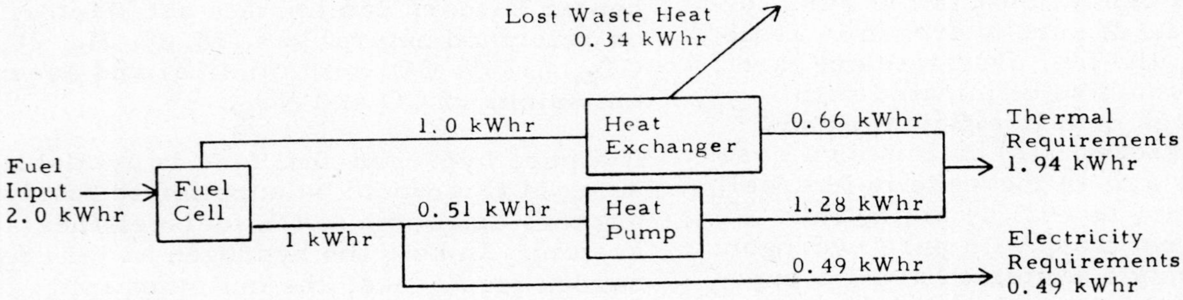

Fig. 5. Annual basis energy balance for a fuel-cell/heat-pump system, based on 1 kWhr of electricity production, hydrogen fuel input (Source: United Technologies Corp.)

with a number of other alternatives to gasoline and concluded that hydrogen was not among the best choices as long as some carbon-bearing energy source remains to be exploited. The primary problems with hydrogen vehicles are economic, not technical, and are related to the fuel delivery systems rather than the equipment on board the vehicle.

Table 4 shows Pangborn's estimates of delivered fuels at the filling station. It can be seen that, although the production costs of hydrogen from coal and from nuclear energy are not unreasonable, the transmission and delivery costs are very high

TABLE 4 Cost of Alternative Fuels on an Energy Basis

Resource Base, Synthetic Fuel	Resource Extraction and Synthesis	Refining or Processing	Transmission and Distribution	Total Cost
	1973 $/10⁶ Btu			
Coal				
Gasoline	1.10 ± 0.15	0.80 ± 0.10	1.10 ± 0.10	3.00 ± 0.35
Distillate	1.10 ± 0.15	0.45 ± 0.05	1.10 ± 0.10	2.65 ± 0.30
Methanol	1.50 ± 0.10		1.75 ± 0.20	3.25 ± 0.30
Liquid (SNG)	1.20 ± 0.25	0.90 ± 0.05 (liq)	1.10 ± 0.25	3.70 ± 0.45
Liquid H_2	1.55 ± 0.35	1.70 ± 0.10 (liq)	2.30 ± 0.20	5.55 ± 0.65
H_2 - Hydride	1.55 ± 0.35	Distributed as gas	3.55 ± 0.13	5.10 ± 0.50
Oil Shale				
Gasoline	0.85 ± 0.15	1.00 ± 0.05	1.10 ± 0.10	2.95 ± 0.30
Distillate	0.85 ± 0.15	0.55 ± 0.05	1.10 ± 0.10	2.50 ± 0.30
Nuclear Energy Water				
Electrolytic H_2				
Liquid - H_2	3.50 ± 0.30	1.70 ± 0.10 (liq)	2.30 ± 0.20	7.50 ± 0.60
H_2 - Hydride	3.50 ± 0.30	Distributed as gas	3.55 ± 0.15	7.05 ± 0.45
Thermochemical H_2				
Liquid - H_2	2.00 ± 0.25	1.70 ± 0.10 (liq)	2.30 ± 0.20	6.00 ± 0.55
H_2 - Hydride	2.00 ± 0.25	Distributed as gas	3.55 ± 0.15	5.55 ± 0.40
Solar Energy Agriculture				
Ethanol (200 proof) $1.00-$2.00/ bu corn	8.65 ± 2.15	0.30 ± 0.05	1.40 ± 0.20	10.35 ± 2.40
Methanol $1.15-$1.40/ 10⁶ bu pulp-wood chips	2.45 ± 0.20	0.25 ± 0.05	1.80 ± 0.25	4.50 ± 0.45

compared to conventional liquid fuels, because it was assumed that one could not take advantage of existing natural gas delivery lines in this case.

HYDROGEN AS AN AIRCRAFT FUEL

Hydrogen has been considered as an aircraft fuel for many years because it weighs only one-third of that of a conventional hydrocarbon on an equal energy basis. A hydrogen fueled aircraft flew in 1956 as part of a military research project for a long range mission requirement. More recently, NASA has coordinated a number of studies on the various components of a hydrogen civil aircraft system, and, like the automobile case, the problems turn out to be economic, rather than technical, and are related to ground equipment, rather than the aircraft itself. Witcofski (14) points out that hydrogen only becomes attractive as an aircraft fuel if either very long ranges, supersonic speeds, or absence of carbon-bearing fuel sources are considered.

The NASA studies have shown that liquid hydrogen, produced from coal or by electrolysis, is more expensive and produced less efficiently than either liquid methane or liquid "jet fuel" from coal. At the airport, liquid hydrogen storage and delivery facilities would be expensive but not technically impossible. At the aircraft, the advantages of liquid hydrogen are only seen when long range flights are contemplated. The energy efficiency of carrying passengers using hydrogen exceeds that for jet fuel when ranges of more than 1,000 miles are considered, and at 4,000 miles, the energy savings also compensate for the lower efficiency of making hydrogen, rather than synthetic jet fuel, from coal.

CONCLUSIONS

Hydrogen energy, in principle, is a "universal" energy carrier which, by analogy with electricity can be coupled with many different fossil and non-fossil energy sources and could provide a storable and transportable fuel for many applications, some of which cannot be adapted to electricity.

The economics of using hydrogen as a fuel, or non-fossil hydrogen as a feedstock, do not look attractive when compared to today's conventional alternatives. It is primarily for this reason that hydrogen is not in use today as an energy carrier. The economic advantages of delivering hydrogen, rather than electricity, will depend largely on being able to take advantage of the in-place natural gas delivery system, so that one cannot expect hydrogen and natural gas to serve as energy carriers at the same time.

There are a number of special characteristics of hydrogen that could serve to justify its use as a fuel, especially because of a potentially higher end-use efficiency. The combination of catalytic combustion and the resulting elimination of chimneys and flues make hydrogen appealing as a residential fuel. The prospects for operating fuel cells, either alone or in combination with heat pumps, are encouraged by the promised higher efficiencies compared to conventional equipment; hydrogen as a fuel makes the fuel cell even more superior. The advantages of hydrogen as an engine fuel for either automobiles or aircraft include higher end-use efficiency, but are not enough at this time to outweigh the excessive costs of handling hydrogen at the refueling point.

It appears, therefore, that the time for the widespread introduction of hydrogen as a fuel has not yet arrived. Because of the rapid changes in energy source costs, in energy transmission and delivery costs, and in the need to emphasize fuel conservation, the justification for hydrogen is a rapidly moving target. The use of hydrogen energy is an option that should be kept open, and the "energy community" should now be taking steps to be ready to move in the direction of hydrogen utilization when the right time approaches.

REFERENCES

(1) Gregory, D. P. et al., A Hydrogen-Energy System, 6. Arlington, VA: American Gas Association, Cat. No. L21173 (1973).

(2) Kelley, J. H. et al., "Hydrogen Tomorrow", Report of the NASA Hydrogen Energy Systems Technology Study, JPL 5040-1, Jet Propulsion Laboratory, Pasadena, CA(1975).

(3) Konopka, A. J., and Wurm, J., Transmission of Gaseous Hydrogen, Presented at the 9th IECEC, San Francisco, CA (1974).

(4) Pangborn, J. B., and Donakowski, T. D. Hydrogen Automotive Fuel: Production and Delivery, Presented at the 1976 Combined Fuels and Powerplant Meeting of the Society of Automotive Engineers, St. Louis, MO.

(5) ERDA Ad Hoc Advisory Committee on the Use of Hydrogen as a Supplementary Fuel, unpublished communication (1977).

(6) Escher, W. J. D., Hydrogen-Fueled Internal Combustion Engine, A Technical Survey of Contemporary U.S. Projects, ERDA TEC 75/005, Escher Technology Associates, St. John, MI (1975).

(7) Biederman, N., Darrow, K., Jr., and Konopka, A., "Utilization of Off-Peak Power to Produce Industrial Hydrogen," Final Report No. EPRI 320-1 for the Electric Power Research Institute, Chicago, IL (1975).

(8) Sharer, J. C., and Pangborn, J. B., "Utilization of Hydrogen as an Appliance Fuel." Paper presented at The Hydrogen Economy Miami Energy (THEME) Conference, Miami Beach, FL (1974).

(9) Wurm, J., and Panniker, G. P. K., "Evaluation of Engine-Driven Heat Pump Systems of Small Capacities." Paper presented at the Symposium on Space Heating and Water Heating Organized by the International Gas Union, Paris (1976).

(10) Handley, L. M., "The Target and FCG-1 Programs," 5-8. Paper presented at the ERDA/EPRI Fuel Cell Seminar, Palo Alto, CA (1976).

(11) "National Benefits Associated with Commercial Application of Fuel Cell Powerplants," ERDA Order No. WA-76-3405, United Technologies Corp., Power Systems Division (1976).

(12) Fickett, A. P., "An Electric Utility Fuel Cell: Dream or Reality?" in Proceedings of the American Power Conference, 37, 1072-81, Chicago, IL (1975).

(13) Gillis, J. C., Pangborn, J. B., and Fore, J. G., Synthetic Fuels for Automotive Transportation. presented at the Spring Meeting of The Combustion Institute, Madison, WI (1974).

(14) Witcofski, R. D., "The Thermal Efficiency and Cost of Producing Hydrogen and Other Synthetic Aircraft Fuels from Coal. Proceedings of 1st World Hydrogen Energy Conference, 5C-3 to 5C-30, Miami Beach, FL (1976).

END USE APPLICATION OF CHEMICAL SYSTEMS FOR ENERGY TRANSPORT (NUCLEAR LONG-DISTANCE ENERGY)

R. Schulten, H.F. Niessen, M. Röth-Kamat[+], M. Walbeck [++]

[+]Institut für Reaktorentwicklung

[++]Programmgruppe Systemforschung und Technologische Entwicklung

Kernforschungsanlage Jülich GmbH, 5170 Jülich, Germany

INTRODUCTION

For the future energy supply, such energy-carriers, which meet requirements of sufficient certainty of supply, economy, environmental pollution and ease of utilization, will be of special significance. Here nuclear energy plays an important role. However, nuclear power plants have been utilized only for the generation of electricity up to now. On the other hand, however, approximately 3/4 of the end energy in the FRG is consumed in the form of heat, whereby the amounts for room-heating and process heat are nearly equal.

The application of the light-water reactors prevalent today in the heat market sector is limited to the region of low-temperature heat. A part of the process heat required there could be made available by thermal power coupling from nuclear power plants. It has to be considered here, however, that next to a limitation of the temperature region, the transport distance plays a considerable role. Due to technical and economic reasons, distances of 30 km should not be exceeded for the transport of hot water for long-distance heating. Considerably shorter distances result for steam (approx.10 km).

The system of nuclear heat transport via chemically bound energy described in the following is also suitable for application of heat at high temperature levels and an economic and low-loss transport of nuclear heat from high-temperature reactors is rendered possible.

PRINCIPLE OF NUCLEAR LONG-DISTANCE ENERGY

Nuclear plants, which aim at the production of long-distance energy, can be situated far away from the place of heat consumption, e.g. densely populated areas, if the nuclear energy is transported by means of "cold" chemical energy. This can be achieved by the combination of an endothermic chemical reaction taking place at the heat source and an exothermic chemical reaction occurring at the region of heat utilization (Ref.1). For such energy transport systems, a plurality of chemical reactions is conceivable in which the following conditions should hold good with respect to their possible application:

(1) reversibility of the chemical reaction system, i.e. no loss of reactant through irreversible subsidiary reactions;

(2) sufficiently large reaction enthalpy and as high a conversion as possible so that high energy densities result for the products to be transported;

(3) favourable temperature region for the forward reaction (i.e.for the endothermic reaction, up to 850 ^{o}C).Combined with (2), this leads to sufficiently high conversions for the exothermic back reaction in the interesting temperature region > 300 ^{o}C;

(4) the required catalysts should be available in sufficient amount and at low costs;

(5) use of strongly corrosive or toxic substances should be avoided; and

(6) availability of the utilized substances in large amounts and at low costs.

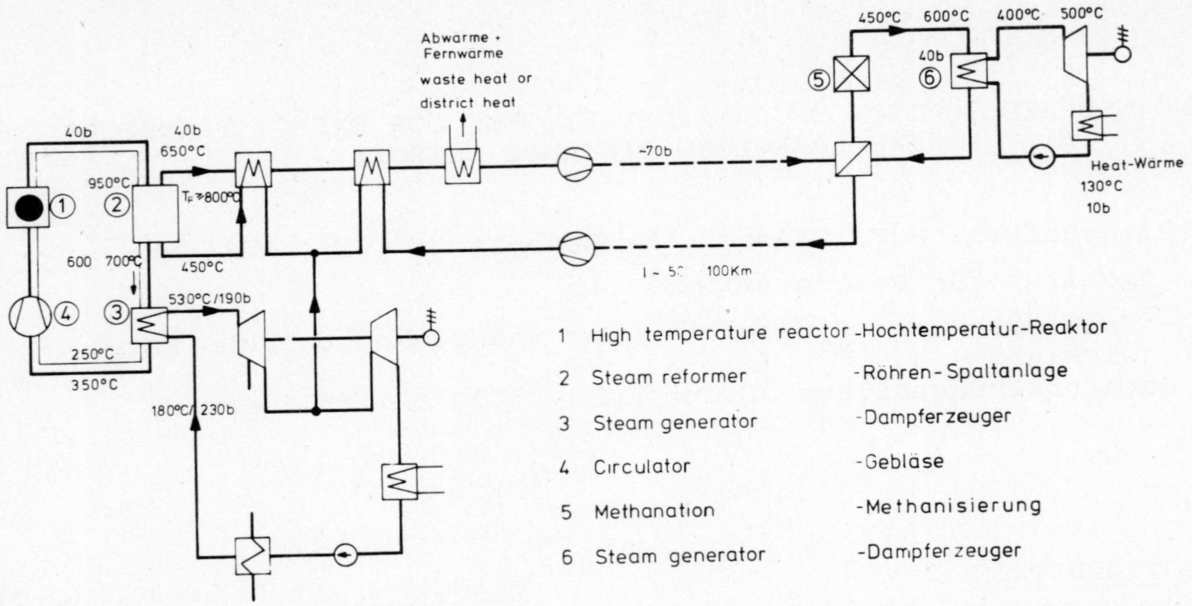

Fig. 1: Principle underlying nuclear long-distance energy (closed circuit system)

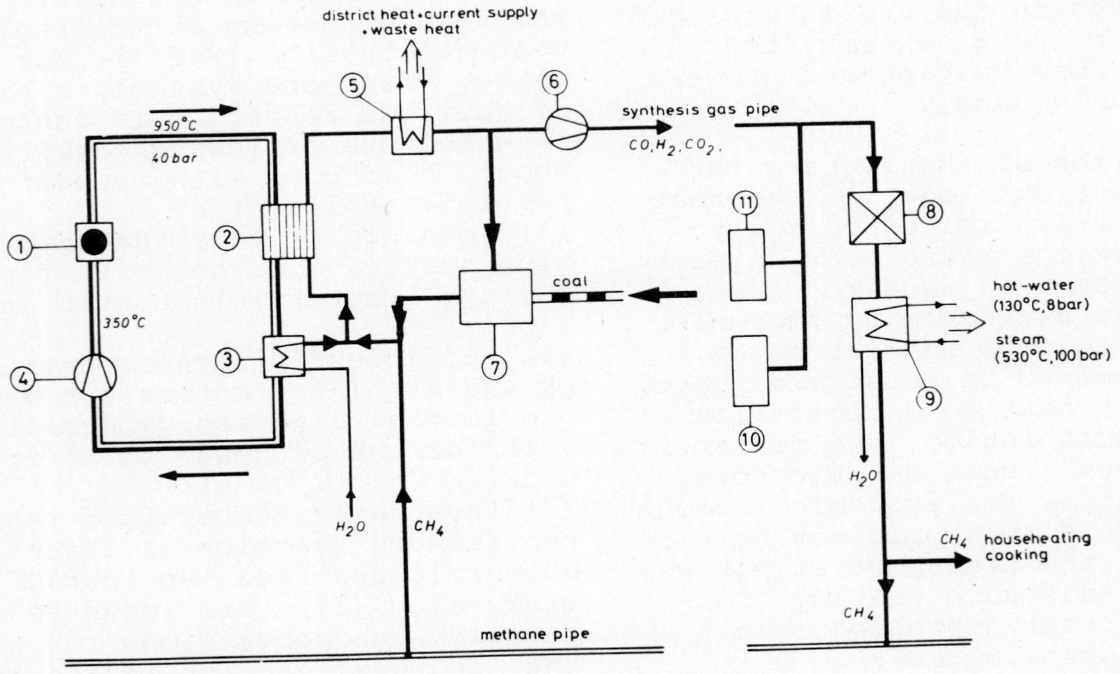

Key: 1.high-temperature reactor; 2.steam reformer; 3.preheater; 4 circulator; 5. waste heat recovery; 6.H_2, CO, CO_2 compressor; 7. coal gasification; 8. methanation; 9. heat exchanger; 10. direct reduction of iron ores; and 11. methanol synthesis

Fig. 2: Principle underlying nuclear long-distance energy (open circuit system)

A technically and economically interesting method of fulfilling the above conditions consists of a combination of the endothermic steam reforming of methane to "reformer gas" (H_2,CO,CO_2) and the exothermic methanation reaction (Ref. 2). This system described as "Nuclear Long-Distance Energy" has the additional advantage of being accomodated by the German Developmental Programme of Nuclear Process Heat, where steam reformers heated by HTR helium are required for hydrogenating coal gasification.

For the conversion of methane by steam, both the following parallel reactions have to be considered:

1) $CH_4+H_2O \rightarrow CO+3H_2 \quad \Delta H = 206$ kJ/mole

2) $CH_4+2H_2O \rightarrow CO_2+4H_2 \quad \Delta H = 165$ kJ/mole

The principle of this heat transport is illustrated by Fig. 1. The system consists of the following parts: Energy production (reactor), energy input (steam reformer), transport pipeline and energy output (methanation). The following detailed process steps occur:

1. Input (of the nuclear heat produced) by the endothermic chemical reaction of steam reforming of methane to synthesis gas.
2. After the usual waste heat utilization, transport of the cold energy-laden synthesis gas over large distances to consumer centres.
3. Output of the chemical energy in the vicinity of consumers by the exothermic chemical reaction of methanation of the synthesis gas. The reaction heat released is utilized for heating purposes, for industrial heat supply as well as for the generation of electricity.
4. After condensation and separation of the water formed in the reaction, transport back so that the cyle is closed.

Additional application possibilities can be attained when the cycle is tapped and synthesis gas is removed from it and used for chemical processes (e.g. methanol synthesis) or as reducing gas in the iron ore reduction plants. In this case, that portion of the synthesis gas which has been used up has to be replaced or substitution takes place by using a coal gasification plant (Fig. 2).

Apart from the above-mentioned extensive application possibilities of the system described and additionally of advantageously substituting fossil raw materials by nuclear heat with a view of providing for long-term energy planning, nuclear long-distance energy displays the following additional advantageous features considering the limiting conditions known today:
(1) considerable lessening of contamination emission in the production and consumption of energy;
(2) by means of a gas interconnecting system it is possible to ensure supplies and reserves;
(3) possibility of covering demands of day-time peaks by pipelines;
(4) already existing infra-structure of district heating can be utilized.
(5) heat provision known and tested at the consumer; and
(6) system easily expandible.

DESCRIPTION OF THE SYSTEM (Ref.3)

Energy Production (High-Temperature-Reactor) and Input (Steam Reforming)

As the described overall system is to be used for the transport of energy, the choice of the most important reaction parameters like pressure, temperature and the initial feed ratio of water/methane should be made according to the requirements of maximum energy utilization, and not according to the product gas composition as is the case in conventional steam-reforming plants. The known concept of the high temperature reactor with a helium outlet temperature of 950 °C and a circuit gas pressure of 40 bar supports the design of a steam-reforming plant with an operation pressure of 40 bar and a temperature of approximately 825 °C. Taking into account the maximum long-distance energy obtainable, an optimum feed ratio of water/methane seems to be 1.5-3 mole H_2O/mole CH_4. A typical example is given in Fig. 3

The energy required for the steam-reforming process is supplied to the gas mixture in the plant at a high temperature level. Helium is cooled from 950 to 600 °C due to the reforming reaction and superheating of the reactant gases from 450 °C to the maximum process temperature of 825 °C. Heat is also given up to the reactant gas in the reformer tubes from the inner gas return tubes.

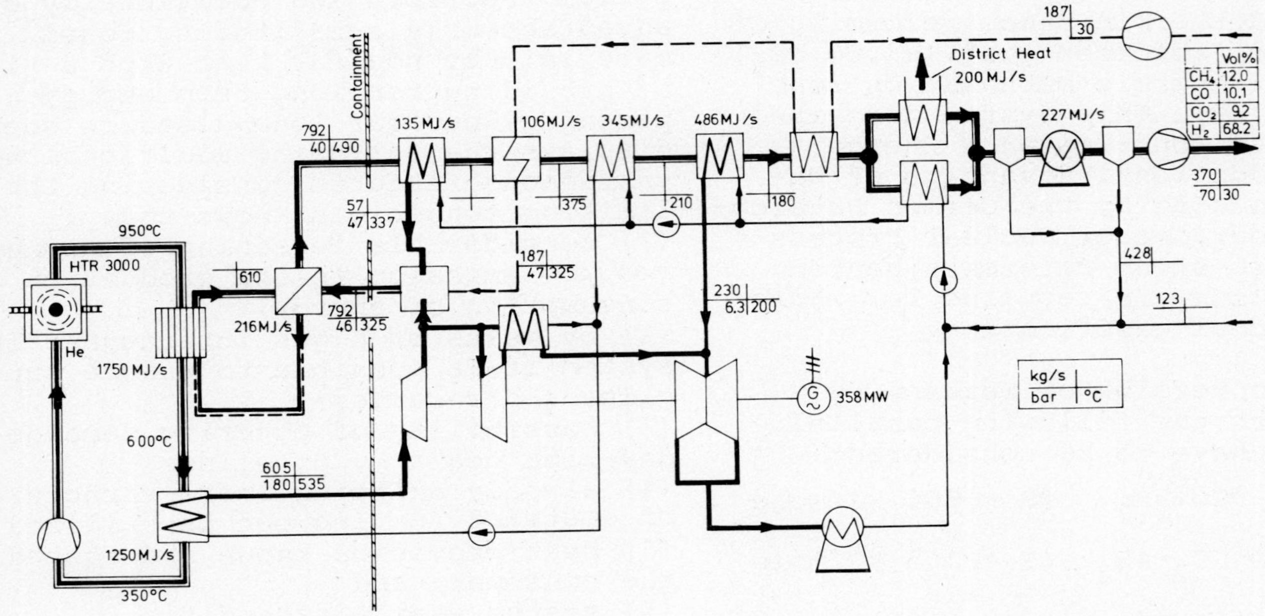

Fig. 3 Flow sheet showing principle underlying the production of nuclear long-distance energy, district heating and electricity

Thus the sensible heat between 825 °C and 610 °C is regained for the process in situ. The enthalpy of helium between 600 and 350 °C is utilized for the production of steam (180 bar, 535 °C).

This steam ist led to the high-pressure part of the turbine for electricity production and ≈ 90 % is led to the steam-reforming plant as process steam after mixing with methane and preheating to 450 °C by utilizing a part of the waste heat of the reformer gases. The remaining part of the steam is led to a condensation turbine with an intermediate superheating of low-pressure steam with medium-pressure steam. The sensible heat of the reformer gases in the temperature region of between 490 and 100 °C is used for the generation of process steam, for methane preheating, for the generation of low-pressure steam, for feed water preheating and for the generation of hot water for district heating. After that the reformer gases are cooled to 40 °C. The condensate is used as feed water for steam generation, whereas the gases are freed from this water. The product gas composition given here is for a steam/methane dilution of 3/1, a reaction pressure of 41 bar and a maximum process temperature of 825 °C considering an approach of 10 °C to equilibrium.

The overall data of the plant is:

High-temperature reactor	3000 MW
Nuclear long-distance energy	1730MJ/s
Gross electric current supply	358 MW
Internal consumption of helium circulator, feed water, compressors for pressure losses, etc.	108 MW
Net current supply	278 MW
District Heating	200MJ/s

The nuclear long-distance energy portion and the overall efficiency can be increased if the helium inlet temperature into the reactor would be raised from 350 °C to 400 °C or even to 450 °C.

Energy Transport

For the example shown here (Fig. 4), it is assumed that the amounts of reformer gas and methane considered are transported from the reactor site to the consumer centres. There the gas streams are split up according to the various large-scale consumers. These methanation stations can be utilized for the production of hot water for district heating or for process steam production. With respect to nominal diameters and transport capacity, and regarding the pressure in the case of the methane return pipe-line,the system corresponds to today's natural gas pipe-lines. As the density is about half that of natural gas, a higher transport pressure for the reformer gas pipe-lines could be optimum. However, the transport pressures usual today of approx.40-70 bar probably are not far removed from the optimum. For a mean transport distance of 70 km, a compression energy of approx. 70 MW is obtained for the amounts according to the article on energy production and input.

Energy Output (Methanation)

The cold reformer gas is transported to the methantion plants, where it is converted into methane and steam according to the back reactions of 1. and 2. The reaction efficiency, which is defined as the ratio of moles CH_4 in the product gas of methanation to the moles C in the reformer gas just before methanation, is shown as a function of reaction temperature and pressure in Fig. 5. Principally, the methanation reactors consist of adiabatic reactors with subsequent heat exchangers and cooled reactors. The design of the overall plant depends on the product desired, whereby the limiting conditions of the catalyst such as starting-up temperature, permissible maximum temperature and space velocity have to be fulfilled. According to today's technical status, maximum temperatures of 600 to 700 °C are permissible for starting-up temperatures of 250 to 300 °C. To attain a high reaction efficiency in the case of adiabatic reactors, generally a multi-step procedure is employed as is shown in Fig. 6. Cooled reactors (heat exchangers in the catalyst bed) are specially appropriate for the production of saturated steam.

As the heat is utilized mostly for the production of hot water or steam, the above-mentioned maximum temperatures are sufficient, especially when one considers that for a certain energy at a high temperature level, a corresponding amount of energy at a low temperature level always results. Thus a higher maximum temperature results in a larger temperature difference for heat transfer in the superheater, however, the average temperature difference for all the heat exchangers can certainly be smaller than for a lower maximum temperature, if it is assumed that there are the same number of successive methanation reactors. Fig. 6 illustrates a possible flow scheme of a methanation station. In the variant shown here,3 adiabatic methanation reactors are connected successively. The cold feed gas is preheated, and after being mixed with recycled product gas of the second step, it is then led to the first methanation reactor. The amount recycled is so chosen, that the desired temperature of 620 °C is obtained after the first methanation step. The product gas of the first step is cooled in the superheater and evaporator to 330 °C and led to the second step, where a temperature of 470 °C is obtained at the outlet. After being cooled in the evaporator, a separation of the recycle stream takes place. The remaining gas is led to a third methanation reactor, where is also reacts and is heated to 330 °C. The sensible heat of the product gas (mainly methane and water) is utilized for preheating feed water and feed gas. After cooling to 30 - 40 °C and separation of the condensate, the product methane is fed into the methane pipe-line and then led back to the energy production plant (steam reformer). The flow scheme chosen here is a good example of the various application possibilities and gives importance to a large flexibility. The following products can be obtained:

1) Net electricity production only	50 MW
2) Electricity production and long-distance heating	33 MW 88MJ/s
3) Net electricity production and low-pressure process steam (3,2 bar, 150 °C)	33 MW 144 t/h
4) Net electricity production and medium-pressure process steam (8 bar, 235 °C)	27 MW 144 t/h
5) Net electricity prod.& med.-pressure process steam(17 bar,315 °C)	21 MW 144 t/h
6) High-press.proc.steam(80bar,535 °C)	165 t/h

It has been emphasized that somewhat better values can be obtained for specific applications, as the plant described here shows all the application possibilities simultaneously and optimization has not been carried out. Similar optimization considerations have to be made regarding the inlet and outlet temperatures, the methanation reac-

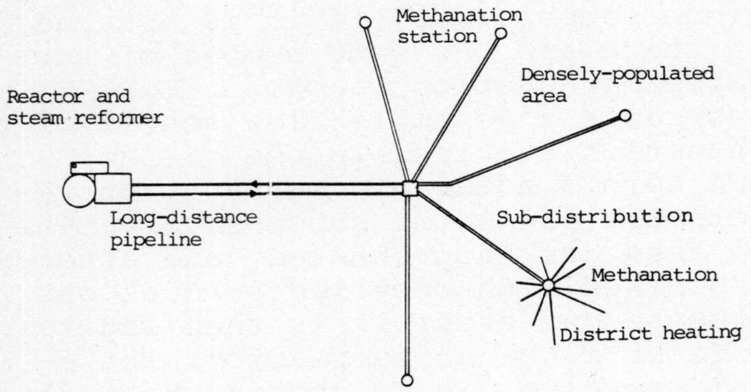

Fig. 4: Transport model for nuclear long-distance energy (closed circuit system)

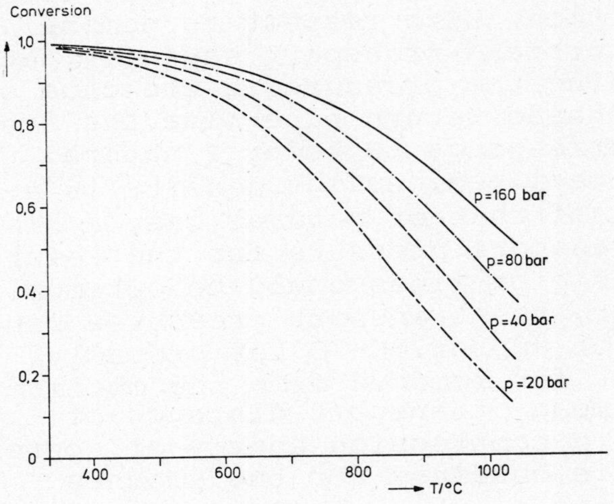

Fig.5: Dependence of reaction efficiency of methanation on reaction pressure and temperature

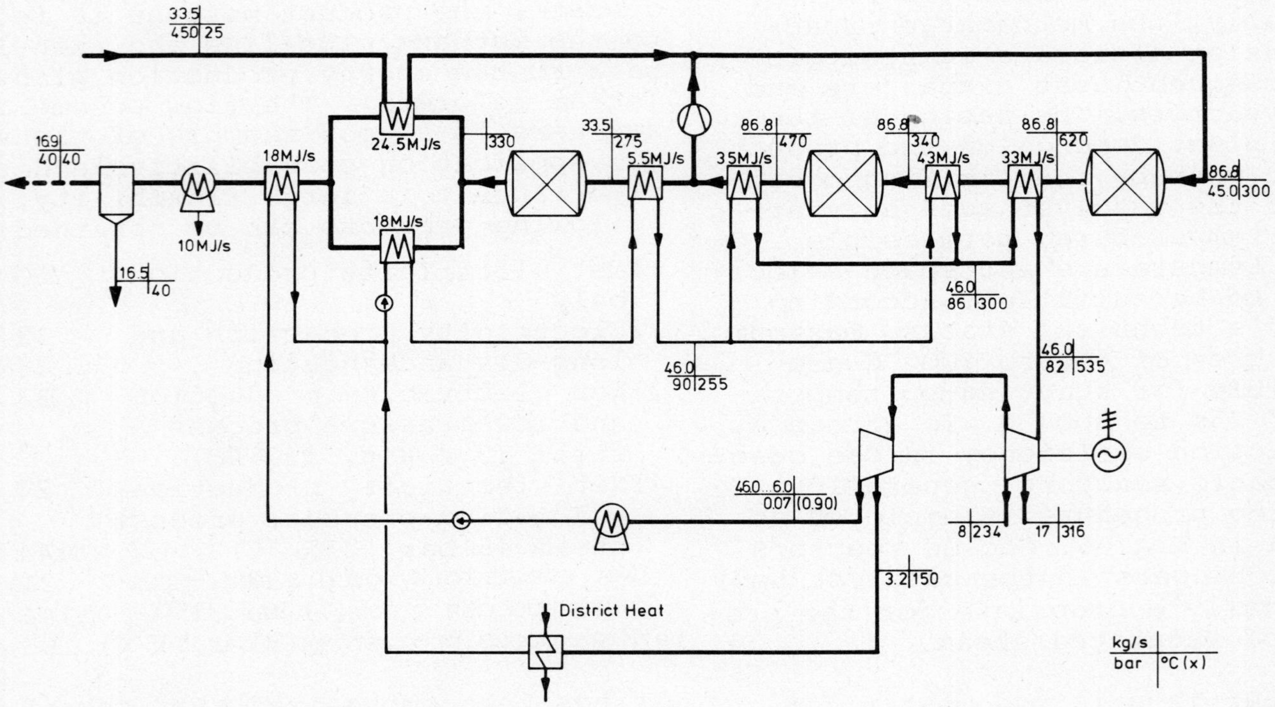

Fig.6: Flow sheet showing principle underlying a methanation station

tors as well as the position of the recycling.

APPLICATION POSSIBILITIES OF NUCLEAR LONG-DISTANCE ENERGY IN THE ENERGY MARKET

General Considerations

In investigating such a system, it has naturally to be tested whether the products can be disposed of in the market and whether the whole system can be introduced.

The products of the system are:

(1) Low-temperature heat (hot water): at the reactor site and at the methanation station.

(2) Process steam: at the methanation station.

(3) Electricity: at the reactor and methanation sites.

For the open circuit system, there additionally results:

(4) Synthesis gas: at the reactor (coal gasification) and, when the separation is undertaken by the customers, at the optional sites in the vicinity of the synthesis gas pipeline.

The customers for these products are:

(1) Long-distance heating enterprises: for low-temperature heat (long-distance heating) and partly process steam.

(2) Local electricity enterprises: local electricity production and long-distance heating.

(3) Individual industrial customers: for all products.

Corresponding to the production or service offered respectively by these customers their demands differ considerably chronologically. Parallel to the electricity demands, the room-heating demands in the long-distance heating supply drop also. A comparability of balance of load can be achieved by decoupling the heating and electricity consumption peaks by an overnight loading of the hot water network and interruption of the heat delivery to the hot water network when an electricity consumption peak occurs. This type of operation is practised in the FRG e.g. in Berlin, where there is no link with the electricity network. This principle could be interesting, however, also for other enterprises which do not have such an isolated "island" situation. Similar problems are posed in the annual demand characteristics as shown in Fig.7. In countries which have a considerable amount of air-conditioning, as e.g.USA, such a summer low does not occur.

The demand characteristics of industrial process heat customers are strongly dependent on the branch, as shown in Fig.8. It is remarked that the chemical industry, which is not marked in the figure, shows practically no annual fluctuations; at the most a conjunctural load could lead to fluctuations.

For a capital intensive system like the nuclear long-distance energy, attention has to be paid that the degree of utilization is as high as possible during the whole year.

A balance can be attained for the annual extreme fluctuations in the heating demands, when e.g. half the narrow pass capacity is met by already existing plants on a fossil basis (s.Fig.9). For the peak loads, the heat is obtained from fossil sources as is represented by the shaded area in Fig.9. As the peak loads occur for only a short time, they amount to only approx. 10 % of the overall heat consumption, although the installed capacity amounts to half the total capacity. This fossil coverage of the peak load enables the average utilization time of the nuclear plant portion to be increased to 5,400 h/a.

Investigation of a Supply Region

In order to find out whether such a system can be introduced and there is a large enough market for it, a study is being undertaken at present by the Kernforschungsanlage (KFA) Jülich commissioned by the Fed.Ministry for Research and Technology (BMFT) for a concrete region. A further goal of this study is to examine the economic aspects.

The region Cologne/Frankfurt was chosen for the study. It was selected because it contains all the main features of the other densely populated areas of the FRG. The results obtained here can therefore be extrapolated without much difficulty. In the Cologne region there are large-scale process heat customers in the chemical industry. Along the route of the future pipe-line, there are administration centres and in the Frankfurt region, there is a good mixture of large-scale and medium industries in various branches, service firms and residential districts. The special local

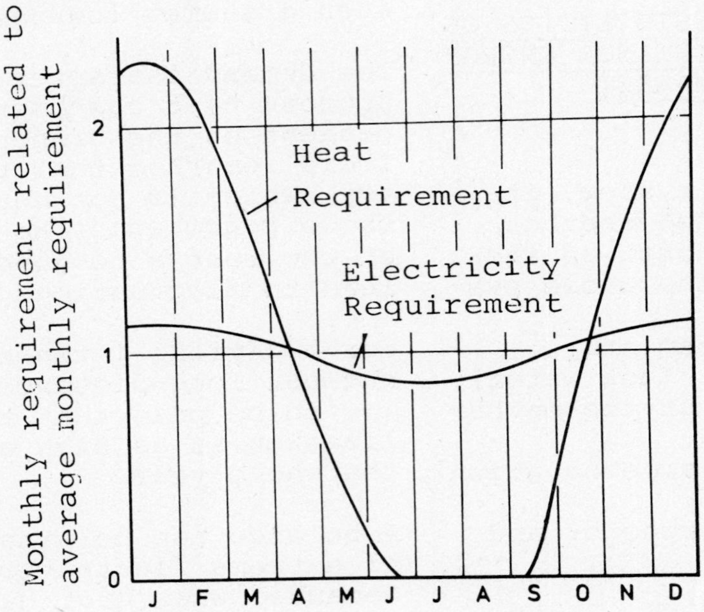

Fig. 7: Relative annual load curves for electricity and heat demands

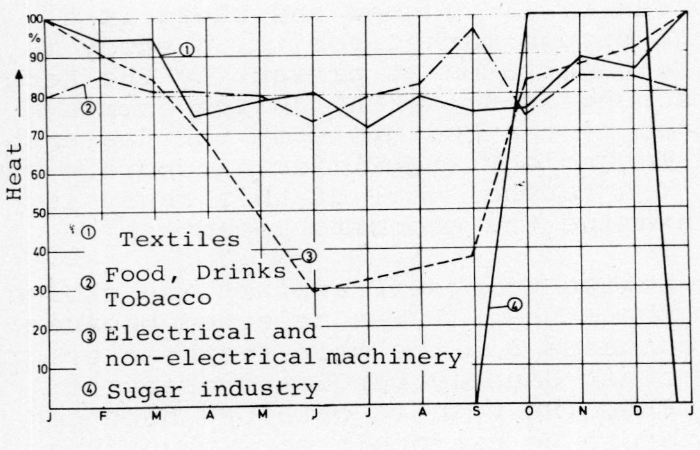

Fig.8: Consumption characteristics for various branches

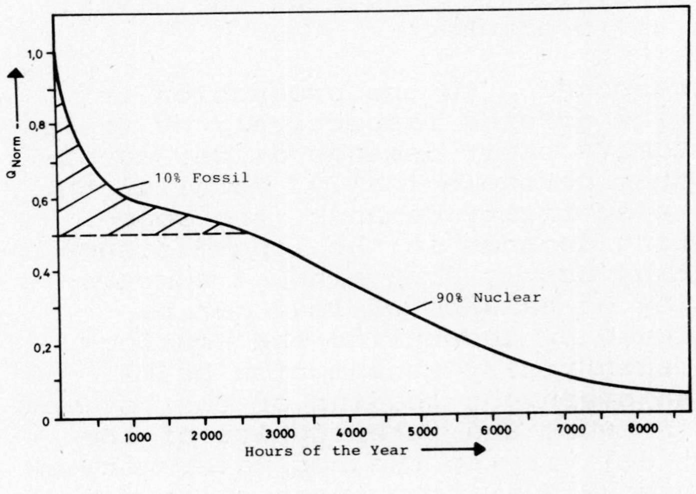

Fig.9: Annual characteristics for specific heat demands

features of the region are considered and recorded as individual information, so that they can be transmitted to other regions of the FRG with a definite aim in view.

Some preliminary results obtained in this study are given in the following. For a capacity installation of 50 % nuclear and 50 % on a fossil basis for the peak and reserves, the potential obtained in 1990 for room heating and water for consumption is:

Cologne	approx.	550 MJ/s
Bonn/Koblenz	approx.	400 MJ/s
Frankfurt region	approx.	850 MJ/s
room heating potential		1800 MJ/s

The results for the local electricity potential have not been obtained yet; however, from experience an increase of about 25 - 30 % could be expected so that an overall potential of 2200 to 2400 MJ/s is obtained for the local municipal region.

The values for the room-heating potential are limited mainly by the network enlargement rate (approx.5%/a). It appears possible to double the potential in 15-20 years'time. The values are also increased when a greater readiness to be connected with the pipe-line is attained by administrative measures (e.g. rational energy utilization, environmental pollution measures to keep the atmosphere clean), as the long-distance heating potential depends strongly on the readiness to be connected or expressed more exactly, the economy depends on the density of the potential consumers willing to be connected.

For process steam the following potential is obtained in the investigated region, if is is assumed that about 50 % of the willing large-scale enterprises are supplied:

Cologne	4400 t/h	($\approx$ 2000 MJ/s)
Frankfurt	1500 t/h	($\approx$ 700 MJ/s)

Furthermore, as is being tested in the study, a part of the long-distance energy gas can be released for material consumption. This variant leads to a further diversification of the possibilities under nearly indentical utilization of the same infra structure for supply and production. A demand of approx. $1.8 \cdot 10^9$ Nm^3/a synthesis gas is to be expected in Cologne region alone for non-energetic purposes.

The process steam or synthesis gas values respectively assume that the residual oils are further combustible and that hydrocracking is not economically interesting.

On the whole the global estimations show:
(1) It is possible to operate several high-temperature reactors, each with a power of 3000 MJ/s, in the investigated region.
(2) There is no necessity of deviating from the base load region ($\leqslant$6000 h/a) to enlarge the potential.
(3) Even for careful assumptions of the process steam potential, it is larger than the room-heating potential for economically interesting heating densities.

DEVELOPMENT

The development of the nuclear long-distance energy upto the technical maturity stage takes place in a research and development project "Nukleare Fernenergie" (Nuclear Long-Distance Energy), which is being jointly carried out by the Kernforschungsanlage Jülich GmbH and the Rheinische Braunkohlenwerke AG. The status of development can be characterized as follows (Ref.5).

Steam reforming is developed worldwide in a large number of conventional ammonia-, methanol- and hydrocracking plants. By the operation of the pilot plant EVA (since 1973), utilizing a single full-sized tube, the following could be shown:
(1) The helium-heated process is feasible.
(2) A maximum temperature of helium of 950 °C is sufficient to obtain heat fluxes through the tube walls similar to that in conventional plants (50 000-60 000 kcal/m^2h).
(3) The process is only limited by the heat transfer possibilities and not by the reaction velocity. The theoretical equilibria are fulfilled.
(4) The inner gas return tube has advantages in the direction of outlet temperature and energy consumption.

While the work done until now has shown that the helium-heated steam-reforming process itself is feasible, we now have to demonstrate the feasibility of a steam reformer bundle of several tubes and the direct coupling with a methanation.

On the one hand, the pilot plant EVA will therefore have a corresponding methanation. On the other hand, a new large-scale experiment SUPEREVA (with 30 helium-heated tubes) is being constructed. In this new experiment also the steam reformer will be directly connected with a methanation plant.

We hope that from the operation of this plant a lot of technical know-how can be gained:
(1) The mechanical behaviour of different types of reformer bundles (arrangement of tubes, helium ducting, support, operation behaviour, simulation of accidents).
(2) Measurement of hydrogen permeation rates.
(3) Metals in helium with impurities (10^4 h).
(4) Change of catalyst.
(5) Testing of the methanation plant.

Kinetic experiments on methanation catalysts are additionally being carried out in order to find out the optimum conditions for each catalyst respectively.

ACKNOWLEDGEMENT

The work described in this paper is being performed within the framework of the project "Nukleare Fernenergie" between the Kernforschungsanlage Jülich GmbH, Jülich and Rheinische Braunkohlenwerke AG., Köln, and in a study by the Kernforschungsanlage Jülich GmbH, Jülich, "Planstudie über halboffene Fernenergie-Versorgung insbesondere für den Raum Frankfurt a.M. mit Heißwasser-Fernwärmeversorgung für den Raum Köln", both sponsored by the Federal Minister for Research and Technology, FRG.

REFERENCES

(1) H.W. Nürnberg, G. Wolf, Energy Conversion Method, USA-Patent, 3 55 8047, Jan. (1971); DT-OS 160 1001 (1967)
(2) F. Hilberath, H. Teggers, DT-PS 1298233, Jan. (1968)
(3) K.Kugeler, H.F.Niessen, M.Röth-Kamat, D.Böcker, B.Rüter, K.A.Theis, Transport of nuclear heat by means of chemical energy (Nuclear long-distance energy), Nucl.Engg. Des. 34(1),65 (1975)
(4) Hauptbericht 1972 über die Fernwärmeversorgung im Bundesgebiet einschl.Berlin (West), Fernwärme International, Zeitschrift der Arbeitsgemeinschaft Fernwärme e.V. bei der VDEW, Sonderdruck Nr. 2597
(5) K.Kugeler, M.Kugeler, H.F.Niessen, R.Harth, Development of helium-heated steam reformers, in Gas-cooled Reactors with Emphasis on Advanced Systems, IAEA Vienna (1976).

DETERMINING THE ENERGY COST OF CONSTRUCTION ACTIVITIES

Barr Z. Segal

Center for Advanced Computation
University of Illinois, Urbana, IL 61801

ABSTRACT

Total (direct and indirect) energy requirements of the construction industry for 1967 were determined in order to examine the potential for energy conservation. An energy input/output model was expanded to include a detailed breakdown of the construction industry. Total primary energy intensities and total energy requirements were calculated for nearly 50 construction sectors, including both new and maintenance categories. Results were used in several specific analyses of energy use in construction, with particular emphasis on new buildings.

INTRODUCTION

Construction and operation of building and non-building facilities requires a huge amount of energy and accounts for a substantial portion of the nation's energy use. Conservation of energy in these activities could significantly reduce our overall demand for energy. Most efforts in evaluating potential energy savings in constructed facilities have focused on reducing operational energy consumption.* As we increase operational efficiency, however, the energy cost of initial construction activities becomes more important.

For this reason, researchers at a major university and a private architectural consulting firm* teamed up in order to determine just how much energy construction activities consume and how much could be saved not only in these activities alone, but also through tradeoffs between initial construction energy costs and lifetime operational energy costs. This paper describes how an energy input/output model (Ref. 1), augmented with detailed data on the construction industry, enabled us to analyze the total energy consumed in a variety of construction sectors, ranging from residential homes to highways. The model provided a "snapshot" of energy flows throughout the U.S. economy in 1967, the most recent year for which comprehensive data was available.

The phrase "total energy" used above has a special meaning. The energy consumed in any activity assumes two basic forms:

Direct Energy - fuels such as coal or refined petroleum consumed by the activity.
Indirect Energy - fuels consumed in other activities related in some way to the production of goods or services necessary for the given activity.

The total energy cost of an activity consists of both the direct and indirect energy consumed.

For example, in construction, the use of refined petroleum products for operating equipment or electricity to light a contractor's office represents direct energy consumption. The use of a steel beam, on the other hand, represents

---

*The Energy Conservation and Production Act of 1976 (PL 94-385) has stimulated this area by recognizing the significant operational energy cost of buildings and encouraging development of "energy conservation performance standards" for new and existing buildings at all levels of government.

*The Energy Research Group of the Center for Advanced Computation, University of Illinois at Urbana-Champaign, and Richard G. Stein and Partners of New York, New York, jointly conducted this research, which was sponsored by the U.S. Energy Research and Development Administration. The author is currently Research Coordinator of the Energy Research Group.

indirect energy consumption. The embodied energy of the beam consists of the energy consumed at each stage of the process leading to its own production. This includes not only the energy used directly by the steel manufacturer, but also that required for mining and extraction of raw materials, production and transportation of intermediate goods, maintenance of factories and offices, and all other activities which were related in some way to production of the steel beam.

The total energy concept is particularly important in examining the construction industry. We have determined that the construction industry consumed more than ten percent of the total U.S. energy requirement in 1967. Less than one-fourth of construction energy use, however, was direct. The great bulk of energy for construction was consumed indirectly, through materials and services used in the construction process. (Although these relationships are based on 1967 data, they are probably valid today as well, since the construction industry has not massively altered its mix of energy and non-energy inputs.) It is evident, therefore, that in order to understand how the construction industry uses energy and how such use might be reduced, both direct and indirect energy consumption must be considered. Hence, we used a model capable of determining both direct and indirect energy requirements for numerous economic activities, including many construction categories.

Our study consisted of two phases. The first, which this paper examines, involved altering an already existing energy input/output model to include detailed construction categories, and then producing some basic results not only for those categories, but also for the construction industry as a whole. The second phase made use of these results to examine some specific questions with regard to energy use in construction. (Although this paper focuses on the first phase and only briefly mentions some of the analyses conducted in the second phase, the complete study is described elsewhere (Ref. 2).)

THE EXPANDED ENERGY INPUT/OUTPUT MODEL

In order to examine total energy use in the construction industry, we augmented an energy input/output model (Ref. 1) with 1967 dollar flow data for 49 construction "sectors." Table 1 displays these sectors, which include 32 new and 17 maintenance construction categories. The categories were taken from a highly disaggregated breakdown developed by the Bureau of Economic Analysis (BEA), U.S. Department of Commerce. Their inclusion resulted in a model containing a total of 399 industrial sectors, which we called the "expanded energy input/output model," (or simply, the "expanded model").

A very brief review of the basic concepts of input/output modeling is in order before continuing. (For more details, see Ref. 3 and Ref. 1.) An input/output model is developed in three stages:

1) Transactions:
   Direct dollar transactions between industrial sectors are assembled for a given year. For each sector, total dollar inputs equal total dollar outputs.

2) Direct Requirements Coefficients:
   Inputs to each sector are normalized by the output of that sector. Thus if sector j purchased some of sector i's goods, the direct requirements coefficient gives the value of sector i goods required by sector j per dollar of sector j's output.

3) Total Requirements Coefficients:
   Using mathematical techniques based on classical input/output economic theory (but beyond the scope of this paper) the total (direct and indirect) requirements of each sector are determined per dollar of the consuming sector's output. Thus, a total requirements coefficient would reveal the direct and indirect input of sector i goods required by sector j per dollar of sector j's output.

The energy input/output model mentioned previously works in a similar way, but uses British Thermal Units (Btu) instead of dollar values for sales of five energy sectors: coal, crude petroleum, refined petroleum, electricity and natural gas. We compute total energy requirements coefficients, which we also denote as total energy intensities, as in basic input/output modeling. For each sector we obtain five such intensities, one for each energy type.* We then form a total primary energy intensity for each sector by combining the coal and crude petroleum intensities with the hydro

---

*Note that the use of total dollar requirements coefficients for energy sector inputs would be misleading, since energy is priced differently both by type (a Btu of natural gas generally costs less than a Btu of refined petroleum) and by consumer (large industrial users get price reductions). The use of physical units (Btu) for sales of energy sectors removes this problem and insures meaningful energy intensities.

TABLE 1 Construction Industry Sectors of Expanded Energy I/O Model

New Construction

1 Residential single family housing, non-farm
2 Residential two-four family housing
3 Residential garden apartments
4 Residential high-rise apartments
5 Residential alterations & additions
6 Hotels & Motels
7 Dormitories
8 Industrial Buildings
9 Office Buildings
10 Warehouses
11 Garages & Service Stations
12 Stores & Restaurants
13 Religious Buildings
14 Education Buildings
15 Hospital Buildings
16 Other Non-farm Buildings
17 Telephone & Telegraph Facilities
18 Railroads
19 Electric Utility Facilities
20 Gas Utility Facilities
21 Petroleum Pipelines
22 Water Supply Facilities
23 Sewer Facilities
24 Local Transit Facilities
25 Highways
26 Farm Residential Buildings
27 Farm Service Facilities
28 Oil & Gas Wells
29 Oil & Gas Exploration
30 Military Facilities
31 Conservation & Development Facilities
32 Other New Non-Building Facilities

Maintenance & Repair Construction

33 Residential
34 Other Non-Farm Buildings
35 Farm Residential
36 Farm Service Facilities
37 Telephone & Telegraph Facilities
38 Railroads
39 Electric Utility Facilities
40 Gas Utility Facilities
41 Petroleum Pipelines
42 Water Supply Facilities
43 Sewer Facilities
44 Local Transit Facilities
45 Military Facilities
46 Conservation & Development Facilities
47 Highways
48 Oil & Gas Wells
49 Other Non-Building Facilities

and nuclear portions of the electricity intensity. The total primary energy intensity of a sector thus gives the direct and indirect primary energy required per dollar of output.

(For the energy sectors themselves, total energy intensities give the direct and indirect energy requirements per Btu of output. Thus, the "energy cost of energy," that is, the energy consumed in order to deliver energy, is accounted for. Electricity is the most intensive energy sector requiring 3.8 Btu of total primary energy for every Btu delivered, due to power plant and transmission losses; coal is least intensive, requiring 1.007 Btu per Btu delivered.)

Returning to our construction energy use project, a crucial step in developing the expanded energy input/output model was the conversion of 1967 direct dollar transactions for the construction sectors into direct energy transactions. This involved collection of detailed information on the prices various construction sectors paid in that year for different types of fuels. We computed direct energy purchases by combining price data with total dollars spent on each type of energy by each construction sector. Table 2 summarizes the results for some aggregated construction categories. The construction industry purchased 1484.7 trillion Btu of direct energy in 1967, the great bulk of which took the form of Refined Petroleum Products. (Construction sectors made no direct coal or crude petroleum purchases in 1967.) The numbers in parentheses in Table 2 are percentages of row, or energy use, totals. They indicate that the pattern of direct energy use varies between building and non-building segments of the construction industry.

The incorporation of these direct energy results into the expanded model allowed computation of total energy intensities for the construction sectors. Table 3 shows the ten most energy intensive construction sectors in terms of total primary energy intensity. Most intensive are New Construction of Petroleum Pipelines (147,197 Btu/$) and New Construction of Gas Utilities (140,038 Btu/$; this sector also involves pipeline construction). This is probably due to the use of heavy construction and welding equipment and large amounts of energy intensive raw materials (steel, pipe, etc.). Table 3 also displays overall average energy intensities for New Construction (74,122 Btu/$) and Maintenance Construction (56,182 Btu/$), indicating the significantly

TABLE 2 Direct Energy Purchases by Construction Sectors - Aggregate Categories (1967, Trillion Btu)

(Numbers enclosed in parentheses are percent of row totals)

	ENERGY TYPE[a]			
	REFINED PETROLEUM	ELECTRICITY	NATURAL GAS	TOTAL[b]
NEW CONSTRUCTION:				
Buildings	415.15 (96.6)	4.28 (1.0)	10.39 (2.4)	429.78 (100.0)
Non-Buildings	785.27 (99.1)	2.18 (0.3)	5.28 (0.6)	792.72 (100.0)
MAINTENANCE CONSTRUCTION:				
Buildings	61.85 (96.6)	.63 (1.0)	1.58 (2.4)	64.06 (100.0)
Non-Buildings	197.09 (99.4)	.54 (0.3)	.54 (0.3)	198.16 (100.0)
TOTAL[b]	1459.36 (98.3)	7.64 (0.5)	17.71 (1.2)	1484.71 (100.0)

[a]There were no direct purchases of coal or crude petroleum by construction sectors in 1967.

[b]Rows and columns may not sum exactly to totals due to round-off.

higher cost of New Construction activity. These results served as the basis for a more detailed investigation of energy use by the construction industry.

TABLE 3 Ten Most Energy Intensive Construction Sectors in 1967

SECTOR	TOTAL PRIMARY ENERGY INTENSITY (Btu/$)
New* Petroleum Pipelines	147,197
New Gas Utilities	140,038
New Highways	123,745
Maintenance** - Petroleum Pipelines	117,158
New Oil & Gas Wells	116,895
Maintenance - Oil and Gas Wells	109,103
Maintenance - Farm Service	96,288
Maintenance - Conservation and Development	92,963
New Oil and Gas Exploration	92,941
New Other Non-Building	89,466
WEIGHTED*** AVERAGES:	
All New Construction (32 sectors)	74,122
All Maintenance Construction (17 sectors)	56,182

*Stands for "New Construction."

**Stands for "Maintenance and Repair Construction."

***Total energy intensities are weighted by Gross Domestic Output of each sector (Ref. 1).

APPLICATIONS OF EXPANDED MODEL RESULTS

To begin this investigation we combined energy intensities of the construction sectors with their 1967 dollar outputs (from BEA) to determine the total primary energy required by each category. Table 4 shows the ten construction sectors which required the most total energy in 1967. Also shown is the percent of each requirement which consisted of direct energy consumption. New Highway Construction required the largest portion of energy: 1035.87 trillion Btu, with nearly 40% of this for direct energy. Interestingly, New Residential 1-Family Construction was second, requiring 780.98 trillion Btu, but with less than 10 percent for direct energy. Note that this residential sector did not even appear in Table 3 - its energy intensity is only 55,511 Btu/$. However, due to the large amount of new residential construction activity in 1967, this category appears in Table 4 as a major "energy user."

Overall the construction industry required 7235.55 trillion Btu in 1967, representing 10.8 percent of the total U.S. energy requirement for that year. As noted earlier, less than 25 percent of the construction industry energy requirement was direct.

The results shown in Table 4 reflect the energy required for production of the total output of construction activities in 1967. In general, total output includes sales both to industrial sectors and to various final demand categories. Considering only final demand energy requirements, the construction industry accounted for 9.4 percent of U.S. energy use in 1967. New construction is defined as selling all of its output to final demand, so the small difference between final demand energy requirements and energy required for production of total output is due to inter-industry sales by maintenance construction sectors. Since such activity (repair of a steel mill roof, for example) is in fact construction, we allocated corresponding energy requirements to the construction industry, thus arriving at the higher energy requirement figure. An approach such as this, which would produce duplication in a study of several different industries, is sound for our analysis, in which only the construction industry is considered.

TABLE 4 Ten Construction Sectors Requiring the Most Total Energy In 1967

SECTOR	TOTAL ENERGY REQUIREMENT (Trillion Btu)	PERCENT DIRECT
New* Highways	1035.87	39.60
New Residential 1-Family	780.98	9.94
New Industrial Buildings	463.38	8.23
New Education Buildings	437.36	15.48
Maintenance** - Other Non-Farm	356.33	10.53
Maintenance - Residential	313.70	7.27
New Electric Utilities	303.94	12.69
New Residential Alterations & Additions	261.85	2.87
New Office Buildings	258.66	17.80
New Oil & Gas Wells	235.54	30.56
ALL CONSTRUCTION SECTORS	7235.55***	20.52

*Stands for "New Construction."

**Stands for "Maintenance & Repair Construction."

***Represented 10.82% of total U.S. energy requirement in 1967.

To set the stage for further analysis of energy use by the construction industry, each sector's total energy requirement was allocated among its direct purchases from all other sectors in the model and corresponding input energy fractions were developed. The resulting tables are too huge (nearly 40,000 figures) to include here. They do however, allow for relatively easy identification of the major contributors of embodied energy to the construction industry.

Table 5 gives an example of this type of information for new 1-family residential construction. The table shows major inputs to this sector ranked by the total primary energy embodied in their sales, and displays the fraction (and cumulative fraction) of this construction sector's total energy use which each direct input represented. As Table 5 indicates, refined petroleum products accounted for the largest portion of the energy required by new residential 1-family construction in 1967 - more than 11 percent. Interestingly, the top 21 energy contributing inputs accounted for more than 70 percent of this sector's total energy use. Adding another 24 input sectors to the ranked list in Table 5 would account for more than 90 percent of the total. Thus a great deal can be deduced about energy use in new single family residential construction by examining less than 50 direct inputs to the process.

A similar analysis of the remaining construction sectors revealed that building sectors and non-building sectors had different energy use patterns in 1967 with non-building sectors using much more direct fuels. We decided to focus our attention on new building construction for the remainder of our study. (In addition, we noted that buildings could serve ideally for some of our specific analyses.) We, therefore, combined certain construction categories (sectors 1 through 16, 26 and 27 in Table 1) into an aggregate New Building Construction Sector. Table 6 gives basic energy use results for this aggregate category, which required 62,671 Btu for every dollar of activity and accounted for over five percent of the total 1967 U.S. energy requirement.

These results enabled us to conduct several inquiries into the energy cost of construction activities. Brief summaries of a few of these analyses are included here in order to display the power and versatility of the energy input/output approach.

Energy Cost per Unit Building Material

We calculated total primary energy costs per unit for a wide variety of inputs to the

TABLE 5 Total 1967 Energy Requirements of New One-Family Residential Construction Ranked by Direct Input Sectors

INPUT SECTOR	TOTAL PRIMARY ENERGY TO CONSTRUCTION SECTOR (Trillion Btu)	INPUT FRACTION	CUMULATIVE FRACTION
Refined Petroleum	88.7	.114	.114
Ready Mix Concrete	67.5	.086	.200
Sawmill Products	53.5	.068	.268
Retail Trade Activities	35.3	.045	.313
Bricks	34.0	.044	.357
Asphalt	24.2	.031	.388
Prefab. Wood Structures	23.6	.030	.418
Wholesale Trade Activities	23.6	.030	.448
Sheet Metal Work	21.5	.028	.476
Veneer Plywood	21.0	.027	.503
Millwork	19.5	.025	.528
Misc. Professional Services	18.7	.024	.552
Cement	18.5	.024	.576
Gypsum Products	17.0	.022	.598
Railroad Transportation	15.3	.020	.618
Metal Doors	14.0	.018	.636
Concrete Blocks	13.2	.017	.653
Misc. Plastics	11.1	.014	.667
Steel Products	11.1	.014	.681
Building Paper	10.6	.014	.695
Paint Products	9.8	.013	.708
TOTAL (all input sectors)	780.98	1.000	1.000

TABLE 6 Summary of 1967 Energy Use in New Building Construction Aggregate (Sectors 1-16, 26, 27)

Direct Energy Use:	429.78 trillion Btu (96.6% Refined Petroleum)
Total Primary Energy Intensity:	62,671 Btu/$
Total Energy Requirement	3,421.6* trillion Btu (12.6% direct)

*Represented 5.1% of total U.S. energy requirement in 1967.

building construction process. By combining 1967 prices from the Census of Manufactures (CM), U.S. Department of the Census, with energy intensities from the expanded model we developed energy cost per unit figures for materials such as lumber, paint, roofing, glass, brick, concrete, steel products and many others.

In most cases, CM categories corresponded with those of the expanded model and represented products meaningful to building construction processes. In certain cases, however, the CM categories did not match well with the model, or did not sufficiently describe the input being investigated. We found the latter to be the case for certain wood window units and performed special "hybrid analyses" (combinations of process and input/output analyses) on such units (Ref. 4). This involved isolating the basic components of each unit, applying energy intensities from the expanded model, and adding in factors for energy required to assemble the unit and deliver it to the construction job site. Our investigation of several wood window units revealed wide variation in energy costs even though these products were all classified under the same CM category.

Overall, we found our results quite comparable to other studies of the energy costs of building materials, although no other studies appeared to have been as comprehensive as ours.

Energy Cost of Building Assemblies

Our knowledge of the energy embodied in many basic building materials enabled us to take our investigation a step further. We were able to determine the energy costs of particular building assemblies composed of some of the basic materials we had already considered. For example, by examining three typical floor bay systems of the same size and with the same performance criteria, we were able to show that proper selection of such assemblies could significantly reduce the energy embodied in the overall structure. Ultimately, we hope to extend this type of analysis in order to determine the energy cost of an entire building.

Life Cycle Energy Cost

Having noted the significant operational energy demand of building facilities, we hypothesized that an initial increment in construction energy cost in some structures might significantly reduce operational cost and lead to overall reduction in lifetime energy consumption. We examined a prototypical residential home, considering the effect of thicker windows and additional insulation on total lifetime energy costs. Assuming a 20-year lifetime for the structure, we found that the initial increase in construction energy embodiment caused by the extra glass and insulation was more than offset by the resulting reduction in lifetime energy demand. Application of this life-cycle approach in building construction could, therefore, significantly reduce the nation's overall demand for energy.

LIMITATIONS OF THE EXPANDED MODEL

The energy input/output approach has several limitations. One problem is that the most recent available input/output data was collected in 1967 by the Bureau of Economic Analysis. Data collection has been performed for 1972, but results will not be available for some time. Several methods exist for updating the 1967 study by focusing on the energy inputs to the various sectors in the model, but these are especially difficult at the 400-industry level. Fortunately, many results of the 1967 energy input/output model remain valid today because the mix of energy and non-energy inputs to the various sectors has not changed radically since 1967. The construction industry has been especially stable in this respect.

Another limitation of our approach is the fact that the input/output model is based on overall average inputs to each industrial activity. Resulting energy intensities, therefore, mask the different technologies used within many sectors. For example, although there were several major steel making technologies in 1967, the model can only reflect the overall mix of these methods in the production of steel nationwide. Analyzing these methods separately might reveal wide variance in their energy intensities.

Similarly, because the input/output model consists of dollar flows based on national averages it masks out regional differences. For example, determination of the energy cost of an actual or planned building would require careful consideration of regional characteristics such as the need for transporting materials from other regions. A detailed and thorough hybrid analysis (Ref. 4) would be essential for such an investigation.

CONCLUSIONS

In spite of the limitations mentioned above, we found the model to be an excellent tool for our investigation. Its ability to determine both direct and indirect energy requirements as well as its high level of detail (399 industries, including 49 construction categories) allowed us to develop some answers to two broad questions:

1) What role does construction activity play in the national consumption of energy?
Our basic expanded model results showed that construction activities accounted for over ten percent of the total U.S. energy requirement in 1967 with less than 25 percent of this for direct energy. We also determined that although construction of new petroleum pipelines was the most energy intensive construction category in 1967, highways and 1-family residences were actually the biggest energy users due to the volume of economic activity in these areas.

2) What are some approaches toward conserving energy in the building construction industry in particular?
Basic model results allowed us to calculate the energy cost of many common building materials. We were then able to analyze energy tradeoffs between different building assemblies with the same performance criteria, and to determine the tradeoffs between construction and operational energy costs of prototypical structures with different performance criteria.

Naturally, there are many additional applications for the expanded energy input/output model. We are continuing our computations of total energy costs for more basic materials and assemblies. We are also considering a life cycle analysis of a more complicated structure such as an office building, and an analysis of the energy tradeoffs (construction vs. operation) of solar home heating systems.

We are confident that our approach to energy conservation in construction activities can lead to effective measures to reduce total national consumption of energy.

REFERENCES

(1) R. A. Herendeen and C. W. Bullard, "Energy Cost of Goods and Services, 1963 and 1967," CAC Doc. No. 140, Center for Advanced Computation, University of Illinois at Urbana-Champaign, November 1974.

(2) B. M. Hannon, R. G. Stein, B. Z. Segal, D. C. Serber and C. Stein, "Energy Use for Building Construction - Final Report," CAC Doc. No. 228, Center for Advanced Computation, University of Illinois at Urbana-Champaign, March, 1977. (Will also be available as NTIS report.)

(3) "The Input-Output Structure of the U.S. Economy," Survey of Current Business, 54, No. 2, Bureau of Economic Analysis, U.S. Department of Commerce, (1974).

(4) C. W. Bullard, P. S. Penner, and D. A. Pilati, "Energy Analysis Handbook," CAC Doc. No. 214, Center for Advanced Computation, University of Illinois at Urbana-Champaign, October 1976.

ENERGY EMBODIED IN BUILDINGS AND BUILDING COMPONENTS

Diane Serber, AIA

New York State ERDA, Project Manager
Formerly Associate, Richard G. Stein and Partners

ABSTRACT

The derivation and expansion of an Energy Input/Output matrix with which to examine energy use in the construction industry has been described elsewhere. (1). This paper describes in greater detail the application of the matrix to the construction industry in 1967 and to specific building materials and components.

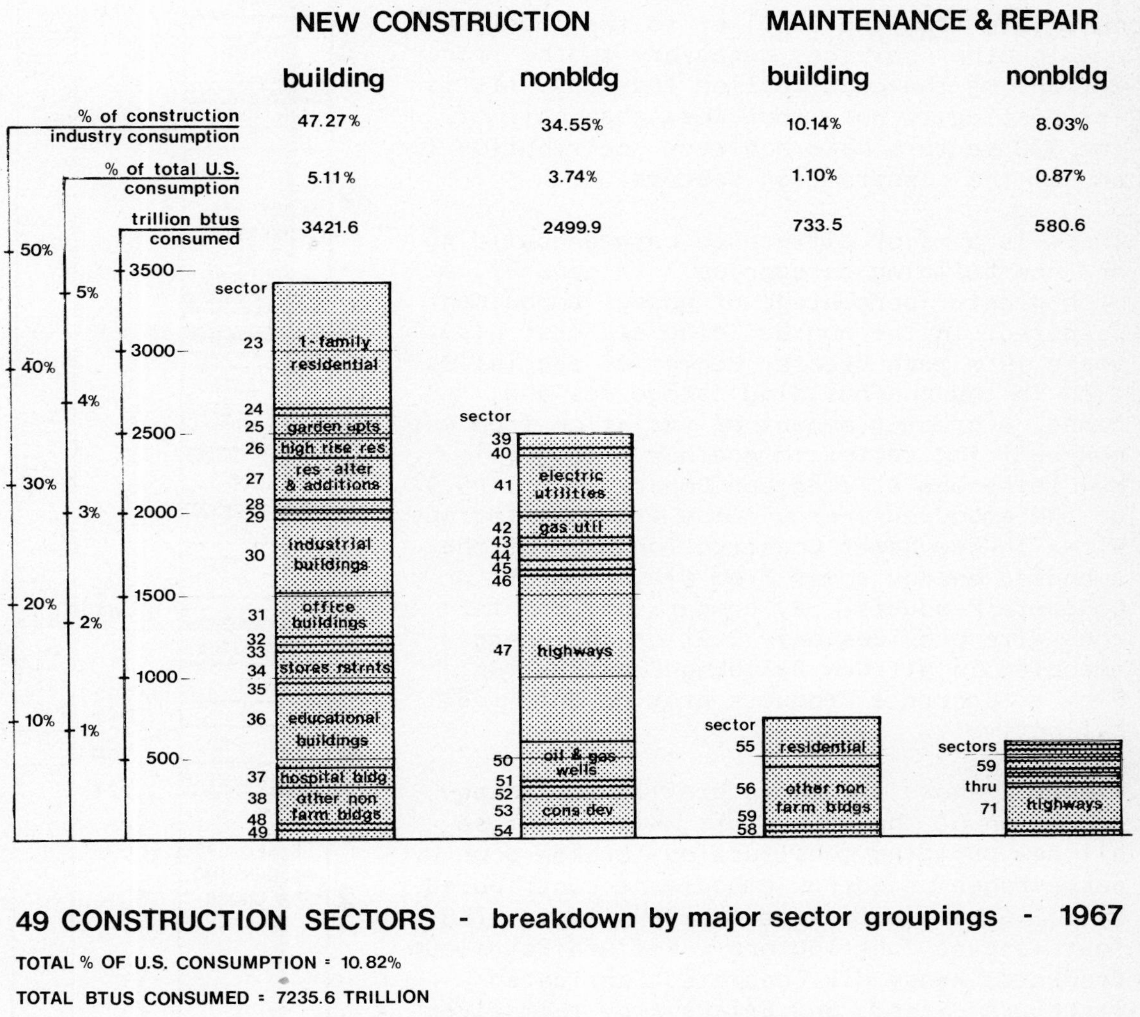

Fig. 1. Energy in construction in 1967

ENERGY IN BUILDING CONSTRUCTION

Overall Energy Patterns

Using an Energy Input/Output Model as a basis, (2) a detailed analysis of the energy used in the construction industry has been made (3). Although we have considered all areas of the industry, our main emphasis has been on new building construction.

Our model (which describes energy flows through the U.S. economy in 1967, the most recent year for which complete data are available), divides the economy into 399 sectors, 49 of which represent construction activity. Figure 1 outlines the gross energy consumption statistics of the industry. The construction sectors are arranged in four basic groups: New Building Construction (18 sectors); New Non-building Construction (14 sectors); Building Maintenance & Repair (4 sectors); and Non-building Maintenance

& Repair (13 sectors). Each column is subdivided into individual sectors, the largest of which are identified.

Further investigation shows varying patterns of materials and services use typical of individual sectors and groups of sectors. These are discovered through the breakdown of the energy embodied in each construction sector into its immediate contributors. i.e., the energy embodied in the direct energy (refined petroleum, natural gas, and electricity) purchased by the contractor and used by him in his office and at the jobsite; energy embodied indirectly in the materials required from all other sectors of the economy which supply products to the construction industry; and the energy embodied in the transport of materials from the supplier to the jobsite and in other services necessary to the transactions of the construction industry. It is interesting to note that less than half of the 399 sectors make a direct contribution to any of the construction sectors.

There is a major difference between building and non-building categories. In general, a much greater percentage of energy embodiment is direct in the non-building sectors; also, there is a much greater degree of specialization in the non-building categories and, hence, a greater amount of variation from one non-building sector to another. (e.g., in New Telephone & Telegraph Construction, 45.9% of the embodied energy comes from Non-ferrous Wire; in New Sewer Construction, 30% of the embodied energy comes from Clay Products or Concrete Products. By comparison, Non-ferrous Wire provides only 2.2% of the energy embodied in all New Building Construction; Clay or Concrete Products provide only 1.8% collectively.

Figure 2 shows the basic profile for an aggregation of the 18 sectors which comprise all new building construction, broken down by percentages of energy embodiment contributed by all other sectors at the 399 level. The four largest contributors - Refined Petroleum Products, Ready Mix Concrete, Fabricated Structural Steel, and Bricks - by themselves account for nearly a third of the total energy embodied.

While there is a great similarity among the various building sectors, there are important differences which must be noted as well. For example, New 1-Family Residential Construction, which was the largest energy consumer among the New Building sectors, is less than half as energy intensive in terms of Btu embodied per square foot of construction (Btu/SF) as New Dormitory Construction, which accounted for a smaller total embodiment. (1-Family Residential = 702,000 Btu/SF vs. 1,431,000 Btu/SF for Dormitories.) Examination of the embodiment profiles for the two sectors (Fig. 2) shows a greater emphasis on the use of low energy intensive materials, such as those

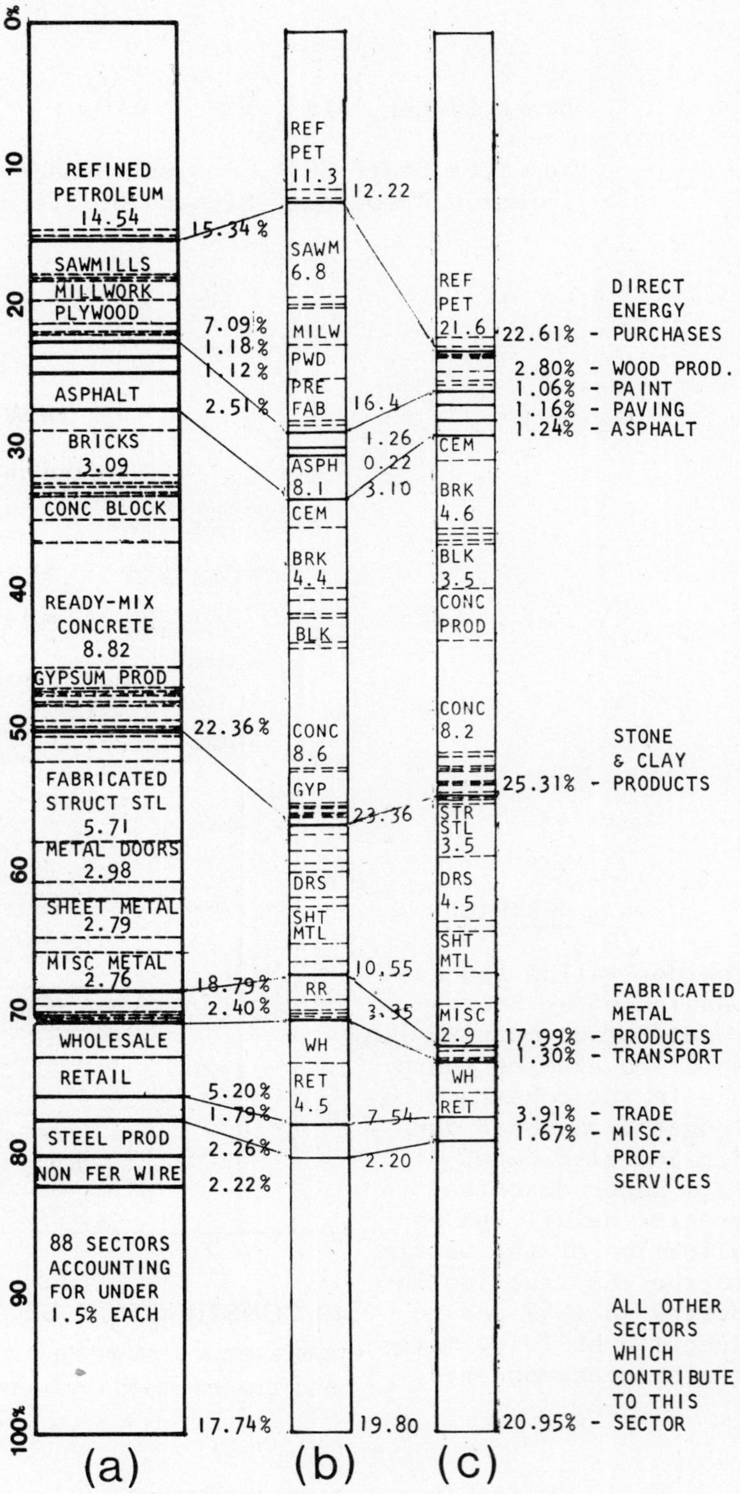

Fig. 2. Energy profiles for:
(a) All New Building Construction
(b) New 1-Family Residences
(c) New Dormitories

found in the various wood products sectors, in 1-Family Residential Construction, and conversely, a greater emphasis on high-intensity materials, such as those found in the fabricated metals sectors, in Dormitory Construction.

An examination of similar profiles for all building types gives an indication of which areas will be fruitful for application of conservation strategies, whether these are the substitution of less energy intensive materials in place of energy-expensive ones; use of less material through more precise structural design and construction techniques; or application of conservation procedures to the production of key materials.

Substitution of Materials

The key tool for any effort to conserve energy through substitution of materials is an energy estimating handbook. Similar in format to a standard cost estimating manual, such a listing will give values for the energy embodied in building components rather than for their dollar cost.

By application of $/unit figures derived from data in the Census of Manufactures (4) to the energy intensities (in Btu/$) available from the Energy I/O matrix (3), we have established the methodology and reporting format for such a handbook and have begun to compile a series of useable energy values. A sample listing is given in Table I below:

TABLE I Energy Embodied in Typical Building Materials

Material	Unit	Embodied Energy (Btu/Unit) After Delivery to Jobsite
Dressed Softwood Lumber	Bd Ft	7,859
Wood Shingles & Shakes	Sq Ft	7,315
Wood Casement Windows	Each*	1,190,349
Wood Doors - Flush Solid Core	Each*	1,191,182
1/2" Softwood Exterior Plywood	Sq Ft	7,705
1/2" Gypsum Wallboard	Sq Ft	6,920
3 1/2" Mineral Wool Insulation	Sq Ft	6,860
Window Glass - Double-Strength	Sq Ft	15,430
3/16" Polished Plate Glass	Sq Ft	48,031
Common & Face Brick	Each	14,283
Ready-Mix Concrete	Cu Yd	2,594,338
Carbon Steel Hot-Rolled Bars & Shapes	Lb	18,736
Carbon Steel Reinforcing Bars	Lb	15,664
Fabricated Structural Steel	Lb	22,707

*An investigation of the units within this category indicated that an average casement window is approximately 3'-0" wide by 4'-0" high, and an average door is 3'-0" wide by 6'-8" high.

It is not possible to derive the energy embodiments of all materials and components in this simple manner. In the case of sectors such as Ready-Mix Concrete, where the entire output of the sector is expressed basically in one product, this method is valid. Similarly in sectors such as Primary Iron & Steel, where the price differential among a multitude of products appears to be a function of the amount of energy needed for their production, the respective costs per unit of product, multiplied by an average Btu/$ for the entire sector, produces a series of energy embodiments which we feel are valid. In other cases, however, a sector includes many disparate products, which are produced by very different processes, and which are reported in non-equatable units. Also, in some cases, there is no definition of the average unit.

An example of this is the Millwork sector, which includes prefabricated wood windows of various types and sizes; wood doors of various types, compositions and sizes; and so forth. In a case of this type, it is necessary to perform a hybrid analysis to arrive at an energy embodiment for a given unit and also to estimate the nature of the "average" unit to which average costs and energy embodiments can refer.

We performed a sample hybrid analysis for wood casement windows, breaking up the window into its major component parts: wood molding and double-strength window glass, and assigning to these components their proper energy embodiments. To the total embodiment of the components we added factors for assembly of the unit; for transport of the components to the window factory; for the overhead operations of the window manufacturer; and for transport of the finished window to the jobsite. Our sample window unit was 3' wide by 4' high - a dimention which, in our experience, is an average unit. The result of our analysis was a total embodiment of 1 million Btu for the window with single glazing and 1.24 million Btu for the same unit double-glazed. These totals exclude hardware, caulking, and plastic components (themselves subjects suitable for hybrid analyses on their own) and are generally in accord with the somewhat higher figure of 1.19 million Btu which may be derived directly from Census of Manufacturs average data. Similar analyses developed values for other sizes.

Sample Building Components

An informed choice in materials selection can reduce building energy use appreciably. A sample analysis of interchangeable floor systems based on similar loading conditions and fire-safety ratings, demonstrates that the production of a reinforced concrete structure will use less than 60% of the energy needed to produce a comparable standard steel structure. As an example we computed the energy embodied in a section of floor slab 30' x 30' square, typical of contemporary high-rise office buildings. Two interchangeable structural systems - steel and reinforced concrete - were investigated (Fig. 3). In spite of their names, each uses both steel and concrete in varying proportions. In standard steel construction the floor deck is typically concrete, designed to be strong enough to span between the beams on which it rests. It is shown poured over a corrugated metal deck, which acts as formwork for the concrete. In concrete construction a great deal of steel (in the form of reinforcing bars) is used to take care of tensile stress. Overall, however, there is less steel (by weight) in a concrete structure, and the steel which is used here is all in reinforcing bars, which have a lower energy embodiment per pound than does fabricated structural steel (15,664 vs 22,698 Btu.) Even so, 55.5% of the energy embodied in the concrete system is due to reinforcing steel.

Although columns have not been considered in this analysis, a preliminary investigation indicates that the proportional difference in embodied energy between steel and concrete systems carries through the entire structure in spite of the fact that the concrete slab is approximately twice as heavy as the steel one. This is a conservative estimate, not taking into consideration lateral loads on tall buildings or certain aspects of the building code which would penalize steel structures to a greater extent than concrete ones and which would pertain to the design of any specific building (5).

It is interesting to note that although the costs of the two systems will fluctuate with market conditions, labor conditions, and location, in general, on large projects, the two

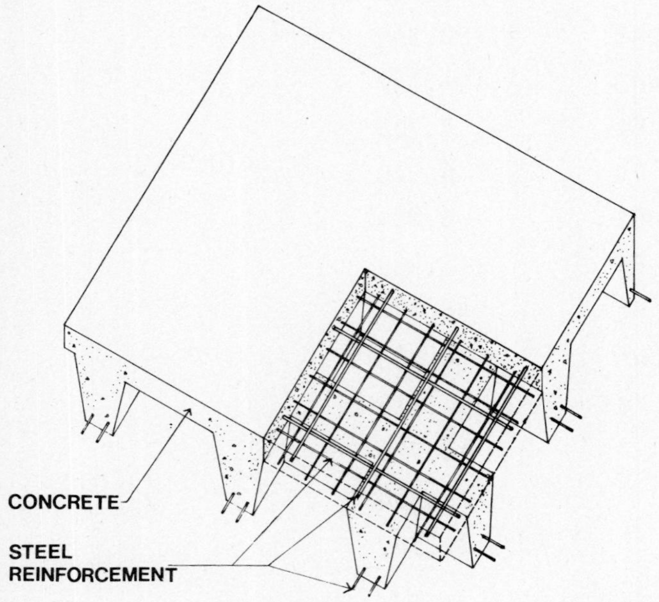

Energy Embodiment: 154,818,749 Btu per 30' x 30' bay
172,021 Btu per SF

Concrete Waffle Slab

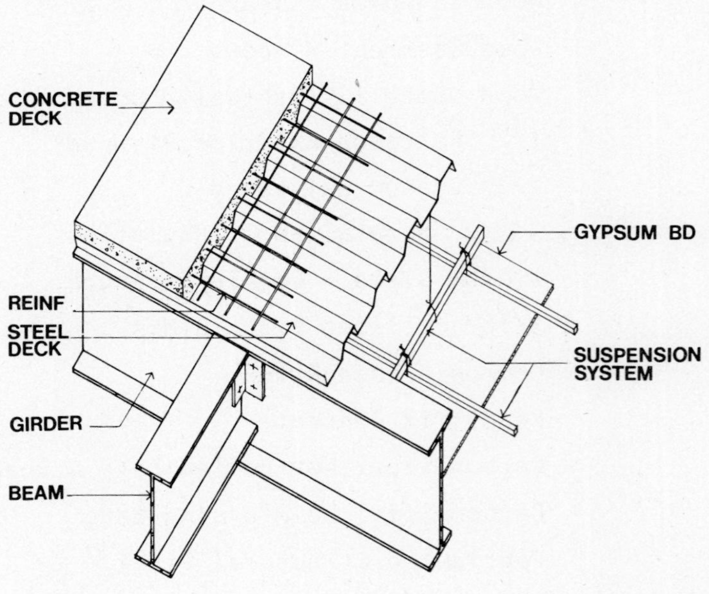

263,450,334 Btu per 30' x 30' bay
292,723 Btu per SF

Standard Steel Construction

Fig. 3. Alternate floor slabs

systems will cost approximately the same. While the cost to the contractor of the materials which make up the steel system is very much higher than the cost of materials of the concrete one, concrete construction uses a great deal more labor on site than does steel. At the moment, in this case, the higher materials cost on the one hand appears to balance the higher labor cost on the other, and the choice of one system over the other is made for reasons other than first cost. e.g., if the depth of the floor structure is critical, a concrete system will be preferred; if scheduling of the work of the various trades within the construction period is expected to be complicated, the steel structure will be chosen. Often the reason for choice is a matter of preference on the part of designer or contractor for one material over the other. Conditions vary sufficiently from one project to another; from one location to another; and from one time to another to invalidate generalizations. If energy embodiment is to be the primary criterion, concrete will obviously be the system of choice.

Although the difference in embodiment appears to be negligible in our example (120,000 Btu/SF), it takes on significance when applied to an entire building category. We have estimated that 157.5 million square feet of new office building construction were built in 1967 (3). Assuming that half of this area had been constructed originally using concrete framing and the other half had been steel (a conservative estimate), a shift to concrete construction for all office buildings in that year would have resulted in a reduction of 9.5 trillion Btu <u>from floor slabs alone</u>. At 6.3 million Btu/barrel, this reduction is equivalent to the saving of 1.5 million barrels of No. 6 fuel oil.

Comparative Wall Sections

Another set of comparisons was made with regard to two wall systems typical of 1-Family residential construction. We have estimated that in 1967, a representative year, about 1.1 billion square feet of 1-family residences were constructed. In that year, this sector accounted for 30% of the square footage of all buildings constructed and 23% of the energy embodied in all new building construction (3). These are significant percentages of the building industry, and a change in the way the buildings in this sector are built can have an important effect on all energy use.

In this building sector, and in other small residential building, 2 x 4 wood stud construction is typical. The examples we chose to study are typical variations of this construction which have a similar thermal performance. Both examples have a basic framework of 2 x 4 studs on 16' centers, gypsum wallboard inside and plywood sheathing and building paper outside. They differ only in their exterior finish: One wall has wood shingles, and the other has brick veneer with a 1" air space between the brick and the sheathing. Thermal performance, that is, the rate at which heat will flow through a given assembly, is expressed in terms of a "U-factor." The U-factors for the shingle and brick walls are .25 and .24 respectively. Adding up the energy embodied in the components of one square foot of each wall yields total embodiments of 25,426 Btu/SF of shingled wall and 119,566 Btu/SF of stud wall with brick veneer. The addition of 3 1/2" of mineral wool batt insulation (at 6,860 Btu/SF) will raise the total embodiments to 32,286 Btu/SF (shingle) and 126,426 Btu/SF (brick) and will lower the U-factor for both alternatives to .085.

The wide gap between the energy embodiments of these two walls is a function of the low energy intensity of wood and the high energy intensity of brick. Brick veneer is a common material in this building type and in other low-rise construction as well. Figure 2 shows brick contributing 4% of the energy embodied in 1-Family Residential Construction. (Similar profiles for each of the 49 construction sectors show brick contributing 4% of the energy embodied in 2-4 Family Residences and 3.4% of the energy in Garden Apartments.) All told, 1-Family Residences alone accounted for 780 trillion Btu in 1967 (see Fig. 1) of which the energy embodied in bricks (4.4%) is equal to 34.32 trillion Btu. At 85,698 Btu/SF (6 bricks/SF of wall at 14,283 Btu/brick) this represents 400 million square feet of brickwork. If the comparison between the brick veneer and the wood shingle construction assemblies shows a difference of 94,140 Btu/SF (119,566 - 25,426), it accounts for a differential of a total of 37.66 trillion Btu for the entire square footage above. At 150 thousand Btu/gallon, this amounts to 251 million gallons of No. 6 oil (or nearly 6 million barrels.) It is apparent that a significant saving in energy consumption could be effected if brick and other energy intensive materials were limited to those uses where their inherent qualities made them most desireable.

In this part of our study we compared only two facing materials. A complete study, which would be necessary for a truly informed choice of materials to be made, would also include asbestos shingles, asphalt shingles, aluminum siding, cement-asbestos board, etc.

Energy Life-Cycle

The figures quantifying the energy embodied in various materials and assemblies are thrown into better perspective when they are considered together with the energy needed to operate buildings.

For example, in a study prepared for New York State, two government office buildings in Albany, New York, were studied for the purpose of reducing the energy needed to operate them (6). As designed, in the mid-1960's, the buildings, with a combined gross area of 1.19 million square feet, required a total of 28 million kwh of electricity and 626 thousand gallons of No. 6 oil annually. At 3,412 Btu/kwh and 150,000 Btu/gallons of oil, this is equal to 190 billion Btu per year for building operation measured on site. Dividing by the gross area, this is equal to 159,664 Btu/SF. According to information derived from the Energy Input/Output matrix, in 1967 (a comparable construction year), the energy embodied in office buildings averaged 1.6 million Btu/SF or ten times the energy needed for one year's operation in this case.

As building owners become more energy conscious, and as buildings begin to be built and operated in energy conservative modes, this ratio can be expected to change. The Albany Study is a good example of this. At the time the study was initiated, New York State had been operating the buildings in accordance with its own conservation program, consisting in the main of selective delamping. The result was a 12% reduction in energy consumption and an equivalent annual energy demand of 183 billion Btu or 153,781 Btu/SF (10 1/2 years' operation energy needs). As a result of the study, further conservation measures, including more delamping and adjustment and rebalancing of the HVAC systems in the two buildings resulted in a projected lowering of the annual operating energy 81.9 billion Btu or 68,824 Btu/SF.* Since no material was to be added, the embodied energy would remain the same, and at this point embodiment would equal nearly 24 years' operation energy.**

The recent GSA Standard for new Federal Government office Buildings (7) sets 55,000 Btu/SF per year as maximum on-site energy to operate all services in new buildings. The two existing Albany office buildings could be brought into conformance with this standard by replacing the existing 120 thousand SF of single glass with double glazing. At 48,800 Btu/SF for the additional 120,000 SF of glass, this would add 5,760 million Btu or 4,800 Btu/SF to the buildings' embodiment. At this point, the annual end use energy demand for both buildings was estimated at 61 billion Btu or slightly over 50,000 Btu/SF. The buildings' energy embodiment would have risen to 1.605 million Btu/SF, which amounts to about 32 years of operation.***

As a result of a concerted effort to conserve operational energy, the energy embodied in construction itself is now close to the amount of energy needed to maintain the building throughout its lifetime.

It is evident that life-cycle costing is as important in energy estimating as it is in cost estimating. The above example, examined in terms of energy life-cycle, projects annual savings of about 16,000 Btu/SF at point of use (or about 18,000 Btu/SF at source) at an energy embodiment cost of 800 Btu/SF. Energy payback, that is, when operational energy saved equals added embodied energy, will occur in three to four months.

The building shell is of particular importance in life-cycle costing because its components affect not only the total embodiment of the building, but also the amount of energy the building will require in its operation. We therefore extended our study of typical residential wall assemblies as previously described to see what the life-cycle energy consequences might be when the thermal transfer characteristics of the residential building shell (walls, roof and windows) were improved and the demand for heating was consequently reduced.

The amount of energy demanded by a given construction is a factor of the rate at which heat will flow through it (expressed in terms of its

---

* These measures are presently being implemented by New York State.

** These figures are based on on-site energy measurement. If the energy consumed by the two buildings is measured at source, the value for electricity will be raised from 3,412 to 11,380 Btu/kwh (based on Niagara Mohawk Power Corporation's heat rate of 10,803 Btu/kwh + 5% transmission losses. Using these figures, the annual energy consumption of the two buildings will be 350,115 Btu/SF for the buildings as designed and 164,175 Btu/SF after implementation of the proposed energy conservation modifications. Energy embodiment for the two buildings would be equal to 4.7 and 10 years of operating energy consumption respectively.

***Based on source energy, this would be about 11 years.

"U-factor") and the average temperature pattern of the location of the building (expressed in terms of "degree days.") The number of Btu which will flow through one square foot of a given assembly in an average heating season is expressed by the formula:

U-factor (Btu/hr) x Degree Days x 24 hrs=Btu

While the addition of insulation to an existing building will be restricted by the amount of interior space which the building occupant will willingly give up, a new building can be designed to accommodate whatever wall thickness is considered to perform best. It need not be restricted in depth to whatever space there may be within an existing wall.

Table 2 tabulates the characteristics of 1 SF of wood stud walls with wood shingle exterior and gypsum wallboard interior. The walls vary in nominal thickness and in amount of mineral wood insulation they contain. For walls over 8" thick, rather than continue to increase the depth of the studs, we have assumed a double 2 x 4 stud wall.

It is evident that while each additional increment of insulation provides an additional savings, the amount of savings is progressively smaller. Increasing insulation from 3 1/2" to 5 1/2" will increase the embodied energy of the wall by 2,384 Btu/SF and will save nearly 4,000 Btu/SF/year in New York City. Energy payback will occur in a little over half a heating season. Increasing insulation from 11 1/2" to 13 1/2", which will increase the embodied energy of the wall by 2,100 Btu/SF, will save only 370 Btu/SF/year. Energy payback would take five to six years.

However, the decision to build a wall of a given thickness is not made in progressive stages but, rather, once and for all. To increase insulation from 3 1/2", which is now typical, to 13 1/2" would increase the embodied energy of the wall by 12,088 Btu/SF and would save 7,329 Btu/SF/year. Energy payback in this case would be in 1.64 years.

These figures are not inconsequential - 7,300 Btu multiplied by the total number of square feet of 1-Family Residential walls constructed in a given year would result in a substantial cumulative effect. (New York City, with an average of 4,848 heating degree days per year is close to the national average of 4,734 (7).

According to the 1971 Statistical Abstracts, there were 844,000 1-Family Residential housing starts in 1967. Assuming that 1,280 SF of exterior wall (as used in the example which follows) is average, this amounts to a total of 1.08 billion square feet of walls built in one year. The total annual saving represented by adding 10" of insulation to these surfaces, as-

Nominal Wall Thickness	Type of Framing	Insulation	U-Factor	Embodied Energy (Btu)	Annual Demand (Btu)
4"	2 x 4 @ 16"	0"	.250	25,426	29,088
4"	2 x 4 @ 16"	3½"	.085	32,286	9,889
6"	2 x 6 @ 24"	5½"	.051	34,670	5,934
8"	2 x 8 @ 24"	7½"	.043	38,074	4,889
10"	(2) 2 x 4 @ 24"	9½"	.032	40,174	3,770
12"	(2) 2 x 4 @ 24"	11½"	.025	42,274	2,932
14"*	(2) 2 x 4 @ 24	13½"	.022	44,374	2,560

*14", which is at first glance a startlingly large dimension for a wood stud wall, was chosen as the upper limit for this investigation because it is nominally equal in thickness to the brick veneer alternative mentioned earlier. Compared to a brick veneer wall with 3½" of insulation, a 14" all-wood assembly would save 82,052 Btu/SF in embodied energy. Assuming again a total of 400 million SF of brick wall constructed in 1967 for 1-Family Residences, this differential per SF is equal to a total of 32.8 trillion Btu or 5.2 million barrels of No. 6 fuel oil.

TABLE 2 Comparison of Energy Embodiment and Annual Operational Energy Demand for Heating Imposed by 1 SF of Shingled Wood Frame Wall With Varying Thicknesses of Insulation. (In NYC: 4,848 deg days)

suming the national average heating degree days of 4,734, will be 7,158 Btu/SF. Multiplied by 1.08 billion square feet of wall surface, the total annual savings becomes 7,730 billion Btu - equivalent to 1.2 million barrels of No. 6 fuel oil saved per year as a result of a simple structural modification of one year's new housing stock. Energy payback for the additional embodiment - 10 billion Btu - would be in less than two heating seasons.

Extention of this analysis to a similar consideration of single versus double glazing and flat roofs with 3 1/2" or 5 1/2" of mineral wool insulation has allowed us to make a general comparison between the outer shells of two typical 1,500 SF 1-family residences, located in New York City, either of them in accordance with today's construction practices in their thermal performance.

Table 3 outlines the general structural and energy characteristics of both building shells. House A, which has 3 1/2" of insulation in the wall and roof and single glazing, would have an embodied energy value of 168.9 million Btu. Operational energy demand by the shell would be 67.3 million Btu per year for heat lost through thermal transmission. House B, which has 5 1/2" of insulation in both roof and walls and double glazing, would have an embodied energy value of 178.8 million Btu (5.9% more than House A) and an operational energy demand due to thermal transmission losses of 35.2 million Btu/year (48% lower than the first example.)

In addition, either building would require a further input of operational energy to counteract heat lost through infiltration and the opening of doors and windows. We have estimated this increment of demand to be 41.8 million Btu/year (3). Total energy demanded for heating, therefore, would be 109.1 million Btu/year for House A and 77 million Btu/year for House B.

General Characteristics:

Floor Area:	1,500 SF = 30' x 50'
Roof Area:	1,500 SF
No. of Stories:	One
Walls:	10' high x 160' perimeter = 1,600 SF
Openings:	23 windows @ 3' wide x 4' high = 276 SF
	2 doors @ 3' wide x 6'-8" high = 40 SF
	Total Area of Openings = 318 SF
Area of solid wall:	1,282 SF

Specific Characteristics

	House A	House B
Walls:	Wood shingles on 2 x 4 studs @ 16" 3 1/2" mineral wool insulation 1/2" gypsum board interior	Wood shingles on 2 x 6 studs @ 24" 5 1/2" mineral wool insulation 1/2" gypsum board interior
Roof:	2 x 12 wood rafters @ 16" built-up roofing on a plywood deck 3 1/2" of mineral wool insulation 1/2" gypsum board ceiling below	As in House A but with 5 1/2" of mineral wool insulation
Windows:	Single-glazed	Double-glazed

Energy

	House A Embodied	House A Demand	House B Embodied	House B Demand
Walls:	62,286 Btu/SF	9,889 Btu/SF	34,670 Btu/SF	5,934 Btu/SF
Roof:	68,073 Btu/SF	8,726 Btu/SF	70,003 Btu/SF	4,189 Btu/SF
Windows:	1,070,652 Btu/EA	131,477 Btu/SF	1,242,852 Btu/EA	67,484 Btu/SF
Doors	346,502 Btu/EA	131,477 Btu/SF	346,502 Btu/EA	67,484 Btu/SF

TABLE 3 Structural and Energy Characteristics of the Shells of Two Typical 1-Family Residences Located in New York City

Thus, an extra energy embodiment of approximately 10 million Btu would net over 32 million Btu saved annually, and energy payback time would be about 1/3 heating season. Thirty-two million Btu is equivalent to over 51 barrels of No. 6 fuel oil saved per year. Multiplied by 844,000 private 1-family housing starts recorded for 1967, the potential savings inherent in a relatively simple set of structural adjustments applied to one year's new housing would amount to 4.29 million barrels - approximately 1% of our 1975 refinery output (8).

It is evident that, although in both houses the energy embodied in the outer shell is a small percentage of the energy which either house will demand over its lifetime, the choice of materials of construction will have a significant effect.

Examination and Modeling of Energy Flow

Another approach to conservation of energy within a large area of the economy such as construction, is to apply conservation practices to its major contributing industries.

From the Energy Input/Output matrix, it is possible to establish immediately:

a) The total amounts of each primary energy resource (coal, crude petroleum, non-fossil electricity, and imports) required due to the direct or indirect demand from New Building Construction.

b) The amounts of energy embodied directly or indirectly in the materials and services provided to New Building Construction directly for incorporation into the completed buildings.

Connecting these two poles are a yet-to-be determined number of inter-industry transactions which, when quantified and graphically represented, will complete the flow diagram. The diagram will show: the transactions between primary energy resources and those sectors which consume primary energy resources directly (e.g. the steel industry, which purchases coal, or the oil refining industry, which purchases crude petroleum); transactions between the energy industries which use energy resources directly and the energy industries which sell the energy product to some non-energy industry (e.g., that portion of the electrical industry which uses refined petroleum for generation); and, finally, the flow of energy embodied in materials, machinery, and administrative and service functions, through inter-industry transactions, leading finally to the products which are sold directly to the building contractor.

The purpose for establishing this network is not only to assess the impact on energy embodied in construction of conservative strategies within contributing industries, but also to permit the identification of nodes in the flow which may become control or limiting points for alternative material strategies. Some nodes, such as machine tool production, may not appear in industries selling to final demand, but may nevertheless be large energy users, with their energy appearing as embodied energy in Fabricated Structural Steel, Ready-Mix Concrete, and other products depending on plants and sophisticated equipment.

Developed as a dynamic computer model, the flow diagram would permit changes at any point in the eventual delivery of goods and services to New Building Construction to be evaluated with respect to the impact which they will have at any other critical point.

Conclusions

The energy required by the construction industry, which has been referred to speculatively up to now, can be analyzed against a sound data base. The category is sufficiently large that simple, available means can be employed to reduce total national energy consumption by millions of barrels of oil annually.

The information which we have assembled so far makes evident the need for additional information and detail to be developed, however, in order to achieve maximum savings predictably and with an understanding of the larger implications of the chosen strategies.

Among the important areas for further research are:

- an amplification of the energy per unit of building material accounting and the incorporation of the complete list in an energy estimating handbook and/or computer estimating program.

- completion of the intermediate steps in the energy flow diagram and translation of the static graphic diagram into a dynamic computer model.

- expansion of the basic research into other conservation paths which appear potentially fruitful, such as:

  - changes in field operation methods

  - more exact structural design, leading to a saving in quantity of material used.

  - the impact on national energy use

of incentives to renovate old buildings rather than demolish them and build new ones.

a study of the transportation margins for materials going into new building construction, broken down by region, to form the basis for a strategy of conservation of transportation energy through greater use of regional materials.

A target reduction of 20% seems feasible to us. Twenty percent of the 5.11% of national energy consumption represented by new building construction in 1967 is equal to 1% of the national energy consumption in that year, or 669 trillion Btu. Assuming an average heat rate of 10,500 Btu per kwh per KW capacity (9), this represents the fuel needed annually to run sixteen or seventeen 1000 MW generating plants.

Or, to use another comparison, at 5.8 million Btu/barrel, 669 trillion Btu is also equivalent to 115.3 million barrels of crude oil, an amount equal to nearly half of the oil we imported from Saudi Arabia (and greater than our total imports from Iran) in 1975 (8).

REFERENCES

(1) B. Z. Segal, Determining the Energy Cost of Construction Activities, paper presented at the International Conference on Energy Use Management, Tucson, AZ, October, 1977

(2) R. A. Herendeen and C. W. Bullard, Energy Cost of Goods and Services, 1963 and 1967, CAC Doc. No. 140, Center for Advanced Computation, University of Illinois at Urbana-Champaign, November, 1974

(3) B. M. Hannon, R. G. Stein, B. Z. Segal, D. Serber and C. Stein, Energy Use for Building Construction - Final Report, ERDA Doc. No. COO-2791-3, Center for Advanced Computation, University of Illinois at Urbana-Champaign, December, 1976. (Will also be available as NTIS report.)

(4) U.S. Dept. of Commerce, 1967 Census of Manufactures, Vols. I and II, U.S. Government Printing Office, Washington, DC, 1971

(5) Structural analysis and computations by Robert Silman, P.E., and Ding Carbonelle, P.E., of Robert Silman Associates, New York City

(6) Pope Evans & Robbins, Inc. and Richard G. Stein & Associates., Energy Conservation Study - State Office Building Campus, Buildings 8 & 12, Albany, N.Y., for New York State Office of General Services, Albany, NY, June, 1975

(7) Data from the National Climatic Center, Asheville, N.C. Average regional weather data (for period of July 1931 through June 1976) weighted in accordance with regional population characteristics.

(8) U.S. Bureau of the Census, Statistical Abstract of the United States, 1976, Washington, DC, 1976

(9) Battelle Research Report: An Input/Output Analysis of Energy Use Changes from 1947-1958 and from 1958-1963, Battelle Institute, Columbus, OH, June, 1972

ENERGY SAVING IN STRUCTURAL MATERIALS FROM CHANGES IN ANALYTICAL AND BUILDING METHODS

Robert Silman, P.E.
President, Robert Silman Associates, P.C., Consulting Engineers
New York, N. Y.

INTRODUCTION

Since the beginning of the Industrial Revolution engineers have been conditioned to produce building structures which are the most "economical". This has usually meant delivering to the Owner a building which adequately fulfills the program and promises to endure for its proposed life at the minimum expenditure of dollars. The simple criterion of lowest ultimate cost has been the guideline of design and construction technology.

But as we now reassess our energy inventory, other criteria have begun to emerge which may supersede the old ways of thinking. New definitions of "cost" and "economy" will have to be developed. Attitudes toward design and construction will have to be remolded and priorities shifted. And perhaps most revolutionary of all--owners will have to accept the reality that the bottom line will be higher and their buildings will cost more.

Engineers have been traditionally trained to design buildings so that the total ultimate dollar cost is as low as possible. Techniques for estimating normally involve taking off quantities of all materials, figuring costs of labor, and adding indirect costs. The body of information which has accumulated over the years has led to a well developed and highly competetive construction market with enormous pressures being exerted to save pennies wherever possible. In terms of achieving its goals, this attitude has succeeded quite remarkably. Experience has counted heavily in being able to deliver the lowest cost structure and in evaluating market conditions, geographical factors, sequencing of operations, etc.

But nowhere in this process has value been placed on conservation of energy resources, unless there is a direct monetary return to be reaped. It is now time to add one more factor into the equation--minimum use of energy.

This paper will explore several areas of potential energy savings in structural materials and methods which can be achieved through more refined design techniques, improved building materials, re-evaluation of code constraints and alternate construction techniques. These savings presume that the designer and contractor are familiar with the relative amounts of energy required to construct a particular system. Such information is only now finding its way into the literature of the industry. Other papers presented at this Conference will deal with techniques for evaluating the intensity of energy required to produce and install a given quantity of a certain material. In discussion of potential techniques for saving energy it is assumed that data on comparative energy expenditure is available in evaluating any construction system.

SAVINGS RESULTING FROM MORE REFINED DESIGN

Energy could be saved by using less material simply by conducting more exact structural analyses of buildings. Engineers often tend to perform approximate analysis because it is faster and more economical for them. It is well known that in many cases a more theoretical or exact analysis will save material. An example of this is in reinforced concrete design for continuous beams or slabs in which bending moments may be calculated approximately by the method of coefficients or more precisely by a rigorous mathematical analysis (e.g. moment distribution or slope deflection). In most cases the latter process yields lower values for bending moments and thus will require less material to resist these forces.

Computers have been a great asset to engineers in performing more precise calculations. Almost any degree of accuracy desired can be stipulated without a great deal of difference in computer costs. In addition the computer can be programmed to furnish ancillary output which might not normally be calculated manually in the design process. Thus items such as deflection, sidesway, bearing stress and secondary stresses can be made to appear as part of the standard output. Members can be selected which satisfy the design criteria exactly instead of being selected based on arbitrary empirical approximations, such as span-to-depth ratios for deflection.

Other areas which can be made more exact in the design process include accumulation of dead loads and specifying of allowable stresses. In the case of dead load analysis the designer should take the time to carefully list all materials and their actual weights. Due allowance must be made for future additions of load where applicable. As for specifying allowable stresses of materials, only the level of stress actually required for each element or group of elements should be called for. There is nothing wrong with specifying various strengths of concrete for various functions in a building. Steel design ought to use only the strengths needed; a hybrid design can be most effective.

There has been a tendency over the past number of years to show many Typical Details on drawings and then apply them to as many conditions in the project as possible. Any particular Typical Detail must be satisfactory for the worst case; thus it is over designed for most other conditions. It would be much more economical to detail each case individually. A good example of this is the stipulation that shear connections of steel beams be designed to resist the full tabulated load in the AISC Handbook. This is a very conservative approach for, in fact, most beams carry far less than the maximum load. It would be much more efficient to furnish reactions for each beam and to provide only that connection material which is actually required. Computer design helps achieve this by furnishing the steel shop a copy of the output showing reactions of each member; but they could also be shown on the drawings.

One of the underlying reasons for designing using Typical Details has been the credibility gap between the engineer in the office and the construction worker in the field. The engineer tends to characterize the construction man as a non-thinking robot who cannot follow any but the simplest of directions. The construction worker on the other hand often feels that the engineer is simply throwing in lots of extra material which isn't really necessary. Caught in the middle is the field inspector who must insure that the job is built in accordance with the contract documents. If all parties assumed a healthy respect for one another there could indeed be a savings of material. The engineer must feel confident that there will be adequate site supervision and a cooperative attitude from tradesmen so that he can stipulate a greater number of more specific details rather than the broader but more wasteful Typical Details.

In the past decade or so new design techniques have been explored which attempt to rationalize certain patterns of structural behavior. The most notable of these is the emergence of ultimate strength design in reinforced concrete. This method has now almost totally replaced working strength theory. Other techniques such as plastic design, limit design and yield line theory have been accepted to some degree or other. Practically all of these lead to savings of material simply by asserting that the design is more rational or more accurate.

Designs can also be refined to utilize

more interaction between various elements. In reinforced concrete structures or in continuous welded steel frames, there is a certain three dimensional aspect which assists in reducing stresses and therefore lowering material requirements. Composite design, in which concrete slabs are made to interact by serving as top flanges for steel beams by means of mechanical shear connectors, has gained great popularity and saved enormous quantities of steel. Other approaches such as orthotropic plate design and stressed skin analysis can promise savings. Engineers should look carefully at the so-called non-structural elements of a building such as walls and partitions sheathing, ceilings, mullions and slabs on grade to determine if they can be made to interact with structural components adjacent to them. Why can't an exterior wall be connected to a column to create a tee beam to resist wind forces? Why can't window mullions be made load bearing and thus reduce the size of spandrel beams? Interaction between building comonents can significantly reduce the amount of total material. The whole can be greater that the sum of the parts!

Another area of refinement in structural design is the exploration of more efficient forms. These forms find particular application in long span roof designs but elements can be incorporated into smaller buildings. Lightweight structures such as cable or air supported roofs come to mind. These often utilize membranes for their coverings. Other structural forms which derive strength from their intrinsic shape include monocoque construction, membranes, tapered members and geodesic domes.

Design engineers will look back over the thoughts presented in this section and undoubtedly comment that their office design costs will be dramatically increased if they have to spend time refining their calculations and drawings. If we are to achieve savings in material we may have to first spend on design labor. Allowances and adjustments will have to be made to engineering fees to permit the required precision and detailing. Part of this added cost will however be earned back immediately through savings in material.

SAVINGS RESULTING FROM IMPROVED PRODUCTS

If material can be saved by improving building products then a net savings in energy can certainly be realized. There are two main areas for product improvement--upgrading existing products and developing new materials.

More efficiency can be achieved through better quality control. This element is particularly important in the production and installation of concrete. Simple controls, such as water-cement ratio, proper placing and consolidating and adequate curing, can dramatically improve the performance of concrete for both strength and durability. This can result in a lower cement factor which ultimately reduces the amount of energy input required for the finished product. Many concrete batching plants are being automated which helps to control quantities of dry materials. But there is still a good deal of work to be done in order to accurately regulate total water content and mixing time.

Another area of structural work where quality control can reduce energy input is welding. This highly energy intensive activity can greatly benefit from more uniform procedures. If consistent results could be achieved, then higher allowable stresses would be possible and smaller quantities of weld would be the end product.

Much research and development is presently beign conducted in order to create new products. The emphasis to date has not been on developing energy saving materials or procedures. If this goal is established, we might find some interesting developments. For certain applications, steel alloys can save materials. In concrete work, admixtures can improve the properties of concrete and reduce cement factors. Mortar can be fortified with additives. Other products can be refined or purified to increase their basic strength.

The manufacturing processes can be improved to require less energy input. The simple operation of chemical analsis for each melt of steel in a basic oxygen furnace requires five or six minutes out of a total of 45 minutes. If this analysis could be performed internally in the ladle, up to 10% of the energy required to fire the furnace might be saved. Similarly, heat recovery installations in steel, cement, brick and other manufacturing operations may lower the total energy required to produce a unit of material.

The use of lighter weight materials reduces the dead load of the structure and thus requires less material in the support system. Decreased dead loads are important not only in the structural frame but also in the finishes and enclosure of a building. The most notable weight-saver in buildings has been the curtain wall which replaced the heavy masonry spandrel of earlier decades. Of almost equal importance is the utilization of lightweight stud interior partitions which have replaced block and plaster. The actual structural frame can show a reduced dead load if lightweight aggregates are employed in the concrete mix. If expanded clays, shale or slag are used as coarse aggregate a weight saving of about 30% can be achieved. Other methods of reducing the weight of concrete include the use of lightweight materials for fine aggregate and the development of structural concrete using foaming agents to decrease density.

Plastics present a tremendous future potential as lightweight building materials, both structural and nonstructural. Their high strength-to-weight ratio makes certain plastics very attractive. Problems such as creep and resistance to ultraviolet light need to be overcome but this seems within our technical capabilities at this time.

An additional area of structural product development is that of adhesives. High strength bonding agents could revolutionize composite and stressed skin design. Imagine developing the required bond between beams and slabs using glue! Adhesives can also be visualized as replacing devices now used in making framed connections such as welds and bolts.

Some of the proposed improvements or new materials may not prove to be energy efficient. Each would have to be evaluated to determine whether the extra energy input could be justified in terms of total energy savings. It does seem clear however that significant energy reductions in structures could result from improved products.

SAVINGS RESULTING FROM FEWER CODE CONSTRAINTS

Building codes have developed along several different lines over the years. Most codes are highly prescriptive and leave almost nothing to the engineer's imagination. Codes which deal with specific structural materials (e.g. ACI Building Code for Reinforced Concrete or AISC Specifications for Structural Steel for Buildings) tend to reflect the latest agreed-upon state of the art by professionals and leaders in industry. Generalized regional or municipal codes (e.g. BOCA, UBC, state and city codes) usually embody the product codes and specifications published by recognized trade associations. Both types of codes are quite definitive in their stipulation of loads, safety factors, design techniques and simplifying assumptions.

Codes change very slowly but certain aspects can lead to savings of material and these ought to be investigated immediately. One area which is gaining code recognition is higher allowable stresses for materials which are inspected or controlled during installation. The latest masonry codes allow stresses twice as high for inspected masonry as for uninspected. There are many other conditions under which an increase in supervision could result in a concomitant increase in allowable stress or reduction in safety factor.

Encouragement and reward can be given to standards of high performance. Instead of prescribing that all producers use the same quantity of materials, an alternate approach would be to demand that everyone meet the same end result. An example is the requirement in many codes that concrete mixes for a given strength have a minimum cement factor. Codes in recent years have abandoned this rigid criterion in faveor of performance records and statistics of previous experience. Thus a concrete producer will be rewarded if he can demonstrate high quality control standards; he will be permitted to use less cement in his mixes. Such incentives can save as much as a full bag of cement per cubic yard of concrete.

Building codes are very specific in defining live loads to be used for various occupancy categories. We have taken many of these standard loadings for granted over the years. It is time to carefully assess all loadings and to re-evaluate where necessary. There should be room within the confines of a building code to make an engineering study of potential loads. This is particularly true for occupancy types which do not fit neatly into the limited listings of live load. Loads in industrial and storage buildings are particularly critical and should be accurately evaluated.

Lateral loads on buildings are presently specified in codes based largely on empirical evidence. A great deal of research needs to be performed to determine accurately the true forces due to wind and seismic effects. Laboratory tests have been conducted in which these loads are simulated on models, such as in wind tunnels, and then a force diagram is prepared. Such tests are very expensive and are only justifiable for large scale projects. Very few existing buildings have been studied in their full scale using instrumentation to assess magnitude, duration and area of loading as well as the response of the building and its foundation. Perhaps it is time to make an investiment of public funds in order to accurately study lateral loads.

Building codes of the future may utilize a statistical or probabilistic approach to failure in establishing factors of safety. This will lead to a proportioning of safety factors for different types of loading as well as for various elements of the building structure. The American Concrete Institute has recognized some of these elements by stipulating various values for load factors according to the type of load: 1.4 for dead load, 1.7 for live load, etc. This same Code has also specified different values for the capacity reduction factor (ø): 0.9 for bending, 0.85 for shear, 0.7 for compression.

A good example of probabilistic approach to failure is the issue of progressive collapse. Since the Ronan Point disaster in 1968 in England a good deal of thought has gone into prevention of progressive collapse in large panel precast concrete structures and in bearing wall masonry buildings. The latest HUD Minimum Property Standards for Multi-Family Housing require that all buildings be resistant to progressive collapse. An analysis should be made of potential causes for progressive collapse--gas explosion (as in Ronan Point), vehicle impact, sabotage, etc. If one or more causes appears likely then a certain probability should be accounted for. If none is likely due to absence of gas or of nearby vehicles, the statistical approach might lead to different conclusions. For instance the area of support which is assumed to be lost in a progressive collapse might vary for each case. Or the requirements for supports at corners and ends, which are difficult to accomplish compared to the interior, might be different.

If the authorities who write the codes can be made more aware of the energy implications of some of their requirements perhaps a savings in materials would be possible. Of course all of these possibilities place a greater burden on the engineer because the less that is specified in the code the more he must supply. But this is precisely the point; if each case is evaluated individually savings can often be realized.

SAVINGS RESULTING FROM ALTERNATIVE CONSTRUCTION METHODS

Perhaps the most difficult change to accept is the overall pricing of a framing system in terms of its energy use. We aren't prepared for seeing the bottom line in BTU's instead of dollars. But if we can adjust to this method of estimating then we can begin to adopt systems which have clear advantages in energy savings in materials.

A perfect example of this sort of comparison is the classic battle over framing systems for midrise buildings. Should they be steel or concrete? If energy is not considered then other factors such as basic material cost, speed of erection, local technology, weather conditions, sizes of members, etc. will determine which to choose. If however the selection of a system is based on a comparison of energy-intensive use of material, then one method may show a clear cut advantage. Studies recently conducted making a comparison between a concrete and steel frame for a typical midrise office building show that the concrete is considerably less energy intensive.

There are many other areas in which systems which are comparable in structural performance may be vastly different in energy expenditure. In the past decade, there has been an increased use of light-gage steel studs and joists in lieu of wood framing. The production of steel of course uses a tremednous amount of energy in comparison with wood, a renewable organic resource whose energy costs are in milling and transporting only, not in production of the basic material.

We now accept certain structural systems as "standard" procedure. This complacency needs to be shaken up if energy considerations are to be promoted. A perfect example is the use of corrugated steel deck as permanent formwork and reinforcing for concrete slabs in steel framed buildings. Prior to the development of metal deck, slabs were formed with re-usable wood members supported from the steel skeleton--a great difference in energy consumption for the formwork material. Of course, the old way used more labor and was therefore superseded by a method which minimized field labor costs. Clearly, a return to some of the older, more energy-economical methods of construction implies an increase in the cost of labor.

The entire industry needs to be re-educated if energy savings are to be a major criterion in selecting a structural system. New priorities mean new approaches to estimating and pricing. All of the techniques developed so carefully over the years may have to be scrapped in favor of energy calculations.

CONCLUSION

Who will take the leadership role in changing the established ways of thinking and in emphasizing energy considerations? Technologists have a unique opportunity at this time to seize the initiative for promoting efficient energy use management in terms of building structures. By adjusting the old rules which determined what was the most "economical" structure, so as to now include intensity of energy input, we can make a very real contribution.

Techniques for saving energy will often mean an increase in labor in the total structure. This should not be classified as retrogressive. Let us rather glorify the element of human energy and craftsmanship. There is no doubt that a positive aspect can be derived from an increase in human involvement in the building process.

The implications of some of the changes proposed in order to minimize energy input may seem far reaching. They will require strong leadership to initiate and great cooperation to implement. No doubt government controls or incentives will be required; voluntary compliance is unlikely in the competitive marketplace. Undoubtedly labor costs will be higher in terms of design, installation and supervision. But the choices are almost nil. We must opt for energy savings in building structures and the sooner, the better.

ANALYSIS AND REDUCTION OF ENERGY USE IN THE BUILDING INDUSTRY

Richard G. Stein, FAIA

588 Fifth Avenue, New York City 10036

ABSTRACT

The building industry, producing 10 percent of the Gross National Product and requiring 15 percent of the energy used by industry, has not been examined for energy conservation because of the more obvious opportunities for reduction of energy use in the 33 percent of energy use in the operation of buildings. Although the industry is fragmented and highly complex in its organization, there are numerous opportunities for energy reduction based on a detailed understanding of the industry. Methods include substitution, industrial efficiency, regionalism, structural refinement and extended building use.

ENERGY USE IN BUILDING

The term "Building Industry" is in common use, yet we seldom consider its performance as an industry when we think of the various groupings that use energy. We have more and more evidence that industry in general performs at an efficiency level far below what it might. We must now examine the Building Industry in detail since in its entirety it is the largest single industry in the country. As will be developed by other papers dealing with the specific analysis of the industry, there are important ways through which tens of millions of barrels of oil can be saved through more efficient industry-wide performance.

The building industry is a complex amalgam of material producers, manufacturers, designers and assemblers. It includes the mining and lumbering sectors as well as the steel mills, the sawmills, the aluminum smelters, the cement factories and the brick kilns. It includes the glass factories, the curtain wall manufacturers and the millwork plants. It includes the lumber yard and the electrical and plumbing jobbers. And it includes the contractors, the architects and all the building tradesmen and subcontractors. The production of a single building is dependent on the coordinated marshalling of the products and participation of this enormous and diverse series of participants. None of them even as industries, represent more than a few percent of the entire product of the building industry. Yet as an industry, its statistics are reported to government agencies in recognized categories of the Standard Industrial Classification, and its cohesiveness as an industry can be confirmed by looking at the precision of trade union jurisdictions on a construction job. Its very variety has made it an elusive target.

In the past the stress on buildings and their relationship to energy use has been almost entirely in the energy required to operate these buildings. This is understandable and important since it represents probably 33 percent of all national energy use. There has been an underestimation, however, of the energy required to build, alter, and maintain the buildings that will be required in the future. These, according to a study we conducted jointly with the University of Illinois Center for Advanced Computation, require 6-1/4 percent of all energy use for all purposes. Together with non-building construction, it totals over 10 percent. There are enormous patterns of variation in what is required to build buildings, what is required to produce the materials for buildings, what the relationship is between capital energy - that is, the energy to build the buildings and the operational energy, the energy to permit them to serve their purpose through their useful lifetimes. In order to understand these variations and to be ready to go to the next step which is to improve the efficiency of energy in the building process, the nature of the building process and the way in which energy is embodied in the components of the building must be understood and given proper quantitative value.

The building industry is highly fragmented and its product is an assembly of thousands of parts produced by a wide range of industries and brought to the jobsite or to the assembly plant, shaped for compatability to all of their neighboring products and finally, assembled as buildings, no two of which are identical. The number of different manufacturers, trades, skills, artisans and procedures that are called into play in order to produce buildings is enormous. The assembly process is equally diverse, depending on skilled craftsmen and unskilled labor. As a result, it is a difficult process to understand; and after one understands what the true nature of this process is, a formidable problem if one seeks to change it. In addition, since the breakdowns of the production of buildings reveal the enormous number of contributors to the process, each dealing with a small percentage of the total end product, the first reaction is one of despair that anything can be done to introduce more rationality into this highly diversified group of contributors. There are factors working the other way, however. Probably the most important is the enormous size of the building industry. It is responsible for about 10 percent of the Gross National Product (GNP) with an end value of over 100 billion dollars annually. Its energy use, however, is considerably greater than its proportionate share of the GNP, largely because it is a major user of great quantities of energy-intensive materials such as steel, aluminum, cement, bricks and glass, bulk materials whose initial manufacture requires large amounts of energy. For example, there may be 10 times the weight of material per dollar in a building as there is in an automobile and 100 times the weight of material per dollar that there is in an office machine. As a result of this high energy use, even what appears to be a small fraction of the entire industry, is an important user of energy.

Further, the largest category of use in the breakdown of the percentage contributions by different sectors is direct energy expended on the jobsite for the construction of buildings, a use that represents about 15 percent of the entire 6-1/4 percent. It becomes apparent that the method of building offers significant possibilities of energy reduction. In the future, either through the expense of the fuel expended in this part of the job operation or as the result of a rationing of fuel and electricity, a great deal of the wastefulness in field use of energy will be removed. Examples of temporary heating and lighting can be cited that use excessive energy on the jobsite. Buildings under construction are provided with temporary heat, enclosed only with pieces of plastic or tarpaulin which are flapping on the outside while propane heaters burn up large amounts of gas to maintain above freezing temperatures. Most of the heat is lost to the area outside the building. Temporary lighting is usually done with large wattage, inefficient incandescent lamps, casually strung throughout the work area of the building under construction and characteristically turned on at the start of the building day and turned off at its conclusion. The proper selection, placement and use of these fixtures could reduce this aspect of energy use. While temporary heat and light have been noted, energy is also used to run construction equipment, trucks, hoists and pumps. Since efficiency of fuel use has never been a consideration, there are undoubtedly worthwhile savings that can be expected.

In addition, the energy efficiency of major industries would have its effect on the energy embodied in construction products incorporated in buildings. Improved efficiency in the steel industry would result in a lower embodied energy in steel products.

At the moment the most carefully documented summary of energy use has been developed through the Energy Input-Output matrix at the Center for Advanced Computation of the University of Illinois, using 1967 as the data base. (1) This is the last year for which sufficiently detailed information was assembled to determine energy use by the construction industry on a yearly basis. The information originally recorded in dollar terms has been translated into British Thermal Units per building type, per building component and per unit of material. There have not been sufficiently significant changes in the building industry's production of material to invalidate the units derived from this study. There would be modifications in the totals. Where we have a figure of 702,000 Btu/sf of single family residences applied to 844,000 single family residences in 1967 (according to Statistical Abstract of the United States, 1971, page 668), the Btu/unit would be applied to 896,000 single family residences in 1975 (Statistical Abstract of the United States, 1976, page 737). The average cost in energy for lumber, bricks, and similar materials has not changed materially during that eight year interval. The figure reported to the F.W. Dodge Co. for total square feet of single family residential construction was 1.051 billion square feet in 1967. Dividing this figure by 884,000 units, gives us a per unit size of just under 1,200 square feet, which is a reasonable figure to use today. In other words, the general information derived from the Energy Input-Output Matrix, using this 1967 benchmark year, gives us information which is sufficiently accurate to make policy decisions based on what their impact will be on the whole construction industry. From a policy-making point of view, the limitation in using the information is that the figures are average figures, reported industry-wide across the entire United States, and with average intensities for what are called the margins in the Input-Output terminology. These margins represent the transportation, the pro-rated cost per plant and equipment, the overhead and office costs and the costs of doing business such as the advertising costs and the profits. Since all of these are average costs, in order to determine how susceptible these are to change, one would then have to know first, what the upper and lower limits are which contribute to that average, and what kind of performance standards each of these upper and lower limits provide. On that basis, knowing what constitutes a satisfactory lower limit of expenditure, one can assume that if this were to become the average, or, allowing for a certain human fallibility, if a somewhat higher figure were used as the new average, one could make a projection of the kind of savings that might be achieveable in the construction industry. Going beyond this, if one could then institute more efficient methods of building, of producing components, of estimating buildings and of planning, one could then begin to lower the average to something that may represent a new level of efficient material and energy use for our building industry.

Parallel investigations can be made in improving the energy efficiency in the manufacturing process. For example, reports comparing energy use in Sweden and West Germany with American industry indicates surprisingly large margins, suggesting that we can anticipate reducing energy use here. For example, in comparison with Sweden, our steel industry uses 1.293 times as much. (2). In comparison with West Germany's steel industry, it uses 1.558 times as much. (3). After materials have been efficiently produced there is also a margin of reduction that can be achieved by using these efficiently produced materials closer to their strength capability. The difference in wood use, for example, in the wood hull of a sailboat in comparison with the wood shell of a vacation house is striking, even though the boat hull is subject to much greater stresses and strains in movement than is its landbound counterpart. Where spaces have been designed for mobility, as in ships, trains and trailers, both material use and space standards are subject to a more rigorous analysis. One can compare a stairway in an early New England house built by a ship's carpenter with the more extravagant space allocations of a typical speculative house builder today. One can also compare the intensive and often very efficient space use in a trailer home designed for mobile vacations with the less efficient space use in tract houses. Or the efficient arrangement for washrooms developed for the Pullman car and the air transport in comparison with the space allotments in our standard construction technology. By a selective borrowing from and cross-referencing of decisions among these various space builders, we can expect that there will be a further reduction in the amount of energy necessary to construct the space needs for our various activities.

CONCLUSION

In summary, while the building of new buildings requires about one fifth as much energy as is expended in operating buildings on an annual national basis - that is, the energy to operate our entire stock of buildings for a year is five times greater than the energy to build one year's worth of new buildings - the picture is quite different when one looks at the constructional and operational energy of individual buildings. As operating energy becomes more efficient, the ratio of operating to capital energy changes. Based on the Federal General Servides Administration energy budget of 55,000 Btu per square foot per year for on-site energy use (the equivalent of about 110,000 Btu of source energy) and an average figure of 1,650,000 Btu to build a square foot of high-rise office building, it will take 5 years of operation to equal the amount of capital energy required to build the building. While the proportion varies from building type to building type and from one individual case to the next, nevertheless, as the effort toward more efficient energy use in all areas of the nation's activities becomes more necessary, the opportunities for reduced energy use in construction will constitute a worthwhile target. The degree of success in achieving it will result from a complex series of apparently small actions including:

- *selection of the least energy intensive materials or assemblies to achieve predetermined performance levels
- *improved methods of construction to reduce on-site energy use.
- *more accurate engineering to determine precise amounts of required material.
- *extended use of worthwhile existing structures to avoid unnecessary new building.
- *improved production methods to reduce embodied energy per unit of material.
- *greater use of regionally available materials to reduce unnecessary transportation.

Other presentations will examine some of these opportunities and their implications in more detail.

REFERENCES

(1) B. M. Hannon, R. G. Stein, B. Z. Segal, D. Serber and C. Stein, Energy Use for Building Construction - Final Report, ERDA Doc. No. COO-2791-3, Center for Advanced Computation, University of Illinois at Urbana-Champaign, December, 1976

(2) Schipper & Lichtenberg, Efficient Energy Use and Well-Being: The Swedish Example, Science, 3 December 1976, Vol. 194, p. 1001

(3) R. K. White & R. Goen, Comparison of Energy Consumption Between West Germany and the United States, Stanford Research Institute, Menlo Park, CA, June, 1975

It should be borne in mind that there are difficulties in making comparisons since process and reporting techniques differ. For example, there is a greater use of scrap metal by the West German steel industry. While this results in a lower per ton energy requirement, it is also a reflection of an economy that still characterizes American industrial methods. There are the beginnings of greater economy in energy use on an industry-wide basis, with reports of savings of 5% and more in steel and other industries.

ENERGY CONSERVATION IN THE BUILT ENVIRONMENT
ENERGY ACCOUNTABILITY SYSTEM

Norman I. Turner, Manager
Architectural and Building Research
Southwest Research Institute
8500 Culebra Road
San Antonio, Texas 78284

ABSTRACT

The construction and occupancy of the built environment uses from thirty to forty five percent of the total annual energy consumed in the United States. Those responsible for the design, materials, selection, construction, occupancy operating programs etc. for this built environment with its annual energy demand are the design professionals; architects, engineers, and planners. They, through their design implementation have the greatest influence in energy conservation over the greatest single segment of the total energy demand. Thus, to afford the greatest opportunity in achieving maximum cost-effective energy conservation solutions to each built environment program, the design professional must have available a rationale or system which will effect an accountability of resources and energy for the design, construction and occupancy of the built environment.

Such a system with its associated rationale, aids, accountability, evaluation mechanisms, and recording processes does not exist. The creation and implementation of the system should be given top priority to insure continuance of design freedom, to insure competitive development within industry, and afford an accountibility of energy-resources within the professions as well as the regulatory authorities locally and nationwide. Three essential segments of the system requiring development are (1) Energy Budget Concept: EB: the annual energy per building use category, per climatic environment, per floor area unit that would be allocated to the building owner under operating conditions; (2) Energy Resource Number: ERN:-an index assigned to each construction material item on a unit basis which identifies quantitatively the energy consumed to produce it; (3) Energy Accountability Assessment Network: EAN:-a network and rationale to establish the flow of information from records of energy consumption, projected and operational, from the design professions through regulatory and review authorities from local or municipal, to region to state to national level: to be available for assessment on a periodic basis of the current energy situation against any planned overall energy goals.

ENERGY-FREE COOLING SYSTEMS FOR HOUSES IN DESERTS

Gideon Golany

The Pennsylvania State University, University Park, Pennsylvania

Because of our current energy crisis and the need for housing and feeding the world's ever-increasing population, arid regions have become important for cultivation and habitation. Housing in such regions must protect people from extreme dryness, the wide daily range of temperature, and intense radiation reflected by the ground. A typical modern solution to these problems is to air condition all buildings, an energy consuming process. Its cost can be prohibitive where the climate is hot most of the year. This chapter will describe possible ways of eliminating the need to expend large quantities of energy to cool buildings and of designing cool houses that even low-income people may afford.

We wish to propose an integrated system that combines three existing methods of alleviating the effects of a desert climate: a subterranean habitat of a type used in ancient China and North Africa (see Fig. 1); ventilation systems used in housing in ancient Mesopotamia and modern Iran; and humidification methods used in modern Israel. We are integrating these three systems to suggest a technologically advanced system that may be more efficient than the three individual ones.*

The climate of the immediate environs of a house is a result of direct solar radiation, reflected radiation, soil composition and its effect on reflected raditation, winds, and the relative humidity of the air, vegetation, and topography. The microclimate of a house is a combination of the climate of its site, the size and number of windows, the thickness of the exterior walls and their composition, the size of the shadowed area outside the house, the design of the openings, and finally, its orientation to the sun.

BUILDING IN THE GROUND

Design may significantly affect these microclimates. Any such house may include a subterranean section, a ground level, possibly a first story, and a roof. This basic pattern may be realized as detached, attached, row, or clustered houses. For our convenience in discussing the design, we will use a detached house in the northern hemisphere as an example (see Fig. 2).

In order to minimize the effects of hot dry environs upon a house and to maintain a comfortable temperature within it, the house could be constructed at least partially below ground level. The ground usually maintains a nearly constant temperature during a season and retains water, which by capillary action will move from the outer to the inner walls and will evaporate and thus moderate its microclimate.

Soil structure is, of course, an important factor in building and operating such a house. Dry soil will conduct some heat to subterranean walls and also radiate heat to them above ground causing a constant evaporation of ground moisture. It may be necessary to water the walls and adjacent ground to keep moisture in the lower part of the house.

Soil such as loess retains moisture for a relatively longer time than some other types under the same conditions. The villagers of some ancient civilizations, such as the Chinese located upon the loess soil of northern China, recognized this wisdom and

---

*I am aware of the need to justify, clarify, and prove the applicability of the assumptions of this concept through research, construction, and operation of an experimental system. The valuable comments of Professors Eliezer Kamon and Raniero Corbelletti, both of The Pennsylvania State University, are gratefully acknowledged. This paper is reprinted with permission from Innovations for Future Cities, ed. Gideon Golany (NewYork: Praeger, 1976).

constructed some of their villages below ground level. (See Bernard Rudofsky, Architecture Without Architects [New York: The Museum of Modern Art, 1964], figures 14-18.) In a dry climate loess soil develops a thick crust and retains moisture. In soil that is subject to quick evaporation, narrow zones of crust formed on the soil adjacent to the walls (.5 to 1 meter in depth) could help retain moisture.

For any shape of house it is necessary to have an extended roof to shade the walls. Because rain is rare in deserts, roofs are usually flat for sleeping. We may decrease the sun's effect on such roofs by building sidewalls for shading or by using a sloping roof with two controlled openings opposite each other for ventilation. Trees around the house may also decrease direct radiation on the walls. Trees without bottom and middle branches and with sparse tops may shade the house without obstructing its view. Such plants should be close enough to the house to provide shade but distant enough to minimize their absorption of water from the subterranean walls. It is also possible to reduce indirect solar radiation by planting ground cover around the house. It is also necessary to make the external walls of the house, or any covering materials, of a light color, which will reflect most solar energy.

The interior of the house may be entirely closed and compact, one floor above the other. This type has the advantage of full control of the entire space of the house. It may have a central, vertical open shaft from the basement to the upper story. (See Fig. 3.) The windows at the top floor must open to allow warm air to flow from the house.

A second type of interior is defined by a large central courtyard extending upward, not from the basement but from ground level. This is commonly known as an oriental or Mediterranean house: in which rooms on all floors face three sides of the courtyard, leaving the northern side without rooms. (See Fig. 3.) The top of the courtyard could have a removable cover to block direct sunlight during the day. At night, the cover could be removed to allow the house to cool.

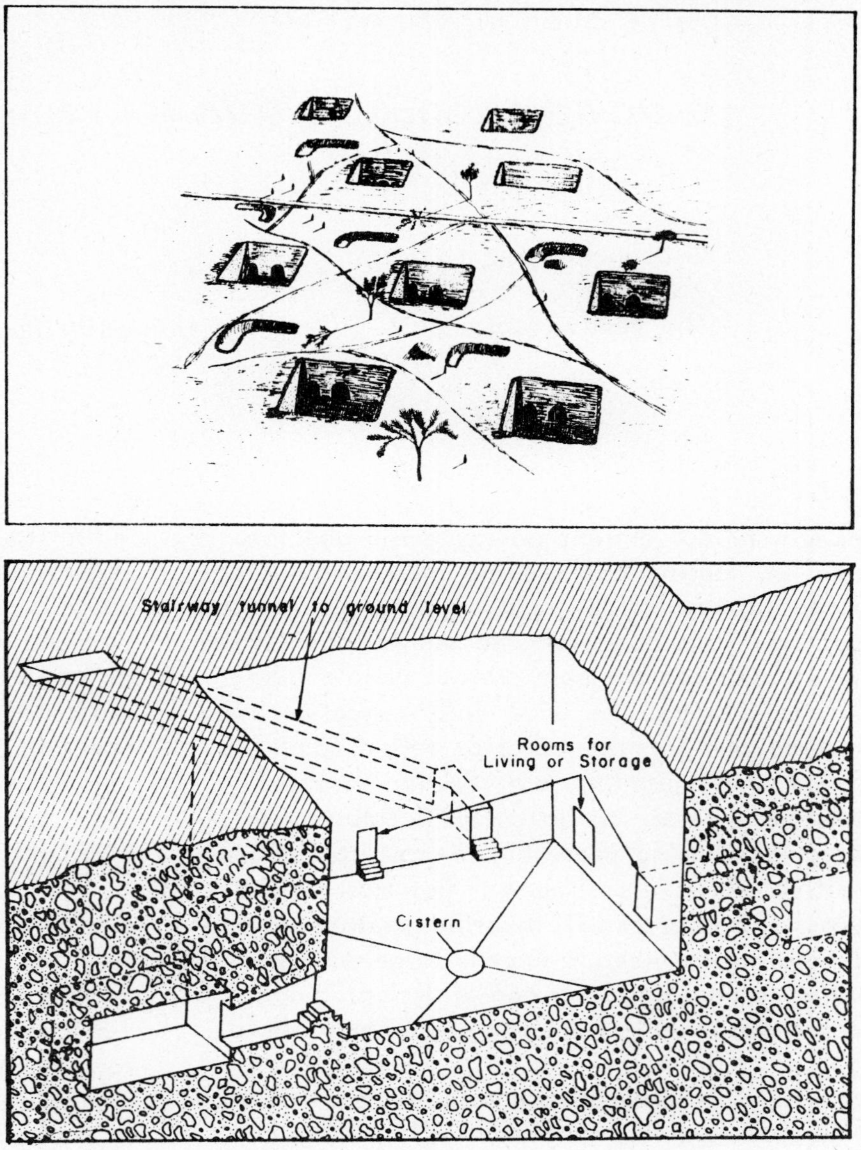

Fig. 1. Subterranean Houses in a Chinese Village

Source: Bernard Rudofsky, Architecture Without Architects (New York: The Museum of Modern Art, 1964), p. 15.

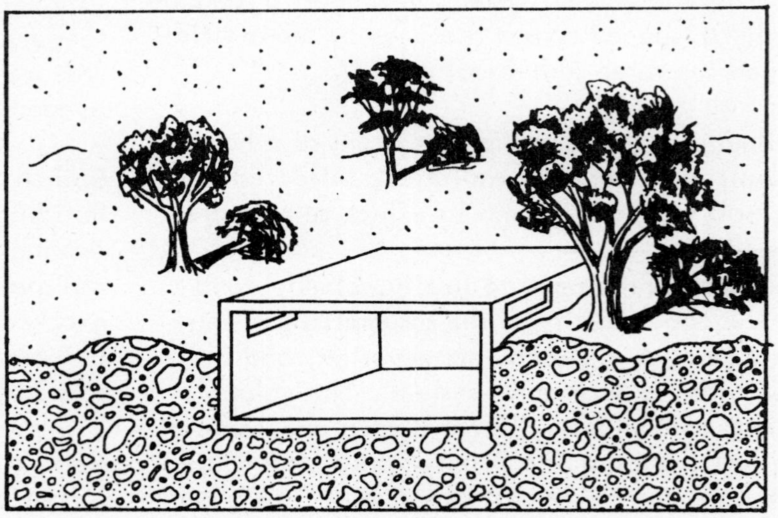

Fig. 2. A Detached House with a Subterranean Floor

Source: Compiled by the author.

VENTILATING

A second feature of design that may moderate these houses' microclimates without expending costly energy is a system for ventilation. One possibility is a system built vertically through the walls at frequent intervals. Rectangular or elliptical tunnels could be accommodated within the walls, either centrally between their sides or asymmetrically, close to their interior sides. (See Fig. 4.) Tunnels could be prefabricated in separate standardized units to be joined for proper length. If the walls were cement, the tunnels might be placed first and then the walls constructed around them.

On the roof the tunnels should open toward the north. Walls facing north would cast a shadow during daylight hours, which would also reduce reflected radiation. Consequently these shadowed areas would be cooler than those exposed to the sun. Regardless of the wind's direction, the openings of tunnels directed to the north, especially those built at the top of a high wall, would catch wind after it had crossed the shadowed area. An opening wider than the tunnel would catch air, which would be forced to channel through the tunnel.

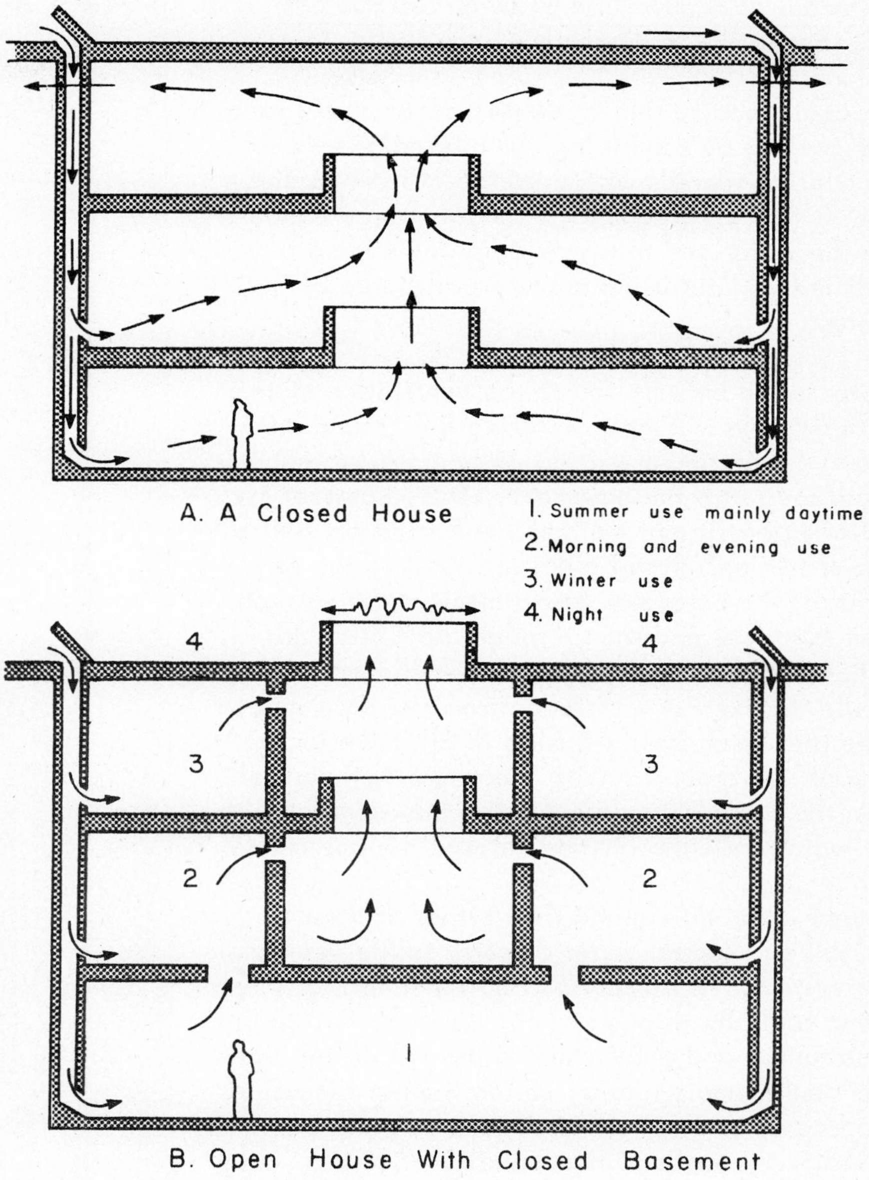

Fig. 3. Two Possible Housing Plans: A Closed, Compact House and an Open House with a Closed Basement

Source: Compiled by the author.

The house should also have an extended top for shadowing. (See Fig. 4). If the house is mostly subterranean, the openings should be well elevated to decrease the amount of dusty air that may be caught by the tunnel in stormy weather and to catch air that is cooler than that immediately above the ground. To avoid insects or even dust from entering a tunnel, a special filter might be built. Both ends of each tunnel should have controlled shutters to regulate the air flow to particular areas of the house. In the daytime, regulation would be most useful in the afternoon hours and sometimes before noon. The tunnels would need to be closed entirely at night, early morning, or sometimes at dusk. See Fig. 5 for an illustration of the pattern of circulation within a two-story house.

HUMIDIFYING AND COOLING

A third feature of design for cooling and humidifying a house is not new. Ancient Mesopotamians in cities along the lower Tigris River constructed sections of their houses below ground level and used them for living in the summer, especially in the afternoon. The uptide water of the river, supplementing the water table, increased the humidity of the soil and therefore improved the cooling system of the basements of the houses. The Mesopotamians combined this system with a channeled system for ventilation in order to lower temperatures within the sections of the house above ground and to cool water in big ceramic jars, which were placed beside the outlets of the ventilation system. (See Fig. 6). Hot air flowed past the jars, evaporated the water on them, and consequently decreased not only the temperature of the water in the jar but also of the air. (See Askar Reuther, Das

Wohnhaus in Bagdad Und Anderen Stadten Des Irak [Berlin: Ernst Wasmuth A.-G. Berlin, 1910].)

We could modify this crude method of humidifying and cooling by combining our internal system of ventilating tunnels with a source of water. The hot air flowing through the tunnels would evaporate the water and thus would be cooled and humidified without using an external source of energy.

Water could be supplied to the ventilation system in three ways. Simply watering the ground adjacent to the house would supply moisture that could seep into the basement. This, however, is not a very efficient method. A horizontal belt of one or two perforated pipes surrounding the exterior of the basement would allow water to seep from the pipes and then through the walls. In either of these methods, nonpermeable materials should be fixed at a distance from the basement to keep the water from diffusing in all directions through the ground. With these two methods, air from the tunnels would not come into contact with the water until the air had entered the basement.

A more efficient method than either of these would be to supply water directly to the ventilating system. In this humidification-cooling system water could be dripped on the filter at the top of each tunnel and small water pipes within the tunnels could supply a mist, which the hot air would evaporate, or a system of perforated pipes could humidify the walls of the tunnels. (See Fig. 7.) The water pipes, of thin but strong permeable material, could run vertically through the walls in the space between the tunnels. Pipelines could be flush with the walls so that water, when run through the perforations, would immediately hit the outside of the tunnels. (See Fig. 7.) This position might minimize water consumption. The air would become increasingly cooler, more moist, and heavier. This would cause it to sink through the tunnel. By the time it emerged from the lowest opening of the interior tunnel of the basement, the air would be cool and relatively humidified. Because there would be an opening in the upper walls to allow hot air to escape, the cool air would circulate upward through the house to change its microclimate. (See Fig. 5.) The oriental design, however, would not allow this system to work effectively beyond the basement level, and may require an independent humidification-cooling system for each floor. (See Fig. 3.)

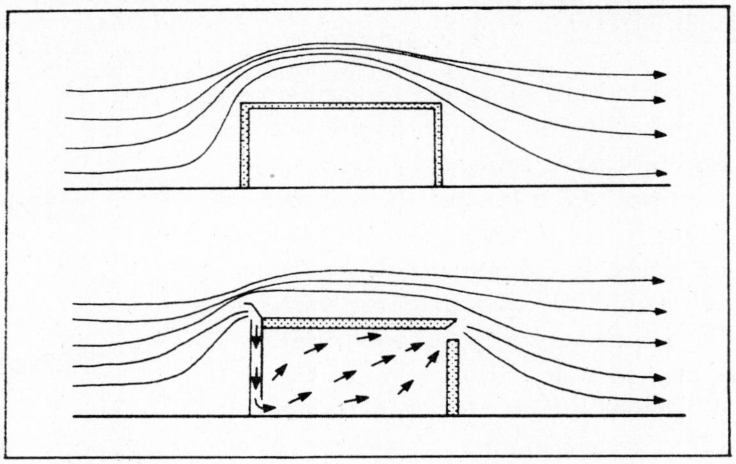

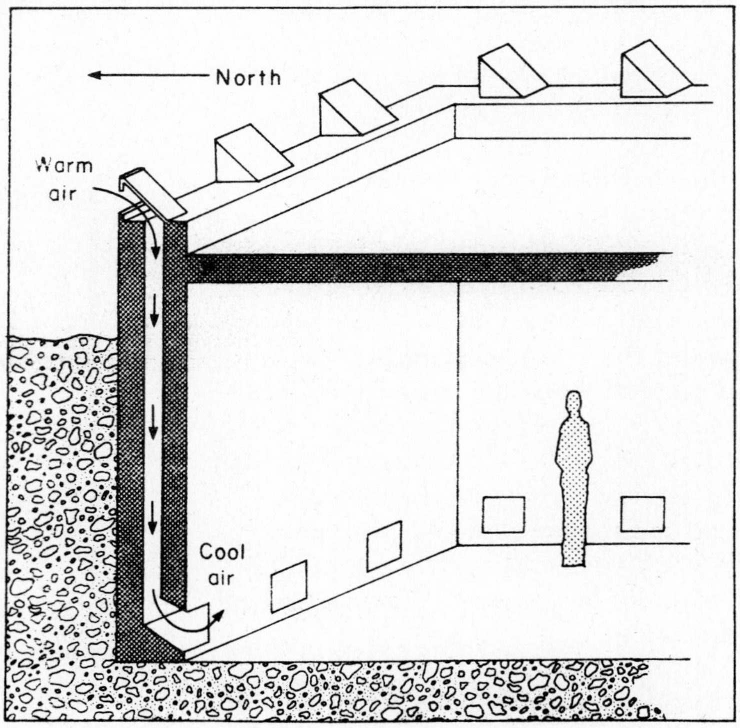

Fig. 4. Ventilation System within the Walls of the House

Source: Compiled by the author.

Although the proposed system has not yet been tested and there is no estimate of the amount of water required to humidify the air, we may anticipate that the amount required will be small. Water consumption rates would also differ among the three alternative methods. Houses built partly below ground would require less water than those with walls partially or entirely above ground and exposed to sunshine and winds.

This house is proposed for deserts that have rivers, but in addition to such local sources, water may be collected from dew on roofs, run-off stored from rare rains, or that recycled from household use. Water with a high concentration of salt should not be used because it may cause damage to the walls over a period of time. In any case, water should enter the house through an underground system of pipes between the foundation and the surface of the ground, to keep the water relatively cooler than the air. Also, water could be cooled at night by exposure to the sky.

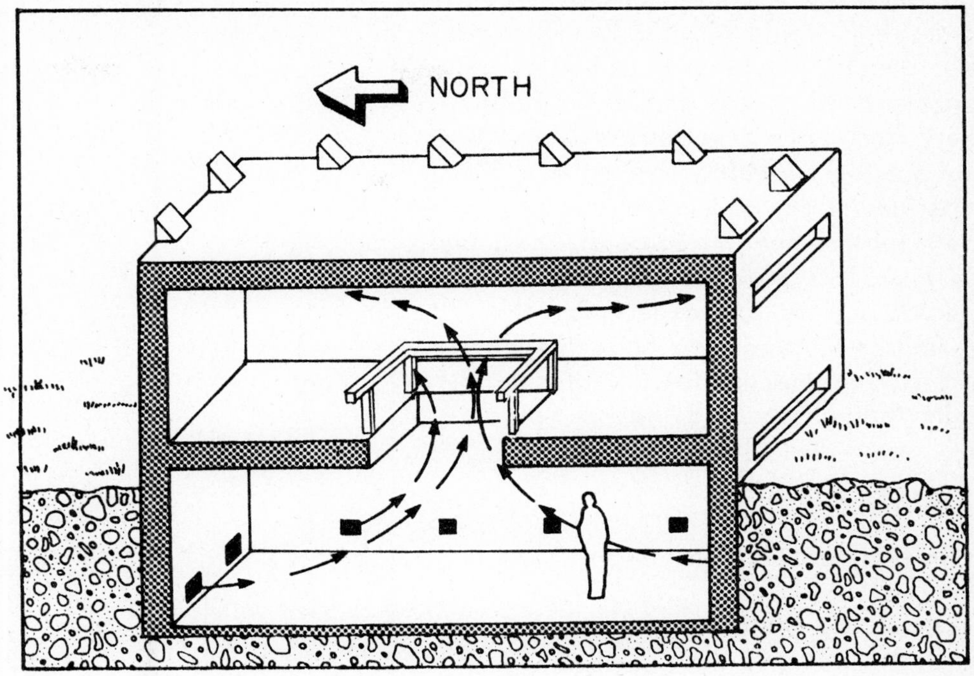

Fig. 5. Possible Design of a Two-Story House and its Pattern of Circulation

Source: Compiled by the author.

To make the best use of this humidification-cooling system, the planner/designer of a house should carefully consider its building materials, particularly those for the roof and walls. Materials have specific properties that affect heat conductance and capillary action and its duration. Limestone, for example, may conduct water less readily than concrete does. It is necessary, therefore, that the humidification-cooling system within the walls be of porous materials with low heat conductance. Thick walls made of stone and gravel would retain humidity for a relatively long time, especially in the summer. Cement, building blocks, gravel, and limestone should allow effective capillary action within the proposed system. If building blocks were to be used, their holes could match each other to form a tunnel, which would eliminate the expense of building special air tunnels; sections of perforated plastic or other porous pipes, such as unglazed clay, might be built into the block close to the air tunnel. (See Fig. 8.)

Both the tunnels and the pipes must have regulators, because when the outdoor temperature dropped below a certain point it would be unnecessary to operate the systems. Shutters at the tops of the tunnels would keep wind from entering them. Valves regulating the water intake should be placed in such a way that they can block the major entrance to the pipes and also each pipe individually.

Using the system in the early morning would not be necessary, since the house and its air would have cooled during the night. The timing of opening and shutting the two systems might well differ. The tunnel system would react immediately to being shut or opened, but the humidification system would not respond as quickly, which means that turning the system on would bring more immediate results than shutting it. This is simply because water penetration would take less time once the system is activiated than the drying process takes when the system is stopped. This difference in timing between the two systems may be regulated after preparation of a schedule of operation based on experimentation or on the residents' experience with the system and the unique nature of the climate of the site and house.

There would, of course, be a few problems to be faced in the construction and maintenance of this humidification-cooling system. Over time, some deterioration of the walls would take place; this decrease in the quality of the walls might be caused by overconcentration of salt from recycled water or by deterioration of the materials making the walls, such as stone, cement, building blocks, bricks, stucco, or clay.

The system would require that every air tunnel have its own independent control of the end shutters and that every one or two water pipes within the wall have separate controls. A careful integration of those items with the walls would be necessary to make them aesthetically pleasing.

Since they would be installed close to the tunnels, the pipelines would require a good control system to keep them from letting in too much moisture and possibly flooding the walls. The pipelines might also degenerate, and their replacement might be difficult and costly. This problem may require special designs and experimentation. It may be necessary to have the pipes installed as independent, replaceable units through spacing that is constructed especially for them.

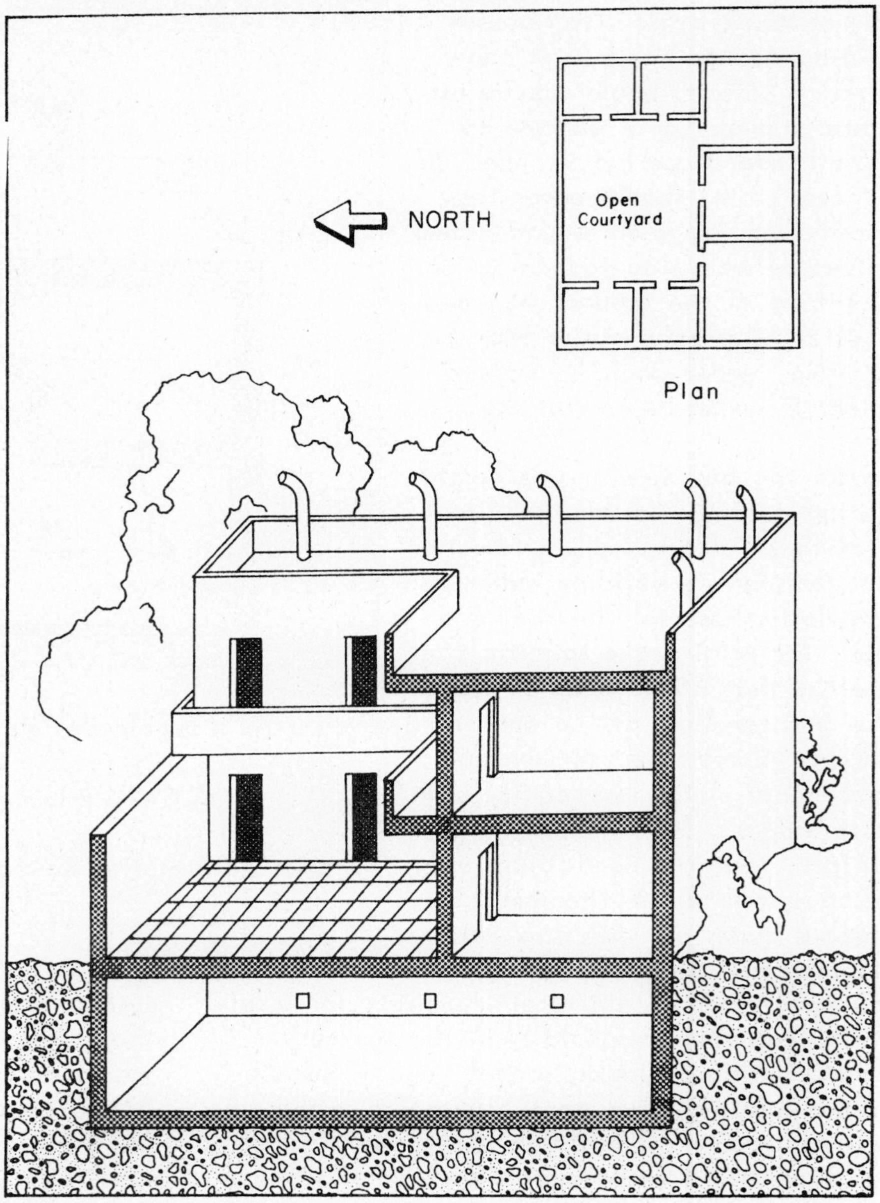

Fig. 6. Plan of an Ancient Mesopotamian House
Source: Compiled by the author.

USING THE SYSTEM

Optimal use of the proposed system depends upon the careful planning and use of each horizontal level of the house. The advantages of each level may vary by the season or by the hour of day or night.

The basement or cellar would be most used during the afternoons, which have the highest dry temperature, since a basement provides the type of microclimate that combines cooled air with the highest humidity in the house. (See Fig. 3.)

Circulation to the upper parts of the house would also begin from here. The basement should have a fairly constant, conditioned microclimate when the humidification-cooling system is efficiently operated. This section should cover the entire base of the house and might extend beyond the foundation. If the extended sections are lighted, they may be cooler than the central basement. A basement rising above ground level may have windows on the northern and eastern sides to allow natural light to enter. This basement may also provide sublevels for different accommodations.

The first floor may be used the most before noon and in the evening in a compact house. In the open, oriental house, this floor would not be comfortable for use in the afternoon, but might still be useful in the morning and early evening. In the open house, use depends upon the orientation of the house, the intensity of solar energy entering the courtyard, and the duration of the of the energy. Plants around this floor may moderate its microclimate somewhat and increase its use during the day. However, the proposed humidification-cooling system might be directed to the rooms located in this section or the courtyard itself to enhance ventilation.

An upper floor in an open house with a courtyard may be useless during the summer, since hot air from the courtyard would move upward toward this floor during the day. However, this would be the best floor in the open house for use in winter. In the compact house, this story might have the humidification-cooling system directed to it to modify its microclimate. The ground and upper floors might be cooled by a direct connection with the humidification-cooling system from the roof down or by channels of the system into the basement and then to each floor. This channeling might effectively introduce humidified air, since air will travel a long way through the humidified tunnels. If a direct connection is used, shorter tunnels might provide less humidity.

The humidification-cooling system might also be applied to high-rise buildings, although its effectiveness would probably be less than for a detached house with a basement. For a high-rise building, a builder cannot rely on trees for shade to cool the walls, but the effect of ground radiation should decrease on each successive floor.

Also, wind for ventilation should increase with the building's height because there would be fewer obstacles blocking the wind, giving the high-rise a ventilation advantage over single housing units. We may also expect that at night the temperatures will be lower on the upper floors of the building than on those closer to the ground.

CONCLUSION

Once the system is demonstrated, it may be commonly used by residents of an arid area, if modular units of the system combined with the walls are prefabricated. We may anticipate that the investment in the construction of such a house would be slightly higher than for a standard house of the same materials and size. This difference in cost might be reduced once the techniques of modular unit production become common.

The cost of maintenance, if any, would be small. Moreover, the cost of system maintenance will be lower than for other systems that have been developed to achieve the same results

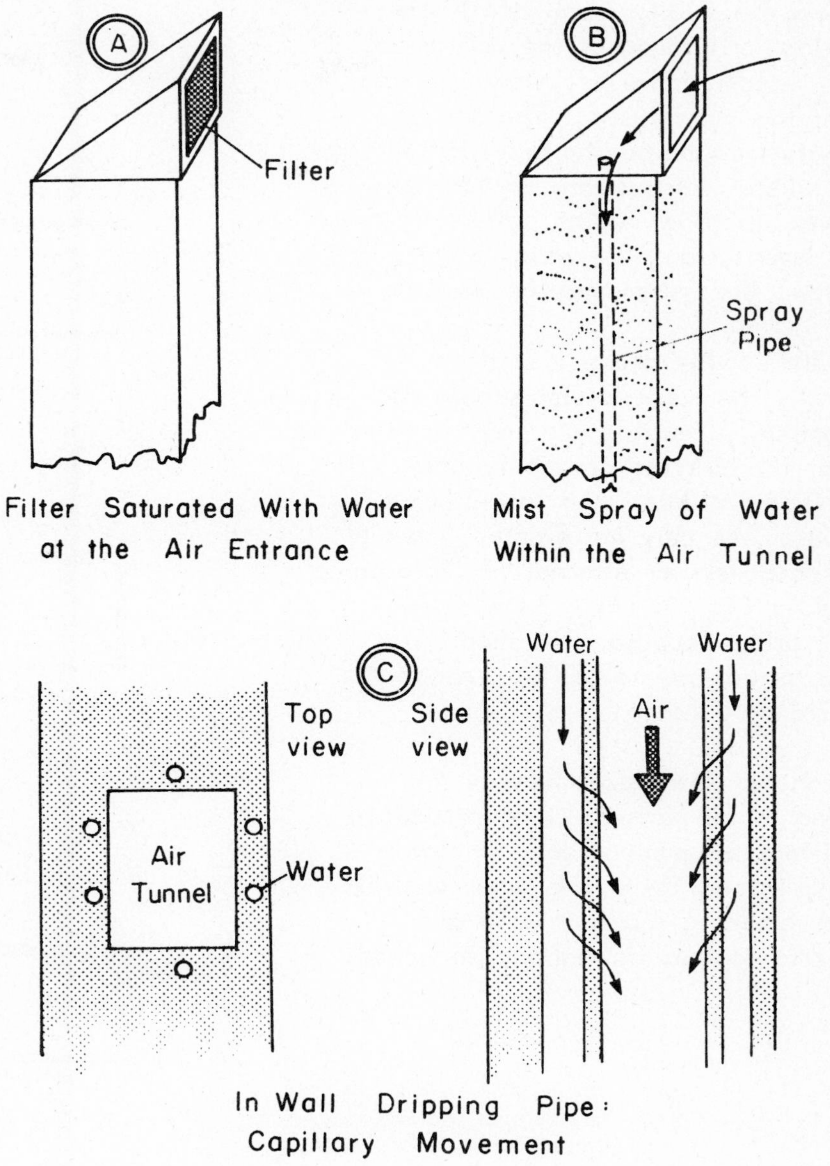

Fig. 7. Three Alternative Methods for Humidification Within Ventilating Tunnels

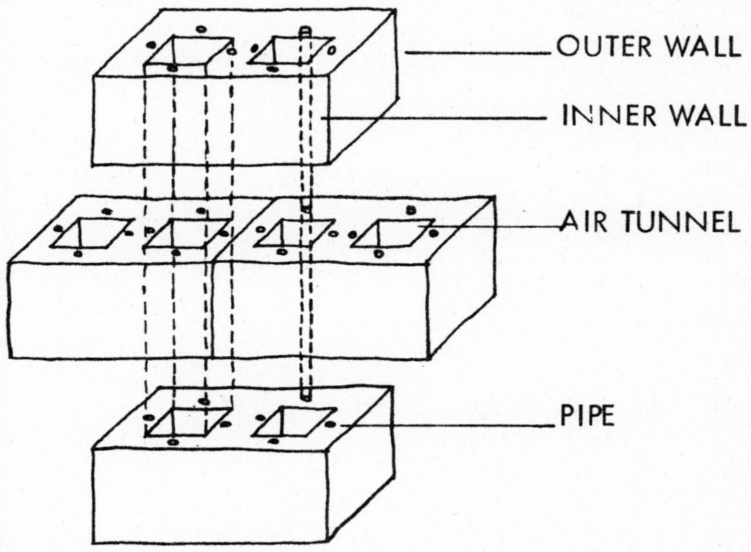

Fig. 8. Building Blocks with Holes Matched to Form Air Tunnels and Built-in Pipes for Water Supply

by using electricity, such as air conditioning. The cost of the operation of the system, however, would be the cost of the water used. The use of the system might be limited in arid areas where water is rare or very expensive. In other regions where rivers cross deserts, as in Arizona, Egypt, or Pakistan, the system may be feasible.

Regardless of the prime investment in research, the house design should be economically feasible for construction so that it could be commonly constructed for middle- and low-income families. We may anticipate, however, that the no-system alternative (Alternative A in Fig. 9) would have the lowest price because of its simplicity, and its price may not differ from that of a standard house.

The concept introduced here is not assumed to be perfect; it is introduced to initiate the project and stimulate its development in a time when urbanization is taking place in arid and semi-arid regions throughout the world.

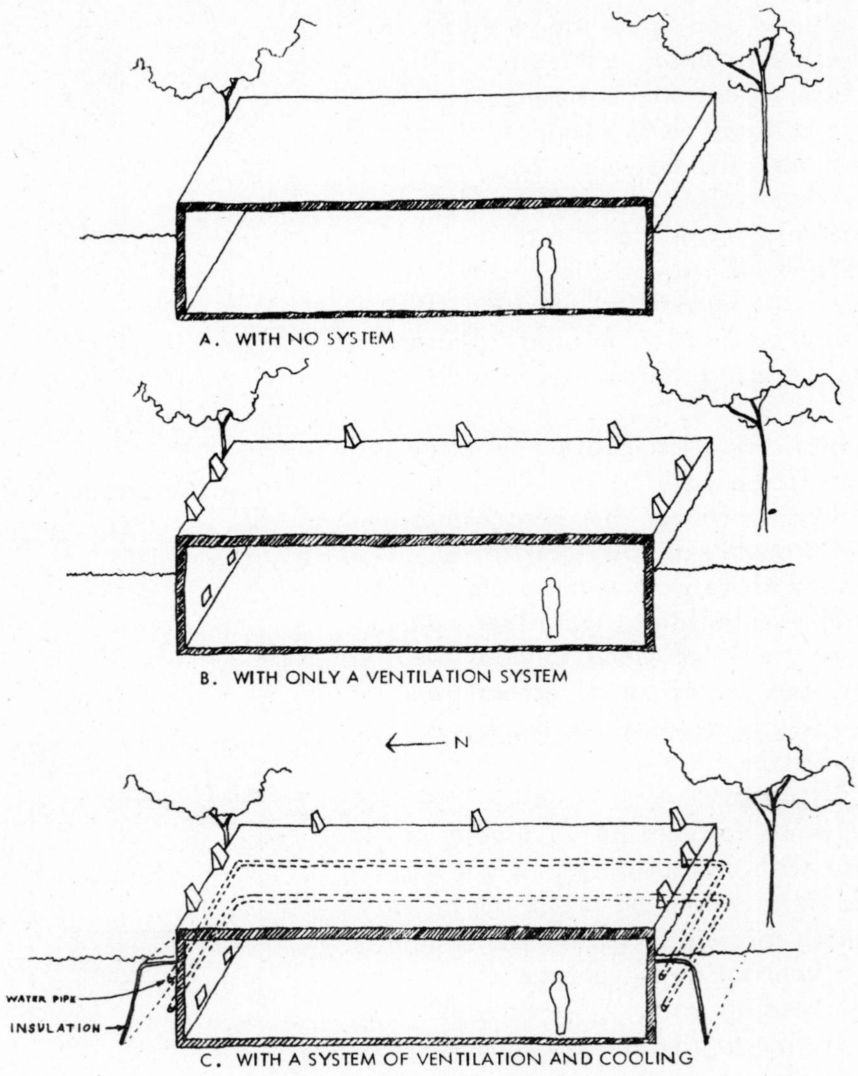

Fig. 9. Three Alternatives of a Below-Ground House Differing in Investment and in Cooling Systems

Source: Compiled by the author.

APPROPRIATE TECHNOLOGY AND ENERGY CONSERVATION: A CANADIAN EXPERIENCE

Brian McCloskey

Associate Professor, University of Petroleum and Minerals, Dhahran, Saudi Arabia

ABSTRACT

Many developing societies still have, yet are rapidly losing environmental attitudes and 'technologies' which are appropriate to their context. Inherent in such attitudes, and particularly manifest in the siting and construction of indigenous habitat, is a strong feeling for energy conservation. The project described herein was an attempt, in part, to identify indigenous technologies and energy-conserving building methods so as to develop, with modifications and additions, a form of housing appropriate to Indian reserves in Northern Quebec.

INTRODUCTION

In fall 1974 and spring 1975, research grants were awarded to the Shelter Systems Group by the Donner Canadian Foundation, the Federal Department of Indian and Northern Affairs and the Ministry of State for Urban Affairs as part of its 'Urban Demonstration Program' for the U.N. Habitat Conference. Four partially solar-heated demonstration houses were to be constructed for, in and by Quebec Indian communities. The design of each of the units was based to some extent on general housing needs/aspirations as expressed in 1973 and 1974 National Indian Brotherhood policy papers and in the Indians of Quebec Housing Study done by Montagnais Indians for Central Mortgage and Housing Corporation in 1974. These papers and reports helped us prepare general objectives relating to Indian desire for more autonomy in administration of housing policy/funds and training Indians in building trades. Earlier studies by our group, of indigenous Indian housing and its appropriateness climatically and technologically, in comparison with more recent and more urban reserve housing, led us into cooperative work with the Brace Research Institute, and considerations of solar and wind energy and northern sanitation problems. As visits were made to selected Indian communities to discuss early ideas for the four buildings, a solar hot air collector prototype was built to test the first Canadian application of the 'Trombe' wall. Flat rock (assembled without mortar) and loose rounded rock were used as heat storage media. After poor results were obtained due to excessive night heat losses from the 1.80 high x 1.80 wide wall through the double acrylic cover, 5 cm. polystyrene insulating sheets were added each night and removed in the morning, thus obtaining better results. It was hence decided that the 'Trombe' system and variations of it would be appropriate for use in all prototype houses. The vertical south wall collector would be much easier for our unskilled labor to build than a tilted collector; surprisingly little radiant energy would be lost as the upright collector would receive added radiation from snow reflection. Yet a manually-operated insulating curtain system for night cover of the collector wall would be required.

As the above solar-heating techniques were being developed, locations for three of the houses were finalized and house plans completed in consultation with Manitou Community College, the client for two of the houses. Models of the third house, at Mistassini, were being discussed with the owners, Anne and Philip Awashish. Soon the fourth house at Waswanipi was nearing design completion with the help of Alan Kitchen, the Band Manager, and future occupant.

It was finally recognized that all houses, given the costs and unreliability (in more isolated and more northern locations) of water systems, should use hot-air type systems, and could be expected to provide between 25% and 60% of the house heating requirement. Back-up heat would be provided by high-efficiency wood-burning space heaters and cook stoves in the more remote Mistassini and Waswanipi locations, and by electrical baseboard heaters at Manitou College. Building systems, like the heating systems, were designed to suit the specific site and client, with log building and rough-cut jackpine framing systems favored. Each house was to have a composting toilet.

By early May, 1975, the Manitou College building team had been assembled and excavations done on two houses. Mistassini began in June and Waswanipi in early July. The following is a description of each of the houses constructed.

The Cote Nord House (46°24'N;74°47'W)

Located at Manitou Community College, an indigenous vocational college, 170 km. north of Montreal, this house was designed by Guy Courtois, an Indian architect in collaboration with the Brace Research Institute and our group at McGill School of Architecture. It was designed to accommodate a small family and to be partially solar-heated using a hot-air, fan-assisted screen-type collector on a 90° south exposure and a rock storage system below the floor in the rear section of the building. Walls were constructed of 5"x5" jackpine squared logs with double tar paper horizontal seals. The north wall, earth-bermed on the outside to 1.20 is insulated on the inside with 10 cm. pressed peat-moss insulation sheets, which were developed specifically for the project by the Centre de technologie de l'Environement at Universite de Sherbrooke. Both jackpine and peat (sphagnum) moss are materials indigenous to most Quebec Indian communities. To reduce heat loss, windows on the north, east and west exposures are minimized, while the roof is shaped to maximize the solar collector on the south exposure. All windows are sealed double glass over which heavy curtains are drawn when light is not required.

The house was built by semi-skilled Montagnais Indians and supervised by Guy Courtois and Richard Kerr. It was occupied in February 1976 and is continually monitored by the occupants in cooperation with the Institut de Recherche de l'Hydro-Quebec. (1)

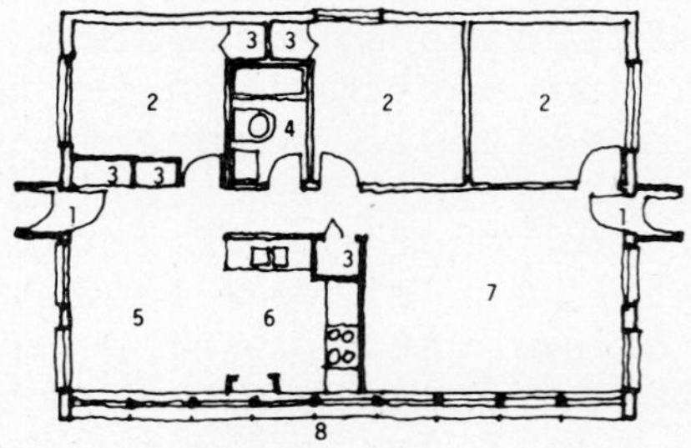

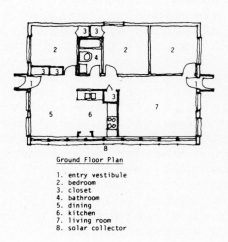

Ground Floor Plan

1. entry vestibule
2. bedroom
3. closet
4. bathroom
5. dining
6. kitchen
7. living room
8. solar collector

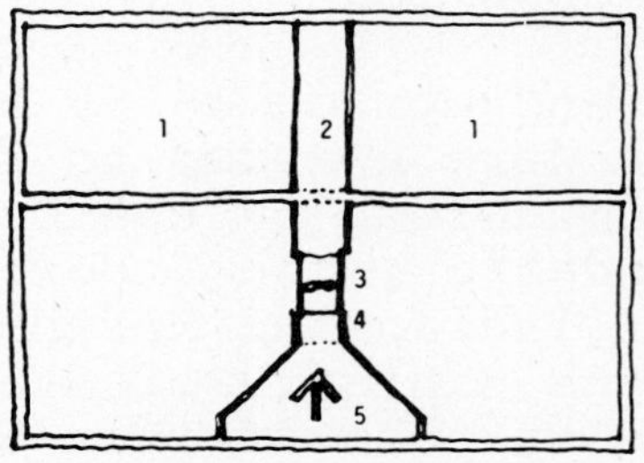

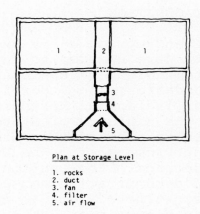

Plan at Storage Level

1. rocks
2. duct
3. fan
4. filter
5. air flow

Fig. 1 Cote Nord House Plans

Floor area = $93m^3$
Net collector area = $42.5m^2$
Rock storage volume = $22m^3$
Air-flow rate = $0.72m^3$/sec.

The Pte. Bleu House (46°24'N;74°47'W)

Also located at Manitou College and developed by the same team, this unit also accommodates a small family (3 BR.), but is two stories. Again, a solar-assisted electrical heating system is used. Here, the system is based on Felix Trombe's 90° south wall hot-air collector, relying on natural convection for solar heat distribution and a 25cm. solid concrete block wall for heat absorption and radiation. The system also uses a 7.5cm. fibreglas insulating curtain drawn manually on a horizontal track in front of the block wall at night and behind the wall during the day as shown in Fig. 4. The two-storey house is 3.60 deep, facilitating the operation of the natural convection solar heating system. The house was built by the same team as the Cote Nord house.

Floor area = 100s.m.
Exterior surface area = 258s.m.
Volume = 235c.m.
Net collector area = 41s.m.

The Mistassini House (50°25'N;73°53'W)

Located on an Indian reserve 320 km. north of Montreal this house was designed for a particular extended family who formed an essential part of the design team. It was developed to be partially solar-heated with wood heat back-up. Like the Pte. Bleu house, it uses a hot air, screen-type collector and circulates air using natural convection as shown in Fig. 5. The storage medium is brine-filled oil barrels, stacked vertically on the south wall, protected at night by a 3" fibreglas insulating curtain. Pressure-impregnated wood foundations (10-15% cheaper than concrete or concrete block)(2) were used, and all framing and cladding done using rough-cut jackpine. All windows are triple-glazed to reduce heat loss and, as in the LaMacaza houses, kept small on windward exposures.

The house was built by unskilled and semi-skilled local Cree workers.

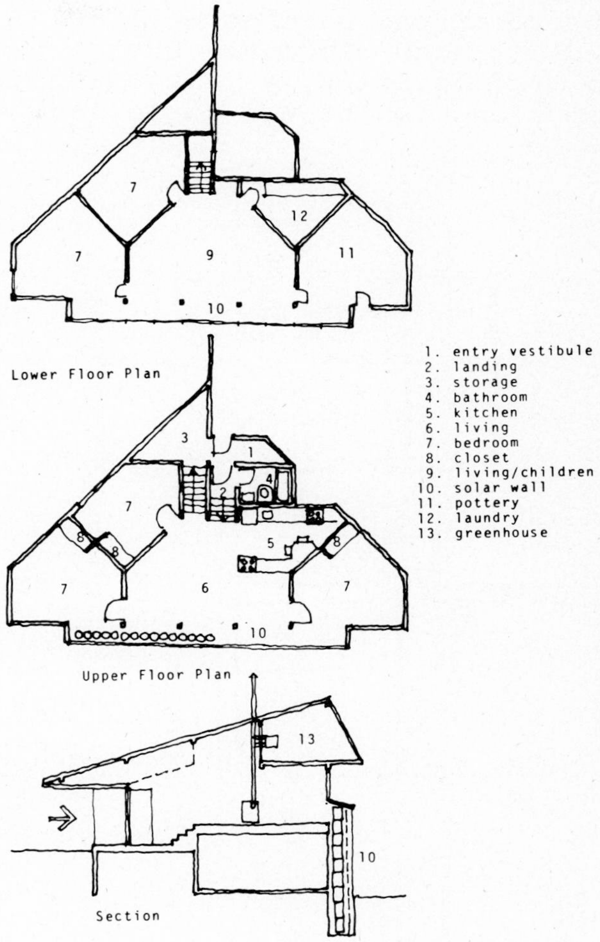

Fig. 2 Mistassini House Plans and Section

Floor area = 231s.m.
Exterior surface area = 431s.m.
Volume = 503c.m.
Storage volume brine = 9092 lit.
Net collector area = 39s.m.

The Waswanipi House (49°42'N;75°57'W)

About 160 km. west of Mistassini, this was the first house on a new Indian reserve. It combines a Band office (ground floor) with living accommodation (upper floor) for the Band Manager and his family. It was designed to be partially solar-heated with wood heat back-up, using a hot-air screen type collector and is 'passive' in mode of operation, as shown in Fig. 6. 1600 water-filled polyethylene bottles placed on a platform between the two floor levels constitute the heat storage medium, absorbing heat from horizontal warm air movement through the system, and radiating heat, primarily to the upper level, and to a lesser extent to the lower floor. Like Mistassini, rough-cut lumber was used as well as triple-glazed windows. Due to its isolated location, a 'Clivus-Multrum' composting toilet was installed.

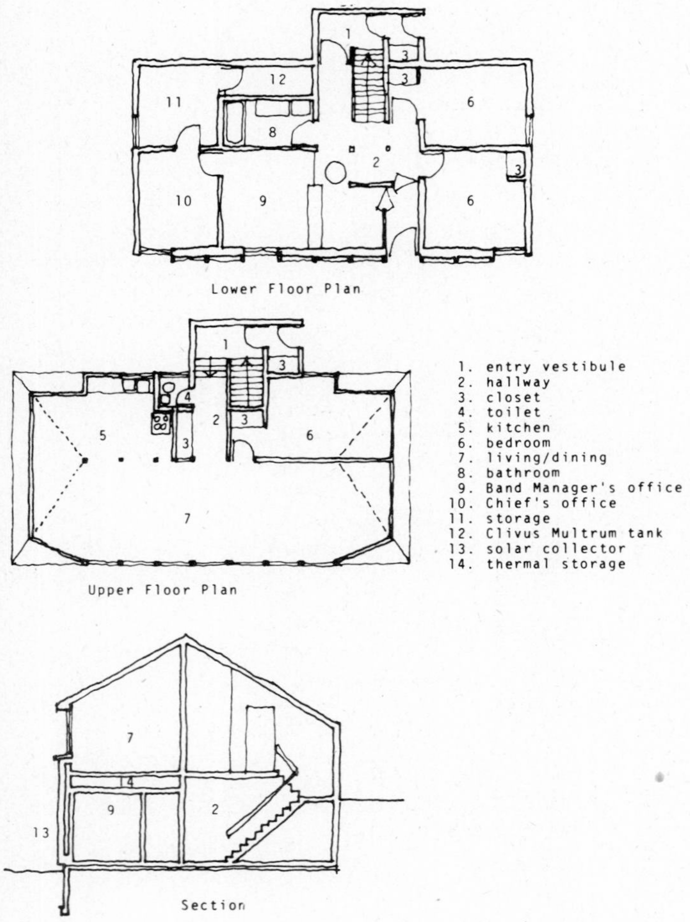

Fig. 3 Waswanipi House Plans and Section

Floor area = 203s.m.
Exterior surface area = 283s.m.
Volume = 613c.m.
Storage volume water = 6819 lit.
Net collector area = 35s.m.

RETROSPECTIVE

While it is felt that the Cote Nord and Pte. Bleu units at LaMacaza to some extent satisfied an expressed Indian desire for a more flexible and larger house, (the 'norm' had been 71.4s.m. regardless of family size)(3) both units lacked a 'permanent' client. Hence evaluation on space use/architectural terms has been difficult. It is felt, on the technical side, however, that the 'appropriate' technologies used for heating, building and sanitation have been successful in the short term. The Cote Nord solar-heating system is being monitored by Hydro-Quebec instrumentation, and to date indicates a very reasonable performance.(4) The cost of this system in relation to that of the house is low ($5000/ $35000)(5) and affordable by Indian owner-builders with an appreciation of operational, rather than 'first-cost' savings. The use of pressure-treated wood foundations for the Mistassini and Waswanipi units also proved to have cost advantages over conventional masonry foundations. Rough-cut framing lumber applications did not, in the course of the project, prove to have cost advantages over dressed lumber, due to transport costs and more important, the small scale of our operation. Yet here, economies of scale and local sawmills could present cost advantages. The pressed peat-moss insulation application in the Cote Nord house (6) will only have advantages to the Indian Community in the case that Indian entrepreneurs take up the development of a small-scale production system with the University of Sherbrooke who have the technical know-how. The 'Humus' ('Mull-Toa') composting toilet, tested in the Cote Nord unit, has repeatedly failed and has had to be replaced several times. Finally and reluctantly a flush toilet was installed. Unfortunately, the 'Clivus-Multrum' composting toilet, considered much superior to the electrically-operated 'Humus' (7), was never installed as planned in the Waswanipi unit.

In spite of some technical achievements, shortcomings were encountered on the 'human' side. Labor-intensive building systems were felt to be desirable and appropriate by the Indians of Quebec Association (8) and ourselves; yet, in practice, we were very pressed for time, and for reasons related to the precision required in the solar-collector construction, chose a more than 'semi-skilled' building team. Labor-intensive building, involving similar costs but more people and more time, however, should be possible in implementing a larger and more repetitive housing project, yet some unskilled and semi-skilled builders were given the benefit of some building training during the project. We also found that the occupants of the Cote

Nord house were always interested and enthusiastic, yet while most Indian people we met were appreciative of our technical follow-through on Canadian Indian Brotherhood policy recommendations, very few, not surprisingly, were prepared to take any individual initiative in building solar houses. As in our project, this will likely again depend upon cooperative ventures between private groups or individuals and either federal government agencies or interested manufacturers.

REFERENCES

(1) Hamilton, B. and McConnell, R. An Experimental Evaluation of a Solar House-Heating System in Quebec Sharing the Sun Conference, Winnipeg, August (1976).

(2) Hamilton, B. Combined Use of Preserved Wood Foundations and a Fibreglas Drainage Layer to Improve Below-Grade Living Space, McGill School of Architecture (1975).

(3) Indians of Quebec Association Indians of Quebec Housing Survey, Village Huron, (1974).

(4) Kerr, R., Turaga, M., Shapiro, M. and McConnell, R. A Performance Evaluation of a Solar Heating System International Solar Energy Society Conference, Miami, June (1977).

(5) McCloskey, B. et al Appropriate Building and Energy Systems for Quebec Indian Communities McGill School of Architecture, December (1976).

(6) Cossette, Marcel, Insulating Panels from Waste Peat Moss Appendix 2, Appropriate Building and Energy Systems for Quebec Indian Communities, McGill School of Architecture, December, (1976).

(7) McKernan, J. and Morgan, D. "Experiences with the Clivus-Multrum and Mull-Toa Toilets in Northern Manitoba" Dept. of Northern Affairs, Gov't. of Manitoba, August (1976).

(8) National Indian Brotherhood First Draft Position Paper on Indian Housing Ottawa (1973).

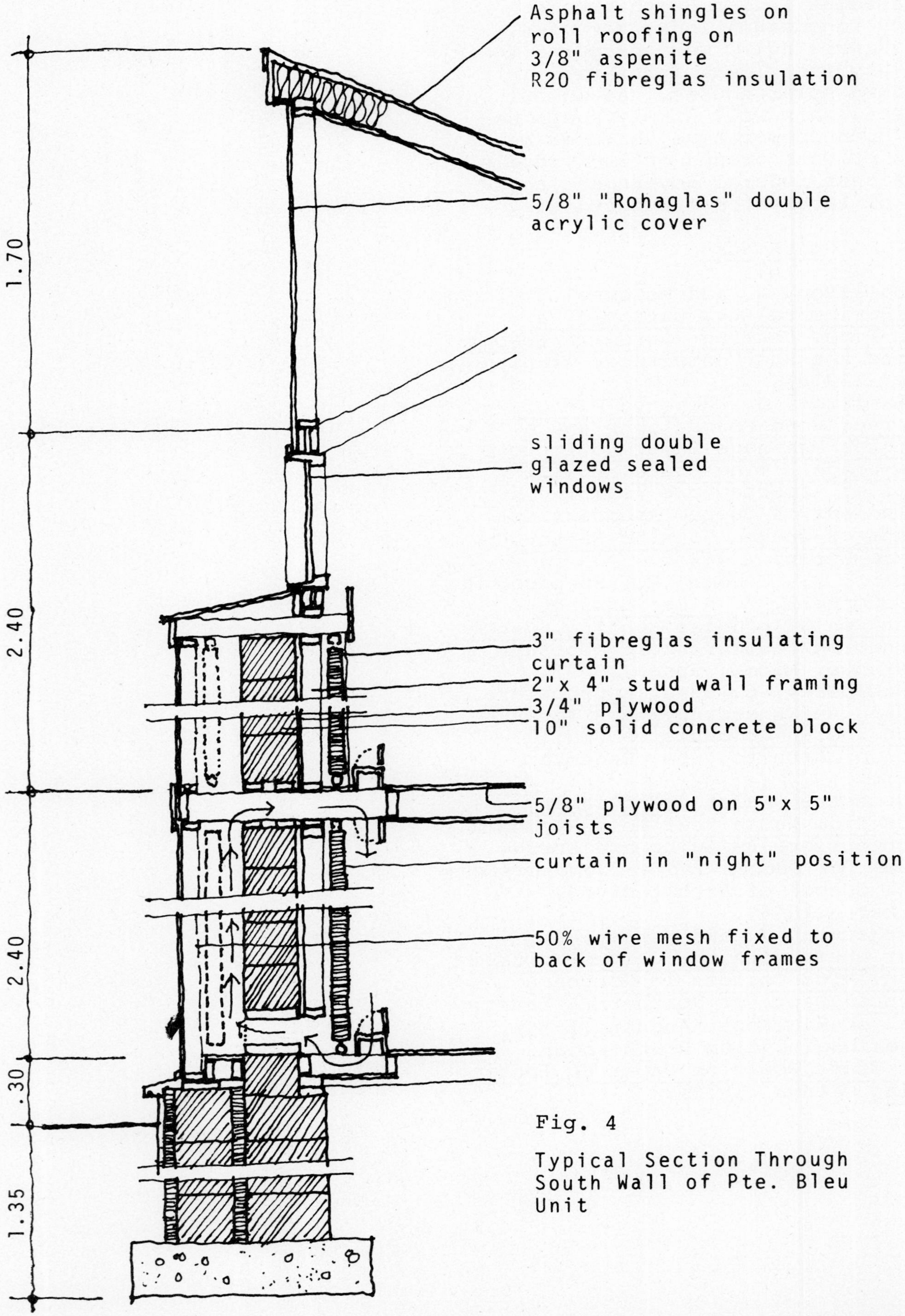

Fig. 4

Typical Section Through South Wall of Pte. Bleu Unit

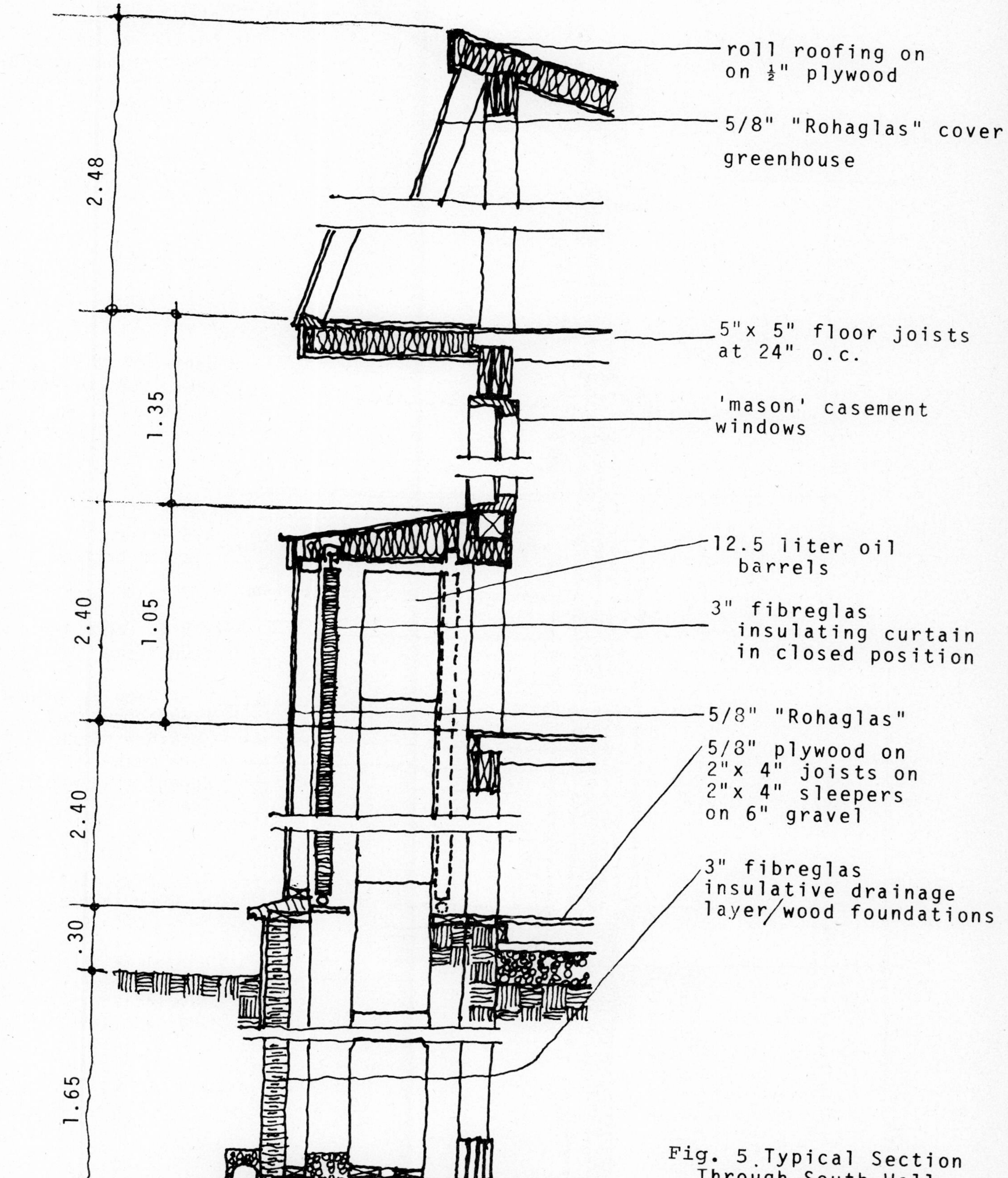

Fig. 5 Typical Section
Through South Wall
of Mistassini Unit

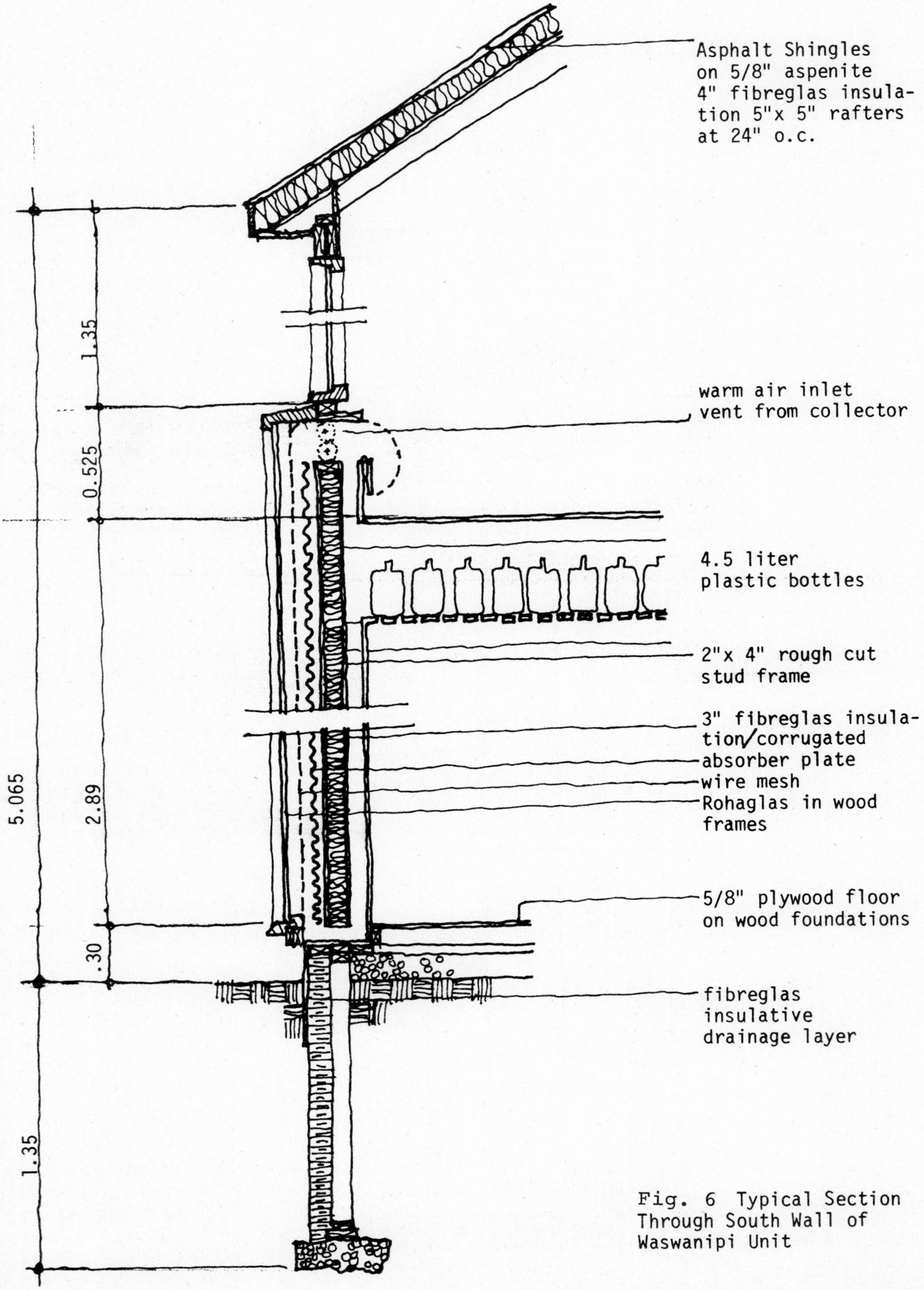

Fig. 6 Typical Section Through South Wall of Waswanipi Unit

SEEKING A BALANCE OF MAN AND NATURE IN REGIONAL DESIGN: THE FLORIDA EXPERIENCE

Don Young
State of Florida, Talahassee, Fla.

ABSTRACT

This paper reports the principal findings and recommendations of a major research study known as the South Florida Environmental Study. The study was co-sponsored by the Florida Division of State Planning and the U.S. Department of the Interior and was conducted by Dr. Howard T. Odum and associates of the Center for Wetlands, University of Florida. Its purpose was to recommend guidelines for future energy, land, and water use within a large regional watershed of South Florida which would help to assure the goal of long range economic vitality for the region.

The choice of competitive resource management strategies for the years ahead requires that they be considered in the context of newly emerging energy and economic realities. Over the past three decades, Florida's economy, population, and resource demands have consistently expanded. The idea of inevitable, continued growth became widely accepted by planners, politicians, state agencies, and others affected the decision-making process at state and local levels.

An abundance of inexpensive, easily accessible resources allowed this rapid growth to occur. Today, however, resources no longer seem limitless; fossil fuels, raw materials, land, and water appear to be in short supply. The physical basis for growth is becoming more costly, growth has slowed, and forecasts for growth in Florida to continue at its previous pace may no longer be valid.

To gain an understanding of likely growth futures for South Florida, the South Florida Study attempts to identify, assess, and integrate the interrelationships between energy, environmental management, and economic issues. It is hypothesized that the growth potential, or carrying capacity of a region is directly dependent on the availability of physical resources, both local natural resources and those imported by means of economic exchange. Energy, land, and water are identified as vital resources whose relative availability sets limits on the region's carrying capacity.

Because energy is central to the issue of regional carrying capacity, an overview of world and national energy supply patterns, including the net energy potential of major existing and proposed sources, is provided. The results indicate that: 1) the net energy return, or "energy profit," of today's principal energy resources (oil, gas) is declining, and 2) the net energy returns of most, if not all, proposed alternative sources (coal, nuclear, solar) are less than today's fossil fuels. If accurate, the results indicate the nation is entering an extended period characterized by less energy availability.

Because of this trend, economic growth will become much more difficult as the costs to import fuels and goods and services into the region continue to rise. Thus energy conservation, or in the language of this conference--energy end use management, assumes even greater importance as a necessary long-term planning objective. The nature of any region's response to declining energy will depend on both the magnitude of energy costs originating outside the region and the internal characteristics of the regional economy. It becomes imperative that state, regional, and local planners begin to examine the direct and indirect energy requirements of their respective economies and the energy costs of present resource management practices.

Examination of the South Florida regional economy reveals a very narrow economic base supported by industrial activities which are highly energy intensive (e.g. agriculture, building construction) or highly dependent on continued national affluence and prosperity (e.g. tourism, retirement living). Thus Florida and South Florida, in particular, emerge as areas which may quickly respond to declining energy conditions and therefore warrant watching as possible bellwethers of national growth and development patterns.

The South Florida Study suggests that the chances of retaining a long-term stable economy can be greatly enhanced by making more effective use of the region's inherent capacity to provide vital services through its natural ecosystems maintained by solar energy. Numerous policy recommendations are given in the areas of energy conservation, land use, and water management. Their common goal is the support of a regional carrying capacity which offers a stable economy and a high quality of life.

SOCIETAL NEEDS AND ENERGY USE IN THE DESIGN OF HUMAN SETTLEMENTS

Harry der Boghosian

College of Architecture, University of Arizona, Tucson, Arizona

ABSTRACT

This paper will discuss some observed changing patterns of society, lifestyles and energy use and the potential implications useful in the design of human settlements.

In an era of total and almost religious reliance on technology as the major problem solver, coupled with an increasing awareness of its doubtful efficacy in view of the economic and environmental costs, small wonder that we sense a need to examine a comprehensive view of energy use management. The magnitude of expected changes in energy use to occur in the remainder of this century will set a pattern for the manner in which the people of the United States will function in the future. These developments respect no boundaries and will evolve into a global dimension.

However a equally potent change is already occurring and we are just now beginning to become aware of and understanding its potential influence, that is the dramatic changes in societal and personal values. Societal needs and energy use may be developments that arise from different causes, but their consequences are closely intertwined. To design and develop new technologies related to energy sources and use, without examining some of the major changes in social and personal values will ultimately lessen the effectiveness of such programs.

Lifestyles

Technological advances in the area of education, social and human planning has brought about the realities for the increased use of individual and family planning techniques and methods. Economic uncertainties, social unrest, multi-earners in the family, and environmental pollution have helped to slow the trend to early marriage and the large family unit. Although the divorce rate is increasing to one in three marriages, the institution of marriage is not declining in need as is expressed in the increasing frequency of re-marrieds. The decreasing family size with its flexibility to alternatives in living and job arrangements, appear to be working a more dramatic change in the home than could all the futuristic labor saving devices of technology. We are thus finding more evidence of this expanded awareness in our lives and desire for individual personal growth and freedom in people of all ages and lifestyles.

Ethics and Values

Among the basic needs of man, self preservation has been recognized as one of the strongest drives in human beings. It was first thought of in terms of conditioned response only, but we have come to recognize its main component as human adaptation. The single most important determinant of our potential as individuals and as a nation to survive and grow in the future, will not depend on our natural resources, wealth, defense, or technologies, but on our capacity for adaptation. This ability combined with our changing social and cultural attitudes and sensitivity toward the natural environment will greatly influence and help formulate the future direction and quality of our lives. A few of the observed shifts in cultural values appear to be forming a concept of a caring society, where we are moving from personal goals of:

1. Achievement to self-actualization
2. Self control to self expression
3. Independence to interdependence
4. Endurance stress to capacity for joy
5. Full employment to full lives
6. Mechanistic forms to organic relationships

7. Competitive relations to collaborative relations

Work and Leisure

The changes in cultural values will stimulate all aspects of our lifestyles and may provide new meanings to the traditional concepts of activities called "work", "living," and "leisure." Aristotle said, "that educational programs must prepare the citizen for the use of leisure, for a state that cannot use its leisure properly will collapse.

The history of non-industrial society showed no distinction between work and leisure and everyday activities consisted of an intertwining attitude of thought and action. Today, due to the high premium on the economic rate of return for labor over a limited period of time, there are created, divisions of activities called work, living and leisure. Changing concepts of work are appearing, such as the humanization of work experiment in the Scandanavian countries. It has encouraged the move towards democratic forms of work organization, which may ultimately abolish the concept of "job." This concept employs the relevant techniques of job rotation, job enlargement and enrichment and autonomous work groups. There are also indications that the single educational experience towards a life time career will probably become more flexible and individuals will flow into related fields and others will have a variety of careers. Therefore as our evolving ethics and social values help to shape our lifestyles, so to, will our attitude towards the natural and man made environment be dramatically altered.

Man and Nature

Much attention has been focused on the importance of ecology as an integrative science related to the natural environment. Humankind has traditionally attempted to dominate the environment instead of harmonizing with it. This prevailing attitude was accelerated by the Industrial Revolution resulting in a general deterioration of the quality of urban life and concomitant degradation of the natural environment. In view of this, it is imperative that we increase our understanding and develop tools to improve the interrelation of man-made urban elements with natural systems, towards the ultimate goal of creating a mutually beneficial interaction of these in an urban ecosystem.

There is precedent of such balanced ecologies in historic and indigenous settlements of the pueblos in the United States, villages in Greece, North Africa and the Middle East. As an example in arid regions, humankind has collaborated with the natural environment through architectural adaptation to the major climatic factor, thermal overloading Villages are built of mud brick masonry, a low energy consumption material that advantageously offsets the wide temperature variation of hot days and cold nights. The urban texture of narrow streets, inward facing buildings with outdoor courts and gardens, shaded by landscaping, all contribute to the achievement of an advantageously modified microclimate about the buildings.

Today, our contemporary urban settlements are not considered to be an efficient system of energy and material utilization or for the support of its inhabitants. Our substantial reliance on the automobile has contributed its share of polluting by products which has negatively effected the climatology of our cities.

It has been estimated that buildings now utilize as much as 34% of all the energy consumed in the United States, and as much as 50% of that energy is wasted. Countries such as West Germany use half as much energy per capita and Switzerland uses less than one-third. It is obvious that energy conscious human settlements of the future will seek a more symbiotic relationship between the societal needs of the users and the conditions of the environment to lessen waste without lowering our living standards. Designing urban environments ecologically in balance that utilize solar energy can potentially lower the total energy demand for power, lighting, heating and cooling, reduce pollution and be directed toward optimizing our utilization of energy and natural resources.

Climate and Architecture

Let us investigate the concept of climate as a modifying factor in the design of the living environments. All living things whether plant, animal or man and their built forms interact with their environment and have always had to contend with climate to survive. One notable variance is that man's built environment is the resultant unconscious or unintentional effect of human civilization on microclimate. Winston Churchill's quote that "we shape our dwellings and they in turn shape us," must be altered to "and they shape our microclimate as well." That the variables of climate are profoundly effected by the physical characteristics and human activities of a city is an ancient thesis that is now more intensively explored by the expanding discipline of urban climatology. The phenomena of visual pollution, dust domes and heat islands and noise pollution are obvious to all of us.

Lessons From Nature

The concept of orientation in climate in order to adapt to the sun's energy is worthy of further discussion. As an introduction, the word orientation derives from Orient, meaning the east as opposed to Occident or west and has its origin in Jerusalem. The significance is old, and like most other things in antiquity, it has religious overtones involving several religions.

The importance of orientation or the effects of differing amounts of sunlight on micro-environment can be illustrated by an example of nature here in Arizona. The enormous Saguaro cactus which is generally located only on southern slopes, always blooms first on the southwest side, no buds develop on the northeast side. The concentration of sap is also greatest in the cells on the southwest side, its growths restricted on its southwest side by the greater dryness there, so that it gradually curves toward the southwest and finally topples. After a period of one to two hundred years.

One of the greatest land structures in history, as compared in size with their builders, are also basic demonstrations of the importance of orientation. They are the immense buildings of the compass termites in Australia, which are blade shaped termite cities that orient on a north-south axis. The east and west exposures help to secure an equable temperature through the large mass of earth. The towers are large, approximately eight to twelve feet high, reaching 400 times their body length, which translated into human terms would equal 2,400 feet.

Lessons From The Past

It may be beneficial to review some historic examples of man's effort to integrate his societal needs in the creation of built environments designed with nature. These historic uses of natural energy in a blend of time, place and cultural values by different societies may provide us with some insights from the past.

The emergence of man on the earth and his creation of shelter evolved on a modest scale based on needs related to survival. Paleolithic man as mobile hunter found shelter when needed and utilized the natural topography and the underside of cliffs. Neolithic man as farmer requiring more permanence constructed his shelter from easily available materials of wood and stone. Stonehenge, the megalithic monument on England's Salisbury Plain is very early evidence of humanity's cognizance of its dependence upon the sun as the earth's primary source of energy. The Egyptians were the first known civilization to produce a sun dried brick and utilized the sun to illuminate, heat and ventilate their buildings. The Greeks introduced the concept of interior courtyards with pools of water obtained by run-off, providing natural lighting and evaporative cooling. They generally built on three sides of the atrium with the fourth open to the south sun. Hippocrates "air light treatise," declared that "cities should be sited so that buildings and streets would be penetrated by the sun and wind for hygienic purposes." In the third century A. D. Vitruvius presented his ideas on the layout of towns and of individual buildings according to various climatic influences. A further indication of his thinking is expressed in the following quotation; "for in one part the earth is oppressed by the sun in its course; in another part, the earth is far removed from it; in another, it is affected by it at a moderate distance, therefore, since, in the sun's course through the inclination of the zodiac, the relation of the heavens to the earth is arranged by nature with varying effects, it appears that in like manner the arrangements of buildings should be guided by the kind of locality and the changes of climate."

The pueblo settlements of southwestern United States built over a thousand years ago demonstrate further the varying cultural approaches of man's attitude to environment and response to climate. The everyday activities of the Indians expressed an integrated attitude of their individual and collective world into a traditional belief system passed on from generation to generation. The various activities for survival such as hunting, farming, fishing, trading and cultural events such as ceremonials and celebrations and spiritual attitudes towards the sun and the natural environment, were fused into a whole concept of their universe.

The Longhouse pueblo at Mesa Verde, Colorado adapts to its environment primarily by location; settled into south facing cave at the base of cliff. The cave is 500 feet long, 130 feet deep and about 200 feet in height. It is shaded by the high position of the summer sun, yet exposed to the warmth of the low angle winter sun.

The Acoma pueblo suggests a more involved architectural adaptation to environment through its design and structural arrangement. Three rows of dwellings extending about 1,000 feet in an east-west orientation are located on a inaccessible mesa 400 feet over the valley for protection. The arrangement of the groupings extend vertically from one to three story units, with no house shading another. The use of thick load bearing walls of rubble stone covered with adobe on the lower untis become thinner and changes to adobe covered brick on the upper units. The summer sun falls primarily on horizontal surfaces with low storage capacity of timbers and mud and the winter sun falls more on vertical surface with high heat retention characteristics of rubble stone and adobe.

Pueblo Bonito offers a combination of location (site) and adaptation (design) for survival and comfort. The main concept consists of a semi-circular plan facing south at the base of sheer rock cliffs, one mile above sea level. The normal hot and arid climate is influenced to a greater extent by the winters of higher elevation. Here again the summer sun falls primarily on horizontal surfaces with low storage capacity and the winter sun falls more on vertical surfaces with high heat retention characteristics. Consequently the total design adapts by allowing a higher percentage of heat gain in the winter and a minimal exposure to the sun in summer.

Contemporary Vernacular Architecture

An example of a cohesive social structure expressing itself into a highly organized architectural structure is found in the Greek vernacular architecture of its villages. The history of Greece reveals the continued scarcity of resources and adversity of piracy, foreign occupations, wars, revolutions, earthquakes and droughts. The continued presence of these pressures over protracted periods of time and generations has imprinted a social need to secure mutual help and maximize their effectiveness towards solidarity for survival. The highly organized systems with subsystems represent autonomy and interdependence structured in a hierarchal order. In the typical village unit, we find the individual belonging to the extended family, the family is an integral part of the in-group system and the in-group is one of the subsystems of the larger village unit. The law that governs all the systems from individual to village is the concept of interdependence. Competition for limited resources pulsates in intensity and is reflected in the individuals need for growth and creativity while simultaneously fulfilling the group needs towards a dynamic equilibrium. It is analogous to an organism with fairly defined area that is allowed to expand and contract its shape and still maintain a constancy of integrity of cellular order. Thus the social cohesiveness of the society as a major determinant is able to express itself in the architecture and planning discipline of Greek settlements. The earlier needs for defense, resulting in contiguous housing, narrow winding streets minimal openings has evolved into individual spaces and shared spaces for the in-group like courtyards. Open doors and the warm and dry climate encourages the openess of entrys and courtyards which facilitates daily contacts and visits. The street artery or the space between the housing groups does not separate the people but serves at various times during the day as an opportunity for playing, conversation, knitting, and other social or commercial activities. The main street as outdoor space serves as a physical and social unifier of people and encourages contacts between its inhabitants. The main street continues and enters the domain of the Greek plazas, of varying sizes, from cluster level to larger village plaza. Here is a center or congregation of the communal level of commercial and social activities and interaction for the whole village. It is undeniable that the visual impact of forms that Greek vernacular architecture evokes, can by itself, be a stimulating experience. However, what insights can we derive from this pattern of a socially cohesive settlement form and what does it suggest for a pluralistic society such as the United States? The clue may be more a ecumenical need of humankind than we suspect even though it expresses itself in varied ways in different cultures. I would like to suggest the concept of the individuals need for a balanced state of freedom and security in his physical, social and psychological domain. The question then posed may become of import to our society; that is, how can we provide the diversity of options for individual expression and integration within the collective structure of society? Consequently, our concern for improving environmental quality and reducing dependence on non-renewable sources of energy will depend to a great extent on our ability to integrate societal needs to form a powerful modifying factor in the design of responsive human settlements.

Knowledge/Leisure Society

Finally, what are some potential interpretations of societal needs and energy use concepts in the design of human settlements? Perhaps, a historical perspective may be in order. It is suggested that the first century of industrial advance in England from 1776 to 1876 was based on the transport revolution of the railways and on steam power. The second century in the United States of 1876-1976, being based on mass production assembly lines and the automobile. The third century of 1976-2076 may be based on the transport revolution of telecommunications allied to knowledge and leisure. This potential force towards encouraging decentralization of population has shown indications of reversing the rural to urban migration trend of the 1950's. We are now experiencing a urban out migration that improved communications such as cable T.V., videophones, satellites and data networks will stimulate and reinforce this dispersal trend.

However, this is not to say taht the concept of centrality will completely disappear in our cities. More important and perhaps paradoxically there seems to be two trends occuring simultaneously, one of centralization and decentralization. In one instance we have the concept of centrality or collective places, where the city centers will continue to survive, not for production of goods, but as meeting places, spectator events, or mass participation and ritual. The other aspect is for more decentralization or individual places for search and discovery experiences. Are these two potential trends in conflict or can they mutually fulfill the diverse needs of the citizens? I would like to suggest that the need for and viability of the latter situation, where the diverse activities can occur in a hierarchal range of individual places to collective places to suit the varied needs of the inhabitants.

In contemporary Japan due to high land costs because of scarcity of available land, heavy inheritance taxes, a housing shortage, all have contributed to fragmentation and high intensity land use. This has encouraged the Japanese designer to practice their particular skill in small scale environment manipulation. In Tokyo, there are many small districts each with its own district ambience and function, even a building such as a hotel has a changing multiple function. Today they are places for conferences, fashion shows, wedding ceremonies, swimming and physical exercise, and family outings. The department store sells more than goods, they serve as galleries for art exhibitions and places for adult education. Even the most basic human activities are being merchandised such as taking children to leisure land because there are no parks for picnicking, the lack of entertainment facilities at home, so people are forced to dine out.

In spite of the absence of "community" there exists many spaces for "territories" with which to identify. On a more personal scale similar to English pubs, and our American neighborhood bar, there are the coffee shops where business takes place, entertainment, renting rooms for group meetings and organizations. This ability of architects in creating small places for people to identify with may suggest a clue for an identifiable scale for the individual to comprehend.

Future Settlement Patterns

Conceptually, therefore, the city may then become a great gathering place comprising many small places. More emphasis will be placed on the comprehensive view of the design of environments for people with attention placed on the interrelatedness of the activities called work, living, recreation, shopping, entertainment and cultural. The settlement pattern may represent hierarchal groupings of diverse activity centers varying in scale and time use related to housing. This concentration of facilities into groups throughout a city of kindred frequencies of use contributes to a marked reduction of energy required for their use by the residents, compared to a complete dispersal of facilities or a concentration in a center. A complimentary concept would encourage sharing of facilities, interior and parking areas into a twenty-four hour utilization period.

Communications

There will be closer living to work, schools and shopping area relationships in order to spend less time commuting. Thus the almost total reliance on the automobile will be lessened and will be used for longer trips and with more passengers where it becomes a more energy efficient means of transport. Persons of all ages could get about within the community by walking, bicycle, para-transit or public transit. Another transportation feature would be deployment of efficient rapid rail transport in the major cities over existing rights of way to connect urban centers of cities.

The concept of para-transit as a travel alternative that can hopefully meet the needs of such groups as the elderly and the handicapped. The various modes include the following: carpool, rental cars, specialized commuter bus services provided on a subscription basis, as well as taxicab and related services like dial-a-ride, jitney and limousine. The problems of transportation are not uniquely a contemporary phenomenon, even in Roman times, Caesar promulgated a set of laws preventing people from coming into the city with a chariot at certain times of the day and the restriction of trade wagons at other times.

Integration of Activities

The evolutionary design of living environments may include further integration of commercial, shopping, places of work, community services in a horizontal or vertical arrangement limited to three or four stories. Many European cities are familiar with this concept and even some of our own cities in the United States showed a propensity for multi-use living environments. Accessibility to nature via walking and bicycling to an open space system from the dwelling, cluster and neighborhood, is also an important consideration to reinforce the interrelationships.

Conceptually the design of individual habitat and that of settlements are more alike than different. The immediate differences are obvious, complexity, number of activities, size and scale however human settlements do reflect and share as macrocosm a similar range of needs expressed in the single habitat. We still require activity spaces, interior and exterior for individual privacy and expression, recreational, entertainment, group socialization, cooking facilities, dining spaces, habitat maintenance systems, health and sanitation systems and communication systems. Admittedly this interpretation of habitat appears overly compartmentalized, it could stimulate our comprehension of human settlements not only as the sum of its parts but as a holistic approach of man living with nature.

Need For Diversity

There is a perceived shifting toward the provision of physical and spatial opportunity for people to carry out the widest range of their desired daily behaviours and activities. What is vital is the creation of environmental diversity, a stage setting where many actors can act out their lives in the play of their choice. Therefore one of our roles is to provide for a broad spectrum of possibilities and options for the individual and ultimately, to show us what a society is, but also what it can be.

As Charles Eames stated, "A block of granite, tells you what you can do or not, it imposes itself. You can't do anything bad, but plasticene is free of restraints and there is no way of telling what to do." "Today, we live in a plasticene world, the restraints or past are being broken down. The greatest task today, is to rediscover the restraints of earlier cultures, not for revival, but to create new restraints appropriate to our time."

Selected References

Aronin, J. (1953) Climate and Architecture, Reinhold, New York

Bell, G. Work and Leisure, Ekistics 236 July, 1975

Bennis, W. (1969) Organization Development Addison Wesley, New York

Carpenter, J. Organizational Techniques and Humanization of Work Ekistics 236 July, 1975

Doumanis, O.B. and M.O. Doumanis Fit and Form in the Greek Village, Design & Environment, Winter 1975

Doxiadis, C. Energy and Ekistics, Ekistics 225 August, 1974

Dozier, E. (1970) The Pueblo Indians of North America, Holt, Rinehart and Winston, New York

Dubos, R. Biological Basis of Urban Design Athens Ekistics Month 1972

Fitch, M. (1972) American Building II, Houghton Mifflin, Boston

Geiger, R. (1965) The Climate Near the Ground Harvard University Press, Cambridge

Goldfinger, M. (1969) Villages in the Sun Praeger, New York

Gottmann, J. The Evolution of Urban Centrality Ekistics, 233 April, 1975

Kirby, R. Para-Transit: Experience and Potential in the U.S.A. Ekistics, 248 July, 1976

Knowles, R. (1974) Energy and Form M.I.T. Press, Cambridge

Maki, F. Some Observations on Urbanization and Communication in Japan Ekistics 226 September, 1974

McHale, J. Beyond the Exponentials Ekistics 250 July, 1976

Olgyay, V. (1963) Design With Climate Princeton University Press, Princeton

Pike, D., Muench, D. (1974(Anasazi American West, Palo Alto

Rapoport, A. (1969) House Form and Culture Prentice-Hall, Engewood

Salk, J. (1973) The Survival of the Wisest Harper/Row, New York

Stearns, W. and Tom Montag, (1974) The Urban Ecosystem Dowden, Hutchinson Press, Stroudsburg

Villecco, M. Architecture as Energy Design Quarterly 103

SOLAR ENERGY UTILIZATION AND ITS EFFECT UPON LAND USE AND URBAN PLANNING

Victoria A. Carella

University of California at Irvine, 2335 Elden #3, Costa Mesa, California, USA.

ABSTRACT

Solar energy utilization will have an effect upon land use and urban planning. This paper considers the implications and effects of large and small scale solar energy development on legislation, comprehensive plans, zoning, local codes, and personal and community attitudes towards building designs, siting and aesthetics. Regional impacts are also considered.

The large land areas required for the large-scale power plants are discussed. Also considered are the growth inducing aspects of such systems along with the environmental effects. Solar energy on a smaller scale is also discussed. The small-scale development will be viewed from the building of new communities with solar energy incorporated from the beginning to the retrofitting of older communities. Sun rights and solar easements will be considered along with other controls to aid a landowner in gaining access to sunlight, such as local regulations of building codes, height regulations, setback and landscape provisions, and general zoning. The paper concludes with a consideration of land use planning of larger areas and their interrelationships.

INTRODUCTION

By the year 2000 a good portion of our Nation's energy needs can be met directly and indirectly by the sun. Project Independence Solar Energy Task Force, in 1973, concluded that "Solar energy could contribute from 15-30% of the nation's total energy requirements by the turn of the century". (Ref. 1) There is a large influx of capital being directed into solar energy development, both from the federal government and private industry. It is almost inevitable that many solar energy power plants will be appearing nation wide. As more and more attention is being directed towards solar use it is time to consider land use planning and solar energy. Every effort should be made to fully consider all the implications and effects of solar energy development.

When considering land use strategies in conjunction with solar power plants, one must look at the plants in their two most outstanding and viable forms: the centralized, large-scale power stations and the decentralized, small-scale power stations. Each of these concepts has their own peculiar attributes and their own distinctive land use considerations.

LARGE-SCALE SYSTEMS

When considering large-scale power plants, the most prominent feature which must be dealt with is the large plant and collector areas required. Because solar energy is so dilute and fluctuates with the weather, a large capacity requires not only a large collection area, but also a large storage area. Following is a general idea as to the land usage required for different types of large-scale systems:

- a thermal station, with conventional collectors, to generate 1000MW: plant area of 30km^2. Thermal stations have the added hazard of thermal pollution and require water for cooling ponds.

- a high pressure steam turbine generator for 1000MW: land area of 17km^2.

- a photovoltaic station producing 1000MW: land area 100km^2. The present low efficiency of the solar cells calls for larger land areas.

- a photovoltaic system which would produce hydrogen for

1433MW: land area 220 square miles.

- a one axis cylindrical parabolic used as an energy concentrator for a 1000MW continuous output plant: land area 4 km^2.

- a two axis concentrator for a 1000MW plant: land area of 17km^2.

- a solar space station which would transmit its energy to a microwave receiving station: land area require for receiving station 36 square miles.

- the extremely large solar farm which would cover a proposed area of 13,000 square miles. Such an idea is not far fetched, for to produce just half of the projection for consumption of electrical energy in the U.S. in the year 2000, 20,000 square miles will be needed.

(Ref. 2, 3, 4, 5)

The large-scale plants also require transmission paths which would necessitate the need for land acquisition, right-of-way, and utility easements.

One may question the necessity for concern about the land use issue and large-scale stations because they will be situated in the desert, which is of little interest to some people. It should be realized that energy related development generates growth in the form of industrial, commercial, and residential development. A site providing energy will attract energy consumers. In the building of the plant, construction workers will be needed who, in turn, will require housing. Once the station is built there will be a need for a maintenance crew who will require housing, schools, police and fire protection, health services, recreational facilities, water, and utilities. There will also be the development of commercial supporting services. All this will put stress on the existing communities to finance and provide services. If there is no existing community, the construction of a solar plant will necestate the development of a new community. The plant will require roads, grading, and filling sites. This will provide vehicular access to large areas of virgin desert. The public should be concerned with the desert ecological maintenance, site access, land acquisition, the effects of transmission paths, and the effects on surrounding land usage, development and values.

When evaluating solar power plants in relation to land use, the environmental effects should be included to provide a total picture of large-scale development. Some of the local environmental effects would include waste heat and affected albedo. In the desert about 65% of the incident sunlight is absorbed and 35% is reflected. With collectors, up to 95% of the incident sunlight would be absorbed. When dealing with large areas of desert being covered, this could alter the local climate. (Ref. 3). Large-scale development could also have visual and aesthetic effects. All of these are minor when compared to the effects on air pollution from fossil-fueled power plants or radioactive releases from nuclear power plants. Large-scale systems also have the redeeming factor that after the plant has gone through its life, the land could be reused or returned to its original use with barely any reclamation programs needed. (Ref. 6)

Consequently, when siting a solar power station a suitable or candidate siting area would be an area which will provide maximum social and economic opportunities to accomodate a power plant, with general minimized environmental effects.

SMALL-SCALE SYSTEMS

Solar energy also lends itself to small-scale development which is potentially decentralizing in its nature. One moves in from the desert land into existing urban areas. It is at this local level that solar energy utilization will be actualized. The overall contribution of solar energy will be based on the successful implementation at the local level.

Small-scale development can be viewed from the building of new communities around the power source or the retrofiting of older communities. A neighborhood could have its own community power and storage systems.

Small-scale development of solar power would consist of locally situated stations which would support the energy needs of a subset of the total community. The stations would be neighborhood-based, or, in the case of commercial areas, industrial-park-based. Each community subset would then become independent of the regional station and would aid in taking the load off of the major power generating stations.

There are several distinct land use problems which arise when considering small-scale development. Solar energy will have to be considered from the inception of the planning process. When considering the use of solar energy within the community, one has to be aware of the total planning picture, zoning, and building requirements. Sunlight will have to be viewed as a resource of the property it strikes. At present, a landowner has no legal right, except for easements, to the free flow of light, even at the common law level. The problem in gaining access to solar energy will involve providing legal protection against unanticipated shadows which adversely affect a solar collector. (Ref. 7)

The more significant controls of access to light stem from the complexities of local regulations comprised of building codes, height regulations, setback provisions and general zoning patterns. These have been shaping our cities since the beginning of the 20th century when the United States Supreme Court, in Village of Euclid, Ohio v. Ambler Realty Co. (272 U.S. 365), affirmed the constitutionality of zoning ordinances. These local laws can severely limit the introduction of this new technology.

The number one impediment is the right to sunlight. The law in the United States is now well established that a surface owner has a right to receive light from the sky directly above his property, but not to receive it across his neighbor's land. Adjacent landowners can construct buildings and fences or plant trees which can block the sunlight and cast shadows over potential or actual collector areas and the owner of the solar device has no legal recourse.

A few states have laws concerning sunlight access for solar energy systems. Effective May 20, 1975, Oregon amended four sections of its planning and zoning code with the result that: (1) comprehensive plans and zoning ordinances now must include consideration of solar energy utilization; (2) planning commissions may recommend ordinances to county governing bodies which are to protect and assure access to solar energy; (3) planning commissions may also recommend public incentives for overall energy conservation; and (4) city councils may consider solar exposure. This "enabling legislation" makes no rules directly affecting the public, but it passes on the rule-making authority to the lower governmental units. In directing city planners to consider solar energy, it gives the owner of a solar collector legal recourse in case a neighbor hinders its use. Colorado, in a statute effective July 8, 1975, provides that solar energy access easements be in writing and subject to the same provisions as are other easements to property. They also require that the written record of such an easement state "the vertical and horizontal angles, expressed in degrees, at which the solar easement extends over the real property subject to the solar easement" and that it list any provisions for compensation in regards to interference with or maintenance of the easement, such as the cost of trimming a tree. (Ref. 7)

The local regulations could also bar the proper placement of a collector. Side and rear yard limits may restrict the placement of a solar device. To take advantage of the maximum amount of solar insolation, landowners may have to place the collector in the front yard, which may be restricted by local ordiances. A solar collector mounted in its own building may not be considered a proper accessory use and might, therefore, be prohibited.

Solar energy, in this decentralized form, will have an effect on community development and psychology. There will be definite aesthetic and visual effects which will require a change in attitudes towards community and building design. Aesthetic controls on uniform rooftops may preclude the placement of a solar collector on a roof. It might be required to have major rezonings in areas to allow for less density, variation in building heights and solar easements. In this respect, the utilization of solar energy in a community can be seen as both growth inducing and a no-growth tool. It would be growth inducing because of the availibility and of the desireability of energy. It can be viewed as a no-growth mechanism and an exclusionary tool because of the restrictions placed on the development of the community, in particular, low densities, low building heights and easements which require good spacing between housing. All this could result in higher costs which will be prohibitive to lower economic groups. Taking, which is prohibited by the Fifth Amendment of the United States Constitution, could also become an issue as the community might require certain landowners to forfeit the use of their land to allow for the placement of communal solar collectors.

The solving of the planning impediments to solar energy utilization will clear one of the major paths for those wishing to employ solar energy in their every day energy usage. These include not only homeowners, but concerned developers and builders as well.

Cities could require their general plans to contain an energy element which promotes solar energy. The comprehensive plans could then help guide local officials in their policy making. The major control is zoning. Zoning ordinances have to be examined and broadened to insure that the controls do not unduly inhibit solar energy use. There is a need for flexibility and performance standards. Zoning can be used to protect solar insolation rights by condemning for solar skyspace. Solar rights could be insured by an envelope or a three dimensional zone, which considers the vertical as well as the horizontal angles, which would be attached to the deed of the house. Solar easements would also guarantee access to sunlight. This will require a written agreement between property owners, with payment for this right, to set back buildings to allow the maximum amount of sunlight to fall on the collector area. (Ref. 7) Included in this might also be an agreement about landscaping near the easement. The question as to who will carry the cost for pruning, etc., will have to be settled amongst the property owners and should probably be included in the solar easement record to avoid future disagreements. The community could work from the zoning concept to control height of buildings, lot sizes, bulk, setback provisions and landscape provisions. Solar devices could be made exempt from the floor area ratio. A permit system could be established to preclude subsequent land use planning decisions that would affect solar rights. Enlargement or modification of existing density configuration regulations could be made so as to induce solar energy use. A comprehensive ordinance to promote solar energy could be established. A change in codes to reflect the possible use of solar energy will be required. Through the official map laws, undeveloped areas could be mapped for future use to maintain low density development and to protect solar rights. A state agency could delegate authority to local agencies to issue Solar Collection System Protection Certification in conformance with state standards. A property owner would pay a fee to the local agency for a permit to install a solar energy device. Adjacent landowners would be notified for comment. If, after the device is installed, changes are allowed to occur on nearby property which affects the efficiency of the collector, the owner of the device could sue for damages. (Ref. 7)

In subdivision, the community could require that the developer provide group or individual collectors. In the site design review process, the developer would show what he is doing in his development to utilize, either passively or actively, solar energy. If he is doing nothing, he could be docked points. To create an ideal design, the developer must consider location in terms of compass orientation, site design and layout. The city could require developers to show proof of efficiency in siting.

Solar energy use could be stimulated through mechanisms which encourage open space. For example, the developer could place collectors on a building at the north end of a north-south oriented lot and then deed the southern end of the lot to the public authority for a solar park. This could be accomplished by an incentive of receiving a tax write off, for the developer, in excess of the actual market value of the land. Another possibility would be providing density bonuses, through the process of transfer of development rights, to developers who use solar energy devices in their development. Land donation could also be required of developers for parks where solar devices would be housed. Specific mandates could require solar energy use in new construction. These methods need a balancing of private and public advantages of solar energy use against economically useful changes in land use. No one particular group, the city, the developer or the landowner, should have to carry the total economic costs of implementing this technology.

This small-scale development of solar power could be integrated into communities by developing the communities around the power source, with the cluster type of development most suitable. In this manner, the station could be placed in a central park with the community hubbed around. Landscaping and building height would be important. Each separate house or apartment could also be equipped with solar collectors. Houses would be aligned to take full advantage of the sun. The placement of structures and vegetation on the lot should maximize the access to sun light.

Retrofiting of established communities will have its own problems. When retrofiting older communities, there will be the need for land purchasing. The neighborhood would collectively buy the land. There might be a problem with the land if it already has housing upon it. This might eventually lead to the community to request the local government to exhert its power of eminent domain if the landowner is uncooperative. All land would have to be purchased at fair market value to avoid the taking of property. As the land goes up in value, those of lower economic status may be forced to relocate. Adjacent land usage and setting will be of concern and land use planning will have to be broadened to view larger areas and their interrelationships.

More state and local legislation will have to begin to address the following issues to promote growth in use at the individual building levels:

1) Improving access to solar insolation. At present no one has any legal right to the sunlight across another person's property. Access to sunlight will have to be made legal, reliable and protected.

2) Optimizing the location of solar energy collectors.
A poor location will be detrimental to the solar energy system. Landowners utilizing solar energy will require more flexible regulations so that they can locate their collectors to maximize the amount of sunlight used.

3) Allocating rights to solar insolation. One's optimal placement of a solar collector may block another's optimal location. Rules will be needed for the allocation of rights to solar insolation.

4) Increasing the number of solar energy systems.
Government may need to require the use of solar energy systems. (Ref. 8)

Solar energy can be encouraged through local land use controls by removing presently existing impediments and by requiring solar access rights and solar systems in old buildings, new buildings and new developments through innovative land use laws and regulations. Solar energy systems can be designed around some existing laws, but it will be necessary to redesign many of the laws and regulations to accomodate solar energy.

Changes in legislation governing local zoning, subdivision control, comprehensive plans and building codes will have to be made. There should be state laws and guidance for local ordinances with compulsory provisions to increase the use of solar energy, in the form of passive or active systems, without unreasonable burden on the builder or user of the land. Laws could be enacted that would require a builder to examine his site carefully and to consider ways to increase the use of the natural forces of the sun. Specific mandates could require solar stubbing in new construction. State agencies should encourage municipal solar energy programs. Another important factor which should be stressed is the education of the general public on solar energy utilization. It will be important that laws and regulations are made specific enough to survive judicial scrutiny. When considering legislation and land planning inducements to encourage solar energy utilization, the comprehensive energy, resources and environmental needs of society must be reconciled with the needs of the private landowner so that both can achieve maximum benefits.

REGIONAL EFFECTS

When planning for the implementation of solar energy utilization, both large-scale and small-scale, the public must recognize the possibility of regional effects. These effects could be created by the extensive use of solar power. Those areas that most likely would be affected are those areas whose economic base is dependent upon coal extraction and the extraction of radioactive materials. If solar energy was used widely, it would cut back on the need for coal and uranium. This could have serious economic impact on the people whose livelyhood is in these areas. A cut back in the need for fossil-fuels and radioactive materials, though, would lessen the need for strip mining and other environmentally negative methods of extraction. The development of particular types of solar energy methods will require the necessary raw materials. This would create jobs in the areas where silicon and other minerals would be mined. Again, this mining would have its environmental impact. The availability of energy in a region would also have a growth inducing impact on that region and on its surrounding area. Thus, land use planning must broaden its outlook to include the regional effects which might arise out of local land use decisions.

CONCLUSION

In summary, solar power is a viable source of energy and, as energy production and consumption become more important, should be included at the inception of the planning process. The different types of solar plants, both large-scale and small-scale, will require their own distinctive land use considerations. The process of incorporating solar power into our daily energy usage will include changes in legislation, comprehensive plans, local zoning, subdivision controls, building and local codes. There should be state laws and guidance for local ordinances with, perhaps, compulsory provisions to increase the use of solar energy. States and municipalities should encourage solar energy programs. There should be public education programs on solar energy utilization.

Local land use controls are going to have to change to facilitate solar energy use. Solar energy can be encouraged by innovative land use laws and regulations. In order for solar energy to make a useful contribution to the overall energy use patterns of this Nation, the impediments to its use must be removed and replaced with viable inducements.

Not only should the public concern itself with the local short and long term effects of solar energy use, but should also consider the total, regional effects. Land use planning must evolve from the small, local viewing of problems and possiblities into a holistic approach where causes and effects from many angles are considered.

REFERENCES

1) David Morris, Solar energy is really power for the people, Planning 8, 6, 16-20 (1976).

2) Charles Beck, Solar energy development, unpublished paper from Energy Projects Information Center, California (1976).

3) Fowler, John M. (1975) Energy and the Environment, McGraw-Hill, San Francisco.

4) Walter Morrow, Jr., Solar energy: its time is here, Technology Review 76,2 (1973).

5) Williams, J. Richard (1974) Solar Energy. Technology and Applications, Academic Press, San Francisco.

6) Gadi Kaplan, For solar power: sunny days ahead?, IEEE Spectrum December (1975).

7) Arnold Reitze, Jr. and Glenn Reitze, Solar laws. Protecting a place in the sun, Environment 18, 5, 2-4; 18, 6, 6 (1976).

8) Richard Robbins, Law and solar energy systems: legal impediments and inducements to solar energy systems, Solar Energy 18, 371-379 (1976).

TRANSPORTATION ENERGY CONSERVATION THROUGH DIRECTED URBAN GROWTH

Robert L. Peskin *
Peat, Marwick, Mitchell & Co.

Joseph L. Schofer
Department of Civil Engineering
Northwestern University

INTRODUCTION

Difficult decisions must be made regarding the most appropriate policies for conserving the use of scarce energy resources in transportation, while producing the least undesirable consequences for the quality of life. Among the most complex are those choices related to urban passenger travel, which is currently so heavily dependent upon the petroleum fueled automobile that a relatively minor aberration, the fuel shortage of 1974, created serious problems in U.S. cities.

Much effort is going into studies of new technologies for providing power to old vehicles, new fuel for old power plants, and even totally new transportation concepts. Yet, little concern has been expressed for energy-conserving options which are not so dependent upon changing technologies. Affecting the patterns of urban growth to encourage more efficient travel has been shown in earlier research (1) to be a potentially important long-term approach toward reducing transportation energy consumption.

There are good reasons for considering non-technology options for conserving energy in urban passenger travel. In particular, actions related to rule changing, transport network and service improvement, and land-use control, are the kind of steps open to local and regional government. Technological change is much less controllable at the local level and in the public sector. Also, the potential for energy savings through systematic behavioral change may be very important, for such shifts may influence the amount of travel people need to consume, not simply the amount of energy consumer per passenger mile.

Finally, planners and decision-makers at the local level seem to have some intuitions regarding what actions do and do not save transportation energy; they are taking those actions--based on intuitions--on a daily basis. It would be desirable to confirm, or deny, these intuitions so that all actors can work effectively together toward conserving energy used in transportation.

Short of full-scale experimentation--which in this case is too risky, costly and time-consuming--analytic studies using models offer the only systematic way to explore the efficacy of various energy-conserving land use and transportation planning actions. This report describes a computer simulation study of the impacts likely to arise due to such energy conservation policies for urban areas. The simulation system used in this effort has evolved in several stages over the past three years (2, 3), and has been applied to many hypothetical cities in search of the most promising energy-conserving actions and policies.

The policy-assessment tests described here are presented as results of an interesting simulation tool, not as certain indicators of what will happen if particular actions are taken. They may provide some clues as to how cities and their transportation system may function under certain conditions. They may support intuition, or they may deny it. If they do the former, one might believe the results and

* This paper is based on research performed while Dr. Peskin was a graduate student in the Department of Civil Engineering at Northwestern University.

deny the need for such models. If they do the latter, one might deny the results and decry the utility of models. This study is based on the assumption that such models, representing an aggregation of useful components of knowledge, can offer helpful guidance in policy search and evaluation. That is, both simulation studies and intuition, together, are needed for good planning. Each sharpens the other, challenges the other, supports the other.

THE SIMULATION SYSTEM

The simulation system for this study was designed to capture the relationships--or what is known about them--between transportation systems, travel behavior and land use development. This serves two purposes: first, it permits the testing and evaluation of energy-conserving actions and policies affecting any one of several of these three key functional systems. Second, the simulation system was constructed to allow comprehensive assessment of the impacts of a particular change--that is, to allow exploration of the effects of changes at one point in the system on the other points in that system.

The model attempts to recognize all of the interactions relevant to passenger travel that occur in an urban system. It cannot succeed in doing this, of course; nor does it promise to capture all of the relationships in completely realistic form. These limitations are important and highly restrictive; they mean that the simulation results are by no means reliable. They are due, in part, to our inability to deal with the complexity of the real world in an efficient and resource-constrained model. They are also due to our strictly limited knowledge of real-world processes. Thus, any effort to translate the simulation results into real-world implications requires both care and good judgment.

The system itself comprises a collection of component sub-models, developed and tested by others, assembled and interlinked in this study to provide a realistic simulation of certain transportation/land use relationships salient to the assessment of energy policies. Given quite general specifications for a city and its transportation system, the simulation system provides a rather complete description of plausible land use and travel patterns, performance of the transportation system, and an estimate of energy consumed in passenger travel.

The heart of this system is an urban land use model developed by Lowry (4) and improved by Garin (5) and Batty (6). Based on an exogenously specified distribution of basic (manufacturing and other land-intensive or exporting) employment, the Lowry model simultaneously allocates service and residential land uses and defines home-to-work trips. Beginning with an initial estimate of free-flow travel times and costs on the automobile and transit networks, the percent of work trips by automobile is determined using a calibrated mode-split model (7).

Using exogenously specified auto-occupancy levels, vehicle trips are computed and assigned to the highway network using a capacity-restrained algorithm (8). It is this explicit consideration of congestion which defines the land use/transportation feedback linkage. The congested travel times and costs (including gasoline costs as a function of average link speed) are used to redefine relative accessibilities to the zones in the city, values which are used as inputs to the Lowry model. The land use allocation and determination of work trips is then repeated based on these revised travel costs. This iterative process continues until convergence is reached--that is, until the land use and transportation systems are in equilibrium.

The distributions of population and employment (both service and basic) are used to determine non-work trips. These include trips from home to service site, home to social purpose, and non-home-based trips. It is assumed that congestion affects only work trips and that most non-work trips occur during the off-peak periods, when congestion on the highway network is slight; thus, a free-flow assignment algorithm (9) is used to determine automobile travel times and costs for these trips. Again, the mode-split model is applied to determine the percent of trips by automobile.

Based on the final assignment of trips to the highway and transit networks, statistics summarizing the passenger travel for a single day are computed. These include total trips by trip type by auto and transit, total vehicle miles travelled, average trip length, average travel time, and an index of congestion (a weighted average of the ratio of congested to free-flow travel time on each link). Having computed the number of vehicles on each link of the transportation network and the speeds at which they travel, the energy required for travel can be determined. Energy consumed by public transit is estimated similarly.

This simulation approach does not yet consider all travel in a city. All trips for the movement of goods and vehicle trips for the purposes of serving a passenger and changing modes (e.g., transit kiss-n-ride and park-and-ride) are ignored. Trips beginning and/or ending outside of the city are also ignored. Thus, results may be valid only in a relative sense, and certainly are systematically incorrect in an absolute sense.

EXPERIMENTAL DESIGN

The nature of the Lowry land use model, the core of the simulation system used in this research, is that it attempts to create an equilibrium city, simulating the entire historical growth process in order to replicate the state of a city at some particular point in time. In this process, the model assumes that policies and conditions affecting urban structure do not change during the growth period. Of course it is obvious that such policies and conditions are changing all the time in a real city.

To test the interactive land use/transportation implications of various energy-saving policies using this simulation system, it makes no sense to assume that such policies were in effect during the entire growth period of the test city. Instead, a base simulation run is made, to define the structure of a pre-energy-policy city; then, this city is "grown" further using the simulation system, with the second-round growth occurring under the energy policies to be tested.

Specifically, the test city is created using the simulation model and a set of policies and conditions defined as "normal," ignoring any concern for energy conservation. Subsequently, a 25% increment of population growth is simulated under the particular energy-saving policies or actions to be tested. The result, then, is a simulation of a city which grew up in an unconstrained era, and then experienced an additional growth increment under energy constraints.

The comparisons presented in this discussion are between a simulated city which experienced its 25% growth increment without any policy changes--known as the "standard incremental run"--and 24 other simulated states of that city having grown under various urban development policies. While the research effort examined several city shapes (concentric ring, one-sided, and polynucleated), this discussion reports only the results of tests with the concentric ring city (10). This urban form is typical of many contemporary cities and should illustrate the general nature of the results of this investigation. Table 1 describes the parameters of the standard incremental run.

Table 1 Description of Concentric Ring City (Experiment 145 - The Standard Incremental Run)

Population:	125,000 (growth of 25% from 100,000 base)
Land Area:	100 mi^2 (256 km^2)
Zonal Arrangement:	52 zones - CBD and three surrounding rings zone size ranging from 0.25 mi^2 (0.64 km^2) to 4.0 mi^2 (10.24 km^2)
Highway Network:	grid arterial street pattern-spaced at intervals of 0.5 mi (0.8 km) in CBD to 1.75 mi (2.8 km) in outermost ring average free-flow speed ranging from 15 mph (24 kph) in CBD to 30 mph (48 kph) in outermost ring
Transit Network:	12 radial bus routes oriented toward CBD all zones served by at least one route average speed ranging from 10 mph (16 kph) in CBD to 20 mph (32 kph) in outermost ring bus frequency of 6 buses per hour
Location of Incremental Basic Employment:	CBD - 53% of added employment Ring 1 - 47% of added employment
Price of Gasoline:	60¢ per gallon

Table 1 (continued)

Transit Fare: 35¢ with 10¢ transfer

Parking Cost: CBD - $2.50/day for work, $1.00 for non-work trips
Ring 1 - $1.25/day for work, $0.50 for non-work trips
Rings 2 and 3 - free

Average Automobile Occupancy:
work - 1.2 persons/car
service - 1.6 persons/car
non-home-based - 1.5 persons/car
social - 1.8 persons/car

EXPERIMENTAL RESULTS

Earlier research (11) based on a less comprehensive simulation approach has provided a strong indication of the potential for reducing transportation energy consumption by promoting certain land use development patterns. In particular, certain multi-centered or polynucleated forms have been found to be more energy efficient than the more typical, sprawled concentric ring city. The general advantages of nucleation and higher density are principally due to shortened trip lengths.

While it is not feasible to shift existing cities quickly to such radically different development patterns, many decisions made in the public sector can and do influence land use trajectories; these include choices about zoning modifications, building and driveway permits, fewer connections, real estate assessments and tax rates, and the construction of public infrastructure. The simulation results presented below are intended to identify some promising land use/transportation, and energy futures.

This analysis begins with a comparison of centralized versus sprawled growth. General additional experimental excursions, in which promising land use and transportation actions are explored in further detail, are then described.

Centralized Versus Sprawled Urban Growth

The first land use policy experiment, summarized in Table 2, is a comparison of the impacts of locating all of the incremental basic employment in the CBD (experiment 154) and spreading it uniformly throughout all the zones (experiment 189). The decreased energy consumption, compared to the standard incremental run (experiment 145), associated with CBD growth seems to be due to reduced congestion, shorter work trips, and increased transit usage. By locating basic employment in the CBD, where the level of transit service is the highest, not only are work trips shorter, but many are diverted to transit, thus reducing traffic volume on the CBD links and allowing the longer automobile trips from the outer zones to travel under less congested, and thus more energy-efficient, conditions. A uniform distribution of basic employment (experiment 189) encourages growth in zones not well served by transit, resulting in more and longer automobile trips on a more congested network.

Table 2 Comparison of Centralized versus Sprawled Urban Growth

Run No.*	Description	Total Energy (gallons/person)
145	Std. Incr. Run	1.083
149	Diagonal Freeways	0.889
154	CBD Growth	0.993
193	CBD Growth with Diagonal Freeways	0.834
189	Uniform Growth	1.131
194	Uniform Growth with Diagonal Freeways	0.915

Run No.*	Average Work Congestion Index	Average Auto Work Trip Length (miles)	Percent Work Trips by Transit
145	1.511	2.837	13.29
149	1.280	2.916	12.91
154	1.386	2.803	16.49
193	1.221	2.758	15.95
189	1.574	2.908	7.69
194	1.291	2.872	8.04

Run No.*	2nd Moment of Population (person - $mi^2 \times 10^2$)
145	11608
149	11206
154	11451
193	11886
189	13012
194	13227

*Run numbers refer to experiments discussed in reference 10.

The favorable impacts of CBD growth are reinforced when combined with the imposition of diagonal freeways to the rectangular grid of arterial streets (experiment 193). Indeed, even uniform growth becomes much more energy-efficient when such a network improvement is added (experiment 194). One adverse impact is that the increased dispersion of population that occurs when diagonal freeways are added alone (as measured by the second moment of population) is also observed when combined with different orientations of directed urban growth. Ultimately this increased sprawl tendency could be costly in terms of many resources. Thus, addition of diagonal freeways needs to be coupled with actions to discourage sprawled growth and increased travel due to considerably improved access to the central city.

There is also a large increase in total automobile vehicle miles of travel (VMT) whenever diagonal freeways are added. This seems to be due principally to longer non-work automobile trips. There is clearly a need to investigate alternatives for directed urban growth which can both control sprawl and shorten the distance between home and work. Also, it is interesting to find that total energy consumption does not vary directly with VMT, a commonly held assumption.

It is interesting to note that the energy saving contribution of diagonal freeways is due, to a significant extent, to the structural change in the grid network imposed by the additional of diagonal links. This is illustrated in Figure 1 in the comparison of experiment 149 (diagonal freeways) with experiment 191 (diagonal arterials). The similar low fuel consumption levels of these experiments indicates that structural change is more important than the level of service (capacity) differences between arterials and freeways for reducing energy consumption.

Nodal Growth Imposed on the Concentric Ring City

A series of experiments was conducted in which alternative nodal growth patterns were imposed on the concentric ring city in an attempt to imitate some of the characteristics of the polynucleated city, which was found to be quite energy-efficient, in a more realistic framework. These alternative nodal growth patterns are ranked in terms of total energy consumption in Fig. 1. The first major finding is the apparent need to limit high density development to those areas where the transportation network can accommodate greatly increased traffic volumes. The most energy-intensive growth pattern is in experiment 185 in which all of the incremental basic employment is placed in a single on-diagonal zone. The tremendous increase in the congestion index is due to overloading most of the arterial highway links near that zone. Adding a single freeway to the CBD in experiment 186 did much to reduce congestion and resulted in a decrease in energy consumption compared to experiment 185.

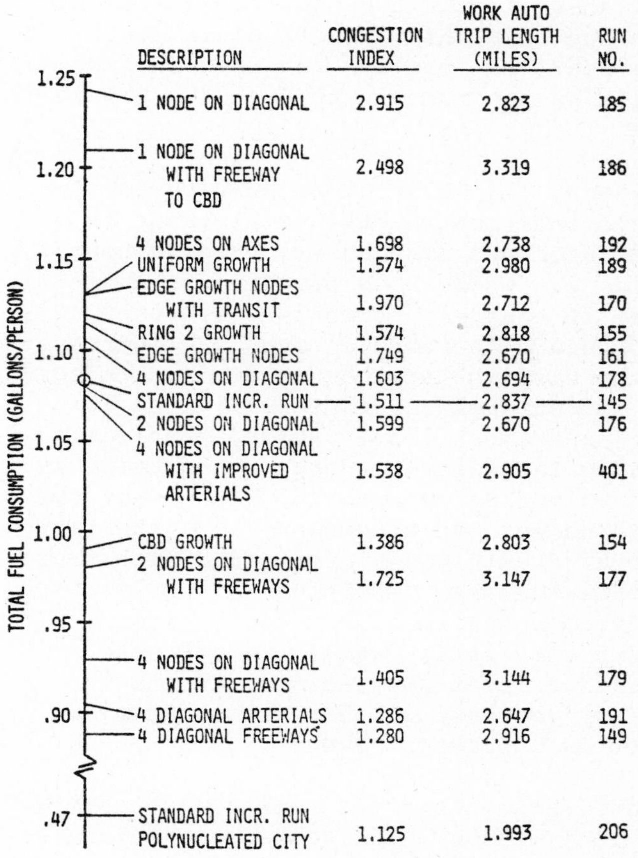

Fig. 1. Effectiveness Ranking of Alternative Nodal Growth Patterns Imposed on Concentric Ring City.

Experiment 161 is an alternative single growth node approach in which incremental basic employment is located east of the CBD, spread among three zones in the outer ring of one city. The resulting "edge" growth node has much more energy-efficient travel than the single-zone growth node in experiment 185 but the congestion index is still high, even though the growth pattern does permit many of the existing arterial highways to serve as direct routes to the CBD. Surprisingly, providing better transit service to the edge growth node, in experiment 170, resulted in more congestion in the adjacent highway links, primarily because this action concentrated more growth in the node. This is the sole case of improved transit service affecting land use to a large degree in this research.

It appears that a series of separated growth nodes is required if energy consumption is to drop below that of the standard incremental run. Dividing the incremental basic employment between two diagonally opposite zones in experiment 176 achieved this goal, although just barely. Maintaining this growth pattern and adding four diagonal freeways to the grid street network, in experiment 177, resulted in a large energy savings. Most of this seems to be due to a shifting of population to zones near the freeways, even to the two facilities not running through the growth nodes.

Continuing this excursion, in experiment 178 the incremental basic employment is split into four diagonally opposite zones in ring 2. While this resulted in an increase in energy consumption, compared to the standard incremental run, the increase is only half as large as when basic employment is uniformly added to all zones in ring (experiment 155). Again, the reason for this increase in energy consumption is the overloading of arterial links near the growth nodes. In experiment 179, the four diagonal growth nodes are paired with four diagonal freeways resulting in the lowest energy consumption of any nodal growth pattern simulated. While the average congestion index was reduced, the now familiar increase in the automobile trip length is observed, however.

Improved grid arterials replaced the freeways in experiment 401 to assess the degree to which the higher level of service and potentially shorter paths (in time) offered by the diagonal freeways influenced the energy consumption of the four diagonal node growth pattern. The energy consumption was only slightly better than the standard incremental run, primarily because the arterials could not reduce congestion as much as the freeways which had higher free flow capacities.

The placement of all incremental basic employment in the four axial nodes in ring 2 (experiment 192) results in a large increase in energy consumption. In this case, congestion is raised by vehicles being forced onto the single arterial routes connecting the growth nodes to the CBD. The diagonal growth node arrangement (experiment 178) seems to allow two rectilinear paths to be chosen to the CBD for each node and results in lower congestion.

The basic conclusion from this series of experiments is that it is desirable to cluster growth, although not so much as to overload the arterial street network. Clustering land use alone, however, is shown not to be a sufficient action to result in any truly meaningful reduction in energy consumption. It is only through the coordinated implementation of directed growth and improvements in the highway network that large energy saving result.

Corridor Growth

Encouraging radial development corridors with supporting highway and transit improvements has potential for reducing energy consumption since it would direct growth across many zones and may avoid the problem of overloading local highway links. Since the establishment of four diagonal growth nodes combined with diagonal freeways seems so promising (experiment 179), a series of experiments conducted to explore the impacts of establishing corridors of growth along these diagonals is presented in this subsection. To accomplish this, the incremental basic employment is placed uniformly in the diagonal zones extending from the CBD of the concentric ring city. Alternatively, growth is encouraged in four axial corridors in a similar manner. Fig. 2 shows a ranking of these experiments in terms of total energy consumption.

The establishment of four diagonal corridors alone (experiment 195) results in a slightly higher energy consumption than the standard incremental run. In a manner similar to imposing four diagonal growth nodes, this is due to increased congestion caused by overloading highway links near the corridors. Experiment 195 also resulted in a noticeable trend toward decentralization of population away from the CBD, which, as a long-term trend could be disadvantageous.

In experiment 196, diagonal arterials are added to the highway network to support corridor growth. This would be expected to define the corridors more sharply (in terms of population concentration) and result in a decrease in energy consumption. A large energy savings does occur, due primarily to a decrease in congestion. In addition, compared to experiment 195, there is a strong shift of population back toward the CBD. Apparently, the accessibility of the CBD is greatly increased by the addition of diagonal arterials to the rectangular grid network.

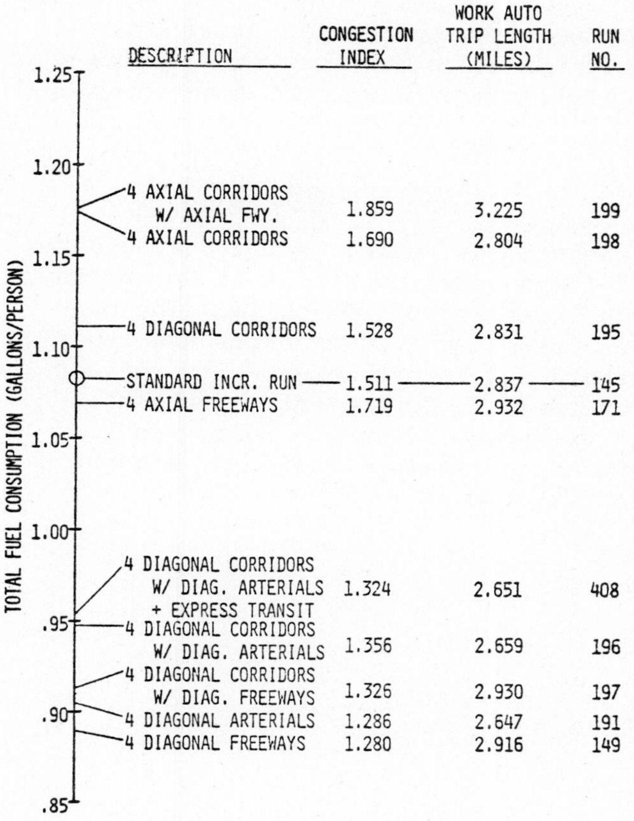

DESCRIPTION
CONGESTION INDEX
WORK AUTO TRIP LENGTH (MILES)
RUN NO.
TOTAL FUEL CONSUMPTION (GALLONS/PERSON)
1.25
1.20
1.15
1.10
1.05
1.00
.95
.90
.85
4 AXIAL CORRIDORS W/ AXIAL FWY. 1.859 3.225 199
4 AXIAL CORRIDORS 1.690 2.804 198
4 DIAGONAL CORRIDORS 1.528 2.831 195
STANDARD INCR. RUN 1.511 2.837 145
4 AXIAL FREEWAYS 1.719 2.932 171
4 DIAGONAL CORRIDORS W/ DIAG. ARTERIALS + EXPRESS TRANSIT 1.324 2.651 408
4 DIAGONAL CORRIDORS W/ DIAG. ARTERIALS 1.356 2.659 196
4 DIAGONAL CORRIDORS W/ DIAG. FREEWAYS 1.326 2.930 197
4 DIAGONAL ARTERIALS 1.286 2.647 191
4 DIAGONAL FREEWAYS 1.280 2.916 149

Fig. 2. Effectiveness Ranking of Alternative Corridor Growth Patterns Imposed on Concentric Ring City.

Experiment 197 continues this excursion by replacing the diagonal arterials with diagonal freeways. This results in slightly less congestion and lower energy consumption confirming the notion that the structural change to diagonal routes is more important in terms of energy savings than is the level of service differential between freeways and arterials. While this action does more strongly define the corridors, it forces more persons toward the extreme corners of the city. This may explain why the average automobile work trip length increases while the congestion index remains the same as in experiment 196. This indicates that while it may be beneficial to coordinate corridor growth with a spinal arrangement of higher capacity transportation facilities, caution must be applied when providing extremely high level of service radial highways, since these have significant potential for encouraging sprawled growth.

To determine if this consideration extends to transit, diagonal corridor growth is combined with diagonal arterials and diagonal express transit service in experiment 408. While the energy consumption is no better than in experiment 196, population growth in the corridor is more sharply concentrated. Apparently, corridor growth coordinated with both improved transit and highway service is advantageous.

The simulation of axial corridors in experiment 198 results in increased congestion and fuel consumption compared to the standard incremental run. The population shift seems to indicate that the sprawl would continue past the edge of the city if another ring of zones were available for development. Replacing the axial arterials with freeways in experiment 199 does not change energy consumption and actually increases congestion by diverting trips from zones adjacent to the corridors, encouraging longer trips. There is a strong shift in population back to the CBD compared to experiment 198. This is opposite of the trend observed when diagonal freeways are added to diagonal corridors. The reason for this difference seems to be due to the structure of the highway network. Diagonal corridors seem to allow for more paths of lower travel time to the CBD than the axial corridors, which force most trips to follow the axial arterials to the CBD.

CONCLUSIONS

A number of general guidelines for urban land use policy development can be drawn from the experimental results. All of these must be considered tentative, however, since they are based on a limited set of experiments on hypothetical cities, using a simple simulation system. The strengths of the findings, in terms of their logic and intuitive appeal, seems to indicate, though, that they are not generally due to biases built into the model.

One final caveat regarding the conclusions is that this research was concerned only with energy requirements for urban passenger travel, ignoring changes in consumption in the residential, commercial, and industrial sectors which might result when urban growth is altered. While a more comprehensive approach has been used in some analytic (12) and empirical (13) studies, this was avoided in this research in order to concentrate on the transportation sector alone and to examine a large set of specific actions.

Particularly when coordinated with the existing transportation network or improvements to that network, directed urban growth can result in more energy-efficient urban forms than sprawled development. The imposition of nodal or corridor growth policies on existing urban forms seems to hold significant promise for energy conservation. The desirability, from an energy perspective, to encourage people to live closer to work was demonstrated. The most appropriate manner in which to accomplish this goal was not identified but it may lie in the realm of tax incentives, zoning ordinances, the appropriate application of urban development funds, and broadly focused efforts to educate people regarding the true cost of transportation.

The potential energy saving benefits of social and physical redevelopment programs for inner city areas is also implied by the results. If older, more central residential locations, now made unattractive by urban blight, could once again become prime residential areas, significant savings in work trip energy consumption might result. Furthermore, persons living in medium to high density areas could be offered better quality transit service more easily, further reducing transportation energy consumption. Such urban development must be coordinated with programs to encourage employers to locate in the CBD, however. Simply locating more population in the the central city may actually increase energy costs if future employment centers locate in the suburbs.

Care must be taken not to overload existing adjacent highway links when implementing directed urban growth, since the resulting congestion could significantly increase energy consumption. Directed urban development balanced with transportation network capacity, perhaps through improved or additional links, can result in lower energy consumption. For example, corridor growth, when combined with diagonal arterial or freeway links on a rectangular highway network, was shown to be among the most favorable of the alternative development patterns simulated.

However, increasing CBD-focused highway service beyond the level needed to control congestion promises to encourage a sprawled growth pattern, and this may increase energy consumption in the long run. Thus, care is needed to limit the capacity of such facilities, either by design or through access controls.

This research has shown the value of comprehensive land use-transportation planning in energy conservation. Piecemeal, incremental urban development is not nearly as energy-efficient and could result in land use patterns with long term adverse transportation energy consumption characteristics. Analytic tools need to be developed to assess the regional and subregional transportation network and energy impacts of medium and large scale development proposals such as industrial parks, shopping centers, and new residential communities. This research has shown that this can be done with a minimum of theoretical model development.

ACKNOWLEDGEMENTS

This research was funded by the U.S. Department of Transportation Office of University Research, contract DOT-OS-50118. The conclusions presented in this paper are those of the authors and do not necessarily reflect the views of the U.S. Government. Technical support was provided by the Energy and Environmental Systems Division of Argonne National Laboratory.

REFERENCES

1. Jerry L. Edwards and Joseph L. Schofer, "Relationships Between Transportation Energy Consumption and Urban Structure: Results of Simulation Studies," Transportation Research Record 599, Transportation Research Board, Washington, D.C., 1976.

2. Ibid.

3. Larry A. Bowman, Daniel H. Goetsch, and Steven E. Polzin, "A Model for Evaluating the Energy Conserving Potential of Transportation and Land Use Policies: Development and Preliminary Application," unpublished M.S. research report, Department of Civil Engineering, Northwestern University, May 1976.

4. Ira S. Lowry, A Model of Metropolis, memorandum RM-4035-RC, the RAND Corporation, Santa Monica, California, August 1964.

5. R. A. Garin, "A Matrix Formulation of the Lowry Model for Intra-Metropolitan Activity Location," Journal of the American Institute of Planners, Vol. 32, No. 6, June 1966, pp. 361-364.

6. Michael Batty, "Recent Developments in Land-Use Modelling: A Review at British Research," Urban Studies, Vol. 9, No. 2, June 1972, pp. 151-177.

7. Charles River Associates, A Disaggregate Behavioral Model of Urban Travel Demand, final report prepared for Federal Highway Administration, contract DOT-FH-11-7566, March 1972.

8. Larry J. LeBlanc, Mathematical Programming Algorithms for Large Scale Network Equilibrium and Network Design Problems, Ph.D. dissertation, Transportation Center at Northwestern University, Evanston, Illinois, 1973.

9. T. C. Hu, "A Decomposition Algorithm for Shortest Paths in a Network," Operations Research, Vol. 16, No. 1, January-February 1968, pp. 91-102.

10. Robert L. Peskin and Joseph L. Schofer, The Impacts of Urban Transportation and Land Use Policies on Transportation Energy Consumption, final report prepared for U.S. Department of Transportation Office of University Research, contract DOT-OS-50118, 1977.

11. Edwards and Schofer, op. cit.

12. T. Owen Carroll, et al, Land Use and Energy Utilization, interim report prepared by the Brookhaven National Laboratory/State University of New York Land Use-Energy Utilization Project for the Office of Conservation and Environment, Federal Energy Administration, BNL Report 20577, October 1975.

13. Real Estate Research Corporation, The Costs of Sprawl, report prepared for U.S. Department of Housing and Urban Development and Environmental Protection Agency, April 1974.

URBAN PLANNING AS AN IMPEDIMENT TO ENERGY CONSERVATION: AN EXAMINATION OF POTENTIAL CONFLICTS BETWEEN EXISTING PLANNING REGULATIONS AND ENERGY CONSERVING SITE PLANNING ALTERNATIVES

John Tschanz

Energy and Environmental Systems Division
Argonne National Laboratory, Argonne, Illinois 60439

ABSTRACT

Energy inefficiencies in urban areas are, in part, a legacy of the layout of streets and buildings during land development. It seems almost inevitable that this process must be more systematically carried out if the energy performance of cities is to be improved in the future. Systematic analyses of the performance of cities and suggestions for achieving public objectives, such as energy conservation, are within the realm of urban planning. Implementation of energy conserving ideas at the site planning level should be particularly effective. As the concepts for energy conserving site planning become more widespread, a situation that might frequently occur is the public review of energy conserving site plans for private development projects. This paper is a report of our examination of conflicts that might arise if existing planning regulations were applied in such reviews.

Although our purpose was not an investigation of the energy consequences of site planning elements, we reviewed some of the research literature that is frequently cited in discussions of energy conservation and site planning. From this literature a representative set of energy conserving site design options was chosen. When compared with criteria that are common in zoning and subdivision regulations, many of these options contain features in conflict with long-established elements of the regulations. Flexible and performance-based regulations hold brighter promise for accommodating innovative design concepts, but experience with their application is limited and their review procedures often turn out to be unacceptably lengthy.

In spite of the conflicts that can be identified between details of the regulations and features of the design options, the overall objectives of the regulations appear generally to be met by the options. It would seem, therefore, that the conflicts can and probably will be removed from the regulations as soon as the case supporting the effectiveness and feasibility of the design options is adequately demonstrated.

INTRODUCTION

Cheap and plentiful energy supplies in the past have permitted several decades of urban and suburban development to ignore the energy consequences of site design and land use decisions. Consumers, land developers, planners, and elected officials have not had to seriously consider energy consumption as a major criterion in development decisions in the past. But the opportunities for energy savings in project design are many. Some of the most obvious examples of energy saving design options are discussed below. In addition to reducing the energy needs of the intended users, a planned approach to total project design will also facilitate the siting of energy system components within development projects.

Although the need to conserve energy has become clear, present institutional structures might hinder the public acceptance of community energy systems and energy-efficient design concepts. By considering the workings of public institutions governing land development and utility operations, we have tried to find preliminary indications of impediments to the commercialization of design innovations in the answers to several questions. Will planners and other public administrators support these concepts, or will they resist them? How are elected officials likely to respond? Do existing laws governing land development and utility operations accomodate such innovations, or will changes in these laws be necessary?

ENERGY CONSERVING SITE DESIGN OPTIONS

Building types, orientations, and densities, street layout, and the choice of land use mix are site design elements that have an effect on energy consumption. They are clear candidates for modification to achieve improved energy efficiency.

Building Types and Densities

The variation in energy consumption with changes in building types and densities can be illustrated by the residential sector which accounts for approximately one-fifth of the national energy use. The most predominant form of housing in this country -- the single-family detached house -- is the least efficient user of energy, both in terms of space conditioning and transportation needs. A number of studies have pointed out that multi-family units tend to have smaller floor areas and suffer from less heat loss in winter and heat gain in summer, than do single-family units. Common wall construction creates less surface area per unit exposed to outside heat and cold.[1] Cluster and higher density development can help to conserve energy through reducing space conditioning and driving needs.

With no increase in overall density, clustering of single-family detached units in conjunction with innovative street design can reduce street lengths by 25%.[2] Because streets are shorter, automobile trips within those projects are shorter and may be more likely to be travelled by foot or bicycle. Through a flexible approach to site planning a developer can also do a better job of adapting a project to topographic and microclimatic features; building orientation for solar utilization with its energy savings is easier. It has been estimated that the total energy savings from this type of planned development, even at low densities, can reach between 8 and 14%.[3]

While energy efficiency in residential development can be increased through clustering and innovative design without increasing overall density or changing housing types drastically, increases in density through multi-family housing and apartment units offer much greater energy savings. Like clustered, detached single-family development, high-density development reduces street lengths. It also conserves energy in space conditioning, especially heating, since less wall surface is exposed to the outside environment. Higher density types of housing such as townhouses, and low- and high-rise apartments are both smaller and more efficient. A typical high-rise apartment unit is 42% smaller and uses 34% less energy per square foot than a single-family detached one-story house.[4]

Building Siting and Orientation

The physical site of a home, a commercial structure, or an entire development has associated with it a variety of characteristics; one is the mini or microclimate of a lot or a project site. The microclimate of a development site with its characteristics of wind, sunlight, and air drainage directly influences the amount of energy necessary to regulate the interior environment of any building in that location. If the site is particularly cold, more heating fuel would be needed. If a site is warm, more electricity for air conditioning would be required.

Many environmental factors determine microclimate; topography is one of the most important. As elevation, slope, and the direction that a slope faces vary, so do the amount of sunlight, velocity of wind, and amount of cool air drainage received by the site. In most of the country a site that is topographically protected from winter winds, oriented towards the sun (i.e., primarily south), and located away from low "cold pockets" is warmer in winter.[5] In hot areas of the country, sites exposed to cooling breezes are most advantageous from an energy standpoint.

In addition to topographic location, the orientation of a building on its site can help to utilize natural warmth and coolness and reduce space conditioning needs. Orienting the

---

[1]Hittman Associates, Inc. *Residential Energy Consumption in Single-Family Housing* (March 1973) and *Residential Energy Consumption in Multi-Family Housing* (October 1972), U.S.G.P.O. Washington, D.C.

[2]Urban Land Institute, *Innovations vs. Traditions in Community Development: A Comparative Study in Residential Land Use,* Technical Bulletin No. 47 (1963), p. 67.

[3]James S. Roberts, *The Costs of Sprawl,* (1974) U.S.G.P.O. Washington, D.C., p. 15.

[4]W. Curtis Priest, et al., *An Overview and Critical Evaluation of the Relationship Between Land Use and Energy Conservation, Vol. 1, Main Report,*" Technology and Economics, Inc. (1975) p. 100.

[5]Olgyay, V., *Design with Climate: Bioclimatic Approach to Architectural Regionalism,*" Princeton University Press (1963) p. 52.

principal facade of a building toward the south means the greatest warmth in the winter and the greatest coolness in the summer. If the building faces east or west, maximum warmth will occur in summer and maximum coolness in winter, when both are least desirable. Optimum southern orientation varies somewhat depending on the climatic region of the country.[6]

In addition to the sun, winter winds and summer breezes have an influence on space-conditioning requirements. Buildings that are positioned perpendicular to the wind direction receive its full force; at 45° the velocity is reduced by 50%.[7] Houses arranged in a row protect each other from wind effects while staggered houses make use of the natural flow of the wind to receive optimum wind velocity. In winter the former arrangement is desirable; in summer the latter. Because winter winds and summer breezes usually come from different directions (e.g., in the midwest region prevailing winds are from the west in winter and southwest in summer), one pattern can provide optimum wind protection and breeze utilization.[8] Sometimes optimum orientation for wind may be different than optimum orientation for solar utilization. In such cases the relative influences of wind and sun in space conditioning needs should be weighed to choose orientation.

Street Layout

Insofar as it affects housing orientation, the orientation of streets also can contribute to the energy efficiency of buildings. In many subdivisions the houses face the street (i.e., the largest number of windows face the street or the back yard). If street orientation is primarily from east to west, either the front or the back of the homes will face south. The advantages of a southern orientation as discussed above will decrease the space conditioning energy demands of those houses.

In 1970 transportation accounted for 25% of all energy consumed in the United States. Fifty-five percent of that energy was burned in cars. To reduce transportation energy use requires more efficient autos and fewer vehicle miles travelled (VMT). The latter depends on reducing the frequency of use of automobiles and the distances of trips travelled. It is here that community design plays an important role. If automobile use for trips within a neighborhood or a subdivision can be made shorter and less frequent, significant savings in fuel consumption could be realized.

Discouraging automobile use and encouraging walking and bicycling within a subdivision can both be accomplished by a number of means. Built-in footpath and bicycle path systems with adequate protection from the elements will facilitate and encourage biking and walking. As mentioned earlier, clustering development can reduce street length and the distance between residences and activity centers in a project. Traditional grid type subdivision and street layout produces maximum street length. A variety of circular design formats that often use cul-de-sacs can decrease street length considerably while serving the same number of dwelling units as a grid type layout.[9] Cul-de-sac streets inhibit through traffic and encourage people to walk or bike to destinations. Reducing street length and width saves initial construction materials and activity, and the energy needs associated with both. The reduction in asphalt and concrete surface areas also minimizes the rise in air temperature from sunlight absorption, thus, additional energy may be conserved through lower air-conditioning requirements.

Development projects of any type -- residential, commercial, or industrial -- can be designed to facilitate maximum use of existing mass transportation systems. If bike or footpaths are included in a subdivision, a plan that links transit stops to residences via these paths will encourage use of mass transit. If a subdivision includes clustering or areas of varying densities, those areas with the highest density might be closest to existing bus stops or transit terminals. Other heavy traffic generators, such as commercial facilities, can also be planned to take advantage of existing services. Such amenities as enclosed walkways, benches, and bike parking facilities near terminals will further encourage use of public transportation.

Mixed Land Use

Mixed use developments offer potential for creative combinations of different land uses within a single project such as including commercial, office, and recreational facilities in residential apartment complexes. Such integration of land uses would reduce the need to

---

[6] Olgyay, V., p. 61

[7] Olgyay, V., p. 100

[8] Olgyay, V., p. 101

[9] Urban Land Institute, *Innovations...*, p. 24

travel, plus offer opportunities for increased efficiencies in energy systems through the utilization of waste heat and other resources.

Some of the most minor automobile trips made involve running errands to grocery stores for small purchases. Supermarkets are located generally within commercial districts distant from residential neighborhoods. Smaller, convenience type grocery stores, too, are often excluded from residential developments and require use of a car to reach. The number and distance of auto trips necessary for minor purchase could be reduced by incorporating the stores that can be supported[10] into the planning of a project, locating them centrally, when possible, to serve the greatest convenience.

In addition to using an automobile for small grocery purchases residents must often drive to other commercial and professional establishments. Doctors' and dentists' offices, drugstores, beauty parlors, and other stores and offices usually are restricted to commercial or high-density residential/commercial development areas. Such uses, if carefully planned, could be compatible with primarily residential development. Depending upon density, some development projects could support a small or moderately sized store/office complex. Not only would it provide goods and services to residents within a short distance, but also would make some jobs available that could be filled by people who lived nearby. Thus, auto use for errands and for some workplace commuting could be cut.

Separation of uses has meant that many buildings are used for only one purpose, often during only part of the day or night. Heavy industrial buildings are used around the clock but commercial, recreational, and educational and social buildings often sit vacant for large amounts of time. These buildings although vacant must be heated, cooled and lighted to some extent. "Doubling up" of uses within one structure could save energy used in space conditioning and an additional amount of energy and materials that would otherwise be necessary in building two structures instead of one. Multiple use buildings have become relatively common in urban areas as large mixed-use shopping, office, and residential complexes are developed.[11] At the subdivision scale there are fewer opportunities for multiple use buildings, but they exist. Developments that have planned both a sales office and activity center could combine the two within one structure. Separate areas within such a building would be required, but its energy efficiency would exceed that of two separate buildings. Other examples might include use of school recreational facilities for evening, adult use or use of a professional or office building for community meetings.

PUBLIC PLANNING CONTEXT

The public planning context for most development proposals is primarily a local one. States are assuming a more active role in some aspects of planning, particularly those related to such resources as environmental quality, coastal zones, wetlands, mineral resources, wild and scenic rivers, and other critical areas, as well as activities such as the siting and construction of large, region-serving facilities. The level of state involvement varies greatly throughout the nation, however,[12] The incorporation of innovative combinations of energy system components within a development could bring it under the control of state utility regulations which are universal, if not uniform from state to state. In this paper we consider only the traditional planning tools that have accompanied the delegation of most planning responsibilities to local authorities: development policy-making and review functions and the establishment and administration of zoning, subdivision, and other land use regulations.

Public Planning

Most planning functions in local government are organized around a planning commission -- a body of appointed laymen (although sometimes elected) who review and approve long range planning policies for community growth and development, and also review specific development proposals. Depending on the size of governmental jurisdiction involved, most planning commissions are served by professional staff who provide technical support --

[10] George Nez, *Standards for New Urban Development - The Denver Background*, Urban Land 20 (1961), p. 6

[11] Urban Land Institute, *Mixed-Use Developments: New Ways of Land Use*, Technical Bulletin No. 71 (1976)

[12] William Toner, et al., *Information Analysis for Selected Land Use Programs*, report prepared for Energy and Environmental Systems Division, Argonne National Laboratory (1975).

conduct studies, monitor programs, review development proposals, and draft plans, policies, and legislation. Planning commissions are generally advisory, referring their recommendations on policy changes to legislative bodies for final action. Citizen participation occurs at a number of points in the planning process, both formally and informally.

Since the planning process is inherently political, with final authority for decisions resting in the hands of elected officials, its ultimate products tend to reflect the interests of its constituents to a large extent, or more accurately, those constituents and private interests who make themselves clearly heard. A conservative attitude is likely to greet radically new energy technologies proposed for implementation at the community level. This does not mean outright rejection of new ideas, but it does mean that the superior merits of new technologies and concepts must be convincingly demonstrated in a variety of community settings.

Assuming consumer demand for new energy technologies and design options is forthcoming, and major technical performance problems are resolved (such as eliminating noise, odor, fumes, safety hazards, etc.), local planning institutions can be expected to respond positively to these concepts themselves. Working in favor of energy innovations is the fact that many energy-efficient design options and related implementation tools have been promoted by planners for years in efforts to achieve objectives besides energy conservation (objectives such as more environmentally sensitive site design and less monotony in development patterns).

Sympathy or even endorsement of these concepts does not necessarily mean that actual implementation will run smoothly. Several major, practical problems may present serious hindrances to the implementation of integrated community energy systems (ICES) and energy-efficient design concepts.

Land use policy changes. It appears that ICES and energy-efficient design concepts can be implemented without the creation of new regulatory devices. Changes in existing zoning and subdivision regulations can be made to accomodate these innovations. Some of the zoning amendments necessary, however, would amount to significant changes in public policies embodied in zoning law, and may be debated at length in many communities. Locating energy production facilities in residential developments close to energy consumers involves a major policy change in conventional use restrictions. Moreover, relaxing the currently high degree of land use segregation found in most zoning ordinances to permit more mixed-use developments and more integration of land uses in general, will also require major policy decisions. Such policy changes are frustrated further by the fact that the political time horizons of elected officials often do not coincide with the benefits of change itself. Initial risks must be taken by one administration, with long-term payoffs not likely to become visible until some later administration is in office.

Information for decision-making. Lack of facts on which to base public policies is also a potentially serious constraint. Energy consciousness is currently very low among public institutions despite the recent energy crisis and continually rising energy prices. Public decision-makers currently are constrained from making rational energy-planning decisions because of a lack of specific facts and figures demonstrating the benefits of various design options so that choices among energy alternatives as well as trade-offs with other community goals can be made.

Not only are data scarce on the relative energy-efficiencies of various design options, but comprehensive energy consumption and consumer behavior data are almost nonexistent for specific communities. Energy-efficient, land-use planning on anything above the scale of a single development can quickly become overly complex because of (1) the high number of variables that must be considered, and (2) the lack of baseline data from which to begin. (Current ERDA research programs are beginning to address this need.)

As was the case with the environmental movement, considerable time lag can be expected for an energy ethic to become institutionalized. Commercial success of new energy technologies and concepts which must be marketed to a highly fragmented number of local communities and public officials will depend, in part, on a massive public education and information program. Even more important, general energy consciousness and improved baseline data will be necessary for planners and public officials to develop sound energy-planning policies.

Fragmentation of public institutions. Another potential hindrance to the commercialization of energy conserving designs in the highly fragmented nature of public institutions at the community level. Each local planning jurisdiction is slightly different in character and governed by a slightly different body of planning law and regulatory techniques. Efforts to standardize the homebuilding industry on a national basis were hampered for years by the fragmented nature of the public institutional and regulatory process at the local level. This fragmented nature of public institutions means that developers most likely will tailor project designs, and possibly ICES hardware also, to specific communities.

Administration

The administrative processing of development proposals is one of the potentially most serious barriers to the commercialization of ICES and energy-efficient design concepts. Regardless of legality or public support, the constraints imposed upon developers during public review and approval procedures are becoming an increasingly critical variable in the developer's success. Both ICES and related energy-efficient design concepts lend themselves best to implementation through flexible regulatory techniques (such as planned unit developments (PUD), cluster provisions, performance zoning, and mixed use districts). Unfortunately, it is these types of highly discretionary and flexible land use controls which result in the highest administrative costs.

Administrative implications of flexible regulations. As a general rule, flexible zoning techniques require some form of administrative review and approval procedure. PUD typically requires a rezoning application coupled with a two or three stage site plan review procedure. A three stage rezoning application procedure might have built-in time estimates totalling approximately 12 to 19 weeks. These review times are relatively low. Review case loads and the complexity and controversial nature of some development proposals have resulted in lead times of up to a year and more. Flexible regulatory techniques are generally perceived to result in higher administrative review times when compared to conventional development procedures, as shown in Table 3.1. Increased review times translate directly into higher front-end carrying costs for such development expenses as interest on land options. This is especially crucial for developers operating close to the margin.

Other complications raised by flexible regulatory procedures include costs of preparing applications and impact statements, the necessity for more sophisticated staff to manage projects (both in public agencies and in private development firms), and more negotiation and discretion exercised by public officials, which in turn can result in more costly demands on developers for design changes, extra amenities, and so on.

All of these constraints imply higher levels of uncertainty and higher costs for developers. In a financially tight market, many developers go back to the basics of conventional subdivisions and the single-family home, avoiding fancy, flexible devices which they can no longer afford to get bogged down in.

Interdepartmental coordination. Development proposals typically involve other public line agencies with single-purpose responsibilities for such services as police and fire protection, streets and roads, water and sewer, etc. In a typical rezoning application procedure, all applications automatically go to the engineering and traffic departments for review and comment. As more single-purpose agencies become involved in administrative review procedures, the process tends to become more complex, costly, and time-consuming. Also, public service line agencies have legitimate concerns about innovations in development. Narrower and more complex street systems in residential developments could inhibit the delivery of police and fire protection services, through restricting emergency vehicle access. Problems are compounded when veto authority over development applications is vested in several agencies. Facilitating smooth, streamlined interdepartmental review procedures may be essential to the success of community system projects.

TABLE 3.1 Estimated Comparisons in Processing Time between PUD and Conventional Development

	Planners % of Responses	Developers % of Responses
PUDs are processed significantly faster	2.2	6.8
About the same	64.7	24.4
PUDs are processed significantly slower	33.1	68.8
Total Responses	100.0	100.0

Source: Planned Unit Development Ordinances, by Frank S. So, David R. Mosena, Frank S. Bangs, Jr., Chicago, IL: American Society of Planning Officials. May 1973.

Local Land Use Regulations

Land use regulations include zoning, subdivision regulations, and aesthetic controls. Several aspects of each of these will affect the acceptability of energy conserving development options.

Zoning. Zoning is a land use control tool which divides a community into use districts, and spells out standards and limitations for each. Permitted uses, lot sizes, yard and setback requirements, and height and bulk restrictions are established for different types of districts. The zoning ordinance is usually directly related to a comprehensive community plan that spells out growth and development goals and identifies where in the community certain uses should occur. Commercial, industrial, and residential districts are generally separated from each other; some specific zoning districts may include a mixture of two or three of these uses.

In many ways the intent of zoning (i.e., separating uses and maintaining relatively low residential development) is in conflict with efforts toward energy conservation. Restrictive districting and maintaining low density, single-family development promote energy intensive community development patterns. There are a number of provisions found in zoning ordinances that could inhibit the use of ICES and energy-efficient design options for energy conservation. They include use restrictions that encourage separation of uses, minimum lot area provisions that determine density, and yard or setback requirements that necessitate detached type housing.

Use Districting. For each district designated by a zoning ordinance is listed a number of permitted uses, permitted by right or subject to some form of administrative approval. In this way the ordinance creates zoning of primarily residential, commercial, industrial, or mixed use. Residential districts tend to be the most restrictive; many exclude virtually all commercial or industrial use. Even in progressive ordinances, innovative housing design concepts such as clustering and PUD are usually permitted only as special uses in residential districts, subject to an administrative review procedure. Many ordinances do not make such provisions, and those options are not possible at all. The result of this type of restrictive zoning is residential neighborhoods of single-family detached housing that have no commercial facilities for use by residents.

The underlying point of use districting, and of all zoning for that matter, is to protect various land uses from the negative impacts of other uses. It is necessary to separate residential areas from noisy, dirty, heavy industry or school and playgrounds from the heavy traffic of expressways. This reasonable concept, however, has been carried too far, in some cases. The availability of the automobile has made possible large expanses of residential neighborhoods, that are located miles from shopping centers or even small stores. Use districting could be relaxed when gross problems with mixing uses are not present. This would permit a tighter integration of compatible though different land uses, such as residential with neighborhood commercial facilities. The planning experience with planned unit development in the past, however, has indicated that the administrative review necessary in most cases to build a PUD discourages developers from doing so. If permitted by right, there would be a greater incentive to build those types of efficient housing.

In addition, ICES may require technologies that under present district definitions might be classified as commercial or industrial uses. These technologies would then be excluded from the residential neighborhoods where they were needed. The light industrial uses that may be part of new energy systems could be regulated in restricted zoning districts, subject to site plan review. If the performance of a community energy system is tolerable, with acceptable levels of noise, pollution, glare or visual disturbance, they could be included in primarily residential districts through the use of performance controls. It has become common practice to regulate industrial use through performance controls that set maximum levels for the variables mentioned above. Specific uses are not listed, but rather any use may be permitted as long as it meets the standards set forth for a particular district. The concept could also be used to permit commercial facilities. Performance controls for this type of use might include maximum traffic generation, maximum floor area of a store, or hours of operation.

Lot Area Provisions. Provisions in the zoning ordinance for minimum lot size determine the density of a particular district. Large minimum lot size means low density. In residential districts density is traditionally low, with single-family detached units the most common form of housing. In the unincorporated areas of Cook County, Illinois, 96% of the land designated for residential use is zoned at a net density of 4.4 units per acre or lower.[13] Where clustering of dwelling units is permitted as a special use, the cluster project usually must meet the density requirements of the district. If a PUD is permitted as a special use, it might not be strictly subject to the density and lot requirements of the district in which it is built. However, to be granted an increase in density, a PUD would be required to provide tangible benefits -- exceptional amenities, design excellence, etc. -- to the neighborhood in which it is located. A significant density increase may require a rezoning. The PUD in itself requires a special permit, but if its density substantially exceeds that of the district, it would require administrative approval including review by the local government council.

Although it is not yet clear what levels of density will be required for efficient performance of ICES, low density development tends to be energy intensive; new community design and utility systems for energy-efficiency may require higher densities. In addition, high density would require less extensive infrastructures for servicing by an ICES than a more spread out, low density development. In the event that these become priorities, minimum lot requirements must be revised or done away with altogether. Performance controls that set maximum lot coverage and thus preserve open space could take the place of lot area requirements. For example, a townhouse apartment project would not be subject to a minimum lot size, but rather would have to provide a percentage of the site area involved as open space.

Yard and Setback Requirements. Yard or setback requirements in zoning ordinances set minimum distances between buildings and between buildings and the street. In specifying minimum distances from front, side, and rear lot lines, yard and setback requirements result in the centering of homes in the middle of their lots. These requirements also mean that building lines must follow lot lines; the shape of the lot determines the axis of the house.

For community energy systems yard and setback requirements could be a barrier to innovative building orientation and solar collector technology. Full-scale clustering requires flexibility in lot size requirements but the limited form of clustering of duplex housing could be prohibited by yard requirements. Side yard specifications would prohibit common wall construction. In order to orient buildings for maximum solar use, it may be necessary to position their southern facades a minimum distance from other buildings. Especially in developments with small lot sizes, this would require some type of zero lot line zoning, allowing buildings to abut lot lines. Rigid yard and setback requirements would make this impossible.

In order to facilitate energy conserving designs, the yard and setback requirements discussed above must be flexible. They could take the form of a performance standard that specified a minimum total lot area without dictating the distribution of yard space on the lot. In this way building orientation, solar collectors, and duplex housing would be possible.

Subdivision regulations. Subdivision regulations control the process by which land is divided into developable lots. They regulate layout of lots and streets, design standards, construction of public utilities (streets, sewers, etc.) and the provision of open space.

The provisions of subdivision regulations that could have the greatest influence upon the introduction of energy conserving designs are those that deal with the layout and design standards for streets. These provisions have a direct influence upon site and building orientation and upon future automobile use.

Street Layout Requirements. Subdivision regulations set forth recommendations for the design of circulation systems. In the past these provisions have specified street layout that was primarily based on the grid system. Many communities have recently amended subdivision regulations to encourage curvilinear street design and use of cul-de-sac streets. Grid type street layout is encouraged in new developments through subdivision specifications that new streets be in alignment with existing streets. "As far as practicable, all proposed streets shall be continuous and in alignment with existing streets."[14] In this way, both the pattern of street layout and the directional orientation of streets are dependent upon the established circulation system. In addition, the use of non-through streets

[13] Cook County Zoning Ordinance (1976) Article 4, pp. 11-48

[14] Ordinances, Regulations, and Laws Pertaining to the Maryland-National Capital Park and Planning Commission as Amended Through December, 1974, p. 379.

like cul-de-sacs while not expressly prohibited is accorded somewhat less acceptance than conventional streets.

Street design specifications in subdivision regulations are not so rigid as other types of provisions. This area of subdivision regulations is not so much a disincentive to energy-efficiency, but rather an opportunity for incentive that could be taken in the future. Guidelines that recommend total system approach to street design with the incorporation of bicycle and pedestrian elements could be a significant incentive to provide efficient street layout. The issue of directional orientation for use of the sun could also be raised in subdivision regulations to further encourage innovation.

Street Width Standards. Subdivision regulations set very specific standards for the design and construction of project streets. Maximum grade, radii of curves, and street width specifications are among those set forth. Width standards vary depending upon the type of street and its traffic volume. Many communities may be using street width standards that are excessive for adequate performance. Reducing street standards to an acceptable width for safety and adequate ease of circulation would save significant energy in the form of construction materials and by discouraging automobile use to some extent. Developers would welcome this reduction because it would save them a great deal of money; residents would in turn benefit by having to pay less for homes.

Subdivision regulations also contain provisions requiring the construction of sidewalks on both sides of the streets. Like large street width standards, requiring sidewalks on both sides in most of the development in a community may be unnecessary for acceptable levels of performance. Especially for cul-de-sac streets that have very little traffic, only one sidewalk may be adequate, and there may be times when none are needed, without discouraging walking or biking.

In reviewing subdivision standards for street width and sidewalks, the performance approach is once again a promising means of regulation. Many subdivision regulations simply give one width figure for each general type of street, ignoring the substantial variation in traffic volume within each category. Street width standards based on projected automobile use or neighborhood density would mean custom tailored and less wasteful street construction.

Aesthetic controls. As of 1968, at least 72 communities had some form of architectural controls.[15] In these towns and cities, an ordinance or review board regulates architectural design of new developments to some extent. Design standards in an ordinance or review guidelines for a board address building materials, architectural style, building color and other aspects of building design. The purpose of such regulations may be to prevent monotonous development or to preserve the character of a historic district or to promote a unified aesthetic style in a neighborhood or community.

Aesthetic controls can serve to limit flexibility in building and site design. Community energy systems may require innovative design to accomodate new technology or building design for solar utilization. For example, buildings oriented toward the south for maximum heat in the winter will require overhangs to keep the house cool in the summer. In the event that architectural controls prohibit this type of design, they would require revision to facilitate this design option. In addition special apparatus or technology necessary to community energy systems such as solar collectors or central heating plants could be prohibited by architectural controls. Ordinance provisions and review criteria could be expanded and made more flexible to allow for the new technologies and design irregularities necessary to energy efficient community systems.

CONCLUSIONS

A general conclusion of this brief survey is that an energy conserving community is not likely to be in serious conflict with the intent of most local planning regulations and policies. In fact, many features that could be included in it have been suggested as desirable for environmental and aesthetic reasons. Several items might be useful, however, to reduce the possibility that energy conserving development proposals will be inadvertently hindered by the existing planning process. A public education and information program based on improved baseline data can raise the general energy consciousness and help planners and public officials develop sound energy-planning policies. Model policies and regulations that overcome stumbling blocks such as those outlined here would provide valuable guidance. Performance-based controls are one potentially valuable approach. To make it all work, both developers

[15] ASPO estimate.

and local planners need the assistance of proven methods and techniques that will facilitate the evaluation of energy conserving alternatives, particularly when implementing flexible controls.

ACKNOWLEDGMENTS

This paper was prepared from a report supplied to the Argonne Community Systems Commercialization Program by the American Society of Planning Officials (ASPO). The ASPO report was written by David Mosena, Duncan Erley, Effraim Gil, and Patricia Slovak. The work has been supported through a contract with the Community Systems Branch, Division of Buildings and Community Systems, USERDA.

ENERGY DILEMMA IN AN URBAN DESERT: PHOENIX

Jeffrey Cook
College of Architecture, Arizona State University, Tempe

ABSTRACT

One of the largest new cities in the U.S.A., Phoenix, is described as a low-density, post-industrial city. Its dispersed settlement pattern is a development built directly over irrigated agricultural land which forms the world's largest man-made oasis. The present eighteen separate municipal governments provide a diffuse network of foci for a population of 1.3 million. An arid conurbation, Phoenix has a typical desert climate with abundant sunshine. Although water supplies are available that could support up to eight times the population, the city is totally dependent on the automobile, and few resources are local.

Although historic desert cultures have developed distinctively appropriate urban patterns, none of these has influenced Phoenix. Alternatively, several distinctive residential patterns have emerged based on the resources/energy, and socio-economics of late Twentieth Century American circumstances. These settlement patterns are examined against the increased scarcity of fossil fuels and projected to even more distinctively appropriate future possibilities.

ENVIRONMENTAL GUIDELINES FOR FOSSIL ENERGY SITE SELECTION

R.L. Scott,[1] Y.H. Amin,[2] J.F. Metzler,[3] and J.W. Mulvihill[4]

U.S. Energy Research and Development Administration
Pittsburgh Energy Research Center
4800 Forbes Avenue
Pittsburgh, PA 15213

ABSTRACT

The Council of Environmental Quality's guidelines state that environmental impact assessments must include an analysis of the probable impact of the proposed action on the overall environment, including impact on ecological systems, land use and development patterns, community and social organizations, and relevant quality of life indicators. At the Pittsburgh Energy Research Center, ERDA, we are involved with the development of environmental guidelines for fossil energy site selection. Fossil fuel conversion plants have the potential to adversely affect the natural as well as the socio-political-economic environment. Each fossil energy conversion process studied has its own specific resource requirements which must be integrated within a region in a manner that does not negatively affect that region's resource requirements and demands. The successful integration of current resource availability and demand with process feedstock requirements will satisfy both the resultant users' energy needs and greatly decrease the adverse impacts associated with these programs.

The paper will discuss how fossil energy plants affect the region's land use patterns and the subsequent urban development which surround the plant. Case studies exhibiting the successes and failures of various types of planning criteria will be reviewed. A plan that suggests possible solutions to the issues will be presented.

---

[1]Chief, Environment Branch, Environment & Conservation Division, PERC

[2]Architect, Environment Branch, Environment & Conservation Division, PERC

[3]Architect, Urban Planner, Energy Conservation Branch, Environment and Conservation Division, PERC

[4]Manager, Environment & Conservation Division, PERC

SUPERBIA - A RESOURCE-CONSERVING URBAN DESIGN THAT MEETS HUMAN NEEDS

Oliver S. Yu
Electric Power Research Institute
Palo Alto, California

ABSTRACT

This paper presents an efficient, flexible, yet people-oriented urban design concept. In this concept, the central business district is enclosed in an integrated megastructure and surrounded by tightly-knitted conventional suburban neighborhoods. For each neighborhood, service facilities such as stores, schools, and clinics, are also centrally located in enclosed structures. With this compact design, energy and material resources are conserved through integrated planning and the elimination of the need for personal autos. Furthermore, residents can simultaneously enjoy the excitement of an urban center and the tranquility of natural wilderness, all within walking distance. Illustrative design lay-outs and supporting technical data are included.

STRATEGIES FOR ENERGY EFFICIENT WATER SUPPLY AND WASTEWATER SYSTEMS.

Leopoldo Baruchello

Balfour-Italia, Consulting Engineers - Rome - Italy

A) CONSIDERED SCOPE, ITS SIZE AND FORESEEABLE FUTURE

If we consider the most probable energy consumption subdivision in a middle-grade industrialized Country such as Italy and assume for a moment that Italy be already in that ideal state corresponding to an aqueduct and a sewerage system with relevant treatment works being available for each urban resident citizen, (such a state being a starting rather an arrival point in the progress of the water systems) we can easily see that nearly the 10% of the Kwh produced in Italy could be used for operating water supplies, sewerages and the relevant infrastructures, including the possible recycling of the treated effluents. This value or-better-figures between 6% and 10% - may be considered valid also for many other countries and thus enable us to state that the size of the field under criticism to which we refer should be of approximately some hundreds of billions of Kwh per year all over the world.These percentages are likely to correspond, in a near future of this Country to 10/15 billions Kwh by rounding off largely to 150 billions Kwh the national power production (which could be in part exported in the alpine if not European level interconnection). The national production is nowadays already of that size and shows an increasing trend, thus imposing to the State Energy Agency (ENEL) a difficult research and look out for new energetic sources, or the building of new conventional or nuclear plants which in any case are likely to affect the already in-deficit situation of the national foreign accounts.

A more detailed analysis of this raw datum is therefore not deprived of interest both in my Country as in many others,since there is a well established hope to find out, within each one of the typical demands, some amounts of waste or excess consumptions which are liable to be drained towards other uses for which a rationalisation seems not possible in the same size.

Indeed, one could easily say that the huge section of the industrial demand, as well as that of the transportations and that of the public and private users of any other kind could be also submitted to similar analyses, being theoretically able to give some optimizing possibilities.However, (especially for the private users which are at present the less rationalized ones) they require a capillary convincing work, or alternatively an action on the norms, for giving some sort of incentives and dissuasions, having such a size to require, for the related costs, most of the savings which could outcome from the effects. All the above, of course, if the procedures for convincing someone by force, such as the rationings or the strong increases of the rates, are excluded.

When on the contrary the counterpart section is formed by the Local Agencies, the industrial Boards and other administrative organizations, which are homogeneous as to their institution and operating way, it could be easier and less burdensome for the State Energy Supply Agency to suggest some particular norms for the already operating plants. This could be done also by imposing for the new plants some energetic norms aiming at a global saving, which ultimately corresponds to a better use of the actual available resources and a possible delay of the future investing requirements.

Therefore, the local Institutions and all those who manage aqueducts, sewage systems and processing plants can review their enpower utilization procedures, although the Supplying Agency is principally the one which aligns itself to the idea of an action for compressing some sonsumptions through rationalization. This will be for the advantage, on both sides, either of the energy availability of other sectors less suitable for a progress in this sense, and either of the national interests for new investments, since, without this action , to be extended to all sectors of the users, any investment would be destined to an increase of the wastes instead of carrying out any saving. The first role therefore is given to the State Energy Supply Agency, since less immediate results should be expected from the spontaneous action of the citizens and of the administrators.

Even if we limit ourselves, as we should for the purpose of this note, to the assumed value of 10% of the global Italian production, we find out that we are considering an intervention scope of a not negligible size, being of the magnitude of 0,30% of the gross national product. It should be very clear that this sector is considered not only because it can be ruled more easily through the Administrations, by the Supply Agency, but also because this is one of the service sectors of the greatest importance, wherein all acquisition of more advanced procedures is first a source of improvements in the service itself and then of a compression of operating costs, a matter which may cause nothing but advantages for the common interests even within the frame of the sector itself. Therefore, even here one should act in two directions : to the inside for a better consumption proportioning within the system, and to the outside for a global saving in the sector as a whole.

TAB. 1

I	II	III	IV	V
water supply, network and processing	- River or Lake intakes and treatment	30%	40%	5
	- idem from springs	30%	20%	-
	- idem from the water table	40%	30%	3
	- idem from recycling	-	10%	2
	- almost completely gravity works	10%	15%	-
	- almost completely pumping networks	90%	85%	3
	TOTAL			13
sewerage, networks and treatment	- gravity networks	30%	40%	-
	- networks with intermediary and final pumpings	70%	60%	3
	- treatment plants with pumpings	95%	80%	4
	- mainly gravity processing plants	5%	20%	-
	TOTAL			7

I Sector

II Subsector

III Probable percentage in Italy nowdays

IV Probable percentage in the 2001 year

V Consumptions, billions of Kwh at 2001

NOTE: With a population in the year 2001 of 70 millions inhabitants and a foreseeable consumption of

$$\frac{200 \times 10^9}{70 \times 10^6} = 2.857 \text{ Kwh/inhab./yr.}$$

the incidence of the above indicated consumptions corresponds approximately to 285 Kwh/inhabitant/year, equalling at the actual average energy prices, $ 6,00 per inhabitant per year or the 0,5% of the actual national average individual income.

In the above indicated Table 1 we did gather in the synthetic way the results of a provisional statistic sampling of the present Italian situation both for the water distributing scope and for that of the sewerage and sewage processings and their recycling. The present situation in Italy could be summarized as follows: little or no recycling , a scarce consideration for the possibilities offered by the hilly configuration of the territory (and therefore mainly pumped networks), many times ill calculated networks, in spite of the possibilities for energetic savings offered by the topographic configuration of the sites. The foreseeable situation may be hypothetically thought only when we credit a reasonable degree of selfconsciousness by the technicians, the Administrations and the citizens, about the desirability of a better power rule. For this development the symptoms are not lacking, whereas we lack any dependable and substantial evidence.

However there is some reason of hope, when we consider that in the recently published rule for integrating the Law n° 319 on the management of the waste water, there is for the first time an invitation to adopt gravity flow solutions in the processing plants still to be built.

Therefore the data of the Table suggest a substantial presence of recycling supplies, as well as a remarkable increase of the gravity operating plants and networks. It will be obviously a matter of new plants mainly , which are still to be built nowadays (and they are for us still the major part), as well as extensions and improvements of already existing networks and infrastructures. According to it, in our opinion no substantial changes in the consumption structure could be previewed, however it is reasonable to assume that the total consumption evaluated by us for the year 2001 is reliable, taking into account a sizable degree of recoveries and self productions , which could be foreseen already now.

Within the re-evaluation of an individual life less bound to the increasing technological interlockings, of which the first symptoms can be seen already now, the recovery of the wastes and of the unexpensive available resources, as well as the valorization of the less expensive natural resources are included. Should it be possible to subdivide the energetic requirements in two classes, one being that of the "unavoidable needs" and the other the "reducible" or "perfectible" needs, we could see that this second class is very large.

In the rest of this note we shall try to resume some concrete possibilities for acting, to improve the management of the energetic resources in the subject sectors, which have a not negligible impact on the effective enhancement of the quality of life. We have some reason to believe that at least one part of these considerations may be considered a "common denominator" by many Countries.

B) POWER CONSUMPTION; SOME FEATURES WITHIN THE CONSIDERED UTILITIES.

B/1 General Matters

Within the group of services consisting of the two considered sections it is possible to notice some not negligible differences in operating modes of the plants and networks as a consequence of different service request features; the whole domestic water infrastructure section is characterized by a basic regular pattern and by the short frequency irregular features which are characteristic of all human activities; the whole being connected to a carrying wave which is apparently subjected to a progressivev variation, both in the sense of an increase of the domestic and total consumptions, as well as towards more reasonable consumptions in the industrial field.

There are some correspondences between the two sections for some common basic parameters (total quantities per day and inhabitant, difference between day and night, consumption peaks corresponding to some particular activities such as the nourishment in the home section and the working time in the industrial section). Such correspondences face some not negligible differences existing between the water supply service and the infrastructures relating to the waste waters. According to what we are concerned with, for the first one there are often, above all, some regulating capabilities by means of accumulation, which do not exist in the second one, where the only regulations are offered by the flooding volume of the network and by the network itself as well as by the capacity of the treatment plant, if any.

This difference is the cause that in the water supply networks the energetic offer is very often proportioned not according to the maximum peaks of the demand, but on lower values as allowed by the accumulation capabilities, whereas in the waste systems the energetic supplies are designed, in the networks, according to the maximum flow rates or to values which are not very far from those. We may say substantially that the aqueducts are more "intelligent" than the waste systems. Indeed, the first ones do pipe purified water with high speed and under pressure through little diameter pipes, whereas the second ones move waste waters, to be purified only downstream at the ends of the piping system, at a low speed and with large diameter cross sections.

Moreover, in the great networks (which are those with a larger influence on the total consumptions) (see Table 1) there is a time staggering between the water supply peaks upstream of the users and the waste discharge flow rates at the input of the plants; this happens just because of the high displacement speeds of many water supply pressure networks in contrast to the corresponding low speeds in the drain systems; the effect of this staggering is not completely negative because the global energy requirement peaks, at least, are only partially added together.

As to the treatment works (upstream of the collecting network and downstream of the discharge network) the energy supplies are conditioned in a favorable sense in the first case by the existing regulating capabilities, whereas in the second case only the energetic uses placed within the processing cycle make use of the existing regulating capacities; the input flows can't. For the drain and sewage networks and relevant plants it is necessary to consider the two basic conditions, one of them being that of the combined networks, the other those decombined into drains and sewerages. The combined networks require somewhat heavy energy supplies, if any, since they must be designed upon the maximum flow rate peaks as possible in the networks, whereas the combined plants are generally made free of their rain flow rate peaks upstream and by skimming, or otherwise they gather them into the so-called "storm water reservoirs", even if they cannot completely avoid that a part of the affluents passes into the plant with, as a consequence, an increase of the power requirements installed within the plant.

TAB 2.

I	II	III	IV	V
from 50 000 inhab. on	30% 17/18 mill.	A= 100% F= 60%	(A + F)/100 1,6	Kwh x 10^9 4,56
up to 50.000 inhab.	30% 17/18 mill.	A=100% F= 30%	1,30	3,70
Rural and spread houses	40%	A= 50% F= 10%	0,60	1,74

A = Water Supply F ≙ Sewers

I users siting

II percentage on total population

III Equipped places %

IV Index of energy consumption

V Energy consumption

NOTE : present total consumption of 10 x 10^9 Kwh per year (instead of 10-15 assumed at the beginning of this paper) for the two sections is assumed.

There are, then, further differences in the energy requirements of the two service systems considered. They consist of an overload of quality and quantity tests and controls in the aqueducts and a marked deficiency of same in the sewage and drain systems, which may cause a decrease in the exact designing of the plants. Due to the above mentioned reasons there are higher energy consumptions for maintenance to be noticed in the sewage and drain systems and finally the energy consumptions in the treatment of the sewage waters could depend in part upon the required recycling condi - tions or upon the features of the receiving body of water, with relevant questionable predetermination. This situation is unknown in the aqueducts, whose water quality could be considered a constant feature requirement.

B/2 Regulation range and capacity. Use of the exceeding energy and alternative energy sources. Desirable quality jumps in the technologies.

Taking into account the above differences and analogies, the main features of the problem are clear. They consist of a need for completely using all available topographic possibilities in order to reduce to a minimum the energy waste by means of the possible gravity downflow; a need for more extensively and accurately regulating the flow rates, in order to allow that lower electric powers, operating upon the base of the ascertainment of the irregular features of the requirements, are installed ; the desirability that the plants, if possible, are designed in such a way as to find out an integration with the other utilities of a particular territory. This integration will enable to find out some "internal energy resources" through which the impact of the energy requirement of the sector on the general energy availability will be reduced.

As a summary then, the main aspects of the matter are two and the prevailing objects are as follows : the first one consists of equalling the energy demand during the day or a longer period, aiming to obtain at least the displacement of the demand peaks to some less overcharged hours and to use if possible the exceeding energy (night production or anyway not peak period production) and the second one consists of the use of energy sources to be drawn within the operating sector of the services or in largely similar sectors. On the second object the maximum research and coordinating effort at any level should concentrate because the guidelines actually given by the present technology are not sufficient for allowing some hope in a short period of time. It is therefore necessary to operate in the technical research level in order to find out, by means of the use of the solid or semi-solid waste products, the maximum possible energy quota for managing the plant; and in the juridical level of the norms as well, in order to obtain that the underlying problems of this sector are not seen separately, but as a whole in order to make use of all the mutual integration possibilities.

We shall come back to this point later; here it is sufficient to point out that both the main objects are of equal interest for the two sectors, even if the first one is more advanced and is therefore situated as a higher energy consumption item, whereas the second one is offering greater possibilities to the purposes of the above mentioned objects; for the first of them there is a longer way to run, whereas the second one is that for which greater integration possibilities are to be found and used.

The two sectors therefore are clearly complementary as it is indicated in the following Table 3.

TAB. 3

Systems	Energy consumption	regulation degree	Energy available	progress possib. degree
water supply	max	max	min	min
waste water	min	min	max	max

B/3 Main targets for the existing water systems.

It seems quite clear that the interventions of which a mention has been made are easier to perform in the new construction plants , whereas all the operating infrastructures are less prone to some in-principle change operations. This is due to their position as service equipments for which a stopping and a modification are difficult and bur - densome, and also because it is really difficult to insert well localized modification which can become, at last, some well func - tioning devices. We have seen that in the aqueducts it is possible to give even more importance to the regulating capacities. However it is above all necessary to better interlock the whole system, with its regulating and operating devices, and particularly those with manual control, in such a way as to avoid that, as is often the case, the valves and the pumps have two positions only, the "all closed" (or flow rate zero) and the "all open" (or maximum flow rate). At least the most important controls should be gradually interlocked in the most flexible manner, with the purpose not to supply more than the required energy. Of course a better policy of the domestic and industrial consumptions could have its due weight, in a sense aiming to avoid all wastes, thus reducing also the relevant energy correlations.

This action for reviewing the consumptions from the bottom, if we may say so,is already operating in some foreign Government Agencies and Agencies of our Government in Italy ; however it may have a more effective push if the Energy supply Agencies and the aqueduct Institutions will develop an action in this sense. Several times we have been told of "ambulant teach-in" for the good use of water and this enterprise should not be considered a minor one. We could probably obtain a compression of the industrial and home consumptions by 20-30% , which could correspond to the 10-15% on the total water consumptions and to a power economy which corresponds approximately to the above mentioned data.

As the waste water is concerned we did see that the lack of a regulating capacity along the network puts a heavy liability on the pumping energy required. It's hard to insert in an operating sewerage scheme new regulating capacities to this purpose. It should not be excluded, however, that the importance of pre-processing basins (and even of septic tanks) at the origin of each contributing effluent, may be increased; it should not even be excluded to use in a better way with the minimum and middle flow rates, the bigger size free surface cross sections, by means of diaphragms to be operated from the existing manholes which could allow a better balancing of the daily maximum flow rates; thus rendering it possible to reduce installed power and tending to reverse in a sense directed to the waste energy hours the "natural" peaks without any regulation.

Moreover, leaving apart the pumpings inside the network, one should take into account these requirements in the plants,by adopting larger manholes for the pumpings and by making use of the maximum possible regulating capabilities.

Even in the existing major sewerage systems , an advantage should be had by using, for the pumpings and the treatment works operation , the energy to be produced from the combustion heat of the collected solid wastes in the area and/or the excess sludges or even of the use of the gases outcoming from the sludge digestion. To consider these not negligible technical options also for already existing layouts, should become, to a certain extent, compulsory.

B/4 Targets to be proposed for the new installations.

In the new water supply and waste water networks the controlling Agencies, and even better the Energy Supply Agency, should impose that some maximum energy saving criteria are already introduced into the design and carried out accordingly. It is however impossible to consider any local level action without coordinating with each other at least the sanitary and hygienic sector problems plannings. Therefore a better use of the energy availabilities of the sector (solid wastes and/or excess sludges), correct siting of the autonomous energy production etc. to meet the sector uses should be obtained.
All the above requires a national and regional

legislative action, and an energy policy which seems still to be conceived. Only after having evaluated the integrating possibilities of the available quantities with autonomous means, shall it be possible to design the accumulating and regulating section of the water supply networks as well as of the waste water systems.

The placing of suitable storage capacities within the waste water systems will help to reduce the costs of the final treatment at the end of the system, in the sense that the storage capacities could become the seat of some pre-treatment directed to limit the standards of the foul waters at the input of the final plant.

In the combined networks, as already mentioned, the alternate use of the basins prearranged for the "storm waters" could be studied very carefully for the ordinary sewage water, in those months where a local greater energy request is present, in order to limit or displace in any suitable way the quantities required by the service.

C) POSSIBLE STRATEGIES FOR OPTIMIZING THE PROBLEM.

C/1 General matters

All the subjects set forth would suggest a first classification of the possible intervention strategies, which could be subdivided into class (α) or maximum priority cases and class (β) or lesser priority cases. We are going now to study them from the points of view of the interventions at the various levels, by starting from the highest, where the intervention power is greater and more decisive, and coming down to the domestic or industrial individual user. The theoretical technical times for obtaining a good start of the above strategies should be as reduced as possible at the Government level, and longer for the other two; as a matter of fact, in our Country, the contrary is often more likely to happen because the selfconsciousness and the relevant concrete actions begin at the private level and are then coordinated and accepted in a legal frame. It is maybe useless to say that this is not, of course, the best way.

C/2 At the Government level.

Once a selfconsciousness at the political level for an action intended to obtain some real economical advantages for the Country by means of a rationalizing of the energy consumption sector has been developed, the competent Ministries should have to plan the class α interventions consisting of the issue of norms for coordinating and setting a frame for the regional, national and intermunicipal plans as to the energy rationalization in the sectors of the sanitary hydraulics, especially for what the new and the already operating systems are concerned. To the class α belongs also the action developed in order to get a better policy for the water consumption in the industrial and domestic field. To the class β belongs the coordination and control of the actions mentioned under point C/3 taken as an enterprise by the National Energy Supply Agency.

C/3 At the electric energy supply level.

Even where no α actions are present, the Agency could study and adopt its own set of standards in order to obtain, when negotiating the contracts with the various users, the supervision of the new projects and the consulting task for the already performed projects in operation. However, the most desirable is a higher incidence of the consulting Authority of the Agency besides and as a help to the Local Institutions in order to carry out a better energy policy. Therefore, the first priority of the Agency is to complete the studies and researches for supporting the Local Institutions in their decisions on the integration of plans and the alternative energy sources; and in addition, all the studies and researches to be made for an intervention on the new project and on the integrations of the existing plants with proposals and suggested guidelines really aswering the purpose and tested even on model networks and on mathematical models, are also of class α. On the contrary, all the other coordination and detail measures, placed on the outside of the frame of the above mentioned points, should have pripirity β only. Finally the Agency could make use, towards the individuals, of some incentives and dissuasions in the rate and contract dealings, in order to reduce the rates for the progressive users and discourage the less rational ones, always in the specific sector, thus addressing them towards a quick updating. It is evident that such an action requires in the Agency a lot of foresight, of which nowadays little trace can be found. The Agency could also make use, when evaluating the possibility of greater energy savings in a sewage project network with the appurtenant treatment, of diagrams according to that indicated by the writer in his study " Considerations about the installing of processing plants for servicing existing sewage networks, as a particular case within the integrated design of the gathering network and processing plant systems ". (XV Hydraulic Convention, Rome, Oct. 1976) -

C/4 At the individual and enterprise levels.

Obviously this level should receive from the top the norms, standards and guidelines, and should obey immediately. It is on the contrary conceivable that the guidelines are delayed and that the private user - industrialist or private citizen - takes the start by rationalizing his own water and energy sector consumptions, by installing additional regulation capacities in the water gathering and for the waste waters upstream of network; he begins to organize the subdivided gathering of the solid wastes in order to make it easier for the recycling and reuse of the energy to be collected from them, and, in other words, let gradually increase a "tide" of an interests and opinion movement pushed by many elementary actions added together, without any action, or still awaiting an action from "the moon" which could decide to perform, from the upper level, the relevant pushing movement. It is clear that the Industries and the Industrial Consortia (here to be considered as "private persons") as well as the residential communities and the joint property estates and even the best advised Municipalities could proceed to rationalise their plants and to a more precise layout of their new projects even without awaiting suitable and particular sets of norms. This could be done in a quicker way if the C/3 action will be timely efficient.

C/5 Costs of the strategies and performing times.

All the points hereabove resumed should not request - on the politic level - to settle difficulties of any kind, the matter being to issue some simple concept and execution measures which could be dealt with at the level of a decree-law in the frame of the purchasing difficulties, at competitive prices, of the oil. All political parties could easily give full powers on the matter. It is easy to demonstrate that for carrying out a plan of the herein defined priority interventions (class α) some specific investments equalling approximately 1/5 of the savings to be drawn from same within the first three years after it comes into force, are required. It is more difficult to determine the costs and the positive consequences of the lower priority interventions (class β) since they are merged within the context of rather complex situations. As repeatedly mentioned before, any action having successful prospects should start from the Energy supplying Agency itself, if it is really solicitous of the common good, free to move in its own field and not too concerned with the forming of precedent cases whose difficult evaluation may hamper the carrying out of new plants. Since all these uncertainties do result really as certainties in the negative sense, we cannot be optimistic, in Italy, on the real possibilities to proceed quickly on the way of a program in the form of the one which has been herein indicated.

ECONOMIC ANALYSIS OF INTEGRATED COMMUNITY ENERGY SYSTEMS

Allen S. Kennedy

Energy and Environmental Systems Division, Argonne National Laboratory

Isaiah O. Sewell

Energy Research and Development Administration, Office of Energy Conservation

Roy J. Faddis

Energy and Environmental Systems Division, Argonne National Laboratory

Chinmei Lee

University of Illinois, Chicago Circle Campus

1. INTRODUCTION

Events in recent years have created public demand at the community as well as the national level for energy supply systems that are energy-conserving, safe, environmentally acceptable, reliable, and "price stable" -- that is, consumers can expect that their energy expenditures will be a relatively constant share of their total budgets. The Integrated Community Energy Systems (ICES) Program of the ERDA Office of Energy Conservation is designed to develop community-scale energy systems with these characteristics. These systems will represent an integration of community design planning and energy technology concepts and will help achieve the national goal of conserving energy, and, in particular, of conserving scarce fuels. This paper documents a case analysis of alternative ICES concepts applied to a major metropolitan development complex.

2. SCOPE OF CASE STUDY

The scope of the Integrated Community Energy Systems (ICES) case study is outlined in Table 2.1. Fox Valley Center (FVC), located 30 miles west of Chicago's loop, is one of the world's largest enclosed-mall shopping centers and provides a real world case for analysis. The Fox Valley Villages (FVV) is in the planning stages and will be developed over a 10-year build-out period; therefore, the opportunity to influence energy decisions exists. The case analysis is, therefore, *retrospective* in the sense that it asks what type of ICES could have been applied to FVC and *prospective* in that it asks which ICES could be applied to FVV as it continues to develop in the future. The systems chosen for study were considered representative of three characteristic states of technology: (1) existing, (2) emerging, and (3) advanced.

At present, the following intensity of development is contemplated at Fox Valley Villages:

Use	Development Intensity
Residential	3,226 dwelling units
Office	1,112,000 ft^2
Commercial (Excluding Fox Valley Center and Town Center)	1,088,000 ft^2
Fox Valley Center	1,709,000 ft^2
Town Center Commercial	206,000 ft^2
Industrial	280,000 ft^2
Public and Institutional	33,000 ft^2

The current site plan indicating the distribution of these uses is shown in Fig. 1.

3. CASE STUDY METHODOLOGY

The case study methodology is outlined schematically in Fig. 3.1. The development site planning information was used to estimate design-day (winter and summer) conditions and load profile estimates for the base case of existing or currently planned energy systems.

A survey of existing, emerging, and advanced technologies resulted in a selection of ICES systems and subsystems to be studied in the context of Fox Valley Villages. Systems and components were identified and sized to meet estimated loads in each zone of the Villages subject to lifecycle costs and fuel and thermal efficiencies. In general, the objective was to replace grid-based systems with site-based systems and subsystems to the maximum possible extent. Fig. 3.2 is a schematic of the general design philosophy, and Table 3.1 lists the alternative ICES systems that were analyzed.

When each system and applicable subsystem had been selected and sized to balance electrical and thermal loads, annual energy consumption was estimated, based on a statistical distribution of weather data for the Chicago area. To complete the engineering analysis, a system and subsystem cost estimation was performed for each alternative. Included in the estimate were purchase and installation costs, operating and fuel charges, and maintenance costs. The final step in the study was to perform a financial analysis of each system to determine whether the systems made economic sense from a investment point of view.

In general, from the developer's perspective, a preferred energy system is one that combines (1) low initial costs, (2) guaranteed fuel supplies at stable, predictable prices, (3) low operating costs, and (4) low owner and tenant charges. Certainly, tradeoffs in these factors will have to be made in all realistic applications. The criteria developed for use in this study are to measure the extent to which an ICES alternative, applied to FVC and FVV, can be considered a viable investment opportunity. Therefore, the analysis method adopted is outlined as follows:

1. Postulate the existence of owners and operators of the various types of buildings in each zone of the community as illustrated in Table 3.2.

2. Assign, for each alternative system, "reasonable" inbuilding systems that the owners and operators of each type of building could be expected to purchase and maintain. For the base case, the existing and planned equipment allocation is shown in Table 3.2.

3. Compare for each alternative to the base configuration, the first cost and annual costs to those which each owner would incur if he were employing the base configuration. Amortize the equipment first costs over the life of the equipment (10%, 25-year life) and sum over all owners and zones. This will result in either a net balance savings or cost that the owners as a whole are either better off with the alternative than under the conventional case if the result is a net savings and vice versa if the result is a net cost.

4. Postulate the existence of a small utility company and assign equipment that it would own and operate under each alternative. Compute the annual revenue requirement that this utility would have to earn under "normal" circumstances to be a going concern.

5. Compute the total net annual community savings (or cost) as the difference between the annual revenues required under the base case and the revenue requirements of the small utility plus the sum of the differences in owner costs.

It should be noted that if the net community savings is exactly zero, all individuals can be made just as well off economically as they were under the base case and a comparison between systems can be made on the basis of energy conservation alone. If the net savings is positive, the community is better off economically by adopting the ICES alternative, and the amount of the savings is available for redistribution to the community. If the savings function is negative, however, energy conservation must be achieved by making some owners worse off economically than they were under the conventional system probably necessitating some form of subsidy to achieve energy conservation.

4. RESULTS

This study was initiated to explore various ICES options applied to a large metropolitan development. Three quantitative measures were computed for each option: (1) net energy conserved, (2) net scarce fuels (oil and natural gas) conserved, and (3) economic viability. The third quantitative criterion -- economic viability -- is a necessary condition for the implementation of any ICES concept. This criterion has been shown to be inextricably linked to: (1) the costs of installing, operating, and maintaining competitive systems; (2) the costs of energy systems to the individual building owners and occupants; (3) the amount, type, and relative prices of fuels and electric rates; and (4) the access of developers to capital markets. A relationship was derived that explicitly relates these variables so that the sensitivity of the economic viability of a particular ICES option could be tested over a range of fuel price, capital cost, and prime mover installed cost scenarios. The fourth criterion -- a qualitative measure -- is the degree to which each option fits into the developer's strategy for planning, constructing, and marketing his ultimate product which is, of course, enclosed, air-conditioned space for various activities and uses.

The systems analyzed fall into two categories: (1) decentralized, inbuilding, customer-owned systems connected to grid energy, and (2) centralized, on-site, energy generation and distribution facilities disconnected from grid energy except for fuel supplies. In each category, existing and emerging technologies have been analyzed, and comparisons of options among categories have been made. Table 4.1 summarizes the results of applying the four criteria to the systems selected.

In general, developers are accustomed to dealing with fairly complex, decentralized, in-building energy systems that are dependent on grid energy. This type of system is modular in that future buildings at the same location do not have to depend on a commitment to a central facility (other than grid energy which is assumed to be always available at some price). Therefore, inbuilding systems, including new and emerging technologies such as heat pumps, are suitable to the developer strategy after such systems have been proved to be cost-effective and reliable. The development industry can be expected to experiment with new, decentralized energy systems even before the economic markets have adjusted fuel prices because of expectations of price increases and curtailments of conventional fuels. To some extent, this is already occurring with solar, storage, and heat pump systems. However, most of the industry will continue to use conventional technology and fuel supplies until actual price signals begin to appear. Therefore, the analysis would indicate that a switch-over to heat pumps in large buildings should be under active consideration; whereas, small-scale, residential heat pumps will be a novelty until gas prices rise significantly or fuel supplies to new customers are curtailed.

Alternatively, centralized systems do not appear to readily fit today's developer strategy. A centralized ICES, of the scale explored in this study, would border on utility service, albeit to a small area. Questions of regulation would arise, and developers do not want to enter into this new area. Basically, developers simply are not equipped to assume a utility operation. Historical evidence from total energy systems also would seem to verify this conclusion.

This discussion is not intended to rule out the developer as a primary actor in the energy planning process. On the contrary, he should be incorporated more fully as an active participant in planning energy services for his developments. However, what is implied is that the developer cannot be relied upon to be an initiator and lobbyist for ICES on a large scale, even if such systems prove to be economically viable. Although he is concerned about energy conservation, it would appear that the developer should be presented with clear-cut energy options from which to choose -- options that are familiar to him and compatible with his overall mode of operation and method of doing business. He has neither the time, the expertise, nor the incentives to delve deeply into initiating on-site ICES systems.

However, the criterion of suitability with developer strategy cannot singly be used to rule out the concept of a certralized on-site ICES. The overall energy savings of ICES as reflected in Table 4.1 cannot be denied. It is true that the concepts studied here do not necessarily conserve more "scarce" fuels but this merely implies that emphasis should be placed on non-scarce-fueled ICES options. Furthermore, if scarce fuels are going to be used, it is significantly better from an economic and energy-conservation viewpoint, to use them in an ICES rather than in central grid plants. This might indicate that peaking could be done in a distributed fashion using ICES plants rather than central station plants. Thus, grid-connected ICES options become an avenue for further exploration and possible implementation.

Finally, this analysis has shown that, unless electric grid rates can be held down as oil and natural gas prices increase, an ICES makes attractive economic sense, particularly for corporations or institutions with access to low-cost capital markets. In the case of fuel cells, gas utilities are logical candidates for initiating the concept as they did for gas-fueled total energy plants. A return to oil plants seems to lack this central organizing mechanism. Diesel engine manufacturers were the prime proponents of oil-fueled total energy plants, but they could not sustain the momentum gained in the 1960s. To rekindle active interest in oil-fired systems, some organizing mechanism that combines equipment manufacturers and oil distribution networks would appear to be required. Such a mechanism should also include small-scale municipal systems that would appear to be prime candidates for oil-fueled ICES systems because of their low-cost capital factor.

Regardless of the mechanism used, oil-fueled systems with waste-heat recovery still appear economically attractive and energy conserving under certain scenarios of oil price, electric rates, and assumed fuel availability. This analysis has also shown that natural gas fuel cells will not be economically competitive with conventional diesel oil sytems unless installed costs can be brought down significantly below current projections or unless fuel cell systems can be developed that use fuels other than natural gas.

TABLE 2.1 Scope of the Case Study

I. Development to Examine
   A. Fox Valley Center - 115-Acre Regional Shopping Center
   B. Fox Valley Villages - 725-Acre New Community

II. Systems to Examine
   A. Existing Technology - Diesel Prime Mover with Waste Heat Recovery
   B. Emerging Technology - Unitary and Applied Heat Pump Systems
   C. Advanced Technology - Fuel Cell Prime Mover with Waste Heat Recovery
   D. Selected Subsystems
      1. Thermal storage with diesel and fuel cell prime movers

III. Products
   A. Description of Development and Design Load Patterns
   B. System Conceptual Designs
   C. Life-Cycle Costs
      1. Installed costs
      2. Fuel costs
      3. Maintenance costs
   D. System Performance Calculations
      1. Define operating rules
      2. Calculate system performance (i.e., efficiencies, temperatures, etc.)
      3. Energy savings
      4. Maintenance requirements and maintenance strategies
   E. Financial Analysis
   F. Problems of Commercialization Identified

TABLE 3.1 Alternative ICES System Concepts for Fox Valley Center and Villages

Decentralized Systems

1. Base System (Currently Existing or Planned Grid-Based Systems)
2. All Electric System with Resistance Heat
3. All Electric System with Limited Heat Pump Application
4. All Electric System with Extensive Heat Pump Application

Centralized Systems

5. Diesel Prime Mover with Heat Recovery and Limited Heat Pump Application
6. Diesel Prime Mover with Heat Recovery, Thermal Storage, and Limited Heat Pump Application
7. Fuel Cell Prime Mover with Heat Recovery and Limited Heat Pump Application
8. Fuel Cell Prime Mover with Heat Recovery, Thermal Storage, and Limited Heat Pump Application

TABLE 3.2 Fox Valley Villages

Equipment Ownership Assumptions by Building Type and Zone

Base System

Bldg/Zone	Square Footage	Energy System Components
Department Stores/A	893,000	Compression Chillers Resistance Heat
Mall and Tenant/A	816,000	Compression Chillers Resistance Heat
Low Rise Commercial-Office/B	100,000	Rooftop Multi-Zone Cooling/Heating
High Rise Commercial-Office/B	925,000	Compression Chillers Resistance Heat
Low Rise Commercial-Office/C, D, E	821,000	Rooftop Multi-Zone Cooling/Heating
High Rise Commercial-Office/C, D, E	560,000	Compression Chillers Resistance Heat
Low Rise Public/D	100,000	Rooftop Multi-Zone Cooling/Heating
High Rise Public/D	23,000	Compression Chillers Resistance Heat
Townhouses/D, E	1,650 D.U.	Individual, ducted gas furnaces and electric air conditioner
Garden Apartments/D, E	1,110 D.U.	Individual, ducted gas furnaces and electric air conditioner
Mid-rise Apartments/D	466 D.U.	Compression Chillers Resistance Heat
Industrial/F	280,000	Rooftop Multi-Zone Cooling/Heating

TABLE 4.1 Summary Comparison of ICES Concepts

CRITERION	DECENTRALIZED SYSTEMS (COMPARED TO BASE CASE)				CENTRALIZED SYSTEMS (COMPARED TO "BEST" DECENTRALIZED SYSTEM)	
	ALL ELECTRIC	SMALL-SCALE HEAT PUMP	LARGE-SCALE/ SMALL-SCALE HEAT PUMP	LARGE-SCALE HEAT PUMP/SMALL-SCALE NATURAL GAS	DIESEL ICES WITH STORAGE	FUEL CELL ICES WITH STORAGE
NET ENERGY CONSERVATION (10^{12} Btu)	-.45 (-20%)	-.20 (-9%)	0 (0%)	+.19 (+8%)	+.89 (43%)	+.95 (+46%)
NET SCARCE FUEL CONSERVATION[a] (10^{12} Btu)	+.05 (+6%)	-.20 (+14%)	+.18 (+21%)	+.05 (+6%)	-.37 (-46%)	-.31 (-39%)
ECONOMIC VIABILITY	NOT VIABLE UNTIL GAS PRICES INCREASE 600%	NOT VIABLE UNTIL GAS PRICES INCREASE 350%	LARGE-SCALE HEAT PUMPS VIABLE - SMALL-SCALE HEAT PUMPS NOT VIABLE AT CURRENT GAS PRICES	VIABLE	VIABLE UNDER CERTAIN CONDITIONS OF INSTALLED COSTS AND FUEL PRICES RELATIVE TO ELECTRIC RATES	VIABLE UNDER CERTAIN CONDITIONS OF INSTALLED COST AND FUEL PRICES RELATIVE TO ELECTRIC RATES
DEVELOPER STRATEGY	SUITABLE	SUITABLE	SUITABLE	SUITABLE	UNSUITABLE FOR DEVELOPER INITIATION	UNSUITABLE FOR DEVELOPER INITIATION
OVERALL POTENTIAL	NOT PROBABLE TO SIGNIFICANT LEVEL OF PENETRATION	PROBABLE IF GAS PRICES RISE SIGNIFICANTLY RELATIVE TO ELECTRIC RATES	PROBABLY BECAUSE OF NATURAL GAS CURTAILMENTS TO LARGE-SCALE CUSTOMERS AND UNCERTAIN OIL AND GAS PRICES	PROBABLE IF GAS PRICES DO NOT RISE SIGNIFICANTLY RELATIVE TO ELECTRIC RATES	UNLIKELY UNLESS POLICIES REGARDING SMALL-SCALE ENERGY PLANTS AND OIL SUPPLIES ARE CHANGED - NO INSTITUTIONS TO IMPLEMENT ON A LARGE-SCALE.	UNLIKELY UNLESS FUELS OTHER THAN NATURAL GAS CAN BE USED - GAS UTILITIES COULD BE AN IMPLEMENTING MECHANISM. IF OIL IS USED, WILL ENCOUNTER SAME PROBLEMS AS DIESEL SYSTEM.

[a] ASSUMES 30% "SCARCE" (OIL AND NATURAL GAS) FUEL MIX ON CENTRAL GRID

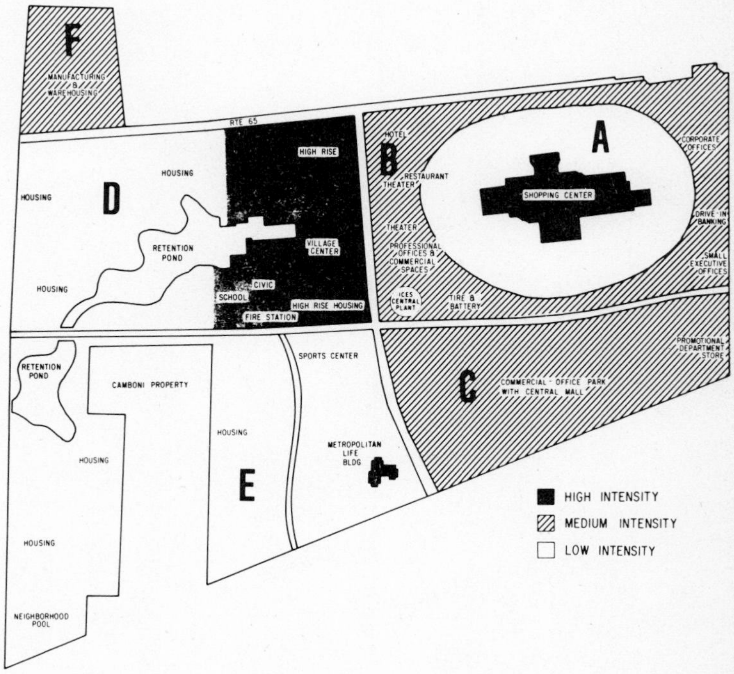

Fig. 2.1. Fox Valley Villages Site Plan

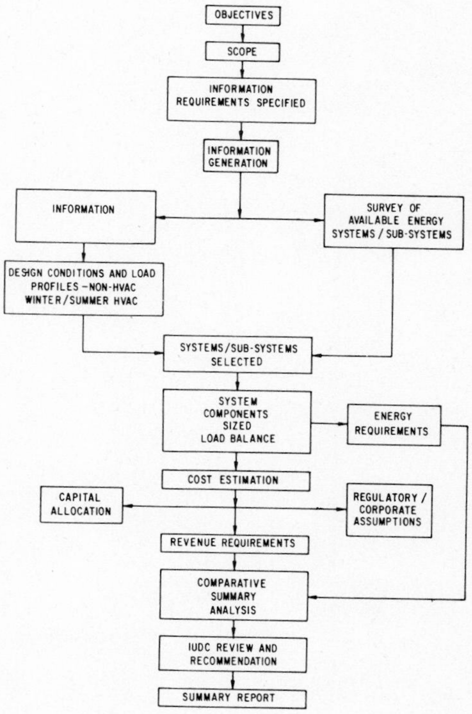

Fig. 3.1. Fox Valley Villages Case Study Methodology

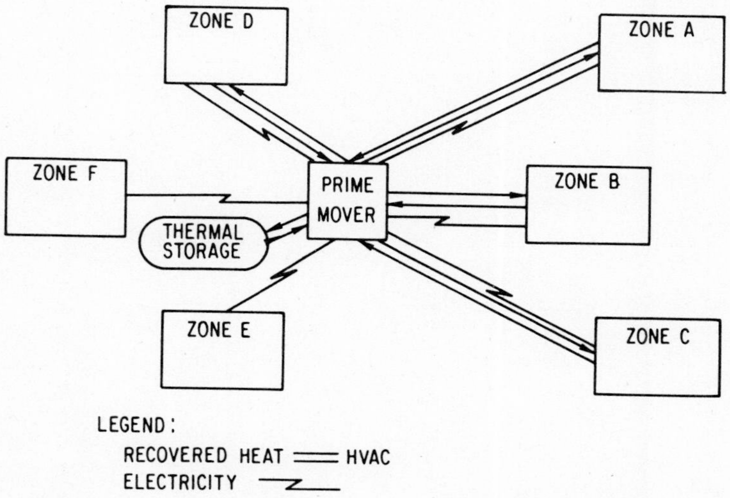

Fig. 3.2. Fox Valley Villages System Design Schematic

THE IMPACT OF SETTLEMENT PATTERNS ON LOW TEMPERATURE HEATING SUPPLY SYSTEMS, TRANSPORTATION AND ENVIRONMENT

Ueli C. Roth

Director, Office for Physical Planning, Environmental Research, Town Planning and Architecture, Zurich, Switzerland and Lecturer for Urban Planning, Federal Institute of Technology (ETH) Zürich, Switzerland

SCOPE OF THE PROBLEM / SUMMARY

The question is: which kind of urban form, of which size, density, mixture of dwellings and working places and structure requires more or less energy, and what kind of energy and which causes more or less environmental problems due to energy production and consumption?

Or vice versa: which energy supply system fosters which kind of urban form, size, density, grain and structure?

Answers to these questions should allow the proper preparation of the urban setting as a "plug" into which advantageous energy supply systems properly fit and v.v; thus such answers should serve as a basis for a coordinated fulfilling of the following five major goals:

- provision of a safe energy supply, e.g.:
  - as independant from other countries as possible
  - diversified in order to reduce the effects of shortage and of technical breakdowns through substitution
- provision of a long-term economic energy supply
- saving of natural resources by limiting end-energy use and by the reduction of waste
- protection of the ecosphere by limiting the emission of thermal and chemical waste and of dust
- securing of full employment under good working conditions

Thus, a coordinated urban energy system should allow to reach the same goals the Energy-Program of President Carter was aimed at.

Energy policy without a simultaneous urban policy will not reach these goals!

With this I do however not wish to be misunderstood: the goals of urban development are not only related to a desirable energy supply system; there are independant goals as well, such as social and general economic as well as other technical ones - but hitherto the energy goals have been largely neglected. They will not be in the future: dwindling resources and increasing pollution as well as the rising risks of energy production will take care of that!

In this paper we are dealing with that part of the energy supply system which has the strongest interdependance with urban systems: low temperature heat supply and transportation, which represents the strongest indirect link between energy supply and the urban system.

Not all too surprisingly, this paper comes to the conclusion that tendentially - and much more research has to be accomplished in this field - decentralized energy supply systems relate rather closely to decentralized urban systems and that it makes little sense to generate energy in gigantic energy parks in order to supply dispersed settlements. If strongly centralized metropolitan areas and uninterrupted megalopoli seem desirable for social, general economic and other technical reasons,a centralized energy supply seems more adequate. But this latter alternative may probably turn out to be more costly, less safe, to cause more pollution and lower the general quality of life - at least in Europe based on the conditions on which this paper was written. Some of the findings may hold true for the US and other continents too.

Correlations between energy supply and settlement patterns can be shown not only in low energy consumption activities, where temperatures under 200°C are required to supply mainly space heating and hot water, but also in high temperature process energy activities in large and small industries and in energy use in general.

Table 1 shows the proportional distribution of end-use energy according to energy supplier and energy consumer groups in the Federal Republic of Germany in 1971. The share of space heating and hot water totalled appr. 43% of the total energy consumption, 48% of which are apportioned to private households, 38% to small industry, trade and service industry and 14% to large industry.

This presentation does not however provide any information on the relative significance of energy supply- and settlement pattern relationship.

In other industrialized countries similar distributions may be found but in small industrialized countries lacking in indigenous raw materials, the share of low temperature and energy use is somewhat higher due to the relativly low consumption of process heat. In Switzerland for example, statistics show that 50% of the total energy consumption falls within the area of low temperature heat[1]

The theories on the changes taking place in the specific energy consumption in transport as a function of settlement density, the number of settlement centres and the distribution pattern of work/residence location, are more or less firmly established, e.g.by Rittel[2] and the American Real Estate Research Corporation[3].

It is equally well known that settlement patterns influence the mode of transport, especially with regard to the division into public- or private transport and consequently also the system of power supply.

To what extent the volume of process energy is influenced by the location of industry is difficult to establish but it may be possible to show a link between site and the choice of a given energy supply system. Such a choice is determined by the limitations imposed by emission standards, as for example in high density areas.

The main object of the following exposition is to examine in detail that proportion of the total energy consumption, (appr. 40-50%) which households and small consumers use for their low temperature requirements.

The influence of settlement patterns on transport energy requirements will be treated separately.

THE INTERRELATION BETWEEN ENERGY SUPPLY AND SETTLEMENT PATTERNS

To what extent are settlement patterns responsible for the volume and type of energy consumption and what impact on the environment do they have?

This question could of course be inverted to read:"What impact do certain energy supply systems have on settlement patterns?" or again: "Which type of settlement pattern ensures the greatest saving in energy and causes the least environmental stress by its energy consumption?"

The relationship between settlement pattern and energy supply has a reciprocal effect. Like a two-way mechanism, 'cause and effect' are interrelated: From settle-

TABLE 1 Consumer groups and energy use in the Federal Republic of Germany 1971 (Enduse)

	Space Heating	Hot Water	Process Heat	Light Power	Total
Households	18.5 %	2.0 %	-	1.0 %	21.5 %
Gcal	293 188 000	31 696 000		15 848 000	340 732 000
Gcal/I	4.80	0.51		0.25	5.58
Small Consumer	15.5 %	1.0 %	3.5 %	0.5 %	20.5 %
Gcal	245 644 000	15 848 000	55 468 000	7 924 000	324 884 000
Gcal/E-E Ind	15.25	0.98	3.44	0.49	20.16
Industry	5.5 %	-	29.5 %	4.0 %	39.0 %
Gcal	87 164 000		467 516 000	63 392 000	618 072 000
Gcal/E Ind	8.39		45.01	6.10	59.50
Transport	0.5 %	-	-	18.5 %	19.0 %
Gcal	7 924 000			293 188 000	301 112 000
Gcal/I	0.12			4.80	4.92
Total	40.0 %	3.0 %	33.0 %	24.0 %	100.0 %
Gcal	633 920 000	47 544 000	522 984 000	380 352 000	1 584 800 000
Gcal/I	10.39	0.77	8.57	6.23	25.96
MWh/I					30.11
KW/I					3.44

Population 1970	61 001 200		
E_I	1 991 000		
E_{II}	12 957 000	E_{Ind}	10 386 000
E_{III}	11 546 000		
E	26 494 000		
E without E_{Ind}	16 108 000		

(Gcal/I = Gcal per capita of population)
(Gcal/E = Gcal per capita of employment) consumption time-unit per
(Gcal/E_{Ind} = Gcal per capita of employment in industry)
MWh/I = MWh per capita of population)
KW/I = KW per capita of population) power

Calculations based on: Federal Ministry for Research and Technology: "Einsatzmöglichkeiten neuer Energiesysteme"(Possibilities of application of recently developed Energy Systems, parts IV and V), Bonn, 1975

ment to energy supply and from energy supply to settlement.

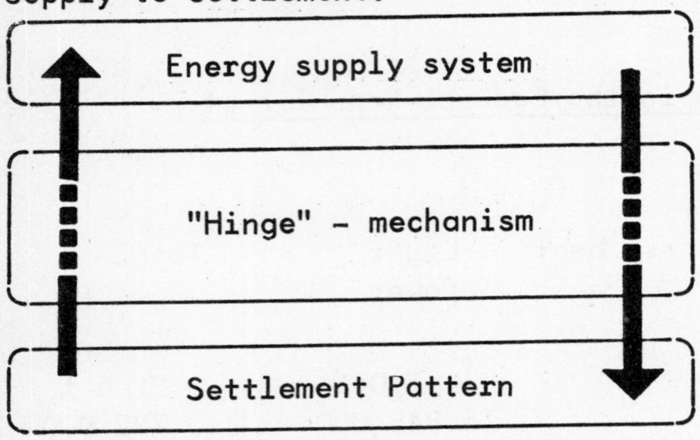

Fig. 1 "Hinge" - mechanism between energy supply and settlement pattern

This illustration provides a very convenient device for demonstrating the relationship between energy supply and settlement pattern.

In the following Chapter a definition of the parameters determining settlement patterns and their impact on energy consumption and energy supply will be provided. The Chapter following describes the more important low temperature energy supply systems. Another paragraph is inserted to analyze in additional the main relationship between energy and settlement patterns with regard to the intermediary link - transportation.
The second-to-last Chapter treats of the problem of pollution of the environment through energy supply systems. The final Chapter contains conclusions and recommendations for a coordinated planning policy on settlement patterns and energy systems within a legislative and administrative framework.

PARAMETERS AND PATTERN OF SETTLEMENT

Parameters facilitate the identification of the pattern of settlement. They characterize on the one hand:

. the distribution of inhabitants

. their place of employment and

. its structure (e.g. their belonging to a particular economic branch, requiring a certain type of building)

on the other, the consequences of these primary elements in terms of amenities provided, such as a transport system, public buildings and parks. Here, a differentiation must be made on at least two levels: Land use at supra regional - or national- level and at regional level, for settlements, agglomerations, towns and villages.

Characteristics of the distribution of the primary elements: Inhabitants (I) and place of employment (E), (the latter being practically identical with employed) are the density, its composition or ratio of gainfully employed $\frac{E}{I}$, the scale of mixture of this ratio (its so called 'grain' the area-spread or size and the pattern of connected settlement centers.

Density

Density usually refers to inhabitants (I) or places of employment (E) per surface unit. Density figures as such are of no value unless they are related to the absolute size of the area measured: A population density of 1000 inhabitants per km^2 would be considered low for a city but extremely high for a continent.

The importance of inhabitant density (I/km^2) and employment density (E/km^2) lies firstly in the diverse building constructions with their varying consumption of heat energy and secondly in the possibilities for cost-effective installation of energy supply systems, e.g. district heating; available literature on the latter set the lower limits for heat consumption density at 20-30 $Gcal/km^2h$.[4)]

The individual heat consumption per m^3 of building volume and the particular type of heat loss varies according to building construction, even by similar insulation ratios. This is due to different volume/surface ratios: it is highest in small detached one-family houses and lowest in large square units.

Data available for the Federal Republic of Germany indicate for one-family houses built after 1949, an average consumption of 110 kcal/m^{2}h for space heating and hot water. Multifamily units consumed 15% less, i.e., 95 kcal/m^{2}h.[5] Case Studies carried out in Switzerland arrive at similar conclusions[6].

The causes of heat loss from constructions and thus their possible remedy vary also according to the type of building, as may be seen from Fig. 2:

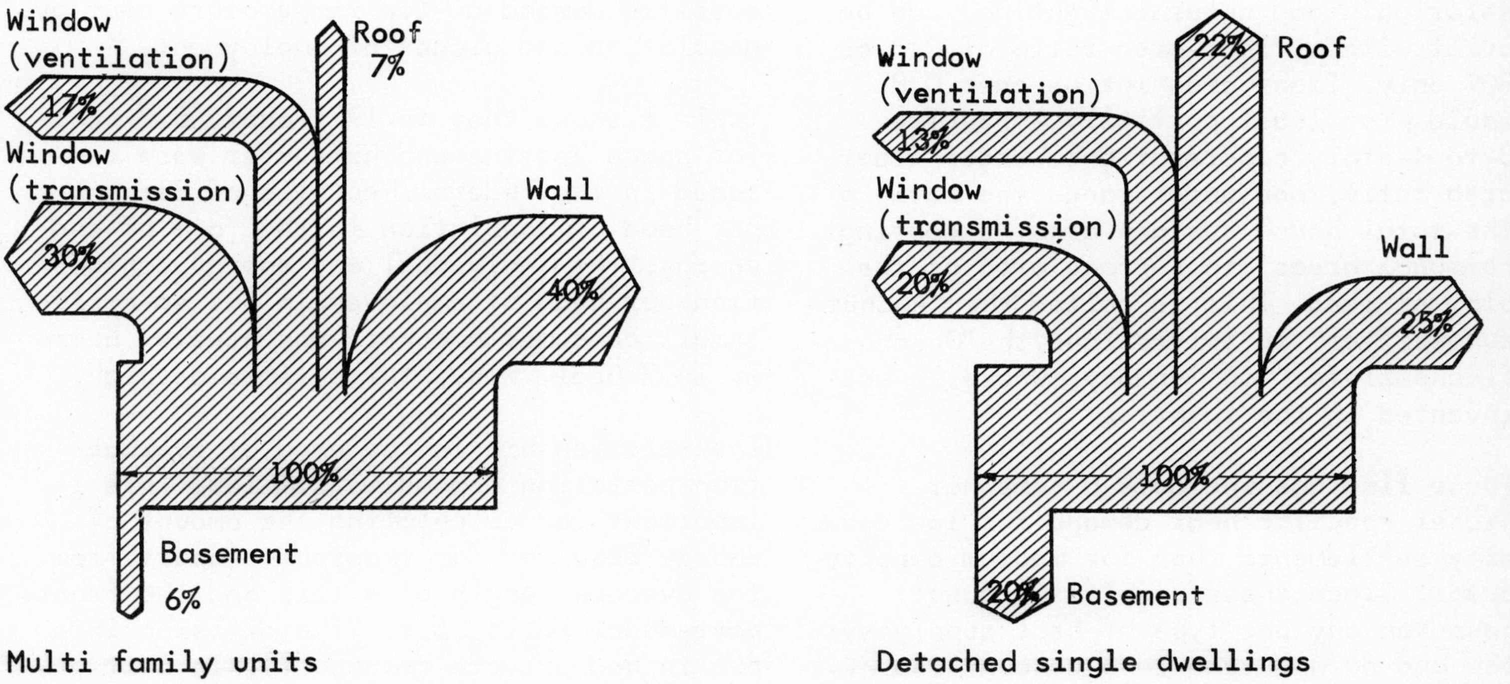

Fig. 2 Heat escape from one- and multi-family structures. Total heat consumption ~ 100%; the standard of construction complies with average demands in the Federal Republic of Germany.
Source: Winkens, H.P.: "Vergleichende Betrachtung von energiesparenden Massnahmen" ("Comparative Study of energy-saving measures", published in International District Heating, Annual No. 5 Public.No. 2, 1976)

An Austrian Study shows the following distribution:

TABLE 2 Heat losses by individual parts of building

Proportion per m^2 utility surface		One- and two-family dwellings	Multi family units
Exterior Wall	m^2	0.89	0.705
Ceiling near roof	m^2	1.02	0.643
Ceiling near basement	m^2	1.02	0.643
Window surface	m^2	0.18	0.163
Exterior door-surface	m^2	0.04	0.02
Length of window-joints	m	0.666	0.603

Ref. Winkens: Oesterreichisches Institut für Bauforschung: "Reduzierung des Energieverbrauchs in Wohnungen" (Austrian Institute for Building Research: "Reduction of energy consumption for dwellings": Case Study Report 117/1)

An additional 50% of total space heating losses are due to ventilation where no heat-recycling installation exists.

Detached one-family houses in the median-latitude regions, taking into account solar gain and natural lighting, can be built with a floor-area ratio of 0.3 or 30% only, floor area ratios over 0.8 would - at least partially - require 3-to 4-story constructions. Under floor area ratio, one understands the ratio of the total house surface area, excepting communal areas, to surface area of the plot of land. This is called the "Ausnützungsziffer" in Switzerland, the "Geschossflächenziffer" in Germany, where it was invented in the twenties.

These findings indicate a somewhat higher specific heat demand for low density settlements than for medium density areas. Since these results were not based on any one type of heat supply system and do not take into account the efficiency rate of conversion of energy, this may vary according to the system chosen for a particular settlement density.

The Employment Ratio

The ratio $\frac{E}{I}$ (ratio of gainfully employed/ number of employment places or employed relative to population figure) is typical for settlements within agglomerations : in European industrial countries it is on the average 0.4; in France it is somewhat less, in Germany somewhat more. In the suburbs, with their typically high number of commuters, the ratio is well below average, somewhere between 0.1 and 0.2, in metropolitan agglomerations well above with from 0.8 to 1.5.

A further differentiation in employment quota is made by classifying places of employment into the three economic sectors as a further structural characteristic (E_I: Primary sector (agriculture, mining, forestry); E_{II}: Secondary sector (industry); E_{III}: Tertiary sector (services); $\frac{E}{I} = \frac{E_I}{I} + \frac{E_{II}}{I} + \frac{E_{III}}{I}$)

One may differentiate still further according to economic-branches, -groups and -types.

The importance of the employment ratio for the energy industry lies in the varying, specific demand of low temperature heat by population and places of employment.

Table 1 shows that in 1971 appr. 5.3 Gcal for space heating and hot water were consumed in the Federal Republic of Germany per head of population in the form of end energy. The individual end energy consumption per head of employed in the so-called 'small consumer bracket' was however nearly 16.2 Gcal, i.e., three times as high.

The relationship in the area of average transportation demand is similar. This is important for calculating the amount of energy required for transport (apart from the average length of a trip and the transport-modal split, i.e., the division into public and private transport). In Switzerland for example, it has recently been established that one inhabitant will generate an average of two daily trips, as against five, generated by a place of employment[7].

The importance of a differentiated employment ratio lies in the impact it has on both the supply of low temperature heat and on process heat, since specific energy demands vary considerably within the individual branches of industry. The two following tables 3 and 4 which are based on studies carried out in Switzerland, provide some examples of these conditions.

Employment ratios are thus conclusively valuable as settlement parameters which, combined with their specific energy consumption values, will be indispensable for the assessment of energy consumption in settlement areas.

TABLE 3 Specific secondary energy demand for heating and hot water for Industry, Trade and Service Industry for 1970

	Specific low temperature heat requirements per employ and year (Gcal/E.a)
Mining, Quarries, Pits	0,1
Food- and beverage industry	15,0
Textile industry	7,7
Clothing and equipment industry	4,3
Wood, Cork, Toys and Sports equipment	9,7
Paper industry	16,2
Printing	7,1
Leather	7,4
Rubber and Synthetic-industry	8,0
Chemical industry	17,9
Stonemasonery, Clay and earth industry	12,7
Metal-industry and -trade	8,4
Machine industry and Machinery	8,6
Watch making and Jewelry trade	4,3
Building industry	3,6
Power, gas and water supply industry	16,0
Trade (Wholesale & Retail)	12,6/6,0
Banks and Insurance Companies	5,4
Transport and Public Transport	18,0
Catering Industry	10,1
Health and Hygene and Cleaning Industry	7,5
Other Services	15,0
Public Administration including Schools	4,0÷19,3

Source: EKONO-Gruner: "Energiekonzeptstudie für den Kanton Basel-Landschaft" (Energy concepts for the Canton Basle-Country, Phase 1) Basle, 1974

TABLE 4 Secondary Process-energy demand for selected Branches of productive Trades.
Source as in Table 3

	Gcal/E.a
Food and Animal Feed	6
Textile industry: Clothing, Linen, Shoes, Bedding	3
Industry producing Wood- and Cork articles, Toys and Sports equipment	1
Paper industry	235
Leather	10
Chemical industry	52
Stone and Earth	90
Metal	6
Machines, Appliances, Vehicles	1
Watch making, Building, Service industry	(0)

The Distribution Pattern

The scale of mixture of the ratio $\frac{E}{I}$, i.e. the "grain" is of particular importance with regard to the economics of transport energy, since, under "fine-grain conditions", work/residence location being in close proximity to one another, travelling distances tend to be shorter. Whether modern mobile society, in as far as it enjoys constitutionally assured freedom of choice of abode will, with given access to efficient transportation, avail itself of the opportunity of closer work/residence location, is another question (Fig. 3).

A typical example of the 'coarse grained' type of settlement pattern is the well known 'city-effect': the inhabitants sleep in the suburbs but work in city centers.

The influence of the grain on the low temperature demand or on the system which will cater to this demand (type of energy supply) is difficult to trace.

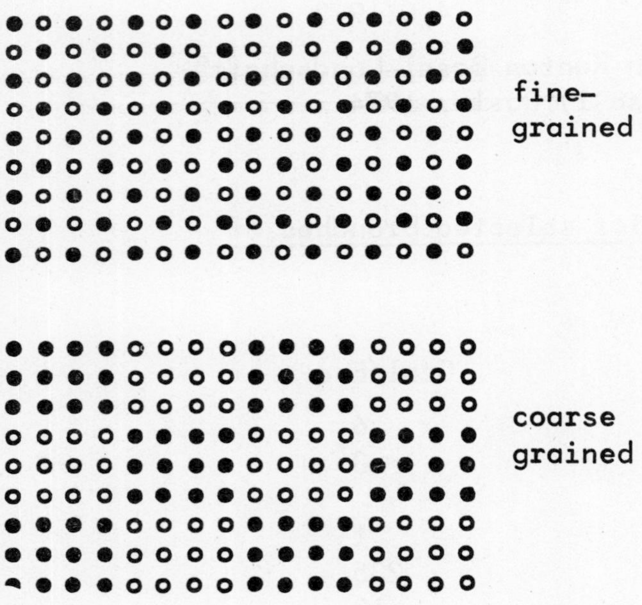

● Residence location or inhabitants respectively

○ Employment location or gainfully employed resp.

Fig. 3 The grain of the urban system

The Size of settlements

The expanse or size of cohesive settlements has immediate impact on the volume of heat required within the supply area. This is of considerable economic and technical importance where connected energy supply systems such as district heating and to some extent also gas and electricity, are to be installed for the provision of low temperature heat. In the case of district heating supplied by a nuclear power station a prerequisite is that the supply area be cohesive as long transmission lines must be installed for security reasons and further, that the area have a population of several 100,000. Only under such conditions will district heating provided by this form of energy, i.e. with compound heat-energy production, be cost-effective. To achieve optimal exploitation of usable waste heat from a 1,000 MW_e nuclear power station, settlement density must be in the region of a 1 Million population figure and 400,000 places of employment[8].

Major national nuclear power station programmes should therefore be supported by land-use planning. They should aim at as many agglomerations as possible with at least 1 Million inhabitants, high settlement concentration spread over limited expanse[9].

The size of a settlement is determined by the settlement pattern of a whole country. In the European context the rule tends to be that the stronger the concentration of a given population on a few high density areas, the larger the expansion of each individual agglomeration area is and therefore the fewer the smaller agglomerations are, the greater its expanse is, even if the total population figure of inhabitants living in agglomerations remains the same (Fig. 4). From the economic and technical viewpoint, the choice of energy supply systems is therefore restricted, producing the so-called 'energy-modal-split', in favour of large and concentrated technical supply systems such as nuclear power stations.

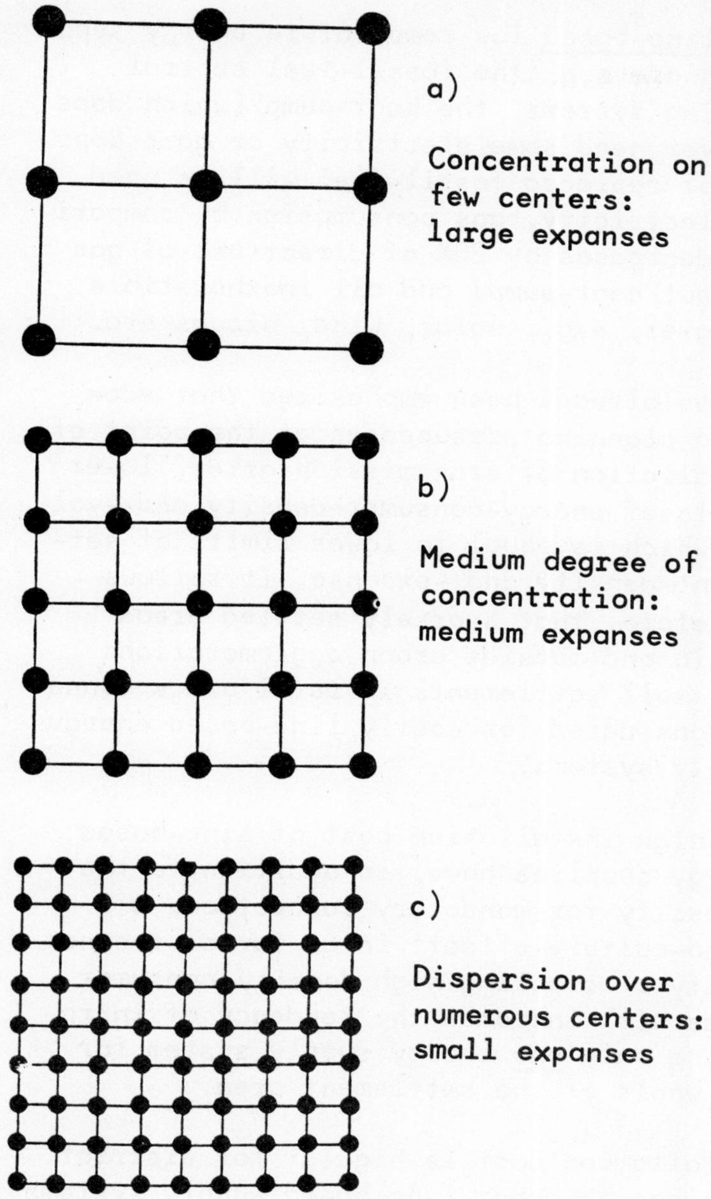

Fig. 4 Concentration level of settlements

The Form of Settlement Patterns

The form of connected settlement patterns, whether mono-centric, circular, starshaped or in a center-orientated pattern or again, in sattelite form has, like the distribution pattern, its greatest impact on the energy cost profile via the transport sector. However, the fact that energy supply systems based on ducts, pipelines and electric mains - we shall call them "line-based" - have a different grid for each of these settlement forms, must also be taken into account. (Fig. 5)

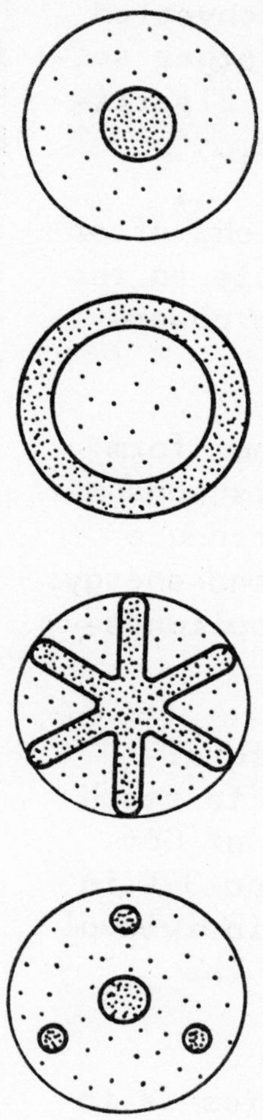

Fig. 5 Types of agglomerations

ENERGY SUPPLY SYSTEMS FOR SPACE HEATING AND HOT WATER

In the previous chapter the major urban parameters were described and their influence on settlement patterns, potential effect on the volume of energy consumption, and on the choice of energy supply systems pointed out. In this chapter a brief summary is given of low temperature supply systems and their correlation with settlement parameters.

Available statistics in Europe are on the whole consistent with a division of 'low temperature heat' consumption of 90% for space heating and 10% for hot water. One distinguishes basically between the following energy types:

P Primary energy is that form of energy, thct nature provides: Oil, Natural Gas, Uranium, Hydro-Power, Solar Heat etc.

S Secondary energy is energy converted from primary energy or from other secondary energy: Nuclear Heat, Electricity, Hydrogen, City-gas etc.

E End-energy is the energy which, after final conversion, is available to the consumer: Petrol, Power from power-points, Natural Gas etc.

N Utility energy finally is that form of energy, for which there is consumer demand and which consumers produce themselves with the aid of end-energy: Heat, Cooling, Hot Water, Appliance-operation, Light etc.

The efficiency rate $\eta = \frac{N}{P}$ of total-energy supply systems is known to reach appr. 40-50% (Federal Republic of Germany 53% in 1973[10]; Switzerland 39% in 1970[2]) and is composed of the individual degrees of efficiency $\frac{S}{P} \cdot \frac{E}{S} \cdot \frac{N}{E}$ of the separate conversion stages.

The differences between countries is to some extent due to the varying proportions of energy supply systems and the resulting variations in the conversion efficiency rate. A large proportion of electricity generated by nuclear power reduces the over-all efficiency rate due to the low efficiency rate in nuclear power stations. Switzerland has currently the largest proportion of nuclear power electricity in Europe. Some of the differences are also due to statistical discrepancies between the countries.

In an assessment of the impact of different energy supply systems on land use and settlement patterns, it is useful to differentiate between line-based and non-line-based energy supply systems.

Line-based energy supply systems comprise: electricity, district heating, district energy, gas. All nuclear or fossil secondary energy suppliers are line-based. Some of the impacts on land use have already been mentioned; they occur at the construction stage, during operation and its consequences.

Non-line-based low temperature energy suppliers are e.g. the fossil-fuel central heating systems, the heat-pump (which does however need some electricity or gas; appr. 20% of replaced fossil-fuel will be used as electricity, gas consumption by comparison decreases by 20% of direct use of gas without heat-pump) and all inexhaustible energies, e.g., solar, wind, biogas etc.

It has already been emphasized that economic planning presupposes at the point of installation of transmission grids, lower limits of energy-consumer-density and -volume which is equal to lower limits of settlement-density and -expanse. It follows therefore, that sparsely settled areas within and outside urban agglomerations and small settlements in rural areas cannot be considered for costly line-based energy supply systems.

The high installation cost of line-based energy supplies have, in addition to the necessity for mandatory connection, a 'mono-culture-effect' in as far as the necessity of creating high density consumer areas will increase the tendency of introducing only one energy supply system for the whole of the settlement area.

Installation cost is highest for district heating. As other line-based supply systems have less obvious bearing on the settlement/energy supply relationship, district heating is used as a model to demonstrate some of the more typical characteristics of the provison of district heating versus non-line-based heat supply systems. These are the fossil-fuel central heating and the heat pump system. The source for this is a recently published German study by Winkens[11] which is likely to be most influential, published on behalf of the Federal Republic Ministry for Research and Technology; another study undertaken by the 'work group district heating' has been Mr. Winken's source as well as Austrian publications based on research on the reduction of energy consumption in dwellings[12].

The calculations for economic viability carried out for this study are based on 1976 prices. From these one may conclude that:

. for new constructions with fossil-fuel central heating, energy saving building designs will prove economical, for those with heat-pumps, such designs will be absolutely essential.

. in areas supplied by district heating by a combined system (heat/power combination), which is already cost-effective, an energy saving building design will be of little economic advantage, even in new constructions.

. in areas already supplied by district heating or in areas where it is planned, no marked change in heat requirements can be expected from heat-insulating measures carried out on existing buildings.

. Where heat-insulating measures are carried out 'exceeding economically acceptable levels', the supply area for district heating could be considerably restricted as a result of lowered heat density.

The consequences of these findings for land use and town planning will be the necessity to effect an early and, in effect, permanent separation into district heating and non district heating perimeters. Given the lifespan of district heating of 50-100 years, these will then be valid interminably. This procedure would be analogeous to the known methods of coordination between sewage systems and settlement and land use planning, (so called 'canalization-perimeters).

The demarcation of these perimeters must be established according to specified heat-insulation values; any subsequent deviation from these values could prove uneconomical. Only in areas beyond such perimeters, where individual or collective heating is in force, will improved heat insulation be economical.

The calculations which are decisive in the choice of suitable district heating installation perimeters are complicated by their great sensitivity to the cost of various energy supply systems[13].

THE "HINGE"-FUNCTION OF THE TRANSPORT SYSTEM

A "hinge" mechanism of a particular nature functions in the transport sector. The correlation to the settlement structure has already been mentioned. It claims 20% of the total energy consumption of European countries, 80% - 90% of which is for urban traffic.

Energy consumption varies per passenger-km or per ton-km of goods according to the mode of transport. Apart from different traffic volumes according to type of settlement, the diverse transport types are important and must also be considered in view of their varying energy consumption.

Between traffic volume - defined as the product of the number and average length of trips - and the settlement pattern, there exists, according to Rittel[2], conclusive evidence. Traffic volume increases proportional to $\sqrt{E^3}$; i.e., given equal conditions, two agglomerations will generate only appr. 70% of the traffic volume of a single agglomeration spreading over an area of equal proportions. Inversely, according to the same author, increasing settlement density 'd' reduces traffic volume in the relation $\sqrt{\frac{I}{d}}$, e.g., a four times higher density will reduce the volume of traffic by half.

From the evidence presented it can be concluded that:

When there is a concentration of population in relatively few urban agglomerations and therefore in clusters of a large size, traffic is reduced between these agglomerations and is concentrated on a few peak-performance axes. However, within each settlement, the specific volume of traffic will increase. Thus, with an increase of traffic within, rather than between, settlement centers the overall volume of traffic will increase. According to the report by the Federal German Council of environment experts[14], the trip-performance within German agglomerations is appr. 8-17 times higher than the Federal average.

TABLE 5 Specific energy consumption according to mode of transport

Passenger transport	Mcal/100 pers·km	Energy supply
Rail transport		
Electric locomotive	15,4	electricity
Diesel locomotive	28,7	diesel
Sub-Way	11,2	electricity
Urban Traffic		
Bus	18,2	diesel
Trolley bus	44,8	electricity
Automobiles	51,8	petrol
Air traffic	166,6	aviation fuel
Goods transport	Mcal/100 tons·km	
Rail transport		
Electric locomotive	15,4	electricity
Diesel locomotive	30,8	diesel
Urban Traffic		
Trucks	89,6	diesel
Electric trucks	229,6	electricity
Other Traffic		
Inland navigation	9,1	diesel
Pipeline	0,7	electricity

Source: Seidenfuss, H.St. Published in "Energie und Verkehr" ("Energy and Transport"; Institute of Transport Research) Münster-University, Vol.77, Göttingen, 1975

In contrast, a pattern of decentralized, smaller and more numerous settlements will increase the inter-community traffic volume as small city-centers are less self-sufficient; intracommunity traffic is reduced. As transport demand is higher within a settlement than for inter-settlement traffic, total transport demand is expected to reduce with the increase in number of settlement centers in a given region.

ENVIRONMENT POLLUTION CAUSED BY ENERGY SUPPLY AND SETTLEMENT

Heat pollution

Energy required for human activities is almost totally released again into the environment as heat. The man-made turnover of heat on the earth's surface and in the atmosphere, currently totalling appr. 0.012 W/m^2, amounts to only 1/10,000 of the total natural solar heat. On the whole it is unlikely, that the human release of energy will, for the time being, affect climatic conditions. This is not so for large agglomerations or high density areas: Compared to the appr. total of 125 W/m^2 of solar heat on the earth's surface at medium latitude, the 630 W/m^2 in Manhattan, 127 W/m^2 in Moscow or 20 W/m^2 in Los Angeles reach equal proportions.
As a result of the so-called 'heat island effect' in large settlements, climatic changes have been noted in these areas, i.e. increased showers, increased cloud formation. There exists therefore a connection between man-made energy conversion and regional and local climate.

TABLE 6 Energy consumption density in selected areas during the late sixties

	km^2	W/m^2
Manhattan, New York City	59	650
Moscow	880	127
West-Berlin	234	21,3
Los Angeles	3500	21
North-Rhine Westphalia (Ruhr)	10300	10,2
Los Angeles Basin	10000	7,5
North-Rhine Westphalia	34000	4,2
Benelux	73000	1,66
Great Britain	242000	1,21
14 US East States	932000	1,11
Central Western Europe	1665000	0,74
the World	510 000 000	0,012

Source: Report of the Study on Man's impact on climate; Inadvertent Climate Modification; MIT Press, Cambridge, Mass. 1971

It is evident that regional climatic changes are the more drastic,

- the higher the energy consumption per inhabitant and per employed,
- the larger the uninterrupted settlement areas (i.e. concentration of total population of a country of a continent ina few high density areas, refer to Fig.4a),
- the lower the efficiency rate of the energy system as a whole $\eta \frac{N}{P}$, (as for a given rate of energy consumption more primary energy is used and higher conversion losses result and are released into the atmosphere) and,
- the higher the unrenewable part of the total energy consumption

Energy sources with regeneration potential are: Solar heat, Wind, Water and Biogas; these energies are being converted in any case and their exploitation by man cause only negligible changes in natural heat reflection due to changes of the earth-surface absorption- and reflection characteristics.

Where negative impacts due to energy consumption are feared to affect climatic conditions, the following targets should be set:

- an optimal efficiency rate of total energy systems. Land use and settlement structures with end-energy demands kept as low as possible and preference for energy supply systems of high efficiency rate.
- a large proportion of renewable energy sources. Land-use- and settlement structures which favour employment of such systems.
- decentralization of settlement in numerous, not too large agglomerations (so-called decentralized concentrations, refer Fig. 4c).

Pollution through noxious substances

Apart from heat, most of the energy conversions also release noxious substances. This applies in particular to the fossil energy sources coal and oil, which contain sulphur. The release of Nitric oxide (NO_x) caused by burning natural gas, may in the future become of increasing importance, since its elimination, contrary to the successful elimination of Sulphur, has as yet not been technically achieved.

Urban traffic contributes to pollution mainly through carbon-monoxide (CO) release.

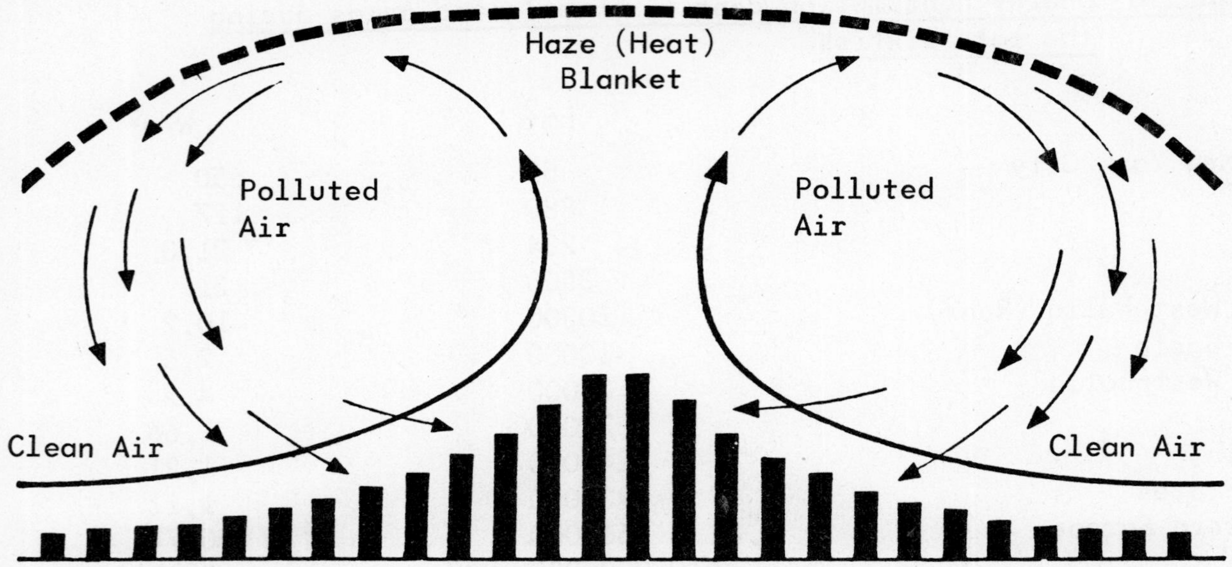

Fig. 6 The 'heat island' - effect: Air rises from heated city centre, clean air is sucked up from surroundings absorbing pollutants on the way to the centre. These noxious substances are carried upward with heat and disperse over the total area of the agglomeration. Symptoms of the procedure are a 'haze blanket', or 'smog-bell'.

The man-made heat blanket hovering over large agglomerations (heat-island effect) causes an inflow of polluted air in the direction of settlement centers. (Fig. 6)

This effect is the more concentrated, the higher the respective energy consumption is and the more concentrated the pollutants in heating and transport fuel are and, finally, the greater the expanse and interconnection of the agglomeration area is.

For this reason, the same targets as are valid for reduction of heat emission apply to the elimination or reduction of chemical and dust immissions caused by heat release through energy conversion. Additional preferences are to be given to the least polluting energy supply systems. Applied to transportation, this would translate into a reduction, as far as possible, of the volume of individual transportation and a maximum support of public transport systems. Land-use and settlement structures should be adapted to this end (refer to preceding chapter).

Conflicting Goals

A particular problem arises with the district heating system in respect to its environmental acceptability. Any reinforced measures to introduce thermal insulation are apparently contradictory to district heating interests. Although such polluting emissions as on-site collective heating systems produce, are eliminated with the district heating system, the problem of heat emission still exists. The factor in favour of district heating is that its objectives of dense settlements coincide with those for public transport schemes. The disadvantage of district heating from the environmental protection point, is the demand for large, connected settlements which are required for district heating supplied by nuclear power stations.

The fact that, inspite of realistic settlement and land use planning, conflicts over a common goal may occur, (which can only be resolved by studying real-life examples and by drawing on the aid of quantitative evaluations,) may be demonstrated by the example of solar energy exploitation for space heating:

As pointed out in the beginning, the density of settlements is strongly reduced where large collecting surfaces are needed by extensive exploitation of solar energy for hot water and space heating. Surveys conducted yielded as an upper limit, a floor area ratio of appr. 0.3. If such a settlement were totally independent of power supplies and free of transportation requirements, the consideration of saving energy and protecting the environment would dictate that nowhere should there be denser settlements. In reality, such utopian settlements cannot exist. In- or outgoing services have to be provided: Electricity, water, sewerage, telephone, television and transportation. This applies also to areas where renewable supply systems are extensively employed. Low density settlements are costly since the connection of services is over longer distances and they complicate the use of the mode of public transport which is environmentally preferable.

Except where there is access to hydropower, the generation of electricity is the most cost-effective and least polluting (efficiency rate!) when generated from fossil fuel or nuclear power and when used in the heat/power combination. For these reasons, the interest of economy and environmental protection centers on land use and settlement structures which can convert the waste heat from electricity generation into district heating. The basic premises for the realization of this are a number of high density settlements.

THE FORMULATION OF COORDINATED URBAN AND ENERGY SYSTEMS. CONCLUSIONS

The question was: what pattern and structure used for the organization of the accommodation of a given size of population and number of employed in terms of:

- large scale spacial arrangement and
- settlement (within agglomerations)

produces:

- high or low energy consumption,

demands:

- a certain type of end-use energy provision,

and has consequently:

- which impact and hence which level of consumption of primary energy?

The most significant contribution contained in this inquiry is the largely qualitative description of the mechanism at work between energy production and distribution on the one hand and settlement on the other hand since they have an impact on the problem. The previous chapter points out that global answers based on qualitative considerations are inadequate, in as far as partly conflicting measures may have to be employed to achieve a final goal. In general it may be assumed with some certainty that major technical and concentrated, line-based energy supply systems tend to promote or presuppose the building of fewer large agglomerations. This is substantiated by the recently concluded study by the Swiss group 'ur' carried out for the Federal German Ministry for physical planning, building and town-planning[9]. Major technical energy supply systems are those which are supplied mainly by nuclear energy.

But it must be stipulated that for each actual case, each country, region and agglomeration, <u>an accord is reached on energy- and settlement concepts</u> because of the correlation which does without doubt exist. To neglect their bearing on the issue will lead to serious conflicts regarding the optimal goal and hence to energy waste and unnecessary pollution.

Just as at present the coordination of water- and sewage supply services and transportation and settlement planning has become a standard rule, so must there be in the future, a systematic, legislated and administratively organized coordination of supply of energy and settlement planning. The emphasis on coordination has led, in some countries, to water protection laws and even to mandatory connection. One such consequence, for example, will be the segregation of district heating supply perimeters; the measures to be prescribed will be for less thermal insulation within but for more outside the limits of perimeters. Other measures should include mandatory connection within the perimeter and competition from other energy suppliers should be suppressed.

The quantitative relationship of energy industry and settlement planning, as well as legislation on the implementation of coordinated concepts is as yet a postulation rather than an accomplished fact in European countries. Legislative planning and technical interference with individual consumer choice of the type of energy supply system may not be the most desirable measure but it is the more important, the greater the growth rate of energy consumption becomes, since here are the primary causes of ecological and technical imbalances.

BIBLIOGRAPHY

1) Hohl, R.:"Einwirkungen der Energieerzeugung auf die Umwelt; Betrachtungen zur Gesamtenergiekonzeption" Switzerland "Bauzeitung" Annual 92, Publication No.17, Zurich, 25.4.1974

2) Rittel, H.W.: "Gesellschaftliche Alternativen im Berufsverkehr" ("Transport alternatives available to a commuting society" in: Project: Future. Standard of Living. Vol. 3 Transportation). Frankfurt a M. 1973

3) Real Estate Research Corporation for the Council of Environmental Quality etc.: "The Cost of Sprawl", Washington, April 1974

4) Dietrich, G.: Kaiser, U.; Solfian, W.: "Fernwärmeversorgung mit Kernkraftwerken" ("District Heating from Nuclear Power Plants" in: International District Heating - FWI, Annual 4) (1975, Publicat. No. 5)

5) German Federal Ministry for Research and Technology: "Auf dem Wege zu neuen Energiesystemen" ("In search of new Energy Systems", Part V), Bonn, 1975

6) Fachkommission für regionale Energieversorgung und Energiekonzeption Basel-Stadt und Basel-Landschaft/GRUNEKO AG. (Work Committee for regional energy supply and energy concepts Basle/City and Basle/Country/GRUNEKO AG. Hectographed Interim report, unpublished), Basle 1975

7) Refer diverse Studies within the framework of the total concept of transport in Switzerland (GVK CH)

8) Muhlhauser, H.; Helbling, W.: "Heizwärme aus Kernenergie; technische und wirtschaftliche Aspekte" ("Heating from Nuclear Power; Technical and Economic aspects"); Technische Rundschau No. 15 8.4.1975

9) Refer to Study of the Swiss Group 'ur', Zurich: "Auswirkung von Entwicklungen im Energiesektor auf die Raum- und Siedlungsstruktur" ("Consequences of developments in the energy sector on space- and settlement- structure". Research project for the German Ministry for Physical planning, Building programmes and Townplanning) Zurich/Bonn 1976

10) Calculations based on: German Federal Ministry for Research and Technology: "Einsatzmöglichkeiten neuer Energiesysteme" ("Application possibilities of recently developed Energy Systems, Part IV"), Bonn 1975

11) Winkens, H.P.: "Vergleichende Betrachtung von energiesparenden Massnahmen" ("Comparative Studies of energy-saving measures", published in 'Fernwärme international' FWI, Year 5, 1976

12) Austrian Institute for Building Planning: "Reduzierung des Energieverbrauchs in Wohnungen" ("Reduction of energy consumption in dwellings" Research Project 117/1)

13) Winkens, H.P. and others: "Die wirtschaftlichen und technischen Ausbaumöglichkeiten der Fernwärmeversorgung in der Bundesrepublik Deutschland - Bericht über die Ergebnisse der Gesamtstudie" ("The economic and technical possibilities for development of the district heating system in the Red.Rep. of Germany-Report on the conclusions of the entire study" ref) 'Fernwärme international' - FWI, year 5 (1976) No. 5

14) The Council of Experts on environmental matters: "Auto und Umwelt" ("Automobiles and environment"; Assessment), September 1973

DISTRICT HEATING AND COOLING UTILIZING TEMPERATURE DIFFERENCES OF CHICAGO WATERS

Danilo J. Santini

Arthur A. Frigo

George T. Kartsounes

Wyman Harrison

Energy and Environmental Systems Division, Argonne National Laboratory

ABSTRACT

The feasibility of using cold water from Lake Michigan and waste-heat water from Commonwealth Edison Company's Fisk Generating Station to cool and heat the buildings of a redevelopment project in Chicago, the South Loop New Town (SLNT) project, has been investigated. The study proposes using modular water-to-water heat pumps for cooling and heating the project. The modular heat-pump system is very attractive for SLNT from both an economic and energy-savings point-of-view. The heat-pump system offers an initial reduction in both the required capital investment for the heating, ventilating, and air-conditioning (HVAC) system when compared to a conventional system and also a significant reduction in annual operating expenses. The heat-pump system, however, requires a significant additional capital investment for the water supply and return system. Even with this investment, the heat-pump system savings in annual operating expenses and reduced capital investment for the HVAC system could pay for the cost of the water supply and return system in less than five years. Added to the economic savings would be an energy savings equivalent of about 450,000 barrels of fuel oil per year when compared to a conventional system.

INTRODUCTION

The SLNT project is a general concept for community development that involves a tract of land adjacent to the southern end of Chicago's central business district. The area covered by the project tract (Fig. 1) includes 335 acres. Plans as of 1976 call for a total of 13,150 dwelling units to be built, housing 30,000 to 35,000 people. The first phase of the development, currently in the design stage, is to contain 3000 dwelling units. Most of these dwelling units will be in high-rise buildings of ten floors or more. The eventual date of completion of all 13,150 units is to be before the year 2000.

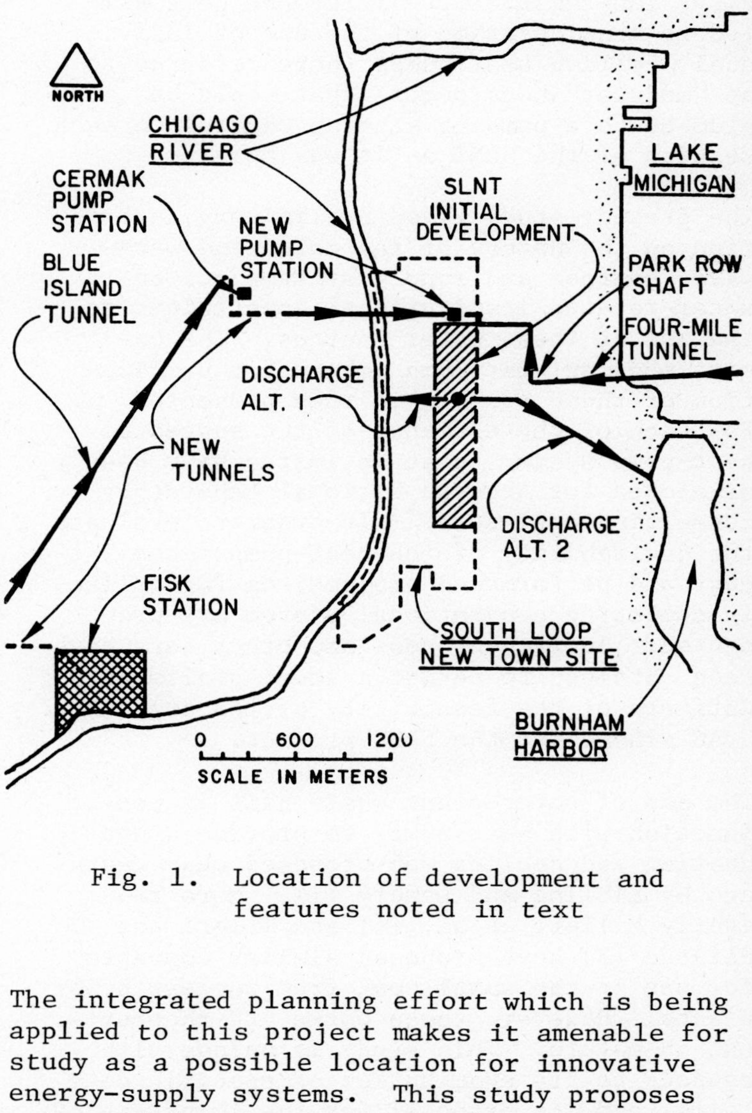

Fig. 1. Location of development and features noted in text

The integrated planning effort which is being applied to this project makes it amenable for study as a possible location for innovative energy-supply systems. This study proposes using modular water-to-water heat pumps for heating and cooling of the SLNT project. Lake Michigan's relatively cool water is to be used as a water source for the heat pumps when the load is predominantly cooling. Condenser waste-heat discharge water from a coal-fired electrical generating station is to be used as a water source when the load is predominantly heating.

The general concept for the project HVAC system was considered likely to be feasible because of the availability of adequate-size abandoned tunnels in the vicinity of the SLNT project. These tunnels could be used to convey waste-heat discharge water from Commonwealth Edison Company's Fisk generating station to the project site for winter heating and cool water from Lake Michigan for summer cooling. The locations of these tunnels, the site, the lake, and power plant are shown in Fig. 1. Attention then turned to consideration of using a high-capacity, low-temperature-difference heat pump for project HVAC and, in particular, linking such a heat pump to the aforementioned water sources. Finally, the relatively high initial capital costs of a single, high-capacity, low-temperature-difference heat pump led to consideration of the use of individual dwelling heat pumps (here referred to as "modular" heat pumps) that could be plumbed to a pumping station central to each section of the SLNT as it was built.

The present study gives preliminary information on the nature of the cold- and warm-water sources and supply systems and on water-related institutional constraints to the use of these water sources. The on-site modular heat-pump scheme for utilization of these sources is then presented. For each of the elements of the suggested heat-pump system, cost estimates have been developed for Argonne National Laboratory by consulting engineers.[+] In order to evaluate the desirability of the heat-pump scheme, cost and performance projections for a likely competing conventional system are presented. Sufficient cost and other data have been obtained to permit a sound preliminary estimate of the feasibility of the suggested HVAC scheme for the SLNT project.

The use of power-plant waste heat in conjunction with heat pumps to provide urban heating and cooling was proposed many years ago by Eibling and Landry (1). More recently Miller, et al. (4) and Illeri and Reistad (3) have proposed similar concepts for use of the waste heat from nuclear power plants. However, these works are theoretical in nature. This study is unique with respect to its examination of specific details, and its proposal for the joint use of power-plant waste heat for winter heating and a natural water body for summer cooling.

[+]Participants were Keifer and Associates, Inc. and Globe Engineering Company.

THE COOL-WATER DISTRIBUTION SYSTEM

The major factors affecting the feasibility of the proposed system are the costs involved in construction of the water-distribution system from sources to the site. Fortunately, there are numerous existing abandoned tunnels which can be used as part of the distribution system. In this section a discussion is presented which examines the use of the existing tunnel network and the construction of new tunnels for the supply and discharge of cool water at the site. In the following section, a similar discussion of warm-water distribution is presented.

Cool-Water Supply

The existing Four-Mile Water Intake Tunnel terminates east of the immediate project area. This 6400-m brick conduit was abandoned for use as a water source by the City of Chicago in 1974. The Lake Michigan cool water would flow west through this Four-Mile Tunnel to the existing Park Row Shaft (Fig. 1). At this point the water would flow north until it comes to the existing 8th Street Shaft. It would then flow west again through the existing Polk Street Tunnel and to the site itself.

The only major refurbishing to the existing tunnel system would be the removal of a bulkhead in the Four-Mile Tunnel just east of the Park Row Shaft. In order to prevent a potential cross connection between the raw Lake Michigan cooling water and the treated Lake Michigan potable water that is pumped at the Cermak Pumping Station, a bulkhead would be constructed. The total cost for removing the bulkhead near the Park Row Shaft and constructing a new bulkhead near the Cermak Pump Station is estimated at \$125,000 - \$85,000 for bulkhead removal, \$15,000 for installing a new concrete bulkhead, and \$25,000 for engineering and contingencies.

Within the site a pumping station and shaft would be constructed to deliver water from the Polk Street Tunnel to the SLNT cooling system. The design capacity of the pumping station is 3.8 m^3/s, based on the peak cooling-water demands for the entire SLNT development. The total power requirement is 1865 kW, based on a total dynamic head of 42.7 m (gage pressure of 3.45×10^5 N/m^2 in the main line of the cooling system and a 7.6-m static head). The total cost for the cooling-water pump station is estimated at \$1,125,000. This cost includes construction of a 6-m diameter shaft which would house the pumping facility. The shaft would be constructed to a depth 23 m below the ground surface to connect to the Polk Street Tunnel.

Cool-Water Discharge

Two alternative systems have been investigated for discharge of the spent cooling water from the study area. Alternative No. 1 would involve the construction of a sewer outfall to the Chicago River. As shown in Fig. 1, the sewer would be constructed from a central area in the site. This alternative would constitute a diversion of Lake Michigan water. The cost of constructing this outfall is estimated at $655,000. It is assumed that the outfall would be constructed in the same manner as a typical storm sewer.

Alternative No. 2, which would not divert Lake Michigan water, consists of tunneling east from the study area under railroad land and constructing an outfall at the north end of Burnham Park Harbor. Cost for discharge via this Lake Michigan outfall is estimated to be $4,315,000.

Table 1 summarizes the costs associated with the two alternatives. An analysis of this table indicates that these cooling-water supply alternatives differ in cost by a factor of about three.

TABLE 1 Cost Summary of Alternatives (Dollars)

Cost Item	No. 1 Discharge to River	No. 2 Discharge to Lake
Bulkhead	125,000	125,000
Renovate Crib	10,000	10,000
Pumping Station	1,125,000	1,125,000
Discharge Cost	655,000	4,315,000
TOTAL	1,915,000	5,575,000

Cool-Water Institutional Constraints

There are at least four institutional problems which would have to be addressed before construction of the cool-water distribution system.

(1) If discharge Alternative No. 1 is implemented, it will constitute a diversion of Lake Michigan water. The annual average diversion due to the SLNT project would be less than 0.6 m^3/s. This amount would have to come from the present 90 m^3/s allowable annual average diversion allotted to Illinois. It might also come from a proposed future increase such as the five-year temporary increase in diversion of 280 m^3/s that is now in effect.

(2) The City of Chicago has an ordinance prohibiting use of lake water for once-through air conditioning. This ordinance may need interpretation to see if it applies to the proposed project and to determine if a variance could be obtained, if one is needed.

(3) State of Illinois' standards limit new sources of cooling-water discharges to no more than 2.9×10^4 kW. An exemption would probably not be required for the proposed system.

(4) The State has a policy of no new cooling-water discharges along the Chicago shoreline. This policy potentially affects discharge Alternative No. 2. Even if the State of Illinois allowed an exemption, U.S. EPA regulations also affect cooling-water discharges into Lake Michigan.

THE WARM-WATER DISTRIBUTION SYSTEM

Commonwealth Edison Company's Fisk Generating Station is located on the Chicago River, 2.1 km downstream from the SLNT site. There are two units at Fisk, a 320- and 150-MW unit. This coal-fired plant uses a once-through cooling system which discharges into the Chicago River. Data on average seasonal river flows, plant condenser-water discharge, heat rejection, and discharge-water temperatures are presented in Table 2. The condenser-water discharge is greater than the river flow in some cases, indicating condenser-water recirculation. River flows are taken as maximum values of available flow when the condenser discharge is greater than the river flow.

TABLE 2 Characteristics of Fisk Station Cooling-Water Discharge (Seasonal Averages)

Season	River Flow (m^3/s)	Fisk Condenser Flow+ (m^3/s)	Discharge Temp.+ (°C)
Winter	16	17.1	10.0
Spring	27	18.1	22.2
Summer	53	18.7	27.8
Fall	21	17.9	19.6

+Commonwealth Edison data for 1975

The Fisk Generation Station is a reliable source of warm water year round. The Fisk plant is an important part of the Commonwealth Edison system; each generating unit is in operation 65-86% of the year with generally noncoincident down time spread throughout the year (as based on data from 1975 and part of 1976).

Warm-Water Supply

The distance between the SLNT site and the power station can be traversed by constructing two new tunnels to connect to the existing abandoned Blue-Island Tunnel. New tunnels totalling 1520 m in length, 910 m for the section from the Fisk station to the Blue-Island Tunnel and 610 m for the section from the site to the Blue-Island Tunnel, are required. The cost would be between $4,000,000 and $4,500,000. This estimate includes the cost of 2.4-m diameter tunnels (at $1000/m), a construction shaft for each tunnel, dewatering and repairing the Blue-Island Tunnel, a gate at the SLNT site pumping station, bulkheads, the connection to the Fisk plant, engineering and contingencies. The cost of refurbishing the abandoned tunnel is a small portion of the total cost amounting to only $450,000 for dewatering, bulkhead construction, and repairs.

Warm-Water Discharge

Regardless of how the warm water arrives at SLNT, it will be returned to its point of origin, the Chicago River. If cooling-water discharge Alternative No. 1 were used, then the same outfall would be used for both cool and warm water. If cooling water were returned to Lake Michigan, then the warm-water outfall to the Chicago River would be an added cost, identical to the earlier cool-water outfall estimate of $655,000. There is no diversion of Lake Michigan water with the warm-water distribution system.

Warm-Water Institutional Constraints

The use of waste-heat water from the Fisk Generating Station is limited by (1) continued operation of Fisk by Commonwealth Edison, and (2) water-quality laws of the State of Illinois regarding secondary-contact waters and indigenous aquatic life. The planned retirement date for Fisk is 1994. It is assumed that Commonwealth Edison Company will build new units at the Fisk site at that time. The retirement of old units and construction of new ones at the same site is common utility practice. An examination of the State of Illinois' water-quality laws indicates that the SLNT warm-water discharge would not violate water-temperature or effluent-concentration standards. Thus water-related institutional constraints do not appear to be a problem in the case of use of power-plant waste-heat water to heat the SLNT development.

THE MODULAR HEAT-PUMP SYSTEM

The construction of the cool- and warm-water supply systems permits the use of a water-to-water heat-pump system on the SLNT site. Warm water from the power plant is combined with cool Lake Michigan water to maintain a relatively constant temperature of 10 to 20°C. Due to this conditioned source of water, the water-to-water heat pump for the SLNT can operate with a COP of 3.2 to 3.5. Conventional cooling systems operate at a COP of about 1.9 to 2.1. The proposed heat-pump system would, therefore, be 1.7 to 3.4 times as efficient.

The SLNT system uses lake water when the load is predominantly cooling and condenser water from the power plant when heating is predominant. There are times in the spring and fall when the needs for cooling and heating are nearly in balance. For this situation, a combination of the water sources can be used.

Most central heat pumps operate in either the cooling or heating mode. A problem exists in apartment complexes because some spaces require heating while others require cooling. This problem can be overcome, at the expense of using two conditioned water-distribution systems to points of use, by installing modular units so that selection of mode of operation can be achieved.

The approach for the SLNT involves sending a single water supply to booster stations located strategically throughout the site. Each apartment is served from one of these stations. This scheme permits each apartment to have individual and independent temperature selectivity. In addition, the discharge gas from the individual compressors can be used to provide part of the domestic hot-water needs of each apartment.

THE CONVENTIONAL SYSTEM

In order to make a valid analysis of the economic feasibility of the proposed modular heat-pump system for the SLNT project, it is necessary to select a proper competing alternative. Often, when attempting to justify an alternative, a biased analyst will select a system which makes the one under consideration appear desirable. On the surface it might seem that this has been done here since the competing system is an electrical system using resistance heating. For this system, the heating/cooling needs of both the initial and ultimate developments would be supplied from a system of six central stations strategically located throughout the site. Cooling towers located at ground level near the central stations would be used as a heat sink for the cooling system. It is estimated that cooling towers will require considerable site work to satisfy esthetic and environmental conditions. Such costs are not included in the cost estimates presented below. Conventional heating with high-voltage electrode boilers would be used for supplying all space-heating requirements for this system.

A combination of factors gives credence to the selection of an electrical system as the likely conventional HVAC method. One of the biggest factors is the strong shift to electrical heating in the residential sector resulting from current and anticipated shortages of natural gas. In the North-Central region, electric space-heating systems were installed in 49% of new multi-family units constructed in 1975, a 10% increase over 1974 (Ref. 5). In the first place, these trends are likely to continue and second, these figures understate the proportion of electrical systems installed in high-rise dwelling units such as found on the SLNT site. In Chicago, current gas-utility policy is to require individual metering of dwelling units. This requirement eliminates gas-fired central systems from high rises. Economics and engineering eliminates decentralized gas-fired systems from high rises. Political considerations and uncertainty rule out oil. A coal ban in Chicago rules out coal. Thus, an all-electric conventional system is a reasonable choice for comparison in the case of the SLNT development.

COSTS OF THE COMPETING SYSTEMS

The economic comparisons of the two systems are based on fairly detailed cost estimates. Heating and cooling demands were broken down by location and building function from the preliminary site plans for the project. Distribution networks within the site were laid out. The HVAC systems themselves were broken down into seven or more categories of capital expenditure and five categories of annual operating expenses. Capital expenditure breakdowns included, as appropriate, cooling towers, central stations, chilled-water distribution, hot-water distribution, ventilation and exhaust, space air conditioning, engineering, lake- and river-water distribution, mixed-water distribution, miscellaneous and contingencies. Annual operating expenditure breakdowns included cooling (electrical power), heating (electrical power), operating labor, maintenance, and miscellaneous. Detailed estimates are presented by Harrison, et al. (Ref. 2). These estimates are further broken down into costs for the initial 3000 unit phase of the development and the remaining 13,150 units of the ultimate development. A summary of the estimated costs for the two alternative systems is presented in Table 4.

TABLE 4 Summary of System Cost Estimates (Thousands of Dollars)

	Initial Development	
	Capital	Annual Operating
Conventional System	9,465	1,635
Modular Heat-Pump System		
On Site	8,771	738
Water Supply and Return	6,415[1] or 10,075[2]	
Difference	-5,721[1] or -9,381[2]	+897
	Remaining Development	
	Capital	Annual Operating
Conventional System	39,596	7,870
Modular Heat-Pump System		
On Site	36,800	3,459
Water Supply and return	Previously Completed	
Difference	+2,796	+4,411

[1]Alternative No. 1

[2]Alternative No. 2

ECONOMIC FEASIBILITY OF THE HEAT-PUMP SCHEME

Table 4 illustrates that the trade-off between the two systems involves an increase in initial cost for the modular heat-pump system to obtain a small reduction in on-site capital cost and a substantial decrease in operating cost. Note that operating savings are annual savings and are repeated every year after capital equipment is installed.

In order to evaluate the economic feasibility of the modular heat-pump system in place of the conventional system, the differences in annual costs of the two projects were compared assuming a twenty-year project lifetime with no salvage value for either alternative. Two twenty-year schedules of construction were developed in order to demonstrate the benefit of rapid project construction. In one case the project is completed in 20 yrs and in the other it is completed in 10 yrs and maintained 10 yrs more. The sensitivity of the modular heat-pump scheme's economic feasibility to underestimates of tunnel costs was tested by assuming up to 200% increases in the initial added capital cost of the heat-pump system rather than the conventional system.

The constant dollar rates of return on the added investment in the modular heat-pump scheme were always positive and ranged up to 35.6% if discharge Alternative No. 1 is adopted with a 10-yr construction schedule. In fact, discharge Alternative No. 1 earns a rate of return of 14.8% if only the initial 3000-unit development is constructed. Even with an initial year cost overrun of 200%, discharge Alternative No. 1 earns 14.9% on the 10-yr construction schedule and 11.3% on the 20-yr schedule. For a 10-yr construction schedule, discharge Alternative No. 2 earns 24.6% without cost overruns and 13.9% with a 100% initial year cost overrun.

CONCLUSIONS

The modular water-to-water heat-pump scheme can save an equivalent of 450,000 barrels of fuel oil per year when compared to the conventional HVAC alternative. The preliminary economic feasibility analysis indicates that the added investment for the heat-pump scheme can be easily justified, even if the cost estimates presented in this paper are substantially in error.

Impediments to implementation of the heat-pump scheme are primarily institutional. In addition to the water-related institutional problems discussed, there are others which are mainly related to current utility regulatory structures. In view of the energy-conservation potential and the economic attractiveness of this scheme, it is felt that a more thorough cooperative study by SLNT and utility planners is warranted.

ACKNOWLEDGMENTS

The authors thank Sarah J. La Belle, Argonne National Laboratory, and Floyd H. Davis, Globe Engineering Company, for their technical contributions and support in the original study (Ref. 2) which formed the basis of this paper.

REFERENCES

(1) Eibling, J. A., and B. A. Landry, The steam generating station as a source and sink for the heat pump, Mech. Engineering 73, 554 (1959).

(2) Harrison, W., A. A. Frigo, G. T. Kartsounes, S. J. La Belle, D. J. Santini, and F. Davis, District heating and cooling utilizing temperature differences of local waters: Preliminary feasibility study for the Chicago 21, South Loop New Town development, Argonne National Laboratory Report ANL/WR-77-1, Argonne, Ill. (May, 1977).

(3) Illeri, A., G. M. Reistad, and W. E. Schmisseur, Urban utilization of waste energy from thermal-electric power plants, Jour. Engineering for Power, 309 (July, 1976).

(4) Miller, A. J., H. R. Payne, M. E. Lackey, G. Samuels, M. T. Heath, E. W. Hagen, and A. W. Savolainen, Use of steam-electric power plants to provide thermal energy to urban areas, Oak Ridge National Laboratory Report ORNL-HUD-14, Oak Ridge, Tenn. (Jan. 1971).

(5) U.S. Department of Commerce, Bureau of the Census, Characteristics of new housing, U.S. Dept. of Commerce, BOC, HUD Report C25-75-13 (1975).

PROJECT SPONSORSHIP

The work described in this article was partially supported by Argonne National Laboratories' Energy and Environmental Systems Divisions' Advanced Technology-Mix Energy Systems program.

DIRECT SEAWATER/LAKEWATER COOLING SYSTEMS

Jules Hirshman
Tracor Marine, Incorporated, Fort Lauderdale, Florida

ABSTRACT

Those urban areas with a large cooling demand located near sources of natural cold water can, under favorable circumstances, cool buildings directly with this resource. Advances in offshore pipeline technology and growing acceptance of district heating/cooling concepts make this a near-term option for energy conservation.

Several studies to evaluate the technical and economic feasibility and the energy savings potential of this concept were performed by Tracor Marine for ERDA. The first was a preliminary examination of four sites near the sea, followed by a site-specific study and a preliminary design for Miami Beach. A further phase of work examined the potential use of fresh water sources, such as lakes, reservoirs, rivers and groundwater for this purpose. The U.S. Navy Civil Engineering Laboratory independently examined this concept for possible use in Naval Facilities, and then had Tracor Marine perform a preliminary design, economic and energy analysis, and an environmental impact assessment for a demonstration project at a small USN facility.

The results of these studies indicate that:

(a) 70% to 80% of the electrical energy used for air conditioning can be saved with this method.

(b) Lifecycle costs for the locations studied are lower for direct cooling systems than for conventional air conditioning.

(c) Technology is available for near-term construction of intake pipelines in deep lakes and ocean sites such as Miami Beach.

(d) Prospective projects of this type must be individually evaluated in terms of specific characteristics of the site.

INTEGRATED UTILITY SYSTEMS- A CONCEPT FOR UTILITY ENERGY MANAGEMENT

Dale W. Kirmse and Edwin F. Coxe
University of Florida

ABSTRACT

The Integrated Utility System (IUS) concept considers the interaction of the five utility subsystems needed by a complex of buildings. The subsystems are: (1) Electric Power Service; (2) Heating-Ventilation-Air Conditioning and Hot Water Service; (3) Solid Waste Handling; (4) Liquid Waste Handling; and (5) Potable Water Service. By and large, current utility subsystems are operated as separate entities. "Integration" of the subsystems is a design approach which seeks optimum overall system performance through interaction and mutual support of subsystems.

The essence of an operationally and financially successful IUS is on-site electrical power generation utilizing the rejected heat, normally wasted by conventional power generation, to provide space heating and cooling. Full integration of the subsystems can be achieved through incineration of solid waste for its heat content and reuse of treated waste water to displace the use of potable water for process makeup water and irrigation.

While optimization is site specific, the multiplicity of subsystem combinations available to maximize the benefits from the application of the IUS concept are addressed. The conceptual designs and results of feasibility studies at two universities, for which utility operating cost savings of over 40 percent have been estimated, are presented.

AN OVERVIEW OF URBAN SYSTEMS TECHNOLOGY OPTIONS

T.J. Marciniak and R.E. Holtz
Energy and Environmental Systems Division
Argonne National Laboratory, Argonne, Illinois 60439

ABSTRACT

The urban environment, because of its high population and energy demand densities, offers a significant opportunity for the application of a variety of energy-related technologies in an energy-conserving, cost-effective manner. Due to the recent, sharp increases in petroleum fuel costs with the expectation of further price increases in the foreseeable future, there is a growing interest in technologies which are: (a) energy-conserving, and (b) which use renewable energy sources, such as solar, "free" sources such as solid waste or less scarce fuels such as coal or nuclear. These technologies are designed to help an urban community meet its electrical, space conditioning and miscellaneous energy needs.

Technologies which are, and will be, available include: (a) prime movers, including diesel engines, fuel cells and perhaps Stirling cycle engines; (b) central heating and cooling components and systems; (c) solid waste utilization technologies based on pyrolysis techniques or incineration with heat recovery; (d) large central heat pumps or dispersed heat pumps; (e) thermal and photovoltaic solar collectors; and (f) coal utilization technologies.

Some, or all, of these technologies can then be used in urban energy systems such as:

1. district heating and cooling systems,
2. total energy or selective energy systems,
3. heating and cooling systems using natural bodies of water, and
4. energy systems based on non-scarce fuels with waste heat recovery and solid waste usage.

This paper will center on an overview of the technologies which are applicable to urban energy systems. These will be structured into a time frame including: (a) technologies which are emerging from the developmental stage to commercialization by 1985; and (c) advanced technologies which will be commercially available after 1985.

THE PERFORMANCE OF OPERATING TOTAL ENERGY PLANTS

D. Wulfinghoff, Energy Consultant
G. McClure, Charles J. R. McClure and Associates, Inc.
3936 Lantern Drive, Wheaton, Maryland 20902
2016 S. Big Bend Boulevard, St. Louis, Missouri 63117

ABSTRACT

Despite growing interest in the use of total energy and its variations as a means of conserving energy in providing utilities to community-type facilities, very little comprehensive information has previously been available about the large number of total energy systems now operating. In order to exploit the experience gained in these plants, detailed analysis of the operation of (23) total energy plants was conducted in mid-1977. This paper summarizes the results of the study.

The total energy systems included in the study serve hospitals, university campuses, (shopping centers), and apartment complexes. (One is a utilities system for a small town.) The prime movers used in most of these systems were diesel or natural gas reciprocating engines. A few used steam turbines or gas turbines, alone or in combination with reciprocating engines.

Surveys of the total energy plants were based on groups of questions, relating to, respectively, the facility site, the technical description of the plant, utilities statistics, operational efficiency, maintenance and reliability, the makeup of the operating staff, competing public utilities, fuel supply, the history of the design process, financing and accounting procedures, related energy conservation activities, and management's view of total energy. From the extensive data provided by the surveys, an attempt was made to correlate the successful operation of total energy plants to the details of operation provided by the data base.

THE ROLE OF NATURAL ENERGIES IN PLANNING FOR THE ECONOMIES OF MAN

Mitchell J. Lavine
Cornell University, Ithaca, N.Y.

ABSTRACT

This paper discusses some theories about energy flow used in research* on environmental assessment of transportation projects. The research analyses include measuring energy flow not only in transportation and related economic systems, but also in related natural systems. The energy-flow theories may help us understand how to use the energies of the environment to gain maximum effect for us.

Energy is a necessary resource for doing work, and work is the means by which all systems operate. Therefore energy is a necessary resource for operating any system. If there is energy or potential energy embodied in all resources, then energy flow is a measure of operational capacity of any system. Thus we

*National Cooperative Highway Research Program, Project 20-11B, "Toward Environmental Benefit/Cost Analysis: Energy-Flow Analysis (Manual)" at the Center for Environmental Research, Cornell University, Ithaca, New York.

may study energy flow in ecosystems to learn not only how they operate, but also how we may manage those ecosystems. To the extent that we may alter energy flow in ecosystems, we may control their operation. Similarly, to the extent that natural processes may alter the energy flow in human economic systems, those natural processes may affect the operation of our economies.

The natural environment includes many energies that we may use. To a great extent, modern industrial systems have operated and grown using energies from fossil fuels. As it becomes more costly, in both monetary and energy terms, to use the energy of fossil fuels, it may become increasingly advantageous for us to make greater use of other energy resources from nature. Energy-flow theories are used in this paper to devise strategies for getting maximum effect by mixing renewable natural energy flows with those already operating in our economic systems.

ENERGY CONSERVATION WITH INTERFACE ECOSYSTEMS

William J. Mitsch, Ph. D.
Illinois Institute of Technology

Natural ecosystems, run by solar energy, can accomplish many functions for man that otherwise may cost large investments of fossil fuels. Identification of these interface ecosystems is an important contribution to energy conservation. Examples of such systems are given for wastewater treatment in north central Florida and flood control and water conservation in southern Illinois. Comparisons are made in economic and energetic terms with technological alternatives for achieving the same tasks.

ELECTRIC GENERATION: ISSUES OF SCALE AND POLITICAL AUTHORITY

Marc Messing

Environmental Policy Institute, Washington, D.C.

QUANTITATIVE MEASURE OF ELECTRIC GENERATION

Early History of Electric Generating Technology

Electric generation emerged toward the end of the 19th century as the synthesis of science and technology. The phenomena of magnetism and electric currents had been observed since early Greek science (1) and had been the subject of Gilbert's magnum opus at the turn of the 17th century (2). Steam, as a source of motion (though not as a source of work) had been known to the Greeks by the 3rd century (3) and had been applied to the mechanics of the industrial revolution by the 17th century (4). In November, 1831, Faraday demonstrated the the production of continuous electric current from mechanical induction and paved the way for the development of dynamos, magnetos, and alternators-- all of which were to be developed within the next fifty years (5). By January, 1880, Edison had completed work on a marketable incandescent light bulb and had conceptualized the possibility of complete centralized electric generating systems, utilizing four steam-pwered reciprocating engines for the production of electricity from six dynamos, distributed through a transmission network capable of supplying light bulbs across a one-half mile square area (6). On January, 1880, Edison filed 32 patents for lamps, 7 patents for distribution components, 6 patents for dynamos, and 5 patents for auxiliary components (7). In 1882, the Pearl Street station went into operation and by the end of the year the six dynamos were providing electricity for almost 4,000 electric lamps in 193 buildings (8).

During the next two decades, from 1882 to 1902, virtually all the technological and institutional mechanisms of the modern electric generating system were developed. Pressure-staging turbines were introduced in 1884, alternators and induction motors were introduced in 1887, condensing turbo-generators in 1891, superheated steam in 1893, (cathode-ray tube in 1897), and tandem turbo-generators in 1900 (9). Both direct current and alternating current systems competed for markets, and both steam-generation and hydro-electric systems were installed throughout the country (10). Institutionally, both General Electric and Westinghouse were incorporated and dominated the market for electric supplies and generating equipment; both publicly and privately-owned systems competed for franchise areas; securities companies were founded to aid in the capital formation necessary for electric system development (the United Electric Securities Company had been formed in 1890 and the General Electric Securities Corporation was formed in 1904); and electrification had been applied not only to lights and motors, but also to cars, trains, toasters, roasters, hair curlers, space heaters, irons, blankets, rocks, caves, lakes, fens, dens, bogs, and, by 1909, to medical diagnosis (11).

<u>Turbine-generators</u>. "The arrival of the steam turbine..." one commentator has

noted, marked "a complete revolution, for with it came a drastic change not only in the design and size of generators, but also in the layout of power stations and in the whole philosophy of the generation of electricity." (12). Steam turbines, which were introduced between 1884 and 1900, provided substantial improvements in the operating efficiencies of electric powerplants from the outset and made possible further increases in the operating temperatures and pressures of increasingly-larger units, thus leading to greater economies of scale, accelerated development of the industry, and increases in the scale of electric generation and electric utility systems.

The first pressure-staging turbines had capacities of 7.5 kw; by 1903, a 5,000 kw unit had been installed in Chicago (13); and, by 1931, the largest generating unit in the United States was 200 MWe (14). The maximum unit sizes of turbine generators increased modestly to about 300 MWe in 1950 but then increased sharply from 1955 to 1970 because of metallurgical developments leading to rapidly increasing steam pressures and temperatures (15). During the ten year period from 1955 to 1965, maximum unit size increased more than three-fold from 300 MWe to 1,000 MWe and, by 1970, the maximum unit size had increased to more than 1,300 MWe (16).

With the introduction of nuclear reactors in the late 1950's, the trend toward maximum unit sizes accelerated. By the late 1960's, U.S. production of commercial nuclear turbine generators had already been scaled up to 400 MWe, and, by the mid-1970's, most new nuclear units were planned in the 800-1,200 MWe range (17). During this time, boiler sizes, steam pressures and temperatures, and the thermal efficiencies of new generating units also continued to increase. Figure 1 indicates the growth in maximum unit sizes for steam boilers and turbine generators from 1900 to 1970 plus projections to 1990; Fig.2 indicates maximum temperatures and pressures in turbines for the same period; Fig. 3 indicates the rate of growth of nuclear turbine generators planned for introduction from 1960 through 1990; and Fig. 4 indicates trends in unit efficiencies.

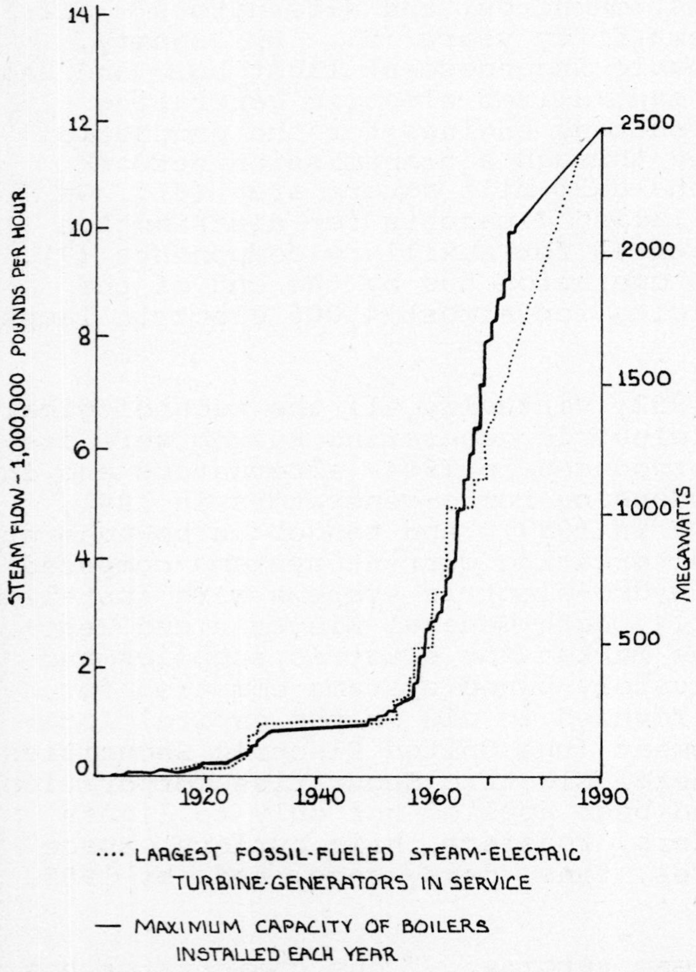

Fig. 1. Maximum unit sizes

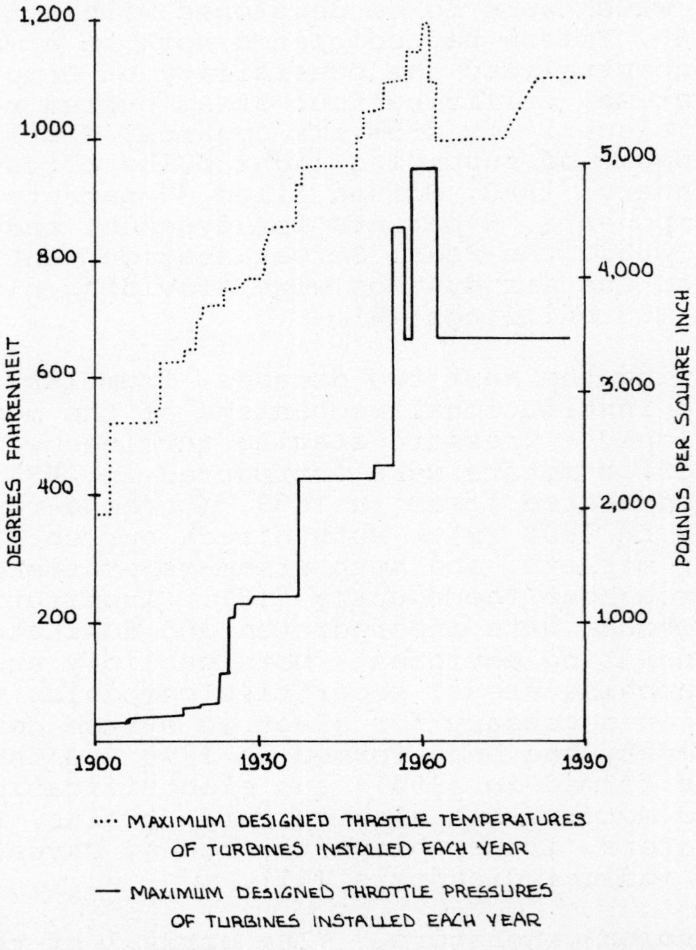

Fig. 2. Maximum steam temperatures/pressures

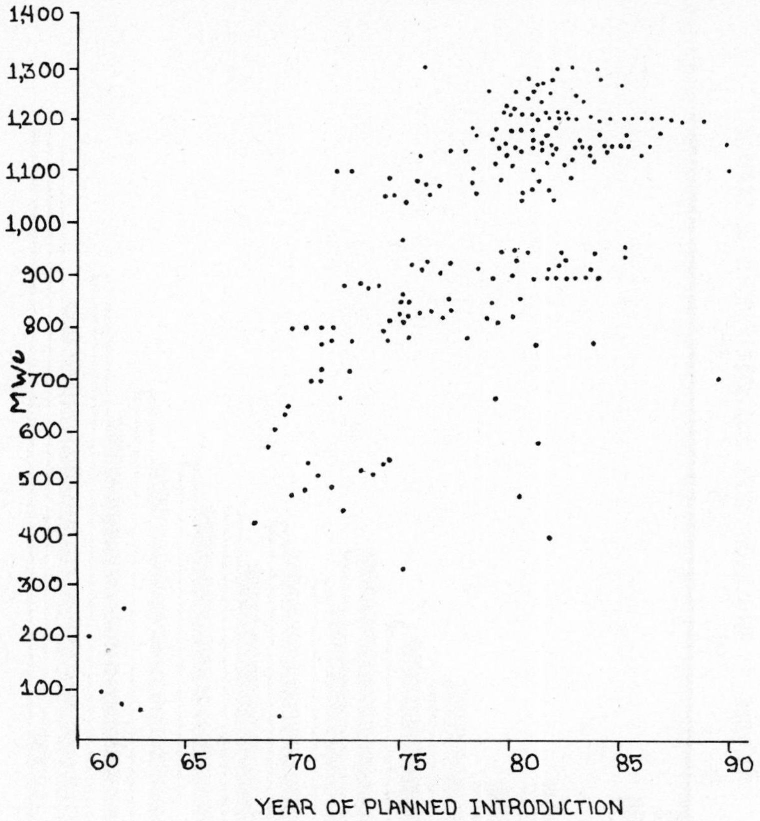

Fig. 3. Nuclear turbine generators

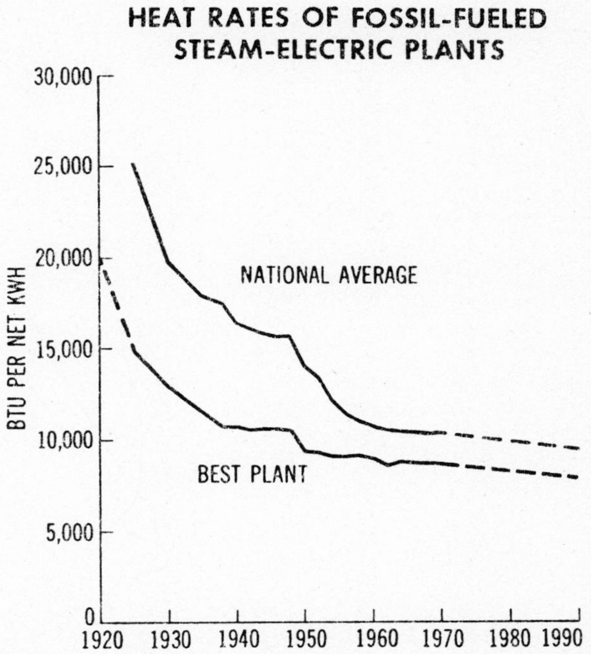

Fig. 4. Heat rates of fossil-fueled powerplants

Trends in Average Unit and Plant Sizes

The continuous introduction of larger and more efficient generating units and the gradual trend toward the construction of multiple units on single sites have resulted in a continuously accelerating rate of increase in the average size of generating units and plants.

In the 1930's the average unit size in the U.S. was about 20 MWe, and more than 95 percent of all units in operation were smaller than 50 MWe (18). During the '30's and '40's average unit sizes continued to increase as the result of larger new units gradually displacing smaller older units, and average unit sizes increased from 26 MWe in 1938 to 30 MWe in 1947 and 49 MWe in 1957 (19). The affects of multiple unit siting can be seen through a comparison of these unit figures with increases in the national averages of plant sizes: average plant sizes increased from 22 MWe in 1938 to 35 MWe in 1947 and to 96 MWe in 1957 (20). From 1957 to 1973 average plant size increased almost five-fold (21). Whereas there were, prior to 1950, only two powerplant sites in the United States with outputs of more than 500 MWe (both located in New York City), by 1974 there were fifteen plants greater than 2,000 MWe, and by 1975 this figure had risen to nineteen (22). Figure 5 compares the rates of growth of average unit size, plant size, and new unit size from 1938 through 1975. Figure 6 indicates the number of fossil- and nuclear-powered units over 300 MWe introduced between 1960 and 1975. Table 1 indicates the relative growth of various size powerplants between 1960 and 1975.

Site Size	1960	1970	1975	Increase 1960-1975	Percentage Increase 1960 - 1975
300 Mwe	1043	1228	1270	228	22%
300 Mwe 2000 Mwe	215	301	392	177	82%
2000 MWe	4	8	23	19	475%

Table 1. Relative growth of small, medium and large powerplant sites: 1960-1975

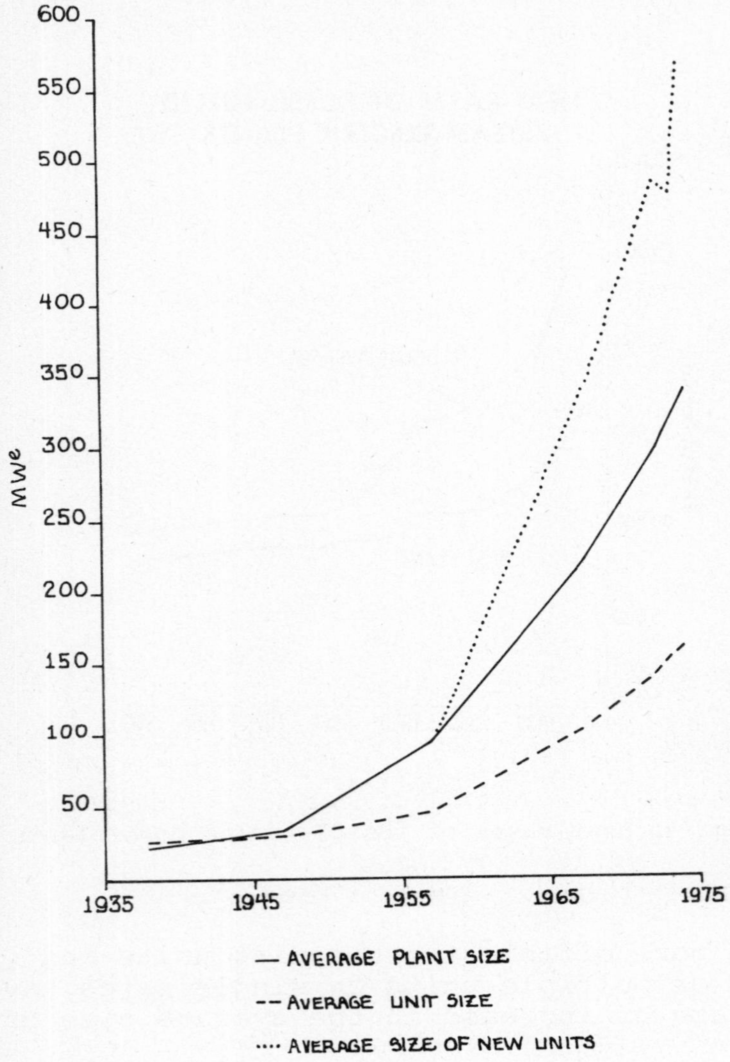

Fig. 5. Average unit size, plant size and new unit size: 1938 - 1975

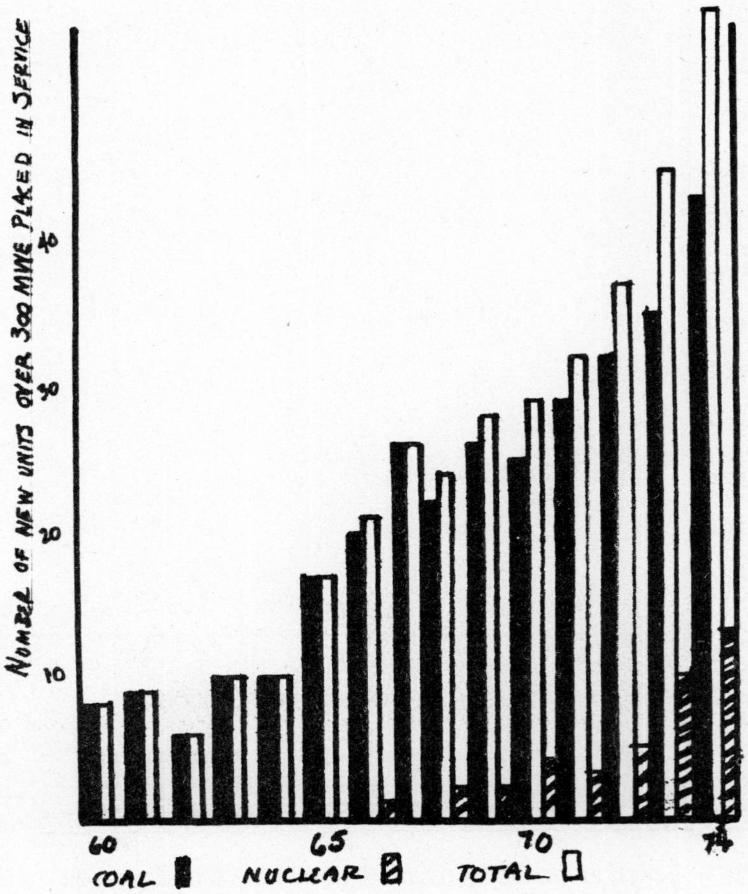

Fig. 6. Large fossil and nuclear units

Electric Production and Consumption

Energy consumption in America has increased at a steady rate since at least the middle of the 19th century, and electric energy has followed a similar pattern at twice the rate since its introduction. From 1900 to 1930, total electric energy consumption increased from approximately 2.5 BKWh to more than 91 BKWh; and, by 1960, this had increased to 753 BKWh (23). By 1970 electric energy consumption had increased to 1.391 TKWh, and government projections through the end of the century predicted an additional six-fold increase (24). Although population growth has been an important factor in these trends, more significant has been the rate of increase in per capita consumption levels. In 1930 per capita electric energy consumption was approximately 740 KWh; in 1960 it was more than 4,000 KWh; in 1970 it was approximately 6,500 KWh; and current projections estimate levels between 15,000 and 32,000 KWh in the year 2,000 (25).

THE ELECTRIC UTILITY INDUSTRY: INSTITUTIONS AND REGULATION

Profile of the Electric Utility Industry

The electric utility industry, which is responsible for the growth of electric energy in the United States, is generally described as pluralistic and diversified. In fact, it is probably more monolithic than pluralistic and more fragmented than diversified. In 1970, there were approximately 3,500 systems in the United States involved in the production, transmission, and generation of electricity, including municipally-owned systems, state-owned systems, federally-owned systems, and privately-owned systems (26). Of these systems, however, more than two thirds are involved solely in the transmission and distribution of power; and, almost 80% of the power generated is produced by approximately 280 investor-owned utilities (27). Less than 150 of these companies account for approximately 90% of all revenues, and 28 are holding companies with almost 100 subdivisions (28). In 1970, 35 private utilities (including 20 independent companies and 15 holding companies) represented the 70 largest utility companies in the U.S. and controlled approximately 70% of all assets (29).

Moreover, it is important to recognize that the electric utility industry represents the single most capital intensive industry in the country, the single largest issuer of securities, and the single largest industry in terms of capital assets (30). Also, perhaps most importantly, the electric utility industry itself is integrally related to the two component suppliers (G.E. and Westinghouse) which control approximately 70% of the U.S. and world nuclear markets and 75% of the domestic turbine market, and which dominate much of the domestic market in the supply of electrical components related to both the supply of and the demand for electric energy (31).

Throughout the 1970's, the growth of the electric utility industry was justified on the basis of several axiomatic beliefs: first, that energy growth and economic productivity were inextricably linked; secondly, that historical patterns of energy consumption and increased electricification of our total energy budget would continue; and, thirdly, that nuclear energy would provide virtually unlimited resource availability (32). Additionally, it was generally understood by economic forecasters that the construction of electric generating facilities and the production of electricity contributed to the GNP index and therefore contributed directly to continuing economic growth and prosperity.

However, by the mid-1970's, the abrupt discontinuities precipitated by the OPEC price increases tended to increase awaremess of the capital intensity of the electric utility industry (33), of the diseconomies of scale of large, new generating facilities (34), of growing capital shortages (35), of the rising capital demands of electric utility companies (36), and of the wide range of energy alternatives available (37). Analysts could now argue that: 1) the cost of additional energy production exceeded the costs of conservation (38), 2)capital investment in non-electric energy sectors would produce more jobs and economic benefits than electric facility construction (39), and 3) increased standards of living could be achieved on reduced rates (or even levels) of energy consumption (40).

Electric Utility Regulation

The earliest electric systems were private corporate ventures franchised either by Edison, who held patent rights for direct current systems, or by Westinghouse, who held patent rights for the competing A.C. systems. By 1902, there were already more than 3600 electric utility systems in the United States of which 2,905 were privately-owned, 815 were municipally-owned, and many were involved in competing jurisdictions. In the city of Chicago, alone, there were 29 electric utility franchises, 3 of which were city-wide (41).

The regulation of electric utility franchises was initially limited to existing local or municipal codes such as building codes and fire permits. State regulation of these utilities began in 1907 with their inclusion under the jurisdiction of state public service commissions in New York, Wisconsin, and Georgia. Such commissions, established for the purpose of regulating railroads and gas, water, and telephone utilities, acquired authority to regulate rate-making, equipment acquisition, financing, the provision of services, and the integrity of corporate operations. Electric utility regulation was thus absorbed within this framework.

From 1907 through 1917, the utility industry continued to proliferate as the number of new systems far outpaced the consolidation of existing systems, so that by 1917 a peak of approximately 6500 separate utility systems was reached. However, by the 1930's, this trend had reversed, and, in 1932, more than 70% of the nation's generating output was controlled by only eight large holding companies. Federal regulation of the electric utilities began during this era.

The federal regulation of the industry increased with the passage of a variety of acts. Under the commerce clause and under the Federal Power Act of 1935, the Federal Power Commission was empowered to regulate the "rates and services of public utilities selling electricity in interstate commerce at wholesale prices" as well as the issuance of securities and the merger or consolidation of companies. Citing Constitutional provisions granting the federal government the power to provide for the common defense, to regulate interstate and foreign commerce, and to control U.S. property, the Congress, in 1946, further reserved for the federal government jurisdiction over all matters related to nuclear energy, and, in 1954, amended this preemption through the Atomic Energy Act which provided for industry participation in the development of peaceful uses of nuclear energy but which did not provide any role for state government in either the planning or the regulation of nuclear energy. Finally, during the period from 1969 to 1975, the Congress passed the National Environmental Policy Act and the Coastal Zone Management Act, and amended the Clean Air Act and the Water Quality Control Act to expand their provisions for the planning and regulation of activities having significant environmental impacts.

Under regulations issued pursuant to the Atomic Energy Act, population guidelines were established which limited the population densities permitted in the vicinity of nuclear powerplants; under standards of the Clean Air Act, state air implementation plans were requested of all states to meet ambient air quality standards; under the Water Quality Control Act, standards for thermal effluents were established; under the Coastal Zone Management Act, coastal areas were recognized to be of special value and planning requirements were mandated for their development; and, under the National Environmental Policy Act, all major federal actions were required to include thorough evaluations of environmental impacts and examinations of available alternatives.

State roles. Since 1973, approximately 23 states have enacted powerplant siting laws which increase the state role in the planning and review of electric generating systems.* Of these, 15 include provisions for long-range planning (extending from 5 to 25 years); and 21 of the 23 states consolidated the procedures for facility certification. In 10 states, the responsibility for siting decisions was assigned to the PUS or PSC, and, in 8 states, the jurisdiction was assigned to specially-created siting boards.

Electric Utility System Planning

Despite the complexity of federal regulations and the variety of local regulations associated with electric utility construction and operation, the authority for planning utility systems is traditionally a state authority which has been delegated to the utilities as regulated public services (42). Accordingly, and notwithstanding recent state initiatives to oversee utility planning, the principle responsibility for demand forecasting and system planning resides in the utility companies and can be divided into two stages: 1) the planning stage--including demand forecasting, technical evaluation of state of the art designs, candidate site selection, and site acquisition; and 2) the regulatory stage--involving permit approvals on the local, state, and federal level. Somewhere between these two stages, there is also the provess of co-ordinated planning with other utility systems and, presumably, with governmental planning units.

---

*The 23 states are: Arizona, Arkansas, California, Connecticut, Florida, Iowa, Kansas, Kentucky, Massachussetts, Minnesota, Montana, New Hampshire, New Mexico, New York, North Dakota, Ohio, Oregon, South Carolina, Vermont, Washington, Wisconsin, and Wyoming.

Without discussing utility system planning in detail, a number of observations are in order on the following subjects.

Demand forecasting. The first is that, historically, utility forecasting has been something between a primitive art and a casual exercise. Given the historical record of growth in demand and the increased efficiency and lower costs of production, new facilities traditionally could be planned on the basis of extrapolated future demand; and gaps between installed capacities and actual demand could be narrowed by market incentives for increased consumer demands. With the increasing costs, complexity, risks, and lead times associated with large new facilities, in general, and large nuclear facilities, in particular, demand forecasting now requires greater sophistication and precision. Table 2 indicates the range of forecast variables contained in the rather sophisticated model prepared by Northeast Utilities.

TABLE 2. FACTORS AFFECTING DEMAND FOR ELECTRICITY*

Economic Factors
- Price elasticities
- Alternate boiler fuels
- Gross economic trends

Social and Demographic Factors
- Population growth
- Number of households
- Appliance saturation
- Promotional activities
- Conservation efforts
- Rooms per household
- Single vs multiple unit dwellings
- Per capita consumption
- Government efforts to control demand
- Standards for building efficiencies

Technical Factors
- New technologies requiring electricity
- New technologies improving efficiencies
- New technologies substituting for electricity

Policy Factors
- Conservation programs
- Fuel restrictions
- Environmental Standards

Consumer Classes
- Railroads
- Residential
- Industrial
- Commercial
- Streetlighting

*Adapted from Northeast Utilities' Forecast Model

Diseconomies of Scale: Once beyond the immediate problem of demand forecasting utility planners face problems of equipment selection and of conflicting evidence regarding the economies of large new units. Throughout the long period of continous technological development and declining fuel prices, economies of scale favored large new units which could be added to utility systems below average costs. However, changing economic conditions and technological limitations seem to have exhausted the economies of increased unit sizes by the mid-sixties or early seventies (43), and problems of increased reserve margins as a function of the decreasing number of units on a system contributed further diseconomies (44). This problem was in turn exacerbated by the failure of large new units to achieve designed performance factors throughout the late 1960's and early 1970's. In June, 1974, the FPC recognized this problem when it stated that "the trend toward larger unit sizes, and the relatively poor availability records of many large units, operate to increase reserve margin requirements over that which is needed when smaller, mature units predominate on a system," (45) and, the FPC noted, as a result, utility reserve margins planned for the next decade tended toward "the upper end of the 15 to 25 percent band currently observed." (46) In a comprehensive analysis of powerplant performance, Komonoff's statistical analysis verified this problem in December, 1976 (47).

Site Selection and Acquisition: Historically, the process for the selection of new generating sites was a simple and straightforward problem based on a limited number of criteria which could easily be quantified and expressed in monetary terms (48). Sites were selected on the following basis: 1) close proximity to load centers, 2) close relation to existing transmission facilities, 3) easy access to fuel transportation, 4) access to water supplies, 5) suitable access for construction, 6) meeting prerequisite geological and load bearing characteristics (49).

Following the selection of candidate sites, parcels of land could be assembled surreptitiously to prevent land speculation or early public opposition (50). More recently, however, as the result of federal regulatory requirements, state planning requirements, and greater public attention to such matters, the site screening and acquisition process has become substantially more complicated*.

External Corporate Obligations: Beyond demand forecasting and site screening, utilities functions as regulated monopolies with prescribed corporate obligations and procedures. Although controversial, we would argue that the financial structure of utilities based on guaranteed rates of return on capital investment, has biased utility planning toward increased capacity expansion rather than the most resource efficient management options; and this bias has been supplemented by a narrow interpretation of the utilities obligation to provide power upon demand without regard to load management or community planning alternatives (51).

Powerpools, Reliability Councils, and Regional Planning: Finally, it should be noted that utility systems have been interconnected through transmission networks since 1920, and in the early 1950's these systems were greatly expanded. Economically, interconnected systems allow for lower overall generating capacities and associated costs (52). Following the great northeast utility failure of 1965 the FPC issued recommendations for increased system coordination and reliability, including the establishment of regional reliability councils and a national electric reliability council to improve regional planning and coordination (53). Accordingly the reliability councils submit planning data on such things as projected capacity additions, retirements, and construction plans, and the data is collated by the FPC for public use; but these arrangements are only indirectly related to the planning activities of the private utilities. The reliability councils comprise a voluntary association and the data which is collected is collected without the force of law (54).

Additionally, powerpools exist throughout the country as the result of private agreements between individual utilities, for the purposes of coordinating planning activities and insuring greater reliability.

POLITICAL AUTHORITIES OF LOCAL GOVERNMENT

There are approximately 78,000 governments in the United States today, including the federal government, 50 state governments, about 2 dozen regional, 3,000 county governments, and 75,000 other jurisdictional entities. Of these, the 3,106 county governments (including boroughs, parishes, districts, 23 consolidated city-counties, and 39 independent cities located outside any other county area) represent 93% of the U.S. population and comprise the most convenient genus of local government. (55)

Administratively, approximately 90% of all counties are governed by county councils, 380 have appointed county administrators, 57 have county charters, and 46 have elected county administrators. Functionally, county governments vary in jurisdictional authority insofar as they are all subdivisions of state government with authorities delegated by the states. But, generally, their authorities include an identifiable range of planning and service functions, including election administration, tax assessment, judicial administration, zoning, and code enforcement. (See Table 3). Additionally, urban-type counties have assumed responsibility for solid waste collection, transit, housing, and air and water quality control functions.

With the increasing urbanization of America, more counties are currently

* One state, Maryland, has a state land-banking procedure, and several states have attempted comprehensive state-wide site surveys, but none of these have met with unqualified success to date.

assuming urban-type functions, and in recent years approximately 23 counties have consolidated into city-county governments to more easily provide these social services on a regional (county) basis. In fact, county government regards itself as the fastest-changing government in the U.S. today, and the National Association of Counties notes that major reorganizations are in progress in approximately 44 states (56).

TABLE 3 SERVICE FUNCTIONS OF COUNTY GOVERNMENT

Finance	Public Utilities
Property tax assessment	Water supplies
Property tax collection	Sewage treatment
Police Protection/Corrections	Power supply
Judicial Administration	Natural Resources
Legal Services and Administration	Coastal zone planning
Public Safety Services	Air pollution control
Fire protection	Water quality control
Emergency medical services	Energy conservation/management
Disaster preparedness programs	Solid waste collection/disposal
Public Health	Land Use
Home health	Comprehensive land use planning
Mental health	Zoning control/open space
Alcoholism/drug abuse	Growth management
Social Services	Community Development
Human resource planning	Public housing programs
Job training	Industrial development
Public service employment	Parks and Recreation
Transportation	Park acquisition and development
Road construction/maintenance	Recreation programs
Mass transit	Miscellany
Education	

But, despite the fact that there are 3,106 counties in the U.S. today and 3,581 electric utility systems, and, despite the fact that both county government and the electric utilities exist as instruments of state government, the geographic boundaries and political jurisdictions of electric utilities and county governments do not coincide; and and hiatus of planning authority exists between them.

It is apparent from comparing the electric demand factors considered in utility demand forecasting (see Table 2) with the planning and service functions for which local governments are responsible (see Table 3) that overlapping authorities fall into two categories: 1) electric demand factors affected by community planning and growth management and 2) community service functions affected by electric system development. Tables 4 and 5 compare these functions.

TABLE 4 ELECTRIC DEMAND FORECASTING FACTORS AFFECTED BY LOCAL GOVERNMENT PLANNING

Population Growth
Number of Households
Single vs. Multiple Dwellings
Building Code Standards
Conservation/Management Programs (Energy)
Industrial Development
Appliance/Technology Saturation

TABLE 5 SERVICE FUNCTIONS OF LOCAL GOVERNMENT AFFECTED BY SITING DECISIONS

Public Health	Natural Resources
Public Safety	Land Use
Social Services	
Human Resources	
Transportation	
Education	
Utility Systems	

Impacts of Large-scale Development on Small-scale Communities

The impacts of large-scale development on small-scale communities are not unique to powerplants: other large-scale industrial developments and other forms of energy facilities, in particular, may have similar impacts on community structures. But, as we have seen, the problem is characteristic of the technological trend in powerplant development toward larger unit and plant sizes, and it has been exacerbated in recent years by the move toward isolated areas in response to regulatory criteria. Restrictions on the radiation exposure of populations have led to population density restrictions on the siting of new power plants; the onus of meeting primary and secondary air quality standards consistent with state air implementation plans has (in the absence of strict non-degradation standards) prompted utilities to site large new fossil-fueled units in remote areas; and increased political opposition to new facility development has provided an incentive for the development of large sites (thereby minimizing separate site review problems for new unit additions) in remote areas (thereby minimizing political opposition). Thus, the problem of large-scale powerplant siting in small communities, has become an increasingly common focus in regional system planning. Table 14 indicates the distribution of base-load powerplants which began operation during the period from 1965-1975 according to plant size and community size.

TABLE 6 DISTRIBUTION OF POWER PLANTS WHICH BEGAN OPERATION, 1965-75

Site Size	County Size by Population: Small (less than 35,000)	Medium (more than 35,000 less than 200,000)	Large (more than 200,000)	Total
Small (less than 300MWe)	15	20	7	41
Medium (more than 300 less than 2000MWe)	35	35	21	91
Large (more than 2000MWe)	56	61	34	151

The socio-political problems associated with this form of development are clear: the construction of these facilities involves the deployment of large work forces with high percentages of specialized skills which must be accomodated to the social infrastructures of the impacted communities, whereas, the ultimate operation of the modern, highly-automated facilities requires only small work forces. Thus, small communities must accomodate both the sudden influx and then the sudden out-migration of work forces which are disproportionate to the size of the existing community. The results are boom-bust cycles

Nuclear Energy Centers

As an extension of the trend toward the remote siting of large-scale electric facilities and toward nuclear energy development, and as an attempt to address the special safety problems associated with nuclear powerplants and fuel cycles, the concept of Nuclear Energy Centers was first proposed several years ago. In the past five years, numerous studies (including one major study mandated by Congress (57)) have examined the technical and the institutional issues of NEC development and, to a lesser extent, of non-nuclear energy center development. As generally defined, nuclear energy centers would agglomerate 10,000 to 40,000 MWe on a single plot of contiguous sites, possibly requiring 40 to 80 square miles of land.

After several years of deversified study, the socio-economic affects associated with NEC deveoopment are now beginning to come into focus. Rob Brenner has written, "A regional population increase of approximately 45,000 will compel local governments to make substantial investments in the expansion of public services such as education, recreation, waste disposal, sewage treatment, police and fire protection, transportation facilities, health care, administrative capacities, and social assistance of various types." (58) The economic problems associated with this development may be more complex but have also benefitted from sophisticated regional economic analysis during the past several years. Less closely studied have been the political issues associated with large-scale regional energy development. In June, 1976, two independent workshop panels evaluated the political impacts of NEC development as part of the NRC's Nuclear Energy Center Site Survey.

It was the conclusion of these workshop panels that: 1) "a decision to go to NECs would tend to result in an upward shift in the locus of decision-making authority with a tendency toward greater federal involvement than is the case with dispersed siting...(2)...the probably-increased role of the federal government in many aspects of power generation, with the attendant transfer of decisions away from more local, presumably more responsive government, was considered disadvantageous from the point of view of public perception as well as state and local interests...(3)the question of the right of federal pre-emption is basic to the practicality of the center concept...(4) the NEC being a multistate concept leads naturally to a perception of diminished state power, which is increased by the predominant federal nuclear role...(5) (it was) generally agreed that NECs would involve federal preemption of state authority, with some members of the panel feeling very strongly that this preemption is particularly undesirable with respect to site certification...(6) NEC sites might constitute among themselves a kind of network for political and economic purposes." (59)

Accordingly the NRC concluded that although "the jurisdictional and institutional context within which NECs will be considered...is a fluid one, where the trend is a movement toward more State and local involvement in decision-making ...the sheer magnitude of a NEC will tend to move the focus of the decision-making back to the national level." (60) However, neither the workshop panels nor the NRC itself identified the specific political authorities (beyond site certification) of state and local government which would be affected by this development: rather it was accepted as a tenet of political theory that the centralization of electric generating systems would necessarily diminish the authorities of local and state governments.

Decentralized Energy Systems

In contrast to the increased centralization of electric generation a variety of technologies currently exist which tend to be smaller in scale and are inherently decentralized in form. Such decentralized technologies can be generally characterized by three elements: 1) the nature of energy conversion is geared to end-use requirements (61), 2) increased energy efficiencies are achieved through proximity of conversion to end-use applications or multiply integrated technologies, and 3) economic feasibility is not dependent upon large-scale unit applications. Such systems include, at a minimum, industrial co-generation (62), district heating (63), modular integrated utility systems (64), resource recovery systems (65), and solar thermal systems (66).

CONCLUSIONS

It is clear that the historical growth of the electric utility industry has been characterized by the technical development of large-scale systems, to the extent that the functional authorities of the electric utility companies effectively transcend the political jurisdictions and geographic boundaries of state and local government. The result has been jurisdictional conflicts between utilities and units of state government, both of which originally derive jurisdictional authorities from state government, and both of which have been superceded in recent years by federal authorities.

The primary social impacts of these jurisdictional conflicts have been manifest in the boom-bust cycles of large-scale energy development in small towns and have been underscored by the prospect of a further continuation of these trends toward large-unit development in geo-politically remote areas.

From a technical standpoint it is unclear whether a continuation of this trend is justified, but it is apparent that alternative smaller-scale decentralized technologies do exist which would mitigate the adverse impacts of large-scale development and eliminate the problems of jurisdictional conflict. In any event, the development of electric generating facilities and energy systems must be integrated with the development of the social systems which require energy, and it would appear that either the jurisdictional authorities for community planning, growth, and development, must be transferred from local government to the utilities or their federal surrogates; or technologies appropriate for meeting energy demands on a decentralized basis compatible with traditional authorities for socio-political growth and land-use planning, must be development

* *

REFERENCES

(1) Aristotle discusses briefly both the forces of lodestones and electric eels; but explanations of magnetism also occur in Empedocles and references date back to Thales.

(2) De Magnete, published in 1600 by William Gilbert, systematically reviewed presious theories on "attraction and repulsion" and introduced experimentation into the inquiries on magnetic forces.

(3) -----------

(4) -----------

(5) P. Dunsheath, A History of Electric Power Engineering, M.I.T. Press, Cambridge (1962).

(6) W. Clark, Energy for Survival, Anchor Books, Garden City (1974).

(7) Dunsheath, op. cit.

(8) Ibid.

(9) Ibid.

(10) Clark, op. cit.

(11) Ibid.

(12) Dunsheath, op. cit.

(13) Ibid.

(14) Federal Power Commission, National Power Survey, 1970.

(15) J. Halvorsen, Improved Energy Conversion, Status Report on Energy Resources and Technology, Atomic Industrial Forum, N.Y. (1975).

(16) FPC, op. cit.

(17) M. Messing, Economies and Diseconomies of Scale in Nuclear Turbine Generators, Environmental Policy Institute, Washington, D.C. (1975)

(18) FPC, op.cit.

(19) Federal Power Commission, Steam Electric Plant Construction Cost and Annual Supplement (1974).

(20) Ibid.

(21) Ibid.

(22) FPC, op. cit.;

(23) FPC, op. cit.

(24) M. Messing, The Need for Energy Facility Sites in the United States,EPI (1975).

(25) Ibid.

(26) FPC, op. cit. (14)

(27) Ibid.

(28) Electricity and the Environment, Report of the Bar Association of the City of N.Y.,N.Y. (1972)

(29) Ibid.

(30) C.R. Ross, Testimony before the U.S. Senate Committee on Judiciary, Hearings on Competitive Aspects of the Energy Industry (1970).

(31) M. Messing, Report on Financial Interests of Electronics and Electric Industry in Electric Energy Growth, EPI (1975).

(32) Messing, op. cit. (24).

(33) FPC, Ad Hoc Task Force on the Financial Outlook of the Electric Power Industry (1975).

(34) D. Huettner, Technological Change: Economics, Managements, and Environment, Pergamon Press (1975).
(35) New York Stock Exchange, Capital Needs and Savings Potential of the U.S. Economy (1974).
(36) Federal Energy Administration, Project Independence Report, November (1975)
(37) W.R.Z. Willey, Energy Conservation, Capacity Expansion, and Investment Finance, Environmental Defense Fund, Berkeley (1977).
(38) Ibid.
(39) Seattle City Light, Energy 1990 Seattle (1976).
(40) Ibid.
(41) The Electric Century: 1874-1974, Electric World (June, 1974).
(42) M.Messing, Report on the Jurisdictional Authorities of State and Local Government, National Academy of Public Administration (1977).
(43) C. Komonoff, Powerplant Performance, Council on Economic Priorities, N.Y. (1976).
(44) D. Huettner and J. Landon, Electric Utilities: Economies and Diseconomies of Scale, Working Paper,(1975).
(45) FPC, News Release No. 21369 (May, 1975).
(46) Ibid.
(47) Komonoff, op. cit.
(48) New England Regional Commission, Power Facility Siting Guidelines in New England, Boston (1976).
(49) Ibid.
(50) Ibid.
(51) Messing, op. cit. (42)
(52) Huettner and Landon, op. cit.
(53) FPC, Prevention of Power Failures (July 1967).
(54) Messing, op. cit (42)
(55) National Association of Counties, County Year Book 1976, Washington,D.C. (1976).
(56) Ibid.
(57) Nuclear Regulatory Commission, Nuclear Energy Site Survey-1975, Washington, D.C. (1976).
(58) R. Brenner, Socioeconomic Impacts of Nuclear Energy Centers, National Academy of Public Administration, Washington, D.C. (1977).
(59) M. Messing, The Political Authorities of Local Governemnts, American Nuclear Society (1977).
(60) NRC, op. cit. (57).
(61) A. Lovins, Energy Strategy: The Road Not Taken, Foreign Affairs (October 1976).
(62) P. McCracken et al., Energy Industrial Center Study, prepared for the National Science Foundation (June 1975).
(63) Use of Steam-Electric Power Plants to Provide Thermal Energy to Urban Areas, Oak Ridge National Laboratory, Oak Ridge, TN. (January 1971)
(64) Dept. of Housing and Urban Development, Evaluating Integrated Utility Systems, National Academy of Science, Washington, D.C. (1974).
(65) Metcalf and Eddy, Inc., Generation of Steam from Solid Wastes, NTIS,(1972).

Figures

1. Maximum Unit Sizes: Source: FPC, National Power Survey, 1970.
2. Maximum Steam Temperatures and Pressures; Source: FPC National Power Survey, 1970.
3. Nuclear Turbine Generators; Source: EPI (see 17 above)
4. Heat Rates of Fossil-Fueled Powerplant; Source: FPC, op. cit.
5. Average Unit Size, Plant Size, and new Unit Size; EPI, 1977.
6. Large Fossil and nuclear units; EPI, 1977.

Tables

1. Relative Growth of small, medium and large powerplant sites; EPI, 1977.
2. Factors affecting Demand for Electricity; EPI, 1977, adapted from Northeast Utilities model (see 48 above).
3. Service Function of County Government: From America's Counties Today, 1973
4. Electric Demand Forecasting Factors Affected by Local Government Planning;EPI,1977.
5. Service Functions of Local Governemnt Affected by Siting Decisions; EPI, 1977.
6. Distribution of Powerplant which began operation, 1965-1975; EPI, 1977.

ENERGY REQUIREMENTS FOR POLLUTION CONTROL AT STATIONARY SOURCES

R. W. Serth and R. S. Hockett

Monsanto Research Corporation, Dayton Laboratory
Dayton, Ohio 45407

ABSTRACT

Estimates of energy requirements for pollution control at stationary sources in the United States, as compiled from the literature, are presented and discussed. The data are analyzed to determine the distribution of energy requirements among economic sectors and among pollutant types. Alternative methods of pollution control which are potentially less energy intensive and still capable of meeting environmental regulations are also discussed.

INTRODUCTION

Fuel shortages in recent years have given rise to questions concerning the compatibility of national goals for a clean environment and goals for energy self-sufficiency. These questions have, in turn, given rise to a growing number of studies related to the energy cost of pollution control. The results of these studies are scattered throughout the literature, and many are in the form of draft reports which are not readily accessible. This paper attempts to summarize and integrate the available results in order to obtain as broad and as accurate a perspective as possible on the problem as it relates to stationary sources of environmental pollution.

Generally speaking, the energy requirements for pollution control given in this paper can be interpreted as energy required to meet all currently enacted federal regulations, after the legal granting of exemptions has been taken into account. Although different studies are not entirely consistent in the regulations that are assumed to be met, uncertainties associated with other assumptions tend to overshadow the differences due to assumed regulations. Exceptions to this statement are noted in the text.

The energy requirements given below are based on the use of "presently available" control technology. This statement must also be interpreted somewhat loosely to accommodate the results, but again, a rigorous interpretation is not warranted by the accuracy of the data.

ESTIMATES OF NATIONAL ENERGY REQUIREMENTS FOR POLLUTION CONTROL

Estimates of nationwide energy requirements for pollution control at stationary sources obtained from the literature are presented in the following subsections. All energy values have been converted to primary thermal energy equivalents using the following conversion factors:

- Electricity: 10,666 Btu primary/kWh electrical, corresponding to a conversion efficiency of 32%
- Oil: 6×10^6 Btu primary/barrel oil
- Coal: 24×10^6 Btu primary/ton coal

Comparison of energy estimates from different literature sources is complicated by the following factors: (1) The estimates are for different years; (2) compliance with different sets of regulations is assumed in different studies; (3) some estimates are for total energy required for control while others are for incremental energy required for compliance with specific regulations; (4) energy accounting is incomplete in most studies, i.e., all types of energy (e.g., electrical, fuels, energy for production of treatment chemicals, energy for maintenance of equipment, etc.) are not taken into consideration; (5) methods of calculation range from gross, cursory estimates to detailed computer simulations. The assumptions involved are often numerous and the manner in which they affect the results is difficult to determine unless a sensitivity analysis was performed in the study; and (6) in some cases, insufficient information is given to permit proper interpretation of the results.

Space does not permit discussion of the methodology and assumptions employed in each individual study. This information is available in summary form in Reference 1. References to each individual study should be consulted for complete details.

Electric Power Plants

Thermal pollution control. Closed-cycle cooling systems for controlling thermal pollution require energy beyond that required to operate once-through cooling systems. In a closed system, the cooling water from the condenser is passed through a cooling device (such as a cooling tower) in which heat is transferred to the atmosphere, and is then returned to the condenser. Additional energy is required for the operation of pumps and blowers, and to compensate for the loss in thermal efficiency of the power plant. The loss in efficiency is due to an increase in condenser temperature, which results in an increased turbine back-pressure.

Estimates of energy requirements for thermal pollution control are listed in Table 1. For comparison, the total United States energy requirement (ref. 2) is also given for each year. A major factor in these estimates is the assumption concerning the number of plants that will receive exemptions under Section 316(a) of the Federal Water Pollution Control Act of 1972. Depending on the assumed number of exemptions, the results can differ by as much as a factor of six (ref. 1, 11).

Air pollution control. Energy requirements for air pollution control at electric power plants are associated primarily with the control of sulfur oxides and particulates. Present standards for nitrogen oxides can be achieved by combustion modification techniques, such as low-excess-air firing and staged combustion, which incur little or no energy penalty, and may in fact increase boiler efficiency by up to 2% (ref. 12). Hence, energy requirements for NO_x control can be considered negligible at present.

Estimated energy requirements for air pollution control at electric power plants are presented in Table 2. The first three studies listed in the Table represent the most thorough analyses of the problem (ref. 1). The three estimates agree to within 40% when corrected for differences in the energy penalty assumed for SO_x scrubbing (ref. 1).

TABLE 1 Operating Energy Requirements for Thermal Pollution Control at Electric Power Plants

Source	Year	Primary energy required, 10^{15} Btu per year: Thermal Pollution Control	Total U.S. Consumption[a]
DSI (ref. 3)[b]	1977	0.086	78
	1983	0.20	95
ERT (ref. 4)[b]	1983	0.22[c]	95
Michigan (ref. 5)	1983	0.17	95
	1985	0.2	101
RPA (ref. 6)	1980	0.27	86
Cywin (ref. 7)	1980	0.13	86
Temple, Barker & Sloane (ref. 8)[b]	1980	0.0[d]	86
	1985	0.2	101
Hirst (ref. 9)	1970	0.16	-
Economics of Clean Water (ref. 10)	1977	0.43	78
	1983	0.79	95
NCWQ (ref. 11)[b]	1983	0.045 to 0.29	95

[a]U.S. Government estimates (ref. 2).

[b]Draft report subject to revision.

[c]Includes fossil-fuel steam electric plants only. If nuclear plants are assumed to make approximately the same contribution, the total energy requirement is roughly 0.4×10^{15} Btu (ref. 3).

[d]Indicates value is less than 0.1.

TABLE 2 Operating Energy Requirements for Air Pollution Control at Electric Power Plants

Source	Year	Primary energy required, 10^{15} Btu per year				
		SO_x scrubbers	Fuel oil desulfurization	ESP's	Total	Total U.S. consumption[a]
DSI (ref. 3)[b]	1977	0.065	0.118	0.009	0.19	78
	1983	0.211	0.104	0.008	0.32	95
ERT (ref. 4)[b]	1983	0.77[c]	-	0.064	0.83	95
Michigan (ref. 5)	1985	0.51	0.15	0.01	0.80[d]	101
RPA (ref. 6)	1980	0.21	-	-	0.21	86
MacDonald (ref. 14)	1975	0.32	-	-	0.32	75
Cywin (ref. 7)	1980	-	-	-	0.32	86
Temple, Barker & Sloane	1980	0.2	-	0.0[e]	0.2	86
(ref. 8)[b]	1985	0.3	-	0.0[e]	0.3	101
Hirst (ref. 9)	1970	-	-	-	0.84[f]	-
Huffman (ref. 15)	1974	0.062				73

[a]U.S. Government estimates (ref. 2).
[b]Draft report subject to revision.
[c]Total for SO_x scrubbing and fuel oil desulfurization.
[d]Includes 0.13×10^{15} Btu/yr for transportation of low-sulfur western coal.
[e]Indicates value is less than 0.1.
[f]Total for power plants and industry.

Industry

Estimates of energy requirements for pollution control in the industrial sector are presented in Table 3. Due to the paucity of available information, rather crude methods have been used to obtain these estimates.

The first four studies listed in Table 3 are the most credible (ref. 1). The first two studies are based on incremental (due to federal regulations) investment schedules for pollution control equipment estimated by the Council on Environmental Quality (CEQ). The resulting energy values are related to an increasing baseline value, and cannot be interpreted as total energy requirements for pollution control (ref. 1).

The Edison Electric Institute (EEI) value is based on a survey of electric

TABLE 3 Operating Energy Requirements for Industrial Pollution Control

Source	Year	Primary energy required, 10^{15} Btu per year			
		Air	Water	Total	Total U.S. consumption[a]
DSI (ref. 3)[b]	1977	0.50	0.23	0.73	78
	1983	0.51	0.28	0.78	95
Michigan (ref. 5)	1985	0.40	0.55	0.95	101
EEI (ref. 15)	1977	-	-	0.88	78
NCWQ (ref. 11)[b]	1977	-	0.38	-	78
	1983	-	0.82	-	95
Cywin (ref. 7)	1980	0.27	0.09	0.36	86
RPA (ref. 6)	1980	-	0.09	-	86

[a]U.S. Government estimates (ref. 2).
[b]Draft report subject to revision.

utilities, and includes electrical energy requirements only (ref. 15). The survey result has been corrected for the fact that the survey covered only 20% of total electric utility sales to industrial customers (ref. 1).

The National Commission on Water Quality (NCWQ) estimate for water pollution control was obtained by extrapolation of analyses of nine major industries (ref. 11).

Clearly, none of the above estimates is very satisfactory. In particular, none of the above values represents the total energy required for industrial pollution control.

Municipal Wastewater Treatment Plants

Estimates of energy requirements for municipal wastewater treatment are presented in Table 4. Each value is noted as being either incremental energy required for compliance with federal regulations or total energy required for wastewater treatment.

Municipal Solid Waste Disposal

Since solid waste disposal is potentially an energy-producing operation, it should not strictly be included in the present context. However, if energy recovery is not practiced (typically the present situation in the U.S.), then energy is required for collection, transportation, landfilling, and incineration of solid waste. Estimates of energy requirements for these operations in the disposal of municipal solid waste are given in Table 5. Estimates of potential energy recovery from municipal solid waste incineration and from recycling materials are also listed.

Capitalization Energy Requirements

A complete accounting of the energy required for pollution control must include the energy expended in the fabrication and installation of pollution control equipment. A national estimate of the capitalization energy required to meet federal regulations was made in the Development Sciences Incorporated (DSI) study (ref. 3). The average annual energy requirement over the period 1972-82 was found to be 0.2 quad[a] for all stationary sources.

[a] 1 quad = 10^{15} Btu.

TABLE 4 Operating Energy Requirements for Municipal Wastewater Treatment Plants

Source	Year	Primary energy required, 10^{15} Btu	Comment	Total U.S. consumption,[a] 10^{15} Btu
DSI (ref. 3)[b]	1977	0.036	Incremental	78
	1983	0.25	Incremental	95
Michigan (ref. 5)	1981	0.26	Incremental	89
EEI (ref. 18)	1971	0.053	Total, electrical energy only	-
	1977	0.19	Total, electrical energy only	78
RPA (ref. 6)	1980	0.055	Incremental above 1968 baseline	86
	1980	0.084	Total	86
Cywin (ref. 7)	1974	0.04	Total	73
	1977	0.06	Total	78
	1980	0.10	Total	86
Huffman (ref. 15)	1968	0.029	Total, from 1968 inventory of municipal plants	-
	1974	0.18	Total, tertiary treatment of all wastewater	73
Hirst (ref. 9)	1970	0.29	Total, secondary treatment of all wastewater	-
NCWQ (ref. 11)[b]	1973	0.15	Total, excluding chemicals production	-
	1990	0.35	Total, excluding chemicals production	-

[a] U.S. Government estimates (ref. 2).

[b] Draft report subject to revision.

TABLE 5 Energy Requirements for Solid Waste Disposal

Source	Year	Primary energy required, 10^{15} Btu per year[a]: Collection, transportation, and landfilling	Electricity for incineration	Electricity production from incineration	Recycling
Hirst (ref. 9)	1970	0.075	0.027	(0.27)	(0.44)
RPA (ref. 6)	1973	0.087	-	-	negligible
	1980	negligible[b]	-	(0.28)	(0.15)
Cywin (ref. 7)	1980	-	-	(0.44)	(0.077)

[a] Values in parentheses represent energy credits.

[b] Incremental energy requirement above the 1973 value due to federal standards. It is assumed that improved collection practices will offset any additional energy demand due to stricter standards for municipal waste management.

DISTRIBUTION OF ENERGY REQUIREMENTS AMONG SECTORS AND POLLUTANTS

The relationships that exist among the data presented in the previous section are obscured by the scatter in the estimated energy requirements and by the diversity of assumptions upon which the estimates are based. Hence, a single set of relatively consistent data was extracted from the above information for purposes of comparison. A combination of the DSI (ref. 3) data for 1977 and the EEI (ref. 15) data was selected as representing the best combination of consistency and accuracy.

The following adjustments were made to the data (ref. 1): (1) the EEI (ref. 15) value of 0.88 quad for the industrial sector was rounded to 1.0 quad to account for energy forms other than electrical. This total was apportioned between air and water pollution control to agree with the air/water ratio given in the DSI study (ref. 3) and the NCWQ (ref. 11) result for water pollution control; (2) for power plant thermal pollution control, a baseline value of 0.06 quad based on the Environmental Research and Technology (ERT) study (ref. 4) was added to the 1977 DSI (ref. 3) incremental value of 0.09 quad; (3) the EEI (ref. 15) value for municipal wastewater treatment was corrected to include fuel and chemical energy in addition to electrical energy, yielding a total of 0.27 quad; and (4) the values listed in Table 5 for landfilling solid waste were rounded to 0.1 quad as an extrapolation from 1970-73 to 1977.

Distribution Among Sectors

Estimates of the energy required for pollution control in 1977 are listed by sector in Table 6.

The total energy required for pollution control from all sectors is found to be 1.7 quad, with estimated error bounds of 0.8 quad to 3.4 quad. In general, these bounds are believed to be quite conservative (ref. 1). The nominal

TABLE 6 Energy Required for 1977 Pollution Control by Sector[a]

Sector	Energy required, 10^{15} Btu: Nominal value	Estimated error bounds	Percent of total energy requirement for pollution control at stationary sources
Industry			58
Air	0.65	-	
Water	0.35	-	
TOTAL	1.0	0.5 to 2.0	
Power plants			20
Air	0.19	0.10 to 0.38	
Thermal	0.15	0.07 to 0.29	
TOTAL	0.34	0.17 to 0.68	
Municipal wastewater treatment	0.27	0.09 to 0.54	16
Municipal solid waste disposal	0.1	0.05 to 0.15	6
TOTAL	1.7	0.8 to 3.4	100
Percent of total U.S. energy consumption in 1977	2.2	1.0 to 4.3	

[a] Part of the information contained in this table is based on draft reports which are subject to revision.

value of 1.7 quad represents approximately 2% of total U.S. energy consumption, with a range of 1% to approximately 4%. These values represent operating energy only; an additional 0.2 quad would be required for fabrication and installation of pollution control equipment.

A comparison of pollution control energy requirements for 1985 is presented in Table 7. These data are less consistent and less reliable than the corresponding data in Table 6. In particular, no satisfactory estimate of energy required for industrial air pollution control is available. The value of 1.0 quad was obtained by extrapolation from the 1977 value of 0.65 quad (Table 7) assuming an annual growth rate of 5%. The very speculative nature of these estimates notwithstanding, the data indicate that energy requirements for pollution control at stationary sources in 1985 will amount to between 2% and 3% of total U.S. energy consumption in 1985.

Distribution Among Pollutants

The above energy requirements for 1977 are regrouped according to pollutant type in Table 8. An additional calculation was required to apportion the energy for industrial air pollution control among the various pollutants. The DSI (ref. 3) breakdown of energy requirements by control device was used to obtain the following split: SO_x, 58%; particulate matter, 26%; others (NO_x, hydrocarbons, carbon monoxide), 16%.

From Table 8, the major energy requirements are for chemical and biological water pollution control and control of sulfur oxides, each of which accounts for about 35% of the total for pollution control at stationary sources.

Distribution Within the Industrial Sector

Energy requirements for industrial pollution control in 1977 are given according to the two-digit Standard Industrial Classification (SIC) scheme in both the DSI (ref. 3) and EEI (ref. 15) studies. Although these two sets of data were obtained by entirely different methods, seven of the top ten categories are identical in the two studies. These seven categories are listed in Table 9 together with the corresponding percentages of the total 1977 industrial pollution control energy requirement.

It can be concluded that the industries within these seven SIC categories account for approximately 80% of the total energy required for pollution control within the industrial sector. The primary metals category alone accounts for 36% of the industrial total, or

TABLE 7 Energy Required for 1985 Pollution Control by Sector[a]

Sector	Energy required, 10^{15} Btu	Percent of total energy requirement for pollution control at stationary sources
Industry		65
Air	1.0	
Water	0.8	
TOTAL	1.8	
Power plants		22
Air	0.4	
Thermal	0.2	
TOTAL	0.6	
Municipal Wastewater Treatment	0.35	13
Municipal Solid Waste Disposal	0.0[b]	0[b]
TOTAL	2.8	100
Percent of total U.S. energy consumption in 1985	2.7	

[a]Part of the information contained in this table is based on draft reports which are subject to revision.

[b]It is assumed that energy recovery from incineration and recycling will offset the energy required for collection, transportation and landfilling.

TABLE 8 Energy Required for 1977 Pollution Control by Pollutant[a]

Pollutant type	Energy required 10^{15} Btu	Percent of total energy requirement for pollution control at stationary sources
Water (chemical and biological)		36
Industry	0.35	
Municipal treatment plants	0.27	
Power plants	Negligible	
TOTAL	0.62	
Sulfur oxides		33
Industry	0.38	
Power plants	0.18	
TOTAL	0.56	
Particulate matter		10
Industry	0.17	
Power plants	0.01	
TOTAL	0.18	
Thermal		9
Power plants	0.15	
Industry	Negligible	
TOTAL	0.15	
Other air pollutants		6
Industry	0.10	
Power plants	Negligible	
TOTAL	0.10	
Municipal solid waste	0.10	6
TOTAL	1.7	100

[a] Part of the information contained in this table is based on draft reports which are subject to revision.

TABLE 9 Distribution of Energy Required for Pollution Control in the Industrial Sector

SIC	Industry name	Percent of 1977 energy requirement for pollution control in industrial sector: DSI (ref. 3)[a]	EEI (ref. 15)
33	Primary metals	36.0	36.5
28	Chemicals and allied products	15.4	15.0
26	Paper and allied products	5.1	12.6
29	Petroleum and coal products	7.2	8.7
34	Fabricated metal products	2.0	4.3
32	Stone, clay, and glass products	4.4	3.6
20	Food and kindred products	11.3	2.1
	TOTAL	81.4	82.8

[a] Draft report subject to revision.

0.36 quad. Other data (ref. 1) indicate that the Iron and Steel Industry accounts for about 70% of the total in the primary metals category.

ALTERNATE POLLUTION CONTROL TECHNOLOGY

The energy requirements considered above have been based on the use of "presently available" pollution control technology. A number of potentially less energy intensive control methods are summarized in Table 10. The methods considered are restricted to those which could have a significant impact in the period 1985-90.

Intermittent systems for SO_x control at combustion sources are designed to meet ambient air quality standards and are not capable of meeting all regulations. Other methods of SO_x control, such as oil desulfurization, coal cleaning, and use of low-sulfur western coal, are at least as energy intensive as flue gas scrubbing. Thus, fluidized-bed combustion (FBC) and more energy efficient scrubbers represent the main opportunities for near-term reduction of energy requirements for SO_x control.

First-generation fluidized-bed power plants (now in the demonstration phase) are expected to have overall thermal efficiencies comparable to conventional plants equipped with scrubbers. However, later-generation FBC systems are projected to have significantly higher efficiencies (ref. 1). The greatest potential energy savings are in electric power plants and large industrial boilers for which pressurized FBC (as opposed to atmospheric FBC) is likely to be economical. Another advantage of FBC is that NO_x emissions are also controlled, so future NO_x standards would be met without an additional energy penalty.

Spray ponds and cooling ponds are about half as energy intensive as forced-draft cooling towers for thermal pollution control. Natural-draft cooling towers are a less energy intensive alternative for industrial sources, but they would save only about one-sixth of the energy saved by installing spray ponds or cooling ponds. These methods have the drawbacks of large land requirements and capital investment costs.

The waste heat rejected from electric power plants and industrial processes represents a substantial energy resource. It is estimated that use of waste heat from electric power

TABLE 10 Less Energy Intensive Alternatives to Present Pollution Control Technology

Pollutant controlled	Alternative
Sulfur oxides	Fluidized-bed combustion of coal
	Intermittent control systems (fuel switching, load shifting, tall stacks)
	More energy efficient scrubbers
Thermal pollution	Spray ponds and cooling ponds as opposed to cooling towers
	Waste heat utilization, e.g., for space heating or wastewater treatment
Municipal wastewater	Recovery of sludge gas
	Trickling filter as opposed to activated sludge for secondary treatment
Municipal solid waste	Energy recovery via pyrolysis or incineration
	Recycling of metals, glass, paper
Industrial air and water pollution	Process modifications to reduce number and size of streams requiring "end-of-pipe" treatment

plants for space heating could save up to 5 quads annually in the U.S. (ref. 16). Thus, integrated systems for the utilization of waste heat in space heating, agriculture, aquaculture, sewage treatment, etc., represent the least energy intensive method of thermal pollution control.

The energy required for municipal wastewater treatment could be reduced through utilization of the gas produced by anaerobic digestion of organic sludge. Sludge gas can be used to fuel internal combustion engines, which can be directly coupled to air blowers and water pumps, or it can be used to drive electrical generators. It is estimated that all of the electrical energy requirements for primary treatment plants or approximately two-thirds of electrical energy requirements for activated sludge plants could be supplied in this manner (ref. 1).

For secondary wastewater treatment, trickling filter plants require up to 50% less energy than activated sludge plants (ref. 1).

Energy recovery via incineration or pyrolysis of municipal solid waste constitutes a much less energy intensive alternative to landfilling. Most processes for energy recovery are in the development or demonstration stages. However, waterwall incinerators constitute a proven technology with a high energy recovery efficiency (ref. 1).

Recovery and recycling of scrap metals, paper, and glass in solid waste is also less energy intensive than landfilling these materials.

CONCLUSIONS

The energy required to meet government regulations for pollution control at stationary sources in 1977 amounts to about 2% of total U.S. energy consumption, with a range of 1% to approximately 4%. Projections for the mid 1980's indicate that this figure will increase only slightly to between 2.5% and 3% of projected total national energy consumption.

Pollution control in the industrial sector accounts for approximately 60% of energy requirements for control at stationary sources. Energy requirements are concentrated in the following industrial categories: Primary Metals, Chemicals, Paper and Paper Products, and Petroleum and Coal Products.

Industrial and municipal wastewater treatment and control of sulfur oxides, primarily from industrial and utility boilers, account for approximately 70% of the energy required for control at stationary sources. It follows that efforts to reduce energy requirements for stationary source pollution control should be directed most heavily toward these two areas.

ACKNOWLEDGEMENT

This work was supported by the Power Technology and Conservation Branch of the U.S. Environmental Protection Agency under Contract No. 68-02-1320, Task No. 22.

REFERENCES

1. R. S. Hockett and R. W. Serth, Energy consumption from present pollution control technology, Draft Final Report submitted to U.S. Environmental Protection Agency, Cincinnati, Contract No. 68-02-1320, Task 22 (1976).

2. A national plan for energy research, development and demonstration: creating energy choices for the future, ERDA-48, vol. 1, U.S. Government Printing Office, Washington, D.C. (1975).

3. Development Sciences Incorporated, First-order estimates of potential energy consumption implications of federal air and water pollution control standards for stationary sources, Draft Final Report submitted to U.S. Environmental Protection Agency, Washington, Contract No. 68-01-2498 (1975).

4. J. P. Mahoney et al., Energy consumption of environmental controls: fossil fuel, steam electric generating industry, Environmental Research and Technology, Inc., Draft Report prepared for U.S. Department of Commerce, Office of Environmental Affairs (1976).

5. J. Davidson et al., Energy needs for pollution control, Chapter 7 in: The Energy Conservation Papers, R. H. Williams, ed., Ballinger, Cambridge, Mass. (1975).

6. H. Baily, P. Cushman and A. Steinberg, A brief analysis of the impact of environmental laws on energy demand and supply, Resource Planning Associates, NTIS publication PB 245656 (1974).

7. A. Cywin, Energy impacts of water pollution control, P. 143-149 in: Energy, Agriculture, and Waste Management, W. J. Jewell, ed., Ann Arbor Science, Ann Arbor, Mich. (1975).

8. Temple, Barker & Sloane, Inc., Economic and financial impacts of federal air and water pollution controls on the electric utility industry, U.S. Environmental Protection Agency, Washington, EPA-230/3-76-013 (1976).

9. E. Hirst, Energy implications of several environmental quality strategies, Oak Ridge National Laboratory, Oak Ridge, Tenn., ORNL-NSF-EP-53 (1973).

10. The Economics of Clean Water - 1973, U.S. Environmental Protection Agency, Washington (1973).

11. Staff Draft Report, National Commission on Water Quality, Washington (1975).

12. D. G. Lachapelle, J. S. Bowen and R. D. Stern, Overview of Environmental Protection Agency's NO_x control technology for stationary sources, Paper presented at 67th Annual Meeting of the American Institute of Chemical Engineers, December 4, 1974.

13. B. I. MacDonald, Alternative strategies for control of sulfur dioxide emissions, Journal of the Air Pollution Control Association, 25, 525 (1975).

14. T. W. Bendixen and G. L. Huffman, Impact of environmental control technologies on the energy crisis, News of Environmental Research in Cincinnati, U.S. Environmental Protection Agency, Cincinnati, January 11, 1974.

15. Power needs for pollution control, Electrical World, 155, 66. (1973).

16. J. Karkheck, J. Powell, and E. Beardsworth, Prospects for district heating in the United States, Science, 195, 948 (1977).

THE EFFECT OF NO_x INPUT ON TROPOSPHERIC OZONE

Donald H. Stedman and Shaw C. Liu
Atmospheric and Oceanic Science Department
The University of Michigan
Ann Arbor, Michigan, USA, 48109

ABSTRACT

Experimental studies of smog chambers and ambient air demonstrate a complex relationship between NO_x emission and ozone formation. These experimental results are discussed together with several model results, and indicate that it is probable that NO_x emissions contribute significantly to high levels of ozone in downwind areas. Several aspects of the importance of this ozone formation are described.

INTRODUCTION

Any combustion process using air as the oxidant tends to emit oxides of nitrogen, because the equilibrium $N_2 + O_2 \rightleftharpoons 2NO$ is driven to the right at the high temperatures in flames. On cooling, the reverse equilibration takes place too slowly for the NO to fully return to N_2.[1] Thus, automobiles and stationary sources emit NO in their effluent at concentrations of the order of 100 ppm. US wide NO_x ($NO + NO_2$) emission sources are shared about 60/40, stationary/mobile.[2]

Natural sources of NO_x include forest fires, lightning[3] and oxidation of natural ammonia emissions. Although the natural source rate is very poorly known, it is fairly clear that in populated regions man is by far the major source.[4] Although emitted as NO, the ultimate fate of all NO_x is to be oxidised. This oxidation takes place slowly by $2NO + O_2 \rightarrow 2NO_2$, and more rapidly by $NO + O_3 \rightarrow NO_2 + O_2$, where the ozone is available from the natural tropospheric ozone background of 20-40 ppb. The NO_2 is further oxidised by other processes to HNO_3 (nitric acid and perhaps some HNO_4, peroxynitric acid) which is then washed out and adds significantly to the acidity of rain.[5] The time constant for this whole process is of the order of 7-14 days. Thus, regional pollution by NO_x emissions may take place during a stagnation episode, although longer range global NO_x pollution is not probable.

Although NO_x as NO_2 is a toxic substance in its own right, and there is an applicable ambient air quality standard,[6] the standard is rarely violated, and even future increased combustion seems unlikely to cause major NO_x violation areas. More significantly NO_x is involved in the generation of ozone. This paper will show that increased regional NO_x emission will probably lead to increased regional ozone concentrations, above the ambient air quality standard of 0.08 ppm for 1 hr. If toxicological studies show that this rural ozone represents a serious problem, then some consideration of further NO_x control may be needed.

CHEMISTRY AND MEASUREMENTS

Ozone in the troposphere arises from downward diffusion from the stratosphere and local photochemistry.[7-10] Concentrations exceeding the 80 ppb standard are frequently observed, particulary in the cloudless but hazy air just after the passage of a summer high pressure system.[11,12] It has been suggested that this haziness also has a man-made source, and may be effecting the local climate.[13]

The photochemistry starts with the photolysis of NO_2, and continues with two further fast reactions thus:

$$NO_2 + h\nu \rightarrow NO + O \qquad j(NO_2)[NO_2]$$

$$O + O_2 + M \rightarrow O_3 + M \qquad k_2[O][O_2][M]$$

$$O_3 + NO \rightarrow NO_2 + O_2 \qquad k_3[NO][O_3]$$

where $j(NO_2)$ represents the solar photolysis rate of NO_2 (maximum value $\sim$0.01 sec^{-1} corresponding to a lifetime $\sim$100 sec),[14] and M represents any third body which takes away excess collision energy. The overall effect of these three reactions is that their forward and reverse rates are approximately equal. Thus $j(NO_2)[NO_2] = k_3[NO][O_3]$, or rearranging, $[O_3] = j[NO_2]/k_3[NO]$. (1) This equation (1) shows that ozone depends not on the $[NO_x]$ but on the ratio of $[NO_2]/[NO]$. This formalism demonstrates why rural ozone events do not obey mass balance, dilution necessarily lowers $[NO_x]$, but may not lower $[O_3]$ since the reservoir for O_3 (normal oxygen) is essentially infinite.

The important ratio $[NO_2]/[NO]$ in turn is controlled by slow reactions of hydrocarbon species (both man-made and natural) which produce oxidizing free radicals such as $HO_2^{\bullet}$ and $RO_2^{\bullet}$. These then oxidise NO to NO_2, thus effectively increasing ozone. Equation (1) demonstrates why the effect of NO_x emission on ozone is complex and non-linear. At night $j(NO_2) = 0$, therefore either O_3 or NO must be zero, and enough NO emission reduces $[O_3]$ to zero. This effect is well known from smog chamber studies,[15] and is illustrated by Fig. 1, which is some data obtained from around Grand Rapids, Michigan. The solid line is the ozone concentration in the air mass around Grand Rapids. The dashed line, the data from the Michigan State DNR Grand Rapids ozone monitor. The Grand Rapids measurements clearly

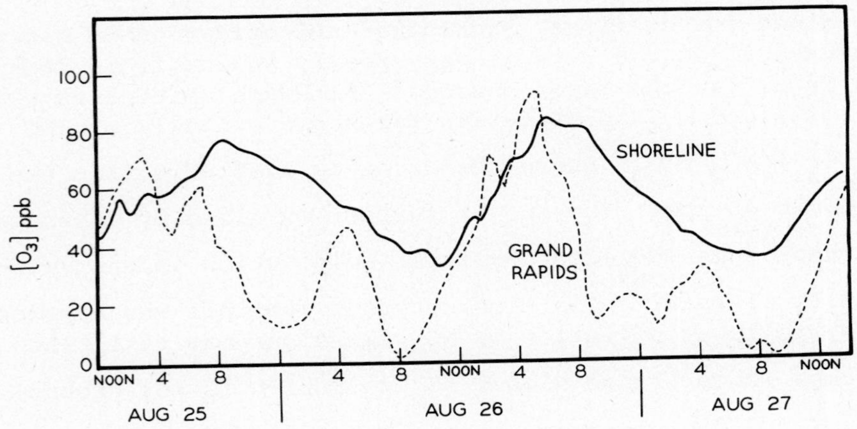

Fig. 1. Variation of ozone in and around Grand Rapids, Michigan. The solid line is the average of several rural ozone stations thought to roughly represent the regional ozone air mass. The dotted line shows the city monitor data.

represent their air mass pollution, except for their evening and morning rush-hour activity which injects NO into the low level nocturnal inversion and depletes the ozone concentration. By contrast, in smog chamber studies the injection of NO_x and hydrocarbons leads eventually to high ozone concentrations. In fact we measured the rural NO_x concentration together with this rural ozone and show a rural NO_x peak just preceeding the ozone peak. This is what would be expected if the air mass has been subject to photochemistry, following a distant injection of NO_x and hydrocarbons. We thus see a negative effect on ozone of local NO_x emissions and a positive effect of distant NO_x emission (the air mass in this case was shown by trajectory analysis to have been fairly stagnant over the Ohio Valley. Studies by Quickert et al[16] have shown no correlation of high ozone in Ottawa with gaseous NO_x, however good correlation with particulate sulfate and nitrate was observed.

Similarly on the remote shoreline of the Upper Peninsula of Michigan, ozone levels exceed 80 ppb frequently for 24-hour periods, partly since local NO_x sources are absent and partly because the shoreline station is always subject to good ventilation from aloft.

These observations can be summarized thus: urban emission of NO and HC lead to a local ozone reduction. If this air mass is then diluted and undergoes photochemistry in its subsequent transport, then high ozone can be formed. As this air mass undergoes further transport the high ozone remains and the NO_x becomes particulate nitrate.

The epidemiology on which the ozone standard is based is largely from urban areas in which the high ozone levels are correlated with intermediate oxidised organic and nitrogeous products such as aldehydes and PAN. Thus it is not clear that toxic effects are directly relatable to rural ozone, wherein these intermediates are both oxidised to particulate matter (haze), and very dilute. Despite this caveat, rural air masses of high ozone over a large region may cause worse pollution in downwind cities because of faster and farther oxidation of emitted NO to NO_2, thus allowing earlier PAN buildup.

If control of the rural ozone concentrations is necessary then either NO_x, or HC, or both must be controlled. Hydrocarbon control on solvent emissions, gasoline combustion etc. is being implemented. However, for rural ozone formation, uncontrollable natural hydrocarbon emissions may be important. We therefore consider the effect of NO_x control alone on clean and on slightly polluted air.

MODEL RESULTS

Figure 2 shows the results of four models of the effects of NO_x addition on otherwise clean air. Since some are steady state, and include stratospheric ozone transport, others are purely photochemical, the models are not strictly comparable, however from Fig. 2 one can observe that there is general agreement over an ozone increase for the first few ppb NO_x. The global background [NO_x] may be a few tenths of a ppb.[17,18] The rest of the graph is very model

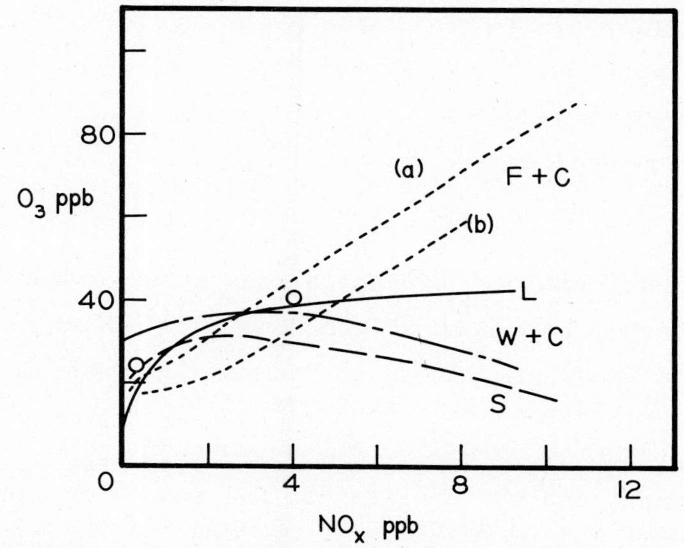

Fig. 2. A comparison of several models of the effect of NO_x addition on ozone in clean air. L Ref. 8; W+C, Ref. 10; F+C, Ref. 19; with faster and slower rate constants for HO_2 + NO, S, Ref. 9; O Ref. 7.

dependent, and clearly subject to some uncertainty as to whether further increase of NO_x alone can cause high ozone. We therefore looked at models with a small amount of hydrocarbon pollution. Figure 3 shows the results of these models, and Table 1 shows a possible relationship between the hydrocarbon pollution levels in these models by comparing the rate OH + HC for

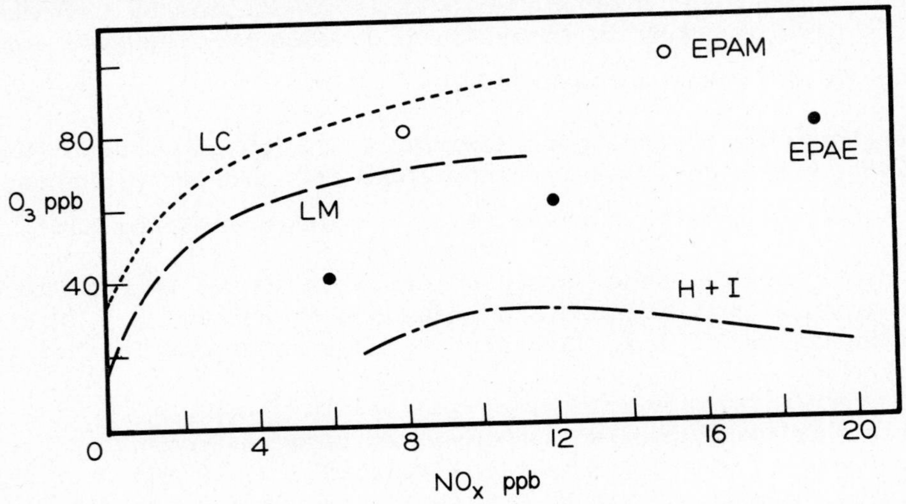

Fig. 3. The effect of NO_x increase at the low hydrocarbon concentrations shown in Table 1. L C, L M Ref. 8; o EPAM the EPA model results of Ref. 21; EPAE, Ref. 22 as reported in Ref. 21; H and I, Ref. 20.

the various studies. Again they are not strictly comparable models, and make different assumptions, although in general they treat the same phenomenon. Apart from the Hesstvedt et al curve[20] which is for a four-hour irradiation at 40°N, all other models show that the ozone produced can readily exceed the 80 ppb standard and that for a given low level of hydrocarbon, NO_x increase from 5 to 15 ppb can be the difference between ozone violation and minimal pollution.

TABLE 1 Photochemical Hydrocarbon Pollution Compared to the Relative Rate of OH + HC.

Model	Contaminant	Rate OH + HC Relative to CH_4
L M	7.5 ppm CH_4	100
L C	3.0 ppm CO	80
EPAM	0.2 ppm C butane propylene mix	260
EPAE	40 ppb isepentane	176
H + I	10 ppb ethylene	38

The maps produced by EPA[21] of estimated NO_x concentrations around cities show a considerable portion of the eastern USA expected to exceed 20 ppb NO_x. If combustion without good NO_x control is increased then these areas are likely to increase, and rural oxidant, which then also impacts on many downwind cities will also increase.

The chemical lifetime for ozone depends on the atmospheric temperature, water vapor concentration, stratospheric ozone content, and the position of the sun. For instance, the lifetime is only 2 to 4 days in the summer but is about 10 to 30 days for spring and fall conditions at mid-latitude. The formation rate of ozone depends on the concentrations of HO_2 and RO_2 and the reaction rate of

$$HO_2(RO_2) + NO \rightarrow NO_2 + OH(RO)$$

which is highly uncertain, 2×10^{-13} $cm^3 sec^{-1}$ was recommended by Hampson and Garvin[23] but Howard and Evenson[24] now report 8×10^{-12} $cm^3 sec^{-1}$. In the models shown in Fig. 2 and Fig. 3 values close to the smaller rate are used. If the faster rate is used the ozone increase due to NO_x emission will be three to four times larger and faster. However, uncertainties in other HO_2 and RO_2 reaction rates are also large. They affect the ozone formation rate because they determine the HO_2 and RO_2 concentrations.

Despite these generalities, there is clearly some room for improvement in models of regional ozone pollution, and certainly in models of NO_x in clean tropospheric air. Also the medical need for control of O_3 levels 0.1 ppm in rural areas is not yet well documented. Perhaps the more toxic local pollution products, such as PAN are the more important health hazards. Again, NO_x control is implicated in PAN control [25] but dilution is more effective in reducing Pan concentrations than it is for ozone. If an urban area is embedded in a high ozone concentration regional air mass, then its NO emission is oxidised to larger concentration of NO_2 sooner, thus leading to increased local PAN generation.

ACKNOWLEDGMENTS

This research was supported by the Atmospheric Research Section, National Science Foundation. The assistance of the State of Michigan, Department of Natural Resources, particularly Mr. D. Armbruster is gratefully acknowledged.

REFERENCES

1. Ya, B. Zel'dovich, P. Ya Sadunikov and D.A. Frank-Kamenetskii, Oxidation of Nitrogen Combustion, Original publication by Publishing House of the Academy of Sciences of USSR, 1947. Translated by M. Shelef.

2. Cleaning our Environment. The Chemical Basis for Action. The American Chemical Society, 1969.

3. Chameides, W.L., D.H. Stedman, R.R. Dickerson, D.W. Rusch and R.J. Cicerone, NO_x Production in Lightning, J. Atmos. Sci., 34, 1, (1977).

4. Liu, S.C., R.J. Cicerone, T.M. Donahue and W.L. Chamedies, Sources and Sinks of Atmospheric N_2O and the possible ozone reduction due to industrial fixed nitrogen fertilizers, Tellus in press (1977).

5. Likens, G.E., Acid Precipitation, Chem. and Eng. News, Nov. 22, (1976).

6. Air Quality Standards, Federal Register, 1975.

7. Chameides, W.L. and D.H. Stedman, Tropospheric Ozone Coupling Transport and Photochemistry, J. Geophys. Res., 82, 12, 1787 (1977).

8. Liu, S.C., Possible effect on the tropospheric O_3 and OH due to NO emissions, submitted to Geophys. Res. Letts.

9. Stewart, R.W., S. Hameed and J.P. Pinto, Photochemistry of Tropospheric Ozone, J. Geophys. Res., in press.

10. Weinstock, B. and T.Y. Chang, Methane and Nonurban Ozone, J. Air Poll. Contr. Assoc., in press.

11. Decker, C.E. et al., Formation and Transport of Oxidants Along the Gulf Coast and in Northern U.A., EPA-450/3-87-033, Aug. 1976.

12. Samson, P.J. and K.W. Ragland, Ozone in the Mid=West: The Evidence for Large Scale Transport, J. Appl. Met., in press.

13. Bolin, B., and R.J. Charlson, On the Role of the Tropospheric Sulfur Cycle in the Shortwave Radiative Climate of the Earth, Ambio, 5, 47 (1976).

14. Stedman, D.H., W. Chameides, and J.O. Jackson, Comparison of Theoretical and Experimental Values for j(NO_2), Geophys. Res. Lett., 2, 22 (1975).

15. Pitts, J.N. Jr., A.M. Winer, K.R. Darnall, G.J. Doyle, J.M. McAfee, Final Report of Contract 3-017, California State ARB, July 1975.

16. Quickert, N. and L. Dubois, Characterization of an Episode with Elevated Ozone Concentration, The Sci. of the Tot. Env., 5, 79 (1976).

17. Noxon, J.F., NO_2 in the stratosphere and troposphere by ground based absorption spectroscopy, Science, 189, 547-549, (1975).

18. Noxon, J.F., Paper presented at the Symposium on Non-Urban Tropospheric Composition, Hollywood, Florida, November 10-16, 1976.

19. Fishman, J., and P.J. Crutzen, A numerical study of tropospheric photochemistry using a one dimensional model, J. Geophys. Res., in press (1977).

20. Hesstvedt, E., O. Hov and I.S.A. Isaksen, Photochemistry of mixtures of hydrocarbons and nitrogen oxides in air, Geophys. Norveg., 31, 6 (1976).

21. Effectiveness of Organic Emission Control Programs as Functions of Geographic Location, Office of Air Quality Planning and Standards, U.A. EPA, January 1977.

22. Sickels, J.E., Ozone precursor relationship of NO_2, isopentane and sunlight under selected conditions, Ph.D. Thesis, Univ. of N. Carolina, 1976.

23. Hampson, F.R. and D. Garvin, Chemical Kinetic and Photochemical Data for Modelling Atmospheric Chemistry, NBS Technical Note 866, 1975.

24. Howard, C.J. and K.M. Evenson, paper presented at AGU Meeting, Washington, D.C., May 3, 1977.

25. Pitts, J.N., Jr., Keys to photochemical smog control, Env. Sci. Tech., 11, 456, (1977).

IMPACTS OF ENVIRONMENTAL REGULATIONS ON THE DEVELOPMENT OF COMMUNITY ENERGY SYSTEMS

Richard R. Cirillo
Michael Senew
Edward J. Croke
Energy and Environmental Systems
Argonne National Laboratory

ABSTRACT

The emergence of federal and state environmental regulations will directly impact the design, siting and economics of community energy development. A combination of air, noise and water control regulations place specific constraints on emissions not only from large central grid power systems, but also from many smaller community systems. Proposed federal and state regulations, depending upon local environmental quality, may prohibit the construction of certain systems or require extensive environmental control on others.

This study reviews existing and proposed environmental regulations affecting community energy systems. Regulatory impacts on large central grid stations are also examined in the context of an alternative approach to community energy production.

A spectrum of designs are included in the study from small developments using gas turbines to large community developments using coal, oil, and solid waste fueled steam turbines. The study is regional in nature considering local air quality and fuel availability.

Potential costs to community energy system developers to comply with environmental regulations are assessed. These costs are compared with alternative costs associated with large centralized grid systems. Fuel availability, control costs, siting restrictions and other constraints imposed by environmental regulations are summarized to assess future impacts and trends relating to community energy development as an alternative to central grid systems.

TRACE SULFUR COMPOUND REMOVAL FOR COAL GASIFICATION

James W. Mulvihill
Director, Environment and Conservation Division
U.S. Energy Research and Development Administration
Pittsburgh Energy Research Center
4800 Forbes Avenue
Pittsburgh, Pa. 15213

ABSTRACT

Numerous coal gasification processes are being developed to utilize our huge coal deposits and to help offset our dependence on foreign imports of crude oil and LNG. Coal can be gasified by addition of steam and oxygen/or air to produce a high BTU-pipeline gas; a low BTU gas suitable for direct combustion; or for use in combined cycle power generation; or a synthesis gas for use as a petrochemical feedstock.

Catalytic methanation is a key process for our future high BTU-pipeline gas energy plans. However, before catalytic methanation can be used for the commercial production of high BTU-pipeline gas, the problem of catalyst deactivation must be solved. This deactivation is caused by trace impurities of sulfur and other compounds in the synthesis gas and/or deposition of carbon on the catalyst surface. This could result in prohibitive operating costs for catalyst replacement or regeneration and could make the process economically impractical.

The U.S. Energy Research and Development Administration (ERDA) is sponsoring research at EXXON Research and Development Company to determine the optimum pre treatment process for trace sulfur compound removal from synthesis gas prior to methanation. The study is aimed at obtaining sufficient experimental data to produce an engineering and economic analysis of the trace sulfur gas clean-up systems as they apply to the SYNTHANE coal gasification process being developed by ERDA at Bruceton, Pa.

Bench scale adsorption breakthrough curves are presented for H_2S, COS, thiophene and methyl mercaptan in dry simulated synthesis gas using metal impregnated activated carbon.

PRELIMINARY ASSESSMENT OF THE ENERGY IMPACT OF AIR AND WATER POLLUTION CONTROLS ON COAL CONVERSION

W.C. Peters,[1] L.L. Lorenzi,[2] D.L. Schwartz,[3] and J.W. Mulvihill[4]
U.S. Energy Research & Development Administration
Pittsburgh Energy Research Center
4800 Forbes Avenue
Pittsburgh, Pa. 15213

ABSTRACT

The enactment and enforcement of air and water pollution control regulations on the proposed coal conversion industry may have a substantial impact on the energy consumption requirements. Environmental protection and energy consumption optimization trade-off guidelines and criteria must be developed to assure a commercially viable technology.

The U.S. ERDA, Pittsburgh Energy Research Center made a preliminary assessment of the energy requirements of environmental control technologies as part of a much broader study involving environmental impact and control technology assessment activities in coal conversion process technology. Energy penalties associated with specific pollution control equipment are presented. The effects on overall coal conversion efficiency are discussed. The energy consumption data are extrapolated to commercial scale applications and the projected energy impacts of proposed environmental standards are discussed in terms of the viability of the standards and adequacy of the energy efficiency of currently available control technologies.

---

[1]Chief, Energy Conservation Branch, Environment & Energy Conservation Division

[2]Environmental Scientist, Environment Branch, Environment & Energy Conservation Division

[3]Mechanical Engineer, Energy Conservation Branch, Environment & Energy Conservation Division

[4]Director, Environment and Energy Conservation Division

MATHEMATICAL MODELING OF THE INFLUENCE ON ENVIRONMENT OF DEVELOPING COUNTRIES IN A WORLD ECONOMY DOMINATED BY INDUSTRIAL NATIONS

H.J. Plass, Jr.
University of Miami
Department of Mechanical Engineering
Coral Gables, Florida

ABSTRACT

By extending to the entire world a simplified mathematical modeling technique developed by the author for the U.S. energy economic-environmental system Refs. (1), (2), (3), (4), it is possible to determine approximately the influences on available resources, the economics, and the environments for developed and developing countries competing for the same resources. The world model, for the purpose of simplicity of analysis, is represented by two domains (countries) which can make independent decisions for their own respective well-being. One domain is that of a developed industrialized country with high energy consumption and high material standard of living; the other domain is a country with low material standard of living but with rapid development toward an industrialized higher standard status. Internal, as well as export-import dynamics are included in the model. Environmental degradation resulting from wastes rapidly dumped into the natural systems of each domain is studied in detail. Possible ways to reduce the degradation through international or national policies in each domain are studied. Similar studies are made of the total energy consumption for each domain when these different policies are applied. Results are obtained for various assumed growth rates in each domain.

References

(1) H.J. Plass, Jr., "A Simplified Equilibrium Model of the U.S. Energy-Economic System and its Use in Comparing Alternatives" Proc. First World Hydrogen Energy Conference, University of Miami, Clean Energy Research Institute, Vol. III (1976)

(2) H.J. Plass, Jr., "How Might Synthetic Fuels from Coal Affect Natural Resources and Environment?" Proc. Second Annual University of Missouri-Rolla Conference on Energy (1975)

(3) H.J. Plass, Jr., "A Detailed Analysis of the Environmental Effects of Energy Utilization in the U.S. Economy" Proc. Third Annual University of Missouri-Rolla Conference on Energy (1976)

(4) H. J. Plass, Jr., "Dynamic Modeling of the Economic-Natural-Human System for the U.S." In preparation for publication.

REGULATORY INCENTIVES FOR EFFICIENT ENERGY USE

Walter J. Cavagnaro
California Public Utilities Commission
455 Golden Gate Avenue, Room 5201
San Francisco, California 94102

INTRODUCTION

The regulatory agency which is responsible for establishing the level of utility rates is in a particularly unique and effective position for promoting efficient energy use. The California Public Utilities Commission's policy of lifeline rates for the small user and the inverted gas rate designs above the lifeline quantity encourages conservation. Allocation of the cost of more expensive new natural gas to industrial users instead of the residential and commercial users is compatible with new federal administration policies. California's lifeline electric rates and leadership in developing time-of-use electric rate designs encourage overall energy conservation and particularly influence use of electricity at the time of peak demand.

Other conservation programs in effect in California include building and appliance standards, voltage regulation, load management, and utility retrofit programs covering home insulation, intermittent ignition devices, gas and electric water heating and many others. The regulatory agency offers financial incentives to the utilities in adopting procedures for rapid recovery of utility costs and added consideration in allowable fair rate of return. Incentives to customers for energy conservation are also encouraged and allowed in rate fixing by the regulatory agency. Cost effectiveness is an essential test in planning conservation expenditures and should be measured by comparing the cost of a conservation program to the marginal cost of new supplies.

GAS AND ELECTRIC RATES

The California Public Utilities Commission has initiated many new utility rate reforms. The first major changes were made in gas offset rate proceedings when the Commission ordered rate increases be permitted principally for monthly usages above 200 therms.1/ The Commission indicated that supply circumstances no longer supported a declining commodity charge based on consumption. The Commission found that the highest rates should be paid by the lowest priority interruptible users, because the highest priced gas is for their benefit -- without that gas those users would have to use alternative fuels. Uniform commodity rates were established by general service and interruptible customers. Subsequently, rate increases were permitted for usages above 75 therms and the general service rate schedule was inverted.

Later in 1975, the Commission issued a decision in PG&E's general rate increase proceeding for the first time implementing lifeline domestic electric

1/ Decision No. 84721, dated July 29, 1975, PG&E, Application No. 55687.

rates and further revising gas rates.2/ The effect of such changes were rate reductions for electric users of less than 300 kwhrs in Zone 1 and 500 kwhrs in Zone 5, while rates for larger users were substantially increased. Similar rate modifications have been put into effect for other California utilities.

In September of 1975, the State Legislature enacted the Miller-Warren Energy Lifeline Act which required the Commission to establish lifeline quantities to cover the minimum energy needs of the average residential user for space and water heating, lighting, cooling, and food refrigerating. After extensive hearings, the Commission established the following quantities for single-family residencies:3/

	Electric Kwhrs	Gas Therms
Basic Residential Use - Lighting, Cooking, Refrigeration	240	
Cooking and Water Heating		26
Water Heating	250	
Space Heating November-April	550 - 1,420	55 - 140

The seasonal allowances for space heating cover four climatic zones ranging from below 2,500 degree-days to areas above 7,000 degree-days. The basic residential uses were combined as appropriate for electric and gas to recognize appliance saturation data and to reduce the complexity of administration. Under the terms of the legislative requirement, lifeline rates cannot be increased until the average system rate has increased 25% or more over the January 1, 1976 rate level. Under this concept and the Commission's policies, it is anticipated that residential rates will be inverted -- the rate for usages above the lifeline quantity will exceed the rate for lifeline usages.

Another major step in gas rate design will result from a policy decision in April of 1977.4/ In examining the use of natural gas for swimming pool heating and other non-lifeline uses it was determined that a multi-tier inverted rate design would be adopted. By continuing the seasonal variations established by the lifeline decision it will be practical to design inverted rates with levels appropriate to the character of residential use. Customers will see a considerable natural gas rate incentive to utilize solar systems for heating swimming pools and for reducing their use of gas for other non-lifeline uses. The level of commercial, small industrial and interruptible rates will be established at appropriate tiers. The replacement cost of energy will be given full consideration in setting rates above the lifeline quantities.

California has also been introducing time-of-day rates in a further effort to reform rate structures. The Commission has required the major electric utilities to file time-of-day rates for all customers with requirements exceeding 4,000 kw.5/ Metering equipment has been required which will permit the extension of this type of rate to 500 kw requirements in the near future.

Additionally, the Commission has required extensive experimentation with load management equipment and rates to test the effectiveness of such devices and metering on all classes and sizes of electric customers.

2/ Decisions Nos. 84902 and 84959, dated September 16, 1975 and October 7, 1975 -- Applications Nos. 54279, 54280, and 54281.

3/ Decision No. 86087, dated July 13, 1976, in Case No. 9988.

4/ Decision No. 87192, dated April 12, 1977, in Cases Nos. 9581, 9642 and 9884.

5/ Decision No. 85559, dated March 16, 1976, in Case No. 9804.

The Commission has joined with the Energy Commission and four major California electric utilities in submitting this experimental program to the Federal Energy Administration. It has been approved for federal funding assistance and is in the process of being implemented.

The Commission has also indicated that marginal cost will be an important element in electric rate design. California, like most of the major states and the Federal Power Commission, bases revenue requirements on reasonable operating expenses and a historical cost rate base. Even though we utilize a future test-year rate base for setting rates, incremental costs substantially exceed average costs. For example, on a system such as PG&E's, having the advantage of substantial hydroelectric resources, the average energy cost was only about 1¢/kwhr in 1975. Incremental energy cost based on current prices of low sulfur oil approximates 2.5¢/kwhr for fuel only. A differential of this magnitude considerably exceeds the difference between average and incremental capacity costs.

With differences of substantial magnitude between average and marginal costs, it can be anticipated that further rate reforms will be involved in balancing portions of the rate structure based on marginal cost with the utilities' allowed revenues based on average costs.

VOLTAGE REGULATION FOR ENERGY CONSERVATION

On February 16, 1977, the California Public Utilities Commission requested that all utilities in California establish a schedule for regulating their service voltages between 114 and 122 volts as measured at the customers' meter. The new service voltage range is 3% or 4 volts below the former upper limit of 126 volts and the lower limit remains unchanged.

While the former service voltage range was 114 to 126 volts, common utility practice was to regulate voltages at the upper end of that range and this caused many customers to receive between 124 and 126 volts during period of light loading. This resulted in wasted energy, increased bills to the customer, and a dramatic reduction in the life of their incandescent lamp bulbs.

For a while we were receiving critical inquiries about motor efficiency with the use of our new voltage range. We have since clearly demonstrated to most critics, including one of our largest electric utilities, that motor efficiency decreases as voltage rises above nameplate voltage (115 volts). Thus, with motors there is also a substantial energy savings to be gained under our "Voltage Regulation" program.

The following test data collected by Southern California Edison Company illustrates the energy savings which occur as voltage is reduced from 126 to 110 volts on a 115-volt 1/2 horsepower Leland Faraday 1725 RPM Motor.

Line Voltage (Volts)	Line Amperes (Amps)	Input Power (Watts)	Speed (RPM)	Output Horsepower (HP)	Efficiency (P.U.)
Normal Operation					
126	9.3	655	1762	0.503	.572
124	9.05	640	1760	0.503	.586
122	8.9	627	1760	0.503	.598
120	8.75	625	1760	0.503	.600
118	8.65	620	1758	0.502	.604
116	8.55	615	1755	0.501	.608
114	8.5	615	1750	0.500	.607
112	8.35	600	1750	0.500	.622
110	8.3	600	1748	0.499	.621

Minimum Start Voltage 76 Volts; Stall Voltage 72 Volts

A typical kitchen refrigerator also uses far less energy at or slightly below nameplate voltage, as indicated by the following test data collected by Pacific Gas and Electric Company on a Dwyer kitchen refrigerator.

Voltage	Current	Phase Angle	Watts
110.0 VAC	2.10 amps	49°	151.5
115.5 "	2.13 "	51°	154.8
117.0 "	2.15 "	51.5°	156.6
118.5 "	2.175 "	52°	158.7
120.0 "	2.20 "	51°	166.1
124.0 "	2.58 "	48°	214.1

Currently, at the last count, Pacific Gas and Electric Company had reduced voltages to within the new range on 924 circuits representing 33% of its system and full implementation is expected by June 30, 1977. With implementation proceeding at a reasonably fast pace we have yet to receive any significant customer complaints from the implementation of this program. We are also pleased to report that the municipal utilities in California are joining us in implementing the "Voltage Regulation" program.

LOAD MANAGEMENT

Electric utilities, as well as major commercial and industrial customers, are developing a significant interest in load management hardware. In the electric utility side, a good deal of the interest relates to the automatic control of peak demands on the system. Many individual utilities are undertaking experimental use of such devices as temperature controlled demand limiters, radio controlled water heaters, ripple control of various electric loads, as well as the use of bi-directional communication systems. An important area for additional research would be in the combined use of communication circuits for various utility control and metering purposes, as well as for non-utility service.

In addition to utility demand management systems, major electric users are applying automatic (computerized) energy management systems both to reduce maximum demand and to reduce energy use. In some cases, even greater opportunities exist for reduction in energy use than for reduction in maximum demand. Customer energy management systems could be coordinated with utility systems to maximize the efficient use of energy.

Further, the Commission has enunciated a policy that in California conservation ranks at least equally with supply as a primary commitment and obligation of a public utility. Therefore, the cost of new supplies will be measured against the alternate costs of load management.

The Federal Energy Administration and others have emphasized the potential for electric energy as a major vehicle for conserving scarce fossil fuels by permitting energy demands to be shifted toward a coal-nuclear base. This concept also appears to be part of the impetus to time of use rates with emphasis on higher peak load rates and lower off-peak rates. However, realistic estimates of the marginal cost of off-peak energy are also imperative. The efficient utilization of off-peak energy is equally important. Increases in total energy use in order to reduce peak load are not necessarily justified. Therefore, the energy efficiency of off-peak electric heating or cooling storage systems are critical to a cost effective development of this technology.

The direct thermal uses of energy -- water heating, space heating, and air conditioning are a very important point of focus in connection with reviewing the end uses of electricity. Research and development in completing systems is imperative to develop the most efficient overall energy utilization. One area in which our Commission and the California Energy Resources Conservation and Development Commission are greatly interested in is solar utilization. The two Commissions are conducting a joint investigation this year to study all aspects of solar utilization. Preliminary investigation by our staff indicates considerable promise for solar assisted water heating systems. All thermal uses including the off-peak electric heating and cooling storage systems should be analyzed on a comparative basis with solar systems. It would appear that many combinations of solar energy utilization supported with other energy forms and various storage systems will merit detailed examination. Improved technology will increase the economic advantage of combined solar and energy storage systems to result in much more efficient systems than are presently available.

The Commission has also directed an extensive review of waste heat utilization. A limited number of cooperative arrangements are in effect between utilities and commercial and industrial customers. These include utilization of waste heat for space conditioning, paper manufacture, refinery processes, and food production. The cement industry has sought Commission and utility support for developing a program for utilization of industrial waste heat to generate electricity. The Commission has directed Southern California Edison Company to submit proposals for joint development of such opportunities. It would appear that the substantial increase in efficiency which could result from widespread waste heat recovery application requires overcoming economic and institutional problems rather than technological difficulties. Nonetheless, improved technology will further facilitate and greatly improve the efficient use of energy in this sector.

UTILITY INCENTIVES FOR CONSERVATION

In its major formal energy decisions between 1975 and 1977 the Commission has established clear and convincing policies setting forth Energy Conservation as a number one priority for California utilities. The utilities were directed to review their priorities to be sure that conservation ranks at least equally with supply as their primary commitment and responsibility. The Commission indicated that it would consider the effectiveness of a utility's conservation efforts in determining the utility's allowed fair rate of return.

The Commission has recognized the reduction in anticipated revenues resulting from conservation in determining future test year revenue requirements.6/ It has also established an offset rate procedure providing for rapid recovery of the utilities' expanding costs in conservation programs.7/

CUSTOMER INCENTIVES FOR CONSERVATION

The incentives provided by the regulatory agency in developing a conservation oriented rate structure are important to utility customers. Through vastly increased retrofit home insulation and other conservation efforts California gas customers are saving substantial quantities of gas. This conserved gas is equivalent to a new supply which is being made available to new customers and reducing the curtailment of existing commercial and industrial customers.

As part of its program of developing additional gas supplies through conservation, California is currently examining a major expansion of the utility's ceiling insulation retrofit programs.8/ Proposals are being considered for utility cash and low interest financing incentives to customers who insulate. Tariff conditions requiring insulation or permitting utility charges for insulation are also being considered. Ceiling insulation and water heater insulation are clearly cost effective when compared to the cost of new gas supplies. Since most utility rates to the customer are at less than marginal cost, it is reasonable to permit the utility to include the payment of incentives to customers in the costs upon which its rates are established. It is the policy in California to

6/ Decision No. 86794, dated December 21, 1976, in Application No. 54946, Southern California Edison Company.

7/ Decision No. 86940, dated February 8, 1977, in Application No. 56845, Pacific Gas and Electric Company.

8/ Case No. 10032, dated January 13, 1977.

require approval by the Commission in advance of payments under any major incentive programs carried out by the utilities.

In summary, we recommend that all regulatory agencies set energy conservation as a high priority activity to be undertaken by the utilities they regulate. Since this country will never again have abundant, cheap, clean energy for all possible uses, we must conserve what we have. All regulatory agencies are urged to follow the leadership established by the California Public Utilities Commission in its Energy Conservation and Energy Management activities and its rate setting policies which serve to further advance these conservation activities.

The energy, use, conservation and supply equation can simply be stated as follows:

Use + Conservation = Supply

As is evident, use plus conservation on one side of the equation equals supply on the other. Thus conservation activities can stretch existing supplies to buy time to solve the very difficult problems of obtaining new supplies.

THE ROLE OF GOVERNMENT IN CAUSING ENERGY END USE EFFICIENCY - AN OVERVIEW

Laurence H. Martin

Martin & Martin
Modesto, California

INTRODUCTION

Objective - The purpose of this paper is to provide an insight to the manner in which government does, or in the future may, participate in global effort to cause energy use efficiency.

Scope - Every effort is made to focus this paper on the subject of end use efficiency. Although heavy emphasis is placed on U.S. programs and policies in most of the sections, the basic strategies identified as well as the problems in developing and implementing such strategies parallel the situation which is found in most other energy consuming nations of the world.

SUBJECT MATTER OF THE PAPER - AN OVERVIEW

Set forth below are the subjects addressed by the paper.

I. Background/Status of Energy Policy in U.S.

II. Review of Key Agencies and Congressional Committees at the Federal Level - U.S.

III. Analysis of Pertinent Regulations and Governmental Policies - U.S.

- A. Methods Used by Government to Cause Energy End Use Efficiency-An Overview
- B. Direct Regulatory Action
  1. Emergency Measures and Federal Management Program
  2. Voluntary Programs Leading to Direct Regulation
  3. Regulation through Gradual Elimination of Inefficient Products or Structures
- C. Use of Financial Pressure
  1. Overview
  2. Overhaul of Tariff Structure
  3. Use of Taxing System
  4. Loans, Direct Financial Aid, or Discontinuance or Refusal Thereof
- D. Educational Programs
  1. Big Three Program
  2. Small Business Conservation Program
  3. Public Schools Energy Conservation Service Programs (PSECS)
- E. Use of Power as a Major Customer
- F. Grant of Special Exceptions to or Exemptions from Existing Laws or Governmental Policies
- G. Establishing Priorities for Research and Development
- H. Summary of Strategy Used by Government to Cause Energy End Use Efficiency/Application in the Future

IV. Need to Eliminate or Modify Existing Laws and Governmental Policies

V. International Energy Management Programs and Policies

- A. General Approach
- B. Programs within Existing Structures
- C. Creation of New International Agencies for Government Interaction
  1. Creation of New Agency/Background
  2. Establishment of International Guidelines for the Development of Energy Use Efficiency Causing Policies and Regulation/Application
  3. Strategy Used to Cause More Efficient Use of Energy - An Overview

D. Use of New International Agencies to Involve the Private Sector

VI. Conclusion - Testing the Merits of Energy Management Stragegies

Section I - Background/Status of Energy Policy in U.S.

Government involvement in the nation's energy system has developed gradually over many years. This evolution has occurred through legislative action resulting in a multitude of acts, codes and ordinances, through administrative actions such as Presidential Executive Orders and Proclamations and judicially through creative interpretations of existing law. Current laws and bodies implementing these laws have their roots in wide and divergent areas of governmental responsibility, which may be summarized as follows:

- Public Land Management from which has come the Mineral Leasing Act of 1920 and such agencies as the Bureau of Land Management;

- Water Policy from which has come the Rivers and Harbors Act of 1899, Reclamation Act of 1902, and Federal Water Power Act and such agencies as the Bureau of Reclamation; Federal Power Commission; and the Bonneville, Alaska, Southeastern and Southwestern Power Administrations;

- Economic Regulation and Redevelopment from which has come the Utility Holding Company Act, Hepburn Act of 1906, and the National Gas Act and such agencies as the Rural Electrification Administration and which greatly expanded the duties of the Federal Power Commission;

- Tax Policies such as the depletion allowance once applicable to taxes on income from oil and gas wells;

- National Security from which has come come the Defense Act of 1916, Atomic Energy Act of 1946, Trade Agreement Extension Act of 1955 now amended to the Trade Expansion Act of 1962, and such agencies as the Tennessee Valley Authority and Atomic Energy Commission;

- Health, Safety, and Environmental Management from which has come the Federal Coal Mine Health and Safety Act of 1969, National Environmental Policy Act of 1970, and, such agencies as Environmental Protection Agency and Council on Environmental Quality; and,

- Research and Development from which has come the National Science Foundation Act of 1950 and much of the recent legislation and such agencies as the Bureau of Mines, Office of Coal Research, and recently the Energy Research and Development Administration.[1]

Regrettably, the evolutionary process leading to regulations and regulatory bodies affecting the energy system was not the result of any conscious "energy policy" or objective, but rather was a by-product of governmental action designed to achieve other and mainly unrelated objectives.[2] This long term, uncoordinated, piecemeal development of energy-related rules and regulations is the unstable foundation on which the U.S. commenced to build a true national energy policy. Much of the recent legislation proposed or passed has been devoted, in substantial part, to undoing the historical structure and building an organized and coordinated basis on which to proceed in the future.[3]

In 1971, the U.S. government's involvement in the energy field rapidly accelerated. On June 4, 1971, and for the first time publicly, the President of the United States acknowledged that the U.S. had an energy problem and suggested a program to help alleviate it. This initial program consisted of the following goals:

- to facilitate research and development for "clean" energy;

- to make available the energy resources on federal lands;

- to assure a timely supply of nuclear fuels;

- to use energy more wisely, with particular reference to a new Federal Housing Administration standard requiring additional insulation in new federally-insured homes;

- to develop and publish additional information on how consumers can use energy more efficiently and to pursue other areas of energy conservation;

- to balance environmental and energy needs including a system of long-range open planning of electric power plant sites and transmission line routes with approval by state or regional agencies before construction; and,

- to organize federal efforts more effectively, particularly by creating a single structure and uniting all important energy resource development programs.[4]

The concepts advanced by the President in 1971 remain virtually unchanged today and until recently, little has been done to solve the fundamental problem of disorganization and duplication of action.[5] By the summer of 1973, there were approximately 46 agencies of the Federal Government that administered programs or effected policies that had specific impacts on the energy systems. There were an additional 18 agencies that had jurisdiction over programs that were not intended to be energy-oriented but which nevertheless had an impact on the energy system. These agencies were proliferated among nine executive departments and 15 independent agencies in the Executive Office of the President. At the state and local levels, matters were no different.

In an effort to eliminate the chaos at the federal level, largely motivated by the sudden and serious fuel shortages in mid-1973, the administration created an Energy Policy Office (EPO) to coordinate the energy related activities of various agencies having an impact on fuel production and consumption. EPO also had jurisdiction over the voluntary oil allocation program and later the voluntary allocation program for middle distillates. In the fall of 1973, in response to a need created by the Administration's decision to impose mandatory allocations on more petroleum products, the Interior Department formed an Office of Petroleum Allocation (OPA) to take charge of the program.[7]

In December 1973, a Federal Energy Office (FEO) was created to take over the responsibilities of both EPO and OPA which were phased out as separate entities. FEO also assumed new powers (mandated by the Emergency Petroleum Allocation Act) vested in the Executive Branch for the mandatory allocation of crude oil and all refined petroleum products and the petroleum price control authority formerly wielded by the Cost of Living Council. FEO, however, was considered to be an embryonic agency that would remain in operation only until statutory authority was provided for a more encompassing Federal Energy Administration (FEA). On June 27, 1974, FEO was actually abolished and FEA became the Federal Government's line agency for petroleum allocation and pricing, energy date collection and analysis, fuel conservation (shared with the Department of Commerce), and energy independency planning.[8]

On October 11, 1974, the President signed a bill (effective 120 days from that date) that abolished the Atomic Energy Commission and created an Energy Research and Development Administration (ERDA) to handle the vast majority of federal research and development projects and a Nuclear Regulatory Commission (NRC) to regulate the nuclear power industry. The FEA and ERDA are sister organizations, both technically reporting directly to the President, but theoretically under the policy control of the Energy Resources Council (ERC) established October 11, 1974. ERC, which was established to insure communication and coordination among federal agencies in energy matters and to set and implement national energy policy, is charged with overall control of the national energy effort.[9]

On the federal level, therefore, the development of a foundation for U.S. energy policy began with the creation of the FEA, ERDA, and the Energy Resources Council. In practice, however, these agencies have not yet caused any meaningful impact onthe problems discussed above and, at present, Congress has extended the FEA life only to December 31, 1977.[10]

On March 1, 1977, the President of the U.S. sent to Congress proposed legislation to reorganize the Federal Government's energy agencies and programs. This bill would establish a new Cabinet-level Department of Energy by combining the functions of the three major Federal energy agencies along with energy-related functions of six other executive and independent regulatory agencies. This new Department would provide the organizational base and the programmatic authorities needed to develop and implement overall Federal energy policies. Among the major responsibilities of the Department will be:

- conservation;
- regulation;
- research and development;
- resource and development and production; and,
- data management.

In submitting the proposed legislation, the President made the following pertinent comments regarding the state of U.S. energy policy:

- nowhere is the need for reorganization and consolidation greater than in energy policy;

- all but two of the Executive Branch's Cabinet departments now have some responsibility for energy policy, but no agency, anywhere in the Federal government, has the broad authority needed to deal with our energy problems in a comprehensive way; and,

- the legislation will bring immediate order to this fragmented system.

The changes called for by the proposed legislation may be summarized as follows:

- abolishes the Federal Energy Administration (FEA), Energy Research and Development Administration (ERDA), and the Federal Power Commission (FPC);

- combines conservation programs which are now split between FEA and ERDA and creates an Assistant Secretary for Conservation, who will be personally responsible for seeing that the conservation program is carried out;

- places under one roof the powers to regulate fuels and fuel distribution systems, powers which are now shared by the FEA and FPC along with the Securities and Exchange Commission and the Interstate Commerce Commission;

- transfers to the new department several energy related authorities and programs, such as the:

  - building thermal efficiency standards from Housing and Urban Development; and,

  - voluntary industrial compliance program from Commerce;

- provides for consultation between the Energy Department and the Department of Transportation on auto fuel efficiency standards;

- establishes a role for the Energy Secretary in the REA loan program at Agriculture;

- abolishes the Energy Resource Council (ERC);

- transfers certain parts of the Interior Department--those concerning fuel data collection and analysis and coal mine research and development to the new Department; and,

- will bring together all energy data gathering and analysis capabiliities currently done by more than 20 executive departments and agencies who collectively operate more than 250 energy data programs (100 of these are FEA, ERDA, FPC, and Interior Department programs).[11]

Many states have initiated similar steps to develop and implement a coordinated energy policy and some, such as New York, California, and Wisconsin, are considerably ahead of the federal government in terms of actual progress.[12] An excellent example of such state level legislation is California'a Warren-Alquist State Energy Resources Conservation and Development Act. This progressive act - which became effective January 7, 1975 - addresses itself specifically to the subject of electrical energy and, in that regard, consolidates for action all major areas of present concern, i.e., emergency measures, short-term conservation programs, and long-term research goals.

Section II - Review of Key Agencies and Congressional Committees at the Federal Level - U.S.

The current U.S. administration (President Carter) has proposed a significant reorganization of the existing system for implementing U.S. Energy policy. However, until such a proposal is accepted by Congress, the existing system stands and hence deserves careful evaluation.

At present, most federal and many state and local agencies have already devised and implemented energy efficiency plans and have additional proposals in the formative stages.[13] However, these programs, whether in existence or in the development stage, are generally restricted to the agency's or governmental entity's specific area of responsibility.[14] For example, the Federal Housing Administration (HUD), General Services Administration (GSA), and the Veterans Administration (VA) just to name a few, are quite active in the area of improving the efficiency of new and existing commercial and residential structures insofar as energy use is concerned. As another example, the

Federal Power Commission (FPC), Tennessee Valley Authority (TVA) and other agencies are active in the area of rate design to achieve more efficient energy use.(15) Therefore, it is important to determine which agencies have responsibility over a specific area of interest since regulations might progress faster and in a more comprehensive fashion in one area than in another. Table 1 lists the great majority of Federal agencies that are now directly involved in, or at least have a substantial impact on, the energy system.(16)

It does appear, however, that certain Federal agencies and Congressional committees at the Federal level of government have played a dominant and perhaps controlling role in the attempt to structure a true national energy policy.(17) It is important to be aware of the make up of these agencies and committees and their basic functions since they furnish a wealth of diverse up-to-date information. These agencies and Congressional committees are described below.(18)

Federal Agencies

- Energy Resources Council (ERC)-- This group was established to insure communication and coordination among federal agencies involved in energy matters and to set up and implement national energy policy. The Energy Resources Council is a Cabinet-level body charged with the overall control of energy efforts. The primary agencies reporting to or dealing with this Energy Resources Council are the Federal Energy Administration and the Energy Research and Development Administration.

- Federal Energy Administration (FEA) Like all its predecessors, the FEA is expected to be a transitional agency, it being authorized in existence only until December 31, 1977. As presently construed, the FEA has jurisdiction in the following areas: fuel allocation and petroleum pricing regulations, energy data collection and analysis, energy planning for making the nation less dependent on foreign sources of oil and energy, energy conservation and strategic petroleum reserves. In addition, FEA has the authority to gather information from persons or firms engaged in any phase of energy supply or major energy consumption and information to the extent that if it is not confidential, it must be made public. The upper eschelon of the FEA is comprised of an administrator, two deputy administrators, a general council, and six assistant administrators, each for the following categories:

  - strategic petroleum reserves;
  - energy information and analysis;
  - resource development;
  - operation, regulation, and compliances; and,
  - international affairs.

  Of particular interest here is the department addressing itself to conservation and environment. Presently, this department has the general mission to reduce the rate of energy demand growth, to implement energy conservation programs and to promote the efficient use of energy resources. Specific functions include:

  - developing and effecting new governmental programs to reduce energy demand in the transportation, building, industrial, and utility sectors;
  - providing leadership for end use of energy conservation and demonstration activities throughout the Federal Government;
  - reducing federal energy consumption through the Federal Energy Management Program;
  - developing and conducting public education and multimedia campaigns in support of energy conservation programs;
  - providing technical and program support on energy conservation to states and localities; and,
  - analyzing and responding to environmental issues affecting proposed energy programs or goals and carrying out FEA's responsibilities under the National Environmental Policy Act.

- Energy Research and Development Administration (ERDA)

  This agency is considered the backbone of the Federal Government's effort to develop nuclear, solar, geothermal, and other alternatives to an increased reliance on foreign oil. ERDA's functions include:

- exercising central responsibility for policy planning, coordination, support, and management of research and development with respect to all energy sources and utilization technologies;
- encouraging and conducting research, development, and demonstration for extraction, conversion, storage, transmission, and utilization phases;
- engaging in supporting environmental, biomedical, physical, and safety research;
- participating in and supporting cooperative research and development projects; and,
- developing, collecting, and distributing scientific energy information.

ERDA has assumed the following areas of responsibility from other agencies:

- from the Interior Department, ERDA has absorbed the Office of Coal Research which was charged with developing new and more efficient methods of mining, preparing, and utilizing coal as a clean fuel;
- from the National Science Foundation, ERDA took over programs formerly conducted by that agency for solar heating and cooling development and for geothermal power production;
- from the Department of the Interior's Bureau of Mines, ERDA assumed the responsibility for fossil fuel research; and,
- from the Environmental Protection Agency, ERDA absorbed the responsibility for the development of alternative power systems except those applicable to assessment or monitoring for regulatory purposes.

ERDA is comprised of an administrator, a deputy administrator and six assistant administrators. For each of the six assistants, the functional jurisdiction is:

- fossil fuels;
- nuclear energy;
- environment and safety;
- conservation;
- solar, geothermal, and advanced energy systems; and,
- national security.

With particular reference to the fuel conservation program, ERDA is concerned with and has the responsibility for conservation research and development programs including automotive power systems, end-use consumption technologies, and improving energy efficiency.

- Department of Commerce (DOC)--DOC involvement in the field of energy is geared mainly towards conservation and efficiency by industry and business. Under the auspices of the Department, the Office of Energy Programs has an energy planning division that is responsible for identifying and stimulating industry's voluntary implementation of energy conservation practices. A second division of the office, the energy resources divison, handles oil policy committee responsibilities of the Department as well as a variety of other functions pertaining to energy policy information. Specialists provide data projections and analysis of the implications, foreign and domestic, on energy policy for the U.S. economy as a whole and commercial enterprises in particular.
- Environmental Protection Agency (EPA)--Since energy-producing facilities and activities normally produce air emissions, water effluents, and other production residuals, nearly all phases of the power generation are subject to regulations under the pollution control programs administered by the EPA. In its air pollution control program, EPA establishes national ambient air quality standards and the states enforce limits on emission to achieve these standards. If a state fails to set such limits or if the state limits are considered inadequate, the EPA is authorized to set the required emission limits for the state. In addition, EPA establishes uni-

form national air pollution control standards for new plants and factories, emission limitations for hazardous air pollutants, and motor vehicle standards. EPA administers a similar national program on water pollution control.

- National Science Foundation (NSF)
The Office of Energy Research and Development Policy is part of the National Science Foundation. The office, established August 10, 1073, is charged with the following functions within the National Science Foundation:
    - providing analysis of specific issues and selective programs related to energy research and development including energy supply technology, energy demand and conservation, and energy-related areas on environmental and sociologic research;
    - developing a general systems framework for the evaluation of energy research and development programs and developing appropriate criteria for assessing the merits of individual technological approaches;
    - identifying and recommending to the Executive Office of the President critical needs in energy research and development;
    - identifying and evaluating significant research findings that could affect energy programs or policies;
    - providing an individual assessment of environmental health and safety standards and identifying additional research to improve standards;
    - maintaining an awareness of current plans and viewpoints of industry and associations in energy matters; and,
    - determining ways in which universities and other research organizations can make the most effective contributions to better energy development.
- Treasury Department--An Office of Natural Resources and Energy has been established within the Treasury Department to monitor energy resources, supply and demand as well as the implications of the United States policies with respect to selected natural resources. The work of this office includes contact with other government agencies concerned with resources and energy on one hand and on the other with the various parts of the Treasury concerned with tax policy, trade policy, economic policy, and international and monetary affairs. Recommendations from the office will be made directly to the deputy secretary regarding the most appropriate policies to produce a favorable U.S. supply-demand balance for energy and other selected resources including coal, oil, natural gas, copper, and other minerals of importance to the balance-of-payments. With respect to tax policies, the Office will advise on such matters as depletion allowances, accelerated amortization, investment credits, and production incentives. Tariffs, quotas, financial and investment policies, price controls, loans and guarantees as they relate to energy and natural resources will also be a part of the on-going studeis and analysis of this new office.

Congressional Committees of Significance (19)

Senate Committees

- Energy and Natural Resources--Jurisdiction includes all proposals relating to energy policy, regulation, conservation and research, including those affecting solar power, non-military nuclear power, the Alaskan Naval Petroleum Reserve, oil and gas production and distribution, mineral extraction from the oceans, energy-related aspects of deepwater ports, hydroelectric power generation, coal production and utilization, mining lands and mineral conservation, and mining education and research. The Committee also has jurisdiction over bills relating to public lands, natural forests, and the territorial possessions of the United States.

- Interior and Insular Affairs--The jurisdiction of this committee includes lands (generally), mineral resources, petroleum and radium conservation.

- Armed Services--Jurisdiction in energy areas includes conservation, development and use of naval petroleum and oil shale reserves, and strategic and critical materials necessary for common defense.

- Finance--Jurisdiction generally includes revenue measures.

- Commerce, Science and Transportation--Jurisdiction includes interstate and foreign commerce generally and specifically covers the regulation of oil and gas pipelines, navigational aspects of deepwater ports, science and engineering research and development, nonmilitary aeronautical sciences, and the transportation aspects of matters relating to the Outer Continental Shelf.

House Committees

- Interstate and Foreign Commerce--Jurisdiction includes inland waterways, oil compacts, petroleum and natural gas, except on public land, and regulation of transmission of power (except installation of connections between government water projects).

- Science and Technology--Jurisdiction covers energy research and development including activities of the Energy Research and Development Administration.

- Interior and Insular Affairs--Jurisdiction includes mineral lands (generally), mineral resources, petroleum and radium conservation.

- Government Operations--Jurisdiction includes conservation, development, and use of the naval petroleum reserve and oil shale reserve.

- Small Businesses--Jurisdiction includes matters dealing with small businesses.

- Outer Continental Shelf--Jurisdiction covers all matters concerning the Outer Continental Shelf, including oil and gas leasing.

Alphabetical Listing of Agencies Affecting Energy Use

Agriculture Department

Alaska Power Administration

Bonneville Power Administration

Commerce Department and particularly the National Bureau of Standards (DOC and NBS)

Defense Department

Department of Housing and Urban Development (HUD)

Energy Research and Development Administration (ERDA)

Energy Resources Council (ERC)

Environmental Protection Agency (EPA)

Executive Office of the President (EOP)

Federal Energy Administration (FEA)

Federal Housing Administration (FHA)

Federal Maritime Commission (FMC)

Federal Power Commission (FPC)

Federal Trade Commission (FTC)

General Accounting Office

General Services Administration (GSA)

Interior Department--Geological Survey

Interior Department--Bureau of Land Management

Interior Department--Bureau of Reclamation

Interstate Commerce Commission (ICC)

Justice Department

Labor Department

National Aeronautics and Space Administration (NASA)

National Science Foundation (NSF)

Nuclear Regulatory Commission (NRC)

Oil Policy Committee (OPC)

Securities and Exchange Commission

Small Business Administration (SBA)

Solar Energy Coordination and Management Project

Southeastern Power Administration

Southwestern Power Administration

State Department

Tennessee Valley Authority (TVA)

Transportation Department (DOT)

Treasury Department

Water Resources Council

Section III - Analysis of Pertinent Regulations and Governmental Policies - U.S.

A. Methods Used by Government to Cause Energy End Use Efficiency - An Overview

Historically, government (at all levels) has promoted or attempted to initiate a desired type of citizen behavior by use of the following methods:

- direct regulation;
- use of financial pressures designed to cause action or inaction;
- educational programs;
- use of power in the marketplace as a major customer;
- selective granting of exceptions to or exemptions from existing law or procedures; and,
- establishment of major research and development programs with well defined priorities.[20]

The activity by government to cause energy end use efficiency on the part of citizens closely follows the above-described pattern and, as a result, can be effectively analyzed in that context.

B. Direct Regulatory Action

Direct regulation of end users to cause more efficient use of energy is a method which has been slow in developing. However, the following distinct patterns of action can be identified:

- maintenance and further sophisitication of "emergency regulations";
- sponsoring voluntary conservation programs involving the private sector which gradually become mandatory in character and directive in nature; and,
- a gradual process of regulating out of existence structures and products of all kinds which do not meet government-imposed efficiency standards.

The so-called "emergency" or "standby" measures and certain aspects of the current Federal Management Program (FEMP) (applicable to Federal employees and facilities only) constitute clear examples of direct regulation of end user behavior, i.e., take out light bulbs, turn down or up thermostat, etc. On the other hand, controlling the end user's buying habits by forcing efficient products, structures, etc., upon them through a program designed to gradually eliminate all inefficient products or structures is no less of a direct regulation of behavior even through it is more subtle in application. Examples follow.

1. Emergency Measures and Federal Energy Management Program

The Federal Government and most, if not all, states and local governmental entities have "emergency" type laws which contain stringent and detailed regulations with reference to how, when, and in what manner end users of energy may or must use it.[21] These laws may now be dormant, but they are nevertheless "on the books" and can be activated swiftly when events dictate the need to do so. Furthermore, studies are continuing in most states regarding how to develop better programs to meet emergency needs.[22]

In the event other more gradual fuel conservation programs and legislation fail to control the existing energy problem, it is quite likely these "emergency measures" will be activated and used on a long-term if not permanent basis.[23] An excellent example of such strict directive regulation is the Los Angeles City Emergency Energy Curtailment Plan. This plan is one of the broadest and most restrictive in the nation and, in fact, has been used as a model by the Federal Government, the Public Utilities Commission in California, and such states as Washington and New York in connection with the development of their own programs.[24]

Federal agencies and many state and local governmental agencies have programs designed to cause their employees to make more efficient use of energy. These programs are important to note since they too are a good indicator of the course of future regulation of the general public. An excellent example of this type of activity is the Federal Energy Management Program which involves sixteen of the most energy-intensive departments and agencies. These 11 cabinet departments and five large agencies make quarterly reports to the FEA, identifying energy savings in two major categories--(1) buildings and facilities, and (2) in the use of fuels in motor vehicles, aircraft, ships and equipment. Energy conservation strategies of the program fall into three general areas, to wit: building operations, transportation, and Federal employee activities. Details of these areas are set forth below:

- Building Operations. This includes:
  - reducing illumination levels by delamping and practicing regular cleaning and replacement of lighting fixtures;
  - maintaining temperature settings during the heating season at a maximum of 65 to 68°F and improving maintenance of cooling

equipment;

- reducing the operating time of machines; and,
- changing from nighttime to daytime cleaning.

- Transportation. This includes:
  - cutting overall travel budget and specifically reducing funds available for air travel, thus causing shift to less energy-intensive modes (train or bus);
  - using the telephone or other telecommunications in lieu of travel whenever feasible;
  - encouraging official travel in off-peak periods by common carrier whenever possible;
  - reducing mileage of federal vehicles (now up to 15 percent);
  - buying and leasing smaller cars; replacing large sedans and limousines;
  - substituting bicycles for automotive vehicles for use at federal facilities;
  - requiring tuneups on all government vehicles every 12,000 miles (or once a year) for a maximum operating efficiency;
  - establishing a maximum of 50 mph for motor vehicles (now 55 mph), and reducing cruising speeds for aircraft and ships;
  - substituting simulators for many flight activities and excusing certain flight personnel from proficiency flying; and,
  - reducing ship steaming time.

- Federal Employee Activities. This includes:
  - encouraging carpooling and use of public transit;
  - encouraging employees to ride bicycles to work and providing bicycle racks at federal facilities;
  - conducting employee awareness programs through printed and other in-house media to encourage energy conservation practices in the office, at home, and in their communities; and,
  - establishing employee suggestion systems and awards programs to induce greater involvement in energy conservation.[25]

2. Voluntary Program Leading to Direct Regulation

In addition to the above-described existing laws and governmental programs now providing some degree of direct regulation, there exist several "voluntary" programs sponsored by the Federal Government and some state governments which likely will lead to direct and very comprehensive regulations of end-user behavoir. Examples of such programs together with an explanation of their purpose and function follow.

- Voluntary Industrial Energy Conservation Program [26]

  This program began in early 1973 and is co-sponsored by the Federal Energy Adminstration (FEA) and the Department of Commerce (DOC). The program consists of two parts:

  - an Energy Management Program which is designed to benefit industry as a whole and individual companies and plants within that industry; and,
  - a reporting system which furnishes data to the Federal Government to enable government policy makers to develop rational energy policies and programs. One of the major aspects of the program is the monitoring and periodic reporting of energy conservation progress.

- State Energy Conservation Program[27]

  Recognizing the potentially important role of States in stimulating energy conservation, the Federal Energy Administration (FEA) in early 1975 began developing a program to provide a framework for State/Federal cooperation in the furtherance of national energy conservation goals. During these early stages, FEA worked closely with the National Governor's Conference and, in the spring of 1975, FEA and representatives of five States worked together in drafting the initial program plan. On June 11, 1975, the National Governors' Conference meeting in New Orleans endorsed this plan for a State/Federal energy conservation program. Initiation of the program was then announced simultaneously by Governor Salmon of Vermont and Chairman of the National Governor's Conference Committee on Natural Resources and Environmental Management and Frank G. Zarb, Administrator of FEA. The voluntary State/Federal program, as was then envisioned, relied heavily on the participation of the States in a data classification system to be used as a base for energy forecasts, on a State-by-State basis, to the year 1985. For the purposes of this program, a State is defined as any one of the 50 States, the District of Columbia, Puerto Rico, Guam, American Samoa, the Virgin Islands, and the Trust Territory of the Pacific. After the data system was in place, the States could establish energy conservation goals and develop specific programs to meet those goals. FEA would provide technical assistance to help the States in their efforts.

On December 22, 1975, then President Ford signed into law the Energy Policy and Conservation Act (EPCA). Title III, Part C of the Act authorized FEA to establish guidelines for the development and implementation of State energy conservation plans. The program is in many respects similar to the voluntary program originally initiated by FEA. It calls for the establishment of energy conservation goals for the reduction by 5 percent of actual State Energy consumption in 1980 from projected consumption. The program is voluntary; States are not penalized if they elect not to participate.

There are, however, some important differences between the original program and the program established by EPCA. First, Section 365(d) authorizes $50 million for each of fiscal years 1976, 1977, and 1978. States can only receive these funds if they commit themselves to the objectives of the program. Second, the provisions of EPCA require that every State plan include at a minimum of five specific energy conservation actions, which are:

- mandatory lighting efficiency standards for non-Federal public buildings;
- programs to promote the availability and use of carpools, vanpools, and public transportation;
- mandatory standards and policies relating to energy efficiency to govern the procurement practices of a State and its political subdivisions:
- mandatory thermal efficiency standards and insulation requirements for new and renovated non-Federal Buildings; and,
- traffic laws or regulations which to the maximum extent practicable consistent with safety, permit motor vehicles to turn right at a red light after stopping.

The Act also lists the following additional actions that the States may (not mandatory), include in their plans for energy efficiency:

- restrictions governing the hours and conditions of operation of public buildings;
- restrictions of the use of decorative or non-essential lighting;
- transportation control;
- programs of public education to promote energy conservation; and,
- any other appropriate methods or programs to conserve and improve efficiency in the use of energy.

The FEA has recently proposed that three more mandatory energy conservation actions be added to the five now required, which are:

- inter-governmental coordination in the area of energy use efficiency;
- a Public Education and Awareness program; and,
- an energy audit program for building and industrial plants.

- <u>State Level Programs (California)</u> (28)

  Both the California Public Utilities Commission (PUC) and the State's Energy Resources, Conservation and Development Commission (ERCD) have programs aimed at encouraging utilities to institute and maintain aggressive programs designed to cause energy use efficiency at the end user level. Although participation in the programs is to a large degree voluntary in nature, reporting requirements regarding progress are mandatory and the degree of cooperation and results achieved receive heavy weight at agency hearings regarding applications for the siting of new power plant facilities or rate increases.

The above-described voluntary programs and others like them will likely lead to comprehensive regulations requiring and directing that:

- energy efficiency "report cards" be submitted to the FEA or like governmental agency on a regular basis;
- a prescribed level of energy use efficiency be maintained; and,
- a precise and detailed procedure be adhered to in maintaining the efficiency standards.

Already this evolutionary process can be seen in the industrial sector in the form of mandatory reporting requirements imposed by the Energy Policy and Conservation Act combined with stringent, albeit still voluntary, efficiency targets which are to be developed by the Federal Energy Administration. Furthermore, once the state's energy conservation programs are finalized (22 now under submission to the FEA for approval) they will cause much direct regulation of end users at the state level.

3. Regulation through the Gradual Elimination of Inefficient Products or Structures

Government action (all levels) in this area presently emphasizes the development of:

- efficiency-descriptive labels on all things consuming energy;
- efficiency-causing building codes for new and existing structures; and,
- regulations requiring motor vehicles to meet specified fuel consumption standards.[29]

At the moment, the Federal government's energy efficiency labeling program is in a state of flux. Initially, the Environmental Protection Agency developed mileage labels for 1974 model automobiles and encouraged car makers to affix these labels to their products along with the price stickers. Later the Federal Energy Administration joined the EPA in this endeavor, and the two agencies administered the voluntary program through mid-March 1976. On March 21, 1976, however, the automobile labeling program became mandatory under the terms of the Energy Policy and Conservation Act, and the EPA recodified its voluntary regulations and made them applicable to all cars manufactured in or imported into the U.S. (the voluntary program had had a 95 percent compliance rate, and that part of the industry that participated in the voluntary program was basically unaffected by the mandatory rules).[30]

As for the labels for other items, the Commerce Department currently administers a voluntary program for household appliances, a program that will be taken over and revamped by the Federal Energy Administration under another mandate contained in the Energy Policy and Conservation Act. FEA has been directed to establish test procedures, labeling rules, and energy efficiency standards for the following products: refrigerators and refrigerator-freezers, freezers, dishwashers, clothes dryers, water heaters, room air conditioners, home heating equipment, not including furnaces, television sets, kitchen ranges and ovens, clothes washers, humidifiers and dehumidifiers, central air conditioners, furnaces, and any other product that is determined to be likely to use more than 100 kW/hrs (or Btu equivalent) a year.[31]

While FEA goes about setting up its program, the Commerce Department's voluntary labeling regulations and specifications remain in effect. Commerce, prior to enactment of the energy policy law, had labels in effect for general guidance, air conditioners, refrigerators, refrigerator-freezers, and freezers, and had proposed guidelines for labeling water heaters and clothes dryers.

In addition to the Commerce Department's voluntary program, and prior to the new FEA labeling authority, the Federal Trade Commission proposed a set of mandatory labeling specifications for air conditioners. That proposal was later withdrawn by FTC.[32]

Eventually, FEA plans to have a comprehensive double-edged program which will include:

- the labeling of virtually everything manufactured which will be aimed to better inform the consumer; and,
- efficiency standards aimed at industry and requiring that products made comply with such standards.

In the case of automobiles, this approach has already been implemented. With reference to other products, a recent statement of current administration policy indicates that:

- the present appliance efficiency program will be strengthened;
- voluntary targets will be replaced by mandatory standards on certain

home appliances, such as air conditioners, furnaces, water heaters, and refrigerators as soon as possible; and,

- the program under existing law to develop test procedures and to establish labeling requirements for appliances will be continued.(33)

Many states and municipalities are developing or already have voluntary labeling programs of their own. In some states, mandatory efficiency standards are either a matter of law or under serious consideration. For example, in California, minimum efficiency standards are now set by law for refrigerators, refrigerator-freezers, and air conditioners and soon the law will be expanded to cover water heaters, televisions, clothes dryers, washing machines, cooking appliances, dishwashers, plumbing fixtures, gas appliances and space heaters. Furthermore, California law will ultimately prohibit continuous burning pilot lights in favor of intermittent ignition devices.(34) It appears, however, that the Federal programs will be expanded and dominate the labeling and efficiency standards areas so that a consistency of action and legislation can be insured.(35)

Much work has been expended at the Federal level to develop a basic national building code, but such a result has not yet been achieved. Generally, government at all levels has been studying both prescriptive and performance standards aimed at causing new residential and commercial structures to use energy more efficiently. The inclusion of such prescriptive or performance standards into design is being pursued through one of the three following approaches.

- by promoting their acceptance by professional organizations as standard operating procedures;
- by causing their inclusion in state and local building codes; or,
- by indirectly forcing their use through some conditioned mechanism in the institutional processes of the building industry.

In August of 1976, the Energy Conservation Production Act (ECPA) was passed. By this act the Department of Housing and Urban Development (HUD) has been charged with the responsibility of formulating national building standards by 1981. The standards developed by HUD must be performance rather than prescriptive in nature. In the meantime, the FEA is making every attempt to cause states to adopt building standards consistent with those propounded by the American Society of Heating, Refrigerating and Air Conditioning Engineers in their work, commonly known as ASHRAE-75. The ASHRAE standards are prescriptive rather than performance-oriented. Aiding the FEA in their effort to cause acceptance by the states of ASHRAE-75 standards, is a provision of the Energy Policy and Conservation Act (EPCA) to the effect that a state's participation in the State Energy Conservation program is conditioned on it adopting into law efficiency standards at least as restrictive as those contained in ASHRAE-75.(36)

The inclusion of prescriptive standards, such as the Department of Housing and Urban Development (HUD) "51 (B)" changes in the "Minimum Property Standards," in conditions for residential financing, is an example of the use of conditioned mechanisms in institutional processes.

At the state level, the development of codes designed to cause the building of energy efficient structures has progressed faster than at the Federal level. As early as March of 1976, the following state progress had been achieved:

- Building Energy Regulatory Authority:
  - In statewide code - 17 states;
  - In separate law - 4 states;
- Regulations:
  - Promogulated - 10 states;
  - Under consideration - 14 states;
- Authority being considered:
  - By legislature - 5 states; and,
  - By staff or study groups - 3 states.(37)

Furthermore, twenty-two states have submitted programs to the FEA under the State Energy Programs.(38)

For the purposes of this discussion regarding direct regulation of end-user behavior, it is important to note the following typical provisions which exist or will exist in most of the new building codes:

- upon completion of the structure or a portion thereof, a card certifying conformance with the requirements of the regulations must be completed, signed by the builder, and approved by a representative of a responsible governmental agency;
- the card denoting approval of the structure (or part thereof) must be posted in a conspicuous location on or in the structure; and,
- until the above procedure has been complied with, the final inspection and ultimately, occupancy itself, cannot be commenced.(39)

The provisions set out above clearly demonstrate how building codes can and will effectively inhibit or prohibit action by end-users.

In addition to rapidly advancing growth of building codes in the "new" construction area, several retrofit programs are gaining momentum. For example, the Federal Energy Administration (FEA) and Department of Commerce (DOC) are aggresively pursuing the development of such a program for national use.(40) This area will most certainly lead to laws requiring that owners (and hence energy users) of structures make certain modification to existing structures so as to meet prescribed energy efficiency standards as a condition of continued use and/or transfer of legal title.

C. Use of Financial Pressure

1. Overview - Over the years, government has substantially influenced behavior through the use of direct or indirect financial pressure. Sometimes the pressure has taken the form of a reward, i.e., tax rebate, and sometimes the form of a penalty, i.e., tax surcharge. In the energy field, numerous proposals exist but little law has evolved as yet to effectively cause energy use efficiency through financial pressure.(41)

Although the precise manner in which financial pressure will be exerted is still the subject of much debate, the basic approaches by which government will exert the pressure appear clear and may be summarized as follows:

- by use of a new utility rate (tariff) design that more accurately communicates the true value of energy to the end-user;
- by use of the taxing system to encourage efficient and discourage inefficient action;
- by way of making loans or granting direct financial aid to persons and business to:
  - encourage voluntary efficiency causing expenditures; or,
  - enable compliance with regulations requiring such expenditures to be made;
- by way of refusal or the threat of refusal of loans or aid; and,
- by way of termination or the threat of termination of existing lending, aid, or related funding programs.(42)

2. Overhaul of Tariff Structure - As distinguished from direct curtailment measures which at present are designed for emergency use only, this rate-design approach is one based on "cost analysis."(43)With the possible exception of those utilities under the jurisdiction of the Federal Power Commission (FPC), the majority of all proposals in this area contemplate that the actual tariff adjustment or restructuring be accomplished locally or regionally through appropriate legislation or utility rulings rather than by some "Federal Grand Design."(44)It has been determined that due to the unique problems existing in different areas of the country any national uniform-rate structure or procedure would be totally impractical and in fact unworkable.(45)

The first clear cut expression of federal policy on this subject is found in Title II of the Energy Conservation and Production Act passed into law on August 14, 1976. The stated purpose of the Title II provisions is to require the Federal Energy Administration to:

- develop proposals for the improvement of electric utility rate design and transmit such proposals to Congress;
- fund electric utility rate demonstration projects;
- intervene or participate upon request, in the proceedings of utility regulatory commissions; and,
- provide financial assistance to State offices of consumer services to

facilitate presentation of consumer interests before such commissions. (46)

The Title II provisions go on to provide that:

- proposals shall be designed to encourage energy conservation, minimize the need for new electrical generating capacity, and minimize costs of electric energy to customers and shall include (but not be limited to) proposals which provide for the development and implementation of:
  - load management techniques which are cost effective;
  - rates which reflect marginal cost of service, or time of use of service, or both;
  - ratemaking policies which discourage inefficient use of fuel and encourage economical purchases of fuel; and,
  - rates (or other regulatory policies) which encourage electric utility system reliability and reliability of major items of electric utility equipment;
- the proposals prepared shall be transmitted to each House of Congress not later than 6 months after the date of enactment of the Act, for review and for such further action as the Congress may direct by law.(47)

A more recent expression of federal philosophy on this subject can be seen in the current administration's proposed energy policy, pertinent portions of which are set forth below.

- Conventional utility pricing policies discourage conservation. The smallest users commonly pay the highest per unit price due to practices such as declining block rates. Rates often do not reflect the costs imposed on society by the actions of utility consumers. The result is waste and inequity. The President will therefore submit legislation which contains the following provisions:
  - State Public Energy Commissions must require their regulated electric utilities to phase out and eliminate promotional, declining, and other rates for electricity that do not reflect cost incidence;
  - to shift energy use from peak to non-peak periods, electric utilities would be required to offer daily off-peak rates to each customer who is willing to pay metering costs and to offer lower rates to customers willing to have their power interrupted at times of highest peak demand;
  - master metering for electricity would generally be prohibited in new structures;
  - State Public Utility Commissions would require gas utilities to eliminate declining block rates and to implement such rules as FPC may prescribe with respect to master metering, summer-winter rate differentials and interruptible rates; and,
  - by amendment to the Federal Power Act, the Federal Power Commission would be authorized to require interconnection and power pooling between utilities even if they are not presently under FPC jurisdiction, and to require "wheeling" (the transmission of power between two non-contiguous utilities across a third utility's system.(48)

3. Use of Taxing System - At present, there is no question that this device will be a major tool in the government's effort to cause consumer efficiency in the use of energy. The area of uncertainty is over how this tool will be used. The following proposals have been under consideration, for some time:(49)

- a windfall profits (income) tax on utilities designed to absorb extraordinary profits that may be generated through substantial modification of "base rate" computations;
- a tax credit or deduction for improvements to existing homes (retrofitting) to increase thermal efficiency;
- the investment tax credit for utilities increased so as to be comparable with that granted other businesses;
- a preferred stock dividend deduction, aimed primarily at the utilities, for the purpose of reducing the cost of capital and stimulating equity rather than debt financing;
- a tax credit or deduction for expenditures in installing a new energy system, i.e., solar energy equipment,

having a fuel consumption rate that complies with government-established efficiency standards;

- a refundable tax credit for certain retrofit activities;
- an excise tax on new products purchased which fail to meet government established energy use efficiency standards;
- an excise tax on use of electricity beyond a predetermined base; and,
- a real property tax exemption relative to the installation of a new energy system, or substantial retrofit activity. For example, the state of North Dakota already has such a law directed towards the installation of machinery and equipment systems that utilize solar energy for heating and cooling in new and existing buildings. The North Dakota law allows the property owner to deduct annually (for five years following installation) from the assessed valuation of the property a sum equal to the lesser of:
    - the remainder of the assessed valuation of the property with the solar system included, <u>minus</u> the valuation of the property without the solar system; or,
    - $2,000.

Many of the above-listed tax proposals are likely to be combined in legislation so as to be more effective in achieving a desired goal. An excellent example of the use of a combination of various tax-related devices to accomplish an energy efficiency causing objective is the current administration (President Carter) energy plan submitted to Congress on April 20, 1977. Among the President's proposals were the tax-oriented devices described below.

- An exemption from Federal and State public utility regulations would be available to industrial cogenerators.
- Tax increases of 7 to 11 cents on aviation fuel and an end to a two-cent-a-gallon rebate on gasoline used in motor-boats.
- Removal of a 10 percent excise tax paid by passengers who ride on intercity bus lines.
- A 10 percent tax credit for businesses that invest in energy conservation.
- Returning increased oil and gasoline taxes to consumers through a federal tax rebate, likely starting at $15 and rising to $25. The rebate is intended to compensate low income families hit hard by sharply higher energy costs.
- A standby gasoline tax increase, starting at five cents per gallon and rising to 50 cents after 10 years if gasoline consumption fails to decline.
- A tax rising to a maximum $2,488 by 1986 on each gas-guzzling car, accompanied by a rebate of up to $493 for fuel-efficient cars.
- Tax credits of up to $410 for homeowners who insulate their houses or take other steps to make their residences more energy-efficient.
- Tax credits of up to $2,000 for homeowners who install solar heating and cooling devices.
- Imposing a new tax on crude oil in addition to the hike in gasoline taxes. The crude oil tax could raise the price of gasoline another seven cents per gallon.(50)

4. <u>Loans, Direct Financial Aid, or Discontinuance or Refusal Thereof</u>

For some time, agencies of the Federal Government have been conducting lending and related programs designed to aid those suffering from energy shortages and those who desire to make expenditures designed to cause energy-use efficiency.(51) An example of such a program is that sponsored by the Small Business Administration (SBA) to assist small firms adversely affected by a fuel shortage. The SBA loans can be used for working capital, to pay existing financial obligations, to refinance debts, or to convert operations to a different fuel source.(52)

On August 14, 1976, Congress passed the Energy Conservation and Production Act (Public Law 94-385), which greatly accelerated the use of this tool as a means to cause or encourage energy use efficiency. Fundamentally, EPCA provides for:

- FEA guarantee of a loan made by a conventional lending institution for:
  - retro-fitting with energy causing devices; or,
  - retro-fitting with renewable resource devices, i.e., solar, geothermal or wind power generating sources; and,
- a Weatherization Assistance program to be administered by the states, or if they fail to do so, local governmental entities, where such entities may apply to the FEA for a grant to be used to weatherize low-income housing.(53)

The current Administration suggests a more extensive use of this strategy as may be seen from the summary of current proposals set forth below.(54)

- A Federal grants program will assist public and non-profit schools and hospitals in installing conservation measures, funded at the rate of $300 million per year for three years.
- The President has requested that the national 55 mph speed limit be vigorously enforced by States and municipalities. The Secretary of Transportation may, if he finds it necessary, withhold highway trust fund revenues from States not enforcing the limit.
- The Secretary of Housing and Urban Development will advance by one year, from 1981 to 1980, the effective date of the mandatory standards required for new residential and commercial buildings by the Energy Conservation and Production Act, with funds to be made available to States to help them in this effort.
- By reducing gasoline consumption, State revenues from gasoline taxes would also be reduced. These funds are used by the States for repair and maintenance of highways. The Administration will develop a program which will reduce their hardships and, to insure adequate highway maintenance, will compensate them for this loss through sources such as the Highway Trust Fund.
- The Federal Government will remove the barriers to opening a secondary market for residential energy conservation loans through the Federal Home Loan Mortgage Corporation and the Federal National Mortgage Association. This action should help to ensure that capital is available to homeowners at reasonable interest rates for residential energy conservation throuh private lending institutions.
- Funding for the existing low-income residential conservation program (weatherization) will be increased to $130 million in fiscal year 1978; and $200 million in FY 1979 and in FY 1980 (budget).
- The Secretary of Labor will take all appropriate steps to ensure that recipients of funds under the Comprehensive Employment and Training Act (CETA) will supply labor for the residential conservation program. The CETA program's employment levels, as proposed by the Administration, would meet the labor requirements of the program.
- The Secretary of Agriculture will vigorously implement a rural home weatherization program in cooperation with the Nation's 1,000 Rural Electric Cooperatives, with loans provided through the Farmer's Home Administration.
- State Public Commissions will be required to direct utilities to offer their customers a residential energy conservation service performed by the utility and financed by loans repaid through monthly utility bills.

D. Educational Programs

Government at all levels is actively involved in the collection and dissemination of data in an effort to cause end-use efficiency of energy. The private sector, particularly regulated utilities, has also been required to participate with governmental bodies.[55] Examples of government activities in this area are described below.

1. Big Three Program[56]

   This program, started in late 1976, consolidated several existing FEA administered educational programs including:

   - lighting and thermal operations for commercial buildings;
   - vanpooling; and, to a degree,
   - the voluntary industrial energy conservation program.

   Basically, the program involves a series of seminars in major metropolitan areas of the U.S. The initial seminars are to involve chief executive officers of high intensity energy using firms of all types, where the general concept of conservation will be discussed from both an ideological and economic viewpoint. There will be 100 seminars held throughout the country of this category. Following the chief executive seminar, there will be three other seminars which is the reason this program received the name "Big Three." These three additional seminars are to be directed to the following problem areas:

   - Transportation - 190 seminars will be hald to discuss such things as vanpooling and other energy use efficiency causing techniques;
   - General building maintenance and light retrofitting projects - 200 seminars will be held; and,
   - Industrial - 190 seminars will be held to discuss such things as equipment energy use efficiency.

   There will be approximately five million dollars diverted to this program.

2. Small Business Conservation Program

   Administered by the FEA, this program is designed to serve a similar service to the small business people that the "Big Three" program serves for the large energy users.[57]

3. Public Schools Energy Conservation Service Program (PSECS)

   This program is administered by Educational Facilities Laboratories, Inc., in conjunction with the FEA. The Public Schools Energy Conservation Service (PSECS), is a product of eighteen months of research, development and testing, involving more than 200 school districts and 1500 elementary schools. By taking advantage of recent advances in engineering and computer technologies, PSECS is able to establish energy use guidelines for electricity and fuel for most elementary school plants. These guidelines take into consideration such factors as climate, pattern of use, and special physical characteristics of the facility. By comparing the actual use with the guideline, the plant's potential for both energy and dollar savings is determined. The Service has been designed to work interactively with school and district staff members, and provides modification information on two levels. On level one, the operating level, the PSECS Self-Audit Program provides suggestions for bringing the plant down to guideline levels with little or very modest expenditures. Experience to date indicates that savings of 25 to 30 percent can usually be achieved if the PSECS operating suggestions are followed. At level two, capital modifications requiring larger investments are analyzed, using simple life cycle cost procedures to obtain a selection of cost effective modifications specific to each school. This computer analysis is intended to indicate those modifications that have sufficient merit to justify the employment of design professionals to carry out a more detailed analysis.[58]

   - All existing or proposed voluntary programs sponsored by the public and private utilities, Department of Commerce, or other agencies designed to increase the efficiency of energy use.
   - FEA, GSA, and DOC Guidelines dealing with improving the energy-use efficiency of new and existing buildings.
   - FEA sponsored "Conservation Award" which will be a highly publicized recognition of energy efficiency causing efforts.

- Current FEA efforts to disseminate information regarding fuel conservation and use efficiency through advertisements on television and in newspapers throughout the U.S.
- FEA intervention in selected legal actions or related utility rate increase application hearings at the state level in order to advance its policies regarding utility tariff rates and related subjects.

The current administration is continuing to encourage the use of educationally orienated programs to cause energy use efficiency as can be seen from a review of the proposals for new action, set forth below.(59)

- The Federal Government will spend up to $100 million over the next three years to add solar hot water and space heating to suitable Federal structures to help demonstrate the commercial potential of such measures.
- The utilities should be required to inform customers of all available residential conservation programs and how to obtain financing, materials and labor to perform residential conservation themselves. Other fuel suppliers would be encouraged to offer similar programs, with the help of their State Energy Offices.

E. Use of Power as a Major Customer

This is emerging as an effective method to expedite "market penetration" of energy efficiency causing hardware. Already this approach has had significant impact through the Federal Energy Management Program and like endeavors. In the immediate future, predictable action in this area will include:

- the purchase of only that equipment which complies with the government's energy use efficiency standards;
- an express prohibition on the use of equipment that does not comply with energy use efficiency standards;
- the requirement that all structures occupied by the government meet government energy use efficiency standards;
- the requirement that only equipment or building and design techniques which comply with the government's energy use efficiency standards be used or in connection with construction or like projects for or financed by the government; and,
- by way of a long-term commitment to buy or lease in large quantities, providing financial support of private industry in the development of products which use energy more efficiently.(60)

One of the more recent examples of how this approach may be used can be seen in the current administration's energy plan advanced by the President April 20,1977. Details of the President's proposals are set forth below.

- The President will direct Federal agencies to alter their auto purchasing practices so that new cars purchased by the Government will, on the average, exceed the average fuel economy standard under the EPCA by at least 2 mpg in 1980 and thereafter.
- The Secretary of Commerce will encourage State and local governments to include items that will contribute to energy conservation in their proposals under the Department's Local Public Works program.
- The President will direct all Federal agencies to adopt procedures which aim at reducing energy use per square foot by 1985 by 20 percent from 1975 energy consumption levels for existing Federal buildings and by 45 percent for new Federal buildings. Investments which are not cost-effective would not be funded under the program. The Director of the Office of Management and Budget and the Administrator of the Federal Energy Administration will implement this program.
- Legislation will be proposed to initiate a Federal vanpooling program. This program will demonstrate the energy conservation and pollution control potential of this form of commuter transportation by the largest employer in the Nation.

About 600 vans will be purchased by the Government and made available for use by the Federal employees. All costs of the program will be repaid to the Federal Government by the riders.

F. Grant of Special Exceptions to or Exemptions from Existing Law or Governmental Policies

Often government has influenced behavior by excepting certain actions from the application of existing regulations. A classic example of this approach is in the tax field where deductions, exemptions, and similar devices abound. In this regard, the current administration has made the following proposals in its energy plan unveiled on April 20, 1977:[61]

- removal of a 10 percent tax paid by passengers who ride on inter-city bus lines; and,
- an exemption from Federal and State public utility regulations would be available to industrial co-generators.

In the area of pricing of energy, the use or proposed use of this approach is most evident. The current administration has stated recently that prices should generally reflect the true replacement cost of energy and that U.S. citizens are only cheating themselves if they make energy artificially cheap.[62] The problem is that energy in the U.S. is made artificially cheap, by various regulations, particularly when U.S. pricing is compared with world market level energy costs.[63] Since deregulation of prices is not a viable solution politically in the U.S. at this time, the current Administration is proposing to address the problem more conservatively by use of the process of selective exceptions, exemptions and related devices.[64] Examples of how the current administration would like to use this technique to address the artificially low prices are set forth below.

- Regarding Oil Pricing[65]
  - Continue indefinitely the current price ceilings of \$5.25 and \$11.28 per barrel for previously discovered oil, subject only to escalation at the general rate of inflation.
  - Define newly discovered oil as oil from a well drilled more than 2-1/2 miles from an existing on-shore well as of April 20, 1977, or more than 1,000 feet deeper than any well within any 2-1/2 mile radius. New oil offshore will be limited to oil from lands leased after April 20, 1977.
  - Allow newly discovered oil to rise over a three-year period to the current world price (adjusted for the rate of inflation); thereafter, newly discovered oil would continue to be priced at the 1977 world price with adjustments for domestic increases in the general rate of inflation.
  - Incremental tertiary recovery from old fields and stripper oil would be free of price controls.
- Regarding Natural Gas Pricing [66]
  - Subject all new gas, sold anywhere in the United States, to a price limitation of the Btu equivalent of the average refiner acquisition cost (before tax) of all domestic crude oil. That price limitation would be approximately \$1.75 per Mcf at the beginning of 1978; the interstate-intrastate distinction would disappear for new gas.
  - Define new natural gas using the same standards as are used to define newly discovered oil (2-1/2 miles, 1000 feet, new leases).
  - Guarantee price certainty at current levels for currently flowing gas, with adjustments to reflect inflation.
  - Authorize the establishment of higher incentive pricing levels for specific categories of high cost gas.
  - Allow gas made available at the expiration of existing interstate contracts or by production from existing reservoirs in excess of contracted volumes to qualify for a price no higher than the current \$1.42 Mcf ceiling adjusted for inflation; gas made available under the same circumstances from existing intrastate production would qualify for the same price as new gas; i.e., \$1.75 per Mcf at the beginning of 1978.

- Allocate the cost of the more expensive new gas to industrial users, not to residential and commercial users.
- Extend Federal jurisdiction to SNG facilities guaranteeing them a reasonable rate of return.

- Other Oil and Gas Measures[67]

  - Inclusion of North Slope Oil in the domestic composite price under the current provisions in EPCA would introduce a degree of unnecessary uncertainty into domestic crude oil pricing. The $5.25, $11.28 and new oil pricing tiers, adjusted for inflation, would be substituted for the composite average limitation. Alaskan Oil would be subject to an $11.28 wellhead ceiling price, but would be treated as foreign oil for purposes of the entitlements program. New Alaskan oil finds would be subject to the new oil wellhead price.
  - Because of the high risks and costs involved in shale oil development, shale oil will be entitled to receive the world price of oil in the United States.

Another area where this technique has been used is in the patent field. In October 1973, a program was announced aimed at encouraging more rapid development and commercialization of energy-related inventions by according applications for patents on such inventions a "special" status that will permit them to be process in much less time than is normally required. Any patent application for an invention which materially contributes to the discovery or development of energy resources, or the more efficient utilization and conservation of energy resource may, if the inventor wishes, be included in the program. Examples of inventions would be those relating to further developments in fossil fuels (natural gas, coal and petroleum), nuclear energy, solar energy, inventions relating to the reduction of energy consumption systems, industrial equipment, household appliances, etc. In the first year of the program, the Patent Office afforded special processing to over 100 applications.[68]

The likelihood of additional government action in this area is good, particularly in its effort to strike a balance between the developing fuel conservation laws and existing environmental protection rules and regulations.

G. Establishing Priorities for Research and Development

Although this area in part involves financial pressure in the form of large expenditures of money for predetermined research and development priorities, the expression of governmental emphasis alone is a potent "cause and effect" device. ERDA acknowledges the existence of this tool in its current plan (ED-76-1) where it states that the planning process is a useful mechanism because the Federal Government can use such an approach as one context for its own actions and as a way to promote consensus on the Nation's approach to energy RD&D. Many states and utility companies have their own plans for research, development, and demonstration which are equally influential in directing the development and market penetration of both supply and conservation technologies.[69]

H. Summary of Strategy Used by Government to Cause Energy End Use Efficiency/Application in the Future

Up to the present time, the above described regulatory and policy-making activities have been used primarily to expedite market penetration of end users, energy efficiency-causing devices or programs in the areas of:

- motor vehicles;
- major appliances, i.e., refrigerators, freezers, dryers, air conditioners, etc.;
- building standards with particular emphasis in the area of insulation, lighting, and design;
- heat pumps and related devices;
- utility rate reform;
- voluntary energy use management in the commercial and industrial sectors; and,

- miscellaneous small hardware for use (voluntarily) by end users (particularly in the residential area) which expedite energy use efficiency, i.e., light bulbs, watt watchers, light dimmers and regulators, water flow regulators and restrictors, air deflectors, weather stripping materials, thermostats, air circulating devices (thermocycler), filter clog control devices for air conditioners, night set-back devices, and flow control devices.

Although vigorous governmental support for the above-described activities continues in the form of existing and proposed regulations and policy, a significant increase in pressure is emerging to expedite the development and implementation of coordinated energy conservation technologies at a level where end use efficiency can be effectively monitored and controlled and which have an acceptable impact on reducing the energy use growth rate.(70) This emphasis appears to be based in part on the factors described below.

- Causing significant voluntary participation by the mass population, particularly in the residential and small commercial areas, is quite difficult because the long-range success of such a program is dependent on the co-existence of the following factors:
    - people must be convinced there is a serious problem which affects them personally;
    - technology must provide tools for people to solve the problem;
    - the tools developed must have market penetration capability, i.e., people must be able to afford them;
    - when applied, the tools must have a visible and curative impact on the problem; and,
    - acceptance by people that the problem exists must continue for a period long enough that attempted solutions can achieve success.(71)
- Reduction in the growth rate of utility load demand and, if possible, overall energy use must be achieved to significantly control the need for new high cost generating facilities.(72)

With the passage of the Energy Policy and Conservation Act (EPCA) in December of 1975 (PL 94-163), the federal government placed the development of conservation technologies on a par with supply technologies. The EPCA has as its stated purpose to "reduce domestic energy consumption through the operation of specific voluntary and mandatory conservation programs." ERDA's current plan reflects the EPCA emphasis on the development of conservation technologies. Many states, including California, have passed energy legislation which places considerable emphasis on the research and development of conservation technology. California is significant to watch since its regulatory programs are advanced and accepted by the Federal Energy Administration (FEA) as models for other states to follow in its State Conservation Program.(73)

Set forth below are anticipated future governmental activities which will have a significant influence on programs designed to develop or improve energy conservation technologies.

- The Federal Government will look to the states and utilities to develop and implement conservation technologies consistent with certain minimum goals. (74)
- The state utility regulating agencies will start, or continue to instruct the utilities in their jurisdiction to develop and implement vigorous energy efficiency-causing programs at the "end user" level which will require:
    - elaborate reporting procedures regarding their programs and the progress thereof;
    - stepped-up research and development programs;
    - mass media educational programs;
    - various market penetration programs for efficiency-causing systems and hardware, i.e., building insulation. (75)
- Permission to utilities for rate increases and new plant construc-

tion will be conditioned on the existence of acceptable energy efficiency-causing programs. In California, the Public Utilities Commission (PUC) and CERDC clearly intend to pursue this type of approach. (76)

- Programs designed to reduce the utility load growth rate can be expected. For example, a study conducted by consultants to the California Assembly Committee on Resources, Land Use and Energy indicated that if all programs anticipated by the California Energy Resource and Development Act (Publ. Rec. Code 25,000 et. seq.) were implemented, there would be very little impact on the load growth over the next ten years.(77)
- Pressure is developing to cause at least limited decentralization of generating facilities. In California, legislation is now pending which would permit "wheeling" by major users (AB-4069-Warren) in order to meet growth needs. This concept has significant support from the large industrial users of electricity.(78)
- Consumer protection is receiving high priority. Solar equipment and home appliances are two areas of immediate emphasis. (79)
- Mandatory efficiency standards for all significant manufactured products, i.e., automobiles, home appliances, and buildings will be developed.(80)
- Various utility rate reform plans are developing and, in some states, have been implemented. (81)
- Increase in government financial assistance or pressure in the lending and tax areas can be expected to expedite development of energy conservation technologies.(82)
- National minimum energy efficiency performance standards for all new buildings will be developed, combined with more emphasis on the thermal and lighting efficiency standards of existing structures.(83)
- Use of funds available for research will be used more effectively to implement policy designed to cause energy use efficiency.(84)
- Increased national emphasis will be placed on a combination of price/taxing strategies.(85)
- Use of monetary rewards for efficiency oriented consumers and products and like penalties for those that are wasteful will increase.(86)
- Numerous programs designed to increase utilization of coal, solar, and nuclear sources of energy will be pressed.(87)

Section IV - Need to Eliminate or Modify Existing Laws and Governmental Policy

One of the most important tasks for lawmakers in developing energy use efficiency programs will be to identify and cause the repeal or appropriate modification of existing laws and underlying policies which directly or indirectly prohibit or inhibit the expeditious implementation of proven technical solutions. Areas which need immediate attention include:

- environmental laws in general;
- procedures governing coal extraction and utilization;
- utility rate restructuring and procedures involved in computing rates;
- siting requirements and procedures for nuclear power plants;
- zoning, land use and related laws, with particular reference to their present application to the use of solar energy facilities or other new and exotic systems for creating energy;
- taxing laws with particular reference to the elimination of de-incentives;
- anti-trust laws with particular reference to the allowance of the pooling of information and technology and elimination of wasteful duplication of action; and,
- safety oriented laws, i.e., OSHA regulations, with particular reference to thermal and lighting standards.(88)

Studies have commenced in the indicated areas, but the development of meaningful conclusions will progress at a slow pace.(89) At this point in time, the great majority of action centers around the conflict between laws and policies designed to protect the environment and laws or proposed laws designed to solve the energy shortage problem.(90)

In 1970, the National Environmental Policy Act of 1969 (NEPA) was enacted to issue a mandate to all Federal agencies to consider the environmental impact of their actions. NEPA requires Federal agencies to prepare environmental statements on major Federal action (including proposals for legislation significantly affecting the quality of human environment). NEPA also established the Council on Environmental Quality (CEQ). Thereafter, the Clean Air Act was passed, designed to eliminate, among other things, air pollution and pollution causing devices. The Environmental Protection Agency (EPA) now coordinates the overall program in conjunction with the Federal Energy Administration (FEA). States and local municipalities followed suit by establishing comparable laws and agencies and the courts, through interpretation of laws, further expanded and strengthened regulations in this area.

Problems have now developed which center on:

- environment-protecting devices causing more inefficiency in energy use; and,
- environment-protecting regulations prohibiting or greatly inhibiting the use of a fuel, i.e., coal, or establishment of an energy creating installation, i.e., a nuclear power plant.

Steps to resolve this environmental protection/energy shortage conflict have already begun in the form of the Energy Supply and Environmental Coordination Act (ESECA) of 1974 and the Federal Non-Nuclear Energy Research and Development Act of 1974 (FNERDA). ESECA causes certain amendments to the Clean Air Act but expressly states its purpose to be to provide for a means to assist in meeting the essential needs in the U.S. for fuels in a manner which is consistent to the fullest extent practical with existing national commitments to protect and improve the environment. FNERDA also purports to manifest the policy of Congress to develop, on an urgent basis, the technical capabilities to support the broadest range of energy policy options through the use of domestic resources by socially and environmentally acceptable means. Most certainly, the two above-described acts are indicative of how future laws, at all levels, will develop in the continuing effort to resolve the environmental protection/fuel shortage conflict.

In addition to the above-described legislation (ESECA and FNERDA), a methodology has been developed by the Federal Energy Administration (FEA) for assessing the environmental consequences of all national energy policies. As touched on above, this FEA function is carried out pursuant to Section 102 (2) (c) of the National Environmental Policy Act of 1969 (NEPA) and Executive Order 11514 (35 FR 4247). The FEA approach is a multi-step process in which the relationship between the following elements are examined:

- proposed energy policy;
- economic implications on energy users and suppliers;
- impact on the production and consumption of the various fuels in the short and long term;
- impact on the rate of development of energy facility construction and application of pollution controls; and,
- resultant environmental impacts.

The environmental impact of three national energy policies has already been assessed by the FEA in the following statements:

- <u>Draft Environmental Impact Statement--Energy Independence Act of 1975 and Related Tax Proposals</u> March 1975);
- <u>Draft Environmental Impact Statement--Mandatory Oil Import Program</u> (June 1975); and,
- <u>Draft Environmental Impact Statement--Electric Power Facility Construction Incentives Act of 1975</u> (July 1975)

The FEA is now assessing each of the state energy plans submitted to it in connection with the State Energy Conservation Program.(91)

There is mounting evidence indicating how policy makers intend to reconcile the conflict between laws designed to protect citizens and their environment and those which advance a worthy cause, i.e., energy supply and conservation technologies, but which might tend to threaten the safety of citizens or the quality of their environment. Fundamentally, the approach involves the requirement that in the process of propounding or implementing laws designed to protect citizens and their environment, the economic and employment impact must be carefully considered and given significant weight.(92) Pertinent examples on proposed laws addressing the conflict are set forth below.

- Toxic Substances Control Act-- (93) Employment effects must be evaluated in implementing this Act, including reduction in employment or loss of employment from threatened plant closures. Any employee can cause an investigation to be made and public hearings to be had if faced with actual or threatened discharge or lay-off or is otherwise faced with adverse or threatened adverse effects on employment.

- Resource Conservation and Recovery Act of 1976 (94)--Under the provisions of this Act, the administrator (Environmental Protection Agency) must conduct continuing evaluations of potential loss or shifts of employment which may result from the administration or enforcement of the provisions of the Act and applicable implementation plans, including, where appropriate, investigating threatened plant closures or reductions in employment allegedly resulting from such administration or enforcement. Every employee who is discharged, or laid-off, threatened with discharge or lay-off, or otherwise discriminated against by any person because of the alleged results of such administration or enforcement may request the administer to conduct a full investigation of the matter including the holding of public hearings.

- HR 6161 (April 6, 1977) Proposed Amendment to the Clear Air Act. (95) This bill calls for an amendment to the Clean Air Act by adding a new section requiring that an "Economic Impact Statement" be made before publication of notice of proposed rulemaking with respect to any standard or regulation covered by the section. The required statement must contain an analysis of the following factors with respect to any standard or regulation:

  - the costs of compliance with any such standard or regulation, including the extent to which the costs of compliance will vary depending on:

    - the effective date of the standard or regulation; and,

    - the development of less expensive, more efficient means or methods of compliance with the standard or regulation;

- the potential inflationary or recessionary effects of the standard or regulation;

- the availability of capital to procure the necessary means of compliance with the standard or regulation;

- the direct and indirect effects on employment of the standard or regulation;

- the effects on competition of the standard or regulation, particularly the effects on small business;

- the effects of the standard or regulation on consumer costs, including costs especially affecting economically vulnerable segments of the population;

- the effects of the standard or regulation on energy use or availability;

- the impact of the standard or regulation on productivity;

- the impact of the standard or regulation on the Nation's balance of payments;

- the economic impact of postponing the standard or regulation or of not promulgating such standard or regulation;

- alternative methods to such standard or regulation for achieving equal or greater degree of emission reduction (or health or environmental protection) at lesser economic costs;

- comparative expenditures required to achieve incremental levels of reduction of emissions (or enhancement of health or environmental protection); and,

- any possible alternative for minimizing or eliminating part or all of any adverse economic impacts of such standard or regulation.

Section V - International Energy Management Programs and Policies

A. General Approach

In evaluating U.S. efforts to cause more efficient use of energy, it is helpful to view such activity from an international perspective. Internally, U.S. interaction with other countries of the world is coordinated by the Office of

International Energy Affairs within the Federal Energy Administration. This office performs the following basic functions:

- identifies, analyzes, develops, proposes, and coordinates U.S. international energy policies to assure appropriate interface between domestic and foreign energy entities as they relate to international industries, producer and consumer countries, their associations, and the relationships between and among these entities and the U.S. Government;
- provides assessments of the international availability of all types of fuels as well as projections of the international environment in which the United States will seek to meet its future energy requirements;
- works worldwide to develop U.S. positions in cooperation with other major energy importers in the areas of demand restraint, emergency sharing, transportation of energy materials, and maritime and environmental questions; and,
- evaluates the adequacy of the following:
  - energy resources in physical terms;
  - stability of contractual arrangements for their acquisition;
  - the firms acquiring such resources for the United States; and,
  - the collateral logistics and refining systems.(96)

At the international level, efforts to resolve the many worldwide energy related problems appear to be developing the following definitive patterns of action:

- formation of energy management and related programs among countries that already have close economic ties, i.e., Common Market countries, within existing agencies;
- formation of new international agencies which involve high government level cooperation in developing and coordinating the implementation of energy policies; and,
- use of new international agencies to cause the private sectors of the participating countries to join in common energy-related programs.

A more detailed discussion of these areas follows.

B. Programs within Existing Structures

The program advanced by the energy commission of the European Communities (EC) is a good example of this approach. Pertinent aspects to the Commission's program submitted to the Council of Ministers are set forth below.

- Objectives for 1985 should be to:
  - keep energy usage 10 percent below pre-energy crisis forecasts;
  - increase electricity use by 10 percent to reach 35 percent of total use;
  - use nuclear energy for 50 percent of electricity production;
  - maintain internal solid fuel (coal lignite, peat) production at its present level;
  - raise natural gas internal output and imports; and,
  - restrict oil consumption to specific uses such as auto fuel and as a raw material.
- To expedite efforts to achieve the above-described policy guidelines for each major energy source.
- By the end of the century, nuclear energy and gas should be the predominant sources of the Communities' energy. By the year 2000, nuclear energy could cover at least 50 perent of total energy requirements while natural or synthetic (oil- or coal-based) gas cover about one-third of total needs. Coal therefore would account for only one quarter of EC needs by that time. The Commission does not expect non-conventional energy sources such as solar and geothermal energy to account for more than a small portion of EC energy sources.
- In the area of electricity, the following specific points were submitted:
  - Expansion of the use of electricity will depend largely on:
    - adequate financing and insured economic stability of nuclear power station;

- more profitable and rational use of available power station; and,
- adoption of appropriate price measures to ensure continued demand during non-peak periods.

- To prevent higher electricity production from increasing the demand for oil, oil burning stations would be restricted to using heavy residue oil from refineries, and would eventually be used only as medium and peak-load plants. Additionally, construction of oil-fired base load plants would be authorized only in exceptional cases. These measures would reduce oil-based electricity production from the current 30 percent to less than 20 percent by 1985. Fuel would be available for power plants only when supplies were interrupted or for economics, technical, or environmental reasons. Coal on the other hand, would be given a larger share of the power station markets.
- The EC's total nuclear power station capacity will have to exceed 200 Gigawatts if nuclear energy is to supply 50 percent of energy needs by the end of the century. To reach that goal, the Commission will draw up proposals to: enable EC industry to build the needed stations, protect EC public health and environment, and guarantee adequate nuclear fuel supplies.
- The EC will act to encourage development of new and better breeder reactors. Work is already underway on the fast breeder reactor, which would increase nuclear fuel supplies, and on the high temperature reactor, which could also be used as a source of process heat for industry.
- The creation of incentive devices would speed up investment and adoption of fiscal measures to prevent excess profiteering from low cost energy sources. The Community budget should be drawn on only if member states or EC action was essential for moderating or encouraging immediate energy related developments.(97)

C. Creation of New International Agencies for Government Interaction

1. Creation of New Agency--Background

The Organization for Economic Cooperation and Development (OECD) was set up under a Convention signed in Paris on December 14, 1960, which provides that the OECD shall promote policies designed:

- to achieve the highest sustainable economic growth,and employment and a rising standard of living in member countries, while maintaining financial stability, and thus to contribute to the development of the world economy;
- to contribute to sound economic expansion in member as well as non-member countries in the process of economic development;
- to contribute to the expansion of world trade on a multilateral, non-discriminatory basis in accordance with international obligations.(98)

The members of OECD are Australia, Austria, Belgium, Canada, Denmark, Finland, France, the Federal Republic of Germany, Greece, Iceland, Ireland, Italy, Japan, Luxemborg, the Netherlands, New Zealand, Norway, Portugal, Spain, Sweden, Switzerland, Turkey, the United Kingdom, and the United States. The Socialist Federal Republic of Yugoslavia is associated in certain work of the OECD, particularly that of the Economic and Development Review Committee.(99)

The International Energy Agency (IEA) was established by decision of the OECD Council on November 15, 1974 as an autonomous body within the framework of the Organization. On November 18, 1974, the sixteen members of the organization then participating in the Agency entered into an agreement on an International Energy Program (IEP). The IEP is implemented through the IEA. The principal aims of the program are:

- development of a common level of emergency self-sufficiency in oil supplies;
- establishment of common demand restraint measures in an emergency;
- establishment and implementation of measures for the allocation of available oil in time of emergency;
- development of a system of information on the international oil market and a framework for consultation with international oil companies;

- development and implementation of a long term cooperation program to reduce dependence on imported oil, including:
  - conservation of energy,
  - development of alternate sources of energy,
  - energy research and development, and,
  - supply of natural and enriched uranium;
- promotion of co-operative relations with oil producing countries, and with other oil consuming countries, particularly those of the developing world.(100)

The countries that participate in the IEA are: Austria, Belgium Canada, Denmark, Germany, Greece, Ireland, Italy, Japan, Luxembourg, the Netherlands, New Zealand, Norway, Spain, Sweden, Switzerland, Turkey, the United Kingdom, and the United States.(101)

2. Establishment of International Guidelines for the Development of Energy Use Efficiency Causing Policy and Regulation/Application

Since its creation, one of the major purposes of the IEA has been the promotion of strong energy use efficiency causing programs. Within the IEA is a Conservation Sub-Group (CSG) on which all member countries are represented. The CSG leads the Agency's conservation efforts. In 1975, the CSG reviewed and evaluated each country's program and submitted a report to the Agency's Governing Board. After the 1975 review, it became apparent that a more uniform approach was needed in the member's conservation programs, hence the CSG developed an "Indicative Test" of conservation measures and urged all members to consider adopting these measures. The suggested measures are set forth below.

- All energy priced at world market levels. This does not necessarily mean thermal equivalent pricing.
- Significant taxes on certain fuels to reinforce the effects of market prices where these prices are judged for national reasons to be inadequate signals (e.g., gasoline taxes).
- Changes in utility marketing practices and price structures to reward conservation by final consumers.
- Comprehensive public education programs with a conservation message, including programs specifically directed at schools.
- Specialized energy conservation education and/or training for such personnel as architects, engineers and building contractors, supervisors, and inspectors.
- Permanent full-time government conservation staff of adequate size.
- Consultants and other government staff spending a significant proportion of their time on energy conservation.
- Programs to increase use of waste heat from electrical generation and from industrial processing.
- Priority for government funding of energy efficient public transport (e.g., rail, bus) over funding for less energy efficient modes (e.g., air travel, highway construction).
- Programs to intensify government energy conservation R & D (excluding those encompassed within IEA R & D activities).
- Thermal and lighting efficiency in new commercial and public buildings and new residences through changes in building codes and standards.
- Incentives to increase retrofitting of existing residences and commercial buildings to improve thermal efficiency (e.g., loans or grants for insulation).
- An exemplary and effective effort to reduce all central government and local government energy use.
- Energy efficiency labeling for all major consumer appliances (e.g., water heaters, air conditioners, refrigerators, freezers, automobiles).
- Programs to improve the efficiency of heating/cooling devices and major appliances.
- Speed limits (of e.g., 90-110 km/hr) on all highways, including superhighways.
- Programs to increase automobile efficiency in countries where average new car efficiency is low (e.g., fuel economy standards, weight, horsepower or displacement taxes).
- Programs to increase load factors on transportation modes with excess capacity (e.g., car pools, public transit).

- Programs to stimulate energy efficiency in industrial production (e.g., target setting, loans for energy improvements, tax credits, rapid depreciation allowance, energy audits of individual companies, provision of information for small companies).
- Policies and programs to improve the efficiency of electrical generation such as Peak Load pricing, Ripple Load Controls, Thermal Storage and other Load Management Techniques.(102)

In 1976, the review process was repeated, this time based on the three criteria described below.

- To determine and measure energy use efficiency within a given country, the actual conservation results in each country over the past two years (1975/1976) were compared, including
    - changes in demand for energy relative to what would have been expected if 1968-1973 consumption trends had continued; and,
    - the relative change in the energy consumed/Gross Domestic Product (GDP) ratio.
- A comparison was made of specific energy efficiencies of major services or products within each energy use sector, including such items as:
    - energy used per passenger kilometer of automobile travel;
    - energy per ton of crude steel produced; and,
    - the capacity factor and conversion efficiency within a country.
- Each member's energy program was compared with the indicative list to assess the comprehensiveness and strength of the measures adopted to promote energy use efficiency.(103)

Table 1 shows the overall conclusions drawn by CSG from its 1976 review based on the above-described criteria.(104)

Obviously, the program for each member country of the IEA could readily justify a lengthy detailed analysis which is here inappropriate. However, a brief analysis of the programs is in order to better provide the reader with an overall view of the effort of government to cause energy use efficiency from an international perspective. To this end, consider the program summaries for Austria, Belgium, Canada, Denmark, Germany, Ireland, Italy, Japan, The Netherlands, New Zealand, Norway, Spain, Sweden, Switzerland, Turkey, United Kingdom and the United States, set forth below.(105)

- Austria--The main element of the program is the pricing of fuel at or above world market levels. Other aspects of the program include:
    - a progressive tax on engine size of automobiles;
    - interest-free loans for new and existing structures conditioned on compliance with mandatory efficiency standards; and,
    - 100 percent loans at subsidized interest rate to companies making energy use efficiency causing investment.
- Belgium--The main element of the program is energy priced at world market levels combined with substantial energy use taxes. Other aspects of the program include:
    - grants to industrial firms that demonstrate a 12 percent reduction in energy use per unit of output from the time assistance is requested;
    - energy use efficiency causing standards for new buildings; and
    - 25 percent grant program for the retrofitting of existing residences.
- Canada--The program is still largely voluntary and maintains prices on oil and gas below world market levels. Other aspects of the program include:
    - regulations requiring a doubling of automobile efficiency by 1985;
    - a voluntary target setting and reporting system for industry has been established which includes:
        - accelerated capital cost allowance to encourage energy efficiency causing changes; and,
        - efforts to encourage waste heat recovery;
    - a building code for new structures subject to acceptance by

TABLE 1 Summary of Conservation Efforts of IEA Member Countries

Member Country	Actual Conservation Results	Specific Efficiencies in Industry	Specific Efficiencies in Transportation	Status of Program
Austria	above average	above average	below average	Adopted, but still important gaps
Belgium	above average	average	below average	Adopted and comprehensive
Canada	below average	below average (poor)	below average (poor)	Adopted program--needs strengthening
Denmark	above average (high)	above average (high)	above average (high)	Adopted--excellent and comprehensive
Germany	average	above average	above average	Adopted and fairly comprehensive
Ireland	average	above average	above average	Adopted and fairly comprehensive
Italy	above average	above average (high)	above average (high)	Adopted and fairly comprehensive
Japan	above average	above average	above average	Adopted and fairly comprehensive
Netherlands	above average (high)	below average	above average	Adopted--needs improvement in areas
New Zealand	above average	below average	below average	Adopted but has gaps--not comprehensive
Norway	average	above average (high)	above average (high)	Adopted--many strong elements
Spain	below average (low)	above average	above average	Adopted--needs improvement
Sweden	below average (low)	below average	below average	Adopted and comprehensive
Switzerland	below average (low)	no information	below average (poor)	No overall plan as yet
Turkey	below average	no information	no information	Adopted but needs substantial improvement
United Kingdom	above average	below average (low)	above average (high)	Adopted but needs improvement in areas
United States	below average	below average	below average	Adopted but needs improvement in areas such as pricing/taxes and buildings

each Province prior to implementation; and,

- development of minimum efficiency standards for major household appliances, combined with energy efficiency labeling.

- Denmark--The program is based on a strong price/tax policy. Other aspects of the program include:
  - taxes that increase with engine size on both new and old automobiles;
  - monetary incentives to public transit combined with reduced funding for both highway and airport construction;
  - loans and grants to industries wishing to invest in energy efficiency causing equipment;
  - 25 percent grants to homeowners for retrofitting residences;
  - mandatory standards for new buildings have been further improved; and,
  - a well-developed district heating program (10 percent of homes).

- Germany--The basis for the program is energy pricing where world market levels apply except in coal which is held above international levels. Other aspects of the program include:
  - high taxes on gasoline and heating oil;
  - large research and development expenditures;
  - 7.5 percent tax allowance for industry and utility investments for dual use of power and other combined heat recovery techniques;
  - assistance program for small businesses;
  - progressive tax on cars to improve efficiency;
  - a new conservation law applicable to buildings and building appliances is in effect by virtue of which:
    - building codes will be upgraded for new buildings;
    - efficiency standards are to be set for appliances including air conditioning, furnaces and water heating;
    - mandatory annual maintenance is required for all building heating units including residences; and,
  - extensive use of peak load pricing and other load management techniques.

- Ireland--The program is premised on maintenance of pricing at world market levels, a strong educational campaign and an improved data operation. Other aspects of the program include:
  - fuel efficiency surveys for industry, one-third of which is paid for by the government;
  - a grant of up to 35 percent for industries that wish to make energy efficiency causing improvements;
  - a new building insulation code has been mandated;
  - homeowners can obtain a grant of two-thirds of the cost, subject to a maximum of £400 for renovating their residences;
  - progressive taxes on automobiles according to engine size; and,
  - elimination of declining block rates combined with extensive use of peak load pricing by utilities.

- Italy--Major elements of this proprogram are prices and taxes. Hydro-carbon fuels are priced at world market levels and all major refined petroleum products are taxed. Other aspects of the program include:
  - a high gasoline tax;
  - a prohibition of automobile use in some urban areas;
  - high annual license fees which increase sharply with engine displacements;
  - a new energy law passed in April, 1976, which:
    - regulates performance of existing heating systems by holding them to 20 °C;
    - establishes standards for new heating systems; and,
    - sets thermal insulation standards for new and renovated buildings;

- a comprehensive mandatory reporting system for industry;
- a major program of building pumped storage plants to deal with peak demand; and,
- utility tariffs which dramatically increase with higher consumption in the domestic sector.

- Japan--Pricing of energy at world market levels is the main area of emphasis in this program. Other aspects of the program include:
  - major use of rail transportation;
  - low "advisory" speed limit;
  - favorable tax treatment for the smaller and lighter automobiles;
  - automobile efficiency testing and labeling procedures are under development;
  - special bus lanes;
  - large funding for public transit;
  - ride sharing utilization of taxis;
  - major industrial program including:
    - a system of administrative guidance and cooperation;
    - accelerated depreciation (one-third first year write-off) for energy use efficiency-causing investments;
    - loans for improving energy efficiency provided by the government bank at interest rates slightly below commercial rates;
    - significant expenditures for research and development; and,
  - increase in utility rates and a gradual elimination of declining block rates.
- The Netherlands--The program is still in the formative stages, at least in terms of specific proposals. Energy prices have been raised to world market levels except for natural gas where the price is moving upward more slowly. Other aspects of the program include:
  - progressive tax on car weight;
  - increased subsidies for public transit;
  - 33 percent grant for owners of residential and commercial buildings who want to improve the efficiency of their structures;
  - a plan to reinsulate all homes in ten years is under development;
  - mandatory building codes exist and have been greatly strengthened;
  - voluntary appliance labeling has been adopted; and,
  - peak load management techniques are in use.
- New Zealand--Until recently this program was retarded by the regulation of prices significantly below international market levels. At present, gasoline, manufactured gas and electricity are priced and/or taxed at high levels, however, coal and natural gas are still sold at below world market prices. Other aspects of the program include:
  - an elimination of declining block rates;
  - a graduated tax on engine size;
  - efficiency labeling for automobiles is under consideration;
  - establishment of an advisory board to work for more efficiency in the industrial sector; and,
  - interest-free loans to homeowners for retrofitting residences.
- Norway--The program relies heavily on pricing to cause energy use efficiency which is set and controlled by the government at or near world market levels. Other aspects of the program include:
  - a heavy tax on gasoline;
  - a 90 km/hour speed limit;
  - a 100 percent import tariff on automobiles;
  - low interest loans for energy efficiency-causing efforts in industry;

- a strong national building code including strict insulation standards;
- loan program for retrofitting existing buildings; and,
- strict controls on growth of electricity use.

- Spain--Pricing at above world market levels combined with heavy emphasis on the reduction of imports provides the basis for this program. Other aspects of the program include:
  - a horse-power tax on automobiles;
  - a high parking tax in urban areas;
  - the building of new subways in several cities;
  - establishment of an institute to provide technical assistance to industry;
  - a non-subsidized line of credit is available to companies to fund investments in new energy efficient technologies;
  - a supply restriction which limits households to 80 percent of 1973 consumption;
  - new building standards designed to cause energy use efficiency; and,
  - a voluntary labeling program for appliances which is gradually becoming mandatory.
- Sweden--Pricing and taxes are providing the foundation for this program. Other aspects of the program include:
  - a 10 to 20 percent tax on electricity;
  - a goal to limit energy use growth to 2 percent per year through 1990 and zero thereafter;
  - a progressive tax on automobiles;
  - speed limits;
  - prior government approval of major energy intensive industrial facilities;
  - a new upgraded building code including both strict insulation standards and a requirement that new homes be individually metered;
  - grants and loans for improving efficiency in existing residences;
  - a study of appliance labeling and efficiency-causing regulations; and,
  - significant increase in funding of research and development projects.
- Switzerland--At present there is no federal government program. Energy use efficiency causing policies and regulations do exist, however, and include:
  - high energy prices and taxes; and,
  - progressive tax on automobiles.
- Turkey--This program is based on pricing, with energy held to world market levels. Other aspects of the program include:
  - revised building insulation standards; and,
  - mandatory quality standards for new boilers and a training program for boiler operators.
- United Kingdom--Pricing energy at international levels combined with a strong and effective public information campaign provides the basis for this program. Other aspects of the program include:
  - government paid energy use audit schemes for industry;
  - a non-subsidized loan program to industry that makes energy use efficiency causing investments;
  - strict mandatory thermal insulation requirements for new and altered or extended buildings;
  - renovation grants for existing homes conditioned on the achievement of a reasonable standard of roof insulation; and,
  - adoption of peak load pricing.
- United States--This program still emphasizes public education and is greatly inhibited by low taxes on all fuels and by regulations holding oil and gas prices below world market prices. Other aspects of the program include:
  - mandatory automobile efficiency standards to double efficiency by 1985;

- the mandatory labeling of automobiles regarding efficiency;
- increased public transit funding;
- voluntary efficiency target setting for the ten most energy intensive industries along with mandatory reporting schemes for the largest firms;
- a reasonably rapid elimination of declining block rates; and,
- sponsorship of innovative load management rate demonstrations.

3. International Strategy Used to Cause More Efficient Use of Energy - An Overview

Overall a number of conclusions were reached and observations made by the Conservation Sub-Group (CSG) after the 1976 review regarding patterns of governmental action in the international effort to cause more efficient use of energy. These conclusions and observations are set forth below.(106)

- Regarding the use of taxes and pricing
  - While crude oil prices vary country by country, the prices of gasoline and gas oil lie in a similar price range for many IEA countries. Particularly in the case of gasoline, only three countries prices are outside the range of 30 to 40 cents per litre.
  - In general, gasoline and motor diesel are taxed at higher rates, while other fuels are rarely taxed except for a value added tax. Further, lower rates of value added tax are applied to those fuels than those applied to gasoline and motor diesel in some countries.
  - The price increases in fuel oil are much smaller than those in gasoline. In particular, it is noted that fuel oil prices in real terms were reduced in at least three countries during this period (1976).
  - In spite of fairly sharp price increases in nominal terms for both fuels, the real price increases are surprisingly modest even for gasoline in many countries. For many IEA countries, the price of energy in real terms decreased between 1974 and 1975. In particular, gasoline prices in real terms decreased in all countries but three. This is a result of the massive increase in the general price level.
- Regarding the transportation sector
  - Automobile efficiency is a critical concern since autos by far are the dominant transport mode in both urban and rural use, and, thus the dominant transport fuel user. High national gasoline prices and/or taxes have promoted the manufacture and purchase of relatively efficient autos notably in Western Europe, and low gasoline prices/taxes have led to large inefficient autos, notably in the United States and Canada. These two countries have established mandatory programs to upgrade the efficiency of their new cars, by standards and labeling. Other countries continue to rely on price effects to maintain already high efficiencies.
  - The auto is used for the majority of urban passenger travel in all countries reporting, even those with cheap and efficient urban mass transit, except for Japan. Load factors for urban trips are uniformly low, all around 30 to 40 percent. Yet only the United States has anything approaching an organized, comprehensive program to promote carpooling or ride sharing. This is a major and surprising weakness of IEA conservation efforts, especially since effective ride sharing will not come about as a result of price effects alone; institutional organizational efforts are needed.
  - Severe speed limits (e.g., 90 to 110 km/hr) are in force in roughly half of the IEA countries, and have resulted in energy savings variously estimated to be between three and five percent, as well as savings in death, injury and damage from accidents avoided that are in some cases quite remarkable. There seems, however, to be no discernible move among the other countries to adopt these lower limits.
  - Urban public transit, both bus and rail, has steadily lost ridership in every IEA country except Japan since World War II. This appears largely to be the

result of increased suburbanization and increased incidence of personal automobiles. As a result, transit is everywhere subsidized, both capital and operating costs, and many measures are suggested, both incentives (such as lowering fares) and disincentives (such as parking surcharges and auto free zones), to increase transit ridership. Yet the real problem of residential housing patterns and general urban planning does not seem to be addressed anywhere effectively.

- Trucks have encroached on rail freight markets in virtually every IEA country in the last 15 years, despite the fact they are on average less energy efficient per ton-kilometre of freight moved. The resulting decline in rail traffic has in general led to national rail subsidies, but few other policies are in place or being considered to promote rail freight.

- Regarding the industrial sector
    - Energy intensive industries such as iron and steel, aluminum and other non-ferrous metal fabrication, cement and pulp and paper account for a great part of national energy consumption in many countries although the number of firms is fairly small. Some countries including the U.S., Canada, Japan, and the United Kingdom, pay special attention to energy conservation potential in these industries by introducing reporting/auditing schemes, and working out sectoral conservation targets.
    - Some countries pay particular attention to small and medium sized firms. These firms as a whole consume large amounts of energy although many of them do not belong to energy intensive industries. They are usually lacking in the knowledge of where energy waste takes place and how it can be reduced economically.
    - In several countries, voluntary or compulsory reports of energy conservation performance by industry are made. In Japan, the selection of a "heat manager" is compulsory.
    - Most countries provide some financial incentives to industry to invest in conservation, although in some countries these incentives are given only to investment for improvement of building insulation, heating and lighting.

- Regarding the domestic/residential sector
    - Most IEA Nations have some form of insulation building codes. However, the force and direction of the codes, the degree of enforcement, and the responsibility for administering the standards varies widely.
    - Most IEA Members have initiated incentive programs to insulate existing buildings. The bulk of the programs work by a direct grant of from one-fourth to one-third of the total cost. A few programs accomplish the same results vis-à-vis tax reductions for insulation, etc. Most of the programs however are limited to a few years and/or have limits on the amount of financing available.
    - The most common form of member governments' internal conservation programs involve the use of circulars and directives to all departments. The circulars instruct the agencies to undertake conservation programs regarding lighting, temperature settings, insulation for public buildings, etc. Where the effects of the programs are known, energy savings of from 9 percent to 24 percent have been achieved.
    - The only country to have actually initiated a program on energy efficiency labeling of appliances is the United States. However, most countries are reviewing such a program. In Canada, a program similar to that of the U.S. will be introduced quite soon.
    - Most IEA Members have yet to initiate a program in the area of appliance and heating efficiency. The U.S., under the Energy Policy and Conservation Act (EPCA), is developing efficiency improvement targets for all major appliances. Other nations have programs of public education on furnace servicing, etc. Germany has standards for industrial boilers, and Canada has a

mandatory minimum furnace efficiency program.

- Regarding the electric utility sector
    - Most members have instituted some form of peak load management;
    - A small number of the members (approximately 5) are using ripple load controls;
    - The use or development of more sophisticated storage mechanisms is found in a majority of the member countries;
    - Techniques involving confined power-heat production exist or are the subject of research and development projects in a majority of the member countries;
    - About half the member countries have instituted or are in the process of developing other load management techniques such as:
        - development of interregional electrical connections;
        - superconduction for electricity transmission;
        - MHD power generation;
        - studies as to how transmission and distribution losses can be reduced;
        - audio-frequency power line carrier control;
        - international electrical inter-connection;
        - central dispatching control; and,
        - grid lineage with neighboring countries with significant exchanges of power.

Certainly, the IEA has made significant progress in developing an international program to cause energy use efficiency which program has, in turn, expedited the development and implementation of similar programs in the Member countries. Expansion of the IEA membership and the Agency's involvement in energy affairs can be anticipated.

1. Use of New International Agencies to Involve the Private Sector

Not only are governments in industrialized nations joining together to resolve energy related problems, but they also are using the newly created international agencies, i.e., IEA, to encourage and aid the private sector of participating countries to likewise cooperate and work towards common goals. By way of example, on March 28, 1975, final approval by the Attorney General of the U.S. was given to the "Voluntary Agreement and Program Relating to the International Energy Program." This program was developed by the Federal Energy Administration and Department of State.(107)

Participation by the U.S. in this voluntary program is premised on the realization by its government that it must be prepared to cooperate with other nations in the distribution of available supplies of fuel or sources of fuel on a rational and equitable basis in order to utilize them with maximum efficiency during any future supply interruption.(108)

The program evolved from a general policy of participating countries to reduce their dependence on foreign oil and to obtain the greatest quantity of supplies during an oil emergency. The U.S. and certain other members of the Organization for Economic Cooperation and Development (OECD) first signed an agreement on November 18, 1974 on an International Energy Program (IEP) pursuant to which the International Energy Agency (IEA) was established as an autonomous institution within the OECD. It is a premise of the IEP Agreement that consultation and cooperation between oil companies and the IEA is essential to the effective functioning of the IEP, and thus to the solution of economic, strategic, and national security problems facing oil-importing nations. Accordingly, the President of the United States requested that this voluntary agreement and program be entered into by the U.S. and that the participating oil companies undertake the actions contemplated thereby in order to further the objectives of the IEP and to implement the related policies and procedures of the IEA. The President's request was premised on the belief that such participation and action would be in the public interest and contributes to the national defense of the U.S.(109) Authorities, with respect to the IEP, are now well defined in the Energy Policy and Conservation Act (P.L. 94-163) passed December 22, 1975.

The basics of the IEP are set forth below.

- It is an entirely voluntary program, consistent with the purpose and scope of the Defense Production Act of 1950.
- It provides immunity from the antitrust laws and the Federal Trade Commission Act with respect to acts or omissions by participants, and such of their affiliates as it may

have been designated in accordance with its terms (Section 9) which are required to implement the objectives of the IEP.

- It contemplates that such acts by the participants will include:
  - membership in standing groups, working parties, advisory bodies or other bodies established at the request of the IEA;
  - consultations, planning, and individual and joint actions to implement the international allocation of petroleum pursuant to the IEP directives;
  - the furnishing by participants of data and information, consultations, and planning in respect thereof; and,
  - membership in ancillary industr groups established by the U.S. government. However, unlike those formed by the IEA, operations of these groups, if established by the State Department, would be governed by the provisions of the Federal Advisory Committee Act, 5 U.S.C. App I (1973 Supp) and if established by the Federal Energy Administration, they would also be subject to the special provisions of Sections 17 of the Federal Energy Administration Act of 1974, 15 U.S.C.A. 776.
- The procedure for an oil company to become a member generally involves:
  - a request to join initiated by the Administrator (FEA) after approval by the Attorney General upon a finding that such participation is in the public interest and contributes to the national defense; or,
  - an oil company asking the Administrator to request that it become a participant which then triggers the above-described clearance procedures.

Notices of all requests and acceptances shall be published in the Federal Register.

- Any participant may withdraw from the Agreement upon at least 30 days notice to the Administrator (of the FEA) subject to fulfillment of the obligations incurred under this agreement prior to the date of such a notice except where emergency measures have been taken. When emergency measures have been undertaken, the effective date of the withdrawal may be postponed up to 60 days.
- Participation by the U.S. in the program may be terminated at any time by the Administrator (FEA) after consultation with the Secretary of State upon notice by letter, telegram, or publication in the Federal Register. In no event, however, shall the program continue beyond June 30, 1985.(110)

The above-described program involving oil companies is significant because it provides a "blue print" for future participation of the private sector in attacking the energy problem on a world-wide front.

Section VI - Conclusions/Testing the Merits of Energy Management Strategies

Engineers, architects, scientists, urban planners, equipment and process designers, personnel at utilities, and other similar disciplines necessarily must play a vital role in the development and implementation programs designed to cause energy use efficiency. This participation will most certainly require the actual drafting of portions of legislation and comprehensive recommendations as to how and in what manner existing laws should be modified or repealed in order to enhance the success of new programs. Numerous existing private technical, design, and similar groups or societies have already recognized the need for their input and are actively participating in the development of policy and legislation. One example is the American Society of Heating, Refrigerating and Air Coniditioning Engineers (ASHRAE) which for several years has played a significant role in the development of building codes designed to cause more efficient use of energy. Other such groups are the American Institute of Architects (AIA) and the International Conference of Building Officials. Groups or societies such as those mentioned above have been and will continue to be a creative and moving force in the formation and implementation of government policy and future legislation designed to cause more efficient use of energy.(111)

Furthermore, teamwork between the legislative bodies (many members of which are lawyers) and members of the technical and related progessional fields is critical. Both groups must have at least a fundamental understanding of the basic principles of governing the other. Any proposed energy use efficiency causing strategy must be tested against answers to such socio-political or legally oriented questions as:

- Where does the economic burden of pursuing the strategy lie?
- Who bears the burden of the extra "bother" which the strategy may cause?
- What alternative modes of behavior are people likely to adopt to accommodate themselves to the changed conditions which the strategy induces and can the energy consequences of those alternative modes of behavior be evaluated?
- Will the strategy require additional government or private sector manpower to implement?
- Is the strategy difficult to enforce?
- Are there any incentives built into the regulations to encourage compliance?
- How will it be financed?
- Has an appropriate agency been assigned responsibility for implementing and enforcing the strategy?
- Does the agency have a constituency which will enable it to resist attempts to change its mission?
- Are there inherent conflicts within the agency itself?
- Is there relevant expertise and power available to enforce actions?
- Are the administrative structures fair?
- Where, if at all, should judicial review of administrative action play a role?
- Does the law create an unnecessary bureaucracy?(112)

The answers to questions such as those set out above could determine the success or failure of any given energy efficiency causing policy, regulation or technology.

Insofar as the citizenry at large is concerned, (particularly non-business "end-users"), knowledge of and support for energy use efficiency programs and laws passed to implement them is essential. The present inability of the government to expedite well-organized and meaningful legislation programs is in large part attributable to a lack of public concern premised on the mistaken belief that the "energy crisis" has passed.(113) A continuation of this "head-in-the-sand" philosophy will only make the adjustment to energy use efficiency causing rules and regulations more difficult.

REFERENCES

1. CCH, Energy Management, Chapters 2502-2510; and Senate Committee on Interior and Insular Affairs, Federal Energy Organization, Serial No. 93-6(92-41), (1973), as cited by ibid.

2. "Presidential Energy Message," (4 June 1971), reprinted in CCH, Energy Management, Chapters 403-414; and CCH, Energy Management, Chapters 2503.

3. "Energy Reorganization Act of 1974," reprinted in CCH, Energy Management, Chapters 10, 721-10, 755; "Federal Non-nuclear Energy Research and Development Act of 1974," reprinted in CCH, Energy Management, Chapters 981-997; "Federal Energy Administration Act of 1974," reprinted in CCH, Energy Management., Chapters 10, 550-580; and "California Warren-Alquist State Energy Resources and Conservation and Development Act." Energy Policy and Conservation Act, reprinted in CCH, Energy Management, Chapters 10, 850-10, 968.

4. "Presidential Energy Message," (4 June 1971), reprinted in CCH, Energy Management, Chapters 403-414.

5. "Presidential State of the Union Message," (15 January 1975), summarized in the White House State of the Union Fact Sheet, reprinted in CCH, Energy Management, Chapters 692-698; US Federal Energy Administration, proposed Energy Independence Act of 1975 and Related Tax Proposals, Chapters 12; and, Presidential Energy Message Detailed Fact Sheet (April 2, 1977).

6. CCH, Energy Management, Chapters 2501.

7. CCH, Energy Management, Chapter 2511; and Presidential Executive Orders, Nos. 11712 and 11726, reprinted in CCH, Energy Management, Chapters 11,001 and 11,021.

8. CCH, Energy Management, Chapters 2511; and Presidential Executive Orders, Nos. 11743, 11748, 11775, and 11790, reprinted in CCH, Energy Management, Chapters 11,031, 11,041, 11,071, and 11,081.

9. CCH, Energy Management, Chapter 2511; Presidential Executive Orders, Nos. 11814 , 11819, and 11834, reprinted in CCH, Energy Management, Chapters 11,091, 11,100, and 11,101; and the "Energy Reorganization Act of 1974."

10. Interview with Dr. Douglas C. Bauer, Associate Assistant Administrator, Utilities Programs, Federal Energy administration Office of Conservation and Environment, Washington, D.C., 18 June 1975; CCH, Energy Management, Chapter 2501.

11. Presidential Message on Department of Energy (March 1, 1977), reprinted in CCH, Energy Management, Chapters 723-730.

12. Interview with Grant P. Thompson, Institute Fellow, Environmental Law Institute, Washington, D.C., 18 June 1975; Warren-Alquist State Energy Resources, Conservation and Development Act (California, 1975); Energy Conservation Project Report No. 2, October, 1975, published by the Environmental Law Institute, Washington, D.C.

13. Interview with Grant P. Thompson, Institute Fellow, Environmental Law Institute, Washington, D.C., 18 June 1975; Warren-Alquist State Energy Resources, Conservation and Development Act (California, 1975); Energy Conservation Project Report No. 2, October 1975, published by the Environmental Law Institute, Washington, D.C. Interviews with Dr. Douglas C. Bauer, Robert R. Jones, David Rosoff, Ted Farfaglis, and Dr. Melbin H. Chiogioji, Federal Energy Administration Office of Conservation and Environment, Washington, D.C., 18 July 1975; Phone conference with Robert R. Jones and David Rosoff, 30 June 1975; Interviews with Walter J. Cavagnaro, Chief Electrical Engineer, and Rufus G. Thayer, Jr., Esquire, Counsel, both of the California Public Utilities Commission, San Francisco, California, 7 June 1975; Interview with Sharon Sellars, Federal Energy Administration, Energy Conservation Division, Region IX, San Francisco, California, 27 April 1977.

14. Ibid.

15. Ibid.

16. CCH, Energy Management, Chapters 2533-2601.

17. Ibid; Interview with Grant P. Thompson, Institute Fellow, Environmental Law Institute, Washington, D.C., 18 June 1975; Interviews with Dr. Douglas C. Bauer, Robert R. Jones, David Rosoff, Ted Farfaglia, and

REFERENCES (Continued)

Dr. Melvin H. Chiogioji, Federal Energy Administration Office of Conservation and Environment, Washington, D.C. 18 July 1975; Phone conference with Robert R. Jones and David Rosoff, 30 June 1975; and Interviews with Walter J. Cavagnaro Chief Electrical Engineer, and Rufus G. Thayer, Jr., Esquire, Counsel, both of the California Public Utilities Commission, San Francisco, California, 7 June 1975.

18. CCH, Energy Management, Chapters 2533-2535, 2540, 2541, 2552, and 2553.

19. CCH, Energy Management, Chapters 2904-2907, 2921-2827.

20. Interview with Grant P. Thompson, Institute Fellow, Environmental Law Institute, Washington, D.C. 18 June 1975; and a National Plan for Energy Research and Development - Creating Energy Choices for the Future - 1976 (ERDA-76-11).

21. Ibid; CCH, Energy Management, Chapters 9676-0689, 9690; Interviews with Dr. Douglas C. Bauer, Robert R. Jones, David Rosoff, Ted Farfaglia, and Dr. Melvin H. Chiogioji, Federal Energy Administration Office of Conservation and Environment, Washington, D.C., 18 July 1975; Phone conference with Robert R. Jones and David Rosoff, 30 June 1975; Phone conference with Steven Powers, Counsel for the Los Angeles Department of Water and Power, Los Angeles, California, 30 June 1975; Energy Policy and Conservation Act (P.L. 94-1963); Energy Conservation Project Report No. 2, October, 1975, published by the Environmental Law Institute, Washington, D.C.

22. Ibid; and "California Warren-Alquist State Energy Resources Conservation and Development Act."

23. CCH, Energy Management, Chapters 9676-9689, 9690; Interview with Grant P. Thompson, Institute Fellow, Environmental Law Institute, Washington, D.C. 18 June 1975; Interviews with Dr. Douglas C. Bauer, Robert R. Jones, David Rosoff, Ted Farfaglia, and Dr. Melvin H. Chiogioji, Federal Energy Administration Office of Conservation and Environment, Washington, D.C., 18 July 1975; Phone conference with Steven Powers, Counsel for the Los Angeles Department of Water and Power, Los Angeles, California, 30 June 1975; Energy Policy and Conservation Act (P.L. 94-163); and Energy Policy and Conservation Project Project No. 2, October, 1975, published by the Environmental Law Institute, Washington, D.C.

24. "Staff Study of Impact of Energy Shortages on Los Angeles," prepared for the Permanent Subcommittee on Investigations of the Senate Committee on Government Operations, Februarly 15, 1974, GPO Stock No. 5270-02211, as reported in CCH, Energy Management, 9758; and Chapter XIII, Los Angeles Municipal Code.

25. White House Fact Sheet, "Presidential Energy Message," (8 October 1974), reprinted in CCH, Energy Management, Chapters 655-679; Interviews with Dr. Douglas C. Bauer, Robert R. Jones, David Rosoff, Ted Farfaglia, and Dr. Melvin H. Chiogioji, Federal Energy Administration Office of Conservation and Environment, Washington, D.C., 18 July 1975; Phone conference with Robert R. Jones and David Rosoff, 30 June 1975; US Federal Energy Administration, Office of Conservation and Environment, Federal Energy Management Program, First Annual Report, Fiscal Year 1974, (Washington, D.C.: December 1974) and Interview with Sharon Sellars, Federal Energy Administration, Energy Conservation Division, Region IX, San Francisco, California, 3 May 1977.

26. US Federal Energy Administration, Office of Conservation and Environment and Office of Industrial Programs, Fact Sheet, "Voluntary Industrial Energy Conservation Program," (Washington, D.C.: June 1975); Interview with Fred King, Federal Energy Administration, Energy Conservation Division, Region IX, 6 June 1976 and Sharon Sellars of that agency on 3 May 1977; description of major programs, Office of Energy Conservation and Environment, Federal Energy Administration, Washington, D.C. November, 1975; and U.S. Department of Commerce, Energy Conservation Programs, April, 1975.

27. Federal Energy Administration, State Energy Conservation Program Fact Sheet; Energy Policy and Conservation Act (EPCA); Interview with Sharon Sellars, Federal Energy Administration, Energy Conservation, Division, Region IX, San Francisco, California, 4 April 1977.

28. Interviews with Robert Watkins, Assistant Division Chief, and Bruce Rogers, Manager of Licensing and Siting Division, Energy Resources, Conservation and Commission, on 2 July 1976; Interviews with Walter Cavagnaro, Rufus Thayer and George Amaroli of the

REFERENCES (Continued)

California Public Utilities Commission on 25 June 1976 and 26 June 1976.

29. Interview with Grant P. Thompson, Institute Fellow, Environmental Law Institute, Washington, D.C., 18 June 1975, Interviews with Dr. Douglas C. Bauer, Robert R. Jones, David Rosoff, Ted Farfaglia, and Dr. Melvin H. Chiogioji, Federal Energy Administration Office of Conservation and Environment, Washington, D.C., 18 July 1975; Phone conference with Robert R. Jones and David Rosoff, 30 June 1975; and Interviews with Walter J. Cavagnaro, Chief Electrical Engineer, and Rufus G. Thayer, Jr., Esquire, Counsel, both of the California Public Utilities Commission, San Francisco, California, 7 June 1975. Energy Conservation in the International Energy Agency, 1976 Review. Interviews with George Amorli, California Public Utilities Commission, 6 June 1976. Interview with Warren Osborn, Energy Conservation Division, Federal Energy Administration, Region IX, 6 June 1976. Interview with Bert Gauger and Craig W. Hoellwarth, Conservation Division, California Energy Resources Conservation and Development Commission, 2 July 1976; Energy Policy and Conservation Act (42 U.S.C. 6201 et. seg.); the Motor Vehicle Information and Cost Savings Act, (15. U.S.C. 1901 et. seg.).

30. CCH, Energy Management, paragraphs 3901, 3902, 3905, 3907, 3909, 3911, 3913, 3914, 4201, and 4301; Interview, Sharon Sellars, Federal Energy Administration, Energy Conservation Division Region IX, San Francisco, California, 27 April 1977; Energy Policy and Conservation Act (P.L. 94-163); and, the White House Detailed Fact Sheet, The President's Energy Program, April 20, 1977.

31. Ibid.

32. Ibid.

33. Interviews with Dr. Douglas C. Bauer, Robert R. Jones, David Rosoff, TEd Farfaglia, and Dr. Melvin H. Chiogioji, Federal Energy Adminstration Office of Conservation and Environment, Washington, D.C., 18 July 1975; Phone conference with Robert R. Jones and David Rosoff, 30 June 1975; Interview with Sharon Sellars, Federal Energy Administration, Energy Conservation Division, Region IX, San Francisco, California, 27 April 1977; Energy Policy and Conservation Act (P.L. 94-163); The White House Detailed Fact Sheet, the President's Energy Program, April 20, 1977.

34. Interview with Marshall F. Johnson, Conservation Division, Energy Resources Conservation and Development Commission, Sacramento, California, 2 July 1976; And phone conversation with Jon Leber, Conservation Division Energy Resources Conservation and Development Commission, Sacramento, California, 4 May 1977.

35. Energy Conservation Project Report No. 2, October, 1975, published by the Environmental Law Institute, WAshington, D.C.

36. Interview with Sharon Sellars, Federal Energy Administration, Energy Conservation Division, Region IX, San Francisco, California, 27 April 1977; Energy Conservation and Production Act (P.L. 94-385); Energy Policy and Conservation Act (P.L. 94-163).

37. Building Energy Authority and Regulations Survey, State Activity, Office of Building Standards, Code Services, National Bureau of Standards, March, 1976.

38. Interview with Sharon Sellars, Federal Energy Administration, Energy Conservation Division, Region IX, San Francisco, California, 27 April 1977.

39. California's proposed "Energy Conservation Standards for New Nonresidential Buildings"; California Administrative Code, Title 25, Chapter 1, Sub-chapter 1, Article 5, Section 1094, "Energy Insulation Standards"; and City of Cerritos, California, Ordinance No. 475, "An Ordinance of the City of Cerritos Amending the Environmental Performance Standards of the Municipal Code by Incorporating a Section on Energy Conservation in Residential Dwelling Units."

40. Interview with Sharon Sellars, Federal Energy Administration, Energy Conservation Division, Region IX, San Francisco, California, 27 April 1977; Energy Policy and Conservation Act (P.L. 94-163), And Federal Energy Administration Fact Sheet, State Energy Conservation Program. (Undated.)

41. Interview with Walter J. Cavagnaro, Chief Electrical Engineer, and Rufus G. Thayer, Jr., Esquire, Counsel, both of the California Public Utilities Commission, San Francisco, California, 7 June 1975; And CCH, Energy

REFERENCES (Continued)

Management, Chapters 4503 et. seg.; White House Detailed Fact Sheet, The President's Energy Program, April 20, 1977.

42. Ibid.

43. Interviews with Dr. Douglas C. Bauer, Robert R. Jones, David Rosoff, Ted Farfaglia, and Dr. Melvin H. Chiogioji, Federal Energy Administration Office of Conservation and Environment, Washington, D.C., 18 July 1975.

44. Ibid.

45. Ibid.

46. Energy Conservation Production Act (P.L. 94-385).

47. Ibid.

48. The White House Detailed Fact Sheet, The President's Energy Program, April 20, 1977.

49. CCH, Energy Management, Chapters 4503 et. seg.; Joint Economic Committee Staff Study, The 1975 Budget: An Advanced Look, 3 December 1973, reprinted on CCH, Energy Management, Chapter 9790; and CCH Energy Management, Chapter 9657.

50. The White House Detailed Fact Sheet, The President's Energy Program, April 20, 1977.

51. Small Business Administration, Fact Sheet, No. 12, September 1974, reprinted in CCH, Energy Management, Chapter 9653.

52. Ibid.

53. Energy Conservation and Production Act (P.L. 94-385) as reprinted in CCH, Energy Management, Chapters 10, 450, et. set.; and an interview with Sharon Sellars, Federal Energy Administration, Energy Conservation Division, Region IX, San Francisco, California 21 April 1977.

54. The White House Detailed Fact Sheet, The Presidential Energy Program, April 20, 1977.

55. Interview with Robert Watkins, Conservation Division, Energy Resources Conservation and Development Commission, Sacramento, California, 2 July 1976; Interviews, Rufus Thayer, Esq., George Amaroli and Walter Cavagnaro, California Public Utilities Commission, San Francisco, California 25 June 1976; and Interview with Sharon Sellars, Federal Energy Administration, Energy Conservation Division, Region IX, San Francisco, California 27 April 1977.

56. Interview with Stacey Swor, Conservation Division, Federal Energy Administration, Region IX, San Francisco, California, 24 June 1976; Interview with Sharon Sellars, Federal Energy Administration, Energy Conservation Division, Region IX, San Francisco, California, 27 April 1977.

57. Interview with Sharon Sellars, Federal Energy Administration, Energy Conservation Division, Region IX, San Francisco, California, 27 April 1977.

58. Federal Energy Administration Fact Sheet, Public Schools Energy Conservation Service (undated).

59. The White House Detailed Fact Sheet, The Presidential Energy Program, April 20, 1977.

60. Interview with Grant P. Thompson, Institute Fellow, Environmental Law Institute, Washington, D.C., 18 June 1975; Interviews with Dr. Douglas C. Bauer, Robert R. Jones, David Rosoff, Ted Farfaglia, and Dr. Melvin H. Chiogioji, Federal Energy Administration Office of Conservation and Environment, Washington, D.C., 18 July 1975; Interview with Sharon Sellars, Federal Energy Administration, Energy Conservation Division, Region IX, San Francisco, California, 27 April 1977.

61. The White House Detailed Fact Sheet, The President's Energy Program, April 20, 1977.

62. The White House Detailed Fact Sheet, The President's Energy Program, April 20, 1977.

63. Energy Conservation, In the International Energy Agency, 1976 Review.

64. The White House Detailed Fact Sheet, The President's Energy Program, April 20, 1977.

65. Ibid.

REFERENCES (Continued)

66. Ibid.

67. Ibid.

68. Report by the U.S. Department of Commerce published in 38 Federal Registry 29629 (26 October 1973); CCH, Energy Management, Chapter 9696; and Department of Commerce "Energy Conservation Programs" Fact Sheet, April 1975.

69. Interview with Michael De Angelis, California Energy Resources, Conservation and Development Commission, 2 July 1976; Interview with Walter Cavagnaro, Rufus Thayer, Esquire, and George Amaroli of the California Public Utilities Commission, 25 June 1976; Interview with Fred King, Conservation Division, Federal Energy Administration, Region IX, 24 June 1976; Proceedings of an EPRI Workshop on Technologies for Conservation and Efficient Utilization of Electric Energy, prepared by Applied Nucleonics Co., Inc., Los Angeles, July, 1976; A National Plan for Energy Research, Development and Demonstration: Creating Energy Choices for the Future, 1976, Vol. I, The Plan (ERDA, 76-1).

70. Interview with Emilio Varanini, Comissioner, California State Energy Commission, Sacramento, California 2 July 1976. The White House Detailed Fact Sheet, The President's Energy Program, April 20, 1977.

71. Ibid.

72. Ibid.

73. Ibid.; Interview with Stacey Swor, Conservation Division, Federal Energy Administration, Region IX, San Francisco, California, 3 May 1977.

74. Ibid.

75. Interview with Robert Watkins, Energy Resources, Conservation and Development Commission, Sacramento, California 2 July 1976; Interview, Walter Cavagnaro, Rufus Thayer, Esquire, and George Amaroli of the California Public Utilities Commission, San Francisco, California, 25 June 1976.

76. Ibid.

77. Interview with Emilio Varanini, Commissioner, California State Energy Commission, Sacramento, California 2 July 1976. The White House Detailed Fact Sheet, The President's Energy Program, April 20, 1977.

78. Ibid.

79. Interview with Robert Watkins, Energy Resources, Conservation and Development Commission, Sacramento, California 2 July 1976; Interview, Walter Cavagnaro, Rufus Thayer, Esquire and George Amaroli of the California Public Utilities Commission, San Francisco, California 25 June 1976.

80. Ibid., Environmental Policy and Conservation Act (Public Law 94-163); CCH Energy Management, Chapters 3901-4306; And phone interview with Jon Leber, Energy Resources Conservation and Development Commision, 4 May 1977. The White House Detailed Fact Sheet, The President's Energy Program, April 20, 1977.

81. Interview with Robert Watkins, Energy Resources, Conservation and Development Commission, Sacramento, California 2 July 1976; Interview, Walter Cavagnaro, Rufus Thayer, Esquire, and George Amaroli of the California Public Utilities Commission, San Francisco, California, 25 June 1976. The White House Detailed Fact Sheet, The President's Energy Program, April 20, 1977. EPRI Energy Workshop Proceedings, prepared by Applied Nucleonics Company, July 1976.

82. Ibid.

83. Interview with Sharon Sellars, Federal Energy Administration, Energy Conservation Division, Region IX, San Francisco, California, 27 April 1977. Energy Conservation and Production Act (P.L. 94-385); Energy Policy and Conservation Act (P.L. 94-163). The White House Detailed Fact Sheet, The President's Energy Program, April 20, 1977.

84. Interview with Robert Watkins, Energy Resources, Conservation and Development Commission, Sacramento, California, 2 July 1976; Interview, Walter Cavagnaro, Rufus Thayer, Esquire and George Amaroli of the California Public Utilities Commission, San Francisco, California, 25 June 1976. The White House Detailed Fact Sheet, The President's Energy Program, April 20, 1977. EPRI Energy Workshop Proceedings, prepared by Applied Nucleonics Company, July, 1976.

85. Ibid.

86. Ibid.

REFERENCES (Continued)

87. Ibid.; Energy Conservation in the International Energy Agency, 1976 Review.

88. Interview with Grant P. Thompson, Institute Fellow, Environmental Law Institute, Washington, D.C., 18 June 1975; Interviews with Dr. Douglas C. Bauer, Robert R. Jones, Davis Rosoff, Ted Farfaglia, and Dr. Melvin H. Chiogioji, Federal Energy Administration Office of Conservation and Environment, Washington, D.C., 18 July 1975; Phone conference with Robert R. Jones and David Rosoff, 30 June 1975; Interviews with Walter J. Cavagnaro, Chief Electrical Engineer, and Rufus G. Thayer, Jr., Esquire, Counsel, both of the California Public Utilities Commission, San Francisco, California, 7 June 1975; Interview with Sharon Sellars, Federal Energy Administration, Energy Conservation Division, Region IX, San Francisco, California, 27 April 1977; And interview with tom Jones, Energy Coordinator, Environmental Protection Agency, Region IX, San Francisco, California, 29 April 1977.

89. Ibid.

90. Ibid.

91. Interview with Tom Jones, Energy Coordinator Environmental Protection Agency, San Francisco, California, Region IX, 29 April 1977.

92. Ibid.; Interview with David R. Andrews, Esquire, Regional Council, Environmental Protection Agency, Region IX, 29 April 1977; H.R. 6161 (Bill to amend Clean Air Act), dated April 6, 1977; Toxic Substance Control Act, October 11, 1976 (P.L. 94-469); Resource Conservation and Recovery Act, October 21, 1976 (p.L. 94-580).

93. P.L. 94-469, October 11, 1976, 15 U.S.C. 2601.

94. P.L. 94-580, October 26, 1976, 42 U.S.C. 6901, amends the Solid Waste Disposal Act (42 U.S.C. 3251).

95. House of Representatives (U.S.) bill to amend the Clean Air Act and for other purposes, April 6, 1977 (pending).

96. United States Government Manual, 1976-1977, published by The Office of Federal Register, National Archives, General Services Administration.

97. European Community Information Service, "A New Energy Policy for the European Community," Background Note No. 13/1974, (15 July 1974), reprinted in CCH, Energy Management, Chapter 9697.

98. Energy Crisis: Strategy for Cooperative Action," Address by Secretary of State Henry Kissinger in Chicago, Illinois, 14 November 1974, reprinted in CCH, Energy Management, Chapter 9644; And, Energy Conservation in the International Energy Agency, 1976 Review.

99. Ibid.

100. Ibid.

101. Ibid.

102. Ibid.

103. Ibid.

104. Ibid.

105. Ibid.

106. Ibid.

107. Ibid.; "Voluntary Agreement and Program Relating to the International Energy Program," reprinted in CCH, Energy Management, Chapter 10,822; Letter of the Administrator of the Federal Energy Administration to the Administrator of the General Services Administration, 6 March 1975, reprinted in CCH, Energy Management, Chapter 10,822; and Letter of the Administrator of the General Services Adminstration to the Administrator of the General Services Administration, 6 March 1975, reprinted in CCH, Energy Management, Chapter 10,822; and Letter of the Administrator of the General Services Administration, 28 March 1975, reprinted in CCH, Energy Management, Chapter 10,822.

108. Ibid.

109. Ibid.

110. Ibid.

111. Interview with Grant P. Thompson, Institute Fellow, Environmental Law Institute, Washington, D.C., 18 June 1975. Interviews with Dr. Douglas C. Bauer, Robert R. Jones, David Rosoff, Ted Farfaglia, and Dr. Melvin H. Chiogioji, Federal Energy Administration Office of Conservation and Environment,

REFERENCES (Continued)

Washington, D.C. 18 July 1975; and Interview with Fred King, Conservation Division, Federal Energy Administration, Region IX, San Francisco, California, 24 June 1976.

112. Interview with Grant P. Thompson, Institue Fellow, Environmental Law Institute, Washington, D.C., 18 June 1975; and Environmental Law Institute, "Energy Conservation Project," summary of a project being conducted under a grant from the National Science Foundation, (Washington, D.C.: n.d.).

113. Statement by Donald Craven, Assistant Administrator for Resource Development, Federal Energy Administration, before the Texas United Press International Editors Association, San Antonio, Texas, June 1975, as reported in Commerce Clearing House (hereafter to be referred to as CCH), Energy Management, "Weekly Bulletin," (18 June 1975).

THE SWEDISH ENERGY CONSERVATION CAMPAIGN

Lee and Agneta Bendsjö Schipper

I. INTRODUCTION

The energy nightmare in Sweden began much as in the United States, under the optimistic assumption of unlimited resources. During the 1960's energy consumption grew at a rapid rate, faster than in the U.S. and faster than the economy as a whole. Even so, by the early 1970's consumption in Sweden, per dollar of gross national product, was still far less than in the United States. Unlike the U.S., however, Sweden imported nearly 3/4 of all energy used, mostly in the form of oil. Domestically, Sweden relied on inexpensive hydropower for much of her electricity, expecting to continue to expand electricity supplies by relying on then cheap nuclear and oil-based power.

In 1973, however, the situation changed drastically. The dramatic rise in world oil prices jarred the oil-based Swedish economy. The transition from hydropower to increasing reliance on thermal power plants assured that electricity prices would rise. Finally the Swedish public and government questioned the future of nuclear power in Sweden. This would not have been so surprising had not Sweden counted on more nuclear power per capita than any other country by the late 1980's. Many foreign observers, especially the nuclear industry, have tried to play down this defacto challenge to nuclear power. Nuclear critics, disappointed that Sweden has not closed the nuclear door completely, also complained.

What neither side has appreciated is that Sweden, faced with higher bills for imported oil, higher costs for electricity, and the risks from over-dependence on nuclear power, has sought a productive resolution of the energy dilemma. That resolution rests upon making energy work harder -- energy conservation. Conservation in Sweden--and in the U.S. --means reducing the energy requirements of a ton of steel, a kilometer of auto travel, or a winter of warmth. To a lesser degree conservation also means changes in behavior and preferences. How far we travel, how warm we heat, how close to work we live, how we spend our free time, ultimately affect energy needs, too.

II. ENERGY CONSERVATION - THE SWEDISH WAY

Parliamentary Decision 1975

In the Spring of 1975, the Swedish parliament reached a decision concerning Sweden's energy future. Among other things, it was decided that the yearly rate of growth in use should be reduced from an average of 4.5% to 2% until 1985. Thereafter, the rate of growth will decrease to an average of 1% in order to reach zero-growth in the beginning of 1990. While the final "Energy Policy" is not due until 1978, the first steps in Sweden have been more dramatic - and steady - than in the U.S. These steps mean that Swedes will have to learn to conserve energy. Unfortunately, the Swedish government is not always consistent in defining what it means by "conservation" or in its use of figures to measure conservation. The use of a target growth figure may lead to problems too, since the full impact of conservation is felt in the long term.

Energy Conservation Committee

In the fall of 1974, even before the parliament's energy policy decision, the government appointed a committee with the task of carrying on a campaign for voluntary energy conservation during the winter of 1974-75. After the parliamentary decision the committee's assignment was expanded:

- to push energy conservation campaigns and negotiations in order to bring about voluntary energy conservation.

- to examine the possible conditions necessary to allow introduction in certain areas of direct restrictions or curtailments on energy use.

- to continuously follow the development of energy use and render an account of the committee's judgemental forecasts for the coming 12 month's period.

The committee is made up of representatives of the five parliamentary parties, interests from business and real estate as well as experts from many agencies and organizations.

The committee will assure that parliament's goals concerning increases in energy consumption will not be surpassed. The parliamentary decision allows industry to increase its energy use faster than the other sectors, that is by 3% per year. The transportation sector is allowed to increase with 2%, while the residential and commercial sector must remain at the same level of consumption as in 1973. This means that energy use for space-heating of homes must actually fall by 0.9% per year. In the U.S.A., leaders have avoided-- perhaps wisely-- specifying consumption targets, but it is widely held that energy use in America need only grow at half the rate of the economy itself.

The Campaign for Voluntary Savings

During the winter of 1976-77 the committee led a campaign directed to the public which emphasized the importance of everyone's role in saving energy.

Means such as billboards, advertisements, matchboxes, TV and radio have been used. Normally, advertisements on radio and TV are not allowed in Sweden, but authorities can invoke allowing air time for important information concerning citizen's rights and duties. It is not clear in the U.S. or Sweden to what degree the government (like commercial advertisers) should augment information with exhortations, especially when life styles, rather than insulation or other technologies, are in question.

The Committee's work has more and more focused in on special groups, e.g., homeowners, apartment managers, auto owners and elected officials in local governments.

Local Government

The Committee is trying to influence the decision makers to take energy saving measures by giving information about the possibilities of saving energy in local governments. The Committee views it important that administrations set a good example. A handbook giving hints how to save energy for local governments will be published during the spring of 1977.

Within governmental administration, there are special instructions regarding the maximum permissible indoor temperature. A campaign affecting all government employees has begun. A captain, who is responsible for distributing information to staff, has been selected for each agency. This could cause friction between management and employees who resent being told what temperature is "best," but the energy savings in Sweden (and the U.S.) from the thermostat adjustments could amount to a few percent of total consumption. Are the savings worth the extra involvement of "big brother"? That decision must be made ultimately by the voters.

The Work Place

The largest part of energy use at work goes into production. In Sweden, as in the U.S. most of the energy is used for heat. Disappointingly little of this heat (in Sweden) is then recycled to heat factories. "Statens Planverk" has worked out an energy conservation folder for different sectors. As a complement to this, the Committee offers the businesses campaign material for energy conservation. The material has been prepared in collaboration with members of the labor market. One special campaign includes carpooling to and from work. Unfortunately, the past decade has seen a rise in Sweden of the "American way of commuting" e.g., one person to one car.

Schools

During the last couple of years, the committee has developed educational materials about energy conservation. This year all junior high and high schools are offered materials consisting of a conservation book and a work book.

Households

"Konsumentverket" (A federal "Consumers Union") has among other things examined which methods of doing the laundry and the dishes are energy-thrifty. During the Spring of 1977 the committee will inform the households of the results. Many people scoff at such details, yet the simple changes - washing in a basin instead of under running water or using a machine with water saving features reduces hot water use considerably.

Agriculture

Information concerning the possibilities of saving energy in agriculture had been prepared in connection with Department of Agriculture and Farmers' Union.

III. ADMINISTRATIVE MEASURES

If voluntary targets are not met, administrative measures have to be taken. The committee has, therefore, started to investigate possibilities of direct restrictions of curtailments of certain energy use. In the Spring of 1977, the committee suggested obligatory inspection of furnaces in order to improve boiler quality. The proposal is for the present being treated by the Department of Housing. This measure shows the ambiguous use of the word 'voluntary' - True mandatory inspections infringe upon individual nights, but these do not prohibit consumers from using as much heat as they please.

Buildings

In cooperation with "Statens Planverk" the Committee has investigated what different possibilities there are to reduce energy use in existing buildings. The investigation is trying to use already known techniques, e.g., insulation, in order to reduce the amount of energy needed for heating. The report was presented in the Spring of 1977. Goals hover around a 20-25% reduction in usage by investing in extra insulation, renovation of windows, etc. An investigation of energy use in second homes has started.

Further steps being investigated to reduce energy use include a maximum temperature in all buildings except homes, and recycling of heat from refrigerators and freezers. Investigations are also being made of electricity use in sports arenas and warm air drying of grains.

In the case of space heating and other uses in buildings, parliament has counted on zero-growth between 1973-85. During the embargo, great conservation measures were taken, but consumption climbed quickly again in 1975. Holding consumption at the 1973 level will depend on what degree society can stimulate property owners to adopt conservation measures. The Committee thinks that it should be possible to maintain 1973 consumption level through various kinds of incentives, perhaps complemented by administrative rules. On the other hand, it is unrealistic to count on a meaningful reduction in this sector's consumption before 1985. The new building codes will not greatly affect consumption before then. These standards will reduce heat losses by nearly 40% compared to existing buildings.

Industry

Industry is responsible for a good 40% of the total energy consumption. According to the parliamentary decision this sector is allowed an average increase of 3% per year. During 1973-76 their consumption decreased. The reason for the industry's low energy consumption ought to be the economic slowdown within the iron/steel and pulp/paper industries. These sectors consume half of all energy used in industry. Only after a long recovery can one for certain judge industry's needs for energy for the future. Moreover, careful observers remind us that the real results in this sector only appear after new factories and machines replace old. Authorities in Stockholm (and Washington) must, therefore, be patient.

Transportation

The transportation and automobile sector has grown much faster than 2% per year. Unfortunately, all prognoses show that the rate of increase will persist. Here additional steps are being considered to slow the rate of increase. However, a recent report suggests that use of more efficient autos can limit the overall increase in energy use in Sweden.

How the authorities will deal with the real problem-- increased ownership of vehicles-- is not clear. In the U.S. this factor, while nearly 40% higher than in Sweden, is much closer to saturation. And the gas-guzzling characteristics of American cars (13.MPG in 1971 vs. 24 in Sweden) make dramatic improvements in the U.S. more likely.

Some Criticisms

To some observers in the U.S. and Sweden, the accelerated conservation programs of these governments appear to entail sacrifices. While the economic benefits of more efficient energy use are well known, the transition can be rough. Consequently, the Swedish government has made available a total of nearly one billion dollars, over four years, in loans and grants to homeowners, municipilities, and industries, in an effort to stimulate investment in more efficient heating systems, insulation, and industrial plant. The alternative, of course, is even greater investment by industry, government, and ultimately the people, in equivalent amounts of energy supplies. If the U.S. could and would mount an investment campaign comparable to that in Sweden, tens of billions of dollars would be involved in addition to sums already earmarked by consumers and businessmen for conservation.

What are the pitfalls of the Swedish campaign? First, the line between lifestyle and investment as factors in conservation is too often ignored. The distinction between voluntary and "administrative" is also unclear, public acceptance of "administrative" measures that involve sensible investments depends upon public-- and governmental understanding of the economic rewards for these outlays.

The second hazard in the Swedish campaign is the stated emphasis upon yearly growth targets. Why not just do everything possible and reasonable to make energy use more efficient? Use taxes or other "administrative measures" that affect lifestyle only if it is decided that "efficient" energy use demands too much energy, particularly nuclear power. This may be justified since it is widely recognized that the market place alone will never represent all of the costs of using energy. But the Swedes must not be disappointed if their ambitious goals are not quite fulfilled: The Swedish plan must, therefore, not be measured by an all-or-nothing outcome. This is the third problem with the Swedish plan - the lack of unambiguous yardsticks that measures success or failure.

As these words were written, the Carter administration was readying its "April 20 Energy Plan." The parallel between the President's energy conservation proposals and those policies now in effect in Sweden is striking: Taxes, direct aid, thermal efficiency standards, cogeneration. The administration ought to take a careful, critical look, therefore, at the progress made in Sweden over the past few years. We may be able to learn from their mistakes!

What is important about the Swedish Program? From an American viewpoint, certain technologies, especially in the buildings sector and in industry look attractive for use here. These include heat recovery and cogeneration in industry, and increased thermal integrity in structures. Of the policies, direct aid for conservation investments is attractive. Taxes on gasoline and autoweight should also be considered.

It is not necessary for America to adopt all Swedish methods, nor is it expected that Sweden is an ideal model for the U.S. On the other hand, it will be important for all to see how much-- and how fast-- determined government, institutional, and private initiatives can be marshalled to improve the effectiveness of energy use. In both Sweden and the U.S. the cost of not using energy more productively will be far greater, in the long run, than the short-run price of adjusting technologies and perhaps lifestyles-- to increasingly scarce and inexpensive energy supplies.

Lee Schipper is information specialist with the Energy and Resources Group, Univ. of California, Berkeley; and a member of the "Study on Nuclear and Alternative Energy Systems" of the National Academy of Sciences. Agneta Bendsjo Schipper is studying Scandinavian Languages at U.C. Berkeley. Lee Schipper will continue his investigation into Swedish energy use and conservation with support next year from the American Scandinavian Foundation and Fulbright - Hayes travel grant, as well as from the Lawrence Berkeley Lab. Information contained herein was adapted from press releases from the Energi Spar Komitteen in Stockholm and "Efficient Energy Use: The Swedish Example" (Science 194, 3 Dec., 1976) Opinions are strictly those of the authors.

This essay was originally prepared at the request of the Swedish Information Service, New York, NY; in conjunction with a new series of information and opinion sheets offered by them.

TOWARD A CALIFORNIA ENERGY STRATEGY

Emilio E. Varanini, III
Commissioner
California Energy Resources Conservation and Development Commission
1111 Howe Avenue
Sacramento, California 95825

INTRODUCTION

The 1973 "Energy Crisis"--and After

Until a few years ago, energy was beyond the consciousness of most Americans and taken for granted by most public agencies. It took the OPEC embargo of the winter of 1973-74 to change all that. This crisis produced a consensus that something had to be done but little agreement as to exactly what.

The Nixon and Ford administrations responded with a policy aimed at ending dependence on foreign petroleum imports by 1985. Initially, this policy appealed to patriotism and self-denial. Later, the strategy consisted of several policies designed to stimulate the development of domestic energy resources and to relax environmental regulations that might hinder the extraction or the utilization of these resources.

During this same period, the Americans were continuously exposed to oil company advertising explaining the source of the energy crisis and why they needed high prices and high profits. In response, environmentalists, consumer groups, and congressmen charged collusion and deceit.

In this atmosphere of discord the administration's program floundered. Bills to implement subsidies for coal and uranium development were stalled in Congress; programs of accelerated leasing of federally owned coal and oil resources were effectively opposed by the affected states and local governments; attempts to relax environmental standards succumbed to legal challenges.

The acrimonious and unproductive energy policy debate was dysfunctional in another important way. While federal efforts and public debate were focused on the oil problem, developments on two other fronts with a potentially greater importance for energy policy were occurring. First, the nation was running out of domestic natural gas--and natural gas cannot be as conveniently imported as oil. Second, with the onset of the petroleum shortage, the administrative regulations authorized by environmental legislation (especially the Natural Environmental Policy Act) passed in the early 1970s were coming into force. These regulations lengthened the regulatory process, precluded many traditional technologies and, in general, made energy supply more costly.

The unpredictable impact of unusual weather--as in the East and Mid-West cold wave and the drought in the West in 1977--has catapulted the gas and electricity supply problems into the public consciousness. We are not sanguine that the new administration will have easy success in effecting the necessary fundamental policy changes. The reason is simple: there is a deep-seated and fundamental disagreement in this country about the extent of our economic and environmental resources, the costs of various energy supply technologies, and, far more important, the proper objectives for an energy policy. In such an atmosphere of discord and uncertainty, fundamental change in any direction would be very difficult to effect.

As a regulatory body faced with the necessity of making decisions, this Commission is acutely aware of the need for an overall strategy, to lend consistency and coherence to decisions. Our Biennial Report legislative mandate, in fact, calls for such a strategy. However, it is a delusion to believe that anyone can construct a comprehensive energy policy that will not engender serious--and perhaps crippling--opposition from one quarter or another. A practical approach must incorporate an understanding of the conflict of values and beliefs and it must be workable in an atmosphere of discord and conflict.

The Policy Conflict: Are We Entering the Age of Scarcity?:

At present, there is a fundamental argument that makes it impossible to develop a consensus energy policy: Are we entering an era of resource scarcity that will inevitably force changes in the structure of our economy and the pattern of our lives? Or will advances in technology and price increases enable us to continue our present lifestyles and balance energy supply and demand?

At one pole are those who believe that the historical trend of economic activity (with its ever-increasing per capita energy use and environmental pollution) will create profound problems if continued. They argue that we must encourage attitudes and create institutions that will channel economic activity in directions that reduce resource consumption and environmental impact.

The advocates of this view contend that scarce resources and limited waste disposal space will prevent the economic trends of the past from continuing far into the future and that increasing resource costs will encourage a more energy efficient economic system. Our only choice is how we make the transition. If we recognize the inevitable and try to make the transition smooth, we will be better off than if we "go down fighting." Continuing "business as usual" to the point of collapse will do damage, perhaps permanent, to our environment and our social institutions.

At the other extreme are those who believe that we can and must continue (or even accelerate) economic trends. Proponents of this view are little impressed by impending scarcity at the global level. They point out that people have been crying wolf about resources for almost 200 years. It has also been shown that the real cost of resources, both mineral and energy to the U.S. economy has declined for this entire century--presumably technological advance has more than kept abreast of declining quality of resources. The proponents of this view concede in the very long run we might feel the effects of resource scarcity--but there is no compelling evidence that we need worry about it now. Similarly, they play down the effect of industrial development on global geophysical or ecological cycles. Since our understanding of global processes is so poor, they argue that we would be unable to do anything if we delayed or halted every activity that might cause problems.

The proponents of this view see poverty and unemployment as far greater threats to societal well-being than either pollution or resource scarcity. Reducing both should be our major goal. Within a democracy it is very difficult to redistribute income; between rich and poor nations it is virtually impossible. The only way that the world's poor can be made better off is by increasing the total amount of wealth available. At home, they argue that high unemployment is the primary problem and the only practical way to reduce it to acceptable levels is to maintain a high level of investment and economic growth.

These two views are fundamentally different, in their perception of the world and in their hierarchy of social goals.

THE CORPORATE PLANNING RESPONSE

A brief look at California's energy history reveals that:

- Conventional sources of energy for the state (natural gas, hydroelectricity, and onshore oil) are limited or declining.
- Our lifestyles, urban and transportation patterns, buildings, and equipment all reflect the historic cheapness and abundance of energy.
- Federal, state, and local environmental protection agencies and groups are demanding that new energy development prove that it is "needed" and will not harm the environment.

To make up for both the declining onshore oil and gas production and to meet what they perceive as California's increasing demand for energy, the major energy companies and utilities are planning new electricity, gas, and oil supply projects. These projects--nuclear power plants, offshore oil platforms, oil supertankers, oil and coal gasification plants, liquified natural gas tankers and terminals--are of a different character than traditional projects, such as onshore oil and gas pipelines and hydro and gas-fired electricity generation, that have supplied California energy requirements. They are much bigger and more expensive. The risks of environmental damage are of unprecedented magnitude. Other projects affect air or water quality and require many permits from the agencies involved in both energy and environmental regulation. But the energy companies and many others are convinced the projects are needed.

We discuss the corporate plans first because, for the present, at least, these companies do all long-range energy planning; the companies decide how much energy will be necessary 10 or 20 years from now and plan the kinds and numbers and locations of the various facilities. State and local governments only get involved when the companies come to them for the necessary permits--thus acting largely only as a filter for unacceptable projects rather than planning an integrated energy system. In the absence of change, the corporate plans and the government's likely reaction to them are the future.

The energy corporations plan to bring us a future in which we gradually come to rely on a few technologies that are of a very large scale. Because these projects are large, with long lead times, they are likely to face a complex private market and governmental approval process that could lead not only to delay, but even to failure. Because some of the projects may be initially economically marginal and because delays and errors in technical judgment can lead to cost overruns, the projects will be difficult to finance. Finally, the nature of the projects themselves present health, safety, and environmental costs and risks.

While we are stressing the regulatory and administrative problems that face these large supply projects, we are neither approving nor condemning them. But it must be admitted that the current regulatory institutions that fulfill these roles are clearly imperfect, and that the regulatory process could be streamlined. But sensitive, effective public decision-making must allow for interested persons and groups to affect the outcome; it must operate in a plural atmosphere. Because of these necessities, public decision-making in comparison with private actions would be necessarily more complex, more time-consuming, and more uncertain, even with perfect institutions. By their size and impacts, the large projects require more public scrutiny, thus at least part of the regulatory delay and uncertainty is a real and unavoidable cost.

It would be clearly desirable to find some philosophy that combines both new or different energy supply technologies with less environmental and health and safety impacts and thus less intrinsic need for regulation and new or improved regulatory institutions that would streamline the regulatory process.

There is no lack of candidate schemes that purport to satisfy this goal. In the next sections we will examine three especially prominent ones.

THE THREE PHILOSOPHIES

The corporate plans and government's likely response to them will bring us the worst of possible worlds: Many projects will be delayed or defeated, raising the possibility of energy shortages; and those projects that make it to the end of the regulatory gauntlet will be large, risky, and environmentally intrusive. Clearly some change in energy policy is desirable.

We have identified three widely held points of view about the direction the change should take place. These views are labeled:

- Conservation and Appropriate Technology, which was termed "Jeffersonian" because of President Jefferson's philosophy of the citizen farmer, self-reliant and creative;

- A Return to Government Fostering Investment and Development, termed "Greelian" after Horace Greeley, a leading proponent of western development in the United States, circa 1850-70; and

- Better Government Planning and Regulation, termed "Periclean" after Pericles, the leader of the golden age of art, music, commerce, and government in ancient Athens.

The visions of the future from each of the three points of view are quite different, in terms of the kind and amount of energy produced and consumed, the level of environmental control, the distribution of economic power, and the importance of the state in making economic decisions.

These differences in perspective stem from differing attitudes toward resource-induced scarcity.

The Jeffersonian is convinced that resource scarcity will be an omnipresent constraint in the future working of our economy. Moreover, he is convinced that technological progress and increasing scale can no longer mitigate the effects of diminishing returns--and increasing emissions into the environment will actually exacerbate the problem. The Jeffersonian believes that grand, centralized decision-making cannot plan growth so that these problems can be avoided--both because he doubts the ability of the political and administrative systems to act rationally and because our understanding of the economy and global ecosystem is so rudimentary as to preclude rational planning.

Instead, the Jeffersonian argues, we must voluntarily accept a simpler, more decentralized economy--an economy whose ability to co-exist with the environment has already passed the test of time, an economy based ultimately on renewable resources that are immune to increasing scarcity.

The Greelian, on the other hand, believes that resource scarcity is not a fundamental difficulty, at least in the near term. It can be overcome by the historically proven remedies of capital investment and technological progress. The real problem is that senseless and irrational environmental regulations may prevent enough investment, deflecting the economy toward recession, high unemployment, and all the social pathologies that accompany economic stagnation.

The Periclean holds that scarcity is real and must be dealt with. He argues, however, that our understanding of the world is such that if we had the proper planning institutions to rationally guide investment, growth could continue without radical changes in lifestyle. The Periclean sees lots of "slack" in the present system, unnecessary and pollution waste created by short-sighted private decision-making and unnecessary delay and confusion caused by contradictory, uncoordinated public decision-making.

PLANNING IN THE FACE OF UNCERTAINTY

Scarcity and the extent to which our institutions can deal with it are the key factors that determine the choice of energy policy. To break the current policy deadlock, these problems must be resolved.

We would like to be able to say that, in its first 18 months of existence, our Commission has been able to do this. Of course, it has not. Resource scarcity is an enormous, difficult problem, in many ways _the_ problem of our era. While we do not expect to resolve once and for all questions of scarcity and its effects, it does hope to continue a program of research and publication that will elucidate these questions and inform the people of California. Work so far has included a comprehensive assessment of the extent of energy resources, oil, coal, gas, and nuclear fuels available to the state and what we believe is a superior analysis of the costs attributable to air pollution. Work continues on the technological feasibility and economics of various gas and electric supply technologies.

In addition, the Commission intends to investigate several other areas that are crucial to the scarcity question: the impacts of the plutonium economy, the long-run availability of transportation fuels, and the economic, social, and political impacts of increasing resource scarcity.

Toward a California Energy Strategy

The Lack of Consensus. We lack confidence in our ability--or anybody else's--to forecast future events accurately. Furthermore, we are not wholly convinced by any of the three points of view described above. Any attempt to implement one of these three programs would be impossible without widespread consensus about energy and environmental policy. Each of the programs required a substantial departure from the status quo, the dismantling of government programs and institutions, and the wholesale creation of new ones. _We believe that determined opposition from any substantial minority or coalition of minorities could stymie the implementation of any one of the three programs_.

What then should we do? Given uncertainty in demand, technological, risk and policy plurality, how does the California state government choose a responsible, prudent energy strategy? Clearly, we cannot wait until the debate over resource scarcity is resolved academically or politically; we require a practical energy policy for the interim.

The Interim Strategy. We propose several principles upon which to base an interim energy policy. These principles are rooted in the premise that the central problem of energy policy is uncertainty about levels of demand, about costs and availability of energy resources and technologies, and about the effects of energy use on the local and global environment. It is not expected that actions based on these principles to interrupt necessary projects currently in progress.

1. There is security in diversity. Making energy decisions is something like playing roulette, except (fortunately) the game is rigged so that on average the players do tend to win. We believe that the prudent policy bettor spreads the risk by planning to use a diversity of strategies so that problems with one will be balanced by successes with others.

2. There is benefit in flexibility. It seems undeniable that we are more uncertain about energy costs, demand, and so on 20 years from now than we are about the same factors only five years from now. It follows from this that projects that can be easily and cheaply modified as to size of environmental impact are more desirable than those that cannot.

3. When there is a choice, we should stick to what we know how to do--rather than rely on totally new technologies. It is a maxim that the further a particular technology is away from commercialization, the better it looks--a technological version of the old "grass is greener" principle. Recent examples are numerous.

4. To the extent possible, regulation should be avoided. This principle has two faces, light and dark.

   On the light side, it seems that the last five years of energy development have been a history of one project after another foundering on the shoals of environmental regulations. To some extent, we believe that the Greelians are correct that these environmental regulations have been chaotic and irrational.

   The dark side is that federal regulation and the controls on energy availability and prices exercised by foreign governments create uncertainties above and beyond that inherent in the technologies involved.

Based on these principles, we have chosen policies from all three points of view. We feel that these comprise a practical, synthetic set of policies--which in the interim can improve California's energy future.

ELEMENTS OF A PRAGMATIC STRATEGY

We have outlined a pragmatic energy strategy as the appropriate response in an uncertain and contentious energy milieu. Here we present the programs to implement this strategy. Four themes--diversity, flexibility, reliance on tested technologies, and avoidance of unnecessary regulation--will reappear throughout.

Conservation

The keystone of the Commission's program is a practical, aggressive conservation program. There are many conservation measures that save an increment of energy more cheaply than a new increment can be supplied, with many added benefits: conservation is more labor intensive, less environmentally harmful, and much less risky. There are no dangers of supply interruptions caused by the actions of OPEC or federal regulators.

State action is needed to stimulate conservation for several reasons: regulated electric and gas prices do not reflect the cost of society of new supplies; there is a "first cost" bias among consumers; and there is a lack of a market infrastructure to "sell" conservation.

We propose conservation programs that can reduce both electricity and gas use by 10 percent by 1985. For electricity, this amounts to one-third of the growth in electricity demand. The conservation effort should include programs to:

- o Refine and strengthen existing conservation standards for new buildings;
- o Promote the retrofit installation of energy-conserving hardware on existing residential and commercial buildings;
- o Refine and strengthen appliance efficiency standards;
- o Improve efficiency of existing electrical generation, transmission, and distribution systems;
- o Develop and implement an energy conservation curricula for primary and secondary schools;
- o Work with professional licensing boards to reform training and licensing procedures for architects, building

contractors, engineers, and other professional groups whose decisions affect energy use.

Co-Generation

Co-generation is un umbrella term covering a number of technologies that allow waste heat created in generating electricity to be used in industrial processes, thus greatly increasing the efficiency with which fuel is utilized. The potential for co-generation in California in existing facilities has been estimated to be in excess of 100 megawatts.

To develop the potential of co-generation, the Commission proposes to:

- o Begin a research program to better identify the co-generation potential in California;
- o Begin a demonstration program, in cooperation with utilities and industry that will demonstrate the use of new technologies for the clean use of fossil fuels;
- o Work, in cooperation with the ARB, toward an air pollution permitting process that accounts for the net pollution benefits of co-generation;
- o Develop a streamlined siting procedure for co-generation facilities (the current three-year process used for conventional generating plants is well suited to co-generation facilities);
- o Develop, with the PUC, policies governing the availability and price at which utilities provide wheeling services and standby reserves.

Diversifying Electricity Supply

Many factors contribute to uncertainty in future electrical demand. Increased economic growth will increase electricity demand as greater numbers of new office buildings and factories are built. The long-term effect of OPEC-induced price increases, past and future, is still uncertain and the future of natural gas price regulation and end-use allocation will have effects. Other demographic, social, the impact of conservation programs and economic factors will also contribute to the uncertainty.

The key question is: What policies should be adopted vis-a-vis electricity supply in the face of this uncertainty?

The Benefits of Diversity. Anyone wanting to build a thermal power plant over 50 megawatts must apply to the Energy Commission for certification. Without going into the complexities of this adjudicatory process, it is sufficient to say that the Commission can influence the type and location of power plants built in this state. We believe that there are compelling arguments in favor of a diverse supply and, therefore, the Commission should use its siting process to encourage the utilities to incorporate a diverse mix of fuel types and generating technologies in their generation mixes.

There are three parts to this argument:

- o There are risks associated with relying on any ore technology or fuel type to generate electricity.
- o The best way to deal with these risks is by building a diversity of power plants. This follows from the fact that the risks are to a large extent unpredictable and uncontrollable.
- o The benefits of spreading the risk exceed the cost. Work done for the Commission shows that the relative costs of various generation technologies is very uncertain. Thus, the relative economic advantage of any particular mix is also uncertain. However, the benefits of hedging are clearly correlated with fuel type.

Right now, California relies largely on hydroelectric and oil-fired power plants for both base and peaking applications. Thus, our rule suggests that in new siting applications we ought to favor coal from diverse sources, nuclear to the extent that the technology can satisfy the strictures provided by state law, and the so-called clean fuels, which will be discussed below. Nor should a mere mix in capacity suffice. If a system's baseload capacity is nuclear and the remaining capacity gas turbines, a disruption of nuclear service would be serious. There should be a mix of plant types in both baseload and peaking applications.

The Smaller Power Plants' Option. The Council on Economic Priorities, a non-profit group dedicated to watchdogging large corporations, recently published a provocative book on coal and nuclear power plant economics. Their statistical analysis showed that decreasing reliability offset economies of scale, suggesting an optimal plant size of about 600 MW for both coal and nuclear. Moreover, their analysis showed that the penalties of erring on either side are not too large. Since smaller plants are often more efficient and have less environmental impact, these smaller plants could likewise often be desirable.

The utilities have heavily criticized the CEP study. While the end result of this debate is not clear, the possibilities, especially the lack of economic penalty for lessened environmental insult, are provocative. In order to stimulate serious discussion of these possibilities, the Commission should:

- Assess the extent to which smaller power plants might reduce environmental problems and increase efficiency in California.
- In the Notice of Intent proceeding, encourage utilities to propose smaller plants as alternatives and to assess the relative economic and environmental benefits of plants of different sizes.

Lead Times and Power Plant Siting. We have repeatedly emphasized our theme that the demand for electricity in the future is quite uncertain. But to avoid shortfalls, one must be conservative in planning electrical generating facilities and, in fact, tend to plan for the worst case, the high-demand future. Conservatism thus imposes the additional cost of building too much capacity and then letting it stand idle.

This uncertainty cost is higher for technologies with long lead times--simply because uncertainty in demand for electricity 15 years from now is far greater than for only five years from now. However, the plants with longer lead times will probably turn out to be cheaper (though not necessarily--recall the great uncertainty in the relative costs of various plant types, discussed earlier). Thus, the probable cost advantages of long lead-time plants must be balanced against the greater probability of building too much capacity.

This problem of lead times is especially crucial because of the Energy Commission's mandate to site power plants only if they are needed--that is, if they are consistent with the Commission's 10-year forecast. There is no absolute need to initiate long lead-time plants, but rather there is an economic need for such plants. The extent of this depends on the trade-off discussed above.

We plan to make a comprehensive investigation of this matter. It will at the same time require the utilities to submit, along with the usual demand and resource forecasts, a clear discussion of the methods they use in making their capacity expansion decisions, since that relates to the problem of choosing short or long lead-time plants.

Marginal Cost Pricing For Gas

In California, gas is currently priced on a rolled-in basis--meaning that the price charged the consumer is based on the average of the costs of the various sources of gas available to the utility.

Rolled-in pricing has the defect that it sends the wrong price signals to consumers. Saving a thousand cubic feet of gas through conservation, for instance, might be cheaper than producing a new synthetic natural gas (SNG) plant. If this were the case, gas prices should signal consumers to make the conservation investment. Thus, in this simple example, gas should be priced at the cost of SNG--the marginal source of gas. When the new gas sources are more costly than historic ones, rolled-in prices will be lower than marginal ones--prompting the consumer to use too much gas and under-invest in conservation or alternative fuels.

Because public utilities are regulated so as to earn only a fixed rate of return, it is impossible to simply charge every customer the marginal cost--the utilities would make too much money. Several schemes have been proposed that attempt to solve this problem and to preserve most of the advantages of marginal cost pricing. Of these, a steeply inverted block rate structure seems to be the most desirable. We suggest that a program to design and implement such a pricing strategy be started.

Toward a More Flexible Gas Policy

The conventional wisdom is that California and the nation are in the midst of a worsening crisis in gas supply. Many customers have already been curtailed, and it is argued that many are more likely before any new gas supply projects are completed. At first glance, there seems to be no reason for hesitation: we need gas and should be willing to invest in whatever projects are necessary to get it. To the contrary, it is argued that the future demand for high-priced supplemental gas <u>is</u> uncertain.

<u>The Causes of Uncertainty</u>. There are four separate aspects to this uncertainty:

- Consumer response to greatly increased gas prices is uncertain, because the historic price advantage of gas relative to oil is likely to be reversed. Depending on the actions of air pollution regulators, the demand for gas could be greatly reduced.

- The continued availability of gas from conventional sources depends on the unpredictable actions of the U.S. and Canadian governments. Deregulation of interstate gas prices and continued availability of Canadian gas could greatly reduce and possibly eliminate the need for planned supplemental and expensive sources.

- Planned supplemental sources--including Alaskan North Slope gas, liquefied natural gas (LNG), and synthetic natural gas (SNG) from Southwestern coal--face technical, financial, and environmental problems that could delay or preclude them as sources.

- A variety of unconventional sources are not being planned for but may become available. Examples are gas from Elk Hills Naval Petroleum Reserve, SNG from naphtha, methanol, liquid petroleum gas (LPG), ultra-low sulfur fuel, oil and interstate gas that could be made available by a state gas purchase authority, which is proposed below. If energy should become available in substantial amounts from these sources, and they are acceptable to environmental regulatory agencies, the demand for gas from supplemental sources could be reduced.

<u>A Strategic Comparison</u>. The aim of the state's gas policy should be to provide gas to all customers who want to buy or must (because of air pollution requirements) buy it at the incremental price. Given the uncertainty discussed above, the key question for California is: given this uncertainty, how much and what kinds of incremental supplies should we commit ourselves to purchasing?

Figure V-1 compares six attributes of several of the incremental sources. Referring to it, the reader will see that the Resource Base column indicates the extent to which California can obtain these sources without increasing the cost of supply. The Domestic Resource Base column is an index of the immunity from foreign control. The third column indicates whether a fuel is not gaseous; if it is not, it will not substitute for gas in many uses. The next column classifies the varying degrees of regulatory constraint that could preclude the possibility of using the incremental source. Capital intensiveness--shown in the fifth column--is crucial because it is a measure of the inflexibility of a commitment to one of the sources. Next, Current Status is important because it tells us what can be accomplished in the short run. The last column shows the relative costs. The table demonstrates that each of the incremental sources has its advantages and disadvantages but no single alternative stands out as California's choice.

To reduce the risks involved in supplying gas to California, the Commission proposes a package of programs to diversify and expand gas supply:

- <u>LNG</u>: The Governor and the Legislature should coordinate to create a special, one-time siting process so that there is no unnecessary regulatory delay in siting a regasification facility, but with the condition that the facility and shipping lanes are located a safe distance from population centers.

- <u>Gas Facility Siting</u>: A comprehensive gas planning and facility siting authority should be created to enable more rational state decisions on gas facility siting.

- <u>Elk Hills</u>: The California congressional delegation should urge the Navy to make public economic analyses of selling the gas produced at Elk Hills, and in particular should urge that the gas be valued at the incremental (LNG or SNG) price.

- <u>Flexible Sources</u>: Because of the uncertainty in demand for costly supplemental sources, it would be desirable to have flexible, short lead-time gas supplies that could be turned on and off. The two obvious candidates are SNG from naphtha and LPG.

  To make these technologies practical, the Commission recommends:

  -- The Governor and the California congressional delegation should urge

FIGURE V-1

A COMPARISON OF CLEAN FUEL ALTERNATIVES

	Resource Base	Domestic Resource Base	Gaseous Fuel	Regulatory Constraints	Capital Intensiveness	Current Status	Cost $/MMBTU
LNG	large	small	yes	high	high	Two projects well into regulatory process	2.50-4.00
Coal SNG	large	large	yes	high	high	Two projects await loan guarantees	3.50-5.00
Naphtha SNG	moderate	moderate/ small	yes	high	low	SoCal project shelved in 1973	3.50-5.00
LPG	small	miniscule	to some extent	low	low	One terminal completed no propane air plants	2.50-4.50
Methanol	large	small	no	none	high	Conceptual design	2.50-4.00
Ultra low sulfur fuel oil	moderate/ large	moderate	no	low	low	Desulfurization capacity exists	2.50-3.00

the FEA to end the allocation of naphtha;

-- The gas facility siting authority should allow rapid decisions for naphtha SNG plants;

-- The CPUC should encourage the gas utilities to make the appropriate modifications in their system to use LPG safely;

-- The possibility of requiring all LNG terminals to have facilities for unloading LPG should be investigated.

o Offshore Gas: The Governor should encourage the development of Santa Ynez gas from offshore Santa Barbara consistent with coastal protection.

o Gas Purchase Authority: A unit within the energy finance authority (see below) should assess the amount and cost of interstate gas that is available and the possible legal difficulties in gaining access to it.

o Methanol: A standby reserve of methanol should be created for use by electric utilities in the South Coast Air Basin during air pollution episodes.

Air Pollution Regulations

The demand for clean air is a key constraint on energy development in California. The impact of air pollution control is greatly worsened by the particular regulatory path that California and the nation have chosen.

Current federal law provides that the required State Implementation Plan must guarantee that all air basins in the state will meet ambient air quality standards by July 1, 1977. Existing sources were required to use "best available control technology" when permitted. New sources are required to meet the New Source Review (NSR) rules, which provide that no new source can contribute to the continued violation or eventual attainment of an ambient air quality standard. In areas whose air quality is already worse than the standards this could preclude any new sources except that the so-called trade-offs policy allows new sources to arrange for reductions from existing sources, and then considers the net contribution.

The trade-offs scheme is desirable in that it equalizes the cost of controlling existing and new sources--but it has several serious drawbacks:

o It is cumbersome because the builder of a new facility must find a willing firm with the correct amount and kind of emissions;

o It is inefficient since it motivates existing sources to maintain their emissions at the maximum level;

o It favors large existing firms over smaller or new firms;

o Control technologies discovered in a trade-off might be subsequently required to be retrofitted on older existing sources--in effect, depleting the supply of trade-offs.

The Commission recommends that a comprehensive re-evaluation of air pollution regulations be made.

Solar Energy

Three separate kinds of solar energy programs are proposed: a wind power commercialization program, continuing research program in solar electric, and a packages of policies designed to increase the penetration of solar devices into the space and water heating markets.

Wind Power. The technology for producing electricity from the wind is old and relatively well proven. The present "one off" wind machines are not competitive with conventional sources, but estimates of future costs are encouraging. Several factors make the lower limit on the cost of wind-generated electricity difficult to estimate. First, the extent of the wind resource is unclear: current estimates which are based on average wind conditions over large areas, tend to under-estimate the amount of wind power available. Second, the effect of mass production is uncertain. Finally, the costs and required amount of energy storage are unclear.

The Commission will propose a three-step program: a wind survey, the development of a limited number of prototypes, and finally, should the evidence recommend it, a commercialization program aimed at having one percent of installed capacity wind-generated by 1985.

Solar Electric. Solar electric technologies are some distance from commercialization. As part of its planning mandate, however, the Energy Commission is participating with the Southern California Edison Company, the Los Angeles Department of Water and Power, and ERDA in a plan to build the nation's first solar electric-power plant, a 10 MW facility near Barstow.

A Program to Accelerate Solar Space and Water Heating. There is a widespread consensus that the solar space- and water-heating technology is ready for commercialization. When compared with the cost of incremental energy sources, solar is economic in many instances. Solar heating has extra benefits: it is less environmentally intrusive and there is less risk of interruption.

The first part of the Commission program is a two-pronged attempt to neutralize the untoward effect of rolled-in pricing on solar investment. First, an income tax credit is proposed that would make the payback period comparable to that which would exist if incremental pricing were instituted. Second, a program of direct utility investment in solar is suggested because the high initial investment will remain a hurdle to many consumers. The utility would include the investment in the rate base and charge the user a rate pegged to the cost of heating his home with gas. This would provide the utilities with the explicit choice of investing at either the LNG-SNG margin or the solar heating margin.

Several other measures can reduce the institutional barriers to solar investments. The Commission:

- Proposes to convene a task force for the purpose of drafting a comprehensive bill on solar rights;
- Will propose legislation that would empower it to test solar devices to guarantee their performance as advertised;
- Recommends the promulgation of a code for passive solar design, with code conformance making builders of new structures eligible for a pro-rated tax credit;
- Encourages the inclusion of sections on solar energy in the licensing examinations of such professionals as architects and assessors, who will have great influence in the commercialization of solar.

State Energy Finance Authority

One of the striking trends in recent years is that the energy industries are increasingly unwilling to invest in many kinds of new facilities without some kind of government subsidy. There are undoubtedly good reasons for such subsidies: the projects are very large and new technologies and long lead times make the projects extraordinarily risky.

But it seems that equally good, if not better, arguments could be made for subsidizing conservation, solar, wind, and clean fuel-investments. Technological risks and market failures do inhibit investment in these areas, but the public benefit of such subsidies is far more obvious than in the case of uranium enrichment, for example.

To help remedy this situation, the Energy Commission proposes an energy finance authority, with the purpose of directing government monies, both federal and state, into energy development projects that are desirable from a social or environmental point of view but would otherwise be unlikely to receive subsidies.

The energy finance authority should be created as a publicly held corporation, governed by a board composed of ex-officio members of state government, staffed by a small cadre of professionals, not constrained by the severe strictures of civil service hiring.

The finance authority would have several extremely useful functions:

- It could act as a credit-acceptance corporation to compete with energy corporations for federal energy subsidies and redirect the monies to environmentally benign forms of supply. For example, the authority might be able to finance the methanol stockpiling program mentioned above.
- It could administer other state energy-development programs funded by bonds. Such administration is required under existing state bond laws. Programs could include:
  - The provision of low-interest loans for conservation investments or installations of solar space or water heating in residences.

- Commercialization of wind power, as proposed above.

- o The proposed state gas purchase authority could obtain required funds for leasing and perhaps purchase of gas reserves from the energy finance authority.

- o The finance authority, using financial auditing methods, could identify implicit subsidies in current financing arrangements and make appropriate recommendations to regulatory agencies.

- o The finance authority could also provide an alternative to all-events tariffs. Replacing the present ratepayer underwriting of risk with no equity return, the finance authority could provide part or all of the equity and receive an appropriate return. A variety of ways of returning these dividends to the ratepayers has been proposed.

Labor Tax Credit

If we are entering an era of scarcity--one in which resource and energy costs increase more rapidly than the general price level--a key problem for government will be how to ameliorate the effects of scarcity on employment and economic activity.

A recent Brookings Institution study on the effect of the OPEC price increase argues that western governments mistook increasing oil prices for inflation, and at exactly the time their economies needed stimulation, the central banks put the economic brakes on in the form of higher interest rates, precipitating the recession of 1974-75.

Most of the financial instruments that the government uses to stimulate the economy--for example, lowering interest rates or increasing the investment tax credit--encourage investment. This is a problem because energy and capital investment are what economists call complements: if the amount of capital used to produce a unit of output increases, so does the amount of energy used. Thus the short-run need to prevent recessionary impacts of increasing energy prices could well conflict with the long-term need to respond to the higher prices by shifting to a less energy-intensive economy. By encouraging investment--and thus energy use--the eventual reckoning is only postponed. But ignoring the recessionary problems is obviously also undesirable.

One fiscal instrument may provide a way out of this dilemma. The labor tax credit would allow employers a credit against income or wage taxes in an amount proportional to the number of employees. As a counter-recessionary measure the labor tax credit is argued to have several benefits. It is less inflationary than a tax cut since it does not stimulate demand in advance of stimulating production. It is favored by many over public jobs programs since the jobs created are more likely to lead to permanent employment. Finally, it does not favor investment and thus energy use.

Currently, there are no less than four separate such proposals before the Congress, including one by President Carter. The Commission supports the implementation of a labor tax credit at the national level. The Commission also recommends that the Governor and appropriate fiscal agencies investigate the creation of such an institution in California.

TREND OF RECENT REVISIONS TO SULFUR OXIDES REGULATIONS AND FUEL UTILIZATION

J.D. Crenshaw
U.S. Environmental Protection Agency
Research Triangle Park, North Carolina 27711

C.H. Kuo
Mississippi State University
Mississippi State, MS 39762

ABSTRACT

Air pollutants such as sulfur oxides, particulate matter and nitrogen oxides are emitted into atmosphere during combustion of fuels. Among these pollutants, sulfur oxides are most closely associated with public hazard and, at the same time, related to characteristics of fuels. Therefore, the present paper reviews and discusses historical development and recent revisions of regulations governing emissions of sulfur oxides and impact of these regulations on fuel utilization.

In 1971, the Environmental Protection Agency (EPA), under the authority of the Clean Air Act of 1970, promulgated National Ambient Air Quality Standards (NAAQS) for six pollutants including sulfur dioxide, particulate matter and others. Two standards, primary and secondary, were set for each pollutant in this promulgation to protect public health and welfare. To attain and maintain these standards, each state and jurisdiction was required by the Clean Air Act to submit for approval by EPA a plan (known as State Implementation Plan or SIP) specifying strategies for control of these pollutants. In addition to the State Implementation Plans, certain new and modified sources are restricted further by Federal Standards for New Stationary Sources (commonly referred to as new source performance standards or NSPS). Since the initial approval for most states in 1972, revised standards for control of emission of the air pollutants have been proposed by some states and approved by EPA. As required by the Energy Supply and Environmental Coordination Act (ESECA) of 1974, EPA also reviewed each State's applicable implementation plans and reported to the State on whether such plans can be revised in relation to fuel burning stationary sources without interfering with the attainment and maintenance of the NAAQS. For plans which were determined by EPA to be feasible for amendments the states were notified to submit revised plans for consideration.

To meet fuel shortage and to utilize energy resources of the United States, power plants and major industrial installations have been encouraged to use coal as primary energy sources and in some instances, a fuel burning installation may be prohibited by ESECA to burn natural gas or petroleum products. Although reserves and supplies of coal are plentiful, some coal is too high in sulfur content to be burned in compliance with State and Federal regulations for sulfur oxides without the use of emission reduction systems which may be either costly or impractical in some cases. Therefore, many states have been re-evaluating their sulfur oxides regulations to insure that scarce low sulfur fuels are being required only in areas where they are needed to protect public health and welfare.

Revised standards have been approved by EPA in recent months for many states including Alabama, District of Columbia, Florida, Georgia, Indiana, Kentucky Maryland, Massachusetts and New Mexico. In addition, a plan for control of sulfur oxides in Ohio has been promulgated by EPA. Unlike the earlier practice of setting a standard for an entire state or an Air Quality Control Region (AQCR), smaller geological domains such as county often has been chosen as a basis in the recent promulgations and revisions. Results of analysis of monitoring information, inventory of emissions from all sources and modeling of impact of emissions on ambient air quality are among the most important consideration in determining level of relaxation permissible. In areas where air qualities are marginal or inferior, stringent limitations in sulfur content of fuels or allowable emissions of sulfur dioxide have been kept to insure the attainment and maintenance of NAAQS. Relaxation of allowable emission of sulfur dioxide to several times greater than the previous standards and in many cases no emission limit are specified in many countries in recent approved ammendments. Thus, burning of fuels, especially coal, with relatively high sulfur content may be permissible in many areas where air qualities are clean. Consequently, both the need for utilizing coal reserves to meet the national fuel shortages and the national commitments for protection of environment may be fulfilled with efforts and cooperation of national and local governments and the industries.

THE ROLE OF GOVERNMENT IN RESEARCH AND DEVELOPMENT

Maxine Savitz
Director, Division of Buildings and Community Systems
Office of Conservation, US ERDA, Washington, D.C.

ABSTRACT

In defining the role of the government in sponsoring research and development I believe that there are four concepts which warrant consideration:

(a) Excludability - the ability of the private sector to exclude individuals from participating in consumption of services.

(b) Affordability - the ability of the private sector to underwrite the cost of the project.

(c) Time preferences - differences between private and public sector discount rates.

(d) Uncertainty - the willingness of the private sector to accept risks as relates to the potential benefits from product innovations.

Although this generalization is not entirely correct, most government RD&D efforts have been justified on the basis of the first two concepts. Excludability relates to the concept of public goods. A public good may be defined as one in which people who do not pay for the service cannot be excluded from the benefits that result. In such cases the market system will breakdown, and unless the government is willing to provide the service this social want will go unsatisfied. Since there is no benefit to be derived by the private sector, it will neither provide the service nor engage in the research necessary to develop methods of satisfying those needs. The government must therefore either purchase the product directly from the private sector or produce it itself if these needs are to be met. A great deal of the governments' research activities in the areas of defense, aerospace, health, law enforcement, education, and environment fall into this category.

The second concept of affordability. This concept is less difficult to grasp. It is obvious that in many instances the costs of a research program may exceed the ability of any single or even consortium of private firms to undertake the research. The nuclear program is a good example of this type of problem as is the ability of a school board to develop new innovative teaching methods. The question is not so much the size of the investment as it is the ability of a single paricipant to absorb the burden.

I do not wish to engage in a long discourse on either of the above concepts. Most of us are sufficiently familiar with the need and justification for public sector participation in these areas to feel relatively comfortable with the government's role. It is the latter two concepts which I believe need to be debated in open forum and to which I will confine my remarks. Both of these areas are of special concern to ERDA in sponsoring energy research.

Setting aside the regulatory programs which the government is pursuing in the energy arena - although I will come back to this area - most of the research projects currently being funded will require commercialization by the private sector. To capture the external benefits which ERDA was mandated to address itself to these new technologies will have to compete successfully with alternative methods of satisfying consumer needs. This is true for both supply and demand based technologies.

The consumer must perceive these alternatives as clearly superior to existing systems and there must be sufficient benefit to be derived by the private sector to insure commercialization.

In the context of RD&D activities the question of time preferences impacts on both the research projects which the private sector will undertake and actions by consumers. The simple fact that the private sector evaluates projects at its after tax cost of capital is sufficient reason to believe there may be a wide discrepancy between those projects which the public sector may consider justified and those meeting the private sector criteria.

(Provide some simple numerical examples to illustrate the concept:

a. Producers
b. Consumers

The government's role in this instance may be seen as subsidizing the cost of the research and development program in order to insure a better chance of meeting the private sector's decision criteria. It may also be seen as pursuing projects with significantly longer time horizons given that the benefits to the private sector may be seen as marginal.

The final concept I wish to address is uncertainty. This I would define to include commercial, institutional, and technical parameters which weigh heavily in the private sector's project selection criteria.

(This discussion will cover a broad range of topics including:

a. Differences in the public and private sector's willingness to accept risk.
b. Methods by which uncertainty can be integrated in the decision making process.
c. Its relationship to how the conservation program is attempting to interface with private sector research programs.)

<u>Summary</u>

Defining the role of the government in funding energy research and development program is critical to insuring that the most effective utilization of scarce resources. If the government was to fund projects which the private sector would have to pursue independently, the net benefits to the society from our programs would be negligible. Likewise if we select only those projects where the probability of commercial success is extremely low the corresponding energy benefits would also probably be insignificant. ERDA's programs hopefully represent a compatible interface with the private sector in which the potential energy benefits are high while minimizing conflicts with private sector in which the potential energy benefits are high while minimizing conflicts with private sector research opportunities. Only in this way can we be reasonably assured that the projects that are being funded will provide choices for the future.

<u>Length/Graphics</u>

The topic paper will be in the range of 30 pages with the speech requiring approximately 20 to 30 minutes. Graphics will be prepared to help illustrate specific points of emphasis.

INTERCOUNTRY DIFFERENCES IN ENERGY USE: LESSONS FOR THE UNITED STATES?

Joel Darmstadter, Resources for the Future, Washington, D.C.
Lee Schipper, Lawrence Berkeley Laboratory and Energy and Resources Group, University of California, Berkeley

Discussions of energy-conservation potentials in the United States have frequently taken note of the lower levels of per capita energy consumption prevailing in other industrialized countries and have given rise to speculation over whether these foreign examples validate the case for energy conservation in this country. Indeed, the topic is one which, in the last several years, has been vigorously debated by economists, environmentalists, and policy makers. And no wonder. America's per capita consumption of energy resources is considerably higher than it is in a number of other advanced industrial countries--e.g., France, Germany, Sweden--whose per capita incomes do not differ appreciably from that of the United States. (See the accompanying table.) Why is this so? The importance of the question is clear enough. If something close to the American standard of living were found to be compatible with a fraction of the prevailing level of energy use, that finding could have an important impact on the direction of U.S. supply development and conservation strategies.

Unfortunately, we have only very recently developed an empirical basis to explain international energy/GDP variability.* Is such variability primarily the consequence of the energy intensity that characterizes given activities within the respective economies? Or does it arise from the fact that there are important "structural" or "mix" differences among the countries? Aggregated energy/GDP comparisons may obscure structural factors highly relevant to the analysis. Thus it is possible for a country to have a much higher energy/GDP ratio than another due solely to the predominance of inherently energy-intensive activities.

Both structural and energy-intensity characteristics of an economy may be only the surface manifestations of more deep-seated features--e.g., geography, resource endowment, technology, demographic factors, and economic policies--which comparative analysis would want to examine. For instance, a comparative advantage in steelmaking (including the ability to export steel to countries lacking that capacity) may have been conferred upon a country through abundant coal and iron ore deposits--scarcely circumstances betraying energy "profligacy." On the other hand, sharply differing intercountry reliance on, say, large, energy-intensive automobiles will, at least to some extent, reflect differences in economic policy instruments--e.g., fuel and horsepower taxes.

It is also worth noting that energy-intensity differences among countries for a given activity may, but need not, spell efficiency differences in the economic use of <u>all</u> the resources going into that activity. It is only partially revealing, in other words, to point to a country's lower (or higher) energy intensity in a given pursuit without specifying the balance of economic advantage in terms of <u>total</u> resource cost. Unfortunately, such an overall reckoning requires knowledge about the relative requirements for costs of capital, labor, and other inputs; this broader dimension has largely been missing from intercountry comparative studies.

Here, we briefly report on the principal findings of two recently published or completed

---

*GDP=Gross Domestic Product. This is a measure closely related to Gross National Product but it is somewhat more suitable for the intercountry energy comparisons with which we are concerned here.

studies* analyzing intercountry differences in energy consumption patterns. The first is a comparative analysis of the United States and eight other highly industrialized economies. The second is a fairly intensive two-country comparison between Sweden and the United States.

Nine-Country Study**

The principal analytical effort in this cross-sectional study for the year 1972 was designed to explore the extent to which intercountry energy/GDP variations were rooted in the differences in (a) the composition of national output, or differences in (b) the energy intensity associated with given economic processes or activities. Beyond identification of these respective contributory elements in a purely statistical sense, some attempt was made to probe underlying economic and noneconomic factors. The six major conclusions are listed below:

(a) Higher U.S. energy consumption relative to GDP was reflected to some extent throughout the range of energy uses, though much more in some sectors of the economy than others. Of the four major sectors--the combined residential/commercial, transport, industry, and transformation losses--there was least intercountry energy/GDP variability in the industry sector: the (unweighted) average of energy consumption relative to GDP in Western Europe and Japan was only some 10 percent below the U.S. level. In the residential/commercial sector, there was more variability, with a ratio averaging about 25 percent below the United States. Transformation losses showed a similar comparative spread. By far the greatest interconntry differences arise in the transport sector, where European and Japanese consumption (relative to GDP) averaged 60 percent below the U.S. ratio. We find that (on the basis of simple averaging), transportation accounts for approximately 48 percent of the energy/GDP difference between the United States and the other countries (excluding Canada); the residential/commercial/agricultural sector to around 17 percent; energy industries to about 16 percent; transformation loss 15 percent; and nonenergy industries to 7 percent.

(b) In a comparative analysis of the composition of consumer budgets, with primary attention directed to energy-related expenditures, we found that in five other countries considered in this part of the study, the share of the consumer budget devoted to purchases of energy for household operation, and particularly automotive transport, was far below that in the United States. Spending for household fuel and power accounts for 3.2 percent of U.S. consumption expenditures, an average of 2.2 percent for the four West European countries, and only 1.3 percent for Japan. The disparity for gasoline is much wider still. There are wide intercountry differences in the respective proportions of the consumer budget accounted for by such energy-using durable goods as household appliances and cars. On the other hand, expenditures for purchased transportation, which is less energy-intensive than passenger car use, are proportionately much higher--and by a wide margin--outside America. Underlying the lower foreign proportion of purchases of energy and energy-using items is a clearcut pattern of substantially higher relative prices abroad for fuels and power as well as for appliances and autos. In addition, higher foreign prices for automotive operation and lower prices for public transport provide additional encouragement to use less energy-intensive modes. The impact on overall energy/GDP variabilities of proportionately greater U.S. direct energy purchases by household was confirmed also by limited input-output analysis.

(c) After correction for climatic differences, the United States emerges as the highest consumer of space-conditioning energy relative to GDP, with a ratio some 40 percent above that of numerous European countries. It was estimated that about half this difference was attributable to U.S. prevalence of larger, single-family homes. For the other half, poorer U.S. insulation practices seemed to figure as a significant factor only in the U.S. comparison with Sweden and Canada. Otherwise, efficiency of the heating plant being, if anything, higher in the United States, most of the other half of the space conditioning energy/GDP difference between the United States and other countries seemed

*J. Darmstadter, J. Dunkerley, and J. Alterman, How Industrial Societies Use Energy: A Comparative Analysis, Johns Hopkins University Press for Resources for the Future, Baltimore, Fall 1977; L. Schipper and A. Lichtenberg, Efficient energy use and well being: the Swedish example, Science, December 3, 1976.

**The countries were: United States, Canada, France, Germany, Italy, Netherlands, United Kingdom, Sweden, and Japan.

to arise from different heating habits--e.g., the heating of unoccupied rooms and the maintenance of higher temperatures.

(d) The study surveyed the role of differential characteristics of energy use in transportation as an element in aggregate energy/GDP variability. As noted earlier, comparative transport practices emerged as a key reason for the higher U.S. ratio. Not only are American passenger cars about 50 percent more energy-intensive (in terms of passenger-miles per gallon) than European cars. Relative to given income levels, Americans also travel a lot more than Europeans. Indeed this factor is quantitatively more important than automotive energy intensity (fuel consumed per passenger-mile) in explaining the far greater amount of energy devoted to transportation in the United States compared to Western Europe. A third contributory element is the proportionately greater share of (less energy-intensive) public transport modes in the foreign energy mix. We found these differences to be a function not only of the higher prices of acquiring and operating cars abroad; but also such aspects as urban density differentials between the United States and other countries and public-policy measures resulting in highly subsidized public transport.

Freight transport also contributes to the higher U.S. energy/GDP ratio. Interestingly, this comes about exclusively by virtue of the high volume of traffic (relative to GDP) that is generated in the United States compared to the grouping of European countries which was analyzed. Indeed, the U.S. freight model mix is, more than Western Europe's, oriented to such energy-saving forms as rail, pipeline, and waterborne traffic. If one argues that size of country and long-distance haulage of bulk commodities (such as ores, grains, and coal) are inherent characteristics of U.S. economic structure and geography, we have here a case where a relatively high U.S. energy/GDP ratio for freight is in no obvious way reflective of comparative energy "inefficiency."

(e) On the industry-wide level of aggregation, the industrial contribution to higher U.S. energy/GDP ratios happens entirely as a result of more energy-intensive production activity. That is, "structure" is not the cause; indeed, if industrial value added were as high a proportion of U.S. national output as it is in most foreign countries, U.S. energy consumption would be higher. What we observed for the industrial sector as a whole appeared to be the case also in a diverse number of specific manufacturing segments. For example, the United States has a relatively smaller iron-and-steel industry compared to a number of other countries, but a higher consumption of energy per ton of crude steel--a characteristic apparently arising from the prevalance of the (energy-intensive) open-hearth process and relatively low proportion of continuous rolling capacity. Such differences between the United States and other countries in industrial energy intensity need not reflect differences in the overall efficiency of carrying on a given industrial operation. To articulate the relationship between energy intensity and economic efficiency, one would need to explore, in addition to the cost of energy resources, such inputs as labor, capital, and use of nonenergy materials. But the findings, as far as they go, constitute at least a presumptive basis for exploring potentials for more efficient industrial energy utilization in the United States.

(f) In sum, the nine-country study points to a variety of complex and diverse reasons for intercountry differences in energy consumption. Some of these differences, such as automotive patterns, appear to be decisively influenced by relative user costs. Other notable differences, such as suburbanization, housing, and mobility characteristics--each with marked influence on energy use--might arguably owe some of their historic momentum to cheap energy in the United States, and costly levels elsewhere, but are clearly also related to many other forces. New conditions of energy supply and costs, new policies, and new attitudes on conservation and the environment may alter some of these features significantly, such that some of the characteristics of energy use overseas may begin to appear here. However, the stamp of past U.S. characteristics will undoubtedly persist in a very marked way as well.

U.S.-Swedish Study

Sweden and the United States were compared using an accounting scheme in which the major emphasis was on a differentiation between comparative structural key factors and comparative energy-intensity factors. Key findings are listed below:

• This analysis involved many differences than had to be accounted for before energy use could be directly compared. Greater natural distances in the United States was one difference. A larger fraction of the oil used in the United States is refined than is the case for Sweden, where the refining mix is also less energy intensive on an output basis. The "need" for air conditioning in the United States is nonexistent in Sweden, but there the "need" to heat factories is more important than in the United States.

• If one counts the energy embodied in the goods and services making up foreign trade, it is found that the United States is a slight importer of energy, in an amount reaching at most 1 percent of total use in 1973. Whereas Sweden is a net exporter of embodied energy, with exports typically amounting to 8-9 percent (net) of total use.

• The greatest differences in energy use patterns are associated with industrial process heat, electricity use, space heating, and surface passenger transportation. While it may come as a surprise that the living space per capita (excluding second homes) is nearly as large in Sweden as in the United States, space heating in Sweden is strikingly less intensive than in the United States when measured in Btu/sq meter/degree day. On the other hand, households in Sweden generally have fewer and smaller appliances than in the United States, reflecting a different lifestyle and lower after tax incomes. Surprisingly, the energy intensity of apartment heating was nearly as great as that in single family dwellings, largely because of common metering and difficulties in controlling temperatures.

• As was found to be the case in the nine-country comparisons, the most marked difference in energy use occurs for passenger transport. Swedes travel only 60 percent as much as Americans, using only 60 percent as much fuel per passenger mile. Mass transit and intercity rail are more efficient and more widely used in Sweden, while air travel is overwhelmingly larger in the United States. This does not mean that the United States is inefficient vis-à-vis air travel, because distances are greater here than in Sweden. Intracity trucking in Sweden is considerably less energy-intensive than in the United States, but long haul trucks in Sweden use slightly more energy/ton-mile than in the United States. The greater distances in the United States mean that ton-mileages are far greater here, allowing much more rail freight in the United States, lowering energy intensity. In all, the lower energy use for transportation in Sweden can be ascribed to greater efficiency in most modes (with notable exception of long haul trucks), shorter travel distances, and substitution of mass transit and walking for commuting.

• Swedish manufacturing industries actually turn out more energy-intensive raw materials, as a percentage of total dollar output and energy use than is the case of the United States. But product for product, each kind of output is generally less energy-intensive (energy/ton) in Sweden than in the United States. This is because technology is newer, and optimized by virtue of considerably higher fuel prices.

• More specifically on comparative prices, while pre-embargo oil prices in both the United States and Sweden were roughly equal, Americans enjoyed 20-50 percent cheaper natural gas and coal. In the case of electricity, however, the two countries have been radically different at least (up to 1972). Since 75 percent of all electricity generated in Sweden was hydro, the ratio of the cost of electricity to the cost of heat from fuel was only half as great in Sweden as in the United States. Industry in Sweden naturally developed a more electricity-intensive technology base. Thermal generation (using oil) was arranged to combine heat and electricity production in industries or in communities where district heat could be used. This meant that only about 7,000 Btu of fuel were required (in 1971-72) for the thermal generation of a kilowatt hour of electricity. Rises in the cost of nuclear electricity and oil make the continued expansion of combined generation of electricity and heat a certainty.

• The most important reason for the higher energy use efficiencies in Sweden appear to stem from the relatively higher costs for fuel in that country. (Partly, differential prices reflect differential tax policies. Thus, taxes on road fuels and on new autos (proportional to weight) also influence automobile travel and efficiency.) However, several nonprice, policy factors that shape energy use in Sweden were also noted. Weatherization standards for buildings, which, when combined with generous allowances in mortgages for energy conservation investments, allowed for economizing on the use of heat, was one important policy in effect in Sweden. Mass transit, while subsidized for reasons not directly related to energy use, played a role in reducing congestion and vehicle miles traveled in cities and on short trips.

• While the mix of industrial production in Sweden is more energy-intensive than is the case for the United States, the mix of personal consumption in Sweden is slightly less energy-intensive than in the United States, the difference in these two indices being accounted for by the exports of energy embodied in raw materials. Abundant hydropower, smaller national distances, and somewhat different lifestyles enter into the comparison, all tending to lower Swedish energy demands and in some cases intensities somewhat, while climate, and the mix of industrial production tend to raise energy demands relative to the United States. It was concluded that the most important single factor accounting for the differences between energy use per capita in two countries was energy-use efficiency. Perhaps the most important lesson to be learned from this study is the fact that, in the long run, energy consumption required to provide a given set of amenities can be considered to be highly flexible.

With the demonstrated long-run flexibility of energy use in mind, the Swedish government is currently embarked upon an energy conservation campaign whose objective is to win still more well-being for each unit of energy use. However, a characterization of the United States as "energy sloppy" or Sweden as "energy-wise" seems improper, given the differences in energy costs and policies that did exist during the study period (1970-72). But the implication that the United States could exact considerably more well-being from energy use, particularly given proper economic signals, was clear from the comparison with Sweden.

Questions Raised By the Two Studies

Since the two studies differed somewhat in their breadth of coverage and mode of analysis, it is difficult to assert a common, unambiguous set of findings. For example, the Swedish study--the more detailed of the two--emphasized comparative energy use in specific tasks rather than a sectoral decomposition of overall energy/GDP differences. As a result, there were--between the two studies--a few differences of interpretation as to the "causes" of the different energy-use patterns. Sweden stands out among the nine countries as second only to North America, on a per capita basis, in total passenger travel, auto ownership, housing space, and total freight miles. (Sweden is nonetheless much below the United States in its energy/GDP ratio because of the high level of Swedish per capita GDP.) On the other hand, Sweden often fell in the middle of all countries studied when ranked by intensity of energy use, possibly reflecting energy prices somewhat lower in Sweden than elsewhere in Europe or Japan. After the few minor discrepancies between the two studies were resolved, Sweden stood out as generally less energy-intensive than most countries in space heating and a wide range of industrial processes, above European average in automotive energy intensity, somewhat less availing of mass transit than in the case of Europe or Japan, and extreme in climate, use of cogeneration and hydropower, and in the use of "noncommercial" fuels.

Both studies cautioned that energy use is only one part of the yardstick by which economic efficiency is measured; thus conclusions that other economies are more efficient (than the United States) in their use of energy must be accompanied by discussions of prices and policies, as the two studies attempted to do, at least in a limited fashion. Furthermore, both studies showed the need to disaggregate energy use and economic activity in order to separate the effects of activity mix, demography, lifestyle, and natural resource base from differences in energy intensity. Nonetheless, these differences in intensity may be considered a sufficient reason for exploring energy conservation paths for the United States. It is clear that technical, behavioral, and consumer preference differences among countries point to the usefulness of examining possibilities for reducing U.S. demand for energy. Whether these changes would be economically feasible or socially desirable in this country is a question that deserves widespread attention and discussion.

TABLE 1 Per Capita Energy Consumption and Per Capita Gross Domestic Product, Nine Developed Countries, 1972

Country	Per capita		Energy/GDP ratio (barrels oil equivalent per $ thousand)	Index numbers (US=100)		
	GDP (dollars)	Energy (barrels oil equivalent)		Per capita GDP	Per capita Energy	Energy/GDP ratio
United States	5,643	61.0	10.8	100	100	100
Canada	4,728	61.2	12.9	84	100	120
France	4,168	24.2	5.8	74	40	54
Germany	3,991	30.1	7.5	71	49	70
Italy	2,612	17.4	6.7	46	29	62
Netherlands	3,678	34.2	9.3	65	56	86
United Kingdom	3,401	27.8	8.2	60	46	76
Sweden	5,000	38.8	7.8	89	64	72
Japan	3,423	21.2	6.2	61	35	57

Note: The hydro and nuclear component of primary energy consumption is converted into Btu's on the basis of fuel inputs into fossil-fueled power plants, assuming 35 percent efficiency. Foreign GDPs are expressed in dollars, using a real purchasing power basis of comparison rather than market exchange rates.

UNITED STATES AND FOREIGN INDUSTRIAL ENERGY USE

Thomas Veach Long, II
Resource Analysis Group
Committee on Public Policy Studies
The University of Chicago
5735 South Ellis Avenue
Chicago, Illinois 60637

ACKNOWLEDGMENT

The material presented below draws heavily on both the research and ideas of R. Stephen Berry of The University of Chicago, who first suggested that international comparisons of energy use might be revealing. My appreciation is offered to all members of the NATO-CCMS project, who labored hard in compiling some of the data reported. These include:

Douglas Harvey (Chairman)
William M. Porter (Former Chairman)
OFFICE OF CONSERVATION, ERDA;

F. Allen Harris (EPA);

Barry Tunnah (Coordinator and Cement)
Robert Kappner (Steel)
Kenneth Stern (Plastics)
GORDIAN ASSOCIATES

Cement Chairman: Robert McLean (Portland Cement Association);

Steel Chairman: A. Palazzi (Centrale Metallurgicol, Rome, Italy);

Plastics Chairman: L. Biondi (Montedison, Milan, Italy);

Methodology Chairman: Thomas Veach Long, II (University of Chicago).

Many thanks also to Sandra Hebenstreit, of The University of Chicago, for data on the energy requirements for cement production in the United States.

INTRODUCTION

Several projects compiling detailed comparisons of energy requirements for the provision of goods and services in different nations are in progress. In the Resource Analysis Group at The University of Chicago, such analyses have been an active research interest in the past three and a half years, with a concentration on industrial energy use (1).

As one of the non-defense activities of NATO, a program seeking to establish an international industrial data base on energy use was initiated two and a half years ago. This program is being carried out under the auspices of the NATO Committee on Challenges of a Modern Society (CCMS), Pilot Study on the Rational Use of Energy. Finally, broad studies of energy use in several nations, such as the study of the Federal Republic of Germany by Stanford Research Institute (2) and that of the Danish economy by the Niels Bohr Institute (3), are becoming available and will permit further comparisons.

The motivation for making these comparisons is two-fold. First, as will be explored below, we have found the analyses to be sensitive quantitative indicators of differences in technologies, and process technologies exhibit marked variations internationally. Thus, the possibilities for energy husbandry through international and interindustry technology transfer can be explored. Second, they provide us with information about how elements in economic society can respond to higher prices for energy goods. It is generally agreed that price elasticities of demand for energy goods generated by regression analysis of time-series data for the pre-1973 period of relatively stable fuel prices may not be reliably applied to today's volatile energy market. To the extent that sectors in different countries have faced widely different prices for energy goods, cross-national comparisons may yield superior information regarding elasticity responses. Of course, these will correspond most closely to long-run price elasticities of demand.

Methods of Analysis

The method of analysis used by our group at The University of Chicago and by the CCMS project is that suggested by the First Workshop on Energy Analysis sponsored by The International Federation of Institutes for Advanced Study (4).

As an initial step, data must be obtained on all the fuel, electricity, steam and material inputs to the process of interest. These results may be presented in tabular form or diagramatically, as in Fig. 1. In this diagram, both the transport energy and the energy requirement of the stage (invented triangle) are aggregate energy requirements. The individual fuel quantities used, which should also be reported, are aggregated on the basis of their heats of combustion, and electricity and steam are evaluated at their respective generation efficiencies. Below, assumed efficiencies are 33 percent for electricity and 80 percent for steam.

The summary data for a representative process, aluminum production, are shown in Fig. 2 (British data from Peter Chapman, The Open University, Milton-Keynes, England). Two points about this method of analysis should be emphasized. First, a complete material flow normalized to one metric tonne of output is required before the energy requirements can be evaluated. Second, all materials are fuels in the sense that use of their thermodynamic potential may substitute for fuel use. For example, Fig. 2 shows that for every tonne of aluminum produced, one tonne of carbon anode is oxidized. The carbon oxidation is fundamentally equivalent to a controlled combustion. Presumably, the thermodynamic potential released by the carbon reaction would have to be substituted by another energy source if an inert electrode were adopted.

Industrial Energy Requirements

International comparisons of energy use in the aluminum, steel, cement and petrochemical industries are given below. These sectors were chosen because they are large consumers of energy in industrial economies. Thus, energy-saving technological modifications can pay large absolute dividends. Data sources are shown in Fig. 3. All energy requirements are given in gigajoules per metric tonne (GJ/tonne). When no attribution is given for an analysis, it is the product of research at Chicago. The CCMS energy data for steel and polymers is preliminary, a report of work in progress, but the cement study has been completed. Production level data and the sectoral comparison data was gleaned from several sources, and detailed attribution is made in CCMS documents.

Figure 3 provides an overview of basic sectoral energy use in fourteen countries. These are striking contrasts between nations, noting particularly the large percentage use by industry in Luxembourg, Japan, Belgium, and Sweden and the high value for transport in the U.S. Of course, to get a fuller picture we should know the total energy use and the energy use per capita, but this does point to the possibility of intriguing international differences in energy use that must be explained.

Aluminum. Our attention was drawn to international comparisons when we were examining data on aluminum production in the U.K., The Netherlands, and the U.S. In this industry, the technology for the energy-consuming electrolytic step is the well-known Hall-Héroult process in all three countries, and we anticipated that the total energy requirements would be similar. Our anticipations were borne out by the data for the totals in three countries: 253 GJ/tonne, U.K.; 232 GJ/tonne, The Netherlands, and 258 GJ/tonne,U.S.A. (1). However, a careful examination of the energy requirements for subprocesses alerted us to some of the

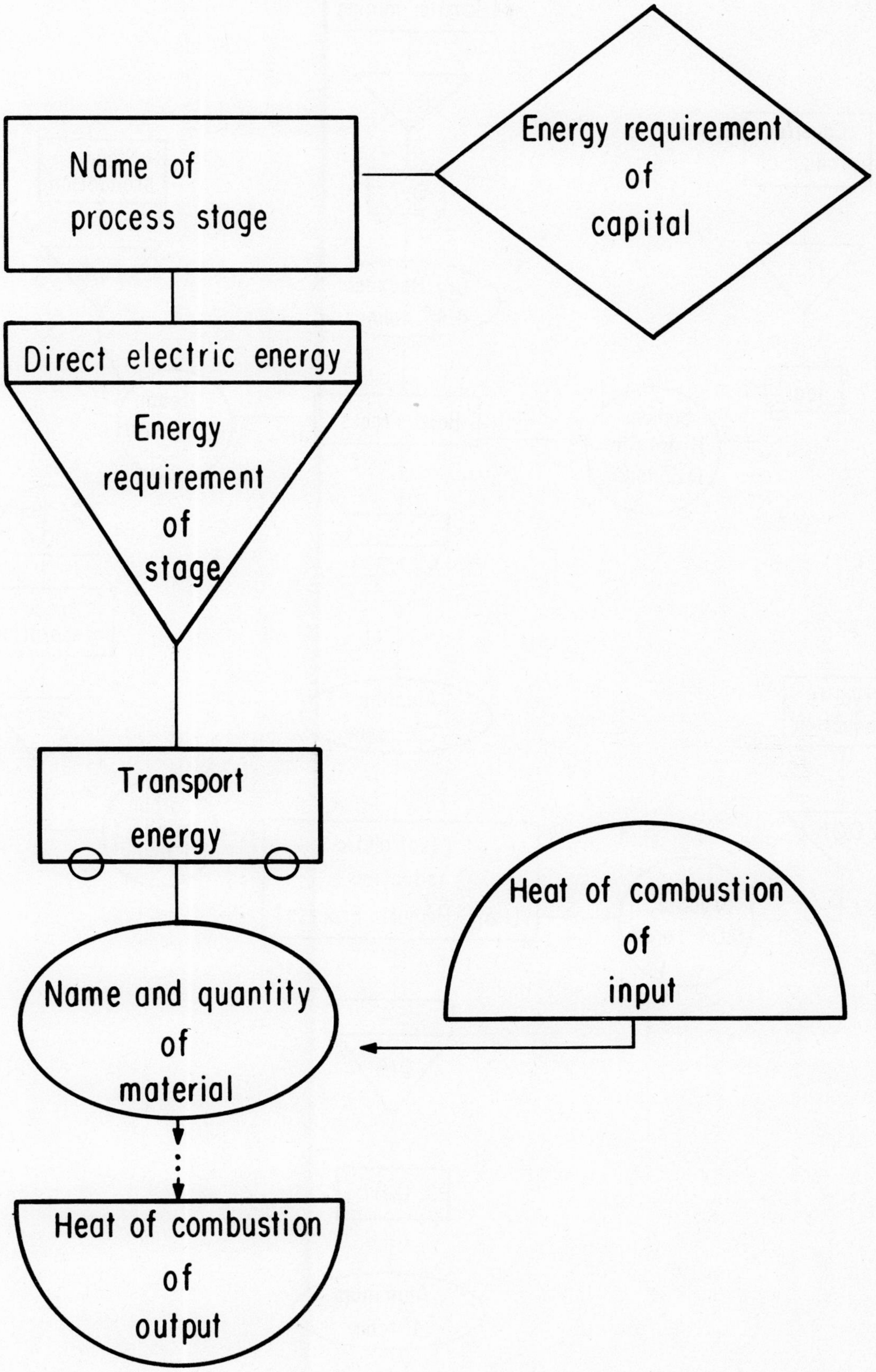
Name of
process stage
Energy requirement
of
capital
Direct electric energy
Energy
requirement
of
stage
Transport
energy
Heat of combustion
of
input
Name and quantity
of
material
Heat of combustion
of
output

FIGURE 1

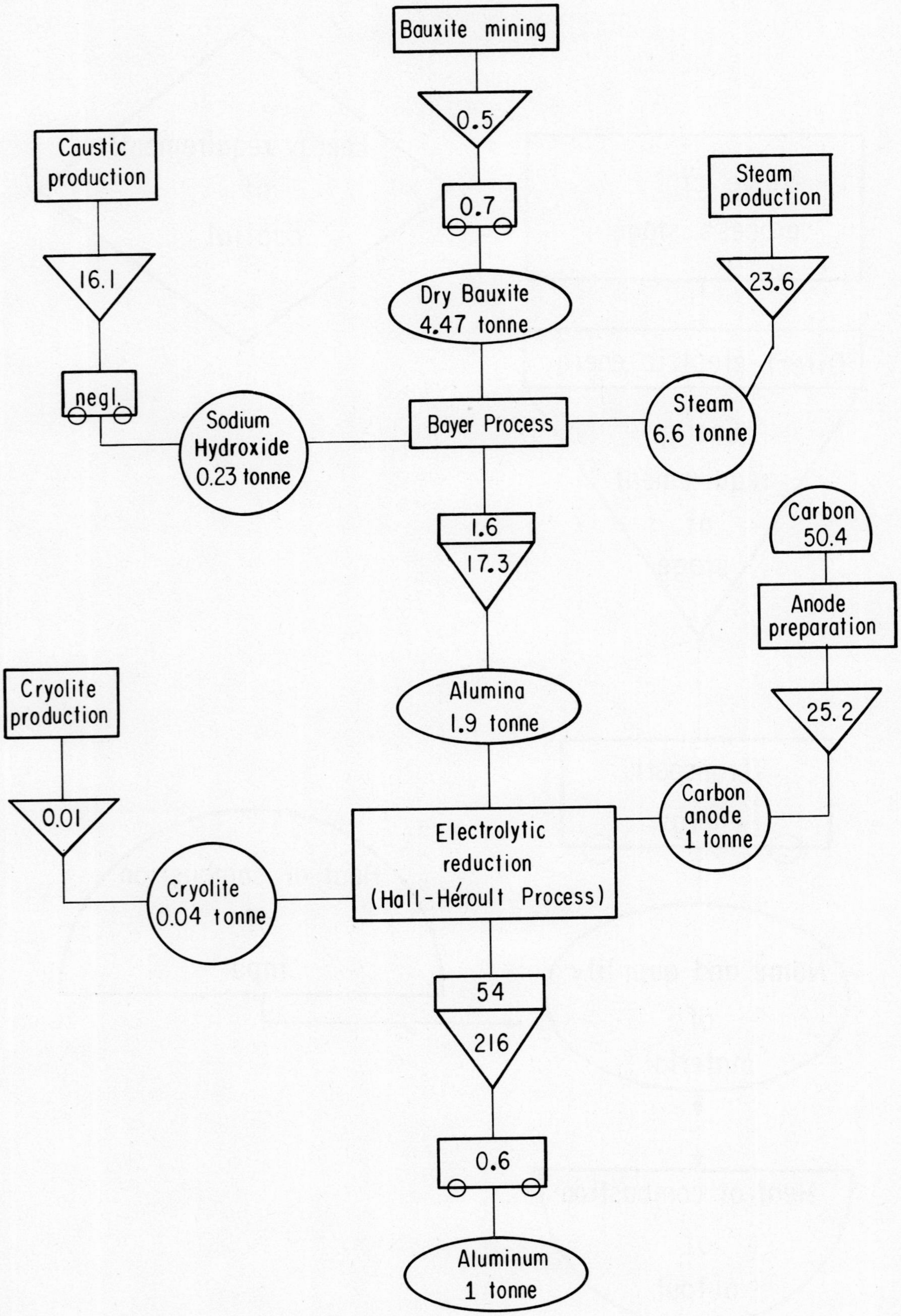
Bauxite mining
0.5
0.7
Caustic production
16.1
negl.
Steam production
23.6
Dry Bauxite 4.47 tonne
Sodium Hydroxide 0.23 tonne
Bayer Process
Steam 6.6 tonne
1.6
17.3
Carbon 50.4
Anode preparation
Alumina 1.9 tonne
25.2
Cryolite production
0.01
Carbon anode 1 tonne
Electrolytic reduction (Hall-Héroult Process)
Cryolite 0.04 tonne
54
216
0.6
Aluminum 1 tonne

FIGURE 2

	Percentage of total energy use in the basic sectors		
	Industry	Transport	Buildings
Belgium	51.7	10.5	37.8
Denmark	26.2	17.2	56.5
France	48.5	16.1	35.4
Germany	41.2	18.3	40.5
Ireland	34.4	18.9	46.7
Italy	49.5	17.2	33.3
Luxembourg	81.2	5.6	13.2
Netherlands	41.3	13.5	45.2
United Kingdom	45.9	14.4	39.7
EUR-8	42.0	16.2	41.8
Canada	31.5	28.6	39.8
Greece	40.0	42.0	18.0
Japan	61.0	17.8	21.2
Sweden	51.8	5.1	43.1
United States	40.0	30.7	29.3

Fig. 3. Sectoral comparison based on end user consumption. CCMS

energy requirements for sub-processes alerted us to some of the pitfalls that one faces in making international comparisons. Referring to Fig. 4. we see that although the total requirements are approximately the same, the requirements for individual steps are substantially different. The variation in energy use in alumina production can be attributed to different technologies and ore grades. The figures for the smelting step were more surprising. Closer examination showed that the smelting number for the U.K. is for their most efficient cells, only, and that the average U.K. value

PROCESS	UK	The NETHERLANDS	USA
Ore extraction	5 (GJ/t)	5(GJ/t)	3(GJ/t)
Alumina production from ore	56	31	13
Aluminum production from alumina	192	196	242
TOTAL	253	232	258

Fig. 4. Energy requirements for primary aluminum production (1) (transportation energy is neglected)

is 226 GJ/tonne. Both the Dutch and U.S. figures are national averages, but the facilities in The Netherlands are of later vintage.

Steel. Represented in the CCMS study are seven nations producing about 40 percent of the total world output (Fig. 5). There are two major energy-using steps in steel making: (1) the production of hot blast furnace metal that is then input into (2) a steel furnace. Blast furnace energy consumption is shown in Fig. 6, and the values range from 21.460 GJ/tonne in West Germany to 30.464 GJ/tonne in the U.S. Note that Italy's value is approximately midway, at 24.509 GJ/tonne.

Turning to Fig. 7, we observe that the total energy input for an average tonne of crude steel ranges from 15.176 GJ/tonne in Italy to 23.767 GJ/tonne in Belgium, without allowing credit for combustible gases after steelmaking. What accounts for the low figure for the energy requirement for crude steel production in Italy, given the intermediate value for the blast furnace step in that country? One explanation could be more modern steelmaking facilities, and this is indeed the case. European facilities, because of reconstruction following World War II, are on the average of more recent vintage.

Another explanation could be differences in steel furnace technology. Figure 8 shows that Italy has a greater percentage of electric furnaces than do other countries, and The Netherlands, Belgium and Germany rely heavily on basic oxygen furnaces. Both of these technologies are reported to be energy efficient.

However, a large part of the explanation for Italy's low crude steel energy requirement arises from the large amount of scrap that is used as feed to the steel furnace. This portion of the feed bypasses the blast furnace step of production. Indeed, the use of electric furnaces makes the injection of a large scrap load possible. In the data summary presented in Fig. 8, scrap is not assigned an "embodied" energy requirement, and, thus, the Italian steel industry appears to be very energy efficient. This is a reasonable analytical assumption to make, and it is appropriate for answering many questions. However, if one asks instead how much energy would be needed per tonne of crude steel if no scrap were available, quite a different ordering would result. Also, the heavy dependence on imported scrap points to a possible strategic fragility for that industry.

Cement. Production data for cement (Fig. 10) indicate that the U.S. turns out more tonnage than West Germany and Italy combined, two other large producers. To illustrate the complexity of a flow diagram even for the relatively simple cement-making process, the analysis for a U.S. wet-process is displayed in Fig. 11. Energy requirements by process and national averages are given in Fig. 12 (5). The most important observation from this data is that replacement of the old capital facilities in the U.S. by modern kilns and processes of the types now operating in West Germany and Japan will yield a large saving in the national energy budget. However, transportation costs of the finished product result in this being a geographically-segmented industry, with little possibility of market penetration outside of a 200 mile radius of a plant. Thus, the large scales of the energy-efficient plants operated in Japan may be inappropriate within U.S. market structure. Nevertheless, there is ample room for improvement in U.S. technology, but this will come slowly because of the substantial capital needs and difficulties in generating cash flow.

Petrochemicals. The most recently-initiated of the CCMS studies is that investigating the petrochemical industry. Data have been particularly difficult to obtain because of proprietary interests. Plyvinyl chloride (PVC) production was chosen for a pilot study because many basic processes are represented. A preliminary analysis of the data is given in Fig. 13. Data for The Netherlands have been made available for public release later this year. However, early analyses of U.S.A. and Dutch PVC production energy requirements (1) are in good agreement with CCMS figures, and data from that study are used for The Netherlands. The primary difference between technologies

	1970	1971	1972	1973	1974
Belgium*	18.1	17.7	20.0	21.4	22.7
France	23.8	22.9	24.1	25.3	27.0
Federal Republic of Germany	45.0	40.3	43.7	49.5	53.2
Italy	17.3	17.4	19.8	21.0	21.1
Netherlands	5.0	5.1	5.6	5.6	5.8
United Kingdom	27.9	24.2	25.4	26.7	22.5
United States	119.3	109.3	120.9	136.8	132.2
Total:	256.4	236.9	259.5	286.3	284.5
Percent of World Production	43.2	40.8	41.2	41.1	40.1

*Includes production data for Luxembourg. In 1972 and 1973, approximately 72% of the combined productin show above was made in Belgium and the balance in Luxembourg.

Fig. 5. Crude steel production data (all data in million metric tons) CCMS

COUNTRY	Total energy consumption for 1 t hot metal GJ	of which for ore preparation Reference: 1 t hot blast furnace metal GJ (%)
BELGIUM	27.779	2.843 (10.2)
FRANCE	25.689	4.449 (17.3)
GERMANY	21.460	2.393 (11.1)
ITALY	24.509	2.209 (9.0)
THE NETHERLANDS	23.688	2.157 (9.1)
U.K.	27.544	2.237 (8.1)
U.S.A.	30.464	4.269 (14.0)

Fig. 6. Energy consumption for one ton of hot blast furnace metal CCMS

Country	Total Energy Input GJ/t	Credit for gas steelmaking GJ/t	Net energy consumption GJ/t	Hot metal ratio, %
Belgium	23.767	n.d.	n.d.	81.5
France	20.812	2.400	18.412	74.6
Germany	16.735	2.252	14.483	71.5
Italy	14.889	2.086	12.803	45.4
The Netherlands	20.139	3.449	16.690	80.7
U.K.	19.589	--	--	53.3
U.S.A.	21.828	1.312	20.516	61.7

Fig. 7. Energy consumption for one average ton of crude steel
CCMS

	PROCESSES			
Country	Thomas Furnace and others	Open Hearth Furnace	Basic Oxygen Furnace	Electric Furnace
Belgium	14.61	1.47	79.90	4.02
France	24.50	13.04	51.76	10.70
Germany	2.94	17.65	69.61	9.80
Italy	---	14.80	43.80	41.40
The Netherlands	---	1.61	92.22	6.17
U.K.	0.80	27.50	43.20	23.50
U.S.A.	---	24.36	55.96	19.67

Fig. 8. Percentage of crude steel made with different processes
CCMS

Country	Total Scrap kg/t	Recycled Scrap kg/t	Purchased scrap kg/t	Imported over Purchased Scrap %	Exported Scrap kg/t
Belgium	305	231	74	25.7	n.d.
France	344	208	136	n.d.	n.d.
Germany	395	200	195	14.0	44.5
Italy	632	230	402	66.7	0.5
The Netherlands	287	212	75	n.d.	n.d.
U.K.	500	200	300	n.d.	18.0
U.S.A.	469	256	213	0.6	59.7

Fig. 9. Scrap indexes and ratios for one average ton of crude steel
CCMS

million tonnes per year

	1970	1971	1972	1973	1974
Federal Republic of Germany	37.5	40.2	42.6	40.9	35.9
Italy	33.1	31.8	33.5	36.3	36.3
Netherlands	3.8	4.0	4.0	4.6	
United Kingdom	17.6	18.1	18.7	21.2	17.8
United States	64.1	68.7	74.5	74.4	73.3

Fig. 10. Cement production data
CCMS

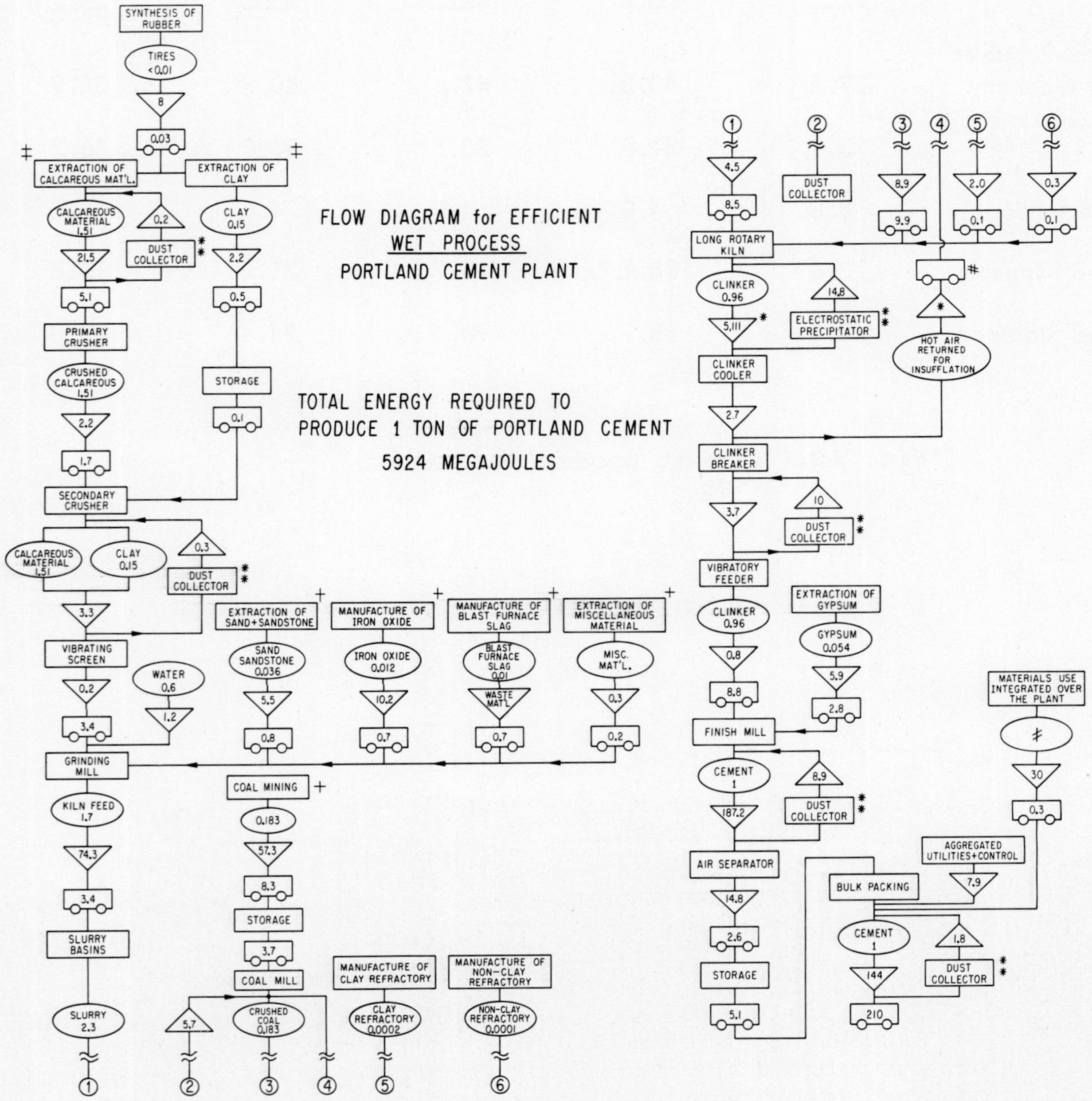
FLOW DIAGRAM for EFFICIENT
WET PROCESS
PORTLAND CEMENT PLANT
TOTAL ENERGY REQUIRED TO
PRODUCE 1 TON OF PORTLAND CEMENT
5924 MEGAJOULES
SYNTHESIS OF RUBBER
TIRES <0.01
EXTRACTION OF CALCAREOUS MAT'L.
EXTRACTION OF CLAY
CALCAREOUS MATERIAL 1.51
DUST COLLECTOR
CLAY 0.15
PRIMARY CRUSHER
CRUSHED CALCAREOUS 1.51
STORAGE
SECONDARY CRUSHER
VIBRATING SCREEN
WATER 0.6
GRINDING MILL
KILN FEED 1.7
SLURRY BASINS
SLURRY 2.3
EXTRACTION OF SAND+SANDSTONE
SAND SANDSTONE 0.036
MANUFACTURE OF IRON OXIDE
IRON OXIDE 0.012
MANUFACTURE OF BLAST FURNACE SLAG
BLAST FURNACE SLAG 0.01
WASTE MAT'L.
EXTRACTION OF MISCELLANEOUS MATERIAL
MISC. MAT'L.
COAL MINING
COAL MILL
CRUSHED COAL 0.183
MANUFACTURE OF CLAY REFRACTORY
CLAY REFRACTORY 0.0002
MANUFACTURE OF NON-CLAY REFRACTORY
NON-CLAY REFRACTORY 0.0001
LONG ROTARY KILN
CLINKER 0.96
ELECTROSTATIC PRECIPITATOR
HOT AIR RETURNED FOR INSUFFLATION
CLINKER COOLER
CLINKER BREAKER
VIBRATORY FEEDER
EXTRACTION OF GYPSUM
GYPSUM 0.054
FINISH MILL
CEMENT 1
AIR SEPARATOR
BULK PACKING
AGGREGATED UTILITIES+CONTROL
MATERIALS USE INTEGRATED OVER THE PLANT

FIGURE 11

	FRG	Italy	Japan	Netherlands	U.K.	U.S.A.
Wet	6.31 - 7.14	7.44	5.73	6.21	7.38	8.18
Semi Dry	4.63 - 5.89	5.51	3.95	----	4.85	---
Dry Sp	4.38 - 5.05	5.53	3.68	3.76	4.91	7.17
Dry Other	4.63 - 6.31	----	----	----	----	----
Shaft Kilns	4.43 - 5.47	6.07	4.18	----	----	----
Average	5.37	5.55	4.52	5.40	6.96	7.77

Note: SP = suspension preheater

Fig. 12. Cement Energy Requirements
GJ/1 tonne clinker
CCMS, PCA, and U. of C.

	Netherlands	Italy	USA Conventional	Transcat
Production of Crude Oil or Natural Gas	0.15	0.15	0.20	0.20
Crude → Naptha	0.7	0.80		
Naptha → Ethylene	12.9	4.79		
Natural Gas → Ethane			9.04	9.33
Ethane → Ethylene			4.45	
NaCl Mining	2.62	2.40	1.97	1.63
NaCl → Cl_2	15.9	13.58	15.95	13.80
Ethane + Cl_2 → VCM				16.42
Ethylene + Cl_2 → VCM	13.1	15.57	14.73	
VCM → PVC	9.1	7.88	14.05	14.05
Subtotal	54.47	45.17	60.39	55.43
Feedstock	28.78	30.46	36.83	35.77
Total	83.25	75.63	97.22	91.20

N.B. CCMS Data for The Netherlands have been withheld from publication until later in 1976 to abide by a proprietary request. Data reported are updated values from University of Chicago research. They differ significantly from Dutch CCMS data only for the naphtha cracking step, where our figure is six times larger--perhaps due to different joint product accounting procedures. If this is taken into account, the Dutch and Italian data are nearly identical.

Figure 13. Energy requirements for PVC
GJ/tonne PVC

in Europe and the U.S. is the use of crude oil as a feedstock in Europe, while natural gas is used in this country. The latter is clearly a more energy-intensive process, and this fact, coupled with natural gas supply shortages is stimulating a rapid conversion to the use of crude oil as a feedstock by the American chemical industry.

Note that the synthesis step for vinyl chloride monomer (VCM) formation requires approximately the same energy in all three countries, and that the figures for electrolytic production of chlorine from sodium chloride are also similar. The Transcat process, developed by the Lummus Company (U.S.A.), directly chlorinates ethane using a circulating molten salt mixture, and this technology appears to offer a possible energy saving if natural gas is used as a feedstock. The smaller energy associated with chlorine production in the Transcat process arises purely because of a smaller chlorine mass input to the reactor per tonne PVC output. No current production facility utilizes this process, and the data are engineering estimates. The data for The Netherlands and Italy do not include any credit for existing cogeneration of steam and electricity which would make these countries appear even more energy efficient in PVC production.

CONCLUSION

The analysis presented above are technological, but the data generated are the sort of information that is at the heart of empirical economics. The energy requirements reported are nothing more than the inverses of average products of energy (or, in some cases, of marginal products). As such, they can be direct inputs into most methods of predicting long-run energy demand for the industries studied. There is a natural interface with the work of Tom Sparrow at The University of Houston, James Griffin at the University of Pennsylvania and others. Economists and physical scientists are finally talking to each other, and I am very glad to see this turn-of-events.

REFERENCES

1. R.S. Barry, T.V. Long, and H. Makino, "An International Comparison of Polymers and Their Alternatives," *Energy Policy*, *3*, 144(1975).

2. R.L. Goen and R.K. White, "Comparison of Energy Consumption between West Germany and the United States," Stanford Research Institute (June, 1975).

3. S. Bjørnholm, "Energien i Danmark, 1990-2005," International Federation of Institutes of Advanced Study and Niels Bohr Institute (April, 1976).

4. First Workshop on Energy Analysis, Workshop Report No. 6, International Federation of Institutes of Advanced Study, Guldswedshyttan, Sweden (August, 1974.

5. U.S. data from Portland Cement Association, "Energy Conservation Potential in the Cement Industry," Federal Energy Administration Conservation Paper Number 26 (1975).

ENERGY USE PROFICIENCY: THE VALIDITY OF INTERREGIONAL COMPARISONS

Dr. Chauncey Starr, President
and Stanford Field

Electric Power Research Institute, Palo Alto, CA.

INTRODUCTION

Conservation of available energy supplies is widely recognized as an important and highly desirable national goal. With respect to the next several years, energy conservation, brought about by both wider use of improved practices and the introduction of more efficient end-use devices, offers the only immediate hope for reducing our growing dependence on foreign oil.

The rapidly changing world-wide energy situation has recently stimulated studies focused on the potential for conservation. Many of these studies compare the energy usage patterns of different countries with those of the United States. While such studies may be useful in identifying specific technical areas that may have high conservation potential, they can be very misleading if used to establish a highly aggregated national energy conservation target.

Variations in energy use patterns between nations occur because of each nation's efforts to employ its total set of human, material and financial resources in an optimal manner. As a consequence of this, practices which may be very efficient in one country may produce a serious misallocation of resources in another. Therefore, any planning based on the aggregate application of all such national energy use practices can result in a large over-estimate of the conservation achievable and may have the dangerous consequence of distorting energy targets.

In the current discussion of national energy conservation, several investigators (1,2) have used comparisons of U.S. Gross Domestic Product* per unit of primary fuel energy (GDP/E) with that of foreign countries to measure the relative "efficiency" of national energy use. The differences in the GDP/E are then used to infer the potential for energy conservation that may exist in the U.S. The analysis in this paper shows that GDP/E is <u>not</u> a valid parameter for this purpose. Further, this paper shows that the GDP/E parameter is a secondary indicator of the mix of economic activities and the real income per unit of labor in any nation. In this paper the term "energy use proficiency" is used to describe the effectiveness with which a country uses its energy resources to achieve its economic and social goals.

Energy and economic comparisons between countries should be expected to be quite complex because of the many different variables involved such as national resource endowments, geography and climate, population density, living styles, relative cost of energy, economic history and structure, social values and criteria, the mix of agricultural, industrial and service activities, and the purchasing value of one currency relative to another. For

---

*Gross Domestic Product is the sum of a nation's economic output measured in terms of expenditures for goods and services by consumers, government, business and foreign countires. Gross National Product is the sum of GDP and earnings from foreign investments.

these reasons, although intercountry comparisons are interesting in delineating differences, the intricacies of such analyses make it exceedingly difficult to draw meaningful inferences about the proficiency of use of energy as an input to a national economy.

The significance of the GDP/E parameter should be most clearly revealed by an analysis of the differences in economic output and energy use among States of the United States where adjustments for different societal characteristics (e.g. currency puchasing powers, relative prices, taxes, investment incentives and living habits) are minimal as compared to international differences. The United States exists as a network of people who are woven into a homogeneous society with similar cultural and living habits, goals, and value systems. The use of a common technology to extensively produce and distribute similar quality goods and services, and the use of a common currency and banking system have contributed heavily to the homogeneity. Further, the relatively free mobility of labor and the absence of interstate trade barriers makes more meaningful any differences between States in their energy use pattern. However, variations in the innate physical and climatic characteristics among the States which comprise the United States, has given rise to individual State economies, and thus distinctive regional agricultural, industrial and service activities.

To explore the relationships among energy, employment and economic output in the United States, an investigation of these variables was undertaken on an individual State basis (for the year 1971) and by individual major industries (for the year 1972). The average GDP/E was $10.85*/million Btu for the U.S. in 1971, derived from 50 States ranging five-fold from 3.6 in Louisiana to 18.4 in Connecticut. These ratios will be compared later with those of other nations, e.g. 14 for Sweden, 17 for West Germany, and 20 for Italy or Japan.

An examination of 87 national industry sectors revealed a seventy three-fold range in economic output (value added) per unit of energy use (VA/E) from 2.5 for water transportation to 183 for real estate activities. Tremendous variations in energy use per employee were found among the 87 industry sectors, varying from only 36 million Btu/year/person in the footwear industry to nearly 500 times more or 17,200 million Btu/year/person for oil and gas extraction. Economic output per employee ranged from $4800/year for footwear to $62,400/year for oil and gas extraction. These figures reveal the wide distribution around the average for the U.S. in 1971 which was 870 million Btu/year/employee for energy use and $9440/year/employee for economic output.

ECONOMIC OUTPUT AND ENERGY USE

The relationship between economic output as measured by GDP and energy use in the United States for the last century is shown in Fig. 1. This relation arises from the elemental fact that in any economy the combination of capital equipment, labor, energy and materials creates our goods and services. Economic output and energy are thus linked, with the relationship determined by the regional economic activities and resources.

Since the turn of the century, the five-fold increase in GDP was accompanied by about a four-fold rise in energy use. The sharp drop in GDP of 30 percent from 1929 to 1933, caused by economic changes unrelated to energy availability, was accompanied by a similar decline in energy use. From 1933 to 1940, as the U.S. climbed out of the Great Depression, GDP rose 60 percent and the energy required to power the rise increased by 50 percent. In the last few years, the data again demonstrate the relatively close linkage of GDP and energy use.

In contrast to the time dependent relationship for the nation shown in Fig. 1, an examination of the distribution of economic output and energy use by States (for the year 1971) reveals a much rougher correlation between the variables

* $ = 1958 dollars

of Gross State Product* (GSP) and energy use (4) as shown in Fig. 2. This chart discloses the wide realm of energy and economic levels which exists among the States. Interspersed among the State points are data for eight foreign countries which will be discussed later.

Even for states with similar economic outputs there is a considerable range in the use of energy as shown in Table 1.

To analyze the differences in GSP and energy use among the 50 states, one should consider a number of important variables some of which are the following.

Relative Cost of Energy and Capital

In some states which are distant from energy sources, the cost of transporting energy has resulted in higher energy prices. For example, in 1971 the average cost of natural gas to the residential consumer in Houston was 93¢/million Btu while in Boston it was $1.57/million Btu. Because of the higher price and the colder climate, the New Englander invested in insulation and double glazing, while the Texan used minimal insulation. Further, the industrial mix of New England was predominantly low energy consuming per unit of economic output. Thus, the existing economic activities are the cumulative result of historical relationships influenced by energy costs and capital costs integrated over time.

Industry Mix

Nearly two-thirds of the energy use in the American economy occurs in the activities that produce our goods and services which are measured by the GDP. The remainder of the energy is used to satisfy final demand, i.e. that purchased directly by final consumers (such as home heating fuels, residential electricity, gasoline, etc. including exports). The energy use for final demand is not converted to the production of marketable goods and services. A summary of energy use by both these categories is shown in Table 2. Thus in 1972, about 62 percent of the total U.S. energy consumption directly contributed to the production of marketable goods and services. The GDP/E ratio therefore, implicitly depends on this split between energy for final demand and for economic output.

The tremendous variations in value added (VA) and energy use (E) for a multitude of economic activities in the U.S. (5) in 1972 is shown in Table 3 and Fig. 3. It can readily be observed that heavy industries such as iron and steel manufacturing, chemicals, oil and gas extraction, petroleum refining and paper require substantially more energy for a specific level of value added than other economic activities such as printing and publishing, finance and insurance business services and real estate.

Thus, the overall economic output and most of the energy use of a nation is determined by summing the mix and types of agricultural, industrial and service activities. In all industries, gross product (primarily value added by manufacturing) is determined primarily by the costs of capital investment, raw materials, labor, and energy. The relative importance of each of these components varies with the nature of the industry and the available technology. Those states such as Texas and Louisiana which house a large proportion of energy-intensive industries such as petroleum refining and petrochemicals manufacturing, and Indiana, Ohio and Pennsylvania where steel manufacturing is prominent, do have much lower GSP/E ratios than those that provide services such as New York which specializes in banking, insurance and real estate industries.

Availability of Natural Resources

The geographic placement of natural resources, the distribution of population and the demand for goods and services in a relatively free economy have led to the development of particular industries in economically attractive areas. For example, in the West South Central census region, (Texas,

---

*Gross State Product was estimated as follows: (3)

$$GSP = \frac{\text{state personal income}}{\text{national personal income}} \times GDP$$

Over the last 75 years, personal income has been 77-86% of GDP.

Oklahoma, Louisiana and Arkansas), one of the world's most prolific sources of oil and natural gas, energy intensive industries such as petroleum refineries and petrochemical plants were built to supply gasoline, home heating oil, industrial fuel oil and chemicals for much of the nation. In contrast, the States in the Northeast developed industries less energy intensive per unit of economic output.

A graphic presentation of GSP/E for each of the states and census regions for 1971 shown in Fig. 4, reveals the distribution of components that make up the U.S. average. The GSP/E varies from 3.6 in Louisiana to 18.4 in Connecticut. GSP and energy use in any state are a result of evolutionary interactions of the previously discussed variables. Because the mix of economic activities varies extensively among regions, the simple GSP/E is not a valid indicator of the comparative proficiency with which energy is used in any State to produce its economic output.

ECONOMIC OUTPUT AND EMPLOYMENT

Thus far the discussion has centered on the linkage between economic output and energy use. However, the major driving force of the economic system in any country is the continuous growth of the population and its labor force which requires the continuous creation of new employment opportunities. Furthermore, most countries are striving for a continuity of gains in economic productivity which may be necessary for steady improvement of societal well-being. A major goal of most countries is to maximize productivity (economic output per employee-hour) by optimizing the use of all of its resources (capital, labor, energy and materials).

In any economy, the level of economic activity as measured by Gross Domestic Product (GDP), is dependent upon (1) the size of the employed labor force, (2) its annual hours of work, and (3) its productivity or ability to produce marketable goods and services. The following identity is descriptive of the relationships:

$$GDP = L \cdot H \cdot P$$

where GDP = currency per year
L = employed persons
H = average hours worked per year by an employee
P = productivity in currency per hour per employee

For the U.S. in 1971, the following quantitative measures were made of the preceding variables:

$$\$745.4\text{x}10^9/\text{yr} = 79.0\ \text{mill. empl.} \cdot 2028\frac{\text{hrs.}}{\text{yr.}} \cdot \frac{\$4.65}{\text{empl.-hr.}}$$

Assuming a constant number of average annual hours worked by an employee, economic productivity (GDP/L) is related to the economic output per unit of energy consumed (GDP/E) and the energy use per employee (E/L) according to the following identity:

$$GDP/L \equiv GDP/E \cdot E/L$$

Thus for the U.S. in 1971:

$$\frac{\$9{,}440/\text{yr}}{\text{employee}} \equiv \frac{\$10.85}{\text{million Btu}} \cdot \frac{870\ \text{million Btu/yr}}{\text{employee}}$$

When this relationship is examined on a State basis by plotting GSP/E vs. E/L as shown in Fig. 5, it may be observed that the 50 states do not deviate greatly from the average GDP/L of $9,440 per year per employee. This may be explained by the relatively free mobility of the U.S. labor force as it equilibrates by moving from place to place in search of higher paying jobs. Thus, for the most part, the average salary paid (reflecting the average economic output) per employee is roughly similar throughout the United States.

Although the relationship in Fig. 5 cannot reveal the comparative proficiency in energy use, it does indicate that the States having a relatively low GDP/E are predominantly the net energy exporters. Many of these energy source States not only installed energy-intensive conversion facilities such as oil and gas extraction plants, but also developed energy-intensive satellite industries such as chemicals, iron and steel manufacturing, and aluminum manufacturing. This pattern exists in many areas of the world. It is now evolving into a secondary stage in Saudi Arabia, for example, as that country progresses from oil and gas extraction operations to additional energy-intensive industries.

The equivalent characteristics (E/L and GDP/L) of several foreign countries are also given in Fig. 5 which shows their levels of economic productivity (GDP/L) are lower than that of the U.S. Assuming a continuation of their national efforts to increase economic productivity, one would expect a future shift of these nations toward the U.S. pattern.

Using the information by industry listed in Table 3 and the parameters of Fig. 5, the distribution of economic activities in the U.S. in 1972 is delineated in Fig. 6. It can be seen that the heavy industries consume a great deal of energy per employee and have relatively low ratios of value added per unit of energy used. As one scans higher VA/E activities, it can be seen that there is a transition to secondary industries such as paper and metal containers, metal fabrication, motor vehicle manufacturing and aircraft manufacturing. At the highest VA/E end of the spectrum, service activities are found such as finance and insurance, printing and publishing, and real estate.

By comparing Figures 5 and 6, it may be inferred that the States with low GDP/E and high E/L are also the States which are weighted toward the heavy industries as a part of their economic mix. Conversely, at the other end of the spectrum are States with high GSP/E and low E/L which are more oriented to light industries and services. Thus, it can be seen that the GSP/E ratio is useful in gaining a gross insight to the mix of industry activities. However, GSP/E comparisons between States clearly cannot disclose differences in the proficiency of energy use between those States, primarily because of the differing mix of industries.

INFERENCES FOR INTERCOUNTRY COMPARISONS

The detailed comparison of Gross State Product and energy use has shown wide differences in GSP/E among States which result from the differences in both the mix of economic activities and the energy consumption per unit of economic output of these activities. Each state has its own economic characteristics which have evolved from its location, natural resource base, population distribution, climate, and other parameters. But more, they have developed as a part of the whole U.S. economic structure with relative free flows of resources, materials, and finished goods among States.

Intercountry comparisons using GDP/E are just as involved as they are for States. Can we compare Sweden or Japan with the U.S. using GDP/E ratios? Each country is as different as each State, and a comparison of their GDP/E cannot reveal the relative proficiency with which they use energy resources.

1971 data for several foreign countries as listed in Table 4 have been superimposed on the State data in Fig. 5. GDP in foreign currencies were converted to dollars not by monetary exchange rates but by using puchasing power parities (6) which have been developed to reflect real purchasing values in each country. In most cases, the foreign incomes were adjusted upward to credit the appropriate countries for purchasing powers higher than the U.S.

Does a comparison of the GDP/E ratio of France with that of the U.S. mean that France is using energy twice as effectively as the U.S. in the production of goods and services? - not likely. As indicated earlier, the reasons for the differences in GDP/E ratios are to be found in the differences attributable to such variables as national resource endowments, geography and climate, population density, living styles, relative cost of energy, economic history and structure, social values and criteria and the mix of agricultural, industrial and service activities.

Although comparisons between nations of specific industries should be more indicative of the relative proficiency of energy resource use, differences in primary fuel availabilities may distort such comparisons. For example, the Schipper-Lichtenberg paper (1) masks the true comparison of the economic value added in manufacturing per unit of energy use as shown in Table 5.

Manufacturing use of electricity is certainly separable from the available mix of generation sources, and this comparison should have been made on equivalent primary fuel energy input (the gross kWh_t described in the heading of Table 9 of Reference 1).

If their table is thus recalculated, as shown in Table 5, then the kilowatt hours of total energy per dollar of value added would show the U.S. at 16.5 kWh_t/$ and Sweden at 21.2 kWh_t/$. Sweden's manufacturing thus appears to be 78% as economically effective in the use of energy as is U.S. manufacturing. However, this should not be interpreted as indicating either Swedish wastefulness or as a potential for conservation -- rather it is undoubtedly the result of the economic optimization of the use of all the resources (capital, labor, materials and indigenous energy sources) available in each country.

With regard to conservation, because of the rapid international transfer of technologies, world-wide differences among nations in the available techniques for effective energy use are likely to disappear in a short time. There are, of course, individual national differences in resource availability, demography, and social structure which influence technology choices. For example, the combination of electricity generation with district heating which is fairly common in Sweden and Northern Europe does represent a more efficient energy use system for these purposes. However, it is not necessarily the most efficient from the point of view of total national resources of all kinds which are needed for such a system in other regions. For this reason, district heating in the U.S. has only rarely been used, even though the technology has been available for decades.

As this study shows, the argument that the United States has a low average GDP/E ratio when compared to other nations, and is therefore energy inefficient, is both simplistic and misleading. One cannot characterize the proficiency of a society's resource use by considering only one of the input variables (i.e., capital, labor, materials or energy). The more significant measure of proficiency is how successfully all these inputs are integrated to establish national well-being.

REFERENCES

1. L. Schipper and A.J. Lichtenberg, "Efficient Energy Use and Well-Being. The Swedish Example." Science Vol. 194, No. 4269, p. 1001, (Dec. 1976).

2. Stanford Research Institute, Comparison of Energy Consumption Between West Germany and the United States, A Report to the Federal Energy Administration, (June 1975).

3. J. W. Kendrick and C. M. Jaycox, "The Concept and Estimation of Gross State Product," The Southern Economic Journal 32, 153 (Oct. 1965).

4. U.S. Department of the Interior, United States Energy Fact Sheets, 1971 (Feb. 1973).

5. University of California, Lawrence Berkeley Laboratory, A Linear Economic Model of Fuel and Energy Use in the United States, A Report to Electric Power Research Institute, (EPRI ES-115), (Dec. 1975), Available from National Technical Information Services.

6. Irving G. Kravis, et al, A System of International Comparisons of Gross Product and Purchasing Power, The John Hopkins University Press (1975).

Figure 1

GDP versus PRIMARY ENERGY USE IN THE UNITED STATES
1875-1975

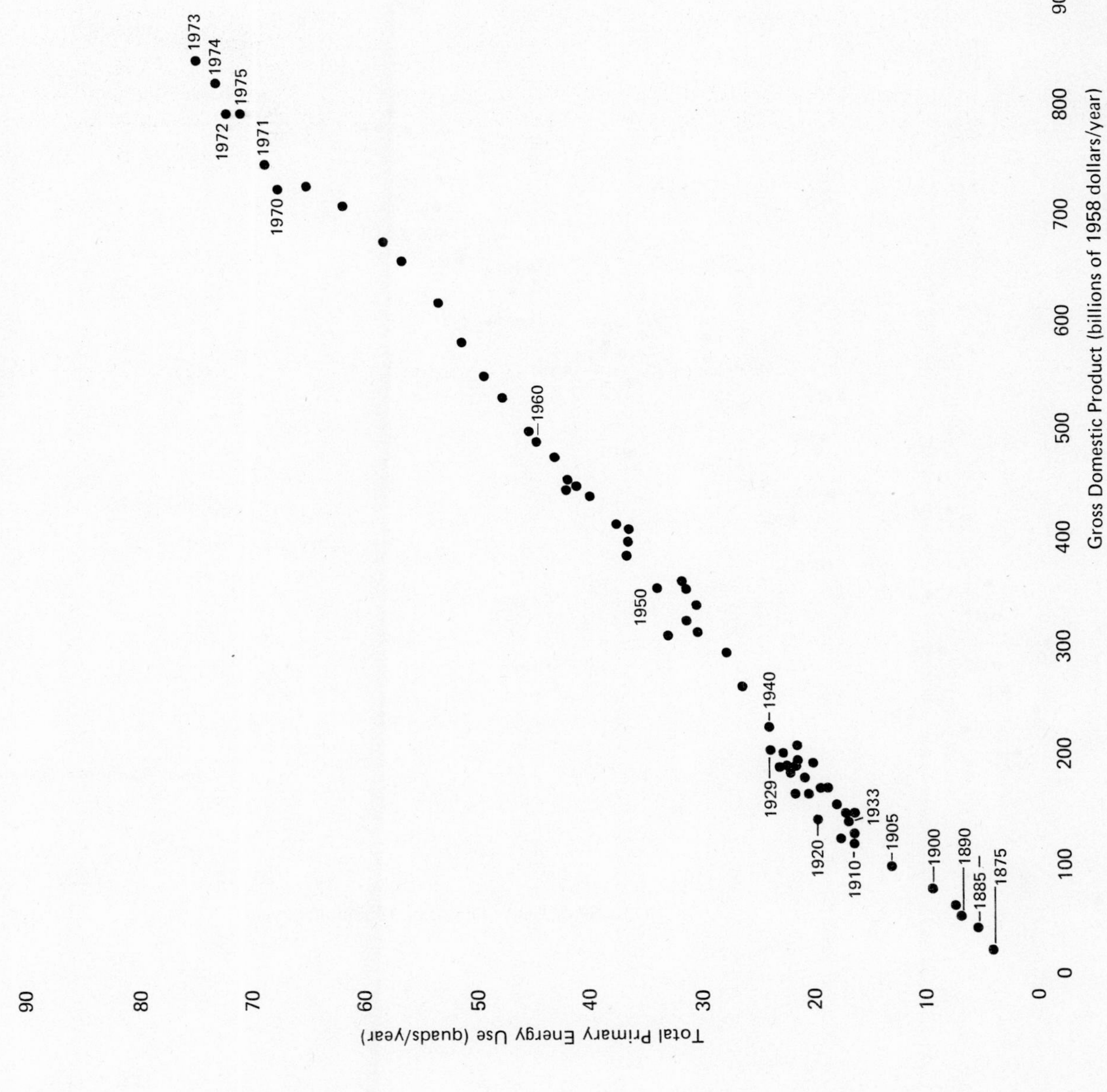

Figure 2

GROSS PRODUCT versus ENERGY USE
1971

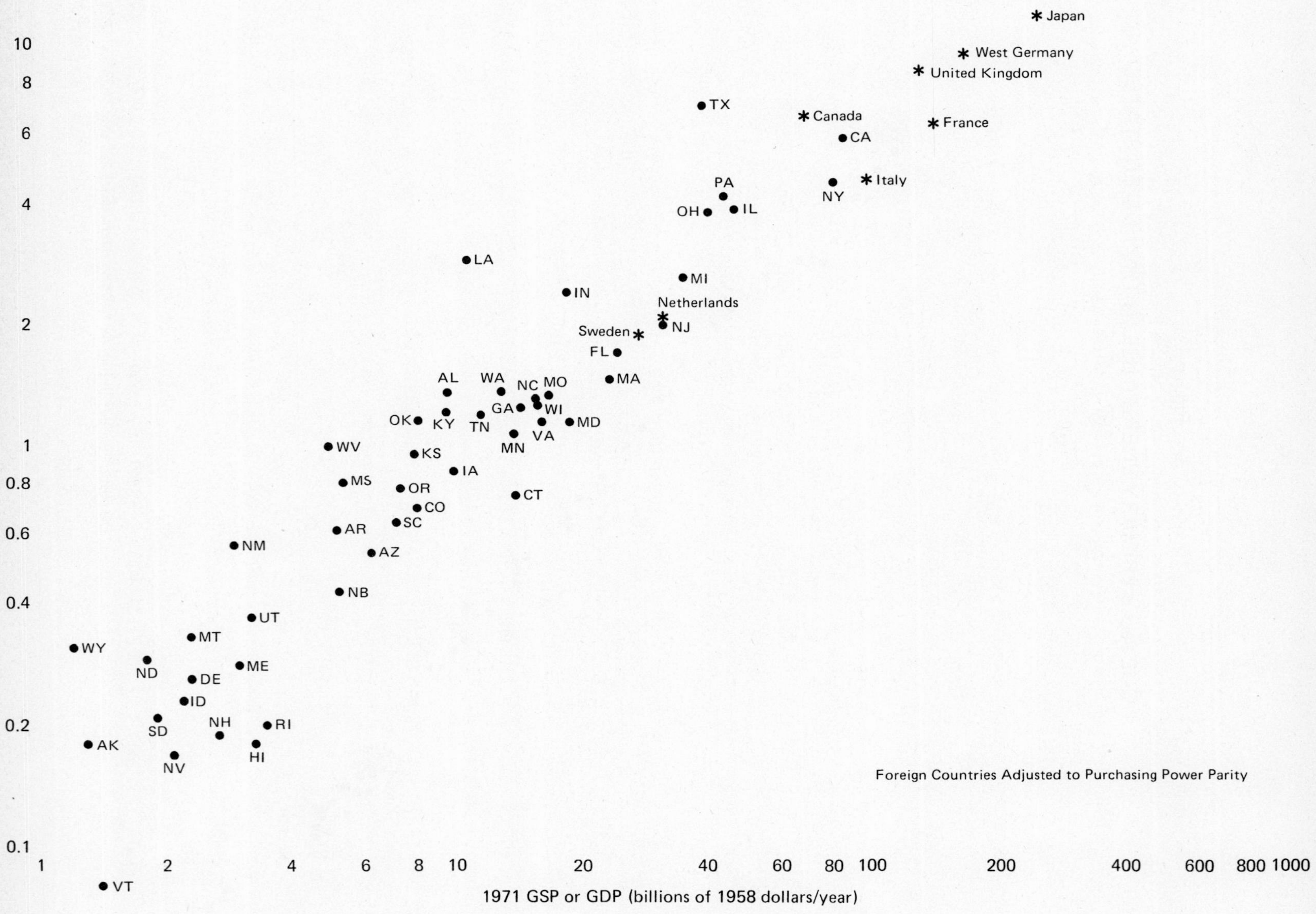

TABLE 1

COMPARISON OF STATE-PAIRS HAVING SIMILAR GSP'S

	Wyoming	Vermont	Louisiana	Connecticut	Texas	New Jersey
GSP ($ Billions/Yr.)	1.3	1.5	10.9	14.3	40.4	31.2
Energy Use (quads**/Yr.)	0.33	0.084	3.07	0.78	7.55	1.98
GSP/E ($/million Btu)	3.9	17.5	3.6	18.4	5.4	15.8

TABLE 2

PRIMARY ENERGY USE IN THE U.S. IN 1972
(quads/year)

	Economic Output Activities	Final Demand	Total Primary Energy
Coal	12.2	2.2 (including 1.8 exported)	14.4
Oil	15.4	18.2	33.6
Gas	15.3	6.1	21.4
Hydroelectric*	1.9	1.0	2.9
Nuclear*	0.4	0.2	0.6
	45.2	27.7	72.9

*converted to primary energy equivalent at the average power generation heat rate of 10,500 Btu/kWh.

TABLE 3

ECONOMIC ACTIVITY, EMPLOYMENT AND ENERGY USE
BY INDUSTRY SECTOR IN THE U.S. IN 1972

(In ascending order of VA/E)

Description of Sector	Value Added* (VA)	Primary Energy Use* (E)	Employment* (L)	VA/L*	VA/E*	E/L*
1. Water Transportation	1,900	777	215	8,840	2.45	3,614
2. Chemical Products	7,442	2,850	434	17,150	2.61	6,567
3. Chemical & Fertilizer Mining	434	164	16	27,130	2.65	10,250
4. Primary Iron & Steel Manufacturing	9,795	3,602	819	11,960	2.72	4,398
5. Nonferrous Metal Ores Mining	445	147	50	8,900	3.03	2,940
6. Pipeline Transportation	601	194	17	35,350	3.10	11,412
7. Paper Products	5,645	1,668	439	12,860	3.38	3,800
8. Petroleum Refining	6,059	1,766	155	39,090	3.43	11,394
9. Air Transportation	5,572	1,562	329	16,940	3.57	4,748
10. Crude Oil & Petroleum Extraction	8,299	2,288	133	62,400	3.63	17,203
11. Primary Nonferrous Metal Manufacturing	4,554	1,215	332	13,720	3.75	3,660
12. Stone and Clay Products	4,785	1,230	473	10,120	3.89	2,600
13. Iron, Ferroalloy Ores Mining	365	90	14	26,070	4.06	6,429
14. Paving Mixtures & Blocks	219	53	10	21,900	4.13	5,300
15. Plastics & Synthetic Materials	2,558	576	208	12,300	4.44	2,769
16. Asphalt Felts & Coatings	213	38	19	11,210	5.61	2,000
17. Glass Products	2,007	354	187	10,730	5.67	1,893
18. Water & Sanitary Services	926	155	94	9,850	5.97	1,649
19. Stone & Clay Mining	965	158	93	10,380	6.11	1,699
20. Gas Utilities	5,959	848	219	27,210	7.03	3,872
21. Forestry & Fishery Products	757	91	79	9,580	8.32	1,152
22. Electric Utilities	13,844	1,660	474	29,210	8.34	3,502
23. State & Local Government Enterprises	5,218	507	449	11,620	10.29	1,129
24. Local, Suburban Passenger Transit	2,309	210	274	8,430	11.00	766
25. Automobile Repairs & Service	4,161	368	557	7,470	11.31	661
26. Textile Goods & Floor Coverings	1,038	91	123	8,440	11.41	740

TABLE 3 (continued)

Description of Sector	Value Added	Primary Energy Use	Employment	VA/L	VA/E	E/L
27. Railroad Services	8,462	676	571	14,820	12.52	1,184
28. Industrial Leather Products	236	18	26	9,080	13.11	692
29. Motor Freight Transportation	11,693	854	1,201	9,740	13.69	711
30. Livestock Products	8,735	623	1,142	7,650	14.02	546
31. Fiber, Yarn & Thread Mills	4,037	288	580	6,960	14.02	497
32. Food Products	19,777	1,319	1,737	11,390	15.00	759
33. Lumber, Wood Products	5,308	332	593	8,950	15.99	560
34. Rubber & Plastic Products	6,512	391	620	10,500	16.65	631
35. Medical & Educational Services & Nonprofit Organizations	35,278	2,033	6,785	5,200	17.35	300
36. New Construction	27,089	1,518	3,153	8,590	17.85	481
37. Paperboard Containers	2,022	102	225	8,990	19.82	453
38. Hotels & Personal Services (e.g. laundry, beauty & barber shops, funeral services)	12,872	634	2,526	5,100	20.30	251
39. Service Industry Machines (e.g. vending, laundry, car wash)	1,484	71	155	9,570	20.90	458
40. Agricultural Products (e.g. wheat, rice)	14,958	687	1,892	7,910	21.77	363
41. Metal Containers	1,002	46	77	13,010	21.78	597
42. Miscellaneous Manufacturing (e.g. jewelry, toys, pens, musical instruments, signs)	3,227	142	417	7,740	22.73	341
43. Fabricated Metal Products (general hardware)	3,616	124	329	10,990	29.16	377
44. Household Appliances	1,595	68	199	8,020	23.46	342
45. Drugs, Cleaning, Toilet Preparations	4,221	174	263	16,050	24.26	662
46. Paint Products	825	32	68	12,130	25.78	471
47. Heating Equipment, Plumbing Fixtures, Structural Metal Products	3,932	152	519	7,580	25.87	293
48. Farm Machinery	1,480	57	125	11,840	25.96	456
49. Wholesale & Retail Trade	133,299	4,759	17,511	7,610	28.01	272
50. Engine & Turbine Manufacturing	1,207	43	109	11,070	28.07	395
51. Miscellaneous Electrical Machinery, Equipment & Supplies	1,071	38	131	8,180	28.18	290
52. Screw Machine Products, Stampings	3,616	124	329	10,990	29.16	377
53. Construction, Mining, Oil Field Machinery	2,200	75	200	11,000	29.33	375
54. Electrical Transmission & Distribution Equipment	3,704	125	387	9,570	29.63	323
55. Electronic Components	2,972	97	323	9,200	30.64	300
56. Household Furniture	2,107	67	373	5,650	31.45	180

TABLE 3 (continued)

Description of Sector	Value Added	Primary Energy Use	Employment	VA/L	VA/E	E/L
57. Motor Vehicles	12,479	393	864	14,440	31.75	455
58. Electric Lighting & Wiring Equipment	1,573	49	222	7,090	32.10	221
59. Radio & T.V. Broadcasting	1,629	49	132	12,340	33.24	371
60. Federal Government Enterprises	5,633	167	691	8,150	33.73	242
61. General Industry Machinery (e.g. furnaces, pumps, bearings)	3,033	89	252	12,040	34.08	353
62. Other Transportation Equipment (e.g. ships, space vehicles)	2,677	76	495	5,410	35.22	154
63. Other Furniture & Fixtures	1,185	32	126	9,410	37.03	254
64. Miscellaneous Fabricated Textile Products	1,053	28	178	5,920	37.61	157
65. Materials Handling Machinery & Equipment	879	23	82	10,720	38.22	280
66. Maintenance & Repair Construction	11,586	292	1,357	8,540	39.68	215
67. Machine Shop Products	1,905	46	229	8,320	41.41	201
68. Special Industry Machinery (e.g. textiles, paper, printing)	2,089	49	160	13,060	42.63	306
69. Coal Mining	1,967	45	148	13,300	43.71	304
70. Metalworking Machinery	3,815	86	278	13,720	44.36	309
71. Professional Sciences, Controlling Instruments & Supplies (e.g. medical, time, engineering)	2,594	57	281	9,230	45.51	203
72. Amusements (e.g. motion pictures, billiards, commercial sports, etc.)	5,272	112	702	7,510	47.07	160
73. Office Machines	2,412	48	237	10,180	50.25	203
74. Optical Equipment	2,646	50	156	16,960	52.92	321
75. Printing & Publishing	10,096	175	1,156	8,730	57.69	151
76. Finance & Insurance	30,186	523	3,146	9,600	52.72	166
77. Communications (e.g. telephone, telegraph, teletype)	17,930	308	984	18,220	58.21	313
78. Ordnance	4,462	76	230	19,400	58.71	330
79. Business Services (e.g. advertising, data processing)	30,274	497	3,089	9,800	60.91	161
80. Aircraft Manufacturing	8,726	137	501	17,420	63.69	273
81. Radio, T.V., Communications Equipment	6,378	99	546	11,680	64.42	181
82. Wooden Containers	241	3	24	10,040	80.33	125
83. Apparel Manufacturing	6,995	80	1,459	4,790	87.44	55
84. Tobacco Products	2,932	27	71	41,300	108.59	380
85. Footwear & Leather Products	1,336	10	278	4,810	133.60	36
86. Agricultural, Forestry, Fishery Services	1,397	9	366	3,820	155.22	25
87. Real Estate Services	82,126	448	980	83,800	183.32	457

* Value Added (\$/Year)
Primary Energy Use (10^{12} Btu/Year)
Employment (10^3 Persons)
VA/L (\$/Year/Employee)
VA/E (\$/million Btu)
E/L (million Btu/Year/Employee)

Figure 3

ECONOMIC ACTIVITIES IN THE UNITED STATES
1972

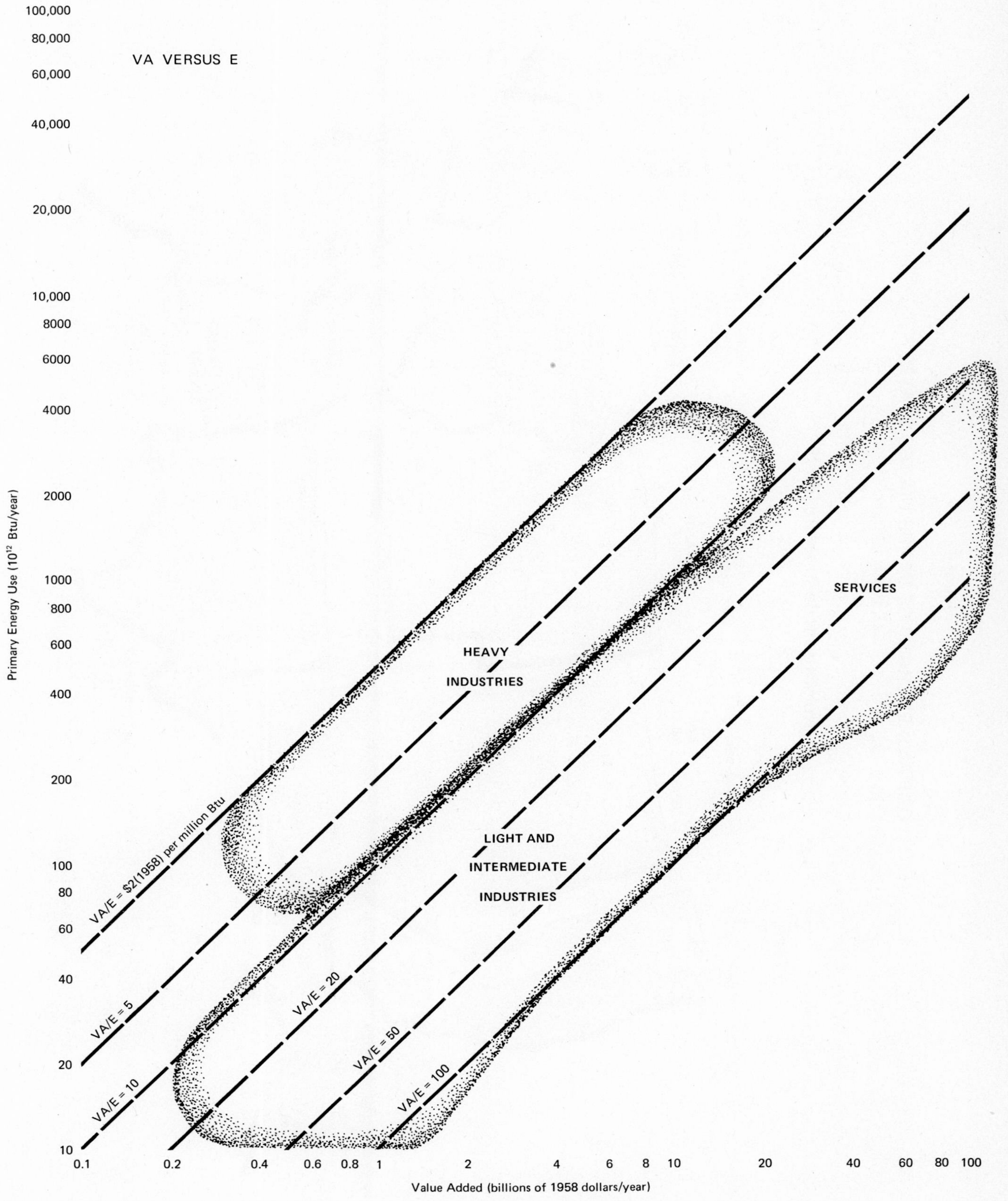

Figure 4

GDP AND ENERGY USE IN THE UNITED STATES
1971

Numbers inside states are GSP/E

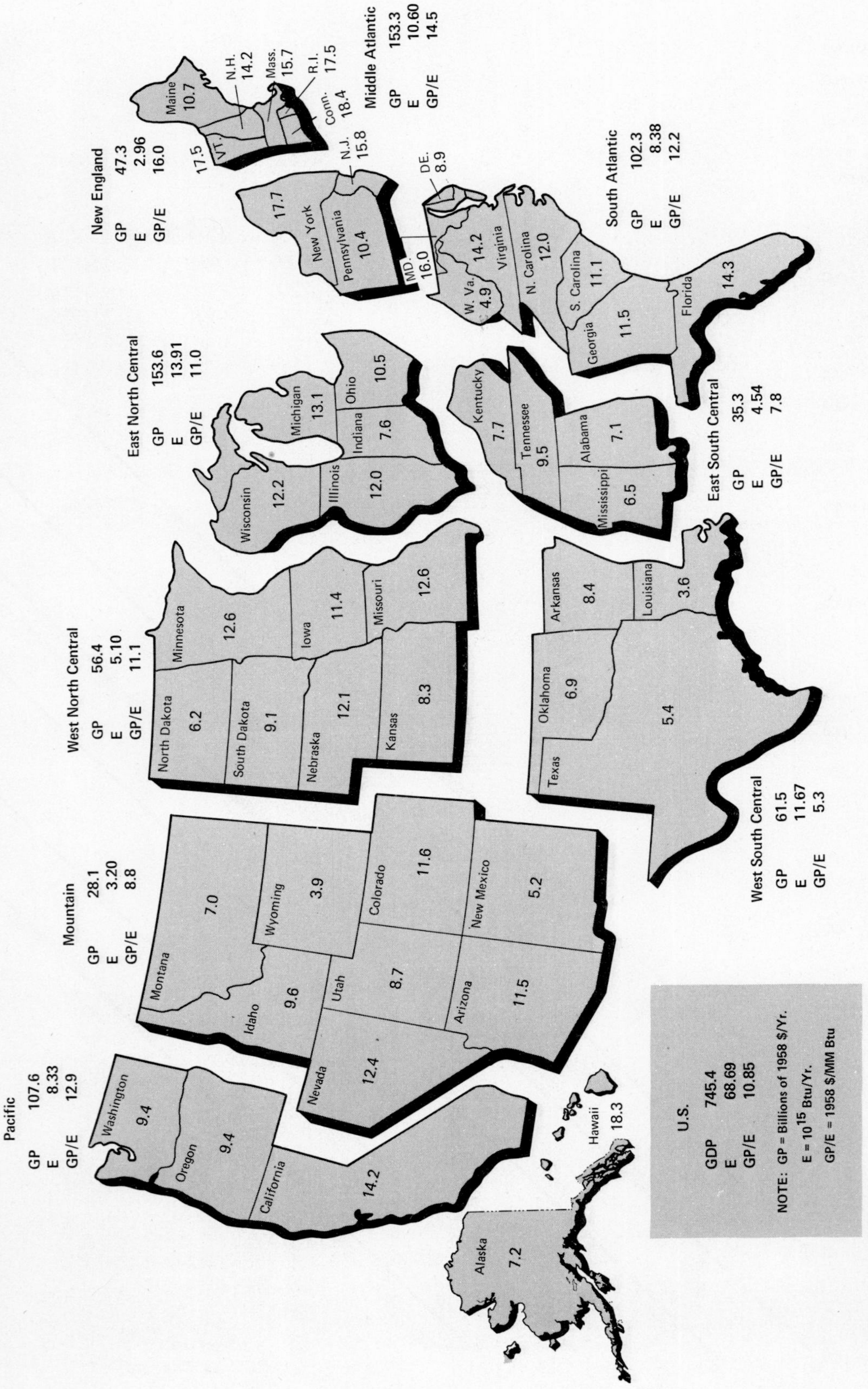

Figure 5

GSP/E or GDP/E versus E/L FOR UNITED STATES
AND FOREIGN COUNTRIES
1971

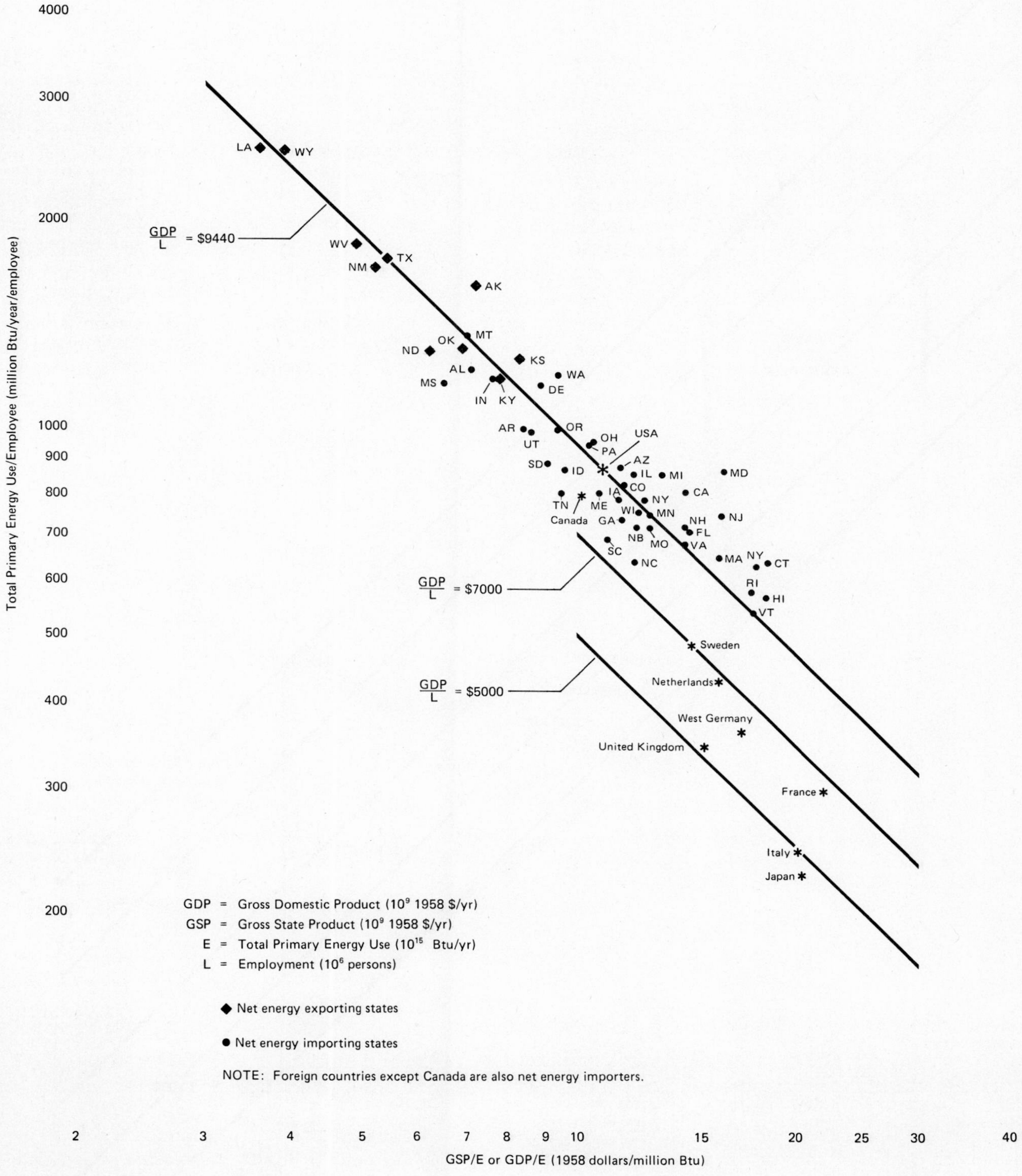

Figure 6

ECONOMIC ACTIVITIES IN THE UNITED STATES
1972

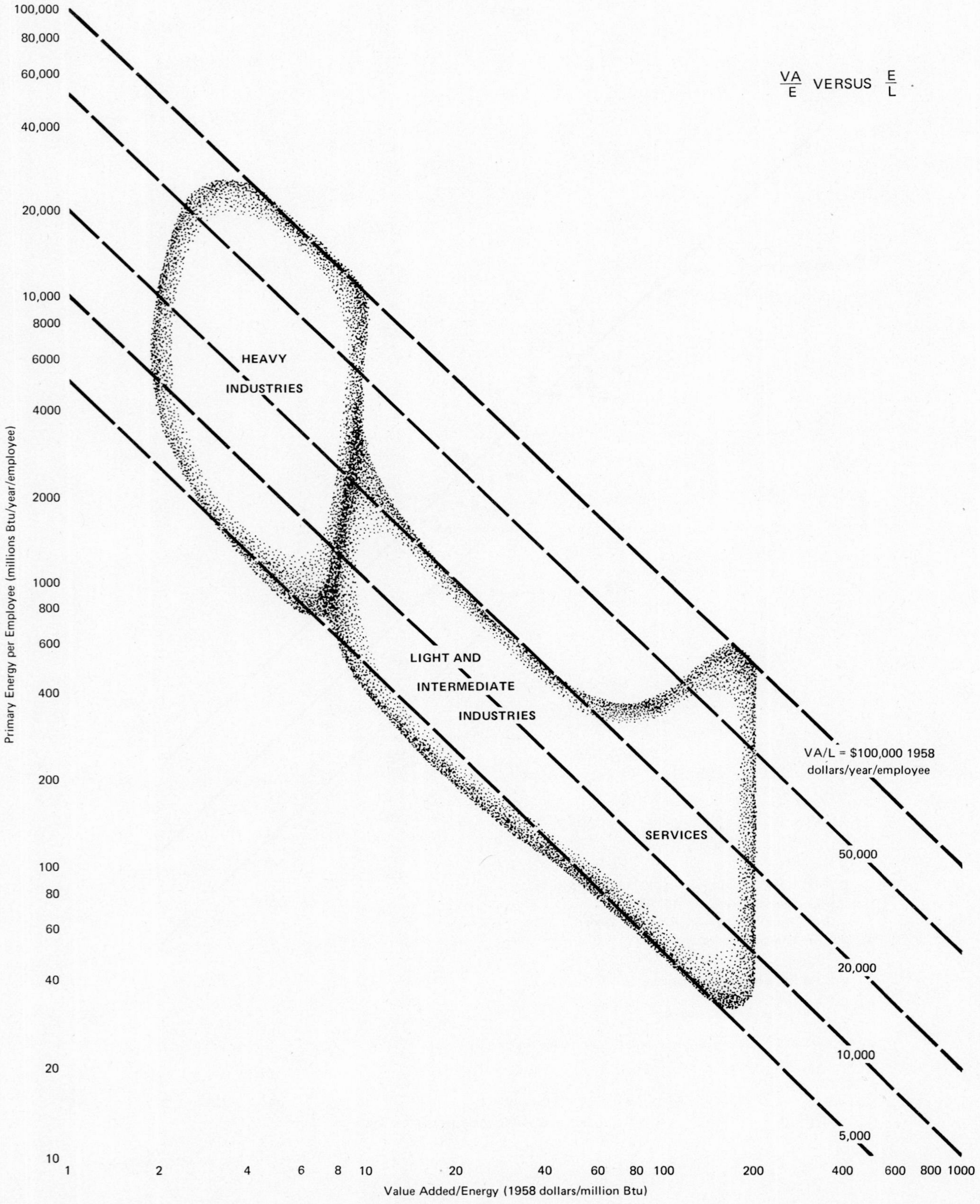

TABLE 4

ECONOMIC OUTPUT, EMPLOYMENT AND ENERGY USE IN 1971
THE U.S. AND FOREIGN COUNTRIES

	GDP/L ($/Yr./employee)	GDP/E ($/million Btu)	E/L (million Btu/Yr./employee)
U.S.A.	$9,440	10.85	870
Canada	$8,110	10.1	800
Sweden	$7,020	14.4	490
Netherlands	$6,740	15.7	430
France	$6,580	22.0	300
West Germany	$6,200	17.0	370
United Kingdom	$5,190	15.0	350
Italy	$4,950	20.2	250
Japan	$4,690	20.6	230

TABLE 5

ENERGY USE IN MANUFACTURING (1971)

(reference: Table 9, Schipper-Lichtenberg, reference 1)

	U.S.	Sweden
Value Added ($/capita)	$1,741	$1,137
Net Energy Basis		
non-electric (kWh_t/capita)	20,455	13,165
electric (kWh_e/capita)	2,750	3,640
total (kWh/capita)	23,205	16,805
energy intensity (kWh/$)	13.3	14.8
Equivalent Primary Fuel Energy Basis		
non-electric (kWh_t/capita)	20,455	13,165
electric (kWh_t/capita)	8,250	10,920
total (kWh_t/capita)	28,705	24,085
energy intensity (kWh_t/$)	16.5	21.2

ANALYSIS AND COMPARISON OF SELECTED ECONOMIC IMPACTS FROM ENERGY CONSERVATION AND ENERGY SUPPLY*

Peter Benenson†

Energy & Environment Division, Lawrence Berkeley Laboratory, University of California, Berkeley, CA 94720

ABSTRACT

The impacts of energy conservation on employment continue to be topic of concern in the formulation of energy policy. In this paper some of those impacts are examined and their relationship to energy policy is discussed. The California energy industry is described quantitatively, analytic tools are developed for estimating direct and indirect income and employment impacts from California energy futures, and the tools are tested on three energy scenarios. The impacts are estimated with a California energy supply model linked to a California input-output model. The three scenarios are characterized by varying combinations of conventional energy technologies, new energy technologies, and specific energy conservation measures. The scenarios are compared in terms of income and employment impacts, energy consumption, and required generating capacity. Also examined are tradeoffs between energy conservation and power plant construction. Finally some policy implications are drawn.

INTRODUCTION

The focus of this paper is a comparison of selected economic impacts from specific energy conservation measures with the corresponding impacts from energy supply strategies. The impacts considered are the labor and material requirements and the energy and generating capacity requirements. Policy recommendations related to the findings are also presented. The material emphasized in this paper is taken from a larger study that was designed to 1) describe quantitatively the California energy industry and its relationship to the California and U.S. economies, 2) provide the analytic capability for determining the direct and indirect employment and income impacts resulting from a given energy future for California, and 3) demonstrate and test the methodology for three scenarios. The three scenarios are characterized by 1) electric utilities' construction plans and published forecasts of oil and gas supply and demand, 2) new electricity supply technologies and decreased availability of natural gas, and 3) a modification of 1) to account for energy conservation from delamping in commerical buildings, residential retrofit ceiling insulation, and passive solar housing construction.

THE PRESENT CALIFORNIA ENERGY INDUSTRY

The present California energy industry is increasingly dependent upon imported energy, both domestic and foreign. The trend over the past decade shows a peak in production of both crude oil and natural gas within California in the latter part of the 1960's. The production of both these fossil fuels has declined since 1968 at annual rates of approximately three and nine percent respecitvely. Meanwhile, in 1970 the sales of electrical energy surpassed the instate production. This differential has since grown larger; in 1975 electricity sales exceeded production by almost 14 percent. Present and potential fuel and energy flows into California are shown schematically in Fig. 1.

FUTURE ENERGY SUPPLY SYSTEMS: AN ASSESSMENT METHODOLOGY

In order to quantify the employment and income impacts of future energy systems, we have developed an energy supply model for California (Ref. 3) and linked it to an input-output (I-O) model of the California economy (Ref. 1). This

---

*From a study entitled "Analysis of the California Energy Industry," J.Sathaye, H. Ruderman, R. Sextro, P. Benenson, L. Kunin, P. Chan, J. Kooser, and Y. BenDov, LBL-5928, January 1977.

†The material selection and policy recommendations are the responsibility of the author. They do not necessarily reflect the views of the other study participants, or the position of the Lawrence Berkeley Laboratory.

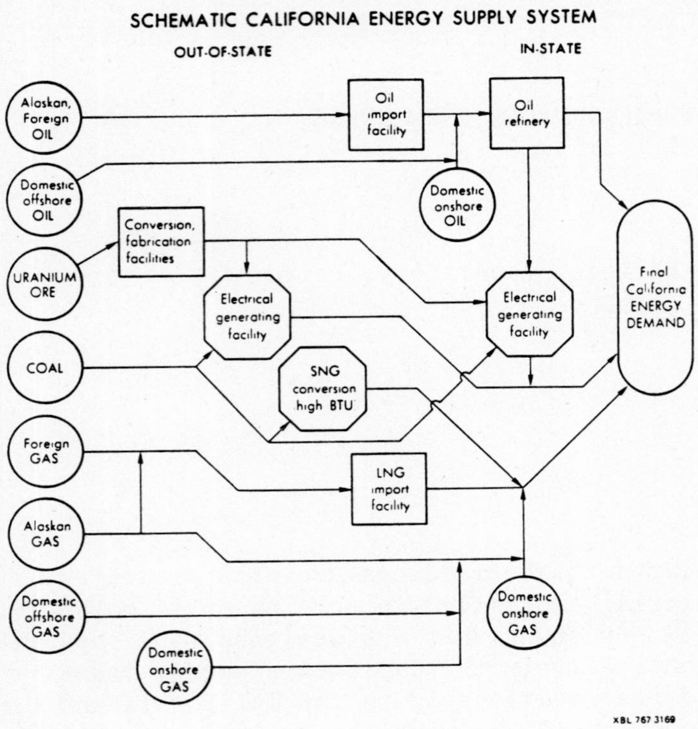

Fig. 1

energy supply model simulates the construction and operation of energy resource extraction, transportation, and conversion facilities. For these facilities it calculates capital and operating requirements, including materials and fuels. For analytical purposes, these are termed the direct effects.* The calculation is made as follows:

Inputs:

- annual schedules are specified for i) gas demand, ii) oil demand, iii) electrical generating capacity, and iv) non-conventional energy supply facilities;
- Gas and oil supply constraints and characteristics are also specified.

Calculations Performed by Model:

- the computer program then calculates the necessary energy facility construction schedules and fuel flows required for these facilities.
- these fuel flows are then converted into transportation facility schedules.

Model Outputs

- The program next calculates the capital, manpower and equipment required to construct and operate these facilities;
- a separate program can compute the environmental pollutants emitted by these facilities.

Output Format

- Finally the output from the two previous steps is reassembled and printed in a tabular and graphical format.

The indirect manpower and income effects are then estimated by transforming these requirements into inputs to the I-O model. The I-O model represents the economy as a matrix of purchases and sales among all producing and consuming sectors. The interdependence among sectors is explicitly accounted for so that the impact of changes in the production of any given sector can be traced to all the other sectors. The I-O model can be described as follows:

$$AX + Y = X$$

where

A is the n x n matrix of technical coefficients derived by forming the ratios of each of the inputs for a given sector to the output of that sector.

X is the 1 x n vector of total outputs where each element is the output of an individual sector.

Y is the 1 x n vector of final demands, the quantity of each output consumed as a final product, in contrast to that used by other industries for intermediate production.

The equation requires that interindustry consumption of output (AX) plus final demand (Y) equal total output produced (X). In this analysis, the Y vector is the output from the California energy supply model, i.e., the estimate of the direct effects. The objective is to solve for X, the indirect effects, i.e., the total production in each sector that is required to produce the estimated amount for final consumption (Y).

Thus

$$Y = X - AX$$
$$= (I-A)X$$
$$X = (I-A)^{-1}Y$$

Once X is solved, an industry by occupation matrix is used to calculate the indirect employment effects (Ref. 2). To summarize, the impacts calculated by these models are fuel requirements, capital and manpower needs for the construction of new energy facilities, the indirect material and manpower required for the

*Direct impacts include all impacts arising directly out of the construction and operation of any facility. Indirect impacts include impacts due to other activities related to the construction and operation of that facility. For example, the manpower required to construct a power plant would be a direct requirement whereas the manpower required to make steel used in constructing the power plant would be an indirect requirement.

supply of construction materials, and some of the costs and manpower associated with the operation and maintenance of these facilities.

SCENARIO DESCRIPTION

To demonstrate this methodology for comparative impact analysis, we have developed, in conjunction with the California Energy Resources Conservation and Development Commission staff, three scenarios of future energy growth in California. The first scenario is based on the utilities' schedule for adding new generating capacity and on published forecasts of oil and gas supply and demand.

The second scenario differs from the first by emphasizing new electricity supply technologies such as wind turbines, solar-thermal and waste-fired power plants, and a decreased availability of natural gas. This scenario also postulates an increased emphasis on geothermal energy rather than nuclear fission for electricity generation, and the deployment of active solar heating devices in residential buildings.

The third scenario was developed based upon the first scenario, with modifications to account for the implementation of three energy conservation measures: 1) delamping in commercial buildings, 2) retrofit ceiling insulation in residential buildings, and 3) a program for housing construction consisting entirely of passive solar houses. These conservation measures are discussed in more detail below.

In a passive solar energy system, the building orientation, insulation, and glazing are designed to take maximum advantage of solar heat gain during the winter and to minimize heat gain during the summer. The building itself serves as the collector and heat storage device. This is in contrast to an active system in which the sunlight heats a fluid that in turn is used for space heating, space cooling, and water heating.

In a recent study, cost data are compared for approximately sixty solar heating, solar heat and cooling and/or hot water systems currently in residential use in Northern California (Ref. 5). System costs for residential space heating and cooling ranged from $200 to $18, 000 with an average cost of $3391; the cost per square foot ranged from $0.91 to $32.29. Although the lifecycle costs for some solar systems are less than those for conventional systems, it is still generally true that when new housing is constructed the first costs are the determining factor in the selection of a heating and cooling system. Therefore, we also consider in the third scenario a cost competitive passive solar house as an attractive alternative to the active system incorporated in the second scenario.

The passive solar housing construction program we hypothesize is predicated upon a change in the state building code. The change would upgrade the insulation and glazing requirement for all new single and multifamily dwellings constructed in California after 1977. The code would require R-19 and R-30 insulation in the walls and ceilings, respecitvely, and double glazed and properly shaded windows. We also assume a minimum of half the window area is on the south wall and that adequate thermal mass is included in the structure for heat storage.

The delamping program we consider entails the removal of a percentage of fluorescent light bulbs from the installed fixtures and the replacement of them by Phantom™ tubes.* This would result in a uniformly lower level of lighting. We assume a delamping program in office buildings, retail stores and schools that would take place in California over a five-year period.

The retrofit insulation program is designed to upgrade the ceiling insulation to an R-19 standard in approximately 4.5 million existing residences. Increased ceiling insulation serves to retard heat loss through the ceiling during the winter, and to limit heat gain through the ceiling during the summer. We decided to limit our estimates to ceiling insulation because it appears most likely to be retrofit. Obviously additional energy savings are possible by retrofit insulating walls also, but this is a more difficult operation.

CAVEATS

As the results are discussed, several important limitations of this study should be kept in mind. The first concerns the accuracy and availability of data. Much of the data on materials and manpower were developed in 1974; the uncertainties range from ± 10 to ± 50 percent. Data on new technologies are even less reliable, since they are based on engineering estimates or experience with pilot plants. The data per-

---

*Most lighting fixtures in office buildings are series-wired multiple lamp fixtures. To reduce the lighting level and maintain uniform lighting throughout the building two options are available: 1) rewire the fixture to take a single tube ballast; 2) install a Phantom™ tube in place of the tube removed. The Phantom™ tube connects a capacitor across the contacts and allows the remaining tube to light. Efficiency is maintained but the total lighting level drops to 30 percent of the original output of the two-tube fixture.

taining to the energy conservation measures investigated here have similar reliability limitations due to the nature of the assumptions. Second, linearity was assumed in both energy supply processes and in economic structural relationships. These assumptions do not take into account economy of scale considerations, or improvements in technologies that may take place over time. Similarly, the input-output model of the economy is based on the structure of the California economy for 1972. It cannot account for structural changes that could take place in the fifteen-year study period. Finally, the boundaries imposed in this study can have an important effect upon the results. Average direct resource requirements could change as the time periods are extended beyond 1990. Similarly, the extent to which indirect effects are included could alter the magnitude of some of these effects.

Scenario Analysis

The estimates are analyzed from two perspectives. First, the scenarios are compared Second, the impacts from each conservation program are analyzed individually to isolate some of the tradeoffs involved in their implementation. In addition to the direct and indirect effects described above, the energy savings and power plant capacity reductions for the three conservation programs were also estimated (Table I).

The twenty-year cumulative energy requirements for the three scenarios are shown in Fig. 2. A direct comparison of the first and third scenarios shows the effect of the energy conservation measures postulated in the third scenario.

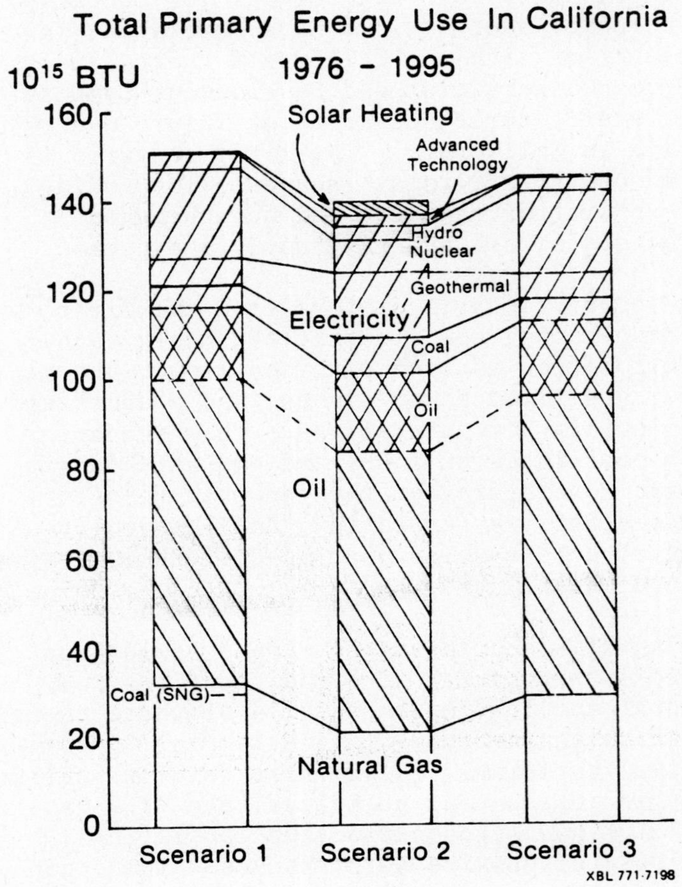

Fig. 2

Table I

Estimates of Energy Savings and Capacity Reduction

	Cumulative Energy Savings						Cumulative Savings in Generating Capacity (MWe)		
	Natural Gas (10^{12}BTU)			Electricity (10^9kwh)					
	to 1981	to 1986	to 1991	to 1981	to 1986	to 1991	to 1981	to 1986	to 1991
Delamping	0	0	0	26.2	78.7	131.3	3080[a]	3080[a]	3080[a]
Retrofit Insulation[c]	248	745	1242	1.1	3.3	5.5	0	0	0
Passive Solar Housing[c]	39	155	349	4.8	19.2	43.1	1400[b]	2800[b]	4200[b]
Total	287	900	1591	32.1	101.2	179.9			

a. Base load.

b. Peak load.

c. These programs, taken together, save natural gas equivalent to the projected output from two synthetic natural gas (SNG) plants (total 141 X 10^9 ft^3/yr).

For nonrenewable resources there is a four percent difference in the total primary energy requirements for the twenty-year period. The major components of this change are a six percent savings in natural gas, a 42 percent drop in coal requirements (accounting for savings in SNG as coal), and about an eight percent decline in nuclear fuel requirements.

The capital and manpower requirements for the three scenarios, aggregated to five-year periods, are shown in Table II. A significant portion of the capital and manpower in the second scenario is expended on active solar heating devices. These devices are installed at a uniform rate over time, making the capital and manpower flows more uniform than those of the first scenario.

In terms of the overall state economy, the capital requirements for the construction of new energy facilities averaged over 15 years is about 14 percent of the 1974 value of Gross Private Capital Formation in California. The difference, on an annual basis, between the first and second scenarios is 500 million dollars, or two percent of the state's Gross Private Capital Formation. A comparison of the first and third scenarios gives a smaller value, amounting to 1.4 percent of the Gross Private Capital Formation.

implementation of the conservation programs would, other things remaining constant, result in reduced electricity and natural gas consumption, and consequently in a somewhat smaller future generating capacity than would otherwise be needed (see Table I).

Cost and manpower data for the conservation porgrams as well as for selected energy facilities are presented in Table III. In the case of the delamping program we have not regarded as net additional costs the cost of producing and installing Phantom™ tubes. If a delamping program of the scope envisaged were undertaken, it is likely that the costs would be comparable to the costs of producing and installing standard fluorescent tubes.

In the case of the retrofit insulation program, we have assumed the insulating material to be fiberglas. Estimates are given for the cost of the insulation, the associated value added (largely wages) and the manpower requirements for installation.

The passive solar housing program assumed would be without net additional capital costs because additional materials going into the construction of passive solar housing would be offset by a reduction in the sizes, and therefore the costs, of heating and air conditioning units.

Table II

Summary of Construction and Operation Requirements (1976-90)

		Capital (10^6 1974 $)		Manpower (Man-years)				
		Construction	Operation	Construction	Operation	Indirect	Total	Average Annual
Scenario 1	1976-80	10,131	7,073	101,900	259,800	189,000	550,700	110,100
	1981-85	22,544	7,906	203,400	274,500	374,600	852,500	170,500
	1986-90	15,239	9,116	177,900	300,000	282,800	760,700	152,140
	Total	47,914	24,095	483,200	834,300	846,400	2,163,900	144,300
Scenario 2	1976-80	19,051	7,470	258,500	298,300	330,700	887,500	177,500
	1981-85	17,712	9,466	224,200	382,700	298,800	905,700	181,100
	1986-90	18,141	10,962	208,000	450,100	302,700	960,800	192,160
	Total	54,904	27,898	690,700	1,131,100	932,200	2,754,000	183,600
Scenario 3	1976-80	9,423	6,948	94,800	258,300	180,600	533,700	106,740
	1981-85	18,283	7,616	172,800	267,800	337,600	778,200	155,640
	1986-90	15,097	8,763	177,000	290,700	284,200	751,900	150,380
	Total	42,803	23,327	444,600	816,800	802,400	2,063,800	137,600

IMPACTS OF ALTERNATIVE INVESTMENTS

We examined some of the tradeoffs between the three energy conservation programs and the energy facility construction programs. The delamping program could result in a savings of 131 X 10^9 kwh of electricity over a fifteen-year period, and in a reduction in capacity requirements of 3080 MWe by the mid-1980's. This

Table III

Comparison of Economic Impacts of Alternative Investments

Investment	Construction: Capital Cost (10^6\$ 1974)	Construction: Direct Manpower (man-years)	Construction: Indirect Manpower (man-years)	Annual Operations: Cost (10^6\$ 1974)	Annual Operations: Manpower (man-years)	Indirect Value Added (10^6\$ 1974)
Nuclear LWR (1100 MWe)	460	6250	8800	5.8	112	164.5
Coal-fired (800 MWe)	260	2970	5700	3.8	125	104.4
Gas turbines (133 MWe)	17	120	330	0.3	7	6.5
Coal Gasification ($250X10^6 ft^3$/day SNG)	750	10780	17100	7.5	590	310.1
Delamping	0	0	0	0	0	0.0
Retrofit Insulation	756	8000	15000	0	0	300.0
Passive Solar Housing (each 5 years)	0	0	3000	0	0	57.0

capacity reduction can be translated into two 1100 MWe nuclear power plants, and one 800 MWe coal-fired power plant that will not be required. These particular units were chosen because they appear in the California utility plans for the coming decade. These power plants would have required \$1.2 billion (1974 dollars)* and 39,000 man-years of labor whereas the delamping program would require no additional resources. The passive solar program reduces the need for peak-load generating capacity by 4200 MWe in 1990, due mainly to a reduction in air conditioning load. A comparison of direct and indirect manpower required to construct new peaking capacity (using gas turbines) with the manpower required for the passive solar heating program indicates that the solar heating program would require 300 fewer man-years annually than the construction of new peaking capacity. The capital savings over this fifteen-year period resulting from the passive solar housing program would be \$510 million. Natural gas savings accruing from both the passive solar housing program and a program of retrofit insulation would amount to 1.6×10^{15} BTU over the next fifteen years. This is slightly more than the projected output of two synthetic natural gas (SNG) plants proposed as one source of additional supply for California. Cancellation of these plants would reduce capital demand by \$1.5 billion (1974 dollars) and would reduce demand for labor by 23,000 man-years.

Obviating the need for a power plant or an SNG facility means that certain amounts of capital and manpower will be available for other sectors of the economy. Whether or not these resources are absorbed elsewhere depends on the level of demand for capital goods, which, of course, is dependent upon the level of aggregate demand. The impact of investment differs in a slack economy compared to one in which resources are fully employed. One cannot therefore make a statement a priori as to what the overall economic impacts of undertaking, or not undertaking, a particular investment project are apt to be.

The problem is further complicated by the fact that capital and labor are combined in varying proportions in different uses, i.e., some avenues of investment are more labor intensive than others. High levels of unemployment combined with relative tightness in capital markets indicate that, other things remaining the same, there are grounds for preferring more labor intensive avenues of investment.

The conservation measures we examined would not increase the demand for labor as much as the power plant construction that would be displaced by these measures. This does not necessarily imply that unemployment will increase in California as a result of energy conservation. First, the labor requirements for other conservation measures may be greater or smaller than the requirements for those measures

*These figures do not include owners' costs. For nuclear facilities these have been estimated to run as high as 40 percent of capital costs; for coal-fired plants, 25 percent. The conservation measures considered do not involve owners' costs of comparable magnitude. Therefore, the estimates of capital savings through implementationof these programs are understated.

we have considered. Second, a projection of the labor required for construction does not imply that an equivalent number of workers are available for employment. If, however, unemployment exists, the financial and real resources that would have been used for power plant construction might be channeled to other sectors of the California economy. This in turn would stimulate income and employment. Expansion in these sectors most probably would involve new construction, so the demand for those occupations directly involved in construction would rise. As 80 percent of the industries are more labor intensive than the energy industries in terms of operation, there appear to be opportunities for net employment gains (Ref. 9).

The analysis has been formulated in terms of quantifiable factors; material and manpower requirements, and fuel and generating capacity expended or reduced. But perhaps the most important impacts from energy conservation are unquantifiable. For example, power plant emissions would be reduced, which in turn reduces the risk of detrimental health effects. Land would be available that would otherwise be used for power plant sites and transmission lines. Requirements of scarce water for fossil fuel extraction and conversion, cooling, and land restoriation would diminish. The negative aesthetic impacts attendant with power plant construction and operation also would be softened. Finally, shipments of radioactive material and wastes would decrease, which lessens the chance of sabotage and reduces the magnitude of an as yet unsolved waste disposal problem.

ENERGY CONSERVATION AND EMPLOYMENT: POLICY IMPLICATIONS

The relationship between energy conservation and employment continues to be a topic of concern. Frequently labor union leaders argue in favor of power plant construction and against energy conservation. It is argued that the decision to cancel construction of a power plant means the loss of specific construction jobs. But this construes the problem too narrowly because of the extremely interrelated nature of the U.S. economy. A large percentage of individual sectors buy from and sell to each other. This interrelatedness, and the fact that energy conservation measures usually require labor and materials for implementation, means that implementing energy conservation measures leads to contractions in some sectors and to expansion in others. The conservation measures examined in this paper illustrate this point. The net impact depends upon the particular conservation measures and the type of power plant construction that are in question. Moreover, the largest total direct and indirect labor requiremets derived from energy production in the scenarios examined account for less than two percent of the California labor force in 1975. The leverage from energy policy on the employment situation is therefore relatively small. Decisions regarding energy conservation and power plant construction should be based on sensible resource allocation and environmental considerations, as they are primarily resource allocation issues, and only secondarily ones of employment.

REFERENCES

1. A national input-output table for 1972 has been constructed by the LBL Energy Analysis staff by updating the 1967 368-order national table published in a supplement to the Survey of Current Business (1974). An LBL report describing the methodology used is being prepared.

2. Deane Merrill, "U.S. Employment for 368 Input-Output Sectors for 1963, 1967 and 1972," Lawrence Berkeley Laboratory Report, UCID-3757, July 1975.

3. Bechtel Corporation, "The Energy Supply Planning Model," Vols. I and II, report submitted to the NSF under contract No. NSF-C867, San Francisco, CA, August 1975.

4. Balcomb, J. Douglas, Simulation Analyses of Passive Solar Heated Buildings - Preliminary Results, Los Alamos Scientific Laboratory of the University of California, 1975.

5. Greene, Barbara, Residential Solar Hot Water Heating and Space Conditioning Systems in Northern California: A Brief Survey, LBL-5229, August 1976

6. TWOZONE is a computer program developed at LBL for calculating thermal loads in buildings. See Deane, E. and Rosenfeld, A.H., "A Model of Natural Energy FLows in Houses," LBL-4411, 1976 and Wall, L. et al., "Preliminary Results from an LBL Residential Computer Model."

7. Lawrence Berkeley Laboratory, "Electrical Energy Consumption in California: Data Collection and Analysis, UCID-3847, July 1976.

8. Ross and Baruzzini, Inc., "Energy Conservation Applied to Office Lighting," Report to FEA, April 1975.

9. Data compiled by Peter Benenson and Robert Weisenmiller from Departments of Agriculture, Commerce and Labor publications, December 1976.

FORECAST OF LIKELY U.S. ENERGY SUPPLY DEMAND BALANCES FOR 1985 AND 2000 AND IMPLICATIONS FOR U.S. ENERGY POLICY

J. F. Gustaferro et al.

Office of Energy Programs, U.S. Department of Commerce, Washington, D. C. 20230

ABSTRACT

This 147 page document projects future U.S. energy balances based on a 1.2% per capita energy growth rate. These balances take into account declining U.S. and world oil and gas reserves and limitations on fuel substitutability. The forecast projects: U.S. energy consumption of 87 Quads in 1985 and 116 Quads in the year 2000; continuing dependence on foreign oil imports for the remainder of this century (40-50% of U.S. petroleum consumption); a continuing shift to electricity (accounting for more than 50% of primary U.S. energy in 2000)-- which will necessitate building the equivalent of one 1100 Megawatt power plant every 16 days to the year 2000; potential disarray in the petroleum-based transportation sector in the 1990-2000 decade. The forecast notes that the U.S. will still be a long way from full transition to non exhaustible energy resources by the year 2000. At that point in time, hydropower and geothermal sources will provide about 7% and solar/biomass/wind only about 4% and the semi-exhaustible fuel sources of coal and fission nuclear will only be supplying 51% of our energy.

PURPOSE

The objectives of this paper are:

1. To develop a forecast of the "most likely" U.S. energy supply/demand balance for 1985 and the year 2000, based on the expected impact of U.S. policies and trends in the key energy areas now in force.

2. To summarize the major implications for U.S. energy policy that emerge from this forecast.

BACKGROUND

This paper presents what is considered to be the most likely U.S. energy picture (supply/demand balances) for 1985 and the year 2000, based on the expected impact of current policy thrusts and socio-economic trends. This is in contrast to the more common "multiple scenario" approach which offers a menu of intervening events, with no indication of what is most likely.

The forecast presentation includes detailed sector breakdowns of both fuel sources and consumption sectors (i.e., residential/commercial, industrial, transportation and electrical generation). These breakdowns take into account the fuel specificity for each energy use (e.g., automobiles run on petroleum as of now, light bulbs operate on electricity, etc.) and the fact that energy forms are not perfectly interchangeable. Such a presentation has special value for policy purposes. Given the constraint that supply and demand must balance, these breakdowns reveal sectoral tradeoffs that must be considered in discussing energy alternatives and are more important for policy evaluation than are aggregate levels of energy consumption and supply.

Finally, the forecast takes special account of the depleting U.S. fossil fuel resource base. That oil and gas particularly are finite and depletable and, towards the end of the century, will be increasingly costly and difficult to recover, has not always been recognized in forecasts of this type.

ASSUMPTIONS

The critical assumption in the forecast

relates to the rate of energy growth.

The forecast assumes an approximate 1.2% per capita rate of energy growth over the forecast period. This is much less than during the decade of the 60's and early 70's and is lower than the rate used in most other energy forecasts. The reasons for this are:

1. Energy consumption patterns have clearly undergone some real change since the 1973-74 crisis and consumption levels have decreased.

2. Government conservation legislation and public awareness programs will have increasing impact on energy consumption.

3. Increasing energy prices will also have a dampening effect on consumption.

4. Technology in the form of energy-saving devices, lighter materials, etc., will contribute to lower per capita energy use.

5. Finally, the U.S. appears to be approaching "energy saturation" on a per person basis under existing patterns of income distribution, e.g., large energy consuming items such as automobiles and air-conditioners are approaching total market penetration.

In short, the forecast asserts that energy growth projections must be viewed in the light of recent events. Thus the forecast projects that per capita primary energy consumption will grow at the rate of about 1.2% per year (as compared to a rate of 1.5% since 1947) and that overall U.S. energy consumption will grow at approximately 2% during this period.

The following additional assumptions it is believed also reflect the political, social and economic realities of America in the near future. Though judgmental, the probability is rated as high and it is realistic to assume that:

1. GNP will continue to grow at the 50 year historical rate of 2.2% per year per capita, or about 3% overall per year in constant dollars.

2. The population growth rate will continue to decrease and in 1985, U.S. population will be about 234,000,000 and in the year 2000, about 262,000,000. (Census Bureau-Series II).

3. There will be no OAPEC embargo; the cartel will hold.

4. The petroleum strategic storage program (EPCA) will be implemented.

5. Until 1985, OPEC will continue to export 28 to 38 million barrels of oil per day (MMBD) at a price equal to or only moderately exceeding the current dollar rate adjusted for future inflation. After 1985, the price increase will accelerate commensurate with the increasing (world-wide) demand and ultimately the decreasing supply (world-wide) of petroleum.

6. Interstate natural gas will approach the price of oil on a Btu basis.

7. The U.S. public will become increasingly conservation-conscious as a result of higher energy prices as well as by national awareness programs conducted by the public and private sectors.

8. EPCA (The Energy Policy and Conservation Act of the 94th Congress) will be enforced and the automobile mileage per gallon targets are technically and economically feasible.

9. There will be some accomodations, on a case-by-case basis, between Clean Air Act requirements and the proposals to utilize coal in generation of electricity. There will also be some lessening of environmental standards as they are applied to automobile NOX emissions.

It should be noted that the assumptions underlying the forecast exclude either "miracles" or "manacles." That is to say dramatic unforeseen technological breakthroughs that might suddenly change the energy picture are not provided for. Neither is it expected that the American public will accept direct controls or direct energy rationing in the absence of a clear and present crisis (e.g., embargo) or as long as the deficit in the energy balance can be made up with foreign oil at a price competitive with the cost of alternate fuels.

Figure A gives the values assumed in the forecast for the various parameters considered.

FIGURE A

Basic Elements of the Forecast

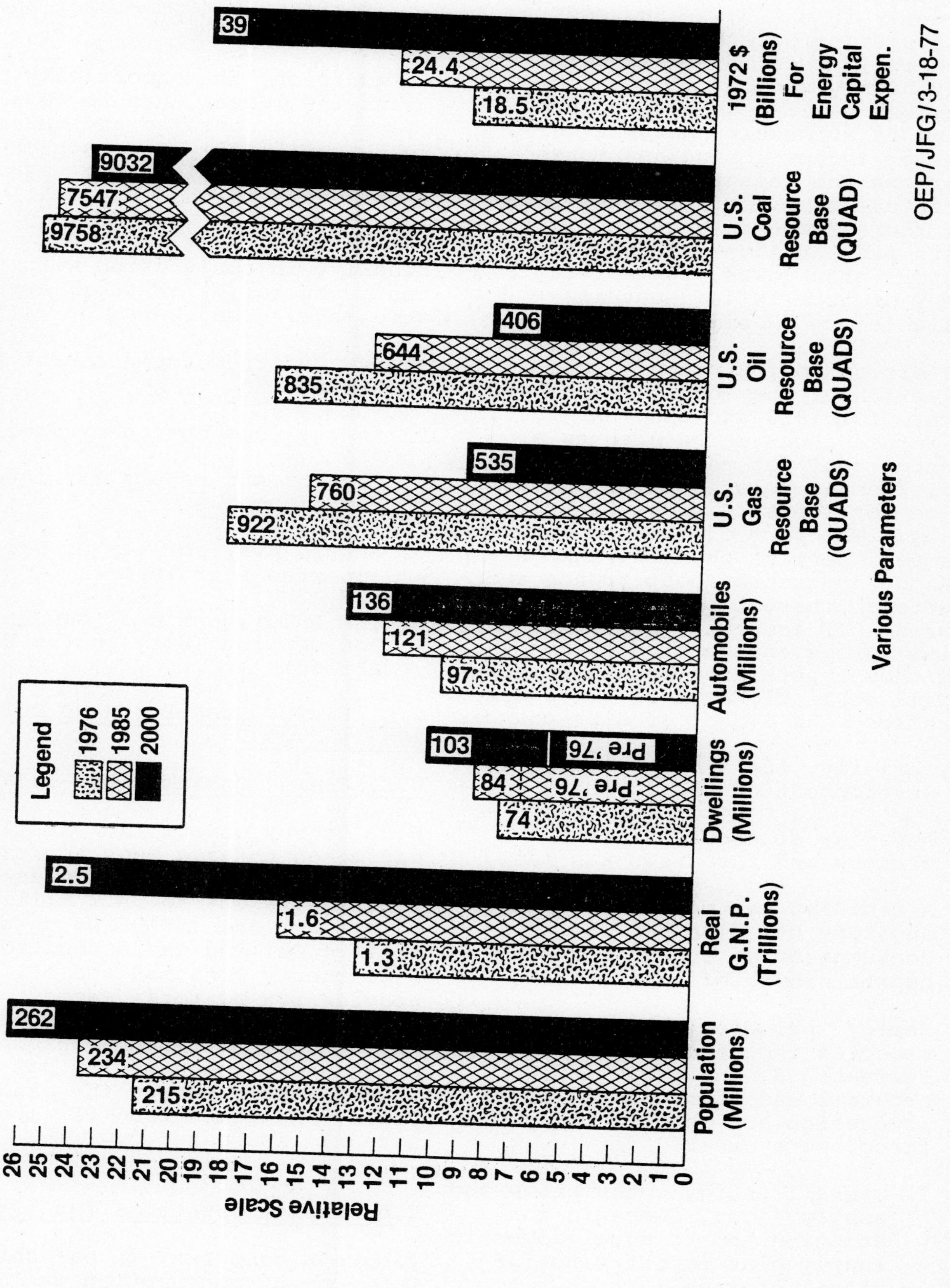

METHODOLOGY

The key element in the forecast has been the judgment applied in assessing the impact of policy thrusts and socioeconomic trends as they relate to energy in the U.S. Judgment is the basic ingredient of all forecasts, of course. Even forecasts which apply highly complex mathematical equations utilize judgments in defining relationships between variables or in developing assumptions regarding the relationship of past and future behavior. In this forecast the assumptions are explicit and in its presentation of detailed "sector" breakdowns, the entire energy picture is laid out in straight-forward simple manner.

In addition to its primary role as a projection of the most likely energy picture for 1985 and the year 2000, the forecast can serve also as a learning tool for understanding energy supply/demand relationships and as a device for identifying key energy issues. The form of the presentation, moreover, permits changing the assumptions or judgments about trends and impacts or current policies, if so desired. If economic, political and social forces change significantly from those postulated in the forecast, it will obviously have to be updated to reflect the impact of such changes.

Very briefly, the steps followed in the development of the forecast were:

a. A review of about 15 other studies and forecasts was made.

b. A decision was made to create an envelope based on per capita consumption of energy and per capita growth of consumption.

c. Energy "balances," based on the expected impact over time of current U.S. energy policies/programs and of trends in energy production and consumption now in force, were created for:

   1. Overall energy consumption and supply (i.e., domestic consumption had to equal domestic supply plus import plus (or minus) storage);

   2. Energy supply components (total supply had to equal the sum of component energies from coal, gas, oil, etc., including imports);

   3. Energy consumption components (total consumption had to equal energy used by the consuming sectors--industrial, residential/commercial and transportation).

In addition, the electricity products, minus the losses, had to equal the electricity consumed and the economical fuel specifically for each energy use had to be satisfied, e.g., gasoline, for the next decade at least, for the automobile.

Finally, in conjunction with the forecast, several brief ancillary studies were made. They were:

a. Per Capita Consumption of Energy by States

b. Energy Consumption by Family Income

c. Present and Future of Liquefied Petroleum Gas

d. Future Availability of Domestic and Imported Petroleum

e. The Relationship of GNP and Energy Consumption (Both in the U.S. and Abroad)

U.S. ENERGY SUPPLY/DEMAND OUTLOOK FOR 1985 and 2000

U.S. Energy Consumption--1976

U.S. energy consumption in 1976 is estimated to have been approximately 74.5 quadrillion BTUs (Quads) or at a rate equivalent to 35.2 million barrels of oil per day (MMBOE/D) on an annual basis. Oil and gas accounted for 74% of U.S. energy consumed and oil imports provide 20% of total energy. These imports, however, constituted more than 41% of U.S. petroleum consumption. While energy consumption has been relatively static during the past three years, the percentage of imports has steadily increased.

U.S. Energy Forecasts for 1985 and 2000

Figure B compares, in bar chart form, U.S. energy consumption in 1976, two projections for 1985 (historical trends and "most probable" cases) and two similar projections for the year 2000.

FIGURE B

Historical and Most Probable U.S. Energy Forecasts for 1985 and Year 2000

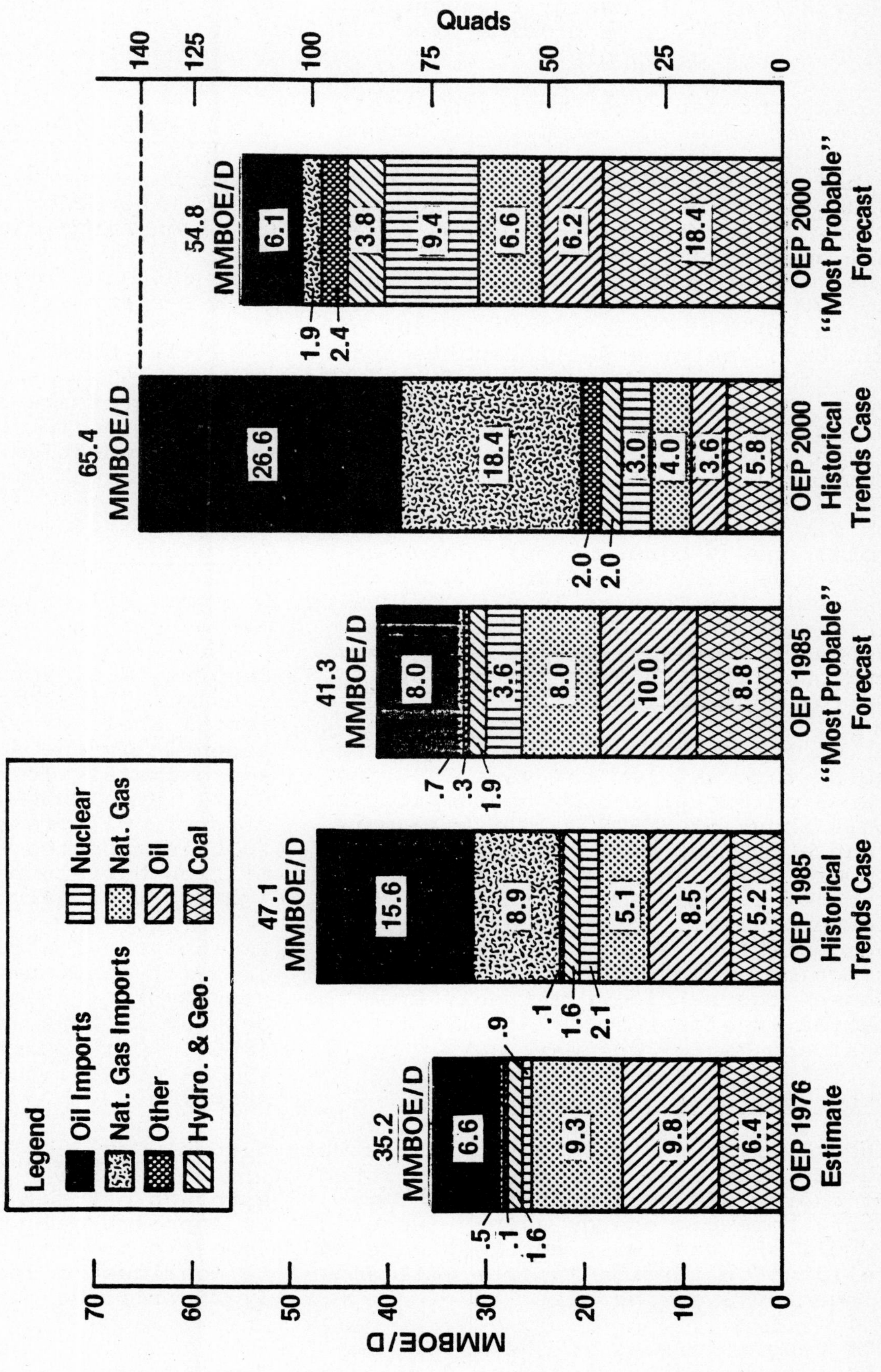

1. The historical trends projection for 1985 indicates an annual energy consumption of approximately 47 MMBOE/D (an increase of 34% over 1976), with oil and natural gas accounting for more than 80% of U.S. energy consumption. Oil and gas imports provide over 50% of U.S. energy supply.

2. In contrast, the "most probable" forecast of energy supply and demand for 1985 (taking into account the likely impact of current energy programs), indicates an annual consumption of approximately 41 MMBOE/D, with oil and gas accounting for some 65% of that amount--considerably less than today. Nevertheless, oil and gas imports still provide more than 20% of all U.S. energy and all imports nearly 45% of U.S. petroleum consumption.

3. For the year 2000, the historical trends case indicates an annual consumption nearly double that of 1976 (65.4 MMBOE/D) and continued reliance on oil and gas (still more than 80% of total energy consumption), which at this time will most likely be at dangerously low reserve levels world-wide. This case also indicates a massive dependency on foreign imports for oil and gas, i.e., nearly 70% of total U.S. energy consumption.

4. On the other hand, the "most probable" forecast for the year 2000 indicates an annual energy consumption of only about 55 MMBOE/D (slightly more than a 50% increase over 1976) and an import dependency of 22% and 50% in the case of the gas and oil using sectors respectively. In this case, coal accounts for 33% of U.S. energy consumption and nuclear power for 17%. These figures differ markedly from the ratios in effect in 1976: 18% for coal and 3% for nuclear.

Other Forecast Highlights

The key finding to be noted at the outset is that even with the lower rate of energy use projected in the forecast, domestic energy supply (especially oil and gas) will not keep pace and reliance on foreign imports will continue and even increase.

The report forsees in the main, however, an orderly adjustment from cheap and abundant oil and gas to limited supplies of oil and gas at much higher prices. The automotive sector will, by the end of the forecast period, have begun to phase out of gasoline internal combustion engines. Heating oil will be in increasingly short supply as well as will natural gas as a source of energy for space heating.

Some specific forecast highlights follow.

U.S. Energy Consumption

Total U.S. energy consumption in 1976 was on the order of 35.2 MMBOE/D (million barrels of oil equivalent per day). U.S. consumption in 1985 and the year 2000 is projected to be 41.3 and 54.8 respectively. Expressed in Quadrillion BTUs this is for 1985--87 Quads; for 2000--116 Quads.

These values are at the low end of the scale in the forecasting community. Any further reduction would require a significant break with past policies and trends.

Petroleum Supply

Domestic oil prices will approach world market prices.

Despite this, domestic oil production, even with Alaskan oil and Outer Continental Shelf development, will remain essentially the same through 1985 (at approximately 10 million barrels per day). Even this will require a major effort. Oil production by the year 2000 is expected to decline to some 7.2 million barrels per day including about 1 MMBOE/D of natural gas liquids.

Extensive reliance on imports of foreign oil will continue (8.0 million barrels per day in 1985 and 6.1 million barrels per day in 2000 as compared to 6.6 in 1976). Oil imports will constitute some 45-50% of U.S. domestic petroleum consumption.

Natural Gas Supply

Although natural gas prices are expected to approach oil prices on a BTU basis, domestic production levels are expected to continue to decline (from approximately 20 trillion cubic feet in 1976 to 17 in 1985 and 13 in the year 2000).

The U.S. will be importing about 1.4 trillion cubic feet of liquefied natural gas (LNG) by 1985 and about 4 trillion cubic feet in the year 2000.

Electricity

The shift to electricity will continue. Electricity will account for 37% of U.S. primary energy consumption in 1985 (as opposed to some 30% in 1976) and for more than 50% in the year 2000.

The use of oil and gas for making electricity will be reduced and ultimately phased out.

Electrical requirements in 1985 will necessitate building the equivalent of 129 additional (new) 1,100 megawatt electrical generation plants (either coal-based or nuclear).

The demand for electrical energy in the year 2000 will necessitate building, somewhere in the U.S., an additional 545 new electrical generating plants of 1,100 megawatt capacity or one every 16 days.

Coal Production

Coal production for electrical generation will expand from 447 million tons in 1976 to some 610 million tons in 1985 and 1,315 million tons in the year 2000.

Overall coal production will increase from 667 million tons in 1976 to 890 in 1985 and 1,860 million tons in the year 2000, accounting for 33% of U.S. energy consumption in that year.

At least 12 full-scale coal to gas or liquid synfuel plants will be built by the year 2000, adding 1 + Quad to U.S. energy supply. If not, the U.S. will be forced to find an alternate fuel to replace this energy, probably electricity.

Nuclear Power

The contribution of nuclear power to total U.S. energy consumption was less than 1 million barrels of oil equivalent (slightly more than 2 Quads) in 1976. This is expected to grow to 3.6 MMBOE/D in 1985 and 9.4 MMBOE/D in the year 2000 (or 17% of primary U.S. energy consumption at that point).

If this expansion of nuclear power capacity is not realized, then the number of coal-fired electrical generation plants and coal gasification plants will have to be increased or power shortages will result.

Solar and Other

A realistic assessment of solar and other non-conventional energy sources (biomass, wind, geothermal, etc.) indicates that they will not play a significant part in our energy future during this century. The contribution of such energy sources is expected to be on the order of .3 MMBOE/D in 1985 and 2.4 MMBOE/D by the year 2000.

Conservation

Projected "savings" from fuel substitution measures (e.g., using coal and nuclear in place of oil and gas) and from the more efficient use of energy will amount to 11.4 MMBOE/D in 1985 and nearly 32 MMBOE/D in the year 2000, compared to historical (pre-1972 rates.

Sector Impacts

U.S. production of petroleum coupled with imports will be insufficient to meet the demand of the transportation sector in the year 2000. Some 26,000,000 automobiles* will have to be powered by something other than gasoline (electricity?) or alternate means of transportation provided.

The industrial sector, by the year 2,000, will have to reduce its oil and gas consumption on a per product basis by some 60%.

Finally, it is worth noting that by the year 2000 the U.S. will be a long way from full transition to inexhaustible energy resources. At that point, coal will provide about 1/3 of our energy, hydro and geothermal sources about 7% and solar/biomass/wind about 4%. The U.S. will still have a considerable way to go (55%) to achieve this transition.

In summary, it should be reiterated that these forecast results represent those which, in the opinion of the forecast team, will occur, given present policy thrusts, trends in U.S. energy production and consumption and most likely events. The path indicated should not be considered an optimum one or even the least disruptive one. Nor should the economic, social and political impacts involved in achieving the energy supply/demand balances indicated be underestimated. In short, even what we have called the "most likely" forecast

*This would be 15-20% of the automobile fleet in 2000.

will still call for major efforts on the part of the Government, industry and the people of the Nation as a whole. On the other hand, it should be obvious that a shift away from the forecast path leads to even more serious consequences and potentially drastic life style changes.

It should also be understood that a "most likely" forecast of this kind needs to be updated on a regular basis in order to keep the anticipated energy supply/demand balances in line with changes in the political, economic, social and technological environment. Key areas to monitor include oil and natural gas domestic production rates, oil and natural gas imports, coal production, electrical generation capacity and experience load factors, population growth and efficiency and use trends in the transportation sector.

IMPLICATIONS OF FORECAST FOR U.S. ENERGY POLICY

U.S. Energy Policy

The core problem to be addressed by U.S. energy policy is the reality of declining domestic reserves and production of oil and natural gas (which account for nearly 75% of current U.S. energy consumption) and the fact that the technology to provide economically and environmentally acceptable substitutes for imports of these energy sources is not yet available. U.S. dependence on foreign sources of energy supply is increasing, as is our vulnerability to supply interruption and to the use of the threat of such interruption as a political weapon.

Current U.S. energy policy for the short and intermediate term has sought to: (a) maintain a special relationship with key oil exporting countries so as to assure continuity of supply; (b) reduce dependence on foreign oil through the establishment of a strategic reserve to help offset the effects of a possible supply interruption (embargo); (c) encourage the discovery and production of additional domestic energy supply--particularly of oil and natural gas and to expand nuclear power electrical generation capacity; (d) encourage the use of and conversion to coal--our most abundant energy resource; and (e) encourage conservation programs to save energy and reduce the rate of energy growth in all sectors, especially in transportation.

U.S. policy for the longer term has emphasized large-scale R&D to develop and enhance the commercialization of fossil-fuel energy substitutes and continued conservation to stabilize energy growth at levels commensurate with available supply.

Policy Implications and Energy Issues

While the energy picture projected in this forecast is considered the "most likely," the trends and policy thrusts on which the forecast is based are not immutable and policy makers can change them.

Among the major policy implications and issues that emerge from this forecast are:

1. Continued U.S. Reliance on Petroleum Imports. The forecast indicates only a partial transition away from oil and gas by the year 2000 and continued extensive reliance on imports of foreign oil. This points up the need both for the maintenance of a strategic petroleum reserve to mitigate the effects of a possible embargo and a foreign posture that allows for the management of oil imports so as to assure continuity of supply.

2. Domestic Energy Resource Development (Oil, Gas and Coal). In the light of forecast results indicating that the U.S. will still be dependent, in large part (38%), on gas and oil to furnish our energy needs in the year 2000, and that oil and gas prices will approach world market prices in the decade of the 80's, a crash effort to immediately maximize oil and gas exploration and development may not be the wisest policy. Steps to bring about an increase in domestic supply should seek to "optimize" oil and gas development and "maximize" coal development and utilization. Optimization of oil and gas production means producing at a level which will recover the maximum amount from domestic reservoirs (over time) rather than seeking to maximize production during the 1976-85 time period.

3. Selective vs. Comprehensive Energy Resource Development.

The forecast indicates that, while there are sectoral and fuel tradeoffs that can be made in dealing with the energy problem as it evolves over the long term, it is clear that no single energy source can be excluded, if the 1985 and year 2000 energy supply/ demand balances are to be achieved. All present forms or sources of energy, in short, will be required.

4. What to do about the Automobile? The key factor in U.S. energy dependence on foreign sources of supply has been, and the forecast indicates will continue to be, the energy requirements for the Transportation Sector--particularly the automobile. The forecast indicates potential disarray in this sector as the year 2000 approaches. As available supplies of oil and gas decline, greater pressure will be exerted on the transportation sector (and particularly the automobile) to phase into non-petroleum burning vehicles. The ability of auto manufacturers to meet transportation demands with high efficiency vehicles will determine the rate of this phase-in. (We see 15-20% of the year 200 fleet utilizing non-gasoline burning automobiles.) If the technology for producing such vehicles is not available by the late 1980's, basic decisions will have to be made regarding alternate modes of transportation for the car-using public or a national commitment of resources necessary for the large-scale synthesization of petroleum form other energy sources (i.e. coal or oil shale).

5. Waste Heat from Electric Generation The forecast indicates an increasing trend toward electrification--30% of total U.S. energy consumption in 1976, 37% in 1985 and 52% in 2000. For the most part, the electrical power generation and distribution system delivers one BTU for every three put into the process. The "waste heat" (or thermal pollution) resulting from electrical generation can provide the country with a valuable energy resource. The barriers to its use now are primarily political and institutional. A strong effort by the Government to bring about conditions which would put to work this so-called "waste heat" (e.g., co-generation and the "district heat" concepts) would result in profound changes to the "most probable" forecast set forth in this paper. Equally as important, when one considers the efficiency of electricity on an end-user BTU basis (much greater than from oil, coal or even natural gas) together with the "savings" resulting from an extensive effort to capture the "waste heat" inherent in electrical generation, there is room for additional growth of GNP (and continued improvement in the standard of living) even at the reduced energy growth rates assumed in this forecast.

6. Coal Gasificaiton. Coal is our most abundant energy resource. An option, and in part a tradeoff for energy deriving from electrical generation based on nulcear power and coal in the year 2000, is the use of this resource for producing high BTU gas, using the already in-place pipeline infrastructure. The forecast estimates that only about 12 coal gasification plants will be operating by the year 2000 even though additional plants could offset--or replace--nuclear or coal fired electrical generating plants on a 2 to 1 basis. ,Although experience and "know how" are on the side of building more electric power plants, it seems desirable to get on with the job of building and testing the first few coal gasification plants as soon as possible. Such an experience base could then be used for evaluating the benefits of such plants vis-a-vis additional coal fired or nuclear electrical generation plants where gas and electricity are substitutable.

7. Shale Oil Exploitation. The arguments cited above for coal gasification also apply to the exploitation of the vast reserves of oil shale in Colorado, Utah and Wyoming. It seems desirable that Government encouragement and support be given to a pilot effort to develop the first large commercial shale oil plant now.

8. Liquefied Natural Gas (LNG) Imports. The forecast projects that LNG imports will reach 1.9 MMBOE/D equivalent by the year 2000 (approximately 4 Quads), twice the current policy ceiling on such imports. The implication is that these LNG import targets will need to be increased

9. Role of Non-conventional Energy Sources. (Solar, Wind Geothermal, Biomass, etc.) Although there is a clear need for a vigorous R&D effort to enhance commercialization and wider

use of these energy sources, the forecast indicates that, given present concepts for end-use application, they will not satisfy a significant portion of U.S. energy demand within the time frame (1976-2000).

10. Conservation. A vigorous conservation effort is assumed in the forecast, which is in part reflected in the quite low rate of per capita energy growth used as an envelope for the projected supply/demand balances. Consideration may have to be given to steps needed to ensure that the per capita rate of energy growth be maintained at a 1.2% (or below) level. In particular, efforts may be needed to induce even greater energy efficiency in the automobile fleet than now contemplated.

TRANSPORTATION AND ENERGY: SOME CURRENT MYTHS

Charles A. Lave
Economics Department
University of California
Irvine, California 92717

In this paper I examine a number of current myths about transportation and energy. They have become myths because they seem self-evidently true, and hence are rarely examined. Rather they are taken as wisdom, and they form the basis for much of the reform position in this area. These myths appeared in speeches by the candidates for the presidential nomination, and one supposes that they will be echoing in Congress as well this year.

Myth # 1: Good Public Transportation Can Attract People Out of Cars

This myth has its origin in a bit of wishful thinking: it is obvious to everyone (and to me) that greater use of public transportation really could solve many of our most pressing urban problems; and since this seems to be the only solution, people then assume that it can be implemented. Somehow. So we have policy dictated out of wish-fulfillment. But how easy is it to produce greater transit usage? It is instructive to examine this question from the viewpoint of the potential user, choosing between transit and his car: transit is not as fast, door-to-door; transit does not depart from his home, nor does it take him directly to his destination; transit is not immediately available the moment he wishes to leave, night or day; transit often cannot provide him a seat at all, much less a private, uncrowded seat; and transit offers less personal security. This is a really formidable list of disadvantages, and there is only one compensating factor to offset them--transit is cheaper to use. We should not then be surprised to learn that transit is used for only 4% of the travel in the U.S., and we also gain some perspective on the long term decline in transit usage: since transit is inferior to the automobile, along every dimension except cost, as user incomes rise over time, transit patronage must decline as people decide to spend part of their new income to buy superior transportation. That is, you can't continue to sell a cheap substitute when income trends are making the real thing affordable to more and more people.

There has been a great deal of research done on the question of what determines the choice of mode (e.g., bus vs. auto) for urban passenger transportation. These studies use statistical procedures to estimate the commuters' sensitivity to the various factors involved in the mode-choice decision, and hence calculate the potential reaction of commuters to possible transit improvements such as lower fares, faster speeds, more frequent service, etc. It is fair to say they have not shown much mode-split sensitivity to cost, the only factor where public transportation has any possibility of comparative advantage. Lest this body of research be dismissed as being somehow inadequate, or auto-biased, it should be pointed out that a number of observable real-world phenomena confirm the public's aversion to "public" transportation: even in cities with good public transportation, only a small proportion of the population uses it; during the OPEC gasoline crises there was very little diversion of people onto transit;[1] and even in European cities with excellent transit, and long traditions of transit use, as family incomes have risen over time, transit use has declined and auto use has increased.[2] Nor have higher gasoline prices changed auto travel significantly: both the computed gasoline price elasticities and observed travel behavior have shown little movement of people from autos to transit.[3] This has also been true for other kinds of price diversion policies such as increased parking charges and increased bridge tolls.

It may reasonably be objected that all of these results apply only to current systems, and that somehow "improved" transit could work. The Urban Mass Transportation Assistance program (UMTA) was organized with this idea in mind, and has devoted more than 11 years, and more than $3 billion to conducting experiments with transit improvements around the United States. These experiments were not narrowly conceived. Any community with an idea for improving its transit service could apply to UMTA and was likely to receive funding. The diversity of the resulting demonstration projects was very broad: they experimented with new vehicles, more frequent scheduling, lower fares (down to zero), more advertising, more coordination between lines, more public information, etc. There were grants available to subsidize operating costs, to subsidize capital costs, and to subsidize training and administrative costs.

An excellent, highly readable evaluation of the UMTA programs has been done by George Hilton.[4] He concludes ". . .the demonstration projects on existing technology have been uniformly unsuccessful. . . [and] None of the new technology which UMTA attempted to produce proved an economic alternative to existing forms of urban transportation . . . [Furthermore] investment in rail rapid transit systems is not an effective way of reducing automobile usage . . . Even to say that [rail transit's] ability to divert drivers is imperceptible . . . overstates [its] positive effect on road use."

We may fervently desire that people leave their cars and switch to transit in order to reduce fuel consumption, congestion, and pollution. But we must recognize that nothing in the way of pure research, applied research, actual demonstrations of transit improvements, or historical evidence gives any expectation that such a diversion can be accomplished.

Myth # 2: Public Transit Saves Energy

Examination of this myth produces a surprising result: the new generation of transit systems appear to be energy wasters. Yet these systems are receiving a great deal of public interest around the United States. There are glamorous new rail transit systems that have recently opened (San Francisco), or are in construction (Washington, Atlanta), or are proposed for construction (seemingly everywhere). Their costs are astronomical: the one that Los Angeles voters just considered, promised to be the largest public works program of all time. Their effects on congestion and pollution are either very small, or even negative. And the amount of subsidy the transit systems require is startling: the BART system in San Francisco currently subsidizes its passengers by about $12 per round trip, the newly opened section of the Washington METRO system subsidizes its passengers by $22 per round trip (though this figure is supposed to drop when the entire system is opened).[5]

Such facts hardly encourage the building of more rail transit systems. But fortunately--for the downtown property owners, who are the principal beneficiaries of these systems--OPEC came riding to the rescue. We now have the energy crunch and it is argued that anything which saves energy, no matter what the dollar cost, is good, and should be promoted. And since it seems self-evident that these rail transit systems save energy, they are good and should be built.

But does rail transit save energy? Let's begin by distinguishing between the energy used to operate the system and the energy used to construct the system. Modern rail systems do use somewhat less energy per passenger mile than the average automobile, for their daily operation. However, the amount of energy used to construct these rail systems is enormous, and completely dominates the daily energy savings.

It required an investment of 164 trillion Btu (British thermal unit) of energy to construct the BART system.[6] As a payback for this investment, enough people are attracted from cars and buses to reduce the need for highways by about 47 lane-miles. Unfortunately, the amount of energy saved by not building these 47 lane-miles is only about 3% of the energy invested in BART, so the offsetting effects on energy investment are trivial; and even doubling or quadrupling BART's patronage would still produce no significant offsetting effect on the net energy invested in the system.

This is a surprising result and is worth restating. If we compare the initial construction energy investment required to transport these commuters by rail versus highway, it takes 34 times more energy to build the rail facilities than it would to build highway facilities to transport an equivalent number of people.

What about the savings in daily operating energy? We find that BART's energy efficiency is intermediate between that of the bus and the car. Hence every time it draws a user away from a car we save energy, but every time it draws a user away from a bus we lose energy. Since only 46.5% of BART's passengers come from cars, the net energy saving is small, only 680 Btu/PM (Passenger Mile). Given BART's patronage (130,000 trips of 13 mile average length, per day) we can calculate how long it will take to save enough energy to repay its construction cost. The answer is 535 years. (Furthermore this figure is not sensitive to the underlying assumptions: let us assume that somehow BART can double its patronage, and divert 75% of its passengers from cars, and somehow operate at a 50% load factor; all three impossible improvements at the same time. Even in such a transit Nirvana it would still take 168 years of operation to reach the break-even energy point.)

In summary, the energy invested in constructing BART is so large, and the operating-energy savings are so small that it will take 535 years to break even on its investment, much less save any energy.

If energy saving were its main rationale, it should never have been built in the first place. Furthermore, because these figures compare BART to a 14 MPG car they are strongly biased toward the effectiveness of rail transit. Congress has already mandated an average auto fuel efficiency of 27.5 MPG by 1985, and such a car is actually 15% more energy efficient than BART. Given that BART is drawing its passengers from more efficient buses and cars, it will actually save energy to simply shut down the system, even ignoring the invested energy.

Is it meaningful to generalize from this one example to other modern rail systems? There are three critical numbers involved: the cost of building the system, its patronage, and its energy consumption. Taking thesse in order, using all of the available comparative data: a) BART's cost per system mile is 7% lower than the average of the systems now being built in Atlanta, Baltimore and Washington; b) BART draws 15% more of its passengers from cars than the average of the Boston MBTA South Shore Extension and the Philadelphia Lindenwold Line; c) BART's operating energy consumption is 14% lower than that of the Lindenwold Line. Since the available statistics on other rail systems are somewhat fragmentary, we cannot make a more detailed comparison than this; but it does seem clear that to the extent that BART is atypical of modern rail systems, it is atypically efficient. These surprising results are apparently not an isolated example of energy waste due to mass transit. In fact, one sees so many overstated projections regarding transit systems, that this practice is worth separate comment.

The Selling of Transit Systems

The "engineering" studies put together to sell a transit system to community voters generally manage to both overstate the number of passengers the system can attract, and to understate the cost of building it. BART again, provides an interesting example of the contrast between reality and forecasts: it cost almost twice as much to build, it draws less than half the passengers, and it uses double the energy that was forecast. The system that was projected to "make a major impact on traffic" actually carries only 2% of the trips in its patronage area.

The Washington METRO system provides another example of the rosy forecasts used to sell transit systems. It has already vastly exceeded its forecasted costs, and it is not yet completed. But its projected patronage seems to be even farther from reality than its costs. The METRO system is 38% bigger than BART, and the population it will serve is 73% bigger, but Washington travel-corridors are far more dispersed than those in San Francisco. On balance we might expect METRO to attract slightly more patronage than BART, perhaps even two to three times as much. But the forecasts used to sell the METRO system fantasize that it will have 11 times as much patronage!

Myth # 3: Transit is More Economical Than Cars

The findings here are the most depressing of any in this article, for they imply that transit's costs are so high as to make it unlikely that transit has any significant future. It is simply much more expensive than our intuitive estimates would indicate: on average, transit costs about two-thirds more per passenger-mile than the private automobile (including all capital and operating costs for the car)[7] but to be attractive to patrons it must charge them less than they would spend by car. That is, transit services are far more expensive to produce than car services, but they earn much less money. The end result has to be enormous deficits: for the U.S. in 1975 the total deficit was $1.7 billion, and has been increasing at an average rate of 59% per year since 1968.

Why are transit costs so high? Labor is the major expense (80% of total costs) in transit systems; transit unions are in a monopoly position with regard to a vital service and the normal discipline of the market is vitiated by the willingness of UMTA (Urban Mass Transit Administration) to provide the necessary subsidies. For example, bus drivers in San Francisco recently rejected, as too small, a pay package that averaged $25,000 per year.

These high labor costs lead to a high unit cost for providing the service. The marginal operating cost of the major rail transit systems (even treating capital costs as sunk, and hence free) is about ten cents per passenger-mile;[8] and this is true for both the traditional systems like New York, and modern systems like BART. (Since BART was designed to minimize marginal operating cost, through substitution of capital for labor, this is especially significant.) That is, their marginal costs are actually greater than the automobile. If we include capital costs as well, the comparison becomes even more surprising, for example the average daily round trip on BART is subsidized by more than $12 (operating subsidy = $2.62; capital subsidy = $9.44, assuming 7% opportunity cost of capital).

Of course these high unit operating costs then imply high deficits. In Boston only 25% of transit expenses are covered by fares, that is, there is a 75% subsidy. In California as a whole, the subsidy is about 60%. In New York, for the coming year, the estimated deficit will be about $350 million.

To understand the economics of public transportation, one must try to understand why we have $10,000/year taxpayers subsidizing $25,000/year bus drivers. It is difficult to see much future for transit in the face of numbers like these.

Furthermore, expansion will not make transit significantly more efficient, as some have predicted. There are possible economies of scale in rail transit, but rail systems are only feasible in a few cities, and even when they are designed for minimum operating cost, as in San Francisco, their marginal costs are still too high to be covered by fares. Bus systems are the most flexible, easiest to expand, and would be the best hope for increasing transit service in most cities, but there are no economies of scale in the provision of bus service.[9]

In summary, transit's current share of travel is only 4%, but it takes a subsidy of $1.7 billion to accommodate even this tiny number of people. The unit cost of providing transit services is already too high and it cannot be reduced. This cost structure has grim implications for the future of transit. At best, taxpayers may agree to meet the increasing cost of maintaining current level of service; but expansion to accommodate a significant amount of urban travel seems highly unlikely.

Myth # 4: The Decline in the Railroads is Due To Federal Subsidies of the Trucking Industry

Seventy-five years ago railroads carried all the overland freight; today they carry only 38% of it. One of the most widely believed explanations for this decline holds that trucks have been able to win a disproportionate share of the freight because their rates are artificially low: railroads have to pay for their own roadbeds, but trucks have the use of a cheap roadway provided by Federal highway subsidies.

Is there such a trucking subsidy? One of the earliest academic studies of this question concluded that trucks in general paid about as much in taxes as they incurred in highway building costs.[10] The railroads have, of course, maintained that this was not true: the United States Railroad Association (USRA) estimates that a five (or more) axle diesel semi-trailer actually causes highway costs of about 1.6 cents per mile more than the taxes it pays. USRA admits that the magnitude of the figure is subject to some controversy but let us go ahead and use it for the moment anyway. Is 1.6 cents per mile an important subsidy? Using a conservative estimate of $1.40 revenue per truck-mile, the alleged subsidy amounts to only 1.1% of truck tariffs. It is difficult to see how a 1.1% price subsidy could be the cause of all that freight diverted from railroads to trucks.

If trucking is not receiving a significant subsidy, and hence charging artifically low prices, what does explain the diversion of freight from railroads onto trucks? The first thing to notice here is that the decline of the railroad freight business began long before there were any significant Federal highway program, and we need look no farther than the method of pricing freight to see why this happened. In the early years of railroading, freight tariffs were set on the value-of-service principle: expensive manufactured goods could afford to pay high tariffs, so the railroads charged them high tariffs; inexpensive bulk goods could only afford low tariffs, so their fees were set low. In essence the railroads set rates based on the value of the service to the shipper, i.e. his ability to pay, rather than being based on the cost of moving the goods. So a manufacturer paid much more for a ton-mile of service than a farmer, and provided a greater share of the railroad's profits. When the Interstate Commerce Commission (ICC) was created, it instutitionalized this value-of-service concept. During the years when railroads were the only source of transportation, the relative exploitation of manufacturers worked to the advantage of the railroads. But this gap between the cost of providing freight service and its sale price was also a tempting opportunity waiting to be exploited when some new form of shipping came along, and it was the trucking industry which first saw this.

Even though it costs a truck more to move a ton-mile of freight than it costs a railroad to do so, the truck's cost were still lower than the artificially high prices charged manufacturers by the railroads. Trucks could provide better service--faster delivery times with greater flexibility--at the same price as the railroads. Thus the ICC regulations which had once worked to the advantage of the railroads now prevented them from lowering their tariffs to give shippers lower rates than trucks, to compensate for the relative quality difference in the services. Naturally the trucking industry prospered and grew, and as a "reward" for its dynamic economic behavior it was eventually placed under the control of the ICC.

Once the trucking industry was under ICC regulation it, in turn, became vulnerable to simple competition. It is ironic that the same ICC regulation which had previously helped the trucking industry now provided the profit opportunity which encouraged the growth of the private, non-regulated trucks. Since under the ICC regulation, the common-carrier trucks had the same value-of-service pricing as the railroads, they were also overcharging high-value goods, and it was not long before some manufacturers did the simple calculations to discover that they could operate their own trucking fleet for less money than the artifically high tariffs that they were paying the common-carrier trucks, even though their private trucks generally had to return home empty. So private trucking prospered, and it now accounts for about one-third of the truck freight.

Unfortunately the ICC was not content to limit its interference to tariff schedules, but also began making direct allocations of freight runs. Thus, for example, it licensed a particular common-carrier to haul one commodity, frozen hush puppies, between a few specific cities: no other commodity, no other cities. A truck could not pick up additional cargo as it dropped off part loads on its outbound trip, and on the way back it returned empty. Obviously such practices were inefficient and cost the freight companies money, but the ICC was not bothered by inefficiency (only by competition, apparently), so they simply allowed the freight companies to raise prices so that they could survive despite their inefficient use of resources. That is, everyone pays higher freight prices to satisfy the ICC's goal that these empty trucks be kept profitable. A number of economists have estiated that the net effect of ICC regulation is an enormous underutilization of capacity: across railroads, common-carrier trucks, and private trucks only about half of total capacity is being utilized.

There is a great deal of evidence which suggests that inefficient use of transportation resources, due to ICC regulation, is the major problem faced by the freight sector of U.S. transportation industry. These regulations cause a complex web of cross-subsidization which misallocates traffic across modes, and produces underutilization of capacity within modes. Thomas G. Moore estimated that the economic cost of ICC interference was 3.6-6.9 billion dollars in 1971, and it is obviously even higher today.[11] That is, about one-quarter of the income generated in transportation is simply wasted (and we all help pay to keep those empty trucks and idle box cars profitable). As Hilton notes, the transportation industry "attracts unspecialized resources from other activities and wastes them in idleness, underutilization and inappropriate uses."

We can now see the reasons for the railroads' decline, and they have nothing to do with alleged highway subsidies. The ICC's value-of-service method of rate setting has become unworkable in the transportation industry because those shippers who have the greatest ability to pay, the high value industries, also have the most alternatives to railroad transportation. And at the same time the excess that must be charged these high value shippers has gotten larger and larger as the railroads acquired an increasing burden of non-profitable services such as rail-passenger, low-density spur line, and small-lot traffic. Meanwhile prices of the bulk, low value goods could not be increased because of the political power associated with these commodities (farmers and mineral interests), and because shipping costs affect demand in these commodities so much more. Thus over time the railroads became burdened with many unprofitable services, could not raise tariffs on the bulk commodities, and every time they raised tariffs on the high value commodities they drove more and more of that business away.

Since the conventional wisdom is that the decline of the railroads is due to highway subsidies, we now have proposals in congress to give "compensating" subsidies to the railroads, either to upgrade their track or to provide them with new rolling stock. But if the above analysis is correct, such grants will have little effect on the economic viability of the railroads. After all, the railroads were losing business back when they still had a sound physical plant, and before there were significant Federal expenditures on highways. Likewise all the current proposals to allow railroads to abandon unprofitable services and reduce track mileage cannot really do much. They increase the short-term profitability of the railroads, but if ICC interference is allowed to continue, then freight will continue to be lost to trucks regardless of increased railroad profitability, and the enormous waste of resources on all modes will continue. Hence the most effective Federal policy for achieving better allocation of transportation resources, reduced energy use, and increased railroad profitability would actually involve no cost to the government. All we have to do is to get the government out of the regulation business.

Myth # 5: Railroads Can Provide Economical Passenger Service

This myth runs something like this: "Railroads can make passenger service viable if they wish to; the figures which the railroads provide about their passenger-service losses are phoney, and result from attributing too much right-of-way cost to passengers and not enough to freight." Unfortunately, recent events have tended to back up the railroads' claims, and it appears that their reported losses on passenger service were an accurate reflection of the difficulty of providing economic passenger service. First, we might note that some of the Eastern railroads with a high volume of passenger service have indeed gone bankrupt. Then, we might note that Amtrak continues to loose money in the passenger business ($325 million in 1975), despite cutting back its service to keep only the high patronage, most profitable runs. Finally, a recent news item from San Francisco provides somewhat whimsical evidence as well.

The Southern Pacific (SP) railroad runs commuter trains up the West side of the Bay, Palo Alto to San Francisco, and has claimed for years that this is a money losing proposition. (Indeed, one might wonder why BART's planners thought a commuter railroad might be more profitable on the other side of the Bay.) In September 1976 the SP made a novel proposal to buy its way out of the passenger business: they offered to give away $6 million worth of passenger vans to anyone who would take one, provided only that the recipients would then run them as van-pools, and hence attract the passengers away from the SP, so it might be permitted to close down. SP's calculations further showed that such van-pool service could offer lower prices than the railroad, as well as faster door-to-door times.

It must be admitted that my approval of the SP plan is somewhat biased since it is a direct descendant of a proposal I made, eight years ago, to solve the passenger transportation problem in Los Angeles. I proposed three simple steps: 1. paint new, narrower lanes on all existing freeways. The new line markings to be a different color from the conventional markings, to make them distinguishable, and the new lanes to be only two-thirds as wide. Hence four conventional lanes become six new narrow lanes, and highway capacity is increased 50% by a simple stroke of the brush. 2. We make one new traffic regulation that says: during the morning and evening rush hours only small cars (say, Hondas) are permitted on the freeways. The narrow lanes are wide enough for these cars to use safely. 3. Since the second step would create a large number of screaming commuters, protesting their inability to drive, we now placate them by giving each and every one a mini-car to drive. Free of charge.

Since these small cars use about one-third the energy of a standard American car, we will obviously make an enormous impact on the energy problem. The combination of narrow lanes and small cars also solves the peak-hour congestion problem. (During other times, any size car may continue to use the freeways.) Furthermore the mini-cars can be parked in half the space needed for a standard car. And finally pollution is reduced because the mini-cars have lower emissions, and because cars can then travel freely, hence reducing the emissions created by stop-and-go congestion. (Electric mini-cars would be even better, of course.)

But can we afford such a solution? The answer is overwhelmingly yes. The ten billion dollars which Los Angeles almost committed to a rail transit system is enough to provide a free mini-car for every commuter in the area; and unlike the proposed transit system, the expenditure would actually do some good. Even if this solution is not taken seriously it should at least provide some perspective on the enormous cost of the transit systems being planned around the United States. Most of us have little concept of what a billion dollars is, but anyone understands the magnitude involved in giving away a free car to every commuter in the city.

Some Final Comments

Although it is discouraging to learn that one's base of "wisdom" on an issue is ill-founded, the conclusions presented here do imply an alternative set of policy measures to deal with transportation problems, and many of these alternative policies promise to be cheaper as well as more effective. First, on the energy question, we need not be discouraged to discover that rail transit is an energy-waster, for the corallary of this is that we are freed of the necessity of building these enormously expensive systems. And once we realize that small cars are more energy efficient than any form of public transportation, we are led toward policies which promote the use of such cars. Hence, rather than continuing to hope for some magical way of convincing people that private autos are bad (when their own experience tells them the opposite), we can simply concentrate on encouraging the use of small cars, either through tax incentives or through mandatory fuel efficiency standards. This is not only easier than trying to sell transit to people who know it is an inferior alternative, but its energy saving is more significant as well: the already legislated 1985 fuel standards will cut U.S. gasoline consumption in half, something which even the most visionary transit advocates cannot promise.

Second, we need not be discouraged to learn that it is ICC regulations, rather than Federal highway subsidies, which have led to the decline of the railroads and the incredible overcapacity and waste of our freight transportation systems. Instead, we should be encouraged to discover that the solution to the problem is simple and cheap, and we should begin phasing out these regulations.

Finally, we ought to adopt a more realistic set of expectations about the possible increases in transit patronage, and the role this might play in solving our various urban problems: yes, transit could solve these urban problems if more people would use it, but they won't; they regard transit as inferior, because it is inferior for most trips. Public transportation is always going to be necessary for some population groups, and we do have to provide it; but this is never going to be a significantly greater number of trips than at present. And short of actual coercion, there is no way of attracting an important share of trips onto transit because there is no way of making transit's service as good as the automobile's (as has been amply demonstrated by the failure of billions of dollars worth of transit improvement projects).

We must drop our wishful thinking and get on with the task of discovering alternative solutions to the transportation problem; and in particular, we ought to give more attention to solutions based on "civilizing" the automobile. Promotion of small, fuel-efficient, clean cars would solve most of the same problems and will be a great deal easier to accomplish than continuing our efforts to convince people that speed, convenience, and privacy are really not important.

FOOTNOTES

1 Mary Stearns, "Social Impacts of the Energy Shortage," Transportation Research Center, USDOT, Cambridge, Mass., April 1975.

2 B. Bruce-Brigs, "Gasoline Prices and the Suburban Way of Life," The Public Interest, No. 37, Fall.

3 Charles River Associates. Energy Impact of Federal Capital Grants Programs for Transportation. Prepared for Federal Energy Administration. 1976.

4 George Hilton, Federal Transit Subsidies: The Urban Mass Transportation Assistance Program, American Enterprise Institute for Public Policy Research, Washington, D.C. 1974.

5 Peat, Marwick, Mitchell. "Transportation and Travel Impacts of BART," P. 174, San Francisco, April 1976.

6 Details of the calculations in this section are contained in: Charles Lave, "Rail Rapid Transit: The Modern Way to Waste Energy." Paper presented at annual meeting of the Transportation Research Board, Washington, D.C.; January 1977.

7 Allen Altshuler, "The Decision Making Environment of Urban Transportation." To appear in Public Policy, Spring 1977.

8 Peat, Marwick, Mitchell, "Transportation Impacts," p. 168.

9 Randall Pozdena and P. McElhiney, Economic Role of the State in Transportation, State Transportation Board, California, 1976.

10 John Meyer, et al. The Economics of Competition in the Transportation Industries. Harvard Press, 1959.

11 T. G. Moore, Testimony in Congressional Record, September 28, SS33610.

HOW SHOULD GOVERNMENTS & INDUSTRY SHARE IN ENERGY INVESTMENTS?

J. Morley English
Professor of Engineering and Applied Science, UCLA
and Econergy, Inc.

ABSTRACT

There are a number of economic conditions that make government participation in new energy investment an imperative. First, a new energy infrastructure may be needed. Characteristically, investment in infrastructure is societal rather than private, although it is shown how the private sector does contribute to the building of infrastructure. Second, because new energy projects are complex and large scale, the investment period is long term. Therefore, the pay-backs appear unattractive to industry which is short term oriented. Finally, alternative cash-flow paths for a petroleum based versus other-than-petroleum based economy are demonstrated to cross some time in the next century. The extra investment needed now to switch from a petroleum based energy economy is shown to be the share that government must contribute in one way or another. While specific formulas for cost charing are not spelled out, a fundamental principle on which such formulas may be based is established.

ASSESSING THE EFFECT OF ENERGY CONSERVATION ON EMPLOYMENT: A METHODOLOGICAL REVIEW

Michael D. Yokell
Energy and Resources Program
University of California
Berkeley, CA 94720

ABSTRACT

A significant reduction in the rate of growth of domestic energy use may entail important changes in the level and composition of domestic employment. These effects depend on the manner and rate at which energy is conserved.

Several methodologies have been utilized to date to predict the employment effects of both deliberate and inadvertant energy conservation. These include "back of the envelope" macroeconomic models, large econometric macro models, production function studies and input-output analysis. This paper is a critical appraisal of the extant methodologies. Although it does not focus on the results, per se, it bears heavily of their validity and points the way towards better analysis in which more confidence can be placed.

Methodology is often relegated by policymakers and the public to the academic stratosphere. In this case, since the predicted employment effects of energy conservation programs depend critically on the methodology used to assess them, methodology should be of general concern. The paper should be of interest to labor leaders, business, energy policy personnel, environmentalists and others vitally concerned with the social and economic effects of energy conservation.

STRUCTURAL DESCRIPTIONS OF ALTERNATIVE ENERGY FUTURES

Clark W. Bullard

Energy Research Group
University of Illinois
Center for Advanced Computation
Urbana, IL 61801

ABSTRACT

Long range energy policy analysis requires a mechanism for characterizing key structural features of a nation's economic system. It is shown that energy analysis techniques are well-suited to such applications, particularly over time horizons beyond the reach of econometric methodologies. Tradeoffs between aggregation and uncertainty are discussed as a function of time horizon. For alternative energy futures, special attention is given to the assessment of impacts on lifestyles, technology, and employment.

INTRODUCTION

Energy use is linked closely to the stock of capital equipment. In fact, capital equipment can be viewed as a "conduit" for channeling energy to perform useful work. Berndt and Wood (1975) presented empirical evidence showing that energy and capital were indeed complementary. The implications for energy demand analysis are significant; short and long-range analyses require substantially different methodologies.

In the short run while stocks of vehicles, buildings and industrial equipment remain essentially unchanged, behavioral factors are the most important determinants of energy use. Reduced travel speed, lowered thermostat settings and other behavioral changes represent only a small part of the overall spectrum of conservation options, but are the only ones available in the short run. From a modeling standpoint, econometric models are best suited to describing such changes in energy demand as a function of price, income, or demographic variables. The parameters of econometric models are regression coefficients derived from observations of time-series and cross-sectional data.* It is implicitly assumed that the basic structure of the economic system is not substantially changed from the period when the parametric data were obtained.

Over the long run, new generations of energy-consuming capital stocks replace the old. The energy-efficiency of the new equipment can vary by at least a factor of two above or below today's level; the range of presently known technology is at least that large. Analyses of long-term energy demand must therefore include an explicit description of the structure of the economic system. Structural parameters (descriptions of technologies) become at least as important as price and income elasticities. Structural models are usually built "from the bottom up," starting at any level of aggregation of production and consumption processes. Constructing such a model forces the analyst to enumerate a series of value judgments, and explicitly deal with problems of internal consistency; he must essentially paint a complete picture of a future state of the system. While this is theoretically true for other methodologies for energy demand analysis, the building-block approach requires that more of these assumptions be explicitly justified.

With a structural model, it is possible to address a variety of physical problems and issues important to energy policy analysis. For example, structural models are needed to assess the energy impact of specific technological options, such as comparing the net energy yield of energy supply and conservation alternatives. The issue of saturation can be addressed; blind extrapolation of econometric data may lead to absurd projections corresponding to people

* This approach has been applied successfully by Klein (1953) and others (1965) for relatively short-term (quarterly and annual) forecasting applications.

spending an inordinate number of hours per day travelling, or to automobiles with less than one occupant. Environmental factors and resource impacts (e.g. labor) can be analyzed directly with structural models through only minor modification.* Finally, explicit representations of technologies and lifestyles (consumer spending patterns) make it possible to relate the effect of these factors on the relationship between energy and GNP.**

In the next section of this paper, a simple linear structural model is described. The third section discusses the application of structural models to analysis of long-term energy demand scenarios.

A SIMPLE STRUCTURAL MODEL

The problem under consideration is to model the flow of energy through a stock of capital equipment at a fixed point in time. For some applications (e.g. net energy) dynamic effects may be important, but at moderate growth rates such problems can usually be handled on an individual basis. The model must be sufficiently detailed to distinguish between technologies having significantly different energy impacts (e.g. coal vs. nuclear, basic oxygen vs. electric steel processes). In practice, this requirement alone creates the need for quantities of data exceeding that available at intervals frequent enough to develop a dynamic model of energy flow in the U.S. economy. Data constraints also limit one to developing a linear model. We are therefore left with the classic Leontief input-output model which is sufficiently disaggregated to permit measurement of key variables in physical units.

This simple linear structural description of the U.S. energy-economic system has been developed by Bullard and Herendeen (1975) at the 360-sector level of detail for 1967 -- the latest year for which data are available.+ Briefly, the model relates the gross output, $\underline{X}$, of the (N = 360) sectors to the demand of final consumers, $\underline{Y}$, through a matrix $\underline{\underline{A}}$ of N^2 coefficients.

$$\underline{X} = (\underline{\underline{I}}-\underline{\underline{A}})^{-1}\,\underline{Y} \qquad (1)$$

The elements of $\underline{X}$ and $\underline{Y}$ are measured in physical units.* The sum of the dollar value of elements of $\underline{Y}$ (prices times quantities) is simply the GNP. A typical coefficient A_{ij} represents the amount of input from sector i needed directly by sector j to produce a unit of output. Thus, $\underline{\underline{A}}$ represents the production technologies of all N sectors, and depends on the stock of energy-consuming capital equipment in those sectors.

If sector 1 is an energy sector, total U.S. energy demand is simply the dot product of the first row of $(\underline{\underline{I}}-\underline{\underline{A}})^{-1}$ with the vector $\underline{Y}$. This first row of the Leontief inverse matrix can be identified as the total (direct plus indirect) energy intensities of goods and services, and is usually denoted $\underline{\varepsilon}$. It is a function soley of $\underline{\underline{A}}$, the production technologies. The elements of $\underline{Y}$ are related more closely to consumers' lifestyles: the market basket of goods and services making up the GNP. Shifts among goods, services or activities of various energy intensities can therefore change total energy demand, even while production technologies remain unchanged.

For long-term energy demand forecasting, it is clearly impossible to specify the N^2 parameters describing production technologies, unless N is very small. The nature of the stocks of capital equipment cannot be known to the degree of detail necessary to specify the elements of $\underline{\underline{A}}$ describing material and energy flows through each sector. As a result, it is necessary to choose a suitable level of aggregation for the model, and to focus attention on those elements of $\underline{\underline{A}}$ to which energy demand is most sensitive.

The advantages of aggregation are clear; the amount of food likely to be consumed per capita in the year 2000 is more easily specified than the amount of olives. The uncertainty associated with thousands of parameters in a detailed model will probably render its results useless for long-term analyses. However, if one can identify those parameters to which energy demand is most sensitive, and select that subset most closely connected to the independent variables, a suitable level of aggregation can be identified.

---

* See for example, Hannon (1977).

**Bullard and Foster(1976).

+ The lag time reflects the complexity of the massive data collection effort needed to support such a simple model. It requires a multi-year, multi-million dollar effort by the Census Bureau and other offices in the U.S. Department of Commerce.

*In theory, physical units should be used for all sectors. Actually, even the 360-sector breakdown is so aggregated that physical units are not applicable in most sectors, so constant dollars are used as a surrogate. Physical units are used in the energy sectors, however.

For the 1967 model, the parameters most important for energy analysis have been identified.* This provided the basis for a 40-sector aggregation scheme employed in the National Academy of Sciences Study of Nuclear and Alternative Energy Systems (1977). Half of these sectors represented various energy production and processing technologies, and the other 20 were energy-consuming sectors. Some sectors such as food were singled out in such a way that fuel substitution options and changes in energy-intensive packaging requirements could be explicitly represented. Other sectors such as "consumer goods" were aggregates of a large number sectors in the detailed model. Since direct energy inputs to "consumer goods" at the final fabrication stage do not vary greatly among various items, aggregation did not present problems. The key parameters from an energy standpoint were the elements of $\underline{A}$ representing inputs of wood, paper, metals and plastics; these changes could be specified at the aggregate level without detailing the various types of goods produced, since the inputs are highly substitutable. As a final example, auto production was singled out as a separate sector so changes in vehicle weight and material composition could be specified explicitly for high-and low-energy scenarios.

Since the model describes the complete network of material and energy flows connecting the N sectors of the economy, indirect energy requirements are automatically included in the energy intensities $\underline{\varepsilon}$. This follows directly from the relationship

$$(\underline{\underline{I}}-\underline{\underline{A}})^{-1} = \underline{\underline{I}} + \underline{\underline{A}} + \underline{\underline{A}}^2 + \underline{\underline{A}}^3 + \ldots \quad (2)$$

showing the equivalence of the input-output approach with a process analysis carried out an infinite number of steps.** All feedback loops are accounted for through the matrix inversion process. Thus the energy required to produce the aluminum and fiberglas for a more efficient automobile is included in the model. The same holds for the energy required to produce insulation sold to the construction industry. This network model is superior to conventional "sectoral" analyses which specify both the technology <u>and</u> the level of production independently. In fact, production levels in one sector are a function of the technology in all sectors (eq. 1), and a network model is needed to assure internal consistency.

Specification of parameters in this structural energy model is a process that deserves some elaboration. It is usually done in the context of <u>scenario analysis</u>, a methodology that will be discussed in more detail in the next section. Regardless of the time horizon, the analyst must use some mental model to assure consistency among the technological and lifestyle changes being specified. In economic terms, this mental model can characterized by a set of relative prices of goods and services; the analyst describes cost-minimizing technical options subject to these prices. This process has been formalized for some applications: for example by Hoffman (1974) for the energy sectors. Specification of the direct energy elements of $\underline{Y}$ has also been formalized by Hirst (1977) for energy use in residential buildings, and by FEA (1974) for transportation fuel using a simple birth and death model to describe turnover of the vehicle fleet. These models are "hybrid" structural/econometric models, a type of model rapidly becoming recognized as best suited for short-mid term energy demand analysis.

LONG-TERM SCENARIO ANALYSIS

The structural modeling techniques described above are suited to short as well as long-term analysis. When combined with an econometric model to account for saturation and other effects, the models can be used in an optimization framework. Driven by cost-minimizing or profit-maximizing algorithms, such models can be used for in a predictive or normative mode for short and mid-term analyses. In this section I will argue that the optimization framework is fundamentally unsuited to analyses of long-term energy demand, and that structural models are better used outside that framework for examining alternative energy scenarios. "Long-term" is meant to represent time horizons of 15-25 years or more, during which substantial shifts in social values can occur.

Based on purly physical considerations, almost any future state of the system can be described, and the simple linear static structural model described above is a useful device for characterizing such a state and its energy requirements. In practice the production technologies and consumption patterns cannot be specified independently. They must be consistent within some value framework, represented by a set of

---

*See Bullard and Sebald (1977). It was shown that uncertainty on the most important 2% of the parameters contributed to 75% of the uncertainty on total demand for electricity.

**See Bullard, Penner and Pilati (1976) for a more detailed discussion.

market prices of goods and services.* The distinction between short and long-term analyses is clear: currently prevailing values (today's relative prices) will simply not be applicable in the distant future.

Harman (1975) cites a long list of changes in societal values that have had significant effects on the structure of the U.S. economy. Among them are attitudes toward family size, the role of women, "reasonable returns" in wages and court settlements, the quality of the physical environment, consumer rights, etc. In the face of such observations we can only conclude that values will change, and so will the ground rules for specifying and evaluating a model of alternative energy futures.

This highlights the "logical trap" of technology assessment. Since values are certain to change, any model of our long-range future based on today's values can only describe states of the economic system that simply would not obtain after changes in values had occurred. Conversely, if model parameters are specified consistent with a different set of values, the results are likely to be rejected because they are inconsistent with today's values as reflected through the political and economic systems which evaluate and select among energy models and scenarios. Faced with this dilemma, where does one find a role for long-range energy demand analysis?

To answer this question one must establish why such long-range modeling is necessary. Most long-range modeling of energy systems is undertaken to support evaluation of research and development priorities in the energy area. Many advanced energy production technologies are uniquely capital-intensive and large-scale; due to both physical and institutional factors they are characterized by exceptionally long lead times. Sometimes these lead times are measured in decades (e.g. the LMFBR) and often are comparable to the time horizon over which societal values may change substantially. The benefits and costs of such technological options must be assessed in the context of the environment in which they will operate (e.g. 15-40 years hence), and models provide a useful framework for such assessments.

I submit that such modeling activities are instruments of social change; that is their appropriate role. The long-range energy demand model is a tool for evaluating scenarios describing alternative futures. The physical nature of energy forces the scenarios to describe in considerable detail the stock of man-made capital equipment (since that is where energy is consumed) and consumer lifestyles (characterized in part by the quantities of goods and services produced). The tool (e.g. the simple linear structural model described earlier) can be used as a starting point of a process for public policy development.

This first step in the process is simply to describe in physical terms several alternative future states of the system. The nature of the model forces the analyst to justify and defend consistency of the model in terms of an explicitly stated set of values. The model thus provides a basis for public debates on the desirability of various alternative future conditions. Many alternatives will be rejected on the basis of this static description alone. After narrowing the range of choices the (possibly more difficult) question "How do we get there from here?" can be addressed. There is a virtually infinite number of paths that could connect the present with any future scenario. Each path has its own implications for compatibility with values, lifestyles, and technologies. Many or all of the paths to a particular future condition might be rejected on these grounds. Moreover, associated with each "feasible" path are millions of possible combinations of policy initiatives that may be necessary or sufficient to achieve the goal. The next level of analysis would be centered on assessing the desirability of policy alternatives.

The analytical sequence of describing alternative futures and then working backward in time is nearly opposite to the process associated with the conventional use of econometric models. It is appropirate, though, for problems of a long-range nature because it first sets out goals and value judgements explicitly for debate, and adoption or rejection. It is not unlike the initial proposal of a Constitutional amendment. Selection of specific paths and policies are debated in only general terms along with the societal values being clarified, but the details are left to be worked out through processes tailored to specific local and contemporary conditions.

The simple static structural model plays a key role in describing the initial set of alternative futures, and other types of models come into play during the process of examining alternative paths and policies over shorter time horizons.

---

*This statement assumes we are interested in describing a future state of a market economy. For a planned economy, the parameters specified must be consistent with the values underlying the plan.

CONCLUSIONS

The process of scenario analysis described above is precisely the one employed by Lovins (1976) in stimulating one of the more provacative discussions of our energy future.* His model is not as complete as the one described in this paper, but in this relatively new area no one has yet struck the optimal balance between rigor and simplicity. It seems clear that the conventional modeling community will be uneasy (at least initially) with the amount of rigor that might have to be sacrificed to achieve the simplicity needed to involve the general public in discussions of the value issues.) The National Academy of Science study** used the model described here to examine a variety of alternative futures all based on essentially the same set of (currently prevailing) values, avoiding the "logical trap" described earlier. The study used the scenarios to illuminate several structural issues, especially the extent to which energy and GNP growth could be "decoupled." Only one scenario in that study dealt with possible consequences of current trends changing societal values.

To be sure, the approach outlined in this paper has many shortcomings. Among the most significant is the inability to foresee and model unanticipated new technologies for consuming energy, or new technological innovations for conserving energy.+ These modeling risks are simply reflections of the real risks inherent in developing long lead-time energy technologies.

By its very nature, the time scale for energy demand analysis is tied to the turnover period for the stocks of energy consuming capital equipment. In addition, energy demand must be analyzed over the same time horizon as energy supplies, so the longer of these two periods is the relevant one from a modeling standpoint. Policies affecting the economic value (hence lifetime) of energy-consuming capital stocks and the choice of supply technologies having various lead-times will also affect the type of modeling approach required.

This returns us to the role of modeling in the policy process: its role as an instrument of social change. If the participants in the modeling process (both analysts and the general public) feel uncomfortable with the degree of uncertainty, the "logical trap," or the prospect of irreversibly imposing their values on future generations,* they may opt for policy options that reduce the risks and uncertainties by shortening the relevant time horizon. Clearly this is the issue underlying the present debate on energy R&D priorities. And to some extent, it may also explain the current Administration's assessment of the political infeasibility of an energy policy dominated by underwriting the risks of long lead-time supply technologies, as opposed to reliance on the relatively short-term conservation technologies tied to the 10-year turnover of the vehicle fleet and the shorter-term retrofit of existing buildings. Value judgements articulated in the Constitution itself prevent one Congress from binding the next, and thereby make it impossible to assure that a particular technology will be economically viable (or even legal to operate) during the 1990-2020 time frame.

ACKNOWLEDGEMENT

This work was supported by the U.S. Energy Research and Development Administration.

*One of the strongest criticisms of Lovins' scenario (Forbes, 1977) is directed at the second phase of the process, challenging the feasibility of paths to the future described.

**NAS (1977)

+For example, would any "reasonable" scenario fornulated in 1935 included the massive growth in air conditioning that occurred in the 1950's and 60's? In an era of rising energy prices (heretofore inexperienced) it may be reasonable to expect significant breakthroughs due to added incentives for private sector R&D in energy efficiency.

* The National Environmental Policy Act of 1969 set up a political process for technology assessment responding to precisely such concerns.

REFERENCES

1) Berndt, E. and D. Wood, "Technology, Prices, and the Derived Demand for Energy," Review of Economics & Statistics, Vol. 57, pp. 259-268 Aug. 1975.

2) Bullard, Clark W. and Craig Z. Foster, "On Decoupling Energy and GNP Growth," Energy Vol 1, pp. 291-300, 1976.

3) Bullard, Clark W. and Anthony V. Sebald, "Effects of Parametric Uncertainty and Technological Change in Input-Output Models," Review of Economics Statistics, Vol. 59, No. 1, pp. 75-81, Feb. 1977.

4) Bullard, Clark W. and Robert A. Herendeen, "U.S. Energy Balance of Trade, 1963-1973," Energy Systems and Policy, Vol. 1, No. 4, pp. 383-390, (1975)

5) Bullard, Clark W., Peter Penner and David A. Pilati, "Energy Analysis Handbook," CAC Doc. 214, Center for Advanced Computation, University of Illinois, Urbana, October 1976.

6) Duesenberry, J. S., Gary Fromm, L. R. Klein, and Edwin Kuh, Brookings Quarterly Econometric Model of the United States, Rand McNally, Chicago, 1965.

7) Federal Energy Administration, "Project Independence Report," Washington, D.C., U.S. Government Printing Office, November 1974.

8) Forbes, Ian, "Energy Strategy: Not What But How," Energy Research Group, Inc., Farmington, MA, March 1977.

9) Hannon, Bruce, "Energy, Labor, and the Conserver Society," Technology Review, pp. 47-53, 1977.

10) Harman, W.W., "The Great Legitimacy Challenge," Vital Speeches of the Day, December 15, 1975.

11) Hirst, E., "Residential Energy Use Alternatives, 1976 to 2000," Science, Vol. 194, No. 4271, pp. 1247-1252, December 1976.

12) Hoffman, K. C., "Unified Framework for Energy System Planning," Energy Policy, Special issue on modeling, IPC Science and Technology Press Ltd., Haywards Heath, Sussex, U.K. 1974.

13) Klein, L. R., A Textbook of Econometrics, Row and Peterson, 1953.

14) Lovins, Amory, "Energy Strategy: The Road Not Taken?" Foreign Affairs, (Oct. 1976).

15) National Acedemy of Sciences, Committee on Nuclear and Alternative Energy Systems, Final Report, Washington, D.C., (forthcoming - July 1977).

ENERGY ANALYSIS: A TOOL FOR EVALUATING THE IMPACT OF END USE MANAGEMENT STRATEGIES ON ECONOMIC GROWTH

Martha W. Gilliland

Science and Public Policy Program, University of Oklahoma, Norman, Oklahoma and Energy Policy Studies, Inc., El Paso, Texas

INTRODUCTION

This conference session on energy analysis has two objectives. First, it is aimed at building a bridge of communications between economists and energy analysts, a bridge which will allow discussion of semantic as well as real theoretical differences. Second, we are reviewing research on what energy analysis has to say about the potential of end use management strategies. This paper focuses on the first objective. It examines the differences between the manner in which economic theory views the production and consumption of goods and services and the manner in which the theory which underpins energy analysis views production and consumption. As such, my focus is at the macro level of planning rather than at the micro level of process analysis. While I view energy analysis as a powerful tool for examining energy flows within and between individual technologies which both produce and consume energy, a tool which is indispensable in analyzing end use management technologies; the quantitative data I will present is at the aggregated level only.

Much of the discussion in the literature which attempts to relate the concept of net energy to economics is controversial. A brief summary of that controversy is given by Gilliland (1). Recognizing that controversy and the fact that more empirical evidence is required to resolve it, I first discuss some energy analysts' view on the relationship between net energy and economic growth and second I compare the net energies achievable via energy production increases with those achievable via end use management strategies.

NET ENERGY AND ECONOMIC GROWTH

The literature which attempts to relate net energy to economic growth generally addresses the question: to what extent is energy growth coupled to economic growth? Current research approaches this question by considering the economic production function as a general case. In considering that, I first compare an economist's perception with an energy analyst's perception of production. Next, I discuss the implications of these two different perceptions for finding solutions to economic growth problems. Finally, I briefly examine the economist's utility theory of value in light of the energy analyst's perception of production.

Production: An Economist's and an Energy Analyst's Perception

Figure 1a represents the flow of money in the economy as it is generally conceptualized in introductory economics textbooks. The real GNP is evaluated as the monetary value of the goods and services produced or the equivalent quantity, the monetary cost of the

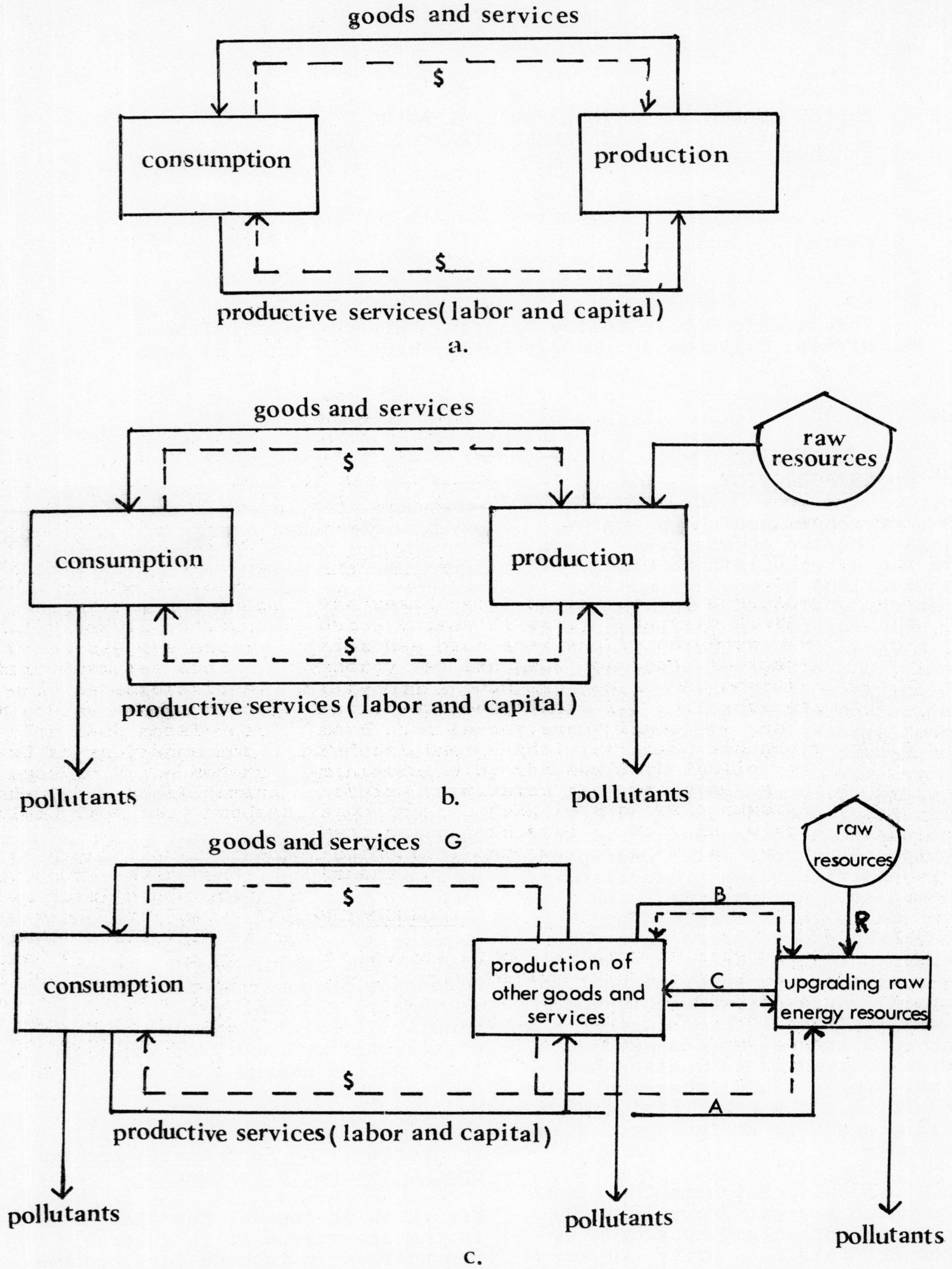

Fig. 1. The production and consumption cycle; solid line is energy flow and dotted line is money. (a) economist's view of the cycle, (b and c) energy analyst's view.

labor and capital required for production. A simple economic production function is one which includes the stock of capital and of labor in order to make predictions of growth. Natural resources, education, technological change and other social factors are occasionally considered in more sophisticated analyses of the production function. In any case, the scheme is one of a closed production-consumption loop, in which the economist's notion of production is one of producing goods and services for final demand. These goods and services have utility to the consumer, where utility includes both the physical attributes of the commodity as well human valuations such as those represented by convenience and status.

In contrast, the energy analyst's thermodynamic view of that same process is shown in Fig. 1b. The difference arises because the energy analyst's notion of production is one of embodying goods and services with energy, the particular quantity of energy embodied and the mix of energy types being determined by the same utility factors as those of the economist. But because the energy analyst views production as an embodiment of energy, he must face the implications of the laws of thermodynamics for that embodiment. The system represented by Fig. 1a when set in energy units violates the laws of thermodynamics; as a result, the energy analyst automatically looks at a larger system. He looks at a system which includes the energy flows that are always present and determined by the laws of thermodynamics. Those laws say: (i) that production requires available work (some people call it potential energy, it is more precisely the thermodynamic quantity known as free energy); (ii) that production is nothing more than increasing the amount of work available in the commodity which is being produced (in comparison to the amount available from the materials used to make the commodity); and (iii) that increasing the available work embodied in the commodity requires that work availability be decreased somewhere else, namely in the ambient environment. Thus, the energy analyst must open the closed loop and show where available work enters the system (natural resources as shown in Fig. 1b) and where the penalty for increasing its availability further must be paid (pollutant outflow to the environment as shown in Fig. 1b). For example, manufacturing an automobile or generating electricity represents an increase in the ability of the energy, which is embodied in the automobile or in the electricity, to do work. One Btu of electricity can do more work than one Btu of coal and an automobile can do more work than the iron, steel, coal, and aluminum which were used to manufacture it. There are a myriad of energy transformations required and inherent in increasing that work availability and each transformation takes place under these thermodynamic constraints.

Since money is not exchanged for either the inflow of raw resources nor the outflow of pollutants, the economist has no problem ignoring them. But why, historically, have economists' predictions of the production function been relatively accurate if he is ignoring both the inflow and outflow of his loop; what difference would it make if they were included?

Implications of Two Perceptions

Historically, accessibility to new sources of raw energy has been easy and pollutant levels have been low. When this is true, conceptualizing production as only part of the system of energy flow does not introduce large errors in predictions of economic production. In fact, it has worked rather well. When the "external" flows associated with accessibility to raw resources and pollution are small compared to the flows which occur within the loop (labor and capital), it makes very little difference whether one excludes or includes these external flows. But when raw resources are difficult to obtain and when pollutant levels become large, ignoring them introduces very large errors in economic predictions. More importantly, it obscures the cause of economic growth problems as well as solutions to them.

While our endowment of raw resources still seems to be enormous when set in the context of our coal and uranium reserves and the seemingly enormous quantities of solar energy available, few would disagree with the fact that the energy required to transform these sources into high quality forms is enormous. In other words, large quantities of energy must be expended to impart to coal or solar energy the same ability to do work as is already present (or was) in Texas oil fields. Furthermore, that fact is thermodynamically determined; no technological breakthrough is going to change it very much. Although technological progress can result in incremental improvements in, for example, synthetic fuels technologies, the amount of improvement is thermodynamically constrained by the quality* of the starting resource.

Probably the most fundamental difference between economists and energy analysts is in this notion of external, physically based constraints on production. The economist does not recognize them while the energy analyst working at the macro level is concerned with precisely those constraints. Just what one perceives is constraining production generally affects the kind of solutions proposed to remove the constraint. This difference in constraint perception is illuminated by an economist's description of the energy shortage. Many economists describe it as a shortage of energy at the going price or as a shortage of income on the part of consumers. This description leads one to solutions which involve manipulating, in one way or another, factors which are perceived to play a role in increasing the flow of goods and services around the loop, so that consumers can make more money with which to buy energy. Such manipulations include increasing the productivity of labor, technological progress, injections of more money from the federal reserve board, or alterations in the distribution of production purchases among those purchased by government, those purchased for personal consumption, and private domestic investment. Regardless of the solution proposed, it is a manipulation of one of the factors operating within the closed loop of production and consumption.

In contrast, the energy analyst looks outside that loop at casual forces, forces which affect the dynamics of production, but which are themselves not affected by that production. To be specific, the quality of raw energy resources in Fig. 1b affects the dynamics of production, but production does not affect the quality of raw resources. This is not a trivial point because many people believe that variables such as technological progress will somehow change the quality of the raw resources. The energy analyst sees a raw resource that, by the laws of thermodynamics, demands increasing amounts of labor, capital, and other resources to upgrade. Fig. 1c should clarify the point; it is similar to Fig. 1b except that the energy producing sector of the production box is disaggregated from the other sectors. Economic production is a measure of the flow labeled G on Fig. 1c, the dollar value of the goods and services provided for final demand. The energy analyst is concerned with the exponentially rising values of flows A and B, the sum of which is the energy cost of upgrading raw resources. The box labeled "Upgrading of Raw Energy Resources" on Fig. 1c represents processes which take raw energy resources and embody them with some of the energy contained in flows A and B; the purpose is to produce a high quality product as represented by flow C, that is, to increase the available work embodied in the raw energy. If the quality of flow C is to remain unchanged (e.g. assume C is electricity or clean gas with a heating value of 1,000 Btu per cubic foot) and if the quality of R is decreasing (e.g. the case of using coal as a source of high Btu gas), then the amount of A plus B must increase and, by the second law of thermodynamics, it must increase exponentially. An increase in flows A and B occurs at the expense of a decrease in flow G (the measure of real GNP). No manipulations within the closed loop such as price floors or price guarantees can change that.

Two different perceptions of "the problem" lead to very different proposed solutions; in fact those different perceptions are causing the energy analysts to look for solutions in

*Quality being defined as the amount of available work in the raw resource naturally.

different places than the economists are looking. Before one possible solution (from the energy analyst's viewpoint) is examined, a final comment is in order on the general misunderstanding that exists concerning the implications of this energy analysis concept for a value theory.

The Perception of Value

The controversy over whether performing energy analyses for the purpose of economic planning automatically means that the analyst subscribes to an energy theory of value usually begins with the following series of statements. Does anyone really care if the energy producing sector produces net energy? The objective is to produce energy with characteristics that have economic value, that is to produce energy at the time and place desired and to produce it in a convenient form. Such a statement represents a utility theory of value. Contrary to what many critics of energy analysis have said, the energy analyst does not argue with a utility theory of value. He argues only that value is ultimately constrained by the physical laws governing energy flow and that those constraints have gone unrecognized because the production-consumption loop is only a part of the system of energy flow. My thesis is that internal factors such as labor, capital, and technological progress have dominated as limiting factors on production since about 1900, but that external forces, namely those associated with the quality of raw energy sources, will limit production at least through the year 2000. Over the long term then, energy flow (including a measure of its quality) can become a value measure.

NET ENERGY INCREASES: PRODUCTION VS. END USE MANAGEMENT

While the net energy dilemna is significant, the potential of end use management to mitigate the problems it creates is also significant. The potential of end use management has gone largely unrecognized because historical solutions, even some involving net energy considerations, have been to increase the amount of gross energy taken out of the ground. Such an approach appears both adequate and appropriate to many policymakers now. In this section, I compare gross and net energy data from 1950 to 1974 and a projection from FEA's National Energy Outlook (2) through 1985. The data is given for U.S. domestic production only. Two projections of net energy totals from 1976 through 1985 are included, one which includes a national emphasis on end use management and one which does not.

Figure 2 gives domestic gross and net energy production beginning in 1950. Note that the net energy calculations involved summing energies of equivalent quality. That is, each Btu of electricity was first multiplied by 3.6 in order to convert it to its coal equivalent; the values were then summed. Many analysts add Btu's of electricity and Btu's of oil, gas, and coal directly; that approach tends to focus on the inefficiency of electric power generation but de-emphasizes the fact that electric power can do more work at its end use point than oil, gas, or coal. In contrast, this approach treats electricity as upgraded oil, gas, or coal. Net energy for a given year, then, was calculated by subtracting the energy cost of producing a given energy resource from the gross amount of that resource produced, where all energy cost measures were first converted to their coal equivalent. The energy cost of producing various energy resource types were derived from a variety of net energy studies carried out over the past five years (3-8). The gross amount of each energy resource type shown on Fig. 2 represents that actually produced domestically for 1950 through 1976 and that projected to be produced in FEA's low energy growth scenario (2) through 1985.

Note in Fig. 2 that the difference between gross amounts of energy produced and net amounts has been increasing. The ratio of gross production to the energy cost of production in 1950 averaged 10 to 1; in 1975 it averaged 6 to 1. The 1985 projection (FEA's low scenario) results in the production of 73 quadrillion Btu's (Q) of domestic energy in 1985 of which 34.2% is oil, 23.3% is natural gas, 27.5% is coal, and 8% is nuclear. Without any conservation techniques or end use management strategies, net production

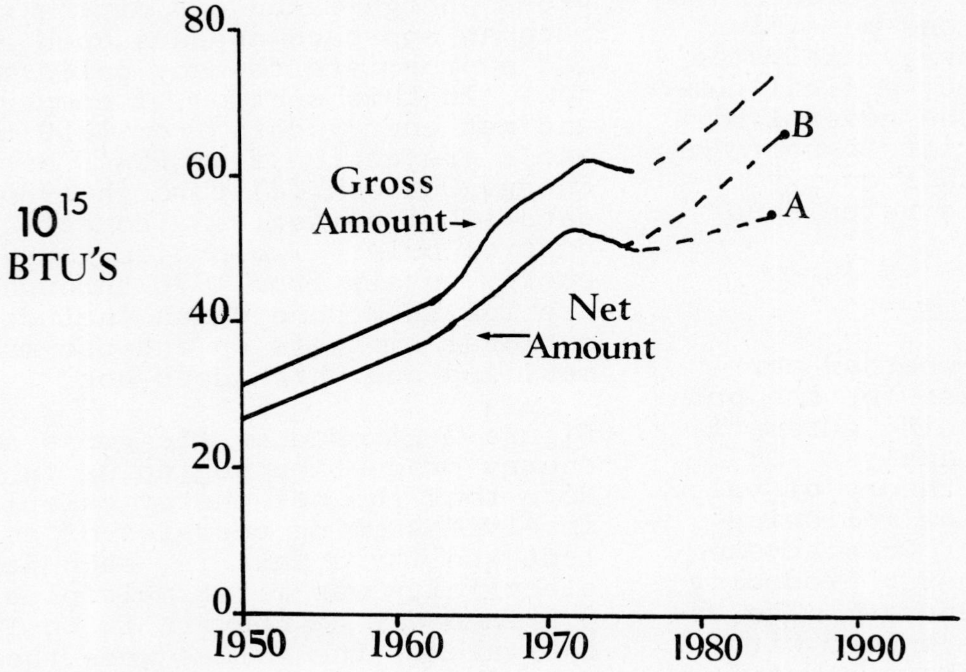

Fig. 2. U.S. domestic energy production: solid line - actual, dotted line - projected. Projected net production without end use management (point A) and with end use management (point B).

in 1985 would be 56Q (point A, Fig. 2). This is an average ratio of gross production to energy cost of production of 4.3:1 [73/(73-56)=4.3].

However, as indicated by the point labeled B on Fig. 2, net energy production can be increased to 65Q by 1985 (a ratio of 9:1) by invoking end use management strategies. This amount of increase does not require any major lifestyle changes, but is based on simple strategies such as insulating and adding storm windows to homes, the use of more efficient home heaters and car engines, industrial use of energy saving capital equipment such as recuperators, some recycling, energy shifts in illumination, the use of energy saving materials in manufacturing and construction, and more efficient home appliances. Beginning with the report of the Energy Policy Project of the Ford Foundation in 1974 (9), the potential of conservation strategies began to be recognized. Since then, many research projects have been directed at analyzing just how much energy could be saved in the end use sectors of the economy (10-15).

Although the reports from these research projects present impressive results concerning the quantities of energy that can be saved, they fail to evaluate the energy cost of replacing "wasteful" devices with energy saving systems. There is energy embodied in insulation and storm windows and energy saving capital equipment is usually energy intensive to make. Energy analysis, itself, is the tool we will have to use to understand the energy conserved by switching from one good or service to an alternative and to evaluate the energy cost of substituting new technologies and new manufacturing processes. Based on what are still preliminary results, it appears that, on the average, for every one Btu of energy invested in a conservation strategy, about four Btu's are saved. I have used this average in calculating data point B in Fig. 2. It should be regarded, however, as a preliminary estimate.

The point is that it is nearly impossible to produce 65Q's of net energy domestically by simply increasing gross production. It appears that it can be done, however, with a strong emphasis on end use management. If that is correct, we can have steadily increasing economic activity without large increases in gross energy growth. My own view is that the thermodynamic constraints on the production of domestic energy are so major that conservation will come about regardless of government programs which either encourage or discourage it. However, the impact of conservation will be felt in the economy more rapidly and more effectively if local, state, and federal government policies encourage it.

REFERENCES

1. Gilliland, M.W. (ed.) (1977) Energy Analysis: A New Public Policy Tool, American Association for the Advancement of Science, Washington, D.C., see specifically the introduction to the book.

2. Federal Energy Administration (1976) National Energy Outlook, Government Printing Office, Washington, D.C.

3. Odum, H.T. and E.C. Odum (1976) Energy Basis for Man and Nature, McGraw Hill, New York.

4. Odum, H.T. (1976) Net Energy Analysis of Alternatives for the United States, Middle and Long-Term Energy Policies and Alternatives, House Committee on Interstate and Foreign Commerce, No. 94-63, Washington, D.C.

5. M.W. Gilliland, Energy analysis and public policy, Science 189, 1051 (1975).

6. Development Sciences, Inc. (1975) A Study to Develop Energy Estimates of Merit for Selected Fuel Technologies, U.S. Dept. of the Interior, Washington, D.C.

7. Colorado Energy Research Institute (1976) Net Energy Analysis: An Energy Balance Study of Fossil Fuel Resources, Golden, Colorado.

8. Pilati, D.A. and R.P. Richard (1975) Total Energy Requirements for Nine Electricity-Generating Systems, Center for Advanced Computation, University of Illinois, Urbana, Illinois.

9. Energy Policy Project of the Ford Foundation (1974) A Time to Choose, Ballinger, Cambridge, Mass.

10. American Physical Society (1975) Efficient Use of Energy: A Physics Perspective, American Physical Society, New York, N.Y.

11. The Conference Board (1974) Energy Consumption in Manufacturing, Ballinger Publishing Co., Cambridge, Mass.

12. Environmental Protection Agency (1973) Energy Conservation Strategies. Office of Research and Monitoring, U.S. EPA, Washington, D.C.

13. Gyftopoulos, E.P., L.J.Lazaridis, and T.F. Widmen (1974) Potential Fuel Effectiveness in Industry, Ballinger Publishing Co., Cambridge, Mass.

14. E. Hirst, Residential Energy Use Alternatives: 1976-2000, Science 194, 1247 (1976).

15. L. Schipper and A.J. Lichtenberg, Efficient Energy Use and Well Being: The Swedish Example, Science 194, 1001 (1976).

USEFUL ENERGY CONSUMPTION: THE DETERMINATION OF RELATIVE FUEL EFFICIENCIES

Jacques GIROD

Institut Economique et Juridique de l'Energie, Grenoble

INTRODUCTION

The total energy consumption of an economic sector or a whole country is generally defined by QE:

$$QE = \Sigma Hi.QAi \quad (1)$$

QAi represents "apparent" final consumption of fuel i, and Hi represents the calorific equivalent. The various fuels are usually divided into four categories -coal, gas, oil, and electricity- and each one of these is measured in a common unit, for example tonne coal equivalent, using conversion factors Hc=1, Hg=1.4 tce/tcal, Hp=1.3 tce/t, He=0.123. 10^{-3} tce/kWh. The product of Hi.QAi defines the QEi consumption measured in the same unit:

$$QE = \Sigma QEi$$
$$= CH+GZ+PE+EL \quad (2)$$

The agregate measurement causes some problems:

. The Hi calorific equivalent is either a theoritical coefficient or one determined by convention. It varies according to the "qualities" of different kinds of gas and oil. As concerns electricity, there must be a clear distinction showing whether the equivalence is of consumption or production.

. Given the substitutions among fuels, and the observed replacement in many countries of low thermal content fuels (coal and manufactured gaz) with high content fuels (gas, oil, electricity), the consumption of the QE aggregate evolves while its volume increases. In other words, QE is not invariant with time; then the quantities of energy effectively used in energy installations are likely to be underestimated.

The first class of problems will not be discussed in this paper. We shall to define another energy aggregate whose property is to be invariant to the proportions of the different fuels in the total energy consumption. To this end, we shall consider the "relative fuel efficiencies": the contribution of each fuel in this particular aggregate is no longer QEi, but a function f(Ri,QEi), with Ri precisely this efficiency coefficient. In this case, a new QU aggregate can be defined.

$$QU = g(f(Ri,QEi)) \quad (3)$$

and it can be considered as a measure of useful energy; the Ri coefficient indicates the relative use efficiency of the various energy installations burning the i fuel.

It is obvious that useful energy figures are more informative than "apparent" energy figures. Though apparent energy is the energy in fact bought by the users, i.e. the energy consumed by their equipment, it does not correspond to the energy they actually need in the physical sense of the therm: temperature of an oven, or moment of a couple. When one kilo of coal is burned, it is presumed that the calorific equivalent is 7000 Kcal; however, the user obtains only 7000.Ri effective calories, with Ri less than the unit.

For each installation, there is other a priori information on Ri, specifically the thermal efficiency supplied by the manufacturers. Like the Hi, it is often simply a theoreticalvalue without respect to variations with equipment age, quality and intensity of maintenance, frequency of start-ups, etc.. If an effort is made to determine the individual efficiency of each installation, great difficulties are met in coming to a meaningful average efficiency for all equipment of a given economic sector. This is specially the case for manufacturing industries having a wide variety of installations. A model linking QEi energy consumptions considered as inputs in processing, to a production volume indicator could be proposed, offering possibilities for the determination of average Ri efficiencies.

It was Adams and Miovic (1) who first explored in this direction. If V is to stand for the volume of production of an industrial sector in a given country, and one takes for granted certain hypotheses (separability, etc.)the model will be written:

$$V = k.QU$$
$$QU = \Sigma Ri.QEi \quad (4)$$

The function f of (3) being multiplicative and the function g additive.(4) is transformed thus into a statistical equation:

$$V = \Sigma Ai.QEi+e \qquad Ai=k.Ri \quad (5)$$

in which, the series (V,QEi)being known,the coefficients Ai of the linear regression may be deduced by adjustment. Hence the Ri may be obtained:

$$Ri/Rj = Ai/Aj \quad (6)$$

Since k may not be calculated directly, it is evident that these figures relate to relative efficiency, the type of fuel j (Coal for example) being considered as a common reference for all types fuel i. This model was estimated by combining observations from 7 Western European countries over a period of 13 years 1950-1962.

Although the results obtained appear plausible (the efficiency of electricity is superior to that of gas and oil, which itself is superior to that of coal), an essential criticism may be opposed to this model: it concerns the place of electricity among other fuels. The economic analysis of past substitutions, backed up by a mass of statistical verification, clearly shows that electricity has had, and doubtless still has, a special role when compared to fossil fuels, that it is impossible to consider it in the same light as the others, in short that the implicit hypothesis of "perfect" substitutability in the plan (CH, GZ,PE,EL) is likely to be untrue.

Inside the aggregate QE, and therefore inside QU, it is necessary to distinguish two groups of fuel which have different physical characteristics and a different economic behaviour: on one side energy which is primarily destined for a thermal use, and on the other, energy destined for a mechanical use. The nature of the equipment and production conditions lead us to assume that coal, gas and oil fall into the first category and electricity into the second (obviously this concerns only the industrial sector). We therefore have expressly:

$$QU = g(TH(CH,GZ,PE),MC(EL)) \quad (7)$$

in which TH and MC denote the consumption of useful energy absorbed by thermal and mechanical use. Moreover these groups differ completely as for as substitution is concerned. In the former, the sources of energy are practically completely substituable for one another, and the winning over from coal of a large part of the thermal market by natural gas and oil bears witness to this. Electricity, on the other hand, has a very specific use throughout the industrial process (primarily for motors) and the possibility of substitution by fossil fuels is still very limited in many countries.

These reasons all argue in favour of a clear separation between (CH,GZ,PE) and EL in the formalization of a model. We must first examine the totality of thermal uses and appraise the importance of substitution between CH,GZ and PE. Then, and only then, may we turn our attention to the relative contributions of TH and MC to the value of the production volume.

1. SUBSTITUTABILITY OF COAL, GAS AND OIL FOR THERMAL USE

The liaison between production and consumption of energy is taken as linear in the plan (V,CH, GZ,PE) hence the expression "perfect substitutability" used above:

$$V = k.TH$$
$$TH = Rc.CH+Rg.GZ+Rp.PE \quad (8)$$

If Ai = k.Ri, we get:

$$V = Ac.CH+Ag.GZ+Ap.PE \quad (9)$$

The formalization adopted by Adams and Miovic reappears here with the important difference that electricity no longer figures in the relation. The measuring of fuel efficiencies takes place only inside the fossil fuel group and, as before, the coefficients are obtained by dividing each of the values Ai by the Aj corresponding to the energy of reference (j). The relation ship Ai/Aj is the marginal rate of substitution between the fuels i and j.

The model(9) was tested for the industrial sectors (steel and energy sector excluded) of 18 countries of the OECD zone during the years 1955-73. The most valid estimation method is to calculate the regression for the totality of the 18 countries for a given year, or better still, to take into account the double variability of space and time and use a covariance model or an error components model.

With these methods it is primordial to postulate a rather high degree of homogeneity in the structures of industrial production, and unfortunately this is not so. We cannot place side by side in the same sample two countries such as Eire and the United States: their structure of energy consumption is unalike and this appears clearly in the light of the results. Even estimations concerning groups of countries considered as relatively homogenous would not improve the results markedly. In spite of the loss of information entailed, we are led to estimate Ai from (9) separately, for each of the 18 countries on the basis of the 19 annual observations. The functions of production considered here are then short-term functions, and it follows that their determination is likely to be affected by variations pertaining to the economic conjuncture.

Table 1 shows the estimates for fuel efficiency

Rcg = Rg/Rc and Rcp = Rp/Rc for the nine bigger countries of the zone. It is to be remarked that these values are expressed in round figures for the relationships $A\hat{g}/A\hat{c}$ and $A\hat{p}/A\hat{c}$.*

. These fuel efficiencies may have different determinations in the different sub-periods considered.

TABLE 1 Relative efficiencies of gas and oil compared to that of coal

		Rcg=Rg/Rc	Rcp=Rp/Rc
France		2	1
Germany		2	1.5
Italy		1.6	0.8
Netherlands	1955-64	2	1
	1965-73	1	1
United States		1	1
United Kingdom		0.5	1.8
Canada		1	1
Japan	1955-64	4	2
	1965-73	5	2
Sweden	1955-60	1.1	0.5
	1961-73	1.1	0.7

In all these estimations we have aimed rather at stability than at precision. Indeed to seek after a precise measurement of fuel efficiencies would be delude oneself. It suffices to obtain approximations which enable us to compare the different countries and to evolve hypotheses for their evolution. The stability of the coefficients, on the other hand, is of prime importance and they have been verified by lengthy cross-checking: the grouping of types of fuels, conditional regressions, the introduction of first differences and of dummy variables in order to distinguish the sub-periods of time etc.. In this way the double determinations of the use efficiencies have only been adopted when the condition of stability required it.

. The differences between the countries are not negligeable, particularly where gas is concerned. This confirms the statement about the heterogeneity of the different countries, and validates the method of estimation.

. Generally the relationship $Rcc=1 \leqslant Rcp \leqslant Rcg$ is verified, and this agrees with the general opinion that gas and oil have a relative fuel efficiency superior (or equal) to that of coal. Where gas alone is concerned, the sole exception is the United Kingdom. This may be explained by the large place retained by manufactured gas until the end of the sixties.

. It is not however without some surprise that we note that the relative efficiency of oil is not as high as one would have supposed. It is equal to 1 for France, the Netherlands, the United-States and Canada. It is even less than 1 for Italy and Sweden. For these two countries, the interpretation of the results is somewhat difficult. Even if the differences between $A\hat{p}$ and $A\hat{c}$ are significative, one may assume that the estimation proceedure for a given time would tend to increase Ac and diminish Ap (The consumption of coal has distinctly dropped, while that of oil has sharply risen). For France, on the other hand, coal still maintains its place, and the interpretation can no longer be only a question of statistics. Other reasons may be advanced.

1) Rules have been laid down by which the efficiencies of the principal energy equipments have been strictly controlled and supervised, and thus have remained close to the optimum. Coal-fired equipment does not seem to have suffered any substantial detriment when compared to oil-fired equipment, although it is usually taken for granted that this latter fuel has a higher calorific content.

2) On another hand, the substitution oil for coal, which should have increased in efficiency was doubtless accompanied by waste, because of the relatively low price of oil. To put it in other terms, the efficiencies of oil-fired equipments could have been superior to what they were and to laboratory measuring, but coal constituted a level of reference below which one must not descend, and above which there was no purpose in rising. This waste, if waste there was, appeared under the different forms that one would expect:absence of regulation, of maintenance, of insulation, of automation, etc..

These two statements are verified from the experiences of boiler manufaturers themselves.

. When we come to the Netherlands (1965-73), the United-States and Canada, we see furthermore that Rcg=Rcp=1; at least, we do not obtain the coefficients $A\hat{c}$, $A\hat{g}$,$A\hat{p}$ significantly different, whatever the estimation methods used. The interpretation of this new result resembles somewhat the interpretation suggested for the case of France, but it would be better to examine this mose closelyand check the conditions of production in each country, the rhythm of the renewal of productive capital, the trend of relative prices, etc..

The case of these three countries enables us to advance the following hypothesis with a certain justification: the use efficiency fac-

*The exact values of these coefficients has been published in the study "Consommation d'énergie et coefficients d'équivalence" commissioned by CNRS.(J.GIROD, P.GODOY, P. RAMAIN).

tors are dynamic, evolving constantly under the influence of changes of equipment and in function of the share of the market alloted to each fuel amongst the fossil fuels; moreover, these factors tend to converge towards a common value (obviously, this concerns only the manufacturing industries).
The example of the Netherlands, with the importance given to natural gas this last decade, justifies in some measure this hypothesis:

. at the time of the breakthrough of a new form of energy onto the thermal market, its relative efficiency increases sharply, its substitution for other fuels entailing a marked improvement in the conditions of use;

. this relative efficiency tends to decrease when the new form of energy has reached its optimum point: the new uses for this energy diminish in efficiency. To this must be added, the obsolescence of the equipments put into service at the moment of the initial breakthrough, and perhaps even a certain waste occasioned by the abundance of supply and favorable price conditions;

. inversely, the other forms of energy, here coal and oil, are earmarked for uses where they conserve high efficiency, almost specific uses.

For the United States and Canada, can we then consider that this phase of progressive leveling of the efficiency coefficients had already been achieved at the beginning of the period studied? Or should we suppose that the laws of supply and demand and the abundancy both in coal and in gas and oil are responsible for this equilibrium? The answer to these questions is quite difficult. Anyway, there is not one single answer because certain regulations intervene which limit the possibility of automatic market adjustment, and secondly, there exist between the United States and Canada differences in consumption habits. Particulary, the relationship (consumption of coal, gas, and oil)/(added value of industrial production) is equal on average during the period 1955-73, to 1.6 tce/1000$ for the United States and to 2.4 tce/1000$ for Canada. The preceeding hypothesis is also confirmed by the following fact: by pooling all countries and if the model(9) is applied to this pool, an equalisation is once more noted for the use efficiencies of coal, gas and oil. The relationships Ri/Rj differ often from one country to another, but when we consider the whole range of particular situations, we find once more the convergence which we have already noted for some countries. It is an application of the law of large numbers and later it will be seen that normal distribution may even be introduced.

The evolution of use efficiencies, principally in those countries where there has been large scale substitution of oil and gas for coal, may be represented as follows:

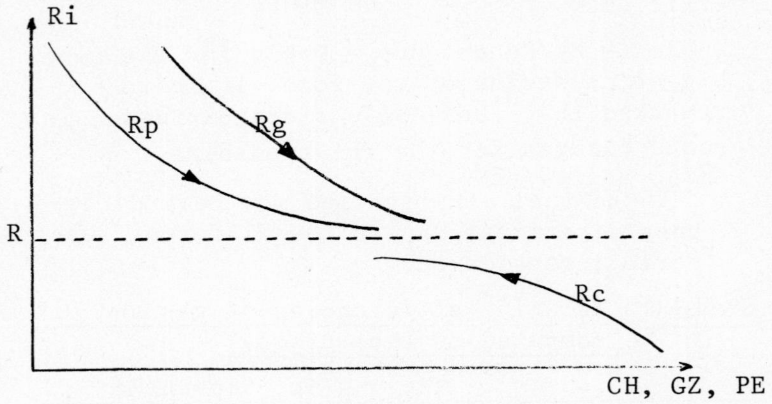

Rc, Rg, Rp evolve progressively towards a common value R. However, it must be noted that:

. this progression may be very slow. It is better here to speak of a tendancy towards an equalisation of use efficiencies;

. the Ri cannot be established in absolute value, only their relationship in pairs.

Finally, if this convergency hypothesis is well founded, one must draw the following conclusion: the calorific equivalents Hi are not only theoritical coefficients and average values, but they have real significance only when convergence is achieved. If Rc=Rg=Rp, then QU is proportional to QE, but the difference between QU and QE is all the greater as the differences between the use efficiencies are significant. This is still the case for a large number of countries and it justifies the need to define the aggregate QU, rather than the aggregate QE, in order to measure the total energy consumption of an economic sector.

2. COMPLEMENTARITY OF ELECTRICITY AND THE OTHER FOSSIL FUELS

The formulation of a general model linking thermal energy TH, mechanical energy MC, and the production value V is more difficult than when thermal energy alone is considered. Here also, there is great diversity between countries, and to the differences in use efficiencies must be added the difference noted when electricity consumption EL is divided by production value V. For the 18 countries, the relationship EL/V varies from 0.084 tce/1000 $ (Denmark 1960) to 0.997 tce/1000 $ (Canada 1960). The study of this relationship completed by that of (FO=CH+GZ+PE)/V leads us to formulate a central hypothesis: the complementarity between fossil fuels (thermal energy) and electricity (mechanical energy). The complementarity is defined as follows: to produce 1 unit of V, it is necessary to introduce, more or less constantly in time, but in variable proportions according to the countries, k units of thermal energy and h units of mechanical energy. In other words, the relationship TH/MC, measured for each country, has not experienced significant variations in time.

This complementarity turns out to be more or less strictly true and, according to intensity, one may propose several models linking V, TH, and MC.

2.1. Production functions

$$\begin{cases} V = k.TH \\ \quad = h.MC \end{cases} \qquad (10)$$

$$\begin{cases} \text{Ln } TH = A'+B'/V \\ \text{Ln } MC = C'+D'/V \end{cases} \qquad (11)$$

$$\begin{cases} \text{Ln } TH = M'+N'/MC \\ \text{Ln } MC = C'+D'/V \end{cases} \qquad (12)$$

Ln:natural logarithm

These three models are characteristic of the functions of production for which there is no possibility of substitution between the inputs. Their economic interpretation is however different. The models (11) and (12) particulary introduce the hypothesis of saturation in the levels of thermal energy consumption and mechanical energy consumption, unlike the model (10) in which the evolution is linear.

This hypothesis of saturation cannot be dismissed:

. the consumption of energy would not remain constantly proportional to the production, neither by sector nor by country, but could follow an asymptotic trend;

. the breakthrough of electricity on the thermal market could also lead to a limit of the levels of fossil fuel consumption.

If MC and UT are defined by

$$\begin{cases} MC = Re.EL \\ TH = Rc.UT \end{cases} \qquad (13)$$

the models (11) and (12) may be immediately transformed into:

$$\begin{cases} \text{Ln } UT=A+B/V \\ \text{Ln } EL=C+D/V \end{cases} \quad (14) \qquad \begin{cases} \text{Ln } UT=M+N/EL \\ \text{Ln } EL=C+D/V \end{cases} \quad (15)$$

in which the coefficients A, B... N, are functions of A', B' ...N', of Rc and Re.

The functions of (14) and (15) are representative of logistic function:

. which passes by an inflection point,

. which has an asymptotic limit when the independent variable increases.

The table 2 shows results obtained for France, the United States, Canada and Sweden. The most precise adjustments are obtained for the relation between EL and V. This example may be used to specify the interpretation and signification of the models (14) and (15).

For the 1955-73 observations, the asymptotic level e^C does not appear in the graph but one may suppose that one is to the right of the inflection point (V=-D/2) and before approaching significatively the asymptotic level.In analogy with innovation diffusion processes, this representative curve may be considered as the envelope curve of a succession of partial curves, each curve corresponding to a given period of time during which the conditions of the use of electricity remain stable, the asymptotic level resulting from a reduction in electricity consumption due to the added value. being hidden by the development of new uses. Through the years, and because of new markets, the asymptotic line is pushed higher and higher. It is therefore interesting to compare e^C (and e^A, e^M) with the values obtained for EL (and UT) in 1973.

2.2. Complementarity rates

The formalizations (14) and (15) are also interesting because they enable us to define certain expressions of the complmentarity rates, and to make comparaisons with the neoclaasical functions of production which, on the contrary, suppose inputs which are completely substitutable. Here, we must limit ourselves to the formulation of essential relations and the presentation of some results.

The complementarity rate Rmt between thermal energy and mechanical energy is defined by the relationship of the derivatives

Rmt = dTH/dMC

From (15), we obtain

$$Rmt = -N/EL.Rc/Re.UT/EL \qquad (-N > 0) \qquad (16)$$

In the same way, with the definitions:

$$Rec=dCH/dEL, \; Reg=dGZ/dEL, \; Rep=dPE/dEL \qquad (17)$$

we obtain, from (8) and (15) the expressions:

$$Rec=-N/EL. \; UT/EL$$
$$Reg=Rec. \; Rc/Rg \qquad Rep=Rec. \; Rc/Rp \qquad (18)$$

. The complementarity rates are proportional to UT/EL (strict complementarity), but inversely proportional to EL: the rate diminishes all the quicker as EL increases rapidly.

. The rates Reg and Rep can be deduced from Rec by multiplying this rate by the use efficiencies of thermal energy: these rates invert therefore the ordered relation established on the Ri.

. The values of Rec, Reg and Rep can be calculated as each of the terms of the expressions (18) is known.

. On the other hand, Rmt cannot be estimated since the relationship Rc/Re is unknown. Between the four complementarity rates, the following relation exists.

$$\frac{Rce.CH+Rge.GZ+Rpe+Pe}{Rc.CH+Rg.GZ+Rp.PE} = \frac{Rtm}{Re} \qquad (19)$$

in which Rce=1/Rec,...,Rtm=1/Rmt. If Rc=Rg=Rp=Re= 1, i.e., if the energy aggregate is QE and not QU and if no distinction is made between TH and MC, then Rec=Reg=Rep=Rmt = 1.

The results of these calculations are to be found in table 3 for France, the United States, Canada and Sweden. These last two countries have been retained because of the importance given to electricity in their industrial consumption. Unlike most countries, electricity had abready been introduced for thermal use and, in the model, that should show in the

TABLE 2 RESULTS OF ESTIMATION FOR MODELS (14) AND (15): DATA 1955-1973

	A	B	M	N	C	D
FRANCE	11.3405 (301.8) R^2=0.97	-23 171 (-25.1) DW=0.56	11.3179 (205.4) R^2=0.94	- 4 240 (-16.7) DW=0.36	9.5347 (577.5) R^2=0.99	- 25 546 (-63.1) DW=1.18
UNITED STATES	13.4270 (209.1) R^2=0.93	-152 733 (-15.3) DW=0.52	13.2390 (202.1) R^2=0.90	-41 431 (-12.2) DW=0.26	12.1487 (185.7) R^2=0.96	-193 532 (-19.0) DW=0.88
CANADA	11.0249 (133.6) R^2=0.89	-9 002 (-11.8) DW=0.42	11.1686 (109.4) R^2=0.87	- 8 768 (-10.9) DW=0.51	9.8839 (188.3) R^2=0.94	-8 163 (-16.8) DW=0.67
SWEDEN	9.2238 (92.3) R^2=0.84	-3 768 (-9.3) DW=0.57	9.1614 (95.6) R^2=0.83	- 1 923 (-9.1) DW=0.73	8.8402 (155.1) R^2=0.96	-4 390 (-19.0) DW=0.45

Figures in parentheses are t-statistics

diminution in the intensity of the complementarity, perhaps even by the appearance of a certain substitution between thermal and mechanical energy.

. The rates Rec are calculated at the mean point of the 1955-73 observations.

. Rec is compared to B/D.($\overline{UT/EL}$) which is the average complementarity rate deduced from (14). Agreement between the two measures of this rate is to be noted.

. It can be seen, above all, that Rec is high for the countries in which TH and MC constitute two separate and complementarity entities of consumption (this is the case for France), and that this rate is all the lower as the substitution between TH and MC has begun (Sweden, Canada).

. Because of the equalities (18), all these conclusions remain valid for Reg and Rep.

TABLE 3 VALUES OF THE RATES Rec

	-N	$\overline{EL}$	$\overline{UT/EL}$	Rec	B/D. ($\overline{UT/EL}$)
FRANCE	4 240	5 471	6.68	5.18	6.06
U.S.	41 431	58 546	4.65	3.29	3.67
CANADA	8 768	8 712	2.88	2.90	3.18
SWEDEN	1 923	2 575	1.71	1.28	1.47

2.3. Complementarity and substitution

The last problem raised by the relationship between thermal and mechanical energy is the passage from complementarity to substitution. The 1955-73 data only enable us to affirm that complementarity exists between these two uses, but we know 1) that electricity producers have envisaged these last few years breaking into the thermal market, 2) that some countries are further advanced in this direction than others. Can mathematical formulation show significantively this modification of electricity's role?

Naturally, we envisage at first introducing the neoclassical production functions, Cobb-Douglas type or better still CES type. It is easy to see that the models (9) and (10) are particular cases of the CES functions (respectively, substitution elasticity infinite and zero). Then, we can show the model (15) may be transformed into a product of two Taylor developments, variables of which are Ln EL and Ln UT, leading to the translog function of n order in which the Cobb-Douglas function is once more a particular case(ref. 2).

Another aspect is worthy of our attention. We can see indeed that for the totality of the countries, the distribution of 18 x 19 data UT/V is approximatively log-normal (mean=0.540, standard deviation=0.339), whereas for the relationship (CH+GZ+PE)/V there are two clearly distinct modes of distribution. This normality may also be verified by the relationship EL/V (-1.206, 0.506). Here one may refer to the work of HOUTHAKKER (3) and JOHANSEN (4) to deduce the "macroeconomic" production function (totality of the 18 countries) from the microeconomic functions (15) estimated for each country separately, these particular functions having zero substitution and the relationship of the inputs at production UT/V and EL/V being distributed log-normally. Its formal execution would doubtless be difficult to carry out completely and would require a simulation. However we know that when the distribution of the input relationships is a Pareto distribution, the macroeconomic production function is a Cobb-Douglas function.

For all these reasons it is worth presenting certain relations obtained with this function, particularly so that they may be compared to the relations (18) and (19). Using the same proceedure as beforehand, the Cobb-Douglas model may be defined:

$$V = k.TH^a.MC^b$$
$$= k'.UT^a.EL^b \quad (20)$$

The marginal substitution rates (negatively noted, so as to avoid confusion with the complementarity rates) may thus be obtained:

$$Rmt = -b/a.Rc/Re.UT/EL \quad (21)$$
$$Rec = -b/a.UT/EL,$$
$$Reg = Rec.Rc/Rg, \; Rep = Rec.Rc/Rp \quad (22)$$

This evidently invites comparison with the relations (16) and (18): the positive relationship -N/EL has been transformed into the constant, negative relationship of the elasticities -b/a. The relation (19) remains valid. Moreover if we take it that the industrial sector acts as a producer seeking to maximise his profits, and if Pc, Pg, Pp and Pe are the prices of the four fuels, a differential calculation leads to the expression:

$$\frac{Pc.CH+Pg.GZ+Pp.PE}{Rc.CH+Rg.GZ+Rp.PE} = \frac{Rtm}{Re}.Pe \quad (23)$$

The passage from (19) to (23) is achieved by means of the three following equalities:

$$Pc=Rce.Pe, \; Pg=Rge.Pe, \; Pp=Rpe.Pe \quad (24)$$

These equalities establish, in the very theoretical case of the maximisation of profits, the proportionality of the prices and of use efficiencies.

Table 4 presents the results of the estimations of the model (20) for the four countries already examined and for the totality of the 18 countries.

The interpretation of these results demands a great deal of caution, despite the relatively high values for R^2 and t. Valid conclusions may only be drawn for comparison 1) between the countries, 2) with regard to the hypothesis of complementarity:

1) the sums a+b are close to the unity for five regressions,
2) the relationships b/a (cf. relation (21)) are all the higher as the substitution between TH and MC seems more rapid (Sweden and Canada).

TABLE 4 AJUSTMENTS OF COBB-DOUGLAS FUNCTION(20)

	Ln k'	a	b	R^2	DW
France*	0.833 (3.3)	0.507 (5.4)	0.476 (5.1)	0.997	1.04
U.S.	0.871 (1.03)	0.582 (3.0)	0.355 (2.3)	0.973	0.99
Canada	-1.294 (-2.7)	0.345 (1.9)	0.778 (3.7)	0.968	0.67
Sweden	1.130 (3.2)	0.104 (0.82)	0.820 (7.0)	0.973	0.60
18 Count.**	0.948 (6.5)	0.636 (9.2)	0.308 (11.6)	0.997	0.87

* 1955-70
** with dummy variable for each country

This caution is due to the very close correlation between the variables UT and EL in (20). However, for the sample of the 18 x 19 observations, the variables UT/V and EL/V are not only log-normal but also independent, and the line of linear regression of UT/V on EL/V is parallel to the axis EL/V. Now the equation of this line may be expressed:

$$\text{Ln } V=\gamma+\alpha.\text{Ln } UT+\beta \text{Ln } EL$$

that is to say (20), in which the coefficients α and β are devoid of statistical sifnificance.

As is often the case for time data, the quality of the estimation of the Cobb-Douglas function is plainly over estimated. The macroeconomic function remains indeterminate.

REFERENCES

(1) Adams, F.G. and Miovic, P., On relative fuel efficiency and the output elasticity of energy consumption in Western Europe, J. Industrial Economics, XVI, 1 (1968).

(2) Girod, J. (1977) La demande d'énergie: méthodes et techniques de modèlisation, 185 p., Centre National de la Recherche Scientifique, Paris.

(3) Houthakker, HS. The Pareto distribution and the Cobb-Douglas production function in activity analysis, The Review of Economic Studies, (1956).

(4) Johansen, L. (1972) Production functions, 274 p., North-Holland, Amsterdam.

Ultimate Limits on the Uses of Energy Analysis

David A. Huettner

University of Oklahoma, Norman, Oklahoma 73019

ABSTRACT

Since the basic objective of energy analysis is to identify the ultimate sustainable limits to human activities, it is appropriate to identify ultimate limits to the usefulness of energy analysis. Human activities can take many forms and any assessment of ultimate limits is not only a forecast of technical progress and resource availability but is also a statement about human values i.e., what activities can or ought to be sustained. Current applications of energy analysis are not merely attempts to find common units of measurement (BTU's rather than dollars) but are attempts to value current human activities in accordance with their believed long run sustainability.

Physical laws (including those of thermodynamics) constrain man's activities but do not dictate values by themself. The problems addressed by energy analysis are of definite social relevance but the science of economics indicates that the objectives of energy analysis are unattainable by the methods employed.

INTRODUCTION

Energy analysts are currently using a variety of concepts to assess energy problems and rank alternatives. These concepts include techniques which might be loosely labeled as net energy analysis, energy analysis, entropy analysis and even economic analysis.[1] While there are various gradations within each of these categories, it is clear that a fundamental conceptual difference exists between analysts using economic measures, such as market prices to value inputs and outputs, and analysts using physical measures, such as energy content.

This cleavage is undoubtedly clearest to economists since they are aware that the direct allocation of resources (i.e., quantities of R & D manpower, etc.) by energy content principles has the same economic effect as an indirect allocation achieved by setting prices according to energy content principles.[2] Indeed, Huettner (7) has shown that resource allocation based on net energy analysis is an energy theory of value and there is little doubt that Odum and possibly Hannon and Gilliland subscribe to an energy theory of value.

---

1. The term net energy analysis is generally used when the energy producing activity or industry is under study. Slesser (16) has defined energy analysis as an assessment of the "energy resource consequences of man's activities." Entropy analysts would simply substitute an entropy measure for BTU's and economic analysts would simply apply economic principles and generally use dollar measures.

2. The duality of resource allocation effects achieved by either price or quantity mechanisms was the subject of much of Leontief's Nobel prize winning work.

Furthermore, it is not true that, given perfect markets, net energy analysis (or energy analysis) and economic analysis "can do the same thing" or "would yield the same perception of the future" as Odum (13) and Slesser (16) have respectively maintained. Energy analysts employing economic principles will generally reach different conclusions from energy analysts using non-economic principles and these differences will remain even if all markets are free or perfect. Claims that one method cuts through confusion or forecasts impending change faster or has more normative content could probably not be proven by proponents of any method since every discipline is rife with examples of poor research to be exploited by the opposition.

Under these conditions, I believe it makes more sense to examine the basic assumptions and logic of a discipline in an effort to determine where it will take us if we let it guide our decisions. It is on this basis that I argue that energy analysis guided by non-economic principles constitutes an energy theory of value. While there is general agreement that the Odum "school" embraces an energy theory of value, Slesser (16) maintains that his "school" does not. Yet if one "values" inputs and outputs in energy terms, ranks or compares alternatives in energy terms, and then acts on this information, I believe that an old adage applies, "when in Italy, all roads lead to Rome." While it is clear that the energy analyses of Slesser, Odum and others guided by non-economic principles do differ, the concern of this study is not with their differences but with their similarities. For this reason, this study will lump all forms of non-economic energy analysis together and simply refer to them as energy analysis.

The basic objective of this paper is to assess whether energy analysis based on non-economic principles is a correct or even a useful way of determining values. This assessment will begin with a review of the basic assumptions and objectives of energy analysis. The next section will review some of the theoretical and methodological issues associated with energy analysis. The final section will review some of the claims of energy analysts and present some concluding remarks.

ASSUMPTIONS AND OBJECTIVES OF ENERGY ANALYSIS

Assumptions

As a former net energy analyst has noted (10), "Net energy analysis began with two reasonable suspicions and an apparently simple method for testing them." The first suspicion was that, as mankind turned to lower quality energy sources, gross energy output would increase but net energy, the amount available to final consumers, would fall and eventually approach zero. The second suspicion was that traditional disciplines, because of narrow viewpoints or the use of prices to measure energy flows, might overlook this ominous possibility. Energy analysis offered, as an alternative, that all energy flows directly or indirectly supporting an energy technology or system be measured in physical units such as BTUs.

The exclusive emphasis on energy and the energy content of inputs in energy analysis rests on the concept of energy as the ultimate limiting factor, since substitutes for other inputs can always be synthesized from it. Energy may be divided into available energy (enthalpy) or unavailable energy (entropy). The second law of thermodynamics tells us that the entropy of a closed system increases continuously and irrevocably toward a maximum. In addition, Gilliland (4) has noted that (i) energy is the only commodity for which a substitute cannot be found, (ii) potential energy is required to run every type of system or production process, and (iii) energy cannot be recycled without violating the second law of thermodynamics.

As society uses up its higher-grade energy resources it will require more and more energy inputs (a larger energy subsidy) to produce a given amount of energy. While gross energy production may increase rapidly over the next few centuries as we consume our remaining fossil fuel resources, net energy will certainly increase less rapidly and may eventually begin to decline, particularly if one views earth, moon and sun as a closed system. Regardless of the exact scenario assumed, however, energy is clearly regarded as the ultimate limiting factor particularly since substitutes for other inputs can always be synthesized from it.

Objectives

The basic objective of energy analysis is to identify the ultimate sustainable limits to human activities (2) and (15). Since energy is viewed as the ultimate limiting factor, pursuit of this objective logically requires that all inputs be valued solely in terms of their energy content. This is particularly true of environmental inputs or services which many energy analysts feel are undervalued by traditional economic techniques.

In fact, the correct valuation of environmental services is essential if ultimate sustainable limits are to be properly assessed. This fact and the importance attached to environmental services by many energy analysts suggest that proper valuation of environmental services be regarded as a sub-objective of energy analysis.

THEORETICAL AND METHODOLOGICAL ISSUES

Theoretical Issues

Turning first to theoretical difficulties, it has been shown (7) that net energy analysis is an energy theory of value. Inputs such as labor, raw materials, machinery and so on are valued according to their direct and indirect energy content alone. More than just appropriate selection of the numeraire is implied, however; prices are formulated as if energy were the only relevant resource constraint, and the relative scarcity of nonenergy inputs becomes a factor only if it leads to a change in the energy content of these inputs. In essence, all nonenergy resources are viewed as transformed energy, and in this one commodity world all derivative products are priced according to their energy content.

In essence, energy analysis assigns values based solely on supply considerations while totally excluding demand considerations. Georgescu-Roegen (3), however, has argued that low entropy (energy supply) is a necessary but not sufficient condition for assigning value. Sufficiency requires that one account for the enjoyment of life (demand factors). Georgescu-Roegen's statement that entropy is the true taproot of economic scarcity means that there are ultimate limits on supply given the state of knowledge assumed but does not imply that value is determined by supply alone. Furthermore, it does not even imply that energy scarcity is the only scarcity that should concern us since we do not live in a one commodity world where perfect substitutes for any given commodity can be synthesized from energy.

At best, net energy analysis can identify only a continuum of possible energy values (a supply curve) and not some unique value or ultimate limit for human activity. Gilliland (5) recognized this problem and argued that "over the long term, low entropy may provide the basis for defining the boundaries of utility and demand." Yet the ultimate, sustainable energy supply curve would undoubtedly stretch over a wide range of energy values (prices) and output levels. Furthermore, the position of this supply curve at any future point would depend on technological progress, hence the range of ultimately sustainable energy output levels and values is expanded enormously.

In fact, given our meager abilities to measure or understand past technological change, let alone forecast future changes (6), it appears safe to conclude that net energy analysis will offer no accurate prediction of the ultimate sustainable boundaries of energy demand (or value).[3] Once this is recognized, it is clear that current energy policy cannot be guided by pronouncements as to ultimate limits. As Leach (10) has noted, "The future is opaque, a dark mirror, and no less to energy analysis than to the rest of mankind. Ultimate limits can wait on more urgent and closer concerns."

A second theoretical issue identified by Yokell (17) is whether an energy

---

3. Note that economic analysis would also fail to define these boundaries.

industry should produce net energy or (by implication) whether one should rank alternatives solely on energy considerations. Since the objective of production is to produce services or products with characteristics useful to humans, society should maximize economic value and not net energy. In addition, society should rank alternatives based on economic value and not energy content considerations.

Economists recognize that renewable and non-renewable resources cannot be treated the same due to the greater and technologically based uncertainties associated with future availability of non-renewable resources.[4] This does not mean, however, that energy or any other non-renewable resource should be selected for conservation and used for ranking of alternatives. Again, it is the value of that resource vis-a-vis other economic values that should be considered i.e., is the current use of this non-renewable resource of greater value to society than postponing its use. Questions of this type (i.e., intergenerational equity) are never easily resolved but energy analysis masks most of the important issues in this regard instead of dealing with them.

Methodological Issues

Given that energy analysis cannot achieve its basic objective of defining ultimate limits to human activity, one can still inquire whether the value weights it produces contain other information useful to decision-makers. Again, however, there appear to be both methodological and theoretical reasons to question the information content of net energy calculations.

Turning first to methodological issues, Leach (10) has identified several of concern.[5] The most important of these is the extent to which external inputs should be counted in net energy calculations, i.e., where should the boundary between supply and demand be drawn? At one extreme, the predominate opinion is to draw the boundary between the energy supply system or facility and the rest of GNP as conventionally defined. For example, the energy used to build and run gasoline stations or new towns for oil shale workers produces "goods" within GNP and is therefore not to be counted as a cost on the energy supply sector. At the other extreme, the prevailing opinion is to capture all possible direct and indirect effects including many remote multiplier and "knock on" effects (12). As noted in (10), however, this would include "the additional energy associated with higher living standards for well paid Alaskan oil workers, the energy to provide all social facilities and infrastructure for new energy developments, and all hidden subsidies provided by natural ecosystem changes."

Clearly, the latter approach, favored by Odum and others, will generate much larger energy subsidies and hence lower net energy calculations than does the GNP approach. Leach concludes (10) that until this boundary problem is settled, "net energy analysis will be arbitrarily inconsistent, uncertain and show large variations . . . (making) . . . Public Law 93-577 virtually unworkable . . . and (suggesting) that net energy analysis has no magic answers to some dilemmas."

A second methodological problem identified by Leach (10) and amplified by Yokell (17) is the problem of valuing physically disparate types of energy. Some energy resources are purchased for their enthalpies hence BTU measures are most relevant yet even this measure omits the fact that a kilowatt hour produced at one place and time of day is not equivalent to another produced elsewhere at a different (or the same) time of day. In addition, some energy resources are purchased for their ability to do electrical or chemical work and for these it is the Gibbs Free

4. For example, see Lee (11).

5. Problems created by uncertainty and joint production are not discussed here. The reader is referred to Leach (10) for treatment of these issues.

Energy (not BTU's) which is relevant. Energy analysis has its apples and oranges problems and one can see how dollar values can circumvent many of these problems.

One can, however, assume that boundary and other problems will be solved[6] and still criticize the resultant calculations as irrelevant for two reasons. First, as has been shown above, these calculations have no information content as to the ultimate limits to human activity (long run policy). Second, if short run rather than long run policy is of concern, what use are these calculations? Should decision makers pay attention to them and in effect allocate resources according to energy content value weights?

As Huettner (7 and 8) has shown, even if we could synethesize perfect substitutes for any input from energy, efficient allocation of resources would require deviations from energy content pricing. Furthermore, an economy with decisions guided by energy analysis would be for different from one guided by economic analysis in terms of level of GNP and types of outputs produced. Indeed, a recent paper by Berry et. al. (1) illustrates the reduction in output accompanying the shift from an economic optimum to a thermodynamic optimum (holding costs constant).

Valuation of Environmental Services

The above paragraphs have argued that energy value weights contain virtually no information of use to decision makers. There still remains, however, the question of whether it can improve the valuation of services provided mankind by the environment. At one end of the spectrum are some economists who argue (incorrectly) that no energy or economic value should be placed on nature's services as long as quantity supplied exceeds quantity demanded.[7]

At the other end of the spectrum are some[8] who argue that the value of environmental services can be measured, in order of increasing preference, as:

1) the energy value of the products and services provided by the environment;
2) the solar energy used by the ecosystem in providing these products and services;
3) the energy that would be required to provide these services by alternate means.

A correct economic analysis would start with the fact that man's demand for services provided by the environment will grow through time and ultimately exceed nature's supply. Beyond that point in time, the value of these services is the cost of providing them by alternate means. Discounting this future stream of costs back to the present yields the present value of future environmental services.

Yet the cost of providing environmental services in the future (and the timing of when they must be provided by man rather than nature) is dependent on how well man protects his environment today. In effect, the ecosystem may be regarded as a complicated machine that provides mankind services through time. By keeping that machine properly maintained in

6. Some of the boundary problem may be solved by eliminating much of the double counting implicit in the "Odum" approach. For example, much of the infrastructure (such as schools) for a new energy development would have been needed anyway assuming that children would require schooling wherever they lived and that the new schools associated with the energy development are offset by school construction postponements in other cities.

7. For example, see (9).

8. For example, see (4) and (14).

the present, mankind can avoid higher maintenance costs in the future. Today's environmental services should not be valued at zero simply because today's supply provided by nature exceeds today's demand. Rather, today's services should be valued at zero only if it is the optimal solution, i.e., it minimizes the discounted present value of the costs of providing a flow of environmental services through time.[9]

While the above argument establishes a rationale for valuing today's environmental services at more than zero (even if the current supply provided by nature exceeds current demand), it does not provide guidelines for placing actual values on these services. Given the difficulties of forecasting man's demand preferences and supply capabilities far into the future, one must conclude that both economic analysis and energy analysis cannot achieve reliable answers. Yet it should also be clear that, even if accurate forecasts were available, energy analysis would not provide mankind the answers it seeks.

CONCLUDING REMARKS

The above sections have argued that energy analysis cannot achieve its primary objective of defining ultimate, sustainable limits to man's activities. In addition, energy value weights were shown to be inappropriate for short run decision making such as resource allocation. Finally, a review of various methodological problems indicated that the actual value weights calculated by energy analysis are arbitrarily inconsistent and highly variable.

Yet proponents of energy analysis frequently argue that economic values are arbitrarily inconsistent and highly variable particularly when compared to energy values. Indeed, Slesser (16) has argued that future energy requirements are more predictable than prices. In part, this notion is due to the mistaken belief that the energy required to produce a good is a purely physical phenomenon and not an economic phenomenon.[10] Yet energy content is clearly affected by the choice of technology or plant location. Prices influences these choices and therefore energy content. Indeed, the isoquant analysis of Berry et. al. (1) shows that cost minimization involves an economically optimal amount of thermodynamic waste.

The greater stability or predictability of energy content values relative to economic values is also questionable in many specific cases (i.e., how accurately would anyone in 1955 have forecasted the energy required to run a computer in 1977). Furthermore, physical laws (including those of thermodynamics) constrain man's activities but do not dictate values by themself. For example, many elements occur as minute fractions of the earth's crust hence are rare yet scarcity (value) can only be determined relative to man's desires for these elements and his ability to extract them. Since these desires and abilities change through time, society's values also change hence so should its cost-benefit ratios or other analytical measures of the desirability of its options. Changing economic values are merely a reflection of these changes and not an inherent defect.

The objective of this paper was to assess whether energy analysis is a correct or even a useful way of determining values. Economics tells us that values are determined by the interaction of supply and demand i.e., market values (perhaps corrected for market imperfections). From this perspective, energy analysis is woefully incomplete since it ignores many supply factors and all demand factors.

The above criticisms are not meant to downgrade the importance of questions regarding ultimate limits nor the vital insights of non-economists. On the contrary, it is clear that the insights and expertise of ecologists,

---

9. One would, of course, have to project future demand for and supply of these services to obtain these future cost estimates.

10. For example, see Slesser (16) and Gilliland (4).

environmentalists and others must be integrated into traditional economic analysis if appropriate values for environmental services and energy resources are to be obtained. The basic purpose of this paper is to indicate that energy analysis is, for various theoretical and methodological reasons, an inappropriate framework for addressing the questions it seeks to answer.

REFERENCES

1. Berry, R.S., Heal, G. and Salamon, P., "On a Relation Between Economic and Thermodynamic Options," Working Paper, University of Chicago, 1977.

2. Chapman, P.F. (1975), Fuel's Paradise: Energy Options for Britain, Penguin, London.

3. Georgescu-Roegen, N., South. Econ. J., 41, 347, (1975).

4. Gilliland, M.W., Science, 189, 1051, (1975).

5. __________, Science, 192, 12, (1976).

6. Gold, B. (1975), Technological Change: Economics, Management and Environment, Pergamon Press, Oxford.

7. Huettner, D.A., Science, 192, 101, (1976).

8. __________, Science, 196, 261, (1977).

9. Langham, M.R. and McPherson, W.W., Science, 192, 8, (1976).

10. Leach, G.L., Energy Policy, 3, 332, (1975).

11. Lee, D.R., Journal of Energy and Development, 1, 291, (1976).

12. Odum, H.T., Ambio, 2, 220, (1973).

13. __________, Science, 196, 261, (1977).

14. Reichle, D.E., Science, 192, 12, (1976).

15. Slesser, M., paper presented at 20th Pugwash Symposium, Arc-et-Senans, France, July (1974).

16. __________, Science, 196, 259, (1977).

17. Yokell, M.D., "Physical Efficiency and Economic Efficiency," Paper presented to the Energy and Resources Seminar, University of California, Berkeley, November 10, 1976.

THE USE AND ABUSE OF ENERGY INTENSITIES

John L.R. Proops

Department of Economics, University of Keele, Staffs., U.K.

ABSTRACT

Energy intensities refer to the energy used by an economy in the production of goods and services. Input-Output Analysis is used to derive two related types of energy intensity and some of their correct and incorrect applications are discussed.

INTRODUCTION

One aspect of energy analysis is the study of the way energy is used by economies in the production of goods and services. This study can be thought of as being directed at three areas:

A. The relationship between the overall amount of industrial energy used by an economy and its GDP.

B. The energy intensiveness of the various manufacturing processes in an economy.

C. The direct and indirect energy costs of producing the various goods and services in an economy.

These three areas are inter-dependent, and their relationship can be seen if we use the standard notation of Input-Output Analysis, introduced by Leontieff (1). This notation allows the inter-relatedness of economic systems to be expressed and analysed, at least to a linear approximation.

APPLICATION OF INPUT-OUTPUT TECHNIQUES

If we consider a simple economy with n producing sectors (i.e. industry types), each selling to both final consumers (*Final Demand*) and other producing sectors (*Intermediate Demand*), then if:

x_i ≡ total output by sector i (*Total Demand*)

y_i ≡ output by sector i going to final consumers (*Final Demand*)

x_{ij} ≡ output by sector i going to sector j (*Intermediate Demand*)

then:

$$x_i = \sum_j x_{ij} + y_i \qquad (1)$$

The units used are arbitrary, and as long as the same units are used *within* sectors, the output of each sector may be expressed in any suitable physical units (litres, tonnes) or financial units ($,£).

If we make the *linear* assumption that doubling the output of a sector requires a corresponding doubling of its inputs, then we can write:

$$x_{ij} = a_{ij}\, x_j \qquad (2)$$

The a_{ij} are known as the *Technical Coefficients*.

Using eqns. (1) and (2) we see:

$$x_i = \sum_j a_{ij}\, x_j + y_i$$

or in vector notation:

$$\underline{x} = \underline{A}\,\underline{x} + \underline{y} \qquad (3)$$

Solving for $\underline{x}$ in (3) gives:

$$\underline{x} = (\underline{I} - \underline{A})^{-1}\underline{y} \qquad (4)$$

Let us now introduce energy into the analysis. Suppose that sector i uses *directly* in its manufacturing processes, in one year, an amount of energy e_i. Then the total energy dissipated by industrial production in the economy in that year is $\sum_i e_i = E_{ind}$. Also, the GDP for this simple economy is just equal to the total quantity of goods delivered to final consumers. If, for simplicity, the y_i are all in \$, then the GDP is given by $\sum_i y_i$. So Area A is about the relationship between $\sum_i e_i$ and $\sum_i y_i$.

Area B is about the energy used by an industrial sector in the production of its total output. i.e. the relationship between e_i and x_i. In particular, we can define an *Energy Intensity* for a particular sector by:

$$c_i = e_i/x_i$$

i.e. c_i = (Total energy used by sector i)/(\$ worth of total output). We should note that this energy intensity imputes all industrial energy use to output to Total Demand. i.e.:

$$\sum_i c_i\ x_i = E_{ind} \quad (\underline{c}\ \underline{x} = E_{ind}) \tag{5}$$

Area C is the trickiest to deal with, and is best approached a step at a time.

Multiplying eqn. (4) throughout on the left by $\underline{c}$ we get:

$$\underline{c}\ \underline{x} = \underline{c}\ (\underline{I} - \underline{A})^{-1}\underline{y}$$

But from eqn. (5):

$$\underline{c}\ \underline{x} = E_{ind}$$

So following Reardon (2), we can define:

$$\underline{c}^* = \underline{c}\ (\underline{I} - \underline{A})^{-1} \tag{6}$$

so that $\underline{c}^*\ \underline{y} = E_{ind}$

i.e. $\underline{c}^*$ imputes all energy use to Final Demand. So we have converted an energy intensity vector referring to Total Demand, $\underline{c}$, to one referring to Final Demand, $\underline{c}^*$.

Now in general $y_i \leq x_i$, so we would expect (and it can be rigorously proved that) $c_i^* \geq c_i$. We can understand this inequality if we think of our economy in terms of a productive system, with a system boundary. If the system boundary is around only one manufacturing sector, sector i, which directly uses energy e_i in the production of total output x_i, then the energy intensity of the output is $c_i = e_i/x_i$. As the system boundary moves out from that manufacturing sector, to include other related industrial activities (i.e. transport, retailing, etc.), the output of good i from the system will decrease, due to the use of the good in other production processes within the system, while the effective energy input increases due to the secondary, indirect, inputs of energy being taken into account. If the system is expanded until it includes all productive activity within the economy, the energy intensity of good i becomes c_i^*.

Embodied Energy

Implicit in the above inclusion of "indirect" energy inputs to production is the notion of *embodied* energy. This allows us to charge for any unit (\$) of good or service purchased in energy terms, the energy charged being equal to the direct plus the indirect energy input to the economy, needed to produce that good. This energy is not physically embodied in the good, most of it simply being dissipated in the course of the irreversible productive processes. The idea of embodied energy simply gives us a consistent way of accounting for the energy used by an economy. We can use the notion of embodied energy to define the energy intensities of goods produced by that economy, by using an *Energy Balance Equation*. For any productive process this equation reads:

Direct Energy In + Embodied Energy In = Embodied Energy Out.

This is the equation used by Bullard and Herendeen (3) and Casper et al (4). If we define an energy intensity for good j as c_j^b, then the energy embodied in good j is the value of j times its energy intensity. So using Input-Output notation we can rewrite the energy balance equation for sector i as:

$$e_i + \sum_j x_{ji}\ c_j^b = x_i\ c_i^b$$

Dividing through by x_i gives:

$$c_i + \sum_j a_{ji} c_j^b = c_i^b$$

i.e.:

$$\underline{c} + \underline{c}^b \underline{A} = \underline{c}^b$$

i.e.:

$$\underline{c}^b = \underline{c} (\underline{I} - \underline{A})^{-1} = \underline{c}^*$$

So the principle of embodied energy also leads to energy intensities which impute all industrial energy use to final demand.

Process Analysis

Process analysis, which uses an engineering approach to the economic "network", is also, at root, an Input-Output technique. Here only physical units are used, and an ideal study of this type would only involve energy use. i.e. instead of using the fundamental equation $\sum_j x_{ij} + y_i = x_i$, we would use $\sum_j e_{ij} + e_i^y = e_i$, where the e_{ij} and e_i^y represent the energy directly used for the production of goods for Intermediate Demand and Final Demand respectively. It has been shown by Proops (5) that such an approach gives energy intensities which, when certain restrictions are imposed upon the energy-financial relationships in the economy, also reduce to $\underline{c}^*$.

So we have seen that there are two basic ways of charging for the energy used in an economy. The first way is to charge to output to Total Demand, which accounts only for the direct energy use. The second way is to charge to output to Final Demand, which accounts for both the direct and the indirect energy use. We should note that neither method involves double counting of energy use, as we have shown that:

$$\underline{c}\ \underline{x} = E_{ind}$$

and

$$\underline{c}^*\ \underline{y} = E_{ind}$$

Further, we can say that the use of these energy intensities must be such that double counting does *not* occur.

Before we proceed, a note on terminology. The term "direct energy input" here refers strictly to the *fuels* used by an industrial sector in producing its output. In the literature one occasionally sees the term "direct energy input" referring to the use of energy costly goods in production processes (e.g. fertilisers in agriculture). In our terminology, though, this is an indirect energy input.

Let us now turn our attention to how energy intensities like $\underline{c}$ and $\underline{c}^*$ can be used to give us some insights into the role of energy in the workings of an economy. We shall also note some methods of study which have been attempted with $\underline{c}$ and $\underline{c}^*$, which are not legitimate applications.

THINGS WE CAN DO WITH $\underline{c}$ AND $\underline{c}^*$

(1) We can attempt predictions of future energy use by an economy (cf. Area A) by simply estimating the size and structure of final demand ($\underline{y}$) and assuming that technologies remain constant, when we can use $\underline{c}^*\ \underline{y} = E_{ind}$. This approach has been used by Bezdek and Hannon (6). If we can estimate technological change, both of energy use within sectors ($\underline{c}$) and also the nature of relations between sectors ($\underline{A}$), then we can predict the change in energy demand, ΔE_{ind}. by differentiating $\underline{c}\ (\underline{I} - \underline{A})^{-1}\ \underline{y} = E_{ind}$.

i.e.:

$$\Delta\underline{c}(\underline{I} - \underline{A})^{-1}\underline{y} + \underline{c}\Delta(\underline{I} - \underline{A})^{-1}\underline{y} + \underline{c}(\underline{I} - \underline{A})^{-1}\Delta\underline{y} = \Delta E_{ind}$$

Also, if we use energy intensities which refer to individual fuels, then the fuel "mix" can also be predicted. This technique is more fully discussed elsewhere by Proops (7).

(2) Most Input-Output studies are static, capital accumulation being included in final demand. So we can use $\underline{c}^*$ to find the energy cost of "extending" the economy by investing in new plant. So we can compare the energy costs of new nuclear reactors, as in Chapman (8), and, say solar energy systems. We should note that using $\underline{c}$ instead of $\underline{c}^*$ here will result in an *underestimate* of the energy cost of such capital projects.

(3) Exports are an element of final demand, so the energy cost of exports can be calculated, using $\underline{c}^*$. However, this is not true of imports, which count as inputs to the economy and therefore cannot be dealt with by this method. This has been performed for the U.S. by Fieleke (9).

(4) We can calculate the direct plus the indirect energy used to produce output of a certain good to final demand. This is simply $c_i^* \, y_i$.

If we are interested in the indirect energy only, this is also simple to calculate. It is just $(c_i^* - c_i)y_i$.

THINGS WE CANNOT DO WITH $\underline{c}$ AND $\underline{c}^*$

(1) We cannot use $\underline{c}$ to calculate the indirect energy input to a manufacturing sector. We might be tempted to write:

Indirect energy input to sector i $= \sum_j c_j \, x_{ji}$

But: Direct energy input to sector i $= e_i$

so: Direct plus Indirect energy input to sector i $= e_i + \sum_j c_j \, x_{ji}$

If this accounting works for one sector in the economy, then it must work for all the sectors, so for the whole economy:

Direct plus Indirect energy input

$$= \sum_i e_i + \sum_i \sum_j c_j \, x_{ji}$$

$$= E_{ind} + \sum_i \sum_j c_j \, x_{ji}$$

$$> E_{ind}$$

So we see that this calculation leads to the double counting of energy use. Even less can one use $\underline{c}^*$ instead of $\underline{c}$ to perform such a calculation, though this is sometimes seen in the literature.

This injunction against adding together direct and indirect energy use may seem to contradict the above use of the Energy Balance Equation. However, we should recall that this equation was used to define energy intensities, which refer to Final Demand, rather than to estimate the direct plus indirect energy use by a sector of the economy.

(2) We cannot use $\underline{c}$ or $\underline{c}^*$ to extend data to include a previously non-included sector. i.e. Suppose we have energy intensities for industrial products ($\underline{c}$ or $\underline{c}^*$), calculated without the inclusion of, say, the agricultural sector in the Input-Output data. Now if we wish to know the energy intensity of agricultural production we might be tempted to solve the energy balance equation for the agricultural sector, independent of the other sectors. So if the agricultural sector is sector a, we might try to solve the equation below for c_a^*:

$$e_a + \sum_j x_{ja} \, c_j^* = x_a \, c_a^*$$

However, the intensities in $\underline{c}^*$ are the solutions of a set of simultaneous equations, and c_a^* thus defined will be a good approximation to its true value only when the agricultural sector is loosely connected to the rest of the economy, both in terms of the outputs from *and* the inputs to the agricultural sector. We can only find c_a^* accurately by reformulating the Input-Output information in the light of the data from the new sector, and thence re-establishing $\underline{c}$ and $\underline{A}$.

To sum up, energy intensities are a potentially powerful tool for the anal-analysis and prediction of the energy use of economies. But enthusiasm for their application must be tempered with caution, so that double counting of energy inputs, and its consequent confusions, is avoided.

REFERENCES

(1) W. Leontieff, (1951) The Structure of the American Economy, 1919-1939, Oxford University Press, London.

(2) W.A. Reardon, (1973) Input-output analysis of U.S. energy consumption, in Energy Modelling (Ed.) M.F. Searl, Resources for the Future, Washington.

(3) C.W. Bullard and R.A. Herendeen, The energy cost of goods and services, Energy Policy 1, 268 (1975)

(4) D.A. Casper, P.F. Chapman and N.D. Mortimer, Energy analysis

of the 'Report of the Census of Production, 1968', Research Report ERG006, Open University, Milton Keynes, U.K. (1974).

(5) J.L.R. Proops, Input-output analysis and energy intensities, Applied Mathematical Modelling 1, 181 (1977).

(6) R. Bezdek and B. Hannon, Energy, manpower and the highway trust fund, Science 185, 669 (1974).

(7) J.L.R. Proops, The output elasticity of energy consumption: an input-output approach (to appear).

(8) P.F. Chapman, Energy analysis of nuclear power stations, Energy Policy 1, 285 (1975).

(9) N.W. Fieleke, The energy trade: the United States in deficit, New England Economic Review May/June, 25 (1975).

A FORECAST OF ENERGY CONSUMPTION AND PEAK LOAD IN THE CONSOLIDATED EDISON RESIDENTIAL SECTOR

Robert K. Weatherwax* and Robert H. Williams

Center for Environmental Studies
Princeton University, Princeton, New Jersey

ABSTRACT

A model of the Consolidated Edison Franchise District residential sector has been formulated and used to forecast yearly electric power usage and seasonally varying peak load, and to evaluate the effects of conservation and load management strategies. The residential sector is disaggregated into 14 end-use categories, and annual usage and seasonally varying demand patterns for each end-use are estimated. Econometric formulas are used to predict end-use saturations. The average unit energy consumption for each end-use for each model year is determined, and an appliance mortality model is used to trace cohort populations. End-use seasonal demand patterns are derived from load studies.

The model predicts significant increase in saturation for nearly every end-use category while total usage grows 1.3%/yr for the period 1975-2000. The peak load grows 2%/yr until 1980 after which it increases .5%/yr. Throughout the period 1970-2000, refrigeration and lighting are the two largest electricity users while room airconditioning and refrigeration are the largest contributors to peak load.

Based upon local weather patterns, it is estimated that air conditioning and space heating usage are respectively 240% and 75% as expensive per kWh as the average of all end-uses. It is concluded that time-of-day rates will not reduce the peak load in the Con. Ed. district. Sensitivity analyses predicted that timely implementation of the current FEA appliance efficiency standards reduce energy usage in 2000 by 25% while higher prices for electricity reduce usage by 15%. By assuming tighter energy efficiency standards after 1980 usage growth can be arrested and the peak load reduced.

*Now with Science Applications, Inc.
2680 Hanover St., Palo Alto, CA.

INTRODUCTION

The residential sector of the Consolidated Edison Electric Franchise District has been analyzed to forecast energy use and peak load, and to identify viable load management and conservation options. (Ref. 1) This work was part of a larger study funded by the Federal Energy Administration and administrated by the New York Public Service Commission that was done in support of a peak load pricing experiment being conducted by Con. Ed. with residential and small commercial customers.

METHODOLOGY

We have developed a computer model to predict the yearly residential usage and peak load in the Con. Ed. district for the period 1970-2000. This modeling is based upon disaggregated projections of usage and usage patterns for 14 end-uses. It utilizes i) econometric formulas to predict appliance saturations for electrical and competing fuels as a function of observed saturation trends modified by the prices of competing fuels and per capita income, ii) a normally distributed appliance mortality model to compute appliance cohort populations by end-use as a function of lifetime and saturation change wherein appliance sales and utility survey data are used to define initial cohort populations, and industry lifetime estimates are used to estimate junkage rates; iii) actual or projected annual energy usage for the average member of each end-use cohort, based upon model year energy usage characteristics as determined by actual sales data, local meteorological conditions and estimated future energy consumption characteristics; and iv) experimentally derived load curves to develop seasonal and hourly variations in end-use load demand patterns. Exogenous variables include price of electricity and competing fuels, rate of household growth, per capita income and end-use energy consumption characteristics. Demand curves for room and central air-conditioning and space heating are determined as a function of weather conditions. Demand for refrigeration, freezers, lighting, waterheating, clotheswashers and dryers, dishwashers, TV households, cooking, and miscellaneous appliances are seasonally determined.

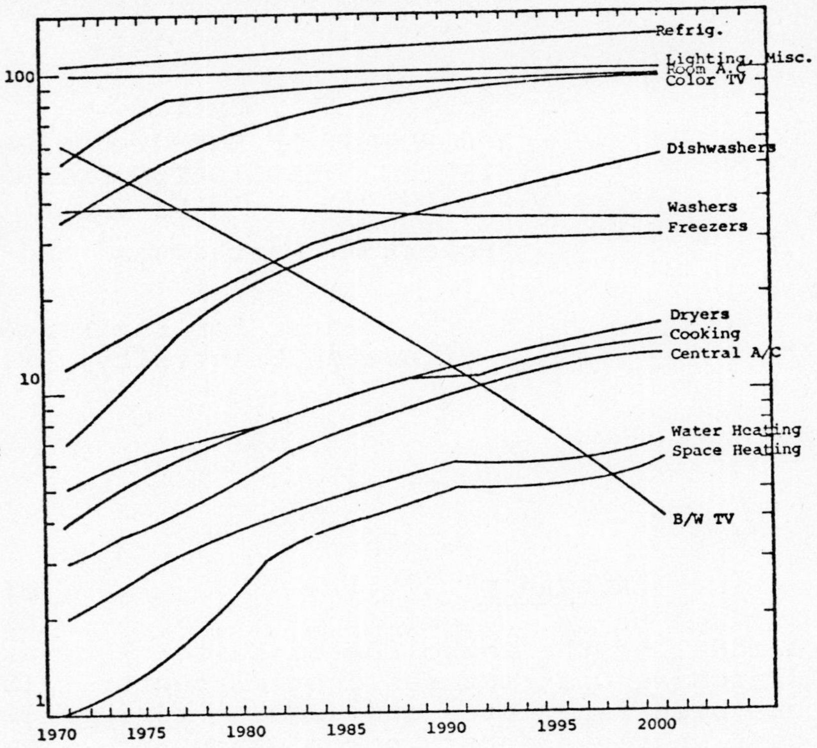

Note: Saturation is given here as the number of appliances per hundred households, except in the case of TV. Color TV saturation is the percentage of households with at least one color TV, while B/W TV saturation is the percentage of households with no color TV's and at least on B/W TV.

Fig. 1 Reference projections of end-use saturation for electrical appliances (%)

RESULTS

Reference electrical appliance saturation and usage projections for the period 1970-2000 are presented in Figures 1 and 2. These projections are based upon an average 2.2%/yr increase in the real cost of electricity between 1975 and 2000, an average growth rate of approximately 0.2%/yr for households in this period, and a 2.5%/yr rate of increase in real per capita income. Also, these projections are based on the assumption that the FEA appliance efficiency standards will be met in stages between 1976 and 1980.

Figure 1 shows that we envision significant increases in saturation in all electrical appliances. We project that there will be over 20% more refrigerators than households by the year 2000. Freezer saturation is arbitrarily limited to 30% (i.e., the approximate percentage of single and double family dwellings). Lighting and miscellaneous appliances are assumed to have a 100% saturation throughout the period.

We estimate that total sector usage will grow at an average rate of 1.3%/yr in the absence of behavior pattern modification or any conservation not directly resulting from the price elasticity of ownership of individual electrical appliances.

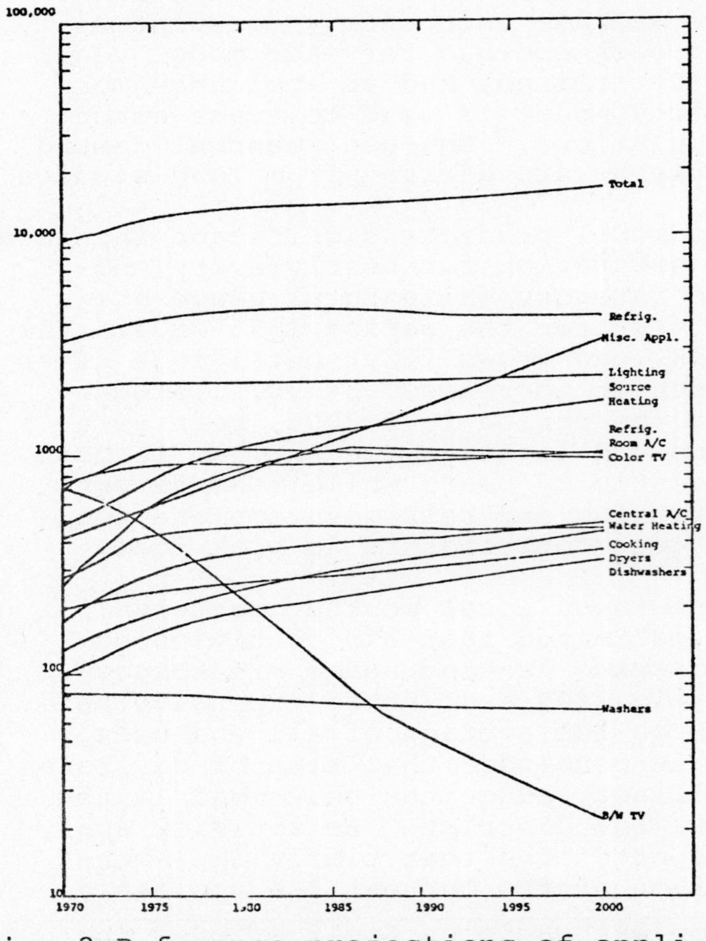

Fig. 2 Reference projections of appliance usage (10^6 kWh)

Our results indicate rising usage of each end-use except black and white TV households. The largest increase in usage is projected for miscellaneous appliances which encompass all non-specified electrical residential end-uses. We arbitrarily assume this end-use is growing at a rate of 7%/yr. throughout the period.

Figure 3 presents the reference projection for growth in residential sector peak load for the years 1970, 1975, 1980, 1985, and 2000 (identical hot summer weather assumed for each year). As can be seen from the figure the peak is projected to shift from 9:00 PM in 1970 to 7:00 PM in 1975 and to grow to nearly 3800 MW by 2000. The predicted growth rate for the peak is 0.6%/yr. from 1975 to 2000, compared to approximately 6%/yr. from 1970 to 1975.

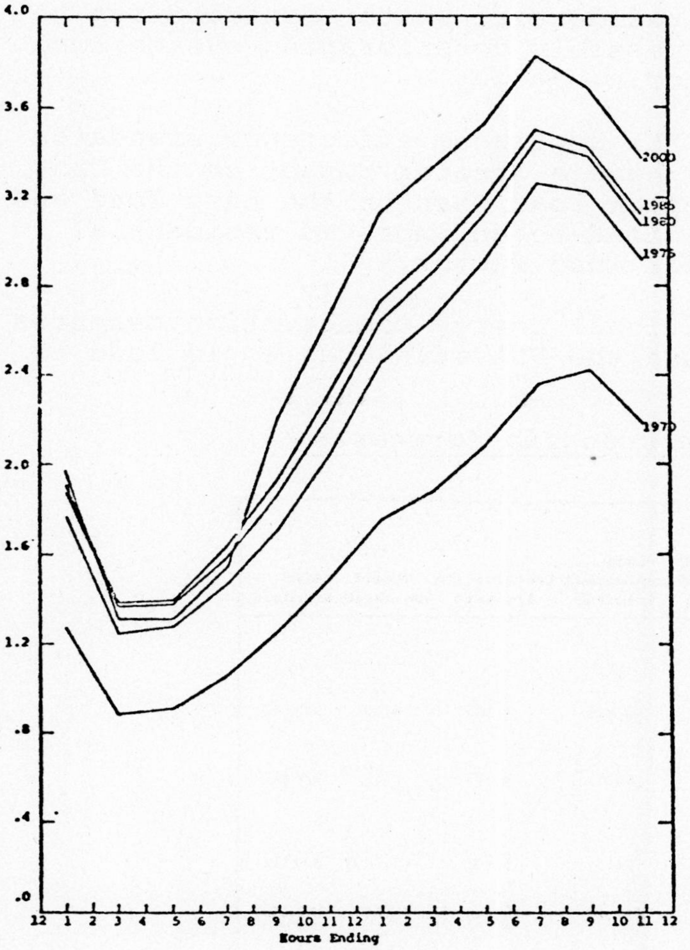

Fig. 3 Reference projections of peak load (MW x 10^3) - hot summer day.

Figure 4 plots the individual end-use demands for the forecast peak day in 1975. The end-use loads are displayed from bottom to top according to descending contribution to the sector peak load. Room air-conditioners and refrigerators comprise respectively 45% and 26% of the peak load.

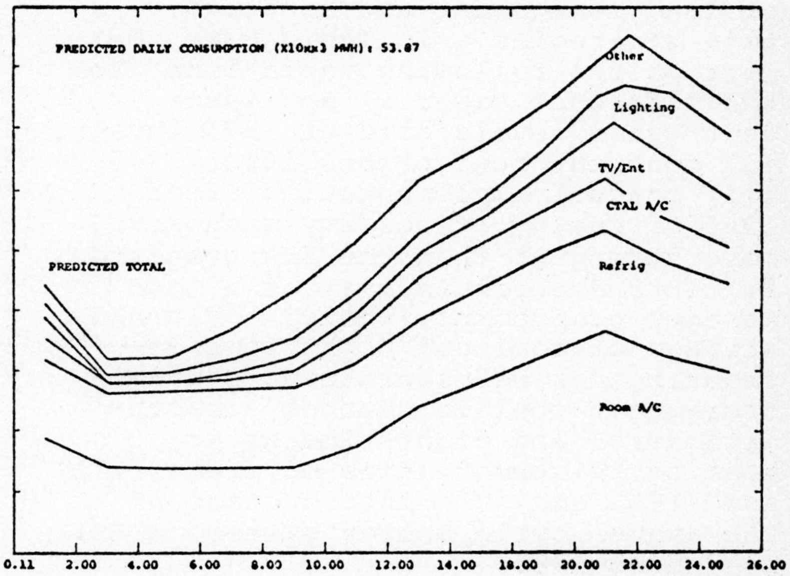

Fig. 4 Load pattern for hot summer day, 1975

Using predicted demand patterns for each end-use throughout the year we calculated the fraction of yearly usage attributable to each end-use during 8 different time periods. These estimates were combined with Schmidt's estimates (Ref. 2) of the marginal cost of electricity generation during each of the eight time periods to arrive at marginal costs for each end-use category in 1975 as shown in Table 1.

Table 1. Marginal cost per kWh consumed by end-use category (1975)

End-Use Category	Marginal cost of Generation (¢/kwh)
Central Air-Conditioning	10.2
Room Air-Conditioning	10.0
Clothes Washers/Dryers	5.0
Miscellaneous Appliances and Dishwashers	4.8
Cooking	4.8
TV/Entertainment	4.5
Refrigerators	4.4
Freezers	4.4
Lighting	4.0
Space Heating	3.3
Total Sector	4.9

The cost variations reflect appliance demand coincident with system peak periods. These results indicate that air-conditioners are approximately 2.5 as costly to the utility and society as are other appliances.

We exercised the model to study the effects of assumptions regarding exogenous variables, and the impact of load management and conservation strategies. In Table 2 we report on the following variations from the reference case: i) appliance unit usage levels held at 1975 levels, ii) constant fuel prices after 1975; iii) gradual implementation after 1980 of energy efficiency standards, more stringent than the EPA standards, iv) introduction in 1981 of a load management program involving diurnal storage of cool and night time defrosting for refrigerators, diurnal storage of cool in central air-conditioners, and night time water heating, v) combination of items (iii) and (IV), and (VI) introduction of the annual cycle energy system (ACES) beginning 1985.

Societal effects are limited strictly to the Con. Ed. district and range from an approximate gross profit of 100$/household/yr in 2000 with a combined conservation and load management scenario to an approximate gross cost of 90$/household/yr in 2000 in the absence of improved efficiency appliances. These calculations do not take into account the costs associated with achieving the alternate scenarios.

CONCLUSIONS AND RECOMMENDATIONS

An across the board time of day rate appears unattractive in the residential sector of the Con. Ed. district since demand for the principal peak load constituents room air-conditioning and refrigeration is not likely to be curtailed on hot summer days. A peak rate structure that could be both effective and practical would be one applied selectively to customers who agree to purchase appropriate storage devices, such as diurnal cool storage for central air-conditioners.

Regardless of fuel price movements it appears that the worst is over for Con. Ed. regarding the deterioration in the residential sector load factor, and the residential sector contribution to the possibly deteriorating system load factor.

The FEA appliance efficiency standards will have a greater impact on the Con. Ed. peak load than on the base load and will lead to an improved residential sector load factor.

Additional energy conservation measures beyond the FEA standards would lead to

Table 2. Comparison of alternate scenario forecasts

Year	Reference Scenario	1975 Appliance Efficiency Characteristics	Const. Fuel Prices, 1976 And Beyond	Load Management Scenario	Conservation Scenario	Hybrid Scenario	ACES Scenario
1985 (2000) Sector Usage Increase (%)**	17(39)	30(66)	20(50)	18(42)	7(10)	8(12)	N/A(40)
Sector Peak Load Increase (%)*	8(18)	24(50)	10(28)	8(35)**	3(-7)	1(5)**	N/A(11)
Sector Load Growth Coincident with System Peak Load (%)*	9(21)	27(55)	12(32)	2(-5)	5(-3)	-2(-22)	N/A(14)
Sector Load Factor (%)	46(51)	45(47)	47(45)	47(51)	45(46)	46(55)	N/A(55)
Marginal Cost (¢/kwh)	4.9(4.6)	5.0(4.8)	4.9(4.8)	4.7(4.1)	4.9(4.7)	4.6(4.1)	N/A(4.6)
Societal Profit (Loss) Relative to Reference Case ($10^6$$)	N/A	-103(-295)	-25(-65)	17(64)	65(279)	61(327)	N/A(4)

*Change relative to reference 1975 forecast which is as follows: total sector usage of 12.2 billion kWh, peak sector load of 3250 MW, peak sector load coincident with system peak load of 2917 MW, sector load factor of 43% and average marginal cost of 4.9¢/kWh.

**Peak occurs at 11:00 PM

considerable societal savings arising from the diminished need for fuel and new capacity in the Con. Ed. district. By 1985 the greatest single source of potential savings would be from the introduction of a screw-in fluorescent light bulb.

Load management initiatives could lead to a further improvement in the Con. Ed. residential sector load factor. Air-conditioning should be given the greatest focus in formulating load management policies because of the higher average marginal cost of providing for new air-conditioning demand.

Diurnal storage of cool stands out as a promising load shifting measure for central air-conditioning systems to both the utility and the individual consumer. In contrast, the benefits of load shifting measures for refrigeration appear to be at best marginal while those for hot water heating are only attractive to the electric utility as long as gas or fuel oil is available.

It appears that in the Con. Ed. district both conservation and load management initiatives could be pursued together in such a way that the load factor is improved while large societal savings result from the reduced need for fuel and new capacity. Pursuing load management options along with conservation would lead to only a slight increase in electricity requirements but would likely result in a net fuel savings, since the load management measures would tend to shift loads from peaking units to more efficient base load generators.

Because the future residential electricity growth rate in the Con. Ed. district will be at most very slow, and many even be zero or negative, if new conservation policies are adopted, serious consideration should be given to permitting the utility to invest in and made a profit from energy conservation and load management equipment used by its customers, wherever the cost of this equipment is less than the cost of the equivalent new supply.

REFERENCES

1. Weatherwax, R.K. and R.H. Williams, "Energy Conservation/Load Management Analysis for the Residential, Small Commercial, and Industrial Sectors of the Consolidated Edison Electric Franchise District (DRAFT)," Center for Environmental Studies, Princeton University, March 1977.

2. Schmidt, S., "Manual for Economic Evaluation of Load Management Techniques," Graduate School of Business and Public Administration, Cornell University, September 1976.

METHODOLOGY FOR ENERGY ANALYSIS

Michael T. Woo, Tetsushi Noguchi, Thomas Veach Long, II, R. Stephen Berry
The University of Chicago
Chicago, Illinois 60637

ABSTRACT

Two methods, input/output (I/O) and process analysis, have been used to evaluate total energy use.[1] Here, we explore the two approaches, with the goal of developing a "better," hybrid method that incorporates the capacity of process analysis to handle disaggregated systems and that of I/O for incorporating information covering the complete chain of indirect inputs. In particular, the I/O method developed at the Center for Advanced Computation at the University of Illinois is examined critically.[2] This report is divided into three sections. First, in order to study the effect of substituting physical energy data for dollar data before matrix inversion in the Illinois model, total sectoral energy requirements thus derived are compared with those derived from the conventional dollar transaction I/O model[3] using a more conventional approach. Large differences occur, indicating that the total energy transaction matrix is sensitive to this basic modification of the direct energy transaction matrix. In the second section, the propagation of data errors in the inversion of the direct matrix is examined by use of power series expansions for the two I/O methods. We thereby examine the sensitivity of the total transaction matrix to variations in the direct transaction matrix. In the third section, we compare the energy requirements for a completely fabricated product, the automobile, calculated with conventional I/O analysis, with the Illinois energy input-output method and with process analysis.[4] The differences are substantial. Following this comparison, I/O data are introduced at successively higher levels of the process analysis to approximate residual indirect energy requirements. If the process method is treated as the benchmark and if an overall accuracy of 10% is to be achieved, then input-output values can only be introduced at the third level.

I. CONVENTIONAL AND ILLINOIS I/O METHODS

Two methods of input-output (I/O) energy analysis are currently used in the U.S. to determine the patterns of demand for energy.[1] The conventional (financial) procedure determine total energy use, E, utilizes Bureau of Economic Analysis (BEA) data tables is expressed in the relation

$$\underline{E} = \underline{e}\,(\underline{\underline{I}} - \underline{\underline{A}}\,(\underline{\underline{T}} - \underline{\underline{P}})^{-1}\,)^{-1} \qquad (1)$$

where

$\underline{E}$ is the vector of total energy intensity coefficients,

$\underline{e}$ is the vector of energy sector conversion factors, e.g. megajoules (MJ) per dollar,

$\underline{\underline{I}}$ is the identity matrix,

$\underline{\underline{A}}$ is the direct transaction matrix,

$\underline{\underline{T}}$ is the diagonal matrix of total sector direct transactions,

$\underline{\underline{P}}$ is the diagonal matrix of imports, introduced to form a self-contained domestic economy.

In this method, the total coefficient, $(\underline{\underline{I}}-\underline{\underline{A}}(\underline{\underline{T}}-\underline{\underline{P}})^{-1})^{-1}$, which is the total energy in dollars, is multiplied by the average energy (MJ*) per dollar value e[3] (i.e., the fuel or electricity unit price) for each energy supply sector.

---

*The HB study used BTU's for their energy units; however MJ are used in the present study as recommended by the International Federation of Institutes of Advanced Study Workshop on energy Analysis and the convention of SI units. One BTU = 1055 joules or, roughly, one BTU is one kilo joules = 10^{-3} MJ.

The Herendeen-Bullard (HB) method, developed at Oak Ridge National Laboratory and the University of Illinois at Urbana-Champaign also utilizes Equation 1, but replaces financial transactions for the energy sectors in the direct transaction matrix, $\underline{A}$, by independently derived energy requirements in physical units (MJ) before inversion.[2] Therefore, in the HB method, $\underline{e}$ becomes a vector of ones for the energy sectors. When Leontief first proposed I/O analysis, he conceived of it in terms of tracing physical flows[5]. Financial data are recorded in the BEA tables because they are more accurate and available, particularly in view of sectorial aggregation problems. In this section, we compare the direct and indirect energy transaction matrices at the level of aggregation of 85 sectors that result from the two methods. We disaggregated the BEA 82 sector data and aggregated the HB 357 sector data to 85 sectors defined by the Survey of Current Business and the separation of the energy sectors.

The direct non-energy transactions are identical in both methods and are entered as dollar transactions. The conventional technique incorporates financial data on direct energy transactions derived mainly from the Census of Manufacturers[6] and the Census of Mineral Industries.[7] The HB method uses these same references as principle sources for direct energy transaction data in physical units, in addition to several others.[8] Thus, the main difference in the two analysis is the attempt by the Illinois group to account for the fact that different industrial users pay different prices for the same fuel (or for electricity). The HB group has also accounted for such potential problems as the use of capitve coal in the steel sector and the allocation of crude oil to the real estate sector.[8] In the Illinois method, energy flows for those sectors for which physical transactions could not be easily assessed, the residual 15% of the non-crude oil energy) is allocated of dollar data. Some sector energies were allocated on the basis of a linear, time-based interpolation. The HB method appears to allocate the direct energy transactions better than the BEA data. However, the accuracy of the HB direct energy allocation is limited by the accuracy of the sources, the residuals, and the two interpolation.

Since direct energy information at the 85-sector level of aggregation is readily available,[6,7] the contribution of alternative methods is not to the understanding of direct energy allocation, but to the assessment of the indirect flows of energy, embodied in purchases of non-energy goods and services, which is not readily available or well-known. The efficacy of an energy analysis method can be gauged by determining the accuracy with which it evaluates the energy attributed to indirect inputs. As noted above, both methods use precisely the same financial BEA data to trace the intersectoral flows of non-energy goods and services. However indirect energy use is the result of upstream direct energy transactions appearing through direct non-energy transactions. For example, the coal used by the steel sector is an indirect energy input to the automobile sector, but a direct energy transaction by the steel sector. Hence, the two formulations of I/O analysis can yield different indirect energies, and even different total non-energy transactions, regardless of the equivalence of their direct non-energy transactions.

Direct energies from the conventional method are generated by multiplying the energy rows of direct transaction matrix, $\underline{A}(\underline{T}-\underline{P})^{-1}$, by the energy sector MJ per dollar values, $\underline{e}$. The conventional indirect energies can be derived by subtracting the conventional method direct energies from the conventional method total energies.

The HB indirect energy transactions are also found by subtracting the direct energy requirement values[8] from the total values.[2] The results of a comparison of the indirect energies calculated using the two methods are given in Table 1 for selected industries. The comparison of indirect energies shows that the HB modifications of direct transactions in the energy secrors of A does indeed lead to large percentage changes in indirect and total energy requirements.

The result indicates possible ambiguities in the use of input-output techniques in evaluating total and indirect energy use for specific industries or technologies. Indirect energy requirements at the 85-sector level are very sensitive to variations in the direct energy transactions. If the financial transactions for non-energy sectors have uncertainties or errors comparable to those of direct energy sectors, then indirect and total energies will have deviations of the same magnitude. Also, to the extent that different industries pay different prices for non-energy goods (as well as for energy commodities), there will be errors carried into the calculated energy requirements. The magnitude of the effect of possible uncertainties in direct non-energy transactions on calculated total and indirect energy requirements is quantified in Section 2.

This comparison of the conventional and HB I/O methods does not indicate which method is more accurate, although one might be prejudiced in favor of the HB method, because it corrects 4 of 85 sectors for supply price inhomogeneity. If the direct transactions matrix were based completely on physical data for all sectors, the results would be as accurate as the precision of the data, taken at that level of aggregation. The HB method transforms a small number of the dollar amounts to physical quantities before inversion. The question is: which will lead to a more accurate evaluation of the indirect energy requirements, a "self consistent" financial method with inaccuracies in the dollar-to-physical-quantity conversions or a mixed method using physical and financial data, with but 4 of 85 of the direct contributions determined on a physical basis? This question will be examined systematically in Section 2 and by example in Section 3.

II. ERROR PROPAGATION ON INVERSION

We now turn our attention to the error propagation in the matrix inversion step of the I/O procedures, the inversion that results in the incorporation of indirect flows to yield total transactions. Errors can be introduced in the direct transaction level in three ways. The first, is through the factors used to convert direct energy transactions given in financial terms to physical energy flow units. In the conventional method, the factors are average MJ per dollar prices for sales by each of the energy supply sectors, as represented by the vector $\underline{e}$. In the HB method, the conversion factors are averaged MJ per physical unit, for example, MJ per ton of coal. The HB conversion factors are multiplied by the physical quantities of fuel or electricity supplied, derived from data sources other than the BEA data, to yield the direct energy transaction in the direct transaction matrix, $\underline{\underline{A}}$. Since the conversion factors are uniform for each energy supply sector, they can likewise be represented in a vector $\underline{e}^*$, which, like $\underline{e}$, does not enter into the inversion. Thus, error in either vector $\underline{e}$ or $\underline{e}^*$, is not propagated by the inversion.

The second way errors are introduced is through inaccurate assessments of direct transactions from energy sectors to other sectors. The conventional direct energy transactions are in financial terms. As described above, the HB method direct energy transaction matrix is given in physical flow quantities; for example, tons of coal.

Finally, errors are introduced through the direct non-energy transactions which are given in financial terms in both methods.

As a guide to understanding error propagation, the inversion was carried out by expansion of the 85-sector matrix[2] in a power series. It is well-known that the inverse of any invertible matrix can be represented by a power series. Let X be the conventional normalized matrix of direct monetary transactions, which corresponds to $\underline{\underline{A}}\ (\underline{\underline{I}} - \underline{\underline{P}})^{-1}$ in Equation (1) and X' is a similar matrix containing additional error terms. The matrices Y and Y' to be obtained are,

$$Y = e(I - X)^{-1}, \text{ and } Y' = e'(I - X')^{-1}, \quad (2)$$

where $e' = e + \varepsilon$ and ε is the row vector of error terms. Expanding these expressions in power series, one obtains

$$Y = e + eX + eX^2 + \ldots, \quad (3)$$

and

$$Y' = e' + e'X' + e'X'^2 + \ldots \quad . \quad (4)$$

Suppose X and X' are related by $X' = (I + \lambda/100)X$ where λ is a diagonal matrix whose elements are given by $\lambda_{ij} = \lambda_i$ if i = j and zero if $i \neq j$, λ_i is the percent error for the ith sector row of the direct transaction matrix. When this expression is substituted for X', the difference between Y' and Y is

$$Y' - Y = \varepsilon (I - X')^{-1} + \left[\sum_{k=0}^{\infty} Z^k\right] \quad (5)$$

where Z^k is given by a recursion relation,

$$Z^{k+1} = (Z^k + Z^k\ \lambda/100 + eX^k\ \lambda/100)^X . \quad (6)$$

The first term in (5) is the error in the conversion factor for dollars to MJ. The convergence of the series for both I-0 methods is quite fast and was carried to the 14th order. The largest 14th order contribution for any sector is smaller than 0.2% of the total.

The results are summarized by taking the average percentage difference of the energy requirements obtained from the matrix Y and that of Y' - Y for various values of error terms. The symbols r,s,t are used below to identify error terms reflecting, respectively, the percentage energy conversion factor error, the percentage energy (supply) transaction error, and the percentage non-energy sector transaction error. Table 2 summarizes the results obtained for various values of r,s,t. U is the percentage mean error in the total primary energy consumed by each of the 85 sectors. Total primary energy is equal to the direct plus indirect consumption of coal, crude oil, natural gas, and electricity generated in hydro and nuclear plants which is evaluated at its fossil fuel thermal equivalence. A one-percent error is arbitrarily assigned to the conversion factor error under the assumption that it would be small.

Two conclusions can be drawn on inspection of Table 2. It is obvious that when all three error factors increase by an order of magnitude the corresponding total percentage mean error, U, in the inverted matrix also increases by an order of magnitude. Furthermore, for a fixed r, U increases monotonically and linearly in both s and t. Indeed, the points lie on a simple plane as shown in Figure 1.

We also calculated the total error using the data and linear least-squares analysis of the equation U = ar + bs + ct + d. However, if r, s and t are equal to zero, U is definition equal to zero, so d is 0. Since u is carried through the process outside of the inversion, any error in the conversion factor u enters in a completely additive manner. Therefore, a = 1. A linear least-squares fit to the equation U = r + bs + cz gives the values b = 1.83 and c = 1.48 with a standard deviation of 0.10

Since this equation establishes the relation between the total error and the errors in the original direct transactions, it is possible to obtain the boundaries for s and t that yield a certain value of U. Figure 2 shows the relation of s and t for U = 10. The figure indicates that if the total error of less than 10% is desired the energy sector transaction error, s, and the non-energy sector transaction error, t, in the direct matrix must lie in the shaded region.

The row vector Z^k in Equation 5 is the absolute error in primary energy introduced in the kth order of the power series. Each element of vector Z^k corresponds to one of 85 sectors. Hence, by summing of the absolute errors of Z^k for the kth and higher terms for each sector, then dividing by the sum of primary energy consumed by k^{th} and higher orders for each sector, one obtains a value for the percentage error for the kth and all higher orders of the primary energy contribution for each sector. An example of a calculation of the average percentage error for k^{th} and higher orders for various values of the order k with the initial percentage error (1,5,5) is shown in Table 3. "Average error" is the average error of all the sectors. The second and higher orders have a particular significance in application to process analysis. The sum of second and higher order terms in Equation 3 corresponds to the indirect energy requirements, so the sum of second and higher order terms of Table 3 indicate the errors introduced in those indirect energy requirements.

III. Optimized Combination of Process and Input/Output Methods

Process energy analysis is often considered to yield more accurate energy requirements for specific processes or products than input-output methods.

This is because aggregation errors are smaller in process analysis and energy flows are derived purely on the basis of physical flows rather than on financial transactions, either partially or entirely. Process analysis, however, faces the inherent difficulty that the amount of data that must be obtained and analyzed increases rapidly as the regression of indirect energy requirements is carried further. Consequently, a process analysis must generally be terminated at a level lower than that of a similar I/O analysis. However as the amount of process grows, the incremental contribution of new data on the total outcomes becomes smaller. Hence it would be desirable to introduce a method faster than process analysis to analyze high-level contributions to the indirect energy. Input-output methods have the value of being comprehensive and available, and it is an attractive thought that they might be wed to process analysis to evaluate high-level terms. In this section I/O data is introduced into at different levels of the analysis. The results are tested using the Berry-Fels (BF) process analysis of the automobile[4] as the benchmark, providing the total energy requirement for an automobile.

Process analysis is carried out using a multi-level analysis as illustrated in Figure 3. In our use of the term here, the first level of analysis includes the energy and non-energy commodities that are used directly in fabrication of the final product. For example, first level inputs to the automobile would include electricity for fabrication, steel for the body, glass for the windshield, and others. Second level inputs consist of the energy and non-energy goods that go directly into producing non-energy inputs of the first level. For the automobile example, some of the second level inputs are the coal and ferroalloys that go into steel making, and the silica that goes into glass production. The third level consists of the energy and non-energy inputs that are used to produce the second level non-energy inputs. Such a regression continues until the effects on the total calculation of adding an additional level to the analysis are the same magnitude as the uncertainties in the largest (or accumulated) terms. (These uncertainties are about 10% in the B-F study.) The transactions are all described in physical units.

The successive terms of a power series expansion of the inverted I/O matrix corresponds approximately to the levels of process analysis described above. There are three major differences.

The first difference is that in practice (but not in principle), process analyses have frequently neglected non-energy inputs to fuels, whereas I/O calculations automatically include them. For I/O analysis of the automobile, direct energy and non-energy inputs for coal are included, just as are the non-energy inputs ferroalloys, for example. The second difference lies in the degree of aggregation implicitly accepted in the two methods. For example, a tonne of carbon steel may be assigned an energy requirement equal to that of the average tonne of steel produced in the United States using I/O techniques, while a value for the particular steel type would be sought by a researcher carrying out a process energy analysis. If a study is to be made of a particular process in an industry, the use of industry-averaged data may introduce significant errors. First level inputs in particular are very sensitive to the aggregation problem. Finally, as mentioned above, transactions are analyzed in financial or in mixed financial-physical terms in I/O analysis,while process analysis utilizes purely physical flows.

Our goal is to see whether one can obtain a useful hybrid of process analysis and I/O analysis. To pursue this hybrid approach, values from both of the I/O method previously discussed were combined with the BF automobile process analysis. The mergers, employing 357-sector I/O tables in each case, were carried out at the first, second and third levels of analysis. To do this, the BF flow diagram was first aggregated so that transactions occurred between sectors that corresponded to I/O sectors. Then it was organized according to the definitions of levels given in the second paragraph of this section. The BF study is basically a three-level study, although the criterion of termination was a total uncertainty of less than 10%.

Because process analysis is assumed to be more accurate, there would be no point to using I/O analysis for the low levels of the calculation and process analysis at higher levels. When I/O results are introduced at a given level, all successive level requirements are entered from I/O data. The I/O energy contribution can be introduced in three ways. For each input identified by process analysis, the I/O data may be used to impute its direct energy requirement, its indirect energy requirement, or its total energy requirement.

The methods developed in Section I are used to derive the direct, indirect and total energy intensities for the conventional and HB I/O methods at a 356 sector level. Primary energy is found by the method explained in Section II. For the rest of this section, the term energy refers to primary energy, with all electrical energy evaluated at its thermal equivalent. It is important to note that the I/O methods differ from the process analyses in that the former treat refined fuel oil, utility gas and electricity (specifically non-hydro and non-nuclear) as purchased goods, rather than as energy inptus at the primary level. They appear as energy only as the fuels and other primary sources from which they are derived. In process analyses, energy inputs are alwasy expressed in terms of their primary sources. Hence, in order to compare the relative accuracy of I/O and process methods and of the introduction of I/O data after successive levels of process analysis, it was (and generally will be) necessary to adjust the entries in the direct I/O matrix to be consistent with the process analysis data.

Information in the BF study is given in physical terms, but the I/O energy intensities are given as megajoules per 1967 producer's dollar. In this analysis, we convert the I/O entries from megajoules per dollar to megajoules per unit of product by multiplying the elements of the I/O table by the cost of the physical unit as given by the unit value shipped, available from either the Census of Manufacturers or the Census of Mineral Industries.* This method has the advantage that it is consistent, and the cost data are available for most commodities. The choice of unit-value-shipped to represent the appropriate price is not the only possible one. Strictly, this price should be the producer's price, which is not a generally available figure. Herendeen's initial study[9] evaluated the producer's price for automobiles using data from the 1971 Statistical Abstract of the United States,[10] based in turn on Ward's Auto Study.[11] Producer's prices evaluated from the Census Bureau's value-shipped and from Ward's auto study differ significantly ($2600 and $2100 per auto, respectively). Since the I/O values of energy requirements are proportional to this figure, the results are quite sensitive to this choice. Note that although the resulting I/O-derived energy intensities in physical units, called conventional HB I/O values, are evaluated by multiplying the producer's price, determined by our adoption of the values shipped, by the conventional and HB I/O data, respectively.

Multiplying the energy intensities by the price, yields energy per unit of physical

*We are grateful to Philip Ritz of the Bureau of Economic Analysis, U.S. Department of Commerce, for a conversation regarding this conversion.

output. The energy intensities in physical units for automobile-related transactions are listed in Table 4. The identifying numbers are assigned as follows: 0 signifies that the entry is the product being examined (automobiles) and entries 1-19 signify first-level non-energy inputs to the product being examined. The letters identify second-level non energy inputs to the automobiles production. Thus, the second-level inputs are first-level inputs for preceding numbered non-energy inputs. The corresponding energy intensities derived from the BF study are also given. With the BF values as reference, the mean percent difference of the conventional I/O energy intensities is -39%. The standard deviation of the percent difference is 53%. With the BF values as reference, the mean percent difference of the HB I/O total energy intensities is -2%. The standard deviation of the percent difference is 49%. A major portion of the difference in the case of the conventional I/O method can be attributed to inaccuracies inherent in the evaluation of direct energy inputs to primary material sectors. Note, also, that these statistics encompass both intermediate production goods and articles delivered to final demand. Thus, one should interpret them in a utilitarian way: if one wishes to substitute an I/O energy requirement in the place of one derived using process analysis, on the average, the HB method gives a more accurate approximation. However, both I/O methods have large (and, on average, similar) variances, relative to process analysis in the energy requiremetns for individual process stages. They are therefore likely to be equally unreliable guides to the energy requirements of individual stages or technologies.

Several implications emerge from Table 4. The direct energy calculated for a product from I/O data is usually lower than the direct energy calculated for that product from process analysis data. For example, the direct energy in fabrication of an automobile is 32,200 MJ as compared with 19,000MJ using the HB I/O method and 4640 MJ using conventional I/O analysis. Also they may stem from different assignments of energy to joint products. This partitioning is made on the basis of relative product dollar values in I/O analysis, while joint-product energy requiremetns are partitioned on the basis of physical units in process analysis. The unit price of the principal product of an industry may be less than the average price of that industry's output because of economics of scale, which would cause the direct energy figure for the material calculated from the I/O financial partitioning to appear lower than that evaluated from process analysis. In the case of the conventional I/O analysis, the direct energy discrepancy unboubtedly reflects the much lower-than average prices for fuels and electricity paid by major consumers under long-term, special contractural agreement.

The indirect energy calculated for a product by I/O analysis is usually higher than the indirect energy calculated by process analysis. This may occur because I/O is more comprehensive and the inversion regression is carried to higher order. At a given level, inputs are evaluated from all sectors of the economy, and each inputs is regressed to an arbitrary number of levels, while practical process analysis encompasses the principal energy-intensive material inputs in a three- or four-level regression.

A contributor to both of these differences may be a problem in allocation of secondary products by the I/O method. I/O sector transaction data is based on firms whose main activity is defined by the sector, plus that output of other sectors which are secondarily engaged in the same activity. The added output is treated as a transaction input from the secondary sector. Recent work by Bullard[12] shows that this leads to inconsistent total energies (for example: total energy less than direct energy). Again , this is one form of aggregation error.

Given the physical transactions and the I/O total energy intensity in physical terms, there is only one method to determine the total energy for a transaction--multiplying the amounts of goods exchanged by the total energy intensity. Two producers can be used to determine the indirect energy if the direct non-energy inputs are given. Procedure one is to multiply the dollar quantity of product of the sector under examination by its I/O indirect energy intensity, i.e. the indirect energy required per unit output of that sector. Procedure two is to multiply the process-analysis-determined quantity of each non-energy input of the examined sector in dollars by the total energy intensities in MJ/$ and sum.

We introduce a convention here to specify how I/O data is introduced: we identify the I/O contribution level with the lowest level whose direct energy includes I/O data explictly. The I/O contribution of procedure 2, which employs the energy intensity data from I/O analysis and evaluates the inputs of non-energy materials and services using process analysis,is well suited to this convention. The direct energies of all previous (lower) levels are determined by process analysis and are referred to as the process analysis contribution. For example, consider an automobile. The indirect I/O-derived-energy requirement for the steel in automobile production is defined as a third level I/O energy contribution to automobiles. However, this convention does not differentiate the two methods of introducting I/O indirect energies mentioned above. Since the first procedure makes use of only the indirect energy contribution for the sector in question and does not explicitly look at all the inputs to that sector, we subtract half a level from the definition of level of initial I/O contribution given above. Thus, procedure 1, which uses financial I/O table data to evaluate both the energy intensities and the flows of non-energy material and service inputs is identified with a "level designation" one half unit lower than that used for procedure 2. Hence if the indirect energy of steel is determined according to procedure 1 by multiplying the physical quantity of steel determined by process analysis, translated into dollar value using producer's prices, by the I/O indirect energy intensity of steel, the product of these numbers is classified as a 2.5th level of initial I/O energy contribution in determining the energy requirement of an automobile. If the indirect steel energy is determined by procedure 2, that is by multiplying the non-energy inputs of steel by their total I/O energy intensities, the product is classified as a 3rd level I/O energy contribution to automobiles. The one-half unit level differences (subtracted) should not be interpreted in any quantitative sense, but as a qualitative reflection of possible accuracy differences in the two procedures. This convention also reflects the fact that more process analysis work has to be done as the introduction of I/O data is deferred to higher levels.

Table 5 shows the different combinations of process energy contributions and I/O energy contributions, as catalogued by the I/O initial contribution conventions, to the total energy intensity of the automobile. The BF value can be thought of being equal to the total, within 10%, of all direct levels. Values for the categories defined by Table 5 are given in Tables 6 and 7, for the I/O energy contribution of the conventional and HB I/O methods, respectively.

The percentage differences from the BF benchmark value for the different levels of contribution of the conventional and HB I/O methods are graphed in Figure 4. Our results show how the combination of I/O and process analysis values converges on the total process result as the I/O data is entered at successively higher levels. The results converge slowly. If we wish to incorporate an error of less than 10% as a criterion, the I/O data should not be entered prior to at least the third level in the auto study. It is interesting that the convergence is not monotonic. The I/O results based on both I/O method oscillate above and below the process study.

The contributions drawn on here involve only a sampling of the 357 I/O sectors for a single product. The findings are, therefore, indicative rather than definitive.

The implication of this section can be summarized as follows: when interfacing I/O and process analysis, at most levels of I/O contribution, the HB I/O appears to be more accurate than the use of conventional I/O techniques. However, for roughly 10% accuracy, HB I/O is not accurate enough to interface with process analysis except at the third level, where it can be used to represent residual terms that account for about 10% of the total energy requirement.

REFERENCES

1) IFIAS Report No. 9, Workshop Report International Federation of Institutes for Advanced Study, Workshop on Energy Analysis and Economics, Lidingo, Sweden (June 22-27, 1975); Thomas V. Long, II, "Net Energy Via Process Analysis," Report of the NSF-Stanford Workshop on Net Energy Analysis, Stanford, California (August 25-28,1975); Clark W. Bullard," Energy Costs, Benefits, and Net Energy," CAC Document No. 174, Center for Advanced Computation, University of Illinois, (August, 1975).

2) R.A. Herendeen and C.W. Bullard, "Energy Cost of Goods and Services, 1963 and 1967," Document No. 140, Center for Advanced Computation, University of Illinois, (November, 1974).

3) "Input-Output Structure of the U.S. Economy: 1967," Survey of Current Businesses, (February, 1974). Also, "Input-Output Structure of the U.S. Economy: 1963," Vol. 1-3, Bureau of Economic Analysis, U.S. Department of Commerce, (1969).

4) R.S. Berry and M.F. Fels, "The Production and Consumption of Automobiles: An Energy Analysis of the Manufacture, Discard and Reuse of the Automobile and Its Component Materials," Report to the Illinois Institute for Environmental Quality, (July, 1972).

5) W. Leontief, Input-Output Economics, Oxford University Press, New York, (1966).

6) 1967 Census of Manufacturers, U.S. Department of Commerce.

7) 1967 Census of Minerar Industries U.S. Department of Commerce.

8) D. Simpson, "Direct Energy Use in the U.S. Economy: 1967, "Technical Memorandum No. 39, Center for Advanced Computation, University of Illinois (January, 1975).

9) R.A. Herendeen, An Energy Input-Output Matrix for the United States, 1863: User's Guide, CAC Document No. 69, University of Illinois, (March., 1973).

10) Statistical Abstract of the United States, 1971 U.S. Department of Commerce.

11) Ward's 1967 Automobile Yearboook, Detroit, (1967).

12) C.W. Bullard and R.A. Herendeen, "Energy Impact of Consumption Decisions," Proceedings of the IEEE, 63, 484, (March, 1975).

INDIRECT ENERGY FROM CONVENTIONAL I/O METHODS

IO CODE	INDUSTRY	COAL	CRUDE	REFINED	ELECTRIC	GAS
24.00	PAPER AND ALLIED PRODUCTS, EXCEPT C	25.4	55.0	23.7	4.2	-1.6
27.00	CHEMICALS AND SELECTED CHEMICAL PRODUCTS	22.7	189.6	40.5	3.2	30.5
28.00	PLASTICS AND SYNTHETIC MATERIALS	24.9	108.9	53.7	3.9	26.5
30.00	PAINTS AND ALLIED PRODUCTS	21.6	87.4	43.2	3.0	20.7
31.02	PAVING MIXTURES AND BLOCKS	17.8	351.2	34.6	2.8	21.4
31.03	ASPHALT FELTS AND COATINGS	18.1	245.6	27.9	2.9	21.6
35.00	GLASS AND GLASS PRODUCTS	15.4	45.6	12.8	2.4	30.4
36.00	STONE AND CLAY PRODUCTS	23.9	54.2	18.2	2.8	24.3
37.00	PRIMARY IRON AND STEEL MANUFACTURING	58.3	40.5	13.4	3.1	25.1
38.00	PRIMARY NONFERROUS METAL MANUFACTURING	21.2	35.4	13.7	4.3	22.1
39.00	METAL CONTAINERS	67.00	31.1	13.7	3.5	22.7

INDIRECT ENERGY FROM HB I/O METHOD*

IO CODE	INDUSTRY	COAL	CRUDE	REFINED	ELECTRIC	GAS
24.00	PAPER AND ALLIED PRODUCTS, EXCEPT C	29.9	100.5	20.0	5.2	31.7
27.00	CHEMICALS AND SELECTED CHEMICAL PRODUCTS	48.4	170.0	18.8	8.1	56.3
28.00	PLASTICS AND SYNTHETIC MATERIALS	39.2	133.0	20.9	9.8	62.9
30.00	PAINTS AND ALLIED PRODUCTS	30.1	83.5	20.2	7.6	45.8
31.02	PAVING MIXTURES AND BLOCKS	18.0	537.9	56.0	4.5	47.4
31.03	ASPHALT FELTS AND COATINGS	24.0	467.8	43.9	5.4	50.1
35.00	GLASS AND GLASS PRODUCTS	19.5	99.5	10.8	3.4	25.2
36.00	STONE AND CLAY PRODUCTS	26.5	110.9	17.7	4.2	29.7
37.00	PRIMARY IRON AND STEEL MANUFACTURING	49.7	91.2	15.7	6.0	32.7
38.00	PRIMARY NONFERROUS METAL MANUFACTURING	49.9	83.9	-16.7	11.9	42.8
39.00	METAL CONTAINERS	59.6	54.9	16.4	6.7	33.8

% COMPARISON OF THE TWO INDIRECT TABLES/CONVENTIONAL I/O METHOD STANDARD**

IO CODE	INDUSTRY	COAL	CRUDE	REFINED	ELECTRIC	GAS
24.00	PAPER AND ALLIED PRODUCTS, EXCEPT C	18.0	82.8	-15.3	25.0	46.6
27.00	CHEMICALS AND SELECTED CHEMICAL PRODUCTS	113.6	-9.8	-53.6	152.9	84.9
28.00	PLASTICS AND SYNTHETIC MATERIALS	57.2	22.1	-61.0	165.8	136.1
30.00	PAINTS AND ALLIED PRODUCTS	39.2	-6.6	-53.2	135.1	120.6
31.02	PAVING MIXTURES AND BLOCKS	1.0	53.1	48.9	62.7	116.3
31.03	ASPHALT FELTS AND COATINGS	32.6	90.5	57.0	89.2	131.6
35.00	GLASS AND GLASS PRODUCTS	26.3	118.1	-15.2	41.5	-16.9
36.00	STONE AND CLAY PRODUCTS	10.7	104.5	-2.9	47.2	8.7
37.00	PRIMARY IRON AND STEEL MANUFACTURING	-14.9	125.1	17.2	92.3	30.5
38.00	PRIMARY NONFERROUS METAL MANUFACTURING	135.2	136.9	21.7	178.9	90.5
39.00	METAL CONTAINERS	-11.1	36.3	19.0	93.6	48.6

*These values are sound through methods described in text, and not directly from BEA or HB data.
**100 x (conventional-HB/conventional)

TABLE 1: INDIRECT ENERGIES FOR SELECTED INDUSTRIES (UNIT = MJ/$)

r,s,t	U	r,s,t	U	r,s,t	U	r,s,t	U
1,1,1	4.3	1,5,5	17.3	1,10,5	26.4	1,20,5	44.7
10,10,10	42.9	1,5,10	24.4	1,10,10	33.5	1,50,10	107
		1,5,20	38.6	1,10,20	47.7	1,50,5	99.5

TABLE 2

THE DEPENDENCE OF TOTAL PERCENTAGE ERROR U IN TOTAL ENERGY ON R,S,T ERROR FACTORS (ALL FACTORS AS PERCENTAGES OF "CORRECT" VALUES).

k (LEVEL)	ERROR (%)
1	17
2	18
3	21

TABLE 3

Table 3: Average error in k^{th} and higher order I/O primary energy contributions for (r,s,t) = (1,5,5) as a percentage of total contribution of those levels.

TABLE 4: PRIMARY ENERGY INTENSITY IN MJ PER PHYSICAL UNIT FOR THE CONVENTIONAL I/O, THE HB I/O, AND THE BF PROCESS ANALYSIS METHODS

			Direct Energy/ Physical Unit			Indirect Energy/ Physical Unit			Total Energy/ Physical Unit		
Name	I/O #	Physical Unit	Conventional I/O	HB I/O	BF I/O	Conventional I/O	HB I/O	BF P/A	Conventional I/O	HB I/O	BF P/A
0 Automobile											
	59.03	auto	19400	23700		13100	161800		150000	185500	
1 Iron Casting	37.02	ton	9800	17600	16300	13100	15100	10100	22900	32700	26400
a) Pig Iron	37.01	ton	8500	10300	23200	6500	5400	1930	14700	15700	25200
b) Ferro Alloy	37.01	ton	20400	25200	61700	15500	13200	19900	35700	38400	81500
2 Steel Casting	37.02	ton	17900	32000	42300	23900	27500	7050	41700	59500	49300
a) Pig Iron	37.01	ton	8300	10300	22700	6300	5400	2830	14700	15700	25500
b) Ferro Alloy	37.01	ton	31600	39100	62900	24100	20400	21000	55800	59500	83900
3 Metal Stamping	41.02	ton	3300	4000	4000	23100	28200	53500	26600	32200	57600
a) Cold Rolled C.S.	37.01	ton	18500	22900	32200	14100	12000	4100	32600	34900	56300
4 Steel Forging	37.03	ton	12200	24800	30700	50500	52200	50200	62700	77000	81000
a) Raw C.S.	37.01	ton	16500	20400	46200	12600	10700	4080	29000	31000	50300
5 C.S. Wire	37.01	ton	32800	40600	61100	25000	21200	4150	57800	61800	65200
Other C.S.	37.01	ton	22800	28200	52200	17300	14700	4100	40100	42900	56300
Alloy Steel	37.01	ton	29900	37000	55000	22800	19300	4090	52700	56400	59100
Stainless Steel	37.01	ton	95400	118000	79500	72600	61600	4300	16900	117600	83800
a) Ferro Alloy	37.01	ft^3	26400	32700	61500	20100	17100	20200	46600	49800	81600
b) Oxygen	27.01	ton	.12			.06			1.77×10^{-1}		
c) Iron Ore Mining	5.00	ton	313	800	900	367	400		680	1220	
d) Limestone Mining	9.00	ton	109	108	200	52	54		157	163	
e) Lime	36.13	ton	7600	11300	200	630	277		8190	11600	
f) Primary Al	38.04	ton	33600	136200	228200	16100	20100	17000	49700	156300	245100
g) Primary Ni	38.05	ton	25400	417600	368600	150000	-15200		177000	402400	
h) Primary Sn	38.05	ton	28800	173700	50400	170000	-17000		198000	456800	
i) Primary Zn	38.03	ton	22100	36000	90000	16800	36800		390000	72800	
6 Pig Iron	37.01	ton	8300	10300	23200	6300	5380	1980	14700	157000	25200
a) Iron Ore Mining	5.00	ton	313	815	.871	367	410		677	1220	
b) Limestone Mining	9.00	ton	109	109	190	52	54		190	163	
7 Al Forging	38.14	ton	56600	92800	24200	130000	213000	240000	177000	305800	267400
a) Primary Al	38.04	ton	22600	230400	219900	27300	34000	20100	83800	264300	240000
8 Al Casting	38.11	ton		44600	47700	51500	130300	178100	74100	174900	226000
a) Primary Al	38.04	ton		166300			24506		60500	170800	
b) Primary Cu	38.01	ton		24200			70800		50300	75000	
9 Al Finished	38.08	ton	17400	32300	27500	56500	178100	240300	73900	210400	267800
a) Primary Al	38.04	ton	40800	166300	220100	19700	24500	20200	60500	190800	240300
10 Primary Al	38.04	ton	38200	155400	220300	18400	22900	20500	56600	178300	240800
a) Carbon Electr.	53.07	not available									
b) Buaxite Minning	6.02	ton	13200	17800	6160	6500	2470	2730	19800	20200	9090
11 Cu Casting	38.12	ton	25000	30000	41500	46000	82600	88600	69500	112500	130200
a) Primary Cu	38.01	ton		33200			96800		68800	130000	
b) Primary Al	38.04	ton		166300			24500		60500	190800	
12 Cu Wire	38.10	ton	9800	14900	13500	51500	92800	101300	61300	107700	114800
a) Primary Cu	38.01	ton	9900	34000	42800	60600	99100	58500	70450	133100	101300
13 Cu Finished	38.07	ton	17300	26300	36600	53800	94600	101110	71100	120900	137700
a) Primary Cu	38.01	ton	8600	29700	42800	52800	86800	58300	61800	116500	101100
14 Zr Die Casting	38.13	ton	25500	34100	31100	57000	117000	61800	82500	151100	92900
a) Primary Zn	38.03	ton		36000			36800		38800	72800	
b) Primary Al	38.04	ton		166300			24500		60500	190800	
15 Zn Finished	38.09	ton	8400	14900	14400	25500	49100	72000	33800	63900	86400
a) Primary Zn	37.03	ton	21900	35800	54600	16700	36600	14400	38900	72400	72000
16 Glass	35.01	ft^2	25	45	95	8	21		3.36	66	
a) none given	not given										
17 Plastics	28.01	ton	15500	35300	28100	55900	52300		71400	84500	
18 Fabrics	17.06	not given	19	24	14	71	91		90	115	
19 Metal Powder	37.04	ton		35600			38300		40100	73700	

LEVEL OF I/O CONTRIBUTION (INITIAL)	DESCRIPTION OF I/O ENERGY CONTRIBUTION	DESCRIPTION OF P/A ENERGY CONTRIBUTION
BENCHMARK	NONE	TOTAL
1	1ST LEVEL TOTAL ENERGY	NONE
1.5	1ST LEVEL INDIRECT ENERGY (METHOD ONE)	1ST LEVEL DIRECT ENERGY
2	2ND LEVEL TOTAL ENERGY (METHOD TWO)	1ST LEVEL DIRECT ENERGY
2.5	2ND LEVEL INDIRECT ENERGY (METHOD ONE)	1ST & 2ND LEVEL DIRECT ENERGY
3	3RD LEVEL TOTAL ENERGY (METHOD TWO)	1ST & 2ND LEVEL DIRECT ENERGY
3.5	3RD LEVEL INDIRECT ENERGY (METHOD ONE)	1ST, 2ND, & 3RD LEVEL DIRECT ENERGY

TABLE 5: PRIMARY ENERGY DESCRIPTION OF I/O - PROCESS ANALYSIS MIX FOR DIFFERENT LEVELS OF I/O CONTRIBUTION

LEVEL OF I/O CONTRIBUTION (INITIAL LEVEL)	I/O CONTRIBUTION (MJ)	(%/TV)	AUTO STUDY CONTRIBUTION (MJ)	(%/TV)	TOTAL VALUE (MJ)	COMPARISON OF TOTAL VALUE TO BENCHMARK VALUE (% DB)
BENCHMARK	0	0	133000	100	133000	0
1	150200	100	0	0	150200	13
1.5	131000	80	32200	20	163200	23
2	57200	64	32200	36	89400	-33
2.5	41200	39	64400	61	105100	-21
3	46500	42	64400	58	11100	-17
3.5	17300	12	126000	88	143300	8

TABLE 6: TOTAL PRIMARY ENERGY AT DIFFERENT CONTRIBUTION LEVELS OF CONVENTIONAL I/O METHOD TO THE BF AUTOMOBILE STUDY (TV = TOTAL VALUE; DB = DIFFERENCE FROM BENCHMARK)

LEVEL OF I/O CONTRIBUTION (INITIAL LEVEL)	I/O CONTRIBUTION (MJ)	(% TV)	AUTO STUDY CONTRIBUTION (MJ)	(% TV)	TOTAL VALUE	BENCHMARK COMPARISON (% DB)
BENCHMARK	0	0	13300	100	13300	0
1	18600	100	0	0	18600	45
1.5	16200	83	17	19400	45	-18
2	7600	70	32200	17	19400	-13
3	45200	41	64400	59	11000	-18
3.5	16300	11	12600	89	142700	7

TABLE 7: TOTAL PRIMARY ENERGY AT DIFFERENT CONTRIBUTION LEVELS OF HB I/O METHODS TO THE BF AUTOMOBILE STUDY (TV = TOTAL VALUE; DB = DIFFERENCE FROM BENCHMARK)

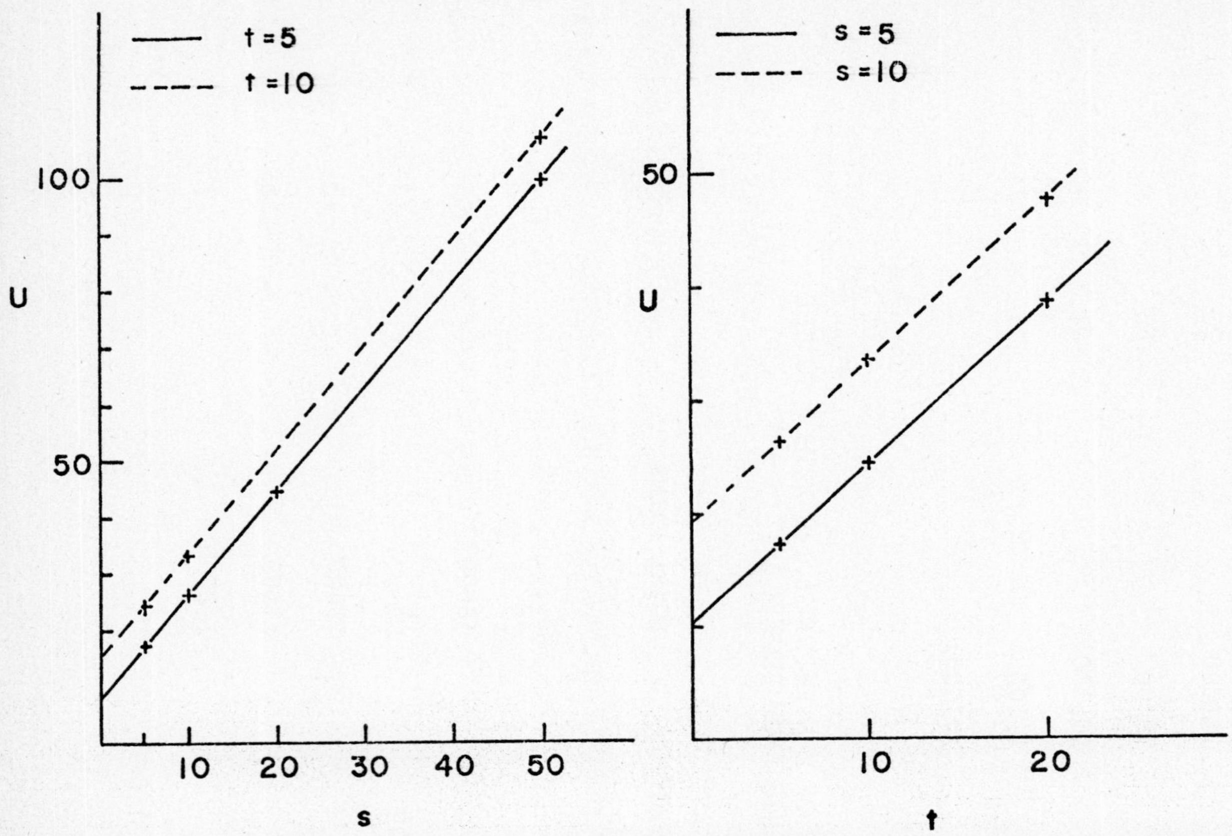

FIGURE 1: TWO VIEWS OF THE PLANE DEFINED BY THE RELATION BETWEEN THE ERROR FACTORS TOTAL ERROR U
(VALUES ARE IN % r = 1)

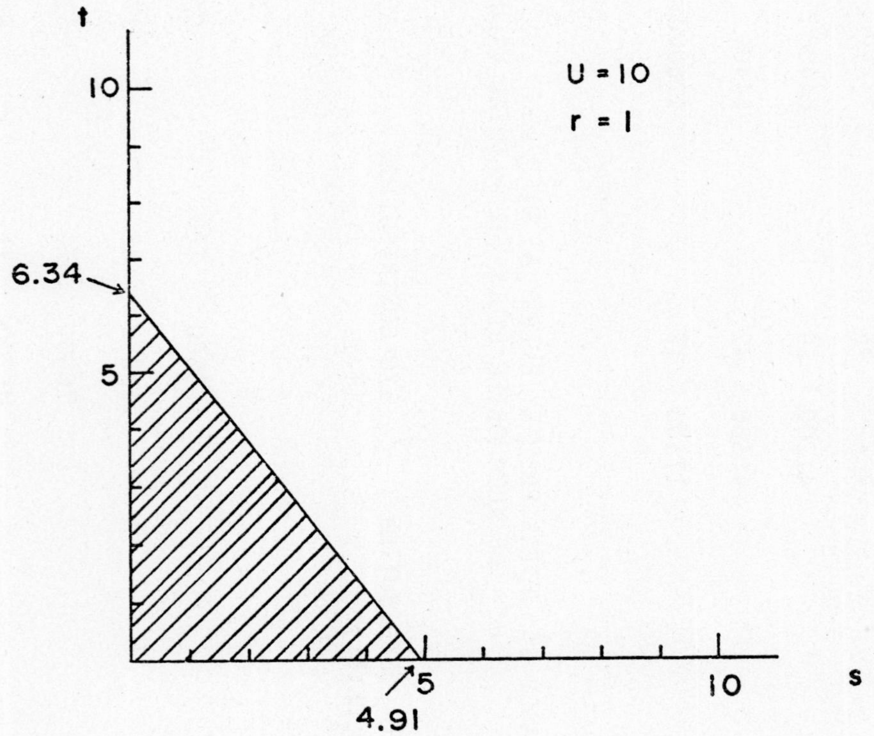

FIGURE 2: LIMITING RELATIONSHIP OF s AND t FOR A FIXED U
(VALUES ARE IN %)

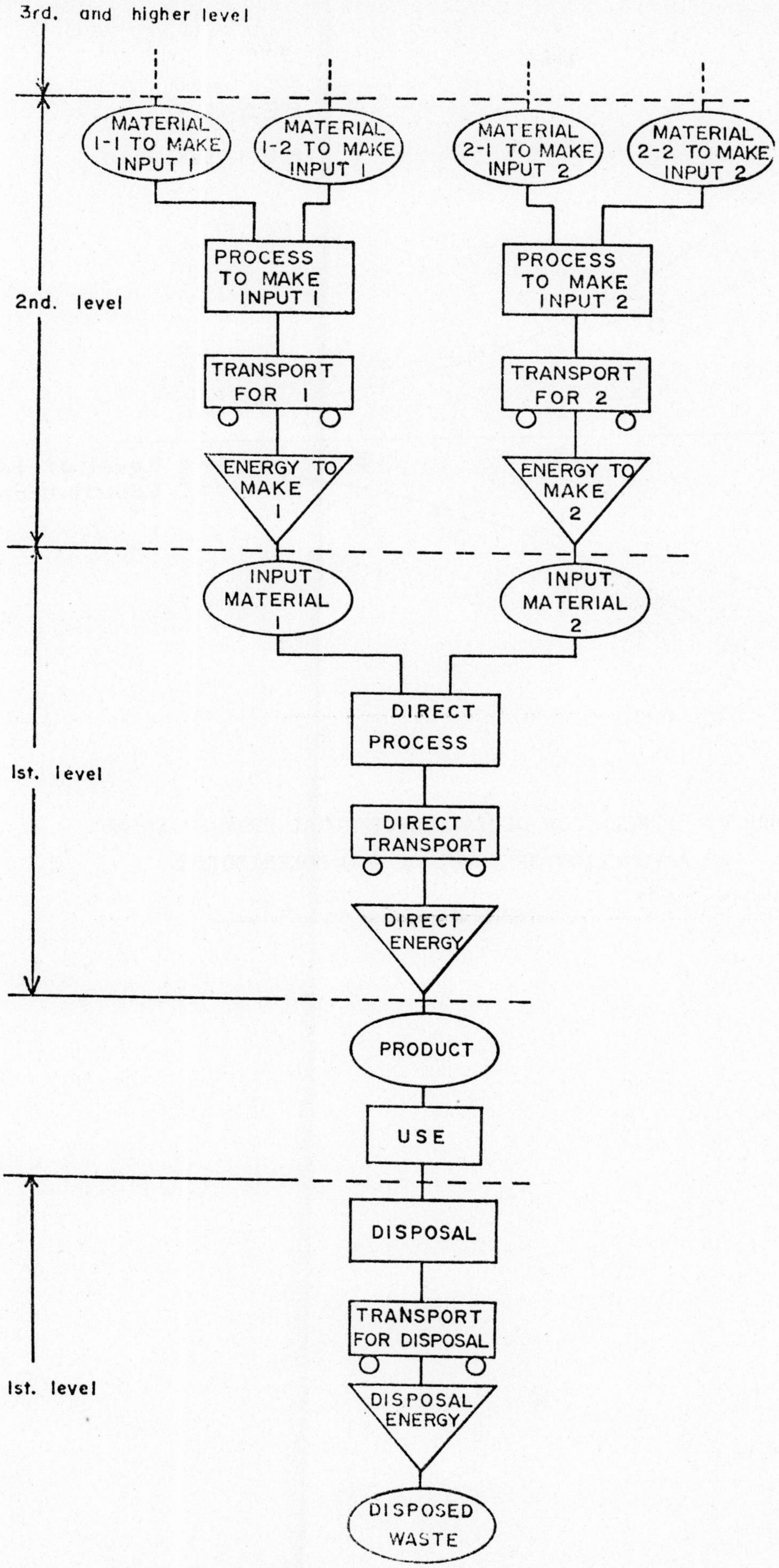

FIGURE 3: PROCESS ANALYSIS AS A LEVEL ANALYSIS

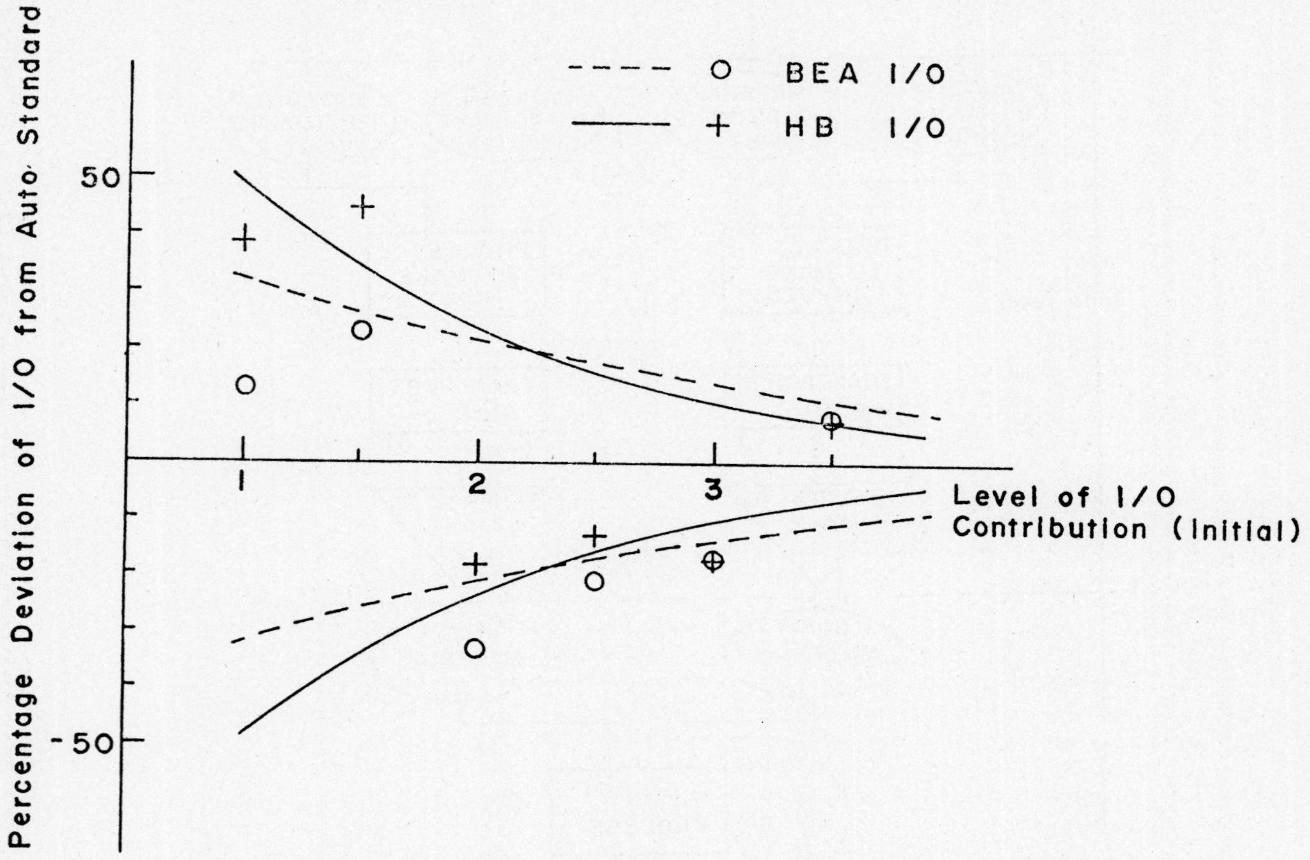

FIGURE 4: PERCENTAGE DEVIATION OF TOTAL PRIMARY ENERGY
AS A FUNCTION OF LEVEL OF I/O CONTRIBUTION

ENERGY PRICE AND CONSUMPTION CHANGES IN THE MID - 70'S

Bernard A. Gelb
The Conference Board, New York

ABSTRACT

Events of the last four years have wrenched historical relationships between energy use and other economic variables, forcing business and government to make energy decisions now and in the future. A major factor in these decisions will be the expectations of future energy demand, which will be partly based on the response of energy users to the recent price and supply developments. The response has, however, been confounded and obscured by the sharp price inflation in non-energy commodities and the deep recession of 1974-1975.

This paper will analyze movements in U.S. energy consumption per unit of output in the major energy using sectors and industries from 1972 through mid-1977 - thus incorporating the more "normal" 1976-1977 period - and relate them to changes in energy prices during the same period. It is hoped that the resulting approximation of the short run price elasticity of demand for energy will be added to the empirical base used by decisionmakers. Preliminary indications are that the long term decline in energy use per unit of output, analyzed in detail for manufacturing by The Conference Board,* accelerated sharply in 1976. The outline is expected to be as follows:

I. Introduction and description of price changes
II. Sector by Sector Analysis
   A. Industry (including 5-7 specific manufacturing industries)
   B. Residential and commercial (using degree-days and floor space as demand data)
   C. Transportation (using passenger car miles and freight ton-miles as output data)
   D. Electric utilities (load factors; heat rate)
III. Entire economy (summary of sectors; effects of aggregation)
IV. Conclusions and considerations
   A. One-time effects and future prospects
   B. Price elasticity and "conservation"

---

*John G. Myers, Bernard A. Gelb, Leonard Nakamura et al., Energy Consumption in Manufacturing (Cambridge: Ballinger Publishing Company, 1974)

PHYSICAL EFFICIENCY AND ECONOMIC EFFICIENCY AS CRITERIA FOR RANKING ENERGY SYSTEMS

Michael D. Yokell
Energy and Resources Program
University of California
Berkeley, California 94720

ABSTRACT

Three forms of energy analysis are compared with economic analysis as techniques for assessing energy conversion technologies. None of the three - net energy, energy payback time, or "present value of net energy" is found to be suitable. Economic analysis is superior to any type of energy analysis, though it is badly flawed. Suggestions are offered for improving economic analysis through the use of shadow pricing techniques which allow environmental and equity considerations to be incorporated into the analysis.

ROADS, TAXES, LIFE STYLES, AND TRANSPORTATION
Running Out of Gas?

Michael B. Barker, AIP

American Institute of Architects, 1735 New York Avenue N W
Washington, D C 20006

ABSTRACT

The new Carter Administration has the opportunity to undertake a comprehensive review of the nation's transportation programs. Central to the discussion will be the fate of the Highway Trust Fund which has been the backbone of U.S. Transportation policy since 1956. The work of the administration and the Congress of this important issue could substantially affect all of our individual life styles for decades to come. Without an integrated federal urban growth policy the outcome of the transportation debate will take place in a policy vacuum.

INTRODUCTION

Question: Can the U.S. continue to ransom its economy for an inefficient petroleum based transportation system? Item: In 1976 the importation of petroleum fuels (45-50% of total consumed) added 38 billions to the U.S. Trade deficit. This represents a substantial loss of U.S. jobs, adds to inflation and adversely effects the standard of living in the United States. This paper indicts existing federal transportation policy as a major cause of our present woes and points to new directions in which to seek solutions.

In June 1956 Congress passed legislation setting up the Highway Trust Fund to finance construction of a nationwide Interstate Highway System and other highways. As adopted, the law (PL 84-627) earmarked certain highway-users taxes for the Trust Fund (see Table 1). The $6 billion plus fund was seen as a remedy to urban traffic congestion and as a stimulus to economic development. In the late 1960's environmentalists attacked the Trust Fund because they felt the nation was becoming overly committed to highways at the expense of other forms of transportation. The International Oil Cartel pricing policies and absolute shortages of petroleum are swelling the ranks of those who question the validity of a petroleum based transportation system.

Highways The Urban Armatures

Throughout history urban form has reflected human ingenuity in using available resources to construct urban settlements. Historically, the form of urban settlements closely reflects economic, physical, and social needs. The structure of urban settlements, largely determined by transportation technology, remained relatively constant in industrialized societies up through the 1920's when there began a revolution in urban form based on motor vehicles and an abundance of petroleum fuels. The classic city forms, concentric, linear and radial corridor, have been superseded by low density urban sprawl. This rapid spreading out of urban form greatly accelerated after World War II. In 1956 the establishment of the Highway Trust Fund was a popular public policy based on a high public demand for automotive mobility, increasing use of trucks to move goods, and the belief that the nation had an inexhaustible supply of inexpensive petroleum.

TABLE 1 List of Highway User Taxes Supporting The Trust Fund

	Rate	Net Yield Fiscal Year 1974 (In Millions)	Percent of Total
Gasoline*	4 cents per gallon	$3,906,614	62.4
Diesel Fuel*	4 cents per gallon	394,681	6.3
New trucks, buses and trailers	10 percent of manufacturers' wholesale price. Vehicle of 10,000 pounds or less gross weight, school and transit buses are exempt	614,132	9.8
Tires	10 cents per pound for highway type 5 cents per pound for others	837,717	13.4
Tubes	10 cents per pound	33,382	0.5
Tread rubber	5 cents per pound if for highway type tires only	24,131	0.4
Heavy vehicle use	$3 per 1,000 pounds annually on total gross weight of vehicles rated at more than 26,000 pounds gross weight	225,193	3.6
Parts and accessories	8 percent of manufacturers' wholesale price of truck and bus parts and accessories	130,455	2.1
Lubricating oil	6 cents per gallon in highway use. All non-highway use refundable	94,005	1.5
	Total	$6,260,310	100.0

*Highway use only

Source: Highway Users Federation

disturbing when you consider that in 1974 alone, the National Safety Council reported that tractor-trailers were involved in 440,000 accidents on the highway. Of these, 3,300 involved fatalities. Wouldn't it have been wiser to encourage substitution of more efficient rail type transportation of goods over the long haul than strenghtening the already subsidized competitive position of trucks? Trucks now haul over 50% of intercity freight. A 1976 study by the Department of Transportation showed that at least 30% of the rail market is now handled by truckers.

Other economic incentives are needed. For example the federal government controls the consumption of alcohol by limiting its use through taxation. Would not a similar approach work for petroleum products. An incremental tax on gasoline could even raise enough revenue to balance the federal budget thereby curbing inflation and reducing dependence on the vicissitude of foreign oil exporters. The Carter administration has proposed an emergency gas tax in its "Energy Program" and a tax incentive for smaller cars. These measures are puny compared to the problem yet are causing The President a substantial drop in popularity and The Congress and the special interests to howl. In the final analysis the transportation-energy problem is not only a transportation problem; rather, it is a land-use and life style problem.

Need for National Growth Policy

Each year the federal government spends tens of billions of dollars on transportation, community development, housing and community services. These activities are all intricately interrelated on the ground but not in the government. There is no National Growth Policy or strategy to coordinate these important public investments.

As an example of how federal policies interrelate, consider the problem of air pollution. Poor air quality is an important side effect produced by our transportation system and suburban life styles. The primary source of air pollution in the United States are automobiles and trucks. Poisonous air used to be the exclusive problem of the central city dweller. New studies by the Environmental Protection Agency point out alarming and increasing rates of air pollution in suburban and rural areas across the country. Detriments to the public health, damage to crops, and degradation of the environment are subsidies we pay as a society to keep our inefficient transportation system operating. For too long these costs have been largely ignored. The Administration and the Congress say it is national policy to clean up the nation's air. Can the air be cleaned up at the same time new roads are being built and their use encouraged? As a nation we can no longer afford to overlook how our public policies interrelate. The very credibility of governmental institutions is at stake.

Solutions - Long Term

Will it be easy to achieve a balanced transportation system which will function within an overall national growth policy? Of course not. Vested interests in transportation systems are exceedingly entrenched. Indeed these interests are institutionalized even in governmental organizations. For example, the committees in the Congress that deal with transportation are not the same as the committees that deal with housing and community development. Yet these two massively funded federal programs are undisputably interdependent. Many people are thinking about how these organizational problems can be solved over time. But in the immediate future two important steps can be taken.

Firstly, The Administration and the Congress can consolidate the transportation modal trust funds into a comprehensive transportation or community development funding mechanism. This would allow state and local governments to use federal assistance as a tool to begin to reshape metropolitan urban regions into more fuel efficient urban systems. Secondly, The Administration working with the Congress should embark upon the task of establishing a National Growth Policy. This work has begun with the issuance of the first three biennial reports the President is required to make to the Congress on National Growth and Development. But this is just a beginning. The excellent report of the National Advisory Committee on National Growth Policy Processes (January 1977) titled "Forging America's Future: Strategies for National Growth and Development" contains some very sound recommendations on how to go about developing a National Growth Policy.

From 1956 to 1975 about 95 cents of every Federal surface transportation dollar was spent on highways. Our great metropolitan areas where 75% of our population lives, have been built on a highway armature which assumes an abundance of inexpensive petroleum fuels and has had a devastating effect on most central cities.

Roads Building and Urban Sprawl

For decades there have been "voices in the wilderness" speaking out for more compact cities which are capable of being serviced by public forms of transportation, are efficient in the distribution and consumption of energy, are accessible to open space and recreational opportunities, and are in a sympathetic relationship to the natural environment. While many recognized the theoretical validity of these arguments, little was done to achieve results. Even the environmentalists liked their single family housing in the midst of sprawl and used their cars to get to work and to the mountains and to the seashore. Besides, gas was plentiful and cheap. Without social and economic incentives to do resource conserving urban planning, the highly speculative land development market spurred by massive road building, produced low density, disorganized, inefficient, energy consuming urban sprawl, in which our society now has a great investment, both socially and economically. It is not exaggerating to say that the metropolitan areas in the U.S. run on gas. The question is should public policy continue to encourage this form of growth, or should public policy be directed toward reducing our dependence on a deminishing resource which is increasingly beyond our means.

Petroleum Shortage

The dire predictions of those early environmental prophets such as Louis Mumford and Buckminster Fuller, who pointed out that the earth could supply only a finite amount of natural resources, are being vidicated by contemporary global events. Petroleum is no longer seen as an inexpensive and inexhaustible source of fuel. As a nation we have only sampled in a small way the consequences of a dwindling supply of this resource. During the Arab oil embargo of 1974, which only slightly constrained the supply, there was a significant amount of social and economic dislocation. This dramatically demonstrated the vulnerability of suburban low density life styles. The form of metropolitan areas will not be changed rapidly. However, opportunities to move these urban systems in the right direction should not be overlooked.

Need for Automobiles and Trucks

In the foreseeable future there will be need for automobiles and trucks and other petroleum consuming vehicles: automobiles for recreational, social, and some work trips; trucks primarily for rural and intra-city frieght; buses for public transportation. The question is one of balance not absolutes. The Highway Trust Fund does not provide the Congress and the Administration the flexibility necessary to "balance" our nation's transportation system. A "balanced" system on the short run could possibly save enough petroleum fuel to maintain our cherished automobile related suburban life styles awhile longer and on the long run reduce hardships and inequities in the transition to more efficient mobility systems.

Economic Considerations

For a moment let us look at the economic considerations.

The highway transportation industry accounts for over 17% of the gross national product of the United States. One out of every six employed Americans or over 13 million jobs are in the highway transportation industry. Approximately 820,000 individual businesses involve motor vehicles and their needs. This tremendous investment, both economically and socially is increasingly vulnerable to severe dislocation because of the growing scarcity of petroleum. Simply trying to increase the supply of petroleum is not enough. While increasing supplies and conservation are important, they cannot solve the problem on the long pull. A gradual shift in the general transportation system is needed to lessen reliance on our petroleum fueled highway dependent system. The actions of the Congress so far on this critical question have been unimpressive. Indeed the recent action of Congress to increase permitted truck weights seems to go exactly in the wrong direction. Doubly

For too long the nation has been employing short term and often shortsighted solutions to "transportation" problems which are much more than transportation problems. The road base transportation system which is supported by the singular Highway Trust Fund is at once a land use problem, a life style problem, an economic problem, a defense problem, and an environmental problem. The recognition of this reality in public policy making is long overdue.

CHANGING LIFESTYLES - OVERVIEW

Albert J. Fritsch, Ph.D.

Center for Science in the Public Interest, 1757 S. St. Washington, DC

Modern lifestyles, especially those in the richer nations of the world, may be characterized by their use of large quantities of energy resources, noteably those of a non-renewable type. Part of this phenomenon has been due to the availability of cheap and readily available oil and natural gas. The cars we use, the homes we heat and cool, the consumer products we purchase and use all require energy to built and operate.

However, the age of cheap, abundant oil and gas is rapidly nearing an end. Scarcity in this country is already being experienced. The OPEC oil prices have been accepted for the past four years as a new factor in the energy economy which must be reckoned with. The money drain due to increased oil imports threaten the stability of numerous nations. Poorer lands do not have the needed fuel to operate irrigation pumps and tractors, which translates into food shortages and possible famine. The possibility of global disaster through resource exhaustion or catastrophic changes in the environment has been raised by respected scientists and cannot be lightly dismissed. In the more consuming lands, electric power plants are being required to convert from low polluting natural gas to coal. In many nations, a shift has been made towards environmentally questionable uranium fuel sources. Many are proposing new or under-used technologies such as solar, wind, geothermal, and tidal energy.

The winds of change are not just blowing in the line of more efficient and renewable energy sources. Today, there is a conscious effort on the part of world leaders to embrace the conservation ethic. As the United States "National Energy Plan" states:

> The attitude and habits developed during the era of abundant, cheap energy are no longer appropriate in an era of declining supplies of America's predominant energy sources. Conservation offers vast opportunities for American creativity and know-how. The challenge of saving energy should galvanize the ingenuity and talents of the American people.*

This cheapest and cleanest source of new energy supply will not be accepted by the American people without some change in their respective lifestyles. To help effectuate this type of national change in consciousness, several steps must be taken:

a) an awareness of how much each person uses in non-renewable energy sources, and how much resource wastes accrue as a result of this consumption;

b) a set of reasons for desiring a change in lifestyle to a conservation ethic;

c) and a series of strategies both individual and community-oriented to help bring this about.

Changes in lifestyle appear to many,

*The National Energy Plan, released by the Executive Office of the President, Wash., DC 20500, April 29, 1977. p 35.

even the emerging middle class and poorer groups of citizens, as somewhat threatening. Hard-earned goods such as automobiles or heated homes or appliances or recreational vehicles will have restrictions placed upon them or will become too costly for the individual consumer. To overcome these psychological pressures to retain and conserve current commercial habits requires effort. However enough research is available to convince sensible people that quality of life might improve with reduction in use of energy resources. Fewer emissions mean a cleaner environment and better health; less energy means conversion to more human service--rather than energy intensive ones; less dependence on foreign energy sources adds to national stability; lower consumer expectations means reduced need to work and earn financial resources needed for maintaining a commercial and consumer economy.

Granted, some quantity of resources is needed by every person to raise them from the level of destitution and supply adequate food, clothing and shelter. Still there is a point- or better, a range depending on the needs of various individuals- where increased quantitative growth in energy use leads to deterioration in the quality of life. Even standards of living (often both qualitative and quantitative in measurement) are not necessarily proportional to energy and natural resources expended on a per capita basis.

People in this country need to be convinced about the need for conservation of resources. A number of appeals- some more self-serving than others- may be proposed. These do not demand retreating from society to the life of a hermit or a rigid turn to the life of the unadorned (no cosmetics, jewelry, colorful artistic work). In fact, simplifying one's life might mean becoming more social and turning to artistic rather than energy-intensive alternative consumer items.

Convincing people to change may be done by the carrot or the stick, the enticement of voluntary methods or the more forceful method of regulations and rules. Maybe the most effective means is a both-and procedure. Whatever the final combination might be, a more effective measure is always one where people are convinced that the change is both necessary and better in the long run. A number of reasons for such changes might include:

a) Ecological--save resources which are not renewable; reduce air and water pollution; reduce ultimate heat losses due to use of fuels.

b) Personal--improved physical and psychic health; less noise, congestion and haste; opportunities for improved human relations and spiritual growth; ability to appreciate and come into harmony with nature.

c) Community--ability to share more often in communal projects and activities; openness to assist others in need especially those who might otherwise be institutionalized.

d) Global--create a more stable economic order; establish better relationships with less fortunate people.

e) Economic--it takes less money to live more simply and thus reduces the need to earn more; increases leisure time.

f) Social--by living simply, people are more free to pursue a rich variety of challenges, including working for social change;people living simply are not tied as tightly to institutions which need reform and thus can work for change.

g) Political inequity--simple living is a political statement when people can exercise some choice over the goods and services they provide for themselves and others rather than rely on the judgment of the multinationals.

Simple lifestyle is thus defined in terms of conservation even though it is really much more. But this minimal attitude requires some points of focus, for it is not enough to buy a gas guzzler and at the same time consider oneself a conservationist because a few papers or cans are being recycled. A conscious effort to make social changes imply taking steps to realize through sacrifice some individual strategies might include:

a) Decreased individualization:

1) Shifting from suburban and low density home living situations to a more communal or apartment type. Residential energy use is quite sensitive to the exploding number of new residences which far exceed population increase. Each of these require separate appliances, furnishing and added fuel bills.

2) Opting more for mass transit in place of individual autos for work and business. One quarter of our national energy budget is in transportation and major savings will accrue through changes to more hopefully efficient and rapid mass transit systems.

3) Changing to less individualized packaging, preserving and preparation of food. Food consumes one-eighth of our energy budget and sizeable savings can accrue through use of such modes as community stores, canneries, and food preparation services.

b) Increasing the lifetime of products:

1) Making and purchasing cars of longer duration. About one-quarter of the amount of energy to fuel cars in a single year is required by the industrial sector in their manufacture. By extending the lifetime of autos only a few years, major savings will result.

2) Refurbishing old and abandoned dwellings instead of building new ones. Many such buildings are better built than newer ones and require only a fraction of the building materials to make them quite liveable.

3) Recycling all items which can be returned to the economy. This includes aluminum, paper, and other resources which require enormous energy resources to process from virgin materials, and also items which can be reused in the original state, such as furniture, appliances, clothing and tools. When possible, source reduction is preferable to recycling.

4) Purchasing goods on a life-cycle costing basis. For example a house that is poorly insulated might have a low initial cost but will be very costly over the long-term because of high heating bills. Life-cycle costing helps factor in the energy sustinence costs of capital goods.

c) Moving to a service instead of goods-oriented economy:

1) Giving personal service to those who otherwise may have to be institutionalized.

2) Selecting recreational options which are less energy-intensive, more participative, less competitive and more person-oriented.

3) Developing arts and crafts and encouraging others to do activities which reflect the whole person in the creative process.

4) Educating others, especially youth, in the value of becoming more attuned to nature.

d) Reducing excessive consumption:

1) Turning down thermostats in winter and up in summer.

2) Reducing travel speed and distance.

3) Using less lighting and unnecessary appliances.

4) Refraining from the use of processed food and over-the-counter drugs.

These are just some of the ways which have been discussed and are being developed as individual strategies towards a conservation ethic. However, each citizen must be willing to move into the social and political arena and make changes through the process of collective action. The citizen must look to other nations which many times have far better experience at conservation than his or her own more wasteful culture; there is a need to overcome institutional hurdles to the use of solar energy as a way for better citizen control of energy sources; the citizen must see that the Highway Trust Fund is used for more ecologically sound transportation measures; financial and sociological strategies must also be thoroughly researched and discussed. Each of the following papers gives us some idea of how we might expand our social options for making profound lifestyle changes and ultimate energy conservation.

LIFESTYLE IMPLICATIONS OF DECENTRALIZED SOLAR ENERGY

Alan Okagaki
Center for Science in the Public Interest, 1757 S St, NW
Washington, D.C. 20009

ABSTRACT

Decentralized applications of solar technologies can be integrated into community structures and lifestyles that are radically different than the conventional American lifestyle. It is argued that for environmental and sociological reasons it would be wise to explore alternative communities and lifestyles based on "appropriate" or "intermediate" technologies such as decentralized solar power. A federal program to support this kind of exploration is proposed.

INTRODUCTION

The federal government, the public, and the business community have grudgingly accepted the fact that energy policy will interact with people's lifestyles. For a variety of cultural reasons, lifestyles have seemed almost sacred in the sense that they should be determined strictly by the individual and never be meddled with in any compulsory way by some outside force. Consequently, government has been reticent to invoke energy policies (particularly in conservation) which might put any curbs on lifestyles. Although government and the public have recognized that some lifestyle change will be necessary, they still define lifestyle in the narrowest possible terms. Lifestyle changes are interpreted to be such actions as turning down thermostats, carpooling, and putting plastic over windows to seal out drafts--all of which are, on balance, very minor changes indeed.

Perhaps the time has come to shed our reticence and deal with lifestyle more directly and in concert with issues more fundamental than energy supply and demand. Perhaps the time has come to ask ourselves "Is the American lifestyle really satisfying to the American people?" Have we become locked into a way of life that is deficient in some respects and have we been unwilling to treat the deficiency? If we were visionaries, if we were to construct utopian social models that we would truly like to live in, what would we do? Although this line of thinking runs counter to any kind of conventional policy orientation, there are sound reasons for adopting this new kind of visionary role.

Although the American way of life is continually proclaimed to be superior for a variety of reasons, we have been receiving warnings about its shortcoming which we should not ignore. A sampling of the shelves of a typical book store indicates that the American lifestyle is dissatisfying to at least some people. Riessen's The Lonely Crowd , Roszak's The Making of a Counter Culture, Fromm's The Sane Society, and many many other works mirror discontent with different facets of the American lifestyle. Our social pathologies reveal in more concrete terms the ways in which our lifestyles fail to fulfill the whole person. Rising alcoholism, drug abuse, violent crime, suicide all attest to the failures of individuals to cope with the American way of life and the American community. Our vocabulary has changed also; words such as "retreat", "escape", and "drop out" have gained new meanings, new connotations. One can remember Thoreau's classic utterance of 130 years ago: that "The mass of people live lives of quiet desperation". His line seems to hold so much more truth today.

Of greater significance are the actions of many Americans who have chosen to reject the standard American lifestyle and adopt an alternative mode of living. The most prominent group to do so in recent years, and the group which the public still identifies most strongly with alternative life styles, was the "hippies who lived in communes". Some 2000 communal groups were established between 1965 and 1970, and undoubtedly far more since then. (1) A number of people belonging to religious organizations have also set up alternative communities. They run the gamut from very conservative Christian groups to Eastern religious organizations considered by many to be on the "fringe". Yet, the most telling fact is told in a set of statistics released by the Census Department which shows that for the first time in many years, the flight from the rural areas to the urban centers has been reversed. More people, having found the trend within our cities towards noisier faster-paced lifestyles not to their liking, have retreated to the land and to simpler lifestyles in rural America.

THE ENVIRONMENTAL PERSPECTIVE

From an entirely different viewpoint, there is another need to investigate different lifestyles. We have come to learn that there is not an environmental crisis, or an energy crisis, or a food crisis, which can be dealt with exclusive of the others. If there is a basic issue, or perhaps more aptly a "theme" which underlies many of the major issues of the day, it would be the protection of the earth as a life-support system. Adlai Stevenson articulated quite profoundly this fundamental concept of earth as life-support system.

> We travel together, passengers on a little spaceship, dependent on its vulnerable resources of air, and soil; all committed for our safety to its security and peace; preserved from annihilation only by the care, the work, and I will say, the love we give our fragile earth.

That our way of life is overtaxing the capacities of our life-support system is abundantly clear. We suffer from poor air and water quality because we dump pollutants into the biosphere in quantities far too large for the environment to absorb. Our civilization will consume in a matter of decades the fossil fuel reserves thattook hundreds of millions of years to form. The world's people are expanding beyond the bounds of agricultural production, and even today, agricultural productivity in the Western world is alarmingly dependent on fossil-fuel based fertilizers and chemicals.

Although energy conservation based on minor lifestyle modification may allow us to balance our demand with our energy supply, we still have no guarantees that our use and abuse of the world's resources are within the carrying capacities of the earth life-support system. Two big problems come immediately to mind: soil fertility and the carbon dioxide content of the atmosphere. For centuries, the most successful agriculture has imitated nature's cycles. A diversity of crops was raised on the same land; livestock was kept on the farm; livestock wastes were returned to the soil along with other organic material; crops were rotated; the soil remained fertile. Such practices have disappeared completely from modern agriculture. Care for the soil and its fertility has been replaced by massive application of synthetic fertilizer. The long-term effects of diminishing soil fertility, especially in light of disappearing feedstock for fertilizer, are nearly impossible to calculate. But the possibilities, using a little bit of imagination, are deadly frightening.

Similarly, the continued combustion of carbon-based fuels holds the possibility of catastrophic alteration of climate. Scientists have already calculated that increases in the carbon dioxide content of the atmosphere could significantly raise average world temperatures with unknown impacts on climate. Major changes in climate could well force massive changes in the eco-system which would quite probably be irreversible. Depending on what these changes would be, the impacts on human civilization could be staggering. We should remember that our responsibilities do not end with ourselves, or with our children, or with our children's children. When we pass the world onto the next generation it must be in condition good enough to support life indefinitely. The key element in our endeavors may well be to set up a way of life which could be sustained indefinitely by the earth, instead of our current one which is based on exhaustion of resources.

SOLAR ENERGY AS APPROPRIATE TECHNOLOGY

Solar energy has become something of a pet energy source for many people and many institutions and organizations. Clean, non-polluting, and technologically fairly simple, it has obvious advantages over the environmentally dangerous nuclear and fossil fuel technologies. In addition, it is a "renewable" energy source; the earth will receive a constant supply of solar energy until the collapse of the solar system. The materials from which solar units are constructed are not consumed in the operation of the units. Thus, most of the materials from solar units could at least in theory be recycled. In other words, solar technologies can be self-sustaining; in economist's terms, they allow us to live on the earth's income rather than on its capital.

There is a subtle, but potentially extremely important war going on over solar energy. The American industrial mainstream has become aware of the possibilities of solar energy and is moving towards mass marketing and distribution of solar equipment. A number of large-sized manufacturing corporations are already selling solar space and water heating equipment and others are conducting research on the more sophisticated solar-electric technologies. Their ranks include Boeing, General Electric, Grumman, Pittsburgh Plate Glass, RCA, and many others. At least three of the seven largest petroleum corporations are active in solar energy including Exxon, Shell and Mobil. The nation's electric utilities are working in solar energy through cooperation with the Energy Research and Development Administration's solar-electric programs and through their own research group, the Electric Power Research Institute.

But there is another entirely separate group which has also taken a powerful interest in solar energy. The members of this group are disparate and for the most part working independently of each other. To identify or classify them is difficult. Perhaps the best way to identify them is through a book. In 1973, the same year as the Arab Oil Embargo, a collection of essays authored by British economist E. F. Schumacher was published in a volume entitled Small Is Beautiful. Although many people were attracted to the title with its allusion to simplicity and scale, the substance of the book was reflected in its subtitle: Economics as if People Mattered. The basic principle behind Small Is Beautiful is that technologies, economic systems, and social organizations should be designed to fit human needs. The book argued forcibly that we have become the prisoners of our technologies and our institutions instead of the masters. Schumacher also stressed that technologies must be made compatible with the environment and inflict as little harm as possible to eco-systems. From the book was born two new phrases, "appropriate technology" and "intermediate technology" which sought to describe this change in relationship between man and his technology.

Despite the lack of organization within the appropriate technology movement, there is a basic consensus of principles and on what appropriate technology really is. John Todd, director of an appropriate technology group called the New Alchemy Institute, gives these characteristics of appropriate technology: (2)

1) they are low in cost
2) they are easily maintained
3) they are labor-intensive
4) they use local materials
5) they protect the environment
6) they are resource conserving
7) they increase the well-being and dignity of man

The vision behind appropriate technology is the self-reliant individual and community. Within practical limitations, AT visionaries want to set up economic structures whereby communities or regions are able to produce as much of their essential goods as possible by themselves. Alternative models of economic institutions are put forward by AT proponents, usually community-level industries run on some kind of cooperative basis. Local governance and decision-making is another important component of the AT vision. Another very important aspect is that through community self-reliance on activities, a genuine sense of community can be built.

Solar energy fits extremely well into this scheme of political and economic decentralization, and thus has become one of the mainstays of the AT movement. A number of reasons support this view of solar energy as appropriate technology. Some of the more important ones are:

1) Solar energy is available everywhere. All people in all areas of the country have access to solar energy. The solar resource itself is so immense that the radiation falling on the 48 continental states is equivalent to 700 times this country's rate of energy consumption. All regions in the country could conceivably draw all their energy from the sun. An AT community could be located anywhere and still have a source of energy.

2) Community owned or individually owned solar technologies are clearly an alternative to dependence on existing energy institutions, especially utility and petroleum companies.

3) Solar technologies themselves are most energy efficient and cost-effective in decentralized, on-site applications. The most promising utilizations of solar technologies are the so-called "total energy systems" where the radiation falling on one collection unit is used for both heat and electricity. In residential structures, photovoltaic cells can be imbedded onto the absorber surface of flat plate collectors. For industrial and commercial uses, photothermal systems, where high temperature heat from focussing collectors would power a turbine and the lower temperature waste heat would be used for heating, might be more practical. By making one collection unit serve double duty, costs are cut and solar energy is utilized more effectively.

4) Solar-electric units have minor if any economies of scale. Unlike conventional power-plants, generating efficiencies of photovoltaic and photothermal systems do not improve greatly with increased size. The limiting factor with photovoltaic units is the efficiency of the individual cell; size of the total unit (i.e.the number of cells in the system) will make little difference on overall operating efficiency. Similarly, efficiencies of photothermal units depend on collector output temperature, which is primarily a function of the concentration ratio of the focussing collector. Concentration ratio is determined primarily by the shape and type of collector, not by its absolute size.

5) Solar energy is extremely versatile. Technologies for solar space, industrial and agricultural heating, solar cooling, and solar electricity are already operational. Solar energy can be used for making hydrogen, thought by some to be the "fuel of the future", by electrolysis or thermal-chemical water cracking. Through bio-conversion, solar energy can be converted into liquid or gaseous fuels. Thus, it is conceivable that solar energy could supply all categories of a community's energy needs.

The technical arguments for stressing decentralized use of solar energy are quite convincing. However, since the entire concept of energy decentralization runs so strongly against conventional thinking, decentralized use of solar energy is down-played by the federal government, not to mention industry. The problem is especially difficult politically, where decentralized solar energy could essentially make big energy industries obsolete. The sum effect is that a basically decentralized energy source is being forced into a centralized mode for political and philosophical reasons. Decentralized solar energy is left for the "fringe". Yet, the possibilities of decentralized solar energy really deserve serious consideration, both on its own intrinsic merit and in the context of an alternative community.

ALTERNATIVE LIFESTYLE COMMUNITIES

Rather than continuing our reluctance to deal constructively with radical life style change, perhaps the logical approach is to actively support experimentation with alternative lifestyles and alternative communities, including those based on appropriate technology. A federal program could be devised, perhaps in the spirit of New Towns program in the late 1960s. Organizations wishing to set up alternative communities could approach the federal program for funding and technical and logistical assistance. They would have to submit a detailed plan as to the size and location of the community, the technologies on which the community could be based, the community's economic organization, activities and structure, the procedures for self-governance, etc. People who shared the philosophy of the proposed community would be recruited. The organization would be granted start-up funds and technical assistance to construct and get the community started. After a fairly short initial period, the community would be financially on its own. Although many of these communities would strive towards self-reliance and autonomy, they would have to maintain some kind of ties to the societal mainstream in order to survive. Hopefully good relationships would continue between the alternative community and the mainstream, diffusing the kind of polarization which often exists today between groups interested in radical change and conventional American institions and culture.

A number of benefits would accrue from an alternative communities program. From a purely technical standpoint, the communities could shake down many of the new appropriate technologies. No one is really quite certain how practical it is to power a community exclusively on alternative energy, or to process all community waste through recycling and composting, or to rely on community-wide subsistence agriculture for food. The most severe critics of AT have denounced appropriate technologies as "toys", playthings which are fun but too insubstantial to build an economic structure upon. With an alternative communities program, we could see how viable these technologies really are. And even the technologies which are a failure will be a stimulus for devising more new technologies which can be more serving of human needs.

This technical information could be extremely valuable if it does become apparent that our civilization has exceeded the life-support capacity of the environment and that it is necessary to drastically reduce the energy, resource, and pollution intensiveness of our way of life. The experience gained in working with low-energy, low-resource, and environmentally sound technologies would enable us to effect a much more carefully thought-out transition.

In addition to the purely technical information, these alternative communities could serve as a test for many sociological and psychological theories about the effects of different communities, working relationships and environments, and modes of self-governance on people's sense of well-being. Concepts of worker alienation, the workings of cooperative ownership and decision-making, alternative methods of education, all aspects of the cultural fabric could be judged in some kind of real way rather than relying on abstract arguments. Two issues are of personal interest to the author. The first relates to labor. A key assumption in the AT movement is that a job becomes more meaningful depending on the control and ownership of the workplace. In other words, people should be more satisfied with jobs in a worker-owned cooperative than with a similar job under traditional ownership. Coupled with this is AT's pre-occupation with labor-intensive technologies, which often translates into menial work. Will people be happier in AT jobs? It is a difficult question. Another question is how can a sense of community be instilled within people and what are the sacrifices which must be made in order to build community? Both of these issues have been discussed to death in the abstract. By testing these principles in real communities, we could have a much better sense of what constitutes a satisfying way of life.

In some respects, an alternative communities program has overtones of a great oppressive social experiment where people are being used and manipulated to answer questions. Such an argument has very little real validity. The thrust of the program will be to provide people a genuine opportunity to develop and evolve their own lifestyles, individually and in community, which could be radically different than typical American lifestyles. It would be a learning experience for the community itself as well as for the nation as a whole. These communities could serve as models for other people contemplating alternative lifestyles; they would be catalysts for more alternative communities by showing people that real opportunities exist for creating and living different lifestyles. In summary, the effect of actively addressing lifestyle issues through an alternative communities program will be to create more real lifestyle choices rather than restricting lifestyles as most people think today.

The long-term impact of such a program is nearly impossible to predict. A possible outcome could be a mainstream American culture along the lines of today's culture, except modified to reflect a conservation ethic and more humane economic values. Tied into the mainstream would be a number of alternative communities, each with its own unique organization, operating principles, and values. One of the answers that will probably come out of the alternative communities experience is that different people will be happiest in different kinds of social organizations. Given a country with a mainstream and satellite alternative communities, people will have a chance to find the community and lifestyle which suits them best.

It would be well to remember that this kind of pluralist society has its precedent. The centralized and largely homogeneous society we live in now is a new phenomenon. The colonization of the New World was carried on by very different groups of people, each wishing an opportunity to create their own communities based on their own principles and lifestyles. To take steps which could lead to cultural diversity is very much a returning to our roots. It would be to re-create the same kinds of opportunities that gave the United States its appeal.

Finally, one comment has to be made about the principles behind the environmental movement. The chief reason behind opposition to the environmental philosophy is that the environmental philosophy is based on the concept of limits and limitations. The world's resources are finite; the world's capacity to support humans is finite; certain technologies should not be developed or deployed because they violate environmental limits. Perhaps a movement towards new communities can be looked upon as a new frontier, a new opportunity to explore. There are real possibilities for visionaries, explorers and people who have desires to expand beyond limitation. They can quite literally create new worlds while preserving the one world which supports life. Such opportunities should not be allowed to be lost.

References

[1]Center for Science in the Public Interest, 99 Ways to a Simple Lifestyle (Garden City, NY: Anchor/Doubleday, 1977) p 311.

[2]John Todd, "The New Alchemists", Co-Evolution Quarterly, Spring 1976, p 63.

EUROPEAN ENERGY CONSERVATION POLICIES : AN EVALUATION

Hazel Ranninger and Jean-Marie MARTIN

Institut Economique et Juridique de l'Energie, Grenoble

ABSTRACT

Comparative analysis of energy conservation policies in 11 European countries reveals important differences in the nature, intensity and rapidity of national government responses to the energy crisis. Reviewing the "mix" of measures adopted in each country suggests that while all conservation policies have encountered major obstacles, certain institutional, economic and social factors are explicative of many cross-national differences.

INTRODUCTION

West European countries were profoundly shaken by the escalation of oil prices at the end of 1973, and far more so because they had become highly dependant on oil imports from North Africa and the Middle East. Reactions to these price changes were initially rapid. During 1974, each national government, as well as the European Economic Community, hastily prepared certain measures through which two complementary objectives were sought :
The first set of measures sought the reorganization of energy supplies to accelerate the substitution of domestic energy resources for imported oil. This reorientation gave a strong impetus to nuclear energy but it touched also coal and lignite production (notably in the United Kingdom and West Germany), the expansion of North Sea oil prospecting, and for a more distant future, the development of geothermics (at low temperature) and solar energy.

A second set of measures was aimed at economizing energy ; ie. to halt increases in consumption, without limiting the growth rate of national economies. What was proposed was the increase in the overall efficiency of the transformation and final utilisation of energy by consuming smaller quantities of primary energy to obtain a given quantity of useful energy. Several countries defined precise objectives in this respect. France, for instance, in the 7th plan which began in 1977, set a target of energy saving at a volume of 45 Mtoes to be attained before 1985.

It is this second set of measures which is the object of our study in the following. In the first section, we compare energy conservation policies in various West European countries (the European zone of the Organisation for Economic Co-operation and Development) attempting to highlight and explain certain basic similarities and differences, and suggesting several explicative hypotheses.

One is however, forcibly struck by the nature of the difficulties encountered without exception by energy conservation policies in all these countries. It appears far easier to produce new quantities of energy, even where technologies as sophisticated as those concerned with nuclear energy are involved, than to economize an equivalent quantity of energy, even though such savings appear perfectly realisable technically and economically. A basic problem therefore is to understand these fundamental difficulties and to evaluate the extent to which they are surmountable. In section 2 we examine some of these obstacles.

I. COMPARISON OF ENERGY CONSERVATION MEASURES

Though the majority of measures were decided at the end of 1974, many came into operation only at the end of 1975 or the beginning of 1976. We lack, therefore a perspective from which precise comparisons can be made. It is still not possible to evaluate with precision the actual costs of conservation measures in order to compare the results in different countries. The measures so far adopted are infact many and heterogeneous.

Firstly, they are aimed at extremely diverse activities ; the transportation of persons and of goods, domestic heating systems, electricity production, industrial energy consumption etc. Secondly, they have taken a variety of forms : information campaigns (based on educational and psychological techniques of persuasion), taxes, subsidies, investment allowances (ie. financial incentives) and finally, statutory regulation and controls (ie. demanding manditory compliance) as well as contracts between public authorities and representatives of industry etc. Finally, though the measures themselves are wellknown far less is known about the conditions surrounding their application. Certain measures have been effective, others much less so. Recognizing these limits and uncertainties, we can however, pin-point a few major trends in European energy conservation policy.

1.1.1. Major Tendencies

In the Table provided in the Appendix, we present graphically an overall view of the intensity and type of measures adopted in each of eleven countries*. We can see at a glance that the "mix" of measures adopted (as of mid 1976) differs substantially from one country to another, as well as the relative weight placed on the various types of measures (i.e. publicity campaigns, financial incentives, statutory regulation etc.) according to each target sector. We must note however, that in so summarizing, in a qualitative way, the vast amount of information analysed for all of these countries, much richness of variation is lost. For the purposes of compiling the table, we have estimated the degree of accent placed on conservation measures as "strong", "medium", "weak" or "non-existant", after analysis of the major energy policy statements and recent legislation in each country. Obviously, the impact of a given measure differs considerably according to a range of circumstances specific to a given country. Where, forinstance, the park of private automobiles is already dominated by small, economical models (as is the case in Denmark) the net impact of measures designed to promote more economical fuel consumption will be less than in countries where larger and more powerful vehicles predominate. Also we must note that in several countries, conservation policy is still at a government-study stage. Though the tardiveness may indicate a lack of urgency attached to energy conservation or indicate differences in the length of governmental decision processes, new measures may of course be introduced in these countries in the near future.

Four broad spheres of intervention can be distinguished.

1.1.2. The provinces of industry and the transportation of goods are marked, in all European countries, as the least touched by energy conservation measures. In no single country have stringent measures, whether fiscal or regulartory been aimed at these sectors. Industry, rather than obligated to respect specific norms is generally merely incited to comply either by means of investment policy, encouraged by public authorities in the form of subsidies and loans, or by the authorisation of accelerated depreciation for conservation investments. Hence, a 7.5 % and 12 % investment grant is available in West Germany and in Belgium respectively. In the Netherlands, a subsidy is earmarked for insulation of industrial premises and in Sweden and France for conservation in industrial processes. In certain countries however, Switzerland, Italy, Austria and Norway, there is a notable absence of measures concerned with industrial energy consumption and with the transportation of goods, a default which is even more paradoxical in view of the fact that in Europe, energy consumption by industry accounts for a larger portion of total consumption than in the United States. Undoubtably, the critical events facing European economies since 1974 have played a significant role but by no means explain the entire situation.

1.1.3. In the sphere of the transportation of persons (intra-and intercity), measures have been more numerous but their objectives still remain partial or negative. They have been aimed far more at the reduction of fuel consumption in private vehicle use than towards the development of other more energy economical modes of transport. In almost all European countries, we find similar measures concerning speed restrictions on major roads and motorways from 80 to 130 km/h. (Only in West Germany is the maximum speed still recommended rather than manditory) : - heavy taxes on petrol to as much as ten times those currently imposed in the United States, (particulary high in Belgium, Italy, France and the Netherlands) - progressive duty on vehicles in relation to weight or power. (The United Kingdom applies a uniform duty on all cars).

Conversely, energy conservation measures aimed at reducing the pre-eminance of the automobile in the total transport system are rare.

---

* Indicative lists of conservation measures adopted in each country was supplied by the International Energy Agency. A detailed comparative analysis of recent legislation and statutory regulations in energy conservation and their modes of application is currently being undertaken at the Institut Economique et Juridique de l'Energie at Grenoble.

Belgium and Sweden are the only countries to have seriously proposed studying the introduction of carpooling. Few measures, with the exception of those in Denmark, have been designed to give real priority to public transport particularly in city areas. A few countries, (the United Kingdom and Italy) have nevertheless instigated a transfer to the railway system of a part of public finances previously allocated to motorway construction. Several countries have introduced traffic lane systems to favour public transport as can be seen from the Table in the Appendix.

1.1.4. The residential use of energy (the heating of individual dwellings, apartment blocks, public and commercial buildings) has been the selected target of all West European conservation policies. Here, can be expected in fact, the largest and most immediate energy savings. It is in this area, far more than in others, that recourse has been made to statutory controls and regulation concerning space-heating, insulation and the operating efficiency of equipment, (notably boilers and central heating units).

Although many governments imposed maximum temperatures in government and administrative offices, France is the sole country to have extended this measure to all public and private, communally and individually owned buildings. (The measure is infact, applicable to collectively owned apartment buildings only in as far as occupants collaborate to respect the 20°C norm.)

Regulatory measures are more systematically applied to the insulation of new builings. All countries have revised, or propose to revise, their building codes but there exist substantial differences between countries in insulation standards*. Obviously, a major and important difficulty is to induce respect of such norms by construction firms for whom insulation materials increase building costs. On the otherhand, such norms are generally not applicable to existing buildings for which most countries have introduced financial incentives in the form of direct grants or tax rebates with the exception of Italy, Switzerland and Norway. In Belgium, Denmark and Ireland, the subsidy available for insulation of existing buildings is up to 25 % of insulation costs. In the United Kingdom, insulation standards are required if renovation grants are awarded. The improvement of lighting standards and controls on public lighting and display signs has so far received little attention by national governments, with the notable exception of France. Measures designed to impose minimum energy efficiency standards for equipment, especially boilers, have also been relatively feeble. Most governments appear reluctant to impose such norms on firms constructing equipment.

1.1.5. A final sphere is the production of electricity in which considerable economies of energy can be made either on the side of the load curve or in promoting the combined production of heat and electricity. The problem of the load curve is so complex that it would be vain to try to compare directly the measures of adjustment taken in each country, the nature of which dependson the special characteristics of each electricity supply system. The difficulties surrounding the extension of combined production are equally clear, but it can be noted that several countries, (Sweden, West Germany and Denmark) have given a far greater priority to this question than others. (France and Italy for exemple). One of the differentiating factors in the development of combined heat-power generation, is obviously the length of the heating season, but other institutional, political and socio-economic factors are highly relevant.

1.2. The differentiating factors

Across these broad tendancies characteristic of all energy conservation policies, several inter-country differences are clearly discernable. A more detailed examination appears to show in fact that by no means all countries reacted to rising oil prices and henceforth to much dearer energy, with the same rapidity or intensity. A large number of factors, difficult to seperate and evaluate, explain these differences but can be sought in essentially two directions.

1.2.1. The conditions determining the supply of energy appear to play an important though not entirely decisive role. Those countries still endowed with relatively abundant and cheap national energy resources (compared with oil prices on the international market), or where there exists the potentiality to exploit such resources in the future, appear to have in general, a lower propensity to economize on energy. This seems to be the case for : (i) Norway, which produces electricity at low cost due to hydroelectric power and where important North Sea oil fields have been exploited ; (ii) the United Kingdom which is still endowed with coal reserves but which hopes, above all, to recover autonomy in energy between 1980 and 1985 through oil and natural gas

* "Evolution of Regulations and Practice on Thermal Insulation of Buildings". Report prepared by EURIMA dec. 1976. (concerns Denmark, France, Germany, Holland, Italy,Sweden and the United Kingdom).

reserves, and (iii) the Netherlands, well endowed with large reserves of natural gas in the Groningue region. In these three countries, average energy prices remain lower than for the rest of Europe, while measures designed to conserve energy seem to have been notably less powerful or slower to appear. National resources levels alone however, fail to explain all differences. Indeed, we note that by no means all countries poor in domestic energy resources are amongst those which have invested in the most vigorous and extensive conservation policies.

1.2.2. A number of social, economic and institutional factors must therefore be taken into consideration as explicative of observed between-country differences.

The existence of a centralized, trained and relatively effective public adminstration has undoubtably played a major role in the rapid development of the French energy conservation programme which is much more highly ambitious than that of countries either richer in energy resources (e.g. West Germany) or poorer (e.g. Italy)★. The lower level of committment of the West German administration may be attributable partly to the federal structure of the country which disperses responsability and partly to the far greater confidence placed upon the sole mechanism energy prices and the market.

Nevertheless, what appears to be a weakness in certain respects may constitute an asset in others. Hence, where the electricity industry is less monolithic than in France (in Sweden and in West Germany for instance), it has undoubtably played a significant role in the development of combined electricity-heat production.
Directly involved in the distribution of electricity, (occasionally even in its production), city authorities are better able to assert themselves in the development of district heating systems. Thus Swedish or German cities have obtained from electricity producers what the single French producer has so far refused to envisage.

Other institutional factors undoubtably intervene in explaining cross-national differences with respect to energy conservation, but their study seems of less interest here than those factors which in all countries actually run counter to a full scale development of energy conservation policy.

2 OBSTACLES TO ENERGY CONSERVATION IN WEST EUROPE

It is, of course still too early to judge the overall effectiveness of energy conservation measures and clearly, a first major difficulty lies in the methods available to estimate energy savings. A recent first-step in this direction, undertaken by the recently created International Energy Agency★ illustrates the problem of reference to GDP as a measure of individual country energy savings. Although most countries appeared to make reductions in the amount of energy consumed for every unit of GDP, one cannot ignore the fact that the structure of the GDP has been considerably modified by the economic crisis. Since each branch of the economy has a different energy content,any change in the proportion between these branches inevitably modifies the energy content of the GDP. Secondly, the fore-mentioned evaluation fails to record the cost of energy savings realized. Such an operation can have significance in economic terms only if the effectiveness of different policies are compared.

We must regonize these difficulties in arriving at an exact evaluation of energy savings, which must await full scale studies, extensive surveys and a detailed country by country analysis. We can however, attempt here to identify certain major obstacles which confront all European conservation policies. Even in France where a new administration has been created and personnel and means put at is disposal, these difficulties are clearly visible. Certain major obstacles to effective energy conservation lie in the institutional and economic structures of society and in life-styles and patterns of social behaviour. The energy savings possible by individual energy users is likely to be seen as infintesimally small in comparison with overall national targets. This poses important problems for adequate and durable incentives, a problem we shall explore in greater detail in the final section. Moreover, life styles, consumption patterns and expectations of social status are not easily modified. They are to a large extent structured by cultural, social and environmental factors beyond immediate individual control such as the distance between residence and work,

★ One of the important functions of the French Agency for the Economies of Energy is to propose to the government new or revised measures (induding new legislation). A similar, though smaller scale agency exists in Belgium : the Monsieur Calorie office, established in Febuary 1975,but no other country has an Agency comparable.

★ Energy Conservation in the International Energy Agency 1976 - Review. OCDE Paris 1976

the adequacy of public transport, the quality of the built environment, etc. We know infact, relatively little about the mechanisms which would promote significant changes in social behaviour towards less energy expensive life-styles, or as to the measures most effective in inducing and maintaining such change. Very few European countries have as yet launched full scale surveys of public responsiveness to energie conservation measures anywhere comparable with American studies ★ or have sponsored in-depth studies to determine the diversity and potential saturation levels of energy needs according to different life-styles and social categories.

2.1. The myopia of energy users

Whatever the nature of measures adopted by public authorities, (statutory controls or financial incentives), energy savings are always, in terms of equivalent returns, the result of an investment : a better insulated house, a more efficiently functioning automobile or heating system, a new transport infrastructure etc. They imply therefore, an arbitrage between immediate additional capital outlay and reduced outlay spread over time. It is the sum total of future discount gains which determine or impede the realisation of the investment. In many cases, this comparison between present and future is badly articulated, mainly because a variety of social and institutional factors intervene. A few examples will illustrate this difficulty.

In many European countries, housing, whether individual or communal, is constructed by developers who must reckon with the high degree of elasticity of housing demand. In order to avoid lowering the volume of sales, they must reduce costs which leads to minimization of costs of insulation and to the choice of the least capital expensive modes of central heating. That such a choice leads finally to high running costs is irrelevant since it is borne by the house-purchaser. We might query whether the purchaser himself is unable to make the elementary economic calculation. In some cases, obviously he is. In other cases, he makes the calculation, but nevertheless chooses the least costly immediate solution because lack of available money constitutes the most pressing restraint. A way out of this difficulty could be sought through a policy of easier credit, but such an option frequently runs counter to other economic policies (notably during a period of inflation) to the extent that it is not generally adopted★. Whatever the case, it seems essential that energy conservation must be conceived as priority by governments. That this is not the case (sometimes quite explicity), we shall see below.

Investment for energy conservation in industry encounters similar obstacles. Industrialists are (or are becoming) more familiar with economic calculation, which ought to lead them, (now that more than three years have lapsed since the drastic changes in energy prices) to numerous investments, particularly in the recovery of waste heat which turns out to be economically perfectly worthwhile, at least in France at the current discount rate (10 %). Why are such investments not being realized ? A recent survey of French industries ★★reveals several explanations for this inertia : firstly, industries frequently simply do not know in detail, their energy consumption, largely because during the long period of low-price energy, the practice of counting therms has been abandonned. Secondly, even though the know-how and means of conservation exist, the investment is frequently renounced because the delay in returns is considered too long : in other words, the profit rate is far superior to the rate of returns from public investments of energy production. Finally, this lacuna is often aggrevated by the dichotomy between financial criteria and the criteria of exploitation, handled by different sectors of one and the same firm.

These obstacles are so important that a recent evaluation of the french energy conservation policy provided the results in the following Table.

★ In 1975, the American Opinion Research Corporation prepared for the Federal Energy Administration, a detailed and extensive survey on Public Attitudes and Behaviour regarding Energy Conservation.

★ We note however, that in Sweden, the Mortgage law of 1957 which gave priority to home builders or buyers who intended to insulate beyond the building code minimums, has been extended in the 1975 programme.

★★ Rapport du Groupe Reflexion sur les investissements Economisant l'Energie dans l'Industrie et leur Financement. (Dec. 1976) Agency for the Economies of Energy. Paris.

TABLE 1

ENERGY-SAVINGS IN FRANCE

Sector	1975	1976
	In Mtoes	
Residential	9.5	10
Industry	0	1
Transport	1.5	1
Transformation of energy	1	1
	12	13

Source : French Agency for the Economies of Energy

In order to remedy the serious deficiency on the part of industry the Agency for the Economies of Energy, in addition to its extensive information and technical services to industry, is negotiating contracts with the directors of the principal branches who are to guarantie certain economies in energy in exchange for state financial aide. Whether or not this kind of approach will actually be effective depends on whether energy conservation is given first priority amongst the sets of government objectives for industry.

2.2. The non-priority nature of energy-saving

The absence of real priority given to energy conservation policies in West European countries is clearly manifest★. In 1976, the majority of countries still lacked any special governmental agency devoted to the problem. Allocated budgets frequently represent no more than several percent of finances devoted to new energy resources, notably nuclear based electricity production. Officially paying lip-service to the need for energy conservation, governments have tended to support or tolerate other wholly contradictory choices, such as the large scale development of all electric domestic heating systems or the resumption of the growth of the automobile industry. These difficulties of giving priority to energy saving are closely related to economic structures and their mode of development in Europe. In other words, energy conservation runs counter to certain interests of powerful industrial groups, for whom a high rate of energy consumption remains indispensible. This is true for both energy producing and energy consuming industries.

★ In the major energy policy statements of several countries reference is made explicity to the fact that the maintenance and improvement of living-standards and social welfare must not be jeopodized by concerns for energy conservation.

The escalation of oil prices has been considered in West Europe as the departure point for a strong development programme for the nuclear industry. This is supported by two mutually re-enforcing processes : the reduction in cost per kwh as a result of the lower cost of installed kw capacity, resulting from large scale production ★ , and the expansion of electricity on the whole energy market, and particularly for thermal uses (approximately 60 % of the market) for electric heating of homes. Thus the greater the penetration of electricity in the market for heating uses, the greater the increase in consumption justifies an extensive nuclear program. This fills the constructors order books who can lower production costs and in so doing facilitates the competitiveness of electricity over other energy resources. Electricity companies find in this high growth-rate the assurance of maintaining a satisfactory level of self-financing (thus independence with respect to the government or banks) and the assurance of preserving future salary increases and promotion of personnel hence limiting tensions and social conflict. Construction firms of nuclear power plants are no less interested in an extensive nuclear program which makes investment profitable and opens the door to export markets. Both parties are able to justify to the government that their interests coincide perfectly with those of the community at large since a strong growth of nuclear energy (i) diminishes the dependance of the national economy on imported oil ... hence lightening the commercial balance of payments, and (ii) leads to the creation of employment, hence effectively counteracts unemployment at the same time favouring the conquest of external markets which has further repercusions on the balance of payments. All social costs related to such a development tend to be systematically overlooked or under-estimated.

Other intersts combine with those of companies and nuclear power plant constructors to forestall the priority ranking of energy policies. The automobile construction industry has not, forinstance, ceased to play upon the power of various models inorder to maintain sales growth. All the persuasion techniques which the industry exploits continue to emphasize the importance of vehicle speed and performance rather than such elements as the lifetime of the

★ It is on this basis that the French electromecanic industry anticipates the construction annually of 8 to 10 reactors of 1000 Mwe : 3 or 4 for exportation, 6 or 7 for the internal market. The effect of mass production is obvious in the case of the proposed super-enrichment reactors where the principal national buyer (Electricity of France) would be able to reduce the rythm of its yearly orders.

vehicle. This fact is in clear contradiction with the measures adopted by governments to limit vehicle speed, but such a contradiction is tolerated by the majority of governments because the automobile industry is creative of employment and contributes to the balance of payments. Other examples could be equally well drawn from the glass and paper industries which oppose any policy of recycling on the grounds that it would limit their growth rate.

2.2.1. We must clearly study a far wider range of social and institutional factors which determine the effectiveness and mode of application of conservation policy. Such factors as the levelof public concern with environmental problems, the degree of law-abidingness of citizens, respect for authority, the disparity between socio-economic categories, may each play a crucial role in promoting or forestalling the degree of anticipated attitude and behaviour change in various countries.

It is thus of prime importance to understand how and for which sectors of the population each type of conservation measure can be supposed to have an impact on behaviour. This kind of knowledge seems indispensible for accurate predictions of energy demand and could provide valuable data for a wider sphere of scientific interest in processes and mechanisms of social change.

In fact, we know surprisingly little as to how and over what time scale certain types of measures "work".★ Whether T.V. advertising campaigns are more successful than home or factory demonstrations of energy economies or prize awards for energy savings is still largely guess-work. We know relatively little about whether financial incentives are more promising in inducing attitude and behaviour change than financial disincentives such as price or tax increases★★ Though the former avoid the element of social injustice of the latter, there is some evidence, particulary from Austrian experience, that state grants available for home insulation etc. are not being used to the full. We might draw some interesting parallels with the apparent reluctance amongst segments of the population to take advantage of financial benefits in such fields as the social services. Those factors which tend towards social apathy or abuse of grant systems are complex and call for extensive study.

Similarly, we still lack a clear understanding of whether and under what circumstances, statutory regulations and controls are more effective if they appear as "advisory" rather than compulsory. Certainly compulsion under certain circumstances leads to wholesale evasion. On the other hand, the public image of the authority issuing, "advise" may be the key to successful compliance. We need, perhaps, to undertake a detailed examination of the relative efficiency of various modes of issuing ministerial instructions (circulaires, white papers, statutory regulations etc.) This is the more vital as the volume and density of such material increases in energy conservation policy as in other fields.

2.2.2. Also of prime importance are cultural differences in the social prestige attatched to energy expensive elements of various life styles which exert pressures forinstance for the ownership of the large private automobile or individual villa residences, secondary residences, labour saving but energy expensive domestic equipment etc. Such social pressures may be far more persuasive than government efforts to incite energy savings in these areas.

Finally, there are clear differences between European countries in the extent to which manditory conservation measures have been backed by effective sanctions. As in all areas of law compliance depends in large measure on the certainty of sanctions, and it is in part this certainty which in the publics mind distinguishes window-dressing measures and folk-crimes from seriously intended legislation. It is certain that several countries have increased sanctions and controls for maximum speed limits and for excessive waste of energy. (In France forinstance advertising of products or equipment excessive in energy consumption is prohibited). But we must await detailed analysis of the between country differences in enforcement methods and practices before this particular element in conservation policy can be evaluated.

★Though there is a rapidly growing body of american survey data on public responses to the energy crisis, we have as yet little comparable survey data for European countries.

★★An American survey illustrates the complexity of factors associated with theimpact of petrol price increases and shortages on private automobile use. The rate of price increases is highly relevant in reducing vehicle use. A sudden and steep price change is more effective in the immediate and longterm, than a gradual increase in price. Robert.L. Peskin, "The immediate impact of Gasoline Shortages on Urban Travel Behaviour". Report prepared for U.S. Department of Transportation. April 1975.

These few reflections based on West European energy conservation policies should not lead us to entirely pessimistic conclusions. Conservation policies are faced with hostile interests and social inertia but have generated other positive aspects, one at least being the progressive uncovering of the major obstacles to effectively enforced policy. Nevertheless such obstacles are likely to remain entrenched, especially in countries where industry, strongly committed to a high growth rate in energy consumption is both highly concentrated and powerful.

REFERENCES

Energy Conservation in the ECC and OCDE member Countries :

1. Collection : les Dossiers de l'Energie. Vol. 2. Documents sur la Politique energétique OCDE.CEE. Ministère de l'Industrie et de la Recherche. Documentation Française, 1975.
2. Energy Conservation in the International Energy Agency. 1976 Review. Organisation for Economic Co-operation and Development. Paris. 1976.
3. Parliament Européen - Documents de séance 1976-1977. Rapport sur le premier rapport périodique de la Commission des Communautés Européennes au Conseil sur le programme d'utilisation rationnelle de l'Energie. (Document 314/76).
4. Tableaux comparatifs des mesures spécifiques adoptées depuis oct. 1973 par les Etats membres dans le cadre d'un programme d'Utilisation Rationnelle de l'Energie (Juillet 1976).
5. Evaluation des répercussions des Economies d'Energie sur la Consommation d'Energie dans la Région de la CEE. Réunion spéciale sur l'économie et l'efficacité en matière d'énergie dans la région de la CEE (Genève 21-23 mars 1977) (ECE/AC.3/R.I)
6. Evaluation de l'incidence des Mesures d'Economie sur la consommation d'énergie dans la Région de la CEE. Note concernant certaines Questions Conceptuelles, Statistiques et Méthodologiques (ECE/AC.3/R./Add.I.)
7. Examen des Travaux effectués ou prévus dans le cadre de la CEE concernant l'économie et l'efficacité en matière d'énergie (ECE/AC.3 R.2)
8. Mesures qui ont été prises ou qui pourraient être prises dans la région de la CEE pour économiser davantage l'énergie et rendre plus efficace son extraction, sa conversion, son transport et son utilisation. Fév. 1975 (E/ECE 883).

Reports available for various countries :

9. Belgium : Les Economies d'énergie et de matières premières. Royaume de Belgique. Ministère des Affaires économiques. Bruxelles. Oct. 1975.
10. Denmark : Energy in Denmark, 1990 : a case study. Report n° 7. The work Group of the International Federation of Institutes for Advanced Study. University of Copenhagen. Sept. 1976.
11. France : Actions des Pouvoirs Publics pour les Economies d'Energie. Bulletin I.F.C.E. N° 157. Janv. 1975.
12. Rapport d'activité 1975. Agence pour les Economies d'Energie . Paris 1975.
13. Rapport de la Commission Energie. Commissariat Général du Plan. Préparation du 7° Plan. La Documentation Française. Paris 1976
14. Germany : First Revision of the Energy Policy Programme for the Federal Republic of Germany. Federal Minister of Economics 1975.
15. Sweden : Energy Planning in Sweden. A presentation of the resolution on energy policy adopted by the Spring Session of the 1975 Riksdag. The Ministry of Industry. Fack. S-103.10. Stockholm.
16. United Kingdom : Energy Act 1976, Ch. 76 : Permanent and reserve powers for energy conservation and control. H.M.S.O. 1976
17. Energy Conservation. First Report from the Select Committee on Science and Technology. H.M.S.O. July 1975.

APPENDIX

TABLE 2 - STRENGTH OF ACCENT ON ENERGY CONSERVATION MEASURES (as of mid 1976)

TARGET SECTORS

RESIDENTIAL

BROAD CATEGORIES OF MEASURES	A	B	D	F	G	I	Ne	No	Swe	Swi[d]	U.K	TYPE OF MEASURE ★★
Persuasive	medium	strong	medium	strong	medium	medium	medium		medium		medium	Publicity campaigns.
	weak	medium		strong					strong		medium	Advice services
Fiscal	weak	strong	medium		medium		weak		strong		strong	Grants
			medium	strong								Tax, Investment incentives
Statutory Control	weak	weak				medium						Advisory
	weak		strong	strong	strong	medium		weak	medium		weak	Manditory

PUBLIC AND GOVERNMENT BUILDINGS

BROAD CATEGORIES OF MEASURES	A	B	D	F	G	I	Ne	No	Swe	Swi	U.K	TYPE OF MEASURE
Persuasive	weak			medium					weak			Information campaigns
	weak		weak						medium			Advice services
Fiscal		weak	weak				weak		weak			Grants
												Investment incentives
Statutory Control		medium	weak		weak	medium						Advisory
	weak	strong		strong	medium				medium	medium	weak	Manditory

TRANSPORT

BROAD CATEGORIES OF MEASURES	A	B	D	F	G	I	Ne	No	Swe	Swi	U.K	TYPE OF MEASURE
Persuasive	medium	medium	medium	strong	weak	medium	medium				medium	Publicity campaigns
	weak			strong							strong	Services to motorist
Fiscal	weak	strong	medium	medium	medium	strong	strong	medium	medium	medium	weak	Increased fuel tax
	weak		weak		weak b	weak					weak	Aid for public transport
Statutory Control		medium	medium	medium		medium			medium		medium	Public transport lanes
	130/100 km/h	120/90	110/90	110/90		120/100	100/80	90/80	110/70	130/100	113/97	Speed Controls Motorways Major roads

INDUSTRY

BROAD CATEGORIES OF MEASURES	A	B	D	F	G	I	Ne	No	Swe	Swi	U.K	TYPE OF MEASURE
Persuasive	weak		weak	strong	medium	medium	weak		medium		strong	Information campaigns
			weak	strong	medium	weak					medium e	Advice/Demonstration
Fiscal			weak						strong c		medium	Grants-insulation
		medium		medium	medium						medium	Investment incentives
Statutory Control					weak							Advisory
				medium a	medium						medium	Manditory

★ Countries : Austria, Belgium, Denmark, France, West Germany, Italy, Netherlands, Norway, Sweden, Switzerland, United Kingdam.

a. Manditory controls relate to maximum space temperature, industrial building-insulation standards and inspection of equipment.
b. Policy based on investment incentives for the transport infra-structure.
c. State grants are available for conservation in industrial processes, the use of waste fuels and the development of prototypes.
d. Introduction of specific measures is still under government review (The Report of the General Energy Commission will be published in July 1977)
e. An Energy Audit Schema and Industrial Thrift Scheme exists for Industry.

★★ [solid black box] strong accent; [hatched box] medium; [dotted box] weak; [empty box] non-existant

SOCIOLOGICAL IMPLICATIONS OF EFFICIENT ENERGY USE

Marcia R. Untracht

Applied Nucleonics Company, Inc., Santa Monica, California

ABSTRACT

The question of using energy more efficiently is basically one of technology, economics, and social change. That is to say, we can have access to new technological developments, understand how to benefit by use of alternate sources of energy, and understand the basic processes involved in using energy more efficiently; but without recognizing the implications of these changes economically and socially, efficient energy use may never become a permanent part of our lifestyle.

Thus, the essential problem concerns *how* to implement more efficient technologies and practices such that they will be integrated in and complement other aspects of our culture. This paper will use a model of social change derived from Thomas LaBelle's description of social determinants of behavior (an extension of Honigmann's model).[1] The model, which proposes that the behavior of a particular culture can be viewed by examining its three basic components -- technology, social organization, and ideology -- will provide us with a basis for understanding the sociological implications of more efficient energy use. As the availability of energy changes, economic factors come into play, new technologies develop, and our attitudes and particular aspects of our social organization will change.

This paper begins with an explanation of the LaBelle model and then applies it to the question of more efficient energy use, particularly in regard to the "American way of life." The result is a framework for change which states that in order for more efficient energy use to become a permanent part of our value system such that it is practically applied by all members of our society, all aspects of our culture must change. The paper will propose specific programs which seek to change the ideology and social organization components such that they coincide with the changes occurring in the technology component.

INTRODUCTION

It has been argued that the way a particular society uses energy can be traced to two major factors: (1) the availability of the energy sources and (2) the price of energy. Thus, an economist might claim that more efficient use of energy is simply a question of supply and demand.

Herein, I will argue that while economics does play a major role in understanding the world's energy problems, the surrounding sociological factors are important as well in order to understand why a particular society appears to be an "energy waster" instead of one that uses its energy sources efficiently.

As 1977 progresses, Americans are finally beginning to realize the implications of the energy problems faced not only by them but by all nations of the world. At this point in time, it is apparent that the economic factors affecting energy use are changing; the world is acknowledging the fact that unless all peoples end wasteful practices and develop new sources of energy, a global energy shortage could occur. Secondly, as a result of the political factors surrounding the question of energy supply, energy prices are rising drastically. The American people, therefore, must come to grips with the realization that we are becoming more "energy dependent" and must act now to make the most efficient use of our energy as we shift the emphasis of our energy use to sources such as coal, nuclear energy, and such solar energy forms that are economic and practical.

Political leaders, for the most part, have made few attempts to renew concern among the American people for finding ways to use energy efficiently; but the question is not solved simply by bringing the situation out in the open. Some leaders propose strict mandatory controls and tremendous price increases, assuming that without them, the inefficient practices that have become integral parts of the "American lifestyle" will not change quickly.

Americans, however, take pride in freedom of choice and the idea of mandatory controls is in direct opposition to their value system. Although some controls and price increases are necessary and inevitable, we must look at the long-term social change strategies which will alter our wasteful attitudes and make more efficient use of energy a way of life that is considered acceptable if not desirable. Efficient energy use does not have to decrease the quality of life nor cause drastic sacrifice. To the contrary, we must learn that using energy more efficienty is necessary and can lead to a stronger and better America.

In the following pages I will view efficient energy use as a necessary social change. I will employ a model of social behavior derived from Honigmann and expanded on by Thomas LaBelle to explain the components of culture. Then, I will look at the changes that must occur in these components to arrive at a new behavior pattern. Finally, practical suggestions for instigating these long-term changes in lifestyle patterns will be offered.

BASIC MODEL OF SOCIAL BEHAVIOR

"Americans still act like the grasshopper rather than the ant of the old fable; we do not have the national discipline to save for the harsh winter; we happily use what we can get our hands on at the moment."[2]

In comparing historical energy use patterns in the United States to those of other nations, it becomes apparent that we have taken little heed in managing our energy use. Schipper and others have done studies which conclude that other societies have not forfeited a comfortable standard of living while they manage their energy efficiently.[3] Even though Starr et al. point out that Schipper's comparisons are simplistic and fail to account for fundamental differences in United States and European economies, the fact remains that Americans can and are improving energy use efficiency substantially.[4]

One way to examine this difference in cultural behavior is to use Honigmann's "heuristic model of culture" or what LaBelle calls the "windmill" model of culture. There are three main components in this model as shown in Figure 1 which are:

- technology;
- social organization; and,
- ideology.

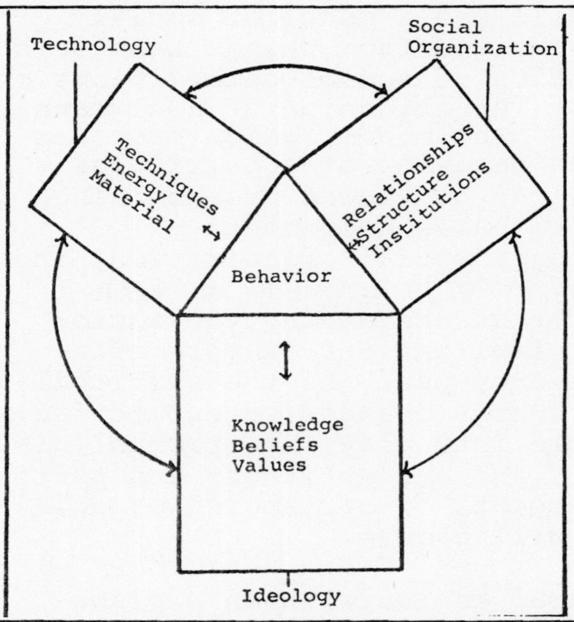

FIGURE 1

HEURISTIC MODEL OF CULTURE

Each of the above components are assumed to be interdependent but as LaBelle states, "any one may be viewed as an independent variable in order to note how it affects the others."[1] For my purpose herein, I will use the term "lifestyle" instead of the term "behavior" as seen in the model. Therefore, I will conclude that the three components of the model all taken together form the basis for the way of life of a particular society. I will now explain each component independently.

A. Technology

Technology refers to both the activities and material objects by which man manipulates his material world. In LaBelle's words, technology is "the manifestation of the available energy sources that condition other factors in a way of life."[1] An example would be the particular type of economy which evolves in a society as a result of the available natural resources and energy supplies. In the United States, for example, we can look back to the birth of the nation and the huge expanse of natural resources and energy which our ancestors found as they arrived on this continent. With the advent of the industrial revolution in the 1860s came the realization that the seemingly endless supply of energy sources (primarily sources including fuel wood, coal, and water power) would permit the development of a highly industrialized society. Thus, the American lifestyle evolved partially as a result of the technological advances that occurred.

B. Social Organization

This component refers to the activities and structures used by humans to interact with other humans. Here, it must be noted that given the availability of certain basic resources such as those noted above, our basic forms of social organization developed as men attempted to fulfill their economic needs vis-a-vis the environment in which they lived. As our technological society developed, our ancestors realized that in order to maintain a peaceful society, the development of specific social organizations to deal with the problems was necessary. A forum for the exchange of ideas and a basic structure to deal with relationships among individuals from various economic sectors, therefore, evolved.

C. Ideology

Ideology refers to the basic knowledge, beliefs, and values which guide our lives and, along with the two components described above, shape our behavior and thus our lifestyle patterns. The particular value system that has evolved in the United States -- the "American way of life" -- is closely related to the availability of natural resources and energy which we have used to build our economy and which shape the type of work which we do in our daily lives. The American value system has been influenced by the variety of job opportunities we have had vis-a-vis the technological developments which occurred in light of our particular environment. Furthermore, our social organization is a manifestation of these values.

In all, from the explanations of each component, it is apparent that one cannot describe one of the components without mention of the other two; thus, LaBelle's argument regarding the interdependency of the components is valid.

VIEWING EFFICIENT ENERGY USE AS A SOCIAL CHANGE

To continue with the basic premise of this paper, I view efficient energy use in the United States as a necessary social change or a basic change in the values which guide our lifestyle. Let us look again at the model of social behavior outlined above and relate it to the energy problems we face. In the technology component it can be noted that for many years scientists and engineers have been studying the question of more efficient use of energy and new sources of energy. The developments in the technological world are numerous in this regard and they continue to progress rapidly. Thus, it is apparent that the changes in the technology component are being made continuously because the scientific community came to the basic realization that we would face an energy crisis years before the Oil Embargo.

Thus, we now arrive at a major question for discussion -- *how do concurrent changes occur in the ideology/social organization components of the windmill model to support the changes in the technological component and finally result in a holistic change in our society, altering our lifestyle patterns permanently?*

I, therefore, have adopted what might be termed a "grassroots" approach; that is to say, while political and economic forces make the immediate decisions and have the power to shape many of our daily activities, the people within our society must come to a basic realization that they must learn to change the lifestyle patterns they have been accustomed to in the past.

To be more specific: first, I see changes in technology which support more efficient use of energy and the development of new sources. Second, I see our political leaders finally coming to grips with the problem which so long has been a concern of the scientific community. Thus, the political community presents the problem to us, announces concern, and deals with it by economic constraints -- raising prices and establishing mandatory controls on energy use. Thus, Americans will be "forced" to use less energy. However, what happens when these constraints are lifted? Will Americans go back to their wasteful habits? I believe the answer is yes *unless* basic changes occur in the ideology and social organization components of our culture. Thus, I have adopted this "grassroots" approach to change which is, in essence, a process of social change.

Going back to the windmill model and looking at it with the energy question in mind, I see more of a progression of interrelated events and can alter Honigmann's model slightly to create a flow chart for change (see Figure 2). The progression is as follows: (1) technology changes; (2) social organization changes; and (3) ideology changes. If all the components had changed at once, our lifestyle pattern would have been altered correspondingly. In order to see a complete, permanent change in the behavior of people, the change, in my opinion, must be holistic. However, I must note that the changes in the components of our society have not been concurrent. Our social organization and ideologies have not kept up with the fast pace of the developments occurring in the technology component and herein lies the problem. Only in the recent past have we seen the beginning of changes in the social organization/ideology components. These are: (1) the government finally coming to realize that change in our energy use practices is essential if our economy is to continue to expand; (2) government plans to establish a new energy department to handle the energy problems which have arisen (and will not disappear) and at the same time to implement economic strategies (i.e., mandatory controls, higher prices on energy, and tax incentives for making various home improvements which lead to more efficient use of energy); and (3) the American public beginning to show signs of "belief" that there is, in fact, a serious crisis and that we must develop a new "energy ethic." Thus, the social organization component begins to change as does the ideology component. The next section will present specific action-oriented suggestions which will expand the development of change in these two components and result in a true lifestyle change which will create an energy efficient America.

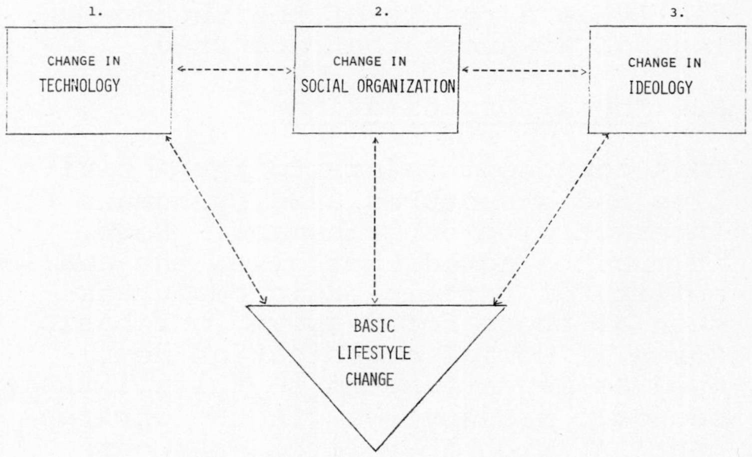

FIGURE 2

AN ALTERATION OF HONIGMANN'S MODEL

CHANGES NECESSARY TO EFFECT LONG-TERM LIFESTYLE MODIFICATIONS RESULTING IN AN ENERGY EFFICIENT SOCIETY

Having completed an explanation of the theoretical concepts which form the basis for this paper, I now arrive at the practical, action-oriented section. As I stated above, changes have already occurred in the technology and social organization components of our society. However, for a total, long-term effect on our lifestyle, change must occur in the ideology component as well. We now require more programs which foster changes in our knowledge, values, and beliefs regarding energy use. Admittedly, the programs suggested below will not only promote change in the ideology component. The social organization and ideology components are so closely related that the suggested programs for change take these components into account simultaneously. Educational courses or the plan for a new energy department to be created in our government structure will certainly assist in altering our value systems while being, in essence, new forms of social organization. However, at this point in time, while it appears that change has progressed rapidly in the technology component, and while changes are beginning to unfold in the social organization component, the ideology component is the one that has changed the least and deserves the most careful study

Thus we must ask, "What must be done to foster long-term changes in our "energy consciousness?"

My approach is "people-oriented," one which seeks to change individuals. Americans must realize that
(1) the energy crisis is real,
(2) as Alfred Fritsch states, "Conservation is not a fad or a temporary belt-tightening...Conservation and preservation of resources are not lifestyles in themselves but the means whereby freedom to choose one's lifestyle is ensured."[5]

Table 1 presents a summary of basic programs and actions which will serve to bring about the realization of points (1) and (2) above. I see the need for continued activity in three areas:

- citizen action,
- educational programs, and
- media activities.

Probably the most influential of these three areas is the media (television, radio, and the press) since media activities touch all Americans. In recent months we have seen an expanded amount of television commercials, newspaper articles and editorials about energy conservation. However, many Americans still take these pleas with a "grain of salt." Part of the problem, in my opinion, is the fact that these announcements are presented in much the same way as commercials which sell products. There is a great necessity for the media to approach the problem in a more sophisticated way. This can be achieved by halting the commercial approach and instead presenting more documentaries, educational programs, and interviews with highly credible experts who are respected by the general public.

It is apparent that we will be faced with higher energy prices and some mandatory controls. Until Americans are convinced that they can effect change and that their actions now will make a difference later, it is necessary to enact programs such as those suggested in the table to run concurrently with the economic actions taken by government. As I see it, this is the best way to attempt to create a permanent social change vis-a-vis more efficient use of energy. People must see that this change will not imply a decrease in the quality of lfe but instead will build a stronger nation.

TABLE 1

SUGGESTED STRATEGIES FOR A NEW "ENERGY CONSCIOUSNESS" IN THE UNITED STATES

I. Citizen Actions

- backing political candidates who speak out on the issue
- lobbying
- volunteer efforts with public interest and ecology groups
- joining citizen coalitions
- making citizen's, professional, and civic organizations more aware of environmental problems

II. Educational Strategies

- development and initiation of courses on energy conservation in the public schools, colleges, and at adult education institutions

- development of energy management campaigns by utilities and large corporations which include free literature handouts and free lectures on energy conservation

III. Media Actions

- television, radio, and press coverage -- documentaries on efficient energy use and interviews with experts

- attempt to present the problems as serious and vital to our well-being now as well as later (avoid over-saturation of the public by the use of announcements which are presented in much the same way as a commercial advertisement)

- educational television networks should plan series on efficient energy use for the residential sector

---

CONCLUSIONS

Changes must occur in all three components of the cultural model in order for a true, permanent change to occur in our lifestyle patterns of energy use. In view of more efficient energy use, the technology component has been seen to change most rapidly while the other two components of the cultural model lag behind. We must, therefore, center our strategies on ways to change our social organization ; ideologies and suggestions in this regard have been pointed out in the preceding section. The strategies for change which have been suggested are supportive, long-term strategies; immediate action must take the form of higher prices for energy, mandatory controls on energy use, and economic incentives for consumers. Only in combination with these economic strategies will the social change strategies such as those presented in Table 1 be affective. This is unfortunate but Americans have been recalcitrant in regard to the so-called "energy crisis" and obviously will have to face some economic constraints as a result.

The situation is not one for despair however. As President Carter stated, *If we fail to act soon, we will face an economic, social, and political crisis that will threaten our free institutions. But we still have another choice. We can begin to prepare right now. We can decide to act while there is still time.*[6]

Furthermore, if we act now, we will soon realize that the changes we make will not decrease the quality of our lives nor will they mean tremendous hardships. Instead, we will become an "energy efficient society" while maintaining the democratic principles which have contributed to our progress as a nation.

REFERENCES

1. LaBelle, Thomas, "Alternative Educational Strategies: The Integration of Learning Systems in the Community," The Conventional and the Alternative in Education, J.L. Goodlad, ed., Berkeley, California McKutchen Publishing Corporation, 1965, pp 165-187.

2. Szulc, Ted, The Energy Crisis, New York, Franklin Watts, Inc., 1974, p.3.

3. Schipper, Lee and Lichtenberg, A.J., "Efficient Energy Use and Well-Being. The Swedish Example." Science, December 3, 1976, Vol. 194, No. 4269, pp. 1001-1013

4. Starr, Chauncey and Field, S., "Energy Use Proficiency: The Validity of Interregional Comparisons," Palo Alto, California, Electric Power Research Institute, March 1977

5. Fritsch, Albert J., The Contrasumers: A Citizen's Guide to Resource Conservation, New York, Praeger Publishers, 1974, p. 156.

6. Address to the Nation, April 18, 1977.

ENERGY CONSERVATION POLICY IN THE U.S.: THE EMERGING ROLE OF THE FEDERAL GOVERNMENT

Timothy A. Hall
Science and Public Policy Program
The University of Oklahoma
601 Elm Avenue, Room 432
Norman, Oklahoma 73019

ABSTRACT

Until the 1973 shortages and price increases, the consumption side of the U.S. energy system received little public attention. However, as domestic energy supplies and demands have become increasingly imbalanced and the nation has become more dependent on foreign oil, energy use has assumed a new significance as a policy-making issue. Public policy options to reduce energy demand generally include: (1) the immediate or gradual removal of price controls on oil and natural gas; (2) financial incentives such as tax credits to encourage energy-saving investments or penalties to discourage overconsumption; (3) mandatory energy conservation measures, for example, mileage standards on new automobiles and efficiency standards for appliances; (4) the development of energy-conserving technologies; and (5) the collection and dissemination of information to help consumers make efficiency and life-style decisions that could alter energy-wasteful habits. This paper summarizes the advantages and disadvantages of each of these options and then reviews recent federal mandates in the direction of reduced demand in the primary energy-consuming sectors of the U.S. economy. Special attention is given to the social, political and institutional barriers to implementing the adopted courses of action.

Although the policy developments discussed here are relatively recent, several observations can be made regarding the federal government's emerging role in energy conservation. First, conservation legislation is currently fragmented and uncoordinated and as yet has not been tied together in a comprehensive national program. Moreover, no "lead" agency for the purpose of reduced demand has been identified. Second, policies and programs already in place have some strong elements, but deliberate action is needed in several crucial areas, such as energy prices, utility rate structures, and tax incentives for the residential/commercial sector. Finally, very little policy is aimed at targets of social change; that is, most of the measures adopted within the past three years rely on regulatory mechanisms which reflect a "technical fix" bias. This last trend raises critical questions with regard to the more behaviorally-oriented energy use problems, especially in the long-term.

FINANCIAL STRATEGIES FOR ENERGY SYSTEMS

Ivan M. Von Zuckerstein
Energy and Environmental Systems
Argonne National Laboratory

Kevin Croke
University of Illinois
Circle Campus

Arthur P. Hurter, Jr.
Northwestern University

David Kelly
Arthur Andersen & Co.

ABSTRACT

The energy shortage has stimulated search for new, more efficient methods of energy utilization that would satisfy the demands for greater fuel efficiency without undue sacrifice of economic considerations and operational characteristics. The ERDA Office of Energy Conservation has launched a program of Integrated Community Energy Systems (ICES) that will develop community-scale energy systems with these characteristics.

The successful commercialization of Community Energy Systems will depend in part on the development, adaptation and use of financial mechanisms that are fitted to the financial environ ent of combined utility application. The unique features of the community energy systems which are envisioned to be participative ventures in varying degrees among the municipalities, developers amd utility companies give rise to the possibility of certain, less conventional financial arrangements that might compound the technological and socio-economic benefits derived from the ICES applications.

Among the financial techniques to be explored in the paper will be a variety of leasing arrangements that would take full advantage of tax incentives, alternate sources of capital and incidental benefits of ownership by one party and operation by the other. In this context, it is planned to review the applicable federal and representative state tax laws pertaining to the tax exempt status of municipal obligations issued for the purpose of financing leased facilities, the options available for allocations of investment tax credits and the tax and regulatory ramifications as related to the use of normalizing or flow-through treatment of accelerated depreciation tax benefits. In addition, a survey will be conducted of the existing federal grant and loan guarantee programs with a view toward identifying specific government agencies/and or programs that have the authority to underwrite a part or the total ICES facility cost and thereby contribute to the lowering of financing charges. In conclusion, the paper will compare and contrast the novel financing approaches with conventional debt/equity financing.

INDEX — VOLUME I

INDEX — VOLUME II